A Dictionary of

Biology

VISIT ONLINE

OXFORD
QUICK
REFERENCE

The most authoritative and up-to-date reference books for both students and the general reader.

Accounting
Agriculture and Land
  Management
Animal Behaviour
Archaeology
Architecture
Art and Artists
Art Terms
Arthurian Literature and Legend
Astronomy
Bible
Biology
Biomedicine
British History
British Place-Names
Business and Management
Chemical Engineering
Chemistry
Christian Art and Architecture
Christian Church
Classical Literature
Computer Science
Construction, Surveying, and
  Civil Engineering
Cosmology
Countries of the World
Critical Theory
Dance
Dentistry
Ecology
Economics
Education
Electronics and Electrical
  Engineering
English Etymology
English Grammar
English Idioms
English Language
English Literature
English Surnames
Environment and Conservation
Everyday Grammar
Film Studies
Finance and Banking
Foreign Words and Phrases
Forensic Science
Fowler's Concise Modern
  English Usage
Geography
Geology and Earth Sciences

Hinduism
Human Geography
Humorous Quotations
Irish History
Islam
Journalism
Kings and Queens of Britain
Law
Law Enforcement
Linguistics
Literary Terms
London Place-Names
Marketing
Mathematics
Mechanical Engineering
Media and Communication
Medical
Modern Poetry
Modern Slang
Music
Musical Terms
Nursing
Philosophy
Physics
Plant Sciences
Plays
Political Quotations
Politics and International
  Relations
Popes
Proverbs
Psychology
Quotations
Quotations by Subject
Rhyming
Rhyming Slang
Saints
Science
Scottish History
Shakespeare
Social Work and Social Care
Sociology
Statistics
Synonyms and Antonyms
Weather
Weights, Measures,
  and Units
Word Origins
World Mythology
Zoology

Many of these titles are also available online at
www.oxfordreference.com

A Dictionary of
Biology

EIGHTH EDITION

OXFORD
UNIVERSITY PRESS

OXFORD
UNIVERSITY PRESS

Great Clarendon Street, Oxford, OX2 6DP,
United Kingdom

Oxford University Press is a department of the University of Oxford.
It furthers the University's objective of excellence in research, scholarship,
and education by publishing worldwide. Oxford is a registered trade mark of
Oxford University Press in the UK and in certain other countries

First published 1985 as *A Concise Dictionary of Biology*
Second edition 1990
Third edition 1996
Fourth edition 2000
Fifth edition 2004
Sixth edition 2008
Seventh edition 2015
Eighth edition 2019

Published in the United States of America by Oxford University Press
198 Madison Avenue, New York, NY 10016, United States of America

British Library Cataloguing in Publication Data
Data available

Library of Congress Control Number: 2018964542

ISBN 978-0-19-882148-9

Printed and bound in Great Britain by
Clays Ltd, Elcograf S.p.A.

Contents

Preface

Biology is one of the most dynamic and fast-moving fields in all science, with vital and growing importance in technological, economic, social, and political spheres. It is hoped that anyone interested in biology, whether student, academic, or layperson, will find this dictionary informative and helpful for their understanding of biological terms and concepts.

This dictionary was originally derived from the *Concise Science Dictionary*, first published by Oxford University Press in 1984 (seventh edition, 2017, retitled *A Dictionary of Science*). It was published in 1985 and consisted of all the entries relating to biology and biochemistry in the *Science Dictionary*, together with relevant entries relating to palaeontology, soil science, physics, and chemistry that are required for an understanding of the physical and chemical aspects of biology (including laboratory techniques for analysing biological material). It also included a selection of terms used in medicine and palaeoanthropology. Subsequent editions saw the addition of more terms relating to human biology, environmental science, biotechnology and genetic engineering, and food technology (among other fields), as well as several chronologies tracing the history of some key areas in biology, and a number of longer feature articles on selected topics.

For this edition many entries have been substantially updated and around 150 new entries have been added in all the major fields. The extensive revisions reflect advances in areas such as molecular systematics, genomics, bioinformatics, and cell biology, as well as the unprecedented challenges for the natural environment and human societies posed by global warming and climate change. There is a new appendix on anatomical terms. Many entries throughout the work have web links to selected sites that provide further information on the topic.

Using the dictionary

An asterisk placed before a word used in an entry indicates that this word can be looked up in the dictionary and will provide further explanation or clarification. However, not every word that appears in the dictionary has an asterisk placed before it. Some entries simply refer the reader to another entry, indicating either that they are synonyms or abbreviations or that they are most conveniently explained in one of the dictionary's longer articles or features. Synonyms and abbreviations are usually placed within brackets immediately after the headword, as are irregular plural forms. Terms that are explained within an entry are highlighted by being printed in bold type.

The more chemical aspects of biochemistry and the chemistry itself will be found in *A Dictionary of Chemistry*; this and *A Dictionary of Physics* are companion volumes to this dictionary.

SI units are used throughout this book and its companion volumes.

Robert Hine
2018

Credits

Editor
Robert S. Hine BSc, MSc

Editors of the fourth, fifth, and sixth editions
Robert S. Hine BSc, MSc
Elizabeth Martin MA

Editor of the earlier editions
Elizabeth Martin MA

Market House Books
Jonathan Law

Advisers
B. S. Beckett BSc, BPhil, MA(Ed)
R. A. Hands BSc
Michael Lewis MA
W. D. Phillips PhD

Contributors
Tim Beardsley BA
Lionel Bender BSc
John Clark BSc
H. M. Clarke MA, MSc
Belinda Cupid MSc, PhD
E. K. Daintith BSc
Malcolm Hart BSc, MIBiol
Robert S. Hine BSc, MSc
Elaine Holmes BSc, PhD
Anne Lockwood BSc
J. Valerie Neal BSc, PhD
R. A. Prince MA
Michael Ruse BSc, PhD
Brian Stratton BSc, MSc
Elizabeth Tootill BSc, MSc

AAV *See* ADENO-ASSOCIATED VIRUS.

A band The region of a striated muscle fibre that contains both thick (myosin) and thin (actin) filaments. It is visible as a dark band with a lighter central zone (*see* H ZONE) in the middle of a *sarcomere.

ABC model A model to explain the genetic control of floral organ determination during flower development. It proposes that the four principal organs—sepals, petals, stamens, and carpels—are specified by organ identity genes belonging to three classes: A, B, and C. The organs are arranged in four concentric whorls, and their identity is determined according to which combinations of genes are expressed in each whorl. Class A genes alone specify the outermost whorl of sepals; the formation of petals depends on activation of both class A and B genes; stamens are specified by class B and C genes acting together; and the innermost whorl of carpels is determined by class C genes alone. The model was based on observations of mutant flowers in the thale cress (*Arabidopsis thaliana*). Mutations in these genes cause the transformation of one flower organ into another (i.e. homeosis); for example, a class B mutation causes sepals to develop instead of petals, and carpels instead of stamens. The genes corresponding to these homeotic mutations are now known to be *MADS box genes that encode transcription factors, capable of activating different target genes depending on what other transcription factors act with them. *See also* FLOWERING.

abdomen The posterior region of the body trunk of animals. In vertebrates it contains the stomach and intestines and the organs of excretion and reproduction. It is particularly well defined in mammals, being separated from the *thorax by the *diaphragm. In many arthropods, such as insects and spiders, it may be segmented.

abductor (levator) A type of muscle whose function is to move a limb away from the body. Abductors work antagonistically with *adductors.

abiogenesis The origin of living from nonliving matter, as by *biopoiesis. *See also* SPONTANEOUS GENERATION.

abiotic factor Any of the nonliving factors that make up the **abiotic environment** in which living organisms occur. They include *edaphic factors and all the aspects of *climate, geology, and atmosphere that may affect the biotic environment. *Compare* BIOTIC FACTOR.

abomasum The fourth and final chamber of the stomach of ruminants. It fills from the *omasum and empties into the small intestine. The abomasum is referred to as the 'true stomach' as it is in this chamber that protein digestion occurs, in acidic conditions. *See* RUMINANTIA.

ABO system One of the most important human *blood group systems. The system is based on the presence or absence of carbohydrate *antigens A and B on the surface of red blood cells and *antibodies against these in blood serum. A person whose blood contains either or both of these antibodies cannot receive a transfusion of blood containing the corresponding antigens as this would cause the red cells to clump (*see* AGGLUTINATION). The table illustrates the basis of the system: people of blood group O are described as 'universal donors' as they can give blood to those of any of the other groups. *See also* IMMUNITY.

The ABO blood group system

Group	Antigens on red cell surface	Antibodies in serum	Blood group of people donor can receive blood from	Blood group of people donor can give blood to
A	A	anti-B	A, O	A, AB
B	B	anti-A	B, O	B, AB
AB	A and B	none	A, B, AB, O	AB
O	neither A nor B	anti-A and anti-B	O	A, B, AB, O

abscisic acid (ABA) A naturally occurring *plant hormone that appears to be involved primarily in seed maturation, stress responses, and in regulating closure of leaf pores (stomata). It thus acts primarily to slow growth, in antagonism to the effects of growth-promoting plant hormones such as auxins and cytokinins. Levels of ABA increase greatly in maturing seeds to prevent premature germination, and germination can only occur when the ABA is removed or inactivated, e.g. by prolonged cold or light. ABA promotes the synthesis of proteins that enable the seeds to survive the dehydration associated with maturation. In leaves, ABA is produced in large amounts when the plant lacks sufficient water, promoting loss of turgor in *guard cells, closure of stomata, and hence reducing further water losses. Levels of ABA increase suddenly in response to various forms of stress, including heating, waterlogging, and chilling. It was formerly believed to play a central role in *abscission; hence the name.

abscission The separation of a leaf, fruit, or other part from the body of a plant. It involves the formation of an **abscission zone**, at the base of the part, within which a layer of cells (**abscission layer**) breaks down. This process is suppressed so long as sufficient amounts of *auxin, a plant hormone, flow from the part through the abscission zone. However, if the auxin flow declines, for example due to injury or ageing, abscission is activated and the part becomes separated. *Ethylene acts as the primary trigger for abscission, inducing

cells in the abscission zone to produce cellulase enzymes that degrade cell walls. Before separation occurs, a protective layer of cork forms on the stem side to prevent invasion by pathogens.

absolute refractory period See REFRACTORY PERIOD.

absorbed dose See DOSE.

absorption The movement of fluid or a dissolved substance across a plasma membrane. In many animals, for example, soluble food material is absorbed into cells lining the alimentary canal and thence into the blood. In plants, water and mineral salts are absorbed from the soil by the *roots. See OSMOSIS; TRANSPORT PROTEIN.

absorption spectrum See SPECTRUM.

abundance (in ecology) The absolute number of individuals of a species in a given sample, community, or area of habitat. An abundant species is one that is locally numerous or occurs at high densities. **Relative abundance** is the proportion of individuals of any one species in relation to the total number of individuals of all species found. For example, suppose a pond contains 100 individuals belonging to four species of amphibian, A, B, C, and D; and the numbers of individuals for each are as follows: A = 40; B = 30; C = 25; D = 5. The relative abundances are: A = 0.4; B = 0.3; C = 0.25; and D = 0.05. The distribution of relative abundance values is described by *species evenness; when these values are similar across a range of species, the community is more diverse (*see* BIODIVERSITY).

abyssal zone The lower depths of the ocean (below approximately 2000 metres), where there is effectively no light penetration. Abyssal organisms are adapted for living under high pressures in cold dark conditions. *See also* APHOTIC ZONE.

Acarina (Acari) A subclass of small arthropods belonging to the class *Arachnida and comprising the mites and ticks. There are about 50 000 described species, with perhaps 20 times this number still unknown, distributed worldwide in a wide variety of terrestrial and aquatic habitats. Many are free-living in soil or on vegetation, feeding on organic matter or preying on other small arthropods, while a significant number are parasites of plants and animals, including domesticated animals and humans. The adult body is generally globular or ovoid, with four pairs of legs. Unlike spiders, there is no 'waist', the abdomen being fused to the more anterior prosoma. At the front of the body the capitulum bears the mouthparts, variously adapted for cutting, crushing, or piercing. The eggs hatch into a three-legged larva, which subsequently moults to a nymph resembling the adult. Ticks (up to 3 cm long) are ectoparasites of vertebrates, feeding on blood drawn through the skin of their host. They transmit a wide range of diseases, including certain forms of encephalitis and Lyme disease. Mites are much smaller (up to 4 mm long) and are parasitic or free-living. They tend to feed on feathers, hair, skin secretions, or skin debris, causing, for example, scabies in humans and mange in domesticated animals. The house-dust mite (*Dermatophagoides* sp.) can provoke allergies or dermatitis. Spider mites are damaging parasites of plants and may infest some arable and greenhouse crops.

⊕ SEE WEB LINKS
• More information about mites and ticks, including news, publications, and societies

acceleration A form of *heterochrony in which, during the course of evolution, the rate of development of an organism is speeded up and new stages are added to the end of the ancestral developmental sequence without prolonging the total development time. The morphological outcome is an example of *peramorphosis, and the developmental sequence (ontogeny) conforms to the theory of *recapitulation.

acceptor **1.** (in chemistry) A compound, molecule, ion, etc., to which electrons are donated in the formation of a coordinate bond. **2.** (in biochemistry) A receptor that binds a hormone without any apparent biological response.

accessory bud A bud that is situated at the side of or above an axillary bud (*see* AXIL).

accessory chromosome (B chromosome; supernumerary chromosome) Any chromosome that is additional to the regular *karyotype of a species. Such chromosomes vary in size and composition and may or may not affect the phenotype. In humans, about one person in every 1500 carries an accessory chromosome, and some of these are associated with mental or physical abnormalities. For example, cat-eye syndrome occurs in people who carry an extra chromosome that partly duplicates chromosome 22. It is characterized by enlarged pupils, skin malformations, cardiac and urinary defects, and, sometimes, mental handicap.

accessory pigment A *photosynthetic pigment that traps light energy and channels it to chlorophyll *a*, the primary pigment, which initiates the reactions of photosynthesis. Accessory pigments include the *carotenoids, *phycobiliproteins, and chlorophylls *b*, *c*, and *d*.

acclimation The physiological changes occurring in an organism in response to a change in a particular environmental factor (e.g. temperature). Thermal acclimation studies reveal how such properties as metabolic rate, muscle contractility, nerve conduction, and heart rate differ between cold- and warm-acclimated members of the same species. These changes occur naturally during *acclimatization and equip the organism for living in, say, cold or warm conditions according to the seasons. Metabolic acclimation is explained mainly by changes in concentration and/or activity of crucial enzymes. For example, different *isozymes may be expressed at different times of the year. Changes in composition of membrane lipids, particularly their degree of saturation, also occur, helping to maintain membrane stability in changing conditions. *Heat-shock proteins help to protect and repair proteins damaged by thermal stress, and their expression increases under such conditions. Plants too show better tolerance of environmental extremes following previous exposure to such conditions (see COLD HARDENING).

acclimatization The progressive adjustments in the biochemistry and physiology of an organism that occur in response to changes in its natural environment, for example seasonal changes in temperature or changes in oxygen concentration at different altitudes. See ACCLIMATION. *Compare* ADAPTATION (sense 1).

accommodation 1. (in animal physiology) Focusing: the process by which the focal length of the *lens of the eye is changed so that clear images of objects at a range of distances are displayed on the retina. In humans and some other mammals accommodation is achieved by reflex adjustments in the shape of the lens brought about by relaxation and contraction of muscles within the *ciliary body. **2.** (in animal behaviour) Adjustments made by an animal's nervous or sensory systems in response to continuously changing environmental conditions.

acellular Describing tissues or organisms that are not made up of separate cells but often have more than one nucleus (see SYNCYTIUM). Examples of acellular structures are muscle fibres. *Compare* UNICELLULAR.

acentric Describing an aberrant chromosome fragment that lacks a centromere. Such fragments are normally lost because they are unable to orientate properly during cell division.

acetabulum (cotyloid cavity) The semicircular cavity in the *pelvic girdle that houses the ball-shaped head of the *femur.

acetate (ethanoate) A salt or ester of acetic acid.

acetic acid (ethanoic acid) A carboxylic acid, CH_3COOH, that is used as a carbon source by certain green algae. Combined with coenzyme A (see ACETYL COENZYME A), it plays a crucial role in the energy metabolism of all organisms. **Acetic acid bacteria** comprise a group of bacteria, e.g. *Acetobacter* spp., that perform incomplete oxidation of alcohols to yield organic acids as end products. When the initial substrate is ethanol, the product is acetic acid.

acetone *See* KETONE; KETONE BODY.

acetylation The introduction of an acetyl group (CH_3CO-) into a compound. The acetylation of *coenzyme A to acetyl coenzyme A is an important stage in the *Krebs cycle: the acetyl group is derived from pyruvate after the removal of a molecule of carbon dioxide and two hydrogen atoms. Acetylation of histone proteins in *chromatin loosens the tightly packed nucleosomes, thereby making the DNA of the chromosomes accessible for gene transcription. *See* CHROMATIN SILENCING.

acetylcholine (ACh) One of the main *neurotransmitters of the vertebrate nervous system. It is released at certain (**cholinergic**) nerve endings and may be excitatory or inhibitory; it initiates muscular contraction at *neuromuscular junctions and controls many physiological functions via the *parasympathetic nervous system. Acetylcholine promotes contraction of smooth muscle in the gut, whereas it has an inhibitory effect on cardiac muscle, reducing heart rate. It also has a role in various brain activities, including attentiveness, aggression, sex drive, and thirst. Acetylcholine receptors (**cholinoceptors**) fall into two main classes, *muscarinic and *nicotinic receptors; hence acetylcholine can have different effects in different tissues, depending on the type of receptor. Once acetylcholine has been released it has only a transitory effect because it is rapidly broken down by the enzyme *cholinesterase.

acetylcholinesterase *See* CHOLINEST-ERASE.

acetyl coenzyme A (acetyl CoA) A compound formed in the mitochondria when an acetyl group (CH_3CO-), derived from the breakdown of fats, proteins, or carbohydrates (via *glycolysis), combines with the thiol group (-SH) of *coenzyme A. Acetyl CoA is a key metabolic building block. It feeds into the *Krebs cycle, plays a role in the synthesis and oxidation of fatty acids, and is a component of steroid hormones, plant hormones, certain pigments, and other substances.

achene A dry indehiscent fruit formed from a single carpel and containing a single seed. An example is the feathery achene of clematis. Variants of the achene include the *caryopsis, *cypsela, *nut, and *samara. *See also* ETAERIO.

aciclovir (acyclovir) A drug used to treat cold sores, shingles, genital blisters, or other lesions caused by herpesvirus infection. It is an analogue of the base guanine and acts by interfering with DNA replication of the virus.

acid A type of compound that contains hydrogen and dissociates in water to produce positive hydrogen ions. The reaction, for an acid HX, is commonly written:

$$HX \rightleftharpoons H^+ + X^-$$

In fact, the hydrogen ion (the proton) is hydrated, and the complete reaction is:

$$HX + H_2O \rightleftharpoons H_3O^+ + X^-$$

The ion H_3O^+ is the **hydroxonium ion**. The strength of an acid depends on the extent to which it dissociates: **strong acids** (e.g. sulphuric acid and hydrochloric acid) are almost completely dissociated in water; **weak acids** (e.g. carbonic acid) are only partially dissociated. *See also* BUFFER; PH SCALE. *Compare* BASE.

acid-base balance The regulation of the concentrations of acids and bases in blood and other body fluids so that the pH remains within a physiologically acceptable range (*see* PH SCALE). This is achieved by the presence of natural *buffer systems, such as the haemoglobin, hydrogencarbonate ions, and carbonic acid in mammalian blood. By acting in conjunction, these effectively mop up excess acids and bases and therefore prevent any large shifts in blood pH. The hydrogencarbonate buffer system is regulated by the kidneys, in which there is net excretion of hydrogen ions, while hydrogencarbonate ions are conserved, and by the lungs, which control the level of carbon dioxide in the blood.

acid growth theory A theory, originally proposed in 1970 by R. Cleland, A. Hager, and co-workers, that describes how auxins stimulate cell expansion in certain plant tissues, notably in coleoptiles of cereals and other grasses, such as oat (*Avena*). It asserts that the auxins induce acidification of the immediate cell-wall environment, thereby activating enzymes called *expansins that loosen load-bearing bonds within the cell wall. This permits expansion of the walls by the cell's internal turgor pressure, and thus enlargement of the cell. It is thought that the auxin binds with *auxin-binding proteins in the plasma membrane and stimulates the synthesis of *proton pumps (also in the cell's plasma membrane), which excrete protons (H^+) into the cell wall from the cytoplasm. The consequent lowered cell-wall pH activates the wall enzymes, while the increased membrane potential enhances ion uptake by the cell, which causes water to enter the cell, thereby increasing turgor pressure. Experimental evidence broadly supports the theory.

acidic stains *See* STAINING.

acidophilic 1. Describing organisms that grow best in acidic conditions (i.e. with a pH less than 7). Such organisms are called **acidophiles**. *Compare* ALKALIPHILIC. **2.** Describing cells or tissues that stain with an acidic dye.

acidosis A condition in which the body fluids become more acidic, i.e. the pH is less than 7.4, and the capacity of the body to *buffer hydrogen ions is diminished. A decrease in the elimination of carbon dioxide from the body gives rise to **respiratory acidosis**, while a deficiency of hydrogencarbonate results in **metabolic acidosis**.

acid protease A protein-digesting enzyme (see PROTEASE) that exhibits maximum activity and stability in acid conditions (pH 2.0–5.0) and is inactivated at pH values above 6.0. Acid proteases have a low *isoelectric point and are low in basic amino acids. Two types are widely used in the food and beverage industries: those from *Aspergillus*, which resemble pepsin; and those from *Mucor*, which resemble rennin.

acid rain Precipitation having a pH value of less than about 5.0, which has adverse effects on the fauna and flora on which it falls. Rainwater typically has a pH value of 5.6, due to the presence of dissolved carbon dioxide (forming carbonic acid). Acid rain results from the emission into the atmosphere of various pollutant gases, in particular sulphur dioxide and various oxides of nitrogen, which originate from the burning of fossil fuels and from car exhaust fumes, respectively. These gases dissolve in atmospheric water to form sulphuric and nitric acids in rain, snow, or hail (**wet deposition**). Alternatively, the pollutants are deposited as gases or minute particles (**dry deposition**). Both types of acid deposition affect plant growth—by damaging the leaves and impairing photosynthesis and by increasing the acidity of the soil, which results in the leaching of essential nutrients. This acid pollution of the soil also leads to acidification of water draining from the soil into lakes and rivers, which become unable to support fish life. Lichens are particularly sensitive to changes in pH and can be used as indicators of acid pollution (see INDICATOR SPECIES).

acinus (*pl.* **acini**) The smallest unit of a multilobular gland, such as the pancreas. Each acinus in the pancreas is made up of a hollow cluster of **acinar cells**, which produce the digestive enzymes secreted in pancreatic juice. Minute ducts from the pancreatic acini eventually drain into the pancreatic duct.

acoelomate Describing any bilaterally symmetrical animal of the group Metazoa that does not possess a *coelom (see also BODY CAVITY). Examples of acoelomate animals are the flatworms.

acquired characteristics Features that are developed during the lifetime of an individual, e.g. the enlarged arm muscles of a tennis player. Such characteristics do not involve any alteration in the base sequence of genes and hence cannot be passed on to the next generation. However, exposure to certain environmental factors, such as malnutrition or toxic chemicals, can cause *epigenetic changes that are transmitted to subsequent generations. Although these can be regarded as acquired characteristics, they are not stably inherited and after a variable number of generations are liable to revert to the original state. See also BALDWIN EFFECT; LAMARCKISM; NEO-LAMARCKISM.

acquired immune deficiency syndrome See AIDS.

acquired immunity See IMMUNITY.

acridine A chemical (see formula) that is capable of causing *frameshift mutations in the DNA sequence. Several derivatives of acridine (such as acridine orange) are used as dyes or biological stains.

Acridine

acrocentric See CENTROMERE.

acromegaly A chronic condition developing in adulthood due to overproduction of (or oversensitivity to) *growth hormone, usually caused by a tumour in the pituitary gland. This leads to a gradual enlargement of the bones, causing characteristic coarsening of the facial features and large hands and feet.

acrosome A membranous sac at or near the front of a sperm that assists in penetration of the egg. The acrosome contains enzymes, which are released when the sperm contacts the egg prior to fertilization. The enzymes break down the outer layers of the egg to permit entry of the sperm. In some invertebrate sperms the acrosome contains actin filaments, which elongate to help gain entrance to the egg. In sea urchins molecules of the protein bindin on the surface of the acrosome attach to receptors on the vitelline envelope surrounding the egg. Binding occurs only between bindin proteins and receptors of the same species, thereby preventing attachment of sperm from a different species.

ACTH (adrenocorticotrophic hormone or adrenocorticotropic hormone; corticotrophin or corticotropin) A hormone produced by the anterior *pituitary gland that controls secretion of certain hormones (the *corticosteroids) by the adrenal glands. Its secretion, controlled by corticotrophin-releasing hormone from the hypothalamus, occurs in short bursts every few hours and is increased by stress. An analogue of ACTH, **tetracosactide**, is given by injection to test adrenal function.

actin A contractile protein found in muscle tissue, in which it occurs in the form of filaments (called thin filaments). Each thin filament consists of two chains of globular actin molecules, around which is twisted a strand of *tropomyosin and interspersed *troponin. Units of muscle fibre (see SARCOMERE) consist of actin and *myosin filaments, which interact to bring about muscle contraction (*see also* SLIDING FILAMENT THEORY). Actin is also found in the *microfilaments that form part of the *cytoskeleton of all cells.

Actinobacteria A phylum of Gram-positive mostly anaerobic nonmotile bacteria characterized by a high *GC content. Many species are fungus-like (hence the former names Actinomycetes and Actinomycota), with filamentous cells producing reproductive spores on aerial branches similar to the spores of certain moulds. The phylum includes bacteria of the genera *Actinomyces*, some species of which cause

disease in animals (including humans); and *Streptomyces*, which are a source of many important antibiotics (including streptomycin).

actinomorphy *See* RADIAL SYMMETRY.

Actinomycetes *See* ACTINOBACTERIA.

action potential A sudden, large, transient change in the *membrane potential of a cell (i.e. a depolarization) that is capable of propagating to other parts of the cell. It therefore can act as means of signalling to adjacent cells. A prime example is the passage of a nerve *impulse, when an action potential travels in a wavelike manner along the *axon of a nerve; it causes a localized and transient switch in electric potential across the membrane, typically from -70 mV (millivolts; the *resting potential) to $+40$ mV in a mammalian neuron. Essentially the change in electric potential is caused by an influx of sodium ions (Na^+) via voltage-gated ion channels, which makes the inside of the cell electrically positive with respect to the outside. Then the sodium channels close and voltage-gated potassium channels open, allowing the outflow of potassium ions (K^+). This has the effect of restoring the resting potential. Nervous stimulation of a muscle fibre also produces an action potential.

Plants too propagate action potentials, although these travel much more slowly than in neurons. They can transmit signals throughout the plant in response to a localized stimulus, such as the progressive folding of the leaflets of a sensitive plant (*Mimosa pudica*) in response to touching just a single leaflet. *Compare* GRADED POTENTIAL. *See also* MYELIN SHEATH.

action spectrum A graphical plot of the efficiency of electromagnetic radiation in producing a photochemical reaction against the wavelength of the radiation used. For example, the action spectrum for photosynthesis using light shows a peak in the region 670–700 nm. This corresponds to a maximum absorption in the absorption *spectrum of chlorophylls in this region.

activated sludge process A common *sewage and waste-water treatment. The sludge produced after primary treatment is pumped into aeration tanks, where it is continuously stirred and aerated, resulting in the formation of small aggregates of suspended colloidal organic matter called **floc**. Floc contains numerous slime-forming and nitrifying bacteria, as well as protozoans, which decompose organic substances in the sludge. Agitation or air injection maintains high levels of dissolved oxygen, which helps to reduce the *biochemical oxygen demand.

activation energy Symbol E_a. The minimum energy required for a chemical reaction to take place. In a reaction, the reactant molecules come together and chemical bonds are stretched, broken, and formed in producing the products. During this process the energy of the system increases to a maximum, then decreases to the energy of the products. The activation energy is the difference between the maximum energy and the energy of the reactants; i.e. it is the energy barrier that has to be overcome for the reaction to proceed. The activation energy determines the way in which the rate of the reaction varies with temperature. It is usual to express activation energies in joules per mole of reactants.

activator 1. A type of *transcription factor involved in assembling proteins to form an initiation complex at the *promoter of a gene in readiness for transcription. Activators can bind to an *enhancer site some distance upstream of the coding region and also to a promoter element near the start site, causing the DNA to form a loop. Some activators also direct remodelling of chromatin by catalysing the acetylation of histones required for unwinding of DNA. This provides access to other transcription factors as a prelude to transcription. *Compare* REPRESSOR. **2.** A substance that—by binding to an *allosteric site on an enzyme—enables the active site of the enzyme to bind to the substrate. **3.** Any compound that potentiates the activity of a drug or other foreign substance in the body.

active immunity *Immunity acquired due to the body's response to a foreign antigen.

active site (active centre) The site on the surface of an *enzyme molecule that binds and acts on the substrate molecule. The properties of an active site are determined by the three-dimensional arrangement of the polypeptide chains of the enzyme and their constituent amino acids. These govern the nature of the interaction that takes place and hence the degree of substrate specificity and susceptibility to *inhibition. Binding of the substrate involves the formation of weak bonds between it and the enzyme, which induces subtle changes in shape of the active site so it fits more tightly around the substrate—a so-called **induced fit** mechanism.

active transport The movement of substances through membranes in living cells, often against a *concentration gradient: a process requiring metabolic energy. Organic molecules and inorganic ions are transported into and out of both cells and their organelles. The substance binds to a *transport protein embedded in the membrane, which carries it through the membrane and releases it on the opposite side. **Primary active transport** uses energy from the hydrolysis of ATP directly, as in the sodium/potassium ATPase *sodium pump. **Secondary active transport** is driven by the movement of ions down a pre-existing concentration gradient established by a separate energy-consuming mechanism, as in *antiporters and *symporters. Active transport serves chiefly to maintain the normal balance of ions in cells, especially the concentration gradients of sodium and potassium ions crucial to the activity of nerve and muscle cells. *Compare* FACILITATED DIFFUSION.

active zone A relatively dense and thickened region of presynaptic membrane within a *synapse where synaptic vesicles fuse with the membrane to release neurotransmitter from the neuron into the synaptic cleft. Active zones contain specific proteins (*see* SNARE), are highly structured, and appear darker than neighbouring regions on electron microscopy.

actomyosin The complex formed from the interaction of the proteins *actin and *myosin during the process of muscle contraction. *See also* SLIDING FILAMENT THEORY.

acute-phase response A mechanism of innate immunity in which the liver markedly increases its synthesis of certain immune proteins, the **acute-phase proteins**, in response to infection. This change in protein output by liver cells is triggered by tumour necrosis factor α (TNF-α), interleukin-1 (IL-1), and IL-6, which are released by macrophages following their activation as a result of contact with bacteria or other pathogens. The main acute-phase proteins are *C-reactive protein and *mannose-binding lectin. These behave like antibodies but are nonspecific and bind to a much broader range of targets. Their binding sites attach to surface components of bacteria and fungi, making them more susceptible to ingestion by phagocytic cells (*see* OPSONIZATION); they also activate the *complement system of immune proteins, which initiates destruction of the targeted cell.

acyclovir *See* ACICLOVIR.

adaptation 1. (in evolution) Any change in the structure or functioning of successive generations of a population that makes it better suited to its environment. *Natural selection of heritable adaptations ultimately leads to the development of new species. Increasing adaptation of a species to a particular environment tends to diminish its ability to adapt to any sudden change in that environment. **2.** (in physiology) The alteration in the degree of sensitivity (either an increase or a decrease) of a sense organ to suit conditions more extreme than normally encountered. An example is the adjustment of the eye to vision in very bright or very dim light.

adaptive immunity *See* IMMUNITY.

adaptive radiation (divergent evolution) The evolution from one species of animals or plants of a number of different forms. As the original population increases in size it spreads out from its centre of origin to exploit new habitats and food sources. In time this results in a number of populations each adapted to its particular habitat: eventually these populations will differ from each other sufficiently to become new species. A good example of this process is the evolution of the Australian marsupials into species adapted as carnivores, herbivores, burrowers, fliers, etc. On a smaller scale, the adaptive radiation of the Galápagos finches provided Darwin with crucial evidence for his theory of evolution (*see* DARWIN'S FINCHES).

adaptive thermogenesis *See* THERMOGENESIS.

adaptor protein A protein that takes part in intracellular signalling pathways by recruiting specific components of the pathway in response to activation of the pathway's receptor molecule. Thus, a particular pathway may have its own specific adaptor(s), designed for key members of that pathway. In general, adaptors lack enzymatic activity themselves but possess at least two types of binding site: one type to bind to the cytoplasmic region of the activated receptor, and others to bind to specific intracellular molecules in such a way that they become activated, for example, through phosphorylation by tyrosine kinases associated with the receptor. *See* SIGNAL TRANSDUCTION.

adductor (depressor) A type of muscle whose function is to pull a limb inwards, towards the body of an animal. *Compare* ABDUCTOR.

adenine A *purine derivative. It is one of the major component bases of *nucleotides and the nucleic acids *DNA and *RNA.

adeno-associated virus (AAV) A member of the parvovirus family that is used as the basis for vectors in certain *gene therapies. A recombinant AAV particle lacking viral DNA is now favoured for delivering DNA for therapeutic applications. The DNA of interest is inserted between a suitable promoter and terminator, and all three sequences are flanked by inverted terminal repeat (ITR) sequences, needed for packaging in the virus particle and replication following entry into the host cell. There the DNA persists as an independently replicating *episome in the nucleus. The engineered virus contains no viral genes, so the likelihood of an immune response by the host is reduced. Also, the protein coat and promoters can be designed so that the virus targets specific tissues in which to deliver its DNA. Naturally occurring AAVs require a 'helper' virus, such as an adenovirus, to replicate. They are harmless to humans and stimulate little or no immune response, so that infected cells are not destroyed and the treatment is potentially long lasting. Also, they can infect both actively dividing and nondividing cells, enabling them to target a wide range of cell types, including the central nervous system, eye, and muscles. However, the small size of the viral genome restricts the size of the inserted gene to about 5 kbp. Patients with Parkinson's disease have been injected with AAVs into the brain to deliver a gene encoding glutamate decarboxylase, the enzyme that catalyses synthesis of the deficient neurotransmitter gamma-aminobutyric acid.

adenohypophysis See PITUITARY GLAND.

adenosine A nucleoside comprising one adenine molecule linked to a D-ribose sugar molecule. The phosphate-ester derivatives of adenosine, AMP, ADP, and *ATP, are of fundamental biological importance as carriers of chemical energy.

adenosine diphosphate (ADP) See ATP.

adenosine monophosphate (AMP) See ATP; CYCLIC AMP.

adenosine triphosphate See ATP.

adenovirus One of a group of DNA-containing viruses found in rodents, fowl, cattle, monkeys, and humans. In humans they produce acute respiratory-tract infections with symptoms resembling those of the common cold. They are also implicated in the formation of tumours (see ONCOGENIC).

adenylate 1. Any anion of adenylic acid (i.e. the phosphoric acid ester of adenosine). 2. Any salt or ester of adenylic acid, especially either of the nucleotides *ATP or ADP.

adenylate cyclase (adenylyl cyclase) The enzyme that catalyses the formation of *cyclic AMP. It is bound to the inner surface of the plasma membrane. Many hormones and other chemical messengers exert their physiological effects by increased synthesis of cyclic AMP through the activation of adenylate cyclase. The hormone binds to a receptor on the outer surface of the plasma membrane, which then activates adenylate cyclase on the inner surface via a *G protein or *calmodulin.

ADH See ANTIDIURETIC HORMONE.

adherens junction (zonula adherens) A type of cell junction, found especially in epithelial cells, that forms a strengthening and interlocking belt encircling the exterior of adjacent cells. It consists of a band of cadherin molecules (see CELL ADHESION MOLECULE) on the outside of the transverse cell surface, which are linked through the plasma membrane to a circumferential belt of actin microfilaments inside the cell. The cadherin bands of neighbouring cells are interlocked, thus contributing to the stability and integrity of the cell layer. *Compare* TIGHT JUNCTION.

adipocyte See FAT CELL.

adipokine (adipocytokine) Any of a group of *cytokines that are secreted primarily by *adipose tissue. They include *leptin, *tumour necrosis factor, *interleukin 6, and adiponectin; they have roles in regulating a range of bodily functions, including energy metabolism, inflammation, cardiovascular function, and fetal growth.

adipose tissue A body tissue comprising cells containing *fat and oil. It is found chiefly below the skin (see SUBCUTANEOUS TISSUE) and around major organs (such as the kidneys and heart), acting as an energy reserve, providing insulation and protection, and generating heat. Secretion of the hormone *leptin by adipose tissue regulates the amount of adipose tissue and adjusts the body's energy balance. *See* BROWN FAT; THERMOGENESIS.

adjuvant A nonantigenic substance (such as aluminium hydroxide) that, in combination with an antigen, enhances antibody production by inducing an inflammatory response, which leads to a local influx of antibody-forming cells. Adjuvants are used therapeutically in the preparation of vaccines, since they increase the production of antibodies against small quantities of antigen and lengthen the period of antibody production.

adolescence The period in human development that occurs during the teenage years, between the end of childhood and the start of adulthood, and is characterized by various physical and emotional changes associated with development of the reproductive system. It starts at **puberty**, when the reproductive organs begin to function, and is marked by the start of menstruation (*see* MENSTRUAL CYCLE) in females and the appearance of the *secondary sexual characteristics in both sexes. In males the secondary sexual characteristics are controlled by the hormone testosterone and include deepening of the voice due to larynx enlargement, the appearance of facial and pubic hair, rapid growth of the skeleton and muscle, and an increase in *sebaceous gland secretions. In females the secondary sexual characteristics are controlled by oestrogens and include growth of the breasts, broadening of the pelvis, redistribution of fat in the body, and appearance of pubic hair.

adoptive immunity Immunity produced by transferring lymphoid cells (e.g. T cells) from an immune donor to a genetically identical recipient. In cancer patients, for example, **adoptive immunization** is performed to boost immunity following intensive radiotherapy, using the patient's own T cells harvested prior to the treatment.

ADP *See* ATP.

adrenal cortex The outer layer of the *adrenal gland, in which several steroid hormones, the *corticosteroids, are produced.

adrenal glands A pair of endocrine glands situated immediately above the kidneys (hence they are also known as the **suprarenal glands**). The inner portion of the adrenals, the **medulla**, contains *chromaffin tissue and secretes the hormones *adrenaline and *noradrenaline; the outer **cortex** secretes small amounts of sex hormones (*androgens and *oestrogens) and various *corticosteroids, which have a wide range of effects on the body. *See also* ACTH.

adrenaline (epinephrine) A hormone (see formula), produced by the medulla of the *adrenal glands, that increases heart rate and *blood pressure, improves the power and prolongs the action of muscles, and increases the rate and depth of breathing to prepare the body for 'fright, flight, or fight' (*see* ALARM RESPONSE). At the same time it inhibits digestion and excretion and stimulates the mobilization of body fat (lipolysis) and glycogen stored in the liver, hence increasing the concentration of glucose in the blood for energy metabolism (glycolysis). Similar effects are produced by stimulation of the *sympathetic nervous system. Adrenaline causes these effects by binding to *adrenoceptors on target cells. It can be administered by injection to relieve bronchial asthma and reduce blood loss during surgery by constricting blood vessels. *See* ADRENOCEPTOR.

Adrenaline

adrenal medulla The inner part of the *adrenal gland, in which *adrenaline is produced.

adrenergic 1. Describing a cell, especially a neuron, or a cell *receptor that is stimulated by *adrenaline, *noradrenaline, or related substances. *See also* ADRENOCEPTOR.

2. Describing a nerve fibre or neuron that releases adrenaline or noradrenaline when stimulated. *Compare* CHOLINERGIC.

adrenoceptor (adrenoreceptor; adrenergic receptor) Any cell receptor that binds and is activated by the catecholamines adrenaline or noradrenaline. Adrenoceptors are therefore crucial in mediating the effects of catecholamines as neurotransmitters or hormones. There are two principal types of adrenoceptor, alpha (α) and beta (β), and various subtypes of each, with differing sensitivities to the catecholamines and to certain drugs. Adrenaline acts equally on both α- and β-adrenoceptors, whereas noradrenaline acts more strongly on α-receptors. All types are coupled to a *G protein, which initiates the signal transduction pathway inside the cell. The **alpha adrenoceptors** fall into two main subtypes: α_1-adrenoceptors, which mediate the contraction of smooth muscle and hence, for example, cause constriction of blood vessels due to contraction of their muscular walls; and α_2-adrenoceptors, which occur, for example, in presynaptic neurons at certain nerve synapses, where they inhibit release of noradrenaline from the neuron. As regards cell signalling, α_1-receptors activate phospholipase C, which causes the synthesis of second messengers, whereas α_2-receptors inhibit adenylyl cyclase, thereby decreasing levels of cyclic AMP. The **beta adrenoceptors** also have two main subtypes: β_1-adrenoceptors, which stimulate cardiac muscle causing a faster and stronger heartbeat; and β_2-adrenoceptors, which mediate relaxation of smooth muscle in blood vessels, bronchi, the uterus, bladder, and other organs. Both act through a G protein that stimulates adenylyl cyclase, thus increasing intracellular levels of cyclic AMP. Activation of β_2-adrenoceptors thus causes widening of the airways (bronchodilation) and blood vessels (vasodilation). *See also* BETA BLOCKER; SIGNAL TRANSDUCTION.

adrenocorticotrophic hormone *See* ACTH.

adsorption The formation of a layer of solid, liquid, or gas on the surface of a solid or, less frequently, of a liquid. There are two types depending on the nature of the forces involved. In **chemisorption** a single layer of molecules, atoms, or ions is attached to the adsorbent surface by chemical bonds. In **physisorption** adsorbed molecules are held by the weaker physical forces. The property is utilized in adsorption *chromatography.

adventitious Describing organs or other structures that arise in unusual positions. For example, ivy has adventitious roots growing from its stems.

aeciospore (aecidiospore) An asexual spore formed in the rust fungi (*see* RUSTS) from the fusion of cells. The nuclei from the two cells do not fuse and the spore is binucleate. Aeciospores develop in a sorus known as an **aecidium**.

aeolian soil A type of soil that is transported from one place to another by the wind.

aerobe *See* AEROBIC RESPIRATION.

aerobic respiration A type of cellular *respiration in which food is completely oxidized to carbon dioxide and water, with the release of chemical energy, in a process requiring atmospheric oxygen. The oxygen performs the role of terminal electron acceptor in the *electron transport chain, and the reaction can be summarized by the equation:

$$C_6H_{12}O_6 + 6O_2 \rightarrow 6CO_2 + 6H_2O + \text{energy}$$

The chemical energy released is stored mainly in the form of *ATP. Most organisms have aerobic respiration (i.e. they are **aerobes**); exceptions include certain bacteria, archaea, and yeasts. Using the *alternative respiratory pathway, many plants can respire aerobically without conserving energy as ATP.

aerotaxis *See* TAXIS.

aerotolerant Describing an anaerobic organism (*see* ANAEROBIC RESPIRATION) that is not harmed by exposure to oxygen, although it cannot use it for respiration. *Compare* OBLIGATE ANAEROBE.

aestivation 1. (in zoology) A state of inactivity occurring in some animals, notably lungfish, during prolonged periods of drought or heat. Feeding, respiration, movement, and other bodily activities are considerably slowed down. *See also* DORMANCY. *Compare* HIBERNATION. **2.** (in botany) The arrangement of the parts of a flower bud, especially of the sepals and petals.

aetiology The study of causes, especially the causes of medical conditions.

afferent Carrying (nerve impulses, blood, etc.) from the outer regions of a body or organ towards its centre. The term is usually applied to types of nerve fibres or blood vessels. *Compare* EFFERENT.

affinity (in cytology and histology) The tendency of a dye, marker, or other substance to bind to a particular cellular component or tissue. *Compare* AVIDITY.

affinity chromatography A type of *chromatography that is dependent on the affinity between specific molecules. The matrix, which can be packed into a column, is made from a material that is able to specifically bind to the molecule under investigation. Antibody purification can be achieved by this method when an appropriate antigen is bound to the matrix. In addition, components of protein complexes can be isolated and purified using this technique, for analysis by mass spectrometry.

aflatoxin Any of a family of toxic compounds produced by some fungi, particularly *Aspergillus flavus* and *A. parasiticus*, that can contaminate certain crops, such as maize and peanuts, under warm, humid conditions. Aflatoxins bind to DNA and prevent replication and transcription. They can cause acute liver damage and cancers: humans may be poisoned by eating stored peanuts and cereals contaminated with the mould or by eating meat and other products derived from animals that have fed on contaminated feed.

AFLP *See* AMPLIFIED FRAGMENT LENGTH POLYMORPHISM.

AFM *See* ATOMIC FORCE MICROSCOPY.

afterbirth The *placenta, *umbilical cord, and *extraembryonic membranes, which are expelled from the womb after a mammalian fetus is born. In most nonhuman mammals the afterbirth, which contains nutrients and might otherwise attract predators, is eaten by the female.

agamospecies A species of organism in which sexual reproduction does not occur, represented typically as a collection of clones. Examples include many bacteria and some plants and fungi. The absence of sexual reproduction means that the *biological species concept cannot be applied, and instead taxonomists must rely on identifying certain diagnostic traits and *molecular systematics to distinguish between closely related asexual lineages. Consequently, the boundaries of agamospecies are often hard to define. *See* TYPOLOGICAL SPECIES CONCEPT.

agamospermy *See* APOMIXIS.

agar An extract of certain species of red seaweeds that is used as a gelling agent in microbiological *culture media, foodstuffs, medicines, and cosmetic creams and jellies. **Nutrient agar** consists of a broth made from beef extract or blood that is gelled with agar and used for the cultivation of bacteria, fungi, and some algae. **Agarose** is a polysaccharide component of agar that in purified form is widely used to form a gel-like matrix in electrophoresis and chromatography. It consists of alternating units of D-galactose and anhydrogalactose.

ageing *See* SENESCENCE.

age polyethism The phenomenon whereby an animal shows different forms of behaviour at different ages. It is best known in colonies of social insects, where certain members of the colony may perform different tasks as they get older. For example, worker honeybees generally perform duties within the hive (such as brood care) up to the age of about 2–3 weeks and then switch to foraging outside the hive for nectar and pollen for the remaining weeks of their lives. The switch is controlled by social cues within the hive, such as pheromones, and is associated with an array of genetic, metabolic, and physiological changes in the insect.

agglutination The clumping together by antibodies of microscopic foreign particles, such as red blood cells or bacteria, so that they form a visible pellet-like precipitate. Agglutination is a specific reaction, i.e. a particular antigen will only clump in the presence of its specific antibody; it therefore provides a means of identifying unknown bacteria and determining *blood group. When blood of incompatible blood groups (e.g. group A and group B—*see* ABO SYSTEM) is mixed together agglutination of the red cells occurs (**haemagglutination**). This is due to the reaction between antibodies in the plasma (**agglutinins**) and *agglutinogens (antigens) on the surface of the red cells.

agglutinogen Any of the antigens that are present on the outer surface of red blood cells (erythrocytes). There are more than 100 different agglutinogens and they form the basis for identifying the different *blood groups. Antibodies in the plasma, known as **agglutinins**, react with the agglutinogens in blood of an incompatible blood group (*see* AGGLUTINATION).

aggressin A toxic substance that is secreted by certain parasitic microorganisms and inhibits the natural defence mechanisms of a host organism.

aggression Behaviour aimed at intimidating or injuring another animal of the same or a competing species, generally as a means of excluding rivals from a resource such as food, mates, or territory. Aggression between individuals of the same species often starts with a series of ritualized displays or contests that can end at any stage if one of the combatants withdraws, leaving the victor with access to the disputed resource or with increased social dominance (*see* DOMINANT). It is also often seen in *courtship. Aggression or threat displays usually appear to exaggerate the performer's size or strength; for example, many fish erect their fins and mammals and birds may erect hairs or feathers. Special markings may be prominently exhibited, and **intention movements** may be made: dogs bare their teeth, for example. Some animals have evolved special structures for use in aggressive interactions (e.g. antlers in deer) but these are seldom used to cause actual injury; the opponent usually flees first or adopts *appeasement postures. Fights 'to the death' are comparatively rare. *See* AGONISTIC BEHAVIOUR; DISPLAY BEHAVIOUR; RITUALIZATION.

agitator A bladelike instrument used in fermenters and *bioreactors to mix the medium continuously in order to maintain the rate of oxygen transfer and to help keep the cells in suspension.

agnathan Any jawless craniate animal. Living representatives are the lampreys and hagfishes. Morphological and molecular evidence now strongly indicates that hagfishes (class Myxini) and lampreys (class Petromyzontida) are sister groups of the same vertebrate clade, *Cyclostomata, with hagfishes deemed to have secondarily lost certain characteristic vertebrate features. Fossil agnathans, covered in an armour of bony plates, are the oldest known fossil vertebrates. They have been dated from the Silurian and Devonian periods, 440–345 million years ago. Fossil chordates resembling hagfishes have been found dating from much earlier times, notably *Myllokunmingia* of the Cambrian *Chengjiang fossils.

agonist A drug, hormone, neurotransmitter, or other signal molecule that forms a complex with a *receptor site, thereby triggering an active response from a cell. *Compare* ANTAGONIST.

agonistic behaviour Any form of behaviour associated with *aggression or submission, including threat, attack, *appeasement, flight, predation, or avoidance of predation. Agonistic behaviour between members of the same species generally involves a mixture of aggressive and submissive tendencies. It is often associated with defence of a territory; for example, a **threat display** by the defending individual is often met with an appeasement display from the intruder, thus avoiding harmful conflict. In contrast, predatory behaviour does not involve conflicting tendencies of aggression and submission: the predator is motivated by hunger, and the behaviour of potential prey is motivated by the imperative to survive.

agranulocyte Any white blood cell (*see* LEUCOCYTE) with a nongranular cytoplasm and a large spherical nucleus; *lymphocytes and *monocytes are examples. Agranulocytes are produced either in the lymphatic system or in the bone marrow and account for 30% of all leucocytes. *Compare* GRANULOCYTE.

agriculture The study and practice of cultivating land for the growing of crops and the rearing of livestock. The increasing demands for food production since the mid-20th century have seen many developments in agricultural technology and practices that have greatly increased crop and livestock production. However, these advances in modern **intensive farming** techniques have had their impact on the environment, particularly with increased use of *fertilizers and *pesticides, as well as greenhouse gas emissions from livestock and farming operations. Concerns about intensification, such as degradation of soil, harm to wildlife, and pesticide residues in food, have led to a growth in organic farming, which adheres to more 'natural' husbandry techniques. The now widespread practice of crop **monoculture** (in which one crop is grown densely over an extensive area) has required an increase in the use of *pesticides, as monoculture provides an ideal opportunity for crop pests. Monoculture also requires vast areas of land, which has meant that natural habitats have been destroyed. *Deforestation has resulted from the clearing of forests for crop production and cattle rearing. Food supply in many less-developed countries relies on **subsistence farming**, in which the crops and livestock produced are used solely to feed the farmer and his family. In such countries a system known as **slash and burn** is common, in which the vegetation in an area is cut down and then burnt, thus returning the minerals to the soil. The area can then be used for crop cultivation until the soil fertility drops, at which point it is then abandoned for a number of years and another site is cultivated. Concerns about greenhouse gas emissions have led to an increased use of arable land to grow *biofuel crops, often instead of food crops,

prompting further controversy about the ethics of farming practices.

The selective *breeding of crop plants and farm animals has had an enormous impact on productivity in agriculture. Modern varieties of crop plants have increased nutritional value and greater resistance to disease, while animals have been selectively bred to enhance their yields of milk, meat, and other products. Developments in genetic engineering have enabled the introduction to commercial cultivation of genetically modified crop plants, such as tomatoes and soya, which contain genes from other organisms to enhance crop growth, nutritional properties, or storage characteristics. Genetic modification can also confer resistance to herbicides, thereby allowing more effective weed control, as well as improved resistance to insects and other pests and to diseases. The application of similar technology to animal production is being researched. *See also* GENETICALLY MODIFIED ORGANISMS (Feature).

agrin A *heparan sulphate proteoglycan released by motor neurons and muscle cells that is crucial for the normal development or regeneration of neuromuscular junctions. As a growing axon makes contact with a muscle cell, the agrin acts as a signal to recruit essential components and start the formation of a synapse between the axon and muscle. Agrin binds to its receptor, called muscle-specific receptor tyrosine kinase (MuSK), in the muscle cell membrane and triggers a cascade of events that results in the accumulation of acetylcholine receptors beneath the muscle cell membrane. Binding of agrin to other receptors, such as *laminin β2, is also required to stabilize the newly forming synaptic structure. Agrin is equally important in the natural regeneration of damaged neuromuscular junctions and also seems to have a role in the differentiation of presynaptic nerve terminals.

Agrobacterium tumefaciens A Gram-negative soil bacterium that infects a wide range of plants and causes tumorous growths (*galls), especially at the root/stem junction (crown gall). It is of interest because the bacterial cells contain a *plasmid, the **Ti plasmid** (tumour-inducing plasmid), a segment of which is transferred to cells of the plant host. This T-DNA (transfer DNA) segment, which comprises the genes responsible for the tumorous growth, becomes integrated into the genome of infected plant cells. Possession of the Ti plasmid has made *A. tumefaciens* an important tool in genetic engineering for the introduction of foreign genes into plant tissue. The tumour-inducing genes are usually replaced with the gene of interest, and a marker gene (e.g. the antibiotic resistance gene) is added to enable selection of transformed cells. *See* GENETICALLY MODIFIED ORGANISMS.

AI *See* ARTIFICIAL INSEMINATION.

AIDS (acquired immune deficiency syndrome) A disease of humans characterized by defective cell-mediated *immunity and increased susceptibility to infections. It is caused by the retrovirus *HIV (human immunodeficiency virus). This infects and destroys *helper T cells, which are essential for combating infections. HIV is transmitted in blood, semen, and vaginal fluid; the major routes of infection are unprotected vaginal and anal intercourse, intravenous drug abuse, and the administration of contaminated blood and blood products. A person infected with HIV is described as **HIV-positive**; after the initial infection the virus can remain generally dormant for up to ten years before AIDS develops. **Antiretroviral therapy (ART)**, which involves a combination of *antiviral drugs drawn from several classes, including reverse transcriptase inhibitors (e.g. zidovudine, lamivudine), protease inhibitors, fusion and entry inhibitors, and integrase inhibitors, can enable HIV-positive individuals to lead long and healthy lives.

((⊕)) SEE WEB LINKS
• Latest initiatives to combat HIV and AIDS, with WHO technical reports and statistics

air bladder *See* SWIM BLADDER.

air pollution (atmospheric pollution) The release into the atmosphere of substances that cause a variety of harmful effects to the natural environment. Most air pollutants are gases that are released into the troposphere, which extends about 8 km above the surface of the earth. The burning of fossil fuels, for example in power stations, is a major source of air pollution as this process produces such gases as sulphur dioxide and carbon dioxide. Released into the atmosphere, carbon dioxide is the major contributor to the *greenhouse effect. Methane, derived from livestock and rice cultivation, is another significant greenhouse gas. Sulphur dioxide and nitrogen oxides, released in car exhaust fumes, are air pollutants that are responsible for the formation of *acid rain; nitrogen oxides also contribute to the formation of *photochemical smog. *See also* OZONE LAYER; POLLUTION.

air sac 1. Any one of a series of thin-walled sacs in birds that are connected to the lungs and increase the efficiency of ventilation. Some of the air sacs penetrate the internal cavities of bones. **2.** A structural extension to the *trachea in insects, which increases the surface area available for the exchange of oxygen and carbon dioxide in respiration.

akinete A nonmotile reproductive cell of certain filamentous cyanobacteria. An akinete is an enlarged resting cell with a thick wall and large amounts of food reserves and DNA. After cell division has occurred within the akinete the cell wall ruptures, releasing a filament of cells.

alanine *See* AMINO ACID.

alarm response (acute stress response; 'fight-or-flight' response) An immediate response to any stimulus that potentially threatens the wellbeing of an organism. It involves the release of *adrenaline and noradrenaline from the adrenal glands, triggered by increased sympathetic nervous activity. These hormones enhance the effects of the sympathetic nervous system (e.g. by increasing heart and breathing rates) and promote glycogen breakdown, which supplies large amounts of glucose for increased respiration and energy release. They also promote the release of glucose by liver cells and of fatty acids from fat cells. *Compare* RESISTANCE RESPONSE.

alarm signal A warning signal given by an animal to other members of its population in response to perceived danger, usually the approach of a predator. This is a form of *altruism, since the animal that has perceived the danger may waste valuable time in giving the signal (or attract the attention of the predator by doing this) and thus reduce its own chances of survival. For example, the alarm signal of a rabbit to a threatening situation involves thumping the ground and then flashing the white of its tail as it runs, which alerts the rabbits nearby.

albinism Hereditary lack of pigmentation (*see* MELANIN) in an organism. Albino animals and human beings have no colour in their skin, hair, or eyes (the irises appear pink from underlying blood vessels). The *allele responsible is *recessive to the allele for normal pigmentation.

albumen *See* ALBUMIN.

albumin One of a group of globular proteins that are soluble in water but form insoluble coagulates when heated. Albumins occur in egg white (the protein component of which is known as **albumen**), blood, milk, and plants. Serum albumins, which constitute about 55% of blood plasma protein, help regulate the osmotic pressure and hence plasma volume. They also bind and transport fatty acids. α-Lactalbumin is one of the proteins in milk.

albuminous cell *See* COMPANION CELL.

alburnum *See* SAPWOOD.

alcohol An organic compound that contains the –OH group bound to a carbon atom. In systematic chemical nomenclature alcohol names end in the suffix *-ol*. Examples are methanol, CH_3OH, and *ethanol, C_2H_5OH. Alcohols that have two –OH groups in their molecules are **diols** (or **dihydric alcohols**), those with three are **triols** (or **trihydric alcohols**), etc.

alcoholic fermentation *See* FERMENTATION.

aldohexose *See* MONOSACCHARIDE.

aldose *See* MONOSACCHARIDE.

aldosterone A hormone produced by the adrenal glands (*see* CORTICOSTEROID); it controls excretion of sodium by the kidneys and forms part of the renin–angiotensin–aldosterone system, which regulates blood volume and blood pressure (*see* RENIN). Aldosterone promotes the reabsorption of sodium ions from the collecting ducts and distal tubules of nephrons in the kidneys, thereby causing enhanced reabsorption of water. *See also* ANGIOTENSIN.

aleurone layer The outer layer of living cells of the endosperm of wheat and other cereal species. It is a single layer of cells that during germination synthesizes α-amylase and proteases. These enzymes are secreted into the starch-filled endosperm to break down the starch into maltose and glucose, and to hydrolyse proteins into amino acids, making them available for the developing embryo. The embryo secretes *gibberellins that diffuse into the aleurone layer, where they trigger the synthesis of the enzymes by switching on their respective genes.

alga (*pl.* **algae**) Any of various unrelated simple organisms that contain chlorophyll (and can therefore carry out photosynthesis) and live in aquatic habitats and in moist situations on land. The algal body may be unicellular or multicellular (filamentous, ribbon-like, or platelike). Molecular studies have confirmed that red and green algae are related to modern land plants in the supergroup Archaeplastida— *see* PLANT (sense 2)—whereas other algal groups, such as brown algae, are protists. Algae are assigned to separate groups based primarily on the composition of the cell wall, the nature of the stored food reserves, and the other photosynthetic pigments present, augmented increasingly by molecular systematics. The main phyla (or groups) are the Bacillariophyta (diatoms); Chlorophyta (most green algae); Chrysophyta (golden-brown algae); Phaeophyta (brown algae); and Rhodophyta (red algae). *See also* STREPTOPHYTE.

The organisms formerly known as blue-green algae are now classified as bacteria (*see* CYANOBACTERIA).

algal bloom The rapid increase in populations of algae and other phytoplankton, in particular cyanobacteria, that occurs in marine and fresh waters, particularly coastal waters and lakes. The density of the organisms may be such that it may prevent light from passing to lower depths in the water system. Algal blooms are often caused by an increase in levels of nitrate, a mineral ion essential for algal and bacterial growth. The source of increased nitrate may be from agricultural *fertilizers, which are leached into water systems from the land, or *sewage effluent. Algal blooms contribute to the eutrophication of water systems. *See also* EUTROPHIC; RED TIDE.

algin (alginic acid) A complex polysaccharide occurring in the cell walls of the brown algae (Phaeophyta). Algin strongly absorbs water to form a viscous gel. It is produced commercially from a variety of species of *Laminaria* and from *Macrocystis pyrifera* in the form of **alginates**, which are used mainly as stabilizers and texturing agents in the food industry.

alien (exotic) A species of organism that is not native to a locality, having been moved there from its natural range by humans or other agents. Some alien species, such as rats, are introduced mainly by accident in cargoes or transport vessels, while others are transferred intentionally, often for their ornamental or economic value. An alien that establishes a self-sustaining wild population is described as **naturalized**, whereas one that depends on continual introduction is termed a **casual**.

alignment (in bioinformatics) The process of matching up base sequences (e.g. of genes) or amino acid sequences (of proteins) to reveal similarities and differences between them. It enables researchers to compare, for example, a newly sequenced gene or protein fragment with well-characterized sequences and is a key step in identifying the nature, possible function, and evolutionary relationships of novel genes and proteins. Alignment is performed by any of various computer programs and makes use of the vast amount of sequence data stored on public databases, which can be accessed via the Internet. In a pairwise alignment, just two sequences are compared, whereas multiple sequence alignment compares three or more. The program compares the sequences and computes the best alignment(s), allowing for gaps and mismatches. There are two main types of sequence alignment—global and local. **Global alignment** extends over entire sequences of a gene, genomic region, or protein and tends to be used to highlight mismatches in otherwise similar sequences, such as between *homologous proteins. This can provide information about the evolutionary relationships of the organisms from which the sequences were obtained. **Local alignment** extends over relatively short sequences and pinpoints regions in which a novel gene or protein sequence (the query sequence) is similar to sequences of genes or proteins whose structure and function are well described. This can provide clues about structural features of the novel protein, such as DNA-binding or protein-binding domains, and hence its possible function. It also helps to identify similarities due to homology with proteins of other organisms. The distinction between local and global alignment strategies is arbitrary, and the two strategies are complementary.

The degree of similarity of alignments is assessed quantitatively by giving each match, mismatch, or gap a score. So, for example, positive scores may be given for matching bases at a given position, whereas mismatches or gaps are penalized with a negative score. Scoring of protein sequence alignments is complicated by the differences in physical and chemical properties of constituent amino acids. A mismatch involving substitution by an amino acid with similar properties (e.g. arginine for lysine) is unlikely to affect the functional properties of the protein and so is penalized very lightly, whereas one in which the mismatched amino acid has radically different properties, and hence is potentially of functional significance for the protein (e.g. alanine for tryptophan), is penalized heavily. The score for every possible mismatch is listed on a **scoring matrix** designed for use with the alignment tool. The overall score provides a means of assessing the biological significance of the alignment. If there is a very low probability that such an alignment score could be obtained purely by chance, then the two sequences are likely to share some meaningful property, such as a functional protein domain or sequence homology derived from a common ancestor. *See also* BLAST.

alimentary canal (digestive tract; gut) A tubular organ in animals that is divided into a series of zones specialized for the ingestion, *digestion, and *absorption of food and for the elimination of indigestible material (see illustration). In most animals the canal has two openings, the mouth (for

the intake of food) and the *anus (for the elimination of waste). Simple animals, such as cnidarians (e.g. *Hydra* and jellyfish) and flatworms, have only one opening to their alimentary canal, which must serve both functions.

alkali A *base that dissolves in water to give hydroxide ions (OH⁻).

alkaline phosphatase A membrane-bound glycoprotein enzyme that catalyses the hydrolysis of phosphoric acid esters under conditions of alkaline pH. In humans the blood levels of alkaline phosphatase are measured as part of the assessment of liver function; in the blood the enzyme also breaks down phosphates required for mineralization of bone. There are four distinct forms of the enzyme (*see* ISOZYME) in humans—intestinal, placental, germ cell, and liver/bone/kidney

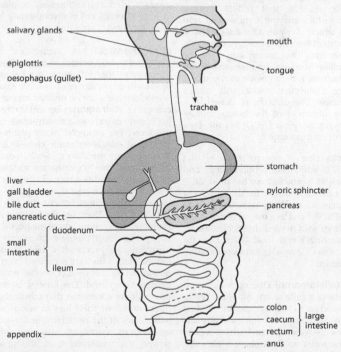

The human alimentary canal

alkaline phosphatases—each encoded by different genes.

alkaliphilic Describing organisms that grow best in alkaline conditions (i.e. with a pH above 7). Such organisms are called **alkaliphiles**. *Compare* ACIDOPHILIC.

alkaloid One of a group of nitrogenous organic compounds derived from plants and having diverse pharmacological properties. Alkaloids include morphine, cocaine, atropine, quinine, and caffeine, most of which are used in medicine as *analgesics or anaesthetics. Some alkaloids are poisonous, e.g. strychnine and coniine, and *colchicine inhibits cell division.

alkalosis A condition in which the body fluids become more alkaline, i.e. the pH is more than 7.4.

alkaptonuria (alcaptonuria) An inherited metabolic disorder that results from a deficiency of the enzyme homogentisic acid oxidase, which is required for the complete breakdown of the amino acids tyrosine and phenylalanine. The accumulation of the intermediate product, homogentisic acid, which imparts a dark colour to the urine, damages connective tissue and causes joint disease. The disorder is caused by a recessive mutation of the homogentisate 1,2-dioxygenase gene (*HGD*) on the long (q) arm of chromosome 3.

allantois One of the membranes that develops in embryonic reptiles, birds, and mammals as a growth from the hindgut. It acts as a urinary bladder for the storage of waste excretory products in the egg (in reptiles and birds) and as a means of providing the embryo with oxygen (in reptiles, birds, and mammals) and food (in mammals; *see* PLACENTA). *See also* EXTRAEMBRYONIC MEMBRANES.

allele (allelomorph) One of the alternative forms of a gene or any other particular site (*locus) on a chromosome. In a diploid cell there are usually two alleles of any one gene (one from each parent), which occupy the same locus on *homologous chromosomes. These alleles may be the same, or one allele may be *dominant to the other

(known as the *recessive), i.e. it determines which aspects of a particular characteristic the organism will display. Alternatively, neither allele may be completely dominant (*see* INCOMPLETE DOMINANCE), or both alleles may be expressed fully (*see* CODOMINANCE). Within a population there may be numerous alleles of a gene; each has a unique nucleotide sequence.

allele frequency (gene frequency) The occurrence of an *allele in a population in relation to all alleles of that gene at the same *locus, expressed as a fraction.

allelic exclusion The mechanism by which a diploid cell expresses only one of the alleles of a particular gene, and so synthesizes just one of the possible protein variants. It occurs in certain cells of the immune system. For example, mature B cells belonging to a particular clone all produce identical antibodies, despite the presence in each cell of several genes determining antibody structure.

allelochemical A substance (*see* SEMIOCHEMICAL) produced by members of one species that influences the behaviour or growth of members of another species. Allelochemicals can be divided into several categories. **Kairomones** benefit the receiving organism but cause disadvantage to the producer. For example, many plants (e.g. cabbages) release aromatic chemicals that attract insect predators, while parasites often exploit the *pheromones released by their hosts to locate a suitable host; certain insect predators detect their prey in a similar way. **Allomones** benefit the producer but have no effect on the receiver. For example, many members of the beetle family Lycidae emit pungent chemicals that warn potential predators of their distasteful nature. Hence they are protected from predation, while the impact on the potential predator is neutral. The flowers of certain orchids emit allomones that mimic the sex pheromones of their bee or wasp pollinator. Males of the respective insect species attempt to copulate with the orchid flower, and pollinate it in the process, thus benefiting the orchid, while the cost to the deceived male insect is minimal.

Synomones are beneficial to both producer and recipient. For example, pine trees damaged by beetles often emit terpenes that attract *parasitoid insects that parasitize the pest beetles. Hence the parasitoid finds a suitable host, and the tree's pests are controlled. *See also* ALLELOPATHY.

allelomorph *See* ALLELE.

allelopathy The secretion by plants of chemicals, such as phenolic and terpenoid compounds, that inhibit the growth or germination of other plants, with which they are competing. For example, the aromatic oils released by certain shrubs of the Californian chaparral pass into the soil and inhibit the growth of herbaceous species nearby. Some plants produce chemicals that are toxic to grazing herbivorous animals.

allergen An antigen that provokes an abnormal immune response. Common allergens include pollen and dust (*see* ALLERGY).

allergy A condition in which the body produces an abnormal immune response to certain *antigens (called **allergens**), which include dust, pollen, certain foods and drugs, or fur. In allergic individuals these substances, which in a normal person would be destroyed by antibodies, react with pre-existing antibodies or trigger responses in primed T cells. A common mechanism is binding of allergen to IgE on *mast cells. This causes the latter to secrete *histamine and other vasoactive agents, leading to inflammation and other characteristic symptoms of the allergy (e.g. asthma or hay fever). This response is a type of *hypersensitivity. *See also* ANAPHYLAXIS.

(⊕) SEE WEB LINKS

• Overview of common allergies, tests, and treatments, with illustrations of some typical symptoms

allochthonous Describing an organism that originates from a place other than that in which it is found. The organism is usually a transient member of a community. *Compare* AUTOCHTHONOUS.

allogamy Cross-fertilization in plants. *See* FERTILIZATION.

allogenic 1. Relating to or caused by a change in the environment or an individual organism brought about by some external factor. For example, the increased predation in a habitat caused by an immigrant predator would be described as allogenic. *Compare* AUTOGENIC. **2.** (**allogeneic**) Describing the variation that exists in the genotypes of different individuals, usually when these belong to the same species.

allograft *See* GRAFT.

allometric growth The regular and systematic pattern of growth such that the mass or size of any organ or part of a body can be expressed in relation to the total mass or size of the entire organism according to the allometric equation: $Y = bx^a$, where Y = mass of the organ, x = mass of the organism, a = growth coefficient of the organ, and b = a constant. *See also* KLEIBER'S LAW.

allomone *See* ALLELOCHEMICAL.

allopatric Describing or relating to groups of similar organisms that could interbreed but do not because they are geographically separated. *Compare* SYMPATRIC. *See* DICHOPATRIC SPECIATION; PERIPATRIC SPECIATION.

allopolyploid A *polyploid organism, usually a plant, that contains multiple sets of chromosomes derived from different species. Hybrids are usually sterile, because they do not have sets of *homologous chromosomes and therefore *pairing cannot take place. However, if doubling of the chromosome number occurs in a hybrid derived from two diploid ($2n$) species, the resulting tetraploid ($4n$) is a fertile plant. This type of tetraploid is known as an **allotetraploid**; as it contains two sets of homologous chromosomes, pairing and crossing over are now possible. Allopolyploids are of great importance to plant breeders as advantages possessed by different species can be combined. The species of wheat, *Triticum aestivum*, used to make bread is an **allohexaploid** ($6n$), possessing 42 chromosomes, which is six times the original haploid number (n) of 7. *See also* AMPHIDIPLOID. *Compare* AUTOPOLYPLOID.

all-or-none response A type of response that may be either complete and of full intensity or totally absent, depending on the strength of the stimulus; there is no partial response. For example, when the membrane potential of a nerve cell exceeds a certain threshold, the cell transmits a complete nervous impulse, but otherwise no impulse occurs; a stinging *thread cell of a cnidarian is either completely discharged or it is not.

allosteric enzyme An enzyme that has two structurally distinct forms, one of which is active and the other inactive. In the active form, the quaternary structure (*see* PROTEIN) of the enzyme is such that a substrate can interact with the enzyme at the active site (*see* ENZYME-SUBSTRATE COMPLEX). In the inactive form the conformation of the substrate-binding site is altered so that interaction with the substrate is not possible. The changes in the binding site are the result of regulatory molecules binding to a second site—the allosteric site—elsewhere on the enzyme. Allosteric enzymes tend to catalyse the initial step in a pathway leading to the synthesis of molecules. The end product of this synthesis can act as a feedback inhibitor (*see* INHIBITION) and the enzyme is converted

to the inactive form, thereby controlling the amount of product synthesized.

allosteric site A binding site on the surface of an enzyme other than the *active site. In noncompetitive *inhibition, binding of the inhibitor to an allosteric site inhibits the activity of the enzyme. In an *allosteric enzyme, the binding of a regulatory molecule to the allosteric site changes the overall shape of the enzyme, either enabling the substrate to bind to the active site or preventing the binding of the substrate.

allozyme Any one of a number of different forms of the same enzyme that are coded by different alleles at the same locus.

alpha adrenoceptor (alpha adrenergic receptor) *See* ADRENOCEPTOR.

alpha diversity (alpha richness) (in ecology) The diversity of species (*see* BIODIVERSITY) within a particular habitat or community. The alpha diversity of component communities combined with the intercommunity *beta diversity gives a measure of regional species diversity, or *gamma diversity.

alpha helix The most common form of secondary structure in *proteins, in which the polypeptide chain is coiled into a helix.

••• hydrogen bond
Ⓡ = amino-acid side chain

Alpha helix

The helical structure is held in place by weak hydrogen bonds between the N–H and C=O groups in successive turns of the helix (see illustration). *Compare* BETA SHEET.

alpha-naphthol test A biochemical test to detect the presence of carbohydrates in solution, also known as **Molisch's test** (after the Austrian chemist H. Molisch (1856–1937), who devised it). A small amount of alcoholic alpha-naphthol is mixed with the test solution and concentrated sulphuric acid is poured slowly down the side of the test tube. A positive reaction is indicated by the formation of a violet ring at the junction of the two liquids.

alternation of generations The occurrence within the *life cycle of an organism of two or more distinct forms (generations), which differ from each other in appearance, habit, and method of reproduction. The phenomenon occurs in some protists and simple multicellular organisms, certain lower animals (e.g. cnidarians and parasitic flatworms), and in plants. The malaria parasite (*Plasmodium*), for example, has a complex life cycle involving the alternation of sexually and asexually reproducing generations. In plants the generation with sexual reproduction is called the *gametophyte and the asexual generation is the *sporophyte, either of which may dominate the life cycle, and there is also alternation of the haploid and diploid states. Thus in vascular plants the dominant plant is the diploid sporophyte; It produces spores that germinate into small haploid gametophytes. In mosses the gametophyte is the dominant plant and the sporophyte is the spore-bearing capsule. *See* INTERPOLATION HYPOTHESIS; TRANSFORMATION HYPOTHESIS.

alternative respiratory pathway A pathway for cellular respiration occurring in many plants as well as some fungi and protists that permits electron transport and reduction of oxygen to water with greatly reduced ATP formation. It involves activity of the **alternative oxidase (AOX)** enzyme in the *electron transport chain (ETC) in mitochondria. This effectively uncouples electron flow through the ETC from ATP

production and allows the cell to regulate the balance between energy generation and utilization. The alternative pathway also has important roles in stress tolerance and in protecting the photosynthetic apparatus. It is not affected by cyanide and other substances that completely inhibit respiration in animals. Alternative oxidase catalyses the oxidation of reduced ubiquinone and transfers electrons directly to oxygen; in so doing it bypasses major sites for proton pumping into the mitochondrial intermembrane space and hence ATP formation by phosphorylation. Thus, instead of being conserved as ATP, energy flowing through the alternative pathway (in the form of electrons) is converted mainly to heat. This provides a mechanism for 'burning off' energy—produced by photosynthesis in the form of carbohydrate—that is temporarily more than the plant can cope with. Some plants exploit this heat-generating system. For example, before pollination the spadix of the skunk cabbage (*Sympocarpus foetidus*) undergoes a temperature rise of about 10°C, which causes the emission of volatile chemicals that attract insect pollinators. The expression of genes encoding alternative oxidase and its enzymatic activity are regulated by numerous factors, including stresses such as drought, cold, salinity, and infection by pathogens. *See also* UNCOUPLING PROTEIN.

alternative splicing The splicing of the RNA transcript of a gene in different ways by the cellular machinery during *RNA processing to create distinct mature messenger RNAs (mRNAs) encoding different proteins. Hence, the same gene can give rise to related but variant forms of a protein at different stages of development or in different tissues. The mechanism involves small RNAs and proteins contributing to the assembly of a *spliceosome at a splice site in the pre-mRNA, which enables splicing to proceed. Selection of splice sites is determined by other proteins and ribonucleoproteins that bind to specific enhancer or silencer sites in the pre-mRNA. In this way, exons can be included or excluded to produce mature mRNAs with different lengths and coding sequences. For example, the gene encoding α-tropomyosin contains 14 exons; alternative splicing in different cell types yields at least nine variants of the protein. About 90% of human protein-coding genes undergo alternative splicing, which accounts for the disparity between the relatively small number of such genes identified, roughly 20 000, and the potentially large number of different gene transcripts, although estimates of this vary widely, from 30 000 to more than 100 000.

altricial species Any species of bird in which hatching occurs at an early stage of development. The offspring of altricial species tend to be born with no feathers and remain in the nest for a comparatively long period of time; they are described as **nidicolous**. *Compare* PRECOCIAL SPECIES.

altruism Behaviour by an animal that decreases its chances of survival or reproduction while increasing those of another member of the same species. For example, a lapwing puts itself at risk by luring a predator away from the nest through feigning injury, but by so doing saves its offspring. Altruism in its biological sense does not imply any conscious benevolence on the part of the performer. Altruism can evolve through *kin selection, if the recipients of altruistic acts tend on average to be more closely related to the altruist than the population as a whole. Some animals perform altruistic acts in the expectation that the 'favour' will be returned in the future. This **reciprocal altruism** often involves unrelated individuals; it is a complex survival strategy that is practised most notably by humans. *See also* ALARM SIGNAL; INCLUSIVE FITNESS.

Alu element (Alu sequence) Any of a group of closely related DNA sequences arising from a single ancestral sequence (*see* GENE FAMILY) and found dispersed and repeated many times within the genome of humans and other primates. Full-length Alu sequences are about 280 bp long, and many are cleaved by the restriction enzyme *Alu*I (hence the name). The diploid human genome contains over 1 million copies of Alu, accounting for over 10% of DNA and constituting a large proportion of the class of moderately *repetitive DNA known as *SINEs. Alu sequences occur in the nontranslated regions (*introns) of genes and between genes and probably have no function. They are *retrotransposons and can thus replicate themselves. The variations in structure and distribution of Alu sequences among primates can be used both to help track the evolutionary history of the genome and as genetic markers (*see* MARKER GENE) for physical mapping.

Alvarez event The collision of a giant meteorite with the earth 66 million years ago that caused catastrophic changes to the earth's climate and environment and a *mass extinction of species, including the dinosaurs. This hypothesis was advanced in 1980 by the US physicist Luis Walter Alvarez (1911–88) and his geologist son Walter Jr, based on the unusually high concentration of the element iridium in a thin layer of clay deposited at the end of the Cretaceous (*see* IRIDIUM ANOMALY). This clay marks the boundary between the Cretaceous period and the more recent Palaeogene (the so-called **K-T boundary**). Subsequently, geologists discovered a possible impact crater, roughly 160 km in diameter, along the coast of eastern Mexico, and other evidence has tended to support the hypothesis. Such a collision would have produced a massive tidal wave and fireball and sent a vast cloud of rock and other debris into the atmosphere. The resulting upheaval in the climate is estimated to have caused the extinction of some 75% of all species.

(⊕) SEE WEB LINKS

• Resources detailing the evidence and main theories regarding mass extinctions and the Alvarez event

alveolates A clade of unicellular eukaryotic organisms, sometimes placed in the phylum Alveolata, that are believed to have common ancestry based on certain morphological similarities and molecular systematics. They comprise the ciliates (*see* CILIOPHORA), the apicomplexans (*see* APICOMPLEXA), and the dinoflagellates (*see* DINOMASTIGOTA). In spite of very diverse lifestyles, they all have flattened vesicles (alveoli) supporting the plasma membrane and forming a flexible *pellicle. The alveolates, along with the *stramenopiles and *rhizarians, are now tentatively regarded as comprising the *SAR supergroup.

alveolus (*pl.* **alveoli**) **1.** The tiny air sac in the *lung of mammals and reptiles at the end of each *bronchiole. It is lined by a delicate moist membrane, has many blood capillaries, and is the site of exchange of respiratory gases (carbon dioxide and oxygen). **2.** The socket in the jawbone in which a tooth is rooted by means of the *periodontal membrane.

Alzheimer's disease A neurological disease characterized by progressive loss of intellectual ability. The disease, which is named after German physician Alois Alzheimer (1864–1915), is associated with general shrinkage of the brain tissue, with deposits of β-*amyloid protein and abnormal filaments composed of tau protein in the brain, and changes in the neurotransmitter systems within the brain that include a loss in the activity of *cholinergic neurons. Some inherited forms are associated with a genetic locus on chromosome 21.

amacrine cell A type of nerve cell found in the *retina of the eye. Amacrine cells receive sensory information from the rod and cone receptor cells in the retina and are able to integrate sensory information before sending this to the brain.

amber A yellow or reddish-brown fossil resin. The resin was exuded by certain trees and other plants and often contains preserved insects, flowers, or leaves that were trapped by its sticky surface before the resin hardened. Amber is used for jewellery and ornaments. It also has the property of acquiring an electrical charge when rubbed (the term electricity is derived from *electron*, the Greek name for amber). It occurs throughout the world in rock strata

from the Cretaceous to the Pleistocene, but most commonly in Cretaceous and Palaeogene rocks.

ambient Describing the conditions and factors that are present in the immediate environment. For example, the ambient temperature is the local temperature in a specific environment.

ameloblast Any of the epithelial cells that secrete *enamel during tooth formation. They die before the tooth erupts, therefore damaged enamel in a tooth cannot be replaced. *Compare* ODONTOBLAST.

amensalism An association between two species that is detrimental to one of the species but has no effect on the other. A common example of amensalism is the release of chemical toxins by plants that can inhibit the growth of other plant species (*see* ALLELOPATHY).

Ames test (*Salmonella* **mutagenesis test**) A test to determine the effects of a chemical on the rate of mutation in bacterial cells, and hence its likely potential for causing cancer in other organisms, including humans. Devised by US biologist Bruce Ames (1928–), it is widely used in screening chemicals occurring in the environment for possible carcinogenic activity (*see* XENOBIOTIC). The chemical is applied to plates inoculated with a special mutant strain of bacteria, usually *Salmonella typhimurium*, that require the amino acid histidine for growth. Cells that mutate back to the wild type are detected by the occurrence of colonies able to synthesize their own histidine and therefore to grow on the medium.

ametabolous Describing insect development in which there is no metamorphosis and immature stages appear very similar to the adults, except that they lack genitalia. It occurs, for example, in silverfish. *Compare* HEMIMETABOLOUS; HOLOMETABOLOUS.

amine Any one of a group of organic compounds derived by replacing one or more of the hydrogen atoms in ammonia by organic groups. **Primary amines** have one hydrogen replaced, e.g. methylamine, CH_3NH_2. They contain the functional group $-NH_2$ (the **amino group**). **Secondary amines** have two hydrogens replaced, e.g. methylethylamine, $CH_3(C_2H_5)NH$. **Tertiary amines** have all three hydrogens replaced, e.g. *trimethylamine. Amines are produced by the decomposition of organic matter.

amino acid Any of a group of water-soluble organic compounds that possess both a carboxyl (-COOH) and an amino ($-NH_2$) group attached to the same carbon atom, called the α-carbon atom. Amino acids can be represented by the general formula $R-CH(NH_2)COOH$. R may be hydrogen or an organic group, which may be nonpolar, basic, acidic, or polar; the nature of the R group determines the properties of any particular amino acid. Through the formation of peptide bonds, amino acids join together to form short chains (*peptides) or much longer chains (*polypeptides). Proteins are composed of various proportions of about 20 commonly occurring amino acids (*see* table). The sequence of these amino acids in the protein polypeptides determines the shape, properties, and hence biological role of the protein. Some amino acids that never occur in proteins are nevertheless important, e.g. *ornithine and citrulline, which are intermediates in the urea cycle. (*See* table.)

Plants and many microorganisms can synthesize amino acids from simple inorganic compounds, but animals rely on adequate supplies in their diet. The *essential amino acids must be present in the diet whereas others can be manufactured from them.

(⊕) SEE WEB LINKS

• Depicts molecular structures of all the amino acids

aminopeptidase Any enzyme that cleaves amino acids from the N-terminus of peptides or polypeptides. For example, membrane-bound aminopeptidases in the small intestine break down peptides and dipeptides into amino acids.

amino sugar (aminodeoxysugar) Any sugar containing an amino group in place of a hydroxyl group. The **hexosamines** are amino derivatives of hexose sugars and include **glucosamine** (based on glucose) and **galactosamine** (based on galactose). The former is a constituent of *chitin and the latter occurs in cartilage.

ammocoete The larva of a lamprey (*see* CYCLOSTOMATA). Ammocoetes have a slender elongate body, typically reaching 150–70 mm long, with nonfunctional eyes, a series of gill slits behind the head on either side, a horseshoe-shaped mouth with a prominent upper lip called the oral hood, and one or two shallow dorsal fins that merge into the caudal fin at the tail. They live in soft silty beds of rivers and streams for several years, filter feeding on microscopic particles by means of a network of tentacle-like cirri in the mouth entrance. Metamorphosis into the adult lamprey may be delayed, sometimes for many years.

ammonia A colourless gas, NH_3, with a strong pungent odour. Ammonia is produced by the *deamination of excess amino acids in the liver. Industrially it is made from its constituent elements by the Haber process for use in the manufacture of nitric acid, ammonium nitrate, ammonium phosphate, and urea (the last three as fertilizers), explosives, dyestuffs, and resins. The participation of ammonia in the *nitrogen cycle is a most important natural process. By means of *nitrogenase enzymes, nitrogen-fixing bacteria are able to achieve similar reactions to those of the Haber process, but under normal conditions of temperature and pressure. The reactions release ammonium ions, which are converted by nitrifying bacteria into nitrite and nitrate ions.

ammonite An extinct aquatic mollusc of the class *Cephalopoda. Ammonites were abundant in the Mesozoic era (225–65 mya) and are commonly found as fossils in rock strata of that time, being used as *index fossils for the Jurassic period. They were characterized by a coiled shell divided

amino acid	3-letter	1-letter	formula
alanine	Ala	A	$CH_3-\overset{\overset{H}{|}}{\underset{\underset{NH_2}{|}}{C}}-COOH$
arginine	Arg	R	$H_2N-\overset{}{\underset{\underset{NH}{\|}}{C}}-NH-CH_2-CH_2-CH_2-\overset{\overset{H}{\|}}{\underset{\underset{NH_2}{\|}}{C}}-COOH$
asparagine	Asn	N	$H_2N-\overset{}{\underset{\underset{O}{\|}}{C}}-CH_2-\overset{\overset{H}{\|}}{\underset{\underset{NH_2}{\|}}{C}}-COOH$
aspartic acid	Asp	D	$HOOC-CH_2-\overset{\overset{H}{\|}}{\underset{\underset{NH_2}{\|}}{C}}-COOH$
cysteine	Cys	C	$HS-CH_2-\overset{\overset{H}{\|}}{\underset{\underset{NH_2}{\|}}{C}}-COOH$
glutamic acid	Glu	E	$HOOC-CH_2-CH_2-\overset{\overset{H}{\|}}{\underset{\underset{NH_2}{\|}}{C}}-COOH$
glutamine	Gln	Q	$H_2N-\overset{}{\underset{\underset{O}{\|\|}}{C}}-CH_2-CH_2-\overset{\overset{H}{\|}}{\underset{\underset{NH_2}{\|}}{C}}-COOH$
glycine	Gly	G	$H-\overset{\overset{H}{\|}}{\underset{\underset{NH_2}{\|}}{C}}-COOH$
*histidine	His	H	$HC=C-CH_2-\overset{\overset{H}{\|}}{\underset{\underset{NH_2}{\|}}{C}}-COOH$ (imidazole ring: N, NH, C–H)
*isoleucine	Ile	I	$CH_3-CH_2-\overset{}{\underset{\underset{CH_3}{\|}}{CH}}-\overset{\overset{H}{\|}}{\underset{\underset{NH_2}{\|}}{C}}-COOH$
*leucine	Leu	L	$(H_3C)_2CH-CH_2-\overset{\overset{H}{\|}}{\underset{\underset{NH_2}{\|}}{C}}-COOH$
*lysine	Lys	K	$H_2N-CH_2-CH_2-CH_2-CH_2-\overset{\overset{H}{\|}}{\underset{\underset{NH_2}{\|}}{C}}-COOH$

(*continued*)

Continued

amino acid	abbreviation 3-letter 1-letter		formula
*methionine	Met	M	
*phenylalanine	Phe	F	
proline	Pro	P	
serine	Ser	S	
*threonine	Thr	T	
*tryptophan	Trp	W	
*tyrosine	Tyr	Y	
*valine	Val	V	

*an essential amino acid
The amino acids occurring in proteins

into many chambers, which acted as a buoyancy aid. The external suture lines on these shells increased in complexity with the advance of the group.

ammonotelic Describing animals that excrete nitrogenous waste in the form of *ammonia. Most aquatic animals are ammonotelic. *Compare* UREOTELIC; URICOTELIC.

amniocentesis The taking of a sample of amniotic fluid from a pregnant woman to determine the condition of an unborn baby. A hollow needle is inserted through the woman's abdomen and wall of the uterus and the fluid drawn off. Chemical and microscopical examination of cells shed from the embryo's skin into the fluid are used to detect spina bifida, *Down's syndrome, or other serious biochemical or chromosomal abnormalities.

amnion A membrane that encloses the embryo of reptiles, birds, and mammals within the **amniotic cavity**. This cavity is filled with **amniotic fluid**, in which the embryo is protected from desiccation and from external pressure. *See also* EXTRAEMBRYONIC MEMBRANES.

amniote A vertebrate whose embryos are totally enclosed in a fluid-filled sac—the *amnion. The evolution of the amnion provided the necessary fluid environment for the developing embryo and therefore allowed animals to breed away from water. Amniotes comprise the reptiles, birds, and mammals. *Compare* ANAMNIOTE.

amniotic egg The type of egg produced by reptiles, birds, and prototherian (egg-laying) mammals (*amniotes), in which the embryo develops inside an *amnion. The shell of the egg is either calcium-based or leathery.

Amoeba A genus of protists (*see also* PROTOZOA) formerly placed in the phylum Rhizopoda but now, on the basis of molecular systematics, classified as *amoebozoans. *Amoeba* spp. have lobe-shaped *pseudopodia, which are used for locomotion and feeding and result in a constantly changing body shape (*see* AMOEBOID MOVEMENT). Most species are free-living in soil, mud, or water, where they feed on smaller protists and other single-celled organisms, but a few are parasitic. The best-known species is the much studied *A. proteus.*

amoebocyte An animal cell whose location is not fixed and is therefore able to wander through the body tissues. Amoebocytes are named after their resemblance, especially in their movement, to amoebas (*see* AMOEBOID MOVEMENT), and they feed on foreign particles (including invading bacteria).

They occur, for example, in sponges and mammalian blood (e.g. some *leucocytes).

amoeboid movement The mechanism of movement demonstrated by *Amoeba* spp. and other cells that are capable of changing their shape (e.g. *phagocytes). The cytoplasm of *Amoeba* spp. consists of a central fluid plasmasol surrounded by a more viscous plasmagel. The plasmagel is converted to plasmasol, which slides towards the front of the cell, forming a *pseudopodium and propelling the cell forward. On reaching the tip of the pseudopodium, this plasmasol is reconverted into plasmagel; at the same time the plasmagel at the rear of the cell is converted into plasmasol and streams forward, thus maintaining continuous movement.

Amoeboid movement is brought about by reversible changes in the *actin filaments of the cell's cytoskeleton. Cross-linking of these filaments by other proteins creates a three-dimensional network with gel-like properties in the plasmagel region. Disassembly of this network causes reversion to the sol state of plasmasol.

Amoebomastigota *See* DISCOMITO-CHONDRIA.

amoebozoans An assemblage of eukaryotic protists, based largely on molecular systematics, that includes the plasmodial (or true) *slime moulds, the dictyostelid cellular slime moulds, the lobose amoebas (e.g. *Amoeba* spp.), and archamoebas, such as the pelobionts (which lack mitochondria and many other features of eukaryotic cells apart from a membrane-bounded nucleus) *See also* UNIKONTS.

amount of substance Symbol *n*. A measure of the number of entities present in a substance. The specified entity may be an atom, molecule, ion, electron, photon, etc., or any specified group of such entities. The amount of substance of an element, for example, is proportional to the number of atoms present. The SI unit of amount of substance is the *mole.

AMP *See* ATP; CYCLIC AMP.

AMPA receptors *See* GLUTAMATE RECEPTOR.

amphetamine A drug, 1-phenyl-2-aminopropane (or a derivative of this compound), that stimulates the central nervous system by causing the release of the transmitters noradrenaline and dopamine from nerve endings. It inhibits sleep, suppresses the appetite, and has variable effects on mood; prolonged use can lead to addiction.

Amphibia The class of vertebrate chordates (*see* CHORDATA) that contains over 7000 known species of frogs, toads, newts, and salamanders. The amphibians evolved in the Devonian period (about 370 million years ago) as the first vertebrates to occupy the land, and many of their characteristics are adaptations to terrestrial life. All adult amphibians have a passage linking the roof of the mouth with the nostrils so they may breathe air and keep the mouth closed. The moist scaleless skin is used to supplement the lungs in gas exchange. They have no diaphragm, and therefore the muscles of the mouth and pharynx provide the pumping action for breathing. Fertilization is usually external and the eggs are soft and prone to desiccation, therefore reproduction commonly occurs in water. Amphibian larvae are aquatic, having gills for respiration; they undergo metamorphosis to the adult form. In recent decades many amphibian populations around the world have suffered sharp declines or even extinction. The causes are often unknown, but pesticide pollution, habitat loss, and climate change are among those implicated.

((⊕)) SEE WEB LINKS
- Hosted by the University of California, Berkeley; gives free access to information about amphibian biology and conservation

amphibolic pathway A biochemical pathway that serves both anabolic and catabolic processes. An important example of an amphibolic pathway is the *Krebs cycle, which involves both the catabolism of carbohydrates and fatty acids and the synthesis of anabolic precursors for amino-acid synthesis (e.g. α-ketogluturate and oxaloacetate).

amphidiploid Describing an organism, cell, or nucleus that contains diploid sets of chromosomes originating from two different species. Crosses between taxonomically unrelated organisms are usually infertile, principally because the chromosomes lack a partner with which to pair during meiosis. However, if there is doubling of the parental sets of chromosomes, *pairing can take place within each set, and meiosis may proceed to produce fertile gametes. For example, the F_1 hybrid resulting from a cross between a cabbage (*Brassica* sp.) and a radish (*Raphanus* sp.) is sterile. Yet such crosses may produce a few seeds that are capable of germinating into fertile F_2 plants. These seeds form following the chance fusion of two unreduced gametes, each of which contain all parental chromosomes. Hence, instead of the 18 chromosomes of the F_1 hybrid (nine from each parent), the F_2 has 36 chromosomes in its somatic nuclei, and thus contains the full diploid sets of both its progenitor species. During meiosis the chromosomes undergo pairing just like a normal diploid plant, and the result is a generally true-breeding hybrid, named *Raphanobrassica*. See also ALLOPOLYPLOID.

amphimixis True sexual reproduction, involving the fusion of male and female gametes and the formation of a zygote. *Compare* APOMIXIS.

amphioxus Another name for the lancelet: *see* CEPHALOCHORDATA.

amphipathic Describing a molecule that contains both polar (i.e. hydrophilic) and nonpolar (i.e. hydrophobic) parts. Many biomolecules are amphipathic, including glycolipids, phospholipids, and cholesterol. They tend to orientate with their hydrophilic part in an aqueous environment and their hydrophobic part shielded from the water, as in *micelles and *lipid bilayers, whose properties are a fundamental feature of cell membranes.

amphistylic jaw suspension A type of jaw suspension seen in certain sharks, in which the upper jaw is braced against the cranium and is also supported by the

cranium

upper jaw

hyomandibular

lower jaw

Amphistylic jaw suspension

hyomandibular (*see* HYOID ARCH). See illustration. *Compare* AUTOSTYLIC JAW SUSPENSION; HYOSTYLIC JAW SUSPENSION.

amphoteric Describing a compound that can act as both an acid and a base. Amino acids, which contain both acidic and basic groups in their molecules, can be described as amphoteric. Solvents, such as water, that can both donate and accept protons are usually described as **amphiprotic** (*see* SOLVENT).

amplified fragment length polymorphism (AFLP) A method of *DNA fingerprinting that is especially useful for analysing samples of DNA where there is little or no knowledge of the organism's genome. It is often employed to determine if two organisms, e.g. two bacterial samples, belong to the same species. Other uses include tracking genetic variation in populations and forensic analysis of human DNA. Essentially the DNA sample is digested with particular restriction enzymes (*see* RESTRICTION FRAGMENT LENGTH POLYMORPHISM) and the resultant restriction fragments are incubated so the ends join to specially designed adaptor sequences. A selected subset of the DNA fragments then undergoes amplification via the *polymerase chain reaction, and the amplified fragments are separated by *gel electrophoresis according to their size and electrical charge. This creates a pattern of bands characteristic for the original DNA sample.

ampulla (*pl.* **ampullae**) **1.** An enlargement at one end of each of the *semicircular canals of the inner ear. Each ampulla contains a group of receptors—sensory *hair cells—embedded in a gelatinous cap (**cupula**), which detects movement in one particular dimension, corresponding to the plane of the canal. Movement of the head causes the cupula (and the hairs within it) to bend in a direction opposite to that of the head movement (see illustration); this stimulates nerve impulses in the receptors, which are interpreted by the brain as movement in a

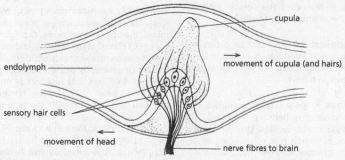

cupula

movement of cupula (and hairs)

endolymph

sensory hair cells

movement of head

nerve fibres to brain

Ampulla

particular dimension. **2.** Any small vesicle or saclike process. **3.** (**ampulla of Lorenzini**) *See* ELECTRORECEPTOR.

amylase Any of a group of closely related enzymes that degrade starch, glycogen, and other polysaccharides. Plants contain both α- and β-amylases; the name **diastase** is given to the component of malt containing β-amylase, important in the brewing industry. Animals possess only α-amylases, found in pancreatic juice (as **pancreatic amylase**) and also (in humans and some other species) in saliva (as **salivary amylase** or **ptyalin**). Amylases cleave the *glycosidic bonds of the long polysaccharide chains, producing a mixture of glucose and maltose.

amyloid Tissue consisting of protein fibrils that may accumulate between cells in various animal tissues, especially in the disorder **amyloidosis**. Amyloid deposits are insoluble and can exert pressure on various vital organs. These deposits are generally detected by staining with the dye Congo red. In **AL amyloidosis**, abnormal plasma cells secrete excessive amounts of antibody proteins called light chains. These are present in the blood and accumulate in tissues, especially the kidney, where they aggregate to form fibrils. A build-up of amyloid tissue in the brain is also a feature of *Alzheimer's disease, *Creutzfeldt-Jakob disease, and *bovine spongiform encephalopathy.

amylopectin A *polysaccharide comprising highly branched chains of glucose molecules. It is one of the constituents (the other being amylose) of *starch.

amyloplast An organelle in plants that stores starch. Amyloplasts are often found in nonphotosynthetic tissue, such as roots and storage tubers.

amylose A *polysaccharide consisting of linear chains of between 100 and 1000 linked glucose molecules. Amylose is a constituent of *starch (the other being amylopectin). In water, amylose reacts with iodine to give a characteristic blue colour.

anabolic steroid Any steroid compound that promotes tissue growth, especially of muscles. Naturally occurring anabolic steroids include the male sex hormones (*androgens). Synthetic forms of these are used medically to help weight gain after debilitating diseases; their use by athletes to build up body muscles can cause liver damage and is banned by most athletic authorities.

anabolism The metabolic synthesis of proteins, fats, and other constituents of living organisms from molecules or simple precursors. This process requires energy in the form of ATP. *See* METABOLISM. *Compare* CATABOLISM.

anadromous Describing the migration of certain fish, such as salmon, that spend most of their lives in oceanic waters before travelling to breed in the upper reaches of rivers and streams. *Compare* KATADROMOUS.

anaemia A condition that arises when either there are too few erythrocytes (red blood cells), the erythrocytes do not contain sufficient amounts of haemoglobin, or the erythrocytes are abnormal in other respects. Anaemia often results from loss of blood or from a deficiency in the factors necessary to synthesize haemoglobin (e.g. iron) or erythrocytes (e.g. folic acid and vitamin B_{12}). Increased destruction of erythrocytes may be induced by certain drugs or severe infection, and an abnormal form of haemoglobin results in sickle-cell anaemia (*see* POLYMORPHISM).

anaerobe *See* ANAEROBIC RESPIRATION.

anaerobic respiration A type of cellular *respiration in which oxygen is not the terminal electron acceptor in the *electron transport chain. Many bacteria and archaea can live in environments in which oxygen is limited or absent by using a range of small molecules or ions that function as the terminal acceptor; they can be chemically reduced and hence enable the generation of ATP by the chemiosmotic mechanism. Examples include the sulphate ion (SO_4^{2-}), as used by certain *sulphur bacteria, and CO_2, which is reduced by *methanogens. *Obligate anaerobes are organisms that cannot use free oxygen for respiration; *facultative anaerobes are normally aerobic but can respire anaerobically or use fermentation during periods of oxygen shortage. Note that anaerobic respiration is distinguished from *fermentation, which generates limited amounts of ATP in some yeasts and bacteria and in muscle tissue when oxygen is absent (*see* OXYGEN DEBT).

anaesthetic Any compound that can render an animal unconscious of painful stimuli. Anaesthetics can be used either at a local level, when sensation is removed from a specific area of the body; or at a general level, when a state of complete unconsciousness is induced.

analgesic A substance that reduces pain without causing unconsciousness, either by reducing the pain threshold or by increasing pain tolerance. There are several categories of analgesic drugs, including morphine and its derivatives (*see* OPIATE), which produce analgesia by acting on the central nervous system; nonsteroidal anti-inflammatory drugs (e.g. *aspirin); and local anaesthetics.

analogous Describing features of very disparate organisms that are superficially similar but have evolved from vastly different origins. The wings of butterflies and birds are analogous organs. *Compare* HOMOPLASY.

anamniote A vertebrate that lacks an *amnion and whose embryos and larvae must therefore develop in water. Anamniotes comprise the agnathans, fishes, and amphibians. *Compare* AMNIOTE.

anaphase One of several stages of cell division. In *mitosis the chromatids of each chromosome move apart to opposite ends of the spindle. In the first anaphase of *meiosis, the paired homologous chromosomes separate and move to opposite ends; in the second anaphase the chromatids move apart, as in mitosis.

anaphylaxis An abnormal immune response that occurs when an individual previously exposed to a particular *antigen is re-exposed to the same antigen. Anaphylaxis may follow an insect bite or the injection of a drug (such as penicillin). It is caused by the release of *histamine and similar substances and may produce a localized reaction or a more generalized and severe one, with difficulty in breathing, pallor, or drop in blood pressure, unconsciousness, and possibly heart failure and death. *See also* ALLERGY.

anaplerotic Describing a metabolic pathway that replenishes an intermediate in one of the main metabolic pathways. For example, some *Krebs cycle intermediates function not only in energy metabolism but are also used as building blocks for the biosynthesis of various compounds. Unless these intermediates are replaced, these crucial pathways will slow down or halt completely. In plants, mitochondrial oxaloacetate, which is a Krebs cycle intermediate used for the synthesis of amino acids, can be replenished by the conversion of phosphoenolpyruvate derived from glycolysis in the cytosol. The cytosolic oxaloacetate is then reduced to malate, which enters the mitochondria, where it is reoxidized to oxaloacetate.

anatomy The study of the structure of living organisms, especially of their internal parts by means of dissection and microscopical examination. *Compare* MORPHOLOGY.

ancestral trait *See* PLESIOMORPHY.

androecium (*pl.* **androecia**) The male sex organs (*stamens) of a flower. *Compare* GYNOECIUM.

androgen One of a group of male sex hormones that stimulate development of the testes and of male *secondary sexual characteristics (such as growth of facial and pubic hair in men). **Testosterone** is the most important. Androgens are produced principally by the testes when stimulated with *luteinizing hormone but they are also secreted in smaller amounts by the adrenal glands and the ovaries. Injections of natural or synthetic androgens are used to treat hormonal disorders of the testes and breast cancer and to build up body tissue (*see* ANABOLIC STEROID).

anemophily Pollination of a flower in which the pollen is carried by the wind. Examples of anemophilous flowers are those of grasses and conifers. *Compare* ENTOMOPHILY; HYDROPHILY.

aneuploid Describing a nucleus, cell, or organism in which one or more chromosomes have been added to or deleted from the complete set, so that the total number of chromosomes is not an exact multiple of the haploid number (n); for example, $2n + 1$ (*see* TRISOMY) or $2n-1$ (**monosomy**). *Compare* EUPLOID.

angiosperms *See* ANTHOPHYTA.

angiotensin Any of three related peptide hormones, two of which raise blood pressure. Angiotensin I is derived, by the action of the enzyme *renin, from a protein (α-globulin) secreted by the liver into the bloodstream. As blood passes through the lungs, another enzyme (**angiotensin-converting enzyme; ACE**) splits angiotensin I, forming angiotensin II. This causes constriction of blood vessels and stimulates the release of *antidiuretic hormone and *aldosterone, which increase blood pressure. Angiotensin III, formed by removal of a single amino acid from angiotensin II, also stimulates aldosterone release by the adrenal gland. Drugs that block the activity of ACE (ACE inhibitors) are widely used to treat chronic high blood pressure (hypertension).

angstrom Symbol Å. A unit of length equal to 10^{-10} metre. It was formerly used to measure wavelengths and intermolecular distances but has now been replaced by the nanometre. 1 Å = 0.1 nanometre. The unit is named after the Swedish pioneer of spectroscopy A. J. Ångstrom (1814–74).

anhydrobiosis *See* CRYPTOBIOSIS.

animal In traditional classifications, any member of the kingdom Animalia, which comprises multicellular organisms that develop from embryos formed by the fusion of haploid eggs and sperm. Unable to manufacture their own food, they feed on other organisms or organic matter (*see* HETEROTROPHIC NUTRITION). Animals are therefore typically mobile (to search for food) and have evolved specialized sense organs for detecting changes in the environment; a *nervous system coordinates information received by the sense organs and enables rapid responses to environmental stimuli. Animal *cells lack the cellulose cell walls of *plant cells. Molecular systematics now includes animals in the *opisthokonts, a eukaryote assemblage that also contains the fungi. For a simplified classification of animals, see Appendix.

animal behaviour The activities that constitute an animal's response to its external environment. Certain categories of behaviour are seen in all animals (e.g. feeding, reproduction) but these activities involve different movements in different species and develop in different ways. Some movements are highly characteristic of a species (*see* INSTINCT), whereas others are more variable and depend on the interaction between innate tendencies and *learning during the individual's lifetime. Physiologists study how changes in the body (e.g. hormone levels) affect behaviour, psychologists study the mechanisms of learning, and ethologists study the behaviour of the whole animal: how this develops during the individual's lifetime and how it evolved through natural selection (*see* ETHOLOGY).

animal pole The point on the surface of an egg cell centred over the **animal hemisphere**—the half of the egg containing the nucleus and the least amount of yolk. It lies opposite the **vegetal pole**, centred over the **vegetal hemisphere**, which contains the bulk of the yolk. The two poles, and the asymmetric distribution of yolk and other substances, define the front–rear axis of the embryo as the fertilized egg divides and determine the pattern of subsequent *cleavage into daughter cells. *See also* CYTOPLASMIC SEGREGATION.

animal starch *See* GLYCOGEN.

anion A negatively charged *ion, such as the chloride ion (Cl⁻). *Compare* CATION.

anisogamy Sexual reproduction involving the fusion of gametes that differ in size and sometimes also in form. *See also* OOGAMY. *Compare* ISOGAMY.

anlage (*pl.* **anlagen**) (in embryology) The earliest developmental stage of an organ or tissue, particularly in an embryo; a *primordium. The name is taken from a German word meaning 'foundation, basis'.

Annelida A phylum of invertebrates comprising the segmented worms (e.g. the earthworm). Annelids have cylindrical soft bodies showing *metameric segmentation, obvious externally as a series of rings separating the segments. Each segment is internally separated from the next by a membrane and bears stiff bristles (*see* CHAETA). Between the gut and other body organs there is a fluid-filled cavity called the *coelom, which acts as a hydrostatic skeleton. Movement is by alternate contraction of circular and longitudinal muscles in the body wall. Traditionally the phylum contained three classes: *Polychaeta, *Oligochaeta, and *Hirudinea. However, recent molecular studies have shown that this taxonomy does not reflect the annelids' evolutionary relationships, although alternative schemes are contentious. One proposal is that the annelids constitute two main clades—the Errantia and the Sedentaria—with each containing representatives from one or more of the traditional classes. Molecular systematics now places the annelids in the clade Lophotrochozoa, along with other wormlike groups and the molluscs.

(((⊕))) SEE WEB LINKS
• A basic round-up of evolution, life history, systematics, and morphology

annotations (in bioinformatics) Descriptive notes that are attached to database entries for, say, chromosomal loci or nucleic acid or amino acid sequences, detailing pertinent features such as transcripts, markers, structure, function, disease associations, and citations to relevant journal articles.

annual A plant that completes its life cycle in one year, during which time it germinates, flowers, produces seeds, and dies. Examples are the sunflower and marigold. *Compare* BIENNIAL; EPHEMERAL; PERENNIAL.

annual rhythm (**circannual rhythm**) The occurrence of a process or a function in a living organism on a yearly basis. Events that display an annual rhythm can include life cycles, such as those of *annual plants; mating behaviour; some kinds of movement, such as *migration; or growth patterns, such as the *growth rings of woody plant stems. *See also* BIORHYTHM.

annual ring *See* GROWTH RING.

annulus (*pl.* **annuli**) **1.** (in botany) **a.** A ragged ring of tissue that remains on the stalk of a mushroom or toadstool. Also called a **velum**, it is formed from the ruptured membrane that originally covered the lower surface of the cap. **b.** The region of the wall of a fern sporangium that is specialized for spore dispersal. It consists of cells that are thickened except on their outer walls. On drying out, the cells contract and the sporangium ruptures, releasing the spores. The annulus springs back into position when the residual water in the cells vaporizes and any remaining spores are dispersed. **2.** (in zoology) Any of various ring-shaped structures in animals, such as any of the segments of an earthworm or other annelid.

anoikis A form of programmed cell death (*apoptosis) induced in certain cell types when they become detached from their extracellular matrix or from a surface. Epithelial, endothelial, and many other tissue cells depend on continual interchange of signals with neighbouring cells and their immediate environment in order to survive. The absence of such signals following detachment from the matrix normally results in death of the cell. However, when tumour cells detach they resist anoikis, allowing them to migrate to new sites and seed new tumours. Hence understanding this process is key to the effective treatment of various cancers. The term is derived from the Greek word for homelessness.

Anoplura See SIPHUNCULATA.

anoxic Lacking or not involving or requiring oxygen. For example, a culture of anaerobic microorganisms is called an **anoxic culture.**

anoxic reactor A *bioreactor in which the organisms being cultured are anaerobes or in which the reaction being exploited does not require oxygen.

ANP See ATRIAL NATRIURETIC PEPTIDE.

ANS See AUTONOMIC NERVOUS SYSTEM.

antagonism 1. The interaction of two substances (e.g. drugs, hormones, or enzymes) having opposing effects in a system in such a way that the action of one partially or completely inhibits the effects of the other. For example, one group of anti-cancer drugs acts by antagonizing the effects of certain enzymes controlling the activities of the cancer cells. See also ANTAGONIST. **2.** An interaction between two muscles, known as **antagonistic muscles,** in which contraction of one prevents that of the other. For example, the *biceps and triceps are an antagonistic pair. See SKELETAL MUSCLE. **3.** An interaction between two organisms (e.g. moulds or bacteria) in which the growth of one is inhibited by the other. Compare SYNERGISM.

antagonist A substance (e.g. a hormone or drug) that inhibits the effect of an *agonist in such a way that the combined biological effect of the two substances becomes smaller than the sum of their individual effects. **Competitive antagonists** act by binding to agonist receptors, while **noncompetitive antagonists** do not bind to the same receptor sites as the agonist. A **functional antagonist** binds to other receptors that elicit an effect opposite to that of the agonist.

antenna (pl. **antennae**) A long whiplike jointed mobile paired appendage on the head of many arthropods, usually concerned with the senses of smell, touch, etc. (see SENSILLUM). In insects, millipedes, and centipedes they are the first pair of head appendages and are specialized and modified in many insects. In crustaceans they are the second pair of head appendages, the first pair (the **antennules**) having the sensory function, while the antennae are modified for swimming and for attachment.

antennal gland (green gland) Either of a pair of ducts (coelomoducts) found in the third segment of a crustacean and opening to the exterior at the base of the second antenna. They function as osmoregulatory organs. For example, in the antennal gland of the freshwater crayfish (Astacus), fluid is filtered from the blood into an end sac and passes through a tubular labyrinth, where ions are reabsorbed, to produce a hypotonic urine that passes via a renal tubule to the bladder.

antenna pigment See PHOTOSYSTEMS I AND II.

antennule See ANTENNA.

anterior 1. Designating the part of an animal that faces to the front, i.e. that leads when the animal is moving. In humans and other bipedal animals the anterior surface corresponds to the *ventral surface. **2.** Designating the side of a flower or axillary bud that faces away from the flower stalk or main stem, respectively. Compare POSTERIOR.

anther The upper two-lobed part of a plant *stamen, usually yellow in colour. Each lobe contains two pollen sacs within which are numerous pollen grains, which are released when the anther ruptures.

antheridium (pl. **antheridia**) The male sex organ of algae, fungi, bryophytes, clubmosses, horsetails, and ferns. It produces the male gametes (**antherozoids**). It may consist of a single cell or it may have a wall that is made up of one or several layers forming a sterile jacket around the developing gametes. Compare ARCHEGONIUM.

antherozoid (spermatozoid) The motile male gamete of algae, fungi, bryophytes, clubmosses, horsetails, ferns, and certain gymnosperms. Antherozoids usually develop in an *antheridium but in certain gymnosperms, such as *Ginkgo* and *Cycas*, they develop from a cell in the pollen tube.

Anthocerophyta (Anthocerotophyta) A phylum comprising about 100 species of simple nonvascular plants, the hornworts (or horned liverworts), found worldwide in temperate and tropical regions on tree trunks, riverbanks, and other damp locations. They resemble thallose liverworts (*see* HEPATOPHYTA), but produce long horn-shaped green sporophytes, which split longitudinally to release the spores. Hornwort cells each contain a single chloroplast inside which, uniquely among plants, is a *pyrenoid, associated with starch production. Some species have separate male and female plants, while others have both types of sexual organ on the same plant. The motile sperm swim through the surface water film to fertilize the female gametes, and the resultant embryo gives rise to young sporophytes. Young gametophytes arise directly from the germinating spores.

Hornworts were formerly classified as a class (Anthocerotae) of the *Bryophyta.

((⊕)) SEE WEB LINKS
• An entry into the world of hornworts, with links to images

anthocyanin One of a group of *flavonoid pigments. Anthocyanins occur in the cell vacuoles of various plant organs and are responsible for many of the blue, red, and purple colours in plants (particularly in flowers). *Compare* BETACYANIN.

Anthophyta (Angiospermophyta; Magnoliophyta) A phylum comprising the flowering plants (angiosperms). The gametes are produced within *flowers and the ovules (and the seeds into which they develop) are enclosed in a carpel (compare CONIFEROPHYTA). The angiosperms are the dominant plant forms of the present day. They show the most advanced structural organization in the plant kingdom, enabling them to inhabit a very diverse range of habitats. The two main clades are the monocots (traditionally placed in the class *Monocotyledoneae, broadly the same as the Liliopsida) with one seed leaf (cotyledon) in the seed; and the eudicots (class *Dicotyledoneae, or Magnoliopsida) with two seed leaves. Other clades are the magnoliids, containing the magnolias, laurels, cinnamon, etc.; the water lilies; and star anise and its relatives. Sister to all these is the genus *Amborella*, represented now by a single species of woody shrub found only on the island of New Caledonia.

((⊕)) SEE WEB LINKS
• Overview of flowering plant diversity and systematics

Anthozoa *See* CNIDARIA; CORAL.

Anthropocene A proposed new epoch in the *geological time scale that marks how human activities in recent decades have become the dominant force in shaping earth history, superseding natural geological, chemical, and climatological forces. It is argued that since the mid-20th century the growth in human population and economic activity has led to impacts of unprecedented scale and speed, such as *climate change, land transformation, and rapid species extinction, which justify the transition to a new epoch from the current *Holocene. Rock strata of any epoch are defined by a 'signal' that is incorporated into geological deposits globally, and several candidates have been suggested to define the Anthropocene. These include radionuclides from atomic bomb tests, microplastic particles, aluminium or concrete particles, and unburned carbon spheres emitted by power stations—all evidence of human activity. The term Anthropocene was coined by scientist Paul Crutzen (1933–2021) in 2000; formal acceptance of the new epoch by the International Commission on Stratigraphy is still awaited.

anthropoid Any member of the clade of *primates that includes the New World and Old World monkeys, gibbons, orangutans, African apes, and humans, and comprises the superfamily Anthropoidea.

antibiotics Substances that destroy or inhibit the growth of microorganisms, particularly disease-producing bacteria and fungi. Antibiotics are obtained from microorganisms (especially moulds) or synthesized. Common antibiotics include the *penicillins, **streptomycin**, and **tetracyclines**. They are used to treat various infections but tend to weaken the body's natural defence mechanisms and can cause allergies. Overuse of antibiotics leads to the development of resistant strains of microorganisms, such as methicillin-resistant *Staphylococcus aureus* (MRSA), with the consequent need to develop new and effective antibiotics.

SEE WEB LINKS

• Information about the use and abuse of antibiotics and the search for new ones

antibody A protein (*see* IMMUNOGLOBULIN) produced by certain white blood cells (*plasma cells) in response to entry into the body of a foreign substance (*antigen) in order to render it harmless. An antibody-antigen reaction is highly specific. Antibody production is one aspect of the immune response and is stimulated by such antigens as invading bacteria, foreign red blood cells (*see* ABO SYSTEM), inhaled pollen grains or dust, and foreign tissue *grafts. Specific *monoclonal antibodies are now used in various types of *immunoassay. *See also* B CELL; IMMUNITY.

anticholinesterase Any substance that inhibits the enzyme *cholinesterase, which is responsible for the breakdown of the neurotransmitter acetylcholine at nerve synapses. Anticholinesterases, which include certain drugs, nerve gases, and insecticides, cause a build-up of acetylcholine within the synapses, leading to disruption of nerve and muscle function. In vertebrates, these agents often cause death by paralysing the respiratory muscles. *See* PESTICIDE.

anticlinal (in botany) At right angles to the surface of an organ or part. In anticlinal cell division the plane of division is at right angles to the surface of the plant body. *Compare* PERICLINAL.

anticoagulant A substance that prevents the formation of blood clots. *Heparin is a natural anticoagulant, which is extracted to treat such conditions as thrombosis and embolism. Synthetic anticoagulants include *warfarin.

anticoding strand (noncoding strand; template strand) The strand in duplex DNA that by convention contains the sequence of bases that is complementary to that of the messenger RNA (mRNA) transcribed from the DNA (except that the RNA has uracil substituting for thymine). It is the strand used as a template for mRNA assembly during transcription, and is complementary to the other strand of the DNA molecule, the *coding strand. *See also* ANTISENSE DNA; NEGATIVE-SENSE.

anticodon A sequence of three nucleotides (trinucleotide) on a strand of transfer *RNA that can form base pairs (*see* BASE PAIRING) with a specific trinucleotide sequence (*see* CODON) on a strand of messenger RNA during *translation. *See also* PROTEIN SYNTHESIS; WOBBLE.

antidiuretic hormone (ADH; vasopressin) A hormone, secreted by the posterior *pituitary gland, that stimulates reabsorption of water by the kidneys and thus controls the concentration of body fluids (i.e. osmolarity) and also blood pressure. ADH is produced by specialized nerve cells in the hypothalamus of the brain and is transported to the posterior pituitary in the bloodstream. Secretion is stimulated by signals from *osmoreceptors in the hypothalamus, which detect increased osmolarity, and inhibited by signals from *baroreceptors in the great arteries, which detect increases in blood pressure. ADH causes increased insertion of *aquaporins in the collecting ducts of the kidneys, thereby allowing more water to pass from the collecting duct fluid into the renal medulla. Deficiency of ADH results in a disorder known as **diabetes insipidus**, in which large volumes of urine are excreted; it is treated by administration of natural or synthetic hormone. *See also* NEUROPHYSIN.

antifreeze molecule Any substance produced by an organism in order to prevent freezing of its tissues or body fluids when subject to subzero environmental temperatures. Many animals living in cold climates adopt a strategy of preventing ice formation in their tissues when subject to freezing conditions. One way of achieving this is to accumulate solutes in their blood, thereby raising the osmotic concentration and so depressing the *supercooling point. Salts and sugars contribute to this, but organisms also produce relatively inert molecules, notably glycerol and other polyhydric alcohols (polyols), such as sorbitol and ribitol, specifically for this purpose. For example, high concentrations of glycerol can enable the survival of certain cold-hardy invertebrates at temperatures as low as −60°C. Certain vertebrates, fungi, plants, and bacteria manufacture **antifreeze peptides** or **antifreeze glycopeptides**, which are effective antifreeze agents at relatively low concentrations. They bind to the edges of ice crystal lattices and prevent the addition of further water molecules, causing a phenomenon termed 'thermal hysteresis', in which the freezing point is depressed well below the melting point—hence these peptides are also called **thermal hysteresis proteins**. Frost-tolerant plants increase the concentration of sugars in their cells in preparation for winter. The sugars help to prevent water loss from the cells when the water content in the cell walls is reduced by freezing. See also CRYOPROTECTANT.

antigen Any substance that the body regards as foreign and that therefore elicits an immune response, particularly the formation of specific antibodies capable of binding to it. Antibodies recognize and bind to specific structures on the surface of the antigen called *epitopes. Antigens may be formed in, or introduced into, the body. They are usually proteins. **Histocompatibility antigens** are associated with the tissues and are involved in the rejection of tissue or organ *grafts (see HISTOCOMPATIBILITY); an example is the group of antigens encoded by the *HLA system. A graft will be rejected if the recipient's body regards such antigens on the donor's tissues as foreign. See also ANTIBODY; B-CELL RECEPTOR; T-CELL RECEPTOR. Compare HAPTEN.

antigenic variation The ability of certain pathogenic microorganisms, particularly viruses, to alter the antigens on their outer surface. This prevents the pathogen from being easily recognized and destroyed by the immune system of the host. See also PHASE VARIATION.

antigen-presenting cell A cell, such as a *dendritic cell or macrophage, that 'presents' foreign antigens to *T cells as part of the immune response. A direct reaction between these lymphocytes and foreign antigens is not common. The antigen-presenting cell takes in the foreign protein, processes it, and then displays peptide fragments on its surface as a complex with *MHC class II proteins.

antihistamine Any drug that inhibits the effects of *histamine in the body and is therefore used to relieve and prevent the symptoms associated with allergic reactions, such as hay fever. Since one of the side-effects produced by antihistamines is sleepiness, some are used to prevent motion sickness and induce sleep.

anti-oncogene See TUMOUR-SUPPRESSOR GENE.

antioxidants Substances that slow the rate of oxidation reactions. Various antioxidants are used to preserve foodstuffs and to prevent the deterioration of rubber, synthetic plastics, and many other materials. Some antioxidants inhibit the oxidation reaction by removing oxygen *free radicals. Naturally occurring antioxidants with this ability include *vitamin E, β-*carotene, *glutathione, and *superoxide dismutase; they limit the cell and tissue damage caused by superoxide radicals, hydrogen peroxide, hydroxide radicals, and other highly reactive substances.

antipodal cells The three haploid cells in the mature *embryo sac of flowering plants that are situated at the opposite end to the micropyle.

antiporter A membrane protein that effects the *active transport of a substance across a cell membrane while transporting ions in the opposite direction. An antiporter is a type of *cotransporter; the ions, typically hydrogen ions (H^+) or sodium ions (Na^+), flow down their concentration gradient, and in so doing provide the energy for the transport of another substance in the other direction. For example, heart-muscle cells have a Na^+/Ca^+ antiporter, which is driven by the inward flow of sodium ions to pump calcium ions (Ca^+) out of the cell. Hence, the energy for antiporters derives ultimately from the energy-consuming mechanisms that establish the concentration gradient of the driving ions. *Compare* SYMPORTER; UNIPORTER.

antipyretic A drug that reduces fever by lowering body temperature. Certain *analgesic drugs, notably paracetamol (acetaminophen), *aspirin, and phenylbutazone, also have antipyretic properties.

antisense DNA A single-stranded DNA molecule that can bind to a complementary base sequence in a particular messenger RNA (mRNA) molecule and so prevent synthesis of the protein encoded by the mRNA. Although antisense DNA thus has the potential to block the expression of a particular gene, this is achieved more effectively by short double-stranded RNA molecules (*see* RNA INTERFERENCE), which are not susceptible to degradation by DNase enzymes. *See also* ANTISENSE RNA.

antisense RNA An RNA molecule whose base sequence is complementary to that of the RNA transcript of a gene, i.e. the 'sense' RNA, such as a messenger RNA (mRNA). Hence, an antisense RNA can undergo base pairing with its complementary mRNA sequence. This blocks gene expression, either by preventing access for ribosomes to translate the mRNA or by triggering degradation of the double-stranded RNA by ribonuclease enzymes. Antisense RNA has therapeutic potential for modifying the activity of disease-causing genes, and drugs based on this mechanism are being developed. Also, genes encoding antisense RNAs can be used in genetic engineering to alter the makeup of organisms. For example, the FlavrSavr tomato was engineered with an artificial gene for antisense RNA that prevented expression of a gene for an enzyme involved in ripening, in order to retard spoilage. In the 1980s it was discovered that double-stranded RNA molecules have a much greater ability to suppress their corresponding genes than single-stranded RNAs, due to the phenomenon of *RNA interference.

antiseptic Any substance that kills or inhibits the growth of disease-causing microorganisms but is essentially nontoxic to cells of the body. Common antiseptics include hydrogen peroxide, the detergent cetrimide, and ethanol. They are used to treat minor wounds. *Compare* DISINFECTANT.

antiserum Serum containing antibodies, either raised against particular antigens, and hence of known specificity, or a broad mixture of antibodies. It is used to provide short-term passive immunity, e.g. against hepatitis A virus, and to treat an infection to which the patient has no immunity. Antisera may be obtained from large animals, such as horses, that have been inoculated with particular antigens, or from pooled donated human serum.

antitoxin An antibody produced in response to a bacterial *toxin.

antiviral Describing a drug or other agent that kills or inhibits viruses and is used to combat viral infections. Several types of antiviral drug are now in use, such as *aciclovir, effective against herpesviruses, and zidovudine (AZT), a *reverse transcriptase inhibitor that is used to treat HIV infection. Protease inhibitors (such as ritonavir and saquinavir), integrase inhibitors (such as raltegravir), and chemokine receptor antagonists (such as maraviroc) are additional classes of antiviral drugs used to treat HIV infection. The body's own natural antiviral agents, *interferons, can now be produced by genetic engineering and are sometimes used therapeutically. However, many antiviral agents are extremely toxic, and viruses evolve rapidly so that a drug's effectiveness can soon be lost.

anucleate Describing any cell that does not contain a nucleus. For example, mature *erythrocytes are anucleate.

anus The terminal opening of the *alimentary canal in most animals, through which indigestible material (*faeces) is expelled.

aorta The major blood vessel in higher vertebrates through which oxygenated blood leaves the *heart from the left ventricle. The aorta branches to form many smaller arteries, which in turn branch many times to supply oxygen and essential nutrients to all living cells in the body. *See also* DORSAL AORTA; VENTRAL AORTA.

aortic arches (arterial arches) Six pairs of arteries in vertebrate embryos, which connect the *ventral aorta to the *dorsal aorta by running between the gill slits. The arches are numbered I to VI from the anterior end. In adult tetrapods arches I and II are lost, III gives rise to the *carotid arteries, IV (one side of which is lost in birds and mammals) becomes the *systemic arch supplying the trunk and limbs, V is lost (*see* DUCTUS ARTERIOSUS), and VI gives rise to the pulmonary arch supplying the lungs (*see* PULMONARY ARTERY). In adult fish four to six arches persist as branchial arteries supplying the gills.

aortic body *See* VENTILATION CENTRE.

aortic valve *See* SEMILUNAR VALVE.

apatite A complex mineral form of the salt calcium phosphate, $Ca_5(PO_4)_3(OH,F,Cl)$, that is the main constituent of the enamel of teeth.

aphotic zone (bathypelagic zone) The region of a lake or sea where no light penetrates; it is situated beneath the *euphotic zone. The aphotic zone contains no algae or phytoplankton, and its inhabitants are exclusively carnivorous animals or organisms that feed on sediment or detritus, all reliant on energy inputs from the euphotic zone. It extends downwards from a depth of about 1000 m, or less in turbid waters, and includes the *abyssal zone.

apical dominance Inhibition of the growth of lateral buds in a plant by the presence of a growing apical bud. It is brought about by a combination of the demand for sugar by the growing bud and the actions of certain plant hormones. While the apical bud is growing, it has a strong demand for sugars, which are thus denied to lateral buds lower down the stem, thereby maintaining their state of dormancy. Moreover, polar flow of the plant hormone *auxin from the shoot tip downwards inhibits the growth of new side shoots. It also triggers the synthesis of *strigolactones, which directly repress bud growth. These effects are antagonized by *cytokinins transported into the stem from the root. Removal of the apical bud allows sugar to be redistributed to nearby lateral buds, which are disinhibited and start to grow, assisted by the decline in the polar flow of auxin and synthesis of strigolactones.

apical meristem A region at the tip of each shoot and root of a plant in which cell divisions are continually occurring to produce new stem and root tissue, respectively (*see* MERISTEM). The new tissues produced are known collectively as the **primary tissues** of the plant. *See* GROUND MERISTEM; PROCAMBIUM; PROTODERM. *Compare* CAMBIUM.

Apicomplexa (Sporozoa) A phylum of parasitic protists (*see also* PROTOZOA) whose members may have a number of different animal hosts. Their complex life cycle involves the alternation of asexual reproduction (multiple fission) and sexual reproduction and the production of resistant spores (see illustration). The phylum includes the agents causing malaria (*Plasmodium*) and toxoplasmosis (*Toxoplasma*). Apicomplexans are now classified as *alveolates.

apocarpy The condition in which the female reproductive organs (*carpels) of a flower are not joined to each other. It occurs, for example, in the buttercup. *Compare* SYNCARPY.

apocrine secretion *See* SECRETION.

Apicomplexa: life cycle of *Plasmodium vivax* (malaria parasite)

apodeme An inward projection from the exoskeleton of an arthropod, to which muscles that enable movement of the limbs are attached.

apoenzyme An inactive enzyme that must associate with a specific *cofactor molecule or ion in order to function. *Compare* HOLOENZYME.

apomixis (agamospermy) A reproductive process in plants that superficially resembles normal sexual reproduction but in which there is no fusion of gametes. In apomictic flowering plants there is no fertilization by pollen, and the embryos develop simply by division of a *diploid cell of the ovule. *See also* PARTHENOCARPY; PARTHENOGENESIS.

apomorphy (derived trait) A novel evolutionary trait that is unique to a particular species and all its descendants and which can be used as a defining character for a species or group in phylogenetic terms. Hence, the possession of feathers is unique to birds and defines all members of the class Aves. An apomorphy that is restricted to a single species is termed an **autapomorphy**. It alone cannot provide any information about the phylogenetic relations of that species, although it can indicate the degree of divergence of a species from its nearest

relatives. An example is speech, which is found solely in humans (*Homo sapiens*) and not in other primates. An apomorphy that is shared by two or more species or groups is termed a *synapomorphy, or a shared derived trait. *Compare* PLESIOMORPHY.

apoplast An interconnected system in plants that consists of all the cell walls, the water that exists in them (the cell wall is composed of cellulose fibres, between which are spaces filled with water), the intercellular spaces, and dead cells such as vessel elements and tracheids. The movement of water (and dissolved ions and solutes) through the apoplast is known as the **apoplastic pathway**. This is the main route by which water taken up by a plant travels across the root cortex to the *endodermis (*see* CASPARIAN STRIP). *See also* PHLOEM LOADING; PHLOEM UNLOADING. *Compare* SYMPLAST.

apoptosis (programmed cell death) The process of cell death that occurs naturally as part of the normal development, maintenance, and renewal of tissues within an organism. During embryonic development it plays a vital role in determining the final size and form of tissues and organs. For example, the fingers are 'sculpted' on the spadelike embryonic hand by apoptosis of the cells between them; the tubules of the embryonic kidney are hollowed out by a similar process. Apoptosis follows an orderly sequence of events, and involves the action of enzymes called *caspases. These are cellular proteins that when activated cleave target proteins within the cell, including other 'executioner' enzymes, which digest the cytoskeleton, DNA, and other cell components. Various signals can result in activation of caspases. Internal signals include irreparable damage to the cell's DNA or the release of cytochrome c and other mitochondrial components into the cytosol following damage to mitochondria. External 'death-signal' molecules, e.g. from neighbouring cells, activate intracellular signal pathways, such as the *Fas signal pathway. The cell shrinks as apoptosis proceeds and components are dismembered, until finally the cell is condensed into fragments called **apoptotic bodies**. These are degraded by scavenging phagocytic cells. Apoptosis is normally suppressed as long as cells continue to receive extracellular survival signals, in the form of trophic factors—an example being nerve growth factor (*see* NEUROTROPHIN). In the absence of such signals, the cell embarks on a 'suicide' programme. Sometimes, other cells, for example immune cells, release specific 'murder' signals, which activate apoptosis in target cells (*see* CYTOTOXIC T CELL; NATURAL KILLER CELL). Cancer is associated with the suppression of apoptosis (*see also* GROWTH FACTOR), which also occurs when viruses infect cells—in order to inhibit the activity of killer cells. Apoptosis differs from cell *necrosis, in which cell death may be stimulated by a toxic substance. Plants also perform apoptosis as a means of removing cells infected by invading fungi or bacteria. *Compare* ANOIKIS.

(⊕) SEE WEB LINKS
• Video of apoptotic cells

aposematic coloration *See* WARNING COLORATION.

appeasement Behaviour that inhibits aggression from another animal of the same species, frequently taking the form of a special posture or display emphasizing the weakness of the performer. Threatening structures (e.g. antlers) and markings are covered or turned away, and vulnerable parts of the body may be exposed. Appeasement is seen in *courtship, in greeting ceremonies, and often (from the loser) after a fight.

appendicular skeleton The components, collectively, of the vertebrate skeleton that are attached to the main supporting, or *axial, skeleton. The appendicular skeleton is made up of paired appendages (e.g. legs, wings, arms) together with the *pelvic girdle and *pectoral girdle.

appendix (vermiform appendix) An outgrowth of the *caecum in the alimentary canal. In herbivores, such as the horse, it is a relatively large, muscular organ that contains symbiotic microorganisms that help digest the plant-based diet. In humans it is a *vestigial organ containing lymphatic tissue and serves no function in normal digestive processes. Appendicitis is caused by inflammation of the appendix.

apposition The addition of layers of cellulose to the inner surface of a plant cell wall, at its junction with the plasma membrane. This type of growth results in thickening and strengthening the cell wall and usually occurs when elongation of the cell is complete. *See also* CELL WALL. *Compare* INTUSSUSCEPTION.

Apterygota A subclass of small primitive wingless insects in which metamorphosis is slight or absent. It includes the orders *Thysanura and *Diplura (bristletails), *Collembola (springtails), and *Protura. *Compare* PTERYGOTA. *See* HEXAPODA.

aquaporin A protein occurring in cell membranes that functions as a water channel. Aquaporins increase permeability of the membrane to water, and can thus accelerate the process of osmosis. They are integral to the plasma membranes of red blood cells and *proximal convoluted tubules, and are inserted into the plasma membrane of the renal collecting ducts in response to *antidiuretic hormone. In plants they occur in the plasma membrane and *tonoplast, where alterations in their abundance and permeability regulate the rate of osmosis across those membranes.

aqueous humour The fluid that fills the space between the cornea and the lens of the vertebrate eye. In addition to supplying the cornea and lens with nutrients, the aqueous humour helps to maintain the shape of the eye. It is produced and renewed every four hours by the *ciliary body.

Arabidopsis A genus of flowering plants of the family Cruciferae (Brassicaceae). The species *A. thaliana* (thale cress) is widely used as a research tool in molecular genetics and developmental biology because it has a small and simple genome (five pairs of chromosomes), over half of which codes for protein, and it can be easily cultured, having a life cycle of only 6–8 weeks. Its full genome sequence was published in 2000. *See*, e.g., ABC MODEL; FLOWERING.

arachidonic acid A polyunsaturated fattyacid, $CH_3(CH_2)_3(CH_2CH:CH)_4(CH_2)_3COOH$, that is essential for growth in mammals (*see* EICOSANOID). It can be synthesized from *linoleic acid. Arachidonic acid acts as a precursor to several biologically active compounds, including *prostaglandins, and plays an important role in membrane production and fat metabolism. The release of arachidonic acid from membrane *phospholipids, particularly diacylglycerol, is triggered by certain hormones. *See* ESSENTIAL FATTY ACIDS.

Arachnida A class of terrestrial *arthropods of the phylum *Chelicerata, comprising around 150 000 described species and including spiders, scorpions, harvestmen, ticks, and mites. An arachnid's body is divided into an anterior *prosoma and a posterior *opisthosoma. The prosoma bears *chelicerae, *pedipalps, and four pairs of walking legs. The opisthosoma may bear various sensory or silk-spinning appendages (*see* SPINNERET). Arachnids are generally carnivorous, feeding on the body fluids of their prey or secreting enzymes to digest prey externally. Spiders immobilize their prey with poison injected by the fanglike chelicerae, while scorpions grasp their prey in large clawed pedipalps and may poison it using the posterior stinging organ. Ticks and some mites are parasitic but most arachnids are free-living. They breathe either via *tracheae (like insects) or by means of thin highly folded regions of the body wall called *lung books. *See also* ACARINA.

((●)) SEE WEB LINKS

• Lively and accessible introduction to major groups of arachnids

arachnoid membrane One of the three membranes (*meninges) that surround the brain and spinal cord of vertebrates. It lies between the *pia mater and the *dura mater. The arachnoid membrane is very delicate and carries *cerebrospinal fluid, which sustains and cushions the nervous tissue.

arbovirus Obsolete name for any RNA-containing virus that is transmitted from animals to humans through the bite of mosquitoes and ticks (i.e. arthropods, hence *ar*thropod-*bo*rne viruses). They cause various forms of encephalitis (inflammation of the brain) and serious fevers, such as dengue and yellow fever.

arbuscular mycorrhiza *See* MYCORRHIZA.

Archaea A *domain of prokaryotic organisms containing the **archaea** (or **archaebacteria**), including the *methanogens, which produce methane; the thermoacidophilic bacteria, which live in extremely hot and acidic environments (such as hot springs; *see* THERMOPHILIC); and the halophilic bacteria, which can only function at high salt concentrations and are abundant in the world's oceans. Archaea are distinguished from the *Eubacteria in that their membrane lipids are ether-linked as opposed to being ester-linked, and they lack *peptidoglycan in their cell walls. However, the archaea are grouped together principally on the basis of molecular systematics, particularly similarities in the base sequence of their ribosomal RNA. This molecular evidence shows the archaea to be phylogenetically distinct and closer to the eukaryotes than the eubacteria. Multiple groups of archaea are now recognized, including the euryarchaeotes, the *TACK supergroup, and the nanoarchaea and other minute organisms constituting the DPANN supergroup.

Archaean The earliest eon of geological time, in which there is the first evidence of life on earth. It follows the *Hadean eon of pregeological time and extends from the time of the earliest known rocks, roughly 3900 million years ago, to the beginning of the *Proterozoic eon, about 2500 million years ago. Rock formations called *stromatolites, dated at 3500 million years old or older, are among the oldest of all fossil remains. They are thought to have been produced by the activities of microbial mats of filamentous purple and green bacteria. These prokaryotes performed photosynthesis anaerobically, perhaps using hydrogen sulphide as an electron donor instead of water. Some of their descendants evolved the ability to use water as an electron donor, producing oxygen as a byproduct, and eventually brought about the change in atmospheric conditions necessary for aerobic life.

archaebacteria See ARCHAEA.

Archaeplastida See PLANT (sense 2).

archegonium (*pl.* **archegonia**) The multicellular flask-shaped female sex organ of bryophytes, clubmosses, horsetails, ferns, and many gymnosperms. Such plants are described as **archegoniate** to distinguish them from algae, which do not possess archegonia. The dilated base, the **venter**, contains the oosphere (female gamete). The cells of the narrow neck liquefy to allow the male gametes to swim towards the oosphere. The archegonium is thus an adaptation to the terrestrial environment as it provides a means for the male gametes to reach the female gamete. *Compare* ANTHERIDIUM.

archenteron (gastrocoel) A cavity within an animal embryo at the *gastrula stage of development. All or part of the archenteron eventually forms the cavity of the gut. It is connected to the outside by an opening (the **blastopore**), which becomes either the mouth, the mouth and anus, or the anal opening of the animal.

archosaur Any member of a clade of reptiles including the crocodilians, *pterosaurs, *dinosaurs, and birds. They are distinguished from other *amniote vertebrates by having a single opening in the skull behind each eye socket, a high narrow skull with pointed snout, and teeth set in sockets.

ardipithecine Any member of a clade of extinct *hominins that contains the earliest known species in the human ancestral lineage. Sparse fossil evidence, mostly from East Africa, indicates a combination of apelike and humanlike features. The oldest known remains belong to *Sahelanthropus tchadensis* and date from between 7 and 6 million years ago (mya). They reveal a small brain and prominent brow ridge, similar to apes, but small canine teeth and an opening for the spinal cord beneath the skull, indicating that the head was held upright on the body and hence bipedal locomotion. Fossilized thigh bones from another ardipithecine species, *Orrorin tugenensis*, also suggest bipedalism; these date from 6.2 to 5.8 mya. The genus *Ardipithecus* contains two species: *Ar. kadabba*, from about 5.8 to 5.2 mya, and *Ar. ramidus*, about 4.4 mya. A partial female skeleton of the latter has adaptations for tree climbing, namely an opposable big toe, as well as a pelvis suggestive of bipedalism. It is estimated to have stood about 120 cm tall.

areolar connective tissue A type of *connective tissue consisting of a gel-like matrix incorporating strands of protein fibres (*collagen and *elastin) and such cells as *fibroblasts, *mast cells, *macrophages, and fat cells. This tissue is found throughout the body under the skin and linking organs and other tissues.

arginine See AMINO ACID.

aril An outgrowth that grows around and may completely enclose the testa (seed coat) of a seed. It develops from the placenta, funicle, or micropyle of an ovule. The aril surrounding the nutmeg seed forms the spice mace. See also CARUNCLE.

arousal A level of physiological and behavioural responsiveness in an animal, which tends to vary between sleep and full alertness. It is controlled in part by the **reticular formation** in the *brainstem, which is crucial in controlling wakefulness and sleep. The level of arousal can be detected by changes in brain electrical activity, heart rate and muscle tone, responsiveness to new stimuli, and general activity.

arrector pili A small muscle in the dermis of the skin that is attached to the base of a hair follicle. Contraction of the arrector pili, in response to cold or fear, pulls the hair into a vertical position, thus trapping an insulating layer of air around the body and causing 'goose flesh' in humans.

arrhenotoky The phenomenon occurring in the reproduction of certain animals in which fertilized eggs give rise to females and unfertilized eggs to males. It is found among certain insect groups, notably the wasps and bees (Hymenoptera), and in mites, as well as rotifers and nematodes. The males are haploid and transmit only the maternal genome—their production represents arrhenotokous *parthenogenesis—whereas the females are diploid. In **pseudo-arrhenotoky** both males and females arise from fertilized eggs and are diploid, but males subsequently become effectively haploid by inactivation of the paternal genome, either in all cells or only in germ-line cells. This occurs in certain scale insects and mites. Compare THELYTOKY.

arrhythmia A disturbance in the normal rhythm of the heart, usually caused by reduced blood flow within the heart or a defect in the *pacemaker (the sinoatrial node) of the heart.

artemisinin An antimalarial drug obtained from the shrub sweet wormwood (*Artemisia annua*) and isolated by Chinese scientists in the 1970s to treat North Vietnamese soldiers suffering from quinine-resistant malaria during the Vietnam War. It acts by inhibiting calcium transport in the malarial parasite *Plasmodium falciparum*.

arteriole A small muscular blood vessel that receives blood from the arteries and carries it to the capillaries. See MICROCIRCULATION.

artery A blood vessel that carries blood away from the heart towards the other body tissues. Most arteries carry oxygenated blood (the *pulmonary artery is an exception). The large arteries branch to form smaller ones, which in turn branch into *arterioles. Collagen and elastin fibres in the walls of arteries enable them to withstand the high pressures of circulating blood. They also create elastic recoil of the artery wall during the nonpumping phase (diastole) of the cardiac cycle, which maintains and 'smooths' the directional flow of blood around the body. Contraction and relaxation of the smooth muscle layer in the wall regulates *blood pressure under the influence of the autonomic nervous system and hormones. The accumulation of fatty deposits in the walls of the arteries leads to **atherosclerosis**, which limits and may eventually block the flow of blood. Compare VEIN.

Arthrophyta See HORSETAILS.

arthropod Any invertebrate animal that, characteristically, possesses an outer body layer—the *cuticle—that functions as a rigid protective exoskeleton; growth is thus possible only by periodic moults (see ECDYSIS). There are over one million described species of arthropods, inhabiting marine, freshwater, and terrestrial habitats worldwide. The arthropod body is composed of segments (see METAMERIC SEGMENTATION) usually forming distinct specialized body regions, e.g. head, thorax, and abdomen. These segments may possess hardened jointed appendages, modified variously as *mouthparts, limbs, wings, reproductive organs, or sense organs. The main body cavity, containing the internal organs, is a blood-filled *haemocoel, within which lies the heart. Although arthropods are often placed in a single phylum, Arthropoda, the origins and relationships of the various groups of arthropods remain uncertain, and they are now usually assigned to several subphyla within the clade *Ecdysozoa, notably *Crustacea (shrimps, barnacles, crabs, etc.); *Hexapoda (insects); Myriapoda (centipedes and millipedes); and *Chelicerata, including the *Arachnida (spiders, scorpions, mites, and ticks). Genetic evidence now indicates that insects evolved from crustacean ancestors and that some crustaceans are more closely related to insects than to other crustaceans. Hence, the insects and crustaceans are now placed in the clade Pancrustacea, with a group of predatory crustaceans, the Remipedia, regarded as the closest relatives to all insects.

articulation The attachment of two bones, usually by means of a *joint. The thigh bone (femur), for instance, articulates with the pelvic girdle.

artificial chromosome A type of cloning *vector that has some features of true chromosomes and is used to clone relatively large fragments of DNA. **Bacterial artificial chromosomes** (BACs) are based on the F (fertility) plasmid found naturally in *E. coli* bacteria (*see* SEX FACTOR). They can accommodate inserts of foreign DNA up to about 300 kilobase (kb) in length. Also included are several bacterial genes necessary for replication of the plasmid by the host cell and a gene (usually for resistance to an antibiotic) that allows selection of BAC-containing cells. Similar to BACs are PACs (**P1 artificial chromosomes**), which incorporate phage P1 elements. Larger DNA fragments are cloned using **yeast artificial chromosomes** (YACs). These are linear vectors derived from a circular plasmid found naturally in baker's yeast (*Saccharomyces cerevisiae*) and capable of accommodating DNA inserts of up to 2000 kb. YACs have a centromere, enabling them to attach to the mitotic spindle of their yeast host and undergo normal segregation during cell division. They are also engineered with *telomeres, the DNA sequences that cap either end of a chromosome. Thus YACs behave like mini-chromosomes. They are used for cloning eukaryotic genes or gene segments, for making *DNA libraries of organisms with large genomes (e.g. mammals), and for studying gene function. **Human artificial chromosomes** are mini-chromosomes derived from fragments of full-length human chromosomes or synthesized *de novo* using elements from BACs or YACs. Their great advantage is that they replicate and segregate normally alongside the cell's other chromosomes during cell divisions without disrupting the other genes in the cell. Hence they are used as vectors for studies of human gene function in experimental animals, and have potential for delivering normal genes to human tissue cells to correct genetic defects.

artificial insemination (AI) The deposition of semen, using a syringe, at the mouth of the uterus to make conception possible. It is used in the selective *breeding of domestic animals and also in humans in some cases of impotence and infertility. It is timed to coincide with ovulation in the female.

artificial selection The modification of species by selective *breeding. Animals or plants with desirable characteristics are interbred with the aim of altering the *genotype and producing a new strain of the organism for a specific purpose. For example, sheep are bred by means of artificial selection in order to improve wool quality. Traditional breeding techniques have been supplemented, and in many cases supplanted, by more recent methods of *genetic engineering, *genome editing, genetic testing, and embryo manipulation. Sequencing the genomes of commercially important animal and plant species has enabled the inheritance of desired genes to be monitored directly by molecular methods, instead of by phenotypic analysis. These methods have simultaneously opened up new approaches to selection and enabled it to become more refined and focused. *See also* DE-EXTINCTION.

Artiodactyla An order of hooved mammals comprising the even-toed ungulates, in which the third and fourth digits are equally developed and bear the weight of the body. The order traditionally includes cattle and other ruminants (*see* RUMINANTIA), camels, hippopotamuses, and pigs. All except the last are herbivorous, having an elongated gut and teeth with enamel ridges for grinding tough grasses. Evidence from molecular studies now indicates that whales and dolphins are closely related to hippos and represent a lineage that split from ruminants about 55 million years ago; pigs are believed to have diverged even earlier. Hence whales plus members of the Artiodactyla are now placed in the superorder Cetartiodactyla. *Compare* PERISSODACTYLA.

ascending tracts *See* SPINAL CORD.

ascocarp The reproductive body of fungi of the phylum *Ascomycota, which contains *ascus cells. An ascocarp may be a closed sphere (**cleistothecium**), a flask-shaped structure (**perithecium**) with a small opening (**ostiole**), or an open cup-shaped structure (**apothecium**). The ascocarp consists of both ascus-bearing and vegetative *hyphae.

ascogenous hyphae The hyphae, in fungi of the Ascomycota, that grow from the *ascogonium after it has fused with the *antheridium. The ascogenous hyphae are made up of binucleate cells containing one nucleus derived from the male antheridium and the other from the female ascogonium. This condition is represented as $n + n$, rather than $2n$, as the cells are not true diploid cells. Asci (*see* ASCUS) develop from the ascogenous hyphae.

ascogonium (*pl.* **ascogonia**) The female gametangium of certain fungi of the Ascomycota, from which *ascogenous hyphae develop.

Ascomycota (sac fungi) A phylum of fungi, formerly classified as a class (Ascomycetes) or a subdivision (Ascomycotina).

It contains about 65 000 described species, including the *yeasts; many some species of filamentous fungi, some of which produce edible fruiting structures (e.g. the morels and truffles); Claviceps purpurea, which causes *ergot in rye; and serious plant pests such as blight fungi and powdery mildews. Many are fungal partners of lichens. Sexual reproduction is by means of ascospores produced within a sac-like *ascus. The asci are usually grouped together in an *ascocarp. The phylum also now includes asexual fungi, such as *Penicillium* and *Candida* species, which were formerly placed in the taxon *Deuteromycota.

ascorbic acid *See* VITAMIN C.

ascospore The spore of fungi belonging to the phylum Ascomycota. *See* ASCUS.

ascus (*pl.* **asci**) A specialized cell in fungi of the phylum *Ascomycota, which contains two haploid nuclei that fuse during sexual reproduction and then undergo meiosis, giving rise to eight **ascospores** contained within the ascus.

asepsis The condition of an environment that is free of pathogens. This is achieved by *sterilization techniques.

asexual reproduction Reproduction in which new individuals are produced from a single parent without the formation of gametes. It occurs chiefly in lower animals, microorganisms, and plants. In microorganisms and lower animals the chief methods are *fission (e.g. in protists), *fragmentation (e.g. in some aquatic annelid worms), and *budding (e.g. in cnidarians and yeasts). The principal methods of asexual reproduction in plants are by *vegetative propagation (e.g. bulbs, corms, tubers) and by the formation of *spores. Spore formation occurs in mosses, ferns, and other plants showing alternation of generations, as a dormant stage between sporophyte and gametophyte, and in some algae and fungi, to produce replicas of the organism. *Compare* SEXUAL REPRODUCTION.

asparagine *See* AMINO ACID.

aspartic acid *See* AMINO ACID.

aspirin (acetylsalicylic acid) A drug that reduces inflammation, combats fever, and alleviates pain. Aspirin works by inhibiting the formation of *prostaglandins, which are major factors in the inflammation process. It also reduces the aggregation of blood platelets, hence its use in maintaining blood flow following heart and circulatory disorders.

assimilation The utilization by a living organism of absorbed food materials in the processes of growth, reproduction, or repair.

assisted reproductive technology (ART) Any of a range of techniques for improving the fertility of animals, including humans, in which eggs and/or sperm are extracted from the respective animal(s) for manipulation in some way. One of the best known is *in vitro* fertilization, in which ova (egg cells) are placed with sperm in a culture medium so that fertilization occurs and embryos start to develop. Healthy embryos are then selected and implanted in the mother's uterus to continue their development. Such techniques are widely used to overcome certain forms of human infertility and in breeding elite livestock. Early-stage embryos can undergo genetic analysis, for instance to screen for certain genetic diseases or (in livestock) to select progeny with desirable genetic traits or a particular sex.

association An ecological unit in which two or more species occur in closer proximity to one another than would be expected on the basis of chance. Early plant ecologists recognized associations of fixed composition on the basis of the *dominant species present (e.g. a coniferous forest association). Associations now tend to be detected by using more objective statistical sampling methods. *See also* CONSOCIATION.

association centre (association cortex) Any part of the brain that links a primary sensory area (a part of the cerebral cortex that receives primary sensory impulses) with other parts of the brain, such as memory and motor areas, and deals with the interpretation and meaning

of the primary sensory inputs. For example, the auditory association area interprets a 'moo' sound as that coming from a cow. It is estimated that some three-quarters of the cerebral cortex is association cortex.

assortative mating A form of nonrandom mating in which individuals select mates with a similar phenotype to themselves (**positive assortative mating**) or with a dissimilar phenotype to themselves (**negative assortative mating** or **disassortative mating**). For example, humans tend to choose mates who are of a similar height to themselves.

aster A starlike arrangement of microtubules radiating from a *centrosome. Asters become conspicuous in animal cells at the ends of the spindle when cell division starts. The astral microtubules help to locate the spindle in relation to the cell's boundaries and to position the cleavage furrow of the cytoplasm when nuclear division is completed.

asthenosphere *See* PLATE TECTONICS.

astigmatism A lens defect in which rays in one plane are in focus when those in another plane are not. The eye can suffer from astigmatism, usually when the cornea is not spherical. It is corrected by using an anastigmatic lens, which has different radii of curvature in the vertical and horizontal planes.

astrobiology The study of life on other planets. Many disciplines—including astrophysics, biochemistry, microbiology, and geology—are combined in looking for signs of extraterrestrial life in the past or present. The work of astrobiologists provides insights into the possible origin of life on earth, for instance whether living organisms were initially carried to earth from space by meteorites (the *panspermia theory). Studies of bacteria that live in extreme conditions on earth (*extremophiles) indicate how organisms might live in the harsh conditions elsewhere in space, and survive as spores for very long periods. Astrobiology also contributes to our understanding of the conditions necessary for life

astrocyte *See* GLIA.

AT content (A+T content) *See* GC CONTENT.

atherosclerosis Obstruction of the arteries by localized deposits (plaques) of fatty material (including *cholesterol) and fibrous tissue on their inner walls, which makes the walls less elastic and predisposes to the formation of blood clots (thrombi). Atherosclerosis is associated with high blood levels of cholesterol, particularly in the form of low-density *lipoprotein; other risk factors include smoking, a sedentary lifestyle, and high blood pressure. It can result in coronary thrombosis and heart attack if it affects the coronary arteries (*see* CORONARY VESSELS).

atlas The first *cervical vertebra, a ringlike bone that joins the skull to the vertebral column in terrestrial vertebrates. In advanced vertebrates articulation between the skull and atlas permits nodding movements of the head. *See also* AXIS.

atomic force microscopy (AFM) A variation of *scanning probe microscopy that measures the force of interaction between a fine-tipped probe and the surface of a sample. Capable of nanometre-scale resolution, it is suitable for imaging the topography of biomolecules such as DNA and proteins, cell surfaces, and cell organelles. Essentially the apparatus consists of a silicon-tipped probe, mounted on a flexible cantilever, which is moved across the sample surface. Deflections of the probe are detected by a laser beam focused onto the back of the cantilever and reflected to a photosensor position detector. The basic technique can operate in various modes. In contact mode the tip of the probe touches the surface, so that it interacts with the attractive (van der Waals) forces exerted by the atoms of the sample surface. The probe (or sample) can be moved very precisely in three dimensions (by a piezoelectric device), and feedback from the position detector enables the probe position to be adjusted and maintained at a constant force or constant distance from the surface. Measures of these adjustments provide the raw data, which are processed by computer and converted into images. In other modes the probe is oscillated very close to the surface but not allowed to touch the sample, or to touch it only intermittently. These can give better resolution with certain types of sample.

(((●))) SEE WEB LINKS
• Succinct account of the technique

ATP (adenosine triphosphate) A nucleotide that is of fundamental importance as a carrier of chemical energy in all living organisms. It consists of adenine linked to D-ribose (i.e. adenosine); the D-ribose component bears three phosphate groups,

ATP

linearly linked together by covalent bonds (see formula). These bonds can undergo hydrolysis to yield either a molecule of **ADP** (**adenosine diphosphate**) and inorganic phosphate or a molecule of **AMP** (**adenosine monophosphate**) and pyrophosphate (*see* ATPASE). Both these reactions yield a large amount of energy (about 30.6 kJ mol^{-1}) that is used to bring about such biological processes as muscle *contraction, the *active transport of ions and molecules across plasma membranes, activation of proteins involved in signal transduction, and the synthesis of biomolecules. The reactions bringing about these processes often involve the enzyme-catalysed transfer of the phosphate group to intermediate substrates, for example by a *kinase enzyme. Most ATP-mediated reactions require Mg^{2+} ions as *cofactors.

ATP is regenerated by the rephosphorylation of AMP and ADP using the chemical energy obtained from the oxidation of food. This takes place during *glycolysis and the *Krebs cycle but, most significantly, is also a result of the reduction-oxidation reactions of the mitochondrial *electron transport chain, which ultimately reduces molecular oxygen to water (*oxidative phosphorylation). ATP is also formed by the light-dependent reactions of *photosynthesis.

ATPase Any enzyme that brings about the hydrolysis of ATP. This results in the cleavage of either one phosphate group, with the formation of ADP and inorganic phosphate (P_i), or of two phosphate groups, with the formation of AMP and pyrophosphate (PP_i); the second reaction yields twice as much energy as the first. ATPase activity is associated with many energy-consuming processes; for example, in muscle contraction it is associated with *myosin when activated by actin. One form of ATPase, *ATP synthetase, can catalyse the synthesis of ATP, for example in the mitochondrial *electron transport chain (*see also* CHEMIOSMOTIC MECHANISM).

ATP synthetase (ATP synthase; F_0F_1 complex) An enzyme complex that catalyses the formation of ATP from ADP and inorganic phosphate. It occurs in the inner mitochondrial membrane and is responsible for *oxidative phosphorylation during respiration. Similar complexes are also found in the *thylakoid membranes of chloroplasts, where they generate ATP in the light-dependent reactions of *photosynthesis, and in the cells of bacteria. Hence, ATP synthetase is of fundamental importance to the vast majority of living organisms. The mitochondrial synthetase complex consists of a membrane-spanning proton channel (the **F_0 portion**), an ATP-synthesizing part (the **F_1 portion**), and a membrane-spanning stator. Each portion comprises multiple polypeptides. According to the *chemiosmotic mechanism, protons (H^+) flowing through the F_0 channel provide the energy for phosphorylation of ADP at catalytic sites in the F_1 portion. The stator guides protons to binding sites on subunits of the F_0 portion; this causes it to act as a rotor, spinning within the mitochondrial inner membrane and turning an internal rod. The spinning rod projects like an axle into the F_1 portion, which is held stationary just inside the mitochondrial matrix. Rotation of the rod changes the shape of the F_1 subunits, exposing catalytic sites that convert ADP to ATP. It is estimated that in humans this mechanism produces ATP equal to a person's body weight every day.

(((•))) SEE WEB LINKS

• Models and animations of ATP synthetase and ATP formation

atrial natriuretic peptide (ANP; atrial natriuretic hormone) A peptide hormone, produced by certain cells in the wall of the atrium of the heart, that promotes the excretion of sodium ions in the urine (i.e. natriuresis). Secretion of ANP is triggered by increased stretch of the atrial wall, due to raised blood pressure or increased blood volume. It acts to inhibit sodium reabsorption in the kidneys and the secretion of *aldosterone by the adrenal glands. Consequently, sodium losses to urine are increased, and water follows by osmosis, thereby decreasing blood volume and blood pressure.

atrioventricular node (AVN) A specialized group of *cardiac muscle fibres situated in the fibrous ring between the right atrium and ventricle of the heart. The AVN is the only pathway between the atria and the ventricles through which electrical impulses can pass. Thus, following the contraction of the atria, the AVN initiates a wave of contraction in the ventricles via the *bundle of His.

atrioventricular valve See BICUSPID VALVE; TRICUSPID VALVE.

atrium (auricle) (*pl.* **atria**) **1.** A chamber of the *heart that receives blood from the veins and forces it by powerful muscular contraction into the *ventricle(s). Fish have a single atrium but all other vertebrates have two. **2.** Any of various cavities or chambers in animals, such as the chamber surrounding the gill slits of the lancelet and other invertebrate chordates.

atrophy The degeneration or withering of an organ or part of the body.

atropine A poisonous crystalline alkaloid, $C_{17}H_{23}NO_3$. It can be extracted from deadly nightshade and other solanaceous plants and is used in medicine to treat colic, to reduce secretions, and to dilate the pupil of the eye.

attenuation 1. (in medicine) A process of reducing the disease-producing ability of a microorganism. It can be achieved by chemical treatment, heating, drying, irradiation, by growing the organism under adverse conditions, or by serial passage through another organism. Attenuated bacteria or viruses are used for some *vaccines. **2.** (in mycology) The conversion by yeasts of carbohydrates to alcohol, as in brewing and wine and spirit production. **3.** (in genetics) A mechanism for regulating gene expression in prokaryotes, observed especially in functional gene clusters (*operons), such as the *trp* genes that encode enzymes responsible for synthesizing tryptophan in *E. coli* bacteria. Attenuation comes into play when the product of the enzymes (in this case tryptophan) is present to excess in the medium; transcription of the operon is drastically reduced, perhaps by as much as 90% of the maximum rate. This

attenuation is caused by formation of stem-loop structures in the initial part of the RNA transcript, encoded by the *trpL* gene upstream of the structural genes. These RNA structures either allow transcription to proceed past the *trpL* gene or cause termination, depending on available concentrations of tryptophan transfer RNA, and hence tryptophan.

atto- Symbol a. A prefix used in the metric system to denote 10^{-18}. For example, 10^{-18} second = 1 attosecond (as).

audibility The state of being perceptible to hearing. The limits of audibility of the human ear are between about 20 hertz (a low rumble) and 20 000 hertz (a shrill whistle). With increased age the upper limit falls quite considerably.

audiometer An instrument that generates a sound of known frequency and intensity in order to measure an individual's hearing ability.

auditory Of or relating to the sense of *hearing. For example, the **auditory meatus** is the canal leading from the pinna to the tympanum (eardrum) in the ear.

auditory nerve The nerve that transmits sensory information from the ear to the brain. See also ORGAN OF CORTI.

auricle 1. See ATRIUM. **2.** See PINNA.

Australian region See FAUNAL REGION.

Australopithecus A genus of fossil hominins that lived roughly 4–2 million years ago, coexisting for some of this time with early forms of humans (see HOMO). They had both apelike and human features; for instance, they walked erect and had teeth resembling those of modern humans, but they also climbed trees and their brain capacity was less than one-third that of a modern human. Various finds have been made, chiefly in East and southern Africa (hence the name, which means 'southern ape'). The earliest belong to the species *A. anamensis* from Kenya and Ethiopia. This may have been an ancestor of *A. afarensis*, which includes the specimen of a female, dubbed 'Lucy', found at Laetoli in Tanzania. In 2008 came the discovery of the first fossil skeleton of *A. sediba*, which lived around 1.9 mya. *Australopithecus* and related genera are known as **australopithecines**.

autacoid (autocoid) A physiologically active substance, especially one that acts locally to regulate the activities of certain cells within a given tissue. Examples include serotonin and histamine.

autapomorphy *See* APOMORPHY.

autecology The study of ecology at the level of the species. An autecological study aims to investigate the ecology of *populations or individuals of a particular species, including habitat, distribution, life cycle, etc. This should enable a full description of the *ecological niche of the organism to be made. *Compare* SYNECOLOGY.

autochthonous Describing an organism that is native to the place in which it is found. *Compare* ALLOCHTHONOUS.

autoclave A strong steel vessel used for carrying out chemical reactions, sterilizations, etc., at high temperature and pressure.

autocrine Describing a type of cell signalling in which the signal molecule influences only the cell that produced it, by binding to receptors in or on itself. Autocrine signals are important in, e.g., embryonic development, tissue differentiation, inflammation, and wound healing; when disrupted, they can lead to the uncontrolled proliferation seen in tumour cells. *Compare* PARACRINE. *See* SIGNAL TRANSDUCTION.

autoecious Describing rust fungi (*see* RUSTS) that pass their life cycle in association with only one host plant. An example is *Puccinia menthae* (mint rust). *Compare* HETEROECIOUS.

autogamy 1. A type of reproduction that occurs in single isolated individuals of ciliate protists of the genus *Paramecium* (*see* CILIOPHORA) The nucleus divides into two genetically identical haploid nuclei, which then fuse to form a diploid zygote. The onset of autogamy is associated with

changing environmental conditions and may be necessary to maintain cell vitality. **2.** Self-fertilization in plants. *See* FERTILIZATION.

autogenic Relating to or caused by a change in the environment or an individual organism due to some *endogenous factor, i.e. one that comes from within the environment or organism. *Compare* ALLOGENIC.

autograft *See* GRAFT.

autoimmunity A disorder of the body's defence mechanisms in which an immune response is elicited against its own tissues, which are thereby damaged or destroyed. Effector cells of the immune system called T cells bind to self antigens and elicit the production of antibodies or a cytotoxic response. For example, insulin-dependent diabetes mellitus is caused by an immune response against proteins in the B cells of the pancreatic islets, thus abolishing the production of insulin. Rheumatoid arthritis, systemic lupus erythematosus, myasthenia gravis, and several forms of thyroid dysfunction are other examples of autoimmune diseases. Autoimmunity arises from a breakdown in immunological *tolerance to self tissues and may also involve a failure of regulatory T cells to suppress immune responses to self antigens.

autoinducer *See* QUORUM SENSING.

autolysis The process of self-destruction of a cell, cell organelle, or tissue. It occurs by the action of enzymes within or released by *lysosomes. *See also* LYSIS.

autonomic nervous system (ANS) The part of the vertebrate *peripheral nervous system that supplies stimulation via motor nerves to the smooth and cardiac muscles and to the glands of the body. It is divided into the *parasympathetic and the *sympathetic nervous systems, which tend to work antagonistically on the same organs (see illustration). The activity of the ANS is controlled principally by the *medulla oblongata, midbrain, and pons of the brainstem, which gives rise to the paired cranial nerves, and by the *hypothalamus.

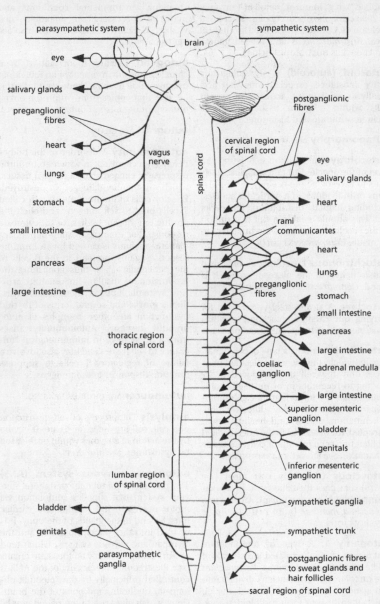

The autonomic nervous system (showing one side of each system of the ANS)

autopolyploid A *polyploid organism in which the multiple sets of chromosomes are all derived from the same species. For example, doubling of the chromosome number during mitotic cell division, possibly induced by *colchicine, gives rise to a tetraploid known as an **autotetraploid**. *Compare* ALLOPOLYPLOID.

autoradiography An experimental technique in which a radioactive specimen is placed in contact with (or close to) a photographic plate, so as to produce a record of the distribution of radioactivity in the specimen. The film is darkened by the ionizing radiation from radioactive parts of the sample. Autoradiography is used to study the distribution of particular substances in living tissues, cells, and cultures. A radioactive isotope of the substance is introduced into the organism or tissue, which is killed, sectioned, and examined after enough time has elapsed for the isotope to be incorporated into the substance. Another common application of autoradiography is the location of radioactively labelled DNA probes or antibodies employed in such techniques as *Southern blotting and *western blotting.

autosome Any of the chromosomes in a cell other than the *sex chromosomes.

autostylic jaw suspension A type of jaw suspension seen in the lungfish (*see* DIPNOI) and tetrapods, in which the upper jaw is connected directly to the cranium (see illustration). *Compare* AMPHISTYLIC JAW SUSPENSION; HYOSTYLIC JAW SUSPENSION.

autotomy The shedding by an animal of part of its body followed by the regeneration of the lost part. Autotomy is achieved by the contraction of muscles at specialized regions in the body. It serves as a protective mechanism if the animal is damaged or attacked (e.g. tail loss in certain reptiles) and is common as a method of asexual reproduction in polychaete worms, in which both new head and tail regions may be regenerated.

autotrophic nutrition A type of nutrition in which organisms synthesize the organic materials they require from inorganic sources. Chief sources of carbon and nitrogen are carbon dioxide and nitrates, respectively. All green plants are autotrophic and use light as a source of energy for the synthesis, i.e. they are **photoautotrophic** (*see* PHOTOSYNTHESIS). Some bacteria are also photoautotrophic; others are **chemoautotrophic**, using energy derived from chemical processes (*see* CHEMOSYNTHESIS). *Compare* HETEROTROPHIC NUTRITION; MIXOTROPHIC.

auxanometer Any mechanical instrument or measuring device used to study the growth or movement of plant organs. One type of auxanometer consists of a recording device that translates any increase in stem height into movement of a needle across a scale.

auxin Any of a group of *plant hormones responsible for such processes as the promotion of growth by cell enlargement, the maintenance of *apical dominance, and the initiation of root formation in cuttings. Auxins are also involved in suppressing the *abscission of leaves, fruit, or other plant organs and in the development of flowers and fruits. Naturally occurring auxin is principally **indoleacetic acid** (**IAA**), which is

cranium —————

upper jaw —————

lower jaw

Autostylic jaw suspension

synthesized in actively growing regions of the plant, particularly shoot tips, from where it is transported in one direction only, from tip to base of the plant—so-called polar transport. Auxin acts by binding to *auxin-binding proteins within target cells. According to the *acid growth theory, auxin stimulates cell expansion by causing acidification of the cell wall and consequent loosening of the cell wall structure. It also activates the expression of genes involved in the cell's growth responses. Various other aspects of a plant's development are controlled, at least in part, by the flow of auxin, including the pattern of branches, the arrangement of leaves, and the organization of leaf veins. IAA is stored in the plant in an inactive form, conjugated (attached) to various compounds, such as *myo*-inositol. Synthetic auxins include *2,4-D, which is used as a weedkiller, whereas the natural auxin **indolebutyric acid** is sold in preparations of 'rooting hormones'.

auxin-binding protein A protein found in plant cells that binds *auxins and thus acts as the initial receptor of the signal transduction pathway for these *plant hormones. A receptor for auxin has been identified as the protein TIR1 (transport inhibitor response 1), found in the nucleus. Binding of auxin to TIR1 leads to the degradation of certain repressors of gene transcription, which allows genes to be 'switched on'. Another, cytosolic, auxin-binding protein called ABP1 has been proposed as a key component of the auxin signalling pathway, although recent research has refuted this role in *Arabidopsis*. See also ACID GROWTH THEORY.

Avery, Oswald Theodore (1877–1955) Canadian bacteriologist who worked at the Rockefeller Institute Hospital in New York (1913–48), where he and his colleagues Maclyn McCarty and Colin Macleod identified DNA as the hereditary material in the cells of pneumococcus bacteria. It had previously been thought that protein was the hereditary material; Avery's work was an important step in leading Watson and Crick to the discovery of the true chemical basis of heredity.

Aves The birds: traditionally a class of bipedal vertebrate chordates (*see* CHORDATA) with *feathers, wings, and a beak, they are now regarded as bona fide reptiles. They evolved probably in the mid to late Jurassic period (165–150 million years ago) from reptilian ancestors—the *theropod clade of saurischian dinosaurs (e.g. *Tyrannosaurus*)—although the identity of their immediate ancestors remains controversial. Modern birds still have scaly legs, like reptiles, but are warm-blooded (*see* HOMOIOTHERMY). The skin is dry and loose and has no sweat glands, so cooling is effected by panting. Their efficient lungs and four-chambered heart (which completely separates oxygenated and deoxygenated blood) ensure a good supply of oxygen to the tissues. Birds can therefore sustain a high body temperature and level of activity necessary for *flight. The breastbone bears a keel for the attachment of flight muscles. The skeleton is very light; many of the bones are tubular, having internal struts to provide strength and *air sacs to reduce weight and provide extra oxygen in flight. Their feathers are vital for flight, streamlining the body, and insulation against heat loss.

Many birds show a high degree of social behaviour in forming large flocks and pair bonding for nesting, egg incubation, and rearing young. Fertilization is internal and the female lays hard-shelled eggs. In some classifications all modern birds are placed in the subclass Neornithes, whereas extinct birds having clawed forelimbs and toothed jaws (e.g. *Archaeopteryx*, considered to be one of the earliest birds) are placed in the subclass Archaeornithes. See also RATITAE.

((⊕)) SEE WEB LINKS
• Global partnership of organizations involved in bird conservation

avidin A glycoprotein component of egg white that binds strongly to the vitamin *biotin. Proteins and nucleic acids can be linked to biotin (**biotinylated**), and the avidin-biotin reaction can then be used in a number of assay methods, such as antigen-antibody reactions or *DNA hybridization. For example, enzymes conjugated with avidin can be used to bind to biotinylated antibodies.

avidity (in immunology) The total binding strength of an antibody molecule to its target site(s), or *epitopes, on the antigen. Avidity is thus determined by both the number of binding sites of a particular antibody that are attached to epitopes and the strength of binding of each individual site (*see* AFFINITY).

Avirulence gene (*Avr* gene) *See* R PROTEIN.

AVN *See* ATRIOVENTRICULAR NODE.

awn A stiff bristle that projects from the tip of a plant part or organ. The bracts in a grass inflorescence commonly bear awns (*see* SPIKE).

axenic culture A *culture medium in which only one type of microorganism is growing. Such cultures are widely used in microbiology to determine the basic growth requirements or degree of inhibition by antibiotics or other chemicals of a particular species.

axial skeleton The main longitudinal section of the vertebrate *skeleton, including the *skull, the *vertebral column, and the rib cage. *Compare* APPENDICULAR SKELETON.

axil The angle between a branch or leaf and the stem it grows from. **Axillary** (or **lateral**) **buds** develop in the axil of a leaf. The presence of axillary buds distinguishes a leaf from a leaflet.

axillary bud *See* AXIL.

axis The second *cervical vertebra, which articulates with the *atlas (the first cervical vertebra, which articulates with the skull). The articulation between the axis and atlas in reptiles, birds, and mammals permits side-to-side movement of the head. The body of the axis is elongated to form a peg (the **odontoid process**), which extends into the ring of the atlas and acts as a pivot on which the atlas (and skull) can turn.

axon The long threadlike part of a nerve cell (*neuron). It carries the nerve impulse (in the form of an *action potential) away from the *cell body of a neuron towards either an effector organ or the brain. *See also* NERVE FIBRE.

axoneme The core of a cilium or eukaryotic *flagellum, consisting of two central *microtubules surrounded by nine other pairs of microtubules. The outer microtubules are associated with the protein *dynein, which is responsible for the movement of the organelle.

BAC *See* (BACTERIAL) ARTIFICIAL CHROMOSOME.

Bacillariophyta In some classifications, a phylum (or group, i.e. the bacillariophytes) of protist *algae comprising the diatoms and belonging to the *stramenopiles. These microscopic marine or freshwater unicellular organisms have cell walls (**frustules**) composed of pectin impregnated with silica and consisting of two halves, one overlapping the other. Diatoms are found in huge numbers in plankton and are important in the food chains of seas and rivers. Past deposition has resulted in diatomaceous earths (kieselguhr) and the oil reserves of these species have contributed to oil deposits. The global abundance of diatoms means that they play a significant role in absorbing carbon dioxide from the atmosphere. This has led to the proposal that encouraging blooms of diatoms by fertilizing the oceans with essential nutrients may be a means of capturing this greenhouse gas and storing it in the bodies of diatoms on the ocean floor.

bacillus (*pl.* **bacilli**) Any rod-shaped bacterium. Generally, bacilli are large, Gram-positive, spore-bearing, and have a tendency to form chains and produce a *capsule. Some are motile, bearing *flagella. They are ubiquitous in soil and air and many are responsible for food spoilage. The group also includes *Bacillus anthracis*, which causes anthrax.

backbone *See* VERTEBRAL COLUMN.

back breeding *See* DE-EXTINCTION.

back cross A mating between individuals of the parental generation (P) and the first

generation of offspring (F_1) in order to identify hidden *recessive alleles. If an organism displays a *dominant characteristic, it may possess two dominant alleles (i.e. it is homozygous) or a dominant and a recessive allele for that characteristic (i.e. it is heterozygous). To find out which is the case, the organism is crossed with one displaying the recessive characteristic. If all the offspring show the dominant characteristic then the organism is homozygous, but if half show the recessive characteristic, then the organism is heterozygous. *See also* TEST CROSS.

background radiation Low-intensity *ionizing radiation present on the surface of the earth and in the atmosphere as a result of cosmic radiation and the presence of radioisotopes in the earth's rocks, soil, and atmosphere. The radioisotopes are either natural or the result of nuclear fallout or waste gas from power stations.

Bacteria (Eubacteria) A domain of life containing a diverse group of ubiquitous

A generalized bacterial cell

microorganisms all of which consist of only a single *cell that lacks a distinct nuclear membrane and has a *cell wall of a unique composition (see illustration). Bacteria constitute the prokaryotic organisms of the living world. However, their classification is a controversial issue. It is now recognized, on the basis of differences in ribosomal RNA structure and nucleotide sequences (see MOLECULAR SYSTEMATICS), that prokaryotes form two evolutionarily distinct domains: *Archaea (the archaea) and Bacteria. Defining characteristics of bacteria include the possession of cell walls containing peptidoglycan, and membrane lipids containing fatty acids in ester linkage to glycerol, whereas archaea lack peptidoglycan and have ether-linked lipids. However, in general parlance, the term 'bacteria' can still, erroneously, encompass both archaea and bacteria.

Bacteria can be characterized in a number of ways, for example by their reaction with *Gram's stain, their *GC content, or on the basis of their metabolic requirements (e.g. whether or not they require oxygen: see AEROBIC RESPIRATION; ANAEROBIC RESPIRATION) and shape. A bacterial cell may be spherical (see COCCUS), rodlike (see BACILLUS), spiral (see SPIRILLUM), comma-shaped (see VIBRIO), corkscrew-shaped (see SPIROCHAETE), or filamentous, resembling a fungal cell. The majority of bacteria range in size from 0.5 to 5 μm. Many are motile, bearing *flagella, possess an outer slimy *capsule, and produce resistant spores (see ENDOSPORE). In general bacteria reproduce only asexually, by simple division of cells, but a few groups undergo a form of sexual reproduction (see CONJUGATION) and *lateral gene transfer is common. Bacteria are largely responsible for decay and decomposition of organic matter, producing a cycling of such chemicals as carbon (see CARBON CYCLE), oxygen, nitrogen (see NITROGEN CYCLE), and sulphur (see SULPHUR CYCLE). A few bacteria obtain their food by means of *photosynthesis, including the *Cyanobacteria; some are saprotrophs; and others are parasites, causing disease. The symptoms of bacterial infections are produced by *toxins.

((⊕)) SEE WEB LINKS
• Online textbook of bacteriology devised by Kenneth Todar, University of Wisconsin-Madison

bacterial artificial chromosome See ARTIFICIAL CHROMOSOME.

bacterial growth curve A curve on a graph that shows the changes in size of a bacterial population over time in a culture. The bacteria are cultured in sterile nutrient medium and incubated at the optimum temperature for growth. Samples are removed at intervals and the number of viable bacteria is counted. A logarithmic growth curve is plotted, which shows various phases (see graph).

In the **lag** (or **latent**) **phase** there is only a small increase in numbers as the bacteria imbibe water, and synthesize ribosomal RNA and subsequently enzymes, in adjusting to the new conditions. The length of this phase depends on which medium was used to culture the bacteria before the investigation and which phase the cells are already in. As the life span (generation time) of the cells decreases, they enter the **log** (or **exponential**) **phase**, in which the cells reach a maximum rate of reproduction and the number of bacteria increases directly with time, giving a straight slope on a logarithmic scale (see EXPONENTIAL GROWTH). For example, the fastest generation time for

Bacterial growth curve

E. coli is 21 minutes. Growth rate can be estimated in this phase. With time, as the population grows, it enters the **stationary phase**, when the nutrients and electron acceptors are depleted and the pH drops as carbon dioxide and other waste poisons accumulate. As the cell's energy stores are depleted the rate of cell division decreases. The **death** (or **final**) **phase** occurs when the rate at which the bacteria die exceeds the rate at which they are produced; the population declines as the levels of nutrients fall and toxin levels increase. *See also* POPULATION GROWTH.

bactericidal Capable of killing bacteria. Common bactericides are some *antibiotics, *antiseptics, and *disinfectants. *Compare* BACTERIOSTATIC.

bacteriochlorophyll A form of chlorophyll found in photosynthetic bacteria, notably the purple and green bacteria. There are several types, designated *a* to *g*. For example, bacteriochlorophyll *a* and bacteriochlorophyll *b* are structurally similar to the chlorophyll *a* and chlorophyll *b* found in plants, but absorb light maximally in different parts of the spectrum. The purple bacteria contain either of these two types of bacteriochlorophyll, depending on the species, located in specialized membrane systems (**chromatophores**) in the form of sheets, tubes, or vesicles arising inside the cell from the plasma membrane. In green bacteria the photosynthetic apparatus is contained in cylindrical structures (**chlorosomes**) underlying the plasma membrane. *Compare* BACTERIORHODOPSIN.

bacteriocin A peptide or protein produced by bacteria that inhibits or kills closely related species or similar strains of the same species. Bacteriocins are more specifically targeted than *antibiotics, serving to enhance the survival of the producing cell at the expense of genetically similar but distinct competitors. They are often encoded by plasmids, and traditionally were named according to the organism producing them. Hence, strains of *Escherichia coli* produce colicins, whereas *Bacillus subtilis* synthesizes subtilisins. Some form channels in the cell membrane of the target organism, allowing leakage of ions, whereas others act as nucleases, cutting DNA or RNA inside the cell. Bacteriocins produced by lactic acid bacteria include *lantibiotics and pediocins; these are useful in the food industry for their activity against food spoilage organisms and pathogenic *Listeria* bacteria.

bacteriology The study of bacteria, including their identification, form, function, genetics, reproduction, and classification. Much attention is focused on the role of bacteria as agents of disease in animals, including humans, and in plants, and on methods of controlling pathogenic bacteria in the food chain and elsewhere in the environment. However, bacteriologists also investigate the many benefits of bacteria, e.g. in the production of antibiotics, enzymes, and amino acids, and in sewage treatment. *See* BIOTECHNOLOGY.

bacteriophage (phage) A virus that is parasitic within a bacterium. Each phage is specific for only one type of bacterium. Most phages (**virulent phages**) infect, quickly multiply within, and destroy (lyse) their host cells. However, some (**temperate phages**) remain dormant in their hosts after initial infection: their nucleic acid becomes integrated into that of the host and multiplies with it, producing infected daughter cells (*see* LYSOGENY). Lysis may eventually be triggered by environmental factors. Phages are used experimentally to identify bacteria, to control manufacturing processes (such as cheese production) that depend on bacteria, and, because they can alter the genetic make-up of bacterial cells, they are important tools in genetic engineering as cloning *vectors. **Phage therapy** is the targeted application of phages to treat bacterial infection. It was first used in 1919 by the French physician Félix d'Hérelle and subsequently developed especially in the Soviet Union and eastern Europe; in the West it was superseded by the introduction of antibiotics in the 1940s. The emergence of antibiotic-resistant strains of bacteria has prompted renewed interest in phage therapy since the 1990s, especially using bioengineered phages and phage-derived lytic proteins capable of destroying pathogen bacterial cells. *See* LAMBDA PHAGE.

bacteriorhodopsin A membrane-bound protein manufactured by the halophilic ('salt-loving') archaeans, such as *Halobacterium salinarum*, that exploits light energy to make ATP without the involvement of chlorophyll-like pigments. When activated by light, it pumps protons out of the cell; this creates a concentration gradient, which enables ATP to be synthesized. Bacteriorhodopsin is composed of seven α-helix segments, which span the membrane and are joined together by short amino-acid chains. It contains the prosthetic group retinal, which is also found in the pigment *rhodopsin in the rod cells of vertebrates.

bacteriostatic Capable of inhibiting or slowing down the growth and reproduction of bacteria. Some *antibiotics are bacteriostatic. *Compare* BACTERICIDAL.

bacterium (*pl.* **bacteria**) A member of the domain *Bacteria.

bacteroid The form adopted by a nitrogen-fixing bacterium when active within the root nodule of a host plant. For example, *Rhizobium* bacteria change into enlarged irregularly shaped branching cells when they infect the root cells of their legume host. These bacteroids become surrounded by a **peribacteroid membrane**, derived from membranes of their host cells, and differentiate to produce the key enzymes and other components of nitrogen fixation, such as *nitrogenase. As the infection process progresses, the bacteroids become housed in a *root nodule, where they are totally dependent on their host for the energy required for nitrogen. In return they supply their host with an assimilable form of nitrogen, i.e. ammonia, which is incorporated into amino acids. *See* NITROGEN FIXATION.

baker's yeast Strains of the yeast *Saccharomyces cerevisiae* that are used in bread-making to enable the dough to rise. Water is added to flour, which activates the *amylase enzymes that hydrolyse the starch in flour to glucose. Baker's yeast is then added, which uses the glucose as a substrate for alcoholic *fermentation. The carbon dioxide produced from the yeast's fermentation causes bubbles to form in the dough; these become larger during heating in an oven, giving bread its typical texture. As baking proceeds, the heating removes the carbon dioxide and alcohol.

balance 1. (in animal physiology) Equilibrium in the posture of the body. In vertebrates balance is sensed and maintained by the *vestibular apparatus of the inner ear. **2.** (in nutrition) *See* DIET.

balanced polymorphism *See* POLYMORPHISM.

balancing selection Any mechanism of *natural selection that maintains genetic diversity in a population. Such diversity better equips the population to adapt to changes in its environment, by producing individuals of varying fitness in respect of, e.g., seasonal changes in climate or spatial variations in habitat. Balancing selection thus maintains genetic and phenotypic polymorphism in the population and prevents loss of rare alleles by *genetic drift. One mechanism of balancing selection is **frequency-dependent selection**, in which the fitness of a particular phenotype, and thus its corresponding alleles, depends on its frequency in the population. In negative frequency-dependent selection, rare phenotypes are favoured over more common variants of the same characteristic. A classic example is self-incompatibility in plants, which prevents male pollen from fertilizing female flower parts (pistils) on the same or a closely related plant, thereby preventing inbreeding. In essence, compatibility is controlled by alleles at a single *S* locus, and if these are the same or similar in male and female tissues, fertilization is prevented. Hence pollen carrying a rare *S* allele is more likely to find a compatible plant and breed successfully than one carrying a common *S* allele; consequently the frequency of the rare allele in the population is likely to increase over time. This type of balancing selection accounts for the often large number of *S* alleles that are maintained in a plant species over the long term. Another mechanism of balancing selection is **heterozygote advantage**, in which individuals carrying different alleles for a particular gene (i.e. heterozygotes) have greater reproductive success than either of the corresponding homozygotes. An example is sickle-cell disease in humans, which is due to a mutant form of haemoglobin causing abnormally shaped red blood cells. Individuals carrying both the mutant sickle-cell allele and the normal allele are more likely to survive in areas where malaria is endemic, compared to individuals who are homozygous for either allele. This results in a balanced polymorphism. *See* POLYMORPHISM. *Compare* POSITIVE SELECTION.

Baldwin effect An idea advanced in 1896 by US psychologist James Mark Baldwin (1861–1934) and others to explain the social evolution of traits that are learned during the lifetime of individuals. Originally called **organic evolution** by Baldwin, it proposes that acquired behavioural traits, such as a thrush learning to smash a snail against a rock, can be transmitted to subsequent generations without contradicting Darwin's theory of natural selection in favour of *Lamarckism. Essentially, the thrush that learns to smash snails improves its life chances and is more likely to breed successfully and produce fertile offspring. What the offspring inherit from their parent is not the snail-smashing behaviour itself but a propensity to acquire behaviours. Hence, there is an increased likelihood that the offspring will themselves acquire the behaviour, for example by imitating their parent, so increasing its frequency within the population. Over successive generations, the behaviour itself influences the evolution of the population as other genetically controlled traits are selected to enhance and consolidate the learned trait, which ultimately itself comes

under genetic control as an instinctive behaviour. There is continuing controversy about the validity and biological significance of the Baldwin effect. Notably, it has been invoked to explain the rapid evolution of complex behaviours, cognitive skills, and language in primates, especially humans. Proponents argue (contentiously) that the emerging mind, through the dramatic advantage it confers in a social environment, effectively accelerates the evolution of the brain to accommodate it. *See also* PHENOTYPIC PLASTICITY.

baleen *See* WHALEBONE.

BALT *See* BRONCHIAL-ASSOCIATED LYMPHOID TISSUE.

barb 1. (in zoology) Any one of the stiff filaments forming a row on each side of the longitudinal shaft of a feather (see illustration). Together the barbs form the expanded part (**vane**) of the feather. *See* BARBULE. **2.** (in botany) A hooked hair.

barbiturate Any one of a group of drugs derived from barbituric acid, which have a depressant effect on the central nervous system. Barbiturates were originally used as sedatives and sleeping pills but their clinical use is now limited due to their toxic side-effects; prolonged use can lead to addiction. Specific barbiturates in clinical use include butobarbital, used to treat insomnia, and thiopental, used as an anaesthetic.

barbule Any of the minute filaments forming a row on each side of the *barb of a feather. In a *contour feather adjacent

Interlocking barbs of a contour feather

barbules interlock by means of hooks (**barbicels**) and grooves, forming a firm vane. Down feathers have no barbicels.

Barfoed's test A biochemical test to detect monosaccharide (reducing) sugars in solution, devised by the Swedish physician C. T. Barfoed (1815–99). **Barfoed's reagent**, a mixture of ethanoic (acetic) acid and copper(II) acetate, is added to the test solution and boiled. If any reducing sugars are present a red precipitate of copper(II) oxide is formed. The reaction will be negative in the presence of disaccharide sugars as they are weaker reducing agents.

bark The protective layer of mostly dead cells that covers the outside of woody stems and roots. It includes the living and dead tissues external to the vascular *cambium, including the secondary phloem and periderm. The term can be used more specifically to describe the periderm together with other tissues isolated by the activity of the *cork cambium. In some species, such as birch, there is one persistent cork cambium but in the older stems of certain other species a second cork cambium becomes active beneath the periderm and further periderm layers are formed every few years. The result is a composite tissue called **rhytidome**, composed of cork, dead cortex, and dead phloem cells.

baroreceptor A *receptor that responds to changes in pressure. Baroreceptors in the aorta and *carotid sinus (in the carotid artery) respond to changes in arterial pressure and relay information via sensory nerves to the cardiac control centre in the medulla oblongata. This regulates *blood pressure and heartbeat via the parasympathetic nervous system.

Barr body A structure consisting of a condensed X chromosome (*see* X INACTIVATION) that is found in nondividing nuclei of female mammals. The presence of a Barr body is used to confirm the sex of athletes in sex determination tests. It is named after the Canadian anatomist M. L. Barr (1908–95), who identified it in 1949.

basal body (kinetosome) *See* FLAGELLUM.

basal ganglia (basal nuclei) Small interconnected clusters of neurons found deep inside the cerebrum of the brain. The main ones are the caudate nucleus, the putamen, and the globus pallidus (the latter two constituting the lentiform nucleus)—one of each occurring in each cerebral hemisphere. They are involved in the subconscious initiation and coordination of voluntary movements, and act in conjunction with the *cerebellum. Information from the *cerebral cortex is processed by the basal ganglia, which relay their output via the thalamus back to the motor cortex. Patients whose basal ganglia are damaged have difficulty in starting movements, and experience slowness and tremor when performing movements. The bundles of myelinated axons ('white matter') that form the internal capsule traverse between the 'grey matter' of the caudate nucleus and lentiform nucleus, giving a striped appearance; hence the three structures together form the corpus striatum ('striped body').

basal group (basal taxon) (in cladistics) A group of organisms that diverges early in the evolutionary history of a clade (*see* CLADISTICS). In a cladogram or phylogenetic tree (*see* PHYLOGRAM) the basal group branches near the root of the tree, which represents the common ancestor of the clade.

basal lamina *See* BASEMENT MEMBRANE.

basal metabolic rate (BMR) The rate of energy metabolism required to maintain a nongrowing, unstressed, endothermic animal at rest. BMR is measured in terms of heat production per unit time and is usually expressed in kilojoules of heat released per square metre of body surface per hour ($kJm^{-2}h^{-1}$). It indicates the energy consumed in order to sustain such vital functions as heartbeat, breathing, nervous activity, active transport, and secretion. Different tissues have different metabolic rates (e.g. the BMR of brain tissue is much greater than that of bone tissue) and therefore the tissue composition of an animal determines its overall BMR. For organisms generally, BMR is proportional to body weight according to Kleiber's law, which is an example of an allometric equation (*see* ALLOMETRIC GROWTH); small animals tend to have a higher metabolic rate per unit weight than large ones. *Compare* STANDARD METABOLIC RATE.

base A compound that reacts with an *acid to give water (and a salt). A base that dissolves in water to produce hydroxide ions is called an **alkali**. For example, ammonia reacts as follows:

$$NH_3 + H_2O \rightleftharpoons NH_4^+ + {}^-OH$$

Similar reactions occur with organic *amines (*see also* NITROGENOUS BASE).

basement membrane A thin sheet of fibrous proteins that supports the cells of an overlying *epithelium or endothelium, separating this from underlying connective tissue. Such membranes also surround muscle cells, Schwann cells, and fat cells, and a thick basement membrane is found in the kidney glomerulus, where it acts as a filter (*see* ULTRAFILTRATION). Basement membranes are components of the *extracellular matrix and help to regulate passage of materials between epithelial cells and adjacent blood vessels. Electron microscopy reveals that a basement membrane consists of two principal layers: an electron-dense **basal lamina**, typically about 30–70 nm thick, which is secreted by the epithelial or endothelial cells and consists largely of the protein *laminin; and an underlying **reticular lamina**, which comprises collagen fibrils and other components and is secreted by connective tissue cells. The laminins bind the basement membrane to neighbouring cells via *cell adhesion molecules.

base pairing The chemical linking of two complementary nitrogenous bases in *DNA and in certain types of *RNA molecules. Of the four such bases in DNA, adenine pairs with thymine and cytosine with guanine. In RNA, thymine is replaced by uracil. Base pairing is responsible for holding together the two strands of a DNA molecule to form

a double helix and for faithful reproduction and reading of the *genetic code. The links between bases take the form of *hydrogen bonds. *See also* WOBBLE.

basic fibroblast growth factor (bFGF) *See* FIBROBLAST GROWTH FACTOR.

basic stains *See* STAINING.

Basidiomycota A phylum of fungi, formerly classified as a class (Basidiomycetes) or a subdivision (Basidiomycotina). Sexual reproduction is by means of basidiospores produced externally on a club-shaped or cylindrical *basidium. Basidia are often grouped together on fruiting structures, such as mushrooms, puffballs, and bracket fungi. Exceptions are the *rusts and *smuts, which do not produce obvious fruiting bodies.

basidiospore The characteristic spore of fungi belonging to the phylum *Basidiomycota. *See* BASIDIUM.

basidium (*pl.* **basidia**) A specialized cell in fungi of the phylum *Basidiomycota, in which nuclear fission and meiosis occur. This results in the formation of four **basidiospores**, which are attached to the basidium by means of **sterigmata** (short stalks).

basilar membrane *See* ORGAN OF CORTI.

basket cell A type of inhibitory interneuron found in various parts of the central nervous system, including the cortex of the cerebellum and cerebral cortex. They are characterized by widely branching dendrites that form a netlike arrangement around the cell body of a connecting neuron. In the cerebellum the basket cells synapse with the cell bodies of the *Purkyne cells.

basophil A type of white blood cell (*leucocyte) that has a lobed nucleus surrounded by granular cytoplasm (*see* GRANULOCYTE). Basophils are produced continually by stem cells in the red bone marrow and move about in an amoeboid fashion. Normally present in very low numbers in the circulation, they may act in a similar manner to eosinophils in killing antibody-coated parasites. They accumulate with eosinophils at sites of allergic reactions and release *histamine.

bast An old name for *phloem.

batch culture A technique used to grow microorganisms or cells. A limited supply of nutrients for growth is provided; when these are used up, or some other factor becomes limiting, the culture declines. Cells, or products that the organisms have made, can then be harvested from the culture.

Batesian mimicry See MIMICRY.

bathypelagic zone See APHOTIC ZONE.

bats See CHIROPTERA.

B cell (B lymphocyte) A *lymphocyte that is derived from stem cells in the bone marrow but does not mature in the thymus (*compare* T CELL); in birds it matures in the bursa of the cloaca (hence *B* cell). Each B cell has a unique set of some 100 000 receptor molecules on its surface, designed to recognize a specific antigen. These *B-cell receptors bind soluble antigen, such as pathogen-derived proteins or toxins, and take it into the cell. Here the internalized antigen is processed and peptide fragments derived from it are bound with *MHC class II proteins, which migrate to the cell surface. These antigen-MHC class II complexes are recognized by CD4-bearing helper *T cells that respond to the same antigen, causing the T cell to adhere to the B cell and stimulate it by various means, including the secretion of lymphokines, notably *interleukin- 4. The B cell is prompted to undergo repeated division to form a clone of cells (i.e. clonal expansion). These mature into *plasma cells, capable of producing large amounts of specific antibody (*see* IMMUNOGLOBULIN), which circulates in the blood and lymph and binds to the corresponding antigen, whether this is free in body fluids or located on the surface of a pathogen. In so doing the antibodies neutralize the pathogen, preventing it from entering body cells, or mark the pathogen

for destruction by macrophages, a process called *opsonization. After a few days of antibody production the plasma cells die. However, some cells from the clone remain in the form of **memory cells** (*see* IMMUNOLOGICAL MEMORY), which initiate a more rapid immune response on subsequent exposure to the same antigen. *See also* CLONAL SELECTION THEORY.

B-cell receptor An assembly of proteins, attached to the plasma membrane of a B cell, that binds specific extracellular antigen. This activates an intracellular signal pathway that ultimately leads to changes in gene expression and hence triggers an immune response by the B cell. Additionally, bound soluble antigen (such as bacterial toxin) or small particles (such as viruses) can be taken into the cell attached to the receptor (i.e. internalized). Here the antigen is processed, bound with *MHC class II proteins, and presented at the cell surface to be recognized by *CD4-bearing effector T cells. The receptor consists essentially of a Y-shaped immunoglobulin molecule with the same specificity as the antibodies that the cell will secrete following activation. Hence, like an antibody, it has paired heavy and light chains and two antigen-recognition sites, but is anchored to the membrane through its heavy chains. The latter are associated with two immunoglobulin signal molecules, Igα and Igβ, whose intracellular portions carry amino acid sequences called **immunoreceptor tyrosine-based activation motifs** (**ITAMs**). When antigen binds to the extracellular recognition sites, these motifs recruit the cytoplasmic protein kinases responsible for initiating the signal pathway inside the cell. The involvement of several *coreceptors is required for maximal activation of the pathway. *Compare* T-CELL RECEPTOR.

B chromosome See ACCESSORY CHROMOSOME.

BDNF Brain-derived neurotrophic factor. *See* NEUROTROPHIN.

Beadle, George Wells (1903–89) US geneticist who, after holding several professorships, went to Stanford University, where he worked with Edward Tatum (1909–75). Using moulds, they deduced that the function of genes is to control the production of enzymes, which in turn control metabolic processes. They found that mutant genes result in abnormal (and non-operative) enzymes. For this 'one gene-one enzyme' theory (*see* ONE GENE-ONE POLYPEPTIDE HYPOTHESIS), they were awarded the 1958 Nobel Prize in physiology or medicine.

becquerel Symbol Bq. The SI unit of activity (*see* RADIATION UNITS). The unit is named after the discoverer of radioactivity, A. H. Becquerel (1852–1908).

bees *See* HYMENOPTERA.

beetles *See* COLEOPTERA.

beet sugar *See* SUCROSE.

behaviour The sum of the responses of an organism to internal or external stimuli. The behaviour of an animal can be either instinctive (*see* INSTINCT) or learned. *See* ANIMAL BEHAVIOUR.

behavioural genetics The branch of genetics concerned with determining the relative importance of the genetic constitution of animals as compared to environmental factors in influencing animal behaviour.

behavioural isolation *See* ISOLATING MECHANISMS.

Benedict's test A biochemical test to detect *reducing sugars in solution, devised by the US chemist S. R. Benedict (1884–1936). **Benedict's reagent**—a mixture of copper(II) sulphate and a filtered mixture of hydrated sodium citrate and hydrated sodium carbonate—is added to the test solution and boiled. A high concentration of reducing sugars induces the formation of a red precipitate; a lower concentration produces a yellow precipitate. Benedict's test is a more sensitive alternative to *Fehling's test.

benthos Flora and fauna occurring on the bottom of a sea or lake. Benthic organisms may crawl, burrow, or remain attached to a substrate. *Compare* NEKTON; NEUSTON; PLANKTON.

beriberi A disease caused by a low intake of vitamin B₁ (thiamine; *see* VITAMIN B COMPLEX), resulting in damage to peripheral nerves and heart failure. Beriberi was once common in regions of the Far East where the diet was based on polished white rice, which lacks the thiamine-rich seed coat, but its prevalence has decreased in recent years with the adoption of more varied diets.

berry A fleshy fruit formed from either one carpel or from several fused together and containing many seeds. The fruit wall may have two or three layers but the inner layer is never hard and stony (as in some drupes). Examples of berries are grapes and tomatoes. A berry, such as a cucumber, that develops a hard outer rind is called a **pepo**. One that is segmented and has a leathery rind, such as a citrus fruit, is called a **hesperidium**. The rind contains oil glands and is lined by the white mesocarp, commonly called **pith**.

beta adrenoceptor (beta adrenergic receptor) *See* ADRENOCEPTOR.

beta blocker (beta-adrenoceptor antagonist) Any of a group of drugs that bind preferentially to beta *adrenoceptors and hence block their stimulation by the body's own neurotransmitters, adrenaline and noradrenaline. Beta blockers, such as propranolol, oxprenolol, and sotalol, are used to treat disorders of the cardiovascular system, including high blood pressure (hypertension), angina pectoris, and irregularities of heartbeat (arrhythmias). They are also effective in treating anxiety and glaucoma (as eye drops) and in preventing migraine. They tend to dampen the effects of exercise or stress on heart rate, heart output, and blood pressure, as well as improving the oxygenation of the heart muscles. The release of the enzyme *renin from the kidneys is also reduced, leading to an overall fall in arterial blood pressure.

••• hydrogen bond
Ⓡ = amino-acid side chain

Beta sheet

beta cells (β cells) *See* ISLETS OF
LANGERHANS.

betacyanin Any of a group of red pigments found mainly in plants of the order
Chenopodiales, which includes the goose-
foot, cactus, and portulaca families. They
are nitrogen-containing glycosylated compounds responsible for the red colour of
beetroot, for example, and are chemically
distinct from the anthocyanins, which
include many red and pink plant pigments.
A group of yellow pigments, called **betaxanthins**, are chemically similar to betacyanins and are restricted to the same families.
Both betacyanins and betaxanthins are
classed as **betalains**.

beta diversity (beta richness) (in ecology) The rate, magnitude, and direction of
change in species diversity (*see*
BIODIVERSITY) between different habitats
or communities. It is thus a measure of
the heterogeneity between the various
habitats. When combined with the *alpha
diversity (i.e. diversity of individual communities) it gives a measure of regional
species diversity, or *gamma diversity.

beta sheet (β-pleated sheet) A form of
secondary structure in *proteins in which
extended polypeptide chains lie parallel to
each other and are linked by hydrogen
bonds between the N–H and C=O groups
(see illustration). Beta sheets occur in many
globular proteins and link polypeptides of
the same type in certain fibrous proteins,
including fibroin (the protein of silk). *Compare* ALPHA HELIX.

betaxanthin *See* BETACYANIN.

bicarbonate *See* HYDROGENCARBONATE.

biceps A muscle that runs along the large
bone of the upper arm (*humerus) and is
connected to the *radius at one end and the
shoulder bone (*scapula) at the other. Contraction of the biceps causes the arm to flex
at the elbow joint (*see* FLEXOR). It works
antagonistically with the triceps, which
contracts to extend the arm (*see*
ANTAGONISM). *See also* SKELETAL MUSCLE.

bicuspid valve (left atrioventricular
valve; mitral valve) A valve, consisting of
two flaps, situated between the left atrium
and the left ventricle of the heart of birds

and mammals. When the left ventricle contracts, forcing blood into the aorta, the bicuspid valve closes the aperture to the left atrium, thereby preventing any backflow of blood. The valve reopens to allow blood to flow from the atrium into the ventricle. *Compare* TRICUSPID VALVE.

biennial A plant that requires two growing seasons to complete its life cycle. During the first year it builds up food reserves, which are used during the second year in the production of flowers and seeds. Examples are carrot and parsnip.

bilateral symmetry A type of arrangement of the parts and organs of an animal in which the body can be divided into two halves that are mirror images of each other along one plane only (usually passing through the midline at right angles to the dorsal and ventral surfaces). Bilaterally symmetrical animals (i.e. **bilaterians**) are characterized by a type of movement in which one end of the body always leads. The two groups of bilaterians are the *protostomes and the *deuterostomes. In botany this type of symmetry is usually called **zygomorphy** when applicable to flowers (e.g. foxglove and antirrhinum flowers are zygomorphic). *Compare* RADIAL SYMMETRY.

bile (gall) A bitter-tasting greenish-yellow alkaline fluid produced by the *liver, stored in the *gall bladder, and secreted into the *duodenum of vertebrates. It assists the digestion and absorption of fats by the action of **bile salts**, which decrease the surface tension of fat droplets so that they are emulsified, hence increasing their surface area exposed to the action of lipase enzymes. It may also stimulate gut muscle contraction (*peristalsis). Bile contains the **bile pigments**, **bilirubin** and **biliverdin**, which are produced by the breakdown of the blood pigment *haemoglobin and subsequently eliminated in faeces.

bile duct The tube through which bile passes from the *liver or (when present) the *gall bladder to the duodenum.

bilirubin *See* BILE.

biliverdin *See* BILE.

binary fission *See* FISSION.

bindin *See* ACROSOME.

binding site An area on the surface of a molecule that combines with another molecule. Binding sites on enzymes can be *active sites or *allosteric sites.

binocular vision The ability, found only in animals with forward-facing eyes, to produce a focused image of the same object simultaneously on the retinas of both eyes. This permits three-dimensional vision and contributes to distance judgment.

binomial nomenclature The system of naming organisms using a two-part Latinized (or scientific) name that was devised by the Swedish botanist Linnaeus (Carl Linné); it is also known as the **Linnaean system**. The first part is the generic name (*see* GENUS), the second is the specific epithet or name (*see* SPECIES). The Latin name is usually printed in italics, starting with a capital letter. For example, in the scientific name of the common frog, *Rana temporaria*, *Rana* is the generic name and *temporaria* the specific name. The name of the species may be followed by an abbreviated form of the name of its discoverer; for example, the common daisy is *Bellis perennis* L. (for Linnaeus). There are several International Codes of Taxonomic Nomenclature that lay down the rules for naming organisms. *See also* CLASSIFICATION; TAXONOMY.

bioaccumulation An increase in the concentration of chemicals, such as *pesticides, in organisms that live in environments contaminated by a wide variety of organic compounds. These compounds are not usually decomposed in the environment (i.e. they are not biodegradable) or metabolized by the organisms, so that their rate of absorption and storage is greater than their rate of excretion. The chemicals are normally stored in fatty tissues. *DDT is known as a **persistent pesticide**, as it is not easily broken down and accumulates along *food chains, so that increasing concentrations occur in individual organisms at each trophic level—a process termed **biomagnification**.

bioactivation A metabolic process in which a product that is chemically reactive is produced from a relatively inactive precursor. *See also* ECOTOXICOLOGY; PRODRUG.

bioassay (biological assay) A controlled experiment for the quantitative estimation of a substance by measuring its effect in a living organism. For example, the amount of the plant hormone auxin can be estimated by observing its effect on the curvature of oat coleoptiles – the concentration of the hormone is proportional to the curvature of the coleoptile.

biocapacity *See* ECOLOGICAL FOOTPRINT.

biochemical evolution (molecular evolution) The changes that occur at the molecular level in organisms over a period of time, particularly in the molecular constitution of the genome. These range from deletions, additions, or substitutions of single nucleotides, through the rearrangement of parts of genes, to the duplication of entire genes or even whole genomes. Such *mutations may result in functional changes to the proteins encoded by the genes, or even the evolution of novel genes and proteins. *See also* MOLECULAR CLOCK.

biochemical oxygen demand (BOD) The amount of oxygen taken up by microorganisms that decompose organic waste matter in water. It is therefore used as a measure of the amount of certain types of organic pollutant in water. BOD is calculated by keeping a sample of water containing a known amount of oxygen for five days at 20°C. The oxygen content is measured again after this time. A high BOD indicates the presence of a large number of microorganisms, which suggests a high level of pollution.

biochemical taxonomy *See* MOLECULAR SYSTEMATICS.

biochemistry The study of the chemistry of living organisms, especially the structure and function of their chemical components

(principally proteins, carbohydrates, lipids, and nucleic acids). Biochemistry has advanced rapidly with the development, from the mid-20th century, of such techniques as chromatography, spectroscopy, X-ray diffraction, radioisotopic labelling, and electron microscopy. Using these techniques to separate and analyse biologically important molecules, the steps of the metabolic pathways in which they are involved (e.g. *glycolysis and the *Krebs cycle) have been determined. This has provided some knowledge of how organisms obtain and store energy, how they manufacture and degrade their biomolecules, how they sense and respond to their environment, and how all this information is carried and expressed by their genetic material. Biochemistry forms an important part of many other disciplines, especially physiology, nutrition, molecular biology, and genetics, and its discoveries have made a profound impact in medicine, agriculture, industry, and many other areas of human activity. *See* Chronology.

(((@))) SEE WEB LINKS
• A virtual library of biochemistry, cell biology, and molecular biology

biocoenosis All the organisms living in a particular place at a particular time. It is the equivalent of a *biome, and is used especially in the ecological literature of eastern Europe.

biodegradable *See* POLLUTION.

biodiversity (biological diversity) The existence of a wide variety of species (**species diversity**) or other taxa of plants, animals, and microorganisms in a natural community or habitat, or of communities within a particular environment (**ecological diversity**), or of genetic variation within a species (**genetic diversity**). The maintenance of a high level of biodiversity is important for the stability of ecosystems. Certain habitats, especially rainforests, have a rich species diversity, which is threatened by the continued destruction of habitats (*see* DEFORESTATION; DESERTIFICATION; GREENHOUSE EFFECT). Such ecosystems

BIOCHEMISTRY

1833	French chemist Anselme Payen (1795-1871) discovers diastase (the first enzyme to be discovered).
1836	Theodor Schwann discovers the digestive enzyme pepsin.
c.1860	Louis Pasteur demonstrates fermentation is caused by 'ferments' in yeasts and bacteria.
1869	German biochemist Johann Friedrich Miescher (1844-95) discovers nucleic acid.
1877	Pasteur's 'ferments' are designated as enzymes.
1890	German chemist Emil Fischer (1852-1919) proposes the 'lock-and-key' mechanism to explain enzyme action.
1901	Japanese chemist Jokichi Takamine (1854-1922) isolates adrenaline (the first hormone to be isolated).
1904	British biologist Arthur Harden (1865-1940) discovers coenzymes.
1909	Russian-born US biochemist Phoebus Levene (1869-1940) identifies ribose in RNA.
1921	Canadian physiologist Frederick Banting (1891-1941) and US physiologist Charles Best (1899-1978) isolate insulin.
1922	Alexander Fleming discovers the enzyme lysozyme.
1925	Russian-born British biologist David Keilin (1887-1963) discovers cytochrome.
1926	US biochemist James Sumner (1877-1955) crystallizes urease (the first enzyme to be isolated).
1929	German chemist Hans Fischer (1881-1945) determines the structure of haem (in haemoglobin).
	K. Lohman isolates ATP from muscle.
1930	US biochemist John Northrop (1891-1987) isolates the enzyme pepsin.
1932	Swedish biochemist Hugo Theorell (1903-82) isolates the muscle protein myoglobin.
1937	Hans Krebs discovers the Krebs cycle.
1940	German-born US biochemist Fritz Lipmann (1899-1986) proposes that ATP is the carrier of chemical energy in many cells.
1943	US biochemist Britton Chance (1913-2010) discovers how enzymes work (by forming an enzyme-substrate complex).
1952	US biologist Alfred Hershey (1908-97) proves that DNA carries genetic information.
1953	Francis Crick and James Watson discover the structure of DNA.
1955	Frederick Sanger discovers the amino acid sequence of insulin.
1956	US biochemist Arthur Kornberg (1918-2007) discovers DNA polymerase. US molecular biologist Paul Berg (1926-) identifies the nucleic acid later known as transfer RNA.
1959	Austrian-born British biochemist Max Perutz (1914-2002) determines the structure of haemoglobin.

1960–61	South African-born British molecular biologist Sydney Brenner (1927–2019) and French biochemist François Jacob (1920–2013) discover messenger RNA.
1961	British biochemist Peter Mitchell (1920–92) proposes the chemiosmotic theory. Brenner and Crick discover that the genetic code consists of a series of base triplets.
1970	US virologists Howard Temin (1934–94) and David Baltimore (1938–) discover the enzyme reverse transcriptase.
1970	US molecular biologist Hamilton Smith (1931–) discovers restriction enzymes.
1973	US biochemists Stanley N. Cohen (1935–) and Herbert Boyer (1936–) use restriction enzymes to produce recombinant DNA.
1977	Sanger determines the complete base sequence of DNA in bacteriophage φX174.
1984	British biochemist Alec Jeffreys (1950–) devises DNA profiling.
1985	US biochemist Kary Mullis (1944–2019) invents the polymerase chain reaction for amplifying DNA.
1986	US pharmacologists Robert Furchgott (1916–2009) and Louis Ignarro (1941–) demonstrate the importance of nitric oxide as a signal molecule in the blood vascular system.
1988	US biochemist Peter Agre (1949–) identifies a water-channel protein (aquaporin) in the plasma membrane of cells.
1998	US biochemist Roderick MacKinnon (1956–) reveals detailed three-dimensional structure of potassium-ion channel in brain cells.
2001	US molecular biologist Harry Noller and colleagues produce first detailed X-ray crystallographic image of a complete ribosome.
2002	First synthetic virus created by Eckard Wimmer and associates, based on human poliovirus.
2004	A team led by David L. Spector produces the first real-time imaging of gene transcription in a living cell, using different fluorescent markers to tag nucleic acids and proteins.
2007	Several research teams produce high-resolution crystal structures of the beta adrenoceptor, a G protein-coupled receptor with important roles in cell signalling.
2011	A high-resolution 3D crystal structure of photosystem II is published, showing for the first time the detail of the catalytic centre.
2014	*E. coli* bacteria engineered to incorporate two artificial nucleotides, X and Y, in their DNA.
2015	Discovery of teixobactin, one of a new class of antibiotics, using the iChip screening tool.
2017	Cryo-electron microscopy used to visualize cellular components, including protein complexes such as spliceosomes.

typically support large numbers of rare species, and population sizes of individual species tend to be small; they are therefore especially vulnerable to habitat destruction. Biodiversity in natural habitats also represents an important pool of species and genetic material of potential use to human societies. For example, wild plants continue to be used as a source of new drugs and other products, and the development of new strains and varieties of crop plants with increased disease resistance usually depends on incorporating genetic material from wild plants. Areas with exceptionally high biodiversity are termed **biodiversity hotspots**. Some 35 such hotspots have been identified; although they represent only 2.3% of the earth's land area, they contain 77% of all endemic plant species, 43% of all endemic vertebrate species, and 80% of threatened amphibians. *See* ALPHA DIVERSITY; BETA DIVERSITY; GAMMA DIVERSITY; CONVENTION ON BIOLOGICAL DIVERSITY.

biodiversity gradient The gradual reduction in biomass and species numbers that occurs with increasing latitude. There are several theories to explain why life is more abundant in the tropics than in cooler regions. The simplest explanation is that the greater surface area of the planet at the equator, compared to the poles, provides more space in which species can evolve. Another theory suggests that the relative environmental stability of the tropics enables species to specialize to a greater extent, so that more can be packed into any given ecosystem. Further, the greater input of solar energy in the tropics increases available resources, resulting in greater biomass and population sizes compared to colder regions.

bioelement Any chemical element that is found in the molecules and compounds that make up a living organism. In the human body the most common bioelements (in decreasing order of occurrence) are oxygen, carbon, hydrogen, nitrogen, calcium, and phosphorus. Other bioelements include sodium, potassium, magnesium, and copper. *See* ESSENTIAL ELEMENT.

bioenergetics The study of the flow and the transformations of energy that occur in living organisms. Typically, the amount of energy that an organism takes in (from food or sunlight) is measured and divided into the amount used for growth of new tissues; that lost through death, wastes, and (in plants) transpiration; and that lost to the environment as heat (through respiration).

bioengineering 1. The application of engineering principles to investigate biological systems, create synthetic systems, or develop materials or devices that supplement or replace natural components. The design of medical implants and prostheses is one important field of bioengineering, involving the use of artificial tissues, organs, and organ components to replace parts of the body that are damaged, lost, or malfunctioning, e.g. artificial limbs, heart valves, and heart pacemakers. Researchers are now developing so-called 'organs on chips', which are microchips containing human tissue cells designed to mimic the microanatomy and functions of organs, such as the heart, lungs, and kidney. They are intended to replace experimental animals, such as mice, in testing new drugs and other treatments. Other areas include work on *biosensors, *biomechanics, and *biomimicry. *See also* SYNTHETIC BIOLOGY; TISSUE ENGINEERING. **2.** The application of engineering knowledge to study the living world.

biofeedback The technique whereby a subject can learn to control certain body functions, such as heart rate or blood pressure, that are usually unconsciously regulated by the autonomic nervous system. It is facilitated by the use of monitoring devices, such as pulse monitors, electroencephalographs, and electromyographs, and can be useful in treating high blood pressure, migraine, epilepsy, and other disorders.

biofilm A colony of bacteria and other microorganisms that adheres to a substrate and is enclosed and protected by secreted slime. Biofilms readily form on virtually any surface, whether nonliving or living, where there is moisture and a supply of nutrients.

They are important components of aquatic and terrestrial ecosystems, typically providing nutrients for small organisms at the base of food chains. Moreover, they form in the microenvironment (*rhizosphere) surrounding plant roots, where they assist the plant in absorbing nutrients from the soil. Cells in such colonies secrete chemical signals that recruit neighbouring cells and also produce a matrix of polysaccharides and proteins that binds the colony members together and to the substrate. Channels through the colony allow cells in the interior access to nutrients, and information about the size of the colony is transmitted by a mechanism called *quorum sensing. Biofilms can also occur in the body and in industrial installations. For example, dental plaque is a bacterial biofilm that forms on the surface of teeth and causes tooth decay, and biofilms on hospital equipment are often resistant to eradication and a persistent source of infection. In industry biofilms are common in water-based processes and can clog up and corrode pipelines, but can also be beneficial, as demonstrated by their role in sewage treatment.

biofuel A gaseous, liquid, or solid fuel that contains an energy content derived from a biological source. The organic matter that makes up living organisms provides a potential source of trapped energy that is increasingly being exploited to supply worldwide energy demand. Biofuels are claimed to have a lower 'carbon footprint' than fossil fuels, and thus contribute less to the *greenhouse effect. However, such claims are often exaggerated, and there is concern that biofuel crops are replacing food crops, with adverse implications for food security, especially in poorer countries. An example of a biofuel is rapeseed oil, which can be used in place of diesel fuel in modified engines. The methyl ester of this oil, **rapeseed methyl ester** (**RME**), can be used in unmodified diesel engines and is sometimes known as **biodiesel**. Other biofuels include *biogas and *gasohol.

(⊕) SEE WEB LINKS

• Highlights the dangers of unsustainable biofuel production

biogas A mixture of methane and carbon dioxide resulting from the anaerobic decomposition of organic waste materials, such as surplus food, manures, and sewage. The decomposition is carried out by methanogenic bacteria (*see* METHANOGEN); these obligate anaerobes produce methane, the main component of biogas, which can be collected and used as an energy source for domestic processes, such as heating, cooking, and lighting; in some cases it can be injected into a pre-existing gas grid to supplement natural gas. The production of biogas is carried out in special **digesters**; these were traditionally used in China and India and have now been developed on a large scale in many parts of the world as a means of generating renewable energy from organic waste. As well as providing a source of fuel, these systems also enable *sewage, which contains pathogenic bacteria, to be digested, thereby removing the danger to humans that could otherwise result from untreated domestic and agricultural waste.

biogenesis The principle that a living organism can only arise from other living organisms similar to itself (i.e. that like gives rise to like) and can never originate from nonliving material. *Compare* SPONTANEOUS GENERATION.

biogenic amine Any amine that is produced by living organisms, especially the physiologically active amines that serve as neurotransmitters in animals. These include *adrenaline (epinephrine), *noradrenaline (norepinephrine), *dopamine, *serotonin (5-hydroxytryptamine), and *histamine.

biogeochemical cycle (nutrient cycle) The cyclical movement of elements between living organisms (the biotic phase) and their nonliving (abiotic) surroundings (e.g. rocks, water, air). Examples of biogeochemical cycles are the *carbon cycle, *nitrogen cycle, *oxygen cycle, *phosphorus cycle, and *sulphur cycle.

biogeography The branch of biology that deals with the geographical distribution of plants and animals. *See* PLANT GEOGRAPHY; ZOOGEOGRAPHY.

bioinformatics The collection, storage, and analysis of DNA- and protein-sequence data using computerized systems. The data generated by *genome projects and protein studies are held in various databanks and made available to researchers throughout the world via the Internet. Many computer programs have been developed to analyse sequence data, enabling the user to identify similarities between newly sequenced material and existing sequences. This allows, for example, predictions about the structure and function of a protein from its amino-acid sequence data or from the nucleotide sequence of its gene. Genome-wide sequence analysis of newly discovered organisms, especially bacteria or protists, indicates the array of proteins they are likely to manufacture, and therefore the kind of lifestyle they are likely to lead. Also, comparisons between genomes of different species provides information about their possible evolutionary relationships. *See* Feature. *See also* GENOMICS.

biolistics A technique for introducing genetic material into living cells, especially plant cells, in which DNA-coated microscopic particles are fired into the cell using a special gun. The microprojectiles, typically gold particles of diameter 0.5 to 1 μm, are accelerated to high velocity by a specially modified small-calibre gun and penetrate the cell walls and plasma membrane with minimal damage. Hence novel DNA can be inserted into intact plant cells (*see* GENETICALLY MODIFIED ORGANISMS). The technique can also be used for delivering DNA through the double membranes of intact chloroplasts and mitochondria.

biological clock The mechanism, presumed to exist within many animals and plants, that produces regular periodic changes in behaviour or physiology. Biological clocks underlie many of the *biorhythms seen in organisms (e.g. sleep-wake cycle, hibernation in animals). They continue to run even when conditions are kept artificially constant, but eventually drift out of step with the natural environment without the specific signals that normally keep them synchronized. Studies in the fruit fly *Drosophila* have revealed the molecular basis of the biological clock, and similar mechanisms are thought to occur in other animals, including mammals, and in plants, fungi, and cyanobacteria. It involves various proteins encoded by 'clock genes', some of which serve as transcription factors for their own genes; examples of clock genes in *Drosophila* and other animals are PER (encoded by the *per* gene) and TIM (encoded by the *tim* gene). These form part of a negative feedback loop in which the concentration of the proteins cyclically rises and falls. The timing of each cycle is determined by the time required for transcription, export of messenger RNA to the cytoplasm, translation, and, crucially, the formation of PER-TIM dimers—the only form in which these two proteins can enter the nucleus. Also, some of the proteins, including TIM, are sensitive to light and are degraded during the day. Hence, the biological clock is entrained to the day-night cycle. Although the clock proteins differ in different organisms, similar mechanisms have evolved independently in diverse organisms. In humans the central oscillator of the biological clock resides in the paired suprachiasmatic nuclei (SCN), lying above the optic chiasma deep within the brain. They receive input from the retina of the eyes to synchronize the 'clock' with the day-night cycle. Genes of the human 'clock' are also expressed cyclically in other tissues, such as the liver, heart, and skin, acting as peripheral clocks. These are synchronized with the central SCN oscillator but may also be entrained by other factors, such as mealtimes. Human equivalents of the *Drosophila* proteins are PER and CRY, which interact with a complex of two other proteins, CLOCK and BMAL1, to regulate expression of their own genes over a roughly 24-hour cycle. In 2017 the US scientists Jeffrey Hall, Michael Rosbash, and Michael Young won the Nobel Prize for physiology or medicine for their work on biological clocks.

(((●))) SEE WEB LINKS

- Introduction to circadian rhythms and other aspects of chronobiology, hosted by the Millar Lab at Edinburgh University

BIOINFORMATICS

Since the 1980s molecular biology has undergone a revolution, with the introduction of automated high-throughput techniques for determining the chemical composition of biomolecules, such as DNA and proteins. Genome projects, notably the Human Genome Project, have generated vast amounts of sequence data, detailing the order of bases throughout entire genomes.

Much of the information is held in various databases hosted by public institutions in different countries. Storage, handling, and analysis of these data have been made possible by advances in computer science and information technology, creating the field of bioinformatics.

Bioinformatics developed on the founding principles of unlimited access without charge or constraint. Fortuitously, the bioinformatics revolution coincided with the rise of the Internet and the free flow of information around the world.

A consequence has been the integration of data from many different sources, with an emphasis on accessibility. The resulting global bioinformatics community has proved enormously beneficial to many areas of research, including medicine, pharmacology, and forensic science, as well as biology.

Making sense of sequences

Stretches of DNA, or entire genomes of organisms, can be sequenced in a matter of hours or days. But what does the resulting string of As, Cs, Gs, and Ts mean?

The usual starting point is to compare a sequence—the query sequence—with existing sequences held on public databases. Such comparisons typically use computing methods (algorithms) and software packages developed and provided as open source (free) by the database-hosting organization.

Finding a similar sequence might provide important clues about possible function. For example, a homologous gene in another organism might have the same function as the query gene. Genes (and proteins) with important functions tend to be highly conserved even in distant relatives.

Computerized analysis of a sequence will also identify such regions of the sequence as open *reading frames, promoters, enhancers, and *repetitive DNA and translate any putative coding sequences into corresponding amino acid sequences. Information about the *epigenetic chemical modifications to DNA and histone proteins that influence gene activity will have a bearing on predicting how a particular gene will function, e.g. in cancer cells or other disease states.

Amino acid sequences can be analysed similarly, to pinpoint possible structural features, such as DNA- or protein-binding domains, and to produce the most likely three-dimensional structures of the assembled proteins.

Comparative genomics

Thousands of organisms from all domains of the living world are now represented in sequence databases. This torrent of information is expanding our knowledge of how genomes are organized and how they work, as well as clarifying the evolutionary relationships between different groups of organisms.

Computerized genome maps are available for many organisms. These make it much easier to compare the order of genes and markers along chromosomes in different species, to establish how the genetic material evolves.

The wonderful world of '-omics'

Genome sequence data are just the start; they can be regarded as the 'instruction manual' available to a particular organism in the form of its genes, and their ability to encode proteins. Bioinformatics also encompasses data accumulating from automated analyses of all aspects of cell metabolism directed by those genes, including:

- transcriptomics—which genes are transcribed into messenger RNA molecules by different types of cells?
- proteomics—what proteins do the cells actually make, when do they make them, and in what quantities?
- interactomics—what combinations of proteins are required for particular tasks?
- epigenomics—how do epigenetic modifications to chromatin affect gene activity?

These data come from such techniques as *microarrays, *RNA-seq, and mass spectrometry. Integrating the data from these sources, although a hugely challenging task, is providing new insights into various aspects of how cells work during embryological development, in health and disease, why they sometimes become cancer cells, and how they respond to drugs.

A long-term goal of bioinformatics is to pave the way for accurate computerized models of entire cells or even organisms (*see* SYSTEMS BIOLOGY).

This graph shows percent identity between human and mouse DNA in the region of the *APOAV* gene, which encodes apolipoprotein A–V (a regulator of triglyceride levels in blood), and the *ZNF259* gene, which encodes zinc finger 259. Black bars indicate exons of genes; arrows indicate direction of transcription. Note increased similarity in exons compared with introns.

```
Human   4   MAAVLTWALALLSAFSATQARKGFWDYFSQTSGDKGRVEQIHQQKMA   50
            MAAV+TWALALL+ F++TQARK  WDYFSQ S  KG + Q   QK+A
Mouse   1   MAAVITWALALLAVFASTQARKSLWDYFSQNSWSKGVMGQ--PQKLA   43


Human   51  REPATLKDSLEQDLNNMNKFLEKLRPLSG--SEAPRL........HSHLGDP   366
            +E   LK S EQDL NMN +LEKL PL G    E P L........HSHL DP
Mouse   44  QE--NLKGSFEQDLYNMNNYLEKLGPLRGPGKEPPLL.........HSHLSDP   363
```

Part of an alignment of amino acid sequences of an isoform of apolipoprotein A–V from human and mouse, using BLAST (Basic Local Alignment Search Tool). It enables comparison of the human sequence with the well-characterized mouse sequence and is a key step in identifying its nature, possible function, and evolutionary relationships. Dashes in human and mouse lines indicate gaps; the line between them indicates shared identical amino acids and conserved residues (i.e. chemically similar amino acids, denoted by +).

biological control The control of *pests by biological (rather than chemical) means. This may be achieved, for example, by breeding disease-resistant crops or by introducing a natural enemy of the pest, such as a predator or a parasite. This technique, which may offer substantial advantages over the use of pesticides or herbicides, has been employed successfully on a number of occasions. Examples include the control of the prickly pear cactus (*Opuntia*) in Australia by introducing the cactus moth (*Cactoblastis cactorum*), whose caterpillars feed on the plant's growing shoots, and the use of the ladybird to prey upon the scale insect (*Icerya*), which kills citrus fruit trees. Insect pests have also been subjected to genetic control, by releasing large numbers of males of the pest species that have been sterilized by radiation: infertile matings subsequently cause a decline in the pest population. This method has been used to control the screw worm fly (*Cochliomyia hominivora*), which lays its eggs in the open wounds of domestic cattle. Biological control is considered to reduce a number of the problems associated with chemical control using *pesticides, but care should be taken to avoid upsetting the natural ecological balance; for example, a particular predator may also destroy harmless or beneficial species.

biological fuel cell (biochemical fuel cell) A device that exploits biological reactions for the conversion of biomass (chemical energy) to electricity (electrical energy). Microbial fuel cells harness the respiratory apparatus of microorganisms to generate electrons, whereas enzymatic fuel cells use enzymes as catalysts. The fuel is oxidized by oxygen from air by the living cells or enzymes, which are in close contact with the anode of the fuel cell. The anode conducts the electrons through a circuit to the cathode. One potential application is the generation of electricity from industrial waste and *sewage. Microbial electrolysis cells can produce hydrogen, a useful fuel. All are classified as bioelectrochemical systems.

biological rhythm *See* BIORHYTHM.

biological species concept (BSC) The concept of a species as a group of populations whose members are capable of interbreeding successfully and are reproductively isolated from other groups. This concept became influential during the late 19th and early 20th centuries, largely replacing the *typological species concept favoured by pioneer naturalists. Central to the concept is the role of sexual reproduction. This maintains the broad uniformity of species' members through genetic recombination and sharing of a common gene pool. *Isolating mechanisms prevent breeding, and hence gene flow, between different groups, thus ensuring genetic divergence between groups. However, the concept cannot be applied to exclusively asexual organisms, such as certain groups of fungi and bacteria. Nor does it account satisfactorily for the many instances in which interspecies mating does occur, especially in plants, fungi, and prokaryotes.

biological warfare The military use of microorganisms, such as bacteria, viruses, fungi, and other microorganisms, including the agents for anthrax and botulism, to induce disease or death among humans, livestock, and crop plants. Though officially banned in most countries, research continues with the aim of developing virulent strains of existing microorganisms, using genetic engineering and other techniques.

biology The study of living organisms, which includes their structure (gross and microscopical), functioning, origin and evolution, classification, interrelationships, and distribution.

(((⊕))) SEE WEB LINKS

• This University of Arizona website contains useful links to various fields, including biochemistry, cell biology, and human biology

bioluminescence The emission of light without heat by living organisms. The phenomenon occurs in glow-worms and fireflies, bacteria, fungi, some dinomastigotes, and in many deep-sea fish (among others); in animals it may serve as a means of protection (e.g. by disguising the shape of a fish) or species recognition or it may provide mating signals. The light is produced during the oxidation of a compound called **luciferin** (the composition of which varies according to the species), the reaction being catalysed by an enzyme, **luciferase**. Bioluminescence may be continuous (e.g. in bacteria) or intermittent (e.g. in fireflies). *See also* PHOTOPHORE.

SEE WEB LINKS

• Comprehensive overview of bioluminescence hosted by the University of California at Santa Barbara

biomagnification *See* BIOACCUMULATION.

biomarker A normal metabolite that, when present in abnormal concentrations in certain body fluids, can indicate the presence of a particular disease or toxicological condition. For example, abnormal concentrations of glucose in the blood can be indicative of diabetes mellitus (*see* INSULIN).

biomass The total mass of all the organisms of a given type and/or in a given area; for example, the world biomass of trees, or the biomass of elephants in the Serengeti National Park. It is normally measured in terms of grams of *dry mass per square metre. *See also* PYRAMID OF BIOMASS.

biome A major ecological community or complex of communities that extends over a large geographical area characterized by a dominant type of vegetation. The organisms of a biome are adapted to the climate conditions associated with the region. There are no distinct boundaries between adjacent biomes, which merge gradually with each other. Examples of biomes are *tundra, tropical *rainforest, *taiga, *chaparral, *grassland (temperate and tropical), and *desert.

biomechanics The application of the principles of mechanics to living systems, particularly those living systems that have coordinated movements. Biomechanics also deals with the properties of biological materials, such as blood and bone. For example, biomechanics would be used to analyse the stresses on bones in animals, both when the animals are static and when they are moving. Other types of problems in biomechanics include the fluid mechanics associated with swimming in fish and the aerodynamics of birds flying. It is sometimes difficult to perform realistic calculations in biomechanics because of complexity in the shape of animals or the large number of parts involved (for example, the large number of muscles involved in the movement of a human leg). *See also* BIOMIMICRY.

biomimicry (biomimetics) The science of applying or emulating designs found in living organisms to make materials, devices, or machines for use in engineering, construction, technology, medicine, and other fields. For example, the properties and shapes of natural surfaces, such as insect wings, birds' feathers, or leaves, have inspired many innovations in materials science and aerodynamics. Moreover, nature typically employs materials that have minimal environmental impact and uses them in a resource-efficient way. Biomimicry is increasingly seen as a source of more sustainable design in many areas.

SEE WEB LINKS

• Fascinating website Ask Nature shows how natural designs can inspire human engineering

biomolecule Any molecule that is involved in the maintenance and metabolic processes of living organisms (*see* METABOLISM). Biomolecules include carbohydrate, lipid, protein, nucleic acid, and water molecules; some biomolecules are *macromolecules.

biophysics The study of the physical aspects of biology, including the application of physical laws and the techniques of physics to study biological phenomena.

biopoiesis The development of living matter from complex organic molecules that are themselves nonliving but self-replicating. It is the process by which life is assumed to have begun. *See* ORIGIN OF LIFE.

bioprospecting The search for economically valuable materials, compounds, or genes from organisms living in their natural state. Humans have sought novel foods, medicines, and fibres from nature since prehistory, and in the process have accumulated a wealth of knowledge about sources of beneficial products in their localities. This local knowledge has often been exploited by foreign bioprospectors with the intention of commercializing it. In the process benefits have frequently been denied to indigenous peoples, and natural resources plundered in an unsustainable manner amounting to **biopiracy**. The pharmaceutical industry in particular has undertaken systematic bioprospecting of biodiverse tropical regions to source many of its drugs from wild plants and animals. The *Convention on Biological Diversity, adopted in 1992, established the principles that states have sovereign rights over their biological resources, that access to those resources requires informed consent, and that benefits should be shared in a fair and equitable way with the country of origin.

biopsy The removal of a small section of tissue from a potentially diseased organ or tissue in a living organism. The biopsied tissue is usually analysed by microscopic techniques in order to identify the nature of the disease.

bioreactor (**industrial fermenter**) A large stainless steel tank used to grow producer microorganisms in the industrial production of enzymes and other chemicals. After the tank is steam-sterilized, an inoculum of the producer cells is introduced into a medium that is maintained by probes at optimum conditions of temperature, pressure, pH, and oxygen levels for enzyme production. An *agitator mixes the medium, which is constantly aerated. It is essential that the culture medium is sterile and contains the appropriate nutritional requirements for the microorganism. When the nutrients have been utilized the product is separated; if the product is an extracellular compound the medium can be removed during the growth phase of the microorganisms, but an intracellular product must be harvested when the batch culture growth stops. Some bioreactors are designed for *continuous culture.

biorhythm (**biological rhythm**) A roughly periodic change in the behaviour or physiology of an organism that is generated and maintained by a *biological clock. Well-known examples are the *annual and *circadian rhythms occurring in many animals and plants. *See also* INFRADIAN RHYTHM; ULTRADIAN RHYTHM.

biosensor A device that uses an immobilized agent to detect or measure a chemical compound. The agents include enzymes, antibiotics, organelles, or whole cells. A reaction between the immobilized agent and the molecule being analysed is transduced into an electronic signal. This signal may be produced in response to the presence of a reaction product, the movement of electrons, or the appearance of some other factor (e.g. light). Biosensors are used in diagnostic tests: these allow quick, sensitive, and specific analysis of a wide range of biological products, including antibiotics, vitamins, and other important biomolecules (such as glucose), as well as the determination of certain *xenobiotics, such as synthetic organic compounds.

biosphere The whole of the region of the earth's surface, the sea, and the air that is inhabited by living organisms: i.e. all of its *biomes collectively.

biostratigraphy The characterization of rock strata on the basis of the fossils they contain. This involves identifying and establishing the distribution and succession of various fossil groups in order to define **biozones**, containing particular fossils or fossil assemblages that can generally be correlated with rock strata of a particular type in different locations. Ideally, a fossil used in biostratigraphical zoning has a limited range over geological time, so its occurrence is restricted to rock strata of a fairly narrow vertical range in the sequence. For example, the succession of numerous different ammonite species provides an important means of zoning rocks of the Mesozoic era throughout the world. Biozones thus form the basic biostratigraphy units. There are several types: for example, an **assemblage zone** is defined by the coincident and overlapping ranges of a particular group of fossil taxa, whereas an **acme zone** is defined by the exceptional abundance of one group or species.

biosynthesis The production of molecules by a living cell, which is the essential feature of *anabolism.

biosystematics *See* SYSTEMATICS.

biota All the organisms living in a particular region, including plants, animals, and microorganisms. *See also* COMMUNITY.

biotechnology The development of techniques for the application of biological processes to the production of materials of use in medicine and industry. For example, the production of antibiotics, cheese, and wine rely on the activity of various fungi and bacteria. *Genetic engineering can modify bacterial cells, plants, or animals to synthesize completely new substances, e.g. hormones, vaccines, *monoclonal antibodies, etc., or introduce novel traits into plants or animals. *See also* BIOENGINEERING; SYNTHETIC BIOLOGY.

biotic factor Any of the factors of an organism's environment that consist of other living organisms and together make up the **biotic environment**. These factors may affect an organism in many ways; for example, as competitors, predators, parasites, prey, or symbionts. In time, the distribution and abundance of the organism will be affected by its interrelationships with the biotic environment. *Compare* ABIOTIC FACTOR.

biotic potential (intrinsic rate of increase) Symbol r. The number of offspring of an individual organism that would survive to reproductive age under ideal conditons. It is a measure of an individual's reproductive potential, although this is seldom fully realized under natural conditons. Organisms with a high biotic

potential undergo r selection. *Compare* REPRODUCTION RATE.

biotin (vitamin B$_7$) A vitamin in the *vitamin B complex. It is the *coenzyme for various enzymes that catalyse the synthesis of fat, glycogen, and amino acids. Adequate amounts are normally produced by the intestinal bacteria in animals although deficiency can be induced by consuming large amounts of raw egg white. This contains a protein, *avidin, that specifically binds biotin, preventing its absorption from the gut. Other sources of biotin include cereals, vegetables, milk, and liver. Symptoms of deficiency include skin inflammation and neuromuscular disorders.

biotope A region that has a characteristic set of environmental conditions and consequently a particular type of fauna and flora (*biota).

biotype 1. A group of individuals within a species that are genetically very similar or identical. **2.** A physiological *race. *Compare* ECOTYPE.

biozone *See* BIOSTRATIGRAPHY.

bipedalism The ability to maintain a standing position, and to move about, using only two legs. Birds and humans are bipedal. The development of bipedalism was an important factor in *hominin evolution.

bipolar cell *See* RETINA.

bipolar neuron A *neuron that has two processes, an axon and a dendron, extending in different directions from its cell body. Many sensory neurons are bipolar neurons. *Compare* MULTIPOLAR NEURON; UNIPOLAR NEURON.

biramous appendage A type of appendage that is characteristic of arthropods of the *Crustacea. It forks from the basal **protopodite** to form two branches, the inner **endopodite** and the outer **exopodite** (see illustration). Each of these branches can be composed of either one or more segments. There are many variations on this generalized structure; the branches

protopodite { coxopodite
              basipodite

epipodite

endopodite

exopodite

A generalized biramous appendage

often possess highly specialized extensions—the **epipodite** (on the coxopodite), the **exite** (on the exopodite), and the **endite** (on the endopodite). *Compare* UNIRAMOUS APPENDAGE.

birds *See* AVES.

birth *See* PARTURITION.

birth control (contraception) The intentional avoidance of pregnancy by methods that do not normally hinder sexual activity. The methods used can be 'natural' or 'artificial'. Natural methods, often used because of religious or moral objections to artificial methods, include the **rhythm method,** in which sexual intercourse is avoided during times when ovulation occurs; and **coitus interruptus,** an unreliable method in which the penis is withdrawn from the vagina before ejaculation. The rhythm method requires a monitoring of the woman's menstrual cycle and may be unsuitable in those women with irregular cycles. Artificial methods use devices or other agents (**contraceptives**) to prevent pregnancy. They include the **condom,** a rubber sheath placed over the penis to trap the sperm; and the **diaphragm,** a rubber cap placed over the cervix. Contraceptives that prevent *implantation include the **intrauterine device** (**IUD**), a metal or plastic coil placed in the uterus by a doctor (which may cause unacceptable bleeding in some women); and the 'morning-after

pill', taken within three days after sexual intercourse. Other *oral contraceptives prevent ovulation. Research is ongoing to develop a safe and effective male contraceptive pill or injection that will suppress sperm production. *Sterilization is usually considered to be irreversible but attempts at reversing the process are possible. For casual relationships, or relationships involving a partner whose sexual history is not known, health workers advise the use of condoms with all other forms of contraception as this provides the safest means of reducing the risk of infection by *sexually transmitted diseases.

birth rate (natality) The rate at which a particular species or population produces offspring. The birth rate of a species is used to measure its fecundity (reproductive capability). It is also an important factor in controlling the size of a population. *Compare* DEATH RATE; REPRODUCTION RATE.

bisexual *See* HERMAPHRODITE.

biuret test A biochemical test to detect proteins in solution, named after the substance **biuret** ($H_2NCONHCONH_2$), which is formed when urea is heated. Sodium hydroxide is mixed with the test solution and drops of 1% copper(II) sulphate solution are then added slowly. A positive result is indicated by a violet ring, caused by the reaction of *peptide bonds in the proteins or peptides. Such a result will not occur in the presence of free amino acids.

bivalent (in genetics) *See* PAIRING.

Bivalvia (Pelecypoda; Lamellibranchia) A class or clade of approximately 30 000 species of aquatic *molluscs (the bivalves) that include the oysters, mussels, and clams. They are characterized by a laterally flattened body and a shell consisting of two hinged shells (i.e. a bivalved shell). The enlarged gills are covered with cilia and have the additional function of filtering microscopic food particles from the water flowing over them. Bivalves live on the sea bed or lake bottom and are sedentary, so the head and foot are reduced.

bladder 1. (in anatomy) **a.** A hollow muscular organ in most vertebrates, also known as the **urinary bladder**, in which urine is stored before being discharged. In mammals urine is conveyed from the *kidneys to the bladder by the *ureters and is discharged to the outside through the *urethra. **b.** Any of various other saclike organs in animals for the storage of liquid or gas. *See* GALL BLADDER; SWIM BLADDER. **2.** (in botany) **a.** A modified submerged leaf of certain aquatic insectivorous plants, such as the bladderwort (*Utricularia*). It forms a hollow with a single opening that is sealed by a valve to trap small aquatic invertebrates after they have been sucked in. **b.** An air-filled cavity in the thallus of certain seaweeds, such as the bladderwrack (*Fucus vesiculosus*).

bladderworm (cysticercus) A larval stage of some tapeworms (*see* CESTODA), consisting of a fluid-filled sac containing an inverted scolex. It develops in the muscle of the secondary host and matures into an adult worm in a primary host that eats this infected tissue.

BLAST (Basic Local Alignment Search Tool) a computational method, or algorithm, that is widely used in *bioinformatics for comparing sequences of bases (e.g. from genes) or of amino acids (from proteins) by the process of local *alignment. It enables researchers to assess the degree of similarity of a particular query (novel) sequence with the vast amount of sequence data held in public databases, amounting to billions of base pairs. This enables the identification of gene family members and potential functional and evolutionary relationships between sequences. The BLAST algorithm uses a feature called **word matching**, which initially identifies short strings of nucleotides or amino acids that match between two sequences. It then attempts to extend the match on either side to produce an alignment, and as it does so assesses the degree of similarity. If it fails to find any initial match, or to extend a match, the comparison is deleted. This avoids the need to compare every base or amino acid in the query sequence with every part of every sequence in the database—a hugely time-consuming operation. The highest scoring alignments are then listed in the result, along with an **E-value** ('expect' value), which indicates the expected number of random alignments with the same score or less. The lower the E-value, the greater the potential biological significance of the alignment. Several BLAST programs are now available for specialized searches.

blastocoel *See* BLASTULA.

blastocyst *See* BLASTULA; IMPLANTATION.

blastoderm The layer of cells that surrounds the central cavity (blastocoel) of a *blastula. In yolky eggs, such as those of insects, the blastoderm forms a layer covering the yolk. *See also* IMAGINAL DISC.

blastomere *See* BLASTULA.

blastopore *See* ARCHENTERON.

blastula The stage of *development of an animal embryo that results from *cleavage of a fertilized egg. This stage generally resembles a hollow ball with the dividing cells (**blastomeres**) forming a layer (**blastoderm**) around a central cavity (**blastocoel**). In vertebrates the blastula forms a disc (**blastodisc**) on the surface of the yolk. In reptiles, birds, and mammals the upper (dorsal) cell layer of the blastodisc is called the **epiblast**; it gives rise to the embryo. In placental mammals it also contributes to the formation of the amnion, one of the membranes enclosing the embryo. The lower (ventral) cell layer of the blastodisc, called the **hypoblast**, gives rise to extraembryonic membranes that will enclose the developing embryo. In mammals the blastula stage develops from the *morula and is known as a blastocyst, comprising an *inner cell mass, blastocoel, and enveloping outer cells of the trophoblast. *See also* GASTRULA.

blending inheritance The early theory that assumed that hereditary substances from parents merge together in their offspring. Mendel showed that this does not occur (*see* MENDEL'S LAWS). In breeding experiments an appearance of blending may result from codominant alleles (*see* CODOMINANCE) and *quantitative inheritance, but close study shows that the alleles retain their identity through successive generations. *Compare* PARTICULATE INHERITANCE.

blind spot The portion of the retina at which blood vessels and nerve fibres enter the optic nerve. There are no rods or cones in this area, so no visual image can be transmitted from it.

blood A fluid body tissue that acts as a transport and homeostatic medium within an animal. It is contained within a blood *vascular system and in vertebrates is circulated by means of contractions of the *heart. Oxygen and nutrients are carried to tissues, and carbon dioxide and chemical (nitrogenous) waste are transported from tissues to excretory organs for disposal (*excretion). In addition blood carries *hormones, acts as a defence system, and is important in maintaining a suitable osmotic and pH environment for tissues. Blood consists of a liquid (*see* BLOOD PLASMA) containing blood cells (*see* ERYTHROCYTE; LEUCOCYTE) and *platelets.

blood-brain barrier The mechanism that controls the passage of substances from the blood to the cerebrospinal fluid bathing the brain and spinal cord. It resides in the special properties of the blood capillaries that supply the central nervous system, the walls of which act like a semipermeable membrane, permitting the passage of solutions but excluding particles and large molecules, including potentially noxious substances. This barrier provides

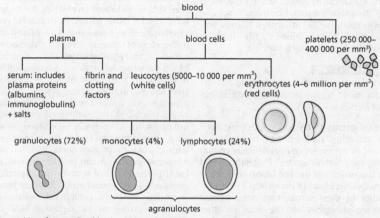

Composition of mammalian blood

the central nervous system (CNS) with a constant environment, while not interfering with the transport of essential metabolites. But it is also a constraint on the effective delivery of drugs to the CNS. However, any breakdown in the blood-brain barrier, e.g. due to a stroke or trauma, is a major factor in the pathology of such diseases. *See also* GLIA.

blood capillary *See* CAPILLARY.

blood cell (blood corpuscle) Any of the cells that are normally found in the blood plasma. These include red cells (*see* ERYTHROCYTE) and white cells (*see* LEUCOCYTE).

blood clotting (blood coagulation) The production of a mass of semisolid material at the site of an injury that closes the wound, helping to prevent further blood loss and bacterial invasion. The clot is formed by the action of *clotting factors and *platelets. The cascade of reactions that culminate in the formation of a blood clot is initiated by *thromboplastin (Factor III), a membrane glycoprotein found in tissue cells. When tissue is damaged, this forms a cell-surface complex that, with phospholipid and calcium ions, converts a plasma glycoprotein, Factor X, to Factor Xa, which in turn converts *prothrombin in the blood to its enzymically active form **thrombin**. Thrombin catalyses the formation of the insoluble protein **fibrin** from soluble **fibrinogen**; the fibrin forms a fibrous network in which blood cells become enmeshed, producing a clot.

(🌐) SEE WEB LINKS

• Overview of blood coagulation and its disorders from the Indiana University School of Medicine

blood groups The many types into which an individual's blood may be classified, based on the presence or absence of certain antigenic carbohydrates (*agglutinogens) on the surface of the red blood cells. Individuals with blood of one group have *antibodies in their serum that react against the agglutinogens on the cells of other groups. These antibodies are produced in

response to exposure to bacteria that have very similar carbohydrate groups on their surface. **Incompatibility** between groups results in clumping of cells (*agglutination), so knowledge of blood groups is important for blood transfusions. In humans, the two most important blood group systems are the *ABO system and the system involving the *rhesus factor.

blood pigment Any one of a group of metal-containing coloured protein compounds whose function is to increase the oxygen-carrying capacity of blood. *See* RESPIRATORY PIGMENT.

blood plasma The liquid part of the *blood (i.e. excluding blood cells). It consists of water containing a large number of dissolved substances, including proteins (e.g. albumin, immunoglobulins, fibrinogen), salts (especially sodium and potassium chlorides and bicarbonates), food materials (glucose, amino acids, fats), hormones, vitamins, and excretory materials. *See also* BLOOD SERUM; LYMPH.

blood platelet *See* PLATELET.

blood pressure The pressure exerted by the flow of blood through the major arteries of the body. This pressure is greatest during the contraction of the ventricles of the heart (**systolic pressure**; *see* SYSTOLE), which forces blood into the arterial system. Pressure falls to its lowest level when the heart is filling with blood (**diastolic pressure**; *see* DIASTOLE). Blood pressure is measured in millimetres of mercury using an instrument called a **sphygmomanometer**. Normal blood pressure for a young average adult human is in the region of 120/80?mmHg (the higher number is the systolic blood pressure; the lower number the diastolic blood pressure), but individual variations are common. Abnormally high blood pressure (**hypertension**) may be associated with disease or it may occur without an apparent cause. Mean arterial pressure (MAP)—the mean of systolic and diastolic pressures—is determined by *cardiac output (CO) and total peripheral resistance (TPR) according to the equation MAP = CO × TPR. It is regulated by various neural and hormonal mechanisms, both local and systemic. CO depends on *stroke volume and heart rate, which are governed chiefly by the *autonomic nervous system, and hormones such as *adrenaline. TPR is determined by the resistance of the arteries and arterioles feeding the capillary beds (*see* MICROCIRCULATION). Contraction and relaxation of smooth muscle in the walls of these vessels alters their internal diameter, and hence resistance to blood flow. These changes may occur in response to local requirements for oxygen or delivery/removal of metabolites, or more general changes. For instance, in the *alarm response noradrenaline binds to receptors in vessels in the gut and other 'non-essential' tissues, causing vessels to constrict. Another hormone, *angiotensin, also causes constriction of arterioles in peripheral tissues and is important in raising blood pressure.

blood serum Blood plasma from which the fibrin and clotting factors have been removed by centrifugation after allowing the plasma to clot. Serum containing a specific antibody or antitoxin may be used in the treatment or prevention of certain infections. Such serum is generally derived from a nonhuman mammal (e.g. a horse).

blood vascular system The tissues and organs of an animal that transport blood through the body. In vertebrates it consists of the heart and blood vessels. *See* VASCULAR SYSTEM.

blood vessel A tubular structure through which the blood of an animal flows. *See* ARTERY; ARTERIOLE; CAPILLARY; VENULE; VEIN.

blue-green bacteria *See* CYANOBACTERIA.

B lymphocyte *See* B CELL.

BMP *See* BONE MORPHOGENETIC PROTEIN.

BMR *See* BASAL METABOLIC RATE.

BOD *See* BIOCHEMICAL OXYGEN DEMAND.

body cavity The internal cavity of the body of an animal, which is present in most invertebrates and all vertebrates and contains the major organs. The body cavity of vertebrates and many invertebrates is the *coelom. In vertebrates the body cavity is divided by a transverse septum just posterior to the heart into the abdominal and thoracic cavities (*see* ABDOMEN; THORAX). In mammals the septum is the *diaphragm.

body fluid Any of the fluids found within animals, including blood, lymph, tissue fluid, urine, bile, sweat, and synovial fluid. Body fluids are generally involved with the processes of transport, excretion, homeostasis, or lubrication. They allow the distribution of oxygen and nutrients to the tissues and organs and the transport of waste products from the tissues, enabling their elimination from the body.

body plan The structural and functional arrangement of the parts of an organism. The 'blueprint' according to which an organism develops a predetermined number, arrangement, and size of body components is embodied and implemented by the organism's genes (e.g. *homeotic genes), but can be influenced by environmental factors, such as malnutrition or exposure to toxic agents (e.g. *teratogens). *See also* BILATERAL SYMMETRY; RADIAL SYMMETRY.

bog *See* HYDROSERE.

Bohr effect The effect of pH on the dissociation of oxygen from haemoglobin, first discovered by the Danish physiologist Christian Bohr (1855–1911). An increase in carbon dioxide concentration makes the blood more acidic and decreases the efficiency of the uptake of oxygen by haemoglobin molecules. This shifts the *oxygen dissociation curve to the right and increases the tendency of haemoglobin to release oxygen (*see* HAEMOGLOBINIC ACID). Thus in actively respiring tissues, where the concentration of carbon dioxide in the blood is high, haemoglobin readily releases its oxygen, while in the lungs, where blood carbon dioxide is low (due to its continual diffusion into the alveoli), haemoglobin readily binds oxygen.

bolus (*pl.* **boluses**) The ball of chewed food bound together with saliva that is formed in the mouth by the action of the tongue. The bolus is shaped to a size that allows it to pass into the oesophagus after being swallowed (*see* DEGLUTITION).

bomb calorimeter An apparatus used for measuring heats of combustion. It consists of a strong container in which the sample is sealed with excess oxygen and ignited electrically. The heat of combustion at constant volume can be calculated from the resulting rise in temperature. The apparatus is used to calculate the *calorific values of foods and the energy content of a sample of biomass, required for constructing a *pyramid of energy.

bone The hard connective tissue of which the *skeleton of most vertebrates is formed. It comprises a matrix of *collagen fibres (30%) impregnated with bone salts (70%), mostly calcium phosphate (hydroxyapatite, $Ca_{10}(PO_4)_6(OH)_2$), in which are embedded bone cells: *osteoblasts (which secrete the matrix) and *osteocytes. Bone generally replaces embryonic *cartilage and is of two sorts—compact bone and spongy bone. The outer **compact bone** is formed as concentric layers (**lamellae**) that surround small holes (*Haversian canals): see illustration. The inner **spongy bone** is chemically similar but forms a network of bony bars. The spaces between the bars may contain bone marrow or (in birds) air

for lightness. Bone is resorbed by cells called *osteoclasts, which work in dynamic interaction with the bone-forming osteoblasts to remodel bone during growth or repair, e.g. following injury or in response to changes in loads. *See also* BONE MORPHOGENETIC PROTEIN; CARTILAGE BONE; MEMBRANE BONE; PERIOSTEUM.

(SEE WEB LINKS)
• Descriptions and images of bone tissue from the School of Anatomy and Human Biology, University of Western Australia

bone marrow A soft tissue contained within the central cavity and internal spaces of a bone. At birth and in young animals the marrow of all bones is concerned with the formation of blood cells: it contains *haemopoietic tissue and is known as **red marrow**. In mature animals the marrow of the long bones ceases producing blood cells and is replaced by fat, being known as **yellow marrow**.

bone morphogenetic protein (BMP) Any of a family of cell-signalling molecules that play key roles in tissue formation in both embryos and adult animals. Their role in the induction of bone formation was first discovered in the 1960s, and since then over 20 BMPs have been identified in various other tissues, instrumental in the development of heart muscle, nerves, and cartilage among others. Recombinant human BMP-2 is used therapeutically to promote bone growth in various interventions, e.g. to promote spinal fusion or to treat nonunion of a fracture. Most BMPs belong to the *transforming growth factor-β superfamily of proteins. *See also* MORPHOGEN.

bony fishes *See* OSTEICHTHYES.

bony labyrinth *See* LABYRINTH.

borax carmine A red dye, used in optical microscopy, that stains nuclei and cytoplasm pink. It is frequently used to stain large pieces of animal tissue.

botany The scientific study of plants, including their anatomy, morphology, physiology, biochemistry, taxonomy, cytology,

Haversian canal

concentric lamellae

calcified matrix

bone cell

single Haversian system
Structure of compact bone

genetics, ecology, evolution, and geographical distribution. *See also* ETHNOBOTANY.

botulinum toxin A nerve toxin produced by the bacterium *Clostridium botulinum*, which can cause fatal *food poisoning. It is the most toxic natural substance known. In minute doses it is used to treat certain conditions involving muscle dysfunction, and in cosmetic medicine 'Botox' treatment is used to reduce skin wrinkles and other signs of ageing.

bouton (synaptic knob) *See* SYNAPSE.

bovine spongiform encephalopathy (BSE) A degenerative disease of the brain that affects cattle and is caused by an abnormal form of a cellular protein (*see* PRION). Known colloquially as 'mad cow disease', it results in a build-up of *amyloid tissue in the brain. The infective agent can be transmitted to other cattle via feed containing offal derived from infected animals. It can also, under certain circumstances, be transmitted to other species. Humans infected by eating contaminated beef or beef products have developed a variant form of *Creutzfeldt-Jakob disease.

Bowman's capsule (renal capsule) The cup-shaped end of a kidney *nephron. Its epithelium contains *podocytes, which facilitate the passage of glomerular filtrate from the blood into the nephron. It is named after its discoverer, the British physician Sir William Bowman (1816–92).

bp 1. Symbol for base pair(s), used as a unit at the molecular level for measuring distances along a duplex polynucleotide and corresponding to the number of paired bases in a particular segment of DNA (or duplex RNA). *See also* KILOBASE. **2.** (in medicine) Abbreviation for blood pressure.

Brachiopoda A phylum of marine invertebrates—the lamp shells. They live in shallow waters, attached to a firm substratum by means of a flexible stalk (**peduncle**), and are protected by a bivalved shell consisting of dorsal and ventral valves. A food-gathering *lophophore protrudes from the shell. Brachiopods thrived in Palaeozoic times but are now much less numerous; living brachiopods include *Terebratella*, with an articulated shell; and *Lingula*, in which the shell valves are held together by muscles only. Brachiopods belong to the clade of protostome animals called *Lophotrochozoa.

bract A modified leaf with a flower or inflorescence in its axil. Bracts are often brightly coloured and may be mistaken for the petals of a flower. For example the showy 'flowers' of poinsettia and *Bougainvillea* are composed of bracts; the true flowers are comparatively inconspicuous. *See also* INVOLUCRE.

bracteole A reduced leaf that arises from the stalk of an individual flower.

bradycardia A decrease in heart rate. *See* TACHYCARDIA.

bradykinin *See* KININ.

bradymetabolism Metabolism that is sustained at a relatively low rate. It is characteristic of 'cold-blooded' animals (*see* ECTOTHERM), and generally excludes any specific heat-generating mechanism. *Compare* TACHYMETABOLISM.

brain 1. The enlarged anterior part of the vertebrate central nervous system, which is

The human brain in transverse section viewed from the left side

encased within the cranium of the skull. Continuous with the spinal cord, the brain is surrounded by three membranes (*see* MENINGES) and bathed in cerebrospinal fluid, which fills internal cavities (*vent-ricles). It functions as the main coordinat-ing centre for nervous activity, receiving information (in the form of nerve impulses) from sense organs, interpreting it, and transmitting 'instructions' to muscles and other *effectors. It is also the seat of intelli-gence and memory. The embryonic verte-brate brain is in three sections (*see* FOREBRAIN; HINDBRAIN; MIDBRAIN), which become further differentiated during devel-opment into specialized regions. The main parts of the adult human brain are a highly developed *cerebrum in the form of two cerebral hemispheres, a *cerebellum, *hypothalamus, and the *brainstem com-prising the *medulla oblongata, *pons, *midbrain, and reticular formation (see illustration). **2.** A concentration of nerve *ganglia at the anterior end of an inverte-brate animal.

(())) SEE WEB LINKS

• Multilevel exploration of brain structure and function, sponsored by the Canadian Institutes of Health Research

---

brain death The permanent absence of vital functions of the brain, which is marked by cessation of breathing and other reflexes (including the *pupillary reflex) controlled by the *brainstem and by a zero reading on an *electroencephalo-gram. Organs may be removed for trans-plantation when brain death is established, which may not necessarily be associated with permanent absence of heartbeat.

brain-derived neurotrophic factor (BDNF) *see* NEUROTROPHIN.

brainstem The part of the brain compris-ing the *medulla oblongata, the *midbrain, and the *pons. Superficially it resembles and is continuous with the spinal cord. The midbrain controls and integrates reflex activities (such as breathing and swallow-ing) that originate in higher centres of the brain. The **reticular formation** (sometimes called the reticular activating system) is a network of neural fibres and cell bodies that extends the length of the brainstem. It receives sensory inputs of all kinds and regulates how much attention we give to them by acting as a sensory filter. It also influences muscle tone, and is thus a crucial part of the system governing arousal, wake-fulness, and sleep. Ascending fibres of the reticular formation connect with neurons in the adjacent thalamus and hypothalamus, which in turn influence the cerebral cortex.

branchial Of or relating to the gills. Most fish possess five pairs of **branchial arches**, skeletal supports for the gills, which are contained in a cavity known as the **bran-chial chamber**. Water is drawn into the branchial chamber through narrow *gill slits (or **branchial clefts**) in the wall of the pharynx and passes over the gills, where gaseous exchange takes place. Blood is transported to and from the gills in afferent and efferent **branchial arteries**, respectively.

Branchiopoda A class of *crustaceans found chiefly in freshwater ponds, lakes, and brackish waters. It includes fairy shrimps (order Anostraca), tadpole shrimps (order Notostraca), water fleas (order Cladocera; *see* DAPHNIA), and clam shrimps (order Conchostraca). Branchiopods are small (typically 0.25–100 mm long) with fused thorax and abdomen and leaflike appendages used variously for swimming, feeding, or respiration. All except the anos-tracans have a platelike carapace enclosing some or all of the body.

brassinosteroid (brassin; BR) Any of a group of steroid derivatives that occur at very low concentrations in plant tissues and are potent *plant hormones. Brassinos-teroids have been found in pollen, leaves, stems, and flowers from various species; the most active form is **brassinolide** (BL). Extracts behave in a manner similar to *auxins, for example stimulating elongation of hypocotyl and epicotyl tissue from seed-lings when applied at concentrations as low as 10^{-12} M. Brassinosteroids have various physiological effects, such as promoting growth by stimulating cell elongation and cell division in shoots, and promoting seed germination and pollen tube elongation. They promote root growth at low concentrations but have an inhibitory effect at high concentrations. Brassinosteroids have been shown to act synergistically with auxins and additively with gibberellins. BL binds to a membrane receptor called BRI1, thereby triggering downstream signal responses and ultimately regulating gene expression.

breathing *See* EXPIRATION; INSPIRATION; RESPIRATORY MOVEMENT.

breed A domesticated *variety of an animal or, rarely, a cultivated variety of plant. Cultivated plants are usually called varieties or *cultivars. Examples of animal breeds are Friesian cattle and Shetland sheepdogs.

breeding The process of sexual *reproduction and bearing offspring. **Selective breeding** of both plants and animals is used in *agriculture to produce offspring that possess the beneficial characters of both parents (*see also* ARTIFICIAL INSEMINATION). *Inbreeding is the production of *homozygous phenotypically uniform offspring by mating between close relatives. Plants that self-fertilize, such as wheat and tomatoes, are inbreeders. *Outbreeding is the production of *heterozygous phenotypically variable offspring by mating between unrelated organisms.

breeding season (mating season) A specific season of the year in which many animals, including mammals and birds, mate, which ensures that offspring are produced only at a certain time of the year. This timing is important as it enables animals to give birth at a time of the year when environmental conditions and food supply are at their optimum. The breeding season of most animals is in the spring or summer. The stimulus to mate is the result of a photoperiodic response (*see* PHOTOPERIODISM), which is thought to be controlled by day length affecting levels of the hormone *melatonin. Differences in the timing of breeding season is one *isolating mechanism whereby closely related species living in the same area can avoid the risk of mating between individuals of different species.

brewing The process by which beer is made. *Fermentation of sugars from barley grain by the yeasts *Saccharomyces cerevisiae* and *S. uvarum* (or *S. carlsbergensis*) produces alcohol (ethanol). In the first stage the barley grain is soaked in water, a process known as **malting**. The grain is then allowed to germinate and the natural enzymes of the grain (the amylases and the maltases) convert the starch to maltose and then to glucose. The next stage is **kilning** or **roasting**, in which the grains are dried and crushed. The colour of a beer depends on the temperature used for this process: the higher the temperature, the darker the beer. In the next stage, **mashing**, the crushed grain is added to water at a specific temperature and any remaining starch is converted to sugar; the resultant liquid is the raw material of brewing, called **wort**. The yeast is then added to the wort to convert the sugar to alcohol, followed by hops, which give beer its flavour. Hops are the female flowers of the vine *Humulus lupulus*; they contain resins (humulones, cohumulones, and adhumulones) that give beer its distinctive bitter taste.

Broca's area A region of the brain located in the frontal lobe of the cerebral hemisphere (*see* CEREBRUM), just anterior to and below the primary motor cortex, concerned with speech generation. It is named after the French neurosurgeon Paul Broca (1824–80), who found that damage to this area caused speech deficits in his patients. Broca's area is one of several regions that are involved in processing language and producing speech; it sends impulses to the primary motor cortex, which in turn instructs the vocal apparatus to articulate words. It also seems to be crucial for learning the rules of language and giving meaning to speech. Broca's area receives inputs via a large bundle of nerves (the arcuate fasciculus) from *Wernicke's area in the temporal lobe, which is concerned more with processing language inputs, whether as speech or written words. These language areas occur most commonly in the left cerebral hemisphere (in 90% of right-handed and 70% of left-handed persons).

bronchial-associated lymphoid tissue (BALT) An organized collection of lymphocytes and lymphoid tissue found in the walls of the airways, where it protects the respiratory epithelium. Each of the numerous BALT aggregations comprises B and T cells, which are crucial in inducing an immune response to invading microorganisms and other inhaled antigens. BALT forms part of the network of *mucosal-associated lymphoid tissue.

bronchiole A fine respiratory tube in the lungs of reptiles, birds, and mammals. It is formed by the subdivision of a *bronchus and in reptiles and mammals it terminates in a number of *alveoli.

bronchus (bronchial tube) (*pl.* bronchi) One of the major air tubes in the *lung. The *trachea divides into two main bronchi, one for each lung, which split into smaller bronchi and then into *bronchioles. The walls of the bronchi are stiffened by rings of cartilage.

brown algae *See* PHAEOPHYTA.

brown fat A darker coloured region of *adipose tissue found in newborn and hibernating animals (in which it may also be called the **hibernating gland**). Adult humans also have small amounts of brown fat distributed around the body, the amounts of which seem to be inversely related to total body mass. Compared to normal white *fat, deposits of brown fat are more richly supplied with blood vessels and have numerous mitochondria (hence the brown colour, due to the high concentrations of cytochrome oxidase). They can also be more rapidly converted to heat energy—a process that takes place in the fat cells themselves—especially during arousal from hibernation and during cold stress in young animals (*see* UNCOUPLING PROTEIN). Since the deposits are strategically placed near major blood vessels, the heat they generate warms the blood returning to the heart. Brown fat may serve to burn excess calories, hence the suggestion

that some types of obesity in humans may be linked to a lack of brown fat. *See* THERMOGENESIS.

Brownian movement The continuous random movement of microscopic solid particles (of about 1 micrometre in diameter) when suspended in a fluid medium. First observed by British botanist Robert Brown (1773–1858) in 1827 when studying pollen grains in water, it was originally thought to be the manifestation of some vital force. It was later recognized to be a consequence of bombardment of the particles by the continually moving molecules of the liquid. The smaller the particles the more extensive is the motion. It can be observed in the particles of a colloidal solution and in the cytoplasm and nucleoplasm of dead cells.

Brunner's glands (duodenal glands) Glands in the submucosa of the duodenum that secrete mucus and an alkaline fluid that contributes to the *succus entericus and neutralizes the acidic *chyme leaving the stomach. They are named after Swiss anatomist J. C. von Brunner (1653–1727).

brush border A region of surface epithelium that possesses densely packed microvilli (*see* MICROVILLUS), rather like the bristles of a brush. This greatly increases the surface area of the epithelium and facilitates the absorption of materials. Brush borders are found in the convoluted tubules of the kidney and in the lining of the small intestine.

Bryophyta A phylum containing some 15 000 known species of simple plants possessing no vascular tissue and rudimentary rootlike organs (rhizoids). They grow in a variety of damp habitats, from fresh water to rock surfaces. Some use other plants for support. Mosses show a marked *alternation of generations between gamete-bearing forms (gametophytes) and spore-bearing forms (sporophytes): they possess erect or prostrate leafy stems (the gametophyte generation, which is *haploid); these give rise to leafless stalks bearing capsules (the sporophyte generation, which is *diploid), the latter being dependent on the former for water and nutrients. Spores formed in the capsules are released and grow to produce new plants.

Formerly, this phylum also included the liverworts and hornworts, now regarded as separate phyla (*see* HEPATOPHYTA; ANTHOCEROPHYTA) and the mosses were classified as a class (Musci) of the Bryophyta. The term 'bryophytes' may still be used, erroneously, to refer to members of all three phyla.

(⊕) SEE WEB LINKS

- A resource devoted to mosses, liverworts, and hornworts from Southern Illinois University Carbondale

Bryozoa (Ectoprocta) A phylum of around 5000 species of aquatic, mainly marine, invertebrates comprising the moss animals and sea mats. Bryozoans live in colonies, 50 cm or more across, which are attached to rocks, seaweeds, or shells. The individuals (**zooids**) making up the colonies are about 1 mm long and superficially resemble cnidarian *polyps, with a mouth surrounded by ciliated tentacles of the *lophophore that trap minute particles of organic matter in the water. Some have a horny or calcareous outer skeleton into which the body can be withdrawn. Bryozoans are placed in the clade of protostome animals called the *Lophotrochozoa. *Compare* ENTOPROCTA.

BSC *See* BIOLOGICAL SPECIES CONCEPT.

BSE *See* BOVINE SPONGIFORM ENCEPHALOPATHY.

buccal cavity (oral cavity) The mouth cavity: the beginning of the *alimentary canal, which leads to the pharynx and (in vertebrates) to the oesophagus. In vertebrates it is separated from the nasal cavity by the *palate. In mammals it contains the tongue and teeth, which assist in the mechanical breakdown of food, and the openings of the *salivary glands.

bud 1. (in botany) A condensed immature shoot with a short stem bearing small folded or rolled leaves. The outer leaves of a bud are often scalelike and protect the delicate inner leaves. A **terminal** (or **apical**) **bud** exists at the tip of a stem or branch while **axillary** (or **lateral**) **buds** develop in the *axils of leaves. However, in certain circumstances buds can be produced anywhere on the surface of a plant. Some buds remain dormant, but may become active if the terminal bud is removed. It is common gardening practice to remove the terminal buds of some shoots to induce the development of lateral shoots from axillary buds. *See also* APICAL DOMINANCE. **2.** (in biology) An outgrowth from a parent organism that breaks away and develops into a new individual in the process of *budding.

budding 1. (in biology) A method of asexual reproduction in which a new individual is derived from an outgrowth (**bud**) that becomes detached from the body of the parent. In animals the process is also called **gemmation**; it is common in cnidarians (e.g. *Hydra* and stony corals) and also occurs in some sponges and other invertebrates. Among fungi, budding is characteristic of the yeasts. **2.** (in horticulture) A method of grafting in which a bud of the scion is inserted onto the stock, usually beneath the bark.

buffer A solution that resists change in pH when an acid or alkali is added or when the solution is diluted. Acidic buffers consist of a weak acid with a salt of the acid. The salt provides the negative ion A^-, which is the conjugate base of the acid HA. An example is carbonic acid and sodium hydrogencarbonate, in which molecules H_2CO_3 and ions HCO_3^- are present. When acid is added most of the extra protons are removed by the base:

$$HCO_3^- + H^+ \rightarrow H_2CO_3$$

When base is added, most of the extra hydroxide ions are removed by reaction with undissociated acid:

$$OH^- + H_2CO_3 \rightarrow HCO_3^- + H_2O$$

Thus, the addition of acid or base changes the pH very little. Basic buffers have a weak base and a salt of the base (to provide the conjugate acid).

Natural buffers occur in living organisms, where the biochemical reactions are very sensitive to change in pH (*see* ACID-BASE BALANCE). The main natural buffers are H_2CO_3/HCO_3^- and $H_2PO_4^-/HPO_4^{2-}$ (*see also* HAEMOGLOBINIC ACID). Buffer solutions are also used in the laboratory (e.g. to keep microscopical preparations at their original pHs in order to prevent the formation of artefacts), in medicine (e.g. in intravenous injections), in agriculture, and in many industrial processes (e.g. fermentation processes).

bugs See HEMIPTERA.

bulb An underground plant organ that enables a plant to survive from one growing season to the next. It is a modified shoot with a short flattened stem. A terminal bud develops at the centre of its upper surface, surrounded by swollen leaf bases that contain food stored from the previous growing season. Papery brown scale leaves cover the outside of the bulb. The stored food is used in the growing season when the terminal bud produces foliage leaves and flowers. The new leaves photosynthesize and some of the manufactured food passes into the leaf bases forming a new bulb (see illustration). If more than one bud develops, then additional bulbs form, resulting in vegetative propagation. Examples of bulb-forming plants are daffodil, onion, and tulip. *Compare* CORM.

bulbil A small bulblike organ that may develop in place of a flower, from an axillary bud, or at the base of a stem in certain plants. If it becomes detached it develops into a new plant.

bulbourethral glands See COWPER'S GLANDS.

bulla (*pl.* **bullae**) The rounded hollow projection of bone from the skull that encloses the *middle ear in mammals.

bundle of His The specialized cardiac muscle fibres in the mammalian heart that receive electrical stimuli from the *atrioventricular node and transmit them throughout the network of *Purkyne fibres. This allows the excitation to reach all parts of the ventricles rapidly and initiates a wave

Development of a bulb

of contraction to expel blood into the aorta and pulmonary artery. The fibres are named after Swiss anatomist Wilhelm His (1831–1904).

bundle sheath cells A layer of cells in plant leaves and stems that forms a sheath surrounding the vascular bundles. In C_4 plants (*see* C_4 PATHWAY) the bundle sheath cells contain chloroplasts and are the site of the *Calvin–Bassham–Benson cycle. The initial fixation of carbon dioxide to form malic acid takes place in the palisade mesophyll cells, which in C_4 plants form a circle around the bundle sheath. This arrangement, known as **Kranz anatomy** or **structure** (after the German *Kranz*, 'wreath'), ensures that the palisade cells are in close contact with the bundle sheath cells so that the malic acid can easily pass to the bundle sheath. It also means that the products of photosynthesis can be quickly transferred from the bundle sheath to the adjacent phloem tissue for transport to other parts of the plant. *See also* HETEROBARIC LEAF ANATOMY.

Burgess shale A fossil-rich deposit of shale and slate dating from the mid-Cambrian period (about 505 million years ago) and located in the Burgess Pass in the Rocky Mountains of British Columbia, Canada. First excavated in 1909, it has since yielded one of the world's oldest assemblages of fossilized marine invertebrates and early vertebrate animals, as well as algae and sponges. The species described include brachiopods, crustaceans, trilobites, and other arthropods, plus a remarkably well-preserved array of fossil worms and other soft-bodied animals. Many forms are of uncertain classification and have been assigned to new phyla. For example, *Wiwaxia* resembles a mollusc

but the body is covered with scalelike elements and has two rows of bladelike spines projecting from the dorsal surface. *Nectocaris* has a kite-shaped body with wide lateral fins and a nozzlelike anterior funnel; two long slender tentacles extend from the head, which also has a pair of stalked eyes. It is now regarded as an early mollusc with affinity to the cephalopods. Another curious design is *Opabinia*, which has a segmented body bearing gills and paired flaplike appendages and a head with five eyes and a long flexible hoselike proboscis. The Burgess shale also contains specimens of *Pikaia*. Originally thought to be an annelid worm, it is now interpreted as a small eel-like chordate with a putative notochord and segmented muscle blocks characteristic of modern lancelets. Some of the organisms found in the Burgess shale have also been unearthed in deposits of the lower Cambrian, notably at the Sirius Passet site in Greenland and the Maotianshan shales in Chengjiang County, China (*see* CHENGJIANG FOSSILS). The great diversity of forms in deposits from this period is seen by some biologists as evidence for a burst of rapid evolution (*see* CAMBRIAN EXPLOSION).

bursa (*pl.* **bursae**) A sac of fibrous connective tissue lined with synovial membrane and filled with synovial fluid. Bursae are found between bones and other tissues, such as skin, ligaments, tendons, and muscles, where they reduce friction when one part moves over another.

butanedioic acid *See* SUCCINATE.

butterflies *See* LEPIDOPTERA.

buttress root *See* PROP ROOT.

cadherin *See* CELL ADHESION MOLECULE.

caecum (cecum) (*pl.* **caeca** or **ceca**) A pouch in the alimentary canal of vertebrates between the *small intestine and *colon. The caecum (and its *appendix) is large and highly developed in herbivorous animals (e.g. rabbits and cows), in which it contains a large population of bacteria essential for the breakdown of cellulose. In humans the caecum is a *vestigial organ and is poorly developed.

Caenorhabditis elegans A soil-dwelling nematode worm that is used experimentally as a model organism in genetics and developmental biology. It was the first multicellular organism to have its genome fully sequenced, in 1998. The genome contains an estimated 17800 genes and amounts to about 97 megabase (Mb). Adults have a fixed number of body cells (959), and the developmental pathway of each cell can be traced. This has yielded valuable insights into mechanisms of development, including genetic control and the role of programmed cell death (*apoptosis), and also the organization of the nervous system.

Cainozoic *See* CENOZOIC.

calciferol *See* VITAMIN D.

calcineurin A phosphatase enzyme that plays a crucial role in relaying intracellular signals following activation of immune cells, particularly lymphocytes (T cells and B cells). It is itself activated by the rise in intracellular concentration of calcium ions that follows stimulation of receptors on the cell surface. Calcineurin is inhibited by the drugs ciclosporin and tacrolimus, which is the basis for their use as immunosuppressants to inhibit T-cell activity, for example in preventing rejection of tissue grafts.

calcitonin (thyrocalcitonin) A peptide hormone in vertebrates that lowers the concentration of calcium (and phosphate) in the blood by inhibiting the action of *osteoclasts, thereby suppressing the release of calcium from bone, and promoting excretion of calcium and phosphate by the kidneys. It operates in opposition to *parathyroid hormone. Calcitonin is produced by the *C cells, which in mammals are located in the thyroid gland. Its precise role in adult humans is unclear, because neither very low nor very high concentrations in the blood seem to have a deleterious effect.

calcitriol The active form of *vitamin D.

calcium Symbol Ca. A soft grey metallic element that is an *essential element for living organisms, being required for normal growth and development. In animals it is an important constituent of bones and teeth and is present in the blood, being required for muscle contraction and other metabolic processes. In plants it is a constituent (in the form of calcium pectate) of the *middle lamella. Calcium ions play a key role in various intracellular signalling pathways and control fusion of vesicles with the plasma membrane in diverse situations, including release of neurotransmitter at synapses, secretion of hormones such as insulin, and activation of the fertilized egg (*see* CALCIUM ION CHANNEL; CALCIUM PUMP; SNARE). *See also* CALMODULIN.

calcium ion channel Any of a family of *voltage-gated ion channels that selectively allow the passage of calcium ions into cells when triggered by a change in the membrane potential of the cell. The resulting influx of calcium ions can have various physiological effects, notably the *contraction of muscle fibres in muscle tissues, including skeletal muscles and heart muscle. Calcium ion channels are also crucial in regulating the tone of smooth muscle, for instance in blood vessels, and generating the action potentials of cardiac pacemaker cells in the sinoatrial node of the heart. Their activity can be modified by drugs—calcium channel blockers—to treat conditions such as angina pectoris, hypertension, and cardiac arrhythmias.

calcium pump (calcium ATPase; Ca^{2+}-ATPase) A membrane-spanning protein that uses *ATP actively to remove calcium ions from cells to maintain a low internal calcium ion concentration. Calcium pumps are especially numerous in the membrane of the *sarcoplasmic reticulum of muscle cells, where they rapidly restore the resting state following the influx of calcium ions that triggers contraction of muscle fibres; hence the calcium pumps allow the fibres to relax in readiness for a further contraction. For each molecule of ATP hydrolysed to ADP, a calcium pump transfers two calcium ions outwards, in exchange for two or three hydrogen ions moving in the opposite direction.

calliclone A culture of cloned tissue cells derived from *callus.

callose An insoluble polysaccharide produced by plants as a temporary cell wall component, for instance in response to damage or stress to seal off affected tissues. It is deposited as drops, plugs, or plates, which stain specifically with aniline blue.

callus 1. (in botany) A tissue consisting of undifferentiated totipotent cells that develops over a cut or damaged plant surface. Hence, callus forms at the cut end of a cutting and can give rise to adventitious roots, either spontaneously or following treatment with hormone 'rooting' powder—a key technique of vegetative propagation. The dedifferentiation of tissue cells to form undifferentiated callus is used as the basis for the commercial cloning of many plant species. By culturing the callus in specialized media containing appropriate nutrients and plant hormones, new genetically identical embryos are formed, which develop into new plants. **2.** (in pathology) A thick hard area of skin that commonly forms on the palms of the hands and soles of the feet as a result of continuous pressure or friction. **3.** (in physiology) Hard tissue formed round bone ends following a fracture, which is gradually converted to new bone.

calmodulin A protein, consisting of 148 amino-acid residues, that is an important regulator of numerous cellular activities, including intracellular signalling pathways. The protein is capable of binding four calcium ions (Ca^{2+}), which causes a conformational change in the molecule, enabling it to interact with various enzymes, including *adenylate cyclase, guanylate cyclase, phosphorylase kinase, and phospholipases. During the contraction of smooth muscle calmodulin binds calcium ions and subsequently activates the enzyme **myosin light chain kinase**. This enzyme phosphorylates the head of the *myosin molecule, enabling it to bind to *actin. Calmodulin may also regulate the functioning of the spindle observed during mitosis.

calnexin A protein, bound to the membrane of the endoplasmic reticulum (ER) of cells, that acts as a *molecular chaperone to newly synthesized proteins of the immunoglobulin superclass. It associates with proteins, including T-cell receptors, MHC class I and II molecules, and immunoglobulins, that are in a partly folded or part-assembled state and retains them in the ER while they undergo further maturation.

calorie The quantity of heat required to raise the temperature of 1 gram of water by 1°C (1 K). The calorie, a c.g.s. unit, is now largely replaced by the *joule, an *SI unit. 1 calorie=4.186 8 joules.

Calorie (kilogram calorie; kilocalorie) 1000 calories. This unit is still in limited use in estimating the energy value of foods, but is obsolescent.

calorific value The heat per unit mass produced by complete combustion of a given substance. Calorific values are used to express the energy values of fuels; usually these are expressed in megajoules per kilogram (MJ kg^{-1}). They are also used to measure the energy content of foodstuffs; i.e. the energy produced when the food is oxidized in the body. The units here are kilojoules per gram (kJ g^{-1}), although Calories (kilocalories) are often still used in nontechnical contexts.

calorimeter Any of various devices used to measure thermal properties, such as *calorific value. *See* BOMB CALORIMETER.

calreticulin A multifunctional calcium-binding protein that acts as a *molecular chaperone for glycoproteins in the endoplasmic reticulum (ER) of cells. It also serves as a storage protein for calcium ions in the ER and may regulate the transcription of certain genes in the nucleus.

Calvin, Melvin (1911–97) US biochemist. After World War II, at the Lawrence Radiation Laboratory, Berkeley, he investigated the light-independent reactions of *photosynthesis. Using radioactive carbon-14 to label carbon dioxide, he was one of the codiscoverers of the *Calvin–Bassham–Benson cycle, for which he was awarded the 1961 Nobel Prize for chemistry.

Calvin–Bassham–Benson cycle (photosynthetic carbon reduction cycle; reductive pentose phosphate cycle) The metabolic pathway of the light-independent stage of *photosynthesis, which occurs in the stroma of the chloroplasts. The pathway was elucidated by Melvin *Calvin, James Bassham (1922–2012), and Andrew Benson (1917–2015), and involves the fixation of carbon dioxide and its subsequent reduction to carbohydrate. During the cycle, carbon dioxide combines with *ribulose bisphosphate, through the mediation of the enzyme ribulose bisphosphate carboxylase/oxygenase (*rubisco), to form an unstable six-carbon compound that breaks down to form two molecules of the three-carbon compound glycerate 3-phosphate. This is converted to glyceraldehyde 3-phosphate, which is used to regenerate ribulose bisphosphate and to produce glucose and fructose. The cycle depends on ATP supplied by the light-dependent reactions, and it generally ceases in the dark.

calyptra (*pl.* **calyptrae**) **1.** A layer of cells that covers the developing sporophyte of mosses, liverworts, clubmosses, horsetails, and ferns. In mosses it forms a hood over the *capsule and in liverworts it forms a sheath at the base of the capsule stalk. **2.** *See* ROOT CAP.

calyptrogen The region within the root *apical meristem that divides to produce the *root cap (calyptra).

calyx (*pl.* **calyxes** or **calices**) The *sepals of a flower, collectively, forming the outer whorl of the *perianth. It encloses the petals, stamens, and carpels and protects the flower in bud. *See also* PAPPUS.

CAM 1. *See* CELL ADHESION MOLECULE. **2.** *See* CRASSULACEAN ACID METABOLISM.

cambium (lateral meristem) (*pl.* **cambiums** or **cambia**) A plant tissue consisting of actively dividing cells (*see* MERISTEM) that is responsible for increasing the girth of the plant, i.e. it causes secondary growth. The two most important cambia are the **vascular** (or **fascicular**) **cambium** and the *cork cambium. The vascular cambium occurs in the stem and root; it divides to produce secondary *xylem and secondary *phloem (new food- and water-conducting tissues). In mature stems the vascular cambium is extended laterally to form a complete ring: the sections of this ring between the vascular bundles comprise the **interfascicular cambium**. *Compare* APICAL MERISTEM.

Cambrian The earliest geological period of the Palaeozoic era. It is estimated to have begun about 542 million years ago and lasted for some 56 million years. During this period marine animals with mineralized shells made their first appearance and Cambrian rocks are the first to contain an abundance of fossils. Cambrian fossils are chiefly of marine animals; they include *trilobites, which dominated the Cambrian seas, echinoderms, brachiopods, molluscs, and primitive *graptolites (from the mid-Cambrian). Trace *fossils also provide evidence for a variety of worms. *See also* CAMBRIAN EXPLOSION.

( ) SEE WEB LINKS
• Illustrated insights into Cambrian stratigraphy and life forms

Cambrian explosion A relatively short interval of rapid intense evolution that supposedly occurred in the early to mid-Cambrian period, some 540 to 520 million years ago. The supposition is based on the sudden appearance in the fossil record from this time of many diverse and novel forms, particularly marine animals, among which can be found representatives of all major modern groups. Notable well-preserved fossil assemblages dating to this period include the *Burgess shale fossils of Canada and the *Chengjiang fossils of China. There is debate about whether such a radiation of forms actually occurred or whether the evidence merely reflects discontinuity in the fossil record Evidence from studies of DNA and recent fossil finds suggests that many animal phyla originated long before the mid-Cambrian, in the Ediacaran period (*see* PROTEROZOIC).

camouflage A high degree of similarity between an animal and its visual environment, which enables it to be disguised or concealed. By blending into the background the animal can elude predators or remain invisible to potential prey. *See also* CRYPTIC COLORATION; MASQUERADE; MIMICRY. *Compare* WARNING COLORATION.

cAMP *See* CYCLIC AMP.

Canada balsam A yellow-tinted resin used for mounting specimens in optical microscopy. It has similar optical properties to glass.

canaliculus (*pl.* **canaliculi**) A very small channel that occurs between the cells of the liver and bone. In the liver the bile canaliculi carry bile to the bile ducts; in bone,

canaliculi connect lacunae, the cavities containing bone cells.

canalization (in evolutionary genetics) A developmental mechanism that limits variation of the phenotype within narrow bounds by repressing underlying genetic variation. It thus maintains a fairly uniform phenotype over a range of different environments in which the organism might normally occur. Canalization is achieved by various genes concerned with development and stress responses. Mutation of these, or exposure to extreme environmental stress, will uncover the genetic variation hitherto hidden by canalization, enabling the population to undergo rapid evolution.

cancer Any disorder of cell growth that results in invasion and destruction of surrounding healthy tissue by abnormal cells. Cancer cells arise from normal cells whose nature is permanently changed. They multiply more rapidly than healthy body cells and do not seem subject to normal control by nerves and hormones. They may spread via the bloodstream or lymphatic system to other parts of the body, where they produce further tissue damage (**metastases**). **Malignant tumour** is another name for cancer. A cancer that arises in epithelium is termed a **carcinoma**; one that arises in connective tissue is called a **sarcoma**. Leukaemia is cancer of white blood cells; **lymphoma** is cancer of *lymphoid tissue; and **myeloma** is cancer of *plasma cells of the bone marrow. Causative agents (**carcinogens**) include various chemicals (including those in tobacco smoke), ionizing radiation, silica and asbestos particles, and *oncogenic viruses. Hereditary factors and stress may also play a role. A key mechanism is mutation of genes in somatic cells; such mutations can, for example, relax normal controls on cell division, circumvent the orderly removal of cells through *apoptosis, or disrupt normal cell signalling pathways. *See* ONCOGENE; TUMOUR-SUPPRESSOR GENE.

( ) SEE WEB LINKS
• Extensive multimedia coverage of cancer biology and different types of cancer, produced by Emory University

cane sugar *See* SUCROSE.

canine tooth A sharp conical *tooth in mammals that is large and highly developed in carnivores (e.g. dogs) for tearing meat. There are two canines in each jaw, each situated between the second *incisor and the first *premolar. In some animals (e.g. herbivores, such as giraffes and rabbits) canine teeth are absent.

capacitation The final stage in the maturation process of a spermatozoon. This takes place inside the genital tract as the sperm penetrates the ovum.

capillarity *See* SURFACE TENSION.

capillary (blood capillary) The narrowest type of blood vessel in the vertebrate circulatory system. Capillaries conduct blood from *arterioles to all living cells: their walls consist of a single layer of endothelial cells, so that oxygen and nutrients can pass through them into the surrounding tissues. Capillaries also transport waste material (e.g. urea and carbon dioxide) to venules for ultimate excretion. Capillaries can be constricted or dilated, according to local tissue requirements. *See* MICROCIRCULATION.

capitulum (*pl.* **capitula**) **1.** (in anatomy) A small rounded head of a bone that articulates with another bone. **2.** (in botany) A type of flowering shoot (*see* RACEMOSE INFLORESCENCE) characteristic of plants of the family Compositae (Asteraceae), e.g. daisy and dandelion. The tip of the shoot is flattened and bears many small stalkless flowers (**florets**) surrounded by an involucre (ring) of bracts. This arrangement gives the appearance of a single flower. **3.** (in zoology) The palps and mouthparts of a tick.

capsid The protein coat of a *virus, which is made up of units called *capsomeres. The chemical nature of the capsid is important in stimulating the body's immune response against the invading virus. In some viruses the capsid is surrounded by an envelope derived chiefly from membranes of the host cell.

capsomere (capsomer) Any of the protein units that make up the regularly organized outer coat (*capsid) of the viruses. The capsid of some viruses contains more than one type of protein molecule.

capsule (theca) 1. (in botany) **a.** A dry fruit that releases its seeds when ripe; it is formed from several fused carpels and contains many seeds. The seeds may be dispersed through pores (as in the poppy), through a lid (as in plantain), or by the splitting and separation of the individual carpels (as in the crocus). Various other forms of capsules include the *silicula and *siliqua. **b.** The part of the sporophyte of mosses and liverworts in which the haploid spores are produced. It is borne on a long stalk (**seta**) and sheds its spores when mature (*see* PERISTOME). **2.** (in microbiology) A thick relatively rigid gelatinous layer completely surrounding the cell wall of certain bacteria (*see* GLYCOCALYX). It appears to have a protective function, making ingestion of the bacterial cell by *phagocytes more difficult and preventing desiccation. **3.** (in animal anatomy) **a.** The membranous or fibrous envelope that surrounds certain organs, e.g. the kidneys, spleen, and lymph nodes. **b.** The ligamentous sheath of connective tissue that surrounds various skeletal joints.

carapace 1. The dorsal part of the *exoskeleton of some crustaceans (e.g. crabs), which spreads like a shield over several segments of the head and thorax. **2.** The domed dorsal part of the shell of tortoises and turtles, formed of bony plates fused with the ribs and vertebrae and covered by a horny epidermal layer. The ventral part of the shell (**plastron**) is similar but flatter.

carbamates Salts or esters of carbamic acid, H_2NCOOH, or their derivatives. They include various insecticides, such as aldicarb, methiocarb, and propoxur. These are active against a wide range of insects upon contact or ingestion; they work by inhibiting *cholinesterase enzymes, which are essential for nervous system function. Hence carbamates are also toxic to other animals, including humans. However, their persistence in the environment is relatively low.

carbamide See UREA.

carbohydrate One of a group of organic compounds based on the general formula $C_x(H_2O)_y$. The simplest carbohydrates are the *sugars (saccharides), including glucose and sucrose, and the *oligosaccharides. *Polysaccharides are carbohydrates of much greater molecular weight and complexity; examples are starch, glycogen, and cellulose. Carbohydrates perform many vital roles in living organisms. Sugars, notably glucose, and their derivatives are essential intermediates in the conversion of food to energy. Starch and other polysaccharides serve as energy stores in plants, particularly in seeds, tubers, etc., which provide a major energy source for animals, including humans. Cellulose, lignin, and others form the supporting cell walls and woody tissue of plants. Chitin is a structural polysaccharide found in the body shells of many invertebrate animals. Carbohydrates also occur in the surface coat of animal cells and in bacterial cell walls and serve as markers for the immune system of potentially pathogenic cells (see EPITOPE).

(()) SEE WEB LINKS

• Summary of carbohydrate nomenclature and classification

carbon Symbol C. A nonmetallic element that occurs in all organic compounds and is therefore fundamental to the structure of all living organisms. It is an *essential element for plants and animals, being ultimately derived from atmospheric carbon dioxide assimilated by plants during photosynthesis (see CARBON CYCLE). The ubiquitous nature of carbon in living organisms is due to its unique ability to form stable covalent bonds with other carbon atoms and also with hydrogen, oxygen, nitrogen, and sulphur atoms, resulting in the formation of a variety of compounds containing chains and rings of carbon atoms.

There are two stable isotopes of carbon (proton numbers 12 and 13) and four radioactive ones (10, 11, 14, 15). Carbon-14 is used in *carbon dating. See also ISOTOPIC DISCRIMINATION; ISOTOPIC SIGNATURE.

carbon assimilation The incorporation of carbon from atmospheric carbon dioxide into organic molecules. This process occurs during *photosynthesis. See CARBON CYCLE.

carbon cycle One of the major cycles of chemical elements in the environment (see BIOGEOCHEMICAL CYCLE). Carbon (as carbon dioxide) is taken up from the atmosphere, or from dissolved carbon compounds in water, and incorporated into the tissues of plants and other primary producers (e.g. phytoplankton) in *photosynthesis. It may then pass into the bodies of animals as the primary producers are eaten (see FOOD CHAIN). During the respiration of plants, animals, and organisms that bring about decomposition, carbon dioxide is returned to the atmosphere. The combustion of fossil fuels (e.g. coal, natural gas, and peat) also releases carbon dioxide into the atmosphere. See illustration. See also GREENHOUSE EFFECT.

carbon dating (radiocarbon dating) A method of estimating the ages of archaeological specimens of biological origin. As a result of cosmic radiation a small number of atmospheric nitrogen nuclei are continuously being transformed by neutron bombardment into radioactive nuclei of carbon-14. Some of these radiocarbon atoms find their way into living trees and other plants in the form of carbon dioxide, as a result of *photosynthesis. When the tree is cut down photosynthesis stops and the ratio of radiocarbon atoms to stable carbon atoms begins to fall as the radiocarbon decays. The ratio $^{14}C/^{12}C$ in the specimen can be measured and enables the time that has elapsed since the tree was cut down to be calculated. The method has been shown to give consistent results for specimens up to some 10 000 years old, but it is less accurate in older material, depending upon assumptions concerning the past intensity of the cosmic radiation. The technique was developed by Willard F. Libby (1908-80) and his coworkers in 1946-7.

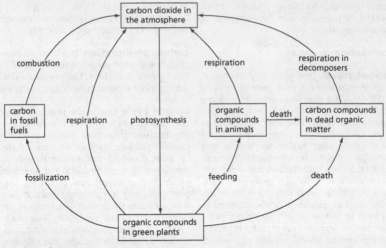

The carbon cycle in nature

carbon dioxide A colourless odourless gas, CO_2, which dissolves in water to give *carbonic acid. It occurs in the atmosphere (0.04% by volume) but has a short residence time in this phase as it is both consumed by plants during *photosynthesis and produced by *respiration, fermentation, and combustion.

The level of carbon dioxide in the atmosphere has increased by some 40% since the Industrial Revolution, mainly because of extensive burning of fossil fuels and the destruction of large areas of *rainforest. This has been postulated as the main cause of the average increase of about 0.94°C in global temperatures over the last 130 years, through the *greenhouse effect. Atmospheric CO_2 concentration, currently nearly 410 ppm, continues to rise in spite of some steps to control emissions, giving the prospect of accelerated global warming and more dramatic *climate change in the foreseeable future.

carbonic acid A weak acid, H_2CO_3, formed in solution when carbon dioxide is dissolved in water:

$$CO_2 + H_2O \rightleftharpoons H_2CO_3$$

The acid is in equilibrium with dissolved carbon dioxide, and also dissociates into hydrogencarbonate and hydrogen ions. These reactions, catalysed by *carbonic anhydrase, take place in the red blood cells when carbon dioxide diffuses into them from the surrounding tissue cells. It is a component in a key natural *buffer system that maintains the pH of body fluids at a certain level. *See also* HYDROGENCARBONATE.

carbonic anhydrase An enzyme, present in red blood cells and kidney cells, that catalyses the reaction between carbon dioxide and water to form carbonic acid, which subsequently dissociates:

$$CO_2 + H_2O \rightleftharpoons H_2CO_3$$
$$H_2CO_3 \rightleftharpoons H^+ + HCO_3^-$$

This reaction is one of the fastest known and controls the elimination of carbon dioxide from the body and the pH of urine. It also facilitates the transfer of carbon dioxide from the tissues to the blood and from the blood to the alveoli (air sacs) of the lungs. *See also* CHLORIDE SHIFT; HAEMOGLOBINIC ACID.

Carboniferous A geological period in the Palaeozoic era. It began about 359 million years ago, following the Devonian period, and extended until the beginning of the Permian period, about 299 million years ago. It is divided into the Mississippian and Pennsylvanian subperiods, dating before and after about 323 mya, respectively. During the Mississippian a marine transgression occurred and the characteristic rock of this division—the Carboniferous limestone—was laid down in the shallow seas. Fauna included foraminiferans, corals, ectoprocts, brachiopods, blastoids, and other invertebrates. The Pennsylvanian saw the deposition of the millstone grit, a mixture of shale and sandstone formed in deltaic conditions, followed by the coal measures, alternating beds of coal, sandstone, shale, and clay. The coal was formed from the vast swamp forests composed of seed ferns, lycopsids, and other plants. During the period fishes continued to diversify, amphibians became more common, insects diversified, and the first reptiles evolved.

carbon monoxide A colourless odourless gas, CO. It is formed by the incomplete combustion of carbon and is present in car-exhaust gases. Carbon monoxide is able to bond with metals; this accounts for its toxicity, which is due to the binding of the CO to the iron in haemoglobin, thereby blocking the uptake of oxygen. In vertebrates small amounts of CO are produced naturally by the enzyme haem oxygenase as a by-product of haem degradation. The gas acts as a neurotransmitter and signalling molecule, and it is also thought to modulate functions of the cardiovascular system and to inhibit the aggregation of platelets. *See* CARBOXYHAEMOGLOBIN.

carbon:nutrient balance hypothesis (environmental constraint hypothesis) A hypothesis that seeks to explain the allocation of a plant's resources to defensive substances in terms of the relative availability of carbon and major nutrients, particularly nitrogen. When carbon (produced by photosynthesis) is in excess, as when a plant is growing in nutrient-poor soil, part of the excess is allocated to the production

of carbon-based secondary metabolites, such as terpenes and phenolics, which act as deterrents to herbivores. When nitrogen is in excess, for example when a plant grows in the shade or when fertilizer is used, the plant uses some of the excess to make toxic nitrogenous metabolites, such as alkaloids. The hypothesis is controversial, and experimental evidence is often contradictory, suggesting that the reality is more complex; some have called for its abandonment.

carboxyhaemoglobin The highly stable product formed when *haemoglobin combines with *carbon monoxide. Carbon monoxide competes with oxygen for haemoglobin, with which it binds strongly: the affinity of haemoglobin for carbon monoxide is 250 times greater than that for oxygen. This reduces the availability of haemoglobin for combination with (and transport of) oxygen and accounts for the toxic effects of carbon monoxide on the respiratory system.

carboxylase Any enzyme involved in the transfer of carbon dioxide or carboxyl groups.

carboxyl group The organic group –CO.OH, present in *carboxylic acids.

carboxylic acids Organic compounds containing the group –CO.OH (the **carboxyl group**; i.e. a carboxyl group attached to a hydroxyl group). Many long-chain carboxylic acids occur naturally as esters in fats and oils and are therefore also known as *fatty acids. *See also* GLYCERIDE.

carboxypeptidase Any enzyme that catalyses the hydrolysis of amino acid residues from the carboxyl terminus of a peptide or polypeptide (*see* EXOPEPTIDASE). Pancreatic juice contains a carboxypeptidase that is secreted into the duodenum.

Carboxylic acid structure

The enzyme is secreted as an inactive precursor, **procarboxypeptidase**, which is activated by another pancreatic protease, *trypsin. *See also* CHYMOTRYPSIN.

carboxysome An organelle found in certain autotrophic bacteria that contains the carbon-fixing enzyme ribulose bisphosphate carboxylase (*see* RIBULOSE BISPHOSPHATE).

carcerulus (*pl.* **carceruli**) A dry fruit that is a type of *schizocarp. It consists of a number of one-seeded fragments (**meri-carps**) that adhere to a central axis. It is characteristic of mallow.

carcinogen Any agent that produces *cancer, e.g. tobacco smoke, certain industrial chemicals, and *ionizing radiation (such as X-rays and ultraviolet rays).

carcinoma (*pl.* **carcinomas** or **carcinomata**) *See* CANCER.

cardiac 1. Relating to the heart. **2.** Relating to the part of the stomach nearest to the oesophagus.

cardiac cycle The sequence of events that occurs in the heart during one full heartbeat. These events comprise contraction (*see* SYSTOLE) and relaxation (*see* DIASTOLE) of the chambers of the heart, associated with opening and closing of the heart valves. When both the atria and the ventricles are relaxed, pressure in the heart is low and blood flows from the vena cava and pulmonary vein into the atria and through to the ventricles. The aortic and pulmonary *semilunar valves, at the junction between the left ventricle and aorta and the right ventricle and pulmonary artery, respectively, are closed; therefore, blood can enter but not leave the heart, which increases the pressure in the chambers. As the pressure in the heart increases, the atria begin to contract, forcing the blood into the ventricles and closing the *tricuspid valve and the *bicuspid valve. A wave of ventricular contraction follows, expelling the blood into the aorta and pulmonary artery to complete the cardiac cycle. At a resting heart rate of, say, 72 beats per minute, the human cardiac cycle lasts approximately 0.85 seconds.

(((∙))) SEE WEB LINKS

• Animation of the cardiac cycle explaining the changes in blood pressure, heart sounds, and ECG trace

cardiac muscle A specialized form of *muscle that is peculiar to the vertebrate heart. Each cardiac muscle fibre is an individual small cell, tapering at either end and containing a single nucleus. The cells branch and interdigitate in a meshlike arrangement, and adjacent cells are joined by intercalated discs and gap junctions to provide extra adhesion and strength to resist the powerful forces of muscular contractions. There are two types of cardiac muscle cells: contractile cells, which are striated and contain numerous myofibrils; and conducting cells, or *Purkyne fibres, which branch extensively and conduct electrical signals throughout the muscle. The muscle itself shows spontaneous contraction and does not need nervous stimulation (*see* PACEMAKER). However, the rate of contractions and muscle contractility (strength of contractions) are regulated by the heart's autonomic nerve supply (*see* HEART).

cardiac output The volume of blood pumped per minute by each ventricle, which is also the total blood flow through the pulmonary circuit. At rest, normal human cardiac output averages approximately 5 litres per minute, which can increase up to fivefold during physical exertion. The cardiac output can be calculated from heart rate (number of beats per minute) and stroke volume (volume of blood expelled from the heart per beat).

cardiovascular centre A cluster of neuron cell bodies in the medulla oblongata of the brain that controls the reflexes affecting heart function and blood vessel diameter. The neurons fall into four functional groups depending on whether they stimulate or inhibit the heart; or dilate or constrict blood vessels. The centre receives nerve inputs from the cerebral cortex, limbic system, and hypothalamus in the brain, and also information about blood pressure (from baroreceptors in the great arteries), blood composition (from chemoreceptors), and physical activity (from proprioceptors in muscles and joints). Output to the heart and blood vessels is via the sympathetic and parasympathetic divisions of the autonomic nervous system, which control the contractility of cardiac muscle and rate of heartbeat.

cardiovascular system *See* CIRCULATORY SYSTEM.

carnassial teeth Molar and premolar teeth modified for shearing flesh by having cusps with sharp cutting edges. They are typical of animals of the order *Carnivora (e.g. tigers, wolves), in which they are the first molars in the lower jaw and the last premolars in the upper.

Carnivora An order of mainly flesh-eating mammals that includes the dogs, wolves, bears, badgers, weasels, and cats. Carnivores typically have very keen sight, smell, and hearing. The hinge joint between the lower jaw and skull is very tight, allowing no lateral movement of the lower jaw. This—together with the arrangement of jaw muscles—enables a very powerful bite. The teeth are specialized for stabbing and tearing flesh: canines are large and pointed and some of the cheek teeth are modified for shearing (*see* CARNASSIAL TEETH).

carnivore An animal that eats meat, especially a member of the order *Carnivora (e.g. tigers, wolves). Carnivores are specialized by having strong powerful jaws and well-developed canine teeth. They may be *predators or carrion eaters. *See also* CONSUMER. *Compare* HERBIVORE; OMNIVORE.

carnivorous plant (insectivorous plant) Any plant that supplements its supply of nitrates in conditions of nitrate deficiency by digesting small animals, especially insects. Such plants are adapted in various ways to attract and trap the insects and produce proteolytic enzymes to digest them. Venus' fly trap (*Dionaea*), for example, has spiny-margined hinged leaves that snap shut on

an alighting insect. Sundews (*Drosera*) trap and digest insects by means of glandular leaves that secrete a sticky substance, and pitcher plants (families Nepenthaceae and Sarraceniaceae) have leaves modified as pitchers into which insects fall, drowning in the water and digestive enzymes at the bottom.

carotene A member of a class of *carotenoid pigments. Examples are β-carotene and lycopene, which colour carrot roots and ripe tomato fruits respectively. α- and β-carotene yield vitamin A when they are broken down during animal digestion; β-carotene has *antioxidant properties.

carotenoid Any of a group of yellow, orange, red, or brown pigments chemically related to terpenes. Carotenoids are responsible for the characteristic colour of many plant organs, such as ripe tomatoes, carrots, and autumn leaves. They also occur in certain algae and other photosynthesizing organisms (such as phototrophic bacteria), in which they function as *accessory pigments in the light-dependent reaction of *photosynthesis; some also have a vital protective function, by absorbing and dissipating excessive light energy *See* CAROTENE; XANTHOPHYLL.

carotid artery The major artery that supplies blood to the head. A pair of **common carotid arteries** arise from the aorta (on the left) and the innominate artery (on the right) and run up the neck; each branches into an **external** and an **internal carotid artery**, which supply the head.

carotid body One of a pair of tissue masses adjacent to the *carotid sinus. Each contains receptors that are sensitive to oxygen and pH levels (acidity) in the blood. High levels of carbon dioxide in the blood lower the pH (i.e. increase the acidity). By responding to fluctuations in pH, the carotid body coordinates reflex changes in respiration rate. *See also* VENTILATION CENTRE.

carotid sinus An enlarged region of the *carotid artery at its major branching point in the neck. Its walls contain many receptors that are sensitive to changes in pressure and it regulates *blood pressure by initiating reflex changes in heart rate and dilation of blood vessels.

carpal (carpal bone) One of the bones that form the wrist (*see* CARPUS) in terrestrial vertebrates.

carpel The female reproductive organ of a flower. Typically it consists of a *stigma, *style, and *ovary. It is thought to have evolved by the fusion of the two edges of a flattened megasporophyll (*see* SPOROPHYLL). Each flower may have one carpel (**monocarpellary**) or many (**polycarpellary**), either free (**apocarpous**) or fused together (**syncarpous**). *See also* PISTIL.

carpogonium (*pl.* **carpogonia**) The female gametangium of the red algae (*see* RHODOPHYTA). Carpogonia, which are found at the tips of the female gametophyte, are flask-shaped structures from which protrudes a slender elongation called a **trichogyne**. The nonmotile male spermatia, released from the antheridia, attach to the trichogyne during sexual reproduction. After fertilization the carpogonium develops into the **cystocarp**, which contains the *carpospores.

carpospore Any of the spores of the red algae (*see* RHODOPHYTA), which are produced in the fertilized *carpogonium. After release the spores develop into a sporophyte (*see* TETRASPORE) or a new gametophyte.

carpus (*pl.* **carpi**) The wrist (or corresponding part of the forelimb) in terrestrial vertebrates, consisting of a number of small bones (**carpals**). The number of carpal bones varies with the species. The rabbit, for example, has two rows of carpals, the first (proximal) row containing three bones and the second (distal) row five. In humans there are also eight carpals. This large number of bones enables flexibility at the wrist joint, between the hand and forelimb. *See also* PENTADACTYL LIMB.

carr *See* HYDROSERE.

carrageenan (carrageen) A naturally occurring polysaccharide isolated from red algae (Rhodophyta). The polymer is composed of D-galactose units, many of which are sulphated. **K-carrageenan** is a gelling agent and stabilizing agent used in foods, cosmetics, and pharmaceuticals.

carrier 1. (in medicine) An individual who harbours a particular disease-causing microorganism without ill-effects and who can transmit the microorganism to others. *Compare* VECTOR. **2.** (in genetics) An individual with an *allele for some defective condition that is masked by a normal *dominant allele. Such individuals therefore do not suffer from the condition themselves but they may pass on the defective allele to their offspring. In humans, women may be carriers of such conditions as redgreen colour blindness and haemophilia, the alleles for which are carried on the X chromosomes (*see* SEX LINKAGE). **3.** (in immunology) A protein that covalently binds to a *hapten to form a hapten-carrier conjugate. Unlike the hapten alone, the conjugate can cause the immune system to produce antibodies against the hapten, against the carrier, and against the hapten-carrier combination. **4.** (in biochemistry) *See* CARRIER MOLECULE; HYDROGEN CARRIER.

carrier molecule 1. A molecule that plays a role in transporting electrons through the *electron transport chain. Carrier molecules are usually proteins bound to a nonprotein group; they can undergo oxidation and reduction relatively easily, thus allowing electrons to flow through the system. There are four types of carrier: flavoproteins (e.g. *FAD), *cytochromes, iron-sulphur proteins (e.g. *ferredoxin), and *ubiquinone. **2.** A lipid-soluble molecule that can bind to lipid-insoluble molecules and transport them across membranes. Carrier molecules have specific sites that interact with the molecules they transport. Several different molecules may compete for transport by the same carrier. *See* TRANSPORT PROTEIN.

carrying capacity Symbol K. The maximum population of a particular species that can be supported indefinitely by a given habitat or area without damage to the environment. It can be manipulated by human intervention. For example, the carrying capacity for grazing mammals could be increased by boosting the yield of their grassland habitat by the application of fertilizer. *See also* K SELECTION.

cartilage (gristle) A firm flexible connective tissue that forms the adult skeleton of cartilaginous fish (e.g. sharks). In other vertebrates cartilage forms the skeleton of the embryo, being largely replaced by *bone in mature animals (although it persists in certain areas). Cartilage comprises a matrix consisting chiefly of a glycosaminoglycan (mucopolysaccharide) called **chondroitin sulphate** secreted by cells (**chondroblasts**) that become embedded in the matrix as **chondrocytes**. It also contains collagenous and elastic fibres. **Hyaline cartilage** consists largely of glycosaminoglycan, giving it a shiny glasslike appearance; this type of cartilage gives flexibility and support at the joints. **Fibrocartilage**, in which bundles of collagen fibres predominate, is stronger and less elastic than hyaline cartilage; it is found in such areas as the intervertebral discs. **Elastic cartilage** has a yellow appearance due to the presence of numerous elastic fibres (*see* ELASTIN). This cartilage maintains the shape of certain organs, such as the pinna of the ear.

(((•))) **SEE WEB LINKS**
• Images and descriptions of different types of cartilage

cartilage bone (replacing bone) *Bone that is formed by replacing the cartilage of an embryo skeleton. The process, called **endochondral ossification**, is brought about by the cells (*osteoblasts) that secrete bone. *Compare* MEMBRANE BONE.

cartilaginous fishes *See* CHONDRICHTHYES.

caruncle A small outgrowth from the testa of a seed that develops from the placenta, funicle, or micropyle. Examples include the warty outgrowth from the castor-oil seed and the tuft of hairs on the testa of the seed of willowherb. *See also* ARIL.

caryopsis (*pl.* **caryopses**) A dry single-seeded indehiscent fruit that differs from an *achene in that the fruit wall is fused to the testa of the seed. It is the grain of cereals and grasses.

casein One of a group of phosphate-containing proteins (phosphoproteins) found in milk; it is also the principal protein of cheese (*see* CURD). Caseins are easily digested by the enzymes of young mammals and represent a major source of phosphorus. *See* RENNIN.

Casparian strip A band of *suberin, an impermeable substance, found in the endodermal cell walls of plant roots; it was named after R. Caspary. Movement of water through the *apoplast pathway is diverted from the cell wall to the cytoplasm, where it then follows the *symplast pathway. The endodermal cells actively secrete salts into the vascular tissue. This results in a low (more negative) water potential, allowing water to move down a water potential gradient from the endodermis into the vascular tissue. The Casparian strip prevents water and solutes from returning to the cortex; consequently a positive hydrostatic pressure is established in the vascular tissue—the phenomenon of *root pressure. It also blocks the entry of potentially toxic substances into the vascular system.

caspase Any of a family of protease enzymes that are central to initiating and carrying out programmed cell death (*see* APOPTOSIS). They are formed, in response to certain triggers, from inactive precursors present in the cell. Proapoptotic signals include the activation of death receptors on the cell surface and bursting of mitochondria due to cell damage. Initiator caspases activate other caspases in a sequential cascade, culminating in effector caspases, such as caspases 3 and 6. Effector caspases cleave crucial cellular components, such as cytoskeletal proteins, and also activate DNase enzymes and inhibit DNA repair enzymes, leading to breakdown of the cell's DNA. Granzyme B, delivered into cells by *cytotoxic T cells, can activate effector caspases directly. Release of cytochrome *c* from damaged mitochondria activates caspases 9 and 3, a process mediated by the formation of a multiprotein complex called an **apoptosome**.

caste A division found in *eusocial insects, such as the *Hymenoptera (ants, bees, wasps) and the *Isoptera (termites), in which the individuals are structurally and physiologically specialized to perform a particular function. For example, in honeybees there are queens (fertile females), workers (sterile females), and drones (males). There are several different castes of workers (all sterile females) among ants. *See* PHENOTYPIC PLASTICITY.

casual *See* ALIEN.

catabolism The metabolic breakdown of large molecules in living organisms to smaller ones, with the release of energy. Respiration is an example of a catabolic series of reactions. *See* METABOLISM. *Compare* ANABOLISM.

catalase An enzyme, found in *peroxisomes, that catalyses the decomposition of hydrogen peroxide, which results from oxidation reactions in the cell (*see also* SUPEROXIDE DISMUTASE), into water and oxygen. High concentrations of catalase are found in the liver. Catalase is one of the fastest known enzymes; it is employed in the rubber industry to break down hydrogen peroxide to form oxygen, thereby converting latex to foam rubber.

catalysis The process of changing the rate of a chemical reaction by use of a *catalyst.

catalyst A substance that increases the rate of a chemical reaction without itself undergoing any permanent chemical change. The catalyst provides an alternative pathway by which the reaction can proceed, in which the *activation energy is lower. It thus increases the rate at which the reaction comes to equilibrium, although it does not alter the position of the equilibrium. *Enzymes are the catalysts in biochemical reactions; they are highly specific in the type of reaction they catalyse.

catalytic activity The increase in the rate of a specified chemical reaction caused by an enzyme or other catalyst under specified assay conditions. It is measured in *katals or in moles per second.

catalytic RNA *See* RIBOZYME.

catecholamine Any of a class of amines that possess a catechol ($C_6H_4(OH)_2$) ring. Including *dopamine, *adrenaline, and *noradrenaline, they function as *neurotransmitters and/or hormones.

category *See* RANK.

catenin Any of a family of proteins that act with other *cell adhesion molecules to bind cells together. Catenins anchor the actin filaments of the internal cytoskeleton to the cadherin molecules that span the space between adjacent cells at *adherens junctions and (in the case of β-catenin) at *desmosomes. Together these components maintain the three-dimensional structure and function of cell layers. They are also important in determining cell fate and organization during development. Moreover, β-catenin functions with *WNT protein in a signal transduction pathway inside cells that regulates cell development, homeostasis, and survival. Binding of WNT to its receptor on the cell surface causes a build-up of β-catenin inside the cell, some of which enters the nucleus where it switches on the transcription of target genes. Mutations in β-catenin are implicated in certain cancers.

cathepsins Protein-digesting enzymes (proteases) that are found in *lysosomes.

cation A positively charged *ion, such as the sodium ion (Na^+). *Compare* ANION.

catkin A type of flowering shoot (*see* RACEMOSE INFLORESCENCE) in which the axis, which is often long, bears many small stalkless unisexual flowers. Usually the male catkins hang down from the stem; the female catkins are shorter and often erect. Examples include birch and hazel. Most plants with catkins are adapted for wind pollination, the male flowers producing large quantities of pollen; willows are an exception, having nectar-secreting flowers and being pollinated by insects.

caudal vertebrae The bones (*see* VERTEBRA) of the tail, which articulate with the *sacral vertebrae. The number of caudal vertebrae varies with the species. Rabbits, for example, have 15 caudal vertebrae, while in humans these vertebrae are fused to form a single bone, the *coccyx.

caveola (*pl.* **caveolae**) A small static flask-shaped depression in the plasma membrane of a cell that is lined by transmembrane proteins called **caveolins** and contains receptors for various extracellular signalling molecules. There is a two-way traffic in and out of the cell via caveolae, and caveolin-containing vesicles are transported from within the cell to the caveolar membrane along microtubules. Caveolae are also involved in endocytosis and play an important role in the cellular uptake of bacterial toxins, viruses, and proteins.

cavitation 1. The formation of a void or gas bubble in a fluid subject to reduced pressure. Cavitation occurs in the water-conducting vessels of plants—the xylem vessels and tracheids—especially when the plant is losing water rapidly by transpiration or soil moisture levels are low. In this situation the water column within the conducting vessels is liable to break, creating a void into which air can diffuse to form an embolism. This causes the flow of water in the affected vessel to cease, thereby reducing the plant's overall conducting capacity. Plants have several mechanisms for overcoming or repairing embolisms. The air bubble is generally contained within a section of vessel, enabling water to bypass the obstruction by flowing laterally into neighbouring vessels. In some cases when transpiration is low, for instance at night, the reduced tension in the water column enables the gas to redissolve in the vessel water. Also, raised root pressure can force the gas to redissolve and the vessel to refill, a mechanism observed in many trees and herbaceous plants. Another mechanism seems to involve an osmotic refilling mechanism that can operate even in water-stressed plants. **2.** (in pathology) The

formation of a cavity, as in the lung due to tuberculosis.

CBD *See* CONVENTION ON BIOLOGICAL DIVERSITY.

C cell (parafollicular cell) Any one of a group of cells in vertebrates that are derived from the terminal pair of gill pouches. In mammals these cells are incorporated into the *thyroid and *parathyroid glands. In other vertebrates they occur mainly in the lung and the paired **ultimobranchial bodies**.

CCK *See* CHOLECYSTOKININ.

CD (cluster of differentiation) Any group of antigens that is associated with a specific subpopulation of human *T cells. The differentiation antigens expressed by a T cell vary with its stage of development and thus with its role in the immune response. Hence, for example, *CD4 antigens are expressed by helper T cells, whereas *CD8 antigens are expressed by cytotoxic T cells. The antigens are glycoproteins and are characterized using *monoclonal antibodies.

CD4 A surface glycoprotein that characterizes *helper T cells. CD4 recognizes foreign peptides bound to *MHC class II proteins expressed on the surface of B cells, macrophages, and other antigen-presenting cells. It acts as a *coreceptor, binding simultaneously with the T-cell receptor to the antigen-MHC II complex and enhancing the T-cell's response. CD4 consists of a single polypeptide chain organized into four immunoglobulin-like domains.

CD8 A surface glycoprotein that characterizes *cytotoxic T cells. It recognizes *MHC class I proteins, which are produced by virtually all cells and form complexes with processed antigen derived from viruses or other pathogens that have invaded the cell. CD8 binds to any antigen-MHC I complex on the surface of an infected host cell and acts as a *coreceptor, helping to activate the T cell and prompting it to kill the infected host cell. The CD8 molecule comprises two distinct chains, α and β, linked by a disulphide bond.

CD40 ligand *See* HELPER T CELL.

cDNA *See* COMPLEMENTARY DNA.

CDR *See* COMPLEMENTARITY-DETERMINING REGION.

cecidium (*pl.* **cecidia**) *See* GALL.

cecropin Any of a group of peptides that have potent activity against Gram-negative and Gram-positive bacteria. First isolated from the haemolymph of the giant silk moth (*Hyalophora cecropia*), cecropins are components of the innate immune response of insects. They disrupt the bacterial cell membranes by forming pores.

CED protein Any of several proteins in the nematode worm *Caenorhabditis elegans* that control programmed cell death (*apoptosis) and are crucial in normal development of the larva. The CED proteins have molecular structures homologous to the *caspases and other apoptotic proteins found in humans and other vertebrates.

cell The structural and functional unit of most living organisms (*compare* COENOCYTE; SYNCYTIUM). Cell size varies, but most cells are microscopic (average diameter 0.01–0.1 mm). Cells may exist as independent units of life, as in bacteria and certain protists, or they may form colonies or tissues, as in all plants and animals. Each is differentiated into *cytoplasm and a *nucleus, which contains DNA, and is bounded by a *plasma membrane, which in the cells of plants, fungi, algae, and bacteria is surrounded by a *cell wall. There are two main types of cell. In **prokaryotic cells** (bacteria and archaea) the nuclear material is not bounded by a membrane and chemicals involved in cell metabolism are associated with the plasma membrane. Reproduction is generally asexual and involves simple cell cleavage. In **eukaryotic cells** the nucleus is bounded by a nuclear membrane and the cytoplasm is divided by membranes into a system of interconnected cavities and separate compartments (**organelles**), e.g. *mitochondria, *endoplasmic reticulum, *Golgi apparatus, *lysosomes, and *ribosomes (see illustration). The shape and internal organization of cells depends on a network of tubules and filaments called the *cytoskeleton. Reproduction can be either asexual (*see* MITOSIS) or sexual (*see* MEIOSIS). Plants and animals consist of eukaryotic cells but plant cells possess *chloroplasts and other *plastids and bear a rigid cellulose cell wall.

⊕ SEE WEB LINKS

• Overview of cell biology, with animations and illustrations, from Cells Alive!

Golgi apparatus
smooth endoplasmic reticulum
mitochondrion
cellulose cell wall
chloroplast
nucleolus
ribosome
plasma membrane
cytoplasm
rough endoplasmic reticulum
lysosome
nuclear membrane
nucleus
cell vacuole
glycogen granules
centrioles
centrosome
mitochondrion
starch granules
fat droplets
rough endoplasmic reticulum
Golgi apparatus
ribosome
Animal
Plant

Generalized eukaryotic cells

CELL BIOLOGY

1665	English physicist Robert Hooke (1635–1703) coins the word 'cell'.
1831	Robert Brown discovers the nucleus in plant cells.
1838	German botanist Matthias Schleiden (1804–81) proposes that plants are composed of cells.
1839	Theodor Schwann states that animals are composed of cells and concludes that all living things are made up of cells.
1846	German botanist Hugo von Mohl (1805–72) coins the word 'protoplasm' for the living material of cells.
1858	German pathologist Rudolf Virchow (1821–1902) postulates that all cells arise from other cells.
1865	German botanist Julius von Sachs (1832–97) discovers the chlorophyll-containing bodies in plant cells later named chloroplasts.
1876–80	German cytologist Eduard Strasburger (1844–1912) describes cell division in plants and states that new nuclei arise from division of existing nuclei.
1882	German cytologist Walther Flemming (1843–1905) describes the process of cell division in animal cells, for which he coins the term 'mitosis'. Strasburger coins the words 'cytoplasm' and 'nucleoplasm'.
1886	German biologist August Weismann (1834–1914) proposes his theory of the continuity of the germ plasm.
1887	Belgian cytologist Edouard van Beneden (1846–1910) discovers that the number of chromatin-containing threadlike bodies (subsequently named chromosomes) in the cells of a given species is always the same and that the sex cells contain half this number.
1888	German anatomist Heinrich von Waldeyer (1836–1921) coins the word 'chromosome'.
1898	Camillo Golgi discovers the Golgi apparatus.
1901	US biologist Clarence McClung (1870–1946) discovers the sex chromosomes.
1911	Thomas Hunt Morgan produces the first chromosome map.
1949	Canadian geneticist Murray Barr (1908–95) discovers Barr bodies.
1955	Belgian biochemist Christian de Duve (1917–2013) discovers lysosomes and peroxisomes.
1956	Romanian-born US physiologist George Palade (1912–2008) discovers the role of microsomes (later renamed ribosomes).
1957	US biochemist Melvin Calvin (1911–97) publishes details of the photosynthetic carbon-fixation cycle (Calvin–Bassham–Benson cycle).
1960–61	South African-born British biochemist Sydney Brenner (1927–2019) discovers messenger RNA, in conjunction with François Jacob (1920–2013) and Matthew S. Meselson (1930–).
1964	US microbiologists Keith Porter and Thomas F. Roth discover the first cell receptors.
1970	US biologist Lynn Margulis (1938–2011) proposes the endosymbiont theory for the origin of eukaryote cellular organelles.

1971	German-born US cell biologist Günter Blobel (1936–2018) proposes the signal hypothesis to explain how proteins are delivered to their correct destinations within cells.
1975	British biologists J. A. Lucy and E. C. Cocking achieve successful fusion of plant and animal cells.
1979	The first 'test-tube baby', Louise Brown, is born in the UK using *in vitro* fertilization.
1982	British cell biologist Timothy Hunt (1943–) discovers cyclins, proteins that control the cell cycle.
	US neurologist Stanley Prusiner (1942–) discovers prions.
1983	A mouse embryo is engineered to include the gene for human growth hormone, creating a 'supermouse'.
1984	Sheep embryos are cloned for the first time.
1986	US cell biologist Robert Horvitz (1947–) identifies genes involved in programmed cell death in the nematode *Caenorhabditis elegans*. First licence granted in USA for marketing a genetically engineered organism.
1993	First successful cloning of human embryos.
1997	Birth of Dolly the sheep, the first mammal to be cloned from adult body cells.
1998	Approval given in USA for therapeutic use of a synthetic skin containing live cultured human tissue cells.
2000	The embryo of a gaur, an endangered mammal, is cloned from skin cells of an adult and develops inside the womb of a cow.
2002	A pluripotent stem cell is isolated from adult human bone marrow. Discovery of new mechanism for regulating gene expression, called a riboswitch.
2004	World's first bank for stem cells opens in north London.
2005	First cloned dog (an Afghan hound called Snuppy) is created, using somatic cell transfer, by Korean researchers led by Woo Suk Huang.
2006	Reprogramming of skin cells to form pluripotent stem cells achieved by a team headed by Shinya Yamanaka (joint winner of the Nobel Prize for physiology or medicine in 2012).
2008	Stem cells used to grow a new trachea for transplant.
2010	The world's first 'synthetic cell' unveiled by the J. Craig Venter Institute; it is based on a bacterial cell containing an artificially constructed chromosome.
2013	Functional human liver tissue buds are derived from induced pluripotent human stem cells implanted into mice.
2014	Yeast cells engineered to synthesize opiates and semisynthetic opioids.
2016	'Artificial' mouse eggs created from tissues cells are successfully fertilized and implanted in a surrogate mother to produce pups.
2017	Approval granted for first cancer therapy based on a patient's own T cells, which are extracted and engineered to express chimeric antigen receptors (CARs) designed to target the patient's specific tumour. The CAR T cells are then reintroduced to the patient to destroy the tumour cells.

cell adhesion molecule (CAM) Any of the proteins in the plasma membrane of animal cells that 'glue' cells to the extracellular matrix (ECM) or to each other, for example by forming *cell junctions. CAMs are also important during growth and development in enabling cells to recognize each other and migrate appropriately within the embryo to ensure correct cell-cell interactions. CAMs typically protrude from the membrane to attach to another molecule, often another CAM, by means of a binding site. There are several families of CAMs, the largest being the **cadherins**, which are glycoproteins found abundantly in junctions between epithelial cells and also in *desmosomes. The **integrins** bind cells to components of the ECM, such as collagens and laminins, and can form cell-matrix junctions, such as hemidesmosomes, which act like 'spot welds' between cells and the ECM. **Selectins** and **intercellular adhesion molecules** (ICAMs) occur on the surface of endothelial cells lining blood vessels, where they help to tether passing leucocytes (white blood cells) at sites of inflammation; **nerve-cell adhesion molecules** (N-CAMs) are important in ensuring proper cell-cell contacts during development of the nervous system and muscle tissue.

cell body (perikaryon; soma) The part of a *neuron that contains the nucleus. The cell processes that are involved in the transmission and reception of nervous impulses (the axon and the dendrites respectively) develop as extensions from the cell body.

cell culture The culturing of cells from plants, animals, or microorganisms on a culture medium *in vitro*. *See* CULTURE.

cell cycle The sequence of stages that a cell passes through between one cell division and the next. The cell cycle can be divided into four main stages: (1) the **M phase**, which consists of *mitosis (nuclear division) and **cytokinesis** (cytoplasmic division); (2) the **G₁ ('first gap') phase**, in which there is a high rate of biosynthesis and growth; (3) the **S phase**,

Cell cycle

in which the DNA content of the cell doubles and the chromosomes replicate; (4) the **G₂ ('second gap') phase**, during which the final preparations for cell division are made.

*Interphase consists of the G_1, S, and G_2 phases, which comprise about 90% (16–24 hours) of the total time of the cell cycle in rapidly dividing cells. The M phase lasts about 1–2 hours. A point is reached in the G_1 phase, known as the **restriction point**, after which the cell becomes committed to passing through the remainder of the cell cycle regardless of the external conditions. (see illustration). The cell cycle is regulated by external and internal signals that determine if the cycle can proceed past several checkpoints. Key to this control mechanism are cyclic fluctuations in cellular concentrations of proteins called *cyclins and in the activity of cyclin-dependent kinases. The timing and frequency of cell division are crucial to normal growth and development and vary according to the type of cell. In a mature human, for example, skin cells, which need constant replacement, divide frequently throughout life, whereas nerve and muscle cells exist in a nondividing state called G0. Many other types of tissue cells are in the G0 state, but some can re-enter the cell cycle if new cells are needed, e.g. to repair damaged tissue.

cell division The formation of two or more daughter cells from a single mother cell. The nucleus divides first; this is followed by division of the cytoplasm (**cytokinesis**) with the formation of a plasma membrane between the daughter nuclei. *Mitosis produces two daughter nuclei that are identical to the original nucleus; in plants a *cell plate forms between the new cells. *Meiosis results in four daughter nuclei each with half the number of chromosomes in the mother cell nucleus. *See also* CELL CYCLE.

SEE WEB LINKS
• Animation comparing mitosis with meiosis

cell fusion The combining of two cells to form a single cell. It occurs in nature, most notably in the merger of gamete cells in sexually reproducing organisms, but also in normal development of certain tissues such as muscle and bone, where multinucleate cells (syncytia) are formed, and in some immune cell lines. It is also implicated in the spread of cancer cells, with evidence that they may fuse with macrophages to create mobile hybrid cells. The fusion of plasma membranes requires proteins called fusogens. Cell fusion is performed experimentally to combine cells from different tissues or species in a cell culture (i.e. **somatic cell hybridization**). The fused cells coalesce (*see* CHEMICAL FUSOGEN) but their nuclei generally remain separate. However, during cell division a single spindle is formed so that each daughter cell has a single nucleus containing sets of chromosomes from each parental line. Subsequent division of the hybrid cells often results in the loss of chromosomes (and therefore genes), so that absence of a gene product in the culture can be related to the loss of a particular chromosome. Thus the technique is used to determine the control of characteristics exerted by specific chromosomes. Hybrid cells (*see* HYBRIDOMA) resulting from cell fusion are used to produce *monoclonal antibodies.

cell junction Any of various kinds of connection between cells. *Tight junctions form a seal between adjacent cells, particularly in epithelia, to prevent the passage of materials between cells. A primarily structural bond between cells is provided by *adherens junctions and *desmosomes, whereas communication between adjacent cells is facilitated by *gap junctions in animal cells and *plasmodesmata in plant cells.

cell membrane Any membrane that is found in a living cell, especially the *plasma membrane, which forms the cell boundary. Other cell membranes include the *nuclear envelope; the *tonoplast, which encloses the vacuole of plant cells; and the membranes of the various cell organelles, such as the endoplasmic reticulum, Golgi apparatus, mitochondria, chloroplasts, and lysosomes.

cell plate The structure that forms in a dividing plant cell at the end of mitosis; it separates the cytoplasm of the two daughter protoplasts. The cell plate is formed from vesicles made by *dictyosomes and arranged by the microtubules of the *phragmoplast in the equatorial region of the *spindle. These vesicles contain pectin, cellulose, and *hemicellulose, which contribute to the *middle lamella and the primary wall of the new cell wall. The cell plate eventually fuses with the cell wall of the parent cell, dividing it into two daughter cells.

cell sap The solution that fills the vacuoles of plant cells. It contains sugars, amino acids, waste substances (such as tannins), and mineral salts.

cell theory The theory that was born of the findings of Matthias Schleiden in 1838 and Theodor *Schwann in 1839, who postulated, respectively, that plants and animals were made up of cells and that these units were basic to the structure and function of all organisms. Previously, in 1665, Robert Hooke (1635–1703), while examining cork under the microscope, had observed that its structure consisted of hollow boxlike units, which he called 'cells'. At the time, however, he did not realize the significance of these units, which were in fact dead cells.

cellular respiration *See* RESPIRATION.

cellulase A carbohydrate-digesting enzyme (a **carbohydrase**) that hydrolyses cellulose to sugars, including **cellobiose** (a disaccharide consisting of two β-(1,4) linked molecules of glucose) and glucose. Cellulase breaks the β-glycosidic links that join the constituent sugar units of cellulose. It plays an essential part in *abscission in plants, while microbial cellulases enable the digestion of plant material by herbivores. *See also* RUMINANTIA.

cellulolytic Able to digest cellulose. For example, cellulolytic bacteria, in the stomach of ruminants, digest cellulose by means of the enzyme *cellulase.

cellulose A polysaccharide that consists of a long unbranched chain of glucose units linked by 1-4 β-*glycosidic bonds (see formula). It is the main constituent of the cell walls of all plants, many algae, and some fungi and is responsible for providing the rigidity of the cell wall. It is an important constituent of dietary *fibre. Cellulose occurs typically as microfibrils, each consisting of long parallel arrays of 50–80 cellulose molecules. The fibrous nature of extracted cellulose has led to its use in the textile industry for the production of cotton, artificial silk, etc.

cell wall A rigid outer layer that surrounds the plasma membrane of plant, fungal, algal, and bacterial (but not animal) cells. It protects and/or gives shape to a cell, and in herbaceous plants provides mechanical support for the plant body. The cell walls in most plants and algae are composed of polysaccharide fibrils embedded in a matrix of other polysaccharides and proteins.

In young plant cells *cellulose microfibrils and matrix components are secreted into the extracellular space outside the plasma membrane, to form a **primary cell wall**. This typically ranges from 0.1 μm to several micrometres in thickness and is perforated by fine cytoplasmic channels called *plasmodesmata. Adjacent cells are joined together by a *middle lamella of sticky pectins. When plant cells have stopped growing, they can strengthen the primary cell wall by secreting hardening substances, or add a **secondary cell wall** inside the primary wall. This consists of further layers of cellulose microfibrils embedded in *lignin, a complex polymer and a major component of wood. This creates a strong, waterproof, protective layer for the plant. The cell walls of fungi consist mainly of *chitin. Bacterial cell walls consist of complex polymers of polysaccharides and amino acids (*see* GRAM'S STAIN; PEPTIDOGLYCAN). *See also* CELL PLATE.

Celsius scale A temperature scale in which the fixed points are the temperatures at standard pressure of ice in equilibrium with water (0°C) and water in equilibrium with steam (100°C). The scale, between these two temperatures, is divided in 100 degrees. The degree Celsius (°C) is equal in magnitude to the *kelvin. This scale was formerly known as the **centigrade scale**; the name was officially changed in 1948 to avoid confusion with a hundredth part of a grade. It is named after the Swedish astronomer Anders Celsius (1701–44), who devised the inverted form of this scale (ice point 100°, steam point 0°) in 1742.

1–4 β-glycosidic bonds

cellulose

cement (cementum) A thin layer of bony material that fixes teeth to the jaw. It covers the dentine of the root of a *tooth, below the level of the gum, and is attached to the *periodontal membrane lining the tooth socket in the jawbone.

Cenozoic (Cainozoic; Kainozoic) The geological era that began about 66 million years ago and extends to the present. It followed the *Mesozoic era and is subdivided into the *Palaeogene and *Neogene periods. The Cenozoic is often known as the **Age of Mammals** as these animals evolved to become an abundant, diverse, and dominant group. Birds and flowering plants also flourished. The era saw the formation of the major mountain ranges of the Himalayas and the Alps.

(((⊕))) SEE WEB LINKS
• Survey of the Cenozoic era by the University of California Museum of Paleontology

centi- Symbol c. A prefix used in the metric system to denote one hundredth. For example, 0.01 metre=1 centimetre (cm).

centigrade scale *See* CELSIUS SCALE.

centimorgan Symbol cM. *See* MAP UNIT.

centipedes *See* CHILOPODA.

Central Dogma The basic belief originally held by molecular geneticists, that flow of genetic information can only occur from *DNA to *RNA to proteins. It is now known, however, that information contained within RNA molecules can also flow back to DNA, for example in the replication of *retroviruses. *See also* GENETIC CODE.

central nervous system (CNS) The part of the nervous system that coordinates all neural functions. In invertebrates it may comprise simply a few *nerve cords and their associated *ganglia. In vertebrates it consists of the *brain and the *spinal cord. The vertebrate CNS contains *reflex arcs, which produce automatic and rapid responses to particular stimuli.

centre (in neurology) A part of the nervous system, consisting of a group of nerve cells, that coordinates a particular process. An example is the respiratory centre in the vertebrate brainstem, which controls breathing movements. The stimulation of a centre will initiate the process, while destruction of the centre will prevent or impair it. *See also* CARDIOVASCULAR CENTRE.

centric fusion A type of chromosomal rearrangement in which the *centromeres of two 'one-armed' (acrocentric) chromosomes fuse to form a single chromosome with two large arms (dicentric). In humans centric fusion involving chromosomes 13 and 14 is found in about 1 in 1000 people; as no genetic information is lost, it has no ill effects.

centrifuge A device in which solid or liquid particles of different densities are separated by rotating them in a tube in a horizontal circle. The denser particles tend to move along the length of the tube to a greater radius of rotation, displacing the lighter particles to the other end.

centriole A structure associated with the *centrosome and found mainly in animal cells. It consists of two short cylinders, orientated at right angles to each other and composed of microtubules. When present, the centriole replicates during the nondividing phase of the cell cycle, and during the prophase of mitosis a centriole migrates with each centrosome to lie at opposite poles of the cell. Centrioles seem to act as an orientational device, or jig, in the assembly of *centrosomes; they are not obligatory for the assembly of the spindle microtubules—they are not present in the cells of most higher plants, and their removal from cells does not affect spindle formation. However, centrioles are essential for the assembly of eukaryotic cilia and flagella. *See also* FLAGELLUM.

centroblast A rapidly proliferating B cell (B lymphocyte) found within the germinal centre of a lymph node follicle. Centroblasts are derived from B cells activated by exposure to a specific antigen in the first phase of a humoral immune response; they seed the germinal centres and divide rapidly. As numbers increase their progeny cells divide and undergo subtle genetic changes (*see* SOMATIC HYPERMUTATION) that fine-tune their specificity and enhance affinity for the particular antigen. After a few days, some differentiate into either *memory cells or antibody-secreting *plasma cells and leave the germinal centre.

centromere (kinomere; spindle attachment) The part of a *chromosome that attaches to the *spindle during cell division (*see* MEIOSIS; MITOSIS). It is characterized by highly repetitive DNA. Spindle attachment is via a platelike structure called the *kinetochore. The position of the centromere is a distinguishing feature of individual chromosomes: a chromosome with the centromere at its centre is described as **metacentric**; one with the centromere towards one end is **acrocentric**; and a chromosome with the centromere at either end is **telocentric**. The centromere usually appears as a constriction when chromosomes contract during cell division.

centrosome (cell centre; centrosphere) A specialized region of all eukaryote cells except fungi, situated next to the nucleus, that organizes the microtubules of the *spindle during cell division. It also serves as the chief *microtubule-organizing centre in animal cells. The centrosomes of most animal cells contain a pair of *centrioles. During *metaphase of mitosis and meiosis, the centrosome separates into two regions, each containing one of the centrioles (where present). The two regions move to opposite ends of the cell and a spindle forms between them.

centrum *See* VERTEBRA.

cephalization The tendency among animal groups for the major sense organs, mouth, and brain to be grouped together at the front (anterior) end of the body. These are usually contained in a specialized cephalic region—the head.

Cephalochordata A subphylum or phylum of chordates that contains only the lancelets, a group of small invertebrate marine fishlike animals. There are 30 known extant species in two genera—*Branchiostoma*

(Amphioxus) and *Epigonichthys* (*Asymmetron*). About 5-15 cm long, lancelets have gill slits and a notochord that persist into adulthood, the notochord extending the whole length of the body and providing the main skeletal support. The pharynx is modified for filter feeding (*see* ENDOSTYLE). Molecular evidence has established the lancelets as the earliest diverging (i.e. most basal) of all chordate animal groups, rather than the tunicates (*see* UROCHORDATA).

Cephalopoda The most advanced class of molluscs, containing the squids, cuttlefishes, octopuses, and the extinct *ammonites. Cephalopods have a highly concentrated central nervous system within a protective cartilaginous case. The eye has a well-developed retina and is comparable to that of vertebrates. All cephalopods are predacious carnivores capable of swimming by jet propulsion; they have highly mobile tentacles for catching and holding prey. Octopuses have a large central brain, highly developed vision, and also groups of neurons ('sub-brains') in each of their eight tentacles that function semi-autonomously, e.g. when sensing and handling objects. They are now regarded as highly intelligent, with short- and long-term memory and capabilities such as problem-solving, tool use, recognition of individual humans, and play.

cephalothorax The fused head and thorax of many crustaceans and of arachnids (*see* PROSOMA), which is connected to the abdomen. *See also* TAGMA.

cercaria (*pl.* **cercariae**) A tadpole-like larval stage of trematode worms (flukes) that develops in the body of a mollusc (the secondary host). It subsequently infects a primary host, in which it matures into an adult worm.

cerci (*sing.* **cercus**) A pair of many-jointed appendages that are present on the last segment of the abdomen in certain insects, such as mayflies, earwigs, and cockroaches.

cerebellum (*pl.* **cerebellums** or **cerebella**) The part of the vertebrate *brain concerned with the coordination and regulation of muscle activity and the maintenance of muscle tone and balance. In mammals it consists of two connected hemispheres, composed of a core of white matter and a much-folded outer cortex of grey matter containing numerous *Purkyne cells, and it is situated above the medulla oblongata and partly beneath the cerebrum. In humans it also receives input from hearing and vision centres elsewhere in the brain, analyses visual signals associated with body movements, and estimates the time intervals between events or when performing movements, e.g. in hand–eye coordination.

cerebral cortex (pallium) The layer of *grey matter that forms the outer layer of the hemispheres of the cerebrum in many vertebrates. It is most highly developed in mammals (*see* NEOCORTEX). The cortex is responsible for the control and integration of voluntary movement and the senses of vision, hearing, touch, etc.; it also contains centres concerned with memory, language, thought, and intellect. *See* CEREBRUM.

cerebral hemisphere Either of the two halves of the vertebrate *cerebrum.

cerebroside Any one of a class of *glycolipids in which a single sugar unit is bound to a sphingolipid (*see* PHOSPHOLIPID). The most common cerebrosides are **galactocerebrosides**, containing the sugar group galactose; they are found in the plasma membranes of neural tissue and are abundant in the myelin sheaths of neurons.

cerebrospinal fluid (CSF) The fluid, similar in composition to *lymph, that bathes the central nervous system of vertebrates. It is secreted by the *choroid plexus into the *ventricles of the brain, filling these and other cavities in the brain and spinal cord, and is reabsorbed by veins on the brain surface. Its function is to protect the central nervous system from mechanical injury, to supply it with nutrients and hormones, and to remove wastes.

cerebrum (telencephalon) The largest part of the vertebrate *brain. It consists of paired left and right **cerebral hemispheres**, which develop from the embryonic *forebrain. The hemispheres have an outer convoluted layer of grey matter—the

Cerebrum: the lobes of one cerebral hemisphere showing functions associated with different areas

*cerebral cortex—which contains an esti-mated ten billion nerve cells. Underneath this is *white matter, deep within which lie the *basal ganglia, responsible for process-ing information from the cortex. The two halves of the cerebrum are linked by the *corpus callosum. The function of the cere-brum is to integrate complex sensory and neural functions. The cerebrum plays a critical role in the process of learning, which involves both short-term and long-term memory. Anatomically, each cerebral hemisphere is divided into four lobes—frontal, parietal, temporal, and occipital lobes—and different functions can be local-ized to certain areas of each lobe (see illus-tration). Just in front of the gulley (central sulcus) dividing the frontal and parietal lobes is a strip of frontal cortex called the **primary motor cortex**. Neurons in this region control muscles in particular parts of the body. The strip of parietal cortex just behind the central sulcus is called the **primary somatosen-sory cortex**. This receives sensory infor-mation from the body, relayed via the thalamus. Apart from motor or sensory functions, each lobe also integrates the mass of information received, whether from inside or outside the body, or from *memory. Higher functions such as speech, reading, writing, planning, prob-lem-solving, and emotion also reside in the cerebrum. For many bodily functions outside the head, the left cerebral hemi-sphere serves chiefly the right side of the body, and the right hemisphere the left side. However, some functions reside pri-marily in one or other hemisphere; for instance, language ability is most often centred in the left hemisphere (see LATERALIZATION). See also NEOCORTEX.

cerumen Ear wax, which is secreted by **ceruminous glands** in the auditory meatus of the outer ear. Cerumen protects the deli-cate lining of the meatus and helps to pre-vent microorganisms from entering the ear.

cervical vertebrae The *vertebrae of the neck. The number of cervical vertebrae var-ies: for example, most mammals (including humans) have 7. Their main functions are to support the head and to provide articulating surfaces against which it can move relative to the backbone. See ATLAS; AXIS.

cervix (*pl.* **cervices**) A narrow or necklike part of an organ. The cervix of the uterus (**cervix uteri**) leads to the vagina. Glands in its walls produce mucus, whose viscosity changes according to the oestrous cycle. During labour, the cervix enlarges greatly to allow passage of the fetus.

Cestoda A class of some 6000 described species of flatworms (*see* PLATYHELMINTHES) comprising the tapeworms—ribbon-like parasites within the gut of vertebrates. Tape-worms are surrounded by partially digested food in the host gut so they are able to absorb nutrients through their whole body surface. The body consists of a **scolex** (head), bearing suckers and hooks for attachment, and a series of **proglottids** (or **proglottides**), which contain male and female reproductive systems. The life cycle of a tapeworm requires two hosts, the primary host usually being a predator of the secondary host. *Tae-nia solium* has humans for its primary hosts and the pig as its secondary host. Mature proglottids, containing thousands of fertilized eggs, leave the primary host with its faeces and develop into embryos and then larvae that continue the life cycle in the gut and other tissues of a secondary host (*see* BLADDERWORM).

Cetacea An order of marine mammals comprising the whales, dolphins, and porpoises; it includes what is probably the largest known animal—the blue whale (*Balaenoptera musculus*), over 30 m long and over 150 tonnes in weight. The fore-limbs of whales are modified as short sta-bilizing flippers and the skin is very thin and almost hairless. A thick layer of blubber insulates the body against heat loss and is an important food store. Whales breathe through a dorsal blowhole, which is closed when the animal is submerged. The toothed whales (suborder Odontoceti), such as the dolphins and killer whale, are carnivorous; whalebone whales (suborder Mysticeti), such as the blue whale, feed on plankton filtered by *whalebone plates.

Molecular systematics now indicates that whales are closely related to cows, camels, deer, and other members of the order Artiodactyla; hence they are classified with them in the superorder Cetartiodactyla.

(⊕) SEE WEB LINKS

• Website for the Whale and Dolphin Conservation Society

Cetartiodactyla *See* ARTIODACTYLA; CETACEA.

CFCs *See* CHLOROFLUOROCARBONS.

CGH *See* COMPARATIVE GENOMIC HYBRIDIZATION.

cGMP *See* CYCLIC GMP.

c.g.s. units A system of units based on the centimetre, gram, and second. Derived from the metric system, it was badly adapted to use with thermal quantities (based on the inconsistently defined *calorie) and with electrical quantities (in which two systems, based respectively on unit permittivity and unit permeability of free space, were used). For scientific purposes c.g.s. units have now been replaced by *SI units.

chaeta (*pl.* **chaetae**) A bristle, made of *chitin, occurring in annelid worms. In the earthworm they occur in small groups projecting from the skin in each segment and function in locomotion. The chaetae of polychaete worms (e.g. ragworm) are borne in larger groups on paddle-like appendages (**parapodia**).

Chaetognatha A phylum of about 180 species of marine coelomate invertebrates, the arrow worms, in which the head bears hooks for catching prey and the trunk and tail support paired lateral and tail fins. They lack organs for excretion, circulation, and respiration and are hermaphrodite. Fossil evidence shows that arrow worms were abundant 500 million years ago but their relationship to other animal groups has long been uncertain. They are now regarded as protostomes, albeit ones with developmental traits ancestral to all triploblastic animals.

Chain, Sir Ernst Boris (1906–79) German-born British biochemist, who began his research career at Cambridge University in 1933. Two years later he joined *Florey at Oxford, where they isolated and purified *penicillin. They also developed a method of producing the drug in large quantities and carried out its first clinical trials. The two men shared the 1945 Nobel Prize for physiology or medicine with penicillin's discoverer, Alexander *Fleming.

chalaza (*pl.* **chalazae**) **1.** A twisted strand of fibrous albumen in a bird's egg that is attached to the membrane at either end of the yolk and thus holds the yolk in position in the albumen. **2.** The part of a plant *ovule where the nucellus and integuments merge.

chalcone Any of a class of *flavonoids having the basic C_6–C_3–C_6 arrangement but in which the middle three carbon atoms do not form a closed ring, in contrast to other flavonoids. Chalcones are important precursors in the synthesis of anthocyanin pigments and other flavonoids in plants; their formation is catalysed by the enzyme **chalcone synthase**.

chalk A very fine-grained white rock composed of the fossilized skeletal remains of marine plankton known as **coccoliths** and consisting largely of calcium carbonate ($CaCO_3$). It is the characteristic rock of the *Cretaceous period. It should not be confused with blackboard 'chalk', which is made from calcium sulphate.

chalone Any of various naturally occurring, tissue-specific inhibitors of mitosis. They are oligopeptides and typically act at very low concentrations in a reversible and nontoxic manner. Experimentally they can induce transient inhibition of cell proliferation in certain cancers and may have therapeutic potential.

chamaephyte A plant life form in Raunkiaer's system of classification (*see* PHYSIOGNOMY). Chamaephytes are essentially low-growing shrubs, in which the overwintering buds are borne above ground but near the surface to minimize exposure to the wind.

channel (in cell biology) A pore formed by a protein molecule in a plasma membrane that aids the diffusion of certain substances into and out of the cell. These substances are usually charged ions or lipid-insoluble molecules. *See* ION CHANNEL; LIGAND-GATED ION CHANNEL; VOLTAGE-GATED ION CHANNEL.

chaparral A type of stunted (scrub) woodland found in temperate regions with little summer rainfall. It is dominated by drought-resistant evergreen shrubs, forming dense thickets, interspersed with dwarfed trees, such as oaks and eucalyptus. It is the typical vegetation found in the western United States and the Mediterranean region (where it is called **maquis**).

chaperone *See* MOLECULAR CHAPERONE.

character (trait) A distinctive inherited feature of an organism. Organisms in a population may display different aspects of a particular character; for example, the A, B, and O human blood groups (*see* ABO SYSTEM) are different aspects of the blood group character. *See also* PHENOTYPE.

character displacement A more pronounced difference in a particular character occurring where two similar species coexist than where their populations do not overlap. A classic example involves two species of Galápagos finch, *Geospiza fortis* and *G. fuliginosa* (*see also* DARWIN'S FINCHES). On islands occupied exclusively by one or the other species, the beak of either species tends towards an intermediate size. But where these two species occupy the same island, they can easily be distinguished by beak size, the beak of *G. fortis* being noticeably larger than that of *G. fuliginosa*. This is an effect of competition between the two coexisting species, which has caused the finches to evolve different-sized beaks to exploit seeds of different sizes more effectively. Other types of character that affect the competitive ability of overlapping species, including physiological and behavioural characters, may also be displaced. This **ecological character displacement** is driven by competition, whereas **reproductive character displacement** reinforces reproductive barriers and prevents mating between closely related species.

Chargaff, Erwin (1905–2002) Ukrainian-born US biochemist who became professor of biochemistry at Columbia University (1952–74). Stimulated by Oswald *Avery's identification of DNA as the basis of heredity in pneumococcus bacteria, he discovered that the constitution of DNA is consistent within a species but that there are as many types of DNA as there are species. However, his discovery that the number of purine bases is always equal to the number of pyrimidine bases, and that a similar relationship exists between adenine and thymine bases and between cytosine and guanine bases—a principle known as **Chargaff's rule**—provided *Watson and *Crick with an important clue in their elucidation of the chemical basis of heredity.

charophyte *See* STREPTOPHYTE.

chela (*pl.* **chelae**) The terminal segment of an arthropod appendage when this can be opposed to the segment that precedes it. The chela is often enlarged and modified to act as a pincer, as in the lobster. Any appendage possessing a chela is described as **chelate**.

chelicerae (*sing.* **chelicera**) The first pair of appendages on the head of arachnids and other arthropods of the phylum *Chelicerata. These appendages take the form of pincers or claws and are used for grasping or tearing food.

Chelicerata A subphylum of arthropods containing the classes *Arachnida (spiders and scorpions), Merostomata (horseshoe crabs), and Pycnogonida (sea spiders). The body of chelicerates is divided into an anterior *prosoma and a posterior *opisthosoma and bears six pairs of appendages, typically comprising *chelicerae, *pedipalps, and four pairs of walking legs.

chemical bond A strong force of attraction holding atoms together in a molecule or crystal. In general, atoms combine to form molecules by sharing or transferring electrons in their outer shells. Typically chemical bonds have energies of about 1000 kJ mol^{-1} and are distinguished from the much weaker forces between molecules. *See also* COVALENT BOND; ELECTROVALENT BOND; HYDROGEN BOND.

chemical control The use of chemicals to kill pests (*see* PESTICIDE). *Compare* BIOLOGICAL CONTROL.

chemical dating An absolute *dating technique that depends on measuring the chemical composition of a specimen. Chemical dating can be used when the specimen is known to undergo slow chemical change at a known rate. For instance, phosphate in buried bones is slowly replaced by fluoride ions from the ground water. Measurement of the proportion of fluorine present gives a rough estimate of the time that the bones have been in the ground. Another, more accurate, method depends on the fact that amino acids in living organisms are L-optical isomers. After death, these racemize and the age of bones can be estimated by measuring the relative amounts of D- and L-amino acids present.

chemical evolution *See* ORIGIN OF LIFE.

chemical fossil Any of various organic compounds found in ancient geological strata that appear to be biological in origin and are assumed to indicate that life existed when the rocks were formed. The presence of chemical fossils in Archaean strata indicates that life existed almost 3500 million years ago, perhaps even as much as 3800 million years ago.

chemical fusogen Any chemical that is used in the fusing of two cells or protoplasts (*see* CELL FUSION). Polyethylene glycol (**PEG**) is used in the formation of *hybridomas, and sodium nitrate in the fusion of plant protoplasts in solution. Naturally occurring fusogens are glycoproteins.

chemical reaction A change in which one or more chemical elements or compounds (the **reactants**) form new compounds (the **products**). All reactions are to some extent **reversible**; i.e. the products can also react to give the original reactants. However, in many cases the extent of this back reaction is negligibly small, and the reaction is regarded as **irreversible**. *See also* ENDERGONIC REACTION; EXERGONIC REACTION.

chemiosmotic mechanism A mechanism, first postulated by the British biochemist Peter Mitchell (1920–92), to explain the formation of ATP in the mitochondrial *electron transport chain. As electrons are transferred along the electron carrier system in the inner mitochondrial membrane, hydrogen ions (protons) are actively transported (by *proton pumps) into the space between the inner and outer mitochondrial membranes, which thus contains a higher concentration of protons than the matrix. This creates an electrochemical gradient across the inner membrane, down which protons move back into the matrix. This movement occurs through special channels associated with *ATP synthetase, the enzyme that catalyses the conversion of ADP to ATP, and is coupled with the phosphorylation of ADP (see illustration at ELECTRON TRANSPORT CHAIN). A similar gradient is created across the thylakoid membranes of chloroplasts during the light-dependent reactions of *photosynthesis (*see* PHOTOPHOSPHORYLATION).

chemoautotroph (**chemolithotroph**) Any organism that is able to obtain its energy by the oxidation of inorganic compounds, including those of iron, nitrogen, and sulphur, and uses carbon dioxide, hydrogencarbonate, or related compounds as a carbon source. Chemoautotrophs are exclusively bacteria or archaea, generally living in hostile environments (*see* EXTREMOPHILES), such as certain *sulphur bacteria. *See* AUTOTROPHIC NUTRITION; CHEMOSYNTHESIS.

chemoheterotroph (**chemoorganotroph**) An organism that obtains its energy by the oxidation of organic compounds.

Animals, fungi, and many prokaryotes and protists are chemoheterotrophs. *Compare* CHEMOAUTOTROPH.

chemokine Any of a large family of small proteins, produced by many types of cells, that attract and guide white blood cells (leucocytes) to sites of infection or influence lymphocyte development and migration in lymph nodes. Hence they are a crucial aspect of innate immunity during the initial response to injury or pathogen invasion and affect lymphocyte behaviour in adaptive immune responses. They fall into four main structural categories: C, CC, CXC, and CX$_3$C, according to the number of cysteine residues (C) or variable amino acids (X) near the amino terminus of the protein. For example, CXCL8 (formerly interleukin 8) is produced by monocytes, macrophages, and other cells and causes neutrophils to migrate from nearby blood vessels to the site of inflammation. CCL2 similarly attracts monocytes from the bloodstream and induces them to change into tissue macrophages. Chemokines act in conjunction with other factors, such as tumour necrosis factor α (TNF-α), which induces the adhesion factors necessary to bring the leucocytes alongside the blood vessel wall before they 'squeeze through' into the tissues.

chemoreceptor A *receptor that detects the presence of particular chemicals and (in multicellular organisms) transmits this information to sensory nerves. Examples include the *taste buds and the receptors in the *carotid body.

chemosynthesis A type of *autotrophic nutrition in which organisms (called **chemoautotrophs**) synthesize organic materials using energy derived from the oxidation of inorganic chemicals, rather than from sunlight. Chemoautotrophs are exclusively archaea or bacteria, including *Nitrosomonas*, which oxidizes ammonium to nitrite; and *Thiobacillus*, which oxidizes sulphur to sulphate.

chemosystematics *See* SYSTEMATICS.

chemotaxis *See* TAXIS.

chemotaxonomy The *classification of plants and microorganisms based on similarities and differences in their natural products and the biochemical pathways involved in their manufacture. *See also* TAXONOMY.

chemotherapy The use of chemicals, especially drugs, in the treatment of disease. The term is often used specifically to denote drug therapy for cancer, as distinct from treatments with radiation (radiotherapy).

chemotropism The growth or movement of a plant or plant part in response to a chemical stimulus. An example is the growth of a pollen tube down the style during fertilization in response to chemical attractants secreted by the synergid cells in the embryo sac.

Chengjiang fossils An assemblage of fossil organisms found in deposits of Maotianshan shale in Chengjiang County, Yunnan Province, China, dating from the lower Cambrian (about 525 million years ago). Discovered in 1984, the Chengjiang fossils represent a great diversity of animal forms, including well-preserved soft tissues, and are comparable in importance with the slightly younger *Burgess shale fossils. Some 250 fossil species have been identified, including numerous arthropods, worms, sponges, cnidarians, and ctenophores, plus several chordates, including a jawless fish named *Myllokunmingia*, discovered by Chinese palaeontologist Degan Shu in 1999. It resembles a hagfish and is one of the oldest known fossil vertebrates to date. Another fossil vertebrate, the eel-like *Zhongjianichthyes*, was found in 2003. UNESCO gave the site World Heritage status in 2012. *See also* CAMBRIAN EXPLOSION.

chiasma (*pl.* **chiasmata**) The point at which paired *homologous chromosomes remain in contact as they begin to separate during the first prophase of *meiosis, forming a cross shape. A number of chiasmata can usually be identified and at these points *crossing over occurs.

chikungunya A togavirus (*see* VIRUS) that is transmitted by mosquitoes and causes disease in humans characterized by fever and severe joint pain. Other symptoms include muscle pain, headache, nausea, fatigue, and rash. Illness generally lasts a few days, before a full recovery, although joint pain may be prolonged in some cases. It was first reported in southern Tanzania in 1952 and has since occurred in over 60 countries in Africa, Asia, Europe, and the Americas. The name derives from the Makonde language and means 'to become contorted', reflecting the stooped posture of infected persons.

chill haze The precipitation that occurs when beer is stored at cold temperatures. Chill haze consists of proteins that can be broken down by using protease enzymes.

Chilopoda A class of wormlike terrestrial *arthropods belonging to the clade or subphylum *Myriapoda and comprising the centipedes. These are characterized by a distinct head, bearing a single pair of relatively long antennae and one pair of poison jaws, and 15–177 body segments, each bearing one pair of similar legs. Centipedes are fast-moving predators found in damp environments. *See also* DIPLOPODA; UNIRAMIA.

chimaera An organism composed of tissues that are genetically different. Chimaeras can develop if a *mutation occurs in a cell of a developing embryo. All the cells arising from it have the mutation and therefore produce tissue that is genetically different from adjacent tissue, e.g. brown patches in otherwise blue eyes in humans. *Graft hybrids are examples of plant chimaeras.

Chiroptera An order of flying mammals comprising the bats. Their membranous wings are supported by very elongated forelimbs and digits and stretch along the sides of the body to the hindlimbs and tail. Whenever bats rest they allow their body temperature to fall, hibernating in winter when food is scarce. Most bats are nocturnal; their ears are enlarged and specialized for *echolocation, which they use to hunt prey and avoid obstacles. Bats feed variously on insects, fruit, nectar, or blood.

SEE WEB LINKS
• Website of Bat Conservation International, dedicated to conservation, education, and research concerning bats

chitin A *polysaccharide comprising chains of N-acetyl-D-glucosamine, a derivative of glucose. Chitin is structurally very similar to cellulose and serves to strengthen the supporting structures of various invertebrates. For example, the exoskeleton of arthropods contains about 30–50% chitin embedded in a protein matrix. It also occurs in fungi, strengthening the cell walls of hyphae.

chitinase An enzyme that catalyses the hydrolysis of *chitin, a major constituent of fungal cell walls and of the exoskeleton of insects and other arthropods. It is synthesized by certain insect-eating animals, such as frogs, and also by certain plants as part of their defence against fungal infection (*see* HYPERSENSITIVITY). Transfer of the chitinase gene between different plant species has been successfully undertaken in an attempt to enhance the resistance of susceptible species to pathogenic fungi. Human macrophages can also secrete a chitinase enzyme, which may help to destroy chitin-containing pathogens, and chitinase-like proteins are expressed in several diseases, including asthma, cystic fibrosis, and other inflammatory conditions.

chlamydospore A thick-walled asexual spore that is produced from a fungal hypha. It is a resting spore, capable of withstanding unfavourable growing conditions.

chlorenchyma *Parenchyma tissue that contains chloroplasts and is photosynthetic. Chlorenchyma makes up the mesophyll tissue of plant leaves and is also found in the stems of certain plant species. *Compare* COLLENCHYMA; SCLERENCHYMA.

chloride secretory cell Any of the cells in the gills of marine bony fish that remove chloride from the blood and secrete it into the sea, thereby maintaining the osmotic composition of the body. The excretion of chloride is accompanied by movement of sodium ions from the blood to the sea, which keeps the electrochemical balance of the fish's body neutral.

chloride shift The movement of chloride ions (Cl^-) into red blood cells. Carbon dioxide reacts with water to form carbonic acid in the red blood cells (*see* CARBONIC ANHYDRASE). The carbonic acid then dissociates into hydrogencarbonate ions (HCO_3^-) and hydrogen ions (H^+). The plasma membrane is relatively permeable to negative ions. Therefore the hydrogencarbonate ions diffuse out of the cell into the plasma, leaving the hydrogen ions, which create a net positive charge; this is neutralized by the diffusion of chloride ions from the plasma into the cell.

chlorocruorin A greenish iron-containing *respiratory pigment that occurs in the blood of polychaete worms. It closely resembles *haemoglobin.

chlorofluorocarbons (CFCs) Compounds obtained by replacing the hydrogen atoms of hydrocarbons by chlorine and fluorine atoms. Their high stability to temperature makes them suitable for a variety of uses, including aerosol propellants, oils, polymers, etc. They are often known as **freons**. Their former widespread use in aerosols and refrigerator coolants led to increased concentrations in the upper atmosphere, where photochemical reactions cause them to break down and react with ozone, which results in depletion of the *ozone layer. They also contribute to the *greenhouse effect. For this reason CFCs have now largely been replaced with less damaging alternatives, such as hydrofluorocarbons (HFCs). *See also* POLLUTION.

chlorophyll Any one of a class of pigments found in all photosynthetic organisms; the most important members are chlorophyll *a* (see formula) and chlorophyll *b*, which occur in all land plants and are responsible for their green colour. Chlorophyll molecules are the principal sites of light absorption in the light-dependent reactions of *photosynthesis (see PHOTOSYSTEMS I AND II). They are magnesium-containing *porphyrins, chemically related to *cytochrome and *haemoglobin. *See also* BACTERIOCHLOROPHYLL.

Chlorophyll *a*

Chlorophyta (green algae) A large phylum (or group, i.e. the chlorophytes) comprising most of the green *algae. They possess chlorophylls *a* and *b*, store food reserves as starch, and have cellulose cell walls. In these respects, and on the basis of molecular studies, they are now regarded as the sister group of the *streptophytes (or charophytes), which includes the land plants and certain freshwater algae, i.e. the stoneworts and choleochaetophytes. Chlorophytes are widely distributed and diverse in form. Unicellular forms may occur singly (sometimes with flagella for motility) or in colonies, while multicellular forms may be filamentous (e.g. *Spirogyra*) or platelike (e.g. *Ulva*).

chloroplast Any of the chlorophyll-containing organelles (*see* PLASTID) that are found in large numbers in those plant and algal cells undergoing *photosynthesis. Plant chloroplasts are typically lens-shaped and bounded by a double membrane. They contain membranous structures called *thylakoids, which are piled up into stacks (*see* GRANUM), surrounded by a gel-like matrix (**stroma**). The light-dependent reactions of photosynthesis occur on the thylakoid membranes while the light-independent reactions take place in the stroma. Chloroplasts contain multiple copies of a DNA molecule, typically 120–200 kbp in size and carrying genes required for chloroplast functions.

chlorosis The abnormal condition in plant stems and leaves in which synthesis of the green pigment chlorophyll is inhibited, resulting in a pale yellow coloration. This may be caused by lack of light, mineral deficiency, infection (particularly by viruses), or genetic factors.

chloroxybacteria (grass-green bacteria; prochlorophytes) Green-pigmented *cyanobacteria that resemble the chloroplasts found as organelles in green plants and green algae. Like chloroplasts, chloroxybacteria perform photosynthesis using both chlorophylls *a* and *b* and carotenoids as pigments, and they lack phycobiliprotein accessory pigments (found in most cyanobacteria). However, this similarity is thought to have evolved independently of plant chloroplasts. The first to be discovered, in the 1960s, was *Prochloron*, a spherical (coccoid) cyanobacterium that lives as a symbiont on the surface or inside the cloaca of certain tunicates. Others include the filamentous *Prochlorothrix*, found free living in certain lakes.

choanae (internal nares) (*sing.* choana) *See* NARES.

cholecalciferol *See* VITAMIN D.

cholecystokinin (CCK; pancreozymin) A hormone, produced by the duodenal region of the small intestine, that induces the gall bladder to contract and eject bile into the intestine and stimulates the pancreas to secrete its digestive enzymes. Cholecystokinin is released by cells of the duodenal epithelium, in response to fat entering the duodenum.

cholesterol A *sterol (*see also* STEROID) occurring widely in animal tissues and also in some plants and algae. It can exist as a free sterol or esterified with a long-chain fatty acid. Cholesterol is absorbed through the intestine or manufactured in the liver. It serves principally as a constituent of blood plasma *lipoproteins and of the lipid-protein complexes that form plasma membranes. It is also important as a precursor of various steroids, especially the bile acids, sex hormones, and adrenocorticoid hormones. The derivative 7-dehydrocholesterol is converted to vitamin D_3 by the action of sunlight on skin. Cholesterol is transported in the bloodstream chiefly as lipoproteins: low-density lipoproteins (LDLs) deliver cholesterol to cells, while high-density lipoproteins (HDLs) retrieve excess cholesterol and return it to the liver. A high ratio of LDLs to HDLs increases the risk of *atherosclerosis and consequent heart attack or stroke.

choline An amino alcohol, $CH_2OHCH_2N(CH_3)_3OH$. It occurs widely in living organisms as a constituent of certain types of phospholipids—the *lecithins and sphingomyelins—and in the neurotransmitter *acetylcholine. It is sometimes classified as a member of the *vitamin B complex.

cholinergic Describing a nerve fibre that either releases *acetylcholine when stimulated or is itself stimulated by acetylcholine. *Compare* ADRENERGIC.

cholinesterase (acetylcholinesterase) An enzyme that hydrolyses the neurotransmitter *acetylcholine to choline and acetate. Cholinesterase is secreted by nerve cells at *synapses and by muscle cells at *neuromuscular junctions. Organophosphorus insecticides (*see* PESTICIDE) act as *anticholinesterases by inhibiting the action of cholinesterase.

Chondrichthyes A class of vertebrates comprising the fishes with cartilaginous skeletons. The majority belong to the subclass Elasmobranchii (skates, rays, and sharks—*see* SELACHII). Most cartilaginous fishes are marine carnivores with powerful jaws. Unlike bony fishes, they have no swim bladder, and therefore avoid sinking only by constant swimming with the aid of an asymmetrical (**heterocercal**) tail. There is no operculum covering the gill slits, the first of which is modified as a *spiracle. Fertilization is internal so the few eggs produced are consequently yolky, large, and well-protected. Some cartilaginous fishes show viviparous development of the young (*see* VIVIPARITY).

chondrin The matrix of *cartilage, which is made up of chondrocytes embedded in chondroitin sulphate.

chondrocyte Any of the cells that make up the matrix of *cartilage.

chordamesoderm Tissue found in the midline of the early chordate embryo that forms the *notochord. Derived from dorsal mesoderm, it acts as an *organizer, producing chemical signals that cause the overlying ectoderm (*neural plate) to differentiate into the neural tube—the precursor of the brain and spinal cord. In vertebrates the notochord is subsequently replaced by the vertebral column.

Chordata A phylum of animals characterized by a hollow dorsal nerve cord and, at some stage in their development, a flexible skeletal rod (the *notochord) and *gill slits opening from the pharynx. There are three major groups: the *Urochordata (sea squirts), *Cephalochordata (lancelets), and *craniates.

((⊕)) SEE WEB LINKS
• Tree of Life web page outlining main characteristics and phylogeny of chordates

chordotonal organs Sensory receptors that detect alterations in the tension of insect muscles.

chorion 1. A membrane enclosing the embryo, yolk sac, and allantois of reptiles, birds, and mammals. In mammals a section of the chorion becomes the embryonic part of the *placenta. See EXTRAEMBRYONIC MEMBRANES. **2.** The protective shell of an insect egg, produced by the ovary. It is pierced by a small pore (**micropyle**) that allows the entry of spermatozoa for fertilization. See also EGG MEMBRANE.

chorionic gonadotrophin *See* GONADOTROPHIN.

choroid A pigmented layer, rich in blood vessels, that lies between the retina and the sclerotic of the vertebrate eye. At the front of the eye the choroid is modified to form the *ciliary body and the *iris.

choroid plexus A membrane rich in blood vessels that lines the *ventricles of the brain. It is an extension of the *pia mater and secretes *cerebrospinal fluid into the ventricles; it also controls exchange of materials between the blood and cerebrospinal fluid.

chromaffin tissue A group of cells in the adrenal medulla (*see* ADRENAL GLANDS) that synthesizes *noradrenaline. Chromaffin tissue also contains the enzyme that converts noradrenaline into *adrenaline.

chromalveolates In some classifications, an assemblage of eukaryotic organisms comprising the *stramenopiles and *alveolates. Some authorities now place them with the *rhizaria in the *SAR supergroup.

chromatid A threadlike strand formed from a *chromosome during the early stages of cell division. Each chromosome divides along its length into two chromatids, which are at first held together at the centromere. They separate completely at a later stage. The DNA of the chromosome reproduces itself exactly so that each chromatid has the complete amount of DNA and becomes a daughter chromosome with exactly the same genes as the original chromosome from which it was formed.

chromatin The substance of which eukaryotic *chromosomes are composed. It consists of proteins (principally histones), DNA, and small amounts of RNA. The DNA molecule is wrapped around the histones to form a series of linked globular *nucleosomes, resembling beads on a string. This is itself coiled to form a highly condensed solenoid arrangement, constituting a form of chromatin called **heterochromatin**, which stains densely with basic stains. The genes in the solenoid can only be transcribed if the solenoid unfolds to some extent, forming an expanded chromatin (**euchromatin**), which is lighter staining. The degree of condensation in any particular region is regulated by reversible acetylation of the histones: the greater the degree of acetylation, the less condensed the chromatin, and hence the greater the availability of genes for transcription.

⊕ SEE WEB LINKS

• Diagram showing levels of chromatin packing

chromatin remodelling Alterations in the degree of condensation of chromatin as a means of *epigenetic regulation of gene expression. A key mechanism is chemical modification of the histone proteins that form the scaffolding around which the DNA is coiled. Methylation of certain amino acids in the 'tails' of the histones, which protrude from the coiled chromatin structure, is associated with silencing of genes. Conversely, acetylation of histones tends to relax the nucleosome and activate gene transcription. Chromatin-binding proteins, such as heterochromatin protein 1 (HP1), bind to the methylated histones and reinforce the silenced state by recruiting histone methyltransferases, which add more methyl groups to the histones. Noncoding RNA molecules are also involved in chromatin remodelling, for example in inactivation of one or more X chromosomes in female mammals (*see* X INACTIVATION). In yeast cells, small interfering RNAs (siRNAs) and other noncoding RNAs coooperate in reforming the highly condensed chromatin of the centromeres following replication of the chromosomes. *See also* DNA METHYLATION.

chromatogram A record obtained by chromatography. The term is applied to the developed records of *paper chromatography and *thin-layer chromatography and also to the graphical record produced in *gas-liquid chromatography.

chromatography A technique for analysing or separating mixtures of gases, liquids, or dissolved substances, such as mixtures of amino acids or chlorophyll pigments. The original technique (invented by the Russian botanist Mikhail Tsvet (1872–1919) in 1906) is a good example of **column chromatography**. A vertical glass tube is packed with an adsorbing material, such as alumina. The sample is poured into the column and continuously washed through with a solvent (a process known as **elution**). Different components of the sample are adsorbed to different extents and move down the column at different rates. In Tsvet's original application, plant pigments were used and these separated into coloured bands in passing down the column (hence the name chromatography). The usual method is to collect the liquid (the **eluate**) as it passes out from the column in fractions.

In general, all types of chromatography involve two distinct phases—the **stationary phase** (the adsorbent material in the column in the example above) and the **moving phase** (the solution in the example). The separation depends on competition for molecules of sample between the moving phase and the stationary phase. The form of column chromatography above is an example of **adsorption chromatography**, in which the sample molecules are adsorbed on the alumina. In **partition chromatography**, a liquid (e.g. water) is first absorbed by the stationary phase and the moving phase is an immiscible liquid. The separation is then by partition between the two liquids. In **ion-exchange chromatography** the process involves competition between different ions for ionic sites on the stationary phase (*see* ION EXCHANGE). *Gel filtration is another chromatographic technique in which the size of the sample molecules is important.

See also AFFINITY CHROMATOGRAPHY; GAS-LIQUID CHROMATOGRAPHY; PAPER CHROMATOGRAPHY; THIN-LAYER CHROMATOGRAPHY.

chromatophore 1. A pigment-containing cell found in the skin of many lower vertebrates (e.g. chameleon) and in the integument of crustaceans. Concentration or dispersion of the pigment granules in the cytoplasm of the cell causes the colour of the animal to alter to match its surroundings. A common type of chromatophore is the **melanophore**, which contains the pigment *melanin. **2.** A membrane-bound structure in photosynthetic bacteria that contains photosynthetic pigments. *See* BACTERIOCHLOROPHYLL.

chromophore Any group of atoms in a molecule that includes an unsaturated chemical group (such as C=C) capable of absorbing *ultraviolet radiation; the chromophore is responsible for the colour of the compound. For example, retinal is the chromophore of *rhodopsin, the light-sensitive pigment in the rod cells of the retina.

chromoplast Any of various pigment-containing *plastids in plant cells. Red, orange, and yellow chromoplasts contain carotenoid pigments and are responsible for the coloration of fruits and flowers. *See also* PLASTOGLOBULUS. *Compare* CHLOROPLAST; LEUCOPLAST.

chromosome A structure consisting of a single DNA molecule and associated proteins. Several to many chromosomes are found in the nucleus of eukaryotic cells (e.g. in plants and animals), whereas prokaryotes (bacteria and archaea) have a single chromosome. Chromosomes are composed of *chromatin and carry the *genes in a linear sequence; these determine the individual characteristics of an organism. When the nucleus is not dividing, individual chromosomes cannot be identified with a light microscope. During the first stage of nuclear division, however, the chromosomes contract and, when stained, can be clearly seen under a microscope. Each consists of two *chromatids held together at the *centromere (*see also* MEIOSIS; MITOSIS). The number of chromosomes in each cell is constant for and characteristic of the species concerned. In the normal body cells of *diploid organisms the chromosomes occur in pairs (*see* HOMOLOGOUS CHROMOSOMES); in the gamete-forming germ cells, however, the diploid number is halved and each cell contains only one member of each chromosome pair. Thus in humans each body cell contains 46 chromosomes (22 matched pairs and one pair of *sex chromosomes) and each germ cell 23. Abnormalities in the number or structure of chromosomes may give rise to abnormalities in the individual; *Down's syndrome is the result of one such abnormality. *See* CHROMOSOME MUTATION.

Bacterial cells contain only a single circular chromosome, typically tethered to the cell's plasma membrane and highly aggregated into a *nucleoid. The organelles of eukaryotes, specifically mitochondria and chloroplasts, also contain multiple copies of very small chromosomes arranged in nucleoids. Viral chromosomes can consist of one or several single- or double-stranded nucleic acid molecules (DNA or RNA). *See also* ARTIFICIAL CHROMOSOME.

(⊕) SEE WEB LINKS

• Overview of DNA, genes, and chromosomes from the Virtual Genetics Education Centre, University of Leicester

chromosome conformation capture (3C) A technique used to investigate the three-dimensional configuration and interactions of chromosomes in cells. It provides information about how various nucleotide sequences, such as promoters and enhancers, physically come together to regulate gene activity, as well as other aspects of chromatin organization.

chromosome diminution (chromatin diminution) A phenomenon seen in the early embryonic development of members of various animal groups, including certain nematodes, crustaceans, insects, and vertebrates, in which some of the genetic material is eliminated from progenitor body (somatic) cells during cell division. In some cases the somatic chromosomes break down, with only a few remaining intact, whereas the integrity of chromosomes in the germ-cell lineage is maintained. It serves to silence or reduce the dosage of certain genes and is thought to have evolved independently in different lineages of organisms.

chromosome jumping *See* CHROMOSOME WALKING.

chromosome map Any plan that shows the positions of genes, genetic markers, or other landmarks along the length of a chromosome. There are essentially two complementary types of map: *linkage maps, which give the relative positions of genetic sites (loci) determined by studies of how frequently recombination occurs between the loci; and *physical maps, which show the arrangement of the chromosomal material, whether it be in the form of banding patterns produced by staining (a type of **cytological map**), the order and distance between restriction enzyme cleavage sites (a restriction map), or the sequence of bases in the DNA. Maps of either type can be constructed in various ways, depending on such factors as the type of organism, the complexity of its genome, and the amount of pre-existing data. Accumulated data for the chromosomes of many species of organism are now held in databases and available freely via the Internet for geneticists and others worldwide.

chromosome mutation A change in the gross structure of a chromosome, which usually causes severely deleterious effects in the organism. Chromosome mutations often occur due to an error in pairing during the *crossing over stage of meiosis. The main types of chromosome mutation include *translocation, *duplication, *deletion, and *inversion. *Compare* POINT MUTATION. *See also* COPY NUMBER; MUTATION.

chromosome painting Any of various techniques based on *fluorescence in situ hybridization (FISH) that uses a palette of fluorescently labelled probes to identify specific chromosomes or chromosomal regions by 'painting' them in different colours. It is used diagnostically in clinical cytogenetics to screen for translocations or other structural aberrations, for example in hereditary diseases and cancer, and in comparative cytogenetics to determine the structural changes in genomes occurring during evolution. *See* SPECTRAL KARYOTYPING.

chromosome walking A technique, used in constructing a *physical map, for selecting contiguous overlapping clones from a DNA library and thus reconstructing the order of genes along a segment of chromosome. For example, one can effectively 'walk' in either direction from a known *marker gene to identify adjacent genes, a technique that is useful in fine-structure mapping of a genome. Essentially, the initial clone containing the marker gene is fragmented and each fragment subcloned for use as a *DNA probe to identify other clones containing adjacent and overlapping segments. In turn, these adjacent segments are fragmented and subcloned and used to probe for further overlaps, and so on. The cloned segments can then be placed in order corresponding to that on the chromosome. In a refinement of the technique, called **chromosome jumping**, only the ends of segments are identified, allowing the investigator to 'jump' over the middle regions. This speeds up the process, and also is a way of bypassing stretches of *repetitive DNA, which are not amenable to chromosome walking.

chrysalis *See* PUPA.

Chrysophyta (Chrysomonada) A large phylum (or group, i.e. the chrysophytes) of mostly freshwater photosynthetic protists, including the golden, or golden-brown, *algae. Now classified as *stramenopiles, they possess *carotenoid pigments (responsible for their colour), in addition to chlorophylls. Their main storage products are oils and the polysaccharide **chrysolaminarin**. Most chrysophytes are unicellular, with two unequal-sized flagella, but some form colonies or are filamentous; one group of marine chrysophytes construct tests of silica. *Mixotrophic species can consume small prey as well as performing photosynthesis.

chyle A milky fluid consisting of *lymph that contains absorbed food materials (especially emulsified fats). Most chyle occurs in the lymphatic ducts (*lacteals) in the *villi of the small intestine during the absorption of fat.

chylomicron Any of the *lipoprotein particles synthesized by intestinal epithelial cells. They consist mainly of triglycerides, in combination with cholesterol and phospholipids, all inside a protein coat to make them water soluble. They pass from the epithelium into the *lacteal of each villus, and thence via the lymphatic system to enter the bloodstream via the thoracic duct. Hence, chylomicrons are the form in which dietary fat is transported in the circulatory system.

chyme The semisolid and partly digested food that is discharged from the stomach into the duodenum.

chymosin *See* RENNIN.

chymotrypsin An *endopeptidase enzyme in pancreatic juice that is secreted into the duodenum. The enzyme is secreted as an inactive precursor, **chymotrypsinogen**, which is activated by another pancreatic protease, *trypsin.

chymotrypsinogen *See* CHYMOTRYPSIN.

Chytridiomycota A phylum of microscopic fungi, the chytrids, that live in soil or fresh water. The body (thallus) is unicellular or a *coenocyte and gives rise to threadlike hyphae or rhizoids; some species form a branching network (mycelium). The cell walls contain chitin, and some also have cellulose. Chytrids feed by secreting enzymes to digest material extracellularly in order to absorb the nutrients. They produce motile stages (zoospores) equipped with a single flagellum. Sexual reproduction is the norm; fusion of the gametes results in a zygote, which either produces motile zoospores or germinates directly into a new thallus.

ciliary body The circular band of tissue surrounding and supporting the *lens of the vertebrate eye. It contains the **ciliary muscles**, which bring about changes in the shape of the lens (*see also* ACCOMMODATION). The ciliary body produces the *aqueous humour.

ciliary feeding A method of feeding used by lancelets and many other aquatic invertebrates. The movement of cilia causes a current of water to be drawn towards and through the animal, and microorganisms in the water are filtered out by the cilia.

ciliary muscle *See* CILIARY BODY.

ciliated epithelium A region of *epithelium consisting of columnar or cuboidal cells bearing hairlike appendages (*see* CILIUM) that are capable of beating rapidly. Ciliated epithelium performs the function

of moving particles or fluid over the epithelial surface in such structures as the trachea, bronchial tubes, and nasal cavities. It often occurs in the vicinity of mucus-secreting *goblet cells.

Ciliophora A phylum containing ciliated *protozoa—ciliates—(including *Paramecium*) that possess two types of nuclei, a micronucleus and macronucleus (*see* NUCLEUS). The cilia are used for feeding and locomotion. Ciliates reproduce sexually by *conjugation. They are included in the *alveolates.

cilium (*pl.* **cilia**) A short minute hairlike structure (up to 10 μm long) present on the surface of many cells, notably in certain protozoans and some types of vertebrate *epithelium. Cilia usually occur in large groups and are shorter than eukaryotic flagella, although both organelles have the same structure (*see* FLAGELLUM). Beating of cilia can produce cell movement or create a current in fluid surrounding a cell. *See also* AXONEME.

 SEE WEB LINKS
• Details of the ultrastructure of cilia and flagella

circadian rhythm (diurnal rhythm) Any 24-hour periodicity in the behaviour or physiology of animals or plants. Examples are the sleep/activity cycle in many animals and the growth movements of plants. Circadian rhythms are generally controlled by *biological clocks. In 2017 the US scientists Jeffrey Hall, Michael Rosbash, and Michael Young won the *Nobel Prize for physiology or medicine for their work on these topics.

circalunar rhythm A *biorhythm that corresponds with the lunar cycle (approximately 29.5 days). The reproductive cycles of many organisms, especially marine organisms, are linked to changing levels of moonlight and the tidal cycle, both of which are governed by the phases of the moon. *See* ULTRADIAN RHYTHM.

circannual rhythm *See* ANNUAL RHYTHM.

circulation The mass flow of fluid (e.g. blood, haemolymph, or lymph) through the tissues and organs of an animal, allowing the transport and exchange of such materials as oxygen, nutrients, and waste products (*see also* VASCULAR SYSTEM; LYMPHATIC SYSTEM). Smaller animals (e.g. arthropods and most molluscs) have an **open circulation**, i.e. the haemolymph is pumped into the body cavity, in which the internal organs are suspended. In open circulatory systems the tissues are in direct contact with the haemolymph, and materials are exchanged directly by diffusion. In a **closed circulation**, found in larger animals, the blood flows in vessels, which usually contain a series of one-way valves to maintain the flow in one direction. *See also* DOUBLE CIRCULATION; HAEMODYNAMICS; MICROCIRCULATION; SINGLE CIRCULATION.

circulatory system (cardiovascular system) The heart, blood vessels, blood, lymphatic vessels, and lymph, which together serve to transport materials throughout the body. *See also* DOUBLE CIRCULATION; SINGLE CIRCULATION; VASCULAR SYSTEM.

cirrus (*pl.* cirri) **1.** A thoracic appendage of a barnacle or other cirripede crustacean, modified as a filtering or food-collecting organ. **2.** A tiny mouth tentacle of a lancelet (*see* CEPHALOCHORDATA). **3.** Any of various slender appendages of annelid worms. **4.** A bundle of cilia that serves as a locomotory organ in certain ciliate protists (*see* CILIOPHORA), such as *Stylonychia*. **5.** A spiny or whiplike tip of a leaf.

cisterna (*pl.* cisternae) Any of the membrane-bound sacs that form the branches of the *endoplasmic reticulum (ER) and *Golgi apparatus. The cisternae of the ER are continuous with each other and with the nuclear envelope; in rough ER the cisternae are flattened sacs, whereas in smooth ER they are more tubular in shape.

***cis-trans* test (complementation test)** A test used in genetics to determine whether two independent mutations, affecting the same phenotypic function, occur within the same gene or in different genes (*see* CISTRON). A pair of homologous chromosomes are brought together in the same cell (normally a bacterial cell). Mutations located on the same chromosome are described as being in the *cis* position; if they occur one on each homologue they are in the *trans* position. If the cell is phenotypically normal for both *cis* and *trans*positions of the mutations, these mutations are considered to be present in different cistrons. If the resultant cell is normal when the mutations are in the *cis* position but mutant when the mutations are in the *trans* position, the two mutations

Cis-trans test

are considered to lie in the same cistron. See illustration.

cistron The unit of genetic function defined by the *cis-trans* test, operationally equivalent to a gene.

citric acid A white crystalline hydroxy carboxylic acid, $HOOCCH_2C(OH)(COOH)CH_2COOH$. It is present in citrus fruits and is an intermediate in the *Krebs cycle in plant and animal cells.

citric acid cycle *See* KREBS CYCLE.

CJD *See* CREUTZFELDT-JAKOB DISEASE.

clade A taxonomic group of organisms comprising an ancestor and all of its descendant forms. *See* CLADISTICS.

cladistics A method of classification in which animals and plants are placed into taxonomic groups called **clades** strictly according to their evolutionary relationships. These relationships are deduced on the basis of certain *homologous characters—shared derived traits known as *synapomorphies that are thought to indicate common ancestry (*see* MONOPHYLETIC). Implicit in this is the assumption that the ancestral line always splits into two descendant lineages, rather than multiple lineages arising at the same point (the latter case is described as an unresolved phylogeny). Also, it requires that truly homologous characters are distinguished from homoplasic features, i.e. ones resulting from convergent

Cladistics: a cladogram showing the relationships of six species (A–F)

evolution (*see* HOMOPLASY). A diagram indicating these relationships (called a **cladogram**) therefore consists of a system of dichotomous branches: each point of branching (called a node) represents divergence from a common ancestor, as shown in the diagram. Thus the species A to F form a clade as they share the common ancestor X, and species A to D form a clade of a different taxonomic rank, sharing the ancestor X_2. Species C to F do not form a clade, since the latter must include *all* the descendants of a common ancestor. *Compare* PHYLOGRAM.

cladode (cladophyll) A flattened stem or internode that resembles and functions as a leaf. It is an adaptation to reduce water loss, since it contains fewer *stomata than a leaf. An example of a plant with cladodes is asparagus. *Compare* PHYLLODE.

cladogram *See* CLADISTICS.

cladophyll *See* CLADODE.

class A category used in the *classification of organisms that consists of similar or closely related orders. Similar classes are grouped into a phylum. Examples include Mammalia (mammals), Aves (birds), and Dicotyledoneae (dicots).

classification The arrangement of organisms into a series of groups based on physiological, biochemical, anatomical, or other relationships. An **artificial classification** is based on one or a few characters simply for ease of identification or for a specific purpose; for example, birds are often arranged according to habit and habitat (seabirds, songbirds, birds of prey, etc.) while fungi may be classified as edible or poisonous. Such systems do not reflect evolutionary relationships. A **natural classification** is based on resemblances and is a hierarchical arrangement. The smallest group commonly used is the *species. Species are grouped into genera (*see* GENUS), the hierarchy continuing up through *tribes, *families, *orders, *classes, and phyla (*see* PHYLUM) to *kingdoms and, in some systems, to *domains. In traditional systems of plant classification the phylum was replaced by the *division. Higher up in the hierarchy the similarities between members of a group become fewer. Present-day natural classifications try to take into account as many features as possible and in so doing aim to reflect the phylogeny of organisms, i.e. the history of their evolutionary relationships (see CLADISTICS). Natural classifications are also predictive. Thus if an organism is placed in a particular genus because it shows certain features characteristic of the genus, then it can be assumed it is very likely to possess most (if not all) of the other features of that genus. See also BINOMIAL NOMENCLATURE; MOLECULAR SYSTEMATICS; TAXONOMY.

clathrin A protein that is the major constituent of the 'coat' of the coated pits and coated vesicles formed during endocytosis of materials at the surface of cells (see ENDOSOME). The clathrin molecules are arranged in a localized polyhedral lattice on the membrane, which subsequently invaginates to form the coated pit and vesicle. When the coated vesicle has delivered its contents within the cell, the clathrin-coated membrane is recycled and returns to the cell surface. A similar process also tranfers materials between membranous organelles within the cell.

clavicle A bone that forms part of the *pectoral (shoulder) girdle, linking the *scapula (shoulder blade) to the sternum (breastbone). In humans it forms the collar bone and serves as a brace for the shoulders.

clay An inorganic constituent of *soils consisting chiefly of clay minerals (mainly hydrous silicates of aluminium) in the form of particles less than 0.002 mm in diameter. See also FLOCCULATION.

cleavage (in embryology) The series of cell divisions by which a single fertilized egg cell is transformed into a multicellular body, the *blastula. Characteristically no growth occurs during cleavage, the shape of the embryo is unchanged except for the formation of a central cavity (the blastocoel), and the ratio of nuclear material (DNA) to cytoplasm increases.

cleistothecium (pl. **cleistothecia**) See ASCOCARP.

climate The characteristic pattern of weather elements in an area over a period. The weather elements include temperature, rainfall, humidity, solar insolation, wind, etc. The climate of a large area is determined by several climatic controls: (1) the latitude of the area, which accounts for the amount of solar radiation it receives; (2) the distribution of land and sea masses; (3) the altitude and topography of the area; and (4) the location of the area in relation to the ocean currents. Weather elements are important *abiotic factors.

climate change Any alteration in the earth's climate whether due to natural causes or human (i.e. anthropogenic) activities. Evidence from the fossil record and geophysical data shows that the earth's climate has undergone periods of warming and cooling throughout geological time, due primarily to shifts in the earth's orbit around the sun. These fluctuations are linked with changes in global glaciation levels, and in some cases *mass extinctions of organisms. In the period since 1850 the earth has been warming at a rate unprecedented in recent geological history, a fact attributed chiefly to anthropogenic emissions of carbon dioxide and other greenhouse gases from the burning of fossil fuels (see GREENHOUSE EFFECT). The average temperature of both the atmosphere and oceans has risen, and energy in the global climate system has increased. Besides melting glaciers, reducing the extent of snow and ice, and raising sea levels, global warming is changing rainfall patterns, with longer dry spells and bouts of more intense precipitation, and this trend is projected to accelerate in future. Average global temperatures are predicted to rise by 1.5–5°C by the end of the 21st century, depending on future carbon emissions. These changes are reflected in natural communities worldwide. Average temperature ranges are shifting rapidly towards the poles, and species that cannot adapt quickly enough must either extend their range, migrate, or die. This particularly

clitellum

Glaciers melting
Polar ice melting
Snow cover reduced
Permafrost melting
Rising sea levels
Flooding of coastal and
low-lying areas
Expansion of deserts
Warming and acidification
of oceans

Ecological impacts
Changes to ecosystems
Shifts in species'
geographical ranges
Species' extinctions
Loss of biodiversity

Human impacts
Food shortages
Water shortages
Damage/loss of homes and
infrastructure
Mass migration
Political instability

Climate trends
Higher temperatures
More/longer dry spells
More intense precipitation
Changing pattern of seasons
More intense/longer droughts
More intense storms

**Impacts on farming
and fishing**
Loss of farmland
Increased salinity of coastal land
Lower crop yields
Changes in cropping patterns
Spread of crop pests
and diseases
Fisheries in decline

Health impacts
Illness due to extreme weather
(e.g. heatstroke)
Air pollution
Pathogens spreading to new areas
Food and water shortages
Stress and mental illness

Impacts of climate change

affects organisms living on islands or mountains, where physical limits to extension of species' ranges are quickly reached. The impact on agriculture is variable: some crops are able to grow in hitherto unsuitable areas, whereas other areas are becoming more prone to droughts, affecting farming practices. Pests also may move into new regions, threatening established agriculture and forestry. Warmer temperatures and increased atmospheric dioxide mean that more carbon dioxide is dissolving in the oceans as carbonic acid. For example, coral organisms are harmed by both warming and acidification, with damaging consequences for the wider marine ecosystems and local fisheries. The Intergovernmental Panel on Climate Change (IPCC) was set up in 1988 by the United Nations Environment Programme and the World Meteorological Association to review the current state of knowledge about climate change and its potential environmental and socioeconomic impacts. It draws information from scientists around the world and periodically publishes Assessment Reports, intended to inform the national policies of its 195 member

countries to reduce carbon emissions and mitigate the effects of climate change. See illustration.

SEE WEB LINKS
• Home page of the Intergovernmental Panel on Climate Change (IPCC)

climax community A relatively stable ecological *community that is achieved at the end of a *succession.

cline A gradual variation in the characteristics of a species or population over its geographical range. It occurs in response to varying environmental factors, such as soil type or climate.

clinostat A mechanical device that rotates whole plants (usually seedlings), so removing the effect of any stimulus that normally acts in one particular direction. It is most often used to study the growth of plant organs when the influence of gravity has been removed.

clitellum (*pl.* **clitella**) A thickened glandular region of the body in certain annelid worms that is prominent when the worm becomes sexually mature. The clitellum possesses long slender chaetae. Goblet cells in the clitellum secrete mucus, which forms the *cocoon in which fertilized eggs develop.

clitoris An erectile rod of tissue in female mammals (and also some reptiles and birds) that is the equivalent of the male penis. It lies in front of the *urethra and *vagina.

cloaca (*pl.* **cloacae**) The cavity in the pelvic region into which the terminal parts of the alimentary canal and the urinogenital ducts open in most vertebrates. Placental mammals, however, have a separate anus and urinogenital opening.

clonal selection The mechanism whereby the cells of the immune system produce large quantities of the right antibody at the right time, i.e. when the appropriate antigen is encountered. There is a pre-existing pool of lymphocytes (*B cells) consisting of numerous small subsets. Each subset carries a unique set of surface antibody molecules with its own particular binding characteristics. If a cell encounters and binds the corresponding antigen it is 'selected'—stimulated to divide repeatedly and produce a large clone of identical cells, all secreting the antibody. The involvement of *helper T cells is essential for activation of the B cell. A similar process generates T cells with specificity towards certain antigens. A form of clonal selection is also invoked to explain the development of immunological *tolerance.

clone 1. A group of cells, an organism, or a population of organisms arising from a single ancestral cell. All members of a particular clone are genetically identical. In nature clones are produced by asexual reproduction, for example by the formation of bulbs and tubers in plants or by *parthenogenesis in certain animals. New techniques of cell manipulation and tissue culture have enabled the cloning of many plants and some animals. A wide range of commercially important plant species, including potatoes, tulips, and certain forest trees, are now cloned by *micropropagation, resulting in more uniform crops. Cloning in animals is more complex, but has been accomplished successfully in sheep, cattle, and various other species. The first mammal to be cloned experimentally from the body cell of an adult was a sheep ('Dolly') born in 1997 after over 200 previous failed attempts. The nucleus containing DNA was extracted from an udder cell (which had been deprived of nutrients) and inserted into an 'empty' egg cell (from which the nucleus had been removed) using the technique of *nuclear transfer. This reconstituted egg cell was then stimulated to divide by an electric shock and implanted into the uterus of a surrogate mother ewe, who subsequently gave birth to a clone of the original sheep. This breakthrough has led to the production of exact replicas of animals with certain genetically engineered traits, for example to manufacture drugs in their milk or provide organs for human transplantation. Cloning may also be used to perpetuate endangered species, where rates of natural reproduction are dangerously low, or even to replicate pet animals for doting owners. **2.** (**gene clone**) An exact replica of a gene. *See* GENE CLONING.

cloning vector *See* VECTOR.

clonotype 1. (in immunology) An attribute or specificity that is unique to a particular clone of cells, especially the structure of the surface receptors of a T cell or of the secreted antibodies of a plasma cell. **2.** (in botany) A plant, or line of plants, that has been propagated by cloning a type specimen.

closed circulation *See* CIRCULATION.

clotting factors (coagulation factors) A group of substances present in blood plasma that, under certain circumstances, undergo a series of chemical reactions leading to the conversion of blood from a liquid to a solid state (*see* BLOOD CLOTTING). Although they have specific names, most coagulation factors are referred to by an agreed set of Roman numerals (e.g. *Factor VIII, Factor IX). Lack of any of these factors in the blood results in the inability of the blood to clot. *See also* HAEMOPHILIA.

clubmoss *See* LYCOPODIOPHYTA.

cluster of differentiation *See* CD.

Cnidaria A phylum of aquatic invertebrates that includes *Hydra*, jellyfish, sea
anemones, and *corals. A cnidarian's body
is *diploblastic, with two cell layers of the
body wall separated by *mesoglea, and
shows *radial symmetry. The body cavity
(**gastrovascular cavity**) is sac-shaped,
with one opening acting as both mouth
and anus. This opening is surrounded by
tentacles bearing *thread cells. Cnidarians
exist both as free-swimming *medusae (e.g.
jellyfish) and as sedentary *polyps. The latter may be colonial (e.g. corals) or solitary
(e.g. sea anemones and *Hydra*). In many
cnidarians the life cycle alternates between
these two forms (*see* ALTERNATION OF
GENERATIONS). The phylum contains the
classes Hydrozoa (e.g. *Hydra*, *Obelia*),
most members of which show alternation
of generations; Scyphozoa (jellyfish), in
which the medusa phase is dominant; and
Anthozoa (corals and sea anemones), in
which medusae are absent. Some authorities recognize a fourth class, Cubozoa,
containing the box jellies (traditionally
regarded as anthozoans). The smallest
cnidarians are the almost microscopic
myxozoans. They live as parasites, typically
with a two-host life cycle involving a fish
and an annelid worm or bryozoan, and are
significant pathogens of fish and amphibians. The traditional classes are now placed
in two clades: the Medusozoa, containing
all members that produce a medusa; and
the Anthozoa, whose members occur only
as polyps. *See also* COELENTERATE.

(⊕) SEE WEB LINKS

• A summary of cnidarian biology and
systematics: part of the website of the
University of California Museum of
Paleontology

cnidoblast *See* THREAD CELL.

CNS *See* CENTRAL NERVOUS SYSTEM.

CoA *See* COENZYME A.

coacervate An aggregate of macromolecules, such as proteins, lipids, and nucleic
acids, that form a stable *colloid unit with
properties that resemble living matter.
Many are coated with a lipid membrane
and contain enzymes that are capable of
converting such substances as glucose
into more complex molecules, such as
starch. Coacervate droplets arise spontaneously under appropriate conditions and
have been suggested as possible prebiolo-
gical systems from which living organisms
originated.

coadaptation The mutual adaptation of
two species that occurs during *coevolution.

coagulation The process in which colloidal particles come together irreversibly
to form larger masses. Coagulation can be
brought about by adding ions to change the
ionic strength of the solution and thus desta-
bilize the colloid (*see* FLOCCULATION). Ions
with a high charge are particularly effective
(e.g. alum, containing Al^{3+}, is used in styptics
to coagulate blood). Alum and iron(III)
sulphate are also used for coagulation in
*sewage treatment. Heating is another way
of coagulating certain colloids (e.g. boiling an
egg coagulates the albumin). *See also* BLOOD
CLOTTING.

coal A brown or black carbonaceous
deposit derived from the accumulation
and alteration of ancient vegetation, which
originated largely in swamps or other moist
environments. As the vegetation decomposed it formed layers of peat, which were
subsequently buried (for example, by marine sediments following a rise in sea level or
subsidence of the land). Under the
increased pressure and resulting higher
temperatures the peat was transformed
into coal. Two types of coal are recognized:
humic (or **woody**) **coals**, derived from
plant remains; and **sapropelic coals**,
which are derived from algae, spores, and
finely divided plant material.

As the processes of coalification (i.e. the
transformation resulting from the high temperatures and pressures) continue, there is a
progressive transformation of the deposit:
the proportion of carbon relative to oxygen

rises and volatile substances and water are driven out. The various stages in this process are referred to as the **ranks** of the coal. In ascending order, the main ranks of coal are: **lignite** (or **brown coal**), which is soft, brown, and has a high moisture content; **subbituminous coal**, which is used chiefly by generating stations; **bituminous coal**, which is the most abundant rank of coal; **semibituminous coal**; **semianthracite coal**, which has a fixed carbon content of between 86% and 92%; and **anthracite coal**, which is hard and black with a fixed carbon content of between 92% and 98%.

Most deposits of coal were formed during the Carboniferous and Permian periods. More recent periods of coal formation occurred during the early Jurassic and Palaeogene periods. Coal deposits occur in all the major continents. Coal is used as a fuel and in the chemical industry; by-products include coke and coal tar. Combustion of coal is a major source of greenhouse gases worldwide, and efforts are under way to develop 'clean coal' technology or to phase out its use.

coated pit *See* ENDOSOME.

coated vesicle *See* ENDOSOME.

cobalamin (vitamin B$_{12}$) *See* VITAMIN B COMPLEX.

cobalt Symbol Co. A light-grey metallic element that is a trace element (*see* ESSENTIAL ELEMENT) required by animals.

It is present in foods of animal origin and is a constituent of vitamin B$_{12}$. Cobalt is also a *micronutrient for plants.

coccus (*pl.* **cocci**) Any spherical bacterium. Cocci may occur singly, in pairs, in groups of four or more, in cubical packets, in grapelike clusters (**Staphylococcus*), or in chains (**Streptococcus*). Staphylococci and streptococci include pathogenic species. They are generally nonmotile and do not form spores.

coccyx (*pl.* **coccyges**) The last bone in the *vertebral column in apes and humans (i.e. tailless primates). It is formed by the fusion of 3–5 *caudal vertebrae.

cochlea (*pl.* **cochleae**) Part of the *inner ear of mammals, birds, and some reptiles that transforms sound waves into nerve impulses. In mammals it is coiled, resembling a snail shell, and is divided by membranes into three parallel canals (see illustration): the middle cochlear duct (scala media) and two outer canals—the vestibular canal (scala vestibuli) and the tympanic canal (scala tympani)—formed from one long canal folded on itself at a bend at the apex of the cochlea. The small opening at this point, where the vestibular and tympanic canals communicate, is called the **helicotrema**. The cochlear duct is filled with a fluid (*see* ENDOLYMPH) and contains the *organ of Corti, which houses the sound receptors. The other two canals also contain a fluid (*see* PERILYMPH).

Diagram of the cochlea (coiling reduced for simplicity)

Sound-induced vibrations of the *oval window are transmitted through the perilymph and endolymph and stimulate *hair cells in the organ of Corti. These in turn stimulate nerve cells that transmit information, via the *auditory nerve, to the brain for interpretation of the sounds.

cockroaches See DICTYOPTERA.

cocoon A protective covering for eggs and/or larvae produced by many invertebrates. For example, the larvae of many insects spin a cocoon in which the pupae develop (that of the silkworm moth produces silk), and earthworms secrete a cocoon for the developing eggs (see CLITELLUM).

codeine A pain-relieving drug that is derived from the plant Papaver somniferum. See OPIATE; ANALGESIC.

coding strand (plus strand; sense strand) The strand in duplex DNA that by convention contains the same sequence of bases as that in the messenger RNA (mRNA) transcribed from the DNA (except that the RNA has uracil substituting for thymine). It is complementary to the other *anticoding strand, which is actually the one used as a template for mRNA assembly during *transcription. See also POSITIVE-SENSE.

codominance The condition that arises when both alleles in a *heterozygous organism are dominant and are fully expressed in the *phenotype. For example, the human blood group AB is the result of two alleles, A and B, both being expressed. A is not dominant to B, nor vice versa. Compare INCOMPLETE DOMINANCE.

codon A triplet of nucleotides within a molecule of messenger *RNA that functions as a unit of genetic coding (the **triplet code**), usually by specifying a particular amino acid during the synthesis of proteins in a cell (see GENETIC CODE). A few codons specify instructions during this process (see START CODON; STOP CODON). The term codon may also loosely refer to any of the corresponding nucleotide triplets of DNA that are transcribed into codons. See also READING FRAME. Compare ANTICODON.

coelacanth A bony fish of the genus *Latimeria*, which was believed to be extinct until 1938, when the first modern specimen of *L. chalumnae* was discovered in the Indian Ocean around the Comoros Islands, off the SE coast of Africa. A second species, *L. menadoensis*, was discovered in 1999 in the Celebes Sea, SE Asia. The coelacanth belongs to the same order (Crossopterygii—lobe-finned fishes) as the ancestors of the amphibians. It is a large fish, 1–2 m long and weighing 80 kg or more, with a three-lobed tail fin. The body is covered in rough heavy scales and the pectoral fins can be used like crutches to help movement across the sea bed. The young are born alive. Fossil coelacanths are most abundant in deposits about 400 million years old and no fossils less than 70 million years old have been found.

coelenterate An obsolete term for invertebrate animals with a radially symmetrical body, a body wall consisting of two cell layers, and a central **gastrovascular cavity**, including *Cnidaria (hydras, jellyfish, sea anemones, and corals) and *Ctenophora (comb jellies).

coelom A fluid-filled cavity that forms the main *body cavity of vertebrate and most invertebrate animals. It is formed by the splitting of the *mesoderm. Ciliated ducts (**coelomoducts**) connect the coelom to the exterior allowing the exit of waste products and gametes; in higher animals these are specialized as oviducts, etc. The coelom is large and often subdivided in annelid worms (in which it functions as a hydrostatic skeleton) and vertebrates. In arthropods it is restricted to the cavities of the gonads and excretory organs, the body cavity being a blood-filled *haemocoel.

coelomoduct See COELOM.

coenobium (cenobium) (*pl.* **coenobia**) A loose association of unicellular organisms that live in a colony and may be surrounded by a common membrane. Often the cells are held together in a jelly excreted by all individuals in the colony. Both algae and bacteria form coenobia. For example, *Volvox* (a green alga) forms a hollow sphere in which 20000 cells may live; some of these are reproductive and others are concerned with photosynthesis. The cells in this coenobium are interconnected by protoplasmic strands.

coenocyte A mass of cytoplasm surrounding many nuclei and enclosed by a cell wall. It is found in certain algae and fungi. *Compare* CELL; PLASMODIUM; SYNCYTIUM.

coenzyme An organic nonprotein molecule that associates with an enzyme molecule in catalysing biochemical reactions. Coenzymes usually participate in the substrate-enzyme interaction by donating or accepting certain chemical groups. Many vitamins are precursors of coenzymes. *See also* COFACTOR.

coenzyme A (CoA) A complex organic compound that acts in conjunction with enzymes involved in various biochemical reactions, notably the oxidation of pyruvate via the *Krebs cycle and fatty-acid oxidation and synthesis (*see* ACETYL COENZYME A). It comprises principally the B vitamin *pantothenic acid, the nucleotide *adenine, and a ribose-phosphate group (see formula).

coenzyme Q *See* UBIQUINONE.

coevolution The evolution of complementary adaptations in two species caused by the *selection pressures that each exerts on the other. It is common in symbiotic associations (*see* SYMBIOSIS). For example, many insect-pollinated plants have evolved flowers whose shapes, colours, etc., make them attractive to particular insects; at the same time the pollinating insects have

evolved sense organs and mouthparts specialized for quickly locating, and extracting nectar from, particular species of plants.

cofactor A nonprotein component essential for the normal catalytic activity of an enzyme. Cofactors may be organic molecules (*coenzymes) or inorganic ions. They may activate the enzyme by altering its shape or they may actually participate in the chemical reaction.

cohesin A multiprotein complex that binds sister chromatids together during the initial stages of nuclear division in cells. It includes members of the SMC (structural maintenance of chromosomes) family of proteins. Cohesin complexes form a series of rings around the chromatids. At the onset of anaphase, when the sister chromatids have already attached to the spindle fibres, formation of the anaphase-promoting complex (APC) causes activation of the enzyme separase, which cuts the cohesin rings and abolishes the cohesion between the sister chromatids. This enables them to separate and move to opposite poles of the cell.

cohesion 1. The force of attraction between like molecules. The hydrogen bonds between water molecules provide the cohesive force that holds up a column of water in the xylem tissue of plants without it breaking. The **cohesion-tension theory** is the most widely accepted explanation for the continual flow of water upwards through the xylem of a plant. Water is removed from the plant by the process of *transpiration, which creates a tension that

site of acetylation
to form acetyl CoA

Coenzyme A

pulls the water in the xylem upwards as a single column held together by cohesive forces. Also, adhesion between water molecules and the walls of the xylem counteracts the downward gravitational force. **2.** (in botany) The union of like parts, such as the fusion of petals that occurs in some flowers.

cohort A group of individuals who share some characteristic, such as being born within a particular period or having a certain disease.

coitus *See* SEXUAL INTERCOURSE.

colchicine An *alkaloid derived from the autumn crocus, *Colchicum autumnale*. It inhibits *spindle formation in cells during mitosis so that chromosomes cannot separate during anaphase, thus inducing multiple sets of chromosomes (*see* POLYPLOID). Colchicine is used in genetics, cytology, and plant breeding research and also in cancer therapy to inhibit cell division.

cold-blooded animal *See* ECTOTHERM.

cold hardening The process of exposing plants to successively cooler conditions over days or weeks to acclimate them to low temperatures. This particularly applies to plants reared indoors, e.g. in heated glasshouses, so as to avoid chill injury when they are planted out. Cold hardening induces certain biochemical and structural changes, such as increasing the proportion of unsaturated fatty acids in cell membranes, which helps maintain membrane fluidity, and thickening the protective cuticle on leaves.

Coleoptera An order of insects comprising the beetles and weevils and containing about 350000 known species—the largest order in the animal kingdom. The forewings are hardened and thickened to form **elytra**, which meet at a precise mid-dorsal line and protect the underlying pair of hindwings and abdomen. The mouthparts are generally modified for biting and in some species assume antler-like proportions. Beetles occur in a wide variety of terrestrial and aquatic habitats; many feed on decaying organic matter, some eat

living vegetation, while others prey on other arthropods. A number of beetles and weevils are economically important pests of stored grain, timber, and crops. The young emerge as larvae and generally undergo metamorphosis via a pupal stage to form the adult beetle.

coleoptile A protective sheath that covers the young shoot of the embryo in plants of the grass family. It bursts open when the first leaves develop. Experiments investigating growth movements of the oat coleoptile led to the discovery of the plant hormone indoleacetic acid (IAA).

coleorhiza A protective sheath that covers the young root of the embryo in plants of the grass family.

coliform bacteria A group of Gram-negative rod-shaped bacteria that are found in the vertebrate gastrointestinal tract; their presence in water is an indicator of faecal pollution. They obtain their energy by aerobic respiration or fermentation; some of them can ferment lactose. Well-known coliform bacteria include *Escherichia coli* and *Salmonella*.

colinearity *See* HOX GENES.

collagen An insoluble fibrous protein found extensively in the connective tissue of skin, tendons, and bone, in the walls of the large arteries, and in the *extracellular matrix. The polypeptide chains of collagen (containing the amino acids glycine and proline predominantly) form triple-stranded helical coils that are bound together to form fibrils, which have great strength and limited elasticity. Collagen accounts for over 25% of the total body protein of mammals.

collectin *See* LECTIN.

collecting duct Any of the ducts in the mammalian *kidney that drains into the renal pelvis, which leads to the ureter. They are the main sites of water reabsorption from the glomerular filtrate, which drains into the ducts from the *distal convoluted tubules of the *nephrons. The cells of the collecting ducts are relatively impermeable to water. However, the influence of *antidiuretic hormone increases the permeability of the collecting ducts, allowing the reabsorption of water and controlling the final urine concentration according to the body's state of hydration.

Collembola An order of small wingless insects, the springtails, less than 10 mm long, which leap by means of a specialized forked organ (**furcula**) that is attached on the underside of the abdomen by a special catch (**retinaculum**) and acts as a spring. The mouthparts are largely concealed within folds of the head. Most springtails are scavengers but some are pests of leguminous plants. Although some authorities place the Collembola in the subclass *Apterygota, with other wingless insects, others regard them as constituting a separate class within the superclass *Hexapoda, most closely affiliated with the *Protura.

collenchyma A plant tissue (*see* GROUND TISSUES) that consists of living cells with additional cellulose thickening in their walls, providing support and protection for young stems and leaves. It is most commonly found in the stem cortex. *Compare* PARENCHYMA; SCLERENCHYMA.

colloblast *See* LASSO CELL.

colloids As originally defined by Thomas Graham (1805–69) in 1861, substances, such as starch or gelatin, which will not diffuse through a membrane. Graham distinguished colloids from **crystalloids** (e.g. inorganic salts), which would pass through membranes. Later it was recognized that colloids were distinguished from true solutions by the presence of particles that were too small to be observed with a normal microscope yet were much larger than normal molecules. Colloids are now regarded as systems in which there are two or more phases, with one (the **dispersed phase**) distributed in the other (the **continuous phase**). Moreover, at least one of the phases has small dimensions (in the range 10^{-9}–10^{-6} m). Colloids are classified in various ways.

Sols are dispersions of small solid particles in a liquid. The particles may be macromolecules or clusters of small molecules. **Lyophobic sols** are those in which

there is no affinity between the dispersed phase and the liquid. An example is silver chloride dispersed in water. In such colloids the solid particles have a surface charge, which tends to stop them coming together. Lyophobic sols are inherently unstable and in time the particles aggregate and form a precipitate (*see* FLOCCULATION). **Lyophilic sols**, on the other hand, are more like true solutions in which the solute molecules are large and have an affinity for the solvent. Starch in water is an example of such a system. **Association colloids** are systems in which the dispersed phase consists of clusters of molecules that have lyophobic and lyophilic parts. Soap in water is an association colloid (*see* MICELLE).

Emulsions are colloidal systems in which the dispersed and continuous phases are both liquids, e.g. oil-in-water or water-in-oil. Such systems require an emulsifying agent to stabilize the dispersed particles.

Gels are colloids in which both dispersed and continuous phases have a three-dimensional network throughout the material, so that it forms a jelly-like mass. Gelatin is a common example. One component may sometimes be removed (e.g. by heating) to leave a rigid gel (e.g. silica gel).

Other types of colloid include **aerosols** (dispersions of liquid or solid particles in a gas, as in a mist or smoke) and **foams** (dispersions of gases in liquids or solids).

colon The section of the vertebrate *large intestine that lies between the *caecum and the *rectum. Its prime function is to absorb water and minerals from indigestible food residues passing from the small intestine, which results in the formation of *faeces. In humans the colon contains the greatest abundance and diversity of all the intestinal microflora, with an estimated 10^{12} bacteria per gram (dry weight) of contents. These constitute a complex *microbiome, with important protective, metabolic, and nutritional functions.

colony 1. (in zoology) A group of animals of the same species living together and dependent upon each other. Some, such as the corals and sponges, are physically connected and function as a single unit.

Others, such as insect colonies, are not physically joined but show a high level of social organization with members specialized for different functions (*see* CASTE). **2.** (in microbiology) A group of microorganisms, usually bacteria or yeasts, that are considered to have developed from a single parent cell. Colonies that grow on *agar plates differ in shape, colour, surface texture, and translucency and can therefore be used as a means of identification.

colony-stimulating factor (CSF) Any of several *cytokines that stimulate development of certain types of blood cells from progenitor cells in the bone marrow and in other tissues. They include **GM-CSF**, a glycoprotein that causes haemopoietic stem cells to develop into mixed colonies of granulocytes and monocytes/macrophages (hence the name); **G-CSF**, which stimulates production of granulocytes only, especially neutrophils; and **M-CSF**, which promotes only monocyte/macrophage cell production. *Interleukin-3 (IL-3) is sometimes called the 'multi-potential CSF' because it stimulates the production of all types of lymphocytes.

colostrum A liquid with a high content of nitrogen, antibodies, and vitamins that is secreted from the mammary glands before and just after giving birth. The change of secretion from colostrum to proper milk takes place gradually during the days after birth.

colour blindness Any disorder of vision in which colours are confused. The most common type is red-green colour blindness. This is due to a recessive gene carried on the X chromosome (*see* SEX LINKAGE), and therefore men are more likely to show the defect although women may be *carriers. It results in absence or malfunctioning of one or more of the three types of cone cell responsible for *colour vision. In **protanopia** the individual lacks cones sensitive to red light; in **deuteranopia** cones sensitive to green light are absent. **Tritanopia** is a rare form of colour blindness in which the individual cannot distinguish between blue and green due to a lack of cones sensitive to blue light.

colour vision The ability of the eye to detect different wavelengths of light and to distinguish between these different wavelengths and their corresponding colours. In the mammalian eye this is achieved by the *cone cells, which are located in and around the *fovea near to the centre of the retina. The cone cells contain the light-sensitive pigment **iodopsin**, which exists in three forms (called **photopsins**), each occurring in a different cone cell and sensitive to a different range of wavelengths of light (originally proposed as the 'trichromatic theory'). The relative stimulation of each type of cone will determine the colour that is interpreted by the brain. *See also* COLOUR BLINDNESS.

Full colour vision is widespread among fishes, amphibians, reptiles, birds, and primates. The *compound eye of certain insects is also capable of colour vision.

columnar epithelium *See* EPITHELIUM.

commensalism An interaction between two animal or plant species that habitually live together in which one species (the **commensal**) benefits from the association while the other is not significantly affected. For example, the burrows of many marine worms contain commensals that take advantage of the shelter provided but do not affect the worm.

communicating junction *See* GAP JUNCTION.

communication An interaction between two organisms in which information is conveyed from one to the other. Communication can occur between individuals of the same species (**intraspecific communication**) or between members of different species (**interspecific communication**). It generally involves the transmission of a signal from one organism to another; signals can be visual, chemical, or tactile or they can take the form of sounds. Visual signals between members of the same species are widely used by animals in such activities as defining and protecting *territories and finding suitable mates (*see* COURTSHIP; DISPLAY BEHAVIOUR; BIOLUMINESCENCE). Chemical and tactile signals also play an important role in these activities (*see* PHEROMONE). Social species rely heavily on all three types of signalling, the classic example being provided by the *dance of the bees, in which information about the distance and direction of a food source is conveyed to other members of the colony. Visual signals, in the form of body coloration, are the principal means of communication between animals of different species (*see* MIMICRY; WARNING COLORATION). Sounds are more effective than visual signals for intraspecific communication over long distances and at night. Certain insects produce sounds by *stridulation, while birdsong and language are sophisticated examples of sound signals in birds and humans, respectively.

Among plants, visual and chemical signals are important in communication. Flowering plants whose flowers are pollinated by insects or other animals depend on the colour, shape, and scent of their flowers to attract suitable pollinating agents. Some plants produce chemical signals to deter competitors and predators (*see* ALLELOPATHY), and certain volatile chemicals may induce defensive mechanisms in neighbouring plants. Trees can transmit chemical signals to their neighbours via the networks of hyphae produced by the mycorrhizal fungi associated with their roots.

community A naturally occurring assemblage of plant and animal species living within a defined area or habitat. Communities are named after one of their *dominant species (e.g. a pine community) or the major physical characteristics of the area (e.g. a freshwater pond community). Members of a community interact in various ways (e.g. through *food chains and *competition). Large communities may be divided into smaller component communities. *See* ASSOCIATION.

community immunity *See* HERD IMMUNITY.

companion cell A type of cell found within the *phloem of flowering plants. Each companion cell is usually closely associated with a *sieve element, with which it communicates via numerous fine cytoplasmic strands (*plasmodesmata) penetrating the adjoining cell walls. The companion cell shares its nucleus and RNA with its adjacent sieve element and regulates the latter's activity. In some cases it may also take part in loading and unloading sugar into the sieve element (*see* PHLOEM LOADING; PHLOEM UNLOADING). In gymnosperms a similar function is attributed to **albuminous cells**, which are found closely associated with gymnosperm sieve elements.

comparative genomic hybridization (CGH) A technique for scanning the genome of an organism or tissue sample for variations in *copy number, i.e. deletions or duplications of stretches of DNA. Such variations can be associated with certain diseases, and CGH is used in medicine as a diagnostic tool, e.g. to assess genetic markers of inherited disease compared to a reference (normal or control) genome. In taxonomy CGH is useful for comparing the genomes from different organisms to determine how closely related they are, e.g. whether two microorganisms are different strains of the same species. The technique creates a set of single DNA strands from each genome, differentially labels them with coloured fluorescent markers, and allows the strands to competitively bind (hybridize) to a complete set of metaphase chromosomes. Variations in the intensity and colour of the bound labels indicate gain or loss of DNA in the original sample. These variations are viewed under a fluorescence microscope, recorded, and analysed. Greater speed and better resolution are obtained by **microarray-based CGH** (or **array CGH**). Instead of a complete set of chromosomes, the labelled DNA samples are applied to a *microarray of DNA probes representative of the genome. A scanner then measures the fluorescent signals and a computer analyses the data.

compass plant A plant that has its leaves permanently orientated in a north-south direction. Such an arrangement enables the plant to take full advantage of morning and evening sun, while avoiding the stronger midday sunlight. An example is the compass plant of the prairies (*Silphium laciniatum*).

compatible solute A highly water-soluble substance that accumulates within the cell(s) of an organism and enables it to maintain its water balance with the external environment. Halophilic archaea, for example, which are adapted to living in very salty water, contain intracellular solutes (e.g. glycine betaine, trehalose) that lower the water potential of their cell contents so that the organism stays in physiological balance with its surroundings. Some plants employ a similar strategy to adjust the water potential of their cells as a means of protecting against damage due to water shortage. Compatible solutes are generally sugar alcohols, sugars, alcohols, amino acids, or their derivatives. Some are synthesized by the organism, whereas others are accumulated from the environment.

compensation point The point reached in a plant when the rate of photosynthesis is equal to the rate of respiration. This means that the carbon dioxide released from respiration is equivalent to that which is taken up during photosynthesis. The compensation point is reached as light intensity increases. If the light intensity is increased beyond the compensation point, the rate of photosynthesis increases proportionally until the point of **light saturation** is reached, beyond which the rate of photosynthesis is no longer affected by light intensity.

competent 1. (in embryology) Describing embryonic tissue that is capable of developing into a specialized tissue when suitably stimulated. *See* INDUCTION; EVOCATION. **2.** (in microbiology) Describing a bacterial cell that is able to take up a molecule of DNA in the process called *transformation (sense 1). Typically, competence is restricted to certain strains of bacterial species, and to certain stages of the growth cycle when the competence-specific proteins are synthesized by the bacterial cells.

competition 1. (in ecology) The interaction that occurs between two or more organisms, populations, or species that share some environmental resource when this is in short supply. Competition is an important force in evolution: plants, for example, become tall to compete for light, and animals evolve various foraging methods to compete for food. There may be a direct confrontation between competitors, as occurs between barnacles competing for space on a rock, or the numbers or fecundity of the competitors are indirectly reduced through joint dependence on limited resources. Competition occurs both between members of a species (**intraspecific competition**) and between different species (**interspecific competition**). Interspecific competition often results in the dominance of one species over another (*see* DOMINANT). Since competition ultimately results in the displacement by one competitor of the others, it is to the advantage of the competitors to avoid one another wherever possible. Thus in time the competitors become separated from each other geographically or ecologically, which promotes evolutionary change. Competition for mates may lead to *sexual selection. **2.** (in neurophysiology) The interaction between multiple nerve endings or synapses during nervous system development that results in the elimination of superfluous connections and establishment of the mature 'wiring' pattern. This **pruning** mechanism occurs, for example, during the formation of a vertebrate skeletal neuromuscular junction. In the embryo, motor axons branch to innervate several muscle fibres, with individual fibres receiving branches from multiple neurons. However, as development proceeds after birth, this polyneuronal innervation is pruned so that each muscle fibre comes to be innervated by just a single axon branch. Similar processes prune synaptic connections in the central nervous system. One mechanism involves competition between nerve endings for a limited supply of *neurotrophin from the target cells, leading to survival of some synapses and degeneration of others. Macrophage cells called glia make and break physical connections between brain cells to help maintain healthy brain function.

competitive exclusion principle A rule, derived by G. F. Gause in 1934, stating that two species that occupy the same habitat cannot also occupy the same *ecological niche. Any two species that occupy the same niche will compete with each other to the detriment of one of the species, which will thus be excluded.

competitive inhibition *See* INHIBITION.

complement A group of proteins present in blood plasma and tissue fluid that aids the body's defences following an immune response; the genes encoding it form part of the *major histocompatibility complex. Following an antibody-antigen reaction, complement is activated chemically and becomes bound to the antibody-antigen complex (**complement fixation**). It can cause lysis of certain types of bacteria, through the formation of a *membrane attack complex, or it can render the target cell more susceptible to phagocytosis—a process called *opsonization. Binding of complement components to their respective *complement receptors on immune cells stimulates various defence mechanisms.

(((•))) SEE WEB LINKS
- An illustrated overview of complement activation and regulation, from the US National Library of Medicine

complemental males The small males of certain animals that live in or on the females and are usually more or less degenerate apart from the reproductive organs. They occur in certain crustaceans (e.g. some barnacles), in which the normal individuals are hermaphrodite but the complemental males have suppressed ovaries, lose their alimentary canal, and lead a semiparasitic existence in the mantle

cavity of the larger partner. This may ensure that cross fertilization occurs.

complementarity-determining region (CDR) A region on the surface of an antibody molecule that contributes to the antigen-binding site (*see* IMMUNOGLOBULIN). Each of the two arms of the Y-shaped antibody comprises a heavy chain and a light chain, both of which carry three **hypervariable loops**. These are stretches of the polypeptide along which the amino acid sequence can vary greatly in different B-cell clones, thereby generating differently shaped antigen-binding sites, and hence antibodies with different binding specificities. The three-dimensional arrangement of each loop constitutes a CDR.

complementary DNA (cDNA) A form of DNA prepared in the laboratory using messenger *RNA (mRNA) as template, i.e. the reverse of the usual process of *transcription in cells; the synthesis is catalysed by *reverse transcriptase. cDNA thus has a base sequence that is complementary to that of the mRNA template; unlike genomic DNA, it contains no noncoding sequences (*introns), and it is the form of DNA that is sequenced in the *RNA-seq technique. cDNA is used in *gene cloning for the expression of eukaryote genes in prokaryote host cells, or as a *DNA probe to locate particular base sequences in genomic DNA. cDNA molecules are inserted into plasmid or phage vectors to create cDNA libraries of expressed genes.

complementary genes Two (or more) genes that are interdependent, such that the dominant *allele from either gene can only produce an effect on the *phenotype of an organism if the dominant allele from the other gene is also present.

complement receptor A protein found on the surface of certain cells of the immune system that binds specifically to complement components coating the surface of a bacterium or other pathogen (*see* OPSONIZATION). There are several types of complement receptor associated with various cell types and binding specifically to different complement proteins. One of the most important is the CR1 receptor found on macrophages, monocytes, polymorphonuclear leucocytes, and erythrocytes. This plays a key role in stimulating the uptake and destruction of complement-coated target cells by macrophages. The CR3 receptor, found on B cells, helps in reinforcing the antibody response when the B cell encounters its specific antigenic target if the target is also coated with the corresponding complement protein.

complexity science The science of complex systems. It draws on a wide range of disciplines, including biology, physics, computer science, mathematics, and engineering, to investigate the behaviour of natural complex systems, such as the brain, the cell, or an ecosystem, as well as human-constructed systems, such as computer networks. Modelling of complex systems may reveal underlying governing principles, enable prediction of emergent behaviour, and allow for better management of existing systems. *See also* SYSTEMS BIOLOGY.

composite fruit A type of fruit that develops from an inflorescence rather than from a single flower. *See* PSEUDOCARP; SOROSIS; STROBILUS; SYCONUS.

compost A mixture of decaying organic matter, such as vegetation and manure, that is used as a *fertilizer. The organic material is decomposed by aerobic saprotrophic organisms, mostly fungi and bacteria. Some decomposition is also carried out by *detritivores. Compost is used mainly on a domestic scale.

compound eye The eye of insects and crustaceans, which consists of numerous visual units, the **ommatidia**. Each ommatidium consists of an outer cuticle covering a lens, beneath which are 6–8 retinal cells surrounding a light-sensitive **rhabdom**. Adjacent ommatidia are separated by pigment cells. The eye is convex, with nerve fibres from the retinal cells converging onto the optic nerve. There are two types of compound eye. In **apposition eyes**, typical of diurnal insects, each ommatidium focuses rays parallel to its long axis so that

each gives an image of a minute part of the visual field, producing a detailed mosaic image. In **superposition eyes**, typical of nocturnal insects, the pigment separating ommatidia migrates to the ends of the cells, so that each ommatidium receives light from a larger part of the visual field and the image may overlap with those received by many neighbouring ommatidia. This produces an image that is bright but lacks sharpness of detail.

(((∴))) SEE WEB LINKS

- A series of magnified images of a bee's compound eyes

compound microscope *See* MICROSCOPE.

concanavalin *See* LECTIN.

concentration The quantity of dissolved substance per unit quantity of solvent in a solution. Concentration is measured in various ways. The amount of substance dissolved per unit volume (symbol c) has units of mol dm^{-3} or mol l^{-1}. It is now called 'concentration' (formerly **molarity**). Osmolarity is the concentration of osmotically active particles expressed as *osmoles per litre of solution. The mass concentration** (symbol π) is the mass of solute per unit volume of solvent. It has units of kg dm^{-3}, g cm^{-3}, etc. The **molal concentration** (or **molality**; symbol m) is the amount of substance per unit mass of solvent, commonly given in units of mol kg^{-1}. **Osmolality** is the concentration of osmotically active particles expressed as osmoles per kilogram of solvent.

concentration gradient (diffusion gradient) The difference in concentration between a region of a solution or gas that has a high density of particles and a region that has a relatively lower density of particles. By random motion, particles will move from the area of high concentration towards the area of low concentration, by the process of *diffusion, until the particles are evenly distributed in the solution or gas.

conceptacle A flask-shaped cavity with a small opening (the **ostiole**) that is found in the swollen tip of certain brown algae, such as *Fucus*. It contains the sex organs.

conception The *fertilization of a mammalian egg cell by a sperm cell, which occurs in the fallopian tube. Conception is followed by *implantation.

condensation reaction A chemical reaction in which two molecules combine to form a larger molecule with the elimination of a small molecule (H_2O in biological systems). *See* ESTERIFICATION; GLYCOSIDIC BOND; PEPTIDE.

condensin A protein that coats chromosomes towards the end of the G_2 phase of the *cell cycle to make them more compact in readiness for mitosis or meiosis. Condensins are also involved in transcription regulation and DNA repair.

conditional response (conditioned reflex) A learned response that develops to an initially ineffective stimulus in classical *conditioning.

conditioning A process by which animals learn about a relation between two events. In **classical (or Pavlovian) conditioning**, repeated presentations of a neutral stimulus (e.g. the sound of a bell or buzzer) are followed each time by a biologically important stimulus (such as food or electric shock), which elicits a response (e.g. salivation). Eventually the neutral stimulus presented by itself produces a response (the **conditional response**, or **conditioned reflex**) similar to that originally evoked by the biologically important stimulus. For example, Pavlov's dogs learned to salivate in response to the sound of a metronome that preceded the presentation of food. In **instrumental (or operant) conditioning** the animal is rewarded (or punished) each time it makes a particular response; this eventually causes the frequency of the response to increase (or decrease). For example, a rat will learn to press a lever in order to obtain food. *See* LEARNING (Feature); REINFORCEMENT.

condyle A smooth round knob of bone that fits into a socket on an adjoining bone, forming a *joint. Such a joint permits up-and-down or side-to-side movement but does not allow rotation. There are condyles where the lower jawbone (mandible) is attached to the skull, which permits chewing movements. *See also* OCCIPITAL CONDYLE.

cone 1. (in botany) A reproductive structure occurring in gymnosperms, known technically as a **strobilus**. It consists of *sporophylls bearing the spore-producing sporangia. Gymnosperms produce different male and female cones. The large woody female cones of pines, firs, and other conifers are made up of structures called **ovuliferous scales**, which bear the ovules. Cones are also produced by clubmosses and horsetails. **2.** (in animal anatomy) A type of light-sensitive receptor cell, found in the *retinas of all diurnal vertebrates. Cones are specialized to transmit information about colour (*see* COLOUR VISION) and are responsible for the *visual acuity of the eye. They function best in bright light. They are not evenly distributed on the retina, being concentrated in the *fovea and absent on the margin of the retina. *Compare* ROD.

confocal fluorescence microscopy A light microscopic technique that produces high-resolution images of fluorescently stained specimens without requiring elaborate preparation of the sample. The fluorescent markers (fluorophores), generally fluorescently labelled antibodies (*see* IMMUNOFLUORESCENCE), are excited by light from a laser focused by the objective lens of the microscope so that it scans a single plane in the specimen, creating an optical section, under computer control. At any moment the stimulating light is focused on a pinpoint and only emitted light from the same (confocal) point forms an image. This eliminates unfocused light and gives much greater resolution. The emitted fluorescent light is captured by a photomultiplier and assembled into digital images by a computer. Serial scanning of, say, an entire cell can thus visualize successive sections through the cell or create three-dimensional, or even time-lapse, images. Numerous fluorescent probes are available for labelling different components of cells or other material. The principles of fluorescence microscopy have been adapted in numerous related techniques so that the dynamic processes of living cells can be followed in space and time or assessed quantitatively. Moreover, computer algorithms now enable **super-resolution microscopy (nanoscopy)** with resolution greater than the limit of 200–300 nm achievable with conventional light microscopy. Hence, macromolecular or even molecular structures can be observed in living as well as dead specimens, avoiding the need for expensive and time-consuming electron microscopy, which necessarily uses dead material.

conformer Any organism whose *internal environment is highly influenced by external factors. Many marine invertebrates are conformers: they have no need to control their internal environment since the external environment is fairly constant in terms of temperature, oxygen tension, and nutrients. *Compare* REGULATOR.

confounding The situation in which an apparent *correlation between two *variables is in fact explained by a third variable that correlates with one or both of the variables under investigation. If confounding is not recognized, it can lead to the false assumption that two variables are directly related (a **spurious correlation**). For example, research might show an apparent association between the growth in insect populations in a certain area and a decline in various plant species: however, both observations might well be accounted for by a rise in temperature (the **confounding** or **hidden variable**).

congenital Present at birth. Congenital disorders of the body may be due to genetic factors, e.g. *Down's syndrome, or caused by injury or environmental factors, e.g. drugs (such as thalidomide), chemicals (such as dioxin), and infections (such as those caused by *Listeria* and *cytomegalovirus).

conidiospore *See* CONIDIUM.

conidium (conidiospore) A spore of certain fungi, such as mildews and moulds, that is produced by the constriction of the tip of a specialized hypha, the **conidiophore**. Chains of conidia may be formed in this way; they are cut off, one at a time, from the tip of the hypha.

Coniferophyta A phylum of seed-bearing plants comprising the conifers, including the pines, firs, and spruces. Conifers have an extensive fossil record going back to the late Devonian. The gametes are carried in male and female *cones, fertilization usually being achieved by wind-borne pollen. The ovules and the seeds into which they develop are borne unprotected (rather than enclosed in a carpel, as are those of the *Anthophyta). Internal tissue and cell structure of these species is not as advanced as in the angiosperms. Conifers are typically evergreen trees inhabiting cool temperate regions and have leaves reduced to needles or scales. The wood of conifers, which is called **softwood** in contrast to the **hardwood** of angiosperm trees, is widely used for timber and pulp. *See also* GYMNOSPERM.

(((●))) SEE WEB LINKS
- Illustrated survey of conifer biology from the Tree of Life web project

conjugation 1. The fusion of two reproductive cells, particularly when these are both the same size (*see* ISOGAMY). **2.** A form of sexual reproduction seen in some algae (e.g. *Spirogyra*), some bacteria (e.g. *Escherichia coli*), and ciliate protozoans. Two individuals are united by a tube formed by outgrowths from one or both of the cells. Genetic material from one cell (designated the male) then passes through the tube to unite with that in the other (female) cell. In bacteria conjugation is initiated and directed by *sex factors.

conjunctiva (*pl.* **conjunctivas** or **conjunctivae**) The delicate membrane that covers the cornea and lines the inside of the eyelid of a vertebrate eye. It is kept clean by secretions of the *lacrimal (tear) gland and the reflex blink mechanism.

connective tissue An animal tissue consisting of a small number of cells (e.g. *fibroblasts and *mast cells) and a large amount of *extracellular **matrix**, including variable amounts of fibres (*see* COLLAGEN; ELASTIN). It is widely distributed and has many functions, including support, packing, defence, and repair. The individual constituents vary, depending on the function of the tissue. Different types of connective tissue include **mesenchyme** in the embryo, *adipose tissue, *areolar connective tissue, blood, lymph, cartilage, and bone.

connexon A hexagonal array of protein subunits, typically 10 nm in diameter and arranged as a short hollow cylinder, that penetrates the plasma membrane of certain cells and forms part of a *gap junction. Connexons of closely apposed cells line up exactly to create a channel connecting the two cells, spanning the narrow gap between cells and permitting the flow of ions and small molecules between cells. In neurons and other excitable cells, connexons often occur in dense clusters and serve as low-resistance pathways along which electric current can flow between cells. The individual subunits of connexons, termed **connexins**, vary in molecular mass according to the type of tissue.

conodont Any of a group of extinct jawless marine animals, known chiefly for their distinctive mineralized mouthparts, which are commonly found as fossils in rock strata dating from the early Cambrian to the end of the Triassic periods (about 525 to 200 mya). Specimens of complete animals are much rarer. The well-preserved *Haikouella lanceolata*, found in the *Chengjiang fossil assemblage in China in 1999, is one of the earliest known chordates (*see* CHORDATA). It had a soft, slender body, about 16–22 mm long, with eyes, a brain, a cartilaginous skeleton, and possibly a tail fin. The hardened cone-like dental elements incorporated calcium phosphate, like bone.

consensual Of or relating to movements that take place through the action of the nervous system independently of the will. These movements generally involve the involuntary reaction being correlated with a voluntary action. For example, the reflex action of both pupils contracting will occur even if only one retina has been stimulated by bright light (*see* PUPILLARY REFLEX).

consensus sequence A sequence of nucleotides found in comparable regions of DNA or RNA, e.g. in the promoter regions (*see* OPERON) of different genes, in which certain bases occur with a frequency significantly greater than that expected by chance. Although such sequences may vary from case to case, it is possible to derive the most likely sequence overall. An example is the *Pribnow box of prokaryote promoters. The term is also applied to sequences of amino acids in polypeptides.

conservation The sensible management of the earth's natural resources in order to avoid degradation and impoverishment of the environment (*see* DESERTIFICATION). It should include the search for alternative food and fuel supplies when these are endangered (as by *deforestation and overfishing); an awareness of the dangers of *pollution; the maintenance of *biodiversity; and the protection of natural habitats together with the creation of new ones (e.g. nature reserves, national parks, and *SSSIs) to compensate for losses elsewhere. *See also* SUSTAINABLE.

conservative replication A hypothesis that envisaged *DNA replication occurring by one DNA molecule initiating the synthesis of a new molecule while remaining intact. The accepted mechanism (*see* SEMICONSERVATIVE REPLICATION) is that the DNA molecule divides to provide two templates, each for synthesizing the other half of the molecule. *Compare* DISPERSIVE REPLICATION.

conserved sequence Any sequences of bases (or amino acids) in comparable segments of different nucleotides (or proteins) that tends to show similarity greater than that due to chance alone. For example, if one position is occupied by the same base in all comparable DNA sequences, then that position is said to be completely conserved. If the same base occurs at a given position in, say, 75% of samples examined, it would be described as partially conserved. By extension, the conservation of other positions in a sequence is assessed in the same way, usually by computer analysis. The degree to which sequences are conserved can indicate the extent of structural and functional similarities between different genes or between different proteins and provides clues to their possible evolutionary relations (*see* HOMOLOGOUS). *See* ALIGNMENT; MOLECULAR SYSTEMATICS.

consociation A climax plant *community that is dominated by one particular species, e.g. a pine forest. *See* DOMINANT. *Compare* ASSOCIATION.

constitutive (in genetics) Describing a gene that is expressed continuously to provide the cell with products essential for its maintenance and basic functions. Such genes, also known as **housekeeping genes**, are normally expressed in all cells of the organism at a fairly constant level. The products—**constitutive proteins**—include numerous enzymes, structural proteins, signal proteins, and proteins involved in gene expression and protein synthesis.

consumer An organism that feeds upon those below it in a *food chain (i.e. at the preceding *trophic level). Herbivores, which feed upon green plants, are **primary consumers**; a carnivore that feeds only upon herbivores is a **secondary consumer**; a **tertiary consumer** is a carnivore that feeds on other carnivores. The consumer at the end of a food chain is known as the **top carnivore**. *Compare* PRODUCER.

contact insecticide Any insecticide (*see* PESTICIDE) that kills its target insect by being absorbed through the cuticle or by blocking the spiracles, rather than by being ingested.

contig *See* PHYSICAL MAP.

continental drift The theory that the earth's continents once formed a single mass and have since moved relative to each other. It was first postulated by A. Snider in 1858 and greatly developed by Alfred Wegener (1880-1930) in 1912. He used evidence, such as the fit of South America into Africa and the distribution of rock types, flora, fauna, and geological structures, to suggest that the present distribution of the continents results from the

(a) 200 million years ago

(b) 135 million years ago

(c) 65 million years ago

Continental drift

breaking up of one or two greater land masses. The original land mass was named **Pangaea** and it was suggested that this broke up into the northerly **Laurasia** and the southerly **Gondwanaland** (see illustration). The theory was not accepted for about 50 years by the majority of geologists but during the early 1960s, the sea-floor-spreading hypothesis of Harry Hess (1906–69) and the subsequent development of *plate tectonics produced a mechanism to explain the drift of the continents.

(⊕) SEE WEB LINKS
• Animation of continental drift during the earth's history

continuous culture A technique used to grow microorganisms or cells continually in a particular phase of growth. For example, if a constant supply of cells is required, a cell culture maintained in the log phase is best; the conditions must therefore be continually monitored and adjusted accordingly so that the cells do not enter the stationary phase (*see* BACTERIAL GROWTH CURVE). Growth may also have to be maintained in a particular growth phase if an enzyme or chemical product is produced only during that phase.

continuous replication *See* DIS-CONTINUOUS REPLICATION.

continuous variation (quantitative variation) The range of differences that can be observed in many characteristics in a population. Characteristics resulting from *quantitative inheritance show continuous variation, e.g. the wide range of foot sizes in an adult human population. *Compare* DISCONTINUOUS VARIATION.

contour feathers *Feathers that are arranged in regular rows on a bird's body, giving the body its streamlined shape. Each has a central horny shaft (the **rachis**) with a flattened **vane** on each side. Each vane is composed of two rows of filament-like *barbs, which are connected to each other by means of hooked *barbules to form a smooth surface. There is often a small second vane, the **aftershaft**, near the base of the feather.

contraception *See* BIRTH CONTROL.

contractile root Any of the modified adventitious roots that develop from the base of the stem of a bulb or corm. The new bulb or corm develops at a higher level in the soil than the old one. The contractile roots shorten and pull it down to a suitable level.

contractile vacuole A membrane-surrounded cavity in a cell that periodically expands, filling with water, and then suddenly contracts, expelling its contents to the cell's exterior. It is thus an organ of *osmoregulation and excretion. Contractile vacuoles are common in freshwater sponges and typical of freshwater protists, such as Amoeba (which has one spherical vacuole) and *Paramecium* (in which a number of accessory vacuoles are attached to a main vacuole).

contraction (in animal physiology) The shortening of muscle fibres in order to exert a force on a tissue or organ of the body. In striated muscle contraction is brought about by interaction of actin and myosin filaments (*see* SARCOMERE; SKELETAL MUSCLE; SLIDING FILAMENT THEORY): it provides a force for *locomotion and plays a role in maintaining the balance and posture of the animal. *See also* SMOOTH MUSCLE.

contralateral Of, relating to, or affecting the opposite side of the body. *Compare* IPSILATERAL.

control 1. The part of an experiment that acts as a standard by which to compare experimental observations. **2.** The natural regulation of biological processes. *See* CONTROL MECHANISM. **3.** *See* BIOLOGICAL CONTROL; CHEMICAL CONTROL.

control element A segment of noncoding DNA that regulates the expression of a gene in eukaryotic organisms. Control elements bind *transcription factors, either activators or repressors, to increase or decrease the rate of expression respectively. **Proximal control elements** are located close to the promotor site of the gene and are sometimes considered part of the promoter. Other control elements, called *enhancers, may be located many nucleotides distant upstream from the promoter. These interact with the promoter by bending of the DNA molecule.

control mechanism Any mechanism that regulates a biological process, such as a metabolic pathway or enzyme-controlled reaction, or that helps to maintain the *internal environment (*see* HOMEOSTASIS). *See also* FEEDBACK; SIGNAL TRANSDUCTION.

conus arteriosus A small thick-walled chamber of the heart of vertebrate embryos that receives blood from the single ventricle and leads to the ventral aorta. It is retained as a distinct structure in adult fishes and (with modifications) amphibians, but in higher vertebrates it is incorporated into the roots of the aorta and pulmonary arteries.

Convention on Biological Diversity (CBD) An international agreement drawn up by the United Nations Environment Programme for the conservation and sustainable use of the earth's biological resources. It was adopted in 1992, came into force in 1993, and had been adopted by 194 countries by mid-2014. The main objectives are: the conservation of biological diversity; the sustainable use of individual species; and the fair and equitable sharing of the benefits of utilizing the genetic resources of the living world. In 2010 a Strategic Plan for Biodiversity for the period 2011–20 was adopted, centred on the Aichi Biodiversity Targets. These comprise five strategic goals and 20 specific targets for achieving those goals. The principal means of implementing the Convention is the requirement that countries each draw up their own National Biodiversity Strategy and Action Plan (NBSAP); 190 countries had complied by 2018.

convergence 1. (in neuroanatomy) The arrangement in which a single neuron receives inputs from numerous receptors or other neurons. For instance, in the retina of the eye, the output of over 100 million photoreceptors converge, via intermediate cells, on just 1 million ganglion cells, whose axons comprise the optic nerve.

Consequently, convergence entails a funnelling of information from the input cells and integration of their signals into new output messages. **2.** *See* CONVERGENT EVOLUTION. *Compare* DIVERGENCE.

convergent evolution The development of superficially similar structures in unrelated organisms, usually because the organisms live in the same kind of environment. Examples are the wings of insects and birds and the streamlined bodies of whales and fish. *Compare* ADAPTIVE RADIATION; PARALLEL EVOLUTION.

convoluted tubule *See* DISTAL CONVOLUTED TUBULE; PROXIMAL CONVOLUTED TUBULE; NEPHRON.

coomassie blue A biological dye used for staining proteins.

cooperation An association between two or more members of the same species (**intraspecific cooperation**), or between individuals of different species (**interspecific cooperation**), in which all members benefit. An example of interspecific cooperation is the relationship formed between ants and aphids: the aphids gain protection by living in the ant colonies, while the ants feed on secretions from the aphids. Interspecific cooperation is a looser association than *mutualism.

coordinate bond *See* COVALENT BOND.

coordination (in animal physiology) The processes involved in the reception of sensory information, the integration of that information, and the subsequent response of the organism. Coordination is controlled by regions of the brain that deal with specific functions, such as locomotion and breathing, and is carried out by the nervous system.

Copepoda A class of crustaceans occurring in marine and freshwater habitats. Copepods are usually 0.5–2 mm long and lack both a carapace and compound eyes. Copepods are important members of plankton: some are free-living, feeding on microscopic organisms; others are parasitic. A familiar freshwater genus is *Cyclops*, so named because the members have a single median eye.

coprophagy The ingestion of faeces as a means of obtaining nutrients. Coprophagous animals include dung beetles, which eat cow dung, and rabbits, which ingest their own faeces.

CO protein (CONSTANS protein) A transcription factor involved in controlling the timing of flowering in the model plant thale cress (*Arabidopsis thaliana*). It is produced in the leaves where it activates expression of the *FT* (*FLOWERING LOCUS T*) gene encoding the plant hormone *florigen, responsible for initiating flowering. In *A. thaliana*, CO protein specifically works late in the afternoon on days when it is still light, thus ensuring flowering at the appropriate season for the plant. Only at such times is the CO protein stabilized through the action of light-sensitive components, including *phytochrome and *cryptochrome photoreceptors. A similar interplay between CO and florigen is thought to regulate daylength-sensitive onset of flowering in other species.

copulation *See* SEXUAL INTERCOURSE.

copy number The number of copies of a particular region of the genome occurring in a cell or an individual. Deletions or duplications affecting segments of DNA greater than about 1 kilobase cause **copy number variation**. These variations can affect up to 20% of the genome in humans. Some occur in stretches of DNA between genes where they may have little or no impact, whereas others can affect gene function and cause disease. For example, a three-nucleotide sequence is often found repeated numerous times in tandem, forming a series of trinucleotide repeats. Expansion of these near to or within certain genes results in several human diseases, such as Huntington disease and fragile X syndrome. *Compare* SINGLE NUCLEOTIDE POLYMORPHISM.

coral Any of a group of sedentary colonial marine invertebrates belonging to the class Anthozoa of the phylum *Cnidaria. A coral colony consists of individual *polyps within

a protective skeleton that they secrete: this skeleton may be soft and jelly-like, horny, or stony. The horny skeleton secreted by corals of the genus *Corallium*, especially *C. rubrum*, constitutes the red, or precious, coral used as a gemstone. The skeleton of stony, or true, corals consists of almost pure calcium carbonate and forms the coral reefs common in shallow waters of tropical seas. Photosynthesizing *dinomastigotes (dinoflagellates) live within the cells of the reef-forming corals, providing them with up to 80% of their energy. In return, the coral provides its endosymbiotic residents with nutrients and a place to live. Coral formations play a key role in marine ecosystems, e.g. as 'nurseries' for juvenile fish and substrates for shellfish. But the majority of coral reefs are being damaged by various factors, including overfishing and poor fishing practices, *pollution, unregulated tourism, deposition of sediment, coral mining, and *climate change. Corals are very sensitive to rising sea temperature, which leads them to expel the dinomastigotes, causing the corals to lose their colour ('coral bleaching') and impacting their ability to survive. Also, ocean acidification tends to dissolve the coral skeletons.

coreceptor A receptor molecule on the surface of a cell that enhances the activity of another receptor following binding of its extracellular ligand. For example, optimal activation of T cells in an immune response depends on the involvement of the relevant coreceptors in a cluster with the *T-cell receptors. The coreceptors are *CD4 or *CD8 proteins, which lie alongside the T-cell receptor in the plasma membrane. Only when both coreceptor and receptor bind the MHC-antigen complex simultaneously is the full set of intracellular effector molecules recruited and the signal pathway inside the cell maximally activated. Similarly, *B-cell receptors also require the contributions of several coreceptors, notably CD19, CD21, and CD81 proteins, for full activation of intracellular signalling pathways. *See also* CD.

cork (phellem) A protective waterproof plant tissue produced by the *cork cambium. It develops in plants undergoing *secondary growth and replaces the epidermis. Its cells, whose walls are impregnated with *suberin, are arranged in radial rows and fit closely together except where the cork is interrupted by *lenticels. Some cork cells become air-filled while others contain deposits of lignin, tannins, and fatty acids, which give the cork a particular colour. The cork oak (Quercus suber) produces cork that can be used commercially.

cork cambium (phellogen) A type of *cambium arising within the outer layers of the stems of woody plants, usually as a complete ring surrounding the inner tissues. The cells of the cork cambium divide to produce an outer corky tissue (*cork or **phellem**) and an inner secondary cortex (**phelloderm**). Cork, cork cambium, and phelloderm together make up the **periderm**, an impermeable outer layer that protects the inner stem tissues if the outer tissues split as the stem girth increases with age. It thus takes over the functions of the epidermis.

corm An underground organ formed by certain plants, e.g. crocus and gladiolus, that enables them to survive from one growing season to the next (see illustration). It consists of a short swollen food-storing stem surrounded by protective scale leaves. One or more buds in the axils of scale leaves produce new foliage leaves and flowers in the subsequent season, using up the food stored in the stem. *Compare* BULB.

cornea A transparent layer of tissue, continuous with the *sclerotic, that forms the front part of the vertebrate eye, over the iris and lens. The cornea refracts light entering the eye onto the lens, thus assisting in the focusing of images onto the *retina. *See also* ASTIGMATISM.

cornification *See* KERATINIZATION.

corolla The *petals of a flower, collectively, forming the inner whorl of the *perianth. It encircles the stamens and carpels. The form of the corolla is very variable. The

Development of a corm

petals may either be free (**polypetalous**) or united to form a tube (**gamopetalous** or **sympetalous**).

coronary vessels Two pairs of blood vessels (the coronary arteries and coronary veins) that supply the muscles of the heart itself. The coronary arteries arise from the aorta and divide into branches that encircle the heart. A blood clot in a coronary artery (**coronary thrombosis**) is one of the causes of a 'heart attack'.

corpus allatum (*pl.* **corpora allata**) *See* JUVENILE HORMONE.

corpus callosum (*pl.* **corpora callosa**) The sweeping band (commissure) of *white matter that provides a connection between the two halves of the cerebrum in the brain. It enables the transfer of information and learning from one cerebral hemisphere to the other.

corpus cardiacum (*pl.* **corpora cardiaca**) Either of a pair of long slender *neurohaemal organs that lie immediately behind the brain in insects. They contain the endings of neurosecretory cells originating from the brain and also their own neurosecretory cells, in which hormones (e.g. *eclosion hormone) are stored for release into the adjacent blood-filled spaces.

corpus luteum (yellow body) (*pl.* **corpora lutea**) The yellowish mass of tissue that forms from the *granulosa cells in the cavity of a *Graafian follicle in the ovary of a mammal after the release of the egg cell. It secretes the hormone *progesterone. Sharks, amphibians, and reptiles also have corpora lutea that secrete progesterone.

corpus striatum (*pl.* **corpora striata**) *See* BASAL GANGLIA.

correlation In statistics, the degree of linear relationship between two or more *variables. Pairs of observations can be plotted as a series of points on a graph. The **correlation coefficient** measures the extent to which the points on the resulting scatter diagram form a straight line. The correlation coefficient varies within the range of +1 (where an increase of one variable is associated with a corresponding increase in the other, and vice versa) to −1 (where an increase of one variable is associated with a corresponding decrease of the other); a coefficient of 0 indicates no linear relationship between the two variables. The statistical technique known as **multivariate analysis** can be used to investigate the relationships between several variables. *See also* CONFOUNDING.

cortex (*pl.* **cortices**) **1.** (in botany) The tissue between the epidermis and the vascular system in plant stems and roots. It is composed of *parenchyma cells and shows little or no structural differentiation. Cortex is produced by activity of the *apical meristem. *See also* ENDODERMIS. **2.** (in zoology) The outermost layer of tissue of various organs, including the adrenal glands (**adrenal cortex**), kidneys (**renal cortex**), and cerebral hemispheres (*cerebral cortex).

Corti cell Any of the *hair cells found in the *organ of Corti in the inner ear. Corti cells play a role in the detection of sound.

corticosteroid Any of several steroid hormones produced by the cortex of the *adrenal glands. **Glucocorticoids** regulate the use of carbohydrates, proteins, and fats in the body and include *cortisol and *cortisone. **Mineralocorticoids** regulate salt and water balance (*see* ALDOSTERONE).

corticotrophin *See* ACTH.

cortisol (hydrocortisone) A hormone (*see* CORTICOSTEROID), produced by the adrenal glands, that promotes the synthesis and storage of glucose and is therefore important in the normal response to stress, suppresses or prevents inflammation, and regulates deposition of fat in the body. Synthesis and release of cortisol is stimulated by *ACTH (adrenocorticotrophic hormone), which is itself regulated by corticotrophin-releasing hormone. The cortisol response is slower than the more immediate stress response triggered by adrenaline and the nervous system, but it lasts longer. Exposure to chronic stress may reduce the ability to 'turn off' the cortisol response, which involves negative feedback by cortisol on ACTH-secreting cells and on the hippocampus in the brain. Cortisol is used as treatment for various allergies and for rheumatic fever, certain skin conditions, and adrenal failure (Addison's disease).

cortisone A *corticosteroid that is itself biologically inactive and is formed naturally in the adrenal gland from the active hormone *cortisol, which is structurally very similar to it. Cortisone is reconverted to the active hormone by metabolism in the liver and other organs. It may be administered therapeutically as an inactive precursor (prodrug) of cortisol.

corymb A type of flowering shoot (*see* RACEMOSE INFLORESCENCE) in which the lower flower stalks are longer than the higher ones, resulting in a flat-topped cluster of flowers. Examples are candytuft and wallflower.

cosmid A hybrid *vector, used in *gene cloning, that includes the *cos* gene (from the lambda bacteriophage). It also contains drug resistance *marker genes and other plasmid genes. Cosmids can incorporate larger DNA fragments than either phage or plasmid vectors alone, typically 35–45 kbp, and are especially suitable for cloning large mammalian genes or multigene fragments.

cosmoid scale *See* SCALES.

cotransmitter A substance that is released from a nerve ending along with a primary *neurotransmitter in order to modify the action of the latter. For example, vasoactive intestinal peptide (*see* VIP) functions as a cotransmitter with acetylcholine at cholinergic synapses.

cotransporter A type of *transport protein that transports two or more substances simultaneously across a cell membrane. Energy is derived by movement of one of the substances (usually an ion) down its concentration gradient; this drives movement of the other substance against its concentration gradient. In *antiporters the two substances move in opposite directions, whereas in *symporters they move in the same direction.

cotyledon (seed leaf) A part of the embryo in a seed plant. The number of cotyledons is an important feature in classifying plants. Among the flowering plants, the monocots have a single cotyledon and the dicots have two. Conifers have either two cotyledons, as in *Taxus* (yews), or five to ten, as in *Pinus* (pines). In seeds without an *endosperm, e.g. garden pea and broad bean, the cotyledons store food, which is used in germination. In seeds showing *epigeal germination, e.g. runner bean, they emerge above the soil surface and become the first photosynthetic leaves.

coumarin Any of a group of organic compounds found widely in plants and derived from the amino acid phenylalanine. They include coumarin itself, which gives newmown hay its characteristic fragrance, and **umbelliferone**, which is characteristic of members of the carrot family. Under certain conditions, coumarin is converted by fungi to the toxic substance **dicoumarol**. If ingested this causes haemorrhaging and other symptoms of bleeding disorder, by disrupting vitamin K metabolism and preventing the formation of prothrombin and certain other clotting factors by the liver. Dicoumarol and another coumarin derivative, *warfarin, are widely used as rodenticides. Cassia cinnamon, derived from the bark of certain trees, contains relatively high levels of coumarin, which can cause liver damage in a small number of sensitive people.

countercurrent heat exchange A *counterflow mechanism that enables fluids at different temperatures flowing in channels in opposite directions to exchange their heat content without mixing. An example of countercurrent heat exchange occurs in the feet of penguins, in which heat from blood in the arteries supplying the feet is transferred to blood returning to the body's core in veins that lie close to these arteries. This helps to maintain the core temperature in freezing conditions.

countercurrent multiplier system An active (energy-requiring) process occurring in the *loops of Henle in the kidney, which

Countercurrent multiplier system in the kidney

is responsible for the production of concentrated urine in the collecting ducts of the nephrons. Sodium and chloride ions are actively pumped from the thick segment of the ascending limb of the loop but water is retained, since the ascending limb is impermeable to water. This creates a concentration gradient in the medulla in which the concentration of sodium and chloride is greatest in the region of the bend of the loop. Fluid passing from the loop of Henle to the distal tubule is less concentrated than that entering the loop, but because of the high osmotic pressure in the medulla water diffuses out of the collecting ducts, producing a concentrated urine. See illustration.

counterflow The flow of two fluids in apposed vessels in opposite directions. In biological systems such an arrangement enables the efficient transfer of heat, ions, molecules, etc., from fluids that are rich in these resources to fluids that are deficient in them.

courtship Behaviour in animals that plays a part in the initial attraction of a mate or as a prelude to copulation. Courtship involves signals that guide the behaviour of the participants and often takes the form of displays that have evolved through

*ritualization; some are derived from other contexts (e.g. food begging in some birds). Auditory and chemical stimuli (*see* PHEROMONE) are also important in many mammals and insects, especially nocturnal species.

As well as ensuring that the prospective mate is of the same species, the male's courtship performance allows females to choose between different males. The later stages of courtship may involve both partners in an alternating series of displays that inhibit *aggression and fear responses and ensure synchrony of sexual arousal. *See also* SEXUAL SELECTION.

COV *See* CROSSOVER VALUE.

covalent bond A type of *chemical bond in which atoms are held together in a molecule by sharing one or more pairs of electrons in their outer shells. For example, in the water molecule (H_2O) each hydrogen atom forms a covalent bond by sharing its only electron with one of the six electrons in the outer shell of the oxygen atom. **Coordinate** (or **dative**) bonds are covalent bonds in which one of the atoms supplies both the electrons. **Single bonds** are those in which one pair of electrons is shared; in **double** or **triple bonds** two or three pairs, respectively, are shared.

Cowper's glands (bulbourethral glands)
A pair of pea-sized glands that lie beneath the prostate gland, named after the English surgeon William Cowper (1660–1709). Cowper's glands secrete an alkaline fluid that forms part of the *semen. This fluid neutralizes the acidic environment of the urethra, thereby protecting the sperm. *See also* SEMINAL VESICLE.

COX *See* CYCLO-OXYGENASE.

coxa (*pl.* **coxae**) The first (basal) segment, attached to the thorax, of the leg of an insect, arachnid, or of certain other arthropods. *See also* FEMUR; TROCHANTER.

coxal glands Paired ducts (coelomoducts) in arthropods that lead from the *coelom to the exterior and are normally involved in excretion. In arachnids one or two pairs of coxal glands in the cephalothorax open at the bases (coxae) of one or two pairs of legs.

C₃ pathway The metabolic pathway followed in the light-independent phase of *photosynthesis by most plants of temperate regions, in which the first product is the three-carbon compound glycerate 3-phosphate. This is formed when carbon dioxide combines with *ribulose bisphosphate in the first reaction of the *Calvin–Bassham–Benson cycle. Plants that follow this pathway are referred to as **C₃ plants**. *Compare* C₄ PATHWAY.

C₄ pathway (Hatch-Slack pathway) The metabolic pathway followed in the light-independent phase of *photosynthesis by tropical plants, such as sugar cane and maize, and by plants that live in arid environments; these plants are known as **C₄ plants**. The initial fixation of atmospheric carbon dioxide occurs with phosphoenolpyruvate (PEP), through the mediation of the enzyme PEP carboxylase, to form the four-carbon compound oxaloacetic acid (see illustration; *compare* C₃ PATHWAY). Oxaloacetic acid is then converted into malic acid; this breaks down to give carbon dioxide, which passes to the *Calvin-Bassham-Benson cycle, and pyruvic acid, which generates more PEP, thereby ensuring a constant supply of this compound for further carbon dioxide fixation. The C₄ pathway allows photosynthesis to occur at very low concentrations of carbon dioxide as PEP carboxylase has an extremely high affinity for carbon dioxide. This pathway also works well at high temperatures and light intensity, enabling efficient photosynthesis in tropical plants. In addition, the malic acid formed can be stored before being used, to be later broken down to carbon dioxide when required in the Calvin-Bassham-Benson cycle. This is important in desert plants, which need to close their stomata during the day to reduce water loss (*see* CRASSULACEAN ACID METABOLISM). The anatomy of the C₄ plant leaf is adapted to suit this metabolic pathway (*see* BUNDLE SHEATH CELLS).

C₄ pathway

CpG island A stretch of DNA, generally several hundred to several thousand bases long, that is relatively rich in dinucleotides containing the bases cytosine and guanine, with the cytosines (C) located 'upstream' (i.e. 5′) of a guanine (G) residue. The 'p' denotes a phosphodiester bond, meaning that the C and G residues are joined along the same strand of DNA, not paired between complementary strands. An example is 5′-CGCG-3′. CpG sequences are generally about five times less frequent than would be expected throughout the genome—a phenomenon called **CG suppression**-but their abundance is much greater in the promoter region near the start site of many eukaryote genes. Methylation of the (normally unmethylated) cytosine residues in the promoter suppresses transcription of the gene; this *DNA methylation, which is transmitted through subsequent cell divisions, is significant in the process of ageing and, by switching off tumour suppressor genes, the formation of tumours. It is also a feature of heritable suppression of paternal or maternal genes (*see* GENE IMPRINTING). CpG islands outside promoters are hotspots for mutation, because methylcytosine undergoes spontaneous deamination to thymidine, which creates a T-G mismatch in the base pairing. The DNA repair machinery cannot resolve which is the correct template base, T or G, so in some cases makes the repair T-A instead of the correct C-G.

cranial nerves Ten to twelve pairs of nerves in vertebrates that emerge directly from the brain. They supply the sense organs and muscles of the head, neck, and viscera. Examples of cranial nerves include the *optic nerve (II) and the *vagus nerve (X). With the *spinal nerves, the cranial nerves form an important part of the *peripheral nervous system.

cranial reflex See REFLEX.

craniates Chordate animals having a cartilaginous or bony skull protecting the brain and major sense organs. The taxon 'Craniata' formerly comprised two major clades: the hagfishes, and the lampreys and jawed vertebrates. However, evidence now indicates that the hagfishes and lampreys are sister groups, and they are placed in the clade *Cyclostomata. The jawed vertebrates comprise the *Gnathostomata; jawless craniates are called *agnathans.

cranium (brain case) (*pl.* **craniums** or **crania**) The part of the vertebrate *skull that encloses and protects the brain. It is formed by the fusion of several flattened bones, which have immovable joints (sutures) between them.

crassulacean acid metabolism (CAM) Photosynthesis by the *C$_4$ pathway in which carbon dioxide is taken up during the night, when the plant's stomata are open, and fixed into malic acid. During the day, when the stomata are closed, carbon dioxide is released from malic acid for use in the *Calvin–Bassham–Benson cycle. This is important for plants that live in arid conditions as it enables them to keep their stomata closed during the day to reduce water loss from evaporation. Crassulacean acid metabolism is common in succulent plants of desert regions, including cacti and spurges, and in certain ferns. It is so named because it was originally studied in plants belonging to the family Crassulaceae, which includes stonecrops and houseleeks.

C-reactive protein (CRP) An acute-phase protein (*see* ACUTE-PHASE RESPONSE) secreted into blood plasma by the liver in response to inflammation and infection. It comprises five identical subunits and binds to the phosphocholine component of lipopolysaccharides in the cell walls of certain bacteria and fungi. Binding of CRP increases susceptibility of the target cell to ingestion by phagocytes and can also activate the *complement cascade, hence triggering destruction of pathogens by this means. Concentrations of CRP in blood are determined to assess the presence of inflammation, and elevated levels have been linked to increased risk of coronary heart disease. It was originally so named because of its ability to form a precipitate with the C polysaccharide of pneumococcus.

creatine A compound, synthesized from the amino acids arginine, glycine, and methionine, that occurs in muscle. In the form of **creatine phosphate** (or **phospho-creatine**), it is an important short-term reserve of energy for muscle contraction. This energy is released when creatine phosphate transfers its phosphate to ADP to replenish supplies of ATP, and is itself converted to **creatinine**, which is excreted in the urine (at a rate of 1.2–1.5 g/day in humans). *See also* PHOSPHAGEN.

creatinine *See* CREATINE.

creationist A proponent of the theory of *special creation.

Cre/loxP recombination A technique for making precise rearrangements in the mouse genome, such as removing a segment of a chromosome or inserting a gene at a particular location. Further, it enables such gene *knockouts or *knockins to be localized to specific tissues or developmental stages. Hence it is a valuable tool in investigations of mammalian gene function and is also used in zebrafish. The Cre protein is a recombinase enzyme originally isolated from the P1 bacteriophage, in which it resolves tangles in the virus's circular DNA. It catalyses site-specific recombination by cutting out DNA flanked by two homologous sequences called *loxP* sites (i.e. 'floxed' DNA) and joining the cut ends. Recombinant mice are bred so that they have one copy of the target site flanked by *loxP* sites, and a copy of the Cre gene linked to a cell-type-specific promoter. Only cells in which the promoter is active, for example in response to a hormonal signal, express the Cre gene and synthesize the Cre protein, which duly excises the target site, thereby causing loss of the corresponding function. Thus the knockout can be confined to those tissues receiving the activating signal, enabling more detailed resolution of a particular gene's function and the effects of abolishing it. *See also* GENOME EDITING.

cremocarp A dry fruit that is a type of *schizocarp formed from two one-seeded

carpels. The carpels remain separate and form indehiscent **mericarps** that are attached to a central supporting strand (**carpophore**) for some time before dispersal. It is characteristic of the Umbelliferae (Apiaceae; carrot family).

crenation The shrinkage of cells that occurs when the surrounding solution is *hypertonic to the cellular cytoplasm. Water leaves the cells by *osmosis, which causes the plasma membrane to wrinkle and the cellular contents to condense.

Cretaceous The final geological period of the Mesozoic era. It extended from about 145 million years ago, following the Jurassic, to about 66 million years ago, when it was succeeded by the Palaeogene period. The name of the period is derived from *creta* (Latin: chalk) and the Cretaceous was characterized by the deposition of large amounts of *chalk in western Europe. The Cretaceous was the time of greatest flooding in the Mesozoic. Angiosperm plants made their first appearance on land and in the early Cretaceous Mesozoic reptiles reached their peak. At the end of the period there was a *mass extinction of the dinosaurs, flying reptiles, and ammonites, the cause of which may be related to environmental changes resulting from collisions of the earth with large meteorites (*see* ALVAREZ EVENT; IRIDIUM ANOMALY).

(((●))) SEE WEB LINKS

• Overview of the Cretaceous period: part of the website of the University of California Museum of Paleontology

cretinism The condition that results from inadequate secretion of thyroid hormones during fetal life or early infancy. The brain and skeleton fail to develop properly, resulting in mental retardation and dwarfism.

Creutzfeldt-Jakob disease (CJD) A disease of humans characterized by dementia and destruction of brain tissue. It is now known to be caused by an abnormal *prion protein and is transmissible, although there is also an inherited familial form. This rare disease typically affects middle-aged and elderly people and leads to rapid mental deterioration and death. The abnormal prion interferes with the structure of normal prion protein in brain tissue, resulting in accumulations of the protein and consequent tissue damage. In most cases the source of infection is unknown. However, it is well established that infection can result, for example, via injections of growth hormone derived from infected human cadavers. During the 1990s a novel form of the disease emerged, called 'variant CJD', which typically affects young healthy individuals. This is thought to be caused by consumption of beef products derived from cattle infected with *bovine spongiform encephalopathy or transfusion of blood from infected donors.

Crick, Francis Harry Compton (1916–2004) British molecular biologist, who in 1951 teamed up with James *Watson at Cambridge University to try to find the structure of *DNA. This they achieved in 1953, using the X-ray diffraction data of Rosalind Franklin (1920–58) and Maurice Wilkins (1916–2004). Crick went on to investigate *codons and the role of transfer *RNA. Crick, Watson, and Wilkins shared a Nobel Prize in 1962.

CRISPR-Cas9 A technique for editing DNA in living cells that can block, repair, or modify specific genes. It is based on a system found in prokaryotes (both bacteria and archaea) that involves a genetic locus called CRISPR (clustered regularly interspaced short palindromic repeats) and CRISPR-associated protein 9 (Cas9). These components function to defend the prokaryote cell against infection with bacteriophages. The CRISPR sequences encode small RNA molecules, called CRISPR RNAs (crRNAs), which recognize and basepair with complementary sequences in incoming viral DNA or RNA. This specific pairing guides CRISPR-associated protein 9 (Cas9) to cleave the virus DNA, thus protecting the cell from further viral infection. This system has been adapted so that the CRISPR-Cas9 mechanism can precisely locate and cut double-stranded DNA in eukaryotes, including humans, enabling target sequences to be cut out and new ones inserted at specific sites. This makes it a valuable tool for *genome editing, e.g. in correcting mutations during gene therapy. A guide crRNA sequence is devised based on knowledge of the target site, and its complementary DNA is incorporated into a plasmid vector that also includes sequences encoding a Cas9 protein plus a reporter gene. Thus, when the vector is introduced to a cell, the crRNA and Cas9 protein are expressed and perform the required editing.

crista (*pl.* **cristae**) **1.** (in anatomy) A crest, e.g. the crista nasalis (nasal crest). **2.** *See* SEMICIRCULAR CANALS. **3.** *See* MITOCHONDRION.

critical group A large group of related organisms that, although variations exist between them, cannot be divided into smaller groups of equivalent taxonomic rank to the parent group. Critical groups are found among plants that reproduce by *apomixis; for example, the 400 or so species of *Rubus* (brambles, etc.) are regarded as a critical group.

critical period (in developmental biology) A phase in the life of an organism during which it has heightened receptiveness to external stimuli that enable it to develop a certain behaviour or learn a certain skill. An example is the period soon after birth when offspring learn the identity of their parents. *See* IMPRINTING.

critical thermal maximum *See* UPPER CRITICAL TEMPERATURE.

'Cro-Magnon man' The name given to fossil hominins found at the cave site of Cro-Magnon, near the village of Les Eyzies in the Dordogne region of France. The remains, discovered in 1868, include the skeletons of four adults and a child and are representative of what are now termed Early Modern Humans. The Cro-Magnon fossils date from about 30000 years ago, but similar modern human fossils date back at least 42000 years in Europe. They display the characteristic traits of modern humans, including a tall rounded skull with a near vertical forehead and lack of a prominent brow ridge. Genetic evidence indicates that these European ancestors lived alongside and to some extent interbred with *Neanderthals until the latter went extinct around 30000 years ago.

crop 1. A plant that is cultivated for the purpose of harvesting its seeds, roots, leaves, or other parts that are useful to humans. *See* AGRICULTURE. **2.** An enlarged portion of the anterior section of the alimentary canal in some animals, in which food may be stored and/or undergo preliminary digestion. The term is most commonly applied to the thin-walled sac in birds between the oesophagus and the *proventriculus. In female pigeons the crop contains glands that secrete **crop milk**, used to feed nestlings.

crop rotation An agricultural practice in which different crops are cultivated in succession on the same area of land over a period of time so as to maintain soil fertility and reduce the adverse effects of pests. Legumes are important in the rotation as they are a source of nitrogen for the soil (*see* NITROGEN FIXATION; ROOT NODULE). In the UK, other crops that may be included in a typical four-stage rotation are wheat, barley, and root crops. However, the use of pesticides enables the monoculture of crops in modern farming systems (*see* AGRICULTURE).

cross 1. A mating between two selected individuals. Controlled crosses are made for many reasons, e.g. to investigate the inheritance of a particular characteristic or to improve a livestock or crop variety. *See also* BACK CROSS; RECIPROCAL CROSS; TEST CROSS. **2.** An organism resulting from such a mating.

cross-fertilization *See* FERTILIZATION.

crossing over An exchange of portions of chromatids between *homologous chromosomes. As the chromosomes begin to move apart at the end of the first prophase of *meiosis, they remain in contact at a number of

Crossing over at two chiasmata in a pair of homologous chromosomes

points (*see* CHIASMA). At these points the chromatids break and rejoin in such a way that sections are exchanged (see illustration). Crossing over thus alters the pattern of genes in the chromosomes. *See* RECOMBINATION.

crossover value (COV) The percentage of linked genes (*see* LINKAGE) that are exchanged during the process of *crossing over during the first prophase of *meiosis. The COV can be calculated by the percentage of offspring that show *recombination and is used to map the genes on a chromosome (*see* CHROMOSOME MAP). A small COV for a given pair of genes indicates that the genes are situated close together on the chromosome.

cross-pollination *See* POLLINATION.

CRP *See* C-REACTIVE PROTEIN.

Crustacea In traditional classifications, a subphylum of *arthropods containing over 35000 species distributed worldwide, mainly in freshwater and marine habitats, where they constitute a major component of plankton. Crustaceans include shrimps, crabs, lobsters, etc. (*see* DECAPODA) and the terrestrial woodlice, all of which belong to the class Malacostraca; the barnacles (class Cirripedia); the water fleas (*see* DAPHNIA), fairy shrimps and tadpole shrimps (*see* BRANCHIOPODA); and the copepods (*see* COPEPODA). The segmented body usually has a distinct head (bearing *compound eyes, two pairs of *antennae, and various mouthparts), thorax, and abdomen, and is protected by a shell-like carapace. Each body segment may bear a pair of *biramous appendages used for locomotion, as gills, and for filtering food particles from the water. Appendages in the head region are modified to form jaws and in the abdominal region are often reduced or absent. Typically, the eggs hatch to produce a free-swimming *nauplius larva. This develops either by a series of moults or undergoes metamorphosis to the adult form. Genetic and other molecular evidence now shows that the insects evolved from crustacean ancestors and that certain groups of crustaceans are more closely related to insects than to other crustaceans. Hence, although the evolutionary relationships of many crustacean lineages are still debatable, both groups are now placed in the clade Pancrustacea (*see* ARTHROPOD).

((()) SEE WEB LINKS

• Overview of crustacean phylogeny at the Tree of Life web project

cryobiology The study of the effects of very low temperatures on organisms, tissues, and cells. The ability of some animal tissues to remain viable in a frozen state enables them to be preserved by freezing for future use as *grafts. *See* CRYOPROTECTANT.

Cryogenian A geological period in the *Proterozoic eon that began about 720 million years ago and ended 635 million years ago. It was characterized by a very harsh climate and extensive glaciation, which may have covered the entire planet (the 'snowball Earth' theory). Life consisted of single-celled microorganisms for much of this period, but in the later stages is thought to have undergone rapid evolution, with the emergence of multicellular life forms following melting of the glaciers and the consequent upheaval in the climate.

cryophyte An organism that can live in ice and snow. Most cryophytes are algae, including the green alga *Chlamydomonas nivalis* and some diatoms, but they also include dinomastigotes, certain mosses, bacteria, and fungi.

cryoprotectant A substance that helps to protect an organism's tissues from the effects of freezing. Cryoprotectants are vital to the survival of those organisms living in cold climates that adopt a strategy of tolerating a limited amount of freezing of their tissue fluids. Such organisms include certain insects, molluscs, and nematodes, and even some frogs, lizards, and turtles. One group of these substances consists of *antifreeze molecules, notably glycerol and related polyhydric alcohols (polyols). Another group of cryoprotectants bind to membranes within cells and prevent the binding of water molecules, so preserving the integrity of cell structures. Examples in insects include proline and trehalose. *See also* ICE-NUCLEATING AGENT.

cryptic coloration The type of colouring or marking of an animal that helps to camouflage it in its natural environment. It may enable the animal to blend with its background or, like the stripes of zebras and tigers, help to break up the outline of its body.

cryptic species *See* SIBLING SPECIES.

cryptobiosis (anhydrobiosis) A state of apparent suspended animation entered by certain invertebrate animals in order to survive desiccation or other extreme stresses. It is best documented among the rotifers, nematodes, collembolans, tardigrades, and other minute inhabitants of mosses and lichens, where the water film essential for active life is transient and sporadic. When the film dries out these animals appear to be dead for periods of days, weeks, or even years until moisture returns, when they 'come back to life' and resume their normal activities. Entering cryptobiosis involves various processes. The animal typically retracts its legs and other appendages, or curls up into a ball to minimize its surface area. Biochemical changes in the cuticle or the secretion of wax ensure that at least some water is retained, although this may be only some 5% of the normal content, and the body becomes contracted and shrivelled. Sugars, such as trehalose, manufactured by the body cells protect the integrity of the plasma membranes and also convert the cytoplasm to a glasslike state. When reverting to its normal state the animal absorbs water, swells, and becomes active in a few hours.

cryptochrome Any of a class of proteins that are sensitive to blue and ultraviolet light and are found in a wide range of organisms, including plants, insects, birds, and mammals. Cryptochromes are bound to FAD, which acts as a *chromophore, and are involved in a variety of light-dependent signalling pathways. They are structurally similar to DNA lyases, enzymes that help to repair light-induced DNA damage in cells. In plants, cryptochromes regulate light-induced aspects of growth and development, such as hypocotyl elongation and onset of flowering. In animals, cryptochromes occur in the retina of the eye and act to entrain circadian rhythms with the light–dark cycle. The activity of cryptochromes is influenced by magnetic fields, and it has been suggested that cryptochromes might form part of an internal magnetic compass for migrating birds and other animals.

Cryptomonada (Cryptophyta) A phylum of unicellular eukaryotic organisms with flattened ovoid cells and two flagella arising from an obliquely situated gullet (**crypt**). Cryptomonads include freshwater and marine algae, which contain chlorophyll and *phycobiliprotein photosynthetic pigments; and heterotrophic protozoa, which may be carnivorous or parasitic (in animal intestines). They are placed in the eukaryotic assemblage called *stramenopiles.

Cryptomycota A phylum of single-celled fungi that live as parasites inside minute aquatic organisms such as diatoms and water moulds. Because of the microscopic and obscure nature of cryptomycotes, their diversity and widespread distribution have emerged only by comparing DNA samples from aquatic microbes with those of known fungal groups and finding the presence of a characteristic fungal 'signature'. Like the related *microsporidia, members of the Cryptomycota lack cell walls during their intracellular feeding stage and have flagellated spores, but they can produce resistant spores with chitin-containing walls.

cryptonephridial system An arrangement of the excretory system in certain beetles and lepidopteran larvae in which the distal blind endings of the *Malpighian tubules are held against the wall of the rectum by a perinephric membrane. A high concentration of ions is established in the tubules, which draws water from the gut by osmosis. This creates a highly efficient mechanism for water conservation in species, such as mealworms (e.g. *Tenebrio* spp.), that feed on very dry materials.

cryptozoic Describing animals that live mainly within soil, litter, or detritus, being rarely seen in the 'open air'. Most are invertebrates, such as earthworms, woodlice, centipedes, and numerous insect larvae, although the term may be extended to include burrowing or hole-dwelling vertebrates, for example some lizards, snakes, and rodents.

crypts of Lieberkühn (intestinal glands) Tubular glands that lie between the finger-like projections (*see* VILLUS) of the inner surface of the small intestine. The cells of these glands (called **Paneth cells**) secrete components of the *succus entericus (intestinal juice) as they gradually migrate along the side of the crypt and the villus; they are eventually shed into the lumen of the intestine. The glands are named after German anatomist J. N. Lieberkühn (1711–56).

crystallin Any of the principal structural proteins of the lens of the vertebrate eye. They are responsible for maintaining the transparency and structural organization essential for the refractive properties of the lens. Minor changes in crystallin structure can lead to cataract or blindness. The term 'crystallin' is also applied to certain transparent proteins found in the cornea.

CSF 1. *See* CEREBROSPINAL FLUID. **2.** *See* COLONY-STIMULATING FACTOR.

CSR strategies A concept, devised originally by British plant ecologist J. Philip Grime, in which plants are classified according to their life-history characteristics as competitive (C), stress-tolerant (S), or *ruderal (R), or some combination of these. It is based on the proposition that habitats are affected by two principal external factors: the level of disturbance (e.g. grazing, disease, trampling) and the degree of stress (e.g. drought, nutrient deficiencies, fire, shading). Plants that are S-strategists tend to flourish under conditions of severe stress but little disturbance; R-strategists are favoured by high levels of disturbance coupled with plentiful resources; and competitors thrive under the crowded conditions that occur when resources are abundant and disturbance is rare. The concept illuminates the analysis of plant communities and helps in predicting their responses to change. It is often depicted as a **CSR triangle**, with the three extreme strategies occupying the angles and intermediate forms (e.g. SR, CS) arrayed within the triangle. *Compare* LEAF-HEIGHT-SEED SCHEME.

ctenidia (*sing.* **ctenidium**) The gills of aquatic molluscs, which are present on both sides of the mantle cavity and are held in place by specialized membranes. The gills are involved in both filter feeding and the exchange of respiratory gases.

Ctenophora A phylum of marine invertebrates that contains the comb jellies (e.g. *Pleurobrachia*) and sea gooseberries. They typically have spherical or oval bodies, although some are ribbon shaped or are equipped with tentacles. The body wall encloses a gastrovascular cavity that communicates with the exterior via a mouth and several anal pores. They possess tentacles armed with *lasso cells, for catching prey, and many hundreds of thousands of cilia, which are fused at their bases and grouped together into longitudinal rows (**comb plates** or **ctenes**). The beating of the cilia enables these animals to swim among the plankton. The ctenophores

were formerly believed to be close relatives of the *cnidaria, but molecular systematics now places them as the sister group of all other animals apart from sponges.

CT scanner (computerized tomography scanner) *See* TOMOGRAPHY.

cuboidal epithelium *See* EPITHELIUM.

cultivar A plant that has been developed and maintained by cultivation as a result of agricultural or horticultural practices. The term is derived from *culti*vated *vari*ety.

cultivation The planting and breeding of crop plants in *agriculture and horticulture. It involves the investigation of new means of increasing crop yield and quality.

culture A batch of cells, which can be microorganisms or of animal or plant origin, that are grown under specific conditions of nutrient levels, temperature, pH, oxygen levels, osmotic factors, light, pressure, and water content. Cultures of cells are prepared in the laboratory for a wide spectrum of scientific research. A *culture medium provides the appropriate conditions for growth. *See also* BATCH CULTURE; CONTINUOUS CULTURE; MONOLAYER CULTURE; ORGAN CULTURE; SUSPENSION CULTURE; TISSUE CULTURE.

culture medium A nutrient material, either solid or liquid, used to support the growth and reproduction of microorganisms or to maintain tissue or organ cultures. *See also* AGAR.

cupula (*pl.* **cupulae**) *See* AMPULLA.

cupule 1. A hard or membranous cup-shaped structure formed from bracts and enclosing various fruits, such as the hazelnut and acorn. **2.** A structure in club mosses (*Lycopodium* species) that protects the gemma (resting bud) during its development. It is composed of six leaflike structures. **3.** The bright red tissue around the seed of yew (*Taxus*), forming the yew 'berry'.

curare A resin obtained from the bark of South American trees of the genera *Strychnos* and *Chondrodendron* that causes paralysis of skeletal muscle. It acts by blocking the action of the neurotransmitter *acetylcholine at *neuromuscular junctions. Curare is used as an arrow poison by some indigenous people of South America and was formerly used as a muscle relaxant in surgery.

curd The solid component produced by the coagulation of milk during the manufacture of cheese. After being pasteurized, milk is cooled down and a culture of lactic acid bacteria is added to ferment the milk sugar, lactose, to lactic acid. The resulting decrease in pH causes casein, a milk protein, to coagulate, a process known as **curdling**. The solid curds are then separated from the liquid component, known as **whey**, and inoculated with different types of microbes to produce different cheeses.

cusp 1. A sharp raised protuberance on the surface of a *molar tooth. The cusps of opposing molars (i.e. on opposite jaws) are complementary to each other, which increases the efficiency of grinding food during chewing. **2.** A flap forming part of a *valve.

cuticle 1. (in botany) The continuous waxy layer that covers the aerial parts of a plant. Composed mainly of *cutin, it is secreted by the *epidermis and its primary function is to prevent water loss. **2.** (in zoology) A layer of horny noncellular material covering, and secreted by, the epidermis of many invertebrates. It is usually made of a collagen-like protein or of *chitin and its main function is protection. In arthropods it is also strong enough to act as a skeleton (*see* EXOSKELETON) and in insects it reduces water loss. Growth is allowed by moulting of the cuticle (*see* ECDYSIS).

cuticularization The secretion by the outer (epidermal) layer of cells of plants and many invertebrates of substances that then harden to form a *cuticle.

cutin A polymer of long-chain fatty acids that forms the main constituent of the *cuticle of epidermal plant cells. The cutin polymers are cross-linked forming a network, which is embedded in a matrix of waxes. The deposition of cutin (**cutinization**) reduces water loss by the plant and helps prevent the entry of pathogens. *See also* SUBERIN.

cutinization The deposition of *cutin in plant cell walls, principally in the outermost layers of leaves and young stems.

cutis *See* DERMIS.

cutting A part of a plant, such as a bud, leaf, or a portion of a root or shoot, that, when detached from the plant and inserted in soil, can take root and give rise to a new daughter plant. It depends on the formation of undifferentiated *callus at the detachment site, which can produce adventitious roots under the appropriate culture conditions. Taking or striking cuttings is a horticultural method for propagating plants. *See also* VEGETATIVE PROPAGATION.

cyanelle *See* GLAUCOPHYTA.

Cyanobacteria A phylum consisting of two groups of photosynthetic bacteria: the blue-green bacteria (formerly known as blue-green algae, or Cyanophyta), which comprise the vast majority of members, and the grass-green bacteria, or *chloroxybacteria. Both groups obtain their food by photosynthesis in a manner very similar to that of green plants and true algae, producing oxygen in the process. All the blue-green bacteria contain the photosynthetic pigment chlorophyll *a*, plus accessory pigments called *phycobiliproteins. The blue colour is caused by one class of these pigments, the phycocyanins; some also have red pigments (phycoerythrins). Blue-green bacteria are unicellular but sometimes become joined in colonies or filaments by a sheath of mucilage. They occur in all aquatic habitats. A few species fix atmospheric nitrogen and thus contribute to soil fertility (*see* HETEROCYST; NITROGEN FIXATION). Others exhibit symbiosis (*see* LICHENS). The chloroxybacteria have been found in marine and freshwater habitats. They differ from the blue-green bacteria in containing chlorophyll *a* and chlorophyll *b*, but no phycobiliproteins—a combination like that found in plant chloroplasts. Fossil remains of cyanobacteria called *stromato-lites date back 3500 million years and are among the earliest evidence of life on earth. According to the *endosymbiont theory, eukaryotic cells originally obtained their photosynthetic organelles by engulfing cyanobacteria.

cyanocobalamin See VITAMIN B COMPLEX.

Cycadofilicales (Pteridospermales; seed ferns) An extinct order of gymnosperms that flourished in the Carboniferous period. They possessed characteristics of both the ferns and the seed plants in reproducing by means of seeds and yet retaining fernlike leaves. Their internal anatomy combined both fern and seed-plant characteristics.

Cycadophyta A phylum of seed plants (see GYMNOSPERM) that contains many extinct species; the few modern representatives of the group include *Cycas* and *Zamia*. Cycads inhabit tropical and sub-tropical regions, sometimes growing to a height of 20 m. The stem bears a crown of fernlike leaves. These species are among the most primitive of living seed plants.

cyclic AMP (cAMP; cyclic adenosine monophosphate) A derivative of *ATP that is widespread in cells as a *second messenger in many biochemical reactions induced by hormones. Binding of the hormone to its receptor on the cell surface activates *G proteins, which in turn activate (or in some cases inhibit) *adenylate cyclase, the enzyme that catalyses the formation of cAMP. cAMP removes a regulatory subunit from the enzyme *protein kinase A, thereby enabling it to phosphorylate (and hence activate) intracellular proteins that mediate the ultimate effects of the hormone on the cell. The stimulatory signal wanes as cAMP is degraded to AMP by a phosphodiesterase, and phosphorylated proteins are dephosphorylated and inactivated. Cyclic AMP is also involved in controlling gene expression

and cell division, in immune responses, and in nervous transmission.

cyclic GMP (cGMP) Cyclic guanosine monophosphate: a derivative of the nucleotide guanosine triphosphate (GTP) that, like *cyclic AMP, acts as a *second messenger in signalling pathways within cells. Formed from GTP by the enzyme **guanylate cyclase**, cyclic GMP activates *protein kinase G, which in turn activates specific intracellular proteins by *phosphorylation. Cyclic GMP is found in most animal cells; one of its key roles is in rod cells of the retina, where it regulates light-dependent opening and closing of ion channels in the rod-cell plasma membrane (see TRANSDUCIN).

cyclic phosphorylation (cyclic photophosphorylation) See PHOTOPHOSPHORYLATION; PHOTOSYSTEMS I AND II.

cyclin Any of a family of proteins that help control the various phases of the *cell cycle. Their concentrations fluctuate in step with the cycle, providing the cues for a progression through various checkpoints in the cycle, from mitosis to the G_1, S, and G_2 phases. They act in conjunction with cyclin-dependent kinases (CDKs), which are proteins that phosphorylate other proteins. For example, in all eukaryotes mitosis (M phase) is initiated by high levels of cyclin B, which combines with a CDK to form the *mitosis-promoting factor (MPF). By the end of the M phase, cyclin B is at a very low concentration; thereafter it rises steadily again to peak just before M phase.

cyclomorphosis (seasonal polyphenism) The occurrence of seasonal changes in the phenotype of an organism through successive generations. It occurs in small aquatic invertebrates that reproduce by parthenogenesis and give rise to several generations annually, such as rotifers and cladoceran crustaceans. Cyclomorphic species of *Daphnia*, for example, undergo changes in the shape of the head during the year; it is rounded from about midsummer to spring, and thereafter becomes helmet-shaped, reverting to the rounded shape at midsummer. Also, summer generations tend to be smaller and more transparent than at other times. Such changes are thought to be caused by the interaction of environmental cues with the organism's genes, thereby altering the course of development. The modifications are associated with improved survival of the organism, for instance by reducing the likelihood of predation.

cyclo-oxygenase (COX; prostaglandin-endoperoxide synthase) An enzyme that catalyses the key reactions in the formation of *prostaglandins from arachidonic acid, linolenic acid, and eicosa-5,8,11,14,17-pentaenoic acid. Nonsteroidal anti-inflammatory drugs, such as aspirin and ibuprofen, inhibit the cyclo-oxygenase enzymes COX1 and COX2, thus blocking the synthesis of prostaglandins and related compounds, such as *prostacyclins and *thromboxane A_2, and relieving the symptoms of fever, inflammation, and pain associated with them. Inhibition of COX1 by aspirin is the basis for its antithrombotic (blood-thinning) properties.

cyclosis *See* CYTOPLASMIC STREAMING.

Cyclostomata A clade of jawless vertebrates—the 'circle mouths'—comprising the hagfishes and lampreys. Hagfishes (class Myxini) have long eel-like bodies lacking fins and a lower jaw. They scavenge for food on the seabed, often feeding on dead fish and whales. The notochord is retained into adulthood, and they have a cartilaginous skull and only vestigial vertebrae; they are deemed to have secondarily lost such characteristic vertebrate features. When threatened, hagfishes secrete copious amounts of slime as a means of defence. Lampreys (class Petromyzontida) are wormlike animals that also lack jaws but possess cartilaginous projections, resembling vertebrae, from the notochord in adult life. Some species feed as parasites on fish, using their toothed circular mouth as a sucker to bore into their host and suck its blood. Free-living lampreys feed only during the larval stage (*see* AMMOCOETE) and die shortly after achieving sexual maturity and breeding. *See* CRANIATES.

cyme *See* CYMOSE INFLORESCENCE.

cymose inflorescence (cyme; definite inflorescence) A type of flowering shoot (*see* INFLORESCENCE) in which the first-formed flower develops from the growing region at the top of the flower stalk (see illustration). Thus no new flower buds can be produced at the tip and other flowers are produced from lateral buds beneath. In a **monochasial cyme** (or **monochasium**), the development of the flower at the tip is followed by a new flower axis growing from a single lateral bud. Subsequent new flowers may develop from the same side of the lateral shoots, as in the buttercup, or alternately on opposite sides, as in forget-me-not. In a **dichasial cyme** (or **dichasium**), the development of the flower at the apex is followed by two new flower axes developing from buds opposite one another, as in plants of the

monochasial cymes   dichasial cyme

1 = oldest flower

Types of cymose inflorescence

family Caryophyllaceae (such as stitchwort). *Compare* RACEMOSE INFLORESCENCE.

cypsela (*pl.* **cypselae**) A dry single-seeded fruit that does not split open during seed dispersal and is formed from a double ovary in which only one ovule develops into a seed. It is similar to an *achene and characteristic of members of the family Compositae (Asteraceae), such as the dandelion. *See also* PAPPUS.

cysteine *See* AMINO ACID.

cysticercus (*pl.* **cysticerci**) *See* BLADDER-WORM.

cystine A molecule resulting from the oxidation reaction between the sulphydryl (-SH) groups of two cysteine molecules (*see* AMINO ACID). This often occurs between adjacent cysteine residues in polypeptides. The resultant *disulphide bridges (-S-S-) are important in stabilizing the structure of protein molecules.

cystocarp *See* CARPOGONIUM.

cytidine A nucleoside comprising one cytosine molecule linked to a D-ribose sugar molecule. The derived nucleotides, cytidine mono-, di-, and triphosphate (CMP, CDP, and CTP respectively), participate in various biochemical reactions, notably in phospholipid synthesis.

cytochrome Any of a group of proteins, each with an iron-containing *haem group, that form part of the *electron transport chain in mitochondria and chloroplasts. Electrons are transferred by reversible changes in the iron atom between the reduced Fe(II) and oxidized Fe(III) states. There are four subclasses, designated a to d, with different chemical configurations of the haem group. Members of each group can be denoted by a suffixed number (e.g. cytochrome b_2) or by the characteristic wavelength (in nanometres) at which they maximally absorb light (e.g. cytochrome c_{555}). *See also* CYTOCHROME OXIDASE.

cytochrome oxidase An enzyme complex comprising the terminal two cytochromes (cytochromes a and a_3) of the respiratory chain in the mitochondria (*see* ELECTRON TRANSPORT CHAIN). It is responsible for the reduction of oxygen, receiving two electrons from cytochrome c (which precedes it in the chain) and combining them with two hydrogen ions and an oxygen atom to form water.

cytogenetics The study of inheritance in relation to the structure and function of cells, in particular changes in the number, structure, and behaviour of the cell's chromosomes. The microscopic observation of chromosomes provides insights into the causes and inheritance of many diseases and can be diagnostic in certain cases. This entails various techniques to obtain a *physical map of the chromosomes showing their characteristic banding patterns or other markers.

cytokine Any of numerous small proteins released from a variety of cell types that affect cell behaviour. Cytokines can influence the cell releasing the cytokine (i.e. autocrine activity), or nearby cells (paracrine activity); in some cases they can enter the bloodstream to influence distant cells (i.e. endocrine activity). Hence, cytokines often behave in a similar manner to growth factors and hormones. Cytokines are crucial to many aspects of cell proliferation, differentiation, migration, and function, and play a central role in immune responses and inflammation. Cytokines produced by lymphocytes are termed **lymphokines**, and those affecting cell migration, particularly of immune cells, form a large group of *chemokines. Cytokines can be classified in several ways, e.g. according to their structure, function, or the type of cell receptor to which they bind when exerting their effect. The main groups include the haemopoietins, some of the interleukins, erythropoietin, *colony-stimulating factors, the tumour necrosis factor (TNF) family, and *interferons. *See also* TRANSFORMING GROWTH FACTOR BETA.

cytokinesis *See* CELL CYCLE; CELL DIVISION.

cytokinin (kinin) Any of a group of *plant hormones chemically related to the purine adenine. Cytokinins are involved in numerous aspects of plant metabolism.

They stimulate cell division in the presence of *auxin and have also been found to delay senescence, overcome *apical dominance, and promote cell expansion. They are produced by growing roots in spring and translocated to the stem, where they antagonize the effects of auxin and strigolactone to stimulate growth of buds. The endosperm of seeds contains high concentrations of cytokinins, which are thought to be involved in development of the embryo. Zeatin is a naturally occurring cytokinin. Carefully formulated mixtures of cytokinins and *auxins are used in *micropropagation to generate cloned plantlets from undifferentiated callus tissue; the auxin:cytokinin ratio controls whether the callus develops shoots or roots, and also the extent of branching in a plant.

cytological map See PHYSICAL MAP.

cytology The study of the structure and function of cells. The development of the light and electron microscopes has enabled the detailed structure of the nucleus (including the chromosomes) and other organelles to be elucidated. Microscopic examination of cells, either live or as stained sections on a slide, is also used in the detection and diagnosis of various diseases, especially *cancer. Fluorescent markers are now commonly used to distinguish particular cellular components (see FLUORESCENT PROTEIN).

cytolysis The breakdown of cells, usually as a result of destruction or dissolution of their outer membranes.

cytomegalovirus A virus belonging to the herpes group (see HERPESVIRUS). In humans it normally causes symptoms that are milder than the common cold, but it can produce more serious symptoms in those whose immune response is disturbed (e.g. HIV patients, cancer patients). Infection in pregnant women may cause congenital handicap in their children.

cytoplasm The contents of a *cell apart from the nucleus. It consists of a matrix (see CYTOSOL) in which the cell's organelles are suspended. The cytoplasm may be differentiated into dense outer **ectoplasm**, which is concerned primarily with cell movement, and less dense **endoplasm**, which contains most of the cell's structures. The cytoplasm of a prokaryotic cell (i.e. of bacteria or archaea) lacks organelles but nonetheless has distinct functional regions.

cytoplasmic inheritance The inheritance of genes contained in the cytoplasm of a cell, rather than the nucleus. Only a very small number of genes are inherited in this way. The phenomenon occurs because certain organelles, the *mitochondria and (in plants) the *chloroplasts, contain their own genes and can reproduce independently. The female reproductive cell (the egg) has a large amount of cytoplasm containing many such organelles, which are consequently incorporated into the cytoplasm of all the cells of the embryo. The male reproductive cells (sperm or pollen), however, consist almost solely of a nucleus, and any small organelles, e.g. mitochondria or mitochondrial DNA, are eliminated from the fertilized egg. Cytoplasmic organelles are thus not inherited from the male parent, and the DNA in plastids and mitochondria follows a pattern of strictly **maternal inheritance**, rather than Mendelian inheritance. In plants, male sterility can be inherited via the cytoplasm.

cytoplasmic segregation The unequal distribution of certain materials—**cytoplasmic determinants**—in the cytoplasm of a fertilized egg cell (zygote), early embryo cell, or stem cell. Following cell division, the daughter cells receive varying amounts of those determinants, which influence their subsequent development. The determinants include mRNAs and regulatory proteins, which activate certain genes and so influence the developmental fate of the recipient cell(s). Cytoplasmic segregation specifies polarity of the zygote, i.e. which end is 'top' and which is 'bottom', and this fundamentally influences future embryonic development, as the cells divide repeatedly to form tissues and structures appropriate for the 'head' or 'tail'. See also ANIMAL POLE.

cytoplasmic streaming (cyclosis) The directional movement of cytoplasm in certain cells, which allows movement of substances through the cell, especially around the cell's periphery. It has been observed most clearly in large cells, such as plant sieve elements and unicellular algae, in which simple diffusion is ineffective as a means of local transport in the cell. The mechanism of streaming involves the interaction of myosin motor proteins (attached to organelles) with actin *microfilaments parallel to the direction of flow and requires energy from ATP. A similar streaming of cytoplasm is responsible for *amoeboid movement.

cytosine A *pyrimidine derivative. It is one of the principal component bases of *nucleotides and the nucleic acids *DNA and *RNA.

cytoskeleton A network of fibres permeating the matrix of living eukaryotic cells that provides a supporting framework for organelles, anchors the plasma membrane and certain cell junctions, facilitates cellular movement, conveys vesicles and other components around the cell, and provides a suitable surface for chemical reactions to take place. Reorganization of the cytoskeleton is also responsible for changes in cell shape and cell migration that occur when organs are formed during embryonic development. The cytoskeletal fibres are composed of *microtubules, *intermediate filaments, and *microfilaments.

(((🌐))) SEE WEB LINKS

- In-depth collection of resources about actin, the cytoskeleton, and related structures, hosted by Edinburgh University

cytosol (hyaloplasm) The semifluid soluble part of the cytoplasm of cells, which contains the components of the *cytoskeleton and in which the cell's organelles are suspended.

cytostome A mouthlike structure of certain protists, through which particulate food is ingested. The cytostome is typically located at the base of an indentation in the cell.

cytotaxonomy *See* TAXONOMY.

cytotoxic Destructive to living cells. The term is applied particularly to a class of drugs that inhibit cell division and are therefore used to destroy cancer cells, and to a group of T cells (*cytotoxic T cells) that destroy virus-infected cells.

cytotoxic T cell A type of T lymphocyte, marked by the *CD8 protein on its surface, that is crucial in recognizing and eliminating body cells that are infected with viruses, bacteria, or other pathogens. Cytotoxic T cells are thus an essential part of the adaptive immune response. Exposure to antigenic fragments of a pathogen presented by *dendritic cells causes naive T cells to become equipped to recognize the same antigen when present on the surface of an infected cell in combination with *MHC class I protein. The *T-cell receptor binds to the specific antigen-MHC complex on the target cell and releases lytic granules onto the target's surface. These granules contain several types of effector molecule, including *perforin. After it has dealt with one cell, the cytotoxic T cell can migrate to other infected cells and repeat the process, while avoiding uninfected bystander cells.

2,4-D 2,4-dichlorophenoxyacetic acid (2,4-dichlorophenoxyethanoic acid): a synthetic *auxin used as a weedkiller of broad-leaved weeds. *See also* PESTICIDE.

dance of the bees A celebrated example of *communication in animals, first investigated by Austrian zoologist Karl von Frisch (1886–1982). Honeybee workers on returning to the hive after a successful foraging expedition perform a 'dance' on the comb that contains coded information about the distance and direction of the food source. For example the **waggle dance**, characterized by tail-wagging movements, indicates the direction and distance of a food source. Performing on the wall of the hive, the bee traces a half-circle in one direction, then a straight run during which it waggles its abdomen, another half-circle to complete the circle. The direction of the dance, and the angle of the straight run to the vertical, indicate the direction of the food source in relation to the sun, while the number of waggles during the straight run indicates the distance from the hive. Other workers, called followers, sensing vibrations and other cues from the dance, follow the instructions to find the food source.

Daphnia A genus of crustaceans belonging to the class *Branchiopoda and order Cladocera (water fleas). *Daphnia* species have a transparent carapace and a protruding head with a pair of highly branched antennae for swimming and a single median compound eye. The five pairs of thoracic appendages form an efficient filter-feeding mechanism. Reproduction can take place without mating, i.e. by *parthenogenesis. Some species exhibit *cyclomorphosis.

dark adaptation The changes that need to take place in the eye when an animal moves from a brightly lit environment to a relatively dark one to enable objects to be seen clearly. On moving to a darker environment, the pupils dilate and *rhodopsin—the pigment in the *rod cells that is broken down in bright light—is regenerated from its constituents.

dark period (in botany) The length of night (or experimentally imposed darkness) that determines how plants respond to the changing seasons (*see* PHOTOPERIODISM). Responses such as the onset of flowering are determined by the length of the period of darkness that occurs between two periods of light, i.e. the critical night length.

dark reaction *See* PHOTOSYNTHESIS.

Darwin, Charles (1809–82) British naturalist, who studied medicine in Edinburgh followed by theology at Cambridge University, intending a career in the Church. However, his interest in natural history led him to accept an invitation in 1831 to join HMS *Beagle* as naturalist on a round-the-world voyage. After his return five years later he published works on the geology he had observed. He was also formulating his theory of *evolution by means of *natural selection, but it was to be 20 years before he published *The Origin of Species* (1859), prompted by similar views expressed by Alfred Russel Wallace. Among his later works was *The Descent of Man* (1871). *See also* DARWINISM.

(((i))) SEE WEB LINKS

• The complete works of Charles Darwin online

Darwinism The theory of *evolution proposed by Charles Darwin in *On the Origin of Species* (1859), which postulated that present-day species have evolved from simpler ancestral types by the process of *natural selection acting on the variability found within populations. This was a fuller account of a paper written jointly by Darwin and Alfred Russel *Wallace and presented to the Linnean Society in 1858. *On the Origin of Species* caused a furore when it was first published because it suggested that species are not immutable nor were they specially created—a view directly opposed to the doctrine of *special creation. However, the wealth of evidence presented by Darwin gradually convinced most people and the only major unresolved problem was to explain how the variations in populations arose and were maintained from one generation to the next. This became clear with the rediscovery of Mendel's work on classical genetics in the 1900s and led to the present theory of *neo-Darwinism.

Darwin's finches (Galápagos finches) The 14 species of finch, unique to the Galápagos Islands, that Charles Darwin studied during his journey on HMS *Beagle*. Each is adapted to exploit a different food source. They are not found on the mainland because competition there for these food sources from other birds is fiercer. Darwin believed all the Galápagos finches to be descendants of a few that strayed from the mainland, and this provided important evidence for his theory of evolution. *See also* ADAPTIVE RADIATION; CHARACTER DISPLACEMENT.

dating techniques Methods of estimating the age of rocks, palaeontological specimens, archaeological sites, etc. **Relative dating techniques** date specimens in relation to one another; for example, stratigraphy is used to establish the succession of fossils. **Absolute** (or **chronometric**) **techniques** give an absolute estimate of the age and fall into two main groups. The first depends on the existence of something that develops at a seasonally varying rate, as in *dendrochronology and *varve dating.

The other uses some measurable change that occurs at a known rate, as in *chemical dating, **radioactive** (or **radiometric**) **dating** (*see* CARBON DATING; FISSION-TRACK DATING; POTASSIUM-ARGON DATING; RUBIDIUM-STRONTIUM DATING; URANIUM-LEAD DATING), and thermoluminescence. *See also* PALAEOMAGNETIC DATING.

day-neutral plant A plant in which flowering can occur irrespective of the night length. Examples are cucumber and maize. *See* PHOTOPERIODISM. *Compare* LONG-DAY PLANT; SHORT-DAY PLANT.

DDT Dichlorodiphenyltrichloroethane; a colourless organic crystalline compound, $(ClC_6H_4)_2CH(CCl_3)$, made by the reaction of trichloromethanal with chlorobenzene. DDT is the best known of a number of chlorine-containing *pesticides used extensively in agriculture in the 1940s and 1950s. The compound is stable, accumulates in the soil, and concentrates in fatty tissue, reaching dangerous levels in carnivores high in the food chain. Restrictions are now placed on the use of DDT and similar pesticides; however, DDT is still used for control of malaria-carrying insects and leishmaniasis in some countries, particularly inside buildings.

deacetylation The removal of an acetyl group ($-COCH_3$) from a molecule. Deacetylation is an important reaction in several chemical pathways, including the *Krebs cycle, and in the reversible condensation of *chromatin.

deamination The removal of an amino group ($-NH_2$) from a compound. Enzymatic deamination occurs in the liver and is important in amino-acid metabolism, especially in their degradation and subsequent oxidation (*see also* OXIDATIVE DEAMINATION). The amino group is removed as ammonia and excreted, either unchanged or as urea or uric acid.

death The point at which the processes that maintain an organism alive no longer function. In humans it is diagnosed by permanent cessation of the heartbeat; however, the heart can continue beating after a large part of the brain ceases to function (*see* BRAIN DEATH). The death of a cell due to external damage or the action of toxic substances is known as *necrosis. This must be distinguished from programmed cell death (*see* APOPTOSIS), which is a normal part of the developmental process and the immune response.

death phase *See* BACTERIAL GROWTH CURVE.

death rate (mortality) The rate at which a particular species or population dies, whatever the cause. The death rate is an important factor in controlling the size of a population. *Compare* BIRTH RATE.

deca- Symbol da. A prefix used in the metric system to denote ten times. For example, 10 hertz = 1 decahertz (daHz).

Decapoda An order of crustaceans of the class Malacostraca that are distributed worldwide, mainly in marine habitats. Decapods comprise swimming forms (shrimps and prawns) and crawling forms (crabs, lobsters, and crayfish). All are characterized by five pairs of walking legs, the first pair of which are highly modified in crawling forms to form powerful grasping pincers. The carapace is fused with the thorax and head forming a *cephalothorax. The antennae are especially long in shrimps and prawns, which also possess several pairs of well-developed swimming appendages (**pleopods**) posterior to the walking legs. Following fertilization by the male, females usually carry the eggs until they hatch. The larvae undergo several transformations before attaining adult form.

decarboxylation The removal of carbon dioxide from a molecule. Decarboxylation is an important reaction in many biochemical processes, such as the *Krebs cycle and the synthesis of *fatty acids. *See also* OXIDATIVE DECARBOXYLATION.

decay *See* DECOMPOSITION.

deci- Symbol d. A prefix used in the metric system to denote one tenth. For example, 0.1 metre = 1 decimetre (dm).

decibel A unit used to compare two power levels, usually applied to sound or electrical signals. Although the decibel is one tenth of a **bel**, it is the decibel, not the bel, that is invariably used. Two power levels P and P_0 differ by n decibels when $n = 10\log_{10}P/P_0$. If P is the level of sound intensity to be measured, P_0 is a reference level, usually the intensity of a note of the same frequency at the threshold of audibility.

The logarithmic scale is convenient as human audibility has a range of 1 (just audible) to 10^{12} (just causing pain) and one decibel, representing an increase of some 26%, is about the smallest change the ear can detect.

deciduous Describing plants in which all the leaves are shed at the end of each growing season, usually the autumn in temperate regions or at the beginning of a dry season in the tropics. This seasonal leaf fall helps the plant retain water that would otherwise be lost by transpiration from the leaves. Examples of deciduous plants are rose and horse chestnut. *Compare* EVERGREEN.

deciduous teeth (milk teeth) The first of two sets of teeth of a mammal. These teeth are smaller than those that replace them (the *permanent teeth) and fewer in number, since there are no deciduous *molars. *See also* DIPHYODONT.

decomposer An organism that obtains energy from the chemical breakdown of dead organisms or animal or plant wastes. Decomposers, most of which are bacteria and fungi, secrete enzymes onto dead matter and then absorb the breakdown products (*see* SAPROTROPH). Many decomposers (e.g. nitrifying bacteria) are specialized to break down organic materials that are difficult for other organisms to digest. Decomposers fulfil a vital role in the *ecosystem, returning the constituents of organic matter to the environment in inorganic form so that they can again be assimilated by plants. *Compare* DETRITIVORE. *See also* CARBON CYCLE; NITROGEN CYCLE.

decomposition (decay) The chemical breakdown of organic matter into its constituents by the action of *decomposers.

de-extinction (resurrection biology) The process of resurrecting species that have gone extinct. One strategy is **back breeding**, as used since 2008 by the Tauros Programme to produce cattle resembling the aurochs, an extinct European wild ox. This involves selective mating of living breeds of cattle having morphological characteristics similar to their extinct relatives, coupled with assisted reproductive technologies such as artificial insemination, embryo transfer, and *in vitro* fertilization. Genetic analysis of the resulting offspring is then compared with the genome sequence of an aurochs, obtained from DNA in fossilized tissue samples. This information directs the breeding programme, which it is hoped will produce a sustainable population of wild cattle called 'tauros' that will occupy the ecological niche once filled by the aurochs (*see* REWILDING).

Another approach is to use *nuclear transfer to clone an extinct species. This requires intact cell nuclei from cells derived from well-preserved tissue samples. For example, in 2009 nuclei from thawed cryopreserved skin cells of the extinct Pyrenean ibex were reprogrammed to develop as embryos and transferred to surrogate Spanish ibex mothers. A single viable Pyrenean ibex kid was born but died shortly after due to malfunctioning lungs.

DNA sequencing coupled with *gene editing now offers the potential to reconstruct the genome of an extinct species by directly manipulating the genome of a close living relative. The extinct North American passenger pigeon is a candidate for such an approach. Just 3% of its genome differs from that of its closest living relative, the band-tailed pigeon, prompting efforts to reinstate the key genetic differences that will produce a passenger pigeon.

De-extinction raises several ethical concerns, including whether it is right to tamper with the course of natural history and whether resurrected species could adapt to the often radically altered natural habitats of today.

defecation The expulsion of faeces from the rectum due to contractions of muscles in the rectal wall. A sphincter muscle, which is under voluntary control, is situated at the end of the rectum (the anus); relaxation of this muscle allows defecation to occur. In babies control of the anal sphincter muscle has not been developed and defecation occurs automatically as a reflex response to the presence of faeces in the rectum.

defective ribosomal products *See* DRIPS.

defensin Any of a class of antimicrobial peptides produced by animals and plants as a means of innate defence against invading microorganisms. They act by disrupting the cell membranes of bacteria and fungi, thereby killing the target cell. In mammals they may also modulate inflammation and other immune responses. The similar structures of defensins in plants and animals suggests that they evolved before these two lineages diverged. *See* HYPERSENSITIVITY (sense 2).

deficiency disease Any disease caused by an inadequate intake of an essential nutrient in the diet, primarily vitamins, minerals, and amino acids. Examples are *scurvy (lack of vitamin C), *rickets (lack of vitamin D), and iron-deficiency *anaemia. *See also* MINERAL DEFICIENCY.

definite inflorescence *See* CYMOSE INFLORESCENCE.

deforestation The extensive cutting down of forests for the purpose of extracting timber or fuel wood or to clear the land for mining, agriculture, or infrastructure. Forests are often situated in upland areas and are important in trapping rainwater. Deforestation in these areas, particularly in India and Bangladesh, has resulted in the flooding of low-lying plains; it has also led to an increase in soil erosion and hence desert formation (*see* DESERTIFICATION), resulting in crop loss and economic problems for local communities. The felling and burning of trees releases large amounts of carbon dioxide, thereby increasing global carbon dioxide levels and contributing to the *greenhouse effect and *climate change. Rainforests, particularly those of South

America, are rich in both fauna and flora; their removal leads to an overall decrease in *biodiversity and the loss of plant species that have potentially beneficial pharmaceutical effects. Despite moves to reduce deforestation, economic pressures ensure that the process continues.

degeneration 1. Changes in cells, tissues, or organs due to disease, etc., that result in an impairment or loss of function and possibly death and breakdown of the affected part. **2.** The reduction in size or complete loss of organs during evolution. The human appendix has undergone this process and performs no obvious function in humans. Degeneration of external organs may cause animals to appear to be more 'primitive' than they really are; for example, early zoologists believed whales were fish rather than mammals because of the degeneration of their limbs. *See also* VESTIGIAL ORGAN.

deglutition (swallowing) A reflex action initiated by the presence of food in the pharynx. During deglutition, the soft *palate is raised, which prevents food from entering the nasal cavity; the *epiglottis closes, which blocks the entrance to the windpipe; and the oesophagus starts to contract (*see* PERISTALSIS), which ensures that food is conveyed to the stomach.

dehiscence The spontaneous and often violent opening of a fruit, seed pod, or anther to release and disperse the seeds or pollen. Examples are the splitting of laburnum pods and primrose capsules; such structures are described as **dehiscent** (*compare* INDEHISCENT).

dehydrogenase Any enzyme that catalyses the removal of hydrogen atoms (*dehydrogenation) in biological reactions. Dehydrogenases occur in many biochemical pathways but are particularly important in driving the *electron-transport-chain reactions of cell respiration. They work in conjunction with the hydrogen-accepting coenzymes *NAD^+ and *FAD^+.

dehydrogenation A chemical reaction in which hydrogen is removed from a

compound. Dehydrogenation of organic compounds converts single carbon-carbon bonds into double bonds. In biological systems it is usually effected by *dehydrogenases.

deletion (in genetics) **1.** A *point mutation involving the removal of one or more base pairs in the DNA sequence. **2.** A frequently lethal *chromosome mutation that arises from an inequality in *crossing over during meiosis such that one of the chromatids loses more genetic information than it receives. *See also* GENOME EDITING; KNOCKOUT.

deme A group of organisms in the same *taxon. The term is used with various prefixes that denote how the group differs from other groups. For example, an **ecodeme** occurs in a particular ecological habitat, **cytodemes** differ from each other cytologically, and **genodemes** differ genetically.

demosponge *See* PORIFERA.

denature To produce a structural change in a protein or nucleic acid that results in the reduction or loss of its biological properties. Denaturation involves unfolding of the polypeptide chains of proteins and of the double helix of nucleic acids, with loss of secondary and tertiary structure; it is caused by heat (**thermal denaturation**), chemicals, and extremes of pH. The differences between raw and boiled eggs are largely a result of denaturation. *Compare* RENATURATION.

dendrite Any of the slender branching processes that arise from the *dendrons of the cell body of a *neuron and transmit nerve impulses to the cell body. Dendrites can be very numerous and form connections (*see* SYNAPSE) with many other neurons or sensory cells.

dendritic cell One of the main types of *antigen-presenting cell in the immune system, responsible for presenting antigen to naive T cells and inducing them to become effective components of the adaptive immune response. Derived from precursors in bone marrow, dendritic cells initially are present in tissues, where they have a highly branched (i.e. dendritic) or spiky appearance. If infection arises they ingest antigen, become activated, and migrate to local lymphoid tissue. Here they activate T cells by expressing the processed antigen on their surface, in association with *MHC class I and *MHC class II proteins, as well as producing various co-stimulatory chemical signals. This combination of signals causes naive T cells to transiently bind to the dendritic cell and stimulates them to proliferate and differentiate into effector T cells that recognize and respond to the specific antigen presented by the dendritic cell.

dendrochronology An absolute *dating technique using the *growth rings of trees. It depends on the fact that trees in the same locality show a characteristic pattern of growth rings resulting from climatic conditions. Thus it is possible to assign a definite date for each growth ring in living trees, and to use the ring patterns to date fossil trees or specimens of wood (e.g. used for buildings or objects on archaeological sites) with lifespans that overlap those of living trees. The bristlecone pine (*Pinus aristata*), which lives for up to 5000 years, has been used to date specimens over 8000 years old. Fossil specimens accurately dated by dendrochronology have been used to make corrections to the *carbon dating technique. Dendrochronology is also helpful in studying past climatic conditions. Analysis of trace elements in sections of rings can also provide information on past atmospheric pollution.

dendrogram A diagram, similar to a family tree, that indicates some type of similarity between different organisms. Dendrograms can be based on *phenetic or *phylogenetic similarities; a **cladogram** shows similarities according to the system of *cladistics.

dendron Any of the major cytoplasmic processes that arise from the cell body of a neuron, apart from the *axon. A dendron usually branches into *dendrites.

Denisovan *See* HOMO.

denitrification A chemical process in which nitrates in the soil are reduced to molecular nitrogen, which is released into the atmosphere. This process is effected by denitrifying bacteria (e.g. *Pseudomonas denitrificans*), which use nitrates as a source of energy in respiration. *Compare* NITRIFICATION. *See* NITROGEN CYCLE.

de novo pathway Any metabolic pathway in which a *biomolecule is synthesized from simple precursor molecules. Nucleotide synthesis is an example.

dense body One of numerous structures found in the fibres of smooth muscle to which the thin filaments are anchored within the muscle cell. Dense bodies are rich in the protein α-actinin; some occur within the cytoplasm while others (also called attachment plaques) are attached to the plasma membrane via intermediate filaments of the cytoskeleton. They are functionally analogous to the *Z lines of skeletal muscle.

density-dependent factor Any factor limiting the size of a population whose effect is dependent on the number of individuals in the population. For example, disease will have a greater effect in limiting the growth of a large population, since overcrowding facilitates its spread. *See also* ENVIRONMENTAL RESISTANCE.

density-independent factor Any factor limiting the size of a population whose effect is not dependent on the number of individuals in the population. An example of such a factor is an earthquake, which will kill all members of the population regardless of whether the population is small or large.

dental caries Tooth decay, which involves the destruction of the enamel layer of the tooth by acids produced by the action of bacteria on sugar. Bacteria can bind to teeth on **dextran**, a sticky substance derived from sucrose. The bacterial cells and other waste attached to dextran gives rise to *plaque. If dental caries is not treated it can spread to the dentine and

2 1 2 3	human	2 0 3 3	rabbit	3 1 4 2	bear
2 1 2 3	(32 teeth)	1 0 2 3	(28 teeth)	3 1 4 3	(42 teeth)

Representative dental formulas

pulp of the tooth, which leads to infection and death of the tooth.

dental formula A representation of the dentition of an animal. A dental formula consists of eight numbers, four above and four below a horizontal line. The numbers represent (from left to right) the numbers of incisors, canines, premolars, and molars in either half of the upper and lower jaws. The total number of teeth in both jaws is therefore obtained by adding up all the numbers in the dental formula and multiplying by 2. Representative dental formulas are shown in the illustration. *See also* PERMANENT TEETH.

dentary A *membrane bone, present in the lower jaw of the vertebrates, that supports the teeth. In mammals the dentary is the sole bone of the lower jaw.

denticle (placoid scale) *See* SCALES.

dentine The bony material that forms the bulk of a *tooth. Dentine is similar in composition to bone but is perforated with many tiny canals for nerve fibres, blood capillaries, and processes of the dentine-forming cells (*odontoblasts). Ivory, the material that forms elephant tusks, is made of dentine.

dentition The type, number, and arrangement of teeth in a species. This can be represented concisely by a *dental formula. *See also* PERMANENT TEETH; DIPHYODONT; MONOPHYODONT; POLYPHYODONT; HETERODONT; HOMODONT.

deoxyribonuclease *See* DNASE.

deoxyribonucleic acid *See* DNA.

deoxyribose (2-deoxyribose) A pentose (five-carbon) sugar, a derivative of *ribose, that is a component of the nucleotides (deoxyribonucleotides) that form the building blocks of *DNA.

depolarization A reduction in the difference of electrical potential that exists across the plasma membrane of a cell, i.e. in the cell's *resting potential. A large, transient depolarization of a nerve-cell membrane occurs during the passage of an *action potential along the axon when the nerve is transmitting an impulse.

derived trait *See* APOMORPHY; SYNAPOMORPHY.

dermal bone *See* MEMBRANE BONE.

Dermaptera An order of insects comprising the earwigs. Earwigs typically have long thin cylindrical bodies with biting mouthparts and a stout pair of curved forceps (**cerci**) at the tip of the abdomen, used for catching prey and in courtship. Some species have a single pair of wings, which at rest are folded back over the abdomen like a fan; others are wingless. Most earwigs are nocturnal and omnivorous.

dermis (corium; cutis) The thicker and innermost layer of the *skin of vertebrates, the other layer being the *epidermis. The dermis consists of fibrous connective tissue in which are embedded blood vessels, sensory nerve endings, and (in mammals) hair follicles, sebaceous glands, and sweat ducts. Beneath the dermis lies the *subcutaneous tissue.

descending tracts *See* SPINAL CORD.

desert A major terrestrial *biome characterized by low rainfall. Hot deserts, such as the Sahara and Kalahari deserts of Africa, have a rainfall of less than 25 cm a year and extremely high daytime temperatures (in some cases up to 50°C). Vegetation is sparse, and desert plants are adapted to conserve water and take advantage of the rain when it falls. The perennials include xerophytic trees and shrubs (*see* XEROPHYTE) and *succulents, such as cacti;

many succulents show *crassulacean acid metabolism. Annual plants are *ephemerals, lying dormant as seeds for most of the year and completing their life cycle in the brief rainy periods. Desert animals are typically nocturnal or active at dawn and dusk, thus avoiding the extreme daytime temperatures. In cold deserts, such as the high plateau of Antarctica, winter temperatures can fall as low as −80°C.

desertification The gradual conversion of fertile land into desert, usually as a result of human activities or *climate change. Loss of topsoil leads to further soil erosion until the land can no longer be used to grow crops or support livestock. A major factor contributing to desertification is bad management of farmland. Overgrazing of livestock removes the plant cover and exposes the soil, making it vulnerable to erosion. Overintensive cultivation of crop plants, especially monoculture (*see* AGRICULTURE), depletes the soil of nutrients and organic matter, resulting in loss of fertility and increasing its susceptibility to erosion. In many less developed countries it is difficult to control the process of desertification as the livelihood of the people often depends on practices that contribute to soil erosion. Another major cause of desertification is *deforestation. Loss of vegetation also results from reductions in rainfall resulting from climate change.

desiccation A method of preserving organic material by the removal of its water content. Cells and tissues can be preserved by desiccation after lowering the samples to freezing temperatures; thereafter they can be stored at room temperature.

desiccator A container for drying substances or for keeping them free from moisture. Simple laboratory desiccators are glass vessels containing a drying agent, such as silica gel. They can be evacuated through a tap in the lid.

desmids Unicellular green algae of the order Zygnematales. Like *Spirogyra*, they have an elaborate chloroplast. The cells of desmids are characteristically split into two halves joined by a narrow neck, each half

being a mirror image of the other. Some form filaments comprising long chains of cells. The outer wall of the cell is patterned with various protuberances and covered with a mucilaginous sheath, which is thought to play a role in the cell's slow gliding movement. Desmids are found mainly in fresh water and belong to the clade *streptophytes (charophytes) that includes land plants.

desmosome (macula adherens) A patchlike junction found between adjacent cells in epithelial tissues that helps strengthen the tissue by binding the cells together while allowing movement of materials within the intercellular space. It consists of a discrete cluster of fibres running across a space, some 25 nm wide, between the two cells. On either side, these fibres are anchored beneath the plasma membrane of each cell within a proteinaceous cytoplasmic plaque, from which more fibres extend into the cytoplasm of the cell. **Hemidesmosomes** are similar to desmosomes but anchor the cell to underlying *extracellular matrix. *See also* CELL JUNCTION.

desmotubule *See* PLASMODESMATA.

desorption The removal of adsorbed atoms, molecules, or ions from a surface.

determinate growth *See* GROWTH.

determined Describing embryonic tissue at a stage when it can develop only as a certain kind of tissue (rather than as any kind).

detoxification (detoxication) The process by which harmful compounds, such as drugs and poisons, are converted to less toxic compounds in the body. Detoxification is an important function of the *liver, where enzymes of the *smooth endoplasmic reticulum modify drugs and poisons, e.g. by adding hydroxyl groups to make the molecules more water-soluble and thus more readily excreted from the body. *See also* PHASE I METABOLISM; PHASE II METABOLISM.

detritivore An animal that feeds on *detritus. Examples of detritivores are earthworms, blowflies, maggots, and woodlice. Detritivores play an important role in the breakdown of organic matter from decomposing animals and plants (*see* DECOMPOSER).

detritus Particles of organic material derived from dead and decomposing organisms, resulting from the activities of the *decomposers. Detritus is the source of food for *detritivores, which can themselves be eaten by carnivores in a **detritus food chain:**

detritus → detritivore → carnivore.

detrusor muscle The smooth muscle of the *bladder wall, which is innervated with sympathetic fibres and contracts as a result of a reflex action when the bladder wall tension has reached a certain level.

deuteranopia *See* COLOUR BLINDNESS.

Deuteromycota A taxon formerly used in some classifications to include all fungi in which sexual reproduction is apparently absent. These fungi, described as 'imperfect fungi' or 'Fungi Imperfecti', are now regarded as ascomycotes or basidiomycotes that have lost the ability to produce asci or basidia, respectively, and in some cases it may be possible to identify close sexual relatives, or even sexual stages of the same species, in either of these groups. For example, *Penicillium*, traditionally classified as a deuteromycote, is now known to have a sexual stage in the form of the ascomycote fungus *Talaromyces*.

deuterostome An animal in which the opening (blastopore) of the embryonic cavity (*see* ARCHENTERON) becomes the anus and the mouth forms as a secondary orifice. The name derives from Greek, meaning literally 'second mouth'. Deuterostomes typically display radial cleavage of the blastula, indeterminate development (the fate of cells is not established until later stages of development), and enterocoelic formation of the coelom (i.e. by outpocketing of the embryonic gut). This contrasts with *protostome development, in which the mouth forms from the blastopore. Most, but not all, animals exhibiting deuterostome development are members of the clade Deuterostomia, which includes the three phyla *Hemichordata, *Echinodermata, and *Chordata.

deutoplasm The nutritional material found in the *yolk of eggs.

development The complex series of changes that occurs in multicellular living organisms as they progress from the initial fertilized egg via an embryo stage to the mature individual. Fundamentally important processes in development include: cell division; the determination of cell fate; *differentiation to form the distinct types of cells; *morphogenesis, whereby cells are arranged as tissues and organs; and *growth, whether by increase in cell number or size. In vertebrate animals there are three developmental stages: (1) *cleavage, in which the zygote divides to form a ball of cells, the *blastula; (2) gastrulation, in which the cells become arranged in three primary *germ layers (*see* GASTRULA); (3) **organogenesis** (or **organogeny**), in which further cell division and differentiation results in the formation of organs. The development of many invertebrates (e.g. insects) and amphibians involves the process of *metamorphosis. In all organisms, development is directed by the coordinated expression of various genes in time and space, regulated by the complex interplay of the gene products (*see* MORPHOGEN). Early development may be influenced by maternal genes (*see* MATERNAL EFFECT GENES) or *cytoplasmic segregation of cell fate determinants, as well as embryonic genes. Genes involved in development are often highly conserved across very diverse organisms, as exemplified by the *Hox* genes that determine the structures along the head-to-tail axis of most animals. Many organisms can modify their development in response to environmental cues. Such **developmental plasticity** enables them to adapt to variations in environmental conditions, such as changes in food supply or climate. For example, in some turtles the incubation temperature determines the sex and other traits of the hatchlings, such as adult body size, a mechanism that maximizes reproductive fitness. *See also* ABC MODEL; HOMEOTIC GENES; MORPHOGENESIS; PRIMARY GROWTH; SECONDARY GROWTH.

Devonian A geological period in the Palaeozoic era that extended from the end of the Silurian (about 419 million years ago) to the beginning of the Carboniferous (about 359 mya). It was named by Adam Sedgwick (1785–1873) and Roderick Murchison (1792–1871) in 1839. The Devonian is divided into seven stages based on invertebrate fossil remains, such as corals, brachiopods, ammonoids, and crinoids, found in marine deposits. There were also extensive continental deposits consisting of conglomerates, red silts, and sandstones, forming the Old Red Sandstone facies. Fossils in the Old Red Sandstone include fishes and the earliest land plants (*see* RHYNIOPHYTES; TRIMEROPHYTES; ZOSTEROPHYLLOPHYTES). Graptolites became extinct early in the Devonian and the trilobites declined.

((()) SEE WEB LINKS)

• The Devonian period as described on the website of the University of California Museum of Paleontology

dextrin An intermediate polysaccharide compound resulting from the hydrolysis of starch to maltose by amylase enzymes.

dextrorotatory Denoting a chemical compound that rotates the plane of polarization of plane-polarized light to the right (clockwise as observed by someone facing the oncoming radiation). *See* OPTICAL ACTIVITY.

dextrose *See* GLUCOSE.

d-form *See* OPTICAL ACTIVITY.

diabetes *See* ANTIDIURETIC HORMONE; INSULIN.

diacylglycerol (DAG) *See* INOSITOL.

diagenesis *See* TAPHONOMY.

diakinesis The period at the end of the first prophase of *meiosis when the separation of *homologous chromosomes is almost complete and *crossing over has occurred.

dialysis A method by which large molecules (such as starch or protein) and small molecules (such as glucose or amino acids) in solution may be separated by selective diffusion through a semipermeable membrane. For example, if a mixed solution of starch and glucose is placed in a closed container made of a semipermeable substance (such as Cellophane), which is then immersed in a beaker of water, the smaller glucose molecules will pass through the membrane into the water while the starch molecules remain behind. The plasma membranes of living organisms are *partially permeable, and dialysis takes place naturally in the kidneys for the excretion of nitrogenous waste. An artificial kidney (**dialyser**) utilizes the principle of dialysis by taking over the functions of diseased kidneys.

diapause A period of suspended development or growth occurring in many insects and other invertebrates during which metabolism is greatly decreased. Some long-lived species may undergo diapause as adults, but for many others the egg or pupa is the diapausal stage. In temperate-zone animals, diapause is typically triggered by changes in day length that presage the onset of winter; and prolonged exposure to low temperatures may be the cue to trigger the resumption of development. In other climatic regions, adverse conditions can also trigger diapause. Both strategies enable the animals to survive unfavourable environmental conditions so that their development may continue in more favourable ones. The peptide, **diapause hormone**, is known to initiate diapause in embryos of the silkworm, *Bombyx mori*, but also to terminate diapause in some other moth species, as does the hormone *ecdysone.

diaphragm The muscular membrane that divides the thorax (chest) from the abdomen in mammals. It plays an essential role in breathing (*see also* RESPIRATORY

MOVEMENT), being depressed during *inspiration and raised during *expiration.

diaphysis (*pl.* **diaphyses**) The shaft of a mammalian limb bone, which in immature animals is separated from the ends of the bone (*see* EPIPHYSIS) by cartilage.

diastase *See* AMYLASE.

diastema The gap that separates the biting teeth from the grinding teeth in herbivores. It creates a space in which food can be held in readiness for the grinding action of the teeth. This space is filled by large canine teeth in carnivores.

diastole The phase of a heartbeat that occurs between two contractions of the heart, during which the heart muscles relax and the ventricles fill with blood. *Compare* SYSTOLE. *See* BLOOD PRESSURE.

diatoms *See* BACILLARIOPHYTA.

diatropism *See* PLAGIOTROPISM.

dibiontic Describing a life cycle in which there is *alternation of generations. *Compare* MONOBIONTIC.

dicarboxylic acid A *carboxylic acid having two carboxyl groups in its molecules. In systematic chemical nomenclature, dicarboxylic acids are denoted by the suffix *-dioic*; e.g. hexanedioic acid, HOOC $(CH_2)_4COOH$.

Dicer A protein, originally identified as the product of the *Dicer* gene in *Drosophila*, that trims double-stranded RNA molecules to form *short interfering RNAs (siRNAs) or *microRNAs (miRNAs). Both siRNAs and miRNAs are produced by eukaryotic cells as a means of suppressing gene activity. *See* RNA INTERFERENCE.

dichasium (*pl.* **dichasia**) *See* CYMOSE INFLORESCENCE.

2,4-dichlorophenoxyacetic acid *See* 2,4-D.

dichogamy The condition in which the male and female reproductive organs of a flower mature at different times, thereby ensuring that self-fertilization does not occur. *Compare* HOMOGAMY. *See also* PROTANDRY; PROTOGYNY.

dichopatric speciation A form of *allopatric speciation in which portions of a pre-existing population become separated because of the formation of a geographical barrier between them. For example, the advance of glaciers at the start of each glaciation caused the separation of previously continuous species into isolated refuges, where each subpopulation subsequently diverged. *Compare* PERIPATRIC SPECIATION.

dichotomous 1. Describing the type of branching in plants that results when the growing point (apical bud) divides into two equal growing points, which in turn divide in a similar manner after a period of growth, and so on. Dichotomous branching is common is ferns and mosses. **2.** *See* KEY.

Dicotyledoneae In traditional classifications, one of the two classes of flowering plants (*see* ANTHOPHYTA), distinguished by having two seed leaves (*cotyledons) within the seed. The dicotyledons usually have leaf veins in the form of a net, a ring of vascular bundles in the stem, and flower parts in fours or fives or multiples of these. Dicotyledons include many food plants (e.g. potatoes, peas, beans), ornamentals (e.g. roses, ivies, honeysuckles), and hardwood trees (e.g. oaks, limes, beeches). Dicots are no longer considered a taxonomically valid group. *See* EUDICOT. *Compare* MONOCOTYLEDONEAE.

dicoumarol *See* COUMARIN.

Dictyoptera An order of insects (sometimes classified as *Orthoptera) comprising the cockroaches (suborder Blattaria) and the mantids (suborder Mantodea), occurring mainly in tropical regions. Cockroaches are oval and flattened in shape; some have a single well-developed pair of wings, folded back over the abdomen at rest, while in others the wings may be reduced or absent. They are usually found in forest litter, feeding on dead organic matter, but some species, e.g. the American cockroach (*Periplaneta americana*), are major household pests, scavenging on starchy foods, fruits, etc. In most species the females produce capsules (**oothecae**) containing 16–40 eggs. These are either deposited or carried by the female during incubation.

dictyosome A cup-shaped array of flattened membranous vesicles found in plant cells. Dictyosomes modify proteins from the endoplasmic reticulum, and may also polymerize sugars to polysaccharides. They then package these materials for delivery to destinations within the cell (e.g. the cell wall), for secretion, or for storage. In animal cells, and rarely in plant cells, numerous dictyosomes associate to form the *Golgi apparatus.

Dicyemida (Rhombozoa) A phylum of around 125 known species of microscopic multicellular animals found as parasites inside the nephridia of cephalopod molluscs. They range in size from 0.5 to 7 mm and consist of a single axial cell surrounded by a jacket of ciliated cells, plus an anterior structure, the calotte, which attaches the organism to the host site. Genetic analysis has shown that dicyemids are members of the *Lophotrochozoa clade and are related to the *Orthonectida.

diet The food requirements of an organism. The foods that constitute the human diet should contain vitamins, mineral salts (*see* ESSENTIAL ELEMENT), and dietary *fibre as well as water, carbohydrates and fats (which provide energy), and proteins (required for growth and maintenance). A balanced diet contains the correct proportions of these *nutrients, which will vary depending on the age, sex, body size, and the level of activity of the individual. An inadequate supply of different food types in the diet can lead to *malnutrition.

dietary fibre *See* FIBRE.

differential-interference contrast microscope *See* NOMARSKI MICROSCOPE.

differentiation The changes by which more specialized cells arise from less specialized cells, so that tissues and organs

with particular functions can form during embryonic *development. Furthermore, certain cells, called *stem cells, retain the ability to differentiate throughout the life of the organism so that tissues can be replaced or repaired (*see* REGENERATION). Plants also have undifferentiated tissues (*see* MERISTEM) that give rise to the new tissues required for growth, a process that typically is lifelong. Whatever the type of organism, the specialization inherent in differentiation depends on sequential changes in the pattern of gene expression as cells undergo successive divisions. The mechanisms involved have been most intensively studied in experimental organisms, notably the fruit fly *Drosophila. In this insect the general body plan is established in the early embryo by proteins called *morphogens, which are encoded by maternal genes of follicle cells and diffuse into the developing embryo. The various morphogens set up a pattern of concentrations that activates genes in different zones of the embryo to different extents, creating the basic pattern of body segments. A class of genes within the embryo itself, the segment genes, further refine this pattern. Within each segment the differentiation of appendages, such as limbs, is controlled by *homeotic genes.

diffuse coevolution (guild coevolution) The evolution in several species of similar traits in response to a set of selection pressures imposed by one or more interacting species. For example, shrubby plant species commonly evolve similar structures such as thorns and unpalatable leaves to protect against grazing herbivores of various species. Also, flowers of different species often have similar features to attract particular pollinating animals, such as bees.

diffusion (passive transport) The random movement of particles (e.g. molecules or ions) from an area of high concentration to an area of low concentration until an even distribution of particles (i.e. uniform concentration) is obtained. Small molecules and ions (such as oxygen and Na^+) can diffuse across a plasma membrane.

diffusion gradient *See* CONCENTRATION GRADIENT.

digestion The breakdown by a living organism of ingested food material into chemically simpler forms that can be readily absorbed and assimilated by the body. This process requires the action of digestive enzymes and may take place extracellularly (i.e. in the *alimentary canal), as is the case in most animals; or intracellularly (e.g. by engulfing phagocytic cells), as occurs in many protists and in cnidarians.

digestive system The system of organs that are involved in the process of *digestion. The digestive system of mammals is divided into the **gastrointestinal tract** (*see* ALIMENTARY CANAL) and accessory structures, such as teeth, tongue, liver, pancreas, and gall bladder.

digit A finger or toe. In the basic limb structure of terrestrial vertebrates there are five digits (*see* PENTADACTYL LIMB). This number is retained in humans and other primates, but in some other species the number of digits is reduced. Frogs, for example, have four fingers and five toes, and in ungulate (hooved) mammals, the digits are reduced and their tips are enclosed in horn, forming hooves.

digitalis A preparation of the dried leaves or seeds of the foxglove (*Digitalis*), used historically as a heart stimulant. Modern clinically prescribed drugs derived from digitalis include **digoxin**, which belongs to a class of drugs known as the cardiac *glycosides. They are used to treat heart failure and some forms of *arrhythmia because of their ability to increase the force of contraction of the heart muscle. Their toxic effects arise from their capacity to disturb the normal rhythm of the heart.

digitigrade Describing the gait of most fast-running animals, such as dogs and cats, in which only the toes are on the ground and the rest of the foot is raised off the ground. *Compare* PLANTIGRADE; UNGULIGRADE.

dihybrid cross A genetic cross between parents that differ in two characteristics, controlled by genes at different loci. Mendel performed a dihybrid cross using pea plants and the characteristics of seed colour and texture: the parental plants had either smooth yellow seeds (*SSYY*)—the dominant characteristics—or wrinkled green seeds (*ssyy*)—the recessive characteristics. All the offspring had smooth yellow seeds, being heterozygous (*SsYy*) for the two alleles. Crossing between these offspring produced an F_2 generation of plants with smooth yellow, smooth green, wrinkled yellow, and wrinkled green seeds in the ratio 9:3:3:1 (see illustration). Mendel used these results as the basis for his Law of Independent Assortment (*see* MENDEL'S LAWS). *Compare* MONOHYBRID CROSS.

dikaryon A cell of a fungal hypha or mycelium containing two haploid nuclei of different strains. The nuclei associate in pairs but do not fuse, therefore the cell is not truly diploid. Dikaryosis occurs in the Basidiomycota and Ascomycota. *See* HETEROKARYOSIS.

dikaryosis The most common form of *heterokaryosis. *See also* DIKARYON.

dilation *See* VASODILATION.

dimethylbenzenes (xylenes) Three compounds with the formula $(CH_3)_2C_6H_4$, each having two methyl groups substituted on the benzene ring. 1,2-dimethylbenzene is ortho-xylene, etc. A mixture of the isomers is obtained from petroleum and is used as a clearing agent in preparing specimens for optical microscopy.

dimictic lake A lake that is stratified by a *thermocline that is not permanent but is turned over twice during one year. The thermocline is disrupted due to seasonal changes in the climate. A **meromictic lake** is one in which there is a permanent stratification.

dimorphism The existence of two distinctly different types of individual within a species. An obvious example is **sexual dimorphism** in certain animals, in which

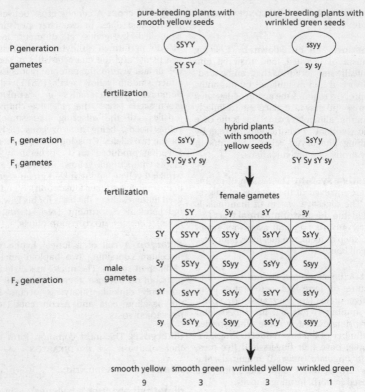

pure-breeding plants with smooth yellow seeds

pure-breeding plants with wrinkled green seeds

P generation — SSYY — ssyy

gametes — SY SY — sy sy

fertilization

F₁ generation — hybrid plants with smooth yellow seeds — SsYy — SsYy

F₁ gametes — SY Sy sY sy — SY Sy sY sy

fertilization — female gametes

F₂ generation — male gametes

	SY	Sy	sY	sy
SY	SSYY	SSYy	SsYY	SsYy
Sy	SSYy	SSyy	SsYy	Ssyy
sY	SsYY	SsYy	ssYY	ssYy
sy	SsYy	Ssyy	ssYy	ssyy

smooth yellow — smooth green — wrinkled yellow — wrinkled green

9 — 3 — 3 — 1

Dihybrid cross

the two sexes differ in colouring, size, etc. Dimorphism also occurs in some lower plants, such as mosses and ferns, that show an *alternation of generations.

Dinomastigota (Dinoflagellata) A phylum of mostly single-celled protists included in the *alveolates. They are abundant in the marine plankton; some are *photoautotrophs, containing brown xanthophyll pigments in addition to chlorophyll, and are important primary producers in marine ecosystems; many of these are *mixotrophs. Approximately half the known species are heterotrophic. Dinoflagellates characteristically have two flagella for locomotion and most have a rigid cell wall of cellulose encrusted with silica. Some species (e.g. *Noctiluca miliaris*) are bioluminescent. Vast numbers of dinoflagellates sometimes accumulate in surface waters as 'algal blooms', colouring the sea typically brown or red—hence the term *red tide. Some species release powerful toxins into the water, which can kill fish, contaminate shellfish, and even affect the breathing of people near the shore. Photosynthesizing dinoflagellates form symbiotic associations with *coral animals and are vital for the health of coral reefs.

dinosaur An extinct terrestrial reptile belonging to a group that arose some 225 million years ago and constituted the dominant land animals of the Jurassic and Cretaceous periods, 190–65 million years ago. There were two orders. The Ornithischia were typically quadrupedal herbivores, many with heavily armoured bodies, and included *Stegosaurus*, *Triceratops*, and *Iguanodon*. They were all characterized by birdlike pelvic girdles. The Saurischia included many bipedal carnivorous forms, such as *Tyrannosaurus* (the largest known carnivore), and some quadrupedal herbivorous forms, such as *Apatosaurus* (*Brontosaurus*) and *Diplodocus*. They all had lizard-like pelvic girdles. Modern birds evolved from saurischian ancestors (*see* AVES) and belong to a clade called theropods, predatory dinosaurs that shared birdlike traits. Many of the herbivorous dinosaurs were amphibious or semiaquatic.

(⊕) SEE WEB LINKS

• A Dino Directory produced by the Natural History Museum, London

dinucleotide A compound consisting of two *nucleotides.

dioecious Describing species, especially plants, in which male and female reproductive parts are on separate individuals. For instance, willows are dioecious because the male and female flowers are on separate plants. *Compare* MONOECIOUS.

dioxin (**2,3,7,8-tetrachlorodibenzo-*p*-dioxin**) A toxic solid formed in the manufacture of the herbicide *2,4,5-T; it was present as an impurity in Agent Orange, used as a defoliant during the Vietnam War. It is the most toxic member of a group of compounds (called **dioxins**) that occur widely as environmental pollutants, being produced during combustion processes (e.g. in incinerators) and as byproducts in various industrial manufacturing processes. Dioxins decompose very slowly and may be concentrated in the food chain; in animals they are stored in fat. Exposure to high levels of dioxins can cause skin disfigurement (chloracne) and may result in fetal defects. Because of their toxicity, many countries have imposed strict controls to reduce industrial emissions of dioxins.

dipeptide A compound consisting of two amino acid units joined at the amino ($-NH_2$) end of one and the carboxyl ($-COOH$) end of the other. This peptide bond (*see* PEPTIDE) is formed by a condensation reaction that involves the removal of one molecule of water.

diphyodont Describing a type of dentition that is characterized by two successive sets of teeth: the *deciduous (milk) teeth, which are followed by the *permanent (adult) teeth. Mammals have a diphyodont dentition. *Compare* MONOPHYODONT; POLYPHYODONT.

diploblastic Describing an animal whose early embryo has only two cell layers, an outer *ectoderm and inner *endoderm. These give rise to the tissues and organs of the mature animal. Diploblastic animals include the cnidarians, ctenophorans, and placozoans. *Compare* TRIPLOBLASTIC.

diploid Describing a nucleus, cell, or organism with twice the *haploid number of chromosomes characteristic of the species. The diploid number is designated as $2n$. Two sets of chromosomes are present, one set being derived from the female parent and the other from the male. In animals, all the cells except the reproductive cells are diploid.

diplomonad Any of a group of small single-celled organisms, typically less than 30 µm long, that have highly reduced mitochondria called *mitosomes, two nuclei, and multiple flagella. Free-living species are typically found in oxygen-deficient environments, such as lake sediments and marshes, while others are anaerobic and live as parasites, including *Giardia intestinalis*, which inhabits the guts of mammals and causes the disease giardiasis. Diplomonads belong to the *excavates assemblage of protists.

diplont An organism that is at the *diploid stage of its life cycle. *Compare* HAPLONT.

Diplopoda A class of wormlike terrestrial *arthropods belonging to the subphylum Myriapoda and comprising the millipedes. Diplopods are characterized by a distinct head, bearing a single pair of short antennae, and 20 to over 60 body segments each bearing two pairs of legs. Restricted to damp habitats, millipedes are slow moving and feed on decaying leaves.

diplotene The period in the first prophase of *meiosis when paired *homologous chromosomes begin to move apart. They remain attached at a number of points (*see* CHIASMA).

Diplura An order of small to medium-sized wingless insects, sometimes known as 'two-pronged bristletails', with elongated bodies, prominent paired tail-like cerci, and slender paired antennae. There are some 800 known species, typically 2–5 mm long, exceptionally reaching 50 mm. They lack eyes, have partially concealed mouthparts, and are found in dark humid places, for example in soil or under bark, living mostly on decaying vegetation, although some species are predators. The number of body segments is fixed during development, and moulting continues throughout life. Although some authorities place the Diplura in the subclass *Apterygota, with the other wingless insects, others regard them as a separate class within the subphylum *Hexapoda.

Dipnoi A subclass or order of bony fishes that contains the lungfishes, which have lungs and breathe air. They are found in Africa, Australia, and South America, where they live in freshwater lakes and marshes that tend to become stagnant or even dry up in summer. They survive in these conditions by burrowing into the mud, leaving a small hole for breathing air, and entering a state of *aestivation, in which they can remain for six months or more. The Dipnoi date from the Devonian era (416–360 million years ago) and share many features with the modern *Amphibia.

Diptera An order of insects comprising the true, or two-winged, flies. Flies possess only one pair of wings—the forewings; the hindwings are modified to form small clublike **halteres** that function as balancing organs. Typically fluid feeders, flies have mouthparts adapted for piercing and sucking or for lapping; the diet includes nectar, sap, decaying organic matter, and blood. Some species prey on insects; others are parasitic. Dipteran larvae (**maggots**) are typically wormlike with an inconspicuous head. They undergo metamorphosis via a pupal stage to the adult form. Many flies or their larvae are serious pests, either by feeding on crops (e.g. fruit flies) or as vectors of disease organisms (e.g. the house fly (*Musca domestica*) and certain mosquitoes).

directed evolution An approach used in biomolecular engineering that employs evolutionary principles to generate molecules with improved or novel properties. Essentially, mutated forms of a target molecule are replicated, and the best-performing ones are selected for further rounds of mutation and selection until the desired outcome is achieved. This may be, e.g. an enzyme with enhanced affinity for a substrate, an antibody with a new specificity for a receptor, or a protein that is stable at higher temperatures. In a typical scenario, a progenitor enzyme with a similar activity to that sought is identified, and its gene is replicated and mutated (e.g. by error-prone *polymerase chain reaction or *site-directed mutagenesis) to create a 'library' of mutants. The mutant genes are then cloned, screened for the desired activity, and the most promising clones selected for further mutation. The random nature of the approach can lead to novel and useful outcomes that molecular modelling would not yield. Moreover, the use of high-throughput screening techniques enables the selection of the few promising candidate genes from the very large numbers of mutants generated. *See also* PROTEIN ENGINEERING.

directional selection *See* POSITIVE SELECTION.

disaccharide A sugar consisting of two linked *monosaccharide molecules. For example, sucrose comprises one glucose molecule and one fructose molecule bonded together.

disassortative mating *See* ASSORTATIVE MATING.

discicristates In some classifications, an assemblage of eukaryotic protists whose members have disc-shaped cristae in their mitochondria and are shown to be related by molecular systematics. They include the Euglenozoa (e.g. euglenids and kinetoplastids) and the Heterolobosea (amoebomastigotes and acrasid cellular *slime moulds). The discicristates belong to the major group of protists called the *excavates. *See also* DISCOMITOCHONDRIA.

Discomitochondria In some earlier classifications, a phylum of protists consisting of motile single-celled organisms characterized by disc-shaped cristae in their mitochondria and the absence of sexual reproduction in their life cycles. There are four classes, all of which are traditionally described as flagellate protozoans. The *Euglenida comprise mainly photosynthetic organisms, such as *Euglena* itself, and have a sculpted outer covering (pellicle). The Kinetoplastida are characterized by a large modified mitochondrion (kinetoplast) at the base of the flagellum and include parasites such as the trypanosomes as well as free-living representatives (e.g. bodos). Members of the third class, the Amoebomastigota, can transform themselves from an amoeboid form into a flagellated form, and back again, depending on the nutrient status of their surroundings. Finally, the Pseudociliata are free-living marine organisms that resemble true ciliates (*see* CILIOPHORA) in having rows of cilia on their surface, but lack the two different types of nuclei found in ciliates. Members of the Discomitochondria are now included, with acrasid cellular slime moulds, in the *discicristates.

discontinuous replication The synthesis of a new strand of a replicating DNA molecule as a series of short fragments that are subsequently joined together. Only one of the new strands, the so-called **lagging strand**, is synthesized in this way.

The other strand (**leading strand**) is synthesized by continuous addition of nucleotides to the growing end, i.e. **continuous replication**. The difference arises because of the different orientations of the parent template strands. The template of the leading strand is oriented in the $3'{\rightarrow}5'$ direction (according to the numbering of atoms in the sugar residues), which means that the leading strand itself is oriented in the opposite $5'{\rightarrow}3'$ direction, providing an –OH group at the $3'$ end for the continual addition of nucleotides by DNA polymerase, which moves forwards as the template strands unwind at the replication fork. However, the template of the lagging strand is oriented in a $5'{\rightarrow}3'$ direction, so the lagging strand itself is oriented in the $3'{\rightarrow}5'$ direction, and hence the DNA polymerase complex must move backwards away from the replication fork. Synthesis of the lagging strand proceeds not continuously, as on the leading strand, but discontinuously in a series of repeated steps. Discontinuous replication produces a series of short DNA fragments (**Okazaki fragments**) complementary to the template strand. These vary in length, being about 100–200 nucleotides in eukaryotes and 1000–2000 nucleotides in prokaryotes. The fragments are then covalently bound together by the enzyme *DNA ligase, forming a continuous chain of nucleotides, thus completing replication of the lagging strand. *See also* DNA REPLICATION; PRIMASE.

discontinuous variation (qualitative variation) Clearly defined differences in a characteristic that can be observed in a population. Characteristics that are determined by different *alleles at a single locus show discontinuous variation, e.g. garden peas are either wrinkled or smooth. *Compare* CONTINUOUS VARIATION.

disease A condition in which the normal function of some part of the body (cells, tissues, or organs) is disturbed. A variety of microorganisms and environmental agents are capable of causing disease, as are mutations of the genes that direct the activities of host cells. The functional disturbances are often accompanied by structural changes in tissue. *Compare* SENESCENCE.

dishabituation The recovery of a habituated response (*see* HABITUATION) in an animal following the application of a different stimulus. For example, if the siphon of the sea slug *Aplysia* is repeatedly tapped, the reflex withdrawal of the siphon is gradually reduced. But if the animal's tail is then given a strong electric shock, further tapping of the siphon will cause a withdrawal response that is nearer normal. *Compare* SENSITIZATION (sense 3).

disinfectant Any substance that kills or inhibits the growth of disease-producing microorganisms and is in general toxic to human tissues. Disinfectants include cresol, bleaching powder, and phenol. They are used to cleanse surgical apparatus, sickrooms, and household drains and if sufficiently diluted can be used as *antiseptics.

disinhibition (in animal behaviour) The tendency to exhibit displacement behaviour at a point of equilibrium in situations where there are two or more conflicting drives. *See* DISPLACEMENT ACTIVITY.

dispersal The dissemination of offspring of plants or sessile animals. Dispersal provides organisms that are not mobile with a better chance of survival by reducing *competition among offspring and parents. It also promotes the colonization of new habitats. Flowering plants produce fruits or seeds that are dispersed by such agents as wind, water, or animals. Specialized structures have evolved in many species to aid dispersal (*see* FRUIT).

dispersion (in ecology) *See* POPULATION.

dispersive replication A form of *DNA replication in which the original DNA chain breaks and recombines in a random fashion before the double helix structure unwinds and separates to act as a template for messenger *RNA synthesis. There is no evidence that it occurs in nature. *Compare* CONSERVATIVE REPLICATION; SEMICONSERVATIVE REPLICATION.

displacement activity An activity shown by an animal that appears to be inappropriate to its situation. Displacement activities are frequently observed when there is conflict between opposing drives or the animal is thwarted in obtaining a goal. For example, birds in aggressive situations, in which there are simultaneous drives to attack and to flee, may preen their feathers as a displacement activity; the situation has caused *disinhibition of the preening activity, allowing it to take place. Preening and feeding are common displacement activities.

display behaviour Stereotyped movement or posture that serves to influence the behaviour of another animal. Many displays in *courtship and *aggression are conspicuous and characteristic of the species; special markings or parts of the body may be prominently exhibited (e.g. the male peacock spreads its tail in courtship). Other displays are cryptic and make it harder for a predator to recognize the displaying animal as potential prey. For example, geometer moth caterpillars, which look like twigs, hold themselves on plant stems with one end sticking into the air. Alternatively, a prey species may deter a predator by means of a *startle display.

disruptive selection *Natural selection that favours the extremes of a phenotype in a population. It often operates when an environmental factor shows distinct variations, for example high temperatures in summer and low temperatures in winter, with no intermediate forms. In this case the population will be variously adapted to withstand both high and low temperatures. *Compare* BALANCING SELECTION.

dissociation constant Symbol K_d. A measure of the affinity of a receptor for a ligand, or of an enzyme for its substrate. It reflects the tendency of the two entities to dissociate under equilibrium conditions. Hence, the lower the value of K_d, the greater the affinity between the two entities.

distal Describing the part of an organ that is farthest from the organ's point of attachment to the rest of the body. For example, hands and feet are at the distal ends of arms and legs, respectively. *Compare* PROXIMAL.

distal convoluted tubule (second convoluted tubule) The part of a *nephron in the kidney that leads from the thick ascending limb of the *loop of Henle and drains into a *collecting duct. The main function of the distal tubule is to 'fine-tune' the composition of the filtrate. For example, potassium ions (K^+) can be either secreted from the distal tubule into the interstitial fluid of the kidney cortex or actively reabsorbed into the tubules, depending on the potassium status of the body fluid. The tubule can also actively transport other ions, such as sodium, chloride, calcium, and hydrogencarbonate ions.

distribution In statistics, the relationship between the values in a data set and the relative frequency with which each value occurs, especially as presented in a table or graph. A **normal** or **Gaussian distribution** is a continuous distribution that is symmetrical around the mean value and in which a large majority (64%) of the observations fall within one *standard deviation of the mean in either direction (and nearly all fall within two standard deviations); such a distribution forms the familiar symmetrical bell curve when plotted on a graph. In a **skewed** or **asymmetrical distribution**, by contrast, the observations are clustered on one side of the mean and spread out over a wider range on the other. *See also* SIGNIFICANCE.

disulphide bridge (sulphur bridge) A covalent bond (S–S) formed between the thiol groups (–SH) of two cysteine residues, usually in the polypeptide chains of proteins. Easily hydrolysed and prone to rearrangement, these bonds contribute to the tertiary structure of *proteins.

diuresis The production of watery urine in large quantities. Diuresis causes the osmotic pressure of the blood to rise and is counteracted by the *antidiuretic hormone.

diuretic A drug or other agent that increases the rate of urine formation and hence the rate at which water and certain

salts are lost from the body. Many diuretic drugs work by decreasing the reabsorption of sodium and chloride ions from the filtrate in the kidney tubules, so that less water is reabsorbed. They are used to treat fluid retention (oedema) arising from disorders of the heart, kidneys, or other organs, and are used in helping to reduce high blood pressure (hypertension). There are several groups of diuretic drugs, with different modes of action. The most powerful are **loop diuretics**, such as furosemide, which act primarily by blocking $Na^+/K^+/Cl^-$ carriers in cells of the *loop of Henle. Another group consists of the thiazides, such as metolazone, which inhibit Na^+/Cl^- transport in the *distal convoluted tubule. Spironolactone exerts its diuretic effect by blocking the binding of the hormone *aldosterone to its receptors. The **osmotic diuretics**, such as mannitol, act by increasing the osmolarity of the filtrate, and hence increasing urine volume.

diurnal Daily; denoting an event that happens once every 24 hours.

diurnal rhythm *See* CIRCADIAN RHYTHM.

divergence 1. (in neuroanatomy) The branching of the terminal axon of a neuron so that it connects with numerous target cells. For example, each horizontal cell in the retina of the eye has a greatly branched axon ending that can transmit signals to many other cells in the retina. **2.** Divergent evolution. *See* ADAPTIVE RADIATION. *Compare* CONVERGENCE.

divergent evolution *See* ADAPTIVE RADIATION.

diverticulum (*pl.* **diverticula**) A saclike or tubular outgrowth from a tubular or hollow internal organ. Diverticula may occur as normal structures (e.g. the *caecum and *appendix in the alimentary canal) or abnormally, from a weakened area of the organ.

division A category used traditionally in the *classification of plants that consists of one or several similar classes. An example is the Spermatophyta (seed-bearing plants). In modern classification systems the *phylum has replaced the division.

dizygotic twins *See* FRATERNAL TWINS.

***dl*-form** *See* OPTICAL ACTIVITY.

DNA (deoxyribonucleic acid) The genetic material of most living organisms, which is a major constituent of the *chromosomes within the cell nucleus and plays a central role in the determination of hereditary characteristics by controlling *protein synthesis in cells (*see also* GENETIC CODE). It is also found in chloroplasts and mitochondria (*see* EXTRANUCLEAR GENES; MITOCHONDRIAL DNA). DNA is a nucleic acid composed of two chains of *nucleotides in which the sugar is deoxyribose and the bases are *adenine, *cytosine, *guanine, and *thymine (*compare* RNA). The carbon atoms of each deoxyribose unit are numbered 1′ (i.e. 1 'prime') to 5′. Attached to each 5′ carbon atom is a phosphate group, via which a covalent bond forms with the 3′ carbon of the ribose of the adjoining nucleotide. The chains run in opposite directions—they are **antiparallel**—with 3′ and 5′ chain ends apposed at either end of the molecule. The two chains are wound round each other and linked together by hydrogen bonds between specific complementary bases (*see* BASE PAIRING) to form a

Detail of molecular structure of the sugar–phosphate backbone. Each deoxyribose unit is attached via its 5′ carbon atom to a phosphate group, and during DNA replication, new nucleotides are added via the 3′ –OH group. Hence, the chain is said to grow in the 5′ to 3′ direction.

Double helical structure of DNA

The four bases of DNA, showing the hydrogen bonding between base pairs

Molecular structure of DNA

spiral ladder-shaped molecule (**double helix**; *see also* SUPERCOILING). Van der Waals forces between adjacent bases on the same strand also contribute to chain stability. See illustration.

When the cell divides, its DNA also replicates in such a way that each of the two daughter molecules is identical to the parent molecule (*see* DNA REPLICATION). Synthesis of DNA from scratch is a key technology of *synthetic biology and particularly crucial to fulfilling the potential of DNA as a medium for storing digital data. *See also* COMPLEMENTARY DNA.

(⊕) SEE WEB LINKS

- An animated primer on basic aspects of DNA, genes, and heredity, produced for the Cold Spring Harbor Laboratory

DNAase *See* DNASE.

DNA barcode Part of an organism's DNA base sequence that serves to identify which species the organism belongs to. The sequence selected for barcoding should ideally show marked variation between species but minimal variation within species. In many animal groups the internationally agreed standard sequence is a 648bp region of the mitochondrial gene for cytochrome oxidase subunit 1 (*CO1*), whereas in plants the chloroplast genes *matK* and *rbcL* are used instead. The barcode is generated by amplifying a specimen of DNA using specially designed primers with the *polymerase chain reaction, and then sequenced. The base sequence is converted into a graphic image of different coloured lines representing the four constituent bases, similar to a barcode on an item in a supermarket. The Barcode of Life Datasystems (BOLD) has compiled a database containing barcodes for nearly 280000 plant, animal, and fungal species (as of early 2018). An investigator can search the database to find a match with a reference barcode and thus identify which species they are dealing with.

(⊕) SEE WEB LINKS

- Website of the International Barcode of Life project

DNA-binding proteins Proteins that are able to bind to DNA in both eukaryotes and prokaryotes and act as either activators or repressors of *gene expression by controlling the binding of *RNA polymerase to DNA during the process of transcription (*see* TRANSCRIPTION FACTOR). They are also involved in *DNA replication: by binding to the nucleotides in the single DNA strands of the template that have unwound, they stabilize these strands so that they do not rewind. DNA-binding proteins do not include the *histones.

DNA blotting *See* SOUTHERN BLOTTING.

DNA chip *See* DNA MICROARRAY.

DNA cloning *See* GENE CLONING.

DNA-dependent RNA polymerase *See* POLYMERASE.

DNA fingerprinting The generation from a sample of DNA of a set of DNA fragments that are characteristic of that sample. DNA fingerprints can thus enable individuals, strains, species, etc. to be distinguished by variations in the base sequence of their genomes. Various techniques can be used to detect such variations, depending on the organism being studied, existing knowledge of the organism's genome, etc. These include analysis of *restriction fragment length polymorphisms (RFLP) and techniques based on the *polymerase chain reaction, such as simple sequence repeat analysis, rapidly amplified polymorphic DNA (RAPD) analysis, and *amplified fragment length polymorphism. *See also* DNA PROFILING.

DNA hybridization A method of determining the similarity of DNA from different sources. Single strands of DNA from two sources, e.g. different bacterial species, are put together and the extent to which double hybrid strands are formed is estimated. The greater the tendency to form these hybrid molecules, the greater the extent of complementary base sequences, i.e. gene similarity. The method is one way of determining the genetic relationships of species. The same principle applies when using *DNA probes to search for particular base sequences in a sample of DNA, e.g. when screening a *DNA library for a particular cloned fragment or in *DNA microarray technology. The technique of **allele-specific oligonucleotide hybridization** is used to test the DNA from individuals to determine whether they are carriers of disease-causing alleles. Here a short single strand of DNA is synthesized to have a base sequence complementary to that of the target sequence from the DNA sample. Binding of the probe to the prepared sample DNA establishes the presence of the target sequence. *See also* COMPARATIVE GENOMIC HYBRIDIZATION.

DNA library (gene library; gene bank) A collection of cloned DNA fragments representing the entire genetic material of an organism. (i.e. a **genomic library**) or just the genes transcribed in particular cells or tissues at a given time (i.e. a **complementary DNA library**, or **cDNA library**). Genomic libraries facilitate the screening and isolation of any particular gene. They are created by fractionating the entire DNA of an organism into fragments using *restriction enzymes and/or physical methods. These fragments are cloned (*see* GENE CLONING) and the host cells containing the recombinant fragments are centrifuged and frozen; alternatively, the phage *vectors are maintained in culture. Screening for individual genes in the library can be done using specific *DNA probes with the *Southern blotting technique or, via their protein products, using *western blotting. Alternatively, the target DNA can be identified with primer sequences and amplified using *PCR. DNA libraries are thus repositories of raw material for use in genetic engineering. A large genome, such as that of humans, is most conveniently cloned using vectors that can accommodate large fragments of DNA, such as yeast *artificial chromosomes, maintained in cell culture. Complementary DNA libraries are created essentially by first isolating the messenger RNA molecules expressed in a particular cell or tissue, and then using reverse transcriptase to produce DNA copies. These are joined to plasmid or phage vectors and cloned to create the library. A cDNA library thus represents only part of the genome, namely that which is being expressed at a particular time by the cell type in question.

DNA ligase An enzyme that is able to join together two portions of DNA and therefore plays an important role in *DNA repair. DNA ligase is also used in recombinant DNA technology (*see* GENETIC ENGINEERING) as it ensures that the foreign DNA (e.g. the complementary DNA used in *gene cloning) is bound to the plasmid into which it is incorporated.

DNA methylation The addition of methyl groups to constituent bases of DNA. In both prokaryotes and eukaryotes certain bases of the DNA generally occur in a methylated form, and in eukaryotes it has a major role in *epigenetic control of DNA transcription. The enzyme DNA methyltransferase catalyses the addition of a methyl group ($-CH_3$) to certain cytosine residues, particularly ones adjacent to guanine residues, which are abundant in the so-called *CpG islands frequently associated with the promoter regions of genes. The pattern of such methylated cytosine residues in the DNA of an individual is called the **methylome**. These $5'$-methylcytosine residues tend to bind repressor proteins, thereby inactivating the gene. Such changes in expression patterns are heritable even though no changes in DNA sequence have occurred. However, many genes become demethylated in sperm, eggs, and early developing embryos. Only later, as cells become specialized for certain functions, does DNA methylation become widespread again as a means of silencing redundant genes. In bacteria this methylation protects the cell's DNA from attack by its own restriction enzymes, which cleave foreign unmethylated DNA and thereby help to eliminate viral DNA from the bacterial chromosome. Methylation is also important in helping *DNA repair enzymes to distinguish the parent strand from the progeny strand when repairing mismatched bases in newly replicated DNA. *See* CHROMATIN REMODELLING.

DNA microarray (DNA chip) A *micro-array containing numerous small DNA molecules, used, for example, to analyse gene transcripts or to detect mutations of specific genes. Oligonucleotide DNA microarrays consist of thousands of short synthetic single-stranded DNA molecules, each comprising 20–25 nucleotides and all with unique sequences designed to complement and bind to specific target nucleotide sequences. Spotted microarrays incorporate much larger DNA fragments obtained from cloned DNA. Such arrays give a quick and convenient method of quantifying gene expression, by determining the total output of messenger RNAs (mRNAs) (i.e. the *transcriptome) of a cell or tissue. This involves adding fluorescent labels to the mRNAs, or in some cases to their single-stranded complementary DNAs (cDNAs), and then incubating the labelled nucleic acids with the DNA microarray. The nucleic acids bind to complementary oligonucleotides on the microarray, and excess nucleic acids are washed off. The microarray is illuminated so that the labels fluoresce, and a computerized scanner measures the intensity of fluorescence at each coordinate on the microarray, and thereby the amount of bound RNA or cDNA. DNA microarrays can also be designed to detect mutations in particular genes, for example the *BRCA* genes involved in hereditary forms of breast cancer. An individual's DNA is denatured, and its binding to a microarray is compared with that of normal (control) DNA on the same microarray. Any disparities between the two binding patterns will pinpoint sequences from the individual with possible abnormalities, enabling closer examination. Because of faster and cheaper DNA sequencing methods, the relative amounts of mRNAs in samples are increasingly being analysed directly using *RNA sequencing, as a preferred alternative to microarrays.

DNA photolyase An enzyme found in bacteria and certain lower eukaryotes (e.g. yeasts) that repairs damage to DNA induced by exposure to ultraviolet (UV) light. In the dark it binds to thymine dimers formed by the UV light, and on subsequent exposure to blue light it absorbs the light energy and

uses it to split the dimer and restore normal structure. This process is called **photoreactivation**; the DNA photolyase of *E. coli* is encoded by the *phr* gene (for 'photoreactivation'). *Compare* CRYPTOCHROME.

DNA polymerase *See* POLYMERASE.

DNA probe A single-stranded DNA or RNA fragment used in genetic engineering to search for a particular gene or other DNA sequence. The probe has a base sequence complementary to the target sequence and will thus attach to it by *base pairing. By labelling the probe with a radioactive isotope or fluorescent label it can be identified on subsequent separation and purification. Probes of varying lengths, up to about 100 nucleotides, can be constructed in the laboratory. They are used in the *Southern blotting technique to identify particular DNA fragments, for instance in conjunction with *restriction mapping to diagnose gene abnormalities or to map certain sequences. *See also* DNA MICROARRAY.

DNA profiling (DNA fingerprinting; genetic fingerprinting) A technique in which an individual's DNA is analysed to reveal the pattern of repetition of certain short nucleotide sequences, called *short tandem repeats (STRs), throughout the genome. This pattern is claimed to be unique to the individual concerned, and the technique is therefore used for identification purposes in forensic science and paternity disputes, and in veterinary science. Sufficient DNA can be obtained from very small samples of body tissue, such as blood, semen, or hair. Usually a person's DNA is analysed to reveal the length of STRs at each of certain genetic loci distributed on different chromosomes. For example, samples for the UK's National DNA Database are based on analysis of a panel of 16 selected STR loci plus a gender identifier. The STR sequences are identified and amplified by the polymerase chain reaction using a commercially available kit, and the lengths of each sequence determined by a DNA sequencer and shown as an output trace of the fluorescence analyser. The results can then be compared with similar data stored on databases. *See also* MICROSATELLITE DNA; RANDOMLY AMPLIFIED POLYMORPHIC DNA; VARIABLE NUMBER TANDEM REPEATS.

DNA repair A variety of mechanisms that help to ensure that the genetic sequence, as expressed in the DNA, is maintained and that errors that occur during *DNA replication, by mutation, are not allowed to accumulate. An error in the genetic sequence could cause cell death by interfering with the replication process. The mechanisms work because DNA is made up of two complementary strands. A damaged section of a strand, or a mismatched base, can be removed by enzymes and replaced by the correct form by DNA *polymerases. The phosphodiester backbone is then sealed by *DNA ligase. *See* EXCISION REPAIR; MISMATCH REPAIR; POSTREPLICATIVE REPAIR; PROOFREADING.

DNA replication The process whereby DNA makes exact copies of itself, which is controlled by the enzyme DNA *polymerase. Replication occurs at rates of between 50 nucleotides per second (in mammals) and 500 nucleotides per second (in bacteria). The hydrogen bonds between the complementary bases on the two strands of the parent DNA molecule break and the strands unwind, forming a Y-shaped **replication fork**. Each strand acts as a template for the synthesis of a new one complementary to itself (*see* DNA-BINDING PROTEINS; PRIMOSOME). DNA polymerases move along the two single strands linking free nucleotides to their complementary bases (*see* BASE PAIRING) on the templates. The process continues until all the nucleotides on the templates have joined with appropriate free nucleotides and two identical molecules of DNA have been formed. This process is known as **semiconservative replication** as each new molecule contains half of the original parent DNA molecule (*compare* CONSERVATIVE REPLICATION; DISPERSIVE REPLICATION). DNA polymerases can add nucleotides only to the 3´ end of the growing DNA strand; hence elongation occurs in the 5´ to 3´ direction, thus preserving the antiparallel orientation of the new strand to the template strand (*see* DISCONTINUOUS REPLICATION). Sometimes mutations occur that may cause the exact sequence of the parent DNA not to be replicated. However, *DNA repair mechanisms reduce this possibility. In bacteria and other prokaryotes, replication starts with the binding of a **pre-replication complex** of proteins, including DNA polymerase, at a specific site (called *ori*, for **origin of replication**) on the chromosome. In eukaryotes the various proteins involved in DNA replication form a **replication complex**, multiple copies of which may be present in the nuclear matrix. Replication is initiated at numerous *ori* sites along each DNA molecule and proceeds simultaneously in both directions at each site.

DNase (deoxyribonuclease; DNAase) An enzyme that catalyses the cleavage of DNA. DNase I is a digestive enzyme, secreted by the pancreas, that degrades DNA into shorter nucleotide fragments. Many other *endonucleases and *exonucleases cleave DNA, including the *restriction enzymes and enzymes involved in DNA repair and replication.

DNA sequencing The process of elucidating the nucleotide sequence of a DNA fragment or of an entire genome (i.e. genome sequencing). Sequencing and analysing genomic DNA (or complementary DNA synthesized from RNA) now lies at the heart of medicine, genetics, and many other fields, providing rapid means of pinpointing genes and their disease-causing mutations, and enabling comparison of gene sequences from different species. A landmark was the publication of the first finished sequence of the human genome in 2003 (*see* HUMAN GENOME PROJECT), since when the genomes of organisms from every domain of life have been unravelled. **First-generation sequencing** relies chiefly on the Sanger method (named after Frederick Sanger), also called the dideoxy method or chain termination method, and introduced in the mid-1970s. This involves synthesizing a new DNA strand using as template single-stranded DNA from the sample being sequenced. Synthesis of the new

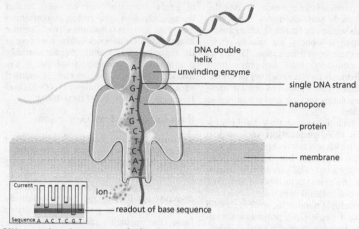

DNA double helix
unwinding enzyme
single DNA strand
nanopore
protein
membrane
Current
ion
readout of base sequence
Sequence A A C T C G T

DNA sequencing using nanopore technology

strand is stopped at any of the four bases by adding the corresponding dideoxy (dd) derivative of the deoxyribonucleoside phosphates; for example, by adding ddATP the synthesis terminates at an adenosine; by adding ddGTP it terminates at a guanosine, etc. The fragments, which comprise fluorescently labelled nucleotides, are subjected to electrophoresis and scanned by a fluorescence detector. The Sanger method can easily be adapted to sequencing RNA, by making single-stranded DNA from the RNA template using the enzyme *reverse transcriptase. This enables, for example, sequencing of ribosomal RNA for use in *molecular systematics. Furthermore, by carrying out electrophoresis in capillaries (instead of on slab gels) and using fluorescent dyes as labels instead of radioisotopes (as originally), the Sanger method has been fully automated. After separation of the fragments, the products of all four reactions are detected by fluorescence spectroscopy and analysed by computer, which gives a printout of the base sequence.

Advances in miniaturization, more sophisticated separation, labelling, and detection techniques, coupled with greater computing power led to the development of **second-generation** (or **next-generation**) **sequencing (NGS)** methods, introduced from around 2005 onwards. These high-throughput methods dramatically speeded up the process and reduced the cost, so that an entire small genome can be sequenced in a day. A key factor is their ability to read millions or even billions of DNA fragments in parallel. Although different approaches to sequencing are employed, they can generally be characterized as 'sequencing by synthesis'. The DNA sample is broken into relatively short fragments, typically several hundred bp long, and adaptor sequences are ligated to the ends of the fragments. This library of fragments is then amplified (e.g. using a form of *polymerase chain reaction) to form clusters of identical fragments immobilized on a substrate, such as a micrometre-sized bead or a 'flow cell'. The fragments in each cluster then serve as templates for the assembly of new DNA fragments having a complementary base sequence, through sequential flooding and removal of known nucleotides. Incorporation of each nucleotide into the growing strand is detected in various ways, for instance by generation of light, fluorescence, or a pH change. A computer program identifies overlapping sequences among the fragment reads and assembles them into contiguous sequences (contigs), after eliminating adapter sequences, low-quality reads, and other extraneous information. These can then be compared with sequences from other individuals (*see* ALIGNMENT) or organisms, enabling features such as open reading frames, regulatory regions, and mutations to be identified.

These techniques are now being supplanted by **third-generation sequencing** methods, which rely not on amplification of short DNA fragments but on sequencing much longer single DNA molecules in real time, in some cases up to 1000 kb long. For example, one method uses microfabricated chips of multiple nanowells, with each well containing a single DNA polymerase molecule. The incorporation of a single nucleotide to a template strand can be detected by a fluorescent signal from that specific well. Another technology rapidly gaining ground is **nanopore sequencing** (see illustration). Double-stranded DNA is denatured and a single strand is made to pass through a nano-scale pore in a membrane, across which a voltage is applied. The ion current through the nanopore changes according to the size of each nucleotide as it traverses the channel. Hence the sequence of bases can be inferred from changes in the current. This simplified approach can be incorporated in small portable devices (the size of a mobile phone) that link to a computer via a USB port. This opens up vast new potential for DNA sequencing in small institutions and in the field. *See also* GENOME PROJECT.

Dobzhansky–Muller model A model that describes how changes at multiple genetic loci can lead to subpopulations of a species becoming unable to interbreed successfully. This reproductive isolation eventually results in the formation of new species. The model assumes that subpopulations of an ancestral species become separated, evolve independently, and over time accumulate new alleles for many different genes. Subsequent breeding between individuals from the different subpopulations will produce sterile hybrids because of incompatibility between the products of each parental set

of genes. Incompatibility between related lineages can also arise owing to rearrangements in the chromosomes (e.g. *centric fusion) within a subpopulation, resulting in the chromosomes' inability to pair normally during sex cell formation (meiosis) in the hybrid offspring. The model is named after Theodosius Dobzhansky (1900-75) and Hermann Joseph Muller (1890-1967).

dodecanoic acid *See* LAURIC ACID.

domain 1. (in biochemistry) A functional unit of the tertiary structure of a *protein. It consists of chains of amino acids folded into *alpha helices and *beta sheets to form a globular structure. Different domains are linked together by relatively straight sections of polypeptide chain to form the protein molecule. Domains allow a degree of movement in the protein structure. *See also* FINGER DOMAIN. **2.** (in taxonomy) The highest taxonomic category, consisting of one or more *kingdoms. Living organisms are divided into three domains: *Archaea, *Bacteria, and *Eukarya (eukaryotic organisms).

dominance hierarchy *See* DOMINANT.

dominant 1. (in genetics) Describing the *allele that is expressed in the *phenotype when two different alleles of a gene are present in the cells of an organism. For example, the height of garden peas is controlled by two alleles, 'tall' (*T*) and 'dwarf' (*t*). When both are present (*Tt*), i.e. when the cells are *heterozygous, the plant is tall since *T* is dominant and *t* is *recessive. Various mechanisms can cause one allele to exert dominance over another in determining phenotype. For example, the recessive allele may simply be defective in some way, and unable to produce any protein product, or at least sufficient amounts of product, so its effects are only apparent when both alleles are recessive (i.e. homozygous recessive). Alternatively, the dominant allele may be aberrant in some way, and hamper the effects of its normal recessive counterpart, e.g. by blocking a metabolic pathway. Another possibility is that the dominant allele codes for a modified protein that evades normal control mechanisms and overrides the normal, recessive allele—a so-called gain-of-function mutation. *See also* CODOMINANCE; INCOMPLETE DOMINANCE. **2.** (in ecology) Describing the most conspicuously abundant and characteristic species in a *community. The term is usually used of a plant species in plant ecology; for example, pine trees in a pine forest. **3.** (in animal behaviour) Describing an animal that is allowed priority in access to food, mates, etc., by others of its species because of its success in previous aggressive encounters. Less dominant animals frequently show *appeasement behaviour towards a more dominant individual, so overt *aggression is minimized. In a stable group there may be a linear **dominance hierarchy** or **peck order** (so called because it was first observed in domestic fowl), with each animal being subservient to those above it in the hierarchy and taking precedence over those below it.

donor 1. (in surgery) An individual whose tissues or organs are transferred to another (the **recipient**). Donors may provide blood for transfusion or a kidney or heart for transplantation. **2.** (in genetics) A cell that contributes genetic material for insertion into another cell, for example to produce a transgenic cell by genetic engineering. **3.** (in chemistry) A chemical species (e.g. a molecule, group, or atom) that donates electrons, atoms, or groups to another chemical species.

dopa (dihydroxyphenylalanine) A derivative of the amino acid tyrosine. It is found in particularly high levels in the adrenal glands and is a precursor in the synthesis of *dopamine, *noradrenaline, and *adrenaline. The laevorotatory form, **L-dopa (levodopa)**, is administered in the treatment of Parkinson's disease, in which brain levels of dopamine are reduced.

dopamine A *catecholamine that is a precursor in the synthesis of *noradrenaline and *adrenaline. It also functions as a neurotransmitter, especially in the brain.

dormancy An inactive period in the life of an animal or plant during which growth

slows or completely ceases. Physiological changes associated with dormancy help the organism survive adverse environmental conditions. Annual plants survive the winter as dormant seeds while many perennial plants survive as dormant tubers, rhizomes, or bulbs. *Hibernation and *aestivation in animals help them survive extremes of cold and heat, respectively. *See also* DIAPAUSE.

dorsal Describing the surface of a plant or animal that is farthest from the ground or other support, i.e. the upper surface. In vertebrates, the dorsal surface is that down which the backbone runs. Thus in upright (bipedal) mammals, such as humans and kangaroos, it is the backward-directed (*posterior) surface. *Compare* VENTRAL.

dorsal aorta The artery in vertebrate embryos that transports blood from the *aortic arches to the trunk and limbs. In adult fish it is a major artery that carries oxygenated blood from the efferent branchial arteries to branches that supply the body organs; in adult tetrapods it arises from the *systemic arch (*see* AORTA). *Compare* VENTRAL AORTA.

dorsal lip *See* SPEMANN'S ORGANIZER.

dorsal root The part of a spinal nerve that enters the *spinal cord on the dorsal side and contains only sensory fibres. The cell bodies of these fibres form the **dorsal root ganglion** (*see* GANGLION), a swelling in the root that lies just outside the cord. *Compare* VENTRAL ROOT.

dose (in radiobiology) A measure of the extent to which matter has been exposed to *ionizing radiation. The **absorbed dose** is the energy per unit mass absorbed by matter as a result of such exposure. The SI unit is the gray, although it is often measured in rads (1 rad = 0.01 gray; *see* RADIATION UNITS). The **maximum permissible dose** is the recommended upper limit of absorbed dose that a person or organ should receive in a specified period according to the International Commission on Radiological Protection.

dot-blot A method for detecting a specific sequence of nucleotides in a DNA or RNA molecule. The nucleic acid sample is adsorbed onto a nitrocellulose filter and an appropriate *DNA probe, which matches the specific sequence being investigated, is added. After a period of incubation any excess probe is washed off and the nucleotide sequence under investigation can be detected by *autoradiography. This technique has been miniaturized and vastly scaled up in *microarray technology.

double circulation The type of circulatory system that occurs in mammals, in which the blood passes through the heart twice before completing a full circuit of the body (see illustration). Blood is pumped from the heart to the lungs and returns to the heart before being distributed to the other organs and tissues of the body. The heart is divided into two separate compartments to prevent oxygenated blood returning from the lungs from mixing with deoxygenated blood from the other parts of the body. *See also* PULMONARY CIRCULATION; SYSTEMIC CIRCULATION. *Compare* SINGLE CIRCULATION.

double fertilization A process, unique to flowering plants, in which two male gamete nuclei, which have travelled down the pollen tube, separately fuse with different female nuclei in the *embryo sac. The

Double circulation in a mammal

first male nucleus fuses with the egg cell to form the zygote; the second male nucleus fuses with the two *polar nuclei to form a triploid nucleus that develops into the endosperm.

double helix *See* DNA.

double recessive An organism with two *recessive alleles for a particular characteristic.

down feathers (plumules) Small soft feathers that cover and insulate the whole body of a bird. In nestlings they are the only feathers; in adults they lie between and beneath the *contour feathers. Down feathers have a fluffy appearance as their *barbs are not joined together to form a smooth vane.

downregulation *See* UPREGULATION.

Down's syndrome A congenital disability due to a chromosome defect in which there are three copies of chromosome no. 21 instead of the usual two (*see* TRISOMY). The affected individual has a short broad face and slanted eyes (as in the Mongolian races), short fingers, and weak muscles. Down's syndrome can be detected before birth by *amniocentesis. It is named after the British physician John Down (1828–96), who first studied the incidence of the disorder.

draft sequence (in bioinformatics) A preliminary assembly of data subsets to form a semblance of a complete base sequence of an entire chromosome or genome. An initial goal in sequencing large genomes is generally the production of a draft sequence, which is subsequently corrected and refined to form the *finished sequence. For example, the draft sequence of the human genome produced by the International Human Genome Sequencing Consortium in 2000 was based on sequencing of over 29 000 cloned fragments of human DNA, representing the human genome many times over. It contained numerous sequence gaps (nearly 150 000), many incorrectly assembled fragments, and averaged one error every 100 base pairs.

dragonflies *See* ODONATA.

DRiPs Defective ribosomal products: peptides produced in cells by enzymic digestion of the natural waste matter of protein synthesis, such as misfolded proteins and prematurely terminated polypeptides arising from errors in the synthetic machinery. DRiPs play a role in the immune system as an important source of self antigen for presentation at the cell surface in combination with *MHC class I proteins.

DRIPs A group of single-celled eukaryotes, also called Mesomycetozoea or Ichthyosporea, that live chiefly as parasites of fish and other aquatic animals. The acronym derives from the initial letters of the names of the four genera known to the discoverers of the group: Dermocystidium, 'Rosette agent' (now given the formal taxonomic name *Sphaerothecum destruens*), *Ichthyophonus*, and *Psorospermium*. The group includes several commercially important parasites and the causative agent (*Rhinosporidium seeberi*) of the human disease rhinosporidiosis, which affects the nose. Molecular systematics shows DRIPs to be among the closest single-celled relatives of the animals, forming a branch in the phylogenetic tree near to the divergence of fungi and animals.

drive The *motivation that results in an animal performing a particular activity. There are two types: **primary drive**, which arises as a direct result of tissue requirements (e.g. the need for food or water); and **secondary drive**, resulting from learned behaviour.

drone A fertile male in a colony of social bees, especially the honeybee (*Apis mellifera*). The drones die after mating with the queen bee as the male reproductive organs explode within the female.

Drosophila A genus of fruit flies often used in genetic and developmental biology research because the larvae possess giant chromosomes in their salivary glands (*see* POLYTENY). These chromosomes have conspicuous transverse bands, which can be studied microscopically to reveal chromosome mutations and gene activity. Fruit flies have a short life cycle and produce a large number of offspring, which also makes them a good model animal for genetic research.

drug Any chemical substance that alters the physiological state of a living organism. Drugs are widely used in medicine for the prevention, diagnosis, and treatment of diseases; they include *analgesics, *antibiotics, anaesthetics, *antihistamines, and *anticoagulants. Some drugs are taken solely for the pleasurable effects they induce; these include *narcotics; stimulants, such as cocaine and *amphetamine; *hallucinogens, such as LSD; and some tranquillizers. Many of these drugs are habit-forming and their use is illegal.

drupe (pyrenocarp) A fleshy fruit that develops from either one or several fused carpels and contains one or many seeds. The seeds are enclosed by the hard protective endocarp (*see* PERICARP) of the fruit. Thus the stone of a peach is the endocarp containing the seed. Plums, cherries, coconuts, and almonds are other examples of one-seeded drupes; holly and elder fruits are examples of many-seeded drupes. *See also* ETAERIO.

dry mass The mass of a biological sample after the water content has been removed, usually by placing the sample in an oven. The dry mass is used as a measure of the *biomass of a sample.

Dryopithecus A genus of extinct apes, fossils of which have been found in Europe and Asia and dated to the mid-Miocene (about 9–13 million years ago). Fossils of *Dryopithecus* and of the similar genus *Proconsul* are often referred to as **dryopithecines**. Dryopithecines are believed to have split into several lines, some of which survived to give rise to the chimpanzees, gorillas, early hominins, and orang-utans.

duct A tube or passage in an organism that is involved in the secretion or excretion of substances (*see* GLAND).

ductless gland *See* ENDOCRINE GLAND.

ductus arteriosus A channel that connects the pulmonary artery with the aorta in the mammalian fetus and therefore allows blood to bypass the inactive lungs of the fetus. Derived from the fifth *aortic arch, it normally closes soon after birth.

duodenum (*pl.* **duodenums** or **duodena**) The first section of the *small intestine of vertebrates. It is the site where food from the stomach is subjected to the action of bile (from the bile duct) and pancreatic enzymes (from the pancreatic duct) as well as the enzymes secreted by digestive glands in the duodenum itself (*see* SUCCUS ENTERICUS), which are required in the breakdown of proteins, carbohydrates, and fats. By neutralizing the acidic secretions of the stomach, the duodenum provides an alkaline environment necessary for the action of the intestinal enzymes.

duplex Describing a biological molecule comprising two cross-linked polymeric chains oriented lengthways side by side. The term is applied particularly to the double-stranded structure of *DNA.

duplication (in genetics) The doubling or repetition of part of a chromosome, which generally originates during the *crossing over phase of meiosis. Occasionally this type of *chromosome mutation may have beneficial effects on a population. For example, a beneficial duplication resulted in the evolution of four types of haemoglobin in humans and apes from a single form. One of these types of haemoglobin (gamma or fetal haemoglobin) has a greater affinity for oxygen and maximizes fetal uptake of oxygen from the mother's blood.

dura mater The outermost and toughest of the three membranes (*meninges) that surround the central nervous system in vertebrates. It lies adjacent to the skull and its purpose is to protect the delicate inner meninges (the *arachnoid membrane and the *pia mater).

duramen *See* HEARTWOOD.

dwarfism *See* GROWTH HORMONE.

dynein A protein, found in many eukaryotic cells, that possesses *ATPase activity and binds to microtubules. Dynein is associated with the microtubules of cilia and *flagella; it moves along the microtubules, using energy derived from the hydrolysis of ATP, causing the cilium or flagellum to bend. Dynein in the cytoplasm of cells causes the movement of organelles along the microtubules.

dynorphin A peptide produced as a neurotransmitter in various parts of the brain and spinal cord that binds to kappa *opioid receptors. The dynorphins are thus classed as endogenous *opioids (with the endorphins and enkephalins), causing analgesia (pain relief) among other effects. They are thought to play a role in depression and certain types of drug addiction.

dysphotic zone *See* EUPHOTIC ZONE.

dystrophic Describing a body of water, such as a lake, that contains large amounts of undecomposed organic matter derived from terrestrial plants. Dystrophic lakes are poor in dissolved nutrients and therefore unproductive; they are common in peat areas and may develop into peat bogs. *Compare* EUTROPHIC; MESOTROPHIC; OLIGOTROPHIC.

E

ear The sense organ in vertebrates that is specialized for the detection of sound and the maintenance of balance. It can be divided into the *outer ear and *middle ear, which collect and transmit sound waves, and the *inner ear, which contains the organs of balance and (except in fish) hearing (see illustration). The term ear is often used for the *pinna of the mammalian outer ear.

eardrum *See* TYMPANUM.

ear ossicles Three small bones—the **incus** (**anvil**), **malleus** (**hammer**), and **stapes** (**stirrup**)—that lie in the mammalian *middle ear, forming a bridge between the tympanum (eardrum) and the *oval window. The function of the ossicles is to transmit (and amplify) vibrations of the tympanum across the middle ear to the oval window, which transfers them to the *inner ear. Muscles of the middle ear constrict the movement of the ossicles. This serves to safeguard the ear from damage caused by excessively loud noise.

earwigs *See* DERMAPTERA.

eccrine secretion *See* SECRETION.

ecdysis (moulting) 1. The periodic loss of the outer cuticle of arthropods and other members of the clade *Ecdysozoa (e.g. nematodes, onychophorans, and tardigrades) to enable growth and metamorphosis within the constraints of a hard body covering. It starts with the release of enzymes from underlying cells that loosen the inner part of the old cuticle, followed by the formation of a new soft cuticle. The remains of the old cuticle then split; the animal emerges and absorbs water or

Structure of the mammalian ear

swallows air and increases in size while the new cuticle is still soft. This cuticle is then hardened with chitin and lime salts. In insects and crustaceans ecdysis is controlled chiefly by three groups of hormones: ecdysteroids, such as *ecdysone; neuropeptides, such as *eclosion hormone and *ecdysis-triggering hormone; and sesquiterpenoids, such as *juvenile hormone. **2.** The periodic shedding of the outer layer of the epidermis of reptiles (except crocodiles) to allow growth to occur.

ecdysis-triggering hormone A peptide hormone, produced by neurosecretory cells in insects, that works in conjunction with other hormones (e.g. *ecdysone and *eclosion hormone) to trigger moulting of the cuticle (*see* ECDYSIS).

ecdysone A steroid hormone, secreted by a pair of **prothoracic glands** in the thorax of insects and by *Y organs in crustaceans, that stimulates moulting (*see* ECDYSIS) and metamorphosis. In insects its release is stimulated by *prothoracicotropic hormone, and its activity is modulated by *juvenile hormone (JH). Ecdysone binds to receptors in epidermal cells, and the resulting ecdysone–receptor complex triggers the expression of genes encoding proteins that digest the old cuticle and form the new one. This gives rise to the next larval stage, provided levels of juvenile hormone remain high. If levels of JH fall, then ecdysone triggers formation of a pupa and metamorphosis ensues. Some plants contain **phytoecdysteroids**, which are structurally similar to ecdysone. They may help to protect the plant by interfering with the moulting cycle of insect pests that consume plant tissue.

Ecdysozoa A clade of protostome animals, based chiefly on molecular systematics, whose members include the arthropods, nematodes, onychophorans, and tardigrades. All are characterized by their habit of periodic moulting (i.e. ecdysis, hence the name of the clade).

e-cell A computer simulation of a living cell based on information obtained from genomics, proteomics, metabolomics, and other data drawn from real cells.

ECG *See* ELECTROCARDIOGRAM.

Echinodermata A phylum of marine invertebrates that includes the sea urchins, starfish, brittlestars, and sea cucumbers. Echinoderms have an exoskeleton (**test**) of calcareous plates embedded in the skin. In many species (e.g. sea urchins) spines protrude from the test. A system of water-filled canals (the **water vascular system**) provides hydraulic power for thousands of **tube feet:** saclike protrusions of the body wall used for locomotion, feeding, and respiration. Echinoderms have a long history: fossils of primitive echinoderms are known from rocks over 500 million years old. Like chordates, they are *deuterostomes. There are five clades of living echinoderms: Asteroidea (sea stars and sea daisies); Ophiuroidea (brittle stars); Echinoidea (sea urchins and sand dollars); Crinoidea (sea lilies and feather stars); and Holothuroidea (sea cucumbers).

((()) SEE WEB LINKS

• Overview of echinoderm characteristics and phylogeny from the Tree of Life project

echolocation A method used by some animals (such as bats, dolphins, and certain birds) to detect objects in the dark. The animal emits a series of high-pitched sounds that echo back from the object and are detected by the ear or some other sensory receptor. From the direction of the echo and from the time between emission and reception of the sounds the object is located, often very accurately.

eclosion hormone A peptide hormone produced by neurosecretory cells in the insect brain that triggers the sequence of events leading to emergence of the adult from the pupa, i.e. eclosion. It is also involved, with other hormones (e.g. *ecdysone), in moulting of the cuticle by immature stages (*see* ECDYSIS).

ECM *See* EXTRACELLULAR MATRIX.

E. coli See ESCHERICHIA COLI.

ecological   equivalents Unrelated organisms that occupy similar habitats and resemble each other. Ecological equivalents result from *convergent evolution. For example, sharks (fish) and dolphins (mammals) live in a marine habitat and superficially resemble each other.

ecological footprint The amount of the environment that a person, business, country, or other entity requires to produce the goods and services it consumes and to deal with its waste products. The concept underlies the Ecological Footprint open source accounting tool, which attempts to quantify the amount of productive land and water needed to supply food, fibres, fish, livestock, timber, and land space for a given population and to absorb its waste, such as carbon emissions and solid waste. The footprint is measured in **global hectares** (gha)—the globally averaged amount of productive land and water available on the planet, which currently equates to about 1.7 gha per person. Hence an individual whose footprint exceeds this is running an ecological deficit. For example, in 2013 it was estimated that individuals in the UK had a footprint of 5.1 gha, while for people in the USA it was 8.6 gha, both far greater than those countries' **biocapacity**, i.e. the available cropland, grazing land, forest land, fishing grounds, and built-up land needed to service the lifestyle of the population. This indicates that such lifestyles are far from *sustainable given current technologies.

ecological niche 1. A summary of the conditions and resources that must be available in order for a species to maintain a population in the long term. A niche therefore is not a place but a conceptual volume with numerous dimensions (an *n*-dimensional hypervolume), such as temperature, humidity, and food supply, each defined by the tolerable range for that particular species. This concept of niche is based on a proposition made in 1957 by British-born US zoologist G. E. Hutchinson (1903–91) and has subsequently been refined and widely adopted. The **fundamental niche** of a species describes its potential hypervolume; this is determined principally by the physiological characteristics of the species and is generally transferable from place to place. However, the **realized niche** is the more limited hypervolume that a species can occupy when the effects of competition, predation, and parasitism by other species are considered. According to the *competitive exclusion principle, two species cannot occupy the same ecological niche in both space and time. **2.** The status or role of an organism in its environment.

ecological pyramid *See* PYRAMID OF BIOMASS; PYRAMID OF ENERGY; PYRAMID OF NUMBERS.

ecology The study of the interrelationships between organisms and their natural environment, both living and nonliving. For this purpose, ecologists study organisms in the context of the *populations and *communities in which they can be grouped and the *ecosystems of which they form a part. The study of ecological interactions provides important information on the nature and mechanisms of evolutionary change. Advances in ecology have placed in harsh focus the effects of human activities on the environment (*see* POLLUTION) and have resulted in a greater awareness of the importance of *conservation and the need to ensure that development is *sustainable.

ecosystem A biological *community and the physical environment associated with it. Ecosystems seldom have discrete boundaries, and they can range in scale from a rotting log to an entire forest. Nutrients pass between the different organisms in an ecosystem in definite pathways; for example, nutrients in the soil are taken up by plants, which are then eaten by herbivores, which in turn may be eaten by carnivores (*see* FOOD CHAIN). Organisms are classified on the basis of their position in an ecosystem into various *trophic levels. Nutrients and energy move round ecosystems in loops or cycles (in the case above, for example, nutrients are returned to the soil via animal wastes and decomposition). *See* CARBON CYCLE; NITROGEN CYCLE.

( ) SEE WEB LINKS
• The latest Millennium Ecosystem Assessment Reports commissioned by the United Nations

ecosystem services The benefits that people receive from ecosystems. These benefits can be classified into several categories. **Provisioning services** include the products that humans derive from ecosystems, such as food, fibres, fuel, drinking water, and medicinal products. **Regulating services** encompass such processes as control of the climate, purification of air and water, amelioration of erosion, reduction of pests and diseases, and provision of pollination mechanisms. **Cultural services** comprise the nonmaterial benefits that accrue from the landscape, such as recreational pursuits, spiritual renewal, and aesthetic experiences. Underlying all these are the fundamental **supporting services** of all ecosystems, such as primary production (i.e. photosynthesis by green plants, algae, etc.), nutrient cycling (e.g. carbon, nitrogen, sulphur cycles), water cycling, and soil formation. The notion that humans depend on their environment, which was deeply entrenched in early societies, came to be widely disregarded with the migration of populations to urban centres and increasing reliance on technology to support those populations. However, in recent decades the widespread degradation of many of the world's ecosystems has seen the concept of ecosystem services acquire greater prominence, chiefly as a means of assessing the economic value of particular ecological benefits and the cost of their destruction.

ecotoxicology The study of the harmful effects of chemicals on the environment and living organisms. Numerous chemicals enter terrestrial and aquatic environments from a wide range of sources, including geological deposits of metals, salts, or hydrocarbons (e.g. coal, petroleum, gas), mining, industries, farming, horticulture, forestry, power generation, vehicles, homes, businesses, and living organisms (i.e. biotoxins). Toxicants individually or in combination interfere with the life processes of microbial, plant, and animal communities and thereby impact on food webs and ecosystems, with possible consequences for human health and society. Ecotoxicologists study the physical and chemical properties of materials,

monitor how they behave in the environment, and use their knowledge to predict how a parent chemical and its metabolites are transferred among organisms; the potential toxic impacts; and also whether clean-up and bioremediation measures may be usefully deployed. They are also concerned with developing 'green chemistry', which seeks to make and use safer compounds that have minimal ecological impact. *See also* BIOACCUMULATION.

ecotype A subgroup of a species that has characteristic genetically determined adaptations to its local environment. In some cases individuals belonging to different ecotypes cannot interbreed, for example where accumulated genetic differences are too great. An ecotype is a broader category than a *biotype.

ectoderm The external layer of cells of the *gastrula, which will develop into the epidermis and related structures (including hair, claws, sweat glands, teeth, etc.) and the nervous system (including the eyes and ears) in the adult. *See also* GERM LAYERS.

ectoparasite A parasite that lives on the outside of its host's body. *See* PARASITISM.

ectoplasm *See* CYTOPLASM.

Ectoprocta *See* BRYOZOA.

ectotherm (poikilotherm) An animal that maintains its body temperature by absorbing heat from the surrounding environment. Most animals are ectotherms, the exceptions being mammals, birds, some fishes and insects, and a few reptiles. Ectotherms are often described as being **cold-blooded** and are unable to regulate their body temperature metabolically. *See* POIKILOTHERMY. *Compare* ENDOTHERM; HETEROTHERM.

ectotrophic mycorrhiza *See* MYCORRHIZA.

edaphic factor An *abiotic factor relating to the physical or chemical composition of the soil found in a particular area. For example, very alkaline soil may be an edaphic factor limiting the variety of plants growing in a region.

eddy covariance technique A micrometeorological technique for simultaneously measuring fluxes of carbon dioxide, water vapour, and energy (heat) above a plant canopy and hence the net exchange of carbon between an ecosystem and the atmosphere. It is now widely used in ecology, for example for determining the amounts of carbon dioxide that are absorbed or emitted by different types of vegetation cover and whether particular sites are sinks or sources of carbon. It involves the installation of a tower that reaches above the tallest vegetation (e.g. above the tree canopy) and carries sensitive monitoring instruments, notably an infrared gas analyser (to monitor carbon dioxide and water vapour concentrations) and an anemometer to detect temperature changes and measure the eddy currents created by the gas fluxes between the atmosphere and the site. Other meteorological data are also collected. Integration of these data shows the extent of carbon and energy fluxes due to photosynthesis and respiration of not just the vegetation but the entire ecological community at the site, including soil organisms. Monitoring can be performed continuously, creating a complete and accurate picture of the overall carbon budget for a site on a daily, monthly, or annual basis. If the chosen site is representative, the information it provides can be extrapolated over a much greater area. Such information is vital not only for estimating levels of productivity and studying other ecological aspects but also for constructing accurate models of how ecosystems interact with the climate and how this might affect weather patterns and climate change.

edge effect (in ecology) Any physical or biological change occurring in the perimeter zone of a habitat related to its immediate surroundings. Such edge effects are particularly marked at artificial boundaries, for instance when fragments of forest remain after surrounding forest is cleared. In this case the edges of the forest fragment are exposed to higher light levels, stronger winds, lower humidity, and greater temperature fluctuations. These factors in turn affect the makeup and diversity of the forest community living on or near the edges; also, species from surrounding areas may colonize the perimeter or act as predators of the forest inhabitants. Edge effects become proportionately greater in extent as the area of the habitat fragment decreases.

Ediacaran fauna *See* PROTEROZOIC.

Ediacaran period A geological period of the Proterozoic eon that began some 635 million years ago and lasted until the start of the Cambrian period around 542 million years ago. It is characterized by the emergence of diverse multicellular soft-bodied organisms, typified by the Ediacaran fossil assemblage found in Australia. *See* PROTEROZOIC.

EDTA Ethylenediaminetetraacetic acid, $(HOOCCH_2)_2N(CH_2)_2N(CH_2COOH)_2$: a compound that acts as a chelating agent, reversibly binding with iron, magnesium, and other metal ions. It is used in certain culture media bound with iron, which it slowly releases into the medium, and also in some forms of quantitative analysis.

EEG *See* ELECTROENCEPHALOGRAM.

effector A cell or organ that produces a physiological response when stimulated by a nerve impulse. Examples include muscles and glands.

effector neuron A nerve cell, such as a motor neuron, that transmits impulses from the central nervous system to an *effector in order to bring about a physiological response to changes in the environment.

effector-triggered immunity (ETI) *See* IMMUNITY (sense 2).

efferent Carrying (nerve impulses, blood, etc.) away from the centre of a body or organ towards peripheral regions. The term is usually applied to types of nerve fibres or blood vessels. *Compare* AFFERENT.

egestion The expulsion from the body of waste food materials that have never left the gut, particularly the expulsion of undigested materials from the gut through the anus (*see* DEFECATION). Egestion should not be confused with *excretion, in which the waste materials are produced by metabolic activity in the body's tissues.

EGF *See* EPIDERMAL GROWTH FACTOR.

egg 1. The fertilized ovum (*zygote) in egg-laying animals, e.g. birds and insects, after it emerges from the body. The egg is covered by *egg membranes that protect it from environmental damage, such as drying. **2.** (**egg cell**) The mature female reproductive cell in animals and plants. *See* OOSPHERE; OVUM.

egg membrane The layer of material that covers an animal egg cell. **Primary membranes** develop in the ovary and cover the egg surface in addition to the normal plasma membrane. The primary membrane is called the **vitelline membrane** in insects, molluscs, birds, and amphibians, the **chorion** in tunicates and fish, and the *zona pellucida in mammals. Insects have a second thicker membrane, also called the chorion. **Secondary membranes** are secreted by the oviducts and parts of the genital system while the egg is passing to the outside. They include the jelly coat of frogs' eggs and the albumen and shell of birds' eggs.

eicosanoid Any of a group of 20-carbon polyunsaturated fatty acids, particularly *arachidonic acid, and their derivatives. The eicosanoids include various biologically important molecules, notably the *prostaglandins, *prostacyclins, *thromboxanes, and *leukotrienes.

ejaculation The propulsion of semen out of the erect penis due to powerful rhythmic contractions of the urethra. An ejaculation coincides with the peak of sexual excitement (**orgasm**) and is accompanied by various physiological effects in the body, such as increased respiration rate and heart rate.

elaiosome *See* MYRMECOCHORY.

Elasmobranchii *See* CHONDRICHTHYES.

elastic cartilage *See* CARTILAGE.

elastic fibres *See* ELASTIN.

elastin A fibrous protein that is the major constituent of the yellow **elastic fibres** of *connective tissue. It is rich in glycine, alanine, proline, and other nonpolar amino acids that are cross-linked, making the protein relatively insoluble. Elastic fibres can stretch to several times their length and then return to their original size. Elastin is particularly abundant in elastic *cartilage, blood-vessel walls, ligaments, and the heart.

electric organ An organ occurring on the body or tail of certain fish, such as the electric ray (*Torpedo*) and electric eel (*Electrophorus electricus*). It gives an electric shock when touched and is used either to stun prey or predators or, in some species, to maintain a weak electric field in the surrounding water that is used in navigation. The organ is composed of modified muscle cells (**electroplate** or **electroplax cells**), nervous stimulation of which greatly increases the potential difference across the cell. The electroplates are in series so a high overall voltage can be achieved.

electric potential Symbol *V*. The energy required to bring unit electric charge from infinity to the point in an electric field at which the potential is being specified. The unit of electric potential is the volt. The **potential difference** (*p.d.*) between two points in an electric field or circuit is the difference in values of the electric potentials at the two points, i.e. it is the work done in moving unit charge from one point to the other.

electrocardiogram (**ECG**) A tracing or graph of the electrical activity of the heart. Recordings are made from electrodes fastened over the heart and usually on both arms and a leg. Changes in the normal pattern of an ECG may indicate heart irregularities or disease.

electroencephalogram (**EEG**) A tracing or graph of the electrical activity of the brain. Electrodes taped to the scalp record electrical waves from different parts of the brain. The pattern of an EEG reflects an individual's level of consciousness and can be used to detect such disorders as epilepsy, tumours, or brain damage. *See also* BRAIN DEATH.

electrogenic Describing an ion transport protein or other cell mechanism that generates an electrochemical potential across a cell membrane, by creating a net redistribution of charged ions. For example, each cycle of the sodium/potassium ATPase *sodium pump causes three sodium ions to leave the cell and two potassium ions to enter the cell, hence creating a net negative charge inside the cell. *See* RESTING POTENTIAL.

electrogenic pump A *transport protein that generates a voltage across a cell membrane, such as a *proton pump or the *sodium-potassium pump.

electrolocation *See* ELECTRORECEPTOR.

electrolyte A liquid that conducts electricity as a result of the presence of positive or negative ions. An example is a sodium chloride solution, which consists of free sodium (Na^+) and chloride (Cl^-) ions. In biology and medicine 'electrolyte' usually refers to the ion itself.

electromagnetic spectrum The range of wavelengths over which electromagnetic radiation extends. The longest waves (10^5–10^{-3} metres) are radio waves, the next longest (10^{-3}–10^{-6} m) are infrared waves, then comes the narrow band (4–7×10^{-7} m) of visible light, followed by *ultraviolet radiation (10^{-7}–10^{-9} m), *X-rays (10^{-9}–10^{-11} m), and gamma rays (10^{-11}–10^{-14} m).

electromyogram (EMG) A recording of the electrical activity of muscle fibres. It is obtained by means of electrodes inserted into the muscle, where they detect the action potentials of individual muscle units. The electrical signals are amplified by an apparatus (**electromyograph**) and displayed on an oscilloscope. Electromyograms are used in experimental muscle physiology and in diagnosing various nerve and muscle disorders.

electron An elementary particle present in all atoms in groupings called shells around the nucleus. When electrons are detached from the atom they are called **free electrons**.

electron flow The transfer of electrons along a series of carrier molecules in the *electron transport chain.

electron microscope A form of microscope that uses a beam of electrons instead of a beam of light (as in the optical microscope) to form a large image of a very small object, such as a cell organelle, a virus, or a DNA molecule. In optical microscopes the resolution is limited by the wavelength of the light. High-energy electrons, however, can be associated with a considerably shorter wavelength than light; for example, electrons accelerated to an energy of 10^5 electronvolts have a wavelength of 0.04 nanometre, enabling a resolution of 0.2–0.5 nm to be achieved. The **transmission electron microscope** (see illustration) has an electron beam, sharply focused by electron lenses (coils producing a magnetic field or electrodes between which an electric field is created), passing through a very thin metallized specimen (less than 50 nanometres thick) onto a fluorescent screen, where a visual image is formed. This image can be photographed. **High-resolution transmission electron microscopy (HRTEM)**, also called phase-contrast TEM, can achieve resolutions below 0.05° nm, enabling direct imaging of atomic structure. The **scanning electron microscope** can be used with thicker specimens and forms a perspective image, although the resolution and magnification

source of electrons

condenser lens

object

objective lens

intermediate image

projector lens

final image

Principle of transmission electron microscope

are lower. It is used particularly for examining surface features of small objects, such as pollen grains. In this type of instrument a beam of primary electrons scans the specimen and those that are reflected, together with any secondary electrons emitted, are collected. This current is used to modulate a separate electron beam in a TV monitor, which scans the screen at the same frequency, consequently building up a picture of the specimen. The resolution is limited to about 10–20 nm. *See also* FIELD-EMISSION MICROSCOPE; FIELD-IONIZATION MICROSCOPE; IMMUNOELECTRON MICROSCOPY.

((( ))) SEE WEB LINKS

- Tutorial featuring virtual scanning electron microscopy, created by the Optical Microscopy Division of the National High Magnetic Field Laboratory

electron transport chain (ETC; electron transport system) A series of molecules that transfers electrons via a sequence of biochemical reduction-oxidation (redox) reactions to effect a stepwise change in free energy. An electron transport chain, also known as a **respiratory chain**, forms the final stage of cellular *respiration. The components of this chain are located in the inner membrane of the mitochondria and consist of multiprotein complexes (labelled I to IV) with tightly bound prosthetic groups (cofactors and coenzymes) essential to catalysing the reactions of the chain. Each component is present in multiple copies in a single

mitochondrion. NADH or $FADH_2$, generated by the *Krebs cycle, transfer their electrons through a chain of electron *carrier molecules, including iron-sulphur proteins, *ubiquinone (Q), and a series of *cytochromes (cyt), that undergo reversible redox reactions, accepting electrons and then donating them to the next carrier in the chain—a process known as **electron flow**. *Cytochrome oxidase combines electrons and hydrogen ions with oxygen—the final electron acceptor in the chain—to form water (see illustration). This process is coupled to the formation of ATP by the *chemiosmotic mechanism (*see also* OXIDATIVE PHOSPHORYLATION). An electron transport chain also occurs in the thylakoid membranes of chloroplasts in *photosynthesis; the carrier molecules include *plastoquinone, *plastocyanin, and *ferredoxin. *See* PHOTOPHOSPHORYLATION.

electro-olfactogram (EOG) A recording of the electrical activity of smell (olfactory) receptors in the nose or other olfactory organ. Electrodes positioned in the olfactory epithelium detect the electrical activity of sensory olfactory cells as they respond to smells (odorants) in the environment. The electrical signals are relayed to an apparatus (**electro-olfactograph**) that amplifies them and displays them on an oscillosope.

electrophoresis (cataphoresis) A technique for the analysis and separation of colloids, based on the movement of charged

The mitochondrial electron transport chain and chemiosmotic synthesis of ATP

colloidal particles in an electric field. There are various experimental methods. At its simplest the sample is placed in a U-tube and a buffer solution added to each arm, so that there are sharp boundaries between buffer and sample. An electrode is placed in each arm, a voltage applied, and the motion of the boundaries under the influence of the field is observed. The rate of migration of the particles depends on the field, the charge on the particles, and on other factors, such as the size and shape of the particles. Electrophoresis can also be carried out using an adsorbent, such as a strip of filter paper, soaked in a buffer with two electrodes making contact. The sample is placed between the electrodes and a voltage applied. Different components of the mixture migrate at different rates, so the sample separates into zones. The components can be identified by the rate at which they move. In *gel electrophoresis the medium is a gel, typically made of polyacrylamide, agarose, or starch.

Electrophoresis, which has also been called **electrochromatography**, is used extensively in studying mixtures of proteins, nucleic acids, carbohydrates, enzymes, etc. In clinical medicine it is used for determining the protein content of body fluids. In modern automated DNA sequencers, electrophoresis is carried out in capillary tubes, typically less than 0.5 mm in diameter and about 48 cm long, containing a gel. *See* DNA SEQUENCING.

electroplax One of the cells found in the *electric organs of certain fish.

electroreceptor An organ specialized for detecting electric currents. Such organs are fairly common in marine fishes, which use them for detecting prey or potential attackers (i.e. **electrolocation**). The best known example is the **ampulla of Lorenzini**, groups of which are embedded in the head of sharks and rays. Each consists of a jelly-filled cuplike structure connected to the skin surface by a duct, sometimes several centimetres long, which is also filled with jelly. Sensory hair cells in the ampulla detect electric currents in the surrounding water, channelled inwards via the external

pore and duct. The apical (outward facing) plasma membrane of each electroreceptor cell has a lower electrical resistance than the basal (inward facing) membrane, so that current moving through the cell causes the basal membrane to become depolarized. This opens voltage-gated calcium channels in the basal membrane, thereby triggering increased release of neurotransmitter by the electroreceptor cell, and hence increasing the frequency of action potentials in the afferent nerves carrying information from the electroreceptor to the brain. The high sensitivity of these organs enables a shark, for example, to sense the very weak electric currents, perhaps just a few microamps, generated by the respiratory muscles of a resting plaice buried in the sand. Sharks and rays also use their ampullae of Lorenzini as *magnetoreceptors to detect the earth's magnetic field.

Similar organs occur in certain teleost fish, for example the marine catfish *Plotosus*. Some fish generate their own weak electric field as an alarm system or as a means of locating objects or communicating with other individuals of the same species. Disturbances in this field are detected by the fish's electroreceptors, warning of possible threats from intruders or receiving signals from conspecifics. In the duck-billed platypus (*Ornithorhynchus*), electrical sensors are distributed in rows over the surface of the bill and used to detect crustaceans, insect larvae, and other prey along the beds of muddy streams and rivers. The sensors—modified mucus glands—are thought to work in conjunction with numerous mechanical sensors, called **pushrods**, also distributed over the bill surface.

electrovalent bond (ionic bond) A type of *chemical bond formed by the transfer of one or more electrons from one atom to another, so that oppositely charged ions are produced. For example, the bond between the sodium and chlorine atoms in sodium chloride (NaCl) is formed by the transfer of an electron from sodium to chlorine, creating Na^+ and Cl^- ions. The electrostatic attraction between these ions provides the bonding in NaCl.

elicitor A substance or other stimulus that triggers the hypersensitive response in a plant (*see* HYPERSENSITIVITY). Most elicitors are polysaccharides, small proteins, or lipids associated with the fungal or bacterial cell wall. However, pectic fragments resulting from microbial damage to the plant's own cell walls may also act as elicitors. *See* IMMUNITY (sense 2).

ELISA (enzyme-linked immunosorbent assay) A sensitive technique (*see* IMMUNOASSAY) for accurately determining the amount of protein or other antigen in a given sample by means of an enzyme-catalysed colour change. Antibody specific to the test protein is adsorbed onto a solid substrate, such as a PVC sheet, and a measured amount of the sample is added; all molecules of the test protein in the sample are bound by the antibody. A second antibody specific for a second site on the test protein is added; this is conjugated with an enzyme, which catalyses a colour change in the fourth reagent, added finally to the sheet. The colour change can be measured photometrically and compared against a standard curve to give the concentration of protein in the sample. ELISA is widely used for diagnostic and other purposes.

El Niño-Southern Oscillation (ENSO) An irregularly recurring climate pattern involving changes in wind direction, sea temperature, and rainfall in the tropical Pacific that has a widespread impact on global weather conditions. There is a warm phase, called **El Niño** (Spanish for 'the little boy'), a neutral or normal phase, and a cold phase, called **La Niña** ('the little girl'). Episodes of El Niño and La Niña typically last 9–12 months and recur every 2 to 7 years. During an El Niño phase, the usual east-to-west trade winds across the Pacific weaken, warm surface water in the western ocean moves eastwards, and upwelling of cold water off the coast of South America declines. The warmer ocean alters the trajectory of the jet stream, affecting temperature and rainfall patterns across the Pacific and further afield. It raises global temperatures and increases the risk of extreme weather, such as drought in tropical Asia and heavy rain in parts of the Americas. During La Niña the pattern is reversed: the trade winds strengthen and push warm surface waters further west, increasing cold water upwelling in the east and cooling the Pacific.

elongation 1. (in protein synthesis) The phase in which amino acids are linked together by sequentially formed peptide bonds to form a polypeptide chain (*see* TRANSLATION). **Elongation factors** are proteins that—by binding to a tRNA-amino-acid complex—enable the correct positioning of this complex on the *ribosome, so that translation can proceed. The key peptidyltransferase reaction required for formation of the peptide bond is catalysed not by a protein but by the large RNA subunit of the ribosome. **2.** *See* TRANSCRIPTION.

elytra (*sing.* elytron) The thickened horny forewings of the *Coleoptera (beetles), which cover and protect the membranous hindwings when the insect is at rest.

emasculation (in plant breeding) The removal of the anthers of a flower in order to prevent self-pollination or the undesirable pollination of neighbouring plants.

Embase (Excerpta Medica Database) A bibliographic database containing citations, abstracts, indexing terms, and codes covering 8500 biological and medical journals from more than 95 countries. It contains over 32 million records, with over 1.5 million new ones added each year, and has a particular emphasis on drug-related fields.

Embden-Meyerhof pathway *See* GLYCOLYSIS.

embedding A stage in the preparation of a sample for examination by microscopy that involves impregnation of the sample with wax or plastic following dehydration. The embedded sample can then be cut into extremely thin sections to reveal cellular and subcellular structure.

embryo 1. An animal in the earliest stages of its development, from the time when the

A developing human embryo

fertilized ovum starts to divide (*see* CLEAVAGE), while it is contained within the egg or reproductive organs of the mother, until hatching or birth. A human embryo (see illustration) is called a *fetus after the first eight weeks of pregnancy. **2.** The structure in plants that develops from the zygote prior to germination. In seed plants the zygote is situated in the *embryo sac of the ovule. It divides by mitosis to form the embryonic cell and a structure called the **suspensor**, which embeds the embryo in the surrounding nutritive tissue. The embryonic cell divides continuously and eventually gives rise to the *radicle (young root), *plumule (young shoot), and one or two *cotyledons (seed leaves). Changes also take place in the surrounding tissues of the ovule, which becomes the *seed enclosing the embryo plant.

((⊕)) SEE WEB LINKS
• A compendium of embryonic development, stem cell research, and regenerative medicine

embryology The study of the development of animals from the fertilized egg to the new adult organism. It is sometimes limited to the period between fertilization

of the egg and hatching or birth (*see* EMBRYO).

((⊕)) SEE WEB LINKS
• This site, hosted by the University of New South Wales, explores human embryology and development

embryo mother cell *See* MEGASPORE MOTHER CELL.

embryonic stem cell *See* STEM CELL.

embryophyte A plant that develops from an embryo that is protected by tissues of the parent plant. This is a shared derived trait (synapomorphy) of the land plants, both nonvascular and vascular, and distinguishes them from the multicellular algae. *See* PLANT.

embryo sac A large cell that develops in the *ovule of flowering plants. It is equivalent to the female *gametophyte of lower plants, although it is very much reduced. Typically, it contains eight nuclei formed by division of the *megaspore mother cell (see illustration). The haploid *oosphere (egg cell), which is associated with two *synergid cells to form the **egg apparatus**, is fertilized by a male nucleus and becomes antipodal cells — embryo sac — polar nuclei — integuments — oosphere ⎱ egg — synergids ⎰ apparatus

Embryo sac

the diploid zygote, which subsequently divides to form the *embryo. The two *polar nuclei fuse with a second male nucleus to form a triploid nucleus that gives rise to the *endosperm. The three remaining nuclei form the **antipodal cells**.

EMG *See* ELECTROMYOGRAM.

emulsification (in digestion) The breakdown of fat globules in the duodenum into tiny droplets, which provides a larger surface area on which the enzyme pancreatic *lipase can act to digest the fats into fatty acids and glycerol. Emulsification is assisted by the action of the bile salts (*see* BILE).

emulsion A *colloid in which small particles of one liquid are dispersed in another liquid. Usually emulsions involve a dispersion of water in an oil or a dispersion of oil in water. Dietary fats are reduced to an emulsion in the duodenum to facilitate their subsequent digestion (*see* EMULSIFICATION).

enamel The material that forms a covering over the crown of a *tooth (i.e. the part that projects above the gum). Enamel is smooth, white, and extremely hard, being rich in minerals containing calcium, especially *apatite. It is produced by certain cells (**ameloblasts**) of the oral epithelium and protects the underlying dentine of the tooth. Enamel may also be found in the placoid *scales of certain fish, which demonstrates the common developmental origin of scales and teeth.

enamel organ A mass of soft pulpy tissue that develops on the surface of a dental *papilla. It consists of a mesh of fibres, fluid *albumin, and *ameloblasts, which

deposit a layer of protective *enamel on the outer surface of the dentine.

encephalin *See* ENKEPHALIN.

encephalization quotient (EQ) A measure of the relative size of the brain of a particular species compared with the expected value for members of the group to which it belongs. It is used to estimate the comparative intelligence of animals, particularly in anthropology. A logarithmic plot of brain mass against body mass for numerous representative species has a slope of roughly 3/4, which provides the baseline for calculating EQ values. The EQ corresponds to the distance above or below this line, and an EQ of 1 equates to the brain mass expected for the group. In mammals and birds the formula for calculating EQ is brain mass$/0.12 \times$ (body mass)$^{0.66}$. Modern humans (*Homo sapiens*) have an EQ of roughly 6, meaning that their brain mass is six times greater than a typical mammal. Estimating EQ values is fraught with pitfalls, particularly when attempting to assess body mass from fragmentary fossil remains.

ENCODE (Encyclopedia of DNA Elements) An international project to describe in detail the functional regions of the human genome. Initiated in 2003 and funded by the US National Human Genome Research Institute, the project involves scientists throughout the world working under the auspices of the ENCODE Consortium. The results are freely available online and provide valuable information about gene function in health and disease. The project aims to annotate not only all protein-coding genes, but also their DNA regulatory elements such as promoters, enhancers, and silencers, functional RNA transcripts and regulatory RNAs, and chromatin structural changes that affect DNA transcription, including modifications of histone proteins and changes in DNA methylation.

(((●))) SEE WEB LINKS

• Encyclopedia of DNA Elements

endangered species A plant or animal species that is in danger of *extinction because its population numbers have reached a critical level or its habitats have been drastically reduced. If these causal factors continue, the species is unlikely to survive. As defined by the IUCN (International Union for the Conservation of Nature and Natural Resources), a **critically endangered species** is at 'extremely high risk of extinction in the wild'; an **endangered species** is at 'high risk of extinction in the wild'; and a **vulnerable species** is one likely to become endangered unless measures are taken to improve its circumstances in the wild. These are all categories of **threatened species**, a list of which is published by the IUCN in its Red Book.

((⊕)) SEE WEB LINKS
• Website of the IUCN

endemic 1. Describing a plant or animal species that is restricted to one or a few localities in its distribution. Endemic species are usually confined to islands and are vulnerable to extinction. **2.** Describing a disease or a pest that is always present in an area. For example, malaria is endemic in parts of Africa.

endergonic reaction A chemical reaction in which energy is absorbed. In an *endothermic reaction the energy is in the form of heat. *Compare* EXERGONIC REACTION.

endocardium *See* MYOCARDIUM.

endocarp *See* PERICARP.

endochondral ossification *See* OSSIFICATION.

endocrine gland (ductless gland) Any gland in an animal that manufactures *hormones and secretes them directly into the bloodstream to act at distant sites in the body (known as **target organs** or **cells**). Endocrine glands tend to control slow long-term activities in the body, such as growth and sexual development. In mammals they include the *pituitary, *adrenal, *thyroid and *parathyroid glands, the *pineal gland, the *ovary and *testis, the *placenta, and part of the pancreas (*see* ISLETS OF LANGERHANS). The activity of endocrine

glands is controlled by negative feedback, i.e. a rise in output of hormone inhibits a further increase in its production, either directly or indirectly via the target organ or cell. *See also* NEUROENDOCRINE SYSTEM. *Compare* EXOCRINE GLAND.

endocrinology The study of the structure and functions of the *endocrine glands and of the *hormones they produce.

endocytosis The process by which materials enter a cell without passing through the plasma membrane. The membrane folds around material outside the cell, resulting in the formation of a saclike vesicle into which the material is incorporated. This vesicle is then pinched off from the cell surface so that it lies within the cell (*see* ENDOSOME). Both *phagocytosis and *pinocytosis are forms of endocytosis. In **receptor-mediated endocytosis**, cells selectively take in substances (e.g. hormones, low-density lipoproteins) that bind to receptors on the cell surface. *Compare* EXOCYTOSIS.

endoderm (entoderm) The internal layer of cells of the *gastrula, which will develop into the alimentary canal (gut) and digestive glands of the adult. *See also* GERM LAYERS.

endodermis The innermost layer of the root *cortex of a plant, lying immediately outside the vascular tissue. Various modifications of the endodermal cell walls enable them to regulate the passage of materials both into and out of the vascular system (*see* CASPARIAN STRIP). An endodermis may also be seen in the stems of some plants.

endogamy The fusion of reproductive cells from closely related parents, i.e. *inbreeding. *Compare* EXOGAMY.

endogenous Describing a substance, stimulus, organ, etc., that originates from within an organism. For example, growth rhythms not directed by environmental stimuli are termed endogenous rhythms. Lateral roots, which always grow from inside the main root rather than from its surface, are said to arise endogenously. *Compare* EXOGENOUS.

endolymph The fluid that fills the membranous labyrinth of the vertebrate *inner ear. *See* COCHLEA; SEMICIRCULAR CANALS. *Compare* PERILYMPH.

endomembrane system The network of membranous compartments that forms part of many of the cell's internal organelles and other functional components. It includes the membranes of the nucleus, endoplasmic reticulum, Golgi apparatus (or dictyosomes), plasma membrane, tonoplast, and outer (but not inner) membranes of the mitochondria and chloroplasts, which form a functional continuum within the cell. New membrane is constantly being made in the *endoplasmic reticulum and transported in the form of vesicles containing cargo (e.g. secretory proteins) along microtubules to their target organelles or to the plasma membrane.

endometrium (*pl.* **endometria**) The mucous membrane that lines the *uterus of mammals. It comprises an upper mucus-secreting layer, which is shed during menstruation, and a basal layer, which proliferates to form the upper layer. *See also* MENSTRUAL CYCLE.

endomitosis The replication of chromosomes in the absence of cell or nuclear division, resulting in numerous copies within each cell. It occurs notably in the salivary glands of *Drosophila* and other flies. Cells in these tissues contain giant chromosomes (*see* POLYTENY), each consisting of over a thousand intimately associated, or synapsed, chromatids.

endonuclease An enzyme that catalyses the internal cleavage of nucleic acids. *See also* RESTRICTION ENZYME. *Compare* EXONUCLEASE.

endoparasite A parasite that lives inside its host's body. *See* PARASITISM.

endopeptidase A protein-digesting enzyme that cleaves a polypeptide chain at specific sites between amino acids. For example, *chymotrypsin cleaves the chain next to aromatic amino acids, such as phenylalanine; *trypsin cleaves the chain next to basic amino acids, such as lysine or arginine; and *pepsin cleaves the chain next to tyrosine and phenylalanine. *Compare* EXOPEPTIDASE.

endophenotype A component part of a broad or complex phenotype. The concept was introduced in 1966 by population geneticists B. John and K. R. Lewis to refer to microscopic and internal phenotypes that are not apparent to the casual observer; since then it has been applied in medicine and psychiatry in efforts to unravel the genetic and physiological basis of multifactorial diseases, such as schizophrenia and heart disease. An endophenotype is thus a measurable and genetically determined aspect of a 'bigger picture'. For example, cortisol secretion is an endophenotype of anxiety. Similarly, in evolutionary biology, complex traits, such as behaviours, can be regarded as the net result of the assembly of various modular endophenotypes.

endoplasm *See* CYTOPLASM.

endoplasmic reticulum (ER) A system of membranes that extends throughout the cytoplasm of plant and animal *cells forming interconnected tubes and flattened sacs (cisternae). It is part of the *endomembrane system, which links many parts of the cell including the plasma and nuclear membranes (*see* CISTERNA). The ER is the site of protein synthesis and modification, and also the synthesis of lipids for new membranes. **Rough ER (RER)** has *ribosomes attached to its surface, giving it a 'rough' appearance under the microscope. Here, newly synthesized proteins enter the interior lumen where they are sorted, chemically modified and folded to attain their tertiary structure, and tagged with a localization signal ('address') that determines their destination. The proteins are then enclosed in vesicles, which pinch off from the surface of the RER to be transported to the *Golgi apparatus for further processing. Membrane proteins and lipids are made in the RER and are integrated into the ER membrane *in situ* to produce new membrane. **Smooth ER** (SER), which is continuous with RER, lacks ribosomes; it is the site of important metabolic reactions, including further protein modification of proteins, synthesis of phospholipids, fatty acids, and steroids, and (in certain animal cells) the breakdown of glycogen. It also chemically modifies and detoxifies molecules that enter the organism from outside, and is especially prominent in liver cells. SER stores calcium ions, release of which into the cytosol acts as a signal to trigger different responses. In muscle cells SER is adapted to form the *sarcoplasmic reticulum.

endopterygote Any winged insect having *holometabolous development, i.e. the eggs hatch into *larvae that are dissimilar to the adults and lack wings. These develop internally before the final moult into a *pupa, from which the adult emerges; metamorphosis is described as **complete**. Endopterygote insects include the beetles (*Coleoptera), ants, bees, and wasps (*Hymenoptera), flies (*Diptera), and butterflies and moths (*Lepidoptera). *Compare* EXOPTERYGOTE.

endoreduplication Repeated duplication of the entire complement of genes within a single nucleus. It occurs in many actively metabolizing plant tissues, where it is evidenced by enlargement of the cell nuclei. The resultant multiplication of genes increases the amounts of messenger RNAs available to the cell, and hence permits the synthesis of large quantities of proteins. *Compare* GENE AMPLIFICATION.

end organ The structure at the end of a peripheral nerve. Examples of end organs are the muscle *end plate at the end of a motor neuron and the *receptor at the end of a sensory neuron.

endorphin Any of a class of three endogenous *opioids—α-, β- and γ-endorphins—found naturally in brain and other tissues that have pain-relieving effects similar to those of morphine. They are released during physical or emotional stress, such as childbirth or strenuous exercise, and can produce feelings of euphoria. All are peptides or polypeptides derived from the precursor pro-opiomelanocortin; for example, β-endorphin is a 31-amino-acid peptide. The endorphins mediate their analgesic effects by binding to *opioid receptors. *See also* DYNORPHIN; ENKEPHALIN.

endoscopy The study of the interior of hollow organs, such as the intestines and the reproductive organs, using a flexible probe (**endoscope**), which contains *optical fibres for transmitting images onto a screen.

endoskeleton A supporting framework that lies entirely within the body of an animal, such as the bony *skeleton of vertebrates or the spicules of a sponge. The function of an endoskeleton is to support the body and in vertebrates it also protects the organs and provides a system of levers on which the muscles can act to produce movement. *Compare* EXOSKELETON.

endosome A vesicle formed within a cell during forms of *endocytosis in which the material to be ingested first binds to *receptor sites on the cell surface. These receptors occur in areas of the plasma membrane called **coated pits**, which are pinched off to become **coated vesicles**, surrounded by a protein coat (*see* CLATHRIN) and containing the ingested material, within the cell. The vesicles then shed their coats and merge with other vesicles to form endosomes. The endosomes fuse with vesicles from the Golgi apparatus to form *lysosomes.

endosperm A nutritive tissue, characteristic of flowering plants, that surrounds the developing embryo in a seed. It develops from nuclei in the *embryo sac and its cells are triploid. In **endospermic** seeds it remains and increases in size; in **nonendospermic** seeds it disappears as the food is absorbed by the embryo, particularly the *cotyledons. Many plants with endospermic seeds, such as cereals and oil crops, are cultivated for the rich food reserves in the endosperm.

endospore The resting stage of certain bacteria, formed in response to adverse conditions. The bacterial cell changes into a partially dehydrated core, enclosed in a multilayered protein coat. On return to favourable conditions the spore germinates and reverts to the normal vegetative form of the organism. Endospores can remain viable for long periods, perhaps millions of years. There are reports of successful germination of endospores obtained from the guts of prehistoric bees preserved in amber for 25–40 million years and from brine isolated in 250-million-year-old salt deposits. Endospores of low G-C content Gram-positive bacteria, such as some members of the *Firmicutes (e.g. *Bacillus subtilis*), are particularly resistant to the usual arsenal of antimicrobial treatments.

endostyle A ciliated groove in the wall of the pharynx of lancelets (Cephalochordata), some tunicates, and larval lampreys that contains mucus-secreting glands. Mucus secreted by the endostyle traps food particles in water drawn into the pharynx by the beating action of the cilia.

endosymbiont An organism that lives inside another organism, often to their mutual benefit. The endosymbiont may reside inside the host's cells (i.e. intracellularly) or in the extracellular spaces. *See* SYMBIOSIS.

endosymbiont theory A theory, devised principally by US biologist Lynn Margulis (1938–2011), that eukaryotic organisms evolved from symbiotic associations between prokaryotic ancestors. Free-living aerobic bacteria and photosynthetic cyanobacteria (*see* CHLOROXYBACTERIA) became incorporated inside larger nucleated prokaryotic cells, where they acted as forerunners of the mitochondria and chloroplasts seen in modern eukaryotes. Such events are held to have occurred on several occasions, producing various lineages of both heterotrophic and phototrophic protists, from which evolved ancestors of animals, plants, and fungi. There is strong evidence for the theory, particularly the finding that mitochondria and chloroplasts have DNA similar in form to that of bacteria, and that they contain prokaryotic-type ribosomes. There is also evidence that the ancestors of some modern groups incorporated eukaryotic cells already

containing primary chloroplast-type plastids, through a process called **secondary endosymbiosis**.

endothelin (ET) Any of several related peptides, all comprising 21 amino acids and first identified in cultured endothelial cells. **Endothelin 1 (ET-1)** is released by the endothelium of blood vessels and is a potent vasopressor, causing constriction of vessels and raising blood pressure. The other endothelins resemble ET-1 structurally but are synthesized in other tissues. The endothelins bind to either of two classes of G-protein-coupled receptors, ET_A and ET_B, and antagonists of these receptors are used to treat pulmonary arterial hypertension.

endothelium (*pl.* **endothelia**) A single layer of thin platelike cells that line the inner surfaces of blood and lymph vessels and the heart. Endothelium is derived from the *mesoderm. *Compare* EPITHELIUM; MESOTHELIUM.

endotherm (homoiotherm) An animal that can generate and maintain heat within its body independently of the environmental temperature. Mammals and birds are endotherms and hence are often described as being **warm-blooded**. Moreover, some fishes, insects, and reptiles can generate heat internally to maintain their body temperature to at least some degree. For example, certain billfishes, tunas, and certain sharks conserve metabolic heat using vascular countercurrent heat exchangers; these are known as **regional endotherms**. *See* HOMOIOTHERMY. *Compare* ECTOTHERM; HETEROTHERM.

endothermic 1. Of or relating to *endotherms. **2.** Denoting a chemical reaction that takes heat from its surroundings. *See* ENDERGONIC REACTION. *Compare* EXOTHERMIC.

endotoxin *See* TOXIN.

endotrophic mycorrhiza *See* MYCORRHIZA.

end plate A slight depression in the plasma membrane of a muscle cell that lies immediately beneath a motor nerve ending at a *neuromuscular junction. It contains transverse folds with a high density of receptors for the *neurotransmitter released by the axon terminals to induce contraction of the muscle fibre.

end-plate potential (EPP) The electric potential, measured in millivolts (mV), produced by depolarization of the *end plate at a *neuromuscular junction in response to stimulation by the motor neuron. The depolarization results from release of excitatory neurotransmitter molecules, which activate receptors in the postsynaptic membrane, causing transient opening of ion channels and allowing cations to flow briefly through the membrane. In vertebrate skeletal muscle a single EPP, typically about 10–20 mV in amplitude, triggers an action potential in the muscle fibre, which propagates through the fibre causing it to contract.

energy A measure of a system's ability to do work. Like work itself, it is measured in joules. Energy is conveniently classified into two forms: **potential energy** is the energy stored in a body or system as a consequence of its position, shape, or state (this includes chemical energy in food substances, etc.); **kinetic energy** is energy of motion and is usually defined as the work that will be done by the body possessing the energy when it is brought to rest.

energy flow (in ecology) The flow of energy that occurs along a *food chain. Energy enters the food chain at the level of the *producers (usually plants) in the form of solar energy. The plants convert solar energy into chemical energy in the process of *photosynthesis. Chemical energy is passed from one trophic level to the next through feeding. Since a large proportion of energy is lost at each trophic level, mostly in the form of heat energy due to respiration, a food chain does not normally consist of more than five trophic levels: the fifth trophic level does not contain enough energy to support further levels. Energy is also lost from the food chain in excretory products and the remains of dead organisms; this is converted into heat energy by the action of

*decomposers. *See also* PRODUCTIVITY; PYRAMID OF ENERGY.

enhancer A cluster of regulatory base sequence in a DNA molecule that can initiate or increase the *transcription of a gene by binding a specific protein called activators (*see* TRANSCRIPTION FACTOR). An enhancer may be some distance, perhaps thousands of nucleotides, from the gene it controls on the same chromosome, and typically interacts with other enhancer and repressor sites to regulate the binding of RNA *polymerase to the DNA molecule and hence expression of the gene. One proposed mechanism involves bending of the DNA molecule so that the enhancer site with its bound activators is brought close to the promoter of the gene, so forming an active initiation complex with other transcription factors and enabling transcription to proceed.

enkephalin (encephalin) Any of a class of endogenous *opioids consisting of five amino acids and found principally in the central nervous system. They bind to *opioid receptors, chiefly in the brain, and their release controls levels of pain and other sensations, for example by blocking the release of *substance P, a neurotransmitter involved in *nociception. Enkephalins themselves can act as neurotransmitters in the brain. *See also* DYNORPHIN; ENDORPHIN.

Ensembl A joint venture between the European Bioinformatics Institute (EBI), part of the European Molecular Biology Laboratory (EMBL), and the Wellcome Trust Sanger Institute that provides annotated sequence data from various eukaryotic genomes and open-source (free) software for researchers mining and handling the data.

enteric nervous system The network of neurons found in the wall of the vertebrate gut that controls and coordinates its muscular activity. In humans it contains an estimated 100 million neurons organized into two layers: a submucosal layer that controls secretions from the gut wall and also has sensory functions, and a second layer between the circular and longitudinal muscle layers, whose contractions it controls. The enteric nervous system operates relatively independently of the central nervous system but transmits signals about the state of the gut to the brain via the vagus nerve. It may play a part in influencing mood and in immunity. The range of enteric neurotransmitters matches that used by the brain and includes *serotonin, which is produced in large amounts in gut tissues.

enterogastric reflex A nervous reflex whereby stretching of the wall of the duodenum results in inhibition of gastric motility and reduced rate of emptying of the stomach. It is a feedback mechanism to regulate the rate at which partially digested food (chyme) leaves the stomach and enters the small intestine. Receptors in the duodenal wall detect distension of the duodenum caused by the presence of chyme and also raised acidity (i.e. low pH) of the duodenal contents due to excess gastric acid. They send signals via the parasympathetic nervous system, causing reflex inhibition of stomach-wall muscles responsible for the stomach emptying.

enterokinase (**enteropeptidase**) An enzyme in the small intestine that activates trypsinogen to *trypsin.

enteron (**coelenteron**) *See* GASTROVASCULAR CAVITY.

enthalpy Symbol H. A thermodynamic property of a system defined by $H = U + pV$, where H is the enthalpy, U is the internal energy of the system, p its pressure, and V its volume. In a chemical reaction carried out in the atmosphere the pressure remains constant and the enthalpy of reaction, ΔH, is equal to $\Delta U + p\Delta V$. For an exothermic reaction ΔH is taken to be negative.

entoderm *See* ENDODERM.

entomology The study of insects.

entomophily Pollination of a flower in which the pollen is carried on an insect. Entomophilous flowers are usually brightly coloured and scented and often secrete nectar. In some species (e.g. primulas) there are structural differences between the flowers to ensure that cross-pollination occurs. Other examples of entomophilous flowers are orchids and antirrhinums. *Compare* ANEMOPHILY; HYDROPHILY.

Entoprocta A phylum of small invertebrate aquatic animals typically having a cup-shaped body bearing tentacles and attached to the substrate by means of a stalk. The entire animal is usually less than 10 mm long. There are nearly 200 known species, widely distributed and almost exclusively marine. Many species are colonial and form mats on seaweeds, rocks, shells, and other surfaces in coastal waters. Each individual has a ring of 4–36 ciliated tentacles (**calyx**) on top of the body, enclosing both the mouth and anus. The tentacles set up a feeding current and trap minute plankton and other particles in mucus, transferring the particles to the mouth. Digestion occurs in the U-shaped gut and waste is discharged via the anus. There is no heart or blood vessels, and the nervous system consists of a single ganglion between the mouth and anus, from which nerves extend to the tentacles, body, and stalk. Dissolved nitrogenous waste is discharged by exocytosis from the stomach wall into the gut, and also collected by a ciliated *flame cell, and discharged via pores. Reproduction is both asexual, by budding, and sexual. Most entoprocts are hermaphrodites, and produce free-swimming larvae that settle on a substrate before undergoing metamorphosis into the sessile adult form. The space between the body wall and the gut is filled with a jelly-like mesenchyme; this has been interpreted by some as a pseudocoelom, prompting the suggestion that entoprocts are allied with other *pseudocoelomate animals, such as nematodes. However, molecular systematics points to a close relationship between entoprocts and bryozoans (*see* BRYOZOA).

entrainment The synchronization of an inherent biological rhythm in an organism with an external cycle of events, such as the daily day-night cycle, the ebbing and flowing of tides, or the seasons. *See* BIOLOGICAL CLOCK.

Entrez An information retrieval service, or browser, provided by the National Center for Biotechnology Information (NCBI), a division of the US National Library of Medicine. It gives access to a collection of databases, both bibliographic and biomolecular, including PubMed, *MEDLINE, and *OMIM.

entropy Symbol S. A measure of the unavailability of a system's energy to do work; an increase in entropy is accompanied by a decrease in energy availability. When a system undergoes a reversible change the entropy (S) changes by an amount equal to the energy (Q) absorbed by the system divided by the thermodynamic temperature (T) at which the energy is absorbed, i.e. $\Delta S = \Delta Q/T$. However, all real processes are to a certain extent irreversible changes and in any closed system an irreversible change is always accompanied by an increase in entropy.

In a wider sense entropy can be interpreted as a measure of a system's disorder; the higher the entropy the greater the disorder. As any real change to a closed system tends towards higher entropy, and therefore higher disorder, it follows that the entropy of the universe (if it can be considered a closed system) is increasing and its available energy is decreasing. This increase in the entropy of the universe is one way of stating the second law of thermodynamics.

environment (in ecology) The physical, chemical, and biological conditions of the region in which an organism lives. *See also* ECOLOGY; ECOSYSTEM.

environmental constraint hypothesis *See* CARBON:NUTRIENT BALANCE HYPOTHESIS.

environmental resistance The sum total of the factors that prevent populations from continually growing and therefore tend to keep populations at constant levels. These factors include predators, disease, and a shortage of any of the various requirements for survival, such as food, water, shelter, and light (which is particularly important for plants). *See also* POPULATION GROWTH.

environmental selection *Natural selection within a population resulting from the influence exerted by the environment. Environmental selection inevitably leads to a change in the composition of genes within a population. For example, a change in the environment resulting in a drop in temperature will favour animals with an increased ability to maintain their body heat and may eventually result in a population of animals with denser fur.

enzyme A protein that acts as a *catalyst in biochemical reactions. Each enzyme is specific to a particular reaction or group of similar reactions. Many require the association of certain nonprotein *cofactors in order to function. The molecule undergoing reaction (the **substrate**) binds to a specific *active site on the enzyme molecule to form a short-lived intermediate (*see* ENZYME-SUBSTRATE COMPLEX): this greatly increases (by a factor of up to 10^{20}) the rate at which the reaction proceeds to form the product. Enzyme activity is influenced by substrate concentration and by temperature and pH, which must lie within a certain range. Other molecules may compete for the active site, causing *inhibition of the enzyme or even irreversible destruction of its catalytic properties.

Enzyme production is governed by expression of the particular gene encoding the enzyme. This expression is in turn regulated by mechanisms depending on a host of factors, such as stage of development of the organism, type of tissue, hormonal and nutritional status, nature of the environment, etc. Enzyme activity is further controlled by pH changes, alterations in the concentrations of essential cofactors, feedback inhibition by the products of the reaction, and activation by another enzyme, either from a less active form or an inactive precursor (*zymogen). Such changes may themselves be under the control of hormones or the nervous system. *See also* ENZYME KINETICS.

Enzymes are classified into six major groups, according to the type of reaction they catalyse: (1) *oxidoreductases; (2) *transferases; (3) *hydrolases; (4) *lyases; (5) *isomerases; (6) *ligases.

The names of most individual enzymes also end in -*ase*, which is added to the names of the substrates on which they act. Thus *lactase is the enzyme that acts to break down lactose; it is classified as a hydrolase.

enzyme inhibition *See* INHIBITION.

enzyme kinetics The study of the rates of enzyme-catalysed reactions. Rates of reaction are usually measured by using the purified enzyme *in vitro* with the substrate and then observing the formation of the product or disappearance of the substrate. As the concentration of the substrate is increased the rate of reaction increases proportionally up to a certain point, after which any further increase in substrate concentration no longer increases the reaction rate (*see* MICHAELIS-MENTEN CURVE). At this point, all active sites of the enzyme are saturated with substrate; any further increase in the rate of reaction will occur only if more enzyme is added. Reaction rates are also affected by the presence of inhibitors (*see* INHIBITION), temperature, and pH (*see* ENZYME).

enzyme-linked immunosorbent assay *See* ELISA.

enzyme-substrate complex The intermediate formed when a substrate molecule interacts with the *active site of an enzyme. Following the formation of an enzyme-substrate complex, the substrate molecule undergoes a chemical reaction and is converted into a new product. Various mechanisms for the formation of enzyme-substrate complexes have been suggested, including the *induced-fit model and the *lock-and-key mechanism.

Eoarchaean (Eoarchean) The earliest era of the *Archaean eon, extending from around 4000 million years ago (mya) to about 3600 mya. It is marked by the accretion of rocks from the molten material that characterized the earth in the preceding *Hadean eon. There is some evidence of

biological life, inferred from isotopic ratios of certain rocks from this time.

Eocene The second geological epoch of the *Palaeogene period. It extended from the end of the Palaeocene epoch, about 56 million years ago (mya), to the beginning of the Oligocene epoch, about 34 mya. The term was first proposed by Sir Charles Lyell (1797–1875) in 1833. Mammals were dominant in the Eocene: rodents, artiodactyls, carnivores, perissodactyls (including early horses), and whales were among the groups to make their first appearance.

EOG *See* ELECTRO-OLFACTOGRAM.

eosin One of a series of acidic dyes, used in optical microscopy, that colours cytoplasm pink and cellulose red. It is frequently used as a counterstain with *haematoxylin for colouring tissue smears and sections of animal tissue.

eosinophil A type of white blood cell (*leucocyte) that has a granular cytoplasm (*see* GRANULOCYTE) and is found chiefly in tissues, with only small numbers in the circulation. Eosinophils attack antibody-coated parasites by releasing from their granules powerful enzymes and free radicals. When activated they also release mediators such as prostaglandins, cytokines, and leukotrienes, which amplify the inflammatory response. Defective regulation of eosinophil activation, as in certain allergies and hypereosinophilia, can cause serious damage to body tissues.

ephemeral 1. (in botany) An *annual plant that completes its life cycle in considerably less than one growing season. A number of generations can therefore occur in one year. Many troublesome weeds, such as groundsel and willowherb, are ephemerals. Certain desert plants are also ephemerals, completing their life cycles in a short period following rain. **2.** (in zoology) A short-lived animal, such as a mayfly.

Ephemeroptera An order of *exopterygote insects that comprises the mayflies, in which the adult stage lasts for only a few hours. The adults have two pairs of wings held vertically at rest, a pair of tail bristles (cerci), and vestigial mouthparts (they do not feed). The nymphs (naiads) live for up to a number of years; they are mainly herbivorous but some possess mandibles for feeding on animal prey.

epiblast *See* BLASTULA.

epiboly A form of coordinated cell movement occurring in a vertebrate embryo during gastrulation—an early developmental stage during which rudimentary tissues form and body axes are defined. It involves the spreading of a sheet of cells from the *animal pole of the *blastula to envelop the vegetal pole. Cells in a single layer can extend by becoming flatter to increase their surface area, whereas multiple layers extend by rearranging into fewer layers covering a larger area, a process called **intercalation**.

epicalyx (*pl.* **epicalyces**) A ring of bracts below a flower that resembles a calyx. It is seen, for example, in the strawberry flower.

epicarp *See* PERICARP.

epicotyl The region of a seedling stem above the stalks of the seed leaves (*cotyledons) of an embryo plant. It grows rapidly in seeds showing *hypogeal germination and lifts the stem above the soil surface. *Compare* HYPOCOTYL.

epidemic An outbreak of a disease (especially an infectious disease) that affects a large number of individuals within a population at the same time. *Compare* ENDEMIC; PANDEMIC.

epidemiology The study of diseases that affect large numbers of people. Traditionally, epidemiologists have been concerned primarily with infectious diseases, such as typhoid and influenza, that arise and spread rapidly among the population as epidemics. However, today the discipline also covers noninfectious disorders, such as diabetes, heart disease, and back pain. Typically the distribution of a disease is charted in order to discover patterns that

might yield clues about its mode of transmission or the susceptibility of certain groups of people. This in turn may reveal insights about the causes of the disease and possible preventive measures.

epidermal growth factor (EGF) The prototype member of a family of proteins that are important in regulating growth, proliferation, and survival of mammalian cells. Other family members include transforming growth factor α (TGF-α) and the neuregulins. EGF mediates its effects by binding to EGF receptors on the surface of target cells, notably in epithelial and mesenchymal (mesoderm-derived) tissue. Recombinant human EGF (rhEGF) is prepared commercially as an agent to promote wound healing. Overexpression or overactivity of the EGF receptor has an important role in certain cancers, such as lung and brain cancers. EGF is commonly added to culture media to promote division of cultured mammalian cells. *See also* GROWTH FACTOR.

epidermis 1. (in zoology) The outermost layer of cells of the body of an animal. In invertebrates the epidermis is normally only one cell thick and is covered by an impermeable *cuticle. In vertebrates the epidermis is the thinner of the two layers of *skin (*compare* DERMIS). It consists of a basal layer of actively dividing cells (*see* MALPIGHIAN LAYER), covered by layers of cells that become impregnated with keratin (*see* KERATINIZATION). The outermost layers of epidermal cells (the *stratum corneum) form a water-resistant protective layer. The epidermis may bear a variety of specialized structures (e.g. *feathers, *hairs). 2. (in botany) The outermost layer of cells covering a plant. It is overlaid by a *cuticle and its functions are principally to protect the plant from injury and to reduce water loss. Some epidermal cells are modified to form guard cells (*see* STOMA) or hairs of various types (*see* PILIFEROUS LAYER). In woody plants the functions of the shoot epidermis are taken over by the periderm tissues (*see* CORK CAMBIUM) and in mature roots the epidermis is sloughed off and replaced by the *hypodermis.

epididymis (*pl.* **epididymides**) A long coiled tube in which spermatozoa are stored in vertebrates. In reptiles, birds, and mammals it is attached at one end to the *testis and opens into the sperm duct (*vas deferens) at the other.

epigamic Serving to attract a mate. Epigamic characters include the bright plumage of some male birds.

epigeal Describing seed germination in which the seed leaves (cotyledons) emerge from the ground and function as true leaves. Examples of epigeal germination are seen in sycamore and sunflower. *Compare* HYPOGEAL.

epigenetic Describing heritable changes that are not the result of changes in DNA sequence. An example of an epigenetic phenomenon is *gene imprinting, in which expression of a gene varies depending on whether it is inherited from the mother or father. The chief mechanisms are *chromatin remodelling, involving changes to the histone proteins that package DNA into chromatin, and *DNA methylation, which causes widespread silencing of genes. In **paramutation**, which has been described in several plant species, certain genes behave as if they have undergone mutation after being in the same genome as other, so-called paramutagenic alleles. The change persists in subsequent generations of plants, even when the paramutagenic alleles are no longer present. Paramutation has also been described in mice; a possible explanation is that microRNAs transmitted from sperm to egg cell are responsible for silencing particular genes. *See also* EPIGENOME; X INACTIVATION.

epigenome The sum of all the chemical modifications to DNA or *chromatin that influence the activity of genes but do not affect the sequence of bases—i.e. changes that are literally 'above the genome'. These *epigenetic tags play a key role in determining which genes are active or inactive in

particular tissues, and they can be inherited by offspring. **Epigenomics** is the study of the epigenome, how it is influenced by environmental and cellular factors, and the role it plays in controlling gene expression and hence cellular function in health and disease. *See also* CHROMATIN REMODELLING.

epiglottis (*pl.* **epiglottises** or **epiglottides**) A flexible flap of cartilage in mammals that is attached to the wall of the pharynx near the base of the tongue. During swallowing (*see* DEGLUTITION) it covers the *glottis (the opening to the respiratory tract) and helps to prevent food from entering the trachea (windpipe), although it is not essential for this purpose.

epigyny A floral arrangement in which the ovary is completely enclosed by the receptacle so that the stamens and perianth arise above it, from the top of the receptacle; i.e. the ovary is *inferior (see illustration). The perianth and stamens are said to be **epigynous** with respect to the ovary, as seen in the daffodil. *Compare* HYPOGYNY; PERIGYNY.

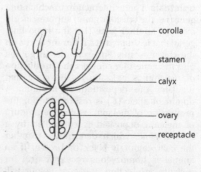

Epigyny

epilimnion (*pl.* **epilimnia**) The upper layer of water in a lake. *Compare* HYPOLIMNION. *See* THERMOCLINE.

epinephrine *See* ADRENALINE.

epipelagic zone *See* EUPHOTIC ZONE.

epiphysis (*pl.* **epiphyses**) The terminal section of a growing bone (especially a long limb bone) in mammals. It is separated from the bone shaft (**diaphysis**) by cartilage. New bone is produced on the side of the cartilage facing the diaphysis, while new cartilage is produced on the other side of the cartilage disc. When the bone reaches adult length the epiphysis merges with the diaphysis.

epiphyte A plant that grows upon another plant but is neither parasitic on it nor rooted in the ground. Epiphytes include many mosses and lichens and some cacti, bromeliads, and tropical orchids. They absorb moisture and nutrients via adventitious or aerial roots, from rain, dew, or surface debris on their host plant. Often the roots are adapted to cling to the host for support and over time these may damage the host.

episome A genetic element that can exist and replicate either independently of its host cell's chromosomes or as an integrated part of the chromosomes. Examples include certain bacterial *plasmids.

epistasis A gene interaction in which one gene (i.e. the epistatic gene) suppresses the effect of another gene (i.e. the hypostatic gene) that is situated at a different *locus on the chromosome. For example, in guinea pigs the gene that controls the production of melanin is epistatic to the gene that regulates the deposition of melanin. A dominant allele (*C*) is responsible for the production of melanin, while the amount of melanin deposited is controlled by a second gene, which determines whether the coat colour is black or brown. If an animal is homozygous recessive (*cc*) for melanin production, the coat colour will be white regardless of the alleles that produce black or brown coloration.

epithelium (*pl.* **epithelia**) A tissue in vertebrates consisting of closely packed cells in a sheet with little intercellular material. It covers the outer surfaces of the body and walls of the internal cavities (coeloms) and is often underlain by a *basement membrane. It also forms glands and parts of sense organs. Its functions are protective, absorptive, secretory, and sensory. The types of cell vary, giving rise to *squamous epithelium; *ciliated epithelium; **cuboidal epithelium**, with cube-shaped cells; and **columnar epithelium**, with rectangular cells (*see also* STRATIFIED EPITHELIUM). Epithelium is derived from *ectoderm and *endoderm. *Compare* ENDOTHELIUM; MESOTHELIUM.

epitope (**antigenic determinant**) The structure on the surface of an *antigen that is recognized by and can bind to a specific antibody. Its shape is generally complementary to that of the antibody's antigen-binding site. Examples are a group of sugar units of the polysaccharide component of a bacterial cell wall and a cluster of amino acids on the exposed surface of a protein molecule. **Conformational** (or **discontinuous**) **epitopes** are composed of parts of a molecule (especially a protein) that are juxtaposed by folding of the molecule into its natural conformation; **continuous** (or **linear**) **epitopes** consist of a single contiguous sequence of a molecule's subunits (e.g. a short peptide from a protein chain).

EPSP *See* EXCITATORY POSTSYNAPTIC POTENTIAL.

Epstein-Barr virus *See* HERPESVIRUS.

EQ *See* ENCEPHALIZATION QUOTIENT.

equator *See* SPINDLE.

equifinality The ability of certain structures or behaviour patterns to form by more than one route in a developing embryo or young animal.

ergocalciferol *See* VITAMIN D.

ergosterol A *sterol occurring in fungi, bacteria, algae, and plants. It is converted into vitamin D_2 by the action of ultraviolet light.

ergot The dark hard-walled mass of hyphae produced by the fungus *Claviceps purpurea* (*see* ASCOMYCOTA) in the grain of the cereals and other grasses that it parasitizes. Ergots are the **sclerotia** of this species—resting bodies that germinate in favourable conditions to produce mycelia or ascocarps. They contain alkaloids related to LSD, which cause blood vessels to constrict and are used therapeutically in the treatment of migraine and haemorrhage. Ingestion of infected grain can result in ergot poisoning (ergotism), with symptoms of gangrene and hallucinations—known as 'St Anthony's fire' in the Middle Ages.

Errantia One of two major clades of annelids proposed on the basis of molecular evidence, the other being *Sedentaria. Errantians typically are highly mobile worm-like animals, either swimming in the sea or crawling or burrowing on the sea floor. However, some are sedentary, such as the tube-dwelling *Platynereis*.

error (in research) Incorrectly rejecting the null hypothesis of no effect when it is true (**type I error**) or failing to reject the null hypothesis when it is false (**type II error**). *See* SIGNIFICANCE.

erythroblast Any of the cells in the *myeloid tissue of red bone marrow that develop into erythrocytes (red blood cells). Erythroblasts have a nucleus and are at first colourless, but fill with *haemoglobin as they develop. In mammals the nucleus disappears. *See* ERYTHROPOIESIS.

erythrocyte (red blood cell) The most numerous type of blood cell, which contains the red pigment *haemoglobin and is responsible for oxygen transport and for transport of some carbon dioxide (as a carbamino group). When mature, mammalian erythrocytes are disc-shaped and lack a nucleus and mitochondria; those of other vertebrates are oval and nucleated. Human erythrocytes are 7–8 μm in diameter and number some 5 to 6 million per cubic millimetre of blood. Each is capable of carrying about 1 billion molecules of oxygen. They survive for about four months and are then destroyed in the spleen and liver. *See also* ERYTHROPOIESIS. *Compare* LEUCOCYTE.

erythropoiesis The formation of red blood cells (erythrocytes), which occurs in the red bone marrow (*see* HAEMOPOIETIC TISSUE). The earliest precursor that can be distinguished microscopically is the **proerythroblast**, which develops from a haemopoietic stem cell and gives rise successively to the **early erythroblast, intermediate erythroblast**, and the **late erythroblast**, in which most haemoglobin is synthesized. In mammals the nucleus is then forced from the cell, which assumes a biconcave shape and is known as a **reticulocyte**. Reticulocytes are released into the blood and develop into mature erythrocytes within two days. *See also* ERYTHROPOIETIN.

erythropoietin A hormone that is released from the kidney (and to a lesser extent from the liver) in response to low concentrations of oxygen in the tissues. It speeds up the process of *erythropoiesis and is the means by which the rate of red cell production is controlled. It is now prepared using genetically engineered cell cultures and administered therapeutically in cases of kidney failure.

Escherichia coli (E. coli) A species of Gram-negative aerobic bacteria that is found in the intestine (*see* COLIFORM BACTERIA) and is also widely used in microbiological and genetics research. The motile rod-shaped cells ferment lactose and are usually harmless commensals, although certain strains are pathogenic and can cause a severe form of food poisoning. Studies of *E. coli* laboratory cultures have revealed much about the biology and genetics of prokaryotes and the bacteriophages that infect them; the species is also frequently used in genetic engineering, particularly as a host for *gene cloning and the expression of recombinant foreign genes in culture.

eserine (physostigmine) An alkaloid, derived from the calabar bean plant, that inhibits *cholinesterase by covalently binding with it (*see* INHIBITION). Eserine is used to treat the eye condition glaucoma, delayed gastric emptying, and the effects of certain poisons, notably atropine.

essential amino acid An *amino acid that an organism is unable to synthesize in sufficient quantities. It must therefore be present in the diet. In humans the essential (i.e. indispensable) amino acids are histidine, lysine, threonine, methionine, isoleucine, leucine, valine, phenylalanine, and tryptophan. These are required for protein synthesis and deficiency leads to retarded growth and other symptoms. Most of the amino acids required by humans are also essential for all other multicellular animals and for most protists.

essential element Any of a number of elements required by living organisms to ensure normal growth, development, and maintenance. Apart from the elements found in organic compounds (i.e. carbon, hydrogen, oxygen, and nitrogen), plants, animals, and microorganisms all require a range of elements in inorganic forms in varying amounts, depending on the type of organism. The **major elements**, present in tissues in relatively large amounts (greater than 0.005%), are calcium, phosphorus, potassium, sodium, chlorine, sulphur, and magnesium (*see also* MACRONUTRIENT). The **trace elements** occur at much lower concentrations and thus requirements are much less. The most important are iron, manganese, zinc, copper, iodine, cobalt, selenium, molybdenum, chromium, fluorine, and silicon (*see also* MICRONUTRIENT). Each element may fulfil one or more of a variety of metabolic roles. Sodium, potassium, and chloride ions are the chief electrolytic components of cells and body fluids and thus determine their electrical and osmotic status. Calcium, phosphorus, and magnesium are all present in bone. Calcium is also essential for cell signalling and nerve and muscle activity, while phosphorus is a key constituent of the chemical energy carriers (e.g. *ATP) and the nucleic acids. Sulphur is needed primarily for amino acid synthesis (in plants and microorganisms), and fluorine maintains healthy tooth structure and combats tooth decay. The trace elements may serve as *cofactors or as constituents of complex molecules, e.g. iron in haem and cobalt in vitamin B_{12}. *See also* MINERAL DEFICIENCY.

essential fatty acids *Fatty acids that must normally be present in the diet of certain animals, including humans. Essential fatty acids belong to the omega-6 and omega-3 series of polyunsaturated fatty acids, the parent compounds of which are *linoleic acid (n-6) and *α-linolenic acid (n-3); all possess double bonds at the same two positions along their hydrocarbon chain (counting positions from the methyl end) and so can act as precursors of eicosanoids, including the *prostaglandins. Deficiency of essential fatty acids can cause dermatosis, weight loss, irregular oestrus, etc. Recommendations for dietary intake of essential fatty acids vary, but both n-3 and n-6 acids should be consumed, ideally in roughly the same amounts.

essential oil A natural oil with a distinctive scent secreted by the glands of certain aromatic plants. *Terpenes are the main constituents. Essential oils are extracted from plants by steam distillation, extraction with cold neutral fats or solvents (e.g. alcohol), or pressing and used in perfumes, flavourings, and medicine. Examples are citrus oils, flower oils (e.g. rose, jasmine), and oil of cloves.

EST *See* EXPRESSED SEQUENCE TAG.

ester An organic compound formed by *esterification. Esters formed from carboxylic acids have the general formula RCOOR′. Esters containing simple hydrocarbon groups are volatile fragrant substances used as flavourings in the food industry. Triesters, molecules containing three ester groups, occur in nature as oils and fats. *See also* GLYCERIDE.

esterification A *condensation reaction of an alcohol with an acid to produce an *ester and water; e.g.

$$CH_3OH + C_6H_5COOH \rightleftharpoons$$
$$CH_3OOCC_6H_5 + H_2O$$

The reaction is an equilibrium and is slow under normal conditions, but can be speeded up by addition of a strong acid catalyst.

ET *See* ENDOTHELIN.

etaerio A cluster of fruits formed from the unfused carpels of a single flower. For example, the anemone has an etaerio of *achenes, larkspur an etaerio of *follicles, and blackberry an etaerio of *drupes.

ethanedioic acid *See* OXALIC ACID.

ethanoate *See* ACETATE.

ethanoic acid *See* ACETIC ACID.

ethanol (ethyl alcohol) A colourless water-soluble *alcohol, C_2H_5OH. It is the active principle in intoxicating drinks, in which it is produced by alcoholic *fermentation of sugar using yeast:

$$C_6H_{12}O_6 \rightarrow 2C_2H_5OH + 2CO_2$$

The ethanol produced kills the yeast and fermentation alone cannot produce ethanol solutions containing more than 15% ethanol by volume. *See also* BREWING.

ethene *See* ETHYLENE.

Ethiopian region *See* FAUNAL REGION.

ethnobotany The study of people's relationships with plants, e.g. how plants are used for food, shelter, medicine, cosmetics, clothing, or weapons, as well as the religious or broader cultural significance of plants. The work of ethnobotanists has contributed to the discovery and identification of many plants used for medicinal purposes by indigenous peoples, particularly those of tropical rainforests. Commonly these plants are assessed by pharmaceutical companies as sources of potential new drugs. Many other economically useful products have been derived from ethnobotanical research. Such work entails stringent ethical obligations (*see* BIOPROSPECTING). Other disciplines, such as anthropology, archaeology, and sociology, also have ethnobotanical dimensions.

ethology The study of the biology of *animal behaviour. Central to the ethologist's approach is the principle that animal behaviour (like physical characteristics) is subject to evolution through natural selection. Ethologists therefore seek to explain how the behaviour of animals, both as individuals and as social groups, in their natural environment may contribute to the survival of the maximum number of its relatives and offspring. This involves recognizing the stimuli that are important in nature (*see* SIGN STIMULUS) and how innate predispositions interact with *learning in the development of behaviour (*see* INSTINCT). Studies of this sort were pioneered by Konrad Lorenz and Niko Tinbergen and have led to the modern disciplines of sociobiology and behavioural ecology.

ethylene (ethene) A colourless gaseous hydrocarbon, C_2H_4, that occurs naturally in plants and acts as a plant hormone in a variety of physiological roles. It is produced in response to stresses, such as drought, flooding, mechanical injury, or infection, and has a key role in senescence, fruit ripening, and leaf abscission. Auxins stimulate tissues to produce ethylene, which diffuses rapidly to trigger responses in surrounding cells. The best known effect is the stimulation of fruit ripening: fruits such as bananas, apples, and avocados naturally produce ethylene during the later stages of ripening, and ethylene gas is used to promote the ripening of fruits, such as bananas, that are picked and shipped 'green'. The orderly programmed cell death (*apoptosis) that occurs during senescence is prompted by a burst of ethylene production, and ethylene triggers cells to produce the enzymes responsible for digesting the abscission layer to allow leaf fall. Ethylene generally suppresses flowering, except in members of the pineapple family (Bromeliaceae)—hence flowering of pineapples may be synchronized by releasing ethylene into the growing crop. Studies have shown varied and often contradictory effects of ethylene on vegetative growth. For example, in rice it acts with *gibberellins to promote stem elongation, while in peas ethylene inhibits root and shoot elongation. Seed germination, bud opening, and root initiation may also be promoted by ethylene.

ETI Effector-triggered immunity. *See* IMMUNITY (sense 2).

etiolation The abnormal form of growth observed when plants grow in darkness or severely reduced light. Such plants characteristically have blanched leaves and shoots, excessively long shoots, and reduced leaves and root systems.

Eubacteria *See* BACTERIA.

eucarpic Denoting a fungus in which the thallus is differentiated into vegetative and reproductive regions. *Compare* HOLO-CARPIC.

eucaryote *See* EUKARYOTE.

euchromatin *See* CHROMATIN.

eudicot Any dicotyledonous flowering plant whose pollen has three apertures (i.e. **triaperturate pollen**), through one of which the pollen tube emerges during pollination. Eudicots contrast with the so-called 'primitive' dicots, such as the magnolia family, which have **uniaperturate pollen** (i.e. with a single aperture). The eudicots form a large clade, but their phylogenetic relationships to other dicots and to monocots remain uncertain.

eugenics The study of methods of improving the quality of human populations by the application of genetic principles. **Positive eugenics** would seek to do this by selective breeding programmes, a strategy that is generally deemed reprehensible. **Negative eugenics** aims to eliminate harmful genes (e.g. those causing haemophilia and colour blindness) by counselling any prospective parents who are likely to be *carriers.

Euglenida In traditional classifications, a class of mostly unicellular protists (including *Euglena*) that move by means of flagella. Most euglenids are photosynthetic and inhabit fresh water, although some are *mixotrophs, engulfing prey when there is insufficient light for photosynthesis. They lack a cell wall, being covered with a proteinaceous *pellicle. Euglenids are sometimes classified as a phylum, Euglenophyta, but are now included in the group Euglenozoa within the clade *discicristates on the basis of their mitochondrial structure (characterized by disc-shaped cristae) and the absence of sexual reproduction.

Eukarya A domain containing all eukaryotic organisms, embracing protists, fungi, plants, and animals. In three-domain classification systems, the other two (prokaryotic) domains are *Archaea and *Bacteria.

eukaryote (**eucaryote**) An organism consisting of *cells in which the genetic material is contained within a distinct nucleus. All organisms except bacteria and archaea are eukaryotes. *See* EUKARYA. *Compare* PROKARYOTE.

Eumetazoa A clade, sometimes given the rank of subkingdom, containing all animals that possess tissues, such as muscle or nerve tissue. Hence it excludes multicellular animals that lack tissues, notably the sponges (Porifera). *Compare* METAZOA.

Eumycota In some earlier classifications, a division, or phylum, containing the 'true fungi', as distinct from slime moulds.

euphotic zone (**epipelagic zone; photic zone**) The topmost layer of a lake or sea in which there is sufficient light for net primary production, i.e. where the energy fixed by photosynthesis exceeds that lost by respiration. The depth varies, depending on such factors as turbidity, supply of nutrients in the water, tidal turbulence, and temperature. For example, high nutrient levels will encourage a greater biomass of phytoplankton near the surface, which causes shading and consequent reduction in depth of the euphotic zone. It typically ranges from 1 m to about 30 m in lakes and coastal waters, and rarely reaches depths of more than 200 m in the open ocean. At depths between 200 and 1000 m blue light may still penetrate sufficiently to allow limited photosynthesis. This is sometimes referred to as the **dysphotic zone** or

mesopelagic zone. Below this is the *aphotic zone, where no light penetrates.

euphyllophyte Any member of a clade containing seed plants, ferns, horsetails, and their allies—i.e. all vascular land plants except the clubmosses and their relatives. Modern euphyllophytes are characterized by possessing true leaves (*megaphylls) and branching stems of unequal length ('overtopping'). In ancestral members the evolution of overtopping led to forms having a main axis with lateral branches. According to the *telome theory, megaphylls subsequently evolved through infilling by photosynthetic tissue of spaces between the twigs of lateral shoots. Fossil evidence shows the first megaphylls appeared in the Devonian period. *See also* MONILOPHYTE.

euploid Describing a nucleus, cell, or organism that has an exact multiple of the haploid number (n) of chromosomes. For example, *diploid ($2n$), *triploid ($3n$), and *tetraploid ($4n$) nuclei or cells are all euploid. *Compare* ANEUPLOID.

eury- A prefix denoting width of range. For example, **euryhaline** aquatic organisms can tolerate a wide range of saline conditions. *Compare* STENO-.

eusocial Describing the most highly developed form of animal societies, such as those of colonial ants, termites, wasps, and bees (*see* HYMENOPTERA). Typically there is extensive division of labour and cooperation, with various *castes each performing particular tasks, such as food-gathering, defence, or tending to the young. Reproduction is undertaken by an elite of fertile individuals, assisted by sterile workers. Moreover, there is an overlap of different generations within the colony at any one time, and continuity of the society is maintained from year to year. Some vertebrate animals may display equivalent levels of social organization. For example, colonies of naked mole rats have a caste system, with nonbreeding adult members that forage, maintain the burrow, or simply keep the other group members warm.

Eustachian tube The tube that connects the *middle ear to the back of the throat (pharynx) in vertebrates. It is normally closed, but during swallowing it opens to allow air into the middle ear, which equalizes the pressure on each side of the *tympanum (eardrum). It was named after the Italian anatomist Bartolomeo Eustachio (?1520–74).

euthanasia The act of ending the life of a person or animal in order to prevent further suffering, e.g. from an incurable and painful disease. This can be achieved by administering a lethal drug or by withholding vital treatment. In human medicine euthanasia is fraught with ethical and legal problems, and is illegal in most countries. Where it is practised, strict safeguards are enforced to ensure that the patient's wishes are determined and adhered to. Euthanasia is widely performed in veterinary medicine.

Eutheria (Placentalia) An infraclass of mammals in which the embryos are retained in a uterus in the mother's body and nourished by a *placenta. The young are thus fully protected during their embryonic development and kept at a constant temperature. Modern placentals are a highly diverse group that occupy all types of habitat in all parts of the world. They include the orders *Artiodactyla, *Carnivora, *Cetacea, *Chiroptera, *Insectivora, *Perissodactyla, *Primates, *Proboscidea, and *Rodentia. Evidence about when placental mammals evolved is contradictory. Studies of the fossil record indicate an origin around 66 million years ago (mya), whereas molecular data suggest a date of around 100 mya, during the Cretaceous period. *Compare* METATHERIA; PROTOTHERIA.

eutrophic Describing a body of water (e.g. a lake) with an abundant supply of nutrients and a high rate of formation of organic matter by photosynthesis. Pollution of a lake by *sewage or *fertilizers renders it eutrophic (a process called **eutrophication**). This stimulates excessive growth of algae (*see* ALGAL BLOOM); the death and subsequent decomposition of these increases the *biochemical oxygen demand and thus depletes the oxygen content of the lake, resulting in the death of the lake's fish and other animals. *Compare* DYSTROPHIC; MESOTROPHIC; OLIGOTROPHIC.

evergreen (Describing) a plant that bears leaves throughout the year, each leaf being shed independently of the others after two or three years. The leaves of evergreens are often reduced or adapted in some way to prevent excessive water loss; examples are the needles of conifers and the leathery waxy leaves of holly. *Compare* DECIDUOUS.

evocation The ability of experimental stimuli (e.g. chemicals or tissue implants) to cause unspecialized embryonic tissue to develop into specialized tissue.

evo-devo Short for evolution of development, or evolutionary developmental biology—a field of study concerned with the interplay of developmental genetics and evolution. It has developed in tandem with advances in molecular biology and genetics that have revealed how evolution of diversity in form and function can arise from mutations that alter the course or timing of development. Central is the concept that all complex animals share a 'genetic toolkit' inherited from a common ancestor dating back perhaps 600 million years. These genes, such as the *Hox* genes, and their signalling pathways are crucial in controlling development and are greatly conserved across a wide range of animal groups. Moreover, embryological development normally must proceed in an orderly sequence of stages (*see* ONTOGENY), disruption of which often leads to death of the embryo. Apart from these constraints on evolution, however, development can also contribute positively to evolution in numerous ways. One is the modular organization of body plans, such as repeated limbs or body segments, which is a universal feature of development. Duplication of such a module may lead to its further adaptation into a novel structure, such as pincers or mouthparts.

Another area involves changes in the timing, location, or level of gene expression (*see* HETEROCHRONY) or alterations in the relative growth rates of different parts of the body. In this way relatively few mutations affecting development can have profound effects on the morphology of the mature organism.

evolution The gradual process by which the present diversity of plant and animal life arose from the earliest and most primitive organisms, which is believed to have been continuing for at least the past 3000 million years. Until the middle of the 18th century it was generally believed that each species was divinely created and fixed in its form throughout its existence (*see* SPECIAL CREATION). *Lamarck was the first biologist to publish a theory to explain how one species could have evolved into another (*see* LAMARCKISM). But it was not until the joint publication in 1858 of the paper by *Darwin and *Wallace outlining a theory of evolution by natural selection, and Darwin's *On the Origin of Species* in 1859, that special creation was seriously challenged. Unlike Lamarck, Darwin and Wallace proposed a feasible mechanism for evolution and backed it up with evidence from the fossil record and studies of comparative anatomy and embryology (*see* DARWINISM; NATURAL SELECTION). The modern version of Darwinism, which incorporates discoveries in genetics made since Darwin's time, remains the most acceptable theory of species evolution (*see also* PUNCTUATED EQUILIBRIUM). *See also* MACROEVOLUTION; MICROEVOLUTION; MOSAIC EVOLUTION. *See* Appendix 8 for further sources of information.

(⊕) SEE WEB LINKS

- The evidence for and mechanisms of evolution, from Understanding Evolution, created by the University of California Museum of Paleontology

evolutionary tree (phylogenetic tree) A diagram depicting the evolutionary relationships between species or groups of organisms as (usually) dichotomously branching lines to form a treelike pattern.

In some trees, changes in the relative abundance of a lineage are indicated by the width of the branch. A *phylogram is a special type of evolutionary tree constructed so that the branch lengths reflect the evolutionary distance separating ancestors and descendants. *See also* CLADISTICS.

exa- Symbol E. A prefix used in the metric system to denote 10^{18} times. For example, 10^{18} metres $= 1$ exametre (Em).

exaptation A morphological or physiological character that predisposes an organism to adapt to a changed environment or lifestyle. For example, a hearing mechanism sensitive to low-frequency sound evolved in crickets, perhaps 250 million years ago, to aid intraspecific communication. When bats evolved, only about 50 million years ago, the cricket hearing mechanism served as a *preadaptation, being modified to perceive the high-frequency sonar emitted by these night-flying predators, in addition to the low-frequency sounds of other crickets.

excavates An assemblage of eukaryotic protists characterized by one or more feeding grooves equipped with flagella that propel food to the mouth. They include, among others, the *diplomonads (e.g. *Giardia*, responsible for giardiasis in humans), other gut parasites or symbionts (retortamonads and oxymonads), trichomonads (such as the human vaginal parasite *Trichomonas*), the jakobids (a group of zoomastigotes), and the *discicristates (euglenids, diplomonads, amoebomastigotes, etc.). Molecular systematics reveals that excavates could be a very early eukaryote lineage, a view supported by their simplified organelles and by the fact that the jakobids have mitochondrial genomes that are remarkably similar to those of bacteria, attesting to the ancient origin of mitochondria as intracellular symbiotic bacteria.

excision repair A form of *DNA repair found in living cells in which damaged or mismatched bases in one of the two strands of DNA are cut out and replaced with the correct bases using the other strand as a template. It is performed by various enzymes. Typically, a **DNA glycosylase** enzyme recognizes the damage and excises the affected base. Then endonuclease enzymes nick the damaged strand either side of the site where the base was removed, allowing an exonuclease to remove the affected region of the strand. The gap is then repaired by a DNA *polymerase, and the new section of the strand is rejoined to the existing strands by a *DNA ligase. *See also* MISMATCH REPAIR.

excitation 1. The act or process of stimulating a cell, especially a nerve cell. **2.** The response of a cell that has been stimulated.

excitatory postsynaptic potential (EPSP) The electric potential that is generated in a postsynaptic neuron during the transmission of a nerve impulse (*see* SYNAPSE). It is caused by *depolarization of the postsynaptic membrane when a *neurotransmitter (such as acetylcholine), released from the presynaptic membrane, binds to the postsynaptic membrane. This will induce an *action potential in the receiving neuron if the EPSP is large enough. *See* FACILITATION. *Compare* INHIBITORY POSTSYNAPTIC POTENTIAL.

excretion The elimination by an organism of the waste products that arise as a result of metabolic activity. These products include water, carbon dioxide, and nitrogenous compounds. Excretion plays an important role in maintaining the constancy of an organism's *internal environment (*see* HOMEOSTASIS). The simplest method of excretion, which occurs, for example, in plants, involves diffusion of waste products from the body, but many animals have specialized organs and organ systems devoted to this function (*see* MALPIGHIAN TUBULE; NEPHRIDIUM). Examples of excretory organs in vertebrates are the lungs (for carbon dioxide and water), and the *kidneys (for *nitrogenous waste and water). In addition, mammals excrete small amounts of urea, salts, and water from the skin in sweat. Removal of excess salt from the blood is accomplished by *rectal glands in elasmobranch fishes, and by *salt glands in marine birds and reptiles.

exercise Increased muscular activity, which results in an increase in metabolic rate, heart rate, and oxygen uptake. Exercise also causes an increase in *anaerobic respiration in order to compensate for the *oxygen debt, which results in a build-up of lactic acid in the tissues. There are two main types of exercise. **Strength exercise** (or **anaerobic exercise**), such as weight-lifting, repeatedly contracts certain muscles to induce the formation of new actin and myosin filaments, and thus increase the maximum force they can exert. **Endurance exercise** (or **aerobic exercise**) sustains a given load over a certain time. This increases the oxidative capacity of the muscle by boosting the numbers of mito-chondria, developing the network of capil-laries that deliver blood to muscle tissue, and increasing the concentration of the oxygen carrier *myoglobin. Consequently, the workload can be maintained for succes-sively longer periods.

exergonic reaction A chemical reaction in which energy is released (*compare* ENDERGONIC REACTION). An *exothermic reaction is an exergonic reaction in which energy is released in the form of heat.

exhalation *See* EXPIRATION.

exine *See* POLLEN.

exocarp *See* PERICARP.

exocrine gland A gland that discharges its secretion into a body cavity (such as the gut) or onto the body surface. Examples are the *sebaceous and *sweat glands, the *mammary glands, and part of the pan-creas. Exocrine glands are formed in the embryo from the invagination of epithelial cells. Their secretions pass initially into a cavity (an **alveolus** or *acinus) and then out through a duct or duct network, along which the secretion may become modified by exchange with the blood across the duct epithelium.

exocytosis The passage of material from the inside of the cell to the cell surface within membrane-bound vesicles. The membranes of the vesicles fuse with the plasma membrane, a process involving various docking proteins in both the vesicle coat and target membrane. The vesicle contents are then released to the exterior. Exocytosis is used both for the removal of waste material from the cell and for secre-tion; for example of mucus by *goblet cells. *Compare* ENDOCYTOSIS.

exodermis *See* HYPODERMIS.

exogamy The fusion of reproductive cells from distantly related or unrelated organ-isms, i.e. *outbreeding. *Compare* ENDOGAMY.

exogenous Describing substances, stim-uli, etc., that originate outside an organism. For example, vitamins that cannot be synthesized by an animal are said to be supplied exogenously in the diet. *Compare* ENDOGENOUS.

exome The protein-coding regions of a genome collectively; in eukaryotes, the exome comprises all the *exons.

exon A nucleotide sequence in a gene that codes for part or all of the gene product and is therefore expressed in mature mes-senger RNA, ribosomal RNA, or transfer RNA. In eukaryotes, exons are separated by noncoding sequences called *introns. Both exons and introns are transcribed into pre-messenger RNA, and subsequently the intronic sequences are removed and the exons spliced together (*see* RNA PROCESSING).

exon shuffling *See* INTRON.

exonuclease An enzyme that catalyses the cleavage of nucleotides from the end of a nucleic acid molecule. *Compare* ENDONUCLEASE.

exopeptidase A protein-digesting enzyme that cleaves amino acids from the ends of a polypeptide chain. *Carboxypep-tidase, which breaks down proteins in the small intestine, is an example of a exopep-tidase. *Compare* ENDOPEPTIDASE.

exopterygote Any winged insect show-ing *hemimetabolous development, i.e. the eggs hatch into young (called *nymphs) that resemble the adults but lack wings;

these develop gradually and externally in a series of stages (*instars) until the final moult produces the adult insect. There is no pupal stage and metamorphosis is described as **incomplete**. Exopterygote insects include the mayflies (*Ephemeroptera), dragonflies (*Odonata), grasshoppers (*Orthoptera), and bugs (*Hemiptera). *Compare* ENDOPTERYGOTE.

exoskeleton A rigid external covering for the body in certain animals, such as the hard chitinous cuticle of arthropods. An exoskeleton protects and supports the body and provides points of attachment for muscles. The cuticle of arthropods must be shed at intervals to allow growth to occur (*see* ECDYSIS). Other examples of exoskeletons are the shells of molluscs and the bony plates of tortoises and armadillos. *Compare* ENDOSKELETON.

exothermic Denoting a chemical reaction that releases heat into its surroundings. *See* EXERGONIC REACTION. *Compare* ENDOTHERMIC.

exotic *See* ALIEN.

exotoxin *See* TOXIN.

expansin Any of a family of proteins that loosen the polymers of plant cell walls enabling expansion of the cells during growth and other developmental processes. They act by breaking the hydrogen bonds between cellulose microfibrils and other polysaccharides, allowing the cell wall components to slide relative to each other without disrupting the cell wall.

experiment A process or trial designed to test a scientific theory.

experimental taxonomy *See* SYSTEMATICS.

expiration (exhalation) The process by which gas is expelled from the lungs (*see* RESPIRATORY MOVEMENT). In mammals, the volume of the thoracic cavity is reduced by contraction of the internal *intercostal muscles and relaxation of the muscles of the diaphragm, assisted by upward pressure of the abdominal organs. As a result,

pressure in the lungs exceeds atmospheric pressure and gas flows out of the lungs, allowing the pressures to equalize. *Compare* INSPIRATION.

expiratory centre *See* VENTILATION CENTRE.

explantation The removal of cells, tissues, or organs of animals and plants for observation of their growth and development in appropriate culture media. The removed parts are known as **explants**. *See also* TISSUE CULTURE; ORGAN CULTURE.

exponential growth A form of *population growth in which the rate of growth is related to the number of individuals present. Increase is slow when numbers are low but rises sharply as numbers increase. If population number is plotted against time on a graph a characteristic J-shaped curve results (see graph). In animal and plant populations, such factors as overcrowding, lack of nutrients, and disease limit population increase beyond a certain point and the J-shaped exponential curve tails off giving an S-shaped (sigmoid) curve.

Graph showing exponential growth of the human population

expressed sequence tag (EST) A short partial sequence, typically 200–400 bp long, of a *complementary DNA (cDNA) clone. Because cDNAs are prepared by reverse transcription of messenger RNA molecules,

ESTs act as markers for genes that are expressed in particular tissues or organs. EST sequence data are held on databases, and researchers use computer programs to search for sequences that correspond to, say, a partial amino-acid sequence for a protein under investigation. The EST sequence can then be used to construct a PCR primer or DNA probe to locate the respective clone from a *DNA library.

expression vector A *vector used in genetic engineering that enables a particular gene to be not only cloned but also expressed in a host cell. The vector is constructed to contain appropriate regulatory sequences, such as a promoter and operator, so that the host-cell machinery can transcribe the gene and translate the resultant messenger RNA to synthesize the corresponding protein. Such vectors are thus essential for the manufacture of, for example, mammalian proteins by bacterial host cells. Expression vectors used in prokaryotes are typically based on plasmids or phages, or plasmid-phage hybrids (**phagemids**). Some eukaryote proteins are extensively modified during or following their synthesis, for example by the addition of carbohydrate groups. Prokaryote host cells are unable to accomplish these modifications, and therefore expression systems based on eukaryotic cells must be used instead. For example, vectors based on baculovirus, a DNA virus of insects, are used successfully in cultures of insect cells. **Secretion vectors** allow both expression and secretion of the novel protein by the host cell, by ensuring that the expressed protein carries a signal peptide that allows it to be transported across the plasma membrane.

extended phenotype The concept, advanced by British evolutionary biologist Richard Dawkins in his 1982 book of the same title, that the phenotype of an organism extends beyond its body to encompass the organism's behaviour and the consequences of that behaviour. Dawkins cites a beaver's lake as an example. This manifestation of the beaver's instinctive dam-building activities is, he argues, an evolutionary adaptation just as much as, say, the beaver's coat, and is likewise subject to natural

selection. Other instances include birds' nests, termite mounds, and spiders' webs.

extensor Any muscle that causes a limb to extend. *See* SKELETAL MUSCLE. *Compare* FLEXOR.

exteroceptor Any *receptor that detects external stimuli. Examples of exteroceptors are the thermoreceptors in the skin, which monitor the temperature of the external environment. *Compare* INTEROCEPTOR.

extinction 1. The irreversible condition of a species or other group of organisms of having no living representatives in the wild, which follows the death of the last surviving individual of that species or group. Extinction may occur on a local or global level; it can result from various human activities, including the destruction of habitats or the overexploitation of species that are hunted or harvested as a resource. Species at the top of a *food chain (e.g. large birds of prey) will be more prone to extinction since they exist in relatively small numbers and will be affected by a deleterious change at any of the levels in the food chain. *See* MASS EXTINCTION. **2.** The termination of a behaviour pattern that is no longer appropriate. For example, dogs can be conditioned to salivate when they hear a bell ring in the absence of a food stimulus (*see* CONDITIONING). However, if the bell continues to be rung in the absence of food the dogs will gradually stop salivating on hearing the bell.

extracellular Located or occurring outside the cell. *Cuticularization is an example of an extracellular process.

extracellular matrix (ECM) The viscous watery fluid that surrounds cells in animal tissues. Secreted by the cells themselves, it is the medium through which they receive materials (e.g. nutrients, hormones) from elsewhere in the body and via which they communicate with other cells. The ECM is the environment in which cells migrate during tissue development and it contains constituents that bind cells together to maintain tissue integrity. The bulk of the matrix consists of *proteoglycans, which associate with water molecules. Other key constituents are *collagens—insoluble fibrous proteins that form various bundles, chains, and other structural components. Also present are **multiadhesive proteins**, which bind to other matrix components and to *cell adhesion molecules in plasma membranes. The ECM is especially prominent in connective tissues, such as bone, cartilage, and adipose tissue, in which it is sometimes called **ground substance**. The development of artificial ECM is an important step in creating three-dimensional scaffolds for *tissue engineering and may also have therapeutic potential as a vehicle for drug delivery and cancer therapy. *See also* BASEMENT MEMBRANE.

(⊕) SEE WEB LINKS

- The extracellular matrix at a glance from the *Journal of Cell Science*

extraembryonic membranes (embryonic membranes) The tissues produced by an animal *embryo for protection and nutrition but otherwise taking no part in its development. The four membranes, which are called **fetal membranes** in humans, are the *chorion, *amnion, *allantois, and *yolk sac.

extrafusal *See* INTRAFUSAL.

extranuclear genes Genes included in the DNA present in organelles other than the nucleus, such as the mitochondria and chloroplasts, some of which code for the synthesis of proteins. The DNA of these organelles is inherited by the offspring via the cytoplasm of the gametes (*see* CYTOPLASMIC INHERITANCE). In organisms in which one gamete is much larger than the other, so that the smaller gamete contains very little cytoplasm, one parent will contribute most or all of the extranuclear genes. For example, human *mitochondrial DNA is passed from generation to generation through the maternal line in the ovum.

extremophile A bacterium or archaean that thrives under extreme conditions, e.g. at very high or very low temperatures, or in very salty or acidic environments. For example, certain archaea, termed **hyperthermophiles**, live in hot springs at temperatures near or even above 100°C. The enzymes of such organisms exhibit great stability and have been extracted for use in laboratory and commercial processes.

eye The organ of sight. The most primitive eyes are the *eyespots of some unicellular organisms. More advanced eyes are the *ocelli and *compound eyes of arthropods (e.g. insects). The cephalopod molluscs (e.g. the octopus and squid) and vertebrates possess the most highly developed eyes (see illustration). These normally occur in pairs, are nearly spherical, and filled with fluid. Light is refracted by the *cornea through the pupil in the *iris and

Structure of the vertebrate eye

onto the *lens, which focuses images onto the retina. These images are received by light-sensitive cells in the retina (*see* CONE; ROD), which transmit impulses to the brain via the optic nerve.

🌐 **SEE WEB LINKS**

• Gross anatomy of the eye, from emedicine

eye muscle Any of the muscles associated with the functioning of the eye, which are divided into two groups. The **intrinsic muscles**, which are involuntary, are situated inside the eyeball and comprise the ciliary muscle (*see* CILIARY BODY) and the iris. The **extrinsic muscles**, which comprise three pairs of voluntary muscles, are inserted on the sclera (outer surface) of the eyeball and control its movements.

eyepiece (ocular) The lens or system of lenses in an optical instrument that is nearest to the eye. It usually produces a magnified image of the previous image formed by the instrument.

eyespot (stigma) 1. A structure found in some free-swimming unicellular algae and in plant reproductive cells that contains orange or red pigments (carotenoids) and is sensitive to light. It enables the cell to move in relation to a light source (*see* PHOTOTAXIS). **2.** A spot of pigment found in some lower animals, e.g. jellyfish. **3.** An eyelike marking on the wings of certain insects, which is revealed in *startle displays.

eye tooth A *canine tooth in the upper jaw.

F₁ (first filial generation) The first generation of offspring resulting from an arranged cross between *homozygous parents in breeding experiments. *See* MONOHYBRID CROSS.

F₂ (second filial generation) The second generation of offspring in breeding experiments, obtained by crosses between individuals of the *F₁ generation. *See* MONOHYBRID CROSS.

Fab fragment One of the fragments produced by partial digestion of *immunoglobulin with the proteolytic enzyme papain. It corresponds to an arm of the Y-shaped molecule and consists of an entire light chain paired with part of a heavy chain, complete with the antigen-binding site. Digestion of an IgG immunoglobulin molecule will thus yield two identical Fab fragments, representing the separated arms of the Y, and an **Fc fragment**, which corresponds to the stem of the Y and consists of the remaining regions of the heavy chains paired via the disulphide bond of the hinge region. *Compare* F(ʌʙ')₂ FRAGMENT.

F(ab')₂ fragment One of the fragments produced by partial digestion of *immunoglobulin with the proteolytic enzyme pepsin. It consists of both arms of the Y-shaped molecule complete with antigen-binding sites and the disulphide bond of the hinge region that joins them together. The remaining heavy chains are cut into small portions, the largest of which is denoted the **pFc' fragment**. *Compare* FAB FRAGMENT.

facilitated diffusion The transport of molecules across the plasma membrane of a living cell by a process that involves a specific transmembrane carrier (*see* TRANSPORT PROTEIN) located within the plasma membrane but does not require expenditure of energy by the cell. The carrier combines with a molecule at one face of the membrane, then changes shape so the molecule is moved through the membrane and released at the opposite face. It enables the diffusion through the membrane of molecules that otherwise could not pass through. For example, fructose is transported from the intestinal lumen into epithelial cells by the GLUT5 transporter in the apical membrane. *Compare* ACTIVE TRANSPORT.

facilitation 1. (in neurophysiology) The effect of successive stimuli on the postsynaptic membrane, which results in the generation of an *excitatory postsynaptic potential (EPSP). Although a single impulse may fail to cross the junction between adjacent nerve cells, the synapse becomes more responsive to the following impulse. **2.** (in ecology) The phenomenon observed during *succession in which the presence of one species increases the likelihood or speed of colonization by a second species. The first species brings about some change in the environment that make it more favourable for the second. For example, a pre-existing plant may provide the germinating seeds and seedlings of another species with vital shelter from the wind or protection from herbivores. Alternatively, it may alter the nature of the soil, for example by changing the pH, sufficiently to permit growth of a new incoming species. *See also* FOUNDATION SPECIES.

Factor VIII (antihaemophilic factor) One of the blood *clotting factors. Factor VIII is a soluble protein that stimulates the activation of Factor X by Factor IXa, which in turn converts *prothrombin to thrombin, thus causing the fibrin matrix of a blood clot to form. *Haemophilia is due to a deficiency or defect of Factor VIII and is treated by administration of blood plasma or plasma concentrate containing the factor. Therapeutic Factor VIII is derived from donated blood plasma or alternatively is obtained from genetically engineered cell cultures, which avoids the risk of contamination with viruses, notably HIV (the AIDS virus).

facultative Describing organisms that are capable of a certain mode of life or function but are not restricted to it. **Facultative parasites** can live as parasites in or on another organism, but can also live independently; e.g. certain types of ringworm fungus are opportunistic parasites of mammals, living in the skin and hair of their host. **Facultative anaerobes** are organisms, such as certain bacteria, fungi, and some internal parasites of animals, that are able to alter their metabolism to grow in either the presence or absence of oxygen. The best-known facultative anaerobe is *Saccharomyces cerevisiae*, the yeast used in brewing. *See* ANAEROBIC RESPIRATION.

FAD (flavin adenine dinucleotide) A *coenzyme important in various biochemical reactions. It comprises a phosphorylated vitamin B_2 (riboflavin) molecule linked to the nucleotide adenine monophosphate (AMP). FAD is usually tightly bound to the enzyme forming a **flavoprotein**. It functions as a hydrogen acceptor in dehydrogenation reactions, being reduced to $FADH_2$. This in turn is oxidized to FAD by the *electron transport chain, thereby contributing to the generation of ATP (1.5 molecules of ATP per molecule of $FADH_2$).

faeces Waste material that is eliminated from the alimentary canal through the *anus. Faeces consist of the indigestible residue of food that remains after the processes of digestion and absorption of nutrients and water have taken place, together with bacteria and dead cells shed from the gut lining.

Fahrenheit scale A temperature scale in which (by modern definition) the temperature of boiling water is taken as 212 degrees and the temperature of melting ice as 32 degrees. It was invented in 1714 by the German scientist G. D. Fahrenheit (1686–1736), who set the zero at the lowest temperature he knew how to obtain in the laboratory (by mixing ice and common salt) and took his own body temperature as 96°F. The scale is no longer in scientific use. To convert to the *Celsius scale the formula is

$$C = 5(F - 32)/9.$$

fallopian tube (oviduct) The tube that carries egg cells from the *ovary to the womb in mammals. The eggs are carried by the action of muscles and cilia. It was named after the Italian anatomist, Gabriel Fallopius (1523–62).

false fruit *See* PSEUDOCARP.

family 1. (in taxonomy) A category used in the *classification of organisms that consists of one or several similar or closely related genera. Similar families are grouped into an order. Family names end in -*aceae* or -*ae* in botany (e.g. Cactaceae) and -*idae* in zoology (e.g. Equidae). The names are usually derived from a type genus (*Cactus* and *Equus* in the examples above) that is characteristic of the whole family (*see* TYPE SPECIMEN). In botany, families are sometimes called **natural orders**. **2.** (in molecular biology) A group of proteins with shared similarities in their amino-acid sequence, and often similarities in function, due to evolutionary divergence from a putative common ancestral protein. For example, the various types and subtypes of *adrenoceptors can be considered as a protein family. *See also* GENE FAMILY.

farming *See* AGRICULTURE.

fascia (*pl.* **fasciae**) A sheet of fibrous connective tissue occurring beneath the skin and also enveloping glands, vessels, nerves, and forming muscle and tendon sheaths.

fascicle 1. (**fasciculation**) A small bundle of nerve or muscle fibres. **2.** *See* VASCULAR BUNDLE.

fascicular cambium *See* CAMBIUM.

fasciculation 1. *See* FASCICLE. **2.** (in development) The process of becoming organized into a bundle, as in the bundling of axons to form a nerve. During nervous system development, the *growth cones of successive axons are guided by a pioneering axon that has already 'navigated' the correct pathway (*see* GUIDEPOST CELL). **3.** A brief involuntary contraction of a group of muscle fibres.

Fas signal pathway An intracellular pathway that initiates the process of programmed cell death (*apoptosis). Fas is a cell surface receptor: when it binds its specific extracellular signal molecule, **Fas ligand**, the intracellular portion of the Fas protein changes shape, enabling it to bind *adaptor proteins and recruit components of apoptosis, notably caspase enzymes, which act as 'executioners' inside the cell. Some effector T cells express Fas ligand on their surface and can kill target cells by binding to Fas and triggering apoptosis. Moreover, after an infection has been cleared, redundant lymphocytes are induced to enter apoptosis by a similar mechanism.

fast green A green dye used in optical microscopy that stains cellulose, cytoplasm, collagen, and mucus green. It is frequently used to stain plant tissues, with *safranin as a counterstain. Unlike **light green**, a similar dye, it does not fade easily.

fast-twitch fibre A type of skeletal muscle fibre that produces relatively rapid, brief, and strong contractions but fatigues quickly. Fast-twitch fibres are characteristic, e.g., of muscles of the hand and eye, where rapid movements are required but are also found together with *slow-twitch fibres in other muscles of the body. They contain a form of the contractile protein *myosin that hydrolyses ATP very rapidly, causing fast interactions of the thick and thin filaments of the muscle fibre and rapid development of muscle tension. But the ATP cannot be replenished fast enough to sustain the rapid muscle contractions; hence fast-twitch fibres are suited for short-term work, as in the biceps of the upper arm. **Oxidative fast-twitch fibres (type IIA)** use aerobic metabolism to manufacture ATP; they contain the oxygen storage protein *myoglobin, are well supplied with blood vessels and mitochondria, and occur as 'red' muscle. In contrast, **glycolytic fast-twitch fibres (type IIB)** rely on *glycolysis to make ATP, have relatively few mitochondria and blood vessels, and make up 'white' muscle. The proportions of fast-twitch and slow-twitch fibres determine the performance characteristics of muscle; athletes with a predominance of fast-twitch fibres are more suited to sprint events, whereas those with more slow-twitch fibres tend to excel at endurance events, such as long-distance running. These proportions in any individual are largely the result of heredity, although training can have some influence.

fat A mixture of lipids, chiefly *triglycerides, that is solid at normal body temperatures. Fats occur widely in plants and animals as a means of storing food energy, having twice the calorific value of carbohydrates. In mammals, fat is deposited in a layer beneath the skin (subcutaneous fat) and deep within the body as a specialized *adipose tissue (*see also* BROWN FAT). The insulating properties of fat are also important, especially in animals lacking fur and those inhabiting cold climates (e.g. seals and whales).

Fats derived from plants and fish generally have a greater proportion of unsaturated *fatty acids than those from mammals. Their melting points thus tend to be lower, causing a softer consistency at room temperatures. Highly unsaturated fats are liquid at room temperatures and are therefore more properly called *oils.

fat body 1. An abdominal organ in amphibians attached to the anterior of each kidney. It contains a reserve of fat that nourishes the gonads during the winter hibernation in readiness for the spring breeding season. **2.** A mass of fatty tissue spreading throughout the body cavity of insects in which fats, proteins, and glycogen are stored as a reserve for hibernation or pupation.

fat cell (adipocyte) Any of the cells of *adipose tissue, in which fats (triglycerides) are stored. Fat cells contain enzymes (lipases) that can break down fat into glycerol and fatty acids, which can be transported in the blood to the liver, where they are used in *fatty-acid oxidation.

fate map A diagram that depicts embryological development, showing the adult structures that develop from the different parts of the embryo. For example, the individual cells (blastomeres) of a blastula can be colour-coded according to which of the three germ layers they will form—ectoderm, mesoderm, or endoderm—and thus ultimately which tissues they give rise to.

fatigue A decline in the level of response of tissues (such as muscle), cells, etc., to nervous stimulation, which occurs after prolonged and continued stimulation of these structures.

fatty acid An organic compound consisting of a hydrocarbon chain and a terminal carboxyl group (see CARBOXYLIC ACIDS). Chain length ranges from one hydrogen atom (methanoic, or formic, acid, HCOOH) to nearly 30 carbon atoms. Ethanoic (acetic), propanoic (propionic), and butanoic (butyric) acids are important in metabolism. Long-chain fatty acids (more than 8–10 carbon atoms) most commonly occur as constituents of certain lipids, notably glycerides, phospholipids, sterols, and waxes, in which they are esterified with alcohols. These long-chain fatty acids generally have an even number of carbon atoms; unbranched chains predominate over branched chains. They may be **saturated** (e.g. *palmitic acid and *stearic acid) or **unsaturated**, with one double bond (e.g. *oleic acid) or two or more double bonds, in which case they are called **polyunsaturated fatty acids** (PUFAs; e.g. *linoleic acid and *linolenic acid). A system of symbols is used for unsaturated fatty acids to denote their chain length (i.e. number of carbon atoms) and the number and position of their double bonds. For example, linoleic acid has 18 carbons and two double bonds, the first of which starts on the sixth carbon from the methyl end; it is denoted as C18:2, n-6 (or omega-6). In contrast, α-linolenic acid, which has 18 carbon atoms and three double bonds, the first starting on the third carbon from the methyl end, is denoted as C18:3, n-3 (or omega-3). See also ESSENTIAL FATTY ACIDS; *TRANS* FATTY ACID.

The physical properties of fatty acids are determined by chain length, degree of unsaturation, and chain branching. Short-chain acids are pungent liquids, soluble in water. As chain length increases, melting points are raised and water-solubility decreases. Unsaturation and chain branching tend to lower melting points.

fatty-acid oxidation (β-oxidation) The metabolic pathway in which fats are metabolized to release energy. Fatty-acid oxidation occurs continually but does not become a major source of energy until the animal's carbohydrate resources are exhausted, for example during starvation. Fatty-acid oxidation occurs chiefly in mitochondria in animal cells, and in *peroxisomes in plant cells. A series of reactions cleave off two carbon atoms at a time from the hydrocarbon chain of the fatty acid. These two-carbon fragments are combined with *coenzyme A to form *acetyl coenzyme A (acetyl CoA), which then enters the *Krebs cycle. The formation of acetyl CoA occurs repeatedly until all the hydrocarbon chain has been used up. See also GLYOXYLATE CYCLE.

fauna All the animal life normally present in a given habitat at a given time. See also MACROFAUNA; MICROFAUNA. Compare FLORA.

faunal region A region of the earth with a distinct and characteristic assemblage of animal taxa. There are six commonly recognized faunal regions used in *zoogeography: the **Ethiopian region**, consisting of Africa south of the Sahara and part of southern Arabia; the **Oriental region**, which covers tropical Asia and associated continental islands; the **Palaearctic region**, comprising Europe and Asia north of the tropics; the **Nearctic region**, consisting of North

America except for tropical Mexico; the **Neotropical region**, consisting of South and Central America including tropical Mexico; and the **Australian region**, covering principally Australia and New Guinea. The regions are described mainly in terms of their terrestrial and freshwater vertebrate fauna, especially mammals and birds, although the invertebrate fauna may sometimes be included. The concept was originated by 19th century naturalists, such as P. L. Sclater and A. R. Wallace, and held as evidence of continental drift (see WALLACE'S LINE). However, another important factor in determining the nature of a region's fauna is the historical migration of species from their centres of origin, a process influenced by the presence of physical barriers, such as mountains. Climate, and hence latitude, is also an important factor.

Fc fragment See FAB FRAGMENT.

FD protein See FLORIGEN.

feathers The body covering of birds, formed as outgrowths of the epidermis and composed of the protein *keratin. Feathers provide heat insulation, they give the body its streamlined shape, and those of the wings and tail are important in flight. Basically a feather consists of a **quill**, which is embedded in the skin attached to a feather follicle and is continuous with the shaft (**rachis**) of the feather, which carries the *barbs. This basic structure is modified depending on the type of feather (see CONTOUR FEATHERS; DOWN FEATHERS; FILOPLUMES). Fossil evidence of feathers can be seen in some of the dinosaur ancestors of the birds, most famously in *Archaeopteryx*, which lived about 150 million years ago (see THEROPODA).

fecundity The number of offspring produced by an organism in a given time. In higher animals, fecundity generally focuses on female offspring, because only they will produce offspring. Normally all organisms, assuming they reach reproductive age, are sufficiently fecund to replace themselves several times over. Darwin noted this, together with the fact that population numbers nevertheless tended to remain fairly constant:

these observations led him to formulate his theory of evolution by *natural selection. *Compare* FERTILITY. *See also* REPRODUCTION RATE.

feedback The use of part of the output of a system to control its performance. In **positive feedback**, the output is used to enhance the input; in **negative feedback**, the output is used to reduce the input. Many biological processes rely on negative feedback. As the population of a species expands, so its food supply per individual is diminished; the result is that the population then begins to fall. Many biochemical processes are controlled by feedback *inhibition. Feedback mechanisms play an important role in maintaining a state of equilibrium within an organism (see HOMEOSTASIS).

feeding See INGESTION.

Fehling's test A chemical test to detect *reducing sugars and aldehydes in solution, devised by the German chemist H. C. von Fehling (1812–85). **Fehling's solution** consists of Fehling's A (copper(II) sulphate solution) and Fehling's B (alkaline 2,3-dihydroxybutanedioate (sodium tartrate) solution), equal amounts of which are added to the test solution. After boiling, a positive result is indicated by the formation of a brick-red precipitate of copper(I) oxide. Methanal, being a strong reducing agent, also produces copper metal; ketones do not react. The test is now little used, having been replaced by *Benedict's test.

female 1. Denoting the gamete (sex cell) that, during *sexual reproduction, fuses with a *male gamete in the process of fertilization. Female gametes are generally larger than the male gametes and are usually immotile (see OOSPHERE; OVUM). **2.** (Denoting) an individual organism whose reproductive organs produce only female gametes. *Compare* HERMAPHRODITE.

femoral Of or relating to the thigh or the femur (thigh bone). For example, the **femoral artery** runs down the front of the thigh.

femto- Symbol f. A prefix used in the metric system to denote 10^{-15}. For example, 10^{-15} second = 1 femtosecond (fs).

femur (*pl.* **femurs** or **femora**) **1.** The thigh bone of terrestrial vertebrates. It articulates at one end with the pelvic girdle at the hip joint and at the other (via two *condyles) with the *tibia. **2.** The third segment of an insect's leg, attached to the *trochanter. *See also* COXA.

fen *See* HYDROSERE.

fenestra (*pl.* **fenestrae**) Either of the two delicate membranes between the *middle ear and the *inner ear. The upper membrane is the **fenestra ovalis** (*see* OVAL WINDOW); the lower membrane is the **fenestra rotunda** (*see* ROUND WINDOW).

fenestration The condition of tissues or other structures within an organism of being perforated by window-like slits. For example, the endothelium of the capillaries in the glomerulus within the kidneys is fenestrated in order to facilitate the filtration of substances.

feral animal Any domesticated animal, such as a dog or a cat, that has returned to live in wild conditions.

fermentation The oxidation of glucose or other food substances by living cells to produce lactic acid or alcohol and energy in the form of *ATP. It occurs in many cell types (e.g. muscle cells) when oxygen is limiting and in certain microorganisms, e.g. yeasts and some bacteria. All fermentations use the series of chemical reactions known as *glycolysis to produce ATP by substrate-level *phosphorylation. This involves reduction of the coenzyme NAD^+ to NADH, and the two types of fermentation are distinguished by the way in which the cell regenerates the NAD^+ so that the glycolysis can continue. In **alcoholic fermentation**, pyruvate (the end product of *glycolysis) is converted to two molecules each of acetaldehyde and carbon dioxide. Each acetaldehyde molecule is then reduced by NADH to yield a molecule of ethanol, in the process regenerating NAD^+. This is the basis of the baking and *brewing industries (*see* BAKER'S YEAST). In **lactic-acid fermentation**, which occurs in many microorganisms and (when sugar is in short supply) in animal cells, the pyruvate is reduced directly to lactic acid by NADH, regenerating NAD^+ without producing carbon dioxide. *Compare* RESPIRATION.

ferns A group of some 12 000 mainly terrestrial vascular plants (*see* TRACHEOPHYTE) traditionally classified in the phylum Pteridophyta. In this broad sense, ferns include clubmosses, *whisk ferns, and *horsetails as well as 'true' ferns. However, it is now recognized that clubmosses and their allies (*see* LYCOPODIOPHYTA) are only distantly related to other ferns and that the latter constitute a distinct clade, the *monilophytes, with closer affinity to the seed plants. The 9000 or so species of true ferns—the so-called **leptosporangiate ferns**—are perennial plants bearing large conspicuous leaves (**fronds**: *see* MEGAPHYLL) usually arising from either a rhizome or a short erect stem. Bracken is a common example. Only the tree ferns have stems that reach an appreciable height. As the young leaves expand into the adult form there is a characteristic uncurling from a structure called a 'fiddlehead', which resembles the scroll at the head of a violin. Reproduction is by means of spores borne on the underside of specialized leaves (*sporophylls).

ferredoxin (Fd) A protein containing iron associated with sulphur that is a *carrier molecule in the electron transport chain in *photosynthesis. It accepts electrons from photosystem I and passes them on to NADP reductase, which donates them for the reduction of $NADP^+$ (*see* PHOTOPHOSPHORYLATION; PHOTOSYSTEMS I AND II). Ferredoxin also takes part in *nitrogen fixation in plants.

ferritin A protein, found predominantly in the tissues of the liver and spleen but also present in nearly all cells of the body, that is used for the storage of iron. A ferritin molecule, which is spherical and has a crystalline core, can store as many as 4000 iron atoms, which can be released when required for *haemoglobin formation. *Compare* TRANSFERRIN.

fertility 1. The potential capability of an organism to reproduce itself. In sexually reproducing plants and animals it is the number of fertilized eggs produced in a given time. For practical purposes this usually cannot be measured, and the only reliable indicators are the numbers of mature seeds produced, eggs laid, or live offspring delivered. However, these measures are strictly referred to as *fecundity, since they exclude fertilized embryos that have failed to develop. **2.** The relative ability of a soil to support plant growth. It consists of both physical factors, e.g. particle size and moisture content, and chemical factors, e.g. concentration and availability of nutrients.

fertilization (syngamy) The union of male and female gametes (reproductive cells) during the process of sexual reproduction to form a **zygote**. It involves the fusion of the gametic nuclei (**karyogamy**) and cytoplasm (**plasmogamy**). As each gamete contains only half the correct number of chromosomes, fertilization and zygote formation results in a cell with the full complement of chromosomes, half of which are derived from each of the parents. In animals the process involves fusion of the nuclei of a spermatozoan and an ovum. In most aquatic animals (e.g. fish) this takes place in the surrounding water, into which the gametes are shed. Among most terrestrial animals (e.g. insects, many mammals) fertilization occurs in the body of the female, into which the sperms are introduced. In flowering plants, after *pollination, the grain pollen produces a *pollen tube, which grows down into the female reproductive organ (carpel) to enable a male gamete nucleus to fuse with the egg nucleus (*see* DOUBLE FERTILIZATION).

In **self-fertilization** the male and female gametes are derived from the same individual. Among plants, self-fertilization (also called **autogamy**) is common in many cultivated species, e.g. wheat and oats. However, self-fertilization is a form of *inbreeding and does not allow for the

mixing of genetic material; if it occurs over a number of generations it will result in offspring being less vigorous and productive than those resulting from cross-fertilization. In **cross-fertilization** (also called **allogamy** in plants) the gametes are derived from different individuals. In plants the pollen comes either from another flower of the same plant or from a different plant (*see also* INCOMPATIBILITY).

fertilizer Any substance that is added to soil in order to increase its productivity. Fertilizers can be of natural origin, such as *composts, or they can be made up of synthetic chemicals, particularly nitrates and phosphates. Synthetic fertilizers can increase crop yields dramatically, but when leached from the soil by rain, which runs into lakes, they also increase the process of eutrophication (*see* ALGAL BLOOM; EUTROPHIC). The inoculation of soil with beneficial bacteria, e.g. nitrogen-fixing bacteria, has potential to increase its fertility. For example, in tropical countries *Azolla* ferns, with their endosymbiotic nitrogen-fixing cyanobacteria (*Anabaena* sp.), are used to increase the fertility of rice paddies.

fetal membranes *See* EXTRAEMBRYONIC MEMBRANES.

fetus (foetus) The *embryo of a mammal, especially a human, when development has reached a stage at which the main features of the adult form are recognizable. In humans the embryo from eight weeks to birth is called a fetus.

Feulgen's test A histochemical test in which the distribution of DNA in the chromosomes of dividing cell nuclei can be observed. It was devised by the German chemist R. Feulgen (1884–1955). A tissue section is first treated with dilute hydrochloric acid to remove the purine bases of the DNA, thus exposing the aldehyde groups of the sugar deoxyribose. The section is then immersed in *Schiff's reagent, which combines with the aldehyde groups to form a magenta-coloured compound.

F_0F_1 complex *See* ATP SYNTHETASE.

fibre 1. An elongated plant cell whose walls are extensively (usually completely) thickened with lignin (*see* SCLERENCHYMA). Fibres are found in the vascular tissue, usually in the xylem, where they provide structural support. The term is often used loosely to mean any kind of xylem element. The fibres of many species, e.g. flax, are of commercial importance. **2.** Any of various threadlike structures in the animal body, such as a muscle fibre, a nerve fibre, a collagen fibre, or an elastic fibre. **3. (dietary fibre; roughage)** The part of food that cannot be digested and absorbed to produce energy. Dietary fibre falls into four groups: **cellulose, hemicelluloses, lignins**, and **pectins**. Highly refined foods, such as sucrose, contain no dietary fibre. Foods with a high fibre content include wholemeal cereals and flour, root vegetables, nuts, and fruit. In human nutrition, a substantial proportion of dietary fibre consists of so-called **non-starch polysaccharides** (cellulosic and non-cellulosic polysaccharides excluding resistant starch). Another distinction is between soluble and insoluble fibre. **Soluble fibre** (e.g. in oats, pulses, fruit, vegetables) is broken down by bacteria in the large intestine to yield short-chain fatty acids, some of which can be absorbed and metabolized by the liver. **Insoluble fibre** (e.g. in wholegrain cereals) is resistant to bacterial attack and, with the bacterial cells, forms a bulky water-retaining mass that promotes gut peristalsis and accelerates the passage of digesta and faeces. Dietary fibre is considered by some to be helpful in the prevention of many of the diseases of Western civilization, such as diverticulosis, constipation, appendicitis, obesity, and diabetes mellitus.

fibre optics *See* OPTICAL FIBRE.

fibril Any small fibre or threadlike structure. *See also* MICROFIBRIL.

fibrin The insoluble protein that forms fibres at the site of an injury and is the foundation of a blood clot. *See* BLOOD CLOTTING.

fibrinogen The protein dissolved in the blood plasma that, when suitably activated, is converted to insoluble *fibrin fibres. *See* BLOOD CLOTTING.

fibrinolysis The breakdown of the protein *fibrin by the enzyme *plasmin (**fibrinase** or **fibrinolysin**), which occurs when blood clots are removed from the circulation.

fibroblast A cell that secretes fibres in the *extracellular matrix of *connective tissue. The cells are long, flat, and star-shaped and lie close to collagen fibres. Fibroblasts are often grown in cell cultures and are used to study genetic reprogramming for cell therapy or stem cell production. *See also* FILOPODIUM; LAMELLIPODIUM.

fibroblast growth factor (FGF) Any of a family of proteins, characterized by a 120-amino-acid core region in the molecule, that influence cell division, migration, or differentiation in a wide range of tissues. They are crucial in embryonic development, the formation of blood vessels (angiogenesis), the differentiation of nerve cells, bone formation, and wound healing. Although originally named for causing the proliferation of fibroblasts in cell culture, FGFs are now known to affect various cell types, with 23 different FGF members identified so far. FGFs bind to heparin or heparan sulphate (as components of proteoglycans) in the extracellular matrix and act on their target cells by binding to one of four types of fibroblast growth factor receptor (FGFR1–FGFR4). Mutations in these receptors can cause developmental abnormalities. **Basic fibroblast growth factor** (bFGF, or FGF-2) promotes the proliferation and organization of embryonic endothelial cells into the tubular precursors of blood vessels, helps determine the fate of nerve cells during development, and triggers fibroblast proliferation and formation of granulation tissue during healing. FGF-2 is also an essential component of the culture medium for human embryonic stem cells.

fibrocartilage *See* CARTILAGE.

fibrous protein *See* PROTEIN.

fibula (*pl.* **fibulas** or **fibulae**) The smaller and outer of the two bones between the knee and the ankle in terrestrial vertebrates. *Compare* TIBIA.

Fick's law of diffusion An equation that describes the rate at which a solute diffuses between two locations in a solution. It has many applications and can be written in various forms, but in biology it is frequently applied to the movement of molecules of respiratory gases, i.e. oxygen or carbon dioxide, during gaseous exchange between an organism's internal body fluids and the external environment. It can be expressed thus:

$$Q = DA \frac{P_1 - P_2}{L}$$

where Q is the rate of diffusion, D is the diffusion coefficient of the solute, A is the area over which diffusion is occurring, P_1 and P_2 are the concentrations of solute at the two locations, and L is the distance between the two locations. The diffusion coefficient depends on the nature of the solute, the medium through which it is diffusing, and the temperature. The equation was derived by the German-born physiologist Adolf Fick (1829–1901).

field capacity The amount of water that remains in a soil when excess has drained away. It is held by capillary forces of the soil pores and reflects the physical nature of the soil.

field-emission microscope (FEM) A type of electron microscope in which a high negative voltage is applied to a metal tip placed in an evacuated vessel some distance from a glass screen with a fluorescent coating. The tip produces electrons by **field emission**, i.e. the emission of electrons from an unheated sharp metal part as a result of a high electric field. The emitted electrons form an enlarged pattern on the fluorescent screen, related to the individual exposed planes of atoms. As the resolution of the instrument is limited by the vibrations of the metal atoms, it is helpful to cool the tip in liquid helium. Although the individual atoms forming the point are not displayed, individual adsorbed atoms of other substances can be, and their activity is observable.

field-ionization microscope (field-ion microscope; FIM) A type of electron microscope that is similar in principle to the *field-emission microscope, except that a high positive voltage is applied to the metal tip, which is surrounded by low-pressure gas (usually helium) rather than a vacuum. The image is formed in this case by **field ionization**: ionization at the surface of an unheated solid as a result of a strong electric field creating positive ions by electron transfer from surrounding atoms or molecules. The image is formed by ions striking the fluorescent screen. Individual atoms on the surface of the tip can be resolved and, in certain cases, adsorbed atoms may be detected.

'fight-or-flight' response *See* ALARM RESPONSE.

filament 1. (in zoology) A long slender hairlike structure, such as any of the *barbs of a bird's feather. 2. (in botany) The stalk of the *stamen in a flower. It bears the anther and consists mainly of conducting tissue. 3. (in cell biology) *See* INTERMEDIATE FILAMENT; MICROFILAMENT.

filoplumes Minute hairlike *feathers consisting of a shaft (**rachis**) bearing a few unattached barbs. They are found between the contour feathers.

filopodium (*pl.* **filopodia**) A slender elongated projection from a cell. Filopodia extend typically from the leading edge of a motile cell, such as a fibroblast, or an elongating cell, such as the growing tip of a nerve fibre (*see* GROWTH CONE). Their finger-like shape is determined by bundles of actin microfilaments extending longitudinally from the leading edge of the filopodium to its base, where it joins the main body of the cell. Assembly or disassembly of these filament polymers causes extension or retraction of the leading edge. These processes are governed by chemical cues in the cell's environment, such as signal molecules, which are detected by receptors in the plasma membrane of each filopodium. Hence, the cell can move forwards in a favourable direction or retreat from unfavourable surroundings. *Compare* LAMELLIPODIUM.

filter feeding A method of feeding in which tiny food particles are strained from the surrounding water by various mechanisms. It is used by many aquatic invertebrates, especially members of the plankton, and by some vertebrates, notably baleen whales. *See also* CILIARY FEEDING; WHALEBONE.

filtrate The clear liquid obtained by filtration.

filtration The process of separating solid particles using a filter. In **vacuum filtration**, the liquid is drawn through the filter by a vacuum pump. *Ultrafiltration is filtration under pressure; for example, ultrafiltration of the blood occurs in the *nephrons of the vertebrate kidney.

fimbria (*pl.* **fimbriae**) **1.** (in anatomy) Any fringelike structure, such as the circle of finger-like extensions surrounding the entrance to a *fallopian tube. **2.** (in microbiology) Another name for a short *pilus (sense 1).

finger domain A finger-shaped structure produced in a protein when a series of the constituent amino acids combines with a metal atom. Finger domains are often found repeated in *transcription factors. *See also* DOMAIN; ZINC FINGER.

fingerprinting *See* DNA FINGERPRINTING; DNA PROFILING; PEPTIDE MAPPING.

finished sequence (in bioinformatics) A high-quality determination of the base sequence of a chromosome or genome that conforms to an agreed level of accuracy. It corrects and refines the *draft sequence using various experimental approaches and data analyses. These steps involve repeated rounds of cloning, sequencing, and assembly until gaps are closed and incorrectly ordered fragments are reassembled. The International Human Genome Sequencing Consortium took a further three years to produce the finished sequence of the human genome, following publication of the draft sequence in 2000. It allows for an average of only a single error in every 10 000 bases.

fins The locomotory organs of aquatic vertebrates. In fish there are typically one or more **dorsal** and **ventral fins** (sometimes continuous), whose function is balance; a **caudal fin** around the tail, which is the main propulsive organ; and two paired fins: the **pectoral fins** attached to the pectoral (shoulder) girdle and the **pelvic fins** attached to the pelvic (hip) girdle, which are used in steering. These paired fins are homologous with the limbs of tetrapods. Fins are strengthened by a number of flexible fin rays, which may be cartilaginous, bony and jointed, horny, or fibrous and jointed.

Firmicutes A phylum of chiefly Gram-positive bacteria characterized by having DNA with a relatively low ratio of G-C base pairs compared to A-T base pairs. It includes bacteria that form resistant *endospores, such as *Clostridium* spp. and *Bacillus* spp., and several important lactic acid-producing species, such as *Lactobacillus* spp., which are used in cheese and yoghurt manufacture. Other significant members include *Streptococcus* and *Staphylococcus*, species of which are common pathogens of humans, and the minute *mycoplasmas, which lack cell walls.

first convoluted tubule *See* PROXIMAL CONVOLUTED TUBULE.

fish *See* CHONDRICHTHYES (cartilaginous fish); OSTEICHTHYES (bony fish); PISCES.

FISH *See* FLUORESCENCE IN SITU HYBRIDIZATION.

fission A type of asexual reproduction occurring in some unicellular organisms, e.g. diatoms, protozoans, and bacteria, in which the parent cell divides to form two (**binary fission**) or more (**multiple fission**) similar daughter cells.

fission-track dating A method of estimating the age of glass and other mineral objects by observing the tracks made in them by the fission fragments of the uranium nuclei that they contain. By irradiating the objects with neutrons to induce fission and comparing the density and number of the tracks before and after irradiation it is possible to estimate the time that has elapsed since the object solidified.

fitness (in genetics) **1.** Direct fitness (absolute fitness). Symbol *W*. A measure of the contribution of an individual to the genetic composition of subsequent generations through its own offspring. Hence, individuals that contribute the most offspring to the next generation are the fittest. Fitness therefore reflects how well an organism is adapted to its environment, which determines its survival. **2.** Indirect fitness. A measure of the reproductive success of an individual's indirect descendants, e.g. siblings and cousins, which share a proportion of the individual's alleles. The sum of direct and indirect fitness is called *inclusive fitness. *See also* SELECTION COEFFICIENT. **3.** Relative fitness. Symbol *w*. The contribution of an individual's genotype or phenotype to a population relative to the contributions of other individuals. Mathematically it can be expressed as the individual's absolute fitness divided by the absolute fitness of the fittest individual, such that the latter has a relative fitness of 1.

fixation 1. The first stage in the preparation of a specimen for microscopical examination, in which the tissue is killed and preserved in as natural a state as possible by immersion in a chemical **fixative**. The fixative prevents the distortion of cell components by denaturing its constituent protein. Some commonly used fixatives are formaldehyde, ethanol, and Bouin's fluid (for light microscopy), and osmium tetroxide and glutaraldehyde (for electron microscopy). Fixation may also be brought about by heat. **2.** *See* NITROGEN FIXATION.

fixed action pattern *See* INSTINCT.

flaccid (in botany) Describing plant tissue that has become soft and less rigid than normal because the cytoplasm within its cells has shrunk and contracted away from the cell walls through loss of water (*see* PLASMOLYSIS).

flagellum (*pl.* **flagella**) **1.** (in prokaryotes) A long slender structure that protrudes from the cell surface of a bacterium. It rotates from its base and propels the bacterium along. Up to several micrometres in length, a flagellum is constructed of numerous subunits of the protein **flagellin**, while at the base a system of rings anchors the flagellum in the cell wall and plasma membrane. Surrounding these rings are paired motor proteins, which impart a rotary motion to the filament, and switch proteins, which can reverse the direction of rotation. Flagella may be attached singly or in groups, for example at the poles of the bacterial cell, or scattered over the cell surface. **2.** (in

nexin protein linking outer microtubules

protein sheath

linked pair of outer microtubules

plasma membrane of cilium

dynein side arms

protein spoke

central microtubules

Cross section of a eukaryotic flagellum

eukaryotes) **(undulipodium)** A slender flexible outgrowth of a eukaryote cell used for locomotion or propelling fluids over the surface of the cell. Many protists and sperm cells swim by means of flagella, and various organisms use them to establish feeding currents or to clear debris from epithelial surfaces. All flagella (and *cilia) have a shaft, about 0.25 μm in diameter, consisting of a longitudinal array of *microtubules, the *axoneme, which is surrounded by an extension of the cell's plasma membrane. The axoneme has two singlet microtubules running down the middle and nine doublet microtubules around the periphery, giving a characteristic $9 + 2$ array. At its base the axoneme connects with a **basal body** (or kinetosome). This is a short cylinder that, like a *centriole, consists of nine triplet microtubules; it organizes assembly of the axoneme microtubules and is part of a complex array of fibres and microtubules forming a **root structure** within the cell. Cilia are typically 1–10 μm long and move by a whiplike power stroke followed by a recovery stroke in the opposite direction. Flagella can be 100–200 μm long and generate successive waves that pass from the base to the tail. In both cases, flexing of the shaft is produced by a sliding motion of the microtubule doublets relative to each other. This involves the successive formation and breakage of molecular bridges between adjacent doublets. The bridges are composed of a protein, *dynein, and their formation requires energy in the form of ATP. *See* TINSEL FLAGELLUM.

flame cells Ciliated cells that form part of the excretory and osmoregulatory system of flatworms (*see* PLATYHELMINTHES), rotifers, and nemertine worms. This system, known as a **protonephridium**, consists of branching tubules that open to the exterior through excretory pores; flame cells occur at the ends of the tubules, into which their cilia project. Fluid, consisting of water, ions, and nitrogenous waste products, filters between cells into the tubules, where it is directed to the exterior by movements of the flame cell cilia, which resemble the flickering of flames. Along its route the fluid is modified by reabsorption and secretion of ions across the tubule walls to produce the urine. Flame cells that possess only one cilium are known as **solenocytes**; these are found in the protonephridia of marine worms of the phylum Priapulida.

flatworms *See* PLATYHELMINTHES.

flavin adenine dinucleotide *See* FAD.

flavonoid One of a group of naturally occurring phenolic compounds many of which are plant pigments. They include the *anthocyanins, **flavonols**, and **flavones**. Patterns of flavonoid distribution have been used in taxonomic studies of plant species. Leguminous plants release flavonoids from their root hairs to attract nitrogen-fixing soil bacteria as a prelude to root nodule formation. Plant-derived flavonoids in foods are strong *antioxidants in their natural state, but are poorly absorbed from the intestine and rapidly metabolized and excreted. However, they may be beneficial to health by modulating cell signalling pathways as well as via their antioxidant activity, and evidence suggests flavonoids may help in preventing cardiovascular disease, diabetes, and cancer. Good sources include vegetables and fruits. *See also* CHALCONE.

flavoprotein *See* FAD.

FLC protein A plant protein encoded by the *FLOWERING LOCUS C* gene that is involved in regulating the onset of *flowering in response to cues such as temperature or growth stage. In certain strains of the model plant species *Arabidopsis thaliana* (thale cress), e.g. it acts as a transcription factor to block expression of the key proteins responsible for initiating flowering, namely FT (*see* FLORIGEN) and FD. Exposure to cold conditions inhibits expression of FLC, thereby relieving repression of the FT–FD pathway and enabling flowering to proceed (*see* VERNALIZATION).

fleas *See* SIPHONAPTERA.

Fleming, Sir Alexander (1881–1955) British bacteriologist, born in Scotland. He studied medicine at St Mary's Hospital, London, where he remained all his life.

In 1922 he identified *lysozyme, an enzyme that destroys bacteria, and in 1928 discovered the antibiotic *penicillin. He shared the 1945 Nobel Prize for physiology or medicine with Florey and Chain, who first isolated the drug.

flexor A muscle that causes a limb to bend by bringing the two parts of the limb together. An example is the *biceps. Flexors work antagonistically with *extensors. *See* SKELETAL MUSCLE.

flies *See* DIPTERA.

flight 1. Any form of *locomotion in air, which can be active or passive (**gliding**). Mechanisms of flight have evolved mainly in birds, bats, and insects: these animals are adapted for flight by the presence of **wings**, which increases the ratio of surface area to body weight. Birds possess powerful flight muscles: the **depressor** muscle runs from the underside of the humerus to the sternum and is responsible for the downstroke of the wing; the **levator** muscle works antagonistically, producing the upstroke. Flight in insects works in a similar fashion but the muscles that control the wing movement are attached to the thorax. A few species of mammals, reptiles, and fish have developed flight to a lesser extent. For example, flying squirrels (order Dermoptera) possess a membrane attached to the limbs that can open and function as a parachute, allowing the animals to glide. **2.** Part of a survival mechanism in an animal that is generated in response to a threatening situation. A potentially dangerous situation can induce the release of *adrenaline, which prepares the animal for 'fight or flight' by increasing the blood pressure and heart rate and diverting the blood flow to the muscles and heart. *See* ALARM RESPONSE.

flip-flop The movement (**transverse diffusion**) of a lipid molecule from one surface of a *lipid bilayer membrane to the other, which occurs at a very slow rate. This contrasts with the much faster rate at which lipid molecules exchange places with neighbouring molecules on the same surface of the membrane (**lateral diffusion**).

flocculation The process in which particles in a colloid aggregate into larger clumps. Flocculation of clay particles in soil can be induced by the addition of calcium salts. Clay particles have an overall negative charge and therefore attract positive ions, such as Ca^{2+}, which form bridges holding the particles together. Flocculation is also often observed in cultures of bacterial and yeast cells.

flora All the plant life normally present in a given habitat at a given time. *See also* MICROFLORA. *Compare* FAUNA.

floral formula A summary of the structure and components of a flower using symbols and numbers. The symbols representing symmetry include $\oplus$ (for actinomorphy) and $\cdot|\cdot$, $\downarrow$, or $\uparrow$ (for zygomorphy). The parts of the flower are represented by K (for calyx), C (for corolla), P (for perianth), A (for androecium), and G (for gynoecium). These are followed by numbers indicating the number of parts in each whorl (e.g. K5 indicates a calyx of five sepals); ∞ indicates an indefinite number of parts (more than 12). If the parts are fused the number is in parentheses; if they are in separate groups or whorls the number is split (e.g. C2 + 2 indicates a corolla of two whorls each of two petals). The symbol G has a line above it to indicate an inferior ovary and a line beneath it to indicate a superior ovary. For example, the buttercup flower has the following floral formula:

$$\oplus\text{K5 C5 A}\infty\underline{\text{G}}\infty,$$

i.e. it has a calyx of five sepals, a corolla of five petals, an androecium of many stamens, and a superior ovary consisting of many carpels.

Florey, Howard Walter, Baron (1898–1968) Australian pathologist, who moved to Oxford in 1922. After working in Cambridge and Sheffield (studying *lysozyme), he returned to Oxford in 1935. There he teamed up with Ernst Chain and by 1939 they succeeded in isolating and purifying

*penicillin. They also developed a method of producing the drug in large quantities and carried out its first clinical trials. The two men shared the 1945 Nobel Prize for physiology or medicine with penicillin's discoverer, Alexander Fleming.

florigen A transmissible signal that transmits stimulus for *flowering from the leaves to apex of the plant in species where flowering is a response to photoperiod (*see* PHOTOPERIODISM). Such a transmissible signal has long been hypothesized, and was termed 'florigen' in 1936; however, its identity in some species has only recently been revealed. For example, in the model plant thale cress (*Arabidopsis thaliana*) the florigen is FT protein, encoded by the *FT* (*FLOWERING LOCUS T*) gene, expression of which is regulated by *CO protein in response to night length. FT protein is synthesized in phloem companion cells and translocates via phloem from the leaf to the apical meristem. Here it binds to FD protein (encoded by the *FLOWERING LOCUS D* gene), and the resultant FD-FT complex acts as a transcription factor, activating genes (i.e. *LEAFY* and *APETALA1*) that prompt the apical meristem to switch from producing leaves to making flowers.

flower The structure in angiosperms (flowering plants) that bears the organs for sexual reproduction. Flowers are very variable in form, ranging from the small green insignificant wind-pollinated flowers of many grasses to spectacular brightly coloured insect-pollinated flowers. Flowers are often grouped together into *inflorescences, some of which (e.g. that of dandelion) are so compacted as to resemble a single flower. Typically flowers consist of a receptacle that bears sepals, petals, stamens, and carpels (see illustration). The flower parts are adapted to bring about pollination and fertilization resulting in the formation of seeds and fruits. The sepals are usually green and leaflike and protect the flower bud. The petals of insect-pollinated flowers are adapted in many ingenious ways to attract insects and, in some instances, other animals. For example, some flowers are adapted to attract short-tongued insects by having an open shallow *corolla tube and nectar situated in an exposed position. Flowers adapted for pollination by long-tongued insects have a long corolla tube of fused petals with nectar in a concealed position. The tongue of the insect brushes against the anthers and stigma before reaching the nectar. Wind-pollinated flowers, in contrast,

Section through a monocarpellary flower at the time of pollination

are inconspicuous. The anthers dangle outside the corolla and the stigmas have a feathery surface to catch the pollen grains.

Some species are adapted for self-pollination and have small flowers, no nectar, and stamens and carpels that mature simultaneously. *See also* ABC MODEL; FLOWERING.

flowering The formation of flowers and their associated reproductive structures in flowering plants. Whether flowering will take place under appropriate conditions depends on many factors, notably the type of plant (i.e. whether an annual, biennial, or perennial), its stage of maturity, and the season. Many plants use day length as a cue (*see* PHOTOPERIODISM) to synchronize individuals within the same population, so that the chances of cross-pollination are maximized. In others flowering is regulated by temperature (*see* VERNALIZATION). The transition from vegetative growth to a flowering state requires that a vegetative apical meristem changes into an inflorescence meristem or floral meristem, which in turn gives rise to the whorls of organs (sepals, petals, stamens, and carpels) that constitute each individual flower (*see* ABC MODEL). Genes controlling this transition have been studied in the model plant *Arabidopsis thaliana* (thale cress). A crucial step is the expression of two **meristem identity genes**, *LEAFY* and *APETALA1*, which encode respective transcription factors. These genes are switched on by the flowering initiation hormone *florigen in conjunction with other proteins, notably FD protein. *See also* CO PROTEIN.

flowering plants *See* ANTHOPHYTA.

fluence A measure, widely used in studies of photosynthesis in plants, of the quantity of light. It is defined as the quantity of radiant energy falling on a small sphere divided by the cross-sectional area of the sphere, and can be expresssed in two ways. **Photon fluence** (units: $mol\ m^{-2}$) measures the number of photons falling on the sphere, whereas **energy fluence** (units: $J\ m^{-2}$) measures the energy incident on the sphere. Hence the corresponding rates of input are the **photon fluence rate** (units: $mol\ m^{-2}\ s^{-1}$) and the **energy fluence rate** (units: $J\ m^{-2}\ s^{-1}$). Many radiation-measuring instruments now used in plant biology record only the part of the spectrum utilized by plants, i.e. radiation of wavelength in the range 400–700 nm; this is termed **photosynthetically active radiation** (PAR), and fluence rates used in practice are often qualified in this way.

fluid mosaic model A widely accepted model, proposed by S. J. Singer and G. L. Nicholson in the 1970s, for the structure of cell membranes. It is based on a *lipid bilayer forming the framework of the membrane, with proteins embedded in it; these proteins may be restricted to either lipid layer (**extrinsic proteins**) or

Fluid mosaic model of cell membranes

they may span both (**intrinsic proteins**), producing an asymmetric structure (see illustration). This structure allows the lateral movement of the constituent phospholipid and protein molecules of the membrane. The proteins may associate in distinct patches to carry out particular functions within the membrane. Moreover, areas of membrane rich in cholesterol and sphingolipids form **lipid rafts**, or lipid microdomains, along with associated proteins. These can diffuse in the plane of the membrane, fold to form flask-shaped structures (*see* CAVEOLA), and rapidly assemble and disassemble. Lipid rafts distribute signalling proteins over the cell surface, and their caveolae can also perform endocytosis.

flukes *See* TREMATODA.

fluorescence in situ hybridization (FISH) A technique in which a DNA probe, labelled with a fluorescent dye, base-pairs (hybridizes) with the complementary base sequence of a target nucleotide. It is used in genetic mapping for locating specific genes or cytogenetic abnormalities within a chromosome set. Another application is for locating particular messenger RNAs (mRNAs) within cells. If just a partial sequence of a gene is known, a DNA probe for the gene can be constructed. A preparation of the organism's entire chromosome complement is then partially denatured to separate the DNA strands and incubated with the labelled probe. The probe binds to its complementary gene sequence, revealing the precise chromosomal location of the target gene. FISH is a valuable diagnostic tool for genetic diseases, and is used in research and clinical medicine to identify chromosomal aberrations such as deletions, duplications, or fusions. *See also* CHROMOSOME PAINTING; SPECTRAL KARYOTYPING.

(((⊕))) SEE WEB LINKS
- FISH factsheet from the National Human Genome Research Institute

fluorescence microscopy *See* CONFOCAL FLUORESCENCE MICROSCOPY.

fluorescent protein Any of various proteins that are used as molecular markers in experimental cell biology. They act as fluorophores, emitting fluorescent light when illuminated by light of a certain wavelength. This enables their abundance, location, and movements to be tracked by fluorescence microscopy. The gene for the fluorescent protein is fused with the gene encoding a protein of interest, and this construct is introduced to the cell. Expression of the target gene inside the cell results in a fusion protein incorporating the fluorescent protein. Hence it is possible to see when and where the target protein is produced, and its fate in the cell. Among the most commonly used are *green fluorescent protein, originally derived from a jellyfish; several mutations giving different colour variants, ranging from blue to yellow, have since been developed. Other fluorescent proteins (e.g. *red fluorescent protein) have been derived from sea anemones and coral animals. Choice of marker is dictated by factors such as its spectral properties, photostability, brightness, and effect on the behaviour of the target protein. Differently coloured markers can be used to track interactions of their respective fusion proteins.

fluoridation The process of adding very small amounts of fluorine salts (e.g. sodium fluoride, NaF) to drinking water to prevent tooth decay. The fluoride becomes incorporated into the fluoroapatite (*see* APATITE) of the growing teeth and reduces the incidence of *dental caries. However, some argue that there is an accompanying risk of fluoride toxicity.

fMRI Functional magnetic resonance imaging. *See* MAGNETIC RESONANCE IMAGING.

focusing (in animal physiology) The process of directing and concentrating light from a source onto the *retina of the eye, by means of the lens, in order to obtain a clear image of objects at a range of distances. *See* ACCOMMODATION.

foetus *See* FETUS.

folacin *See* FOLIC ACID.

folic acid (folacin; vitamin B₉) A vitamin of the *vitamin B complex. In its active form, tetrahydrofolic acid, it is a *coenzyme in various reactions involved in the metabolism of amino acids, purines, and pyrimidines. It is synthesized by intestinal bacteria and is widespread in food, especially green leafy vegetables. Deficiency causes poor growth and nutritional anaemia. Adequate dietary folic acid is essential for pregnant women to reduce the risk of neural tube defects, such as spina bifida, in the developing fetus.

follicle 1. (in animal anatomy) Any enclosing cluster of cells that protects and nourishes a cell or structure within. For example, follicles in the *ovary contain developing egg cells, while *hair follicles envelop the roots of hairs. **2.** (in botany) A dry fruit that, when ripe, splits along one side to release its seeds. It is formed from a single carpel containing one or more seeds. Follicles do not occur singly but are grouped to form clusters (**etaerios**). Examples include larkspur, columbine, and monk's hood.

follicle-stimulating hormone (FSH; follitropin) A hormone, secreted by the anterior pituitary gland in mammals, that stimulates, in female mammals, ripening of specialized structures in the ovary (*Graafian follicles) that produce ova and, in males, the formation of sperm in the testis. It is a major constituent of fertility drugs, used to treat failure of ovulation and decreased sperm production. *See also* GONADOTROPHIN.

follicular phase The phase of the *ovarian cycle during which the Graafian follicles mature and the lining of the uterus thickens under the influence of oestradiol.

fontanelle A gap in the bones of the skull. Newborn babies are born with a fontanelle, which disappears as the bones in the skull fuse; the fusion is complete at around 18 months of age.

food Any material containing *nutrients, such as carbohydrates, proteins and fats, which are required by living organisms in order to obtain energy for growth and maintenance. Heterotrophic organisms, such as animals, ingest their food (*see also* DIET); autotrophic organisms, such as plants, manufacture their food materials. *See also* MIXOTROPHIC.

food additive A substance added to a food during its manufacture or processing in order to improve its keeping qualities, texture, appearance, or stability or to enhance its taste or colour. Additives are usually present in minute quantities; they include colouring materials, sweeteners, preservatives (*see* FOOD PRESERVATION), *antioxidants, emulsifiers, and stabilizers. In most countries the additives used must be selected from an approved list of such compounds, which have been tested for safety, and they must be listed on the food labels of individual products.

food chain The transfer of energy from green plants (the primary producers) through a sequence of organisms in which each eats the one below it in the chain and is eaten by the one above. Thus plants are eaten by herbivores, which are then eaten by carnivores. These may in turn be eaten by different carnivores. The position an organism occupies in a food chain is known as its *trophic level. In practice, many animals feed at several different trophic levels, resulting in a more complex set of feeding relationships known as a *food web. *See* BIOENERGETICS; CONSUMER; PRODUCER; PYRAMID OF BIOMASS; PYRAMID OF ENERGY; PYRAMID OF NUMBERS; TROPHIC EFFICIENCY.

food poisoning An acute illness caused by food that may be naturally poisonous or contaminated by certain types of pathogenic microorganisms, parasites, or toxins from other sources such as pesticides. The most common type of food poisoning in the UK is that caused by the bacteria belonging to the genus *Salmonella, which inhabit the alimentary canal of livestock. Other food poisoning bacteria include Staphylococcus aureus, Clostridium perfringens, Campylobacter jejuni, *Listeria monocytogenes, and pathogenic *Escherichia coli. Norovirus, found in raw shellfish, also causes symptoms of food poisoning, notably diarrhoea and vomiting. Freezing and other types of *food preservation can prevent the growth of the bacteria and thorough cooking will kill the microorganisms before the meat is eaten. However, food poisoning can result if frozen meat is not completely thawed at its centre before cooking, as it may not reach sufficiently high temperatures to kill the bacteria during cooking. Another type of food poisoning, known as **botulism**, is caused by toxins produced by the bacterium *Clostridium botulinum*, which can grow in badly preserved canned foods.

food preservation Prevention of the spoilage of food, which is achieved by a variety of techniques. These aim to prevent bacterial and fungal decay and contamination of food, which can cause *food poisoning. For example, dehydration removes the water from food, which prevents microorganisms from growing. Treating food with salt (salting) causes the microorganisms to lose water due to osmosis. Pickling involves treatment with vinegar (ethanoic acid), which reduces the pH and prevents bacteria from growing. Heating food (**blanching**) to temperatures of 90°C denatures the enzymes that cause the breakdown of food and kills many bacteria. The food is then packed in air-tight containers, such as cans or bottles. Heating milk to high temperatures to kill the bacteria is the basis of *pasteurization. Freezing food prevents the growth of bacteria but does not necessarily kill them; thorough cooking is therefore essential. In *freeze drying, food is rapidly frozen and then dehydrated, usually in a vacuum. Preprepared food can be preserved by the addition of chemicals (*see* FOOD ADDITIVE), such as sodium benzenecarboxylate, propionates, and sulphur dioxide, but some of these may have adverse side-effects. **Irradiation** is a method of food preservation in which the bacteria are killed by irradiating the food with gamma rays.

food production *See* AGRICULTURE.

food reserves Reserves of fat, carbohydrate, or (rarely) protein in cells and tissues that function as an important store of energy that can be released and used in ATP production when required by the organism. For example, in animals *fat is stored in adipose tissue, and carbohydrate—in the form of the **storage compound** *glycogen—is stored in liver and muscle cells. In plants *starch is a major storage compound, being found in perennating organs (*see* PERENNATION) and seeds (in which it is mobilized at germination), and oils are important storage materials in some species (e.g. in the seeds of the castor-oil plant).

food supply (in human ecology) The production of food for human consumption. *See* AGRICULTURE.

food web A system of *food chains that are linked with one another. In a food web a particular organism may feed at more than one trophic level. For example, in a pond food web a freshwater mussel may feed directly on green algae, in which case it is a primary consumer. However, it can also feed on protozoa, which are themselves primary consumers, in which case the mussel is the secondary consumer. A food web does not usually include the decomposers, but these organisms are very important in the flow of energy through a food web (*see* ENERGY FLOW).

footprinting A technique for detecting regions where a protein is bound to DNA; for example, where a specific transcription factor is bound to the DNA of a gene. The protein protects the nucleic acid (i.e. the DNA) from digestion by a nuclease enzyme, so that after such treatment the undegraded DNA, the 'footprint', can be isolated and characterized.

foramen (*pl.* **foramina**) An aperture in an animal part or organ, especially one in a bone or cartilage. For example, the **foramen magnum** is the opening at the base of the skull through which the *spinal cord passes.

forb A nonwoody plant that is not a grass.

forebrain (prosencephalon) One of the three sections of the brain of a vertebrate embryo. The forebrain subsequently enlarges and divides into two regions: the central **diencephalon**, which comprises an upper *thalamus and lower *hypothalamus; and the surrounding **telencephalon**, or *cerebrum. *Compare* HINDBRAIN; MIDBRAIN.

foregut 1. The anterior region of the alimentary canal of vertebrates, up to the anterior part of the duodenum. **2.** The anterior part of the alimentary canal of arthropods. *See also* HINDGUT; MIDGUT.

forest An area of vegetation in which the dominant plants are trees; forests constitute major *biomes. Temperate forests have adequate or abundant rainfall and moderate temperatures. They may be dominated by deciduous trees (such as oak, ash, elm, beech, or maple), often growing together to form mixed deciduous forest, as in temperate regions of Europe, Asia, and North America; or by broad-leaved evergreens (such as southern beech, *Nothofagus*), as in Chile. Cold forests, of northern regions, are dominated by evergreen conifers (*see* TAIGA). Tropical forests include *rainforest, characterized by regular heavy rainfall; monsoon forest, found in SE Asia and having heavy rainfall interspersed with periods of drought; and thorn forest, as in SW North America, SW Africa, and parts of Central and South America and Australia, which has sparse rainfall, is dominated by small thorny trees, and grades into savanna woodland (*see* GRASSLAND) and semidesert. *See also* DEFORESTATION.

form 1. A category used in the *classification of organisms into which different types of a variety may be placed. **2.** Any distinct variant within a species. Seasonal variants, e.g. the tawny brown (summer) and blue-white (winter) forms of the blue hare, may be called forms, as may the different types that constitute a *polymorphism.

forward genetics The traditional approach to genetic investigation, in which the aim is to identify the gene that governs a particular function. Mutant phenotypes provide clues about genetically controlled functions, and co-inherited genetic markers indicate the region of the genome containing the responsible gene. This information enables the gene to be isolated and cloned, for example using *positional cloning, and its base sequence determined. *Compare* REVERSE GENETICS.

fossil The remains or traces of any organism that lived in the geological past. In general only the hard parts of organisms become fossilized (e.g. bones, teeth, shells, and wood) but under certain circumstances the entire organism is preserved. For example, virtually unaltered fossils of extinct mammals, such as the woolly mammoth and woolly rhinoceros, have been found preserved in ice in the Arctic. Small organisms or parts of organisms (e.g. insects, leaves, flowers) have been preserved in *amber.

In the majority of fossils the organism has been turned to stone—a process known as **petrification**. This may take one of three forms. In **permineralization**, solutions originating underground fill the microscopic cavities in the organism. Minerals in these solutions (e.g. silica or calcite) may actually replace the original material of the organism so that even microscopic structures may be preserved; this process is known as **replacement** (or **mineralization**). A third form of petrification—**carbonization** (or **distillation**)—occurs in certain soft tissues that are composed chiefly of compounds of carbon, hydrogen, and oxygen (e.g. cellulose). After the organism has been buried, and in the absence of oxygen, carbon dioxide and water are liberated until only free carbon remains. This forms a black carbon film in the rock outlining the original organism. **Moulds** are formed when the original fossil is dissolved away leaving a mould of its outline in the solid rock. The deposition of mineral matter from underground solutions in a mould forms a **cast**. Palaeontologists often produce casts from moulds using such substances as dental wax. Moulds of thin organisms (e.g. leaves) are commonly known as **imprints**. **Trace fossils** are the fossilized remnants of the evidence of animal life, such as tracks, trails, footprints, burrows, and **coprolites** (fossilized faeces).

The ideal conditions for the formation of fossils occur in areas of rapid sedimentation, especially those parts of the seabed that lie below the zone of wave disturbance. *See also* CHEMICAL FOSSIL; INDEX FOSSIL; MICROFOSSIL; MOLECULAR FOSSIL; TAPHONOMY.

fossil fuel Coal, oil, and natural gas, the fuels used by humans as a source of energy. They are formed from the remains of living organisms and all have a high carbon or hydrogen content. Their value as fuels relies on the exothermic oxidation of carbon to form carbon dioxide $(C + O_2 \rightarrow CO_2)$ and the oxidation of hydrogen to form water $(H_2 + \frac{1}{2}O_2 \rightarrow H_2O)$. Fossil fuels are a major source of the greenhouse gas carbon dioxide; hence their use contributes to the *greenhouse effect and global warming.

fossil hominid *See* HOMINID.

foundation species A species that has a significant physical impact on its environment and thereby creates habitats for other species and underpins the community of organisms. Foundation species facilitate species diversity in a wide range of aquatic and terrestrial environments; they can occupy any trophic level, from primary producers to predators. Many are plants that are *dominant in a community and provide shelter, shade, or substrate for other species. For example, the seagrass *Thalassia testudinum* colonizes coastal sediments, modifying water flow and stabilizing the substrate sufficiently to allow settlement by bottom-dwelling invertebrates, while also protecting the invertebrates from predators. However, foundation species need not be numerically abundant in a community. An example is the beaver (*Castor* spp.), whose dams on rivers have a wide-ranging impact by creating pools upstream. Hence it is also a *keystone species.

founder effect The phenomenon occurring when a population is founded by a small sample of the entire species, perhaps just a handful of individuals. Chance dictates that these founder members will be genetically unrepresentative of the species as a whole, and that the genetic make-up of the new population will differ markedly from the main species population. The founder effect thus increases the likelihood of evolutionary divergence from the main species, and eventually the formation of a new species. *See* PERIPATRIC SPECIATION.

fovea (fovea centralis) (*pl.* **foveae**) A shallow depression in the *retina of the eye, opposite the lens, that is present in some vertebrates. This area contains a large concentration of *cones with only a thin layer of overlying nerves. It is therefore specialized for the perception of colour and sharp intense images. The clarity is enhanced when light is focused on the foveae of both retinas simultaneously. *See* BINOCULAR VISION.

fragmentation **1.** (in physiology) A method of asexual reproduction, occurring in some plants and invertebrate animals, in which parts of the organism break off and subsequently develop into new individuals. Fragmentation occurs naturally in many plants that produce multiple rooted shoots, each of which, when detached, can form a clone of the parent plant. It also occurs in certain cnidarians and annelids. In some, regeneration may occur before separation, producing chains of individuals budding from the parent. **2.** (in ecology) The break-up of an area of habitat into smaller patches. This is most often a consequence of human activity, as when tracts of forest are converted to agriculture, leaving isolated fragments of the original vegetation. Surviving populations of fauna and flora are much reduced in size, and biodiversity tends to decline, particularly in the smallest patches. Remedial steps include provision of wildlife corridors that permit the movement of animals, and even sometimes plants, between fragments.

frameshift An alteration in the sequence of DNA bases read as base triplets during *transcription, caused by the deletion (or addition) of a single nucleotide in the

DNA sequence. The missing (or additional) base results in an abnormal triplet and also causes each subsequent triplet group to be altered. A frameshift mutation therefore produces messenger RNA with corresponding changed *codons and results in the synthesis of an abnormal, and usually nonfunctional, protein. For example, the deletion of uracil (U) in the second codon of the following mRNA sequence:

AUU CAU CGG UAG ACC UGU AUG

results in the frameshifted sequence:

AUU CAC GGU AGA CCU GUA UG

fraternal twins (dizygotic twins) Two individuals that result from a single pregnancy, each having developed from a separate fertilized egg. The two egg cells contain different combinations of *alleles and so do the two sperm that fertilize them. The twins therefore have no more genetic similarity than brothers or sisters from single births. *Compare* IDENTICAL TWINS.

free energy A measure of a system's ability to do work. The **Gibbs free energy** (or **Gibbs function**), G, is defined by $G = H - TS$, where G is the energy liberated or absorbed in a reversible process at constant pressure and constant temperature (T), H is the *enthalpy and S the *entropy of the system. Changes in Gibbs free energy, ΔG, are useful in indicating the conditions under which a chemical reaction will occur. If ΔG is positive the reaction will only occur if energy is supplied to force it away from the equilibrium position (i.e. when $\Delta G = 0$). If ΔG is negative the reaction will proceed spontaneously to equilibrium.

free radical An atom or group of atoms with an unpaired valence electron. Free radicals are formed when a bond is broken without forming ions. Because of their unpaired electrons, most free radicals are extremely reactive. For example, superoxide free radicals, formed in the body during normal metabolic processes and as a reaction to toxins and infections, have damaging effects on cells and tissues. *See also* ANTIOXIDANTS; REACTIVE OXYGEN SPECIES; SUPEROXIDE DISMUTASE.

freeze drying (lyophilization) The removal of liquid from heat-sensitive materials. The material is frozen, placed under a high vacuum, and maintained at a low temperature ($-40°C$ or below). The pressure generated by the vacuum causes the ice to turn from a solid to a gaseous form without passing through a liquid state. This allows the removal of water from the material without otherwise disturbing its composition. Freeze drying is used to preserve tissues (e.g. blood plasma) and foods and to concentrate solutions.

freeze fracture A method of preparing material for electron microscopy that allows the visualization of the interior of plasma membranes and organelles, and tight junctions between cells. Cells are frozen at $-196°C$ and cracked so that the plane of fracture runs through the middle of *lipid bilayers, separating the two halves. The exposed surfaces are then coated with carbon and platinum and the organic material is digested with enzymes (**freeze etching**), leaving a carbon-platinum replica of the fractured surface, which can be examined using the microscope.

frequency-dependent selection *See* BALANCING SELECTION.

Frizzled *See* WNT PROTEIN.

frogs *See* AMPHIBIA.

frond *See* MEGAPHYLL.

frontal lobe The anterior part of each cerebral hemisphere in the brain. It is associated with higher mental functions, such as planning and reasoning, and in humans, speech; it is the seat of an individual's personality. A strip of the posterior frontal lobe, where it meets the parietal lobe, is the primary motor cortex (*see* CEREBRUM). Neurons here control muscle contractions in specific parts of the body.

fructification *See* SPOROPHORE.

fructose (fruit sugar; laevulose) A simple sugar, $C_6H_{12}O_6$, stereoisomeric with glucose (*see* MONOSACCHARIDE). (Although natural fructose is the D-form, it is in fact laevorotatory.) Fructose occurs in green plants, fruits, and honey and tastes sweeter than sucrose (cane sugar), of which it is a constituent. Derivatives of fructose are important in the energy metabolism of living organisms. Some polysaccharide derivatives (fructans) are carbohydrate energy stores in certain plants. Fructose derived from corn starch is now used extensively in food manufacturing as a sucrose substitute, for example in soft drinks, canned fruit, and jams.

fructose 1,6-bisphosphate An intermediate formed in the initial stage of

*glycolysis by the phosphorylation of glucose using ATP.

fruit The structure formed from the ovary of a flower, usually after the ovules have been fertilized (*see also* PARTHENOCARPY). It consists of the **fruit wall** (*see* PERICARP) enclosing the seed(s). Other parts of the flower, such as the receptacle, may develop and contribute to the structure, resulting in a **false fruit** (*see* PSEUDOCARP). The fruit may retain the seeds and be dispersed whole (an **indehiscent fruit**), or it may open (dehisce) to release the seeds (a **dehiscent fruit**). Fruits are divided into

Different types of fruit and methods of seed dispersal

two main groups depending on whether the ovary wall remains dry or becomes fleshy (**succulent**). Succulent fruits are generally dispersed by animals and dry fruits by wind, water, or by some mechanical means. *See* illustration. *See also* COMPOSITE FRUIT.

fruit fly *See* DROSOPHILA.

fruit sugar *See* FRUCTOSE.

frustule The cell wall of a diatom (*see* BACILLARIOPHYTA), which is impregnated with silica. The wall consists of two halves that overlap one another.

FSH *See* FOLLICLE-STIMULATING HORMONE.

FT protein *See* FLORIGEN.

fucoxanthin The major *carotenoid pigment present, with chlorophyll, in the brown algae (*see* PHAEOPHYTA).

fugitive species A species that is able to coexist with a competitively superior species due to its better dispersal capabilities. The fugitive species is quicker to exploit any vacant patches in the environment that become available (for example due to a fire or storm), colonizing and reproducing before the more competitive species occupies the patch and excludes it. For example, the sea palm (*Postelsia palmaeformis*, a brown alga) colonizes bare patches in mussel beds off the northwestern coast of the USA. When a patch is created by wave action, the alga attaches itself to the bare rock, only to be gradually excluded as the surrounding mussels encroach. Although it is eventually driven from each site, the fugitive alga can coexist with the mussels provided sufficient bare patches become available.

fugu *See* PUFFERFISH.

fumaric acid A carboxylic acid, HCOOHC:CHCOOH, that is an intermediate in the *Krebs cycle, being formed by the dehydrogenation of succinic acid.

functional group A group of atoms having specific chemical properties, which it confers on larger molecules to which the group attaches. Common examples in biologically important molecules include hydroxyl, carbonyl (found in ketones and aldehydes), carboxyl, amino, sulfhydryl, and phosphate groups. The number and nature of such groups influences the shape of any large biomolecule, as well as how it interacts with other molecules.

functional magnetic resonance imaging (fMRI) *See* MAGNETIC RESONANCE IMAGING.

functional trait Any characteristic of an organism that influences its performance (e.g. growth rate, competitive ability), and so ultimately affects its ability to survive and reproduce (i.e. its fitness). Examples include leaf photosynthetic rate, leaf size, stomatal density, and herbivore resistance in plants and beak size, feeding rate, and milk yield in animals.

fungi A group of organisms that constitute the kingdom Fungi. Molecular studies have shown that fungi are more closely related to animals than plants, and both fungi and animals are now placed in the clade *opisthokonts. Fungi can either exist as single cells or make up a multicellular body called a *mycelium, which consists of filaments known as *hyphae. Most fungal cells are multinucleate and have cell walls composed chiefly of *chitin. Fungi exist primarily in damp situations on land and, because of the absence of chlorophyll, are either parasites or saprotrophs on other organisms. The principal criteria used in traditional classification are the nature of the spores produced and the presence or absence of cross walls within the hyphae (*see* ASCOMYCOTA; BASIDIOMYCOTA; DEUTEROMYCOTA; GLOMEROMYCOTA; ZYGOMYCOTA). However, comparisons of ribosomal RNA sequences and other molecular studies have revealed other members of the fungi, including the *Chytridiomycota, *Cryptomycota, and *microsporidia. *See also* LICHENS.

(((•))) SEE WEB LINKS
- Tree of Life survey of fungi, including phylogeny, plus many links to other sites

fungicide *See* PESTICIDE.

Fungi Imperfecti *See* DEUTEROMYCOTA.

funicle The stalk that attaches an ovule to the placenta in the ovary of a flowering plant. It contains a strand of conducting tissue leading from the placenta into the chalaza.

furanose A *sugar having a five-membered ring containing four carbon atoms and one oxygen atom.

fusion The combining together of cells, nuclei, or cytoplasm. *See* CELL FUSION; FERTILIZATION.

fynbos The vegetation of the southern and south-western Cape region of South Africa, noted for its diversity of plants. Of more than 7700 species found in the region, some 70% are endemic. These include many popular garden ornamentals, such as ericas, proteas, daisies, freesias, and gladioli. There are few trees or grasses, and vegetation is dominated by reeds, fine-leaved herbaceous species, and *sclerophyllous shrubs with tough leathery leaves, all adapted to the poor soils and Mediterranean climate. Periodic wildfires prevent the growth of trees and help maintain the community by enabling the germination of new plants from seeds buried underground by native ants. The ants feed on special food bodies (elaiosomes) provided with the seeds (*see* MYRMECOCHORY). The spectacular variety of flowers also attracts numerous butterfly species, not to mention ecotourists. Many species unique to the fynbos are threatened with extinction.

GABA *See* GAMMA-AMINOBUTYRIC ACID.

Gaia hypothesis The theory, based on an idea put forward by the British scientist James Ephraim Lovelock (1919–), that the whole earth, including both its biotic (living) and abiotic (nonliving) components, functions as a single self-regulating system or superorganism. Named after the Greek earth goddess, it proposes that the responses of living organisms to environmental conditions ultimately bring about changes that make the earth better adapted to support life; the system would rid itself of any species that adversely affects the environment. The theory has found favour with many conservationists, and its ideas are embodied in the discipline known as earth system science.

galactose A simple sugar, $C_6H_{12}O_6$, stereoisomeric with glucose, that occurs naturally as one of the products of the enzymic digestion of milk sugar (lactose) and as a constituent of gum arabic.

Galápagos finches *See* DARWIN'S FINCHES.

galectin *See* LECTIN.

gall (cecidium) An abnormal growth of a plant tissue or organ elicited by a foreign organism. Galls can take a wide variety of forms, but most frequently occur as swellings or pits in stems, roots, leaves, and buds. Organisms responsible for their formation include bacteria, viruses, fungi, nematodes, mites, and insects. The gall structure can be quite complex, with several distinct cell layers, or relatively simple and undifferentiated, but is typically very distinct from surrounding normal tissue and often is characteristic of the eliciting organism. It can involve cell enlargement (hypertrophy) and/or cell proliferation (hyperplasia). The mechanisms underlying gall formation are known in only a few cases. The bacterium *Agrobacterium tumefaciens*, which is responsible for crown galls, induces a genetic change in infected host tissue by transfer of a plasmid bearing tumour-forming genes. Insects may secrete substances in their saliva that induce gall formation, or in some cases may transmit viruses or other genetic carriers that affect the plant genome.

gall bladder A small pouch attached to the *bile duct, present in most vertebrates. *Bile, produced in the *liver, is stored in the gall bladder and released when food (especially fatty substances) enters the duodenum.

gallery forest A type of forest that grows alongside rivers and streams running through tropical grassland (savanna). The increased soil moisture is sufficient to support tree growth, which may be dense and multilayered, especially on alluvial nutrient-rich soils of flood plains.

gallstone A hard ball of material formed in the gall bladder from surplus cholesterol, which precipitates in the *bile. Gallstones may lodge in and obstruct the bile duct.

GALP *See* GLYCERALDEHYDE 3-PHOSPHATE.

GALT *See* GUT-ASSOCIATED LYMPHOID TISSUE.

gametangium (*pl.* **gametangia**) An organ that produces gametes. The term is usually restricted to the sex organs of algae, fungi, mosses, and ferns. *See* ANTHERIDIUM; ARCHEGONIUM; OOGONIUM.

gamete A reproductive cell that fuses with another gamete to form a zygote. Examples of gametes are ova and spermatozoa. Gametes are *haploid, i.e. they contain half the normal (diploid) number of chromosomes; thus when two fuse, the diploid number is restored (*see* FERTILIZATION). Gametes are formed by *meiosis. Gametes often differ in size, the smaller (usually male) gamete being known as the **microgamete** and the larger (usually female) as the **macrogamete**. *See also* SEXUAL REPRODUCTION.

game theory A branch of mathematics that predicts the outcomes of interactions between 'players' according to the costs and benefits of the strategy each player employs. Although originally developed to elucidate problems in economics, game theory is used in evolutionary biology to study behavioural strategies among animal populations, such as confrontation, cooperation, and altruism, and how these evolve. Organisms expend energy (a 'cost') to acquire resources (a 'benefit') from their surroundings. The net gain or loss of such an interaction is the payoff, and different strategies result in different payoffs. Game theory enables mathematical modelling of such interactions to predict outcomes and how the different strategies might affect reproductive fitness of individuals in the population and through subsequent generations.

gametogenesis The processes involved in the formation of gametes. Gametes are normally formed by *meiosis but sometimes by *mitosis (as in the gametophyte generation of the ferns). In mammals gametogenesis in the female is known as *oogenesis and occurs in the ovaries; in the male it is known as *spermatogenesis and occurs in the testes.

gametophyte The generation in the life cycle of a plant that bears the gamete-producing sex organs. The gametophyte is *haploid. It is the dominant phase in the life cycle of mosses and liverworts, the *sporophyte generation depending on it either partially or completely. In clubmosses, horsetails, and ferns it is the *prothallus. In seed plants it is very much reduced.

For example, in angiosperms the pollen grain is the male gametophyte and the embryo sac is the female gametophyte. *See also* ALTERNATION OF GENERATIONS.

gamma-aminobutyric acid (GABA) An inhibitory *neurotransmitter in the central nervous system (principally the brain) that is capable of increasing the permeability of postsynaptic membranes to chloride ions (*see* INHIBITORY POSTSYNAPTIC POTENTIAL). GABA is synthesized by *decarboxylation of the amino acid glutamate. There are three classes of GABA receptors: $GABA_A$ and $GABA_C$ receptors are *ionotropic receptors whereas $GABA_B$ receptors are *metabotropic receptors. In crustaceans and annelids, GABA is an inhibitory transmitter at synapses influencing muscle activity. Benzodiazepine drugs, such as valium, exert their anti-anxiety effect by allosterically binding to $GABA_A$ receptors and enhancing the action of GABA.

gamma diversity (gamma richness) (in ecology) The overall species diversity (*see* BIODIVERSITY) of a range of habitats or communities within a region. It reflects the diversity of species occupying individual habitats (*alpha diversity) combined with a measure of the heterogeneity between the habitats (i.e. *beta diversity).

gamma globulin *See* GLOBULIN.

gamopetalous Describing a flower in which the petals are fused to form a *corolla tube. *Compare* POLYPETALOUS.

gamosepalous Describing a flower that possesses a calyx consisting of fused sepals. *Compare* POLYSEPALOUS.

ganglion (*pl.* **ganglia**) A mass of nervous tissue containing many *cell bodies and *synapses, usually enclosed in a connective-tissue sheath. In vertebrates most ganglia occur outside the central nervous system; exceptions are the *basal ganglia in the brain. In invertebrates ganglia occur along the nerve cords and the most anterior pair (**cerebral ganglia**) are analogous to the vertebrate brain; invertebrate ganglia constitute a part of the central nervous system.

ganglion cell *See* RETINA.

ganoid scale *See* SCALES.

gap gene *See* SEGMENTATION GENES.

gap junction (nexus) A passage through the lipid bilayers of adjacent plasma membranes that mediates the transfer of small molecules or ions between interacting cells. Gap junctions are abundant in epithelial tissues and cardiac muscle. They consist of hexagonally packed tubes (*connexons), approximately 10 nm in diameter, through which small molecules or ions may directly pass from the interior of one cell to the interior of the other. Gap junctions, together with chemical *synapses (which function through *neurotransmitters), are **communicating junctions** and comprise one of several types of *cell junction.

( ) SEE WEB LINKS
• Description and schematic representation of gap junction

gas bladder *See* SWIM BLADDER.

gaseous exchange The transfer of gases between an organism and the external environment in either direction. It occurs by diffusion across a *concentration gradient and includes the exchange of oxygen and carbon dioxide in respiration and photosynthesis. Successful gaseous exchange requires a large surface area, as is provided by the alveoli of the lungs and the leaves of plants.

gas-liquid chromatography A technique for separating or analysing mixtures of gases by *chromatography. The apparatus consists of a very long tube containing the stationary phase, a nonvolatile liquid, such as a hydrocarbon oil coated on a solid support. The sample is often a volatile liquid mixture (e.g. of fatty acids), which is vaporized and swept through the column by a carrier gas (e.g. hydrogen). The components of the mixture pass through the column at different rates and are detected as they leave, either by measuring the thermal conductivity of the gas or by a flame detector.

Gas chromatography is usually used for analysis; components can be identified by the time they take to pass through the column. It is also used for separating mixtures into their components, which are then directly injected into a mass spectrometer in the technique of gas chromatography-*mass spectrometry.

gasohol A mixture of petrol (gasoline) and a relatively small proportion of alcohol (i.e. typically ethanol at 5% or 10%) that can be used as an alternative fuel for cars and other vehicles without modification of the engine. The percentage of ethanol in the fuel is denoted by the **E number**, e.g. E10. Ethanol-based gasohol has a higher octane rating and burns more completely than conventional petrol, thus lowering some harmful emissions. Often the ethanol is obtained as a *biofuel by fermentation of agricultural crops or crop residues, (e.g. sugar cane waste). Vehicle engines can also be modified to use mixtures containing greater proportions of ethanol. For example, blends with over 20% ethanol have been used in Brazil since the 1970s, and a mixture of 85% ethanol and 15% petrol, called E85, is available for flexi-fuel vehicles in some countries.

gastric Of or relating to the stomach.

gastric gland *See* GASTRIC JUICE.

gastric juice An acidic mixture of inorganic salts, *hydrochloric acid, mucus, and *pepsinogens secreted by **gastric glands** in the stomach lining and important for digestion of food. Oxyntic (parietal) cells in the gastric glands secrete hydrogen and chloride ions, while chief cells secrete the inactive proenzyme pepsinogen. In the lumen of the stomach, the hydrochloric acid converts pepsinogen into the active enzyme pepsin, which catalyses the initial digestion of food proteins.

gastric mill (proventriculus) A type of *gizzard occurring in many crustaceans. It is situated in the anterior region of the stomach and consists of a set of bones (ossicles) and muscles that grind food particles. The food particles are then filtered by bristles in the posterior section of the stomach.

gastrin A hormone, produced by **G cells** in the mucosa of the stomach and first part of the duodenum, that controls the release of gastric juice and stimulates movements of the stomach wall. The secretion of gastrin is stimulated by stretching of the stomach wall due to the presence of food. It is one of the hormones that integrates and controls digestive processes (*see also* SECRETIN). As gastric juice lowers pH in the stomach, negative feedback tends to inhibit further gastrin secretion.

gastrodermis The epithelial lining of the digestive tract of certain invertebrates, including nematode worms, cnidarians, and ctenophores.

Gastropoda A class of molluscs that includes the snails, whelks, limpets, land and sea slugs, and conches. Molluscs have a well-developed head with tentacles, a large flattened foot, and a coiled twisted shell. They occupy marine, freshwater, and terrestrial habitats; in the terrestrial and some freshwater gastropods the *mantle cavity acts as a lung instead of enclosing gills.

gastrovascular cavity The body cavity of a *cnidarian, which has one opening functioning both as mouth and anus. Enzymes secreted into the cavity partly digest prey, and food particles are then taken up by cells lining the cavity, where digestion is completed within intracellular lysosomes.

gastrula (*pl.* **gastrulae**) The stage in the *development of an animal embryo that succeeds the *blastula. The process, called **gastrulation**, is characterized by division and migration of cells to form the primary *germ layers—the ectoderm, mesoderm, and endoderm. The embryo becomes converted to a cup-shaped structure containing a cavity (the *archenteron), with an opening called the blastopore. An area of tissue in the dorsal lip of the blastopore organizes gastrulation and determines the subsequent developmental fate of the germ layers (*see* ORGANIZER). Whether the

blastopore subsequently becomes the mouth or the anus divides triploblastic animals (i.e. all except cnidarians, ctenophores, etc.) into two major evolutionary groups—respectively the protostomes ('mouth first') and the deuterostomes ('mouth second').

GC content (G + C content) The percentage of nucleotides containing the bases guanine (G) and cytosine (C) relative to the total nucleotide base content in a sample of DNA or RNA. The complementary attribute is the AT (adenine+thymine) content—or in RNA the AU (adenine+uracil) content. So a high GC content necessarily means a low AT content. The stacking arrangement of paired G-C bases enhances the physical stability of GC-rich regions of the nucleic acid molecule, particularly in RNA, whereas AT-rich regions are more 'bendable'. Regions of increased GC content in eukaryote genomes are often associated with genes or other functional elements. One form of GC-richness is due to the occurrence of CG or GC dinucleotide repeats along the same strand, and many genes have stretches relatively rich in CG repeats near their start site (so-called *CpG islands). These play a role in controlling expression of the gene. Because GC content varies between different species, it is used in taxonomy, particularly in prokaryote systematics to characterize broad groups (e.g. *see* FIRMICUTES).

G cells *See* GASTRIN.

gel A lyophilic *colloid that has coagulated to a rigid or jelly-like solid. In a gel, the disperse medium has formed a loosely held network of linked molecules through the dispersion medium. Examples of gels are silica gel and gelatin.

gelatin(e) A colourless or pale yellow water-soluble protein obtained by boiling collagen with water and evaporating the solution. It swells when water is added and dissolves in hot water to form a solution that sets to a gel on cooling. It is used in bacteriology for preparing culture media, in pharmacy for preparing capsules and suppositories, and in jellies and other foodstuffs.

gel electrophoresis A technique for separating molecules of DNA, RNA, or proteins according to their size or electric charge. It is used, e.g., to analyse the fragments of DNA obtained following digestion of a DNA sample with *restriction enzymes to detect mutations, or to create a *DNA profile. In essence, the DNA mixture is placed at one end of a chamber containing semi-solid gel made of agarose or poly-acrylamide. An electric field is applied and the negatively charged DNA fragments move through the gel to the positive pole at a speed depending on their size: smaller fragments move faster than larger ones. On completion, a fluorescent dye is applied to visualize the DNA fragments, which form a series of bands in lanes along the gel. Hence the number, size, and relative abundance of fragments in each DNA sample can be determined by comparing the bands with those obtained from a reference sample of known size composition. Protein samples require addition of the detergent SDS to polyacrylamide gel in order to give them the necessary negative charges, in the method termed **SDS-PAGE (sodium dodecyl sulphate-polyacrylamide gel electrophoresis).** In **pulsed-field gel electrophoresis (PFGE)** the voltage periodically switches between three different directions. This enables DNA fragments as large as 10 Mb to be resolved successfully, compared with the maximum of 50 kb for the standard technique. It is used, e.g., to analyse digests of entire chromosomes from bacterial cells to identify pathogenic subtypes. The whole process is integrated and miniaturized in a single instrument, with the gel contained in arrays of capillary tubes, samples applied using a micropip-ette, automatic detection of fluorescence, and computerized data analysis. Gel electrophoresis is an essential part of many techniques in molecular biology. *See also* SOUTHERN BLOTTING; WESTERN BLOTTING.

gel filtration A type of column *chroma-tography in which a mixture of liquids is passed down a column containing a gel.

Small molecules in the mixture can enter pores in the gel and move slowly down the column; large molecules, which cannot enter the pores, move more quickly. Thus, mixtures of molecules can be separated on the basis of their size. The technique is used particularly for separating proteins but it can also be applied to cell nuclei, viruses, etc.

gemmation A type of *vegetative propa-gation in which small clumps of undiffer-entiated cells (**gemmae**) develop on the surface of a plant. These are shed and dis-persed to other areas, where they grow to produce new individuals. Gemmation is found only in certain lower plants, such as mosses and liverworts.

GenBank The genetic sequence database of the US National Institutes of Health. Started in 1982, it now contains over 2 billion annotated sequence entries and over 5 billion whole genome shotgun (WGS) sequences. It can be searched in various ways, for example using the *Entrez Nucleotide browser or the BLAST algo-rithm. GenBank, which is part of the *Inter-national Nucleotide Sequence Database Collaboration, exchanges information with sequence databases maintained by the DNA Data Bank of Japan (DDBJ) and the European Nucleotide Archive, which is funded by the European Molecular Biology Laboratory (EMBL).

gene A unit of heredity composed of DNA. In classical genetics (*see* MENDELISM; MENDEL'S LAWS) a gene is visualized as a discrete particle, forming part of a *chromosome, that determines a particular characteristic. It can exist in different forms called *alleles, which determine which aspect of the characteristic is shown (e.g. tallness or shortness for the characteristic of height).

A gene occupies a specific position (*locus) on a chromosome. In view of the discoveries of molecular genetics, it may be defined as the sequence of nucleotides of DNA (or RNA) concerned with a specific function, such as the synthesis of a single polypeptide chain or of an RNA molecule, corresponding to a particular sequence of the *genetic code. One or more of these **structural genes**, coding for protein, may be associated with other genes controlling their expression (*see* OPERON). Only about 1.5% of human genes encode proteins, although about 75% of the genome is transcribed, at least on some occasions. Much of this gives rise to various types of non-protein-coding RNA, now thought to play crucial roles in controlling the expression of structural genes. *See also* CISTRON.

gene amplification The multiple replication of a section of the *genome, which occurs during a single cell cycle and results in the production of many copies of a specific sequence of the DNA molecule. For example, in the oocytes of amphibians and other animals, in which large numbers of ribosomes are needed, the genes encoding ribosomal RNA are greatly amplified. Viral genes that cause the formation of tumours (*see* ONCOGENE) are amplified in tumour cells.

gene bank *See* DNA LIBRARY.

gene cloning (DNA cloning; molecular cloning) The production of exact copies (**clones**) of genes or DNA sequences using genetic engineering techniques. It is performed to obtain numerous copies of a particular gene so that, e.g., its expression can be investigated or its product isolated in relatively large amounts. Essentially the target gene(s) is inserted into cloning *vectors, such as bacterial plasmids or bacteriophages, which transfer the recombinant DNA to suitable host cells, such as the bacterium *E. coli* or yeast cells. Alternatively, the target DNA can be inserted directly into the host's DNA using the techniques of *genome editing. Inside the host cell the recombinant DNA undergoes replication; thus, a host cell will give rise to a colony of transgenic cells each containing identical cloned DNA fragments. Only a small fraction of host cells produce colonies that successfully incorporate the target gene, and various methods may be used to identify such colonies, e.g. by screening for a *reporter gene that confers resistance to a particular antibiotic or expresses a *fluorescent protein. The initial target gene may be selected from a genomic library or complementary DNA library, which is itself created by cloning (*see* DNA LIBRARY). Alternatively, messenger RNA can be isolated from a tissue or a cell culture, complementary DNA copies made using *reverse transcriptase, and the target DNA amplified using specific primers with the *polymerase chain reaction. Moreover, DNA sequences or even entire genes can be assembled from their component nucleotides using automated synthesizers according to a predetermined base sequence. Gene cloning enables large quantities of a desired protein product to be produced by transgenic host cells (*see* EXPRESSION VECTOR): e.g. human insulin is produced by bacteria containing the cloned insulin gene. *See also* POSITIONAL CLONING.

gene conversion The nonreciprocal transfer of genetic material from one homologous chromosome to another, thereby converting a normal (wild-type) allele to a mutant allele, or vice versa. It is believed to occur as a result of recombination between DNA strands of sister chromatids following a double-strand breakage in one chromatid and subsequent *mismatch repair using the 'wrong' DNA strand as template. This mechanism can account for the non-Mendelian segregation of genetic markers that can sometimes be observed, for example, following meiosis in yeast. Loss of wild-type alleles due to gene conversion is also implicated in the initiation of certain hereditary diseases and cancers in humans.

gene editing *See* GENOME EDITING.

gene expression The manifestation of the effects of a gene by the production of the particular protein, polypeptide, or type of RNA whose synthesis it controls. The *transcription of individual genes can be 'switched on' or 'switched off' according to the needs and circumstances of the cell at a particular time. A number of mechanisms are responsible for the control of gene expression; the *Jacob-Monod hypothesis describes a mechanism operating in prokaryotes (*see* OPERON). This involves the binding of specific regulatory proteins, called *transcription factors, to control sequences in DNA. Other mechanisms operating in prokaryotes include regulation of translation to restrict protein synthesis, and modification or breakdown of transcription factors. Control of gene expression is more complicated in eukaryotes, which possess various control mechanisms not seen in prokaryotes. Eukaryotes have many more transcription factors than prokaryotes, with numerous interacting roles to play in different tissues and at different stages of development. Gene activity is controlled by multiple factors acting synergistically, and sometimes at considerable distance from the gene in question. Moreover, the packaging of DNA into chromatin also plays a part. Reversible chemical modifications of the histone proteins determine the degree of chromatin condensation and whether the genes are accessible for the transcription machinery (*see* CHROMATIN REMODELLING). Also, many protein-encoding genes are regulated by small noncoding RNA molecules that bind to target mRNAs and degrade them or prevent their translation (*see* RNA INTERFERENCE). Methylation of cytosine bases of specific genes in eukaryotic DNA is observed in cells in which the gene is not expressed; if DNA methylation is prevented by the use of inhibitory chemicals, this can cause certain genes to be expressed. In multicellular organisms, expression of the right genes in the right order at the right times is particularly crucial during embryonic development and cell differentiation. This involves subtle and complex interplay of chemical signals with the embryo's genes, in patterns that vary between different types of organism (*see* DIFFERENTIATION). Abnormalities of gene expression may result in the death of cells, or their uncontrolled growth, as in *cancer. See also* TRANSCRIPTOMICS; X INACTIVATION.

gene family (multigene family) A group of genes that have arisen by duplication of an ancestral gene. Such genes show similarities of nucleotide sequence that betray their common origin, and if they have evolved relatively recently they occur close together on the same chromosome. However, more distantly related members of a gene family may be found scattered widely on different chromosomes, reflecting chromosomal rearrangements during the evolution of the genome. Members of a gene family may be functionally very similar or differ widely. For example, members of the histone gene family all produce very similar proteins, whereas the serine protease family contains both trypsin, the proteolytic digestive enzyme, and haptoglobin, a protein that binds globin but has no proteolytic activity.

gene flow The exchange of genetic material by interbreeding between populations of the same species or between individuals within a population. Gene flow increases the variation in the genetic composition of a population, whereas barriers to gene flow can give rise to the emergence of new species (*see* SPECIATION).

gene frequency *See* ALLELE FREQUENCY.

gene imprinting (molecular imprinting) The differential expression of a gene according to whether it is derived from the mother or father. This leads to unequal genetic contributions from the mother and father to their offspring and is essential for normal development. For example, loss of maternal imprinting in mice results in an abnormally large fetus, whereas loss of paternal imprinting leads to a small fetus. In humans only a small proportion of genes normally exhibit imprinting, and these occur chiefly in two clusters, one on the short arm of chromosome 11 and the other on the long arm of chromosome 15. Several human diseases are associated with failures of imprinting, including Angelman syndrome, Prader–Willi syndrome, and certain cancers. A diverse range of organisms, including plants, insects, and mammals, exhibit gene imprinting. It depends on the *DNA methylation of certain bases, particularly cytosine residues, effectively silencing expression of the affected gene by causing inhibitory proteins to bind to methylated residues. Another imprinting mechanism is evident in yeast, fruit flies, and the nematode *Caenorhabditis elegans*.

Here, imprinting is determined by patterns of chemical modifications of the histone proteins that package the DNA into chromatin (*see* CHROMATIN REMODELLING). Such patterns, although reversible, are inherited by being incorporated as the DNA is replicated.

gene knockout *See* KNOCKOUT.

gene library *See* DNA LIBRARY.

gene manipulation *See* BIO-TECHNOLOGY; GENETIC ENGINEERING.

gene mutation *See* POINT MUTATION.

Gene Ontology *See* ONTOLOGY.

gene pool All the *genes and their different alleles that are present in a population of a particular species of organism. *See also* POPULATION GENETICS.

gene probe *See* DNA PROBE.

generation A group of organisms of approximately the same age within a population. Organisms that are crossed to produce offspring in a genetics study are referred to as the **parental generation** and their offspring as the first **filial generation**. *See also* F₁; F₂; P.

generation time The interval between the beginnings of consecutive cell divisions. It may be as short as 20 minutes in bacteria. *See also* INTERPHASE.

generative nucleus One of the two nuclei in an angiosperm *pollen grain (*compare* TUBE NUCLEUS). It divides to produce two male gamete nuclei (*see* DOUBLE FERTILIZATION).

generator potential The local electrical charge that develops in the sensitive part of a receptor cell, known as the **generator region**. The generator potential is caused by *depolarization of the membrane surrounding this part of the cell, brought about by an exchange of ions in response to a stimulus. The strength of the potential is proportional to the strength of the stimulus: when the potential exceeds a certain threshold, it fires an action potential. More action potentials are triggered as long as the generator potential exists.

gene sequencing *See* DNA SEQUENCING; PHYSICAL MAP.

gene silencing *See* CHROMATIN REMODELLING; KNOCKOUT; RNA INTERFERENCE.

gene splicing *See* RNA PROCESSING.

gene therapy The application of genetic engineering techniques to alter or replace defective genes. A defective gene may result from an incorrect sequence of bases in the DNA molecule or an inability of the gene to code for the expression of a particular polypeptide. Cells can be removed from the body to allow their DNA to be corrected and then reinserted; alternatively, the normal 'healthy' gene can be directly targeted to the appropriate cells inside the body. The goal is to make the corrected gene work in a way that is effective, reliable, sustainable, and safe for the patient, but this is often difficult to achieve. *Retroviruses are often used as *vectors for transferring genes into cells as part of the natural retrovirus life cycle involves the insertion of their own genetic material into the chromosomes of their host. Alternatively *liposomes may be used. Another option is to use microRNAs to 'knock out' defective genes in certain tissues (*see* RNA INTERFERENCE). Precise site-specific changes in DNA can now be made using the techniques of *genome editing. These avoid the unpredictable integration of viral vectors and have potential to correct defective genes in tissue cell lines or stem cells derived from a patient. Such cells can then be reintroduced to the patient to treat the condition. *See also* ADENO-ASSOCIATED VIRUS.

genetically modified organisms (GMOs) Organisms whose genomes incorporate and express genes from another species. Genetically modified (or transgenic) individuals are created by genetic engineering, using suitable *vectors to insert the desired foreign gene into the fertilized egg or early embryo of the host. See Feature.

GENETICALLY MODIFIED ORGANISMS

Since the early 1980s developments in genetic engineering have made it possible to produce genetically modified organisms. A gene from one organism is isolated and transferred to cells of another organism, where it is incorporated into the recipient's chromosomes and expressed. Such transgenic organisms can exhibit quite novel characteristics. Commercial applications of this new technology have ranged from the production of human hormones in bacteria and vaccines in yeasts to the development of genetically modified (GM) crop plants. More recently, the techniques of *genome editing offer the potential to make specific targeted mutations that will enhance or suppress expression of certain genes or to introduce DNA sequences that will confer new traits, such as disease resistance.

Techniques

Much of the work with genetic modification of plants involves *protoplasts, cultured spherical cells from which the cell walls have been removed. The Ti plasmid (see illustration) of *Agrobacterium tumefaciens*, the bacterium that is responsible for the tumorous growths of crown-gall disease in plants, has been used successfully as a *vector with certain dicotyledons, including tobacco, tomato, potato, soyabean, and cotton. It works much less well with grasses, cereals, and other monocots. In these plants various other techniques are available, including:

- electroporation—treatment of cells by exposure to an electric field that renders them transiently permeable to DNA fragments;
- microinjection—injection of DNA directly into the cell nucleus;
- biolistics—'shooting' a cell with a DNA-coated tungsten microprojectile, which penetrates the cell wall with minimal damage.

To produce a transgenic animal the novel genes are inserted at a very early stage of development, e.g. the early embryo or the pronucleus of a fertilized egg, typically using microinjection. The recombinant embryos are then transferred to the uterus of a foster mother where they complete their development.

Applications

Plants

- tolerance to herbicides
- improved insect resistance or drought tolerance
- production of materials (e.g. plastics) or drugs
- longer 'shelf life'; improved nutritional qualities (e.g. β-carotene in golden rice)

Animals

- production of therapeutic proteins in milk (e.g. human growth hormone)
- potential for improved growth rates and milk yields
- potential for production of organs for human transplants

Risks

The use of GM organisms in the environment poses certain potential problems. For example, genes for herbicide or insect resistance may spread from crop plants to wild plants, with possible serious consequences for both agriculture and natural ecosystems. Farmers may be faced with new 'superweeds', while insect populations could decline. Moreover, the products of GM crops have to be fully evaluated to ensure that they are safe to eat. Genetic modification of animals often has unforeseen side-effects and raises ethical issues about such treatments.

desired fragment containing gene of interest
cleaved by restriction endonuclease and
isolated by Southern blotting

restriction endonuclease excised T-DNA

fragment inserted into plasmid using ligase

recombinant Ti plasmid
containing foreign gene

plasmid introduced into plant protoplast via host bacterium

recombinant plasmid
chromosome

gene of interest integrated into plant chromosome

transformed plant cell

culture

growth

transformed plant

In plants one of the most useful
cloning vectors is the Ti plasmid of *A.
tumefaciens*. The plasmid is
engineered by replacing its tumour-
inducing genes (T-DNA) with the
gene to be cloned and then returned
to the bacterium, which is allowed to
infect cultured plant protoplasts.
Transformed cells are selected by the
presence of marker genes and
propagated with plant growth
substances so they develop into
complete plants.

genetic code The means by which genetic information in *DNA controls the manufacture of specific proteins by the cell. The code takes the form of a series of triplets of bases in DNA, from which is transcribed a complementary sequence of *codons in messenger *RNA (*see* TRANSCRIPTION). The sequence of these RNA codons determines the sequence of amino acids during *protein synthesis. There are 64 possible codes from the combinations of the four bases present in DNA and messenger RNA and 20 amino acids present in body proteins; also some codons have other functions (*see* START CODON; STOP CODON; WOBBLE). Hence, the code is redundant: i.e. any given amino acid can be encoded by more than one codon. However, it is not ambiguous—each codon specifies only one amino acid. See illustration.

genetic drift (Sewall Wright effect) The random change in the frequency of alleles in a population over successive generations due to sampling error in the gametes. Each new generation differs from its parental generation with regard to allele frequencies simply because of random variation in the distribution of gametes. Over time, this may lead to certain alleles becoming fixed and others being lost altogether. This process is more rapid in smaller populations, following a *population bottleneck, or when the alleles concerned confer no apparent benefit compared with their counterparts. Hence, genetic drift can ultimately cause loss of genetic diversity if there are no counteracting factors.

genetic engineering (recombinant DNA technology) The techniques involved in altering the characters of an organism by modifying its genome. This can mean changing the base sequence of its DNA in some way, e.g. by *knockout or *genome editing, or inserting genes from another organism to create *recombinant DNA. Alternatively, entirely new genes can be assembled from scratch using sequence information and

First base in codon	Second base in codon		Third base in codon	
U	C	A	G	
U	UUU Phe	UCU Ser	UAU Tyr	UGU Cys	U
	UUC Phe	UCC Ser	UAC Tyr	UGC Cys	C
	UUA Leu	UCA Ser	UAA (stop codon)	UGA (stop codon)*	A
	UUG Leu	UCG Ser	UAG (stop codon)	UGG Trp	G
C	CUU Leu	CCU Pro	CAU His	CGU Arg	U
	CUC Leu	CCC Pro	CAC His	CGC Arg	C
	CUA Leu	CCA Pro	CAA Gin	CGA Arg	A
	CUG Leu	CCG Pro	CAG Gin	CGG Arg	G
A	AUU Ile	ACU Thr	AAU Asn	AGU Ser	U
	AUC Ile	ACC Thr	AAC Asn	AGC Ser	C
	AUA Ile	ACA Thr	AAA Lys	AGA Arg†	A
	AUG Met (start codon)	ACG Thr	AAG Lys	AGG Arg	G
G	GUU Val	GCU Ala	GAU Asp	GGU Gly	U
	GUC Val	GCC Ala	GAC Asp	GGC Gly	A
	GUA Val	GCA Ala	GAA Glu	GGA Gly	A
	GUG Val	GCG Ala	GAG Glu	GGG Gly	G

* UGA encodes tryptophan in mitochondrial DNA

† AGA is a stop codon in mitochondrial DNA

The genetic code

DNA synthesizers to code for proteins with particular desired properties (*see* SYNTHETIC BIOLOGY). Bacteria or yeasts containing the recombinant genes serve as 'cellular factories', synthesizing the desired protein. Genetic engineering has many applications, ranging from the commercial production of insulin and other hormones, vaccines, etc., to the creation of *transgenic animals and crop plants in agriculture (*see* GENETICALLY MODIFIED ORGANISMS). *See also* DNA LIBRARY; DNA PROBE; GENE CLONING; GENE THERAPY; MONOCLONAL ANTIBODY.

genetic fingerprinting *See* DNA FINGERPRINTING; DNA PROFILING.

genetic load The extent to which the average fitness (i.e. reproductive success) of all members of a population varies from that of the individual with the greatest fitness. It is denoted by a number from 0 to 1, where 0 denotes that all individuals have genotypes conferring maximum fitness, and 1 indicates that all but one have genotypes conferring zero fitness. In effect genetic load indicates the concentration of deleterious genes within a population.

genetic mapping *See* CHROMOSOME MAP; LINKAGE MAP; PHYSICAL MAP; RESTRICTION MAPPING.

genetic marker *See* MARKER GENE; MOLECULAR MARKER.

genetic polymorphism *See* POLYMORPHISM.

genetics The branch of biology concerned with the study of heredity and variation. **Classical genetics** is based on the work of Gregor Mendel (*see* MENDELISM). During the 20th century genetics expanded to overlap with the fields of ecology and animal behaviour (*see* BEHAVIOURAL GENETICS; POPULATION GENETICS), and important advances in biochemistry and microbiology led to clarification of the chemical nature of *genes and the ways in which they can replicate and be transmitted, creating the field of **molecular genetics**. Since the 1980s, automated DNA sequencing techniques coupled with advances in computerized data handling have transformed genetics, enabling rapid determination of the nucleotide sequences of entire genomes. The *bioinformatics revolution has allowed evolutionary relationships between organisms to be traced at the genome level and gene function to be analysed at the cellular level. Genetics today is thus the starting point for unravelling the complexities of every aspect of metabolism, function, and development. *See* Chronology. *See also* GENETIC ENGINEERING.

genetic screening The process by which the genome of a human or other organism is analysed for genetic markers (*see* MARKER GENE) that indicate the presence of particular genes, especially ones that cause or predispose to certain diseases. Increased knowledge of the human genome (*see* HUMAN GENOME PROJECT) and technological advances have simplified genetic screening in persons with a family history of certain inherited diseases, e.g. certain forms of breast cancer. Clinical gene testing is now used routinely to screen for many different genes, either to assess the risk of disease in susceptible individuals or their offspring or to confirm a diagnosis of inherited disease. When there is a risk of inherited genetic disease, early embryos can be screened to detect the presence of harmful alleles so that normal embryos can be selected for implantation in the mother's womb (*see* PREIMPLANTATION GENETIC DIAGNOSIS). Commercial gene test kits are available to the general population, although claims that these can determine the risk of healthy individuals developing, say, heart disease or cancer should be treated with caution. Such tests have major implications for the insurance industry as well as for medicine. For example, some healthy individuals may be expected to pay a higher premium for life insurance because genetic screening reveals the presence of such genes. *See also* ASSISTED REPRODUCTIVE TECHNOLOGY.

genetic variation *See* VARIATION.

gene tracking A method for determining the inheritance of a particular gene in a family. It is used in the diagnosis of

GENETICS

1866	Gregor Mendel publishes his findings on inheritance in peas and his observation that characters are determined by discrete 'factors'.
1875	German cytologist Oskar Hertwig (1849–1922) describes the process of fertilization and formation of the zygote.
1879–85	German cytologist Walther Flemming (1843–1905) describes the behaviour of chromosomes during cell division, which he terms 'mitosis'.
1886	German biologist August Weismann (1834–1914) publishes a theory of continuity of the germ plasm through successive generations.
1887–92	German cytologist Theodor Boveri (1862–1915), Hertwig, and others describe meiosis, confirming Weismann's prediction of a 'reduction division'.
1900	Hugo de Vries, German botanist Karl Correns (1894–1933), and Austrian botanist Erich von Tschermak (1871–1962) independently rediscover Mendel's work.
1903	US cytologist Walter S. Sutton (1877–1916) describes how the behaviour of chromosomes during meiosis explains Mendel's laws and suggests that genes are located on chromosomes.
1909	Dutch botanist Wilhelm Johannsen (1857–1927) coins the term 'gene'. Frans-Alfons Janssens describes crossing over.
1910	Thomas Hunt Morgan discovers sex-linked traits in fruit flies.
1913	US geneticist Alfred Sturtevant (1891–1970) publishes the first genetic map—of fruit-fly genes.
1916	US geneticist Calvin Bridges (1889–1938) proves the chromosome theory of heredity.
1927	US geneticist Hermann Müller (1890–1967) demonstrates that X-rays can cause mutations.
1930	British statistician Ronald Fisher (1890–1962) publishes *The Genetical Theory of Natural Selection*, a key work in neo-Darwinism.
1941	George Beadle and Edward Tatum begin work with nutritional mutants of bread mould, leading to their 'one gene-one enzyme' hypothesis.
1944	Oswald Avery and colleagues demonstrate that DNA is the genetic material.
1947	Erwin Chargaff establishes the one-to-one ratio of purine and pyrimidine bases in DNA.
1953	James Watson and Francis Crick propose a molecular structure for DNA.
1960	French biochemists Jacques Monod (1910–76) and François Jacob (1920–2013) introduce the term 'operon' for a functionally integrated group of genes.
1961–66	The genetic code is deciphered by US biochemists Marshall Nirenberg (1927–2010), Philip Leder (1934–), and others.
1972	Paul Berg (1926–) creates the first recombinant DNA molecule, based on a lambda phage.

1973	First experimental genetic manipulation of a bacterium takes place.
1977	Techniques for sequencing DNA devised by US biochemist Walter Gilbert (1932–), Frederick Sanger, and colleagues.
1978	Human insulin is produced by genetically engineered bacteria.
1983	US biochemist Kary Mullis (1944–2019) devises the polymerase chain reaction for amplifying DNA. First transgenic plant is created.
1984	British geneticist Sir Alec Jeffreys (1950–) develops DNA (or genetic) fingerprinting.
1988	First patent is awarded for a genetically engineered animal—a cancer prone mouse. Field trials of genetically modified tomatoes take place in the USA. The Human Genome Project begins.
1993	Transgenic sheep are used to produce human proteins in their milk. Genetically modified tomatoes go on sale in the USA.
1997	British geneticist Ian Wilmut and colleagues announce the birth of a lamb ('Dolly')—the first mammal to be cloned from an adult body cell.
1998	The first complete sequence of a genome for a complex animal, the nematode *Caenorhabditis elegans*, is published.
1999	Full DNA sequence of human chromosome 22 is published, making it the first human chromosome to be fully sequenced.
2002	Draft sequences of the mouse and rice genomes published.
2003	Human Genome Project completes DNA sequencing of the human genome, 50 years after publication of Watson and Crick's paper on structure of DNA.
2004	Approval granted in the USA for a microarray-based diagnostic genetic test for use in clinical medicine.
2005	Completion of phase I of the International HapMap Project containing over one million markers of the human genome.
2006	Sequencing of Neanderthal DNA reveals differences with modern humans.
2008	Debut of next-generation DNA sequencing.
2009	Successful gene therapy for X-linked adrenoleukodystrophy using engineered stem cells. First detailed map of the epigenome.
2012	Advent of new genome editing tools TALENS and CRISPR.
2015	First comprehensive map of the human epigenome published.
2016	CRISPR-Cas9 gene editing used to treat a human patient.
2017	Intravenous gene therapy with adeno-associated virus used to insert a new gene in spinal neurons of infants suffering from spinal muscular atrophy 1.

genetic diseases, such as cystic fibrosis and Huntington's disease. Molecular markers, such as single nucleotide polymorphisms or *restriction fragment length polymorphisms (RFLPs) situated in or near the locus of interest, are identified using *DNA probes, and suitable markers selected. These can then be traced through members of the family and used to detect the presence or absence of the disease locus prenatally in future at-risk pregnancies.

genome All the genes contained in a single set of chromosomes, i.e. in a *haploid nucleus. Each parent, through its reproductive cells, contributes its genome to its offspring.

genome editing Any of various techniques for making precise, site-specific changes in the base sequence of the DNA of an organism. Each uses an engineered DNA-binding construct to seek out the target sequence, and a nuclease enzyme to cut the two strands of the DNA double helix at the required site. By exploiting the cellular DNA repair mechanisms, small inserts or deletions can be made at a single site, often causing inactivation (knockout) of the affected gene. Further, by cutting at two sites, larger inversions or deletions can result; and when a donor DNA template or transgene is introduced with the editing module, entire genes can be corrected or replaced. Genome editing is performed chiefly for genetic research, but increasingly is seen as a valuable technique to correct harmful mutations during *gene therapy, and to engineer transgenic organisms, especially in animal and plant breeding. Several different technologies are used, including *CRISPR-Cas9, *zinc finger proteins, *TALENs, and *Cre/*loxP* recombination.

genome project Any undertaking, whether by a single organization or a consortium of scientific institutions, to map and sequence the entire genome of an organism. The bacterium *Haemophilus influenzae* was the first organism to be sequenced; the first eukaryotic genome to be sequenced was that of the budding yeast *Saccharomyces cerevisiae*. A massive international collaboration resulted in the sequencing of the human genome, completed in 2003 (*see* HUMAN GENOME PROJECT). By 2018 the Genomes OnLine Database (GOLD), held details of nearly 200000 sequencing projects, either completed or ongoing, involving species drawn from all groups of organisms.

genomics The branch of genetics concerned with the study of genomes. It has developed since the 1980s, exploiting automated techniques and computer-based systems to collect and analyse vast amounts of data on nucleotide and amino-acid sequences of various organisms, generated by projects such as the *Human Genome Project. There are several distinct but overlapping areas of genomics. **Structural genomics** is essentially about mapping the genome and ultimately producing a complete DNA sequence for any particular organism. However, the term is often extended to include determination of three-dimensional molecular structures of nucleic acids and proteins (*see* PROTEOMICS). **Functional genomics** deals with gene expression and how gene products work. This highly complex area, which involves analysis of transcripts of sets of genes (*see* TRANSCRIPTOMICS), seeks to understand how gene expression is controlled and integrated and how gene functions change under different conditions, such as disease states. **Comparative genomics** identifies regions of sequence similarity between genomes of different species. Knowledge of the functional significance of a particular DNA sequence in one species allows predictions about functions of closely matching sequences in other species. In addition, such comparisons permit inferences about mechanisms of gene evolution and give insights into the evolutionary relationships of different organisms. *Metagenomics collectively analyses the genomes of all microorganisms in an environmental sample, such as a sample of soil or seawater (*see* MICROBIOME). *See also* BIOINFORMATICS; METABOLOMICS; PHYLOGENOMICS.

genotoxicity The condition resulting from the interaction of toxic agents (**genotoxins**) with DNA molecules in genes.

Since the genes are passed down to the next generation, the toxicity induced by genotoxins is heritable. Genotoxins can induce mutations in chromosomes (**clastogenesis**) or in a small number of base pairs (**mutagenesis**). Genotoxic agents include X-rays, natural *carcinogens, some human-made products (e.g. acridine and vinyl chloride), and viruses.

genotype The genetic composition of an organism, i.e. the combination of *alleles it possesses. *Compare* PHENOTYPE.

genotype frequency The percentage of individuals in a population that possess a specific genotype. The genotype frequency can be calculated using the Hardy-Weinberg equation (*see* HARDY-WEINBERG EQUILIBRIUM).

genus (*pl.* **genera**) A category used in the *classification of organisms that consists of a number of similar or closely related species. The common name of an organism (especially a plant) is sometimes similar or identical to that of the genus, e.g. *Lilium* (lily), *Crocus*, *Antirrhinum*. Similar genera are grouped into families. *See also* BINOMIAL NOMENCLATURE.

geochronology *See* VARVE DATING.

geographical isolation The separation of two populations of the same species or breeding group by a physical barrier, such as a mountain or body of water. Geographical isolation may ultimately lead to the populations becoming separate species by adaptive radiation. *See also* ALLOPATRIC; SPECIATION.

geological time scale A time scale that covers the earth's history from its origin, estimated to be about 4600 million years ago, to the present. The chronology is divided into a hierarchy of time intervals: eons, eras, periods, epochs, ages, and chrons. *See* Appendix 4.

geophyte A plant life form in Raunkiaer's system of classification (*see* PHYSIOGNOMY). Geophytes are herbaceous plants in which the perennating buds are below ground, giving rise to corms, bulbs, or rhizomes.

geotaxis (*pl.* **geotaxes**) The movement of a cell or microorganism in response to gravity. For example, certain cnidarian larvae that swim towards the seabed exhibit positive geotaxis. *See* TAXIS.

geotropism (gravitropism) The growth of plant organs in response to gravity. A main root is positively geotropic and a main stem negatively geotropic, growing downwards and upwards respectively, irrespective of the positions in which they are placed. For example, if a stem is placed in a horizontal position it will still grow upwards. This is caused by redistribution of the plant hormone *auxin, which moves to the lower side of elongating cells in the stem or root; in stems auxin promotes cell elongation, whereas in roots it inhibits it. Hence, the accumulation of auxin leads to differential growth of the lower side of the cells in shoots and the upper side in roots, causing shoots to curve upwards and roots to curve downwards. One proposed mechanism involves membrane-bound clusters of starch grains called *statoliths, which are thought to act as tilt 'switches' to regulate the distribution of auxin and ensure that the stem or root attains the correct angle of growth. However, mutant plants lacking statoliths are still capable of geotropism, albeit more slowly, so other gravitational sensors must be present. *See* TROPISM.

germ cell Any cell in the series of cells (the **germ line**) that eventually produces *gametes, especially the first cell in such a series. In mammals the germ cells are the oogonia (in the ovaries) and the spermatogonia (in the testes). *See* OOGENESIS; SPERMATOGENESIS.

germinal epithelium 1. The layer of epithelial cells lining the seminiferous tubules of the testis, which gives rise to spermatogonia (*see* SPERMATOGENESIS). **2.** A misnomer for the ovarian surface epithelium—a layer of epithelial cells on the surface of the ovary that is continuous with the *mesothelium. These cells are involved in repairing defects in the surface of the ovary following ovulation and also are thought to be the source of the granulosa cells of the ovarian follicles. They do not give rise to the ova as was formerly believed.

germination 1. The initial stages in the growth of a seed to form a seedling. Uptake of water into the dry seed by *imbibition causes the seed to expand and the seed coat to rupture. It also triggers metabolic changes that reactivate growth. Enzymes are produced that digest food reserves in the *endosperm or seed leaves (*cotyledons), and the resulting nutrients are transported to the growing embryo. The embryonic shoot (plumule) and embryonic root (radicle) emerge and grow upwards and downwards respectively. *See also* EPIGEAL; HYPOGEAL. **2.** The first signs of growth of spores and pollen grains.

germ layers (primary germ layers) The layers of cells in an animal embryo at the *gastrula stage, from which are derived the various organs of the animal's body. There are two or three germ layers: an outer layer (*see* ECTODERM), an inner layer (*see* ENDODERM), and in most animal groups a middle layer (*see* MESODERM). *See also* DEVELOPMENT.

germ plasm *See* WEISMANNISM.

gestation The period in animals bearing live young (especially mammals) from the fertilization of the egg to birth of the young (parturition). In humans gestation is known as **pregnancy** and takes about nine months (40 weeks).

GFP *See* GREEN FLUORESCENT PROTEIN.

GFR *See* GLOMERULAR FILTRATION RATE.

ghrelin A peptide hormone that is secreted chiefly by cells in the stomach lining when the stomach is empty, and increases hunger by stimulating the release of *neuropeptide Y, a potent stimulant of appetite, from the arcuate nucleus in the hypothalamus. It also stimulates release of growth hormone by binding to receptors (called growth hormone secretagogue receptors) in the anterior pituitary. Ghrelin thus opposes the appetite-suppressive effects of the hormones insulin, *leptin,

and *peptide YY. It is synthesized as a prohormone and processed to yield a 28-amino-acid peptide modified by the addition of n-octanoic acid. The concentration of ghrelin in blood rises during the fasting period before a meal, thereby promoting appetite. Paradoxically, it also appears to suppress the mobilization of fat reserves in adipose tissue. Its role in regulating the body's energy balance has prompted interest in ghrelin as a potential target for anti-obesity treatments.

GI *See* GLYCAEMIC INDEX.

giant chromosome *See* POLYTENY.

giant fibre A nerve fibre with a very large diameter, found in many types of invertebrate (e.g. earthworms and squids). Its function is to allow extremely rapid transmission of nervous impulses and hence rapid escape movements in emergencies.

giant virus *See* NUCLEOCYTOPLASMIC LARGE DNA VIRUS.

gibberellic acid (GA₃) A *plant hormone that is extracted from fungal cultures and is one of the most important commercially available *gibberellins (see formula). It was discovered in 1954.

Gibberellic acid

gibberellin Any of a group of *plant hormones chemically related to terpenes and occurring naturally in plants and fungi. They promote elongation of stems, e.g. bolting in cabbage plants, by enhancing both cell division and elongation. In germinating seeds following imbibition of water, the embryo releases gibberellins, which mobilize food reserves by activating enzymes in the *aleurone layer. They are also influential in inducing flowering and fruit development. For example, gibberellins are sprayed to promote the growth of seedless grapes, to replace the gibberellin normally produced by the seeds. Commercially available gibberellins, such as *gibberellic acid, are used to manipulate the onset of sexual maturity in various species, e.g. to induce cone-bearing in young conifer trees.

giga- Symbol G. A prefix used in the metric system to denote one thousand million times. For example, 10^9 joules = 1 gigajoule (GJ).

gigantism See GROWTH HORMONE.

gill 1. (in zoology) A respiratory organ used by aquatic animals to obtain oxygen from the surrounding water. A gill consists essentially of a membrane or outgrowth from the body, with a large surface area and a plentiful blood supply, through which diffusion of oxygen and carbon dioxide between the water and blood occurs. Fishes have **internal gills**, formed as outgrowths from the pharynx wall and contained within *gill slits. Water entering the mouth is pumped out through these slits and over the gills. The gills of most aquatic invertebrates and amphibian larvae are **external gills**, which project from the body so that water passes over them as the animal moves. Most aquatic vertebrates and invertebrates with gills also use them for excreting nitrogenous wastes and maintaining the ionic composition of body fluids. See CHLORIDE SECRETORY CELL. **2.** (in botany) One of the ridges of tissue that radiate from the centre of the underside of the cap of mushrooms. The spores are produced on these gills.

gill bar A cartilaginous support for the tissue between the gill slits in lower chordates, such as lancelets.

gill slit An opening leading from the pharynx to the exterior in aquatic vertebrates and lancelets. In lancelets they function in *filter feeding. In fish they contain the *gills and are usually in the form of a series of long slits. They are absent in adult tetrapod

vertebrates (except for some amphibians) but their presence in some form in the embryos of all vertebrates is a characteristic of the phylum *Chordata.

gingiva (gum) The part of the epithelial tissue lining the mouth that covers the jaw bones. It is continuous with the sockets surrounding the roots of the teeth.

Ginkgophyta See GYMNOSPERM.

gizzard A muscular compartment of the alimentary canal of many animals that is specialized for breaking up food. In birds the gizzard lies between the *proventriculus and the duodenum and contains small stones and grit, which assist in breaking up the food when the gizzard contracts. See also GASTRIC MILL.

gland A group of cells or a single cell in animals or plants that is specialized to secrete a specific substance. In animals there are two types of glands, both of which synthesize their secretions. *Endocrine glands discharge their products directly into the blood vessels; *exocrine glands secrete through a duct or network of ducts into a body cavity or onto the body surface. Secretory cells are characterized by having droplets (**vesicles**) containing their products. See also RECTAL GLAND; SECRETION.

In plants glands are specialized to secrete certain substances produced by the plant. The secretions may be retained within a single cell, secreted into a special cavity or duct, or secreted to the outside. Examples are the water glands (*hydathodes) of certain leaves, nectaries (see NECTAR), and the digestive glands of certain carnivorous plants. See also SALT GLAND (sense 2).

Glaucophyta A small phylum of single-celled freshwater algae whose primitive chloroplasts (**cyanelles**) resemble cyanobacteria. They are widely regarded as descendants of the most ancient lineage of photosynthetic eukaryotes, which evolved soon after the emergence of the ancestral primary photosynthetic eukaryote (the primary endosymbiont) through acquisition of a cyanobacterium as a permanent cell organelle (see ENDOSYMBIONT THEORY).

Notably, the cyanelles retain a trace layer of *peptidoglycan, the characteristic cell wall material of bacteria, between the inner and outer membranes. Like cyanobacteria and red algae, glaucophyte plastids contain *phycobiliproteins as accessory pigments, and the thylakoids are not stacked, in contrast to their stacked arrangement in green algae and plants. Some members of the three genera are equipped with flagella and are motile. Glaucophytes are members of the supergroup Archaeplastida (*see* PLANT sense 2).

glenoid cavity The socket-shaped cavity in the *scapula (shoulder blade) that holds the head of the *humerus in a ball-and-socket joint.

glia (glial cells; neuroglia) Cells of the nervous system that support the neurons. There are four classes of glial cells: **astrocytes** are star-shaped cells that contribute to the *blood-brain barrier by surrounding the smallest blood vessels in the brain and modulating the endothelial cells of the vessels. This barrier restricts the passage of various potentially noxious substances from the blood to neurons in the central nervous system. Each astrocyte makes contact with numerous synapses, typically over 100 000, and modulates their activity; the astrocytes can themselves take up and release neurotransmitter and thereby regulate communication between neurons. They have stores of glycogen with which they supply energy to the neurons, and they also take part in the repair and regeneration of neurons. **Oligodendrocytes** wrap around the axons of neurons in the brain and spinal cord, covering them with layers of insulating plasma membrane, similar to the role of *Schwann cells, their glia counterparts in the peripheral nervous system. This prevents impulses from 'short-circuiting' between adjacent neurons. **Ependymal cells** line the ventricles of the brain and central canal of the spinal cord, where the beating of their hairlike cilia helps circulate the cerebrospinal fluid through these cavities. They also give rise to the epithelial layer surrounding the *choroid plexus. **Microglia** are the main immune defence cells of the nervous system, acting as *macrophages and able to mount inflammatory responses.

global hectare *See* ECOLOGICAL FOOTPRINT.

global warming *See* CLIMATE CHANGE; GREENHOUSE EFFECT.

globin *See* HAEMOGLOBIN.

globular protein *See* PROTEIN.

globulin Any of a group of globular proteins that are generally insoluble in water and present in blood, eggs, milk, and as a reserve protein in seeds. Blood serum globulins comprise four types: α_1-, α_2-, and β-globulins, which serve as carrier proteins; and γ-globulins (**gamma globulins**), which include the *immunoglobulins responsible for immune responses.

Glomeromycota A phylum of fungi that live as obligate symbionts in association with plant roots, forming endotrophic (or arbuscular) *mycorrhizas. The hyphae lack cross-walls, and they reproduce asexually by producing spores.

glomerular filtrate The fluid in the lumen of the Bowman's capsule of the *nephron that has been filtered from the capillaries of the glomerulus (*see* ULTRAFILTRATION). The glomerular filtrate has the same composition as the plasma except that it does not contain any of the larger components, such as plasma proteins or cells.

glomerular filtration rate (GFR) The volume of fluid (*see* GLOMERULAR FILTRATE) that is filtered from the capillaries of the glomeruli into the kidney tubules per unit time. It is influenced chiefly by blood pressure and by changes in the diameter of arterioles in the glomeruli. In humans average GFR values vary with age and range from about 116 ml/min/ 1.73 m^2 in people aged 20–29 to 60 in those aged 70+. Some 99% of this filtered volume is reabsorbed as it passes along the kidney tubules to form urine. The GFR is usually estimated by measuring the *creatine concentration in a blood sample.

This is then combined with other factors such as age, gender, height, weight, and ethnicity to give an estimate of GFR.

glomerulus (*pl.* **glomeruli**) **1.** A tangled mass of blood capillaries enclosed by the cup-shaped end (*Bowman's capsule) of a kidney tubule (*see* NEPHRON). Fluid is filtered from these capillaries into the Bowman's capsule and down the nephron (*see* GLOMERULAR FILTRATE). **2.** A globular cluster of axons and dendrites found in the olfactory bulb and concerned with relaying signals from the olfactory epithelium (in the nose) to the brain. Each glomerulus receives inputs from numerous olfactory receptors of just one type and combines these signals to send an output via mitral cells to the piriform (olfactory) cortex in the brain.

glottis (*pl.* **glottides**) The opening from the pharynx to the trachea (windpipe). In mammals it also serves as the space for the *vocal cords. *See also* EPIGLOTTIS; LARYNX.

glucagon A hormone, secreted by the α (or A) cells of the *islets of Langerhans in the pancreas, that increases the concentration of glucose in the blood by stimulating the metabolic breakdown of glycogen in the liver. It thus antagonizes the effects of *insulin (*see* ANTAGONISM) and helps maintain normal levels of blood glucose.

glucan Any *polysaccharide composed only of glucose residues, e.g. starch and glycogen.

glucocorticoid *See* CORTICOSTEROID.

gluconeogenesis The synthesis of glucose from noncarbohydrate sources, such as fat and protein. This occurs when the glycogen supplies in the liver are exhausted. The pathway is essentially a reversal of *glycolysis from pyruvate to glucose and it can utilize many sources, including amino acids, glycerol, and *Krebs cycle intermediates. Large-scale protein and fat catabolism normally occurs only in those suffering from starvation or certain endocrine disorders.

gluconic acid An optically active hydroxy-carboxylic acid, $CH_2(OH)(CHOH)_4COOH$. It is the carboxylic acid corresponding to the aldose sugar glucose, and can be made by the action of certain moulds.

glucosamine *See* AMINO SUGAR.

glucose (dextrose; grape sugar) A white crystalline sugar, $C_6H_{12}O_6$, occurring widely in nature. Like other *monosaccharides, glucose is optically active: most naturally occurring glucose is dextrorotatory. Glucose and its derivatives are crucially important in the energy metabolism of living organisms. It is a major energy source, being transported around the body in blood, lymph, and cerebrospinal fluid to the cells, where energy is released in the process of *glycolysis. Glucose is present in the sap of plants, in fruits, and in honey and is also a constituent of many polysaccharides, most notably of starch and cellulose. These yield glucose when broken down, for example by enzymes during digestion. Maintenance of a constant glucose concentration in the blood, primarily through the actions of the hormones *insulin and *glucagon, is vital for cell metabolism. In particular, cells of the nervous system depend almost exclusively on glucose as an energy source, whereas most other cells can switch to fatty acids when levels of glucose fall.

glucuronic acid A compound, $OC_6H_9O_6$, derived from the oxidation of glucose. It is an important constituent of *gums and *mucilages. Glucuronic acid can combine with hydroxyl (–OH), carboxyl (–COOH), or amino (–NH_2) groups to form a **glucuronide**. The addition of a glucuronide group to a molecule (**glucuronidation**) generally increases the solubility of a compound; hence glucuronidation plays an important role in the excretion of foreign substances (*see* PHASE II METABOLISM).

glucuronide *See* GLUCURONIC ACID.

glume Either member of a pair of bracts that are found at the base of a spikelet (*see* SPIKE) of the grasses. *Compare* LEMMA.

glutamate The anion of the amino acid glutamic acid. It functions as a neurotransmitter at excitatory synapses in the vertebrate central nervous system and at excitatory neuromuscular junctions in insects and crustaceans. In plants, glutamate is the initial acceptor molecule in the assimilation of the ammonium ion (NH_4^+), combining with it to form glutamine in a reaction involving ATP. *See also* GLUTAMATE RECEPTOR.

glutamate receptor (GluR) Any receptor protein that binds the neurotransmitter *glutamate as a ligand. Glutamate receptors fall into two main types: **ionotropic glutamate receptors** (iGluRs), which are *ligand-gated ion channels, responsible for fast excitatory transmission; and **metabotropic glutamate receptors** (mGluRs), which work via *G proteins and cause slower longer-lasting effects in postsynaptic cells. Ionotropic glutamate receptors are further subdivided into three classes, each named according to their sensitivity to specific agonists: **NMDA receptors** (named after *N*-methyl-D-aspártate), **AMPA receptors** (named after α-amino-3-hydroxy-5-methylisoxazole-4-propionic acid), and **kainate receptors** (*see* KAINATE). Unusually, NMDA receptors must bind not only glutamate but also glycine, and the membrane must be depolarized, before the ion channel will open. Glutamate receptors mediate changes in the strength of synapses that form the basis of learning and memory. *See* SYNAPTIC PLASTICITY.

glutamic acid *See* AMINO ACID; GLUTAMATE.

glutamine *See* AMINO ACID.

glutathione (GSH) A *peptide comprising the amino acids glutamic acid, cysteine, and glycine. It occurs widely in plants, animals, and microorganisms, serving chiefly as an *antioxidant. Reduced glutathione reacts with potentially harmful oxidizing agents and is itself oxidized, whereupon it is returned to its reduced state by the enzyme glutathione reductase. This is important in ensuring the proper functioning of proteins, haemoglobin, membrane lipids, etc. Glutathione is also involved in amino acid transport across plasma membranes.

gluten A mixture of two proteins, gliadin and glutenin, occurring in the endosperm of wheat grain. Their amino acid composition varies but glutamic acid (33%) and proline (12%) predominate. The composition of wheat glutens determines the 'strength' of the flour and whether or not it is suitable for biscuit or bread making. Sensitivity of the lining of the intestine to gluten occurs in **coeliac disease**, a condition that must be treated by a gluten-free diet.

glycaemic index (GI) An indication of how eating a particular food will affect the concentration of glucose in the blood. GI is determined relative to the effects of a standard test meal of (usually) 50 g of glucose in a fasting individual, which is given the value 100. Foods with a high GI (typically >85), such as white bread, white rice, and potatoes, cause blood glucose to rise rapidly following ingestion, whereas the increase is lower and more prolonged in foods with a low GI (<60), such as apples, beans, and yogurt. The GI of a food is governed by various factors, such as the type of carbohydrate, the concentration and nature of the fibre, the proportions of protein and fat, and the methods of cooking or processing. Lower GI foods are recommended for endurance sports and to optimize control of blood glucose for diabetics.

glycan *See* POLYSACCHARIDE.

glyceraldehyde 3-phosphate (GALP) A triose phosphate, $CHOCH(OH)CH_2O-PO_3H_2$, that is an intermediate in the *Calvin–Bassham–Benson cycle (*see also* PHOTOSYNTHESIS) and glycolysis.

glycerate 3-phosphate A phosphorylated three-carbon monosaccharide that is an intermediate in the *Calvin–Bassham–Benson cycle of photosynthesis and also in *glycolysis. It was formerly known as **3-phosphoglycerate** or **phosphoglyceric acid (PGA)**.

glyceride (acylglycerol) A fatty-acid ester of glycerol. Esterification can occur at one, two, or all three hydroxyl groups of the glycerol molecule producing **monoglycerides**, **diglycerides**, and *triglycerides respectively. Triglycerides are the major constituent of fats and oils found in living organisms. Alternatively, one of the hydroxyl groups may be esterified with a phosphate group forming a phosphoglyceride (*see* PHOSPHOLIPID) or a sugar forming a *glycolipid.

glycerine *See* GLYCEROL.

glycerol (glycerine; propane-1,2,3,-triol) A trihydric alcohol, $HOCH_2CH(OH)CH_2OH$. Glycerol is a colourless sweet-tasting viscous liquid, miscible with water but insoluble in ether. It is widely distributed in all living organisms as a constituent of the *glycerides, which yield glycerol when hydrolysed. Glycerol itself is used as an *antifreeze molecule by certain organisms.

glycerophospholipid *See* PHOSPHO-LIPID.

glycine A sweet-tasting *amino acid that, besides being a component of proteins, is the main inhibitory neurotransmitter for fast synapses in the spinal cord of vertebrates. Glycine is also required for opening of NMDA-type *glutamate receptors.

glycobiology The study of carbohydrates and carbohydrate complexes, especially *glycoproteins.

glycocalyx 1. (cell coat) A layer of carbohydrate on the surface of the plasma membrane of most eukaryotic cells. It is made up of the oligosaccharide side-chains of the glycolipid and glycoprotein components of the membrane and may include oligosaccharides secreted by the cell. It plays a role in cell-cell adhesion and in regulating the exchange of materials between a cell and its environment. **2.** The outermost layer of a bacterium, typically consisting of numerous polysaccharides plus various glycoproteins. The glycocalyx varies in thickness and consistency: in some species it forms a flexible **slime layer** while in others it forms a rigid and relatively impermeable *capsule.

glycogen (animal starch) A *polysaccharide consisting of a highly branched polymer of glucose occurring in animal tissues, especially in liver and muscle cells. It is the major store of carbohydrate energy in animal cells. *See also* GLYCOGENESIS; GLYCOGENOLYSIS.

glycogenesis The conversion of glucose to glycogen, which is stimulated by insulin from the pancreas. Glycogenesis occurs in skeletal muscles and to a lesser extent in the liver. Glucose that is taken up by cells is phosphorylated to glucose 6-phosphate; this is converted successively to glucose 1-phosphate, uridine diphosphate glucose, and finally to glycogen. *See also* GLUCONEOGENESIS. *Compare* GLYCOGENOLYSIS.

glycogenolysis The conversion of glycogen to glucose, which occurs in the liver and is stimulated by glucagon from the pancreas and adrenaline from the adrenal medulla. These hormones activate an enzyme that phosphorylates glucose molecules in the glycogen chain to form glucose 1-phosphate, which is converted to glucose 6-phosphate. This is then converted to glucose by a phosphatase enzyme. In skeletal muscle glycogen is degraded to glucose 6-phosphate, which is then converted into pyruvate and used in ATP production during glycolysis and the Krebs cycle. However, pyruvate can also be converted, in the liver, to glucose; thus muscle glycogen is indirectly a source of blood glucose. *Compare* GLYCOGENESIS.

glycolate pathway *See* PHOTORESPIRATION.

glycolipid Any of a group of sugar-containing lipids, in which the lipid portion of the molecule is usually based on glycerol (*see* GLYCERIDE) or sphingosine and the sugar is typically galactose, glucose, or inositol. Glycolipids are components of biological membranes. In animal plasma membranes they are found in the outer layer of the lipid bilayer; the simplest animal glycolipids are the *cerebrosides. Plant glycolipids are glycerides in which the sugar group is most commonly galactose. They are the principal lipid constituents of chloroplasts.

glycolysis (Embden-Meyerhof pathway)
The series of biochemical reactions in which glucose is broken down to pyruvate with the release of usable energy in the form of *ATP. One molecule of glucose undergoes two substrate-level *phosphorylation reactions and is then split to form two triose-phosphate molecules. Each of these is converted to pyruvate. The net energy yield is two ATP molecules per glucose molecule. In cellular *respiration pyruvate then enters the *Krebs cycle and undergoes partial or complete oxidation, ultimately driving an *electron transport chain to generate more ATP. Alternatively, the pyruvate is converted to lactic acid or alcohol by *fermentation. Other simple sugars, e.g. fructose and galactose, and glycerol (from fats) enter the glycolysis pathway at intermediate stages. *Compare* GLUCONEOGENESIS.

glycomics The study of the entire set of oligosaccharides (i.e. the **glycome**) synthesized by an organism. Many proteins undergo addition of short chains of sugar residues as part of their post-translational processing. This glycosylation can involve different oligosaccharides at various sites on proteins—the totality of which is the **proteoglycome**; it is crucial to effective protein functioning in many areas of metabolism, including cell signalling and cell recognition. The function of glycolipids is also influenced by their component oligosaccharide chains. Glycomics seeks to develop automated high-throughput techniques for identifying and quantifying these oligosaccharides and methods of analysing and storing such data. These data can then be integrated with observed patterns of gene expression (from *genomics) and protein synthesis (from *proteomics) to yield information about how and why the glycome varies during, for example, development or disease.

glycoprotein A carbohydrate linked covalently to a protein. Formed in the Golgi apparatus in the process of *glycosylation, glycoproteins are important components of plasma membranes, in which they extend throughout the *lipid bilayer. They are also constituents of body fluids, such as mucus, that are involved in lubrication. Many of the hormone receptors on the surfaces of cells are glycoproteins. Glycoproteins produced by viruses attach themselves to the surface of the host cell, where they act as markers for the receptors of leucocytes. Viral glycoproteins can also act as target molecules and help viruses to detect certain types of host cell; for example, a glycoprotein on the surface of

The principal stages of glycolysis

*HIV enables the virus to find and infect white blood cells.

glycosaminoglycan Any one of a group of polysaccharides that contain *amino sugars (such as glucosamine). Formerly known as **mucopolysaccharides**, they include *hyaluronic acid and chondroitin (*see* CARTILAGE), which provide lubrication in joints and form part of the matrix of cartilage. The three-dimensional structure of these molecules enables them to trap water, which forms a gel and gives glycosaminoglycans their elastic properties.

glycoside Any one of a group of compounds consisting of a pyranose sugar residue, such as glucose, linked to a noncarbohydrate residue (R) by a *glycosidic bond: the hydroxyl group (-OH) on carbon-1 of the sugar is replaced by -OR. Glycosides are widely distributed in plants; examples are the *anthocyanin pigments and the **cardiac glycosides**, such as digoxin (*see* DIGITALIS) and ouabain, which are used medicinally for their stimulant effects on the heart.

glycosidic bond (glycosidic link) The type of chemical linkage between the monosaccharide units of disaccharides, oligosaccharides, and polysaccharides, which is formed by the removal of a molecule of water (i.e. a *condensation reaction). The bond is normally formed between the carbon-1 on one sugar and the carbon-4 on the other (see illustration). An α-glycosidic bond is formed when the -OH group on carbon-1 is below the plane of the glucose ring and a β-glycosidic bond is formed when it is above the plane.

*Cellulose is formed of glucose molecules linked by 1-4 β-glycosidic bonds, whereas starch is composed of 1-4 α-glycosidic bonds.

glycosuria *See* INSULIN.

glycosylation The process in which a carbohydrate is joined to another molecule, such as a protein to form a *glycoprotein or to a lipid to form a *glycolipid. Glycosylation occurs in the rough endoplasmic reticulum and the Golgi apparatus of cells.

glyoxylate cycle A metabolic pathway in plants, fungi, protists, and bacteria that is a modified form of the *Krebs cycle. It utilizes fats as a source of carbon and enables the synthesis of carbohydrate from fatty acids by avoiding the stages of the Krebs cycle in which carbon dioxide is released. It occurs in tissues rich in fats, such as those of germinating seeds; the enzymes involved in the cycle, which have not been found in animals (except in nematodes embryos, and certain other animal tissues), are contained in organelles (*microbodies) called **glyoxysomes**. The glyoxylate cycle differs from the Krebs cycle in that it utilizes two molecules of acetyl coenzyme A (rather than one), which are derived from *fattyacid oxidation. Isocitrate is converted to succinate (from which glucose can be synthesized in *gluconeogenesis) and glyoxylate (see illustration).

glyoxysome *See* GLYOXYLATE CYCLE.

glyphosate *N*-(phosphonomethyl)glycine: a herbicide, marketed as Roundup, that kills a wide range of plants but shows little

Formation of a glycosidic bond

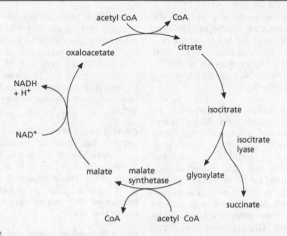

Glyoxylate cycle

persistence in soil and has low toxicity to animals. If applied to the leaves it is rapidly translocated to the rest of the plant, and hence can penetrate the roots of even hardy perennials. It works by blocking the synthesis of aromatic amino acids, so that treated plants are unable to manufacture proteins and other key metabolites. Glyphosate inhibits the activity of 5-enolpyruvyl-shikimate-3-phosphate synthase (EPSPS), a key enzyme in the *shikimic acid pathway, which occurs only in plants and microorganisms. Certain crops, notably soya bean, have been genetically engineered to give them resistance to glyphosate, by inserting genes for an EPSPS enzyme from *Agrobacterium*. These 'Roundup-ready' crops, which can be sprayed with the herbicide without being affected, are now widely grown in North America and elsewhere.

GM-CSF *See* COLONY-STIMULATING FACTOR.

Gnathostomata A subphylum or superclass of chordates consisting of all vertebrates that possess jaws. It contains six extant classes: *Chondrichthyes (cartilaginous fishes), *Osteichthyes (bony fishes), *Amphibia, *Reptilia, *Aves (birds), and *Mammalia. *Compare* AGNATHAN.

Gnetophyta *See* GYMNOSPERM.

gnotobiotic Designating germ-free conditions, especially those in which experimental animals are inoculated with known strains of microorganisms.

goblet cell A goblet-shaped cell, found in the epithelium of the intestine and respiratory system in mammals and in the epidermis of fish, that secretes *mucus. Goblet cells have a wide top and constricted base and possess glycoprotein-containing vesicles.

goitre Enlargement of the thyroid gland resulting in a swelling of the neck. This may be caused by lack of iodine in the diet (which is required for the production of thyroid hormones), as the gland enlarges to compensate for this deficiency; or by *hyperthyroidism.

Golgi, Camillo (1843–1926) Italian cytologist, who experimented with cells and tissues while working as a physician. He later became a professor at Pavia University. He devised a method of staining cells using silver salts, which enabled him to study nerve cells, and identified a type of nerve cell in the brain (later called **Golgi cells**) that made connections with many other nerve cells. This led to the establishment, by the Spanish histologist Santiago Ramón y Cajal (1852–1934), of the *neuron

g

as the impulse-conducting unit of the nervous system. Golgi is also remembered for his discovery of the cell organelle now called the *Golgi apparatus. For his work on the structure of the nervous system he shared the 1906 Nobel Prize for physiology or medicine with Ramón y Cajal.

Golgi apparatus (Golgi complex) An assembly of vesicles and folded membranes within the cytoplasm of eukaryotic *cells that forms part of the *endomembrane system. It modifies proteins and packages them and other materials (e.g. polysaccharides) for delivery to the plasma membrane for secretion or to destinations within the cell. Proteins arrive in vesicles following their assembly in the *endoplasmic reticulum (ER), are processed within a membranous compartment, or cisterna, and ultimately sorted into **Golgi vesicles** for secretion, storage, or transport to lysosomes. As it 'matures' each cisterna gradually moves through the stack. The region of the stack nearest to the ER is called the *cis* **Golgi**, the middle zone is the **medial Golgi**, and the area where Golgi vesicles are released is the *trans* **Golgi**. Plant cells usually contain smaller arrays of Golgi-type vesicles, called *dictyosomes. Some macromolecules are manufactured in the Golgi, such as certain noncellulose polysaccharides in plant cells. The apparatus is named after its discoverer, Camillo Golgi.

(((●))) SEE WEB LINKS

• A succinct account of the Golgi apparatus from a site hosted by Florida State University

Golgi tendon organ A receptor organ (*see* PROPRIOCEPTOR) located at the junction between a skeletal muscle and its tendon that detects contraction of the muscle. Named after Camillo Golgi, it consists of an encapsulated bundle of collagen fibres that penetrate the muscle at one end and the tendon at the other. The organ is supplied by an afferent neuron, whose sensory endings (dendrites) wrap around the collagen fibres. Contraction of the muscle causes stretching of the collagen fibres, which stimulates the sensory neuron. This relays

signals via interneurons in the central nervous system, causing the inhibition of motor neurons supplying the same muscle and hence its relaxation. These signals also lead to the excitation of antagonistic (opposing) muscles. This **inverse myotatic response** protects muscles from tearing or detachment due to excessive contraction forces.

gonad Any of the usually paired organs in animals that produce reproductive cells (gametes). The most important gonads are the male *testis, which produces spermatozoa, and the female *ovary, which produces ova (egg cells). The gonads also produce hormones that control secondary sexual characteristics.

gonadotrophin (gonadotrophic hormone) Any of several hormones, secreted by the mammalian anterior *pituitary gland, that stimulate reproductive activity of the testes or ovaries (the gonads). Pituitary gonadotrophins include *follicle-stimulating hormone and *luteinizing hormone. **Chorionic gonadotrophin** is a hormone produced by the placenta of higher mammals that maintains the *corpus luteum. The presence of large amounts of **human chorionic gonadotrophin (hCG)** in the urine of women is an indication of pregnancy.

Gondwanaland *See* CONTINENTAL DRIFT.

gonorrhoea *See* SEXUALLY TRANSMITTED DISEASE.

G protein Any one of a group of proteins that relay signals in animal cells (*see* SIGNAL TRANSDUCTION). They occur on the inner surface of the plasma membrane and transmit signals from receptors on the outer surface of the cell to intracellular components. G proteins are activated by binding to GTP and become inactive when they bind GDP (*see* GUANOSINE). **Large G proteins** comprise three dissimilar subunits (denoted α, β, and γ: i.e. they are heterotrimeric), and numerous Gα subunits are known. They associate with various types of cell surface receptors and transmit signals to different intracellular pathways. A conformational change in the receptor caused by binding of its ligand enables the G protein to bind to the intracellular part of the receptor. This displaces the molecule of GDP from the G protein and allows GTP to take its place, thereby activating the G protein. The subunits dissociate and interact with other cellular components to trigger the signal pathways. Immediate targets of activated G proteins include *adenylate cyclase and *phospholipase C, which give rise to the second messengers cyclic AMP, inositol trisphosphate (IP$_3$), and raised intracellular calcium ions. The α subunit has intrinsic GTPase activity and subsequently hydrolyses the GTP to GDP, permitting reassociation of the subunits and restoring the G protein to its inactive state. **Small G proteins** comprise a single polypeptide; the best characterized is *RAS protein. The cholera toxin exerts its effects by changing the G protein in the epithelial cells of the intestine so that it is continually activated, which causes an abnormal increase in cellular adenylate cyclase levels. One consequence of this is that sodium ions are actively pumped into the intestine, causing water to follow by osmosis: the result is diarrhoea and dehydration.

G-protein-coupled receptor (GPCR) Any of a superfamily of proteins that are located in cell membranes and act as *metabotropic receptors, relaying signals from the exterior to the interior of the cell via associated G proteins. The latter then activate second messengers, which regulate the activity of the cell. GPCRs function in vision, hearing, smell, and embryonic development, among many other roles, and their malfunction is involved in various diseases. GPCRs are characterized by having seven transmembrane domains, connected by intracellular and extracellular loops, an extracellular amino terminus, and an intracellular carboxy terminus. The second and third intracellular loops and part of the carboxy tail form the binding domain for the G protein. A wide range of substances, including peptide hormones, acetylcholine, adrenaline, noradrenaline, gamma-aminobutyric acid, and glutamate, can bind as ligands to the various types of

GPCR, at binding sites in the extracellular region of the receptor molecule. Ligand binding changes the shape of the receptor allowing the G protein to bind to it, which activates the G protein by causing GTP to replace bound GDP.

Graafian follicle (ovarian follicle) The fluid-filled cavity that surrounds and protects the developing egg cell (oocyte) in the ovary of a mammal. The oocyte is surrounded by *granulosa cells, which secrete hormones that regulate development of the follicle. After the release of the ovum the follicle develops into a *corpus luteum. It is named after the Dutch anatomist Reinier de Graaf (1641–73).

grade A group of organisms that share certain morphological characteristics but are not necessarily related in evolutionary terms. For example, all organisms that possess a coelom can be regarded as a grade. A grade can be contrasted with a *clade, which represents a group comprising an ancestral species and all its descendants.

graded potential A change in the membrane potential of a cell from its *resting potential that varies with the strength of a stimulus reaching the cell. It can result from chemical or physical factors that affect the transfer of ions into and out of the cell via ion channels. Such movements may cause the membrane potential to become more negative (hyperpolarization) or less negative (depolarization). The magnitude of the change reflects the net size of the stimulus and is thus a way of integrating positive and negative inputs. This integration is particularly important in nervous systems, when nerve cells (neurons) receive numerous excitatory and inhibitory signals from other neurons; it determines how the neuron will respond. Graded potentials usually give rise to very small electric currents, which travel short distances from the cell and decay rapidly. But in neurons, if the depolarization reaches a certain threshold, it triggers an *action potential, which travels along the axon as a signal to other excitatory cells. Sensory receptor cells, such as occur in the eye, ear, and other sense organs, typically respond to stimuli by producing a graded potential called a **receptor potential**. This may trigger an action potential in the receptor cell itself, or in a sensory neuron, or change the rate at which the sensory neuron generates spontaneous action potentials.

graft An isolated portion of living tissue that is joined to another tissue, either in the same or a different organism, the consequent growth resulting in fusion of the tissues. (The word is also used for the process of joining the tissues.)

Grafting of plant tissues is a horticultural practice used to propagate plants, especially certain bushes and fruit trees, artificially. A shoot or bud of the desired variety (the **scion**) is grafted onto a rootstock of either a common or a wild related species (the **stock**). This procedure thus combines the desirable characteristics of the scion (e.g. flower form or fruit yield) with those of the stock (e.g. disease or drought resistance), and the scion supplies the stock with food made by photosynthesis. The stock supplies the scion with water and mineral salts and affects only the size and vigour of the scion.

Animal and human grafts are used to replace faulty or damaged parts of the body. An **autograft** is taken from one part of the body and transferred to another part of the same individual, e.g. a skin graft used for severe burns. An **allograft** (**homograft**) is taken from one individual (the **donor**) and implanted in another of the same species (the **recipient**), the process being known as **transplantation**, e.g. a heart or kidney transplant. In such cases the graft may be regarded by the body as foreign (a state of **incompatibility**): an immune response follows and the graft is rejected (*see also* HISTOCOMPATIBILITY).

graft hybrid A type of plant *chimaera that may be produced when a part of one plant (the scion) is grafted onto another plant of a different genetic constitution (the stock). Shoots growing from the point of union of the graft contain tissues from both the stock and the scion.

gram Symbol g. One thousandth of a kilogram. The gram is the fundamental unit of mass in *c.g.s. units and was formerly used in such units as the **gram-atom**, **gram-molecule**, and **gram-equivalent**, which have now been replaced by the *mole.

Gram's stain A staining method used to differentiate bacteria. The bacterial sample is smeared on a microscope slide, stained with a violet dye, treated with acetone-alcohol (a decolourizer), and finally counterstained with a red dye. **Gram-positive** bacteria retain the first dye, appearing blue-black under the microscope; such bacteria have a thick layer of *peptidoglycan in their cell walls. In **Gram-negative** bacteria, the acetone-alcohol washes out the violet dye and the counterstain is taken up, the cells appearing red. The cell walls of these bacteria have an outer layer of lipoprotein overlying a thin layer of peptidoglycan. The more complex cell walls of Gram-negative bacteria can often increase their virulence, giving them greater potential to cause fever or shock in their hosts and more resistance to host defences and to antibiotics. The stain is named after the Danish bacteriologist H. C. J. Gram (1853–1938), who first described the technique (since modified) in 1884.

granulocyte Any white blood cell (*see* LEUCOCYTE) that contains granular material (**secretory vessels**) and *lysosomes in its cytoplasm. *Neutrophils, *basophils, and *eosinophils are examples of granulocytes. *Compare* AGRANULOCYTE.

granulocyte/macrophage colony-stimulating factor (**GM-CSF**) *See* COLONY-STIMULATING FACTOR.

granulosa cells (**granular cells**) A layer of secretory cells in a *Graafian follicle that surround the oocyte. They have a function analogous to that of the *Sertoli cells in the testis, supplying nutrients to the developing oocyte. After ovulation they form the *corpus luteum. The granulosa cells, in conjunction with the outer layer of thecal cells, play a key role in determining which of the several developing follicles reaches maturity and undergoes ovulation in each *ovarian cycle. Testosterone from the thecal cells diffuses into the granulosa cells, where it is converted to oestradiol. This steep rise in oestradiol, along with *follicle-stimulating hormone (FSH), promotes the growth and maturation of the granulosa cells, and also their expression of receptors for FSH and for luteinizing hormone (LH). It is the follicle with the most FSH and LH receptors that matures and ovulates, while other follicles regress. The granulosa cells also secrete the hormone *inhibin, which acts as negative feedback mechanism to regulate FSH levels.

granum (*pl.* **grana**) A stack of platelike bodies (*thylakoids), many of which are found in plant *chloroplasts (each chloroplast contains about 50 grana). Grana bear the light-receptive pigment chlorophyll and contain the enzymes responsible for the light-dependent reactions of *photosynthesis.

granzymes See PERFORIN.

grape sugar See GLUCOSE.

graptolites A group of extinct marine colonial animals that were common in the Palaeozoic era. Graptolites are generally regarded as being related to colonial hemichordates known as pterobranchs (*see* HEMICHORDATA). They had chitinous outer skeletons in the form of simple or branched stems, the individual polyps occupying minute cups (**thecae**) along these stems. Fossils of these skeletons are found in Palaeozoic rocks of all continents; they are particularly abundant in Ordovician and Silurian rock strata, for which they are used as *index fossils. At the end of the Silurian many graptolites became extinct but a few groups continued into the early Carboniferous.

grass-green bacteria See CHLOROXYBACTERIA.

grassland A major terrestrial *biome in which the dominant plants are species of grass; the rainfall is insufficient to support extensive growth of trees, which are also suppressed by grazing animals. Tropical grassland (**savanna**), which covers much of Africa south of the Sahara, has widely spaced trees, such as acacias and baobabs, and supports large herds of grazing animals and their predators. Temperate grasslands, such as the **steppes** of Asia, the **prairies** of North America, and the **pampas** of South America, have few trees and are largely used for agriculture.

gravitropism See GEOTROPISM.

gray Symbol Gy. The derived SI unit of absorbed *dose of ionizing radiation (*see* RADIATION UNITS). It is named after the British radiobiologist L. H. Gray (1905-65).

grazing The consumption of vegetation, usually on *grassland, by animals, particularly cattle and sheep. Overgrazing can lead to *desertification.

green algae See CHLOROPHYTA.

green chemistry See ECOTOXICOLOGY.

green fluorescent protein (GFP) A naturally *fluorescent protein obtained from the jellyfish *Aequorea victoria* and used as a marker for identifying cells containing recombinant DNA or for localizing specific proteins in cells. It absorbs blue light and emits a green fluorescence, and hence the abundance and location of GFP in cells can be visualized microscopically under ultraviolet light. The GFP gene is used as a *reporter gene to pinpoint cells expressing recombinant DNA, by causing GFP to be synthesized and the cell to glow green. Moreover, GFP can tag a protein of interest so that its abundance and distribution within the cell can be assessed. This is achieved by introducing recombinant DNA in which the GFP gene is fused to the target protein gene under the control of its normal promoter. When this fusion gene is expressed, the resultant fusion protein can readily be visualized within the cell and its fate tracked over time. Several colour variants of GFP are now available.

green gland See ANTENNAL GLAND.

greenhouse effect An effect occurring in the atmosphere because of the presence of certain gases (**greenhouse gases**) that absorb infrared radiation. Light and ultraviolet radiation from the sun are able to penetrate the atmosphere and warm the earth's surface. This energy is re-radiated as infrared radiation, which, because of its longer wavelength, is absorbed by such substances as carbon dioxide. The greenhouse effect is a natural phenomenon, without which the earth's climate would be much more hostile to life. However, emissions of carbon dioxide from human activities (e.g. farming, industry, and transport) have increased markedly in the last 150 years or so. The overall effect is that the average temperature of the earth and its atmosphere is increasing (so-called **global warming**). The effect is similar to that occurring in a greenhouse, where light and long-wavelength ultraviolet radiation can pass through the glass into the greenhouse but the infrared radiation is absorbed by the glass and part of it is re-radiated into the greenhouse.

The greenhouse effect is seen as a major environmental hazard. Average increases in temperature are already changing weather patterns, affecting agricultural output (*see* CLIMATE CHANGE), and causing the polar ice caps to melt, with a corresponding rise in sea level. Carbon dioxide, emitted e.g. from coal-fired power stations and car exhausts, is the main greenhouse gas. Other contributory pollutants are nitrogen oxides, ozone, methane, and *chlorofluorocarbons. Many countries have now agreed targets to limit emissions of greenhouse gases, e.g. by switching to renewable energy sources. *See also* POLLUTION.

((⊕)) SEE WEB LINKS
• Summary of the greenhouse effect, from the Live Science website

grey crescent A crescent-shaped band of lightly pigmented cytoplasm visible on the surface of the fertilized eggs of certain amphibians, including frogs' eggs. It develops opposite the point of entry of the sperm, due to rotation of the outer (cortical) layer of cytoplasm relative to the inner cytoplasm. The grey crescent contains cytoplasmic factors (e.g. *catenins) that are essential for

subsequent embryonic development, and certain cells in the grey crescent move inward to form the dorsal lip of the *gastrula, the site of *Spemann's organizer.

grey matter Part of the tissue that makes up the central nervous system of vertebrates. It is brown-grey in colour, consisting largely of nerve *cell bodies, *synapses, and *dendrites. The grey matter is the site of coordination between nerves of the central nervous system. *Compare* WHITE MATTER.

grooming The actions of an animal of rearranging fur or feathers and cleaning the body surface by biting, scratching, licking, etc., which is important for removing parasites and spreading oils over the body surface. In many mammals, especially primates, grooming between individuals (**allogrooming**) has an important role in maintaining social cohesion.

ground meristem The meristem in plant shoots and roots, derived from the *apical meristem, that gives rise to the cortex and pith (the *ground tissues) in stems and the cortex and endodermis in roots.

ground substance The colourless transparent part of *extracellular matrix of connective tissue, in which various cells and fibres are embedded. It consists of *proteoglycans associated with water, which can make up to 90% of the ground substance.

ground tissues All the plant tissues formed by the *apical meristems except the epidermis and vascular tissue. The principal ground tissues are the *cortex, *pith, and primary *medullary rays, and they consist chiefly of *parenchyma. *See also* COLLENCHYMA; SCLERENCHYMA.

group selection A mechanism originally proposed to account for the evolution of *altruism in social groups of animals. It was suggested by the British ethologist V. C. Wynne-Edwards (1906–97) in 1962, and arose from his observations that individual animals often expose themselves to danger (for instance by warning of predators) or forgo reproduction (as with worker bees in a colony) for the greater good of the group as a whole. Hence, groups containing altruistic individuals would have some selective advantage over groups lacking such members. This conflicts with Darwinian orthodoxy, which views natural selection as operating strictly on individuals. More recently, group selection has been reformulated by the biologist E. O. Wilson (1929–) into **multi level selection** theory, which proposes that selection acts simultaneously on many levels, from cells through individuals to groups. However, proponents of the theory of *kin selection regard this as a more plausible explanation of apparently altruistic acts, and the controversy continues.

growth An increase in the dry weight or volume of an organism through cell division and cell enlargement. Growth may continue throughout the life of the organism, as occurs in woody plants (i.e. **indeterminate growth**), or it may cease at maturity, as in humans and other mammals (i.e. **determinate growth**). See also ALLOMETRIC GROWTH; EXPONENTIAL GROWTH.

growth cone The enlarged motile tip of an elongating axon (nerve fibre) that determines the direction of axon growth. It has numerous finger-like projections, called filopodia (see FILOPODIUM), which bear surface receptors. These sense specific signal molecules and other chemical cues in the immediate environment that guide elongation of the axon along its proper pathway so that the neuron connects to its target cell, thereby ensuring correct wiring of the nervous system. Directional growth is achieved by the forward migration of actin filaments inside the filopodia and the elongation of microtubules at the leading edge of the axon.

growth factor Any of various chemicals, particularly polypeptides, that have a variety of important roles in the stimulation of new cell growth and cell maintenance. They bind to the cell surface on receptors. Specific growth factors can cause new cell proliferation (*epidermal growth factor, *insulin-like growth factor, haemopoietic

growth factor—see HAEMOPOIETIC TISSUE) and the migration of cells (see FIBROBLAST GROWTH FACTOR). **Platelet-derived growth factor (PDGF)** is released by platelets in response to tissue damage. It stimulates the proliferation of fibroblasts, which contribute to wound healing. Some growth factors act in the embryonic stage of development; for example, nerve growth factor (see NEUROTROPHIN) stimulates the growth of axons and dendrites from developing sensory and sympathetic neurons. Some growth factors or their receptors are involved in the abnormal regulation of growth seen in cancer when produced in excessive amounts or permanently activated. See also BONE MORPHOGENETIC FACTOR; TRANSFORMING GROWTH FACTOR BETA.

growth hormone (GH; somatotrophin) A hormone, secreted by the mammalian pituitary gland, that stimulates protein synthesis and growth of the long bones in the legs and arms. It also promotes the breakdown and use of fats as an energy source, rather than glucose. Production of GH is greatest during early life, and it triggers the liver to secrete *insulin-like growth factor, which circulates in the blood and promotes the growth of bone and cartilage. GH secretion is controlled by the opposing actions of two hormones from the hypothalamus: **growth hormone releasing hormone (somatoliberin)**, which promotes its release; and *somatostatin, which inhibits it. It is released usually several hours after a meal, or following exercise, in response to falling blood glucose levels. By reducing uptake of glucose by muscles, GH causes blood glucose levels to rise; hence its action antagonizes that of insulin. Overproduction of **human growth hormone (hGH)** results in gigantism in childhood and *acromegaly in adults; underproduction results in dwarfism. **Bovine somatotrophin (BST)** has been used to increase milk and meat production in cattle. Commercially, growth hormone is obtained from bacterial cells containing recombinant DNA expressing the growth hormone gene.

growth ring (annual ring) Any of the rings that can be seen in a cross-section of

phloem
vascular cambium
cortex
pith
medullary ray
cork
cork cambium

autumn wood } growth ring year 3
spring wood

autumn wood } growth ring year 2
spring wood

autumn wood } growth ring year 1
spring wood

Transverse section through a three-year-old woody stem to show the growth rings

a woody stem (e.g. a tree trunk). It represents the *xylem formed in one year as a result of fluctuating activity of the vascular *cambium. In temperate climates pale soft **spring wood**, characterized by large xylem vessels, is formed in spring and early summer. Growth slows down in late summer and a darker dense **autumn wood** with smaller xylem vessels is formed (see illustration). The age of a tree can be determined by counting the rings. Under certain circumstances two or more growth rings may form in one year, giving rise to false annual rings.

growth substance *See* PLANT HORMONE.

GSH *See* GLUTATHIONE.

GTP (guanosine triphosphate) *See* GUANOSINE.

guanine A *purine derivative. It is one of the major component bases of *nucleotides and the nucleic acids *DNA and *RNA.

guanine-nucleotide exchange factor (GEF) *See* RAS PROTEIN.

guano An accumulation of the droppings of birds, bats, or seals, usually formed by a long-established colony of animals. It is rich in plant nutrients, and some deposits are extracted for use as fertilizer.

guanosine A nucleoside consisting of one guanine molecule linked to a D-ribose sugar molecule. The derived nucleotides, guanosine mono-, di-, and triphosphate (GMP, GDP, and GTP, respectively), participate in various metabolic reactions. *See also* CYCLIC GMP; G PROTEIN.

guanylate cyclase *See* CYCLIC GMP.

guard cell Either of a pair of cells that control opening and closing of a leaf pore (*stoma). Each is a sausage- or kidney-shaped cell whose wall varies in rigidity. The wall bordering the pore is thickened and rigid, whereas the outside wall is thin and extensible. As the paired cells absorb water they swell, and the increased turgor pressure causes the thin-walled region to bend outwards, pulling the nonextensible thicker wall with it and opening the pore. Loss of water has the reverse effect, resulting in shrinking of the guard cells and closure of the pore. The guard cells respond to various environmental cues, including sunlight, water availability, and levels of carbon dioxide in the leaf. Movement of water in and out of the cell is controlled by the transport of protons (H^+) and movement of potassium ions (K^+) and other ions into and out of the cell. Sunlight activates a proton pump in the guard cell membrane, which actively pumps H^+ out of the guard cell into the surrounding *apoplast. The resulting electrochemical gradient drives K^+ and other ions into the guard cells, raising their internal ionic concentration (i.e. making the water potential more negative) and causing water to follow by osmosis through pores (aquaporins) in the membrane. Hence the guard cells swell and the stomatal pore opens. In the absence of sunlight, the proton pumps become much less active and K^+ ions diffuse out of the guard cells, causing them to lose water osmotically and shrink. These changes can occur in just a few minutes. If the leaf is stressed due to lack of water, the mesophyll cells release *abscisic acid (ABA), which triggers the guard cells to close the stomata, even in the presence of sunlight, to conserve moisture within the leaf.

guidepost cell A differentiated neuron that influences the development of the axons of other neurons so that they connect ultimately with their target cells, thus ensuring that the nervous system is correctly 'wired'. The guidepost cells have specific signal molecules on their surface, such as nerve-cell adhesion molecules (N-CAMs), which interact with the *growth cone at the tip of the developing neuron and direct it along the correct route. In some cases, axons may form temporary connections (synapses) with guidepost cells while they await the formation or maturation of their target cells. When development is completed, the guideposts may die.

guild (in ecology) A group of species within a community that exploit the same resources in a similar way. Thus, for example, different species of snakes and lizards may belong to a guild by virtue of occupying the same types of underground shelters, and different species of seed-eating birds may constitute a guild. Any particular species can belong to multiple guilds—for example, one based on feeding habit and a second defined by nest-site choice—depending on whether or not it meets the functional criterion for membership.

gullet *See* OESOPHAGUS.

gum 1. Any of a variety of substances obtained from plants. Typically they are insoluble in organic solvents but form gelatinous or sticky solutions with water. Gum resins are mixtures of gums and natural resins. Gums are produced by the young xylem vessels of some plants (mainly trees) in response to wounding or pruning. The exudate hardens when it reaches the plant surface and thus provides a temporary protective seal while the cells below divide to form a permanent repair. Excessive gum formation is a symptom of some plant diseases. *See also* MUCILAGE. **2.** *See* GINGIVA.

gustatory receptors *See* TASTE BUD.

gut *See* ALIMENTARY CANAL.

gut-associated lymphoid tissue (GALT) Any of various lymphoid tissues and structures that protect the digestive tract against infectious organisms and other ingested antigens. GALT includes the palatine *tonsils, adenoids (pharyngeal tonsils), *appendix, *Peyer's patches, and other clustered or diffuse lymphoid cells found throughout the gut wall.

guttation *See* HYDATHODE.

gymnosperm Any plant whose ovules and the seeds into which they develop are borne unprotected, rather than enclosed in ovaries, as are those of the flowering plants (the term gymnosperm means naked seed). In traditional systems of classification such plants were classified as the Gymnospermae, a class of the Spermatophyta, but they are now divided into separate phyla: *Coniferophyta (conifers), *Cycadophyta (cycads), Ginkgophyta (ginkgo), and Gnetophyta (e.g. *Welwitschia). See also* PROGYMNOSPERMS.

gynandromorph An animal that possesses both male and female characteristics because it is genetically a *mosaic, i.e. some of its cells are genetically male and others are female. This phenomenon is found particularly in the insects but also appears in the birds and mammals; it is often due to the loss of an X chromosome in a *stem cell of a female (XX), so that all tissues derived from that cell are phenotypically male. *Compare* INTERSEX.

gynoecium (gynaecium) (*pl.* **gynoecia, gynaecia**) The female sex organs (*carpels) of a flower. *Compare* ANDROECIUM.

gyre A circular rotation of surface water currents in an ocean basin. Gyres are the result of prevailing winds, the earth's rotation, and deflection by land masses, and they occur in all the major oceans. In the northern hemisphere the gyres of the North Pacific and North Atlantic rotate in a clockwise direction, whereas in the southern hemisphere the direction of rotation is anticlockwise. These directional flows exchange warm water from the equator for cooler water from the poles and have a huge impact on the climate of adjoining land masses, as well as regulating temperature, salinity, and nutrients in the oceans.

habitat The place in which an organism lives, which is characterized by its physical features or by the dominant plant types. Freshwater habitats, for example, include streams, ponds, rivers, and lakes. *See also* MICROHABITAT. *Compare* ECOLOGICAL NICHE.

habituation 1. A simple type of learning consisting of a gradual waning in the response of an animal to a continuous or repeated stimulus that is not associated with *reinforcement. *Compare* DISHABITUATION. **2.** The condition of being psychologically, but not physically, dependent on a drug, with a desire to continue its use but not to increase the dosage.

Hadean The earliest eon in the history of the earth, from the time of the accretion of planetary material, around 4600 million years ago (mya), to the date of the oldest known rocks—and hence the beginning of the geological record—about 4000 mya. The young earth was probably a rocky planet with a hot interior and a moist surface with oceans of liquid water. No evidence of life has been found. The Hadean eon precedes the *Archaean eon.

Hadobacteria A group (in some classifications a phylum) of bacteria whose members are able to withstand extreme environmental conditions. For example, some *Deinococcus* species are highly resistant to radiation as well as to cold, dehydration, and vacuum. Another hadobacterium, *Thermus aquaticus*, is an inhabitant of hot springs; it is the source of the very heat-stable enzyme Taq polymerase, which is used in the *polymerase chain reaction.

haem (heme) An iron-containing molecule (see formula) that binds with proteins as a *cofactor or *prosthetic group to form the **haemoproteins**. These are *haemoglobin, *myoglobin, and the *cytochromes. Essentially, haem comprises a *porphyrin with its four nitrogen atoms holding the iron(II) atom as a chelate. This iron can reversibly bind oxygen (as in haemoglobin and myoglobin) or (as in the cytochromes) conduct electrons by conversion between the iron(II) and iron(III) series.

Haem

haemagglutination *See* AGGLUTINATION.

haematophagous Feeding on blood. For example, horseflies and certain leeches are haematophagous.

haematoxylin A compound used in its oxidized form (**haematein**) as a blue dye in optical microscopy, particularly for staining smears and sections of animal tissue. It stains nuclei blue and is frequently used with *eosin as a counterstain for cytoplasm. Haematoxylin requires a mordant, such as iron alum, which links the dye to the tissue. Different types of haematoxylin can be made up depending on the mordant used, the method of oxidation, and the pH.

Examples are **Delafield's haematoxylin** and **Ehrlich's haematoxylin**.

haemerythrin A red iron-containing *respiratory pigment that occurs in the blood of certain worms, especially sipunculids and priapulids, and in lamp shells (*see* BRACHIOPODA). The protein contains two iron atoms to which two oxygen atoms can bind reversibly.

haemocoel The main body cavity of arthropods and molluscs, which forms part of the open circulatory system (*see* CIRCULATION). It consists of a network of spaces (sinuses) that are filled with **haemolymph**, which performs the roles of both blood and tissue fluid. The haemocoel is an enlarged blastocoel (*see* BLASTULA), which greatly reduces the coelom (this is restricted to the cavities of the gonads and excretory organs). The haemocoel can also act as a *hydrostatic skeleton.

haemocyanin (hemocyanin) Any of a group of copper-containing respiratory proteins found in solution in the blood of certain arthropods (e.g. crabs and lobsters) and molluscs. Haemocyanins contain two copper atoms that reversibly bind oxygen, changing between the colourless deoxygenated form (CuI) and the blue oxygenated form (CuII). In some species, haemocyanin molecules form giant polymers with molecular weights of several million.

haemocyte (hemocyte) A multifunctional cell type that is a key component of the immune defences of most invertebrate animals. Haemocytes are found in the coelom and blood vessels of coelomate animals and perform a variety of functions. For example, in insect larvae they can engulf foreign particles by *phagocytosis, secrete opsonins (*see* OPSONIZATION) and other antimicrobial compounds to promote destruction of invading organisms, encapsulate larger pathogens, such as the eggs of parasitic wasps, and attack them by secreting *reactive oxygen species. Haemocytes also promote blood coagulation and repair of damaged tissues and serve various metabolic functions.

haemocytometer A device used to estimate the concentration of cells in blood samples. Typically a haemocytometer consists of a glass slide with a shallow depression divided into a fine-meshed grid; each of the grid compartments is about 0.05 mm square. The slide is placed under a microscope and the number of cells in each square of the grid can be counted, which enables the concentration of blood cells in the sample to be assessed. Haemocytometers are also used in estimating the number of microorganisms present in a sample of water or other fluid.

haemodynamics The study of the flow of blood through the circulatory system (*see* CIRCULATION). Blood flow is influenced by numerous factors, depending on the type of system, for example whether it is open or closed, or single or double. In a closed system, such as that of mammals, flow velocity is determined primarily by the pumping action of the heart and the total cross-sectional area of the vessels through which the blood is being pumped. Hence, blood flows fastest through the aorta and pulmonary arteries, points in the circulation with relatively small overall cross-sectional areas, and slowest through the capillaries, which have the greatest combined cross-sectional area. Other factors affecting blood flow include blood volume and viscosity and the elasticity of blood vessels.

haemoglobin One of a group of globular proteins occurring widely in animals as oxygen carriers in blood. Vertebrate haemoglobin is contained in the red blood cells (*erythrocytes). It comprises two pairs of polypeptide chains, known as α-chains and β-chains (forming the **globin** protein), with each chain folded to provide a binding site for a *haem group. Each of the four haem groups binds one oxygen molecule to form **oxyhaemoglobin**. Dissociation occurs in oxygen-depleted tissues: oxygen is released and haemoglobin is reformed (*see* BOHR EFFECT; HAEMOGLOBINIC ACID; OXYGEN DISSOCIATION CURVE). The haem groups also bind other inorganic molecules, including carbon monoxide (to form *carboxyhaemoglobin). Carbon dioxide can bind to terminal $-NH_2$ groups of the α- and β-chains, forming carbamino groups ($-NHCOO^-$), in which form a significant fraction of carbon dioxide is transported in the red blood cells.

haemoglobinic acid A very weak acid formed inside red blood cells when hydrogen ions combine with haemoglobin. The presence of the hydrogen ions, which are produced by the dissociation of carbonic acid (*see* CARBONIC ANHYDRASE), encourages oxyhaemoglobin to dissociate into haemoglobin and oxygen (*see* BOHR EFFECT). The oxygen diffuses into the tissue cells and the haemoglobin acts as a *buffer for the excess hydrogen ions, which it takes up to form haemoglobinic acid.

haemoglobin S An abnormal form of haemoglobin produced in sickle-cell disease. *See* POLYMORPHISM.

haemolymph *See* HAEMOCOEL.

haemolysis The breakdown of red blood cells. It can be due to the action of disease-causing microorganisms, poisons, antibodies in mismatched blood transfusions, or certain allergic reactions. It produces anaemia.

haemophilia A hereditary sex-linked disease (*see* SEX LINKAGE) in which there is a deficiency or defect of certain clotting factors, causing the blood to clot very slowly. The most common form is haemophilia A, which involves a deficiency of Factor VIII, whereas the much rarer haemophilia B is caused by deficiency of Factor IX. There may be prolonged bleeding following injury and, in severe cases, spontaneous bleeding into the joints and muscles. The disorder is due to a defective recessive allele of the respective gene, both of which in humans are located on the X chromosome. Female carriers of a defective allele are unaffected, whereas all males who inherit a defective allele exhibit the disease. Treatment is by infusion of the deficient clotting factor, obtained either from human donated blood plasma or recombinant tissue culture.

haemopoietic tissue The tissue that gives rise to blood cells in the process of **haemopoiesis**. The haemopoietic tissue of the embryo and fetal stage of vertebrates is the bone marrow, lymph nodes, yolk sac, liver, spleen, and thymus but after birth haemopoiesis occurs in the red bone marrow (*see* MYELOID TISSUE). The different types of *stem cells in haemopoietic tissue that give rise to erythrocytes and leucocytes are all originally derived from **haemopoietic stem cells** (or **haemocytoblasts**). The formation of the different types of blood cell is under the control of **haemopoietic growth factors**, which include hormones and *cytokines. *See also* ERYTHROPOIESIS.

haemostasis The prevention of blood loss following rupture of blood vessels, which is effected by several physiological processes. Initially, bleeding is restricted by constriction of the damaged vessels, whose endothelial surfaces also stick together. The damage to the vessel endothelium exposes collagen, which attracts *platelets to the site. These become sticky and release the vasoconstrictor *serotonin and arachidonic acid, which is converted to *thromboxane A_2. This attracts other platelets to the site, so that a plug forms. Neighbouring undamaged areas continue to release *prostacyclin, which inhibits platelet aggregation and so prevents unnecessary enlargement of the plug. The various stages of *blood clotting then ensue.

hagfish *See* CYCLOSTOMATA.

hair 1. A multicellular threadlike structure, consisting of many dead keratinized cells, that is produced by the epidermis in mammalian *skin. The section of a hair below the skin surface (the **root**) is contained within a *hair follicle, the base of which produces the hair cells. Hair assists in maintaining body temperature by reducing heat loss from the skin. Bristles and whiskers are specialized types of hair. **2.** Any of various threadlike structures on plants, such as a *trichome.

hair cell A cell that is equipped with hairlike cilia and specialized for detecting movement of the surrounding medium, whether water or air; hence, it transduces mechanical stimuli into electrical stimuli, in the form of nerve impulses in associated sensory nerves. Hair cells occur in various types of vertebrate sensory organs, including the mammalian ear and the *lateral-line system of fishes and amphibians. A characteristic feature is the array of cilia projecting from the cell's apical surface. These typically include a single long *kinocilium and 20–300 much shorter *stereocilia. The shafts of the cilia are usually embedded in an accessory structure, such as the cupula in the *ampulla of the inner ear. Movement of the accessory structure causes the cilia to bend, which changes the cell's membrane potential and alters the pattern of impulses in the afferent sensory neuron.

hair follicle A narrow tubular depression in mammalian skin containing the root of a *hair. It is lined with epidermal cells and extends down through the epidermis and dermis to its base in the subcutaneous tissue. The ducts of *sebaceous glands empty into hair follicles.

hallucinogen A drug or chemical that causes alterations in perception (usually visual), mood, and thought. Common hallucinogenic drugs include *lysergic acid diethylamide (LSD) and mescaline. There is no common mechanism of action for this class of compounds although many hallucinogens are structurally similar to *neurotransmitters in the central nervous system, such as serotonin and the catecholamines.

hallux (*pl.* **halluces**) The innermost digit on the hindlimb of a tetrapod vertebrate. In humans it is the big toe and contains two phalanges. The hallux is absent in some mammals and in many birds it is directed backwards as an adaptation to perching. *Compare* POLLEX.

halophyte A plant that can tolerate a high concentration of salt (sodium chloride) in the soil. Such conditions occur in salt marshes and mudflats. Halophytes possess some of the structural modifications of *xerophytes; for example, many of them are *succulents. In addition, they are physiologically adapted to withstand the high salinity of the soil water: their root cells have a higher than normal concentration of solutes, which enables them to take up water by osmosis from the surrounding soil. Moreover, some have salt glands in the leaves that excrete salt. Examples of halophytes are mangrove trees (*see* MANGROVE SWAMP), thrift (*Armeria*), sea lavender (*Limonium*), and rice grass (*Spartina*). *Compare* HYDROPHYTE; MESOPHYTE.

haploid Describing a nucleus, cell, or organism with a single set of unpaired chromosomes. The haploid number is designated as *n*. Reproductive cells, formed as a result of *meiosis, are haploid. Fusion of two such cells (*see* FERTILIZATION) restores the normal (*diploid) number.

haplont An organism that is at the *haploid stage of its life cycle. *Compare* DIPLONT.

haplotype **1.** (in genetics) **a.** A set of linked genes or other genetic markers that are generally inherited together as a unit. This occurs because during meiosis there is little or no *recombination with the corresponding region on the homologous chromosome, and hence shuffling of alleles between the homologous regions is rare. The stretch of DNA containing a haplotype is called a **haplotype block**. For example, the genes of the *major histocompatibility complex in humans are closely linked at the HLA locus on chromosome 6 and behave as a haplotype, with the alleles on maternal and paternal chromosomes generally transmitted to offspring in the same combinations. Levels of recombination differ in different parts of the genome, causing the phenomenon known as *linkage disequilibrium, which means that haplotype blocks are more likely in some genomic regions than in others. **b.** The entire set of genes occurring on a single chromosome, or haploid set of chromosomes. Hence an individual has two haplotypes for each chromosome, one derived from its mother and one from its father. **2.** (in medicine)

a. The antigenic constitution of an individual (i.e. antigenic phenotype) resulting from the inheritance of a particular haploid combination of histocompatibility alleles at the HLA locus. **b.** The set of phenotypic features associated with either the paternal or maternal alleles inherited by an individual. *See also* INTERNATIONAL HAPMAP PROJECT.

hapten A foreign molecule that can bind to an antibody but does not evoke an immune response unless combined with a *carrier protein. In *hypersensitivity reactions haptens bound to endogenous proteins cause the body's immune system to attack both the hapten and the endogenous protein.

haptonasty *See* NASTIC MOVEMENTS.

haptotropism *See* THIGMOTROPISM.

hardwood *See* WOOD.

Hardy-Weinberg equilibrium The balance in the relative numbers of *alleles (*see* ALLELE FREQUENCY) that is maintained within a large population over a period of time assuming that: (1) mating is random; (2) there is no natural selection; (3) there is no migration; (4) there is no mutation; and (5) the population is very large. In such a stable population, for a gene with two alleles, *A* (dominant) and *a* (recessive), if the frequency of *A* is *p* and the frequency of *a* is *q*, then the frequencies of the three possible genotypes (*AA*, *Aa*, and *aa*) can be expressed by the **Hardy-Weinberg equation**:

$$p^2 + 2pq + q^2 = 1,$$

where p^2 = frequency of *AA* (homozygous dominant) individuals, $2pq$ = frequency of *Aa* (heterozygous) individuals, and q^2 = frequency of *aa* (homozygous recessive) individuals. The equation can be used to calculate allele frequencies if the numbers of homozygous recessive individuals in the population is known; this is useful, e.g., for estimating the frequency of heterozygote carriers of a defective allele responsible for a genetic disease in a population. Moreover, the equation can be used to test if evolution is occurring at any particular genetic locus; if it is not, and the above conditions apply, then the calculated allele frequency will agree with that observed in a population for subsequent generations. A disparity in the calculated and observed frequencies might indicate that evolution is occurring. The equation and the equilibrium are named after British mathematician G. H. Hardy (1877–1947) and German physician W. Weinberg (1862–1937).

harvesting 1. The processes involved in gathering in ripened crops (*see* AGRICULTURE). **2.** The collection of cells from cell cultures or of organs from donors for the purpose of transplantation (*see* GRAFT).

Hatch-Slack pathway *See* C_4 PATHWAY. It is named after M. D. Hatch and R. Slack.

haustorium (*pl.* **haustoria**) A specialized structure of certain parasitic plants and fungi that penetrates the cells of the host plant to absorb nutrients. In parasitic fungi haustoria are formed from enlarged hyphae and in parasitic flowering plants, such as the dodder (*Cuscuta*), they are outgrowths of the stem.

Haversian canals Narrow tubes within compact *bone containing blood vessels and nerves. They generally run parallel to the bone surface. Each canal is surrounded by a series of concentric cylinders of bony matrix (**lamellae**) containing *osteocytes within small cavities (lacunae). Each canal and its lamellae forms a **Haversian system**, or **osteon**. Haversian systems are separated from each other by boundaries called glue lines. They are named after the English anatomist Clopton Havers (1657–1702).

hearing The sense by which sound is detected. In vertebrates the organ of hearing is the *ear. In higher vertebrates variation in air pressure caused by sound waves are amplified in the outer and middle ears and transmitted to the inner ear, where sensory cells in the cochlea (*see* ORGAN OF CORTI) register the vibrations. The resulting information is transmitted to the brain via the auditory nerve. The ear can distinguish between sounds of different intensity (**loudness**) and frequency (**pitch**). In insects sound is detected by paired tympanal organs, found in various locations on the body depending on species. Each consists of a tympanum, analogous to the tympanic membrane of the ear, that vibrates in response to changes in air pressure. *See* TYMPANUM.

heart A hollow muscular organ that, by means of regular contractions, pumps blood through the circulatory system (*see* CIRCULATION). The vertebrate heart has a thick wall (*see* MYOCARDIUM) composed of a specialized muscle (*see* CARDIAC MUSCLE); it is surrounded by the *pericardium. Mammals have a four-chambered heart consisting of two atria and two ventricles; the right and left sides are completely separate from each other so there is no mixing of oxygenated and deoxygenated blood (see illustration). Oxygenated blood from the pulmonary veins enters the heart through the left atrium, passes to the left ventricle, and leaves the heart through the *aorta. Deoxygenated blood from the *venae cavae enters the right atrium and is pumped through the right ventricle to the pulmonary artery, which conveys it to the lungs for oxygenation. The tricuspid and bicuspid valves ensure that there is no backflow of blood. The contractions of the heart are initiated and controlled by the sinoatrial node (*see* PACEMAKER); in an average adult human the resting **heart rate** normally ranges from about 60 to 100 beats per minute (bpm), although elite athletes

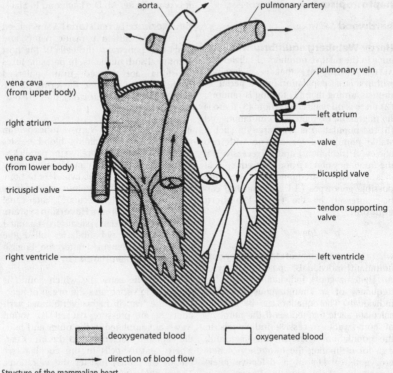

aorta

pulmonary artery

pulmonary vein

vena cava (from upper body)

left atrium

right atrium

valve

vena cava (from lower body)

bicuspid valve

tricuspid valve

tendon supporting valve

right ventricle

left ventricle

deoxygenated blood

oxygenated blood

direction of blood flow

Structure of the mammalian heart

might be in the range 40 to 60 bpm. The rate of contractions (i.e. the heartbeat) and contractility (i.e. strength of the heartbeat) are regulated by the nerves of the autonomic nervous system that supply the heart: the sympathetic nerve fibres of the cardiac nerves release noradrenaline, accelerating heartbeat and increasing contractility, whereas the parasympathetic fibres of the vagus nerve release acetylcholine, slowing the heartbeat and decreasing its strength. These two neurotransmitters thus have antagonistic effects on the pacemaker cells. *See also* CARDIAC CYCLE; CARDIAC OUTPUT.

The hearts of other vertebrates are similar except in the number of atria and ventricles (there may be one or two) and in the degree of separation of oxygenated and deoxygenated blood. Invertebrates, however, show great variation in the form and functioning of the heart.

(⊕) SEE WEB LINKS

• Sequence of diagrams of the human heart from *Gray's Anatomy*

heartwood (duramen) The wood at the centre of a tree trunk or branch. It consists of dead *xylem cells heavily thickened with lignin and provides structural support. Many heartwood cells contain oils, gums, and resins, which darken the wood and provide protection against fungi and wood-boring insects. *Compare* SAPWOOD.

heater cell A type of muscle cell, found in certain fishes (e.g. sailfish and marlin), that is specialized for generation of heat. For example, some cells of the eye muscles of blue marlins and swordfish have evolved as heater cells to provide heat for the retina and brain. Heater cells lack muscle fibrils but retain the sarcoplasmic reticulum. When the cell is stimulated, stored calcium ions are released from the sarcoplasmic reticulum into the cytosol, triggering intense energy metabolism by mitochondria. This releases energy, in the form of heat, at rates of up to 250 W/kg.

heat-shock protein (HSP) Any of various proteins that are synthesized by living cells in response to increased temperature or other forms of stress. They occur in both eukaryotes and prokaryotes, and function mainly as *molecular chaperones, protecting the cell's proteins as they become unfolded due to heating and enabling them to refold correctly. There are several families of heat-shock proteins, named according to their relative molecular mass (in kilodaltons, kDa); the larger ones, including HSP100, HSP90, HSP70, and HSP60, tend to predominate in animal cells. The smaller ones, with molecular masses in the range 17–28 kDa, are more common in plants. Another protein, *ubiquitin, is regarded as a heat-shock protein and acts as a marker for proteins destined for degradation. Many of the heat-shock proteins are also induced by other stresses, both natural and unnatural, including reduced oxygen concentration, changes in osmotic potential, ionizing radiation, and toxins, and hence are also called **stress proteins**.

heavy-metal pollution Environmental *pollution by metals with a high *relative atomic mass, such as lead and mercury. These metals derive from a number of sources, including industrial effluents and leaching of metal ions from the soil into lakes and rivers by *acid rain. They are easily incorporated into biological molecules and exert their toxic effects by displacing essential metals of a lower binding power in biologically active molecules or by acting as noncompetitive inhibitors of enzymes (*see* INHIBITION). However, some plants, known as *hyperaccumulators, can tolerate high concentrations of heavy metals and accumulate them in tissues. *See also* ECOTOXICOLOGY.

hecto- Symbol h. A prefix used in the metric system to denote 100 times. For example, 100 coulombs = 1 hectocoulomb (hC).

Hedgehog protein *See* SONIC HEDGEHOG.

helicase *See* PRIMOSOME.

helicotrema *See* COCHLEA.

heliotropism *See* PHOTOTROPISM.

helix-turn-helix (helix-loop-helix) A structural motif characteristic of certain proteins that bind to DNA. It consists of two *alpha helices connected by a short nonhelical segment called a 'turn'. One of the helical segments recognizes a specific sequence of nucleotides in the DNA molecule and fits into the groove in the DNA double helix, while the other helix stabilizes the bound configuration. This motif is one of several found in DNA-binding proteins (*see also* LEUCINE ZIPPER; ZINC FINGER).

helper T cell (T_H cell) A *T cell, characterized by having *CD4 protein antigens on its surface, that activates other cells of the immune system. CD4 T cells differentiate into several subsets; the two most significant are called T_H1 and T_H2. The **T_H1 cells** have an important role in controlling intracellular viral and bacterial infections. Pathogens can grow within intracellular vesicles of macrophages unless the macrophage is stimulated by binding of a T_H1 cell. A T_H1 cell binds to an *antigen-presenting cell (i.e. a macrophage) that is displaying the specific antigen for the *T-cell receptor. This prompts the T cell to secrete cytokines such as γ-interferon (IFN-γ), which activate the macrophage, causing intracellular vesicles to fuse with lysosomes inside the macrophage, enabling the lysosomal contents to destroy the bacteria. Exchange of cytokines between the antigen-presenting cell and the T cell causes the T cell to proliferate, producing a clone of T cells all with the same antigenic specificity. **T_H2 cells** have a vital role in helping to activate B cells and initiate their proliferation and differentiation into antibody-producing plasma cells. They also coordinate immune responses to large extracellular pathogens. B cell activation occurs when an antigen-specific 'armed' T cell recognizes its antigen presented on the surface of a B cell in association with MHC class II protein. The T_H2 cell becomes attached to the B cell and synthesizes various cell-bound and secreted molecules, notably the **CD40 ligand** (CD40L, or CD154) and interleukin 4 (IL-4). CD40L binds to the CD40 molecule on the B-cell surface, triggering a suite of responses by the B cell. Other subsets of CD4 T cells include regulatory T cells (T_{regs}), which are vital in maintaining homeostasis of the immune system. This involves suppressing immune responses that are no longer needed and ensuring that the system continues to distinguish between self and nonself material.

heme *See* HAEM.

hemicellulose A *polysaccharide found in the cell walls of plants. The branched chains of this molecule bind to cellulose microfibrils, together with pectins, forming a network of cross-linked fibres.

Hemichordata A phylum of bilaterally symmetrical soft-bodied marine invertebrates that comprises two classes: the Enteropneusta (acorn worms), which are cylindrical burrowing animals; and the Pterobranchia, which are colonial forms with vase-shaped bodies. The coelom is divided into three regions and the body into three sections: proboscis, collar, and trunk; hemichordates possess *gill slits, like chordates, but they lack a notochord. Some acorn worms develop via a ciliated larva (**tornaria**), which has some similarities to an echinoderm larva, and molecular studies confirm the hemichordates as the group most closely related to the echinoderms in evolutionary terms.

(⊕) SEE WEB LINKS

- Introduction to the hemichordates, on the website of the University of California Museum of Paleontology

hemicryptophyte A plant life form in Raunkiaer's system of classification (*see* PHYSIOGNOMY). Hemicryptophytes are typically herbaceous perennials, such as grasses, which produce perennating buds at the soil surface, where the buds are protected by leaf or stem bases.

hemimetabolous Describing insect development in which there is incomplete or partial metamorphosis, typically with successive immature stages increasingly resembling the adult. For instance, rudimentary wings are usually apparent, at least in later stages. It is characteristic of the *exopterygote insect orders. The immature stages of hemimetabolous species are called *nymphs. Compare AMETABOLOUS; HOLOMETABOLOUS.

hemiparasite (semiparasite) 1. A parasitic plant that lacks a fully developed root system and forms connections with another plant, from which it obtains some or all of its water and minerals. Such plants have chlorophyll and produce their own food by photosynthesis, and in some cases are capable of limited growth in the absence of the host plant. They tap into the sap-conducting tissue of the host by means of specialized structures called haustoria (see HAUSTORIUM). Some, such as eyebright (*Euphrasia* spp.), attach themselves to the roots of their host and appear like normal plants growing in the soil, whereas others grow on the aerial parts of their host. The mistletoes are well-known examples that colonize the branches of trees. **2.** A facultative parasite (see PARASITISM).

hemipenis (*pl.* **hemipenes**) Either of the paired male sex organs of snakes and lizards (i.e. *Squamata). The hemipenes are contained inverted within the tail and are used alternately during repeated copulations. They are everted by means of erectile tissue, and sperm passes along a seminal groove to enter the female. Some have ridges, hooks, or spines to anchor the hemipenis inside the female for extended periods to ensure fertilization takes place.

Hemiptera An order of *exopterygote insects comprising the true bugs. Hemipterans typically have oval flattened bodies with two pairs of wings, which are folded back across the abdomen at rest. The forewings are hardened, either at their bases only (in the suborder Heteroptera) or uniformly (in the suborder Homoptera). The mouthparts are modified for piercing and sucking, with long slender stylets forming a double tube. Many bugs feed on plant sap and are serious agricultural pests, including aphids, leaf-hoppers, scale insects, mealy bugs, etc. Others are carnivorous, and the order contains many aquatic species, such

as the water boatmen, which have legs adapted for swimming and the exchange of respiratory gases.

hemizygous Describing a diploid organism or cell that contains only a single copy of a particular gene. For example, genes that are not present on the smaller of the sex chromosomes in humans, the Y chromosome, occur only on the X chromosome; hence males are hemizygous for such genes.

Hensen's node (primitive node; primitive knot) A group of cells that acts as an *organizer of development in the early embryo of birds and mammals. It is located immediately anterior to the *primitive streak in a developing gastrula and is analogous to the dorsal lip of the blastopore in the blastula of an amphibian embryo. Named after its discoverer, German zoologist Victor Hensen (1835–1924), it controls gastrulation by secreting diffusible protein signals, initiating a signalling cascade in surrounding cells that regulates the differentiation of such structures as the neural tube and notochord.

heparan sulphate Any of a group of *glycosaminoglycans consisting of alternating residues of glucosamine and hexuronic acid, usually attached to a protein forming a proteoglycan. Heparan sulphates accumulate in a group of inherited metabolic disorders known as mucopolysaccharidoses.

heparin A glycosaminoglycan (mucopolysaccharide) with *anticoagulant properties, occurring in vertebrate tissues, especially the lungs and blood vessels. Heparin salts are administered therapeutically to prevent or dissolve blood clots.

hepatic Of or relating to the liver. For example, the **hepatic portal vein** (see HEPATIC PORTAL SYSTEM) and the **hepatic artery** supply blood to the liver, and the **hepatic vein** carries blood away from the liver.

Hepaticae See HEPATOPHYTA.

hepatic portal system The vein (**hepatic portal vein**) or veins that transport blood containing the absorbed products of digestion from the intestine directly to the liver.

hepatocyte A specialized epithelial cell that is the most abundant type of cell in the *liver. Hepatocytes are arranged around a central vein in units called **lobules**, in close association with branches of the hepatic portal vein, hepatic artery, and bile canaliculi. They are involved with the various functions of the liver, including metabolism, *detoxification, and the production of bile. *See also* KUPFFER CELLS.

Hepatophyta (Marchantiophyta) A phylum comprising the liverworts—simple plants that lack vascular tissue and possess rudimentary rootlike organs (rhizoids). Liverworts occur in moist situations (including fresh water) and as epiphytes on other plants. Like the mosses (*see* BRYOPHYTA), liverworts show marked alternation of generations between haploid gamete-bearing forms (gametophytes) and diploid spore-bearing forms (sporophytes), the latter being dependent on the former for nutrients, etc. The plant body (gametophyte) may be a thallus, growing closely pressed to the ground (thallose liverworts, e.g. *Pellia*), or it may bear many leaflike lobes (leafy liverworts). It gives rise to leafless stalks bearing capsules (sporophytes). Spores formed in the capsules are released and grow to produce new plants. Molecular studies show that the liverworts are the sister clade to all other land plants, including the mosses, with which they were formerly placed (in the class Hepaticae) in the phylum *Bryophyta.

(⊕) SEE WEB LINKS
• Introduction to liverworts, from the Plant Life website

hepatotoxin Any chemical that has adverse effects on the liver. Alcohol (ethanol) is one of the most common hepatotoxins.

herb 1. A *herbaceous plant, i.e. a seedbearing plant that does not form hard woody tissue. *Compare* FORB. **2.** A plant with medicinal or culinary uses. Culinary

herbs are usually plants whose leaves are used for flavouring food; examples are mint and parsley.

herbaceous Describing a plant that contains little permanent woody tissue. The aerial parts of the plant die back after the growing season. In *annuals the whole plant dies; in *biennials and herbaceous *perennials the plant has organs (e.g. bulbs or corms) that are modified to survive beneath the soil in unfavourable conditions.

herbicide *See* PESTICIDE.

herbivore An animal that eats vegetation, especially any of the plant-eating mammals, such as ungulates (cows, horses, etc.). Herbivores are characterized by having mouthparts (e.g. teeth) adapted for shearing, chewing, or grinding plants and alimentary canals specialized for digesting cellulose (*see* CAECUM). **Oligophagous** herbivores are specialized feeders that eat just one or a few plant species, whereas **polyphagous** herbivores eat a wide variety of plants.

herd immunity (community immunity) The general state of immunity occurring when a sufficient proportion of a population has been vaccinated or is naturally resistant so that the proportion of susceptible individuals is too small for the disease to spread.

heredity The transmission of characteristics from parents to offspring via the chromosomes. The study of heredity (*genetics) was first undertaken by Gregor Mendel (*see* MENDEL'S LAWS).

heritability A measure of the degree to which the variance of a particular phenotype is caused by genetic factors. It is given by a value between 0 and 1 and effectively measures the extent to which offspring resemble their parents relative to the population mean. Estimates of heritability are important in applied genetics, especially in agriculture and horticulture, because they enable prediction of the response of a population to artificial selection. The higher the heritability value, the greater the response, although heritability declines after several generations of artificial selection due to increasing homozygosity. The term is used in two different ways. **Heritability in the narrow sense** is the proportion of phenotypic variance due only to the additive genetic effect of all the *polygenes controlling a particular trait; hence it measures the proportion that is transmissible to offspring and therefore amenable to selection. **Heritability in the broad sense** considers other genetic but nontransmissible factors as well, including dominance and *epistasis; it is used, for example, in psychology to quantify genetic and environmental influences.

hermaphrodite (bisexual) 1. An animal, such as the earthworm, that has both male and female reproductive organs. In **sequential hermaphroditism** the individual functions first as one sex, then as the other, as occurs in most sponges. **2.** A plant whose flowers contain both stamens and carpels. This is the usual arrangement in most plants. *See also* INCOMPATIBILITY. *Compare* DIOECIOUS; MONOECIOUS.

heroin (diacetylmorphine) A *narcotic compound that is a synthetic derivative of morphine (*see* OPIATE). The compound is easily absorbed by the brain, due to its lipid-like nature, and is used as a sedative and powerful *analgesic. Highly addictive, it is abused by drug users.

herpesvirus One of a group of complex DNA-containing viruses causing infections in humans and most other vertebrates that tend to recur. The group includes **herpes simplex**, the agent of cold sores; **herpes varicella/zoster**, the virus causing chickenpox and shingles; **Epstein–Barr (EB) virus**, the causal agent of glandular fever and also implicated in the cancer Burkitt's lymphoma; and the *cytomegalovirus.

(((⊕))) SEE WEB LINKS
• Animations of herpes simplex virus infection and replication

herpetology The branch of zoology concerned with the study of nonavian reptiles and amphibians, which are sometimes collectively known as **herps** or **herptiles**.

hertz Symbol Hz. The *SI unit of frequency equal to one cycle per second. It is named after Heinrich Hertz (1857–94), a German physicist.

hesperidium (*pl.* **hesperidia**) *See* BERRY.

heterobaric leaf anatomy An anatomical arrangement in which the smallest veinlets of a leaf span the upper and lower surfaces of the leaf blade and project as ribs onto the leaf surface, forming a network of fine translucent lines. Found in many tree species and in some herbaceous plants, such as sunflower and broad bean, these projections are bundle sheath extensions (BSEs) arising from the *bundle sheath cells. They form a network of barriers within the leaf, separating the mesophyll into numerous compartments (areoles) and inhibiting transverse movements of air. The function of heterobaric anatomy is uncertain, but it has been shown that the translucent nature of BSEs enables them to act as light guides, transferring light to deeper layers of photosynthetic cells within the leaf, thereby permitting a thicker leaf as an adaptation to dry conditions. *Compare* HOMOBARIC LEAF ANATOMY.

heterochromatin *See* CHROMATIN.

heterochrony Any change in the relative rates or timing of development of different cell lines in the body. Such changes can lead to significant alterations in form; for example, if a particular organ develops earlier or faster it may be larger. Hence, mutations causing such changes play an important role in evolution. Heterochronic changes affecting the relative rates of development of germ (reproductive) cells and somatic (body) cells are of particular interest. They fall into four categories—*acceleration, *progenesis, *neoteny, and *hypermorphosis—depending on whether the change in rate (speeded up or slowed down) affects the somatic tissues or the reproductive tissues (see table). However, they have only two possible outcomes: one

Categories of heterochrony

Somatic features	Reproductive organs	Evolutionary process	Morphological result
speeded up	unchanged	acceleration	peramorphosis (by acceleration)
unchanged	accelerated	progenesis	paedomorphosis (by truncation)
slowed down	unchanged	neoteny	paedomorphosis (by retardation)
unchanged	slowed down	hypermorphosis	peramorphosis (by prolongation)

is *paedomorphosis, in which reproduction occurs in an ancestrally juvenile form; the other is *peramorphosis, in which development is extended by the addition of stages to the sequence shown in the ancestral form. *Compare* HETEROMETRY; HETEROTONY; HETEROTYPY.

heterocyst A specialized cell found in nitrogen-fixing filamentous or colonial cyanobacteria. Heterocysts are enlarged cells with thick cell walls and they lack chlorophyll, giving them a colourless appearance. They are the site of nitrogen fixation, for which they produce the enzyme nitrogenase. The lack of photosynthetic activity (which would produce oxygen), together with the thick cell wall, maintain the anoxic conditions in heterocysts that are essential for the activity of nitrogenase. Heterocysts are connected by *plasmodesmata to neighbouring cells, on which they depend for nutrients.

heterodont Describing animals that possess teeth of more than one type (i.e. *incisors, *canine teeth, *premolars, and *molars), each with a particular function. Most mammals are heterodont. *See* PERMANENT TEETH. *Compare* HOMODONT.

heteroduplex DNA Double-stranded DNA in which the two strands are derived from different DNA molecules. Heteroduplex DNA is formed during genetic recombination (*see* HOLLIDAY INTERMEDIATE) and can be produced *in vitro* in *DNA hybridization.

heteroecious Describing a parasite that passes different phases of its life cycle in different host species. For example, adult tapeworms of the genus *Echinococcus* live in the small intestine of dogs and other canids, whereas the eggs develop into hydatid cysts in the lungs or liver of intermediate hosts such as cattle, horses, or humans. Rust fungi are also examples of heteroecious species (*see* RUSTS). *Compare* AUTOECIOUS.

heterogametic sex The sex that is determined by possession of two dissimilar *sex chromosomes (e.g. XY). In humans and many other mammals this is the male sex. The heterogametic sex produces reproductive cells (gametes) of two kinds, half containing an X chromosome and half a Y chromosome. *Compare* HOMOGAMETIC SEX.

heterogamy *Alternation of generations in which parthenogenesis alternates with bisexual reproduction in the life cycle, as seen in certain aphids.

heterogeneous nuclear ribonucleoprotein (hnRNP) *See* HETEROGENEOUS NUCLEAR RNA; RIBONUCLEOPROTEIN.

heterogeneous nuclear RNA (hnRNA) An assortment of RNA molecules assembled around newly transcribed RNA (pre-messenger RNA, or pre-mRNA) in the cell nucleus. The hnRNA associates with proteins to form **heterogeneous nuclear ribonucleoprotein** (hnRNP). After release from the chromosome and processing of the pre-mRNA by the spliceosome (*see* INTRON), the mature messenger RNA remains associated with various proteins, forming a **messenger ribonucleoprotein**

(mRNP), which is then exported from the nucleus. The terms hnRNA and pre-RNA are sometimes used synonymously. *See* RNA PROCESSING.

heterokaryon *See* HETEROKARYOSIS.

heterokaryosis The presence in the same cell of two or more genetically different nuclei. Heterokaryosis occurs naturally in certain fungi, in which it results from the fusion of the cytoplasm of cells from different strains without the fusion of their nuclei. The cell, and the hypha or mycelium containing it, is known as a **heterokaryon**; the most common type of heterokaryon is a *dikaryon. Heterokaryosis can also be induced *in vitro*, to study the interaction between the cellular components from different species (*see* CELL FUSION).

heterokonts *See* STRAMENOPILES.

heterometry (in embryonic development) Literally 'different measure': a mechanism for generating changes in developmental outcomes that depends on the amount of expression of certain genes, and hence on the amount of the proteins they encode. Changes in such control mechanisms can produce significant morphological changes without any mutations in the proteins themselves, and have major evolutionary consequences. For example, the variations in beak shape and size among the different species of Galápagos finches (*see* DARWIN'S FINCHES) are the result of variations in the amounts of two signalling proteins produced during embryonic development, *bone morphogenetic protein 4 (BMP4) and *calmodulin. High levels of calmodulin produce an elongated beak, while high levels of BMP4 produce a wide, deep beak. *Compare* HETEROCHRONY; HETEROTONY; HETEROTYPY.

heteromorphosis (hypermetamorphosis) Development in which there is a radical change in form between successive larval *instars. For example, the tiny insects of the order Strepsiptera have an active first-instar larva (called a **triungulin**), which is succeeded by several inactive, often legless, grublike larval stages. This change in form reflects quite different functions. The triungulin's task is to seek out a suitable larval host, which is then parasitized by the following inactive larval stages.

heteroplasmy The presence within a cell, tissue, or organism of mitochondria with different genetic constitutions. Mitochondrial genomes have a high mutation rate, and mutations probably occur in all individuals. Such mutations can be of two types, either altering the sequence of the nucleotides or varying the total length of the *mitochondrial DNA. They are subject to selection in cells and can be lost or transmitted to progeny cells at cell division. More rarely, heteroplasmy results from inheritance of paternal mitochondria. Generally, an organism inherits its mitochondrial DNA from its mother via the egg cell, with no contribution from the male gamete, and therefore all its mitochondria are genetically identical. However, in some species there is 'leakiness' from the male line. For example, in mice about 1 in 1000 mitochondria originate from the male parent. Exceptionally, in some organisms (such as mussels) mitochondria are apparently inherited equally from both parents. *Compare* HOMOPLASMY.

heterosis *See* HYBRID VIGOUR.

heterosporous Describing a vascular plant that produces two types of spores, a smaller microspore and a larger megaspore. The microspore develops into a male *gametophyte (microgametophyte) stage that produces sperm, while the megaspore develops into a female gametophyte (megagametophyte) that produces eggs. All seed plants exhibit heterospory, and their haploid microspores and megaspores are produced by microsporangia and megasporangia, respectively, inside the parent plant's reproductive organs—cones in the case of gymnosperms and flowers in the case of angiosperms. *Compare* HOMOSPOROUS. *See also* ALTERNATION OF GENERATIONS.

heterospory The condition of producing two types of spore, megaspores and microspores. Heterospory occurs in all seed-bearing plants and in some mosses and ferns.

pin-eyed flower   thrum-eyed flower

stigma

long style

stamens

stamens

stigma

short style

Heterostyly in the primrose (*Primula vulgaris*)

heterostyly The condition of flowering plants in which flowers of the same species have styles of different lengths, so that the stigma is positioned below the anthers in some flowers and above them in others. This ensures that pollinating insects are more likely to transfer pollen from the anther of one flower to the stigma of another, thus promoting cross-pollination. Species showing heterostyly include the primrose (see illustration).

heterothallic Describing species of algae and fungi that reproduce sexually by the *conjugation of cells, thalli, or mycelia from different strains (*mating types), normally referred to as + and − strains. *Compare* HOMOTHALLIC.

heterotherm An organism whose ability to regulate its body temperature is intermediate between an *endotherm and an *ectotherm. Some small birds and mammals—generally endothermic ('warm-blooded') groups—may reduce their metabolic rate during a particular season or even a certain time of day, allowing their body temperature to fall and entering a state of **torpor**. In small animals, for example hummingbirds and certain rodents, this avoids the relatively high cost of maintaining a constant body temperature during periods of food scarcity or other adverse conditions. This strategy is sometimes called **partial endothermy**. At the opposite end of the

spectrum, certain animals that are generally regarded as ectothermic ('cold-blooded') have the ability to generate heat internally for limited periods, temporarily raising their body temperature to enable activity when the environmental temperature is low. Various insects, including bumblebees, are known to shiver in cold conditions in order to attain the body temperature necessary for flight. This may be termed **facultative endothermy**.

heterotony Literally 'different place': a mechanism for generating changes in developmental outcomes that depends on where certain genes are expressed. For example, in the developing hindlimb of a chick the webbing between the toes is lost because of the action of a signal protein called bone morphogenetic protein 4 (BMP4), which stimulates cell death (*apoptosis) of those tissues. By contrast, in the hindlimb of a duck a second protein called Gremlin is also expressed in the webbing cells; this inhibits the action of BMP4, thus preventing apoptosis and resulting in a webbed foot. Changes in the regulatory regions of such genes can alter in which tissues they are expressed, leading to morphological changes that may be of adaptive significance. *Compare* HETEROCHRONY; HETEROMETRY; HETEROTYPY.

heterotrichy The condition occurring in certain filamentous algae in which the algal body is composed of both prostrate filaments and protruding upright filaments.

heterotrophic nutrition A type of nutrition in which energy is derived from the intake and digestion of organic substances, normally plant or animal tissues. The breakdown products of digestion are used to synthesize the organic materials required by the organism. All animals obtain their food this way: they are **heterotrophs**. *See also* INGESTION. *Compare* AUTOTROPHIC NUTRITION; MIXOTROPHIC.

heterotypy Literally 'different form': changes in developmental outcome arising from mutations in key regulatory molecules that control embryonic pattern formation and tissue specification. Such changes can lead to dramatic changes in form during evolution. A classic example is the suppression of limb growth in the abdominal segments of insects. This is thought to have arisen in an ancestral arthropod through mutation of a gene called *Ultrabithorax* (*Ubx*). The mutant gene expressed a Ubx protein capable of repressing the gene *Distalless* required for leg development in the abdomen. Hence, the descendants had legs confined to the thoracic segments, and became the insects. *Compare* HETEROCHRONY; HETEROMETRY; HETEROTONY.

heterozygote advantage *See* BALANCING SELECTION.

heterozygous Describing an organism or cell in which the *alleles at a given locus on *homologous chromosomes are different. An allele that is fully *dominant with respect to its *recessive counterpart at the same locus will solely determine the nature of the corresponding trait (i.e. the phenotype), whereas alleles displaying *incomplete dominance will result in an intermediate phenotype. Heterozygous individuals (i.e. **heterozygotes**) having greater reproductive fitness compared with either homozygote are said to have **heterozygote advantage**. Heterozygous organisms do not breed true. *Compare* HEMIZYGOUS; HOMOZYGOUS.

Hexapoda (Insecta) In traditional classifications, a subphylum (or class) of arthropods comprising over a million known species (of an estimated 5.5 million species). They are distributed worldwide in nearly all terrestrial habitats. Ranging in length from 0.5 to over 300 mm, an insect's body consists of a head, a thorax of three segments and usually bearing three pairs of legs and one or two pairs of wings, and an abdomen of eleven segments. The head possesses a pair of sensory *antennae and a pair of large *compound eyes, between which are three simple eyes (*ocelli). The *mouthparts are variously adapted for either chewing or sucking, enabling insects to feed on a wide range of plant and animal material. Insects owe much of their success to having a highly waterproof *cuticle (to resist desiccation) and, in most species, wings—outgrowths of the body wall that confer the greater mobility of *flight. Breathing occurs through a network of tubes (*see* TRACHEA).

Most insect species have separate sexes and undergo sexual reproduction. In some, this may alternate with asexual *parthenogenesis and in a few, males are unknown and reproduction is entirely asexual. In the wingless insects (subclass *Apterygota) *metamorphosis is slight or absent. In the winged insects (subclass *Pterygota) the newly hatched young grow by undergoing a series of moults. In the more primitive *exopterygotes (including the orders *Dermaptera, *Orthoptera, *Dictyoptera, and *Hemiptera) the young (called a *nymph) resembles the adult. The more advanced *endopterygotes (e.g. *Coleoptera, *Diptera, *Lepidoptera, and *Hymenoptera) undergo metamorphosis, in which the young (called a *larva) is transformed into a quiescent *pupa from which the fully formed adult emerges. Insects are of vital importance in many ecosystems and many are of economic significance—as animal or plant pests or disease vectors or beneficially as crop pollinators or producers of silk, honey, etc.

Some authorities regard the Hexapoda as a superclass comprising the classes *Protura, *Collembola, and *Diplura, all of whose members have mouthparts that are enclosed in folds of the head; and Insecta, which have exposed mouthparts. Genetic and other molecular evidence now indicates that insects evolved from crustacean ancestors. Hence, the insects and crustaceans are now placed in the clade Pancrustacea, with a group of predatory crustaceans, the Remipedia, regarded as the closest relatives to all insects.

(((**◉**))) SEE WEB LINKS

• Overview of insect characteristics and phylogeny, with links to insect orders, at the Tree of Life web project

hexose A *monosaccharide that has six carbon atoms in its molecules.

Hfr (high-frequency recombinant) *See* SEX FACTOR.

hibernation A sleeplike state in which some animals pass the winter months as a way of surviving food scarcity and cold weather. Various physiological changes occur, such as lowering of the body temperature and slowing of the pulse rate and other vital processes, and the animal lives on its reserve of body fat. Hibernating mammals may cool to near freezing and have metabolic rates as much as 20 times lower than normal. They may also briefly undergo short-lived periods of arousal before returning to deep hibernation. Animals that hibernate include bats, hedgehogs, and many fish, amphibians, and reptiles. *See also* DORMANCY. *Compare* AESTIVATION.

hidden variable *See* CONFOUNDING.

hierarchy A type of social organization in which individuals are ranked according to their status or dominance relative to other group members. This affects their behaviour in various ways, e.g. by determining their access to food or to mates. Many vertebrate animals and some invertebrates live in hierarchical social groups.

highly repetitive DNA *See* REPETITIVE DNA.

high-performance liquid chromatography (HPLC) A sensitive technique

for separating or analysing mixtures, in which the sample is forced through the *chromatography column under pressure.

Hill reaction The release of oxygen from isolated illuminated chloroplasts when suitable electron acceptors (e.g. potassium ferricyanide) are added to the surrounding water. The reaction was discovered by Robert Hill (1899–1991) in 1939; the electron acceptors substitute for $NADP^+$, the natural acceptor for the light-dependent reactions of *photosynthesis.

hilum (*pl.* **hila**) A scar on the seed coat of a plant marking the point at which the seed was attached to the fruit wall by the *funicle. It is a feature that distinguishes seeds from fruits.

hindbrain (rhombencephalon) One of the three sections of the brain of a vertebrate embryo. It develops to form the *cerebellum, *pons, and *medulla oblongata, which control and coordinate fundamental physiological processes (including respiration and circulation of blood). *Compare* FOREBRAIN; MIDBRAIN.

hindgut 1. The posterior part of the alimentary canal of vertebrates, comprising the posterior section of the colon. **2.** The posterior section of the alimentary canal of arthropods. *See also* FOREGUT; MIDGUT.

hip girdle *See* PELVIC GIRDLE.

hippocampus (*pl.* **hippocampi**) A part of the vertebrate brain consisting of two ridges, one over each of the two lateral *ventricles. Part of the *limbic system, it is highly developed in advanced mammals (primates and whales) and its function is concerned with forming new memories—by encoding information about events, people, places, etc. so that it can be stored as long-term declarative memory elsewhere in the brain.

Hirudinea A class of freshwater and terrestrial annelid worms that comprises the leeches. They have suckers at both anterior and posterior ends but no bristles. Some are blood-sucking parasites of vertebrates and invertebrates but the majority are predators.

histamine A substance that is released during allergic reactions, e.g. hay fever. Formed from the amino acid histidine, histamine can occur in various tissues but is concentrated in connective tissue. Histamine is one of the inflammatory mediators released from *mast cells in response to antigen binding to IgE antibodies on the surface of the mast cell. It is a short-lived amine that causes the initial symptoms of an allergic response, such as the acute constriction of the airways in an asthmatic attack. It also causes dilation and increased permeability of small blood vessels, which results in such symptoms as localized swelling, itching, sneezing, and runny eyes and nose. The effects of histamine can be countered by the administration of *antihistamine drugs.

histidine *See* AMINO ACID.

histiocyte *See* MACROPHAGE.

histochemistry The study of the distribution of the chemical constituents of tissues by means of their chemical reactions. It utilizes such techniques as *staining, light and electron microscopy, *autoradiography, and *chromatography.

histocompatibility The degree to which tissue from one organism will be tolerated by the immune system of another organism. For any animal, it is essential that its immune system can distinguish its own tissues from foreign cells or tissues, so that only the latter are attacked. This self-recognition is achieved principally by a set of marker molecules, called **histocompatibility proteins** (or **histocompatibility antigens**), which occur on the surfaces of cells. These proteins (in humans also called **human leucocyte antigens**, or **HLAs**) are encoded in vertebrates by a cluster of genes called the *major histocompatibility complex (MHC). Each species has a unique set of MHC proteins, and there is also wide variation within any given species. This explains why in human transplantation it is very difficult to match donor and recipient tissue exactly (*see* HLA SYSTEM). MHC proteins also play a vital role in the immune responses of lymphocytes, notably by enabling *T cells to identify foreign antigens. The MHC encodes two distinct classes of histocompatibility proteins, both of which are glycoproteins (*see* MHC CLASS I PROTEIN; MHC CLASS II PROTEIN).

histocompatibility antigen *See* HISTOCOMPATIBILITY; HLA SYSTEM.

histology The microscopic study of the tissues of living organisms. The study of cells, a specialized branch of histology, is known as *cytology.

histone Any of a group of water-soluble proteins found in association with the *DNA of plant and animal chromosomes to form *chromatin. They contain a large proportion of the basic (positively charged) amino acids lysine, arginine, and histidine, and hence bind tightly to the negatively charged DNA. There are five classes of histones—H1, H2A, H2B, H3, and H4—involved in the condensation and coiling of chromosomes. The DNA winds around clusters of histones forming the spool-like *nucleosomes, the basic units of DNA packaging. Each nucleosome comprises a pair of histones from each of classes H2A to H4 plus a single H1 histone. Chemical modification of histones is a key aspect of suppressing or activating gene activity (*see* CHROMATIN REMODELLING). Histones do not occur in vertebrate sperm cells (*see* PROTAMINE) or in bacteria, although a very similar protein has been found in the genome of the archaean *Thermoplasma*. *See also* EPIGENOME; GENE IMPRINTING.

HIV (human immunodeficiency virus) The *retrovirus that causes *AIDS in humans. It has a specific affinity for the *helper T cells of its host, binding to *CD4 antigens on the cell surface and thereby disabling these cells. The membrane envelope glycoproteins encasing the virus show great variability in their amino-acid sequences, hence the difficulty of preparing an effective AIDS vaccine. Two varieties (serovars) are known: HIV-1 and HIV-2. The latter, which is less virulent, is found chiefly in Africa. HIV is thought to have originated from chimpanzees in central Africa.

HLA system (human leucocyte antigen system) A series of gene loci, forming part of the *major histocompatibility complex on chromosome 6 in humans, that encode the component polypeptide chains of the *MHC class I proteins and *MHC class II proteins. These proteins play crucial roles in the immune system, by binding peptide fragments derived from pathogens and presenting these on the surface of host cells for recognition by effector T cells. There are three gene loci encoding the α chain of class I proteins: HLA-A, HLA-B, and HLA-C (the β2-microglobulin is encoded on chromosome 15). Class II α and β chains are encoded by a further three pairs of loci, designated HLA-DP, HLA-DQ, and HLA-DR, with the latter often having an extra β gene. This range of genes means that any individual can produce three different class I proteins and three or four different class II proteins, giving a variety of antigen-binding properties. These MHC glycoproteins also act as antigens themselves and are important in determining the acceptance or rejection by the body of a tissue or organ transplant (*see* GRAFT). These antigens are one group of the so-called **histocompatibility proteins** (*see* HISTOCOMPATIBILITY). Two individuals with identical HLA types are said to be **histocompatible**. Successful transplantation requires a minimum number of HLA differences between the donor's and recipient's tissues.

hnRNA *See* HETEROGENEOUS NUCLEAR RNA.

hnRNP *See* HETEROGENEOUS NUCLEAR RNA; RIBONUCLEOPROTEIN.

Hogness box *See* TATA BOX.

holandric Describing a trait that is inherited only by the male line. Such traits are determined by genetic loci that occur only on the Y chromosome, and hence are present only in males.

Holliday intermediate An intermediate structure in genetic recombination, proposed by Robin Holliday in 1964, in which two double-stranded homologous DNA molecules are joined by means of a reciprocal crossover involving one DNA strand of each molecule. It is formed when a single strand of DNA from each chromosome is broken and joined to the other strand at the point of crossover. The region in which strands from different DNA molecules are paired (a **heteroduplex** DNA

2 homologous DNA molecules

break in one strand of each molecule
Holliday intermediate

point of cross-strand exchange

Holliday intermediate

point of cleavage of both strands

heteroduplex region extended

recombination products

sequence) is extended and two strands of the Holliday intermediate are cleaved. The breaks in the DNA sequences are then repaired to form the recombinant products (see illustration).

holobiont A host and all its associated symbiotic organisms and viruses. For example, a lichen can be viewed as a holobiont comprising algal and fungal partners. More broadly, a human individual plus the microorganisms contained within the gut or living on the skin surface may also constitute a holobiont. This 'superorganism' is regarded by some as the organizational level at which natural selection acts. *See* HOLOGENOME.

holocarpic Denoting a fungus in which the entire thallus is differentiated into a reproductive sporangium when mature. *Compare* EUCARPIC.

Holocene (Recent) The most recent geological epoch of the *Quaternary period, comprising roughly the past 11700 years since the end of the *Pleistocene up to the present. It follows the final glacial of the Pleistocene and thus is sometimes known as the **Postglacial** epoch. Some geologists consider the Holocene to be an interglacial phase of the Pleistocene that will be followed by another glacial. *See also* ANTHROPOCENE.

holocrine secretion *See* SECRETION.

holoenzyme A complex comprising an enzyme molecule and its *cofactor. Only in this state is an enzyme catalytically active. *Compare* APOENZYME.

hologenome The sum of the genomes of an organism and all its symbiotic microorganisms (i.e. a *holobiont)—for instance, a termite, the cellulose-digesting flagellate protists in the termite's gut, and the protists' own endosymbiotic bacteria. This concept underlies the **hologenome theory of evolution**, in which natural selection is viewed as acting on holobionts, rather than on individual organisms. It is argued that the microbial community of the host (i.e. its *microbiome) can evolve far more

rapidly than a large and complex host, and that the host can achieve selective advantage in just a few microbial generations simply through acquired changes in the makeup of its microbiome. Such a process invokes the *Lamarckian concept of inheritance of acquired characteristics. The microbiome can also on occasion influence mating choices, for instance by altering the scent produced by an animal, which has led to speculation about its possible role in contributing to mating incompatibility and the splitting of species.

holometabolous Describing insect development in which there is complete *metamorphosis and the immature stages, called *larvae, are markedly different from the adults. Transformation of the larvae into the adult takes place during a resting stage called a *pupa. Holometabolous development is characteristic of *endopterygote insect orders. *Compare* AMETABOLOUS; HEMIMETABOLOUS.

holophytic Describing organisms that feed like plants, i.e. that are photoautotrophic. *See* AUTOTROPHIC NUTRITION.

holotype *See* TYPE SPECIMEN.

holozoic Describing organisms that feed by ingesting complex organic matter, which is subsequently digested and absorbed. *See* HETEROTROPHIC NUTRITION; INGESTION.

homeobox A nucleotide sequence containing about 180 base pairs, which are identical or very similar in many eukaryotic organisms, that encodes a 60-amino acid sequence known as a **homeodomain**. Present in many eukaryotic regulatory proteins, this *domain is an important region involved in the binding of transcription factors and other regulatory proteins to the DNA molecule. Homeodomain-containing proteins control the expression of many genes, particularly ones that direct the formation of body structures during embryonic development. A homeobox was first identified in the *homeotic genes of *Drosophila* and has since been found in the homeotic genes of many animals (including humans) and in plants.

homeologous Describing chromosomes or genes that are only partly homologous. *See* HOMOLOGOUS CHROMOSOMES.

homeosis *See* HOMEOTIC GENES.

homeostasis The regulation by an organism of the chemical composition of its body fluids and other aspects of its *internal environment so that physiological processes can proceed at optimum rates. It involves monitoring changes in the external and internal environment by means of *receptors and adjusting the physiological variables, such as the composition of body fluids, accordingly; *excretion and *osmoregulation are important in this process. Examples of homeostatic regulation are the maintenance of the *acid-base balance and body temperature (*see* HOMOIOTHERMY; POIKILOTHERMY). Many variables are continually adjusted so that they remain around a given value or *set point.

homeothermy *See* HOMOIOTHERMY.

homeotic genes A class of genes, including the *Hox* genes, that play a central role in controlling the early development and differentiation of embryonic tissues in eukaryotic organisms, including plants, animals, and fungi. They code for *transcription factors—proteins that bind to DNA and regulate the expression of a wide range of other genes. This binding capability resides in a structural domain of the protein called a homeodomain, encoded by a nucleotide sequence that is characteristic of homeotic genes (*see* HOMEOBOX). These genes were first identified in *Drosophila* fruit flies, through the occurrence of mutations that cause the transformation of one organ into another—the phenomenon of **homeosis**. *Drosophila* flies have two major clusters of homeotic genes: the *antennapedia* complex, which controls development of the head and anterior thoracic segments, and the *bithorax* complex, which governs the development of posterior segments. For example, one mutation of the *bithorax* cluster causes the thoracic segment that

normally bears the halteres (balancing organs; *see* DIPTERA) to be transformed into a segment bearing a pair of wings. In vertebrates there are four clusters of homeotic genes located on separate chromosomes. *See also* ABC MODEL.

((🌐)) **SEE WEB LINKS**

• Description of homeotic genes in Drosophila, based on the poster for the 1995 Nobel Prize for Physiology or Medicine

home range The area in which an animal forages and spends most of its time. The size varies according to the size of the animal and its feeding habits; for example, the home range of African hunting dogs may extend over 4000 km^2, whereas that of a field mouse may be less than 1 ha. Unlike a *territory, the home range is not defended against incursion by other members of the same species, although an animal's territory may lie within its home range. Moreover, the ranges of neighbouring individuals or groups often overlap. Within the range there may be a core area, containing the main feeding and drinking areas, within which the animal's activities are focused.

homing *See* NAVIGATION.

hominid Any member of the primate family Hominidae, which includes humans and their fossil ancestors in the tribe Hominini. The hominid family is now regarded as also including the other extant great apes (chimpanzees, gorillas, orang-utan—formerly constituting the family Pongidae) and extinct groups, such as *Australopithecus* and *Dryopithecus*.

((🌐)) **SEE WEB LINKS**

• Interactive documentary narrated by palaeontologist Donald Johanson and presented by the Institute of Human Origins, Arizona State University

hominin Any member of the clade or tribe (Hominini) that contains modern humans and their recent ancestors in the genus *Homo* plus other closely related extinct bipedal hominids, including members of the genera *Australopithecus*, the *ardipithecines, and *Paranthropus*. Fossil evidence shows that the early hominins diverged from chimpanzees about 7–6 million years ago (mya) in Africa, and evolved the ability to walk upright on two legs. However, some make the contentious claim that the earliest hominin was *Graecopithecus*, which lived in eastern Europe around 7.20 mya. This is based on recent analysis of a fossil tooth and jawbone.

SEE WEB LINKS
• Interactive family tree of hominin evolution from the Smithsonian Institution

Homo The genus of primates belonging to the tribe Hominini that includes modern humans (*H. sapiens*, the only living representative) and various extinct species. The oldest *Homo* fossils are those of *H. habilis* and *H. rudolfensis*, which first appeared in Africa some 2.4 and 1.9 million years ago (mya), respectively. Both species are thought to have used simple stone tools. *H. habilis* appears to have been 1–1.5 m tall and had more human-like features and a larger brain than *Australopithecus*.

H. erectus (sometimes known as *H. ergaster*) appeared in Africa around 1.9 mya and subsequently spread to Asia between 1.8 and 1.5 mya. Fossils of *H. erectus*, which was formerly called *Pithecanthropus* (ape man), include Java man and Peking man. They are similar to present-day humans except that there was a prominent ridge above the eyes and a receded forehead. They used crude stone tools and fire. *H. erectus* may also have given rise to *H. heidelbergensis* (represented by Heidelberg man and Boxgrove man). This latter species now contains all hominin specimens with a mixture of 'erectus-like' and 'modern' characters, dating from at least 700 000 years ago to the emergence of *H. sapiens* around 300 000 years ago (although evidence from Spain, Italy, and England suggests that *H. heidelbergensis* may date back to 1.3 mya). *H. heidelbergensis* is regarded as the ancestral lineage of both *H. neanderthalensis* (*Neanderthals) and *H. sapiens*. In 2003 the first remains of a pygmy-sized human species, *H. floresiensis*, were discovered on the island of Flores, Indonesia. Fossils of this

Homo: timeline for main species

H. habilis

H. rudolfensis

H. erectus (ergaster)

H. heidel-
bergensis

H. flores-
iensis

H. neander-
thalensis
to 400,000 years

H. sapiens

2.0   1.5   1.0   0.5   Present
Million years ago

species date between 100 000 and 60 000 years ago; individuals stood about 100 cm tall; hence their nickname, the 'Hobbit'. A fossil human ancestor, first identified in 2010 from a fragment of finger bone discovered in Denisova Cave, Siberia, is thought to belong to a new species of the genus *Homo* or a subspecies of *H. sapiens*. Sufficient DNA was contained in the bone fragment to reconstruct and sequence nearly the entire genome. The results indicate that so-called Denisovans were more closely related to *Neanderthals than to ancestral modern humans, but interbred with both. *See also* 'CRO-MAGNON MAN'. See illustration.

homobaric leaf anatomy A form of leaf anatomy in which there are few or no internal barriers to lateral movements of air inside the leaf. *Compare* HETEROBARIC LEAF ANATOMY.

homodont Describing animals whose teeth are all of the same type. Most vertebrates except mammals are homodont. *Compare* HETERODONT.

homogametic sex The sex that is determined by possession of two similar *sex chromosomes (e.g. XX). In humans and many other mammals this is the female sex. All the reproductive cells (gametes) produced by the homogametic sex have the same kind of sex chromosome (i.e. an X chromosome). *Compare* HETEROGAMETIC SEX.

homogamy 1. (in botany) The condition in a flower in which the male and female reproductive organs mature at the same time, thereby allowing self-fertilization. *Compare* DICHOGAMY. **2.** Breeding between individuals that are similar in some way or share certain traits; inbreeding.

homoiothermy (homeothermy) The maintenance by an animal of its internal body temperature at a relatively constant value by using metabolic processes to counteract fluctuations in the temperature of the environment. Homoiothermy occurs chiefly in birds and mammals, which are described as *endotherms. However, some degree of homoiothermy is also found in

certain fish and reptiles and in some insects. In homoiotherms, the heat produced by their tissue metabolism and the heat lost to the environment are balanced by various means to keep body temperature constant: 36–38°C in mammals and 38–40°C in birds. The *hypothalamus in the brain monitors blood temperature and controls thermoregulation by both nervous and hormonal means. This produces both short-term responses, such as shivering or sweating, and long-term adjustments to metabolism according to seasonal changes in climate (acclimatization). Endotherms generally possess insulating feathers or fur. Their relatively high internal temperature permits fast action of muscles and nerves and enables them to lead highly active lives even in cold climates. However, in certain animals, homoiothermy is abandoned during periods of *hibernation. *Compare* POIKILOTHERMY.

homologous 1. (in biology) Describing a character that is shared by a group of species because it is inherited from a common ancestor. Such characters, called **homologies**, are used in *cladistics to determine the evolutionary relationships of species or higher taxa. They are divided into two types: a **shared derived homology** (*see* SYNAPOMORPHY) is unique to a particular group and may be used to define a *monophyletic group; a **shared ancestral homology** (*see* PLESIOMORPHY) is not unique to the group, or may not be exhibited by all descendants of the ancestor in which it arose (*see* PARAPHYLETIC). Even though homologous features share the same evolutionary origin, they may have developed different functions. For example the wings of a bat, the flippers of a dolphin, and the arms of a human are homologous organs, having evolved from the paired pectoral fins of a fish ancestor. *Compare* ANALOGOUS. **2.** (in molecular biology) Describing sequences of nucleotides (or amino acids) at corresponding sites of different nucleic acids (or proteins) that show similarity because the molecules are descended from a common ancestral molecule (*see* SEQUENCE ANALYSIS). The term is sometimes used more loosely, but incorrectly, to describe sequences that are merely similar, when no evolutionary relationship is implied or can be established. *See* ORTHOLOGOUS; PARALOGOUS. *See also* CONSERVED SEQUENCE.

homologous chromosomes Chromosomes having the same structural features. In *diploid nuclei, pairs of homologous chromosomes can be identified at the start of meiosis (*see* PAIRING). One member of each pair comes from the female parent and the other from the male. Homologous chromosomes have the same pattern of genes along the chromosome but the nature of the genes may differ (*see* ALLELE).

homoplasmy The presence within a cell or organism of genetically identical mitochondria. *Compare* HETEROPLASMY.

homoplasy The similarity of a particular character in two different, yet often related, groups of organisms that is not the result of common ancestry. Such a similarity may arise due to convergent evolution, parallel evolution, or an evolutionary reversal, and is therefore potentially misleading when examining shared characters in constructing phylogenetic trees (*see* CLADISTICS). For example, wings in bats and birds are a convergent, and therefore homoplasic, character. Hence, all efforts are made to distinguish homoplasic characters from *homologous derived characters (*see* APOMORPHY). *Compare* ANALOGOUS; PATRISTIC.

homosporous Describing a vascular plant in which the sporophyte generation produces a single type of spore. This spore develops into a single type of *gametophyte, which gives rise to both male and female reproductive organs. Most seedless plants exhibit homospory. *Compare* HETEROSPOROUS. *See also* ALTERNATION OF GENERATIONS.

homothallic Describing species of algae and fungi that reproduce sexually by the *conjugation of cells, thalli, or mycelia from the same strain. *Compare* HETEROTHALLIC.

homozygous Describing an organism or cell in which the *alleles at a given locus on homologous chromosomes are identical (they may be either dominant or recessive). Homozygous organisms, which are called **homozygotes**, breed true when crossed with genetically identical organisms. *Compare* HETEROZYGOUS.

hopeful monster A hypothetical new phenotype, or monstrosity, that arises due to mutations that radically alter an individual's developmental pattern. Such an individual, it is claimed, could possess major innovations that equip it for an environment quite different from that of its immediate antecedents. The concept, which was introduced in 1933 by the geneticist R. Goldschmidt, presents a way in which, theoretically, a new group could arise in a single macroevolutionary leap, rather than by the gradual process of natural selection producing many small adaptive changes—the route favoured by orthodox neo-Darwinians. There are some well-documented examples of single-gene mutations causing large morphological changes that appear to be of adaptive value, such as the loss of pelvic spines in freshwater three-spined sticklebacks bred from their marine counterparts. In this light, Goldschmidt's ideas, which were initially ridiculed, have been reappraised.

horizontal cell *See* RETINA.

horizontal gene transfer *See* LATERAL GENE TRANSFER.

hormogonium (*pl.* **hormogonia**) A filament produced by filamentous cyanobacteria that becomes detached from the parent organism and acts as a means of vegetative propagation.

hormone 1. A substance that is manufactured and secreted in very small quantities into the bloodstream by an *endocrine gland or a specialized nerve cell (*see* NEUROHORMONE) and regulates the growth or functioning of a specific tissue or organ in a distant part of the body. For example, the hormone *insulin controls the rate and manner in which glucose is used by the body. Other hormones include the *sex hormones, *corticosteroids, *adrenaline, *thyroxine, and *growth hormone. **2.** *See* PLANT HORMONE.

hornworts *See* ANTHOCEROPHYTA.

horsetails Any of group of *tracheophyte plants belonging to the genus *Equisetum*. Horsetails have a perennial creeping rhizome supporting erect jointed stems bearing whorls of thin leaves. Spores are produced by terminal conelike structures. The horsetails and their extinct relatives have a fossil record extending back to the Palaeozoic with their greatest development in the Carboniferous period, when giant tree forms were the dominant vegetation with the *Lycopodiophyta. Horsetails were formerly placed in a separate phylum (variously called Equisetophyta, Sphenophyta, or Arthrophyta), but recent molecular genetic evidence places them in a clade, the *monilophytes, along with the whisk ferns and true *ferns.

host 1. An organism whose body provides nourishment and shelter for a parasite (*see* PARASITISM) or a *parasitoid. A **definitive** (or **primary**) **host** is one in which an animal parasite becomes sexually mature; an **intermediate** (or **secondary**) **host** is one in which the parasite passes the larval or asexual stages of its life cycle. **2.** An organism that lives in close association with an inquiline (*see* INQUILINISM). **3.** (in genetics) A cell or organism into which foreign DNA is introduced during *gene cloning.

housekeeping gene *See* CONSTITUTIVE.

***Hox* genes** A class of *homeotic genes that control development of structures along the head-to-tail (anteroposterior) axis of a wide range of animals. The *Hox* genes are organized into clusters on certain chromosomes; jawed vertebrates, for example, have four *Hox* gene clusters. In mammals these four clusters are designated *Hox A*, *Hox B*, *Hox C*, and *Hox D*, each on a separate chromosome, with individual genes given numbers, hence, *A1*, *A2*, *B1*, *B2*, etc. Nematodes, arthropods, and cephalochordates have a single cluster. *Hox* genes are highly conserved, showing remarkable similarity of DNA sequence and function; each falls into one of several groups of *paralogous genes, derived by duplication of ancestral genes. Moreover, in embryos of all animals studied, the *Hox* genes show **colinearity**—their sequence of expression in body segments from head to tail reflects their linear arrangement in the homeotic gene clusters.

HPLC *See* HIGH-PERFORMANCE LIQUID CHROMATOGRAPHY.

HSP *See* HEAT-SHOCK PROTEIN.

human chorionic gonadotrophin (hCG) *See* GONADOTROPHIN.

Human Genome Project A coordinated international project, begun in 1988, to map the entire human *genome so that the genes could be isolated and sequenced (*see* DNA SEQUENCING). It involved the production of a *DNA library. The full *draft sequence was completed in 2000 and published in February 2001, and the high-quality *finished sequence was completed in April 2003, two years ahead of schedule. Subsequent work has shown that the haploid human genome contains about 3.2×10^9 nucleotide base pairs, but only around 21 000 protein-coding genes. *Alternative splicing in about 50% of these means that the number of protein products is actually much greater. Moreover, there are many more 'RNA genes', sequences encoding the various types of RNA used for implementation and control of gene expression.

human growth hormone (hGH) *See* GROWTH HORMONE.

human immunodeficiency virus *See* HIV.

humerus (*pl.* **humeri**) The long bone of the upper arm of tetrapod vertebrates. It articulates with the *scapula (shoulder blade) at the *glenoid cavity and with the *ulna and *radius (via a *condyle) at the elbow.

humidity The concentration of water vapour in the atmosphere. The **absolute humidity** is the mass of water vapour per unit volume of air, usually expressed in kg m^{-3}. A useful measure is the **relative humidity**, the ratio, expressed as a percentage, of the moisture in the air to the moisture it would contain if it were saturated at the same temperature and pressure.

humoral Relating to the blood or other body fluids. For example, **humoral immunity** is immunity conferred by the antibodies present in the blood, lymph, and tissue fluids (*see* IMMUNITY).

humus The dark-coloured amorphous colloidal material that constitutes the organic component of soil. It is formed by the decomposition of plant and animal remains and excrement (*see* LITTER) and has a complex and variable chemical composition. Being a colloid, it can hold water and therefore improves the water-retaining properties of soil; it also enhances soil fertility and workability. Acidic humus (**mor**) is found in regions of coniferous forest, where the decay is brought about mainly by fungi. Alkaline humus (**mull**) is typically found in grassland and deciduous forest: it supports an abundance of microorganisms and small animals (e.g. earthworms).

hyaline cartilage *See* CARTILAGE.

hyaloplasm *See* CYTOSOL.

hyaluronic acid (hyaluronan) A *glycosaminoglycan that is part of the ground substance of connective tissue. Hyaluronic acid is a major component of articular cartilage and skin, helping in tissue repair and the lubrication of joints. It also plays a role in cell movement and proliferation and may contribute to the growth of certain tumours. Consisting of around 25 000 disaccharide residues, it is made by a complex of enzymes on the outer cell surface, and absorbs water molecules to form a hydrated gel. Hyaluronan is commonly used as the basis for fillers in cosmetic surgery. *Hyaluronidases break down hyaluronic acid.

hyaluronidase Any of a family of enzymes that break down *hyaluronic acid, thereby decreasing its viscosity and increasing the permeability of connective tissue. Hyaluronidases are used in medicine as local adjuvants to increase the absorption and diffusion of drugs administered by injection or application, such as local anaesthetics, and in cosmetic surgery.

hybrid The offspring of a mating in which the parents differ in at least one characteristic. The term is usually used for offspring of widely different parents, e.g. different varieties or species. Hybrids between different animal species are usually sterile, as is the mule (a cross between a horse and a donkey). *See also* HYBRID VIGOUR; HYBRID ZONE.

hybrid dysgenesis Infertility and an increased incidence of chromosome mutations that is observed in the progeny of laboratory-reared organisms crossed with wild organisms. Hybrid dysgenesis has been best studied in strains of *Drosophila* fruit flies, when laboratory females are crossed with wild males; it is thought to be caused by transposable elements (*see* TRANSPOSON) in the wild-type males, which are activated in eggs from certain strains of laboratory females.

hybridization 1. The production of one or more *hybrid organisms by the mating of genetically different parents. **2.** The production of hybrid cells. *See* CELL FUSION (somatic cell hybridization). **3** *See* DNA HYBRIDIZATION.

hybridoma A type of hybrid cell that is produced by the fusion of a tumour cell (a myeloma cell) with a normal antibody-producing *B cell (*see* CELL FUSION). The resulting hybrid cell line is able to produce large amounts of normal antibody, which is described as monoclonal (*see* MONOCLONAL ANTIBODY) as it results from a cloned cell line.

hybrid vigour (heterosis) The increased vigour displayed by the offspring from a cross between genetically different parents. Hybrids from crosses between different crop varieties (F₁ hybrids) are often stronger and produce better yields than the original varieties. Mules, the offspring of mares crossed with donkeys, have greater strength and resistance to disease and a longer lifespan than either parent. One explanation invokes the phenomenon of *overdominance. Another, called **dominance complementation**, proposes that slightly superior alleles from either inbred parent will complement each other and result in a superior phenotype compared with either parent.

hybrid zone A zone that exists between the populations of two related species, whose geographical ranges overlap, which contain hybrids resulting from matings between the two species. The hybrids, which typically have a reduced fertility, will tend to act as a barrier between further interspecific mating and hence will restrict gene flow between the two species. However, mating between hybrids and individuals of either parent species can also introduce novel genetic variation, which can be of adaptive value in a changing environment. This can lead to stabilizing of the hybrid zone over time or even to enhanced gene flow between the parent species, weakening the reproductive barrier and ultimately causing fusion of the hybridizing species into a single species. Hybrid zones can range in spatial form from simple bands to more complex patterns, dependent on local factors such as topography, vegetation, and climate. *See also* ISOLATING MECHANISMS; PARAPATRIC SPECIATION.

hydathode A pore found in the *epidermis of the leaves of certain plants. Like *stomata, hydathodes are surrounded by two crescent-shaped cells but these, unlike guard cells, do not regulate the size of the aperture. Hydathodes are used by the plant to secrete water under conditions in which *transpiration is inhibited; for example, when the atmosphere is very humid. This process of water loss is called **guttation**.

hydrochloric acid A strong acid, HCl, that is produced by *oxyntic cells in the wall of the stomach and forms a constituent of *gastric juice. Hydrochloric acid is required for the conversion of pepsinogen to pepsin in the lumen of the stomach and kills various microorganisms that enter with food.

hydrocortisone *See* CORTISOL.

hydrogen acceptor *See* HYDROGEN CARRIER.

hydrogen bond A type of electrostatic interaction between electronegative (fluorine, nitrogen, or oxygen) atoms in one molecule and hydrogen atoms bound to electronegative atoms in another molecule. It is a strong dipole-dipole attraction caused by the electron-withdrawing properties of the electronegative atom. Thus, in the water molecule the oxygen atom attracts the electrons in the O-H bonds. The hydrogen atom has no inner shells of electrons to shield the nucleus, and there is an electrostatic interaction between the hydrogen proton and a lone pair of electrons on an oxygen atom in a neighbouring

0.177nm

 oxygen hydrogen

Hydrogen bonds (shown as dotted lines) between water molecules

molecule. Each oxygen atom has two lone pairs and can make hydrogen bonds to two different hydrogen atoms. The strengths of hydrogen bonds are about one tenth of the strengths of normal covalent bonds. Hydrogen bonding does, however, have significant effects on physical properties. Thus it accounts for the unusual properties of water and for its relatively high boiling point. It is also of great importance in living organisms. Hydrogen bonding occurs between complementary base pairs in the two polynucleotide strands of DNA (*see* BASE PAIRING). It also occurs between the C=O and N–H groups in proteins, and is responsible for maintaining the secondary structure.

hydrogencarbonate (bicarbonate) A salt of *carbonic acid in which one hydrogen atom has been replaced; it thus contains the hydrogencarbonate ion HCO_3^-. *See also* BUFFER.

hydrogen carrier (hydrogen acceptor) A molecule that accepts hydrogen atoms or ions, becoming reduced in the process (*see* OXIDATION-REDUCTION). The *electron transport chain, whose function is to generate energy in the form of ATP during respiration, involves a series of hydrogen carriers, including *NAD and *FAD, which pass on the hydrogen (derived from the breakdown of glucose) to the next carrier in the chain. In the light reactions of photosynthesis the key hydrogen acceptor is $NADP^+$.

hydrogenosome A cell organelle found in certain protists and chytrid fungi that live in anaerobic environments, such as Trichomonas vaginalis, a sexually transmitted parasite of the human reproductive and urinary tracts. Hydrogenosomes typically generate hydrogen as a by-product of ATP production; they are thought to be derived from mitochondria and like them are bounded by a double membrane. However, they lack many of the enzymes required for aerobic respiration, such as an electron transport chain.

hydroid A specialized elongated water-conducting cell found in the cortex of certain mosses. Hydroids are aligned end to end; at maturity they lose their cytoplasm and their end walls partially break down, becoming permeable to water and forming a rudimentary vessel.

hydrolase Any of a class of enzymes that catalyse the addition of water to, or the removal of water from, a molecule. Hydrolases play an important role in the construction and breakdown of storage materials, such as starch.

hydrological cycle (water cycle) The circulation of water between the atmosphere, land, and oceans on the earth. Water evaporates from the oceans and other water bodies on earth to form water vapour in the atmosphere. This may condense to form clouds and be returned to the earth's surface as precipitation (e.g. rainfall, hail, and snow). Some of this precipitation is returned to the atmosphere directly through evaporation or transpiration by plants; some flows off the land surface as overland flow, eventually to be returned to the oceans via rivers; and some infiltrates the ground to flow underground forming groundwater storage.

hydrophilic Having an affinity for water. Some hydrophilic substances, e.g. sodium chloride, consist of ions that dissolve in water to form a homogeneous mixture called a *solution. However, other nonionic substances, such as a sugar cube, also dissolve in water due to the presence of ionic and polar regions on their surface.

hydrophily A rare form of pollination in which pollen is carried to a flower by water. It occurs by one of two methods. In Canadian pondweed (*Elodea canadensis*) the male flowers break off and float downstream until they contact the female flowers. In *Zostera*, a marine species, the filamentous pollen grains are themselves carried in the water. *Compare* ANEMOPHILY; ENTOMOPHILY.

hydrophobic Lacking affinity for water. Substances that lack ionic or polar regions cannot form hydrogen bonds with water and hence will not dissolve in water. Their hydrophobic nature is an important property of cell membranes.

hydrophyte Any plant that lives either in very wet soil or completely or partially submerged in water. Structural modifications of hydrophytes include the reduction of mechanical and supporting tissues and vascular tissue, the absence or reduction of a root system, and specialized leaves that may be either floating or finely divided, with little or no cuticle. Examples of hydrophytes are waterlilies and certain pondweeds. *Compare* HALOPHYTE; MESOPHYTE; XEROPHYTE.

hydroponics A commercial technique for growing certain crop plants in culture solutions rather than in soil. The roots are immersed in an aerated solution containing the correct proportions of essential mineral salts. The technique is based on various water culture methods used in the laboratory to assess the effects of the absence of certain mineral elements on plant growth.

hydrosere The sequence of plant communities (*see* SERE) occurring during the change from shallow open water to forest or bog. The process commences with accumulations of silt, enabling initial colonization by submerged or floating plants, such as water lilies and pondweeds, depending on the rate of flow and nutrient status of the water. As silting increases and organic debris is deposited, reeds, sedges, and similar plants begin to appear, forming a **swamp**. Organic matter builds up as *peat, and conditions progressively become drier, creating a stage called a **fen**, dominated by herbaceous species, and then a **carr**, in which shrubs and small trees predominate. Eventually, the substrate is sufficiently stable to support the larger trees of mature forest species. In conditions of high rainfall or low evaporation a different hydrosere may occur. The rate of peat formation may be sufficiently high to create a **bog**, i.e. a permanently waterlogged raised site that receives all its water and nutrient inputs from rainfall or other precipitation.

hydrosphere The water on the surface of the earth. Some 74% of the earth's surface

is covered with water, 97% (or some 1.3×10^{21} kilograms) of which is in the oceans. Icecaps and glaciers contain about 2.6×10^{19} kg, rivers about 10^{15} kg, lakes and inland seas about 2×10^{17} kg, and groundwater (down to 4000 metres) about 1.5×10^{19} kg. Water in the atmosphere contains only about 1.3×10^{16} kg.

hydrostatic skeleton The system of support found in soft-bodied invertebrates, which relies on the incompressibility of fluids contained within the body cavity. For example, in earthworms the coelomic fluid is under pressure within the coelom and therefore provides support for internal organs.

hydrotropism The growth of a plant part in response to a water potential gradient. Roots, for example, grow towards water in the soil; i.e. they show positive hydrotropism. The mechanism is not fully understood, but studies in thale cress (*Arabidopsis thaliana*) show that the elongation zone behind the root tip serves as a moisture sensor and undergoes differential growth to direct the root towards the water. The process involves the plant hormone *abscisic acid. Another study in pea seedlings has shown that roots respond to the acoustic vibrations of water flowing in nearby pipes, suggesting a possible mechanism of long-range detection of water sources by plant roots. Other factors, such as gravity and light, also affect root growth, and how these are integrated for optimal direction of growth remains unclear. *See* GEOTROPISM; TROPISM.

hydroxonium ion *See* ACID.

hydroxyl Denoting the group −OH in a chemical compound or the ion OH⁻.

5-hydroxytryptamine *See* SEROTONIN.

Hydrozoa *See* CNIDARIA.

hygroscopic Describing a substance that can take water from the atmosphere.

hymen A fold of mucous membrane that covers the opening of the vagina at birth. It normally perforates at puberty, to allow the flow of menstrual blood, but if the opening is small it may be ruptured during the first occasion of sexual intercourse.

Hymenoptera An order of insects that includes the ants, bees, wasps, ichneumon flies, and sawflies. Hymenopterans generally have a narrow waist between thorax and abdomen. The smaller hindwings are interlocked with the larger forewings by a row of tiny hooks on the leading edges of the hindwings. Some species are wingless. The mouthparts are typically adapted for biting, although some advanced forms (e.g. bees) possess a tubelike proboscis for sucking liquid food, such as nectar. The long slender *ovipositor can serve for sawing, piercing, or stinging. Metamorphosis occurs via a pupal stage to the adult form. *Parthenogenesis is common in the group.

Ants and some bees and wasps live in colonies, often comprising numerous individuals divided into *castes and organized into a coordinated and complex society. The colony of the honeybee (*Apis mellifera*), for example, consists of workers (sterile females), *drones (fertile males), and usually a single fertile female—the queen. The primary concern of the queen is egg laying. She determines the gender of the egg by either withholding or releasing stored sperm. Unfertilized eggs become males; fertilized eggs become females. Under the influence of pheromones released by the queen, the workers fulfil a variety of tasks, including nursing the developing larvae, building the wax cells (combs) of the hive, guarding the colony, and foraging for nectar and pollen. The single function of the larger drones is to mate with the young queen on her nuptial flight.

hyoid arch The second of seven bony V-shaped arches that support the gills of fish. The dorsal part is specialized as the **hyomandibular**, which is involved in jaw suspension (*see* AMPHISTYLIC JAW SUSPENSION; HYOSTYLIC JAW SUSPENSION). In tetrapods the dorsal part of the hyoid arch has evolved to form the stapes (one of the *ear ossicles). The ventral part forms the **hyoid bone**, which supports the tongue.

hyomandibular *See* HYOID ARCH.

hyostylic jaw suspension A type of jaw suspension seen in most fishes, in which the upper jaw is not directly connected to the cranium. The attachment between cranium and upper jaw is made by a ligament at the front end and by the hyomandibular (*see* HYOID ARCH) at the rear end (see illustration). *Compare* AMPHISTYLIC JAW SUSPENSION; AUTOSTYLIC JAW SUSPENSION.

hyper- A prefix denoting over, above, high; e.g. hyperpolarization.

hyperaccumulator A plant that can accumulate abnormally high concentrations of heavy metal ions, such as lead, cadmium, nickel, zinc, arsenic, and aluminium. For example, alpine pennycress, *Noccaea* (*Thlaspi*) *caerulescens*, grows on spoil heaps and former mine workings and can contain up to 30 000 ppm of zinc compared to a typical concentration of 100 ppm in other plants. Hyperaccumulators are used in *phytoremediation to extract metal contaminants from polluted land.

hyperglycaemia Excessive amounts of glucose in the bloodstream, which occurs in diabetes mellitus due to underproduction of *insulin. *Compare* HYPOGLYCAEMIA.

Hyostylic jaw suspension

Hypermetropia

hypermetamorphosis *See* HETERO-MORPHOSIS.

hypermetropia (hyperopia) Long-sightedness. A vision defect in which the lens of the eye is unable to accommodate sufficiently to throw the image of near objects onto the retina. It is caused usually by shortness of the eyeball rather than any fault in the lens system. The subject requires spectacles with converging lenses to bring the image from behind the retina back on to its surface (see illustration).

hypermorphosis A form of *hetero-chrony in which, during the course of evolution, the rate of development is unchanged but the total development time is extended, permitting the addition of new stages to the end of the ancestral development sequence. The morphological outcome is an example of *peramorphosis. Hence the development (ontogeny) of such organisms conforms to the theory of *recapitulation of their phylogeny.

hyperparasite A parasite that lives in or on another parasite. The most common examples are insects that lay their eggs inside or near *parasitoid larvae, which are themselves parasitizing the tissues of a host, again usually an insect larva. Hyper-parasites, more strictly termed **hyperpara-sitoids**, are found mainly among the Hymenoptera (wasps, sawflies, etc.), although some flies and beetles also adopt this strategy.

hyperplasia Increase in the size of a tissue or organ due to an increase in the number of its component cells. *Compare* HYPERTROPHY.

hyperpolarization *See* INHIBITORY POSTSYNAPTIC POTENTIAL.

hypersensitivity 1. (in immunology) Increased or abnormal sensitivity to compounds, which can elicit a specific immune response accompanied by tissue damage. Hypersensitive reactions include *allergies and *anaphylaxis. *See* IMMUNITY. *See also* HAPTEN. **2.** (in botany) A reaction by the tissues of a plant in response to invasion by pathogenic viruses, bacteria, fungi, or other organisms. Defence responses are induced by binding of *elicitor substances to **resistance proteins (R proteins)** located in the cytoplasm of the plant's cells. The elicitors can be components of the pathogens themselves or products of fungal degradation of the plant cell walls. The reaction varies according to the type of infecting agent, but typically involves the synthesis of enzymes (such as *chitinase) that disarm or degrade the pathogen, the production of *phytoalexins, which inhibit the growth of pathogens, and strengthening of cell walls in the vicinity of the attack by deposition of lignin and other materials. Localized generation of highly reactive oxygen derivatives is also commonly deployed, and various antimicrobial peptides (*defensins) can be synthesized by stressed plant tissues. The final strategy is programmed cell death (*apoptosis) around the infection site, which slows the spread of the pathogen to neighbouring healthy tissues. *See also* IMMUNITY; SYSTEMIC ACQUIRED RESISTANCE.

hypertension *See* BLOOD PRESSURE.

hyperthermophile *See* THERMOPHILIC.

hyperthyroidism Overactivity of the *thyroid gland. This condition causes an increase in *basal metabolic rate, with such symptoms as rapid pulse and weight loss.

It is sometimes associated with a *goitre. *Compare* HYPOTHYROIDISM.

hypertonic solution A solution that has a lower *water potential and a correspondingly higher osmotic pressure than another solution. *See* OSMOSIS.

hypertrophy An increase in the size of a tissue or organ due to an increase in the size of its component cells. Hypertrophy often occurs in response to an increased workload in an organ, which may result from malfunction or disease. *Compare* HYPERPLASIA.

hypervariable loops *See* COMPLEMENTARITY-DETERMINING REGION.

hyperventilation An increase in the amount of air taken into the lungs caused by an increase in the depth or rate of breathing. *See also* VENTILATION.

hypha (*pl.* **hyphae**) A delicate filament in fungi many of which may form either a loose network (*mycelium) or a tightly packed interwoven mass of *pseudoparenchyma, as in the fruiting body of mushrooms. Hyphae may be branched or unbranched and may or may not possess cross walls. The cell wall consists of either fungal cellulose or a nitrogenous compound called *chitin. The cell wall is lined with cytoplasm, which often contains oil globules and glycogen, and there is a central vacuole. The hyphae produce enzymes that in parasitic fungi digest the host tissue, and in saprotrophic fungi digest dead organic matter.

hypo- A prefix denoting under, below, low; e.g. hypogyny, hyponasty.

hypoblast *See* BLASTULA.

hypocotyl The region of the stem beneath the stalks of the seed leaves (*cotyledons) and directly above the young root of an embryo plant. It grows rapidly in seedlings showing *epigeal germination and lifts the cotyledons above the soil surface. In this region (the **transition zone**) the arrangement of vascular bundles in the

root changes to that of the stem. *Compare* EPICOTYL.

hypodermis (exodermis) The outermost layer of cells in the plant *cortex, lying immediately below the epidermis. These cells are sometimes modified to give additional structural support or to store food materials or water. After the loss of the *piliferous layer of the root the hypodermis takes over the protective functions of the epidermis.

hypogeal 1. Describing seed germination in which the seed leaves (cotyledons) remain below ground. Examples of hypogeal germination are seen in oak and runner bean. *Compare* EPIGEAL. **2.** Describing fruiting bodies that develop underground, such as truffles and peanuts.

hypoglycaemia Below-normal levels of glucose in the blood plasma, causing weakness, dizziness, and sweating. *Compare* HYPERGLYCAEMIA.

hypogyny A floral arrangement in which the ovary is *superior, i.e. it arises from the receptacle above the sepals, petals, and stamens (see illustration). The perianth and stamens are said to be **hypogynous** with respect to the ovary, as seen in the tulip. *Compare* EPIGYNY; PERIGYNY.

hypolimnion The lower layer of water in a lake. *See* THERMOCLINE.

hypophysis (*pl.* **hypophyses**) *See* PITUITARY GLAND.

hyporheic zone The region beneath and alongside the bed of a stream or river where the stream water and groundwater mix. Water can flow through the hyporheic zone as a consequence of upwelling of groundwater or seepage of stream water, giving rise to **hyporheic** (or **interstitial**) **flow**. This zone sees continual interchange of material between stream water and groundwater, which can impact on the quality of both, for example in terms of water chemistry and pollutant load. It is also important as a habitat, especially for microbial and invertebrate communities, and for fish spawning.

hypogyny

hypostasis (in genetics) Masking of the phenotype of a gene by the action of another gene. *See* EPISTASIS.

hypothalamus (*pl.* **hypothalami**) Part of the vertebrate brain that is derived from the *forebrain and located on the ventral surface below the *thalamus and the *cerebrum. The hypothalamus regulates a wide variety of physiological processes, including maintenance of body temperature (*see* THERMOREGULATION), the *alarm response to stress, water balance, sleeping, feeding, lactation, and ovarian and uterine cycles, via both the *autonomic nervous system (which it controls) and the *neuroendocrine system. In mammals the central oscillator of the *biological clock is located in the paired clusters of neurons known as the suprachiasmatic nuclei within the hypothalamus. The hypothalamus synthesizes the hormones oxytocin and antidiuretic hormone, which are transported to the posterior pituitary, from where they are released into the bloodstream. Indeed, the endocrine functions of the hypothalamus are largely mediated by the *pituitary gland, to which it is connected by a short stalk. The pituitary responds to releasing hormones produced by the hypothalamus, which thereby indirectly controls hormone production in other glands.

hypothyroidism Underactivity of the *thyroid gland, which is generally due to a deficiency in *thyroid-stimulating hormone. In adults hypothyroidism causes a decrease in *basal metabolic rate, with tissue swelling and weight gain (**myxoedema**). In children thyroid hormone is required for development of the bones and nervous system, and hypothyroidism can result in *cretinism. *Compare* HYPERTHYROIDISM.

hypotonic solution A solution that has a higher *water potential and a correspondingly lower osmotic pressure than another solution. *See* OSMOSIS.

hypoxia A deficiency of oxygen in body tissues, which can result from living in an oxygen-deficient environment, inadequate *inspiration, or deficiency of red blood cells or haemoglobin (required for oxygen transport).

H zone The region of a striated muscle fibre that contains only thick (*myosin) filaments. The H zone appears as a lighter band in the middle of the dark *A band at the centre of a *sarcomere.

IAA (indoleacetic acid) *See* AUXIN.

I band The region of a striated muscle fibre that contains only thin (*actin) filaments. I bands are visible as light bands at either end of a *sarcomere.

ICAM (intercellular adhesion molecule) *See* CELL ADHESION MOLECULE.

ice age A period in the earth's history during which ice advanced towards the equator and a general lowering of temperatures occurred. The last major ice age, that of the Pleistocene epoch (sometimes known as the **Ice Age**), ended about 11 500 years ago. At least four major ice advances (glacials) occurred during the Pleistocene; these were separated by interglacials during which the ice retreated and temperatures rose. Since the end of the Pleistocene the earth has been in an interglacial phase, the Holocene epoch. It has been established that ice ages also occurred during the Precambrian over 600 million years ago (*see* CRYOGENIAN) and during the Permo-Carboniferous (about 250 million years ago).

(((()))) SEE WEB LINKS

• Sponsored by the US Public Broadcasting Service, this site explains why the earth warms up and cools down

ice-nucleating agent Any agent that promotes the formation of ice crystals. Such agents include minute solid particles, such as dust or food particles, and large molecules. Certain organisms living in cold climates, such as some insects, molluscs, and nematodes, may manufacture **ice-nucleating proteins** in order to achieve a slow controlled freezing of their extracellular body water when the temperature falls below freezing. The structure of these proteins encourages the binding of water molecules and prevents sudden and damaging ice formation. However, gut particles can also act as ice-nucleating agents, and hence organisms that cannot tolerate freezing must void their guts at the onset of freezing conditions. *See also* CRYOPROTECTANT.

ichthyosaur Any extinct aquatic reptile belonging to the order Ichthyosauria, which lived during the Mesozoic era (250–66 million years ago). The earliest forms resembled lizards with flippers and probably swam in an eel-like fashion. Their successors evolved to become progressively more fishlike, with thickened bodies, finlike fore and hind limbs, and crescent-shaped tails. Fossilized skeletal remains show that the eyes were unusually large and adapted for hunting in poor visibility. Ichthyosaurs typically lived on a diet of squid, and some species attained lengths of over 15 m.

ICSH *See* LUTEINIZING HORMONE.

identical twins (monozygotic twins) Two individuals that develop from a single fertilized egg cell by its division into two genetically identical parts. Each part eventually gives rise to a separate individual and these twins share identical DNA sequences. However, even identical twins can differ in patterns of methylation and acetylation of their DNA and in the histone proteins that package the DNA, which result in differences in gene expression (*see* GENE IMPRINTING). Such epigenetic phenomena may explain observed differences in looks, personality, etc. *Compare* FRATERNAL TWINS.

identification key *See* KEY.

idioblast A plant cell that differs markedly in structure or function from surrounding tissue cells. Idioblasts occur in clusters or singly and serve various functions, notably the storage of secondary metabolites, such as defensive chemicals that help to deter browsing animals.

idiogram *See* KARYOGRAM.

idiosyncrasy An abnormal reaction to a drug or other foreign substance shown by an individual, which is usually genetically determined. An individual that shows immunological idiosyncrasy is said to be **hypersensitive** to a particular substance, agent, etc.

Ig *See* IMMUNOGLOBULIN.

IGF *See* INSULIN-LIKE GROWTH FACTOR.

ileum (*pl.* **ilea**) The portion of the mammalian *small intestine that follows the *jejunum and precedes the *large intestine. It is a site of digestion and absorption. The internal lining of the ileum bears numerous small outgrowths (*see* VILLUS), which increase its absorptive surface area.

ilium (*pl.* **ilia**) The largest of the three bones that make up each half of the *pelvic girdle. The ilium bears a flattened wing of bone that is attached by ligaments to the sacrum (*see* SACRAL VERTEBRAE). *See also* ISCHIUM; PUBIS.

imaginal disc A group of undifferentiated cells in an insect larva that develops into a specific adult structure. Imaginal discs are derived from the *blastoderm. They do not undergo further development until the pupal stage, when they start to differentiate into adult epidermal structures, such as eyes, antennae, and wings, under the control of the hormone *ecdysone. The developmental fate of the cells in each disc is determined, and different zones of the disc give rise to particular structures. For example, a *fate map of the leg disc of the fruit fly *Drosophila* can be drawn, consisting of concentric circles of cells; the inner zones develop into the distal leg structures (tarsus and tibia), while the outermost zones give rise to the proximal structures (femur, trochanter, and coxa).

imago (*pl.* **imagos** or **imagines**) The adult sexually mature stage in the life cycle of an insect after metamorphosis.

imbibition The uptake of water by substances that do not dissolve in water, so that the process results in swelling of the substance. Imbibition is a property of many biological substances, including cellulose (and other constituents of plant cell walls), starch, and some proteins. It occurs in dry seeds before they germinate and—together with osmosis—is responsible for the uptake of water by growing plant cells.

immersion objective An optical microscope objective in which the front surface of the lens is immersed in a liquid on the cover slip of the microscope specimen slide. Cedar-wood oil (for an **oil-immersion lens**) or sugar solution is frequently used. It has the same refractive index as the glass of the cover slip, so that the object is effectively immersed in it. The presence of the liquid increases the effective aperture of the objective, thus increasing the resolution.

immune clearance The accelerated removal of an antigen from the bloodstream that follows the initiation of an antibody response by the immune system. This leads to the formation of antigen-antibody complexes, which are ingested by macrophages and other phagocytic cells.

immune complex A small soluble cluster of bound antigen and antibody. Immune complexes typically contain small antigens, such as toxins or bacterial cell debris, and are commonly found following infections. Generally they are removed from the circulation after triggering the binding of *complement proteins, which in turn bind to receptors on red blood cells (erythrocytes). These transport the complexes to the liver and spleen, where macrophages remove the complexes and destroy them, leaving the erythrocytes intact. Failure to remove immune complexes results in their deposition in small blood vessels, particularly of the kidney glomeruli, as occurs in the autoimmune disease systemic lupus erythematosus.

immune response The reaction of the body to foreign or potentially dangerous substances (*antigens), particularly disease-producing microorganisms, or to cancerous cells. *See* IMMUNITY.

immunity 1. The ability of an animal to resist infection, to counter the harmful effects of toxins produced by infecting organisms, or to destroy cancer cells. Immunity in vertebrates depends on the presence in the body of a range of defensive cells and substances, notably *antibodies and white blood cells (*lymphocytes), which produce an immune response. **Innate (inherited or natural) immunity** is the body's first line of defence and plays a vital role in controlling invading organisms during the early stages of an infection. The entry of most potential pathogens is blocked by the body's natural barrier defences, primarily the skin and the mucous membranes that line the digestive, respiratory, urinary, and reproductive tracts. Epithelial surfaces secrete mucus, which traps foreign particles, including microorganisms, while tears and saliva help to guard against entry via the eyes and mouth, aided by the enzyme *lysozyme. In the airways a ciliated epithelium effectively sweeps away the mucus and trapped particles. Moreover, any pathogens entering in food must survive the highly acidic environment of the stomach. Pathogens that evade these physical and chemical barriers then face innate cellular defences. Macrophages, neutrophils, eosinophils, and certain other immune cells have cell surface receptors, called *Toll-like receptors, that bind to common surface constituents of bacteria, viruses, fungi, and parasites. Such an encounter can prompt the macrophage to engulf the microorganism and to induce it to secrete an array of *cytokines and *chemokines. These substances attract additional immune cells to the site, including neutrophils and monocytes from nearby blood vessels, and initiate the process of *inflammation. The *complement system of defensive proteins is also activated; these coat target cells with fragments that assist recognition by macrophages. The abnormal surface proteins of virus-infected or cancerous cells may attract the attention of circulating *natural killer cells, which release toxic

chemicals that trigger programmed cell death. This innate response plays a crucial role in promoting **adaptive** (or **acquired**) **immunity.** *Dendritic cells ingest foreign material, such as bacteria and virus particles (*see* ANTIGEN), and carry it to lymph organs, where they signal lymphocytes to carry out a more targeted and versatile repertoire of immune responses; adaptive immunity can persist throughout the lifetime of the individual. **Active immunity** arises when the body produces antibodies against an invading foreign substance (antigen), either through infection or *immunization. **Humoral immunity** is when B lymphocytes produce free antibodies that circulate in the bloodstream (*see* B CELL; B-CELL RECEPTOR); **cell-mediated immunity** is caused by the action of T lymphocytes (*see* T CELL; T-CELL RECEPTOR). **Passive immunity** is induced by injection of serum taken from an individual already immune to a particular antigen; it can also be acquired by the transfer of maternal antibodies to offspring via the placenta or breast milk (*see* COLOSTRUM). Active immunity tends to be long-lasting (*see* IMMUNOLOGICAL MEMORY); passive immunity is short-lived. Invertebrates rely on innate defence mechanisms to combat invading pathogens, particularly the barrier of an external cuticle or exoskeleton, and multifunctional immune cells such as *haemocytes. *See also* AUTOIMMUNITY. **2.** The condition induced in plants in response to infection by pathogens. Two broad lines of immune defence are recognized, both triggered by elicitor molecules derived either from the invading pathogen itself or from the breakdown of plant tissues arising from the infection. **PAMP-triggered immunity (PTI)** is a general immunity that arises from the interaction of transmembrane receptors (called pattern recognition receptors) in plant cells with elicitor molecules that characterize an entire category of pathogens. These are called pathogen-associated molecular patterns (PAMPs), a classic example being the chitin constituent of fungal cells walls. **Effector-triggered immunity (ETI)** is induced by elicitor molecules called effectors, specific to certain species or even strains of pathogens. These bind to *R proteins in the cytoplasm of the plant cell and thereby trigger an immune response, which tends to be stronger than the PTI response; however, many of the effector pathways by which the plant responds are common to both, and the distinction is often blurred. In either case, receptor binding causes signal molecules to enter the cell nucleus and activate genes, resulting in production of defensive molecules that initiate various immune responses. These include the production of reactive oxygen species and nitric oxide, which are toxic to certain pathogens and also contribute to programmed cell death in the *hypersensitive response, by which infected cells are sacrificed to save remaining healthy tissue. The deposition of polymers such as callose and lignin on the inside of cell walls acts to seal healthy cells from adjacent infected ones. *Phytoalexins are nonspecific antimicrobial substances that can kill fungi and bacteria, and are also induced by viral infections and injury. Moreover, in both PTI and ETI a form of long-term immunity, called *systemic acquired resistance, is initiated throughout the plant by the plant hormone salicylic acid.

immunization The production of *immunity in an individual by artificial means. **Active immunization** (**vaccination**) involves the introduction, either orally or by injection (**inoculation**), of specially treated bacteria, viruses, or their toxins to stimulate the production of *antibodies (*see* VACCINE). **Passive immunization** is induced by the injection of preformed antibodies.

immunoassay Any of various techniques that measure the amount of a particular substance by virtue of its binding antigenically to a specific antibody. In **solid-phase immunoassay** the specific antibody is attached to a solid supporting medium, such as a PVC sheet. The sample is added and any test antigens will bind to the antibody. A second antibody, specific for a different site on the antigen, is added. This carries a radioactive label (in **radioimmunoassay**) or fluorescent label (in **fluoroimmunoassay**), enabling its concentration, and thus that of the test antigen, to be determined by comparison with known standards. Variations on this technique include *ELISA and *western blotting. In **counting immunoassay**, polystyrene beads are coated with an excess of the specific antibody. Beads that bind antigen clump together, whereas unbound beads do not and are counted using a cell counter. The proportion of unbound beads is inversely proportional to the concentration of antigen in the sample. The principle of immunoassay has also been employed with certain types of *microarray.

immunoelectron microscopy A form of transmission electron microscopy (*see* ELECTRON MICROSCOPE) in which the specimen is treated with gold-labelled antibodies specific for the protein component of interest. The gold labels are then revealed as electron-dense dark areas on the image. Labelling with different-sized gold particles can enable two different components to be visualized simultaneously. However, because the sections of material used are ultra-thin, only relatively few labels are likely to be bound in any one section.

immunoelectrophoresis An analytical technique used to identify antigens that have been separated by *electrophoresis. The gel medium used for the electrophoresis is irrigated with specific antibodies for the required antigens: when an antigen meets its corresponding antibody a precipitate forms, which can be seen and isolated.

immunofluorescence A technique used to identify specific proteins in tissue samples by applying antibodies that have been labelled with a fluorescent compound (e.g. fluorescein). The labelled antibody combines with the protein under investigation and the resulting complex can be identified under a microscope by its fluorescence (i.e. immunofluorescence microscopy). Proteins and enzymes in tissue samples can be identified and located in this way. In **indirect immunofluorescence** the bound antibody is detected by an 'anti-antibody' fluorescent immunoglobulin. *See also* CONFOCAL FLUORESCENCE MICROSCOPY.

immunogenicity The degree to which an antigen can produce an immune response.

immunoglobulin (Ig) One of a group of proteins (*globulins) in the body that act as *antibodies. They are produced by specialized white blood cells called *B cells and are present in blood serum and other body fluids. An antibody is typically a Y-shaped structure consisting of four polypeptide chains—two heavy chains and two light chains. Each arm of the 'Y' bears an antigen-binding site (*see* COMPLEMENTARITY-DETERMINING REGION). There are five classes (*isotypes), with distinct immunological and physical properties. The most abundant immunoglobulin of blood, lymph, and tissue fluid is **immunoglobulin G (IgG)**. It binds to microorganisms, promoting their engulfment by phagocytes (*macrophages), and to viruses and bacterial toxins, thereby neutralizing them. It also binds *complement, which results in lysis of the target cell, and can cross the placenta to protect the fetus. Mature B cells express IgG on their surface, whereas both immature and mature B cells express **immunoglobulin M (IgM)**, the first antibody to be produced following immunization or infection. Found in blood and lymph, it is an aggregate of five of the basic Y-shaped units, enabling it to 'mop up' microorganisms or other antigens with its ten binding sites. However, its chief role is to activate complement. **Immunoglobulin A (IgA)** is found in saliva, tears, breast milk, and mucous secretions, where its role is to neutralize viruses and bacteria as they enter the body. It also occurs in serum. **Immunoglobulin D (IgD)** is present in serum in very low concentrations, but occurs on the surface of antibody-secreting B cells; it is important in B-cell activation. **Immunoglobulin E (IgE)** also normally has very low concentrations in blood and connective tissues, but it plays a crucial role in allergic reactions. It binds to *mast cells and *basophils, and, when triggered by the presence of antigen, causes histamine release and hence inflammation and other common allergic symptoms. *See also* FAB FRAGMENT; IMMUNITY.

(⊕) SEE WEB LINKS
• Introduction to immunoglobulins

immunologically privileged site Any of various sites in the body where foreign tissue grafts do not induce immune responses. Such sites include the eye, brain, testis, and unborn fetus. So, for example, a corneal graft placed in the eye of an unrelated recipient will generally not be rejected by the recipient's immune system. Although antigens do migrate from these privileged sites, they either induce immunological *tolerance or a nondestructive response. However, damage to such sites often results in autoimmune attack by the host immune system, as in multiple sclerosis.

immunological memory The ability of the immune system to mount a secondary immune response, i.e. to respond faster and more effectively to subsequent exposures to an antigen following an initial, or primary, immune response to the same antigen (e.g. as a result of an infection or immunization). It depends on the formation of a small and persistent population of memory cells, induced by the primary immune response. Memory T cells appear typically only five days following initial immunization, whereas memory B cells may take about a month to reach maximum levels. Populations of such cells persist for the lifetime of the individual and exhibit enhanced affinity for the antigen, especially with repeated exposure to the original antigen. When memory cells encounter their specific antigen, they rapidly divide to produce a clone of effector cells capable of dealing with the target. Memory B cells undergo changes in their immunoglobulin genes that enable them to produce antibodies with greater affinity for the antigen, resulting in its efficient disposal.

immunological synapse The highly organized region of contact between an effector CD4 *T cell and its target *antigen-presenting cell. It involves clustering of T-cell receptors within a local area of the T-cell plasma membrane and of complexes of the antigen-MHC class II protein in the apposed part of the target cell.

With the additional involvement of coreceptors and adhesion molecules, this creates a **supramolecular adhesion complex (SMAC)**. The T cell then becomes polarized, which focuses the secretion of soluble cytokines at the point of contact between the cells.

immunological tolerance *See* TOLERANCE.

immunosuppression The suppression of an immune response. Immunosuppression is necessary following organ transplants in order to prevent the host rejecting the grafted organ (*see* GRAFT); it is artificially induced by radiation or chemical agents that inhibit cell division of lymphocytes. Immunosuppression occurs naturally in certain diseases, notably *AIDS.

immunotoxin A therapeutic agent consisting of a monoclonal antibody linked to a toxin or drug. It is used to target, for example, a particular tumour in a patient. The antibody selectively binds to cancer cells expressing the corresponding antigen and is taken into the cell, where the toxin (e.g. ricin A chain) is cleaved, allowing it to kill the cell.

imperfect (in botany) Describing a flower that has exclusively male or female sex organs, i.e. either stamens or carpels. *Compare* PERFECT.

imperfect fungi *See* DEUTEROMYCOTA.

Imperial units The British system of units based on the pound and the yard. The former f.p.s. system was used in engineering and was loosely based on Imperial units; for all scientific purposes *SI units are now used. Imperial units are also being replaced for general purposes by metric units.

implant Any substance, device, or tissue that is inserted into the body. For example, drug implants and heart pacemakers are typically inserted under the skin.

implantation (nidation) (in embryology) The embedding of a fertilized mammalian egg into the wall of the uterus (womb) where it will continue its development. After fertilization in the fallopian tube the egg passes into the womb in the form of a ball of cells (**blastocyst**). Its outer cells (trophoblast) secrete adhesion molecules and enzymes, which destroy cells of the uterine wall and form a cavity into which the blastocyst sinks. In humans implantation occurs about six days after fertilization.

imprinting 1. (in ethology) A specialized form of learning in which young animals, during a **critical period** (or **sensitive period**) in their early development, learn to recognize and approach some large moving object nearby. In nature this is usually the mother, though simple models or individuals of a different species (including humans) may suffice. Imprinting was first described by Konrad Lorenz, working with young ducks and geese. *See* LEARNING IN ANIMALS (Feature). **2.** (in genetics) *See* GENE IMPRINTING.

impulse (nerve impulse) The signal that travels along the length of a *nerve fibre and is the means by which information is transmitted through the nervous system. It is marked by the flow of ions across the membrane of the *axon caused by changes in the permeability of the membrane, producing a reduction in potential difference that can be detected as the *action potential. The strength of the impulse produced in any nerve fibre is constant (*see* ALL-OR-NONE RESPONSE).

(⊕) SEE WEB LINKS
• Details of nerve impulses and other aspects of neurophysiology

inbreeding Mating between closely related individuals, the extreme condition being self-fertilization, which occurs in many plants and some primitive animals. A population of inbreeding individuals generally shows less variation than an *outbreeding population. Continued inbreeding among a normally outbreeding population leads to **inbreeding depression** (the opposite of *hybrid vigour) and an increased incidence of harmful characteristics. For example, in humans, certain mental and other defects tend to occur more often in families with a history of cousin marriages.

incisor A sharp flattened chisel-shaped *tooth in mammals that is adapted for biting food and—in rodents—for gnawing. In humans there are normally two pairs of incisors (central and lateral) in each jaw. *See* PERMANENT TEETH.

inclusion body 1. Any of various particulate structures formed in plant and animal cells due to viral infection. Inclusion bodies sometimes become embedded in cellular proteins. **2.** A storage granule found in cells of bacteria or archaea. They may consist of energy compounds (e.g. poly-β-hydroxybutyric acid or elemental sulphur) or serve as reservoirs of structural components.

inclusive fitness The quality that organisms attempt (unconsciously) to maximize as the result of natural selection acting on genes that are influential in controlling their behaviour and physiology. It includes the individual's own **reproductive success** (i.e. direct *fitness, usually taken as the number of its offspring that survive to adulthood) and also the effects of the individual's actions on the reproductive success of its relatives (i.e. indirect fitness), because relatives have a higher probability of sharing some identical genes with the individual than do other members of the population. When interactions between relatives are likely to occur (which happens during the lives of many animals and plants) *kin selection will operate.

incompatibility 1. The condition that exists when foreign grafts or blood transfusions evoke a marked immune response and are rejected. **2.** The phenomenon in which pollen from one flower fails to fertilize other flowers on the same plant (**self-incompatibility**) or on other genetically similar plants. This genetically determined mechanism prevents self-fertilization (breeding between likes) and promotes cross-fertilization (breeding between individuals with different genetic compositions). The mechanism is based on chemical interactions between male pollen and female pistil tissues, governed by numerous alleles at the incompatibility *S* locus. This form of genetic polymorphism is maintained by *balancing selection. *See also* FERTILIZATION; POLLINATION.

incomplete dominance The condition that arises when neither *allele controlling a characteristic is dominant and the aspect displayed by the organism results from the partial influence of both alleles. For example, a snapdragon plant with alleles for red and for white flowers produces pink flowers. *Compare* CODOMINANCE.

incubation 1. The process of maintaining the fertilized eggs of birds and of some reptiles and egg-laying mammals at the optimum temperature for the successful development of the embryos. A period of incubation follows the laying of the eggs and precedes their hatching. **2.** The process of maintaining a *culture of bacteria or other microorganisms at the optimum temperature for growth of the culture or of maintaining a reaction mixture at an appropriate or specified temperature. **3.** The phase in the development of an infectious disease between initial infection and the appearance of the first symptoms.

incus (anvil) (*pl.* **incudes**) The middle of the three *ear ossicles of the mammalian middle ear.

indefinite inflorescence *See* RACEMOSE INFLORESCENCE.

indehiscent Describing a fruit or fruiting body that does not open to release its seeds or spores when ripe. Instead, release occurs when the fruit wall decays or, if eaten by an animal, is digested. *Compare* DEHISCENCE.

independent assortment The separation of the alleles of one gene into the reproductive cells (gametes) independently of the way in which the alleles of other genes have segregated. By this process all possible combinations of alleles should occur equally frequently in the gametes. In practice this does not happen because alleles situated on the same chromosome tend to be inherited together (*see* HAPLOTYPE). However, if the allele pairs *Aa* and *Bb* are on different chromosomes, the combinations *AB*, *Ab*, *aB*, and *ab* will normally be equally likely to occur in the gametes. *See* MEIOSIS; MENDEL'S LAWS.

indeterminate growth *See* GROWTH.

index fossil (zone fossil) An animal *fossil of a group that existed continuously during a particular span of geological time and can therefore be used to date the rock in which it is found. Index fossils are found chiefly in sedimentary rocks. They are an essential tool in stratigraphy for comparing the geological ages of sedimentary rock formations. Examples are *ammonites and *graptolites.

indicator species A plant or animal species that is very sensitive to a particular environmental factor, so that its presence (or absence) in an area can provide information about the levels of that factor. For example, some lichens are very sensitive to the concentration of sulphur dioxide (a major pollutant) in the atmosphere. Examination of the lichens present in an area can provide a good indication of the prevailing levels of sulphur dioxide.

indigenous Describing a species that occurs naturally in a certain area, as distinct from one introduced by humans; native.

indoleacetic acid (IAA) *See* AUXIN.

induced-fit model A proposed mechanism of interaction between an enzyme and a substrate. It postulates that exposure of an enzyme to a substrate causes the *active site of the enzyme to change shape in order to allow the enzyme and substrate to bind (*see* ENZYME-SUBSTRATE COMPLEX). This description is generally preferred to the *lock-and-key mechanism.

inducer 1. A signal molecule that influences the fate of cells during embryological development. Inducers activate the expression of specific genes in target cells to bring about differentiation of the latter into specialized tissue cells. **2.** A substance that

relieves repression of a gene and so allows its expression. For example, in the bacterium *E. coli* when lactose (or certain related molecules) binds to the repressor of the *lac operon, the repressor dissociates from the operator region, allowing transcription of the structural genes encoding the enzymes responsible for lactose metabolism. *See* OPERON.

inducible nitric oxide synthase (iNOS) An enzyme, found in activated macrophages and certain other tissues, that catalyses the production of the highly reactive molecule *nitric oxide (NO). Macrophages that have ingested bacteria or other pathogens are prompted by helper T cells (T_H1 cells) to destroy those pathogens. γ-Interferon (IFN-γ) secreted by the T_H1 cell, in combination with tumour necrosis factor α (TNF-α) from the activated macrophage, causes increased expression of the iNOS gene, and hence synthesis of the enzyme, leading to production of NO. This has broad antimicrobial activity: it can be both used inside phagocytic vesicles within the macrophage and released outside the cell, for example to attack large pathogens, such as parasitic worms. Expression of iNOS also occurs more widely during inflammation and wound healing.

induction 1. (in embryology) The ability of natural stimuli (*inducers) to cause unspecialized embryonic tissue to develop into specialized tissue. **2.** (in genetics) Inactivation of repression of a gene or *operon by binding of an *inducer (sense 2) to a repressor. Gene products, such as enzymes, that are synthesized following induction are said to be **inducible. 3.** (in obstetrics) The initiation of childbirth by artificial means; for example, by injection of the hormone *oxytocin.

indusium (*pl.* **indusia**) The kidney-shaped covering of the *sorus of certain ferns that protects the developing sporangia. It withers when the sorus ripens to expose the sporangia.

industrial melanism The increase of melanic (dark) forms of an animal in areas darkened by industrial pollution. The example most often quoted is that of the peppered moth (*Biston betularia*), melanic forms of which markedly increased in the industrial north of England during the 19th century. Experiments have shown that the dark forms increase in polluted regions because they are less easily seen by birds against a dark background; conversely the paler forms survive better in unpolluted areas. *See also* DIRECTIONAL SELECTION.

infection The invasion of any living organism by disease-causing microorganisms (*see* PATHOGEN), which proceed to establish themselves, multiply, and produce various symptoms in their host. Pathogens may invade via a wound or (in animals) through the mucous membranes lining the alimentary, respiratory, and reproductive tracts, and may be transmitted by an infected individual, a *carrier, or an arthropod *vector. Symptoms in animals appear after an initial symptomless **incubation period** and typically consist of localized *inflammation, often with pain and fever. Infections are combated by the body's natural defences (*see* IMMUNITY). Treatment with drugs (*see* ANTIBIOTICS; ANTISEPTIC) is effective against most bacterial, fungal, and protist infections; some viral infections respond to *antiviral drugs. *See also* IMMUNIZATION.

inferior Describing a structure that is positioned below or lower than another structure in the body. For example, in flowering plants the ovary is described as inferior when it is located below the other organs of the flower (*see* EPIGYNY). *Compare* SUPERIOR.

inflammation The defence reaction of tissue to injury, infection, or irritation by chemicals or physical agents. Activated *macrophages and mast cells in the affected tissue release various signalling molecules (cytokines), including *histamine, *serotonin, *kinins, and *prostaglandins. These cause localized dilation of blood vessels so that fluid containing antimicrobial peptides and complement proteins leaks out and blood flow is increased. White blood cells (neutrophils

and monocytes) are also attracted to the site. Overall, these responses lead to the classic signs of inflammation: swelling, redness, heat, and often pain. White blood cells, particularly *phagocytes, digest pathogens and cell debris, and this process leads to an accumulation of **pus**, which is gradually consumed by macrophages. Following this immune response (*see* IMMUNITY) a gradual healing process usually follows. More extensive tissue damage may lead to a systemic inflammatory response in which additional white cells are recruited from the bone marrow. Certain cytokines released by activated macrophages can cause fever, in which the body's thermostat is reset to a higher value. This is thought to speed up chemical reactions and enhance phagocytosis, thereby accelerating the immune response. Septic shock is a life-threatening condition caused by overwhelming systemic inflammation in response to certain bacterial infections.

inflorescence A particular arrangement of flowers on a single main stalk of a plant. There are many different types of inflorescence, which are classified into two main groups depending on whether the tip of the flower axis goes on producing new flower buds during growth (*see* RACEMOSE INFLORESCENCE) or loses this ability (*see* CYMOSE INFLORESCENCE).

infradian rhythm A *biorhythm whose period is more than 24 hours in length. For example, the body temperature of hibernating animals often shows periodic transient increases corresponding to brief periods of arousal. Also, the reproductive behaviour of many animals follows a *circalunar rhythm linked to the 29.5-day lunar cycle. *See also* BIOLOGICAL CLOCK.

infraspecific Occurring within a species. For example, **infraspecific variation** is the *variation occurring between individuals of the same species. *Compare* INTERSPECIFIC; INTRASPECIFIC.

ingestion (feeding) A method of *heterotrophic nutrition in which bulk food is taken into an organism and subsequently digested (*see* DIGESTION). Ingestion is the principal mechanism of animal nutrition. *See also* MACROPHAGOUS; MICROPHAGOUS.

inguinal Of, relating to, or situated in the groin. For example, the **inguinal canal**, which contains the spermatic cord, lies in the abdominal cavity of a male fetus.

inhalation *See* INSPIRATION.

inheritance The transmission of particular characteristics from generation to generation by means of the *genetic code, which is transferred to offspring by genes carried on chromosomes in the gametes. *See also* GENETICS; MENDEL'S LAWS.

inhibin A hormone that plays a role in controlling sperm formation and the maturation of ovarian follicles by acting on the anterior pituitary to reduce secretion of *follicle-stimulating hormone (FSH). In males, inhibin is produced by *Sertoli cells in the testis, whereas in females it is secreted by the *granulosa cells that surround the ovum in a developing follicle. In both cases, the inhibin acts as negative feedback mechanism to regulate FSH levels.

inhibition 1. (in biochemistry) (**enzyme inhibition**) A reduction in the rate of an enzyme-catalysed reaction by substances called **inhibitors**. **Competitive inhibition** occurs when the inhibitor molecules resemble the substrate molecules and bind to the *active site of the enzyme, so preventing normal enzymatic activity. Competitive inhibition can be reversed by increasing the concentration of the substrate. In **noncompetitive inhibition** the inhibitor binds to a part of the enzyme or *enzyme-substrate complex other than the active site, known as an **allosteric site**. This deforms the active site so that the enzyme cannot catalyse the reaction. Noncompetitive inhibition cannot be reversed by increasing the concentration of the substrate. The toxic effects of many substances are produced in this way. Inhibition by reaction products (**feedback inhibition**) is important in the control of enzyme activity. *See also* ALLOSTERIC ENZYME. **2.** (in physiology) The prevention or reduction of the activity of neurons or effectors (such as muscles) by means of certain nerve impulses. Inhibitory activity often provides a balance to stimulation of a process; for example, the impulse to stimulate contraction of a skeletal muscle may be accompanied by an inhibitory impulse to prevent contraction of its antagonist.

inhibitory postsynaptic potential (IPSP) The electric potential that is generated in a postsynaptic neuron when an inhibitory neurotransmitter (such as *gamma-aminobutyric acid) is released into the synapse and causes a slight increase in the potential difference across the postsynaptic membrane (**hyperpolarization**). This makes the neuron less likely to transmit an impulse. *Compare* EXCITATORY POSTSYNAPTIC POTENTIAL.

initial One of a group of cells (or, in lower plants, a single cell) that divides to produce the cells of a plant tissue or organ. The cells of the apical meristem, cambium, and cork cambium are initials.

initiation codon *See* START CODON.

initiation factor (IF) Any of a group of proteins that are required for initiating the *translation stage of protein synthesis. Each ribosome is assembled on the messenger RNA (mRNA) chain in two subunits, one smaller than the other. IFs catalyse the binding of an **initiator tRNA** molecule, which recognizes the *start codon of the mRNA, to the smaller subunit of the ribosome, forming an **initiation complex**. This smaller subunit then binds to the mRNA and the IFs dissociate, allowing the larger ribosomal subunit to bind to the smaller one. IFs also control the rate of translation in some cells.

innate behaviour An inherited pattern of behaviour that appears in a similar form in all normally reared individuals of the same sex and species. *See* INSTINCT.

innate immunity *See* IMMUNITY.

inner cell mass A group of cells that forms during the early development of the fertilized egg in mammals. Evident at the blastocyst stage (*see* BLASTULA), it is destined to form the embryo proper and consists of embryonic *stem cells. The inner cell mass is clustered at one end of the blastocyst cavity (blastocoel), and both are enclosed in the saclike trophoblast.

inner ear The structure in vertebrates, surrounded by the temporal bone of the skull, that contains the organs of balance and hearing. It consists of soft hollow sensory structures (the **membranous labyrinth**), containing fluid (**endolymph**), surrounded by fluid (**perilymph**), and encased in a bony cavity (the **bony labyrinth**). It consists of two chambers, the *sacculus and *utriculus, from which arise the *cochlea and *semicircular canals respectively.

innervation The supply of nerve fibres to and from an organ.

innominate artery A short artery that branches from the aorta to divide into the *subclavian artery (the main artery to the arm) and the right *carotid artery (which supplies blood to the head).

innominate bone One of the two bones that form each half of the *pelvic girdle in adult vertebrates. This bone is formed by the fusion of the *ilium, *ischium, and *pubis.

inoculation 1. See VACCINE. **2.** The placing of a small sample of microorganisms or any other type of cell into a *culture medium so that the cells can grow and proliferate.

inoculum (*pl.* **inocula**) A small amount of material containing bacteria, viruses, or other microorganisms that is used to start a culture.

inositol A cyclic alcohol, $C_6H_{12}O_6$, that is a constituent of certain cell phosphoglycerides; the most important isomer is *myo*-inositol (see formula). It is sometimes classified as a member of the vitamin B complex but it can be synthesized by many animals and it is not regarded as an essential nutrient in humans. **Phosphatidylinositol**, a constituent of plasma membranes, is a precursor of the intracellular *second messenger molecules, **inositol**

Myo-inositol

1,4,5-trisphosphate (IP₃) and **diacylglycerol (DAG)**; these are produced, through the mediation of *phospholipase C, in response to the binding of substances, e.g. serotonin, to their receptors on the cell surface. IP₃ interacts with receptors in the endoplasmic reticulum, causing the release of stored calcium ions into the cytoplasm. Calcium channels in the plasma membrane are also opened, and the marked rise in intracellular calcium triggers the activation of various signal pathways, via the calcium-binding protein calmodulin. DAG stays near the plasma membrane, where it helps to activate members of the *protein kinase C family, which in turn activate their respective signal pathways. These pathways mediate such cellular events as smooth muscle contraction, adrenaline secretion, and histamine secretion.

inquilinism An association between members of two different species in which one, the **inquiline**, lives on or in the other (the **host**), or inside the host's home, obtaining shelter and in some instances taking some of the host's food. For example, certain mosquitoes live and breed in the fluid contained in the pitchers of pitcher plants, benefiting from the protection afforded by the pitcher and also making use of nutrients from prey trapped by the plant. The nest of many social insects harbour inquilines, which have evolved ingenious strategies to gain food or avoid being attacked, as evidenced by the staphylinid beetle *Atemeles pubicollis*. As a larva this lives in ant colonies, adopting the 'begging' posture of ant larvae in order to receive food from adult ants.

INSDC See INTERNATIONAL NUCLEOTIDE SEQUENCE DATABASE COLLABORATION.

Insecta *See* HEXAPODA.

insect growth regulator Any of various substances that interfere with the normal development or growth of insects, some of which are used to control insect pests. The two main categories are distinguished by their modes of action. **Juvenile hormone mimics (juvenoids)** disrupt the hormonal control of larval development so that metamorphosis either fails to occur or any adults that do emerge are weak and sterile. Examples include fenoxycarb, methoprene, and pyriproxyfen. In contrast, the **chitin synthesis inhibitors**, such as diflubenzuron and triflumuron, prevent chitin formation and hence normal replacement of the old cuticle following a moult. Treated insects either fail to moult, or have soft weak cuticles that cannot protect them, and die at or soon after *ecdysis. Both types of growth regulator have low toxicity for vertebrates, but can have devastating effects on populations of beneficial insects, such as pollinators and pest predators, when used in the field. Their most common applications are to control insects in food stores and in houses.

insecticide *See* PESTICIDE.

Insectivora A now invalid order of small, mainly nocturnal mammals that included the hedgehogs, moles, and shrews. They have long snouts covered with stiff tactile hairs and their teeth are specialized for seizing and crushing insects and other small prey. Their evolutionary relationships have long been disputed, and it is now clear that Insectivora was *polyphyletic; hence the term 'insectivore' is restricted to popular nonscientific usage. Most former members are now placed in the order Eulipotyphla, comprising three familes: Soricidae (shrews); Talpidae (moles); and Erinaceidae (hedgehogs). Molecular studies have shown that the golden moles and tenrecs are unrelated, and they now constitute the separate order Afrosoricida.

insectivore An animal that eats insects, especially a mammal of the order Eulipotyphla (hedgehogs, shrews, and moles).

insectivorous plant *See* CARNIVOROUS PLANT.

insertion 1. (of muscles) *See* SKELETAL MUSCLE. **2.** (in genetics) A *point mutation in which an extra nucleotide base is added to the DNA sequence. This results in the misreading of the base sequence during the *translation stage of protein synthesis.

insertion sequence *See* TRANSPOSON.

insight learning A form of learning in which an animal responds to new situations by adapting experiences gained in other contexts. Insight learning requires an animal to solve problems by viewing a situation as a whole instead of relying wholly on trial-and-error learning. Chimpanzees are capable of insight learning. *See* LEARNING IN ANIMALS (Feature).

in silico Describing biological processes or experiments that are simulated by a computer program (from Latin, literally 'in silicon').

inspiration (inhalation) The process by which gas is drawn into the lungs through the trachea (*see* RESPIRATORY MOVEMENT). In mammals the rib cage is raised by contraction of the external *intercostal muscles and the muscles of the diaphragm. These actions enlarge the thorax, so that pressure in the lung cavity is reduced below atmospheric pressure, which causes an influx of air until the pressures are equalized. *Compare* EXPIRATION.

inspiratory centre Either of a pair of collections of neurons (i.e. *nuclei, sense 2) in the *medulla oblongata in the brain that control the regularity of breathing. They are an important subcentre of the *ventilation centre in the brain, and constitute the **dorsal respiratory group**. The neurons receive information from stretch receptors in the bronchial tubes and chemoreceptors in the *carotid bodies and generate impulses that pass down the phrenic nerve to the diaphragm, causing it to contract (*see* INSPIRATION). They alter the basic breathing rhythm in response to the physiological requirement of the body for oxygen.

instar A stage in the larval development of an insect between two moults (ecdyses). There are usually a number of larval instars before metamorphosis. *See* EXOPTERYGOTE; HETEROMORPHOSIS.

instinct An innate tendency to behave in a particular way, which does not depend critically on particular learning experiences for its development and therefore is seen in a similar form in all normally reared individuals of the same sex and species. Much instinctive behaviour takes the form of **fixed action patterns**. These are movements that—once started—are performed in a stereotyped way unaffected by external stimuli. For example, a frog's prey-catching tongue flick is performed in the same way whether or not anything is caught. Some complex instinctive behaviour, however, requires some learning by the animal before it is perfected. Birdsong, for example, consists of an innate component that is modified and made more complex by the influence of other birds, the habitat, etc.

insula (insular cortex) (*pl.* **insulae**) A region of the forebrain lying deep within the cortex of each cerebral hemisphere. The insulae are thought to process information about the state of the body—'how it feels'—and integrate this with parts of the limbic system involved with feeling emotion. They are associated with basic emotions such as pain, joy, sadness, and fear, and with bodily desires, such as cravings for food or drugs. The insulae of humans and great apes have an expanded anterior region and also contain a unique type of spindle-shaped neuron called a VEN (von Economo neuron) after their discoverer, the Austrian neurologist Constantin von Economo (1876–1931). The VENs are thought to account for the higher cognitive functions of the insulae in humans, such as planning and decision-making.

insulin A protein hormone, secreted by the β (or B) cells of the *islets of Langerhans in the pancreas, that promotes the uptake and metabolism of glucose by body cells,

and thereby controls its concentration in the blood. Following a meal, and the consequent rise in blood glucose concentration, insulin secretion enhances energy metabolism, prompts liver cells to convert excess glucose to the storage carbohydrate *glycogen, and in adipose tissue promotes fat synthesis from glucose. Then, as glucose levels fall, so does insulin secretion. However, if blood glucose falls markedly, the pancreas secretes *glucagon, a hormone that counteracts the effects of insulin. Underproduction of insulin results in the accumulation of large amounts of glucose in the blood (**hyperglycaemia**) and its subsequent excretion in the urine in abnormally high concentrations (**glycosuria**). This condition, known as **diabetes mellitus**, can be treated successfully. Type 2, or non-insulin-dependent, diabetes mellitus is caused by failure of target cells to respond to insulin; it can generally be treated by diet alone. Type 1, or insulin-dependent, diabetes mellitus is an autoimmune disease in which the β cells of the pancreas are destroyed; it is more severe and requires treatment by insulin injections. Insulin was the first protein whose amino-acid sequence was fully determined (in 1955). Therapeutic human insulin is now manufactured using recombinant human DNA in cultured bacteria or yeast cells.

insulin-like growth factor (IGF) Any of certain polypeptides (*see* GROWTH FACTOR) that are structurally similar to the hormone insulin and promote division and growth of cells. There are two principal types: **IGF-I (somatomedin C)** and **IGF-II (somatomedin A)**. Growth hormone stimulates the liver to release IGF-I, which circulates in the bloodstream bound to several specific IGF-binding proteins. It exerts its effect by binding to IGF-I receptors on target cells. IGF-I is also produced locally in various tissues, including kidney, muscle, bone, and the gastrointestinal tract, where it can mediate effects by autocrine or paracrine signalling.

integrase An enzyme encoded by retroviruses, such as HIV, that catalyses the integration of the double-stranded DNA copy of the viral genome into the DNA of the host cell. Integrase inhibitors are a class of drugs used to treat HIV and other retroviral infections, usually in combination with other forms of antiretroviral therapy. They block the action of the viral integrase.

integrated pest management (IPM) An approach to controlling insects and other crop pests that combines various physical, chemical, and biological methods in an attempt to reduce reliance on chemical pesticides, and hence minimize pollution and harmful residues in the product. It also helps to counter the threat posed by increasing resistance of many insect pests to conventional insecticides. IPM is now used on certain field crops worldwide, including cotton, rice, alfalfa, and citrus fruits, and with many greenhouse crops. To be effective, IPM requires thorough knowledge of the physiology, ecology and life cycles of not only the pest but also of its natural predators. This enables selection of the most appropriate strategy, drawing from a wide array of control techniques. These include: *biological control, which encourages the pest's natural enemies and may involve the introduction of novel predators in certain situations; improving the resistance of crops to pests, either by conventional plant breeding or by genetic engineering (see GENETICALLY MODIFIED ORGANISMS); the use of agricultural practices that lessen the degree of pest damage, such as mixed cropping or timeliness of planting; and the selective use of insecticides or other chemical agents, such as *insect growth regulators, so that these have the maximum controlling effect with minimum impact on beneficial insects and the environment.

(SEE WEB LINKS)

• Introduction to IPM for homes, schools, and businesses

integration (in neurophysiology) The coordination within the brain of separate but related nervous processes. For example, sensory information from the inner ear and the eye are both necessary for the sense of balance. These stimuli must be integrated by the brain not only with each other but also with various motor nerves, which coordinate the muscles that control posture.

integrin See CELL ADHESION MOLECULE.

integument 1. The outermost body layer of an animal, characteristically comprising a layer of living cells—the *epidermis—together with a superficial protective coat, which may be a secreted hardened *cuticle, as in arthropods, or dead keratinized cells, as in vertebrates (see SKIN). **2.** The outer protective covering of a plant *ovule. It is perforated by a small pore, the *micropyle. Usually two integuments are present in angiosperms and one in gymnosperms. After fertilization the integuments form the *testa of the seed.

intelligence The coordination of *memory, *learning, and reasoning in animals. Intelligence has also been defined as the ability of an animal to form associative links between events or objects of which it has had no previous experience (see INSIGHT LEARNING). In humans intelligence can be expressed as an **intelligence quotient (IQ)**: the mental age of the subject (as measured by standard tests) divided by his or her real age×100. Psychologists have identified different types of intelligence, such as analytical intelligence, creative intelligence, and practical intelligence. Certainly, a high IQ score does not necessarily reflect prowess in other areas such as creativity, empathy, or sociability.

interactome All the interactions that occur between the various molecules produced by an organism, in all its cells and tissues and at all stages of its life. For practical purposes, researchers focus on binary protein-protein interactions (PPIs), because these are the molecules encoded by the organism's genes and are fundamental to all other cellular processes. Determining which proteins form complexes, or bind together in some way, enables researchers to identify the components of the complex pathways and networks that govern different aspects of cellular functioning and potentially play a role in health and disease.

This information can then be correlated with data from genome mapping projects to pinpoint the genes corresponding to proteins of interest. This might allow the development of, for example, a genetic test to identify individuals with susceptibility to a certain condition. With the human genome containing roughly 21000 protein-coding genes, identifying all the protein interactions is a mammoth task. Numerous projects are underway to amass the interactome data, both in humans and in other organisms, including yeast (*Saccharomyces cerevisiae*), the fruit fly *Drosophila melanogaster*, and the nematode *Caenorhabditis elegans*. The data are stored on databases and can be visualized as three-dimensional network maps. Fast high-throughput experimental techniques are essential to identify protein-protein interactions on a large scale. A range of techniques are used; some use X-ray tomography or specialized microscopy to visualize cellular components in situ, whereas others treat the cell with antibodies or chemical reagents to pinpoint protein complexes, which can then be separated out and analysed, e.g. by mass spectrometry. Often a labelled 'bait' protein is tracked to determine which 'prey' proteins bind to it, e.g. by using *yeast two-hybrid screens, in which pairwise protein interactions are linked to expression of reporter genes in yeast cells. Coexpression of proteins, as revealed by *microarrays, can suggest functional relationships; alternatively, query proteins are allowed to interact with pure samples of known proteins on microarray slides to discover their binding affinities.

intercalary Occurring between differentiated tissues. For example, **intercalary meristem**, which is not part of the *apical meristem, occurs in the internodes of grasses (between leaf nodes) and enables longitudinal growth of the stem.

intercellular Located or occurring between cells. *Compare* INTRACELLULAR.

intercellular adhesion molecule (ICAM) *See* CELL ADHESION MOLECULE.

intercostal muscles The muscles located between the *ribs, surrounding the lungs. Comprising the superficial **external intercostal muscles** and the deep **internal intercostal muscles**, they play an essential role in breathing (*see* EXPIRATION; INSPIRATION).

interfascicular cambium *See* CAMBIUM.

interference RNA *See* RNA INTERFERENCE.

interferon (IFN) Any of a number of proteins (*see* CYTOKINE) that increase the resistance of cells to attack by viruses. In mammals, 10 species of interferons have been discovered, falling into three type classes: type I includes α-interferons from white blood cells and β-interferons from connective tissue fibroblasts; type II consists solely of γ-interferon from T cells and *natural killer cells (NK cells); and type III comprises the λ-interferons (also known as interleukins 28A, 28B, and 29). α- and β-interferons induce intrinsic resistance to viral infection in all cells by triggering the expression of genes that encode antiviral proteins. Moreover, they activate NK cells, which selectively kill virus-infected cells, and promote synthesis of *MHC class I proteins by all cell types, thereby protecting uninfected cells from attack by the NK cells. γ-interferon is distinct from the other two classes in that it is not produced directly as a result of virus infection. Its actions include macrophage activation, increasing the expression of MHC molecules, and suppression of T_H2 *helper T cells. It is produced by T_H1 cells and cytotoxic T cells. Interferons are produced commercially for therapeutic purposes using genetically engineered bacteria or human tissue culture. However, interferons also play a role in the routine maintenance of cells; this complicates treatment, which can lead to unwanted side effects.

intergenic suppressor A *point mutation that negates or lessens the effects of another mutation in the same gene. For example, the effects of a primary *frameshift mutation may be suppressed by a second mutation in which the *deletion or *insertion of a nucleotide occurs.

Intergovernmental Panel on Climate Change (IPCC) *See* CLIMATE CHANGE.

interkinesis An abbreviated form of interphase occurring between the first and second nuclear divisions of *meiosis in some organisms. The chromosomes begin to uncoil, and there is at least partial reformation of the nuclear envelope. However, in other organisms meiosis proceeds straight from anaphase I to metaphase II, more or less skipping telophase I, interphase, and prophase II.

interleukin Any of numerous *cytokines that are produced by leucocytes and other cell types and which perform a range of regulatory functions for cells of the immune system, including cell differentiation, proliferation, maturation, migration, adhesion, and death. **Interleukin-1 (IL-1)** is secreted by antigen-activated macrophages and activates liver cells to produce *acute-phase proteins. It has a pro-inflammatory effect. **Interleukin-2 (IL-2)** stimulates the proliferation of T cells, which also secrete it, and also enhances the activity of B cells, natural killer (NK) cells, and macrophages, promoting cytotoxicity, interferon secretion, and antibody production. **Interleukin-3** is a growth factor for lymphoid and myeloid cells, and **interleukin-4** induces T cells and B cells to proliferate and the latter to produce antibodies. **Interleukin-6** is transiently produced at sites of inflammation and in response to infection, and stimulates the *acute-phase response, the production of blood cells (haemopoiesis), and other immune responses. Over 30 interleukins are now known to exist, and some are manufactured using recombinant DNA technology, for use as therapeutic agents.

intermediate filament Any of numerous microscopic protein fibres that form part of the *cytoskeleton of eukaryote cells. With a diameter of about 10 nm, they are intermediate in size compared to the narrower *microfilaments and the wider *microtubules—the other main cytoskeletal components. Intermediate filaments are relatively sturdy, forming a three-dimensional mesh within the cell that gives structural support to the nucleus and other cell organelles. Each filament consists of several twisted strands of protein subunits. Intermediate filaments differ in the nature of their protein subunits, often according to the tissue in which the cell is found. For example, those found in protective skin cells consist of *keratin subunits and form hair, nails, or wool when the cells die. Those of muscle cells consist of **desmin** subunits, while filaments in leucocytes are composed of vimentin.

internal environment The conditions that prevail within the body of an organism, particularly with respect to the composition of the *tissue fluid. The concept of an internal environment was first proposed by the French physiologist Claude Bernard (1813–78), who stated that maintenance of a constant internal environment was necessary for the survival of an organism in a varying external environment. Selective absorption of materials across cell membranes plays a large part in controlling the internal environment of both animals and plants. Animals in addition can regulate their body fluids by the action of hormones and the nervous system. *See* HOMEOSTASIS.

International HapMap Project An international collaborative effort, initiated in 2002 and ended in 2009, to identify and catalogue the common genetic variations among human populations. The information was made freely available to researchers to enable them to identify genes that contribute to human diseases and affect responses to drugs. The project was based on the premise that much of the human genome consists of segments that are inherited more or less intact as *haplotypes. The project sought to identify the 10 million or so *single nucleotide polymorphisms (SNPs) that are scattered throughout the human genome and select those that can be used to tag each haplotype uniquely. It is estimated that 300 000–600 000 SNPs can effectively tag all human haplotypes. Hence a researcher need determine only these tag SNPs to identify the collection of haplotypes in the genome of an individual, rather than all 10 million SNPs, which makes it a lot easier to characterize a person's genotype. This then provides a means of comparing the genotypes of individuals with different traits and of identifying the differences in genetic makeup that could contribute to, for example, the occurrence of high blood pressure or diabetes in certain persons. The HapMap consortium analysed the DNA of 270 individuals taken from four human populations—in Nigeria, Japan, China, and the USA—with African, Asian, and European ancestry, to assess how haplotype frequencies vary according to ethnic origin and determine the most common haplotypes in each population.

International Nucleotide Sequence Database Collaboration (INSDC) A collaborative venture that effectively divides up the task of collecting, updating, and storing the nucleotide sequence data reported by researchers throughout the world. It comprises three databases, which exchange information on a daily basis; these are the National Center for Biotechnology Information (NCBI) in the USA, the DNA Data Bank of Japan (DDBJ), and the Nucleotide Sequence Database of the European Molecular Biology Laboratory (EMBL). Information from the entire collection can be accessed via any one of the partner organizations.

interneuron (interneurone) A relatively short neuron, which in vertebrates is confined to the grey matter of the central nervous system. In the spinal cord interneurons provide links between sensory neurons and motor neurons in *polysynaptic reflexes, while in the brain they have a pivotal role in the integration of sensory inputs.

internode 1. (in botany) The part of a plant stem between two *nodes. **2.** (in neurology) The myelinated region of a nerve fibre between two nodes of Ranvier. *See* MYELIN SHEATH.

interoceptor A *receptor that detects stimuli from the internal environment of an organism. *Chemoreceptors that detect

changes in the levels of oxygen concentration in the blood are examples. *Compare* EXTEROCEPTOR.

interphase The period following the completion of *cell division, when the nucleus is not dividing. During this period changes in both the nucleus and the cytoplasm result in the complete development of the daughter cells. *See* CELL CYCLE; INTERKINESIS.

interpolation hypothesis A hypothetical account for the emergence of a sporophyte generation during the evolution of vascular plants. It postulates that early plants were exclusively haploid gametophytes. At some point a zygote 'germinated' mitotically instead of meiotically, producing a rudimentary diploid sporophyte. Hence, a sporophyte generation was 'inserted', or interpolated, into the life cycle. The sporophyte evolved to become increasingly elaborate and the more prominent form—as in modern seed plants— while the gametophyte became progressively reduced. *Compare* TRANSFORMATION HYPOTHESIS.

intersex An organism displaying characteristics that are intermediate between those of the typical male and typical female of its species. For example, a human intersex may have testes that fail to develop, so that although he is technically a man he has the external appearance of a woman. Intersexes may be produced in various ways; for example, by malfunctioning of the sex hormones. *See also* HERMAPHRODITE. *Compare* GYNANDROMORPH.

interspecific Occurring between members of different species. The term is applied, for example, to some types of *competition and *communication. *Compare* INFRASPECIFIC; INTRASPECIFIC.

interspecific competition *See* COMPETITION.

interstitial cell A cell that forms part of the connective tissue (the **interstitium**) between other tissues and structures, especially any of the cells of the *testis that lie between the seminiferous tubules and secrete androgens in response to stimulation by interstitial-cell-stimulating hormone (*see* LUTEINIZING HORMONE).

interstitial-cell-stimulating hormone See LUTEINIZING HORMONE.

interstitial fauna (meiofauna) Small invertebrate animals that live between individual substrate particles, such as sand grains, in the bed of a sea, lake, or river. *See also* BENTHOS.

intervertebral disc Any of the discs of cartilage that separate the bones of the *vertebral column. The intervertebral discs allow the vertebral column a certain degree of flexibility and they also absorb shock.

intestinal juice See SUCCUS ENTERICUS.

intestine The portion of the *alimentary canal posterior to the stomach. Its major functions are the final digestion of food matter from the stomach, the absorption of soluble food matter, the absorption of water, and the production of *faeces. See LARGE INTESTINE; SMALL INTESTINE.

intine See POLLEN.

intracellular (in biology) Located or occurring within cells. *Compare* INTERCELLULAR.

intrafusal Describing a type of muscle fibre that occurs in the *muscle spindles in skeletal muscles. Two to twelve intrafusal fibres enclosed in a capsule are found in each muscle spindle, lying parallel to the main (**extrafusal**) fibres. Each intrafusal fibre consists of a noncontractile (**equatorial**) region connected to a contractile (**polar**) region at each end of the fibre. The equatorial region is connected to stretch receptors.

intramembranous ossification See OSSIFICATION.

intraspecific Occurring between members of the same species. The term is applied, for example, to some types of *competition and *communication. *Compare* INFRASPECIFIC; INTERSPECIFIC.

intraspecific competition See COMPETITION.

intrinsic factor See VITAMIN B COMPLEX.

intrinsic rate of increase See BIOTIC POTENTIAL.

introgression (introgressive hybridization) The insertion of the genes of one species into the gene pool of another. This can occur when two species interbreed to produce fertile hybrids. These can then back-cross with individuals of one of the parent species. *See also* HYBRID ZONE.

intron (intervening sequence) A nucleotide sequence in a gene that does not code for the gene product (*compare* *EXON). Introns occur principally in eukaryotes, although they have also been found in certain archaea and cyanobacteria and in some viruses. Introns are transcribed into pre-messenger RNA (pre-mRNA) but are subsequently removed from the primary transcript before the mature mRNA transcript moves into the cytoplasm for translation (*see* RNA PROCESSING). In certain cases, removal of the introns is an autocatalytic process—**self-splicing**—whereby the RNA itself has the properties of an enzyme (*see* RIBOZYME). Self-splicing occurs in primary transcripts of some single-celled organisms, such as *Tetrahymena*, as well as chloroplasts, mitochondria, and some viruses. However, splicing of eukaryote primary transcripts produced in the nucleus generally requires the participation of a **spliceosome**, a complex of proteins and RNAs. In the earliest eukaryotes introns may simply have been sequences of *selfish DNA, able to move between different loci within the genome (*see* TRANSPOSON) with no benefit to the host. However, they now serve a range of valuable functions in gene expression and DNA organization. Introns act as 'spacers' for exons and facilitate *alternative splicing to create distinct mRNAs from the same gene. They contain regulatory sequences, such as enhancers and silencers, that control transcription of their 'host' gene. Introns contain genes for non-coding RNAs, such as *microRNAs and *short interfering RNAs, which can suppress translation of the messenger RNA or direct it for degradation. In some cases the transcription of introns may serve as a mechanism to specify the timing between activation of a gene and the appearance of its protein product. Moreover, the placement of introns can influence the way in which chromosomes are packaged into nucleosomes. Of evolutionary significance is **exon shuffling**—the recombination or rearrangement of exons encoding functional domains of proteins that is permitted by the presence of introns. This permits rapid evolution of proteins with novel permutations of functional groups.

intussusception 1. (in botany) The insertion of new cellulose fibres in the space between existing fibres in the wall of an elongating plant cell. This type of growth results in increasing the surface area of the cell wall. *Compare* APPOSITION. **2.** (in medicine) The telescoping of one part inside another, particularly in relation to segments of the intestine.

inulin A polysaccharide, made up from fructose molecules, that is stored as a food reserve in the roots or tubers of many plants, such as the dahlia. Extracted from plants, notably chicory root, it is used in some processed foods as a substitute for sugar, flour, or other ingredients and as a source of soluble fibre.

in utero Describing any event occurring in the uterus of a mammal during pregnancy.

inverse myotatic response *See* GOLGI TENDON ORGAN.

inversion (in genetics) **1.** A *chromosome mutation caused by reversal of part of a chromosome, so that the genes within that part are in inverse order. Inversion mutations usually occur during *crossing over in meiosis. **2.** A *point mutation caused by the reversal of two or more bases in the DNA sequence within a gene.

invertebrate Any animal that lacks a vertebral column (backbone). Invertebrates include all nonchordate animals as well as the more primitive chordates (*see* CHORDATA).

in vitro Describing biological processes that are made to occur outside the living body, in laboratory apparatus (literally 'in glass', i.e. in a test tube). In *in vitro* fertilization, mature egg cells are removed from the ovary of a woman unable to conceive normally and fertilized externally; the resultant blastocyst is implanted into her uterus. *Compare* IN SILICO; IN VIVO.

in vivo Describing biological processes as they are observed to occur in their natural environment, i.e. within living organisms. *Compare* IN VITRO.

involucre A protective structure in some flowering plants and bryophytes. In flowering plants it consists of a ring of *bracts arising beneath the flower cluster of those species with a *capitulum (i.e. members of the dandelion family) or an *umbel (i.e. members of the carrot family). In mosses and liverworts the involucre is a projection of tissue from the thallus that arches over the developing *archegonium.

involuntary (in biology) Not under the control of the will of an individual. Involuntary responses by muscles, glands, etc., occur automatically when required; many such responses, such as gland secretion, heartbeat, and peristalsis, are controlled by the *autonomic nervous system and effected by *smooth muscle.

involuntary muscle *See* SMOOTH MUSCLE.

involution 1. A decrease in the size of an organ or the body. It may be associated with functional decline, as occurs in the ageing process, or follow enlargement, as when the uterus returns to its normal size after pregnancy. **2.** The turning or rolling inwards of cells that occurs during the development of some vertebrate embryos.

iodine Symbol I. A dark violet nonmetallic element that is required as a trace element (*see* ESSENTIAL ELEMENT) by living organisms; in animals it is concentrated in the thyroid gland as a constituent of thyroid hormones. Deficiency causes disruption of thyroid function, with low levels of circulating thyroid hormones leading to enlargement of the thyroid gland (goitre). Seafood and iodized salt are good sources of iodine in areas where the soil, and hence food crops, have low levels of the element.

iodopsin *See* COLOUR VISION.

ion An atom or group of atoms that has either lost one or more electrons, making it positively charged (a cation), or gained one or more electrons, making it negatively charged (an anion).

ion channel A protein that spans a cell membrane to form a water-filled pore through which ions can pass in or out of the cell or cell compartment. Ion channels are found in the plasma membrane and in certain internal cell membranes. They vary in how they open and close and in their selectivity to different ions: some may be specific for one particular ion, whereas others may admit two or more similar ions (e.g. K^+ and Na^+). The electrical and chemical environment inside cells, including the *resting potential, is determined largely by the numbers, types, and activity of the cell's ion channels; they play a crucial role in the excitability of nerve and muscle cells. Ions pass through their respective channels at a rate and in a direction dictated mainly by the electrochemical gradients across the membrane. **Ungated ion channels** are permanently open, whereas **gated ion channels** can open and close. Of the latter there are two main types: *ligand-gated ion channels, which typically open when a signal molecule binds to a receptor region of the channel protein, and *voltage-gated ion channels, which respond to changes in membrane potential. *See* CALCIUM ION CHANNEL; POTASSIUM ION CHANNEL; SODIUM ION CHANNEL.

(⊕) SEE WEB LINKS

• Overview of the mechanisms and functions of ion channels, including diagrams

ion exchange The exchange of ions of the same charge between a solution (usually aqueous) and a solid in contact with it.

The process occurs widely in nature, especially in the absorption and retention of water-soluble fertilizers by soil. For example, if a potassium salt is dissolved in water and applied to soil, potassium ions are absorbed by the soil and sodium and calcium ions are released from it. The soil, in this case, is acting as an ion exchanger. Synthetic **ion-exchange resins** consist of various copolymers having a cross-linked three-dimensional structure to which ionic groups have been attached. An **anionic resin** has negative ions built into its structure and therefore exchanges positive ions. A **cationic resin** has positive ions built in and exchanges negative ions. Ion-exchange resins are used as the stationary phase in ion-exchange *chromatography.

ion-exchange chromatography *See* CHROMATOGRAPHY.

ionizing radiation Radiation of sufficiently high energy to cause ionization in the medium through which it passes. It may consist of a stream of high-energy particles (e.g. electrons, protons, alpha-particles) or short-wavelength electromagnetic radiation (ultraviolet, X-rays, gamma-rays). This type of radiation can cause extensive damage to the molecular structure of a substance either as a result of the direct transfer of energy to its atoms or molecules or as a result of the secondary electrons released by ionization. In biological tissue the effect of ionizing radiation can be very serious, usually as a consequence of the ejection of an electron from a water molecule and the oxidizing or reducing effects of the resulting highly reactive species:

$$2H_2O \rightarrow e^- + H_2O^* + H_2O^+$$
$$H_2O^* \rightarrow .OH + .H$$
$$H_2O^+ + H_2O \rightarrow .OH + H_3O^+$$

where the dot before a radical indicates an unpaired electron and denotes an excited species. Organisms are constantly exposed to background radiation from naturally occurring radioactive elements in soil, water, and air, such as radon gas and cosmic rays. Human-derived sources include nuclear power stations and medical uses, such as X-ray sources in radiography. Radiation damage depends on the *dose received.

ionophore A relatively small hydrophobic molecule that facilitates the transport of ions across lipid membranes. Most ionophores are produced by microorganisms. There are two types of ionophore: **channel formers**, which combine to form a *channel in the membrane through which ions can flow; and **mobile ion carriers**, which transport ions across a membrane by forming a complex with the ion. Examples of ionophores include vancomycin, nystatin, and 2,4-dinitrophenol.

ionotropic receptor A receptor protein that forms part of a *ligand-gated ion channel, so that binding of ligand (e.g. a hormone or neurotransmitter) to the receptor causes opening of the channel, permitting ions to flow through it. *See* GLUTAMATE RECEPTOR. *Compare* METABOTROPIC RECEPTOR.

IP$_3$ *See* INOSITOL.

IPCC (Intergovernmental Panel on Climate Change) *See* CLIMATE CHANGE.

IPM *See* INTEGRATED PEST MANAGEMENT.

ipsilateral Of, relating to, or affecting the same side of the body. *Compare* CONTRALATERAL.

iridium anomaly The occurrence of unusually high concentrations of the relatively scarce metal iridium at the boundaries of certain geological strata. Two such layers have been discovered, one at the end of the Cretaceous, 66 million years ago, and the second at the end of the Eocene, 34 million years ago. One theory to account for these suggests that on each occasion a huge iridium-containing meteorite may have collided with the earth, producing a cloud of dust that settled out to form an iridium-rich layer. The environmental consequences of such an impact, notably in causing a general warming of the earth by the *greenhouse effect, may have led to the extinction of the dinosaurs at the end of the Cretaceous and the

extinction of many radiolarians at the end of the Eocene. *See* ALVAREZ EVENT.

iris The pigmented ring of muscular tissue, lying between the cornea and the lens, in the eyes of vertebrates and some cephalopod molluscs. It has a central hole (the **pupil**) through which light enters the eye and it contains both circular and radial muscles. Reflex contraction of the former occurs in bright light to reduce the diameter of the pupil (*see* PUPILLARY REFLEX); contraction of the radial muscles in dim light increases the pupil diameter and therefore the amount of light entering the eye. Colour is determined by the amount of the pigment melanin in the iris. Blue eyes result from relatively little melanin; grey and brown eyes from increasingly larger amounts.

iron Symbol Fe. A silvery malleable and ductile metallic element that is the fourth most abundant element in the earth's crust. It is required as a trace element (*see* ESSENTIAL ELEMENT) by living organisms. Iron is an important constituent of *haemoglobin and the *cytochromes, being stored in the liver in the form of *ferritin. In animals deficiency of iron results in a form of *anaemia.

irradiation Exposure to any form of radiation; often exposure to *ionizing radiation is implied. *See also* FOOD PRESERVATION.

irrigation The provision of water for crops by artificial methods; for example by constructing ditches, pipe systems, and canals. Irrigation can lead to problems when the water leaches trace elements from the soil; selenium, for example, can be toxic to both local fauna and flora. Irrigation can also increase the salinity of the soil, if diverted rivers are used to provide the water. Evaporation of surface water leaves a crust of salt, which can drain down to deeper layers of the soil.

irritability *See* SENSITIVITY.

ischium (*pl.* **ischia**) The most posterior of the three bones that make up each half of the *pelvic girdle. *See also* ILIUM; PUBIS.

isidium (*pl.* **isidia**) An outgrowth from the thallus of a *lichen, 0.01–0.03 mm in diameter, that includes both algal and fungal cells. Isidia can detach from the main body of the lichen and serve as structures of vegetative propagation. *Compare* SOREDIUM.

islets of Langerhans (pancreatic islets) Small groups of cells in the pancreas that function as an endocrine gland. The *a* (or A) *cells* secrete the hormone *glucagon, the *β* (or B) *cells* secrete *insulin, and the D cells secrete *somatostatin. The islets are named after their discoverer, the German anatomist and microscopist Paul Langerhans (1847–88).

isoelectric point The pH of a medium at which a protein carries no net charge and therefore will not migrate in an electric field. Proteins precipitate most readily at their isoelectric points; this property can be utilized to separate mixtures of proteins or amino acids.

isoenzyme *See* ISOZYME.

isogamy Sexual reproduction involving the production and fusion of gametes that are similar in size and structure. It occurs in some protists, e.g. certain protozoans and algae. *Compare* ANISOGAMY.

isolating mechanism Any of the biological properties of organisms that prevent interbreeding (and therefore exchange of genetic material) between members of different species that inhabit the same geographical area. These mechanisms include **seasonal isolation**, in which the *breeding seasons of the different populations do not overlap; **behavioural isolation**, in which different *courtship behaviour in the populations ensures that mating takes place only between members of the same species; **morphological isolation**, whereby physical differences between species prevent successful copulation or transfer of gametes; and **gametic isolation**, in which, e.g., sperm of one species is unable to survive in the reproductive tract of the other species. These are examples of **premating mechanisms**. **Postmating mechanisms** include hybrid infertility and inviability. *See also* HYBRID ZONE.

isoleucine *See* AMINO ACID.

isomerase Any of a class of *enzymes that catalyse the rearrangement of the atoms within a molecule, thereby converting one *isomer into another.

isomers Chemical compounds that have the same molecular formulae but different molecular structures or different arrangements of atoms in space. **Structural isomers** have different molecular structures, i.e. they may be different types of compound or they may simply differ in the position of the functional group in the molecule. Structural isomers generally have different physical and chemical properties. **Stereoisomers** have the same formula and functional groups, but differ in the arrangement of groups in space. Optical isomers are examples of stereoisomers (*see* OPTICAL ACTIVITY).

isoprene A colourless liquid diene, $CH_2:C(CH_3)CH:CH_2$. The systematic name is **2-methylbuta-1,3-diene**. It is the structural unit in *terpenes and natural rubber and is used in making synthetic rubbers. Isoprene is one of the principal volatile hydrocarbons emitted by plant foliage, especially at high temperatures on sunny days. This scatters sunlight, forming a bluish haze, and also reacts with nitrogen oxides in air, contributing to low-level ozone, which can cause respiratory problems.

Isoptera An order of social exopterygote insects that comprises the termites. These mainly tropical insects have a complex system of *castes, including wingless workers and soldiers and primary and secondary winged reproductive members. A termite colony is founded by a single reproductive pair, the nest comprising an elaborate system of tunnels in wood or soil. Termites rely on gut microflora for cellulose digestion, causing considerable damage to wooden structures if they invade houses.

isotonic Describing solutions that have the same osmotic pressure (*see* OSMOSIS).

isotope One of two or more atoms of the same element that have the same number of protons in their nucleus but different numbers of neutrons. Hydrogen (1 proton, no neutrons), deuterium (1 proton, 1 neutron), and tritium (1 proton, 2 neutrons) are isotopes of hydrogen. Most elements in nature consist of a mixture of isotopes.

isotopic discrimination The uptake or assimilation by living organisms of a particular isotope in preference to another isotope of the same element. A well-known example in nature is the preferential fixation by photosynthetic organisms of the lighter (and vastly more abundant) isotope carbon-12 compared with the heavier carbon-13. The relative abundance of these two stable isotopes in biological material, whether living or dead, differs from that in the atmosphere and according to various factors, including the type of ecosystem, season, and atmospheric conditions. A record of these changes, based on preserved organic material, can provide insights into historical climatic conditions. In plants isotopic discrimination occurs chiefly because carbon dioxide (CO_2) containing ^{13}C diffuses more slowly than lighter CO_2 containing ^{12}C and also because the enzymes involved in photosynthetic carbon fixation discriminate between the two isotopes, particularly *rubisco, the first enzyme to encounter CO_2 in C_3 plants (*see* C_3 PATHWAY). $^{12}C/^{13}C$ isotopic discrimination is less pronounced in C_4 plants, in which CO_2 is first incorporated by the *C_4 pathway. This gives C_3 and C_4 plants characteristically different *isotopic signatures. Determination of $^{12}C/^{13}C$ ratios in plant material using *mass spectrometry is informative about plant physiology and atmospheric conditions. For example, it indicates whether the stomata of a plant were mainly open or mainly closed during its lifetime, and hence whether conditions were likely to have been moist or dry.

isotopic signature The relative proportions of certain isotopes in a sample of tissue, water, air, or other material. In biology the proportions of stable isotopes of common elements in organic matter, especially carbon, nitrogen, and oxygen, can

provide useful clues about growing conditions, lifestyle, and metabolism of the organism from which the material was derived. Differences in the relative tissue abundance of the natural stable isotopes of carbon (i.e. ^{12}C and ^{13}C), oxygen (^{16}O and ^{18}O), and nitrogen (^{14}N and ^{15}N) reflect their differing availability in the environment or in the diet or some form of selective metabolism (*see* ISOTOPIC DISCRIMINATION) by the organism. So, for example, human hair samples with different isotopic signatures can indicate the dietary preferences of the individuals. A relatively high $^{12}C/^{13}C$ ratio indicates that a person has consumed chiefly food derived from *C_3 plants (e.g. wheat, rice, and potatoes), which concentrate the lighter ^{12}C isotope (derived from carbon dioxide in air), whereas a $^{12}C/^{13}C$ ratio closer to that of air indicates a diet dominated by C_4 plants (e.g. maize) or food derived from them (e.g. corn-fed beef). Moreover, the ratio of $^{14}N/^{15}N$ in body tissue changes at successive steps in food chains, with the proportion of ^{15}N increasing at each step. Therefore vegetarians have a greater $^{14}N/^{15}N$ ratio than meat eaters.

isotype 1. (in plant taxonomy) A plant specimen that is a duplicate of or very similar to the *type specimen and can be used as a reference specimen if the type specimen is lost. **2.** (in genetics) A variant product of a gene or gene cluster arising from the translation and assembly of alternative transcripts. For example, the classes of *immunoglobulins (IgM, IgG, IgD, IgA, and IgE) are isotypes, each characterized by a particular constant region of the heavy chain, encoded by one of the immunoglobulin C-region genes.

isotype switching The mechanism by which B cells can change the type of antibody (i.e. isotype) that is produced during the course of an immune response. Early in a response, mature B cells express the immunoglobulins IgM and IgD, but later on can switch to IgG, IgA, or IgE antibodies. This switch is achieved by recombination between stretches of DNA among the genes encoding the constant regions of the immunoglobulin heavy chains, which determine isotype.

isozyme (isoenzyme) One of several forms of an enzyme in an individual or population that catalyse the same reaction but differ from each other in such properties as substrate affinity and maximum rates of enzyme-substrate reaction (*see* MICHAELIS-MENTEN CURVE).

itaconic acid (methylene succinic acid) A product of the fermentation of carbohydrates by strains of the filamentous fungus *Aspergillus*. Itaconic acid is produced on an industrial scale and used commercially as a plasticizer and lubricant in the production of adhesives and fibres and as a source of 'biopolyesters'.

ITAMs Immunoreceptor tyrosine-based activation motifs. *See* B-CELL RECEPTOR.

iteroparity The strategy of reproducing several or many times during a lifetime. Iteroparous organisms include perennial plants and the majority of animals. Organisms that have a specific breeding season and whose breeding population contains individuals of different ages, for example temperate-region trees, are said to show **overlapping iteroparity**. Organisms that breed repeatedly and do so at any time of year, such as humans, exhibit **continuous iteroparity**. *Compare* SEMELPARITY.

Jacob-Monod hypothesis The mechanism postulated by the French biologists François Jacob (1920–2013) and J. Monod (1910–76) in 1961 to explain the control of *gene expression in bacteria (*see* OPERON). Jacob and Monod investigated the expression of the gene that codes for the enzyme β-galactosidase, which breaks down lactose; the operon that regulates lactose metabolism is called the *lac operon.

JAK *See* JANUS KINASE.

Janus kinase (JAK) Any of a family of receptor-associated *tyrosine kinases that play a crucial role in many intracellular signalling pathways, particularly ones involving cytokines. Binding of ligand to the associated receptors brings two JAKs together, enabling them to activate each other and phosphorylate tyrosine residues in the intracellular regions of the receptors. This recruits cytosolic proteins called STATs (signal transducers and activators of transcription), which are phosphorylated by the JAKs, associate in pairs (dimerize), and migrate to the nucleus, where they bind to various gene control sites to alter levels of gene transcription. The name 'Janus' reflects the symmetrical arrangement of the kinase domains in the paired JAK proteins, after the Roman god and gatekeeper of heaven who had two faces, one in front and one behind.

jasmonate Any of a group of *plant hormones that play a vital part in the plant's defensive response against attack by herbivorous insects, fungi, and other pathogens. Substances released by such an attack act as elicitors; they bind to plasma membrane receptors and trigger the hydrolysis of the unsaturated fatty acid *linolenic acid to produce jasmonates, principally **jasmonic acid** and **methyl jasmonate**. These can travel in phloem to undamaged tissue, where they activate genes encoding defensive proteins, e.g. protease inhibitors that interfere with digestion in the insect's gut. They are also believed to act as mobile long-distance wound signals (*see* SYSTEMIC SIGNALLING). Jasmonates modulate the action of many other genes as well, influencing processes ranging from seed germination to nectar production, fruit ripening, root development, and the formation of mycorrhizal symbioses, in conjunction with other hormones.

jaw The part of the vertebrate skeleton that provides a support for the mouth and holds the teeth. It consists of the upper jaw (maxilla) and the lower jaw (mandible; *see also* DENTARY). Agnathans lack jaws.

jejunum (*pl.* **jejuna**) The portion of the mammalian *small intestine that follows the *duodenum and precedes the *ileum. The surface area of the lining of the jejunum is greatly increased by numerous small outgrowths (*see* VILLUS). This facilitates the absorption of digested material, which is the prime function of the jejunum.

jellyfish *See* CNIDARIA.

JGA *See* JUXTAGLOMERULAR APPARATUS.

joint The point of contact between two (or more) bones, together with the tissues that surround it. Joints fall into three classes that differ in the degree of freedom of movement they allow: (1) **immovable joints**, e.g. the *sutures between the bones that form the cranium; (2) **slightly movable joints**, e.g. the *symphyses between the vertebrae of the spinal column; and a ball-and-socket joint (the hip)   a hinge joint (the elbow)

Types of freely movable joint

(3) **freely movable** or **synovial joints**, e.g. those that occur between the limb bones. Synovial joints include the **ball-and-socket joints** (between the limbs and the hip and shoulder girdles), which allow movement in all directions; and the **hinge joints** (e.g. at the knee and elbow), which allow movement in one plane only (see illustration). A synovial joint is bound by ligaments and lined with *synovial membrane.

joule Symbol J. The *SI unit of work and energy equal to the work done when the point of application of a force of one newton moves, in the direction of the force, a distance of one metre. 1 joule$=10^7$ ergs$=0.2388$ calorie. It is named after James Prescott Joule (1818–89).

jugular vein A paired vein in the neck of mammals that returns blood from the head to the heart. It joins the subclavian vein at the base of the neck.

jumping gene *See* TRANSPOSON.

junk DNA DNA that is apparently nonfunctional, insofar as it does not encode proteins or essential RNA molecules, or perform any vital regulatory role, e.g. in controlling gene expression. The term was coined as early as the 1960s and formalized by the Japanese-US geneticist Susumu Ohno (1928–2000) in 1972. Only 1.5% of the human genome codes for proteins, ribosomal RNA, or transfer RNA, with another 5% representing regulatory sequences, roughly 20% comprising noncoding *introns within genes, and a further 15% being noncoding DNA of unique sequence between genes. The remaining 58% is made up of multiple copies of *repetitive DNA and *transposable elements, such as the *Alu family. However, projects such as ENCODE have revealed that around 80% of the human genome is transcribed in at least one cell type, leading to the assumption that much of the 'noncoding' DNA must have some function. But what constitutes a useful function has been questioned; much noncoding DNA accumulates mutations without any noticeable harmful effects, indicating that the base sequence is not critical to its function, or it serves no useful purpose at all. Hence, functional DNA may account for only 8–14% of the total. *See also* SELFISH DNA.

Jurassic The second geological period of the Mesozoic era. It followed the Triassic, which ended about 201 million years ago, and extended until the beginning of the Cretaceous period, about 145 million years ago. It was named in 1829 by A. Brongniart after the Jura Mountains on the borders of France and Switzerland. Jurassic rocks include clays and limestones in which fossil flora and fauna are abundant. Plants included ferns, cycads, ginkgos, rushes, and conifers. Important invertebrates included *ammonites (on which the Jurassic is zoned), corals, brachiopods, bivalves, and echinoids. Reptiles, notably the dinosaurs, dominated the vertebrates and the first flying reptiles—the pterosaurs—appeared. The birdlike dinosaur, *Archaeopteryx*, also made its appearance.

SEE WEB LINKS

- Overview of the Jurassic period: part of the website of the University of California Museum of Paleontology

juvenile hormone (JH) A hormone secreted by insects from a pair of endocrine glands (**corpora allata**) close to the brain. In the larva it modulates the activity of *ecdysone, promoting the formation of larval features and inhibiting metamorphosis, the latter occurring only when levels of JH decline. In adult insects it promotes the development of accessory sex glands, and in females induces synthesis of yolk proteins and promotes egg maturation.

juvenoid *See* INSECT GROWTH REGULATOR.

juxtaglomerular apparatus (JGA) A region of tissue found in each *nephron in the kidney that is important is regulating blood pressure and body fluid and electrolytes. It is located where the distal convoluted tubule passes close to the afferent arteriole supplying the Bowman's capsule, near to the glomerulus, and contains two types of specialized cells. Large smooth muscle cells in the wall of the afferent arteriole form the **juxtaglomerular cells**, or granular cells; these contain granules of the proteolytic enzyme *renin, which are released when the juxtaglomerular cells detect decreased blood pressure in the arteriole. The JGA also includes chemoreceptor cells of the adjacent region of the distal tubule, which form a tightly packed array called the **macula densa**. This detects low concentrations of sodium ions in the filtrate inside the tubule (indicative of reduced plasma sodium levels or reduced *glomerular filtration rate) and triggers release of renin from the juxtaglomerular cells. Release of renin into the bloodstream leads to increased levels of *angiotensins, which raise blood pressure, stimulate thirst, and also stimulate the secretion of *aldosterone from the adrenal cortex and *antidiuretic hormone from the posterior pituitary. Aldosterone promotes reabsorption of sodium ions from the distal tubule, and antidiuretic hormone promotes water reabsorption. The net effect is to raise blood pressure, restore blood volume, and increase the glomerular filtration rate.

kainate (kainic acid) An analogue of glutamic acid obtained from the red alga *Digenea simplex* that is a specific agonist for a subset of *glutamate receptors—the kainate receptors.

Kainozoic *See* CENOZOIC.

kairomone *See* ALLELOCHEMICAL.

kallidin *See* KININ.

karyogamy The fusion of nuclei or nuclear material that occurs during sexual reproduction. *See* FERTILIZATION.

karyogram (idiogram) A diagram representing the *chromosomes of a species, typically in order of size, showing their characteristic features. It is created by analysis and rearrangement of the *karyotype, as seen through the microscope.

karyokinesis The division of a cell nucleus. *See* MEIOSIS; MITOSIS.

karyolysis *See* KARYORRHEXIS.

karyoplasm *See* NUCLEOPLASM.

karyorrhexis A stage of cell death (*see* NECROSIS) that involves fragmentation of a cell nucleus. The nucleus breaks down into small dark beads of damaged *chromatin. This usually leads to further dissolution of the nucleus (**karyolysis**), whereupon the damaged chromatin gradually fades.

karyotype The number and structure of the *chromosomes in the nucleus of a cell. The karyotype is identical in all the *diploid cells of an organism. The procedure of karyotyping involves culturing a sample of tissue cells from the individual concerned, then arresting the cell cycle in metaphase, when the chromosomes are in their most condensed state, by addition of the poison colchicine. The cells are immersed in a hypotonic solution, which causes them to burst; the nuclei are then fixed and stained to reveal the structural features of the chromosomes. The addition of Giemsa dye causes the appearance of the characteristic banding patterns of each chromosome pair. By viewing through a microscope fitted with a digital camera and suitable software, the homologous pairs of chromosomes can be arranged in numerical order. *See* CHROMOSOME MAP; KARYOGRAM; SPECTRAL KARYOTYPING.

katadromous Describing the migration of certain fish that spend most of their lives in fresh waters before swimming to deep oceanic waters to breed. For example, the common freshwater eel (*Anguilla anguilla*), found in European lakes and rivers, travels to the Sargasso Sea in the western Atlantic to breed, from where the young (elvers) may take up to three years to reach their home waters. *Compare* ANADROMOUS.

katal Symbol kat. A non-SI unit of enzyme activity defined as the catalytic activity of an enzyme that increases the rate of conversion of a specified chemical reaction by 1 mol s^{-1} under specified assay conditions.

kb *See* KILOBASE.

kDa *See* KILODALTON.

keel (carina) The projection of bone from the sternum (breastbone) of a bird or bat, to which the powerful flight muscles are attached. The sterna of flightless birds (e.g. ostrich and emu) lack keels.

kelp Any large brown seaweed (*see* PHAEOPHYTA) or its ash, used as a source of iodine.

kelvin Symbol K. The *SI unit of thermodynamic temperature equal to the fraction 1/273.16 of the thermodynamic temperature of the triple point of water. The magnitude of the kelvin is equal to that of the degree Celsius (centigrade), but a temperature expressed in degrees Celsius is numerically equal to the temperature in kelvins less 273.15 (i.e. $°C = K - 273.15$). The unit is named after Lord Kelvin (1824–1907).

keratin Any of a group of fibrous *proteins occurring in hair, feathers, hooves, and horns. Keratins have coiled polypeptide chains that combine to form supercoils of several polypeptides linked by disulphide bonds between adjacent cysteine amino acids. Aggregates of these supercoils form microfibrils, which are embedded in a protein matrix. This produces a strong but elastic structure.

keratinization (cornification) The process in which the cytoplasm of the outermost cells of the mammalian *epidermis is replaced by *keratin. Keratinization occurs in the *stratum corneum, feathers, hair, claws, nails, hooves, and horns.

ketohexose See MONOSACCHARIDE.

ketone Any one of a group of organic compounds that contain the carbonyl group (>C=O) linked to two hydrocarbon groups. The **ketone group** is a carbonyl group with two single bonds to other carbon atoms (–CO–). Examples are acetone (propanone), CH_3COCH_3, and methyl ethyl ketone (butanone), $CH_3COC_2H_5$. See also KETONE BODY.

ketone body Any of three compounds, acetoacetic acid (3-oxobutanoic acid, CH_3COCH_2COOH), β-hydroxybutyric acid (3-hydroxybutanoic acid, $CH_3CH(OH)$ CH_2COOH), and acetone (propanone, CH_3COCH_3), all produced by the liver as a result of the metabolism of body fat deposits. Ketone bodies are normally used as energy sources by peripheral tissues. However, if carbohydrate supply is limited (e.g. during starvation or in diabetics), the blood level of ketone bodies rises and they may be present in urine, giving it a characteristic 'pear drops' odour. This condition is called **ketosis**.

ketopentose See MONOSACCHARIDE.

ketose See MONOSACCHARIDE.

key (identification key) A means of identifying a specimen organism by selecting from a series of choices relating to external features. Each stage in the key presents several character descriptions (two in a **dichotomous key**); each description is followed by a direction to another stage in the key. Selection of the character that matches the specimen leads to another stage, with another two choices, and so on until the specimen is identified.

keystone species A species whose impact on its community is disproportionately large relative to its abundance. This is generally because it alone fulfils some crucial functional role in the community, the continuation of which is essential for the survival of numerous other species. Classic examples are the beaver (*Castor* spp.), whose dam building creates the unique beaver ponds on which many other species depend, and the bison (*Bos bison*), responsible for a mosaic-like grazing pattern that underpinned the biodiversity of the grasslands of North America. *Compare* FOUNDATION SPECIES.

kidney The main organ of *excretion and *osmoregulation in vertebrates, through which nitrogenous waste material (usually in the form of *urine) and surplus water, ions, etc., are eliminated from the body. In mammals there is a pair of kidneys situated in the abdomen (see illustration). Each has an outer **cortex** and an inner **medulla** and is made up of tubular units called *nephrons, through which nitrogenous waste is filtered from the blood, with the formation of urine. The nephrons drain into a basin-like cavity in the kidney (the **renal pelvis**), which leads to the *ureter and *bladder.

killer cell See CYTOTOXIC T CELL; NATURAL KILLER CELL.

vena cava — aorta
right kidney
left kidney
renal artery
collecting duct
cortex
medulla
renal vein
pelvis
ureter
→ direction of blood flow
bladder
urethra — sphincter muscle

The kidneys of a mammal

kilo- Symbol k. A prefix used in the metric system to denote 1000 times. For example, 1000 volts = 1 kilovolt (kV).

kilobase Symbol kb. A unit used at the molecular level for measuring distances along nucleic acids, chromosomes, or genes, equal to 1000 bases (equivalent to 1000 nucleotides or base pairs). *See also* BP.

kilodalton Symbol kDa. A non-SI unit of mass used to express molecular mass, especially for large molecules, such as proteins and polysaccharides. It is equal to 1000 daltons. A dalton is equal to one twelfth of the mass of an atom of carbon-12, or 1.660 33 × 10⁻²⁷ kg. It is named after British chemist and physicist John Dalton (1766–1844).

kilogram Symbol kg. The *SI unit of mass previously defined as a mass equal to that of the international platinum-iridium prototype kept by the International Bureau of Weights and Measures at Sèvres, near Paris; from 20 May 2019 it has been defined in terms of the Planck constant.

kinase (phosphokinase) An enzyme that can transfer a phosphate group from a high-energy phosphate, such as ATP, to an organic molecule. *Phosphorylation is normally required to activate the molecule, which is often an enzyme. For example, kinases activate the precursors of enzymes secreted in pancreatic juice (*see* CHYMOTRYPSIN;

TRYPSIN). *See also* PROTEIN KINASE; TYROSINE KINASE.

kinesin A motor protein, structurally similar to *myosin, that transports cargoes inside cells by interacting with microtubules. The kinesin molecule consists of two heavy chains and two light chains: the heavy chains form a pair of globular 'heads' and a coiled helical tail, to which the light chains attach. The heads bind alternately to a microtubule, effectively 'walking' along the microtubule using energy supplied by the hydrolysis of ATP. The tail region binds to a cargo, such as a vesicle or organelle, which is thereby carried along. Kinesins transport synaptic vesicles inside neurons, from the cell body along the axon to synapses; they are also involved in the assembly of the mitotic spindle and segregation of chromosomes during nuclear division.

kinesis The movement of a cell or organism in response to a stimulus in which the rate of movement depends on the intensity (rather than the direction) of the stimulus. For example, a woodlouse (pillbug) moves slowly in a damp atmosphere and quickly in a dry one.

kinetochore A platelike structure by which the *microtubules of the spindle attach to the centromere of a chromosome during nuclear division. In higher organisms it consists of protein and RNA arranged in three layers closely apposed to the chromatin. It acts as a motor, pulling the centromere along the attached microtubules towards the spindle pole. This process is thought to involve motor proteins, such as *dynein, and disassembly of the microtubule subunits.

kinetoplast *See* DISCOMITOCHONDRIA.

Kinetoplastida *See* DISCOMITOCHONDRIA.

kinetosome (basal body) *See* UNDULIPODIUM.

kingdom In traditional classification systems, the highest category into which organisms are classified. The original two kingdoms, Plantae (*see* PLANT) and Animalia (*see* ANIMAL), were over time supplemented by others to give five kingdoms: Bacteria (or Prokaryotae; *see* BACTERIA),

*Protoctista (including protozoa and algae), Fungi (*see* FUNGI), Plantae, and Animalia. However, the discovery of the *archaea led taxonomists to suggest a superordinate category in the taxonomic hierarchy—the *domain. Modern molecular systematics supports the grouping of organisms into three domains, but the number of kingdoms is much harder to determine. Currently, eukaryotes, for example, are grouped into at least four major assemblages, or superphyla, rendering the traditional concept of kingdom obsolescent.

kinin 1. One of a group of peptides, occurring in blood, that are involved in inflammation and regulating blood pressure. Kinins are formed in response to blood-vessel injury by the splitting of blood plasma globulins (**kininogens**) by the enzyme kallikrein at the site of inflammation. Kinins include **bradykinin** and **kallidin**. They cause local increases in the permeability of small blood vessels and are also potent vasodilators that act to reduce blood pressure. Other roles have been identified in cell proliferation and migration, and in oedema formation. They mediate their effects by binding to two classes of *G-protein-coupled receptors, B_1 and B_2 receptors. Bradykinin is a potent stimulator of pain in the skin. **2.** *See* CYTOKININ.

kinocilium (*pl.* **kinocilia**) The single cilium of a *hair cell that protrudes much further than the other, relatively short, *stereocilia. In cross-section, its shaft has the $9 + 2$ array of microtubules characteristic of motile cilia (*see* FLAGELLUM sense 2). Kinocilia are absent from some hair cells in the mammalian ear; they do not seem to be essential for the cell's sensory function.

kinomere *See* CENTROMERE.

kin selection Natural selection of genes that tend to cause the individuals bearing them to be altruistic to close relatives. These relatives therefore have a higher probability of bearing identical copies of those same genes than do other members of the population. Thus kin selection for a gene that tends to cause an animal to share food with a close relative will result in the gene being spread through the population because it (unconsciously) benefits itself. The more closely two animals are related, the higher the probability that they share some identical genes and therefore the more closely their interests coincide. Parental care is a special case of kin selection. *See* INCLUSIVE FITNESS.

Kleiber's law The relationship whereby metabolic rate (R) varies with body mass (M) of an organism raised to the power 3/4; i.e. $R \propto M^{3/4}$. This example of 'quarter power' biological scaling was first established by Swiss-born animal physiologist Max Kleiber (1893–1976) in 1932 and has since been shown to apply to organisms ranging in size from bacteria to whales, and for plants as well as animals. It is now established that physiological attributes, such as heart rate and respiratory rate, generally change with mass raised to some multiple of 1/4. Kleiber's work contradicted earlier reasoning that R should vary with $M^{2/3}$, on the basis that heat loss varies with surface area, which is proportional to $M^{2/3}$—that is, 'cube root' scaling. In the 1990s, two US biologists, James Brown and Brian Enquist, collaborated with US particle physicist Geoffrey West to construct a theoretical model to explain quarter power scaling. It is based on the geometric properties of fluid-conducting networks, such as the blood circulatory system of animals and vascular tissue of plants, and the constraints these impose on biological designs. To supply, say, double the number of cells, the supply network must more than double in volume. But the maximally efficient network in terms of energy cost is one that occupies a fairly constant proportion of the body's volume. Consequently, metabolic rate must fall as body mass increases, to compensate for a necessarily more sparsely distributed supply network. The mathematics of the model predicts that the log-log relationship between R and M will be a straight line with a slope of 3/4, as indeed is observed experimentally. *See also* ALLOMETRIC GROWTH.

Klinefelter's syndrome A genetic disorder affecting men in which an individual gains an extra X chromosome, so that the usual karyotype of XY (*see* SEX CHROMOSOME) is replaced by one of XXY (*see* NONDISJUNCTION). Symptoms of Klinefelter's syndrome, named after US physician H. F. Klinefelter (1912–90), include underdeveloped testes, infertility, and some female characteristics (such as breast enlargement).

klinostat A device used in experiments to test the influence of gravity on the growth movements of plants (*see* GEOTROPISM). It consists of a motor that slowly rotates a drum inside which seedlings are attached. This prevents any single part of the seedlings from receiving uninterrupted gravitational stimulation and results in horizontal growth of the seedlings.

knee-jerk reflex *See* STRETCH REFLEX.

knockin A technique related to gene *knockout in which a gene is inserted into the genome of a cell, cell line, or organism. Using the technique of *Cre/*loxP* recombination, such an insertion can be targeted to a particular site in the genome and may replace or supplement existing genes. Moreover, it can be induced to switch on only in certain tissues, or at certain stages of development, thus mimicking the normal behaviour of genes.

knockout A technique for inactivating a particular gene or genes within an organism or cell. Both single-celled and multicellular organisms can be genetically engineered so that a normal gene is replaced with a defective homologous gene. Experiments with laboratory organisms (especially mice) treated in this way reveal how defects in particular genes can affect the development and life of the organism. Knockout mice can be engineered by targeting specific chromosomal locations, particular tissues, or certain development stages, using *Cre/*loxP* recombination or *genome editing. An alternative approach involves injecting normal mouse oocytes with defective DNA and selecting transfected embryos that are heterozygous for the defective gene. If adults reared from such embryos are crossed with other similar heterozygotes, 25% of the progeny will be homozygous for the 'knocked out' gene, and hence the impact of the defect on such mice can be assessed. Instead of knockout the phenomenon of *RNA interference can be used to suppress the expression of specific genes in cultured tissue cells—so-called **gene silencing**. The cells are transfected with small double-stranded RNA molecules, tailor-made to bind to specific RNA transcripts inside the cell nuclei, thereby causing degradation of the transcripts or preventing their translation into protein.

Kranz anatomy The arrangement of palisade mesophyll cells in a circle around the vascular bundle of C_4 plants. *See* BUNDLE SHEATH CELLS.

Krebs, Sir Hans Adolf (1900–81) German-born British biochemist, who emigrated to Britain in 1933, working at Sheffield University before moving to Oxford in 1954. Krebs is best known for the *Krebs cycle, the basis of which he discovered in 1937. Details were later added by Fritz Lipmann (1899–1986), with whom Krebs shared the 1953 Nobel Prize for physiology or medicine.

Krebs cycle (citric acid cycle; tricarboxylic acid cycle; TCA cycle) A cyclical series of biochemical reactions that is fundamental to the metabolism of aerobic organisms, i.e. animals, plants, and many microorganisms (see illustration). The enzymes of the Krebs cycle are located in the *mitochondria in eukaryotes, and are in close association with the components of the *electron transport chain. The two-carbon *acetyl coenzyme A (acetyl CoA) reacts with the four-carbon oxaloacetate to form the six-carbon citrate. In a series of seven reactions, this is reconverted to oxaloacetate and produces two molecules of carbon dioxide. Most importantly, the cycle generates one molecule of guanosine triphosphate (GTP—equivalent to 1 ATP) and reduces three molecules of the coenzyme *NAD to NADH and one molecule of the coenzyme *FAD to $FADH_2$. NADH and $FADH_2$ are then oxidized by the electron transport chain to generate three and two molecules of ATP respectively (depending

Krebs cycle

on the values of their respective *P/O ratios). This gives a net yield of 12 molecules of ATP per molecule of acetyl CoA.

Acetyl CoA can be derived from carbohydrates (via *glycolysis), fats, or certain amino acids. (Other amino acids may enter the cycle at different stages.) Thus the Krebs cycle is the central 'crossroads' in the complex system of metabolic pathways and is involved not only in degradation and energy production but also in the synthesis of biomolecules. It is named after its principal discoverer, Hans *Krebs.

krill Any marine crustacean of the order Euphausiacea, members of which constitute a major component of *zooplankton. Krill are small shrimps with semitransparent, often reddish tinged bodies, up to about 6 cm long. They feed chiefly on microscopic phytoplankton and thus act as primary consumers in the marine food chain; they in turn are food for a wide range of other animals, including fish, birds, and whales. Some species congregate in dense swarms, notably the Antarctic krill (*Euphausia superba*), which is one of the most abundant animals, with an estimated population biomass of some 200 million tonnes.

K selection A type of selection that favours organisms with a low rate of reproduction but whose populations expand to the maximum number of individuals that the habitat can support (the *carrying capacity of the habitat). *K*-selected species (or **K strategists**) tend to be highly adapted to their environment and are able to compete successfully for food and other resources. They also tend to inhabit stable environments and have relatively long life spans. *See* SURVIVAL CURVE. *Compare* R SELECTION.

K–T boundary *See* ALVAREZ EVENT.

Kupffer cells Specialized *macrophages that dispose of old blood cells and particulate matter. Kupffer cells, named after Karl Wilhelm von Kupffer (1829–1902), are found in the bloodstream and in the liver, attached to the walls of the *sinusoids.

kwashiorkor *See* MALNUTRITION.

labelling 1. (**isotopic labelling**) The process of replacing a stable atom in a compound with a radioisotope of the same element to enable its path through a biological or mechanical system to be traced by the radiation it emits. In some cases a different stable isotope is used and the path is detected by means of a mass spectrometer. A compound containing either a radioactive or stable isotope is called a **labelled compound** and the atom used is a **label**. If a hydrogen atom in each molecule of the compound has been replaced by a tritium atom, the compound is called a **tritiated compound**. A radioactive labelled compound will behave chemically and physically in the same way as an otherwise identical stable compound, and its presence can easily be detected. This process of **radioactive tracing** is widely used in biology and medicine. **2.** The addition to a target substance of a readily identifiable marker, such as a fluorescent or coloured dye, so that each target molecule is labelled with the same amount of marker. This enables the presence and in some cases the amount of target molecule to be determined by a fluorescence or colorimetric detector. Markers such as fluorescein are now widely used in automated assay techniques, including *DNA sequencing and *microarrays, and *fluorescent proteins are common markers in studies of gene expression and cell metabolism.

labia *See* LABIUM.

labium (*pl.* **labia**) **1.** The lower lip in the *mouthparts of an insect, which is used in feeding and is formed by the fusion of a pair of appendages (the second *maxillae). **2.** Either member of two pairs of fleshy folds that form part of the *vulva. The outer and larger pair, the **labia majora**, are covered by pubic hair and contain adipose tissue; the smaller **labia minora** lack adipose tissue and pubic hair. Both pairs of labia contain sebaceous glands.

labrum (*pl.* **labra**) The upper lip in the *mouthparts of an insect. It is formed from a plate of cuticle hinged to the head above the mouth and is used in feeding.

labyrinth The system of cavities and tubes that comprises the *inner ear of vertebrates. It consists of a system of membranous structures (**membranous labyrinth**) housed in a similar shaped bony cavity (**bony labyrinth**).

lac **operon** The *operon that regulates lactose metabolism in the bacterium *Escherichia coli*. Its form was first postulated in 1961 by François Jacob (1920–2013) and Jacques Monod (1910–76) to explain control of β-galactosidase synthesis, and it is used as a model for the structure and regulation of genes in prokaryotes. *See* JACOB-MONOD HYPOTHESIS.

lacrimal gland (**lachrymal gland**) The tear gland, present in the eyelids of some vertebrates. The fluid (tears) produced by this gland cleanses and lubricates the exposed surface of the eye; it also contains the enzyme *lysozyme, which helps to destroy bacteria. Tears drain into the nose through the lacrimal duct.

lactase (β-galactosidase) The enzyme that breaks down the milk sugar, lactose, to glucose and galactose by catalysing the hydrolysis of galactose residues.

lactation The discharge of milk from the *mammary glands. This generally only occurs after birth of the young and is stimulated by the sucking action of the infants and changes in oestradiol levels. Lactation is under the control of hormones, notably *prolactin, which mediates milk production, and *oxytocin, which induces release, or 'letdown', of milk during suckling of the young.

lacteal A minute blind-ended lymph vessel that occurs in each *villus of the small intestine. Digested fats are absorbed into the lacteals (*see* CHYLE) and transported to the bloodstream through the *thoracic duct.

lactic acid (2-hydroxypropanoic acid) An alpha hydroxy carboxylic acid, $CH_3CH(OH)COOH$, with a sour taste. Lactic acid is produced from pyruvic acid in active muscle tissue when oxygen is limited (*see* OXYGEN DEBT) and subsequently removed for conversion to glucose by the liver. During strenuous exercise it may build up in the muscles, causing cramplike pains. It is also produced by *fermentation in certain bacteria and is characteristic of sour milk.

lactobacillus (*pl.* **lactobacilli**) Any one of a group of rod-shaped Gram-positive anaerobic bacteria of the genus *Lactobacillus*, in which lactic acid is an end product of *fermentation. Lactobacilli are important in food technology for the production of cheese, yoghurt, and other foods from milk.

lactoferrin A glycoprotein present in milk and other *exocrine secretions that contributes to iron transport and the innate immune response. It binds free iron thus withholding it from potentially pathogenic microorganisms; it also helps to break down bacterial cell walls and modulates inflammation. Moreover, lactoferrin has antifungal, antiviral, and anticancer activity and is a factor in the growth and development of various tissues.

lactogenic hormone *See* PROLACTIN.

lactose (milk sugar) A sugar comprising one glucose molecule linked to a galactose molecule. Lactose is manufactured by the mammary gland and occurs only in milk. For example, cows' milk contains about 4.7% lactose. It is less sweet than sucrose (cane sugar). Insufficiency of the lactase enzyme responsible for digesting lactose causes lactose intolerance, where excess lactose is fermented by bacteria in the digestive tract.

lacuna (*pl.* **lacunae** or **lacunas**) A gap or cavity in the tissues of an organism; for example, the hollow centre of certain plant stems or any of the small cavities in bone in which the bone-forming cells are found.

laevorotatory Designating a chemical compound that rotates the plane of plane-polarized light to the left (anticlockwise for someone facing the oncoming radiation). *See* OPTICAL ACTIVITY.

laevulose *See* FRUCTOSE.

lagging strand *See* DISCONTINUOUS REPLICATION.

lag phase *See* BACTERIAL GROWTH CURVE.

Lamarck, Jean Baptiste Pierre Antoine de Monet, Chevalier de (1744–1829) French natural historian. In 1778 he published a flora of France, which included a dichotomous identification *key, and later worked on the classification of invertebrates, published in a seven-volume natural history (1815–22). In 1809 he put forward a theory of *evolution that has become known as *Lamarckism (later rejected in favour of Darwinism).

Lamarckism One of the earliest superficially plausible theories of inheritance proposed by Jean-Baptiste de Lamarck in 1809. He suggested that changes in an individual are acquired during its lifetime, chiefly by increased use or disuse of organs in response to "a need that continues to make itself felt", and that these changes are inherited by its offspring. Thus the long neck and limbs of a giraffe are explained as having evolved by the animal stretching its neck to browse on the foliage of trees. This so-called inheritance of acquired characteristics has never unquestionably been demonstrated to occur and the theory was largely displaced by the genetic theories of Mendel and his successors (*see* MENDELISM). Subsequent attempts to revive Lamarckism were prompted by the now-discredited findings of *Lysenkoism, and the discovery that certain acquired epigenetic changes can be inherited. *See* NEO-LAMARCKISM.

lambda phage A temperate *bacteriophage that infects cells of the bacterium *Escherichia coli*, where it can either exist as a quiescent prophage (in a state called *lysogeny) or undergo replication leading to lysis of the host cell and release of new phage particles. The phage particle consists of an icosahedral head, 64 nm in diameter, and a tail, 150 nm in length. The head contains the double-stranded DNA of the phage genome. Lambda phage has been intensively studied as a model of viral infection and replication and is much used in genetic research and in genetic engineering. Modified lambda phages are used as *vectors in gene cloning, especially for packaging relatively large amounts of foreign DNA.

lamella (*pl.* **lamellae**) **1.** (in botany) **a.** Any of the paired folds of membranes seen between the *grana in a plant chloroplast. **b.** Any of the spore-bearing gills on the underside of the cap of many mushrooms and toadstools. *See also* MIDDLE LAMELLA. **2.** (in zoology) Any of various thin layers of membranes, especially any of the thin layers of tissue of which compact bone is formed.

Lamellibranchia *See* BIVALVIA.

lamellipodium (*pl.* **lamellipodia**) A flat leaflike protrusion of a motile cell, such as a migrating skin cell (keratinocyte) or connective tissue fibroblast, that develops in the direction of movement. It contains bundles of actin microfilaments, which polymerize and extend at the leading edge and become cross-linked to form a supportive network to maintain the shape of the lamellipodium. Lamellipodia often form the 'webbing' between finger-like filopodia (*see* FILOPODIUM).

lamin Any of a group of proteins that are components of nuclear *intermediate filaments and form the nuclear lamina on the inner surface of the nuclear envelope.

Classified into three types—A, B, and C—they help to connect the nucleus to the cytoskeleton and are involved in maintaining the mechanical stability of the nucleus, positioning of chromosomal regions inside the nucleus, and control of gene expression. Mutations in lamin genes cause a variety of diseases—the laminopathies—affecting muscles (e.g. congenital muscular dystrophy), nerves, and metabolism; they have been implicated in certain cancers, possibly through greater nuclear deformability promoting the spread (metastasis) of cancer cells.

lamina (*pl.* **laminae**) **1.** Any thin, flat, platelike or sheetlike structure. **2.** The thin and usually flattened blade of a leaf, in which photosynthesis and transpiration occurs. The bulk of the lamina is made up of *mesophyll cells interspersed by a network of veins (*vascular bundles). The mesophyll is enclosed by a protective epidermis that produces a waxy cuticle. **3.** The leaflike part of the thallus of certain algae, notably kelps. *See also* STIPE.

laminin Any of a family of proteins that are crucial components of the basal lamina, the dense fibrous layer of the *basement membrane underlying most epithelial and endothelial cells. Laminins are large cross-shaped molecules that bind the collagen fibres of the lamina to adjoining cells and to the looser surrounding connective tissue, thus helping to create and stabilize the basic fibrous network. This is accomplished through multiple binding sites on the laminin molecule, each specific for different components. Laminins have vital roles not only in maintaining the integrity of mature tissues but also in organizing cells into tissues during embryonic development.

lampbrush chromosome A type of chromosome found in the oocytes of many vertebrate and invertebrate animals, notably newts and other amphibians, and in the giant single-celled alga *Acetabularia*. It is characterized by having numerous loops arising from the long axis of the chromosomal arms. Under the microscope this gives an appearance reminiscent of a cylindrical brush of the type formerly used for cleaning oil lamps. The loops are regions where the chromosome is uncoiled and the DNA exposed for active transcription by an associated matrix of RNA and proteins. This arrangement enables the transcription of greater amounts of RNA compared with normal prophase chromosomes.

lamprey *See* CYCLOSTOMATA.

lancelet *See* CEPHALOCHORDATA.

Langerhans cells *See* MACROPHAGE.

La Niña *See* EL NIÑO-SOUTHERN OSCILLATION.

lantibiotic Any of a class of antimicrobial peptides produced by Gram-positive bacteria and containing the unusual amino acids lanthionine or β-methyllanthionine. Although the name is a shortening of their original description as 'lanthionine-containing antibiotics', they are not classed as antibiotics; the latter are produced via enzyme-mediated secondary metabolism whereas lantibiotics and other *bacteriocins are encoded in DNA and synthesized ribosomally. Lantibiotics inhibit or kill other Gram-positive bacteria and some, such as nisin, are used in the food industry to inhibit food spoilage microorganisms; they also have potential as antimicrobials in medicine and can be bioengineered to alter their properties.

LAR *See* LEAF AREA RATIO.

large intestine (hindgut) The portion of the alimentary canal of vertebrates between the *small intestine and the *anus. It consists of the *caecum, *colon, and *rectum and its principal functions are the absorption of water and inorganic ions and the formation of faeces ready for expulsion via the anus. In many herbivorous animals the hindgut is the main site for the bacterial digestion of gut contents, while in humans and other nonruminant animals it contains a complex microbial community.

large virus *See* NUCLEOCYTOPLASMIC LARGE DNA VIRUS.

lariat *See* RNA PROCESSING.

larva (*pl.* **larvae**) The juvenile stage in the life cycle of most invertebrates, amphibians, and fish, which hatches from the egg, is unlike the adult in form, and is normally incapable of sexual reproduction (*see* PAEDOGENESIS). It develops into the adult by undergoing *metamorphosis. Larvae can feed themselves and are otherwise self-supporting. Examples are the tadpoles of frogs, the caterpillars of butterflies, and the ciliated planktonic larvae of many marine animals. *Compare* NYMPH.

larvacean *See* UROCHORDATA.

larynx (*pl.* **larynges**) The anterior portion of the *trachea (windpipe) of tetrapod vertebrates, which in amphibians, reptiles, and mammals contains the *vocal cords. Movement of the cartilage in the walls of the larynx (by means of the laryngeal muscles) alters the tension of the vocal cords. This changes the pitch of the sound emitted by the vocal cords when they vibrate. The final voiced sound is modified by resonance within the oral and nasal cavities.

lasso cell (colloblast) A type of cell in comb jellies (phylum *Ctenophora*) that is similar to the *thread cells of cnidarians. Lasso cells are embedded in the tentacles of the organism. Each has two protruding filaments, one straight and one spiralled, with adhesive heads used to catch prey.

latent learning A form of *learning in which there is apparently no immediate reward for the animal, and what is learnt remains 'latent'. The prime example is an animal exploring its surroundings. Learning about the geography of its home area may bring an animal no immediate benefits, but can prove vital in the future when fleeing a predator or searching for food. Many insects learn the details of landmarks near their nest by making orientation flights. This process enables them to locate the nest when returning from distant sites.

latent period (latency) 1. The short time that elapses between the reception of a stimulus and the start of the response. For a contracting muscle the latent period lasts about 0.02 seconds, during which time calcium ions are released from the *sarcoplasmic reticulum. **2.** The period during which a virus persists in a latent state. *See* LATENT VIRUS.

latent virus Any virus that remains in its host organism without undergoing replication. Latent viruses, which include herpes simplex virus (the agent of cold sores), may be induced to replicate and cause cell lysis some time after the initial infection, for example when the host's immunity is reduced.

lateral gene transfer (horizontal gene transfer) The transfer of genetic material 'sideways' between organisms of different species or strains, instead of 'vertically' from parent to offspring. It is common among bacteria, in which genes can be transferred via *plasmids or through *transduction by bacteriophages. An important clinical example is the lateral transfer of genes for antibiotic resistance, which can transform the recipient bacterial cell line and make infection harder to treat. To a lesser extent, eukaryotes also acquire new genes from other species; sometimes these confer selective advantage, in other cases they can cause harm. For instance, the ancestral endosymbiotic prokaryotes that became the mitochondria and chloroplasts of eukaryotes gave marked selective advantage to their eukaryote hosts, and their genes have subsequently mostly been transferred to the nuclear genome and become stably inherited. Infecting viruses integrate into their host chromosomes and can introduce foreign DNA from other species. Bacteria can also transform the genome of cells they infect and may lead to disease, as in the case of plant tumours arising from infection with *Agrobacterium tumefaciens*. Genetic engineering, e.g. using *genome editing, is a way of artificially introducing novel genetic material. *See also* RETICULATE (sense 2).

lateralization Any asymmetry in the function of the nervous system, particularly a functional specialization between left and right cerebral hemispheres of the brain in vertebrates (*see* CEREBRUM). Such asymmetries occur in a wide range of animals, from invertebrates to humans. For example,

studies in humans have shown that language ability resides in the left cerebral hemisphere in around 90% of individuals, while visuospatial processing (e.g. recognition of faces, patterns, and spatial relations) resides more often in the right hemisphere. However, variations in this are common; for example, around a quarter of left-handed people have right-hemisphere language areas, and in some individuals language functions are divided between the hemispheres.

lateral-line system A sensory system in fish and amphibians consisting of a line of sense organs (**neuromasts**) running along either flank of the body from head to tail. Each neuromast contains sensory *hair cells that are sensitive to vibrations or variations in water pressure and are used to detect moving objects, such as predators or members of the same shoal, and to monitor progress through water during swimming. In amphibians and some fishes, the neuromasts lie on the surface, whereas in most teleost and elasmobranch fishes they are located internally to lateral line canals. These are mucus-filled channels that lie just beneath the skin and connect to the exterior via a series of pores. Only the sensitive cilia of the neuromasts project into the canals.

lateral root *See* ROOT.

latex A sticky, milky fluid of mixed composition produced by about 10% of flowering plants. It has no known metabolic function and is thought to serve as a defence against insects and other herbivores that feed on the plant. Latex is produced in specialized cells called **lactifers** and accumulates at the point of damage on the leaf or stem. Insects may become trapped in the sticky exudate or succumb to one of the many bioactive substances that can be found in latex, including alkaloids, cardiac glycosides, phenolics, and terpenes. The latex of some species, notably rubber trees, is collected for commercial purposes.

Laurasia *See* CONTINENTAL DRIFT.

lauric acid (dodecanoic acid) A white crystalline *fatty acid, $CH_3(CH_2)_{10}COOH$. Glycerides of the acid are present in natural fats and oils (e.g. coconut and palm-kernel oil).

Law of Independent Assortment *See* MENDEL'S LAWS; INDEPENDENT ASSORTMENT.

Law of Segregation *See* MENDEL'S LAWS.

LD$_{50}$ Lethal dose 50, or median lethal dose: the amount of a pharmacological or toxic substance (such as ionizing radiation) that causes death in 50% of a group of experimental animals. For each LD$_{50}$ the species and weight of the animal and the route of administration of the substance is specified. LD$_{50}$s are used both in toxicology and in the *bioassay of therapeutic compounds.

L-dopa *See* DOPA.

leaching 1. The transport of plant nutrients in solution by rainwater or irrigation from the upper soil layers to deeper layers where they become unavailable to plants. **2.** Extraction of the soluble components of a solid mixture by percolating a solvent through it.

leading strand *See* DISCONTINUOUS REPLICATION.

leaf A flattened structure that develops from a superficial group of tissues, the leaf buttress, on the side of the stem apex. Each leaf has a lateral bud in its axil. Leaves are arranged in a definite pattern (*see* PHYLLOTAXIS) and usually show limited growth. Each consists of a broad flat *lamina (leaf blade) and a leaf base, which attaches the leaf to the stem; a leaf stalk (**petiole**) may also be present. The leaves of bryophytes are simple appendages, which are not homologous with the leaves

Transverse section through a leaf blade

Simple leaves

Compound leaves

of vascular plants as they develop on the gametophyte generation.

Leaves show considerable variation in size, shape, arrangement of veins, type of attachment to the stem, and texture. They may be **simple** or divided into **leaflets**, i.e. **compound** (see illustration). Types of leaf include: *cotyledons (seed leaves); **scale leaves**, which lack chlorophyll and develop on rhizomes or protect the inner leaves of a bud; **foliage leaves**, which are the main organs for photosynthesis and transpiration; and *bracts and **floral leaves**, such as sepals, petals, stamens, and carpels, which are specialized for reproduction.

Leaves may be modified for special purposes. For example the leaf bases of bulbs are swollen with food to survive the winter. In some plants leaves are reduced to spines for protection and their photosynthetic function is carried out by another organ, such as a *cladode.

leaf area index (LAI) The total surface area of the leaves of plants in a given area divided by the area of ground covered by the plants. In an area of dense vegetation, such as a forest, the LAI will be high.

leaf area ratio (LAR) The total leaf area of a plant divided by the dry mass of the entire plant. Typically measured in square centimetres per gram or square metres per kilogram, it indicates the efficiency with which a plant uses its leaves to produce plant material. It is equal to the product of the *leaf mass ratio (LMR) and the *specific leaf area (SLA).

leaf buttress *See* PRIMORDIUM.

leaf-height-seed scheme (LHS scheme) A scheme proposed by M. Westoby in 1998 to describe the ecological strategy of plant species in terms of three parameters: *specific leaf area (L), height of the plant at maturity (H), and seed mass (S). The strategy of a species is thus depicted by its position in the three-dimensional space bounded by the axes for the three variables. It is argued that, for example, L reflects whether the species is a good competitor or is more able to tolerate stress, and that H and S represent two aspects of coping

with disturbance. The chief advantage is that any plant species can be placed objectively in the scheme by determining the easily measured variables, thus enabling comparisons to be made between different studies and geographical regions. *Compare* CSR STRATEGIES.

leaf litter *See* LITTER.

leaf mass ratio (LMR) The ratio of leaf dry mass to the dry mass of the entire plant, typically measured in grams per gram or kilograms per kilogram. *See also* LEAF AREA RATIO.

LEA protein (late embryogenesis abundant protein) Any of a group of proteins produced in plants, microorganisms, and some invertebrates in response to drought or desiccation. For example, LEA proteins are produced in seeds during maturation as they dry out. They act by stabilizing cell membranes and other protein components to prevent their aggregation when water content falls.

learning A process by which an animal's experience may permanently alter their future behaviour, usually in a beneficial way. Learning allows an animal to respond more flexibly to the situations it encounters: learning abilities in different species vary widely and are adapted to the species' environment. On a physiological level, learning involves changes in the connections of neurons in the central nervous system (*see* SYNAPTIC PLASTICITY). Learning and *memory are intimately linked, because new information can be linked to past experience recorded in memory and new associations formed. Numerous different categories of learning have been proposed, including *habituation, associative learning (through *conditioning), trial-and-error learning, *insight learning, *latent learning, and *imprinting. *See* LEARNING IN ANIMALS (Feature).

lecithin (phosphatidylcholine) A phosphoglyceride (*see* PHOSPHOLIPID) containing the amino alcohol *choline esterified to the phosphate group. It is the most abundant animal phospholipid (being a component of plasma membranes) and

LEARNING IN ANIMALS

An animal's survival prospects are greatly improved if the animal alters its behaviour according to its experience. Learning increases its chances of obtaining food, avoiding predators, and adjusting to other often unpredictable changes in its environment. The importance of learning in the development of behaviour was stressed particularly by US experimental psychologists, such as John B. Watson (1878–1958) and B. F. Skinner (1904–90), who studied animals under carefully controlled laboratory conditions. They demonstrated how rats and pigeons could be trained, or 'conditioned', by exposing them to stimuli in the form of food rewards or electric shocks. This work was criticized by others, notably the ethologists, who preferred to observe animals in their natural surroundings and who stressed the importance of inborn mechanisms, such as instinct, in behavioural development. A synthesis between these two once-conflicting approaches has now been achieved: learning is regarded as a vital aspect of an animal's development, occurring in response to stimuli in the animal's environment but within constraints set by the animal's genes. Hence young animals are receptive to a wide range of stimuli but are genetically predisposed to respond to those that are most significant, such as those from their mother.

Conditioning
The classical demonstration of conditioning was undertaken by Ivan Pavlov in the early 1900s. He showed how dogs could learn to associate the ringing of a bell with the presentation of food, and after a while would salivate at the sound of the bell alone. He measured the amount of saliva produced by a dog, and showed that this increased as the animal learnt to associate the sound of the bell with presentation of food. The dog became conditioned to respond to the sound of the bell.

Such learning is widespread among animals. Pavlov's experiment involved positive conditioning, but negative conditioning can also occur. For example, a young bird quickly learns to associate the black-and-orange markings of the cinnabar moth's caterpillars with their unpleasant taste, and to avoid eating such caterpillars in future.

Trial-and-error learning
This occurs when the spontaneous behaviour of an animal accidentally produces a reward. For example, a hungry cat is placed in a box and required to pull on a string loop to open the door and gain access to food (see illustration). After various scratching and reaching movements, it accidentally pulls the loop and is released from the box. Its behaviour is instrumental in securing a reward. On subsequent occasions, the cat's attention becomes increasingly focused on the loop, until eventually it pulls the loop straightaway on entering the box.

Trial and error learning by a cat

Insight learning: problem solving

Chimpanzees can learn to stack crates or boxes to form a platform or to manipulate poles in order to reach an otherwise inaccessible bunch of bananas. A chimp may apparently solve such a problem suddenly, as if gaining insight after mental consideration of the problem. Such complex learning benefits from previous experience, in this instance by simply 'playing' with crates, boxes, or poles.

Insight learning by a chimpanzee

Imprinting

This is a form of learning found in young animals, especially young birds, in which they form an attachment to their mother in early life, thereby ensuring that they are taken care of and do not wander off. For example, chicks or ducklings follow the first large moving object that they encounter after hatching. This is normally their mother, but artificially incubated youngsters can become imprinted on a wooden decoy, as illustrated here, or even on a human being – as originally demonstrated in goslings and ducklings by Konrad Lorenz Imprinting occurs during a particularly sensitive period of development: the attachment formed by an animal to an imprinted individual or object lasts well into its adult life.

Imprinting in ducklings

also occurs in higher plants, but rarely in microorganisms.

lectin Any of a group of proteins found in plants, animals, fungi, algae, and bacteria that share the property of binding to specific carbohydrate groups. Lectins fall into several categories on the basis of their binding specificities. One large family comprises the **calcium-dependent (C-type) lectins**, which share a highly conserved carbohydrate recognition domain and require calcium ions to stabilize the binding site. The **selectins** comprise a subfamily of C-type lectins that function in immune responses as *cell adhesion molecules. They are expressed by activated endothelium on the inner surface of blood vessels and bind to passing leucocytes, detaining them for subsequent migration through the vessel wall to an infection site. Also classed as C-type lectins are the **collectins** (collagen-like lectins); these include *mannose-binding lectin, an acute-phase plasma protein that binds to bacterial surfaces, and macrophage mannose receptor, which binds certain sugars on the surface of bacteria and viruses, thereby activating the macrophage. **Galectins** form another category of lectins, with specificity for galactose. Plant lectins are a heterogeneous group of molecules involved in a wide range of cellular activities. By specifically binding to carbohydrate groups attached to other cell components, including the cell wall, they are crucial in processes such as cell recognition, growth, and division. In legumes lectins take part in the recognition of suitable bacterial partners for the plant in establishing *root nodule symbioses. Many plant lectins bind to polysaccharides on the surface of invading organisms and are important in the plant's defence against fungi and herbivorous animals. Plant lectins can often bind to animal cells, and some are toxic to animals, including humans. A notable example is ricin, obtained from the seeds of the castor oil plant *Ricinus communis*. Lectins derived from plant seeds, such as **concanavalin A**, can cause cells to clump together by forming cross links between the oligosaccharide groups on cell surfaces. Lectins are widely used for diagnosis and experimental purposes, e.g. to identify mutant cells in cell cultures,

to determine blood groups by triggering *agglutination of red blood cells, or in mapping the surface of plasma membranes.

leeches *See* HIRUDINEA.

legume (pod) A dry fruit formed from a single carpel and containing one or more seeds, which are shed when mature. It is the characteristic fruit of the Leguminosae (Fabaceae; pea family). It splits, often explosively, along both sides and the two halves of the fruit move apart to expose the seeds. A special form of the legume is the *lomentum.

Leishman's stain A neutral stain for blood smears devised by the British surgeon W. B. Leishman (1865–1926). It consists of a mixture of *eosin (an acidic stain), and *methylene blue (a basic stain) in alcohol and is usually diluted and buffered before use. It stains the different components of blood in a range of shades between red and blue. The similar **Wright's stain** is favoured by American workers.

lek An area of ground divided into *territories that are vigorously defended by males for purposes of sexual display and mating during the breeding season. Such a system occurs in various bird species, for example the black grouse and peafowl, and also in some mammals. The most dominant males occupy the smallest territories at the centre of the lek, where they are most likely to attract and copulate with visiting females. The peripheral, and larger, territories are occupied by subordinate males, who have less mating success. Over successive breeding seasons, the younger subordinate males gradually displace older individuals from the most desirable territories and become dominant themselves. The lek territories contain no resources of value to the breeding female, such as food or nesting materials, although males of some species may construct bowers or similar structures used in their display.

lemma (*pl.* **lemmas** or **lemmata**) One of two bracts that protects the floret of grasses. The lemma is situated beneath the second (smaller) bract, known as the **palea**, which it surrounds. *See* SPIKE.

lens A transparent biconvex structure in the eyes or analogous organs of many animals, responsible for directing light onto light-sensitive cells. In vertebrates it is a flexible structure centred behind the iris and attached by **suspensory ligaments** to the *ciliary body. In terrestrial species its main function is to focus images onto the retina. To focus on near objects, the circular muscles in the ciliary body contract and the lens becomes more convex; contraction of the radial muscles in the ciliary body flattens the lens for focusing on distant objects (*see also* ACCOMMODATION). The transparency and refractive properties of the lens are maintained by structural proteins called *crystallins.

lenticel Any of the raised pores in the stems of woody plants that allow gas exchange between the atmosphere and the internal tissues. The pore is formed by the *cork cambium, which, at certain points, produces a loose bulky form of cork that pushes through the outer tissues to create the lenticel.

Lepidoptera An order of insects comprising the butterflies and moths, found mainly in tropical regions. Adults possess two pairs of membranous wings, often brightly coloured and usually coupled together. The wings, body, and legs are covered with minute scales. Adult mouthparts are generally modified to form a long proboscis for sucking nectar, fruit juices, etc. Butterflies are typically small-bodied, active during daylight, and rest with their wings folded vertically; moths have larger bodies, are nocturnal, and rest with their wings in various positions. The larvae (caterpillars) have a prominent head and a segmented wormlike body, most segments bearing a pair of legs. They chew leaves and stems, sometimes causing considerable damage to crop plants. The larvae undergo metamorphosis via a *pupa (chrysalis) to the adult form. In some groups, the pupa is enclosed in a cocoon of silk derived from silk glands (modified salivary glands); others use leaves, etc. to build a cocoon.

leptin A protein hormone, comprising 167 amino acids in humans, that is secreted by adipose tissue; it suppresses appetite and regulates adipose tissue mass and energy balance. It acts on leptin receptors in the hypothalamus and inhibits expression of *neuropeptide Y, thereby countering the appetite-stimulating effects of *ghrelin and inhibiting food intake. It also promotes synthesis of the appetite suppressant *melanocyte-stimulating hormone. Deficiency of leptin or its receptors leads to severe obesity.

leptoid An elongated cell, similar to a sieve cell, that transports photosynthate (sugar) in certain mosses. Mature leptoids lose their nuclei but retain their cytoplasm and have prominent connections with adjacent cells.

leptotene The beginning of the first prophase of *meiosis, when the chromatids can be seen and *pairing begins.

lethal allele (lethal gene) A mutant form of a gene that eventually results in the death of an organism if expressed in the phenotype. Most lethal genes are recessive; for example, sickle-cell anaemia (*see* POLYMORPHISM) results from a recessive lethal gene that causes the production of abnormal and inefficient haemoglobin.

lethal dose 50 *See* LD$_{50}$.

leucine *See* AMINO ACID.

leucine zipper A molecular motif, originally discovered in DNA-binding proteins but also found in other proteins, in which a set of four or five consecutive leucine residues is repeated every seven amino acids in the primary sequence. In a helical configuration this produces a line of leucines on one side of the helix. With two such helixes alongside each other, the arrays of leucines can interdigitate like a zip fastener, thus forming a stable link between the two helices.

leucocyte (white blood cell) A colourless cell with a nucleus, found in blood and lymph. Leucocytes are formed in lymph nodes and red bone marrow and are capable of amoeboid movement. They can produce *antibodies and move through the walls of vessels to migrate to the sites of injuries, where they surround and isolate dead tissue, foreign bodies, and bacteria. There are two major types: those without granules in the cytoplasm, such as *lymphocytes and *monocytes (*see* AGRANULOCYTE), and those with granular cytoplasm (*granulocytes), which include *basophils, *eosinophils, and *neutrophils.

leucoplast Any *plastid in plant cells that contains no pigment and is therefore colourless. Leucoplasts are usually found in tissues not normally exposed to light and frequently contain reserves of starch (in **amyloplasts**), protein, or oil. *Compare* CHROMOPLAST.

leukaemia *See* CANCER.

leukotrienes (LTs) A class of compounds synthesized from *arachidonic acid in a variety of tissues, including the lungs and leucocytes. There are at least six different types of leukotrienes, all of which play an active role in regulating *inflammation. In the immune response, LTs attract neutrophils and eosinophils to sites of infection, increase the permeability of small blood vessels, and cause *vasoconstriction. They play a key role in various inflammatory and allergic diseases such as asthma.

***l*-form** *See* OPTICAL ACTIVITY.

LH *See* LUTEINIZING HORMONE.

LHS scheme *See* LEAF-HEIGHT-SEED SCHEME.

lice *See* MALLOPHAGA (bird lice); SIPHUNCULATA (sucking lice).

lichens A group of organisms that are symbiotic associations (*see* SYMBIOSIS) between a fungus (usually one of the *Ascomycota) and a green alga or a cyanobacterium. The fungal partner (**mycobiont**) usually makes up most of the lichen body and the cells of the alga or bacterium (**phycobiont**) are distributed within it, typically in a layer beneath the lichen's surface. The phycobiont photosynthesizes and passes most of its food to the fungus and the fungus protects its partner's cells. The lichen reproduces by means of *soredia, *isidia, or by fungal spores, which must find a suitable partner on germination. Lichens are slow growing but can live in regions that are too cold or exposed for other plants. Often they are pioneer inhabitants of cleared land or volcanic flows, when they break down the surface and enable colonization by plants. They may form a flattened crust or be erect and branching. Many grow as *epiphytes, especially on tree trunks. Some species are very sensitive to air pollution and have been used as *indicator species. Lichens are classified as fungi, usually being placed in the taxon of the fungal partner; some authorities group them together in the phylum Mycophycophyta.

(⊕) **SEE WEB LINKS**
• Details of lichens, with illustrations, from British Lichens, prepared by Mike Sutcliffe

life cycle The complete sequence of events undergone by organisms of a particular species from the fusion of gametes in one generation to the same stage in the following generation. In most animals gametes are formed by *meiosis of germ cells in the reproductive organs of the parents. The zygote, formed by the fusion of two gametes, eventually develops into an organism essentially similar to the parents. In plants, however, the products of meiosis are spores, which develop into plants (the *gametophyte generation) often very different in form from the spore-forming (*sporophyte) generation. The sporophyte generation is restored when gametes, formed by the gametophyte generation, fuse. *See* ALTERNATION OF GENERATIONS.

life form *See* PHYSIOGNOMY.

ligament A resilient but flexible band of tissue (chiefly *collagen) that holds two or more bones together at a movable *joint. Ligaments restrain the movement of bones at a joint and are therefore important in preventing dislocation.

ligand 1. (in chemistry) An ion, atom, or molecule that donates a pair of electrons to a metal atom to form a type of covalent bond called a coordinate bond. **2.** (in cell biology) A molecule that binds to a protein with a high degree of specificity. Examples are the substrate of an enzyme and a hormone binding to a cell receptor.

ligand-gated ion channel A type of *ion channel that opens when a signal molecule (ligand) binds to an extracellular receptor region of the channel protein. This changes the conformation of the channel protein and hence opens the channel. Many neurotransmitters act as ligands for ion channels. For example, at neuromuscular junctions, *nicotinic acetylcholine receptors form a cation channel that opens when two molecules of acetylcholine bind to the channel's extracellular region. The resultant influx of sodium ions causes depolarization of the muscle cell membrane. Other types of ligand-gated ion channels are found at synapses and in the brain. *Compare* VOLTAGE-GATED ION CHANNEL.

ligase Any of a class of enzymes that catalyse the formation of covalent bonds using the energy released by the cleavage of ATP. Ligases are important in the synthesis and repair of many biological molecules, including DNA (*see* DNA LIGASE), and are used in genetic engineering to insert foreign DNA into cloning *vectors.

light-dependent reaction *See* PHOTOSYNTHESIS.

light green *See* FAST GREEN.

light-harvesting complex *See* PHOTOSYSTEMS I AND II.

light-independent reaction *See* PHOTOSYNTHESIS.

lignin A complex organic polymer that is deposited within the cellulose of plant cell walls during secondary thickening. Lignification makes the walls woody and therefore rigid, enabling plants to grow tall against the pull of gravity. It is the second most abundant biopolymer on earth, after cellulose. *See* SCLERENCHYMA.

lignite *See* COAL.

ligule 1. A membranous scalelike outgrowth from the leaves of certain flowering plants. Many grasses have a ligule at the base of the leaf blade. **2.** A small membranous structure that develops on the upper surface of a young leaf base in certain clubmosses, for example *Selaginella*. It withers as the plant matures. **3.** A strap-shaped extension from the corolla tube in certain florets of a *capitulum. They are termed **ligulate** (or **ray**) **florets**.

limb 1. An appendage of a vertebrate animal, such as the leg or arm of a mammal or the wing of a bird. *See also* PENTADACTYL LIMB. **2.** The expanded upper part of a sepal, petal, or leaf. **3.** The widened upper section of a gamopetalous *corolla.

limbic system A group of regions in the brain that is involved in the expression and control of emotions, mood, instinct, and fear and plays a major role in long-term memory. The limbic system represents the evolutionarily primitive parts of the forebrain and includes the *hippocampus, *hypothalamus, *thalamus, and amygdala. It contains centres for pleasure and pain that play roles in the development of desires and drives and also in learning and emotional memory.

liming The application of lime (calcium hydroxide) to soils to increase levels of calcium and decrease acidity.

limiting factor Any environmental factor that—by its decrease, increase, absence, or presence—limits the growth, metabolic processes, or distribution of organisms or populations. In a desert ecosystem, for example, low rainfall and high temperature will be factors limiting colonization. When a metabolic process is affected by more than one factor, the **law of limiting factors** states that its rate is limited by the factor that is nearest its minimum value. For example, photosynthesis is affected by many factors, such as light, temperature, and carbon dioxide concentration, but on a warm sunny day carbon dioxide concentration will be the limiting factor as light and temperature will be at optimum levels.

limnology The study of freshwater habitats and ecosystems.

LINE (long interspersed element) Any of a class of dispersed moderately *repetitive DNA found in eukaryotes, consisting of numerous copies of relatively long (generally 6–8 kb) sequences scattered throughout the genome. LINEs are *retrotransposons and can spread by reverse transcription: an RNA transcript is formed, then a DNA copy of this, which subsequently undergoes insertion into the genome. However, many have lost this ability to move about the genome and are in effect 'molecular fossils'. In total LINEs account for about 20% of the human genome, of which most belong to the **L1 family**. These are 6 kb long and represent about 17% of the total human DNA. Active L1 LINEs contain two large coding regions plus non-coding flanking sequences. One of these encodes both an endonuclease and a reverse transcriptase.

linear energy transfer (LET) The energy transferred per unit path length by a moving high-energy charged particle (such as an electron or a proton) to the atoms and molecules along its path. It is of particular importance when the particles pass through living tissue as the LET modifies the effect of a specific dose of radiation. LET is proportional to the square of the charge on the particle and increases as the velocity of the particle decreases.

linkage The tendency for two different genes on the same chromosome to remain together during the separation of *homologous chromosomes at meiosis. Linkage can be broken by *crossing over or by a *chromosome mutation, when sections of chromosomes are exchanged and new combinations of genes are produced. *See also* HAPLOTYPE; SEX LINKAGE.

linkage disequilibrium A phenomenon found in linkage mapping of genes whereby a certain combination of alleles at two or more marker loci occurs more frequently than expected. This happens when there is a relatively low level of *recombination in the chromosomal region containing the loci, so that the alleles tend to be inherited as a set, or *haplotype. Linkage disequilibrium is more common in some parts of the human genome than in others, corresponding to recombination 'cold spots'. In these chromosomal regions,

large DNA stretches (haplotype blocks) are transmitted through many generations with the same allelic combinations. This is very useful in mapping disease-related genes because a marker, such as a *single nucleotide polymorphism (SNP), can be used to tag an entire haplotype block containing a particular set of alleles. This makes it much easier to scan the genome for markers that are associated with a disease and is the rationale underlying the *International HapMap Project.

linkage map A plan showing the relative positions of *genes along the length of the chromosomes of an organism. It is constructed by making crosses and observing whether certain characteristics tend to be inherited together. The closer together two allele pairs are situated on *homologous chromosomes, the less often will they be separated and rearranged as the reproductive cells are formed (*see* CHIASMA; CROSSING OVER). The proportion of offspring that show *recombination of the alleles concerned thus reflects their spacing and is used as a unit of length in mapping chromosomes (*see* MAP UNIT). The information obtained from such a **classical linkage map** can be combined with a restriction map, which is a linkage map of sites cleaved by restriction enzymes (*see* RESTRICTION MAPPING), providing a huge number of potential marker sites for genes of interest. Linkage maps provide valuable frameworks for constructing detailed *physical maps giving the base sequence of the chromosomal DNA.

Linnaean system *See* BINOMIAL NOMENCLATURE.

Linnaeus, Carolus (Carl Linné) (1707–78) Swedish botanist. He travelled round Europe and by 1735 had described more than 100 new species of plants. In 1749 he announced his system of *binomial nomenclature, which, with modification, has been used ever since for all organisms.

linoleic acid A liquid polyunsaturated *fatty acid with two double bonds, $CH_3(CH_2)_4CH:CHCH_2CH:CH(CH_2)_7COOH$. Linoleic acid is abundant in many plant fats and oils, e.g. linseed oil, groundnut oil, and soya-bean oil. It is an *essential fatty acid.

linolenic acid A liquid polyunsaturated *fatty acid with three double bonds in its structure: $CH_3CH_2CH:CHCH_2CH:CHCH_2CH:CH(CH_2)_7COOH$. Linolenic acid occurs in certain plant oils, e.g. linseed and soyabean oil, and in algae. It is one of the *essential fatty acids.

lipase Any of a family of enzymes that catalyse the hydrolysis of triglycerides into fatty acids and glycerol. Lipase secreted by the liver digests fats in the small intestine for absorption by the intestinal epithelial cells, whereas lipases in the bloodstream hydrolyse the reconstituted fats in chylomicrons for uptake by muscle cells or adipose tissue. Lipases in adipose cells hydrolyse stored triglycerides when fat is mobilized to re-enter the bloodstream as fatty acids.

lipid Any of a diverse group of organic compounds, occurring in living organisms, that are insoluble in water but soluble in organic solvents, such as chloroform, benzene, etc. Lipids are broadly classified into two categories: **complex lipids**, which are esters of long-chain fatty acids and include the *glycerides (which constitute the *fats and *oils of animals and plants), *glycolipids, *phospholipids, and *waxes; and **simple lipids**, which do not contain fatty acids and include the *steroids and *terpenes.

Lipids have a variety of functions in living organisms. Fats and oils are a convenient and concentrated means of storing food energy in plants and animals. Phospholipids and *sterols, such as cholesterol, are major components of plasma membranes (*see* LIPID BILAYER). Waxes provide vital waterproofing for body surfaces. Terpenes include vitamins A, E, and K, and phytol (a component of chlorophyll) and occur in essential oils, such as menthol and camphor. Steroids include the adrenal hormones, sex hormones, and bile acids.

Lipids can combine with proteins to form *lipoproteins (e.g. in cell membranes). In bacterial cell walls, lipids may associate with polysaccharides to form **lipopolysaccharides**.

(((⊕))) SEE WEB LINKS

• Extensive information about lipids in Lipid Library, compiled by the former head of the Scottish Crop Research Institute

lipid bilayer The arrangement of lipid molecules in *plasma membranes, which takes the form of a double sheet. Each lipid molecule comprises a hydrophilic 'head' (having a high affinity for water) and a hydrophobic 'tail' (having a low affinity for water). In the lipid bilayer the molecules are aligned so that their hydrophilic heads face outwards, forming the outer and inner surfaces of the membrane, while the hydrophobic tails face inwards, away from the external aqueous environment. *See also* FLUID MOSAIC MODEL.

lipid raft *See* FLUID MOSAIC MODEL.

Structure of a lipid bilayer

lipoic acid A sulphur-containing *coenzyme that is synthesized in mitochondria from octanoic acid. It is involved in the decarboxylation of pyruvate by the enzyme pyruvate dehydrogenase. This reaction has to take place before carbohydrates can enter the *Krebs cycle during aerobic respiration. It is also a coenzyme for another Krebs cycle enzyme, alpha-ketoglutarate dehydrogenase, which catalyses the conversion of alpha-ketoglutarate to succinyl-CoA. Various properties of lipoic acid, including antioxidant activity, have been identified using cultured cells *in vitro*, and deficiency in humans has not been described. The value of dietary supplementation remains to be established.

lipolysis The breakdown of storage lipids in living organisms. Most long-term energy reserves are in the form of triglycerides in fats and oils. When these are needed, e.g. during starvation, lipase enzymes convert the triglycerides into glycerol and the component fatty acids. These are then transported to tissues and oxidized to provide energy.

lipoprotein One of a group of compounds consisting of a lipid combined with a protein. Lipoproteins are the main structural materials of the membranes of cells and cell organelles. They also occur in blood and lymph, being the form in which lipids are transported in these media. *Cholesterol is transported in the bloodstream mainly in the form of **low-density lipoproteins** (LDLs) and is removed by means of LDL receptors in cell membranes; the LDLs are bound to the receptors, which are then taken into the cells. LDLs comprise about 50% cholesterol, 25% protein, and 25% lipids. Lack of LDL receptors, occurring as a genetic defect in some individuals, is a cause of high levels of cholesterol in the blood, predisposing to atherosclerosis. Plasma levels of LDL cholesterol can be reduced by dieting and exercise and/or by drugs (such as statins). **Very low-density lipoproteins** (VLDLs) are formed in the liver and consist largely of triglycerides (94%), with just 2% protein and 3% cholesterol. They are linked

to a high risk of cardiovascular disease. **High-density lipoproteins** (HDLs) transport cholesterol from tissues to the liver and contain about 50% protein, 35% lipids, and 15% cholesterol. These, the so-called 'good' lipoproteins, effectively 'scavenge' cholesterol, and their concentration in blood is raised by exercise. A high ratio of LDLs to HDLs in the blood is associated with an increased risk of *atherosclerosis and heart disease. *See also* CHYLOMICRON.

liposome A microscopic spherical membrane-enclosed vesicle or sac (20–30 nm in diameter) made artificially in the laboratory by the addition of an aqueous solution to a phospholipid gel. The membrane resembles a cell membrane and the whole vesicle is similar to a cell organelle. Liposomes can be incorporated into living cells and are used to transport relatively toxic drugs into diseased cells, where they can exert their maximum effects. For example, liposomes containing the drug methotrexate, used in the treatment of cancer, can be injected into the patient's blood. The cancerous organ is heated to a temperature higher than body temperature, so that when the liposome passes through its blood vessels, the membrane melts and the drug is released. Liposomes can also be used as vectors in the *transfection of cells with foreign DNA or RNA. The study of the behaviour of liposome membranes is used in research into membrane function, particularly to observe the behaviour of membranes during anaesthesia with respect to permeability changes.

lipotropin Either of two peptide hormones produced in the anterior pituitary gland that trigger the mobilization of fat deposits and the transfer of lipid components to the bloodstream. β-Lipotropin is formed by cleavage of the precursor proopiomelanocortin, and is itself cleaved to form γ-lipotropin and other active peptides, including endorphins and melanocyte-stimulating hormone.

Listeria A genus of rod-shaped aerobic motile Gram-positive bacteria. Only one species, *L. monocytogenes*, causes disease (**listeriosis**). It is resistant to physical and chemical treatments and can occur as a contaminant in certain foods, in faeces, etc. Listeriosis can take various forms, depending on the site of infection: localization in the central nervous system causes meningoencephalitis, while uterine infection can result in abortion or congenital handicap in the fetus.

lithosphere The earth's crust and upper region of the mantle. It is fragmented into tectonic plates, which move over the semifluid underlying asthenosphere.

litre Symbol l. A unit of volume in the metric system regarded as a special name for the cubic decimetre. It was formerly defined as the volume of 1 kilogram of pure water at 4°C at standard pressure, which is equivalent to $1.000\ 028\ dm^3$.

litter Dead organic matter in the soil that has not yet decomposed. It consists of fallen leaves and other plant remains (**leaf litter**), animal excrement, etc. After decomposition by *decomposers and *detritivores litter becomes *humus.

littoral Designating or occurring in the marginal shallow-water zone of a sea or lake, especially (in the sea) between high and low tide lines. In this zone enough light penetrates to the bottom to support rooted aquatic plants. *Compare* PROFUNDAL; SUBLITTORAL.

liver A large lobed organ in the abdomen of vertebrates that plays an essential role in many metabolic processes by regulating the composition and concentration of nutrients and toxic materials in the blood. It is made up of units called **lobules**, each of which is a roughly hexagonal structure consisting largely of *hepatocytes arranged around a central vein. The liver receives the products of digestion dissolved in the blood via the *hepatic portal vein and its most important functions are to convert excess glucose to the storage product *glycogen, which serves as a food reserve; to break down excess amino acids to ammonia, which is converted to *urea or *uric acid and excreted via the kidneys; and to store

and break down fats (*see* LIPOLYSIS; LIPOPROTEIN). Other functions of the liver are (1) the production of *bile; (2) the breakdown (*detoxification) of poisonous substances in the blood; (3) the removal of damaged red blood cells; (4) the synthesis of vitamin A and the blood-clotting substances prothrombin and fibrinogen; and (5) the storage of iron (*see* FERRITIN).

(((●))) **SEE WEB LINKS**
• Summary of liver anatomy, functions, and diseases, compiled by the British Liver Trust

liverworts *See* HEPATOPHYTA.

living fossil Any organism whose closest relatives are extinct and that was once itself thought to be extinct. An example is the *coelacanth, a primitive fish that was common in the Devonian era, the first recent living specimen of which was discovered in 1938.

lizards *See* SQUAMATA.

LMR *See* LEAF MASS RATIO.

lncRNA *See* LONG NONCODING RNA.

loam A fertile *soil that is made up of organic matter mixed with clay, sand, and silt. Loams differ in their ratios of clay, sand, and silt, which influences which types of plants they can support.

lobopod 1. A blunt-ended *pseudopodium found in certain protozoans. 2. A member of the phylum Lobopodia (*see* ONYCHOPHORA).

lock-and-key mechanism A mechanism proposed in 1890 by Emil Fischer (1852–1919) to explain binding between the active site of an enzyme and a substrate molecule. The active site was thought to have a fixed structure (the lock), which exactly matched the structure of a specific substrate (the key). Thus the enzyme and substrate interact to form an *enzyme-substrate complex. The substrate is converted to products that no longer fit the active site and are therefore released, liberating the enzyme. Observations made by X-ray diffraction studies have shown that the active site of an enzyme is more flexible and dynamic than the lock-and-key theory would suggest. *Compare* INDUCED-FIT MODEL.

locomotion The ability of an organism to move in a particular direction in its environment, which requires a propulsive force acting against a supporting structure. Most animals and many single-celled organisms have powers of locomotion. Some protists possess contractile fibres that exert force on the plasma membrane to change the shape of the cell; this may be combined with *cytoplasmic streaming to bring about locomotion (*see* AMOEBOID MOVEMENT). In many other protists and bacteria the propulsive force is provided by the action of flagella. In animals the force required to initiate locomotion is generated by *skeletal muscles, which act against a supporting framework provided by a *skeleton. *See also* FINS; FLIGHT.

locule (loculus) A small cavity in a plant or animal body. In plants the locule of the ovary is the cavity containing the ovules and the locules of the anther contain the developing pollen grains.

locus (*pl.* loci) The position of a gene or other feature on a chromosome or within a nucleic acid molecule. The alleles of a gene occupy the same locus on *homologous chromosomes.

lodicule Either of two small scales situated at the base of the floret of grasses. They may represent the reduced perianth and be involved in the opening of the floret at the time of pollination.

logarithmic scale A scale of measurement in which an increase or decrease of one unit represents a tenfold increase or decrease in the quantity measured. Decibels and pH measurements are common examples of logarithmic scales of measurement.

log phase *See* BACTERIAL GROWTH CURVE.

lomentum (**loment**) (*pl.* **lomenta**) A type of dry dehiscent fruit formed from a single carpel but divided into one-seeded compartments by constrictions between the seeds. *Legumes (e.g. those of *Acacia*) and *siliquas (e.g. those of wild radish) can be divided in this way.

long-day plant A plant in which flowering can be induced or enhanced by short nights, usually of less than 12 hours duration. Examples are spinach and spring barley. *See* PHOTOPERIODISM. *Compare* DAY-NEUTRAL PLANT; SHORT-DAY PLANT.

long noncoding RNA (lncRNA) Any transcribed RNA molecule that is longer than 200 nucleotides and does not code for a protein. In mammalian cells such transcripts greatly outnumber those from protein-coding genes and their functions are often still unclear. They appear to regulate gene function through a variety of mechanisms. These include epigenetic silencing, by coating gene clusters to make them inaccessible to transcription machinery, as in *X inactivation; blocking or recruitment of transcription factors; activation or trafficking of transcription factors; and posttranscriptional regulation of messenger RNA. Some may be precursors of small RNAs involved in *RNA interference.

long-sightedness *See* HYPERMETROPIA.

long-term depression (LTD) *See* SYNAPTIC PLASTICITY.

long terminal repeat (LTR) A length of double-stranded DNA, typically 250–600 bp, that is repeated at each end of integrated retroviral DNA and certain *retrotransposons. One LTR directs the host cell to initiate transcription of the retroviral DNA, while the other directs cellular enzymes to trim the primary RNA transcript.

long-term potentiation (LTP) *See* SYNAPTIC PLASTICITY.

loop of Henle The hairpin-shaped section of a kidney tubule situated between the proximal and distal tubules in the *nephron. The loop of Henle extends from the cortex into the medulla; it consists of a thin descending limb, which is permeable to water, and a thick ascending limb, which is impermeable to water. In the **descending limb** the filtrate entering from the proximal tubule loses water to the surrounding medulla but not sodium chloride (NaCl) and other solutes; hence its osmolarity increases, reaching a peak at the hairpin bend. In the first part of the **ascending limb**—the thin segment—NaCl diffuses out into the medulla, but water is retained; in the second part—the thick segment—NaCl is actively transported from the tubule into the medulla. These complex movements of ions and water across the walls of the loop, requiring the expenditure of energy for active transport, enable it to function as a *countercurrent multiplier, resulting in the production of concentrated urine in the *collecting duct. It is named after German anatomist F. G. J. Henle (1809–85).

lophophore An organ characteristic of aquatic invertebrates of the phyla Bryozoa, Phoronida, and Brachiopoda that functions in filter feeding. It consists of a ridge of hollow tentacles bearing cilia, which waft food particles into the mouth.

Lophotrochozoa A clade of protostome animals, based on molecular systematics, that includes the molluscs and annelid worms, together with the nemertines, bryozoans, rotifers, flatworms, and other worm-like phyla.

lower critical temperature The minimum body temperature that can be tolerated by an organism. Below this temperature, the biochemical properties of cell structures, especially membranes, are altered, and reactions are slowed such that the organism cannot maintain its usual bodily functions and death may ensue. Also, at subzero temperatures, there is a risk of the body water turning to ice, with consequent physical disruption of cells. The lower critical temperature varies greatly, depending on the 'normal' temperature regime to which the organism is adapted. Hence, plants native to warm regions, e.g. maize and cotton, are much more sensitive to chilling than plants from cooler regions. Many animals have strategies for adjusting to seasonal falls in temperature, often entering a state of torpor in which metabolism is greatly reduced. In this state they may tolerate much lower body temperatures than when

fully active during the summer months. *Compare* UPPER CRITICAL TEMPERATURE.

LSD *See* LYSERGIC ACID DIETHYLAMIDE.

LTR *See* LONG TERMINAL REPEAT.

luciferase *See* BIOLUMINESCENCE.

luciferin *See* BIOLUMINESCENCE.

lumbar vertebrae The *vertebrae in the region of the lower back. They occur below the *thoracic vertebrae and above the *sacral vertebrae. In mammals they bear processes for the attachment of back muscles.

lumen (*pl.* **lumina**) **1.** The space enclosed by a vessel, duct, or other tubular or saclike organ. The central cavity of blood vessels and of the digestive tract are examples. **2.** Symbol lm. The SI unit of luminous flux equal to the flux emitted by a uniform point source of 1 candela in a solid angle of 1 steradian.

lung The *respiratory organ of air-breathing vertebrates. A pair of lungs is situated in the thorax, within the ribcage. Each consists essentially of a thin moist membrane that is folded to increase its surface area. Exchange of oxygen and carbon dioxide takes place between blood capillaries on one side of the membrane and air on the other. The lung is supplied with air through a *bronchus. In mammals and reptiles the membrane of the lung takes the form of numerous sacs (*see* ALVEOLUS) that are connected to the bronchus via *bronchioles (see illustration). The lungs themselves contain no muscular tissue and are ventilated by *respiratory movements, the mechanisms of which vary with the species.

lung book The respiratory organ of some arachnids. Lung books occur in pairs on the ventral side of the abdomen. They consist of leaflike folds of ectoderm sunk into pockets having slitlike openings at the abdominal surface. Gaseous exchange occurs by diffusion through the ectodermal folds.

lungfish *See* DIPNOI.

luteal phase The phase of the *ovarian cycle that occurs after ovulation. During this

The lungs and air passages of a mammal (right lung cut open to show internal structure)

phase the corpus luteum secretes oestradiol and progesterone, which prepare the uterine wall to receive a fertilized egg.

luteinizing hormone (LH; interstitial-cell-stimulating hormone; ICSH) A hormone, secreted by the anterior pituitary gland in mammals, that stimulates in males the production of sex hormones (*androgens) by the *interstitial cells of the testes and in females ovulation, *progesterone synthesis, and *corpus luteum formation. *See* OVARIAN CYCLE.

luteotrophic hormone *See* PROLACTIN.

lyase Any of a class of enzymes that catalyse either the cleavage of a double bond and the addition of new groups to a substrate, or the formation of a double bond.

Lycopodiophyta (Lycophyta) A phylum of *tracheophyte plants containing the clubmosses (genus *Lycopodium*) and related genera—including the spikemosses (*Selaginella*) and quillworts (*Isoëtes*)—as well as numerous extinct forms, which reached their peak in the Carboniferous period with giant coal-forming tree species. Lycophytes have roots and their stems are covered with numerous small leaves. Reproduction is by means of spores; the sporangia are usually grouped into cones. Phylogenetically, the lycophytes are the sister group to the *euphyllophytes, which comprise all other vascular plants.

lymph The colourless liquid found within the *lymphatic system, into which it drains from the spaces between the cells. Lymph (called *tissue fluid in the intercellular spaces) resembles *blood plasma, consisting mostly of water with dissolved salts and proteins. Fats are found in suspension and their presence varies with food intake. The lymph eventually enters the bloodstream near the heart.

lymphatic system The network of vessels that conveys *lymph from the tissue fluids to the bloodstream. Tiny *lacteals (in the small intestine) and **lymph capillaries** (in other tissues) drain into larger tubular vessels that converge to form the right lymphatic duct and the *thoracic duct, which connect with the venous blood supply to the heart. Associated with the lymphatic vessels at intervals along the system are the *lymph nodes. The lymph capillary walls are very permeable, so lymph bathing the body's tissues can drain away molecules that are too large to pass through blood capillary walls. In most vertebrates lymph is pumped by cycles of contraction and relaxation of the lymphatic vessels and also by the action of adjoining muscles. *See also* LYMPH HEART.

(((●))) SEE WEB LINKS

• An illustrated summary of the human lymphatic system: part of the Cancer Research UK website

lymph capillary *See* LYMPHATIC SYSTEM.

lymph heart Any of a series of muscular pumping structures that occur in the major lymphatic vessels of certain lower vertebrates and maintain a flow of lymph around the body. Many animals (including all mammals) do not possess lymph hearts; in these the flow of lymph is maintained by a series of muscle contractions and valves (*see* LYMPHATIC SYSTEM).

lymph node A mass of *lymphoid tissue, many of which occur at intervals along the *lymphatic system. Lymph in the lymphatic vessels flows through the lymph nodes, which filter out bacteria and other foreign particles, so preventing them from entering the bloodstream and causing infection. The lymph nodes also contain *lymphocytes, which can multiply rapidly and mount an immune response if they encounter foreign antigens in the lymph. This leads to swelling of the lymph node. In humans, major lymph nodes occur in the neck, under the arms, and in the groin.

lymphocyte A type of white blood cell (*leucocyte) that has a large nucleus and little cytoplasm. Lymphocytes are formed in the *lymph nodes and provide about a quarter of all leucocytes. They are important in the body's defence and are responsible for immune reactions as the presence of *antigens stimulates them to produce *antibodies. There are two principal populations of lymphocytes: **B lymphocytes** (*see* B CELL), which produce circulating antibodies and are responsible for humoral *immunity; and **T lymphocytes** (*see* T CELL), which are responsible for cell-mediated immunity.

lymphoid tissue The type of tissue found in the *lymph nodes, *tonsils, *spleen, and *thymus. It is responsible for producing lymphocytes and therefore contributes to the body's defence against infection. *See also* BRONCHIAL-ASSOCIATED LYMPHOID TISSUE; GUT-ASSOCIATED LYMPHOID TISSUE; MUCOSAL-ASSOCIATED LYMPHOID TISSUE

lymphokine *See* CYTOKINE.

lymphoma *See* CANCER.

lymphotoxin *See* TUMOUR NECROSIS FACTOR.

lyophilization *See* FREEZE DRYING.

Lysenkoism The official Soviet science policy governing the work of geneticists in the USSR from about 1940 to 1960. It was named after its chief promoter, the agriculturalist Trofim Lysenko (1898–1976). Lysenkoism dismissed all the advances that had been made in classical genetics, denying the existence of genes, and held that the variability of organisms was produced solely by environmental changes. There was also a return to a belief in the inheritance of acquired characteristics (*see* NEO-LAMARCKISM). This state of affairs continued, despite overwhelming conflicting evidence from Western scientists, because it provided support for communist theory.

lysergic acid diethylamide (LSD) A chemical derivative of lysergic acid that has potent hallucinogenic properties (*see* HALLUCINOGEN). It occurs in the cereal fungus ergot and was first synthesized in 1943. LSD acts as an *antagonist at *serotonin receptors.

lysigeny The localized disruption of plant cells to form a cavity (surrounded by remnants of the broken cells) in which secretions accumulate. Examples are the oil cavities in the leaves of citrus trees. *Compare* SCHIZOGENY.

lysimeter An instrument used to measure the loss of water from an area of land covered with vegetation. Water that evaporates from both the soil and from the plants can be determined.

lysine *See* AMINO ACID.

lysis The destruction of a living cell. This may be effected by *lysosomes or *lymphocytes, either as part of the normal metabolic process (as when cells are damaged or worn out) or as a reaction against invading cells (e.g. bacteria). *Bacteriophages eventually cause lysis of their host cells. Experimentally, lysis of cells in culture or suspension is achieved by various methods, including mechanical disruption by a blender or ultrasonic waves in a *sonicator.

lysogeny The relationship between a temperate phage (*see* BACTERIOPHAGE) and a bacterium. A bacterium whose chromosome has been integrated with the DNA of a temperate phage is called a **lysogen**; the viral DNA is known as a *prophage.

lysosome A membrane-bound sac (organelle) found in animal cells and in single-celled eukaryotes. It contains hydrolytic enzymes that degrade aged or defective cell components or material taken in by the cell from its environment, such as food particles or bacteria. The lysosomal enzymes are adapted to work in the acid conditions of the lysosome interior, which has a pH of about 4.8. This means that should the

enzymes escape from the lysosome they are inactivated by the neutral pH of the cell cytosol, and so will not attack the cell contents. **Primary lysosomes** originate from the *Golgi apparatus; they fuse with phagosomes (*see* PHAGOCYTOSIS) containing material for disposal, forming a **secondary lysosome** in which digestion takes place. In plant cells, the *vacuole contains hydrolytic enzymes equivalent to those in the lysosome and can degrade materials in a manner similar to a lysosome.

lysozyme An antibacterial enzyme widely distributed in body fluids and secretions, including tears and saliva. It disrupts the polysaccharide components of bacterial cell walls, leaving them susceptible to destruction.

MAC *See* MEMBRANE ATTACK COMPLEX.

macroevolution Evolution on a relatively large scale, involving, for example, the emergence of entire groups of organisms, such as the flowering plants or the mammals. *Compare* MICROEVOLUTION.

macrofauna The larger animals, collectively, which can be observed without the aid of a microscope (*compare* MICROFAUNA). The macrofauna sometimes includes small soil-dwelling invertebrates, such as annelids and nematodes, but these may be separated into an intermediate category, the **mesofauna**. *Compare* MEIOFAUNA.

macromolecule A very large molecule. Natural and synthetic polymers, including polysaccharides, nucleic acids, and proteins, are macromolecules. *See also* COLLOIDS; MULTIMER.

macronutrient A chemical element required by plants in relatively large amounts. Macronutrients include carbon, hydrogen, oxygen, nitrogen, phosphorus, potassium, sulphur, magnesium, calcium, and iron. *See also* ESSENTIAL ELEMENT. *Compare* MICRONUTRIENT.

macrophage A large phagocytic cell (*see* PHAGOCYTE) that can ingest pathogenic microorganisms (e.g. bacteria, protozoa) or cell debris and forms part of the body's immune system. Macrophages develop from precursor cells (promonocytes) in bone marrow, become wandering monocytes in the bloodstream, and then settle as mature macrophages in various tissues, including lymph nodes, connective tissues (as **histiocytes**), lungs (alveolar macrophages), lining of liver sinusoids (as *Kupffer cells*), skin (as **Langerhans cells**), and

nervous tissues (**microglia**). The sinusoids of the spleen are lined with macrophages that remove worn-out red cells and platelets from the blood and destroy them. Collectively the macrophages make up the **mononuclear phagocyte system**.

Macrophages express surface receptors (*see* TOLL-LIKE RECEPTOR) that recognize and bind to common surface components of bacteria, viruses, fungi, and other pathogens. Such binding activates the macrophage, causing it to engulf microbial pathogens and also to secrete various *cytokines and *chemokines, which attract leucocytes (e.g. neutrophils and *monocytes to the infection site, initiate inflammation, trigger the *acute-phase response, and activate *natural killer cells. Macrophages also recognize and bind to microbial surfaces that have been coated with complement proteins (*see* COMPLEMENT RECEPTOR). Signals from T_H1 *helper T cells play a crucial role in stimulating macrophages to destroy their ingested bacteria, by increasing the expression of *inducible nitric oxide synthase (iNOS). Nitric oxide (NO), superoxide anion (O_2^-), and hydrogen peroxide (H_2O_2) are all toxic effector molecules produced by macrophages to kill engulfed microorganisms. Activated macrophages can also act as antigen-presenting cells and activate T cells, as the first step to an adaptive immune response.

macrophagous Describing a method of feeding in heterotrophic organisms in which food is ingested in the form of relatively large chunks. *Compare* MICROPHAGOUS.

macrophyll *See* MEGAPHYLL.

macropinocytosis A form of 'high-volume' *pinocytosis by which *dendritic cells engulf relatively large volumes of fluid from their surroundings. This enables them to take in antigens nonspecifically in fulfilling their immunological role as antigen monitors.

macula (*pl.* **maculae**) **1.** A patch of sensory *hair cells in the *utriculus and *sacculus of the inner ear that provides information about the position of the body in relation to gravity. The hairs of the cells are embedded in an **otolith**, a gelatinous cap containing particles of calcium carbonate. Movement of the particles in response to gravity pulls the gelatinous mass downwards, which bends the hairs and triggers a nerve impulse to the brain (see illustration). **2.** An area of the *retina of the vertebrate eye with increased *visual acuity. Maculae occur in some animals that lack *foveae and often surround foveae in those animals that possess them.

macula adherens *See* DESMOSOME.

macula densa *See* JUXTAGLOMERULAR APPARATUS.

MADS box gene Any of a superfamily of genes that encode transcription factors having crucial roles in the development of plants, animals, and fungi. The transcription factors are characterized by the MADS box, a sequence of about 56 amino acids that forms part of a DNA-binding domain of the protein. The acronym refers to the founding members of the family: *M*CM1 (from yeast), *A*GAMOUS and *D*EFICIENS (from flowering plants), and *S*RF (serum response factor, from mammals). Gymnosperms, angiosperms, ferns, and mosses all have numerous MADS box genes, which encode proteins involved in ensuring the correct timing and sequence of plant development, including determination of floral organ identity. Some act as homeotic genes, analogous to the *Hox genes of animals, with mutations causing the transformation of one flower part into another (*see* ABC MODEL). Many of these aspects, such as timing of flowering, vernalization, and induction of bolting, are of great significance to commercial plant breeders, which makes manipulation of MADS genes a major avenue of research.

magnesium Symbol Mg. A silvery metallic element that is an *essential element for living organisms. It functions as a cofactor for various enzymes and is a constituent of *chlorophyll in plants. In animals it is involved in the transmission of nerve impulses. Deficiency in plants causes *chlorosis between the veins, especially in older leaves, and is more common on acidic or sandy soils.

magnetic resonance imaging (MRI) An imaging and diagnostic technique based on the phenomenon of *nuclear magnetic resonance. The body or part to be imaged is placed in an MR scanner. This creates a strong magnetic field, which causes the alignment of certain tissue nuclei, notably hydrogen nuclei in water and lipids. Pulses of high-frequency radio waves are generated,

Macula (when the head is positioned at 90° to the direction of the force of gravity)

and images of internal tissues are created by a computer, according to how the radio waves are absorbed and transmitted. These images map tissue 'slices' and can resolve even very similar tissues by detecting subtle differences in their chemical composition. Hence, MRI can detect, for example, a brain tumour buried deep within normal brain tissue. Apart from diagnosing pathological changes, MRI is also used to guide interventional procedures. **Functional MRI (fMRI)** can monitor brain activity and is now commonly used in neuroscience to study brain function. It detects the minute changes in blood flow that accompany increased neural activity. So, for example, it is possible to map regions of the visual cortex that are active when a subject is exposed to different visual stimuli, such as colours, faces, or dictionaries.

🌐 **SEE WEB LINKS**
• Comprehensive online textbook covering all aspects of MRI

magnetite A strongly magnetic mixed iron oxide, Fe_3O_4, postulated to be part of the *magnetoreceptor in some birds and other animals.

magnetoreceptor A device or organ that detects magnetic fields, particularly the earth's magnetic field. Some form of magnetic sense is found in a wide range of animals, including insects, fishes, amphibians, reptiles, birds, and mammals, but the nature of the magnetoreceptor organs remains elusive. Investigators have sought small particles of the magnetic material magnetite, which are postulated to transduce the magnetic field into nervous impulses. But the only confirmed examples of magnetite in organisms are found in magnetotactic bacteria, which use the earth's magnetic field to guide them when swimming. Photoreceptors may also be involved, and there is evidence to support this in various invertebrate and vertebrate species. One theory is that changes in electron spin resonance caused by the earth's magnetic field elicit nerve impulses. The light-sensitive proteins—*cryptochromes—that occur in the retina of the eye have also

been postulated to form the basis of an internal magnetic compass in certain animals. Certain fishes, particularly sharks and rays, may use their *electroreceptors as magnetoreceptors. This is possible because the animals themselves act as conductors moving through the magnetic field, and so generate electric currents, albeit very weak ones, by electromagnetic induction. Seawater is the conductor that completes the circuit and so could permit detection by the animal's own sense organs. Whatever the basis for it, such a sense is exploited by the animals in navigating through the marine environment. It is suggested that electromagnetic induction might underlie a magnetic field-sensing mechanism operating in the semicircular canals of the inner ear in some birds.

Magnoliophyta *See* ANTHOPHYTA.

major histocompatibility complex (MHC) A large gene cluster that encodes various components of the immune system, including the *histocompatibility proteins and components of the *complement system. In humans the MHC contains over 200 genes, is located on chromosome 6, and includes the *HLA system. Other vertebrate species have similar MHC regions. Certain MHC genes can have many variant alleles; this produces an enormous diversity of proteins in a population, each individual possessing a unique set. *See* MHC CLASS I PROTEIN; MHC CLASS II PROTEIN.

malaria A disease caused by the parasitic protozoan *Plasmodium*, which requires two hosts, the bloodsucking female *Anopheles* mosquito and a human, in order to complete its complex life cycle. Symptoms of fever and anaemia in humans are caused by invasion and destruction of the red blood cells during an asexual phase of the life cycle. *See* APICOMPLEXA.

🌐 **SEE WEB LINKS**
• Coverage of all aspects of malaria, from the website of the US Centers for Disease Control and Prevention

MALDI-TOF *See* MATRIX-ASSISTED LASER DESORPTION/IONIZATION-TIME OF FLIGHT.

male 1. Denoting the gamete (sex cell) that, during *sexual reproduction, fuses with a *female gamete in the process of fertilization. Male gametes are generally smaller than the female gametes and are usually motile (*see* SPERMATOZOON). **2.** (Denoting) an individual whose reproductive organs produce only male gametes. *Compare* HERMAPHRODITE.

malic acid (2-hydroxybutanedioic acid) A crystalline solid, HOOCCH(OH) CH$_2$COOH. L-malic acid occurs in living organisms as an intermediate metabolite in the *Krebs cycle and also (in certain plants) in photosynthesis. It is found especially in the juice of unripe fruits, e.g. green apples, and contributes to their sour taste.

malignant Describing a mutant cell or group of cells that proliferates at a faster rate than normal cells and has the capacity to spread to other sites in the body. *See* CANCER.

malleus (hammer) (*pl.* **mallei**) The first of the three *ear ossicles of the mammalian *middle ear.

Mallophaga An order of secondarily wingless insects comprising the bird lice. Bird lice are minute with dorsoventrally flattened ovoid bodies, reduced eyes, and biting mouthparts. They are ectoparasites of birds, feeding on particles of dead skin, feather fragments, and sometimes blood. The eggs hatch to form nymphs resembling the adults.

malnutrition The condition arising due to the lack of one or more of the *nutrients that are required in the *diet to maintain health. Malnutrition can result from a reduced intake of nutrients (**undernourishment**), an inability to use absorbed nutrients, failure to meet a required increase in nutrient intake, or nutrient losses. There are three stages in the process of malnutrition: first, the carbohydrate stores in the body are depleted; secondly, the fat reserves are metabolized (*see* FATTY-ACID OXIDATION); and finally, proteins are broken down to provide energy. Death may result after protein levels have been reduced to half their normal value. **Kwashiorkor** is a type of malnutrition that develops when the diet lacks proteins and hence *essential amino acids. Malnutrition due to reduced absorption of nutrients in the intestine can develop with a cereal-based diet, due to sensitivity of the intestinal lining to gluten, a protein found in cereals. *See also* MINERAL DEFICIENCY.

Malpighian body (Malpighian corpuscle) The part of a *nephron in the kidney that consists of its cup-shaped end together with the *glomerulus that it encloses. It is named after its discoverer, the Italian anatomist M. Malpighi (1628–94).

Malpighian layer (stratum germinativum) The innermost layer of the *epidermis of mammalian *skin, separated from the underlying dermis by a fibrous *basement membrane. It is only in this layer of the epidermis that active cell division (*mitosis) occurs. As the cells produced by these divisions age and mature, they migrate upwards through the layers of the epidermis to replace the cells being continuously worn away at the surface.

Malpighian tubule Any of the organs that are involved in the excretion of nitrogenous waste and osmoregulation in insects and other terrestrial arthropods, including arachnids, and in centipedes. They lie in the haemocoel, and the primary urine is formed by secretion into the lumen of the Malpighian tubules, principally by the active transport of potassium ions (K$^+$) across the tubule walls. The tubules also selectively extract from the haemolymph uric acid, which—together with water and salts—flows from the tubules into the hindgut to be excreted in the faeces. Some reabsorption of water and salts can occur in the tubules, but most occurs in the hindgut, thereby maintaining the ionic concentration of the haemolymph. *See also* CRYPTONEPHRIDIAL SYSTEM.

malt The product of the hydrolysis of starch by β-*amylase that occurs during the germination of barley in *brewing. *See also* MALTOSE.

MALT See MUCOSAL-ASSOCIATED LYMPHOID TISSUE.

maltase (α-glucosidase) A membrane-bound enzyme in the small intestine that hydrolyses the disaccharide maltose into glucose.

maltose (malt sugar) A sugar consisting of two linked glucose molecules that results from the action of the enzyme *amylase on starch. Maltose occurs in high concentrations in germinating seeds; malt, used in the manufacture of beer and malt whisky, is produced by allowing barley seeds to germinate and then slowly drying them.

malt sugar See MALTOSE.

Mammalia A class of vertebrates containing some 5700 species. Mammals are warm-blooded animals (see HOMOIOTHERMY), typically having sweat glands whose secretion cools the skin and an insulating body covering of hair. All female mammals have *mammary glands, which secrete milk to nourish the young. Mammalian teeth are differentiated into incisors, canines, premolars, and molars and the middle ear contains three sound-conducting *ear ossicles. The four-chambered heart enables complete separation of oxygenated and deoxygenated blood and a muscular *diaphragm takes part in breathing movements, both of which ensure that the tissues are well supplied with oxygen. This, together with well-developed sense organs and brain, have enabled mammals to pursue an active life and to colonize a wide variety of habitats.

Mammals evolved from *synapsid ancestors during the Triassic period, and the first true mammals emerged during the Jurassic (201 to 145 mya). There are two subclasses: the primitive egg-laying *Prototheria (monotremes) and the Theria, which includes all other mammals and consists of the infraclasses *Metatheria (marsupials) and *Eutheria (placental mammals).

mammary glands The milk-producing organs (possibly modified sweat glands) of female mammals, which provide food for the young (see MILK; COLOSTRUM). Their number (2 to 20) and position (on the chest or abdomen) vary according to the species. In most mammals the gland openings project as a **nipple** or **teat**. Nipples have a number of milk-duct openings; teats have one duct leading from a storage cavity.

mandible 1. One of a pair of horny *mouthparts in insects, crustaceans, centipedes, and millipedes. The mandibles lie in front of the weaker *maxillae and their lateral movements assist in biting and crushing the food. **2.** The lower jaw of vertebrates. **3** Either of the two parts of a bird's beak.

Mandibulata See UNIRAMIA.

manganese Symbol Mn. A grey brittle metallic element that is a trace element (see ESSENTIAL ELEMENT) for living organisms. It functions as a cofactor for several enzymes, notably the antioxidant *superoxide dismutase, and other enzymes involved in metabolism and maintaining healthy skin, cartilage, and bone.

mangrove swamp A region of vegetation, found along tropical coasts, in which mangrove trees (*Rhizophora* species) predominate. The waterlogged soil is highly saline, and—like other *halophytes—mangroves are adapted to withstand these conditions; they also possess aerial roots (**pneumatophores**) through which gaseous exchange occurs, to counteract effects of the badly aerated soil. Mangroves are not only complex ecosystems but also play a vital role in protecting coastal regions from the effects of tropical storms and high tides.

mannan See MANNOSE.

mannitol A polyhydric alcohol, $CH_2OH(CHOH)_4CH_2OH$, derived from mannose or fructose. It is the main soluble sugar in fungi and an important carbohydrate reserve in brown algae. Mannitol is used as a sweetener in certain foodstuffs and as a *diuretic in medicine to relieve fluid retention, e.g. in acute kidney failure, or to treat swelling inside the eye or around the brain.

mannose A *monosaccharide, $C_6H_{12}O_6$, stereoisomeric with glucose, that occurs naturally only in polymerized forms called **mannans**. These consist exclusively of mannose residues (linear mannans) or a combination of mannose and glucose residues (glucomannans), sometimes with galactose side chains. They are found in plants, algae, and fungi, serving as energy stores in seeds and roots. They are also major components of *hemicelluloses, which play important structural roles in plants, especially gymnosperms. Mannans on the cell walls of *Candida* yeasts are involved in establishing infections and in triggering the body's immune response.

mannose-binding lectin (MBL) An acute-phase protein (*see* ACUTE-PHASE RESPONSE) secreted into blood plasma by the liver in response to infection. It forms part of the innate immune response by binding to bacterial cells, enhancing their susceptibility to ingestion by phagocytic cells, particularly monocytes, and triggering the *complement cascade leading to target cell destruction. The MBL molecule has up to six clusters of carbohydrate-recognition domains, which bind to appropriately spaced mannose or fucose residues on bacterial surfaces. The spacing of the residues is crucial in discriminating bacterial cells from self tissues. *See also* LECTIN.

manometer A device for measuring pressure differences, usually by the difference in height of two liquid columns. The simplest type is the U-tube manometer, which consists of a glass tube bent into the shape of a U. If a pressure to be measured is fed to one side of the U-tube and the other is open to the atmosphere, the difference in level of the liquid in the two limbs gives a measure of the unknown pressure.

mantle The fold of skin covering the dorsal surface of the body of molluscs, which extends into lateral flaps that protect the gills in the **mantle cavity** (the space between the body and mantle). The outer surface of the mantle secretes the shell (in species that have shells).

Maotianshan shale *See* CHENGJIANG FOSSILS.

MAP kinase (mitogen-activated protein kinase; MAPK) An intracellular signal molecule that forms part of a key signal transduction pathway in eukaryotic cells. It enters the nucleus to activate transcription factors that control gene expression. The so-called **MAP kinase cascade** is triggered by a small G protein, such as *RAS protein, and consists of sequentially activating protein kinases, including MAP kinase kinase kinase (MAPKKK) and MAP kinase kinase (MAPKK); the latter doubly phosphorylates MAPK, enabling it to translocate to the nucleus and affect gene expression. This signal pathway is involved in diverse cellular activities and can be activated by a range of surface receptors.

map unit (m.u.; centimorgan) A unit for measuring distance between genes (or other loci) on a chromosome according to the frequency of recombination between due to *crossing over. A distance of 1 m.u.—or 1 centimorgan (1 cM)—corresponds to a recombinant frequency of 1%, i.e. the two genes recombine once in every 100 meioses. Map units are used in constructing *linkage maps; they measure relative genetic distance between loci, not absolute physical distance. They are additive over short distances, but are not reliable for distantly linked genes because of the possibility of multiple crossovers. *Compare* CROSSOVER VALUE.

Marchantiophyta *See* HEPATOPHYTA.

marker gene 1. A gene used to identify a particular bacterial colony or bacteriophage plaque. Such genes are incorporated into cloning *vectors to enable the isolation and replication of colonies containing a desired vector. Typically, marker genes confer resistance to specific antibiotics or produce colour changes. **2. (genetic marker)** A gene that acts as a tag for another, closely linked, gene. Such markers are used in mapping the order of genes along chromosomes and in following the inheritance of particular genes: genes closely linked to the marker will generally be inherited with it. Markers must be readily identifiable in the phenotype, for instance by controlling an easily observable feature (such as eye colour). *See also* MOLECULAR MARKER; REPORTER GENE.

mark-recapture method A method for estimating the size of a population of mobile animals. It involves collecting a random sample of individuals, marking them, and releasing them. After allowing sufficient time for the marked individuals to disperse among the wider population, a second sample is taken. The proportion of marked individuals in this sample is assumed to be the same as the proportion of the original sample to the entire population; hence the size of the population can be estimated. The method assumes that marked and unmarked individuals have the same probability of being captured and that the marked individuals are representative of the population as a whole, e.g. in terms of viability, behaviour, and so forth.

marsupials *See* METATHERIA.

masquerade The resemblance of an organism to some inanimate object in its environment so that it remains effectively 'hidden' from predators. Many insects have evolved particular shapes and colours so they resemble leaves, twigs, or other features of their natural surroundings, making it difficult for predators to detect them visually. For example, some larvae look like bird droppings, certain butterflies have very leaflike wings, while stick insects fully live up to their name. *See also* CAMOUFLAGE.

mass extinction The extinction of a large number of species within a relatively short interval of the geological time scale. The fossil record provides evidence for several mass extinctions, perhaps as many as 20, since the start of the Phanerozoic eon about 570 million years ago. Such extinctions cause radical changes in the characteristic fossil assemblages of rock, which have been reflected in the naming of strata by geologists. Hence, mass extinctions often mark the boundaries between geological strata and between the corresponding geological time intervals. The biggest

mass extinctions occurred at the end of the Permian period (about 248 million years ago), when over 80% of all marine invertebrate genera disappeared (including the trilobites), and at the end of the Cretaceous (65 million years ago), when some 50% of all genera became extinct, including virtually all the dinosaurs (*see* ALVAREZ EVENT). Such cataclysmic changes in the earth's biota have profound effects on the course of evolution, for example by leaving vacant ecological niches into which surviving groups can expand and radiate. Recent decades have witnessed a rapid acceleration in species loss, due largely to human activities. This has prompted calls to designate a new epoch, the *Anthropocene. See* Appendix 6.

mass flow (pressure flow) A hypothesis to explain the movement of sugars in the phloem tissue of plants. At a **source** (site of production) sugars are loaded into *companion cells and thence into the *sieve elements (*see* PHLOEM LOADING), causing water to follow by osmosis. The pressure of water in the tubes (the hydrostatic pressure) causes it to move along the tubes to a **sink** (site of utilization), where the reverse process occurs. Here sugars diffuse or are actively transported from the sieve elements into the companion cells and then into the surrounding tissues, establishing a concentration gradient from source to sink (*see* PHLOEM UNLOADING). Although different solutes can be transported in the phloem in different directions at the same time, it is argued that the mass flow hypothesis can still apply provided transport occurs in different sieve elements.

mass spectrometry (mass spectroscopy) A technique used to determine relative atomic masses and the relative abundance of isotopes and for chemical analysis, as in the identification of metabolites, drugs, and other molecules isolated from biological samples. In a **mass spectrometer** a sample (usually gaseous) is ionized and the positive ions produced are accelerated into a high-vacuum region containing electric and magnetic fields. These fields deflect and focus the ions onto a detector. The fields can be varied in a controlled way so that ions of different types can impinge on the detector. A **mass spectrum** is thus obtained consisting of a series of peaks of variable intensity to which mass/charge (m/e) values can be assigned. For organic molecules, the mass spectrum consists of a series of peaks, one corresponding to the parent ion and the others to fragment ions produced by the ionization process. Different molecules can be identified by their characteristic pattern of lines. Analysis of mixtures can be done by gas chromatography-mass spectrometry (*see* GAS-LIQUID CHROMATOGRAPHY).

Mass spectrometry is commonly used to identify protein components of complex mixtures. One approach involves separating different proteins by chromatography and electrophoresis, extracting the individual protein bands, and cleaving the proteins into small peptide fragments using protease enzymes. The fragments are then analysed by mass spectrometry. Large peptides can also be analysed 'top-down' using **tandem mass spectrometry**, which involves multiple phases or rounds of spectrometry in a single machine. Such a technique can isolate a particular peptide from a mixture, subject it to fragmentation through collisions with gas molecules in a vacuum (collision-induced dissociation), then determine the m/e values of the resulting fragments. The resultant 'signature' of the fragments is compared with a database of known protein fragmentation patterns to establish the identity of the unknown protein. *See also* MATRIX-ASSISTED LASER DESORPTION/IONIZATION-TIME OF FLIGHT; PROTEIN PROFILING.

- The principles and biological applications of mass spectrometry described by Alison Ashcroft, University of Leeds

mast cell A large cell with densely granular cytoplasm that is found in connective tissues, for example around blood vessels and in the skin. Mast-cell granules contain mediators of inflammation (e.g. *histamine) and lipid mediators (such as prostaglandin D_2 and leukotriene C4), as well as

various *cytokines, which cause a local increase in blood flow and attract white blood cells to the area. The granule contents are released from the cell in response to tissue injury or as part of an allergic response. Release is triggered by binding of antigen to a type of antibody (IgE) that is bound to the mast cell. The cell also releases heparin, an anticoagulant that prevents clot formation. *See also* ALLERGY.

mastication The process of chewing food, which involves movements of the jaws and teeth. Mastication breaks up the food into small particles, which provides a greater surface area for digestion and enables the formation of a *bolus, which is small enough to pass through the oesophagus.

mastigoneme *See* TINSEL FLAGELLUM.

mastoid process An outgrowth from the temporal bone of the skull containing air cavities that communicate with the cavity of the middle ear. In humans it is a route through which infection may spread from the middle ear.

maternal effect genes Genes expressed in maternal follicle cells whose products (messenger RNAs and proteins) diffuse into the egg cell to influence its early development. Gradients of the products are established in the egg cytoplasm; following fertilization and subsequent cell division of the zygote, these gradients influence zygotic gene expression and cause regional differentiation of the embryo by sequential expression of *segmentation genes. For example, in many types of embryo, maternal effect genes are responsible for determining polarity, i.e. which end is the 'head' and which is the 'tail'. *See also* PATTERN FORMATION.

maternal inheritance *See* CYTOPLASMIC INHERITANCE.

mating *See* SEXUAL INTERCOURSE.

mating season *See* BREEDING SEASON.

mating type The equivalent in microorganisms, fungi, and algae of the male and female individuals of higher organisms.

Mating types differ genetically but are morphologically the same and therefore difficult to distinguish. For this reason different mating types, of the same species, are designated as + and −. The mating types are distinct strains that are compatible in sexual reproduction; they are able to give rise to a zygote, which can develop into a new individual.

matric potential Symbol ψ_m. A component of *water potential due to the adhesion of water molecules to nondissolved structures of the system, i.e. the matrix, such as plasma membranes or soil particles. It is always negative and is significant only outside living cells in relatively dry systems, for example soils, where much of the water is tightly bound to soil particles.

matrix (*pl.* **matrices** or **matrixes**) (in histology) The component of tissues (e.g. bone and cartilage) in which the cells of the tissue are embedded. *See also* EXTRACELLULAR MATRIX.

matrix-assisted laser desorption/ ionization-time of flight (MALDI-TOF) A form of *mass spectrometry that is used for rapid qualitative and quantitative characterization of samples of biomolecules in a wide range of applications. The test sample is applied to a specially designed matrix on a target plate, which is then irradiated with ultraviolet light from a laser. This heats the matrix, the surface of which vaporizes along with the sample, which is ionized in the process. These ions are directed by electric fields along a flight tube towards a detector. The time of flight depends on the mass-to-charge ratio of each ion—smaller or more highly charged ions move faster. The outcome is a spectrum of the ionized fragments of the test sample. This is compared with a range of standard samples, and the nature of the test sample is established. The technique is a rapid and relatively simple way of identifying bacterial or yeast cells according to their spectra, which are compared with a database of clinically relevant standard spectra. The rapid identification of proteins and peptides is another application.

maturation 1. The process of becoming fully developed, especially the final phase in the development of *germ cells, which renders the egg or sperm capable of fertilization. **2.** Changes in the neuromuscular system as an animal develops that improve coordination regardless of experience.

maturation-promoting factor (MPF) *See* MITOSIS-PROMOTING FACTOR.

maturity 1. The stage in a life cycle that is reached when a developing organism has taken on the appearance of the adult form and is capable of reproduction. **2.** The stage reached in the formation of gametes (*gametogenesis) following meiotic division of precursor cells and their development into fully functional gametes.

maxicell A bacterial cell that has been irradiated with ultraviolet light to intentionally damage the cell's own DNA so that *plasmid genes, which are relatively spared, can be preferentially expressed.

maxilla (*pl.* **maxillae**) **1.** One of a pair of *mouthparts in insects, crustaceans, centipedes, and millipedes. They lie behind the *mandibles and their lateral movements assist in feeding. Crustaceans have two pairs of maxillae but in insects the second pair are fused together forming the *labium. **2.** One of a pair of large toothbearing bones in the upper jaw of vertebrates. In mammals they carry all the upper teeth except the incisors.

maxilliped A paired appendage in crustaceans that contributes to the *mouthparts. There may be up to three pairs of maxillipeds, which are located behind the mandibles and are used to move food to the mouth.

maxillula (*pl.* **maxillulae**) A paired appendage that contributes to the *mouthparts in certain crustaceans. Lying between the mandibles and the maxillae, they may bear hairs or spines designed to move food particles deeper into the mouth.

maximum likelihood In phylogenetics, the statistical principle used to create the phylogenetic tree (*phylogram) most likely to account for the observed data. Methodologies based on this principle are frequently applied in computer analysis of molecular sequence data and require a mathematical model describing how DNA or protein sequences behave. For example, the model will account for how bases change at different rates depending on the particular gene, the position in a codon, and whether the change is a transition (i.e. between two purines or two pyrimidines) or a transversion (i.e. purine to pyrimidine or vice versa). The assumption is that evolutionary histories of organisms adhere to such rules.

maximum parsimony The logical principle that the simplest explanation for the observed facts is preferred over more complex explanations. It is also known as **Occam's razor**, after the English philosopher William of Occam (c. 1285–1349). The principle is often adopted in phylogenetics in construction of phylogenetic trees (*phylograms) showing the ancestral relationships of different groups based on the minimum number of assumed evolutionary changes. When dealing with comparison of DNA or protein sequence data from different taxa, computer programs can construct all possible phylogenetic trees and identify the most parsimonious one that will fit the data—e.g. one that requires the fewest base changes in DNA. However, the principle can also be applied in constructing such trees based on other evidence, e.g. differences in morphological, developmental, or behavioural traits between the taxa.

maximum permissible dose See DOSE.

Mb See MEGABASE.

M band (M line) See SARCOMERE.

M cells See PEYER'S PATCHES.

mean The average value of a set of n numbers, i.e. the sum of the numbers divided by n.

meatus (*pl.* **meatus** or **meatuses**) A small canal or passage in the body. An example is the **external auditory meatus** of the *outer ear in mammals, which connects the exterior opening to the eardrum.

mechanoreceptor A *receptor that responds to such mechanical stimuli as touch, sound, and pressure. The skin is rich in mechanoreceptors, including *Meissner's corpuscles, *Pacinian corpuscles, *Merkel's discs, and Ruffini's capsules. These provide information to the brain about the pressure, texture, and shape of objects in contact with the body. Their density varies according to site; in humans they are most abundant in the skin of the fingertips, palms, soles of the feet, and genitalia. Stretch receptors in muscles, tendons, and ligaments are mechanoreceptors that respond to stretching of muscles and changes in the position of limbs. Their input to the central nervous system is essential for control of posture and coordination of movements (*see* GOLGI TENDON ORGAN; MUSCLE SPINDLE). *Hair cells are found in various sensory organs, including the mammalian ear and the *lateral line system of fishes and amphibians. They respond to movements of the fluid medium surrounding the body, and hence can detect sound waves in air and motion in surrounding water. *See also* BARORECEPTOR.

mechanotransduction The conversion of a physical force into a biological response. Prime examples are the responses of *mechanoreceptors to stimuli such as touch, sound, and stretching of muscles. The physical forces open or close ion channels in the sensory cell, which alters the cell's *membrane potential and triggers the release of neurotransmitter and/or the generation of electrical signals in the form of action potentials. However, the mechanical interactions between individual cells, and between cells and their external environment, affect many aspects of a cell's life, including differentiation, tissue organization and integrity, cell motility, and even tumour formation. The *cytoskeleton, responsible for maintaining the cell's internal structure, is a key component in relaying external forces to the cell's interior. Mechanotransduction is also involved in the colonization of host cells by infecting bacteria.

meconium The first faecal material passed by a newborn mammal, which is usually dark green in colour.

median The middle number or value in a series of numbers or values.

median eye (pineal eye; 'third eye') An eyelike structure, with a rudimentary lens and retina, found on the top of the head of some lizards, frogs, salamanders, sharks, fishes, the Cyclostomata (lampreys) as well as in many fossil vertebrates. The most notable example is found in the tuatara (*Sphenodon*) of New Zealand, a 'living fossil' representative of an order of early reptiles. In adults the median eye is generally covered with skin or scales, and acts as a photoreceptor, detecting changes in light intensity and modifying the physiology and behaviour of the animal, particularly its circadian rhythms. In lizards the median eye develops from part of the forebrain in close association with the pineal sac (which becomes the *pineal gland in animals that have lost the median eye, such as birds and mammals).

median lethal dose See LD$_{50}$.

mediastinum (*pl.* **mediastina**) **1.** A membrane in the midline of the *thorax of mammals that separates the lungs. **2.** The space between the two lungs, which is occupied by the heart and oesophagus.

MEDLINE (Medical Literature Analysis and Retrieval System Online) A bibliographic database administered by the US National Library of Medicine (NLM) that contains over 24 million references to journal articles in biomedicine, health, and related life sciences. Coverage extends to around 5200 journals, with particular emphasis on US journals, and around three-quarters of a million new records are added annually. The articles are indexed using the NLM's Medical Subject Heading (MeSH) thesaurus, which permits searching using both alphabetical and hierarchical approaches. MEDLINE is the major component of the NLM's **PubMed** database and can be accessed via *Entrez.

medulla (*pl.* **medullae**) **1.** (in zoology) The central tissue of various organs, including the adrenal glands (**adrenal medulla**) and kidneys (**renal medulla**). **2.** (in botany) *See* PITH.

medulla oblongata Part of the vertebrate *brainstem, derived from the *hindbrain, that is continuous with the spinal cord. Its function is to regulate the autonomic pathways controlling breathing, heartbeat, blood pressure, digestive movements, and other involuntary processes. Nerve signals from the peripheral nervous system via the spinal cord are relayed by the medulla to the midbrain and forebrain, and the medulla gives rise to many of the *cranial nerves. Dorsal respiratory neurons in the medulla send nerve impulses to control breathing movements of the diaphragm (*see* VENTILATION CENTRE). These neurons receive impulses from chemoreceptors on the surface of the medulla that respond to the pH of their surroundings, which reflects the partial pressure of carbon dioxide in the blood. They also have inputs from oxygen-sensitive chemoreceptors in the major blood vessels, and from higher centres in the brain. The medulla's cardiovascular control centre regulates the autonomic nervous system that governs heart rate and *blood pressure. It integrates inputs from the medulla's pH chemoreceptors and the oxygen sensors in the great vessels, as well as from pressure receptors (*baroreceptors) in the aorta and carotid artery, and from higher brain centres responsible for stress or emotions.

medullary ray (ray) Any of the vertical plates of *parenchyma cells running radially through the cylinder of vascular tissue in the stems and roots of plants. In cut timber they appear as lines radiating from the centre of the cut surface of a log, perpendicular to the growth rings. Each may be one to many cells in width. **Primary medullary rays** occur in young plants and in those not showing secondary thickening; they pass from the cortex through to the pith. **Secondary medullary rays** are produced by the vascular *cambium and terminate in xylem and phloem tissues. Medullary rays store and transport food materials.

medullated nerve fibre A nerve fibre that is characterized by a *myelin sheath, which insulates the axon.

medusa (*pl.* **medusae**) The free-swimming stage in the life cycle of certain members of the *Cnidaria. Medusae are umbrella-shaped, with tentacles round the edge and the mouth in the centre underneath. They swim by pulsations of the body and reproduce sexually. In the Hydrozoa (e.g. *Hydra*) they alternate in the life cycle with *polyps, from which they are produced by budding. In the Scyphozoa, which includes all the common jellyfish, and in the box jellies (Cubozoa), the medusa is the dominant form and the polyp is reduced or absent.

mega- 1. A prefix denoting large size; e.g. meganucleus, megasporangium. **2.** Symbol M. A prefix used in the metric system to denote one million times. For example, 10^6 volts $= 1$ megavolt (MV).

megabase Symbol Mb. A unit of length used to measure the size of polynucleotides (i.e. DNA or RNA) or segments of such molecules. 1 Mb $= 10^6$ bases or base pairs.

meganucleus See NUCLEUS.

megaphyll A type of foliage leaf in ferns and seed plants that has branched or parallel vascular bundles running through the lamina. The megaphylls of ferns are large pinnate leaves called **fronds**. A megaphyll was formerly called a **macrophyll**. *Compare* MICROPHYLL.

megasporangium See SPORANGIUM.

megaspore See MEGASPORE MOTHER CELL; SPOROPHYLL.

megaspore mother cell (**megasporocyte**) A diploid cell in plants that divides by meiosis to give rise to four haploid megaspores (*see* SPOROPHYLL). In flowering plants

the megaspore mother cell (or **embryo mother cell**) is situated in the ovule. One of the megaspores it produces develops into the *embryo sac; the others abort.

megasporophyll *See* SPOROPHYLL.

Megavirus See NUCLEOCYTOPLASMIC LARGE DNA VIRUS.

meiofauna 1. (meiobenthic fauna) Animals inhabiting the bottom of a river, lake, or sea that are just visible to the naked eye; for example, small polychaetes, bivalves, and nematodes. They have dimensions in the range 0.1 to 1 mm, intermediate between *microfauna and *macrofauna. **2.** *See* INTERSTITIAL FAUNA.

meiosis (reduction division) A type of nuclear division that gives rise to four reproductive cells (gametes) each with half the chromosome number of the parent cell. Two consecutive divisions occur. In the first, *homologous chromosomes become paired and may exchange genetic material (*see* CROSSING OVER) before moving away from each other into separate daughter nuclei. This is the actual reduction division because each of the two nuclei so formed contains only half of the original chromosomes. The daughter nuclei then divide by mitosis and four *haploid cells are produced. *See also* PROPHASE; METAPHASE; ANAPHASE; TELOPHASE.

Meissner's corpuscles Receptors for touch that are found in the papillary (upper) region of the dermis of the *skin. Named after German anatomist Georg Meissner (1829–1905), each consists of a mass of dendrites encapsulated by connective tissue. The corpuscles are fast-adapting receptors, meaning that they respond primarily to changes in pressure on the skin, as when an object is rolled between the fingers. They occur in abundance in the palms of the hands, fingertips, and soles of the feet as well as in the nipples and tip of the penis. *Compare* RUFFINI'S CAPSULE.

melanin Any of a diverse group of polymers, derived from phenolic precursors, such as the amino acid tyrosine, that

The stages of meiosis in a cell containing two pairs of homologous chromosomes

cause pigmentation of eyes, skin, and hair in vertebrates, and of various structures in plants and fungi. Vertebrate melanins are produced by specialized epidermal cells called **melanophores** (or **melanocytes**); their dispersion in these cells is controlled by *melanocyte-stimulating hormone and *melatonin. The photo-absorptive properties of melanins are important in protecting the skin against damage by UV light; they also scavenge harmful free radicals, and their pigmentation provides camouflage for many animals. Invertebrates, fungi, and microorganisms also produce melanin pigments: the 'ink' of the octopus and squid is a notable example. Hereditary *albinism is caused by the absence of the enzyme tyrosinase, which is necessary for melanin production.

melanism Black coloration of the body caused by overproduction of the pigment melanin, often as a reaction to the environment. There are several species of melanic moths in industrially polluted areas (*see* INDUSTRIAL MELANISM) and the panther is a melanic form of leopard.

melanocyte-stimulating hormone (MSH) Any of several related peptide hormones that affect skin coloration and, in mammals, also play a role in regulating hunger and metabolism. MSHs are produced from the precursor pro-opiomelanocortin (POMC) and secreted by the anterior or (when present) the intermediate lobe of the pituitary gland. In humans, certain neurons in the arcuate nucleus of the hypothalamus release α-MSH, which is a potent suppressor of appetite and has a key role in regulating energy balance; it acts antagonistically to *neuropeptide Y and has potential as an antiobesity agent. α-MSH acts also as an anti-inflammatory cytokine and as a neuropeptide in the brain. It stimulates sexual activity, is involved in regulation of heart rate and blood pressure, and promotes tanning in people with fair skin by stimulating pigment cells (melanocytes) to produce melanin. In amphibians, fishes, and reptiles, MSH controls the dispersal of melanin in *chromatophores of the skin, causing the skin to darken.

melanophore *See* CHROMATOPHORE.

melatonin A hormone derived from *serotonin and secreted by the pineal gland and retinas of vertebrates. Melatonin secretion by the pineal is linked to the dark-light cycle of the organism's environment, being greatest at night and lowest by day. Production is regulated by signals from the suprachiasmatic nucleus of the hypothalamus, which itself receives input from the retina (*see* BIOLOGICAL CLOCK). The hormone is involved in regulating certain diurnal and seasonal changes in the body, such as the reproductive cycle in seasonally breeding animals. Melatonin also controls pigmentation changes; it triggers aggregation of the pigment *melanin into melanophores in the skin, causing the skin to turn pale. It is used as a drug to treat sleep disorders and symptoms of jet lag.

membrane 1. A thin sheet of tissue or other material that lines a body cavity, forms a partition, or connects various structures. **2.** Any of the various flexible sheetlike structures, composed predominantly of lipids and proteins, that occur in living cells, such as the *plasma membrane forming the cell boundary. *See* CELL MEMBRANE.

membrane attack complex (MAC) An assembly of *complement proteins that creates a pore in the plasma membrane of a pathogen, thereby contributing to its destruction. It is one of the mechanisms of innate immunity that follows activation of the complement cascade. Certain complement components (C5b, C6, and C7) bind to the lipid bilayer of the target cell, insert into it, and then induce polymerization of the C9 component to form a cylindrical channel about 10 nm in diameter. This disrupts the membrane's integrity and electrochemical properties and kills the cell. The MAC is significant for defence against only certain bacterial infections, such as *Neisseria* spp. responsible for gonorrhoea, but is important in certain immune system diseases.

membrane bone (dermal bone) *Bone formed directly in connective tissue, i.e. by intramembranous *ossification, rather than by replacing cartilage (*compare* CARTILAGE BONE). Some face bones, skull bones, and part of the clavicle are membrane bones. Small areas of membrane become jelly-like and attract calcium salts. Bone-forming cells break down these areas forming a bone lattice, which eventually fills in.

membrane potential (transmembrane potential) The difference in electrical charge across the plasma membrane of a cell due to differences in the distribution of charged ions between the interior of the cell and the extracellular fluid. Membrane potentials exist in all living cells, and typically are in the range −20 to −200 mV. Examples include neuron, −70 mV; skeletal muscle cell, −90 mV; smooth muscle cell, −50 mV; plant root or stem cells, −100 to −130 mV. *See* RESTING POTENTIAL.

membranous labyrinth The soft tubular sensory structures that form the *inner ear of vertebrates and are housed within the bony labyrinth.

meme A self-replicating unit of cultural inheritance analogous to a gene. The term was introduced by British biologist Richard Dawkins (1941–) in his book *The Selfish Gene* (1976) to denote a cultural entity, such as a song, a skill, a religion, or a recipe, that is transmitted between individuals and across generations, so that it is inherited and (potentially) can change over time. Such a concept can provide useful insights into the nature of cultural development. More controversial is the supposed role of memes in shaping the evolution of the human brain by a form of sexual selection. Proponents argue that the ability to reproduce memes, such as musical compositions or literary works, is a manifestation of the genetically determined ability to copy them, making that individual more attractive to potential mates. Hence such admired or talented individuals tend to have more offspring, who inherit their parent's meme-imitating, or even meme-generating abilities. Selection for such traits over time would, it is claimed, enhance the size and complexity of the brain.

memory The means by which information is stored in the brain. The exact mechanism of processing and storing information is not known but is thought to involve the construction of circuits of neurons in the cerebral cortex, which are strengthened by repeated use (*see* SYNAPTIC PLASTICITY). Various models of memory have been proposed, with different types of information and levels of processing. One model proposes three principal forms of memory based on duration. **Sensory memory** is the fleeting retention of stimuli perceived by the senses. These memories are almost immediately lost unless attention is paid to them, when they enter **short-term memory** (or 'working memory'). This lasts for up to a minute or so. With effort, for example repetition of a phone number, the information may then enter **long-term memory**, where it can remain for hours, days, or even years. The subsequent conscious recollection of things or facts involves **explicit (declarative) memory**, whereas

implicit memory concerns motor skills, such as the ability to ride a bike, and conditioned reflexes. The transfer of short-term memory to long-term memory involves the *limbic system, in particular the hippocampus. It is thought that connections in the hippocampus are necessary for consolidating and accessing short-term memory but not for recalling long-term memories. The amygdala is instrumental in producing fear and fear memories. Memory is essential to the processes of *learning and recognition of individuals and objects.

(((●))) SEE WEB LINKS
- Comprehensive account of concepts and models of memory, from the *Stanford Encyclopedia of Philosophy*

memory cell *See* IMMUNOLOGICAL MEMORY.

Mendel, Johann Gregor (1822–84) Austrian geneticist, who from 1843 lived as a monk in Brün (now Brno, in the Czech Republic). His fame rests on the plant-breeding experiments he began in 1856, which eventually produced the rules of inheritance summarized in *Mendel's laws. His work was ignored during his lifetime and only rediscovered in 1900 by Hugo de Vries and others.

Mendelism The theory of heredity that forms the basis of classical *genetics, proposed by Gregor Mendel in 1866 and formulated in two laws (*see* MENDEL'S LAWS; PARTICULATE INHERITANCE). Mendel suggested that individual characteristics were determined by inherited 'factors', and when improved microscopes revealed details of cell structure the behaviour of Mendel's factors could be related to the behaviour of chromosomes during *meiosis.

Mendel's laws Two laws summarizing Gregor Mendel's theory of inheritance (*see also* MENDELISM). The **Law of Segregation** states that each hereditary characteristic is controlled by two 'factors' (now called *alleles), which segregate (separate) and pass into separate germ (reproductive) cells. The **Law of Independent Assortment** states that pairs of 'factors' segregate independently of each other when germ cells are formed (*see also* INDEPENDENT ASSORTMENT). These laws are the foundation of classical genetics.

meninges (*sing.* **meninx**) The three membranes that surround the brain and spinal cord of vertebrates: the *pia mater, the *arachnoid membrane, and the outer *dura mater. The pia and arachnoid are separated by the **subarachnoid space**, which contains *cerebrospinal fluid.

meniscus (*pl.* **menisci**) *See* SURFACE TENSION.

menopause The time in a woman's life when ovulation and menstruation cease (*see* MENSTRUAL CYCLE). It normally occurs between the ages of 45 and 55. The effects of the gonadotrophic hormones, *follicle-stimulating hormone and *luteinizing hormone, in the ovaries decrease so that the follicles do not develop properly. There is a change in the balance of the hormones oestradiol and progesterone, secreted by the ovaries, which may be associated with certain physical symptoms, such as weight gain and 'hot flushes', and there may also be mood changes. These symptoms can be treated by long-term **hormone replacement therapy** with oestrogens and progestogens.

menstrual cycle (uterine cycle; sexual cycle) The approximately monthly cycle of events that prepares the uterus to receive the fertilized egg for pregnancy. It occurs in synchrony with the *ovarian cycle and replaces the *oestrous cycle in most primates (including humans). The lining of the uterus becomes progressively thicker with more blood vessels in preparation for the *implantation of a fertilized egg cell (blastocyst). Ovulation occurs during the middle of the cycle (the fertile period). If fertilization does not occur the uterine lining breaks down and is discharged from the body (**menstruation**); the discharge is known as a 'period'. In women the fertile period is 11–15 days after the end of the last menstruation.

mericarp *See* SCHIZOCARP.

meristem A plant tissue consisting of actively dividing cells that give rise to cells that differentiate into new tissues of the plant. The most important meristems are those occurring at the tip of the shoot and root (*see* APICAL MERISTEM) and the lateral meristems in the older parts of the plant (*see* CAMBIUM; CORK CAMBIUM).

meristoderm The outer meristematic layer of the thallus of certain brown algae (Phaeophyta). The cells of the meristoderm divide to increase the width of the thallus.

Merkel's disc A touch receptor, occurring in the outer layer (epidermis) of hairless skin, that responds to sustained light pressure. Named after German anatomist Friedrich S. Merkel (1845–1919), these discs are found in large numbers in the fingertips, lips, and genitalia, where they are important in discriminating textures and shapes. Each consists of a sensory Merkel cell and associated sensory nerve ending, which conveys impulses to a sensory nerve.

merocrine secretion *See* SECRETION.

meromictic lake *See* DIMICTIC LAKE.

Merostomata *See* CHELICERATA.

merozygote A bacterial cell that contains more than the *haploid number of chromosomes but less than the full *diploid number. Merozygotes arise when the genetic material from one bacterial cell is only partially transferred into another cell during *conjugation, *transduction, or *transformation.

mesencephalon *See* MIDBRAIN.

mesenteric artery The artery that delivers blood to the intestines.

mesentery A thin sheet of tissue, bounded on each side by *peritoneum, that supports the gut and other organs in the body cavities of animals. Vertebrates have a well-developed dorsal mesentery that anchors the stomach and intestine and contains blood vessels and nerves supplying the gut. The reproductive organs and their ducts are also supported by mesenteries.

mesocarp *See* PERICARP.

mesoderm The layer of cells in the *gastrula that lies between the *ectoderm and *endoderm. It develops into the muscles, circulatory system, and reproductive organs (excluding the germ cells), and in vertebrates also into the excretory system and skeleton. *See also* GERM LAYERS.

mesofauna *See* MACROFAUNA.

mesoglea The gelatinous noncellular layer between the outer and inner cell layers in the body wall of ctenophores and cnidarians. It may be thin, as in *Hydra*, or tough and fibrous, as in the larger jellyfish and sea anemones. It often contains cells that have migrated from the two body layers but these do not form tissues and organs and the mesoglea is not homologous with the mesoderm of *triploblastic animals.

mesopelagic zone *See* EUPHOTIC ZONE.

mesophilic Describing an organism that lives and grows optimally at moderate temperatures, typically between 10 and 40°C. The vast majority of organisms are mesophiles, occupying all the major biomes of temperate and tropical climates. *Compare* PSYCHROPHILIC; THERMOPHILIC.

mesophyll The internal tissue of a leaf blade (lamina), consisting of *parenchyma cells. In the leaves of eudicots typically there are two distinct forms. **Palisade mesophyll** lies just beneath the upper epidermis and consists of cells elongated at right angles to the leaf surface. They contain a large number of *chloroplasts and their principal function is photosynthesis. **Spongy mesophyll** occupies most of the remainder of the lamina. It consists of spherical loosely arranged cells containing fewer chloroplasts than the palisade mesophyll. Between these cells are air spaces leading to the *stomata.

mesophyte Any plant adapted to grow in soil that is well supplied with water and mineral salts. Such plants wilt easily when exposed to drought conditions as they are not adapted to conserve water. The majority of flowering plants are mesophytes. *Compare* HALOPHYTE; HYDROPHYTE; XEROPHYTE.

mesothelium (*pl.* **mesothelia**) A single layer of thin platelike cells covering the surface of the inside of the abdominal cavity and thorax and surrounding the heart, forming part of the *peritoneum and *pleura (*see* SEROUS MEMBRANE). It is derived from the *mesoderm. *Compare* ENDOTHELIUM; EPITHELIUM.

mesotrophic Describing a body of water, such as a lake, that is intermediate between a *eutrophic lake and an *oligotrophic lake in the amount of nutrients contained within it.

Mesozoic The geological era that extended from the end of the *Palaeozoic era, about 251 million years ago, to the beginning of the *Cenozoic era, about 66 million years ago. It comprises the *Triassic, *Jurassic, and *Cretaceous periods. The Mesozoic era is often known as the **Age of Reptiles** as these animals, which included the dinosaurs, pterosaurs, and ichthyosaurs, became the dominant animal life form; most became extinct before the end of the era.

messenger RNA *See* RNA.

metabolic pathway *See* METABOLISM.

metabolic rate A measure of the energy used by an animal in a given time period. The metabolic rate of an animal is affected by several interacting factors, including temperature and the level of activity. The metabolic rate of a resting animal is known as the *basal metabolic rate (BMR).

metabolic waste The *waste products, collectively, of metabolism.

metabolism The sum of the chemical reactions that occur within living organisms. The various compounds that take part in or are formed by these reactions are called **metabolites**. In animals many metabolites are obtained by the digestion of food, whereas in plants only the basic starting materials (carbon dioxide, water, and minerals) are externally derived. The synthesis (*anabolism) and breakdown (*catabolism) of most compounds occurs by a number of reaction steps, the reaction sequence being termed a **metabolic pathway**. Some pathways (e.g. *glycolysis) are linear; others (e.g. the *Krebs cycle) are cyclic. The changes at each step in a pathway are usually small and are promoted by efficient biological catalysts—the enzymes. In this way the amounts of energy required or released at any given stage are minimal, which helps in maintaining a constant *internal environment. Various *feedback mechanisms exist to govern *metabolic rates.

metabolite *See* METABOLISM.

metabolome The entire complement of metabolites found within a cell under defined conditions, such as a particular physiological or developmental state. The metabolome is determined using various forms of high-throughput mass spectrometry. It excludes nucleic acids and other large molecules, giving a 'snapshot' of the cell's metabolic state. *See* METABOLOMICS.

metabolomics The study of how the pool of metabolites (*see* METABOLOME) of cells changes under various physiological or developmental conditions or in response to genetic modification (e.g. mutation).

metabotropic receptor A cell receptor that, when activated by binding of a ligand, triggers changes in cell metabolism via intracellular second messengers. *See* GLUTAMATE RECEPTOR; G-PROTEIN-COUPLED RECEPTOR. *Compare* IONOTROPIC RECEPTOR.

metacarpal One of the bones in the *metacarpus.

metacarpus (*pl.* **metacarpi**) The hand (or corresponding part of the forelimb) in terrestrial vertebrates, consisting of a number of rod-shaped bones (**metacarpals**) that articulate with the bones of the wrist (*see* CARPUS) and those of the fingers (*see* PHALANGES). The number of metacarpals varies between species: in the basic *pentadactyl limb there are five, but this number is reduced in many species.

metacentric *See* CENTROMERE.

metafemale A fruit fly (*Drosophila* sp.) with three instead of the normal two X chromosomes. In such flies the disproportion of sex chromosomes to autosomes leads to impaired development and often failure to emerge from the pupae. Human females with the 47XXX karyotype have generally normal sexual development and are fertile, although there is an increased risk of learning difficulties and developmental delay. *Compare* METAMALE.

metagenomics The direct genetic analysis of the genomes contained within an environmental sample, such as a sample of soil or seawater. It is used particularly in microbiology to characterize the structure, functions, and species composition of microbial communities (*microbiomes), without the need to isolate and culture species individually. So a single process can reveal the taxonomic and functional relationships of the several thousand different bacterial species in one gram of soil, some of which may be unknown or difficult to culture in the laboratory. If just the species composition is required, then the analysis focuses on one or a few ribosomal RNA genes, which are highly conserved and good indicators of evolutionary relationships among species. To delineate broader functional relationships, the entire DNA (i.e. the **metagenome**) or the transcribed RNA (i.e. the **metatranscriptome**) is analysed. The DNA is extracted from a carefully prepared sample and subjected to *shotgun sequencing using *next-generation sequencing. The resulting multitude of relatively short DNA sequences is then assembled by computer software into successively longer sequences, by identifying overlapping homologous regions, to yield more reliable information. The sequences are then categorized, or 'binned', according to chemical composition or base sequence, by comparing them with several million reference sequences in a database, and assigned to individual strains, species, or closely related groups. Significant features are identified and labelled (*see* ANNOTATION) to establish the taxonomic and functional identities of the organisms within the original sample.

metamale A fruit fly (*Drosophila* sp.) with three (instead of the normal two) copies of each autosome. The disproportion of autosomes to the normal complement of one X chromosome (plus one Y chromosome, which plays little or no part in sex determination) impairs development, and such flies are typically weak and sterile. *Compare* METAFEMALE.

metameric segmentation (metamerism; segmentation) The division of an animal's body (except at the head region—*see* CEPHALIZATION) into a number of compartments (**segments** or **metameres**) each containing the same organs. Metameric segmentation is most strongly marked in annelid worms (e.g. earthworms), in which the muscles, blood vessels, nerves, etc. are repeated in each segment. In these animals the segmentation is obvious both externally and internally. It also occurs internally in arthropods and in the embryonic development of all vertebrates, in which it is confined to parts of the muscular, skeletal, and nervous systems and does not show externally.

metamorphosis (*pl.* **metamorphoses**) The rapid transformation from the larval to the adult form that occurs in the life cycle of lancelets (*see* CEPHALOCHORDATA) and many invertebrates and amphibians. Examples are the changes from a tadpole to an adult frog and from a pupa to an adult insect. Metamorphosis often involves considerable destruction of larval tissues by lysosomes, and is controlled by hormones, such as ecdysone (in insects) and thyroxine (in amphibians).

metanephridium (*pl.* **metanephridia**) Either of the paired excretory organs that occur in the body segments of earthworms and other annelids. Each metanephridium consists of a ciliated funnel-shaped entrance (**nephrostome**) open to the *coelom, a long convoluted tubule, a bladder, and an exit pore (**nephridiopore**) via which urine is discharged to the exterior. The nephrostome of each nephridium lies in the segment anterior to the tubule. Coelomic fluid containing nitrogenous waste in the form of ammonia is swept into the nephrostome by beating of the cilia and passes along the tubule, where substances are exchanged with the blood capillaries that closely associate with the tubule. The distal part of the tubule is enlarged to form a bladder, in which the highly diluted urine collects before discharge.

metaphase The stage of cell division during which the membrane around the nucleus breaks down, the *spindle forms, and centromeres attach the chromosomes to the equator of the spindle. In the first metaphase of *meiosis pairs of chromosomes (bivalents) are attached, while in *mitosis and the second metaphase of meiosis, individual chromosomes are attached.

metaphloem The part of the primary *phloem that develops in a plant after the stem or other part has finished elongating. *Compare* PROTOPHLOEM.

metaplasia The transformation of a tissue into a different type. This is an abnormal process; for example, metaplasia of the epithelium of the bronchi may be an early sign of cancer.

metapopulation A collection of interacting subpopulations of a species. Many species exist naturally as more-or-less discrete subpopulations, each occupying a patch of suitable habitat and separated from neighbouring subpopulations by intervening areas of unsuitable habitat. A pond-dwelling species is a clear example. This pattern of distribution is frequently intensified by habitat loss and fragmentation, for example through human intervention (e.g. deforestation or road building). Individuals may migrate between subpopulations or disperse to unoccupied habitat patches. The sizes and composition of the component subpopulations, and levels of migration and dispersal, are all crucial factors in determining the metapopulation dynamics. Knowledge of this is vital, for example in planning the conservation of a threatened species or setting limits for the sustainable harvest of a species.

metaproteomics The analysis of the proteins synthesized collectively by the microorganisms within an environmental sample. *See* PROTEOMICS.

metastasis (*pl.* metastases) *See* CANCER.

metatarsus (*pl.* **metatarsi**) The foot (or corresponding part of the hindlimb) in terrestrial vertebrates, consisting of a number of rod-shaped bones (**metatarsals**) that articulate with the bones of the ankle (*see* TARSUS) and those of the toes (*see* PHALANGES). The number of metatarsals varies between species: in the basic *penta-dactyl limb there are five, but this number is reduced in some species.

Metatheria An infraclass of mammals containing the marsupials. The female bears an abdominal pouch (**marsupium**) into which the newly born young, which are in a very immature state, move to complete their development. They obtain nourishment from the mother's mammary teats. Modern marsupials are restricted to Australasia (where they include the kangaroos, koala bears, phalangers, and bandicoots) and the Americas (the opossums). Common ancestors of marsupials and placental mammals probably evolved during the early to mid-Cretaceous period, 140–125 mya, and marsupials diverged from placental mammals about 90 mya. In Australia, where the marsupials have been isolated for millions of years, they show the greatest diversity of form, having undergone *adaptive radiation to many of the niches occupied by placental mammals elsewhere. *Compare* EUTHERIA; PROTOTHERIA.

metaxylem The part of the primary *xylem that develops in a plant after the stem or other part has finished elongating (*compare* PROTOXYLEM). The walls of the metaxylem are more extensively lignified than those of the protoxylem.

Metazoa A group (sometimes given the rank of subkingdom) comprising all multicellular animals, including the *Porifera (sponges) and *Placozoa, which in some former classifications were placed in a separate subkingdom, Parazoa. *Compare* EUMETAZOA.

methanogen Any of various *archaea that produce methane; they include such genera as *Methanobacillus* and *Methanothrix*. Methanogens are *obligate anaerobes found in oxygen-deficient environments, such as marshes, swamps, sludge (formed during *sewage treatment), and the digestive systems of ruminants. Mostly they obtain their energy by using carbon dioxide to oxidize hydrogen, with the production of methane:

$$CO_2 + 4H_2 \rightarrow CH_4 + 2H_2O$$

Formate, methanol, or acetate may also be used as substrates by certain types. Methanogenic bacteria are important in the production of *biogas and are responsible for the 'marsh gas' emitted in swamps. They are also major inhabitants of the anaerobic guts of ruminants, such as cattle and sheep, and of termites. Hence they are responsible for the methane produced by grazing ruminants, which is a significant contributor to greenhouse gas emissions.

methionine *See* AMINO ACID.

methylene blue A blue dye used in optical microscopy to stain nuclei of animal tissues. It is also suitable as a vital stain and a bacterial stain.

methylome *See* DNA METHYLATION.

metre Symbol m. The SI unit of length that is equal to 39.37 inches. It is formally defined as the length of the path travelled by light in vacuum during a time interval of 1/299 792 458 of a second. This definition, adopted by the General Conference on Weights and Measures in October 1983, replaced the 1960 definition based on the krypton lamp, i.e. 1 650 763.73 wavelengths in a vacuum of the radiation corresponding to the transition between the levels $2p^{10}$ and $5d^5$ of the nuclide krypton-86. This definition replaced the older (1927) definition of a metre based on a platinum-iridium bar of standard length.

MHC *See* MAJOR HISTOCOMPATIBILITY COMPLEX.

MHC class I protein Any of a class of histocompatibility proteins encoded by the *major histocompatibility complex that are expressed by all nucleated cells. Following synthesis in the endoplasmic reticulum, these proteins bind peptide fragments derived from viruses or other pathogens invading the cytoplasm and are transported to the outer membrane of the cell, effectively flagging the cell for destruction by *cytotoxic T cells bearing *CD8 proteins. Each MHC class I protein consists of two polypeptide chains—an α chain and β_2-microglobulin—and forms a peptide-binding cleft in which a short antigenic peptide, usually 8–10 residues long, becomes stably integrated. Tissue grafts from a mismatched donor are marked for rejection by the recipient's immune system by virtue of the nonself MHC class I proteins carried by all tissue cells.

MHC class II protein Any of a class of histocompatibility proteins encoded by the *major histocompatibility complex whose expression is restricted to certain immune cells, in particular *B cells, *macrophages, other *antigen-presenting cells, epithelial cells of the thymus, and (in humans) activated T cells. These proteins bind antigenic peptide fragments derived from pathogens or toxins that have been taken in and degraded by macrophages or B cells and held in vesicles within the cell. The class II proteins then migrate to the cell surface, where they are recognized by receptors of *CD4-bearing T cells (i.e. *helper T cells), which in turn activate the presenting cell to kill its intracellular pathogens (in the case of macrophages) or to secrete antibodies to neutralize extracellular bacteria or circulating toxins (in the case of B cells). Class II proteins consist of an α chain and a β chain and have a peptide-binding site that is open at both ends and can accommodate longer peptides than MHC class I molecules.

micelle An aggregate of molecules in a *colloid. For example, phospholipids in aqueous solution form micelles—small clusters of molecules in which the nonpolar hydrocarbon groups are in the centre and the hydrophilic polar groups are on the outside. The products of fat digestion are dispersed into micelles by the action of bile salts, which facilitates their absorption in the small intestine.

Michaelis-Menten curve A graph that shows the relationship between the concentration of a substrate and the rate of the corresponding enzyme-controlled reaction. Named after L. Michaelis (1875–1949) and M. L. Menten (1879–1960), the curve only applies to enzyme reactions involving a single substrate. The graph can be used to calculate the **Michaelis constant** (K_m), which is the concentration of a substrate required in order for an enzyme to act at half of its maximum velocity (V_{max}). The Michaelis constant is a measure of the affinity of an enzyme for a substrate. A low value corresponds to a high affinity, and vice versa. *See also* ENZYME KINETICS.

Michaelis-Menten curve

micro- 1. A prefix denoting very small size; e.g. microgamete, micronucleus. **2.** Symbol μ. A prefix used in the metric system to denote one millionth. For example, 10^{-6} metre = 1 micrometre (μm).

microarray A glass slide, silicon wafer, or glass bead on which are deposited biomolecules or other material in a regular micro-scale pattern to enable automated simultaneous multiple assays of target substances or activities. Microarrays are powerful analytical tools with wide-ranging

MICROARRAY TECHNOLOGY

A living cell makes and uses thousands of different RNAs and proteins to perform all its vital functions. Real insight into cellular metabolism requires knowledge of which components are being expressed by different cell types, for example during development, under different physiological conditions, or in disease states, and how their concentrations change over time.

Microarrays were developed in the 1990s for the automated analysis of thousands of different biomolecules simultaneously. They have enabled researchers to study the function of genes and their products (RNAs and proteins) on a genome-wide basis—the field of functional genomics. These techniques generate massive amounts of data, and computerized handling, analysis, and presentation of the data are crucial for valid interpretation.

Examples of microarrays

DNA microarrays ('gene chips' or 'DNA chips')
Contain short oligonucleotides or larger DNA fragments that hybridize with complementary DNA or RNA sequences. Uses include:

- following expression patterns of whole genomes over time;
- identifying chemical changes throughout a cell's chromosomal DNA;
- identifying and measuring minute deletions and duplications of DNA, e.g.in cancer cells.

Protein microarrays
Contain microsamples of purified proteins.
Used to determine protein-protein interactions at a cellular scale.

Tissue microarrays
Contain small thin sections of tissue taken from different tissue locations or pathology specimens.
Used to identify distribution of specific proteins throughout a tissue sample.

Chemical microarrays
Contain chemical compounds.
Used, for example, to select potential drugs with particular chemical binding properties.

Preparation of DNA microarrays

There are two basic types of DNA microarrays.

In situ *oligonucleotide arrays (DNA chips)*
DNA oligonucleotides (20–25 nucleotides) are assembled *in situ* on a quartz wafer or glass slide, using a photolithography process similar to that used to make microprocessor chips.

A single chip may contain up to a million specified oligonucleotides at regularly spaced sites called addresses, or 'features'.

Poor or incorrect hybridization leads to signal 'noise'.

Spotted microarrays
Based on the principle of the dot-blot technique with complementary DNA obtained from clones of the genes of interest.
Assembled by spotting minute amounts of denatured DNA onto a coated glass microscope slide, so that the DNA dries and becomes attached to the slide.
Contain large DNA fragments (hundreds to thousands of base pairs long).
Up to 20 000 features per slide.
Cheaper and easier to make—can be made 'in house' by researchers.

Measuring mRNAs using an oligonucleotide microarray

applications, including the analysis of gene expression and of genetic variation. They can be designed to carry small DNA molecules (*see* DNA MICROARRAY), proteins (e.g. antibodies or antigens), carbohydrates or other organic molecules, or even individual living cells. These reagents are applied to the substrate in a regular microscopic grid pattern, each being identified by its unique coordinate, or address, on the grid. Interaction of a target substance (e.g. an antibody or a complementary nucleic acid) with a particular address on the microarray activates or attaches a label (e.g. a fluorescent dye). The microarray can then be 'read' by a scanner, which automatically assesses the amount of label at each address, and hence the amount of target substance. Even smaller-scale **nanoarrays** have been developed, to increase further the scope and speed of this technology. These are made using a form of high-precision lithography, and employ atomic force microscopy to detect direct interactions between individual molecules with nanometre resolution. No label is required, although some applications do use reporter labelling. Advances in next-generation DNA sequencing mean that it is replacing microarray analysis for some applications, particularly where sensitivity and accuracy are at a premium, e.g. for assessing rare or previously unknown sequence variants (*see* RNA SEQUENCING). *See* Feature.

microbe *See* MICROORGANISM.

microbiology The scientific study of microorganisms (e.g. bacteria, viruses, and fungi). Originally this was directed towards their effects (e.g. in causing disease and decay), but during the 20th century the emphasis shifted to their physiology, biochemistry, and genetics. Microbes are now recognized as important vehicles for the study of biochemical and genetic processes common to all living organisms, and their rapid growth enables their laboratory culture in large numbers for studies in genetics and for biotechnological applications.

(((●))) SEE WEB LINKS
• Home page of the Microbiology Society

microbiome The community of microorganisms associated with a particular site, e.g. a host organism or a habitat. The high-throughput techniques of *metagenomics now enable rapid characterization and analysis of entire microbiomes, which is transforming knowledge of their constituents' significance, whether as symbionts, commensals, or pathogens. For example, it is estimated that the human body has ten times as many microbial cells as body cells. Most of these have not been identified. Hence the establishment of the Human Microbiome Project, the aim of which is to characterize microbial communities found at various sites in the human body and correlate changes in the microbiome with human health. The Earth Microbiome Project, founded in 2010, is a collaborative effort using crowd-sourced samples to characterize microbial life on earth. *See also* HOLOBIONT.

(((●))) SEE WEB LINKS
• NIH Human Microbiome Project

microbody Any of a class of cell organelles that typically are spherical vesicles, 0.2–1.5 μm, bounded by a single membrane. Microbodies contain enzymes responsible for the oxidation of various materials and they originate from the endoplasmic reticulum. They include *peroxisomes and glyoxysomes (*see* GLYOXYLATE CYCLE).

microcirculation The part of the blood circulatory system consisting of the networks of capillaries and associated vessels that supply blood to tissue cells. At its end, each artery gives rise to a microcirculatory bed of increasingly narrow vessels: arterioles, which in turn subdivide to form metarterioles, and then capillaries. The latter are typically 1 mm long and 3–10 μm in diameter, just wide enough for erythrocytes to pass through. The capillaries connect to the venous system via postcapillary venules, which in turn join the larger veins. Blood flow into the capillary bed is controlled by a muscular precapillary sphincter at the end of each arteriole, and blood can bypass the capillaries via a direct arteriovenous connection, or anastomosis.

microclimate The local climate of a small area or of a particular *habitat, which is different from the **macroclimate** of the larger surrounding geographical area.

microdissection (micromanipulation) A technique used for the dissection of living cells under the high power of an optical microscope. It utilizes minute mechanically manipulated instruments, such as needles, scalpels, *micropipettes, and lasers. For example, the instruments may be used to remove a single nucleus from one cell and to implant it in another (*see* NUCLEAR TRANSFER).

microelectrode A *micropipette filled with a physiologically appropriate electrolyte solution (e.g. potassium chloride), whose tip can be inserted through the plasma membrane into cells without major disturbance to cellular function. It can be used to detect changes in electrical potential, for example during the passage of an action potential along a nerve cell. A second electrode is immersed in the extracellular fluid, and both electrodes are connected via an amplifier to an oscilloscope.

microevolution Evolution on a relatively small scale, involving the emergence of new species, or of new groups below the species level, such as races and subspecies. *Compare* MACROEVOLUTION.

microfauna 1. Animals that cannot be seen with the naked eye. They are normally observed with the aid of a microscope. *Compare* MACROFAUNA. 2. The animals that live in a particular *microhabitat.

microfibril A microscopic fibre. Plant cell walls contain microfibrils, about 5 nm in diameter, each consisting of some 50–60 parallel cellulose chains that are associated together to form a rod or a flat ribbon. Cellulose microfibrils are arranged in layers at right angles to each other.

microfilament (actin filament) Any of numerous microscopic protein fibres, typically 7–9 nm in diameter, that form one of the main components of the *cytoskeleton of eukaryotic cells. Each microfilament consists of two helically twisted strands, each with distinct 'plus' and 'minus' ends, and comprising a chain of globular subunits of the protein *actin. They can shorten or extend by the removal or addition of subunits and are linked by crosslinking proteins to form three-dimensional networks. Bundles of microfilaments often occur just beneath the cell surface—**cortical microfilaments**—typically oriented parallel to the long axis of the cell, and some are anchored to the plasma membrane. With the aid of the motor protein *myosin the microfilaments can slide relative to each other, causing contractile movements, as in muscle cells, or other changes in cell shape, such as those occurring in *amoeboid movement. They are also involved in the transport of materials within the cell, and in the peripheral flow of cytoplasm and cell organelles known as *cytoplasmic streaming. Hence microfilaments play a crucial role in the growth of cell extensions, for example the pollen tube that develops from a germinating pollen grain. *Compare* MICROTUBULE. *See also* INTERMEDIATE FILAMENT.

microflora 1. Plants and algae that cannot be seen with the naked eye. They are normally observed with the aid of a microscope. 2. The plants and algae that live in a particular *microhabitat.

microfossil A *fossil that is so small that it can only be studied under a microscope. Microfossils include bacteria, diatoms, and protists and parts of organisms, such as plant pollen and skeletal fragments. Microfossils are important in the correlation of rocks where only small samples are available. The study of microfossils, particularly pollen, is known as *palynology.

microglia *See* GLIA; MACROPHAGE.

microhabitat The local habitat of a particular organism or microorganism. There are normally a number of different microhabitats within a large *habitat (**macrohabitat**), each with its distinct set of environmental conditions. For example, in a stream macrohabitat there will exist different microhabitats, depending on oxygen content, pH, speed of water flow, and other factors in localized areas of the stream.

microinjection A technique for injecting substances into cells, cell organelles, or other microscopic structures (e.g. capillaries) using a *micropipette. It can be used, for example, to introduce dyes to help visualize cell components or to inject drugs or other substances to investigate their effects at the cellular level. In genetics, microinjection is used to insert DNA fragments or entire nuclei into cells.

micromanipulation *See* MICRO-DISSECTION.

micronucleus *See* NUCLEUS.

micronutrient A chemical element required by plants in relatively small quantities. Micronutrients are typically found in cofactors and coenzymes. They include copper, zinc, molybdenum, manganese, cobalt, and boron. *See* ESSENTIAL ELEMENT. *Compare* MACRONUTRIENT.

microorganism (microbe) Any organism that can be observed only with the aid of a microscope. Microorganisms include bacteria, archaea, viruses, protists (including certain algae), and fungi. *See* MICROBIOLOGY.

microphagous Describing the method of feeding of those heterotrophic organisms that take in their food in the form of tiny particles. *Filter feeding and *ciliary feeding are examples of this type of feeding. *Compare* MACROPHAGOUS.

microphyll A type of foliage leaf in clubmosses and horsetails that has a single unbranched midrib. Such leaves are generally no more than a few millimetres long. *Compare* MEGAPHYLL.

micropipette A glass pipette with an ultrafine tip, typically less than 1 μm in diameter. It can be inserted into single cells or other microscopic structures and used, for example, to inject materials (*see* MICROINJECTION). The micropipette is usually held by a **micromanipulator**, a mechanical device that allows precise movement of the tip. *See also* MICROELECTRODE.

micropropagation The *in vitro* propagation of plants by cloning. Typically, this involves culturing excised meristematic tissue on a special medium that encourages axillary bud development. The new shoots are then separated and cultured, and the cycle is repeated until finally the shoots are transferred to a medium that promotes root development, to produce plantlets. Micropropagation is used in agriculture, horticulture, and forestry as special genotypes can be bred and maintained, the process is rapid, and plants can be kept disease-free. It also offers a means of propagating plants that are otherwise difficult to propagate such as orchids, and is useful for conserving rare plants. *See also* CLONE.

micropyle 1. A small opening in the surface of a plant ovule through which the pollen tube passes prior to fertilization. It results from the incomplete covering of the nucellus by the integuments. It remains as an opening in the testa of most seeds through which water is absorbed. **2.** A small pore in some animal cells or tissues; for example, in insect eggs (*see* CHORION).

microRNA (miRNA) A small RNA molecule that is encoded by a cell and can 'silence' the expression of a particular target gene within the cell (*see* RNA INTERFERENCE). miRNAs bind to target messenger RNA (mRNA) molecules and suppress translation of the mRNA into protein. They regulate expression of perhaps half of all human genes and are involved in many aspects of embryological development, cell differentiation, cell death, and cancer. In eukaryotes miRNA genes occur chiefly within the regions between genes or in the introns of protein-coding genes. The primary RNA transcripts are longer precursors (**primary miRNAs**, or **pri-miRNAs**) from which the RNAase-like enzyme Drosha cuts a 70–80-nucleotide region. This forms a double-stranded (ds) hairpin structure (the **pre-miRNA**), which is trimmed by another protein, *Dicer, to produce the miRNA—a single RNA strand approximately 22 nucleotides long. This is then incorporated into a complex of proteins called **RISC** (RNA-induced silencing complex). The miRNA binds to one or more sites with a more-or-less complementary base sequence on the target mRNA, thereby blocking translation of the mRNA in some way. Plant cells lack an equivalent of Drosha, and a Dicer-like RNAase enzyme is able to trim miRNAs directly from long dsRNA precursors. A single miRNA can be involved in regulating the synthesis of numerous proteins within the same cell. *Compare* SHORT INTERFERING RNA.

microsatellite DNA A form of highly *repetitive DNA consisting of large numbers of very short base sequences scattered throughout the eukaryote genome. Each consists of a short base sequence—variously called a *short tandem repeat (STR) or **short sequence repeat** (SSR)—generally 2–5 bp long, that is repeated 5–30 times in a tandem array at any given locus. Hence, for example, one individual might have 17 consecutive repeats of the sequence CA at a given locus, while another individual might have 15 repeats. Overall, the same sequence might occur at many different loci and be repeated tens of thousands of times throughout the genome. The variability in the number of such repeats between individuals provides a valuable set of *molecular markers for characterizing the DNA (*see* DNA PROFILING; PHYSICAL MAP). The number of repeats in any particular region of DNA is determined by using the *polymerase chain reaction (PCR) to amplify the region and assessing the size of the PCR products by gel electrophoresis. The DNA of two individuals with tandem arrays of different lengths at a given locus will yield PCR products of different sizes, which will be revealed as distinct bands on the developed gel. Microsatellite DNA is named by analogy with *satellite DNA, although it does not form visible satellite peaks on centrifugation. *Compare* VARIABLE NUMBER TANDEM REPEATS.

microscope Any device for forming a magnified image of a small object. The **simple microscope** consists of a biconvex magnifying glass or an equivalent system of lenses, either hand-held or in a simple frame. The **compound microscope** uses two lenses or systems of lenses, the second

Compound microscope

magnifying the real image formed by the first (see illustration). The lenses are usually mounted at the opposite ends of a tube that has mechanical controls to move it in relation to the object. An optical condenser and mirror, often with a separate light source, provide illumination of the object. The widely used **binocular microscope** consists of two separate instruments fastened together so that one eye looks through one while the other eye looks through the other. This gives stereoscopic vision. *See* Chronology. *See also* CONFOCAL FLUORESCENCE MICROSCOPY; ELECTRON MICROSCOPE; NOMARSKI MICROSCOPE; PHASE-CONTRAST MICROSCOPE; SCANNING PROBE MICROSCOPY; ULTRAVIOLET MICROSCOPE.

(((⊕))) SEE WEB LINKS

• Accessible and entertaining portal from Microscopy UK

microsome A fragment of *endoplasmic reticulum formed when cells or tissues, e.g.

MICROSCOPY

c.1590   Dutch spectacle-makers Hans and Zacharias Janssen invent the compound microscope.

1610   German astronomer Johannes Kepler (1571–1630) invents the modern compound microscope.

1675   Anton van Leeuwenhoek invents the simple microscope.

1826   British biologist Dames Smith (d. 1870) constructs a microscope with much reduced chromatic and spherical aberrations.

1827   Italian scientist Giovanni Amici (1786–1863) invents the reflecting achromatic microscope.

1861   British chemist Joseph Reade (1801–70) invents the kettledrum microscope condenser.

1912   British microscopist Joseph Barnard (1870–1949) invents the ultramicroscope.

1932   Dutch physicist Frits Zernike (1888–1966) invents the phase-contrast microscope.

1936   German-born US physicist Erwin Mueller (1911–77) invents the field-emission microscope.

1938   German engineer Ernst Ruska (1906–88) develops the electron microscope.

1940   Canadian scientist James Hillier (1915–2007) makes a practical electron microscope.

1951   Erwin Mueller invents the field-ionization microscope.

1978   US scientists of the Hughes Research Laboratory invent the scanning ion microscope.

1981   German physicist Gerd Binning (1947–) and Swiss physicist Heinrich Rohrer (1933–2013) invent the scanning tunnelling microscope.

1985   Gerd Binning invents the atomic force microscope.

1987   James van House and Arthur Rich invent the positron microscope.

1990   Richard Henderson produces the first atomic-resolution images of a protein using cryo-electron microscopy.

1992   Douglas Prasher clones the gene for green fluorescent protein, paving the way for the use of fluorescent proteins in microscopy.

1994–96 Stefan Hell and colleagues introduce the first techniques of super-resolution microscopy, thus breaking the 'diffraction barrier' of light microscopy.

2002   A photoactivable green fluorescent protein is developed that fluoresces only when activated by violet or ultraviolet light.

2006   Further advances in super-resolution microscopy are introduced: photoactivated localization microscopy (PALM) and stochastic optical reconstruction microscopy (STORM).

2008   Live embryo development in zebrafish recorded by Philipp Keller and colleagues using digital scanned laser lightsheet fluorescence microscopy.

2013	Researchers at Stanford University develop CLARITY, a new technique for visualizing intact entire organs by replacing lipids with transparent hydrogel-based mesh to enable successive rounds of staining and antibody labelling.
2016	Cryo-electron microscopy used to visualize the structure of the nucleosome core particle with high resolution.
2018	First ultrathin optical lenses based on metasurface materials constructed to produce full-colour images.

liver, are disrupted. Microsomes can be isolated by centrifugation and are commonly used to investigate the functional properties of endoplasmic reticulum, such as enzymic activity and protein synthesis.

microsporangium *See* SPORANGIUM.

microspore *See* MICROSPORE MOTHER CELL; SPOROPHYLL.

microspore mother cell (microsporocyte) A diploid cell in plants that divides by meiosis to give rise to four haploid microspores (*see* SPOROPHYLL). In flowering plants microspore mother cells are formed within the pollen sacs of the anthers by mitosis; the microspores they produce develop into pollen grains.

microsporidia (*sing.* **microsporidium**) Unicellular fungi that live as parasites inside cells of invertebrates and vertebrates. Many of the 1200 known species cause disease (microsporidiosis) in a range of animal hosts, including humans (chiefly in immunocompromised subjects, such as people with HIV infection). Microsporidia lack cell walls and certain organelles typically found in eukaryotes; notably they have reduced structures called *mitosomes instead of mitochondria. They produce resistant spores with chitin-containing walls and a coiled polar tube used to infiltrate host cells.

microsporocyte *See* MICROSPORE MOTHER CELL.

microsporophyll *See* SPOROPHYLL.

microtome A machine used to cut thin sections (3–5 μm thick) of plant or animal tissue for microscopical observation. There are various designs of microtome, each basically consisting of a steel knife, a block for supporting the specimen, and a device for moving the specimen towards the knife. The specimen is usually supported by being embedded in wax; if a **freezing microtome** is used, the specimen is frozen. An **ultramicrotome** is used to cut much thinner sections (20–100 nm thick) for electron microscopy. The biological material is embedded in plastic or resin, sectioned with a glass or diamond knife, and the cut sections are allowed to float on the surface of water in an adjacent water bath.

microtubule A microscopic tubular structure, with an external diameter of 24 nm and of variable length, found in a wide range of eukaryotic cells. Microtubules are composed of numerous subunits of the globular protein **tubulin**, and can occur singly or in pairs, triplets, or bundles. Each tubulin protein is a dimer consisting of two slightly different polypeptides—α-tubulin and β-tubulin—which gives each microtubule distinct 'plus' and 'minus' ends. Each tubule can add or lose tubulin dimers from its 'plus' end, depending on levels of free GTP-bound dimers. If these are abundant, the microtubule binds more dimers and gets longer; if they are in short supply, dimers dissociate from the microtubule, which gets shorter. This state is called dynamic instability.

Microtubules help cells to maintain their shape by resisting compressive forces (*see* CYTOSKELETON); they also occur in cilia and eukaryotic *flagella (sense 2) and the *centrioles and form the *spindle during nuclear division. A further role is in the intracellular transport of materials and movement of organelles. Formation of microtubules is initiated at *microtubule-organizing centres (MTOCs). *Compare* MICROFILAMENT; INTERMEDIATE FILAMENT.

microtubule-organizing centre (MTOC) A region of a cell from which radiate microtubules of the *cytoskeleton. Most animal cells contain a single MTOC, the *centrosome, which is located near the nucleus and is also involved in organizing the spindle during nuclear division (*see* MITOSIS). In contrast, a plant cell typically has hundreds of MTOCs, producing girdle-like arrays of microtubules running around the cell's inner periphery.

microvillus (*pl.* **microvilli**) One of a number of minute finger-like projections on the free surfaces of epithelial cells. Microvilli are covered with plasma membrane and their cytoplasm is continuous with the main cell cytoplasm. Their purpose is to increase the absorptive or secretory surface area of the cell, and they are abundant on the villi of the intestine, where they form a *brush border.

micturition The discharge of urine from the bladder. It is brought about by reflex contraction of the *detrusor muscle after voluntary relaxation of the sphincter muscle at the junction of the bladder with the urethra.

midbrain (mesencephalon) One of the three sections of the brain of a vertebrate embryo. Unlike the *forebrain and the *hindbrain, the midbrain does not undergo further subdivision to form additional zones. In mammals it becomes part of the *brainstem. The roof of the midbrain is called the tectum (otherwise termed the optic lobe in birds, amphibians, and fishes), while the floor is the tegmentum; between them is a narrow channel, the cerebral aqueduct. The tegmentum contains the substantia nigra ('black matter') and the red nucleus, both of which are involved in controlling voluntary movements. In mammals a pair of anterior prominences in the tectum (the superior colliculi) receive input from the retina of the eyes, as well as other sensory information. These integrate the information to form a visual map of the individual's surroundings, and instigate eye movements, head turning, and other spatial movements of the body. The paired inferior colliculi in the tectum receive input from auditory fibres and project to the auditory relay nucleus of the thalamus, thus serving as a vital relay station in the circuitry responsible for hearing. Maps from different senses are held in different layers of the tectum, but are organized according to the same spatial geometry. In nonmammalian vertebrates the tectum is relatively larger than in mammals, and takes a greater role in controlling body movements.

middle ear (tympanic cavity) The air-filled cavity within the skull of vertebrates that lies between the *outer ear and the *inner ear. It is linked to the pharynx (and therefore to outside air) via the *Eustachian tube and in mammals contains the three *ear ossicles, which transmit auditory vibrations from the outer ear (via the *tympanum) to the inner ear (via the *oval window).

middle lamella A thin layer of material, consisting mainly of pectins, that binds together the primary walls of adjacent plant cells. *See also* CELL PLATE.

midgut 1. The middle section of the alimentary canal of vertebrates, which is concerned with digestion and absorption. It comprises most of the small intestine. **2.** The middle section of the alimentary canal of arthropods. *See also* FOREGUT; HINDGUT.

migration The seasonal movement of complete populations of animals to a more favourable environment. It is usually a response to lower temperatures resulting in a reduced food supply, and is often triggered by a change in day length (*see* PHOTOPERIODISM). Migration is common in mammals (e.g. porpoises and whales), fish (e.g. eels and salmon), and some insects (e.g. butterflies) but is most marked in birds. The Arctic tern, for example, migrates annually from its breeding ground in the Arctic circle to the Antarctic—a distance of some 17 600 km. Migrating animals possess considerable powers of orientation; birds seem to possess a compass sense, using the sun, pole stars (at night), and (in cloud) the earth's magnetic lines of force as reference points (*see* NAVIGATION).

milk The fluid secreted by the *mammary glands of mammals. It provides a balanced and highly nutritious food for offspring. Cows' milk comprises about 87% water, 3.6% lipids (triglycerides, phospholipids, cholesterol, etc.), 3.3% protein (largely casein), 4.7% lactose (milk sugar), and, in much smaller amounts, vitamins (especially vitamin A and many B vitamins) and minerals (notably calcium, phosphorus, sodium, potassium, magnesium, and chlorine). Composition varies among species; human milk contains less protein and more lactose.

milk sugar See LACTOSE.

milk teeth See DECIDUOUS TEETH.

milli- Symbol m. A prefix used in the metric system to denote one thousandth. For example, 0.001 metre = 1 millimetre (mm).

millipedes See DIPLOPODA.

Millon's reagent A solution of mercuric nitrate and nitrous acid used to test for proteins. The sample is added to the reagent and heated for two minutes at 95°C; the formation of a red precipitate indicates the presence of protein in the sample. The reagent is named after French chemist Auguste Millon (1812–67).

mimicry The resemblance of one animal to another, which has evolved as a means of protection or improving predation efficiency. In **Batesian mimicry**, named after British naturalist Henry Bates (1825–92), the markings of certain harmless insects closely resemble the *warning coloration of another insect (the **model**). Predators that have learnt to avoid the model will also avoid good mimics of it. This phenomenon is often found among butterflies. **Müllerian mimicry**, named after German zoologist J. F. T. Müller (1821–97), involves the mutual resemblance of a group of animals, all harmful, such as the wasp, bee,

and hornet, so that a predator, having experienced one, will subsequently avoid them all. In **aggressive mimicry**, a predator or parasite gains some advantage in adopting a resemblance to another species. A notable master of marine disguise is the mimic octopus (*Thaumoctopus mimicus*), discovered in 1998, which can change shape and appearance for either defence or attack. Among its many guises it can imitate a crab in order to approach real crabs so they can be caught more easily. To avoid being predated itself it can mimic a sea snake by extending two tentacles and burying the remainder in the seafloor. *See also* MOLECULAR MIMICRY.

(((⊕))) SEE WEB LINKS
- Summary of the lecture 'Warning colour and mimicry', from the Department of Biology, University College London

Mimivirus *See* NUCLEOCYTOPLASMIC LARGE DNA VIRUS.

mineral deficiency Lack of any essential mineral nutrient, such as nitrogen, phosphorus, or potassium, in living organisms, which can result in mineral deficiency diseases. In humans, for example, lack of calcium causes poor bone development, and lack of nitrogen can cause the disease kwashiorkor, due to a deficiency in protein intake (*see* MALNUTRITION). In plants mineral deficiency results in stunted growth and *chlorosis. A deficiency of trace elements (*see* ESSENTIAL ELEMENT) also leads to diseases; for example, a deficiency of iron can cause anaemia in humans and chlorosis in plants.

mineralocorticoid *See* CORTICOSTEROID.

mineral salts Inorganic salts that need to be ingested or absorbed by living organisms for healthy growth and maintenance. They comprise the salts of the trace elements in animals (*see* ESSENTIAL ELEMENT) and the *micronutrients of plants.

minicell 1. A small roughly spherical cell produced by the abnormal division of a rod-shaped bacterium that contains no chromosomal DNA and is unable to grow or divide. It is used experimentally to express nucleic acids and proteins encoded by DNA introduced to the cell, e.g. in the form of a cloning *vector. **2.** (microcell) An experimentally produced body consisting of a nucleus surrounded by a thin layer of cytoplasm and a plasma membrane and containing variable numbers of chromosomes. It is used as a vector for transferring one or several chromosomes into a normal cell following cell fusion.

minisatellite DNA *See* VARIABLE NUMBER TANDEM REPEATS.

minus 10 sequence *See* PRIBNOW BOX.

Miocene The first epoch of the *Neogene period, extending from the end of the Oligocene, about 23 million years ago, to the start of the Pliocene, roughly 5 million years ago. It saw the radiation of several modern mammal groups, including the ruminants (deer, cattle, and antelopes), certain rodents (beavers, porcupines, and cavies) and the apes. Cooling of the climate during the Oligocene resulted in a continuing shift to deciduous hardwood species, such as oak and maple, at the expense of conifers during the Miocene.

miracidium (*pl.* **miracidia**) The first larval stage of trematode worms (flukes), which hatches from eggs excreted in the faeces of the primary host. Its leaflike body is covered with cilia, enabling the larva to swim towards a secondary host, in which it continues its development.

miRNA *See* MICRORNA.

mirror neuron Any of a class of neurons found in the cortex of the brain of monkeys that fire not only when the monkey performs an action (e.g. picking up a banana) but also when the monkey sees or hears someone else performing the same action, i.e. a 'reflection' of the action. There is evidence that such neurons are also present in parts of the human cerebral cortex. It has been hypothesized that mirror neurons function in understanding the actions of others, and more broadly in various aspects of communication and socialization.

According to some researchers, they may also play a significant role in the development of language, by enabling individuals to imitate gestures and sounds that are goal-oriented, i.e. are associated with a purpose or meaning.

mismatch repair A form of *DNA repair that replaces mismatched bases in newly replicated DNA in living cells. For example, if a guanine (G) on the template strand pairs with thymine (T) instead of the usual cytosine (C), the resulting abnormality in the DNA duplex is detected by the mismatch repair system. A length of strand containing the mismatched T is cut out and the gap is repaired with a new length containing the correct base (C), using the parental strand as template. The mismatch repair system closely follows the Y-shaped fork where the parental DNA unwinds and replication takes place, so enabling immediate correction of replication errors. It accounts for the vast majority of DNA repairs.

missense mutation A type of *point mutation that converts one codon to another, specifying a different amino acid. This results in incorporation of the wrong amino acid in the polypeptide chain during translation and, consequently, a potentially faulty protein.

mites *See* ACARINA.

mitochondrial DNA (mtDNA) A circular ring of DNA found in *mitochondria. In mammals mtDNA makes up less than 1% of the total cellular DNA, but in plants the amount is variable. It codes for ribosomal and transfer RNA but only some mitochondrial proteins (up to 30 proteins in animals), the nuclear DNA being required for encoding most of these. Human mtDNA contains 37 genes, encoding 13 proteins and some RNAs. Mitochondrial DNA is generally inherited via the female line only, although there are exceptions to this (*see* HETEROPLASMY). According to the *endosymbiont theory, mitochondrial DNA is a remnant of the chromosome of the symbiotic prokaryote ancestor of the mitochondrion. *See also* MITOCHONDRIAL EVE.

mitochondrial Eve The hypothetical female claimed by some biologists to be the most recent ancestor of all living humans. Analysis of *mitochondrial DNA (mtDNA) from groups of people throughout the world suggests that mitochondrial Eve lived between about 100 000 and 148 000 years ago, probably in Africa (hence she is also known as 'African Eve'). Mitochondrial DNA is particularly useful for investigating recent genetic history as it mutates quickly (ten times more rapidly than nuclear DNA) and in humans is inherited solely through the female line (therefore it does not undergo recombination by *crossing over). The uniformity of the different samples of mtDNA indicates that modern humans evolved relatively recently from a single region in Africa. This view has been reinforced by studies of Y chromosomes from different groups around the world, which are transmitted only through the male line and originate from a single hypothetical male, dubbed 'Y chromosome Adam', dated to between 120 000 and 156 000 years ago.

mitochondrial theory of ageing See SENESCENCE.

mitochondrion (*pl.* **mitochondria**) A structure within the cytoplasm of eukaryotic *cells that carries out aerobic respiration: it is the site of the *Krebs cycle and *electron transport chain, and therefore the cell's energy production. It also participates in numerous other cellular processes, including calcium buffering, phospholipid biosynthesis, and *apoptosis. Mitochondria vary greatly in shape, size, and number but are typically oval or sausage-shaped and bounded by two membranes, the inner one being folded into finger-like projections (**cristae**); they contain their own DNA (*see* MITOCHONDRIAL DNA). Components of the electron transport chain and, in plants, the *alternative respiratory pathway are located in the inner mitochondrial membrane. They are most numerous in cells with a high level of metabolic activity. Mitochondria are dynamic structures, moving around inside the cell, changing shape, undergoing fission and fusion, and forming a complex network rather than discrete organelles. According to the *endosymbiont theory, mitochondria originated as symbiotic aerobic bacteria living inside the predecessors of eukaryotic cells.

(((●))) SEE WEB LINKS

• Overview of mitochondrial structure and function from BiologyWise

mitogen-activated protein kinase *See* MAP KINASE.

mitosis (*pl.* **mitoses**) A type of nuclear division that results in two daughter cells each having a nucleus containing the same number and kind of chromosomes as the mother cell. The changes during divisions are clearly visible with a light microscope. Each chromosome divides lengthwise into two *chromatids, which separate and form the chromosomes of the resulting daughter nuclei. The process is divided into four stages, *prophase, *metaphase, *anaphase, and *telophase, which merge into each other. Mitotic divisions ensure that all the cells of an individual are genetically identical to each other and to the original fertilized egg. *See also* CELL CYCLE.

mitosis-promoting factor (maturation-promoting factor; MPF) A protein complex responsible for triggering mitosis in somatic cells and for maturation of oocytes into egg cells. Consisting of cyclin B (*see* CYCLIN) bound to a cyclin-dependent kinase (CDK), it catalyses the phosphorylation of proteins that in turn bring about the events of mitosis, including condensation of chromosomes, formation of the mitotic spindle, and breakdown of the nuclear envelope. Levels of cyclins and MPF rise as the cell enters mitosis, reach a peak during mitosis, and then fall during anaphase.

mitosome A very small organelle, thought to be a remnant of a *mitochondrion, found in certain single-celled organisms. Unlike mitochondria, mitosomes lack DNA and the apparatus for energy generation, their only apparent function being the production of iron-sulphur clusters, which are components of certain key metabolic enzymes. Mitosomes occur in several

The stages of mitosis in a cell containing two pairs of homologous chromosomes

groups of anaerobic or parasitic organisms, including the protists *Entamoeba histolytica* and *Giardia intestinalis*, both intestinal parasites of mammals, and the microsporidia, which are microscopic parasitic fungi.

mitral valve *See* BICUSPID VALVE.

mixed function oxidase (mixed function oxygenase) *See* MONOOXYGENASE.

mixotrophic Describing an organism that combines *autotrophic and *heterotrophic nutrition. Certain plants, notably carnivorous species such as Venus flytraps and sundews, supplement their photosynthetic lifestyle by capturing and digesting insects or other small animals as a source of additional nitrogen. However, mixotrophy is far more widespread among protists and bacteria and takes many different forms, with varying degrees of autotrophy and heterotrophy. For example, many photosynthesizing algal species in marine and freshwater plankton can also consume bacteria as prey. Conversely, some heterotrophic dinoflagellates and ciliates ingest algae and retain the chloroplasts of their prey in a functioning state to provide carbohydrate to the host. In some cases more intimate and long-standing symbiotic relationships have evolved, e.g. between *rhizarians and microalgae. It can be argued, however, that the sharing of nutritional roles in such associations is not 'true' mixotrophy.

MLST *See* MULTILOCUS SEQUENCE TYPING.

mmHg A unit of pressure equal to that exerted under standard gravity by a height of one millimetre of mercury, or 133.322 pascals.

model organism A species of organism commonly used for scientific study. Such species are generally chosen because they are relatively easy to work with and maintain in the laboratory and because often they are bred under strictly controlled conditions, with well-documented pedigrees, enabling comparison of results in different centres around the world. They range from the humble thale cress (*Arabidopsis thaliana*) to the long-suffering laboratory mouse (*Mus musculus*). A list of the most common model organisms is given in Appendix 5.

moderately repetitive DNA *See* REPETITIVE DNA.

modern synthesis *See* NEO-DARWINISM.

modifier gene A gene that influences the expression of another gene. For example, one gene controls whether eye colour is blue or brown but other (modifier) genes can also influence the colour by affecting the amount or distribution of pigment in the iris.

molality *See* CONCENTRATION.

molar 1. A broad ridged tooth in the adult dentition of mammals (*see* PERMANENT TEETH), found at the back of the jaws behind the premolars. There are two or more molars on each side of both jaws; their surfaces are raised into ridges or *cusps for grinding food during chewing. In humans the third (and most posterior) molar does not appear until young adulthood: these molars are known as **wisdom teeth. 2.** Denoting that an extensive physical property is being expressed per *amount of substance, usually per mole. For example, the molar heat capacity of a compound is the heat capacity of that compound per unit amount of substance, i.e. it is usually expressed in $J\,K^{-1}\,mol^{-1}$.

molarity *See* CONCENTRATION.

mole Symbol mol. The SI unit of *amount of substance. It is equal to the amount of substance that contains as many elementary units as there are atoms in 0.012 kg of carbon-12. The elementary units may be atoms, molecules, ions, radicals, electrons, etc., and must be specified. 1 mole of a compound has a mass equal to its *relative molecular mass expressed in grams.

molecular biology The study of the structure and function of large molecules associated with living organisms, in particular proteins and the nucleic acids *DNA and *RNA. **Molecular genetics** is a specialized branch, concerned with the analysis of genes and their expression (*see* DNA SEQUENCING).

(((●))) SEE WEB LINKS
● Website of the European Molecular Biology Organization

molecular chaperone Any of a group of proteins in living cells that assist newly synthesized or denatured proteins to fold into their functional three-dimensional structures. The chaperones bind to the protein and prevent improper interactions within the polypeptide chain, so that it assumes the correct folded orientation. This process may require energy in the form of ATP. Other functions include assisting the translocation of proteins across the membranes of cell organelles and binding denatured proteins under stress conditions or in degenerative disease. There are several unrelated families of chaperones, including five classes of *heat-shock proteins—HSP25 (small heat-shock proteins), HSP60, HSP70, HSP90, and HSP100—chaperonins, *calnexin, and *calreticulin.

molecular clock The concept that during evolution the number of nonadaptive, or 'neutral', substitutions in the nucleotides of nucleic acids (DNA or RNA), and hence in the proteins encoded by the nucleic acids, is proportional to time. Hence, by comparing the DNA or proteins of species that diverged a known length of time ago (e.g. determined from fossil evidence), it is possible to calculate the average substitution rate, thereby calibrating the 'molecular clock'. Comparative studies of different proteins in various groups of organisms tend to show that the number of neutral substitutions per site per year is fairly consistent over time, especially if many genes are considered to average out fluctuations in evolutionary rate due to natural selection. Hence, given a fairly constant rate of molecular evolution in comparable sequences of macromolecules, it is possible to date evolutionary events that occurred before any corroborative fossil record. *See* NEUTRAL THEORY OF MOLECULAR EVOLUTION.

molecular fossil A substance found in geological deposits that is indicative of a particular species or group of organisms that lived when the deposits were formed. Such a substance should ideally be uniquely associated with the organisms in question and either resist chemical breakdown over geological time or decay into products that are unambiguous and predictable. Examples include the isoprene residues indicative of ancient archaeans, and oleanane, a defence chemical found only in flowering plants.

molecular imprinting *See* GENE IMPRINTING.

molecular marker Any site (locus) in the genome of an organism at which the DNA base sequence varies among the different individuals of a population. Such markers generally have no apparent effect on the phenotype of the individual, but they can be determined by biochemical analysis of the DNA and are used for a variety of purposes, including chromosome mapping, DNA profiling, and genetic screening. Genetic tools such as restriction enzymes and the polymerase chain reaction plus the ever-growing abundance of DNA sequence data, coupled with automated high-throughput assays, have revealed several classes of molecular markers, including *restriction fragment length polymorphisms (RFLPs), *variable number tandem repeats (VNTRs), *short tandem repeats, *microsatellite DNA, and *single nucleotide polymorphisms (SNPs).

molecular mimicry The close resemblance of a foreign molecule to one present naturally in the body. For example, opiates such as morphine exert their pain-relieving effects by mimicking the body's natural endorphins, which are released to counter stresses such as strenuous exercise. Antigenic molecules of certain invading pathogens can mimic a host tissue molecule. This can cause the host's immune system to produce antipathogen antibodies that will cross-react against self tissue, leading to autoimmune disease. For example, in some forms of rheumatic fever, the infecting streptococcal bacteria induce the formation of antibodies and even T cells that attack not only the bacteria but also various host tissues, including the heart valves, leading to serious long-term consequences.

molecular systematics (biochemical taxonomy; molecular phylogenetics) The use of amino-acid or nucleotide-sequence data in determining the evolutionary relationships of different organisms. Essentially it involves comparing the sequences of functionally homologous molecules (e.g. ribosomal RNA subunits) from each organism being studied, and determining the number of differences between them. The greater the number of differences, the more distantly related the organisms are likely to be. Moreover, since the number of nucleotide substitutions, and hence substitutions of corresponding amino acids, is generally proportional to time, some indication of the time scale involved can be obtained (*see* MOLECULAR CLOCK). This information has proved particularly useful where there are gaps in the fossil record and can be combined with other evidence from morphology, physiology, and embryology to produce more accurate phylogenetic trees. In microbiology molecular systematics has transformed bacterial phylogeny, in particular establishing the view that there are two quite distinct lineages of prokaryotes, the *bacteria and the archaea. There has been an equally radical reassessment of the classification of eukaryotes, which on current molecular evidence form an unrooted phylogenetic tree consisting of four supergroups plus other assemblages. The availability of rapid fully automated *DNA sequencing has led to the growth of comparative *genomics, which enables comprehensive genomic comparisons between different organisms, with computerized analysis of the data lending much greater reliability to inferences about evolutionary relationships.

Molisch's test *See* ALPHA-NAPHTHOL TEST.

Mollusca A phylum of soft-bodied invertebrates characterized by an unsegmented body differentiated into a **head**, a ventral muscular **foot** used in locomotion, and a dorsal **visceral hump** covered by a fold of skin—the *mantle—which secretes a protective shell in many species. Respiration is by means of gills (*ctenidia) or a lunglike organ and the feeding organ is a *radula. Molluscs occur in marine, freshwater, and terrestrial habitats, and there are several clades, of which three are represented by the classes *Gastropoda (snails, slugs, limpets, etc.), *Bivalvia (bivalves, e.g. mussels, oysters), and *Cephalopoda (squids and octopuses).

SEE WEB LINKS

• Website of the Malacological Society of London, dedicated to the advancement of knowledge and research on molluscs

molybdenum Symbol Mo. A silvery hard metallic element that is a trace element required by living organisms. *See* ESSENTIAL ELEMENT.

Monera *See* PROKARYOTE.

monilophyte Any member of a clade of vascular plants (sometimes called Monilophyta) based on molecular genetic analysis, comprising the whisk ferns, horsetails, ferns, and their allies such as adders' tongues (Ophioglossaceae), moonworts, and grape ferns. Like their sister clade, the seed plants, monilophytes have a differentiated main stem and side branches, but they reproduce by spores instead of seeds. *See also* EUPHYLLOPHYTE.

monoamine oxidase (MAO) An enzyme that breaks down monoamines (e.g. *adrenaline and *noradrenaline) in the body by oxidation. Drugs that inhibit this enzyme are used to treat forms of depression.

monobiontic Describing a life cycle in which there is only a single independent generation, i.e. there is no *alternation of generations.

monochasium (*pl.* **monochasia**) *See* CYMOSE INFLORESCENCE.

monocistronic Describing a fully processed mature messenger *RNA that can encode only one polypeptide per RNA molecule. In eukaryotic cells virtually all messenger RNAs are monocistronic. *Compare* POLYCISTRONIC.

monoclonal antibody (Mab) A specific *antibody produced by one of numerous identical cells derived from a single parent cell. (The population of these cells comprises a *clone and each cell is said to be **monoclonal**.) Monoclonal antibodies occur naturally as part of the humoral immune response when activated lymphocytes divide to form a clone of progeny cells all producing identical antibodies. They are also produced experimentally and commercially in cell culture by fusing a normal antibody-producing lymphocyte with a cell derived from a malignant tumour of *lymphoid tissue of a mouse (*see* CELL FUSION). The resulting *hybridoma cell then multiplies rapidly and yields large amounts of antibody. Monoclonal antibodies are used to identify a particular antigen within a mixture and can therefore be used for identifying blood groups; they also enable the production of highly specific, and therefore effective, *vaccines and drugs such as *immunotoxins. Above all, they have transformed medical and biological diagnostics by ushering in a huge range of cheap and convenient kits for identifying and quantifying biological materials (*see* IMMUNOASSAY).

Monocotyledoneae A class of flowering plants (*see* ANTHOPHYTA), distinguished by having one seed leaf (*cotyledon) within the seed. The monocotyledons generally have parallel leaf veins, scattered vascular bundles within the stems, and flower parts in threes or multiples of three. Monocotyledon species include some crop plants (e.g. cereals, onions, fodder grasses), ornamentals (e.g. tulips, orchids, lilies), and a very limited number of trees (e.g. the palms). *Compare* DICOTYLEDONEAE; EUDICOT.

monoculture *See* AGRICULTURE.

monocyte The largest form of white blood cell (*leucocyte) in vertebrates. Monocytes have a kidney-shaped nucleus, are actively phagocytic, and are the precursors of *macrophages.

monoecious Describing plant species that have separate male and female flowers on the same plant. Examples of monoecious plants are maize and birch. *Compare* DIOECIOUS.

monoglyceride *See* GLYCERIDE.

monohybrid cross A genetic cross between parents that differ in the alleles they possess for one particular gene, one parent having two dominant alleles and the other two recessives. All the offspring (called **monohybrids**) have one dominant and one recessive allele for that gene (i.e. they are hybrid at that one locus). Crossing between these offspring yields a characteristic 3:1 (monohybrid) ratio in the following generation of dominant:recessive phenotypes. *Compare* DIHYBRID CROSS.

monolayer culture A type of *culture in which cells are grown in a single layer on a flask or Petri dish containing the culture medium. *Compare* SUSPENSION CULTURE.

monomer A molecule (or compound) that consists of a single unit and can join with others in forming a dimer, trimer, or polymer.

mononuclear phagocyte system (reticuloendothelial system) The network of phagocytic cells, such as *macrophages and their precursors (monocytes), distributed in the body's tissues, particularly the bone marrow, liver, spleen, and lymph nodes. Some monocytes circulate in the bloodstream, along with *dendritic cells, which also are found in a range of tissues, including the lungs, lymphatic system, digestive system, and skin. Monocytes that enter tissues develop into macrophages, some of which are specialized as histiocytes or *Kupffer cells. All have the common purpose of engulfing and destroying microorganisms and other foreign particles as part of the body's innate *immune response. They also serve as *antigen-presenting cells to stimulate further more specific immune responses by lymphocytes.

monooxygenase (mixed function oxidase; mixed function oxygenase) Any of a large group of enzymes that perform oxidation-reduction reactions in which one atom of the oxygen molecule is incorporated into the chemical donor substrate and

A monohybrid cross: the inheritance of stem lengths in garden peas

the other oxygen atom is combined with hydrogen ions to form water. The different classes of monooxygenases are defined according to the type of hydrogen-providing cosubstrate (or coenzyme, e.g. FAD, NADPH) required in the mixed-function oxidation. Such enzymes are commonly involved in the detoxification of harmful substances by tissues, being located in the vertebrate liver and in the hepatopancreas of invertebrates. For example, herbivorous insects employ monooxygenases to neutralize the natural plant toxins (e.g. phenolics, terpenoids, and quinones) they ingest. Such enzymes are vital components of numerous metabolic pathways; *cytochrome oxidase is an example.

monophyletic In systematics, describing a group of organisms that contains all the descendants of a particular single common ancestor. In *cladistics such a grouping is called a **clade** and is the only type of group regarded as valid when constructing classification schemes. Hence the monophyletic grouping Theria contains the marsupial and placental mammals, together with their extinct Mesozoic relatives, all of which share an immediate common ancestor not shared by the more distantly related egg-laying mammals (comprising the Prototheria). Similarly, birds and crocodilians are the living representatives of a monophyletic group, Archosauria (which also includes the dinosaurs), and are more closely related to each other than to other living reptilian descendants. Consequently the grouping 'reptiles', used in many modern classification systems, is not monophyletic but *paraphyletic, since it excludes the birds (and mammals). *Compare* POLYPHYLETIC.

monophyodont Describing a type of dentition that consists of a single set of teeth that last for the entire lifespan of an animal. *Compare* DIPHYODONT; POLYPHYODONT.

monopodium (*pl.* **monopodia**) The primary axis of growth in such plants as pine trees. It consists of a single main stem that continues to grow from the tip and gives rise to lateral branches. *Compare* SYMPODIUM.

monosaccharide (simple sugar) A carbohydrate that cannot be split into smaller units by the action of dilute acids. Monosaccharides are classified according to the number of carbon atoms they possess: **trioses** have three carbon atoms; **tetroses**, four; **pentoses**, five; **hexoses**, six; etc. Each of these is further divided into **aldoses** and **ketoses**, depending on whether the molecule contains an aldehyde group (–CHO) or a ketone group (–CO–). For example glucose, having six carbon atoms and an aldehyde group, is an **aldohexose** whereas fructose is a **ketohexose**. These aldehyde and ketone groups confer reducing properties on monosaccharides: they can be oxidized to yield sugar acids. They also react with phosphoric acid to produce phosphate esters (e.g. in *ATP), which are important in cell metabolism. Monosaccharides can exist as either straight-chain or ring-shaped molecules (see illustration). They also exhibit *optical activity, giving rise to both dextrorotatory and laevorotatory forms.

monosomy *See* ANEUPLOID.

monosynaptic reflex A simple *reflex that involves transmission of information from a sensory neuron to the appropriate motor neuron across a single synapse in the spinal cord. The knee-jerk reflex action is an example of a monosynaptic reflex (*see* STRETCH REFLEX). *Compare* POLYSYNAPTIC REFLEX.

monotremes *See* PROTOTHERIA.

monotypic Describing a species whose members show only minor variation throughout the entire geographical range of the species, so that there are no recognized races or subspecies. *Compare* POLYTYPIC.

monozygotic twins *See* IDENTICAL TWINS.

morph Any of the distinct common forms found in a population displaying *polymorphism.

morphine An alkaloid present in opium (*see* OPIATE). It is an analgesic and narcotic, used medically for the relief of severe pain.

Glucose (an aldohexose)

H — 1C = O aldehyde group

H — 2C — OH

HO — 3C — O

H — 4C — OH

H — 5C — OH

6CH$_2$OH

straight-chain form

α-glucose

6CH$_2$OH

5C — O

H C4 OH H C1

HO C3 C2 OH

H OH below plane of ring

ring forms

β-glucose

6CH$_2$OH

above plane of ring

5C — O

OH 1

H C4 OH H C1

HO C3 C2 H

H OH

Fructose (a ketohexose)

H

H — 1C — OH

2C = O ketone group

HO — 3C — H

H — 4C — OH

H — 5C — OH

6CH$_2$OH

straight-chain form

6HOCH$_2$ O 1CH$_2$OH

5C H OH C2

HO C4 C3 OH

OH H

ring form

Monosaccharides

morphogen A substance that determines the developmental fate of part of an embryo. During development, different morphogens are produced within the embryo or by cells of the surrounding maternal tissue. They diffuse through the embryonic tissue, each one establishing its own concentration gradient, and together they form a chemical pattern on which embryonic development is based (*see* PATTERN FORMATION). The complex interplay of these morphogen gradients regulates the activity of genes in different regions of the embryo and ultimately brings about the *differentiation of the tissues and organs appropriate to the different regions of the embryo, i.e. morphogenesis. In the development of the fruit fly *Drosophila*, for example, the anterior and posterior ends of the egg are 'signposted' by proteins encoded respectively by *bicoid* and *nanos* genes in maternal follicle cells (*see* MATERNAL EFFECT GENES). The messenger RNAs from these genes accumulate at opposite ends of the egg cell (oocyte), and their products subsequently diffuse into the fertilized egg, thereby establishing its polarity at the outset of development. Bicoid protein is a transcription factor that switches on embryonic genes in a concentration-dependent manner. Above a certain threshold level it triggers the transcription of the gene encoding the Hunchback protein; other concentrations of Bicoid switch on genes encoding other proteins. These represent the beginning of a cascade of morphogens that influence gene activity in an increasingly precise manner. *See also* SEGMENTATION GENES.

morphogenesis The development, through growth and differentiation, of form and structure in an organism.

morphology The study of the form and structure of organisms, especially their external form. *Compare* ANATOMY.

mortality *See* DEATH RATE.

morula A solid ball of some 16–32 cells formed during the initial stage of embryonic development in animals, such as humans and other mammals, in which the fertilized egg contains very little yolk and undergoes complete cleavage. It subsequently becomes a blastocyst (*see* BLASTULA).

mosaic 1. An organism made up of cells that have different genotypes but have developed from the same zygote. *See also* CHIMAERA; GYNANDROMORPH. **2.** A viral disease in plants that causes yellow patches to develop on the leaves, giving these a variegated appearance (*see* VARIEGATION). An example is tobacco mosaic, caused by *tobacco mosaic virus.

mosaic evolution The evolution of different parts of an organism at different rates. For example, many aspects of the human phenotype have evolved relatively slowly or not at all since the hominins diverged from their primate ancestors, one notable exception being the nervous system, which has given humans their overwhelming selective advantage. Similarly, at a molecular level, some proteins evolve very rapidly, while others remain unchanged over millions of years. This high degree of evolutionary independence among different aspects of the phenotype permits flexibility; for example, when a population is faced with new selection pressures in a changing environment, only the most crucial components need to evolve, not the entire phenotype.

mosses *See* BRYOPHYTA.

moths *See* LEPIDOPTERA.

motivation The internal conditions responsible for temporary reversible changes in the responsiveness of an animal to external stimulation. Thus an animal that has been deprived of food will accept less palatable food than one that has not been deprived: the difference is attributed to a change in feeding motivation. Changes in responsiveness due to maturation, *learning, or injury are not usually readily reversible and are therefore not considered to be due to changes in motivation. Early attempts to describe motivation in terms of a number of separate *drives (e.g. food drive, sex drive) have not found general favour, partly because drives interact with one another; for example, water deprivation often affects an animal's willingness to feed.

motor cell A type of plant cell that acts like a hinge at joints to enable the movement of plant parts, such as the closing and opening of leaflets in response to light intensity (*see* NYCTINASTY) or the rapid closure of a leaf in a carnivorous plant. Motor cells adjust their internal concentration of potassium ions (K^+) to alter their turgidity, and hence the cell shape. They can accumulate K^+ via potassium channels in the plasma membrane, which promotes the osmotic uptake of water into the cell, making the cell swollen (turgid). Conversely, they can pump K^+ out of the cell, which causes water to leave and the cell to shrink. The movements resulting from the changes in motor-cell turgor are relatively gradual, taking minutes or hours. However, in the case of carnivorous plants, such as Venus' flytrap, a very rapid leaf closure is required to trap insect prey. Here the motor cells along the midrib of the leaf become freely permeable to K^+, which surges out, causing water to follow and leading to near instantaneous collapse of the cells and swift closure of the leaf.

motor neuron A *neuron that transmits nerve impulses from the central nervous system to an effector organ (such as a muscle or gland) and thereby initiates a physiological response (e.g. muscle contraction).

motor protein Any of various proteins that perform mechanical work in cells by reversibly changing shape. Major motor proteins are *kinesin, which moves vesicles along fibres of the cytoskeleton, and *dynein, which causes the bending movements of cilia and flagella. They are powered by energy released by the hydrolysis of ATP.

motor unit A motor neurone and all the muscle fibres it controls. In skeletal muscle, each muscle fibre is innervated by just a single motor neurone, but the axon of a single neurone typically branches to form synapses with multiple muscle fibres. In muscles used for very fine movement, as in the fingers, a single neurone may connect with only one or two muscle fibres, whereas in powerful muscles, such as the quadriceps of the upper leg, a single neurone may form synapses with hundreds of muscle fibres, distributed throughout the body of the muscle. Recruitment of more and more motor units is a key mechanism to increase the force of muscle contraction. Motor units can be categorized according to the number of muscle fibres, the speed at which those fibres contract, and their resistance to fatigue into **slow units**, **fast-fatigable units**, and **fast fatigue-resistant units**; these are deployed according to whether a muscle is used for sustained contraction, e.g. in maintaining posture, or for rapid, forceful exertion, e.g. running.

moulting 1. The seasonal loss of hair, fur, or feathers that occurs in mammals and birds. In mammals the hair growth cycle can be described by four main stages: active hair proliferation (anagen), regression (catagen), rest (telogen), and shedding (exogen). The mechanisms controlling hair loss and renewal are still poorly understood but are regulated by changes in photoperiod linked to the animal's internal *biological clock and variations in the levels of hormones such as *melatonin. So, for example, as days become shorter in the autumn, the summer coat is replaced by a winter coat, which is often denser and longer than the summer coat. In some species, such as the snowshoe hare (*Lepus americanus*), the winter coat is a different colour to provide camouflage against a snowy landscape. Many bird species undergo a 'pre-basic' moult following the breeding season or laying season, and some may have a 'pre-alternate' moult to produce breeding plumage. **2.** The periodic loss of the integument of arthropods and reptiles. *See* ECDYSIS.

mouth The opening of the *alimentary canal, which in most animals is used for the *ingestion of food. It leads to the *buccal cavity (mouth cavity).

mouth cavity *See* BUCCAL CAVITY.

mouthparts Modified paired append- ages on the head segments of arthropods, used for feeding. A typical insect has a *labium (lower lip), one pair each of *man- dibles and *maxillae, and a *labrum (upper lip), although in many the mouthparts are modified to form piercing stylets or a sucking proboscis. Crustaceans, centipedes, and millipedes have one pair of mandibles and two pairs of maxillae used for cutting and holding the food. Crustaceans also have several pairs of *maxillipeds (*see also* MAXILLULA). Arachnids have *chelicerae and *pedipalps.

MPF *See* MITOSIS-PROMOTING FACTOR.

MRI *See* MAGNETIC RESONANCE IMAGING.

mRNA *See* RNA.

MRSA *See* STAPHYLOCOCCUS.

MSH *See* MELANOCYTE-STIMULATING HORMONE.

mtDNA *See* MITOCHONDRIAL DNA.

MTOC *See* MICROTUBULE-ORGANIZING CENTRE.

m.u. *See* MAP UNIT.

mucigel (slime) A mixture of plant secre- tions, bacteria, and soil particles that sur- rounds the tip of plant roots. It contains complex polysaccharides called *mucilages which are secreted by the *root cap and help to lubricate passage of the growing root tip through the soil. However, the root tip also releases proteins, which help to protect the root tip from infection, and amino acids and sugars, which influence microbial activity in the *rhizosphere, thereby helping to mobilize soil nutrients for uptake by the root. *See also* RHIZOSPHERE.

mucilage Any of a large group of complex heterogeneous polysaccharides present in many plants and algae. They occur in the root, bark, seeds, and other parts. Various functions have been proposed, including food reserve, water absorption and storage, and aiding seed dispersal and germination. Mucilages are hard when dry and slimy when wet. Unlike *gums, mucilages are not found on the plant surface as exudates following injury or attack by pathogens. They are a form of soluble fibre in plant-based foods and are also used as stabilizers and thickeners in cosmetics and pharmaceuticals. *See also* GLYCOCALYX; MUCIGEL.

mucin *See* MUCUS.

mucopolysaccharide *See* GLYCOSAMINO-GLYCAN.

mucoprotein A *glycoprotein in which the carbohydrate component is a relatively large polysaccharide. Mucoproteins readily form gels with water and constitute the mucin in *mucus. The mucoprotein chains are joined end to end by disulphide bonds, forming very long mucin strands. When released into water, the network of strands expands rapidly to produce a large volume of mucus.

mucosa (*pl.* **mucosae**) A *mucous membrane, especially that forming the lining of the wall of the mammalian stomach and intestine. In these organs the mucosa includes an outer muscular layer, the *muscularis mucosae, which lies adjacent to the *submucosa.

mucosal-associated lymphoid tissue (MALT) Any of various lymphoid tissues or organs that protect the mucosal surfaces, including the digestive tract (*see* GUT-ASSOCIATED LYMPHOID TISSUE), respiratory tract (*see* BRONCHIAL-ASSOCIATED LYMPHOID TISSUE), eyes, and reproductive tract. This network constitutes the **mucosal immune system** and overall contains about half of all the body's lymphocytes,

emphasizing its importance in immune defence. Plasma cells of MALT produce chiefly IgA-type antibodies, which are found widely in mucosal secretions.

mucous membrane (mucosa) A layer of tissue comprising an epithelium supported on connective tissue. Within the epithelium are *goblet cells, which secrete *mucus onto the surface, and the epithelium often bears cilia. Mucous membranes line body cavities communicating with the exterior, including the alimentary and respiratory tracts. *Compare* SEROUS MEMBRANE.

mucus The slimy substance secreted by *goblet cells onto the surface of a *mucous membrane to protect and lubricate it and to trap bacteria, dust particles, etc. Mucus consists of water, various *mucoproteins (collectively called **mucin**), cells, and salts. Mucus in *saliva is essential for swallowing food, facilitating the smelling and tasting of food, and protecting the mouth and pharynx against abrasion.

Müller cell Any of the cells in the vertebrate retina that support the visual sensory cells, notably the rods and cones. These glial cells (known collectively as **Müller glia**), named after German anatomist Heinrich Müller (1820–64), span the full thickness of the retina, filling the spaces between the retinal neurons. Besides giving physical support to the other cell types, they also regulate the concentrations of certain physiologically important substances within the retina, such as potassium ions, glutamate, and gamma-aminobutyric acid, thereby affecting the activity of other cells. In certain animal groups, such as fish, amphibians, and birds, Müller cells can dedifferentiate following injury and act as stem cells to provide a source of new photoreceptive cells.

Müllerian mimicry *See* MIMICRY.

Muller's ratchet The inevitable and irreversible accumulation of deleterious mutations that occurs in an asexually reproducing population. The mechanism is named after the US geneticist Hermann J. Muller (1890–1967), who likened each successive asexual generation to the one-directional movement of a ratchet. He observed that the evolution of sexual reproduction offered the great advantage of allowing the removal of mutations through recombination. However, factors such as large population size, low mutation rate, and the occurrence of beneficial mutations can all counteract Muller's ratchet and its ultimate consequence—extinction.

multiadhesive   protein   *See* EXTRACELLULAR MATRIX.

multicellular Describing tissues, organs, or organisms that are composed of a number of cells. *Compare* UNICELLULAR.

multienzyme system A complex of enzymes within a cell that form a reaction sequence of a biochemical pathway so that the product of the first enzyme reaction is transferred directly to the next enzyme and immediately undergoes a second reaction, and so on. The rate of an enzyme reaction often depends on the concentration of the enzyme and the substrate, both being required in relatively high amounts. Multienzyme systems, such as those involved in RNA and protein synthesis, help maintain a high rate of cellular metabolism since the intermediate products are transferred directly to the next enzyme and are therefore not required in large concentrations.

multifactorial   inheritance   *See* QUANTITATIVE INHERITANCE.

multilocus sequence typing (MLST) A procedure for characterizing isolates of bacteria or other microbial organisms based on sequencing the DNA of certain internal fragments of seven specific *housekeeping genes that are common to all species. The alleles at each of the seven loci define each isolate unambiguously, enabling the sample profile to be identified by comparing it with a large database of MLST profiles.

multimer (in molecular biology) An aggregate molecule that consists of two or more component molecules, which may or may not be identical, held together by non-covalent bonds. For example, many proteins comprise   distinct   polypeptide   subunits

arranged to give a specific quaternary structure. Multimers can be named variously according to the number of constituent subunits; hence, a dimer has two subunits, a trimer has three, a tetramer four, and so on.

multiple alleles Three or more alternative forms of a gene (*alleles) that can occupy the same *locus. However, only two of the alleles can be present in any individual diploid organism. For example, the *ABO system of blood groups is controlled by three alleles, only two of which are present in an individual.

multipolar neuron A *neuron that has one axon and several dendrons extending from its cell body in different directions. Most vertebrate motor neurons and interneurons are multipolar neurons. *Compare* BIPOLAR NEURON; UNIPOLAR NEURON.

muscarinic Describing one of the two main classes of *acetylcholine receptors, so called because the effect of acetylcholine on them can be mimicked by muscarine, a toxic alkaloid produced by *Amanita muscaria* and certain other gill fungi. Muscarinic receptors occur on target cells innervated by fibres of the vertebrate parasympathetic nervous system, for example in smooth muscle, cardiac muscle, and glands. They are *G protein-coupled receptors, which activate *ion channels via intracellular second messengers. *Compare* NICOTINIC.

Musci *See* BRYOPHYTA.

muscle A tissue consisting of sheets or bundles of cells (**muscle fibres**) that are capable of *contraction, so producing movement or tension in the body. There are three types of muscle. *Skeletal muscle produces voluntary movement (e.g. at joints); *smooth muscle mainly effects the movements of hollow organs (e.g. intestine and bladder); and *cardiac muscle occurs only in the heart.

(⊕) SEE WEB LINKS
• Interactive tutorial identifying the principal muscles of the human body

muscle spindle A *stretch receptor in vertebrate muscle. Muscle spindles are modified muscle cells that run parallel to normal muscle fibres; each consists of a capsule containing small striated muscle fibres (*intrafusal fibres). Muscle spindles respond to stretching of the muscle by sending nerve impulses to the central nervous system via sensory nerves. These signals stimulate motor neurons to increase contraction of the muscle, thereby adjusting the muscle tone. The spindles thus play an important role in the subconscious maintenance of posture and movement. *See also* STRETCH REFLEX.

muscularis mucosae A thin layer of smooth muscle that forms the outermost part of the *mucosa of the mammalian stomach and intestine.

mutagen An agent that causes an increase in the number of mutants (*see* MUTATION) in a population. Mutagens operate either by causing changes in the base sequence of the DNA of the *genes, so interfering with the coding system, or by causing chromosome damage. Various chemicals (e.g. *aflatoxin, benzopyrene in cigarette smoke, *colchicine), forms of radiation (e.g. X-rays, gamma rays), and retroviruses are among the many agents identified as mutagens.

mutant A gene or an organism that has undergone a heritable change, especially one with visible effects (i.e. the change in *genotype is associated with a change in *phenotype). *See* MUTATION.

mutation A sudden random change in the genetic material of a cell that potentially can cause it and all cells derived from it to differ in appearance or behaviour (i.e. in phenotype) from the normal type. An organism affected by a mutation (especially one with visible effects) is described as a **mutant**. **Somatic mutations** affect the nonreproductive cells and are therefore restricted to the tissues of a single organism. These can lead to diseases such as cancer. **Germ-line mutations**, which occur in the reproductive cells or their precursors, may be transmitted to the organism's descendants and cause abnormal development or early embryonic death.

Mutations occur naturally at a low rate but this may be increased by radiation and by some chemicals (*see* MUTAGEN). Most are *point mutations, which consist of changes in the nucleotide sequence of the DNA, but some (the *chromosome mutations) affect the appearance or the number of the chromosomes. An example of a chromosome mutation is that giving rise to *Down's syndrome.

According to the *neutral theory of molecular evolution, the majority of point mutations in DNA are neither useful nor harmful and can spread throughout a population by genetic drift. Mutations that alter phenotypes are generally harmful, but a very small proportion may increase an organism's *fitness; these spread through the population over successive generations by natural selection. Mutation is therefore essential for evolution, being the ultimate source of genetic variation. *See also* COPY NUMBER; MOLECULAR CLOCK; SINGLE NUCLEOTIDE POLYMORPHISM.

(⊕) SEE WEB LINKS
• First of two topics on mutation from the online resource DNA from the Beginning

mutation frequency The average frequency with which a particular *mutation occurs in a population. This frequency can be increased by radiation or by exposure to chemicals, such as mustard gas or hydrogen peroxide. *Compare* MUTATION RATE.

mutation rate The rate at which *mutations occur over unit time. It can be expressed in various ways. For example, the 'time' units may be per replication or cell division, per gamete, or per generation; moreover, the number of mutations can be expressed as an average probability for each mutable site (i.e. per nucleotide), or as a total for the entire genome. Thus, the spontaneous mutation rate for a wild-type strain of the bacterium *E. coli* has been calculated as 1×10^{-3}/genome/generation, based on the mutations accumulated over many generations. Estimates of the mutation rate in humans are generally in the range of 100–160/diploid genome/generation, depending on the method of calculation. In terms of genomic DNA, this equates to a mutation rate of the order of 10^{-9}/base pair/year; similar mutation rates have been found for other organisms. *Compare* MUTATION FREQUENCY.

mutualism An interaction between two species in which both species benefit. (The term *symbiosis is often used synonymously with mutualism.) There are many examples in nature spanning all types of organisms. In some cases—**facultative mutualisms**—the partners can survive on their own, whereas in others—**obligate mutualisms**—one or both partners have evolved to depend on the symbiotic association for their survival. A well-known example of mutualism is the association between termites and the specialized protists that inhabit their guts. The protists, unlike the termites, are able to digest the cellulose of the wood that the termites eat and release sugars that the termites absorb. The termites therefore depend on the association to use wood as a foodstuff, while the protists are supplied with food and a suitable environment. *See also* MYCORRHIZA.

mya *Abbreviation for* million years ago.

mycelium (*pl.* **mycelia**) A network of *hyphae that forms the body of a fungus. It consists of feeding hyphae together with reproductive hyphae, which produce *sporangia and *gametangia.

mycobiont The fungal component of a *lichen.

mycology The scientific study of *fungi.

Mycophycophyta *See* LICHENS.

mycoplasmas A group of bacteria that lack a rigid cell wall and are among the smallest of all living cells (diameter 0.2 μm–2.0 μm). They are either saprotrophic or parasitic and are found on animal mucous and synovial membranes, in insects, and in plants (in which they seem to inhabit sieve tubes). They cause a variety of diseases, including pleuropneumonia in cattle—hence they were formerly known as pleuropneumonia-like organisms (PPLO). Due to their small size and flexible three-layered cell membrane they can pass through a 0.2-μm-diameter filter and they represent a major contaminant of biotechnological products, such as monoclonal antibodies and vaccines, and of other cell cultures, in which they may exist symbiotically with the cells. Eight genera have been described (including *Mycoplasma*) with over 120 species. Some species live as commensals or parasites in the respiratory and urogenital tracts of humans and other vertebrates. For example, *M. genitalium* lives in the respiratory and urogenital tracts of primates; it has one of the smallest known genomes, with only 482 protein-coding genes. *M. pneumoniae* is a causative agent of pneumonia in humans.

mycorrhiza (*pl.* **mycorrhizas** or **mycorrhizae**) The mutually beneficial association (*see* MUTUALISM) formed between fungi and the roots of plants. This is a very common form of mutualism, estimated to occur in over 90% of terrestrial plants. The absorption of mineral ions, especially phosphate, by the plant is greatly enhanced—perhaps up to 1000-fold—by the presence of the network of fungal hyphae penetrating the finest soil pores. In return the fungus receives energy from the plant, in the form of products of photosynthesis. **Ectotrophic mycorrhizas** (**ectomycorrhizas**) form a network of hyphae around the root and grow into the extracellular spaces of the root. They occur in association with about 10% of plant families, mostly woody species such as pine, oak, and birch. The hyphae of **endotrophic mycorrhizas** (**endomycorrhizas** or **arbuscular mycorrhizas**) enter the cortical cells of the host roots, where they form branching structures called arbuscules, which invaginate to fill individual host cells, while remaining enclosed by the host cell's plasma membrane. These are found in over 85% of plant species, including many crops. The vast absorptive surface area of mycorrhizal fungi is vital for optimal growth of most trees and crop plants, and in commercial production steps are taken to ensure the presence of suitable fungal spores in the soil. Indeed, mycorrhizas are likely to have enabled the colonization of the terrestrial environment by the earliest land plants.

There is also evidence that plants transmit chemical defence signals to their neighbours via the network of hyphae. *See also* GLOMEROMYCOTA.

Mycota In older classification systems, a kingdom comprising the *fungi.

myelin A *phospholipid produced by the *Schwann cells of the nervous system. Myelin forms an insulating layer around the nerve fibres (*see* MYELIN SHEATH).

myelin sheath (medullary sheath) The layer of fatty material that surrounds and electrically insulates the axons of most vertebrate and some invertebrate neurons. The myelin sheath enables a more rapid transmission of nerve impulses (at speeds up to 120 m s^{-1}). It consists of layers of membrane derived from *Schwann cells, in neurons of the peripheral nervous system, and from oligodendrocytes in the central nervous system. The sheath is interrupted at intervals along the axon by **nodes of Ranvier**, where there are *voltage-gated ion channels and the axon is exposed to the extracellular fluid. Myelinated sections of axon are called **internodes**. During conduction of a nerve impulse the *action potential effectively 'jumps' from node to node, a mechanism called saltatory conduction, which vastly increases the speed of conduction compared to unmyelinated axons.

myeloid tissue Tissue within red *bone marrow that produces the blood cells. It is found around the blood vessels and contains various cells that are precursors of the blood cells. *See* HAEMOPOIETIC TISSUE.

myeloma *See* CANCER.

Myllokunmingia *See* CHENGJIANG FOSSILS.

myoblast *See* MYOTUBE.

myocardium (*pl.* **myocardia**) The muscular wall of the heart. The thickness of the myocardium varies, reflecting the magnitude of the pressure generated in the heart during contraction: the myocardium is thickest around the left ventricle, which does most of the work in the heart. The inner surface of the myocardium is lined by a layer of *endothelium (the **endocardium**), which is continuous with the endothelium lining the blood vessels.

myocyte A contractile cell, especially a muscle cell.

myofibril *See* SKELETAL MUSCLE.

myogenic Originating in or produced by muscle cells. The contractions of *cardiac muscle fibres and single-unit *smooth muscle cells are described as myogenic, since they are produced spontaneously, without requiring stimulation from nerve cells (*see* PACEMAKER).

myoglobin A globular protein occurring widely in muscle tissue as an oxygen carrier, especially in muscles containing *slow-twitch fibres. It comprises a single polypeptide chain and a *haem group, which reversibly binds a molecule of oxygen. This is only relinquished at relatively low external oxygen concentrations, e.g. during strenuous exercise when muscle oxygen demand outpaces supply from the blood. Myoglobin thus acts as an emergency oxygen store. It occurs in high concentrations in the muscles of diving mammals such as seals.

myometrium The thick layer of smooth muscle that makes up the bulk of the *uterus. The myometrium is lined by the *endometrium.

myopia Short-sightedness. It results from the lens of the eye refracting the parallel rays of light entering it to a focus in front of the retina generally because of an abnormally long eyeball. The condition is corrected by using diverging spectacle lenses to move the image back to the retina.

myosin A contractile protein that interacts with *actin to bring about contraction of muscle or cell movement. The type of myosin molecule found in muscle fibres consists of a tail, by which it aggregates with other myosin molecules to form so-called 'thick filaments'; and a globular head, which has sites for the attachment of actin and ATP molecules. *See* SARCOMERE; SLIDING FILAMENT THEORY.

myotatic reflex *See* STRETCH REFLEX.

myotome One of a series of segmented muscle blocks found in fishes and lancelets. Myotomes are arranged in pairs on either side of the body that work antagonistically (*see* ANTAGONISM) against the backbone (or notochord), providing a means of locomotion by causing the tail to sweep from side to side.

myotube An embryonic precursor of a skeletal muscle fibre. During development in the early embryo, numerous individual precursor cells, called **myoblasts**, proliferate and migrate to sites where muscle formation will take place. There they fuse to form myotubes, each containing numerous nuclei and bounded by a single plasma membrane. The process of differentiation continues as the myotube manufactures the characteristic muscle proteins that will form the contractile myofibrils and eventually matures into a muscle fibre.

Myriapoda In some classifications, a subphylum of arthropods that comprises the classes *Chilopoda (centipedes), *Diplopoda (millipedes), Pauropoda (pauropods), and Symphyla (symphilids). In other classifications the Myriapoda is a class or clade containing only the centipedes and millipedes.

myrmecochory The collection and dispersal of plant seeds by ants. A variety of plant species possess hard seeds that are inedible to ants but are nevertheless gathered by them and taken to the ants' nest.

The ants perform this service because the seeds are equipped with special food bodies (**elaiosomes**). These are variously shaped appendages derived from ovarian tissue and containing proteins, lipids, and carbohydrates. In the nest the ants detach the elaiosomes and feed them to their larvae, discarding the seeds either within or near the nest. The ant benefits by receiving food, while the seed dispersal may benefit the plant in several ways: for example, protection of the seeds inside the ants' nest; reduction of competition from its own seedlings; or removal of the seeds to a suitable germination site.

myxamoeba *See* SLIME MOULD.

myxobacteria A group of *proteobacteria that can live in highly organized swarms that exhibit many characteristics of multicellular organisms, such as cell differentiation, division of labour, intercellular communication, and coordinated movement. When food is scarce, the swarm forms a fruiting body, in which some cells differentiate into resistant spores which can survive adverse conditions, whereas other cells are sacrificed.

Myxomycota *See* SLIME MOULDS.

myxovirus One of a group of RNA-containing viruses associated with various diseases of humans and other vertebrates. **Orthomyxoviruses** produce diseases of the respiratory tract, e.g. influenza; **paramyxoviruses** include the causal agents of mumps, measles, and fowl pest.

myxozoan *See* CNIDARIA.

n *See* HAPLOID.

NAD (nicotinamide adenine dinucleotide) A *coenzyme, derived from the B vitamin *nicotinic acid, that participates in many biological redox reactions (see formula). NAD is characteristically loosely bound to the enzymes concerned. It normally carries a positive charge (NAD^+) and can accept one hydrogen atom and two electrons to become the reduced form, **NADH**. NADH is generated during the oxidation of food, especially by the reactions of the *Krebs cycle. It then gives up its two electrons (and single proton) to the *electron transport chain, thereby reverting to NAD^+ and generating three molecules of ATP per molecule of NADH.

NADP (nicotinamide adenine dinucleotide phosphate) differs from NAD only in possessing an additional phosphate group. It functions in the same way as NAD although anabolic reactions (*see* ANABOLISM) generally use NADPH (reduced NADP) as a hydrogen donor rather than NADH. Enzymes tend to be specific for either NAD or NADP as coenzyme.

nano- Symbol n. A prefix used in the metric system to denote 10^{-9}. For example, 10^{-9} metre = 1 nanometre (nm).

nanoarray *See* MICROARRAY.

NAR *See* NET ASSIMILATION RATE.

narcotic Any drug that induces stupor and relieves pain, especially morphine and other *opiates. Such narcotics are addictive and cause dependence, and their medical use is strictly controlled.

nares (nostrils) The paired openings of the *nasal cavity in vertebrates. All vertebrates have **external nares**, which open to the exterior; in some species these are situated on a *nose. **Internal nares** (or **choanae**) are present only in air-breathing vertebrates (including lungfish) and open into the mouth cavity. In mammals they open posteriorly, beyond the secondary *palate.

NAD

nasal cavity The cavity in the head of a vertebrate that is lined by a membrane rich in sensitive olfactory receptors (*see* OLFACTION). It is connected to the exterior by external nostrils and (in air-breathing vertebrates) to the respiratory system by internal *nares.

nastic movements Movements of plant organs in response to external stimuli that are independent of the direction of the stimuli. Examples are the opening of crocus and tulip flowers in response to a rise in temperature (**thermonasty**), the opening of evening primrose flowers at night (**photonasty**), and the folding up and drooping of leaves of the sensitive plant (*Mimosa pudica*) when lightly touched (**haptonasty or thigmonasty**). *Compare* TROPISM. *See also* MOTOR CELL; NYCTINASTY.

natality *See* BIRTH RATE.

natriuretic peptide Any of several peptide hormones that promote the excretion of sodium ions in the urine (i.e. natriuresis). The first to be discovered was *atrial natriuretic peptide (ANP), which is produced by an upper chamber (atrium) of the heart. ANP is released from atrial muscle cells in response to various factors, including stretching of the atrial walls, raised concentration of sodium in the blood, angiotenism II, and stimulation of β-adrenoceptors. The overall effects are a decrease in blood volume and lowering of cardiac output and blood pressure. Brain natriuretic peptide (BNP, or B-type natriuretic peptide) is released in the brain and also the heart ventricles in response to stretching of the chambers; it has vasodilator and natriuretic properties, and levels increase markedly in cases of heart failure. C-type natriuretic peptide (CNP) is produced more generally in the cardiovascular system and serves as a regulator of vascular tone and growth.

natural group A group of organisms of any taxonomic rank that are believed to be descended from a common ancestor (*see* MONOPHYLETIC). In an ideal natural classification all taxa should be natural groups. *See also* CLADISTICS.

natural history 1. The study of living organisms in their natural habitats. 2. The study of all natural phenomena.

naturalized *See* ALIEN.

natural killer cell (NK cell) A lymphoid cell that recognizes and destroys tissue cells infected with pathogenic organisms and also certain tumour cells. NK cells are important as part of the innate *immune response against infection, before the more antigen-specific mechanisms of the adaptive immune response are mobilized (*see* CYTOTOXIC T CELL), and they play a significant role in combating infections with, for example, herpesviruses. They become activated in response to interferons or the release of cytokines by macrophages, bind to target cells, and release cytotoxic granules onto the surface of their target. The toxic effector molecules penetrate the target's plasma membrane and induce programmed cell death (*see* APOPTOSIS). Receptors on NK cells recognize changes in the expression of surface markers on infected cells, enabling them to distinguish 'stressed' (i.e. infected) cells from uninfected self tissue. By means of various cytokines and chemokines NK cells also undergo adaptive changes, interact with other types of immune cells, and even have a capacity for immunological memory.

natural selection The process that, according to *Darwinism, brings about the evolution of new species of animals and plants. Both Darwin and *Wallace noted that the size of any population tends to remain constant despite the fact that more offspring are produced than are needed to maintain it. They also saw that variations existed between individuals of the population and concluded that disease, competition, and other forces acting on the population eliminated those individuals less well adapted to their environment. The survivors would pass on any heritable advantageous characteristics (i.e. characteristics with survival value) to their offspring and in time the composition of the population would change in adaptation to a changing environment. Over a long period of time this process could give rise to organisms so different from the original population that new species are formed. *See* SELECTION. *See also* ADAPTIVE RADIATION. *Compare* PUNCTUATED EQUILIBRIUM.

nature and nurture The combined effects of inherited factors (nature) and environmental factors (nurture) on the development of an organism. Traits that are determined by numerous genes (*see* QUANTITATIVE INHERITANCE), such as height in humans, tend also to be subject to environmental factors, such as nutrition in this example, i.e. both genes and environment direct the eventual *phenotype, which can show considerable variation. Hence, the genetic potential of an organism will only be realized under appropriate environmental conditions. *See also* PHENOTYPE.

nauplius (*pl.* **nauplii**) The free-swimming larva of many marine and freshwater crustaceans. It has an unsegmented body with a single eye at the front (the **nauplius eye**), mandibles, antennae, and three pairs of limbs.

navigation The complex process that enables animals to travel along a particular course in order to reach a specific destination. Navigation is an important aspect of behaviour in many animals, particularly those, such as birds, fish, and some insects, that undergo *migrations. Landmarks, such as coastlines and mountain ranges, are important reference points for navigation but many animals can navigate successfully without the aid of these, by using the sun, stars, magnetic fields, odours, and polarized light. For example, birds use the sun and stars as landmarks and are sensitive to the earth's magnetic fields, as are many other animals; for example, turtles return decades later to the beach where they hatched by sensing the intensity and inclination of the magnetic field (*see* MAGNETORECEPTOR), while salmon can identify the unique odour of their home river. It is thought that homing pigeons might use very low-frequency sound (infrasound) to build up a topographic map of the area surrounding their loft, as well

as a magnetic sense, to guide their flight back home.

NCLDV *See* NUCLEOCYTOPLASMIC LARGE DNA VIRUS.

ncRNA *See* NONCODING RNA.

Neanderthal A fossil hominin species that lived in Europe and western Asia between about 400 000 and 40 000 years ago, when the climate was much colder than today. Neanderthals are now generally regarded as a distinct species from humans, *Homo neanderthalensis*. The fossil remains indicate that Neanderthals were fairly short, strongly built, and had low brows but that the brain size was the same as, or larger than, modern humans. They were nomadic cave dwellers who buried their dead. Early modern humans in Eurasia overlapped with Neanderthals from about 65 000 years ago until the Neanderthals went extinct, possibly the culmination of competition with the more advanced stone tool technology of humans. Studies have shown that present-day humans in Asia, New Guinea, and Europe each carry about 2.5% of Neanderthal DNA, and collectively may have about 20% of Neanderthal genes, a legacy of interbreeding after humans first spread from Africa. The name is derived from the site in the Neander valley, Germany, where fossils were found in 1856.

(((●))) **SEE WEB LINKS**

• Illustrated account of Neanderthal features and fossil record, from the Smithsonian Institution

Nearctic region *See* FAUNAL REGION.

near point The nearest point at which the human eye can focus an object. As the lens becomes harder with age, the extent to which accommodation can bring a near object into focus decreases. Therefore with advancing age the near point recedes—a condition known as *presbyopia.

necrosis The *death of cells, which can be caused by a variety of chemicals and toxic substances. It often involves denaturation of proteins and may be preceded by a change in the appearance of cells and their contents (such as swelling of the mitochondria and *karyorrhexis) and the appearance of *lysosomes. *Compare* APOPTOSIS.

nectar A sugary liquid produced in plants by **nectaries**, regions of secretory cells on the receptacle or other parts of a flower. It typically comprises a mixture of sucrose, fructose, and glucose, and attracts pollinating insects or other animals, which feed on it.

negative feedback *See* FEEDBACK.

negative selection *See* PURIFYING SELECTION.

negative-sense (in genetics) Describing RNA or a single strand of DNA whose base sequence is complementary to that of the messenger RNA (mRNA) required for translation into protein. By convention, the *anticoding strand of duplex DNA is the negative-sense strand—i.e. the one that undergoes transcription to mRNA. Similarly, negative-sense viral genomes must first be transcribed into a complementary mRNA to replicate inside a host cell. *Compare* POSITIVE-SENSE.

nekton *Pelagic organisms that actively swim through the water. Examples are fish, jellyfish, turtles, and whales. *Compare* BENTHOS; NEUSTON; PLANKTON.

nematoblast *See* THREAD CELL.

nematocyst *See* THREAD CELL.

Nematoda A phylum of *pseudocoelomate invertebrates comprising the roundworms. They are characterized by a smooth narrow cylindrical unsegmented body tapered at both ends. They shed their tough outer cuticle four times during life to allow growth. The microscopic free-living forms are found in all parts of the world, where they play an important role in the breakdown and recycling of organic matter. The many parasitic nematodes of plants and animals may be much larger; they include the filaria (*Wuchereria*) and Guinea worm (*Dracunculus*), which cause serious diseases in humans. Some are

serious crop pests, such as eelworms, which attack plant roots. Nematodes belong to the clade *Ecdysozoa. *See also* *CAENORHABDITIS ELEGANS*.

(((●))) **SEE WEB LINKS**
• Website of the Society of Nematologists; Education Committee link accesses many images of nematodes

Nemertea (Nemertina; Rhynchocoela) A phylum containing some 1200 species of free-living worms found chiefly in marine habitats but also in freshwater and soil. They have flattened bodies, sometimes patterned in striking colours, ranging from less than 1 mm to over 20 m long (hence their popular name of ribbon worms). Unlike flatworms, nemerteans have a complete digestive tract, with mouth and anus, and a blood vascular system. They also typically possess a long hollow muscular proboscis, which lies inverted inside a fluid-filled chamber, the rhynchocoel. The proboscis can be rapidly everted through its own pore and is used to catch prey, such as crustaceans, worms, and molluscs. It can also be used to assist burrowing in sediment or mud. Nemerteans may also crawl or swim by means of muscular undulations of the body. Reproduction can be sexual, with eggs developing directly into adults or via ciliated larvae. Asexual reproduction by *fragmentation also occurs. Nemerteans are included in the clade *Lophotrochozoa.

neocortex (neopallium; isocortex) The part of the *cerebral cortex of the vertebrate brain that has evolved most recently, and is most highly developed in mammals. In humans and other primates it forms a surface layer covering most of the forebrain and comprises six layers of neurons with a multitude of interconnections; it is greatly folded to vastly increase its surface area. The expansion of the neocortex in primates is thought to be correlated with their complex social behaviour. It is the major centre for the coordination of sensory and motor information. *See* CEREBRUM.

neo-Darwinism (modern synthesis; synthetic theory) The current theory of the process of *evolution, formulated between about 1920 and 1950, that combines evidence from classical genetics with the Darwinian theory of evolution by *natural selection (see DARWINISM). It makes use of modern knowledge of genes and chromosomes to explain the source of the genetic variation upon which selection works. This aspect was unexplained by traditional Darwinism. Recent discoveries of *epigenetic inheritance have led some to call for modification, or even replacement, of neo-Darwinian theory, invoking *neo-Lamarckism. But this is still regarded by many as heresy.

Neogene The geological period of the *Cenozoic era, consisting of the *Miocene and *Pliocene epochs, that began 23 million years ago, following the *Palaeogene period, and was succeeded about 2.58 million years ago by the current *Quaternary period.

neo-Lamarckism Any of the comparatively modern theories of evolution based on Lamarck's theory of the inheritance of acquired characteristics (see LAMARCKISM). These include the unfounded dogma of *Lysenkoism and controversial experiments on the inheritance of acquired immunological tolerance in mice. More recently, some theorists have proposed that certain epigenetic phenomena are consistent with Lamarckian principles. They cite the acquired patterns of chemical modification of DNA and histones that alter gene expression and can be replicated in the offspring. Critics retort that such changes are merely a form of *phenotypic plasticity; they do not affect the nucleotide sequence, are not permanent, and can be reversed.

Neolithic The New Stone Age: a stage of human cultural and technological evolution that began in the Middle East in approximately 8500 BC. It was characterized by farming using wild and domesticated crops and herding of livestock, and (in the mid- and late Neolithic) by the making of pottery. Grinding and polishing of stone tools was also practised. The Neolithic is thus not a fixed chronological period, and the emergence and duration of Neolithic cultures varied from region to region. Neolithic cultures developed somewhat later in central Europe and in southern and eastern Asia. The Bronze Age superseded Neolithic cultures from around 3500 BC in the Middle East.

neonicotinoid Any of a class of synthetic pesticides based on the chemical structure of nicotine. They were introduced in the early 1990s for use as insecticides and have relatively low toxicity to humans and other mammals compared to alternatives such as organophosphates. They act by binding irreversibly to *nicotinic acetylcholine receptors in the insect's central nervous system, causing paralysis and death. Various neonicotinoid chemicals are used in horticultural, agricultural, and garden pesticides. Since the mid-2000s there has been growing evidence that these chemicals have harmful effects on pollinating insects, particularly bumblebees and honey bees, and may persist in the soil and vegetation adjacent to treated areas. In 2013 the EU introduced a temporary ban on the use of thiamethoxam, clothianidin, and imidacloprid in flowering crops such as oilseed rape and sunflowers on which bees feed, which remained in place as of 2018.

neoplasm (tumour) Any new abnormal growth of cells, forming either a harmless (benign) tumour or a malignant one (see CANCER).

Neoproterozoic See PROTEROZOIC.

Neornithes In some classifications, a subclass containing all extant species of birds. See AVES.

neoteny The retention of the juvenile body form, or particular features of it, in a mature animal. For example, the axolotl, a salamander, retains the gills of the larva in the adult. Neoteny is thought to have been an important mechanism in the evolution of certain groups, such as humans, who are believed to have developed from the juvenile forms of apes. Most strikingly, adult humans retain a skull shape that closely resembles that of an infant, giving a large brain case, whereas in the closely related chimpanzees, the skull undergoes a dramatic change of shape into adulthood, with a much smaller brain case. *See also* HETEROCHRONY.

Neotropical region *See* FAUNAL REGION.

nephridiopore *See* NEPHRIDIUM.

nephridium (*pl.* **nephridia**) A type of excretory organ found in many invertebrates, consisting of a simple or branched tube, formed by the ingrowth of ectoderm, with cilia at the inner end. Excretory products diffuse into the nephridium and are wafted to the exterior by ciliary action. The simplest type is known as a **protonephridium** and consists of a tubule and its associated *flame cell; networks of protonephridia occur in flatworms (*see* PLATYHELMINTHES) and rotifers. The paired *metanephridia of the earthworm and some other annelids superficially resemble kidney tubules.

nephron The excretory unit of the vertebrate *kidney (see illustration). Many constituents of the blood are filtered from the glomerulus into the *Bowman's capsule at one end of the nephron. The *glomerular filtrate passes along the length of the nephron and some of its water, plus some salts, glucose, and amino acids, are reabsorbed into the surrounding blood capillaries (*see* PROXIMAL CONVOLUTED TUBULE; LOOP OF HENLE; DISTAL CONVOLUTED TUBULE). More water is reabsorbed in the *collecting duct, and the resulting concentrated solution of nitrogen-containing waste matter (*urea in most mammals) plus inorganic salts drains from the collecting ducts of the nephrons and is discharged as urine into the ureter. The kidneys of birds and mammals contain two types of nephrons: **juxtamedullary nephrons**, which have a long loop of Henle extending deep into the kidney medulla, and a glomerulus in the inner cortex; and **cortical nephrons**, which have the glomerulus in the outer cortex and a relatively short loop of Henle. Of the one million or so nephrons in the human kidney, some 85% are cortical

distal convoluted tubule

arteriole

Bowman's capsule

glomerulus

proximal convoluted tubule

collecting duct

ascending limb

loop of Henle

to ureter

descending limb

Structure of a single nephron

nephrons, with the remainder being juxtamedullary nephrons. The latter are essential for producing urine that is hyperosmotic relative to body fluids, and their proportion increases with the need to conserve body water, e.g. in desert-dwelling mammals. Birds also need to conserve water and have juxtaglomerular nephrons in their kidneys, whereas other reptiles have only cortical nephrons, as do freshwater fishes and amphibians. Marine fishes have much smaller kidneys with fewer nephrons, whose main function is to excrete ions taken in by drinking seawater. Osmoregulation is augmented by *chloride secretory cells in the gills of such fishes.

nephrostome *See* NEPHRIDIUM.

nephrotoxin Any toxic substance that targets the kidneys. Common examples of nephrotoxins are mercury salts and certain herbicides, such as *Paraquat.

neritic zone The region of the sea over the continental shelf, which is less than

200 metres deep (approximately the maximum depth for organisms carrying out photosynthesis). *Compare* OCEANIC ZONE.

nerve A strand of tissue comprising many *nerve fibres plus supporting tissues (*see* GLIA), enclosed in a connective-tissue sheath. Nerves connect the central nervous system with the organs and tissues of the body. A nerve may carry only motor nerve fibres (**motor nerve**) or only sensory fibres (**sensory nerve**) or it may be mixed and carry both types (**mixed nerve**). Although the nerve fibres are in close proximity within the nerve, their physiological responses are independent of each other.

nerve cell *See* NEURON.

nerve-cell adhesion molecules *See* CELL ADHESION MOLECULE.

nerve cord A large bundle of nerve fibres, running down the longitudinal axis of the body, that forms an important part of the *central nervous system. Most invertebrates have a pair of solid nerve cords, situated ventrally and bearing segmentally arranged *ganglia. All animals of the phylum *Chordata have a dorsal hollow nerve cord; in vertebrates this is the *spinal cord.

nerve fibre The *axon of a *neuron together with the tissues associated with it (such as a *myelin sheath). The length and diameter of nerve fibres are very variable, even within the same organism. *See also* GIANT FIBRE.

nerve growth factor (NGF) *See* NEURO-TROPHIN.

nerve impulse *See* IMPULSE.

nerve net A network of nerve cells connected with each other by chemical or electrical synapses. Characteristically, stimulation at any point in the net spreads in all directions, although some have fast, through-conducting pathways. The nervous system of certain invertebrates (e.g. cnidarians and echinoderms) consists exclusively of several overlapping nerve nets in the body wall. These show differentiation and adaptation depending on the part of the body and functions they serve. Sensory and motor functions often reside in distinct nets. Some other invertebrates (e.g. flatworms) have nerve nets that connect to the central nerve cords.

nervous system The system of cells and tissues in multicellular animals by which information is conveyed between sensory cells and organs and effectors (such as muscles and glands). It consists of the *central nervous system (in vertebrates the *brain and *spinal cord; in invertebrates the *nerve cord and *ganglia) and the *peripheral nervous system. Its function is to receive, transmit, and interpret information and then to formulate appropriate responses for the effector organs. It also serves to coordinate responses that require more than one physiological process. Nervous tissue consists of *neurons, which convey the information in the form of *impulses, and supporting tissue. During development neurons compete for survival (*see* COMPETITION sense 2), given the limited availability of growth factors such as *neurotrophin. Superfluous neurons and synaptic connections are eliminated—a process called **pruning**—so that only neurons in the correct location and with the appropriate connections become established. In humans this process, which begins in the early embryo, continues into infancy and early childhood. *See also* GROWTH CONE; GUIDEPOST CELL.

(((•))) SEE WEB LINKS
- Basic interactive guide to the anatomy of the human nervous system, from Human Anatomy Online

net assimilation rate (NAR) The increase in plant dry mass per unit leaf area per unit time. Typically measured in kilograms per square metre per day or grams per square centimetre per day, it enables comparisons of the efficiency of production between plants of different sizes.

net primary productivity *See* PRODUCTIVITY.

netrin Any of several related proteins that act as long-range chemical cues to guide the growth of axons in vertebrates. Netrins are secreted and diffuse through the extracellular matrix, creating a concentration gradient that influences how the *growth cone of the developing axon navigates through the tissues. Specific netrin receptors on the growth cone bind the netrin molecules. In some cases the elongating axon grows towards the source of the netrin, whereas in other cases it is repelled by the netrins, depending on the type of nerve cell involved.

neural crest cells See NEURAL PLATE.

neural network See NEURONAL NETWORK.

neural plate A strip of ectoderm, above the notochord, that lies along the central axis of the early embryos (see GASTRULA) of chordates and develops into the central nervous system. Under the influence of *chordamesoderm, the neural plate folds and forms the *neural tube; this process is called **neurulation**. During the stage when the neural tube is forming the embryo is known as a **neurula**. Cells along each side of the neural plate become detached and lie between the neural tube and the overlying epidermis as **neural crest cells**. Subsequently they migrate away to give rise to a variety of cell types, including neurons and other cells of the peripheral nervous system, bones, and connective tissues of the head, and melanocytes in the skin.

neural tube A hollow tube of tissue in the early embryo of vertebrates that subsequently develops into the brain and spinal cord. It forms by folding of the ectodermal *neural plate and has a central canal running through it. Sometimes the folds of the neural plate fail to close properly, resulting in **neural tube defects** (such as spina bifida) in the fetus.

neuregulin See EPIDERMAL GROWTH FACTOR.

neurilemma cell See SCHWANN CELL.

neurite Any projection from the cell body of a *neuron, especially an immature neuron. As the neuron develops, neurites can become either dendrites or axons.

neuroendocrine system Any of the systems of dual control of certain activities in animals by nervous and hormonal stimulation. For example, the posterior *pituitary gland and the medulla of the *adrenal gland receive direct nervous stimulation to secrete their hormones, whereas the anterior pituitary gland is stimulated by *releasing hormones from the hypothalamus. Similarly, development, *metamorphosis, and *moulting in insects are regulated by a neuroendocrine mechanism.

neurofibril Any of the fibres in the cytoplasm of a nerve *axon. Neurofibrils include *neurofilaments and **neurotubules**, microtubules that play a role in the transport of proteins and other substances within the cytoplasm.

neurofilament A type of *intermediate filament found in the axons of nerve cells. Neurofilaments serve as elements of the *cytoskeleton supporting the axon cytoplasm.

neuroglia See GLIA.

neurohaemal organ A discrete body formed by a cluster of neurosecretory-cell terminals, where neurohormones are released into an adjacent blood space or capillary bed. Neurohaemal organs are found widely in both invertebrates and vertebrates. See NEUROSECRETION.

neurohormone Any hormone that is produced not by an endocrine gland but by a specialized nerve cell and is secreted from nerve endings into the bloodstream or directly to the tissue or organ whose growth or function it controls (see NEUROSECRETION). Examples of neurohormones are *noradrenaline, *antidiuretic hormone, and hormones associated with metamorphosis and moulting in insects (see ECDYSONE; JUVENILE HORMONE). Compare NEUROPEPTIDE.

neurohypophysis (pl. **neurohypophyses**) See PITUITARY GLAND.

neuromodulator 1. (in neurophysiology) A substance released by a nerve cell (neuron) that regulates the activity of other neurons but does not directly stimulate nerve impulses. Typically a neuromodulator is secreted into the intracellular space surrounding the target cells, rather than at synaptic junctions, and alters the excitability or other properties of a group of neurons. Neuromodulators thus work more slowly but have longer-lasting effects than *neurotransmitters. However, the distinction between the two can become blurred, especially as some substances can act as both neuromodulator and neurotransmitter, depending on circumstances. Neuromodulators in the brain fall into two main categories: neuropeptides, such as hypocretin, melanin-concentrating hormone, and cortistatin; and the monoamines, such as noradrenaline, dopamine, and serotonin; others include histamine and acetylcholine. The brain's hormonal neurons are located chiefly in the brainstem and central region, but project their axons into many other widespread ('diffuse') parts of the brain to release their neuromodulators. These play important roles in cognition, emotion, and behaviour. **2.** (in medicine) Any drug or device that influences the electrical activity of nerves, e.g. to treat pain or depression.

neuromuscular junction The point where a skeletal muscle fibre comes into contact with a motor neuron carrying nerve impulses from the central nervous system. The impulses travel from the neuron to the muscle fibre by means of a neurotransmitter, in a similar way to the transmission of impulses across a *synapse between two neurons. The neurotransmitter—acetylcholine (ACh) in vertebrates—is released from vesicles at the terminal knob of the motor neuron into a small gap (the cleft), where it diffuses to the *end plate of the muscle fibre. Here the ACh binds to receptors in the muscle cell membrane, causing the *ligand-gated ion channels controlled by the receptors to open. This allows a net influx of sodium ions (Na+) to enter the muscle cell, and some depolarization of the membrane occurs—an *excitatory postsynaptic potential. If sufficient ACh is released by nerve endings (*see* SUMMATION), the depolarization spreads and activates voltage-gated sodium channels in the end plate and elsewhere in the muscle cell membrane, which triggers an *action potential. This is propagated throughout the muscle fibre, causing it to contract. The receptor-bound ACh is subsequently broken down by the enzyme acetylcholinesterase into its components (acetate and choline), which re-enter the presynaptic nerve cell to be reused. During development or regeneration, the formation of a neuromuscular junction depends on signal molecules, such as *agrin, which recruit components required for the synaptic connections. A single axon can have numerous terminal branches, each of which innervates a single muscle fibre.

neuron (neurone; nerve cell) An elongated branched cell that is the fundamental unit of the *nervous system, being specialized for the conduction of *impulses.

Sensory neuron

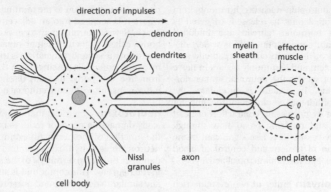

direction of impulses

dendron

dendrites

myelin sheath

effector muscle

Nissl granules

axon

end plates

cell body

Motor neuron

A neuron consists of a *cell body, containing the nucleus and *Nissl granules; *dendrites, which receive incoming impulses and pass them towards the cell body; and an *axon, which conducts impulses away from the cell body, sometimes over long distances. Impulses are passed from one neuron to the next via *synapses. *Sensory neurons transmit information from receptors to the central nervous system. *Motor neurons conduct information from the central nervous system to *effectors (e.g. muscles). *Interneurons form connections between neurons in the brain and ganglia, and play a key role in interpreting and analysing sensory information. (See illustration.) *See also* BIPOLAR NEURON; GROWTH CONE; MULTIPOLAR NEURON; NEURITE; UNIPOLAR NEURON.

neuronal network (neural network)
1. (in biology) A circuit of interconnected neurons. All types of behaviour depend on information, in the form of nerve impulses, being transmitted via synapses between individual neurons (nerve cells) through neuronal networks. Such networks range from the relatively simple connections of a reflex arc (*see* REFLEX), to the highly complex circuits involved in processing information in the brain. However, their activity depends primarily on the number of constituent neurons and the configuration of their connecting synapses (e.g. excitatory or inhibitory). Most networks underlying

complex behaviours include sensory and motor subcircuits that perform specific tasks, for example filtering sensory input or producing repetitive motor output (as in walking or swimming). Moreover, neuronal networks often exhibit plasticity, that is, they can be modified by experience. **2.** (in information processing) A paradigm inspired by the organization and capabilities of neuronal networks, which can detect patterns and trends and exhibit adaptive learning. *See* SYNAPTIC PLASTICITY.

neuropeptide Any of numerous peptides that influence the activity of neurons. Examples include the hypothalamic releasing hormones, antidiuretic hormone, and the gastric peptides (e.g. *VIP) released from cells in the duodenal wall. Neuropeptides may act as neurotransmitters (e.g. *substance P), as *cotransmitters to modify the action of neurotransmitters, as *neuromodulators, or as *neurohormones; some play several roles depending on their location.

neuropeptide Y (NPY) A 36-amino acid neuropeptide that is widespread throughout the animal kingdom and in various body systems and has key roles in energy metabolism and in regulating the activity of the heart and blood vessels. It is a potent stimulant of appetite, being released from cells in the hypothalamus, where it activates NPY receptors, which are *G-protein-coupled receptors. It also regulates secretion of gonadotrophin-releasing hormone from the hypothalamus. Its release is triggered by the gut hormone *ghrelin and inhibited by the hormone *leptin. NPY is widely distributed in the brain and sympathetic nervous system and commonly occurs in nerve cells that secrete the neurotransmitter noradrenaline. In general it modulates the effects of adrenaline and noradrenaline, for example by decreasing contraction of heart muscle and reducing blood flow through coronary vessels. It is also significant in the reduction of anxiety and control of blood pressure, and has antimicrobial activity.

neurophysin Either of two cysteine-rich carrier proteins that bind to *oxytocin (neurophysin I) and *antidiuretic hormone (neurophysin II) in the neurohypophysis of the pituitary and are secreted with them. They have no hormonal activity and are cleaved from their respective hormones following secretion.

neuropil A dense feltlike mass of microscopic threadlike axons, dendrites, and extensions of glial cells within the grey matter of the central nervous system (CNS) or within ganglia outside the CNS. The cell bodies of the neurons that give rise to the axons and dendrites may be dispersed within the neuropil or segregated into distinct zones around it. Numerous highly specific connections (synapses) occur between the branching dendrites and axons of the neuropil, producing the intricate wiring of the nervous system.

neurosecretion The secretion of *neurohormones by **neurosecretory cells**, which possess characteristics of both nerve cells and endocrine cells. They are found, for example, in the hypothalamus, where they receive nerve impulses from other parts of the brain but transmit these signals to the pituitary gland by neurohormones that are released into the blood. Like other neurons, neurosecretory cells consist of a cell body, from which extends a slender axon that ends in a terminal region. The cell bodies typically form a cluster, or nucleus, within the central nervous system. Secretory material is synthesized in the cell body and

passes along the axon to the terminus, from where it is secreted into an adjacent blood space. Several termini may form a distinct body, called a *neurohaemal organ. Discharge of the secretory materials is triggered by action potentials passing down the axon from the cell body, like the discharge of neurotransmitters by conventional neurons.

neurotoxin A chemical that either physically damages a nerve or reduces or alters the function of the nerve. An example of a neurotoxin is 6-hydroxydopamine, which damages the nerve terminals of the sympathetic neurons. This compound is structurally similar to dopamine and noradrenaline and can enter the nerve terminal by existing transport systems.

neurotransmitter (transmitter) A chemical that mediates the transmission of a nerve impulse across a *synapse or a *neuromuscular junction (i.e. excitation), or causes hyperpolarization of a postsynaptic cell making an action potential less likely (i.e. inhibition). Examples are *adrenaline, *noradrenaline, *dopamine, and *serotonin (in adrenergic nerves), *acetylcholine (in cholinergic nerves), *glutamate, *gamma-aminobutyric acid, *substance P, and *nitric oxide. The neurotransmitter is released at the synaptic knob at the tip of the axon into the synaptic cleft. It diffuses across to the opposite membrane (the postsynaptic membrane), where it stimulates receptors and initiates the excitation or inhibition of the postsynaptic cell. At a neuromuscular junction, the neurotransmitter transmits impulses to the muscle-fibre membrane. A single neurotransmitter may exert either an excitatory or an inhibitory action, depending on which type of receptor occurs on the postsynaptic membrane. *See also* COTRANSMITTER.

(((⊕))) SEE WEB LINKS

• Summary of the main neurotransmitters, their biochemistry, and mode of action, from the Medical Biochemistry pages of Indiana University

neurotrophin (NT) Any of several growth factors that promote the development, maintenance, and repair of neurons. Each is a dimer of two identical polypeptides held together by disulphide bonds. The first to be discovered, nerve growth factor (NGF), is essential for the normal growth and survival of neurons of the sympathetic nervous system and also sensory neurons. It is produced by the neurons themselves and by astrocytes, Schwann cells, fibroblasts, and certain other cells. Some cells in the central nervous system (CNS), such as cholinergic neurons in the basal forebrain that supply the hippocampus, are sensitive to NGF, which is also required for the synthesis of neurotransmitter in adult brain cells. Other neurotrophins that are widely distributed in the CNS include brain-derived neurotrophic factor (BDNF) and neurotrophin 3 (NT-3). Both play important roles during development and survival of CNS neurons; they can also influence the activity of brain cells by altering the concentration of ion channels in neurons, which in turn affects synaptic transmission and neuronal excitability. Other neurotrophins are NT-4/5 and NT-6. All neurotrophins exert their effects by binding to *neurotrophin receptors on the surface of target neurons.

neurotrophin receptor A protein on the surface of a cell that interacts with a *neurotrophin molecule outside the cell. There are two types: low-affinity neurotrophin receptors, which are found on both neurons and other cell types; and high-affinity neurotrophin receptors, which are restricted to neurons. The latter are responsible for triggering the characteristic effects of neurotrophins on their target cells; designated as NTRK (neurotrophin tyrosine kinase receptor, or **TRK**), there are three types, NTRK1–3, each of which is the receptor for one or more neurotrophins. For example, NTRK1 is the receptor for nerve growth factor (NGF) and neurotrophin 6 (NT-6). Each neurotrophin binds to a pair of receptors, causing the formation of a receptor dimer, which leads to self-phosphorylation of tyrosine amino acids on the intracellular part of the receptor and activation of intracellular signalling pathways. The NGF-NTRK1 complex is

taken inside the cell and transported along the axon to the cell body of the neuron, where it regulates expression of genes in the cell nucleus. The role of low-affinity receptors is less clear. They may simply bind neurotrophins to provide a local store of these molecules, or they might interact with high-affinity receptors in mediating the effects of neurotrophins. In some cases, binding of neurotrophin to low-affinity receptors can trigger cell death (apoptosis).

neurotubule *See* NEUROFIBRIL.

neurula (*pl.* **neurulae**) *See* NEURAL PLATE.

neuston The organisms that colonize the surface of an aquatic habitat. They are most abundant in freshwater habitats and include pond skaters, certain types of water beetles, and floating plants. *Compare* BENTHOS; NEKTON; PLANKTON.

neuter An organism that does not possess either male or female reproductive organs. Cultivated ornamental flowers that have neither pistils nor stamens are called neuters.

neutral Describing a compound or solution that is neither acidic nor basic. A neutral solution is one that contains equal numbers of both protonated and deprotonated forms of the solvent.

neutral theory of molecular evolution The theory, originally proposed in the late 1960s by Motoo Kimura (1924–94) and others, that most evolutionary changes at the molecular level are due to the random process of *genetic drift acting on mutations, rather than *natural selection. Its proponents, while recognizing the importance of selection in determining functionally significant traits, hold that the great majority of the differences in macromolecular structures observed between individuals in a population are of no adaptive significance and have no impact on the reproductive success of the individual in which they arise. Hence, frequencies of the corresponding mutant alleles are governed by purely random events. This contrasts with the orthodox neo-Darwinian view that nearly all evolutionary changes have adaptive value for the organism and arise through natural selection. For example, many enzymes exhibit *polymorphism with regard to their amino acid sequence, giving rise to morphological variants that are detectable by electrophoresis. However, these variants may apparently perform equally well, and 'neutralists' would argue that evolution consists essentially of random shuffling between them. 'Selectionists' retort that such variants are likely to have subtle differences in function and are susceptible to selective forces, such as minor environmental changes. However, the accumulated sequence data for genomes and proteins largely support the neutral theory. The observed differences in base sequence between species are generally much greater for introns (noncoding sequences) and pseudogenes (nonfunctional or 'dead' genes), and for synonymous base substitutions (i.e. ones that do not change the amino acid in the resulting protein) than they are at sites that are critical for gene function or for nonsynonymous substitutions (i.e. ones that alter the corresponding amino acid and thereby affect the properties of the resulting protein). Because neutral mutations arise on average at a fairly constant rate, they are often used as the basis of a *molecular clock to estimate the time elapsed since the divergence of species.

neutrophil A type of white blood cell (*leucocyte) that has a lobed nucleus and granular cytoplasm (*see* GRANULOCYTE). Neutrophils circulate in the blood, where they engulf bacteria (*see* PHAGOCYTOSIS) and release various substances, such as *lysozyme and oxidizing agents They are attracted to sites of inflammation by cytokines released from injured tissue cells.

newton Symbol N. The *SI unit of force, being the force required to give a mass of one kilogram an acceleration of 1 m s^{-2}. It is named after Sir Isaac Newton (1642–1727).

next-generation sequencing (NGS) *See* DNA SEQUENCING.

nexus *See* GAP JUNCTION.

NFκB *See* NUCLEAR FACTOR κB.

NGF Nerve growth factor. *See* NEUROTROPHIN.

NGS (next-generation sequencing) *See* DNA SEQUENCING.

niacin *See* NICOTINIC ACID.

niche *See* ECOLOGICAL NICHE.

nicotinamide *See* NICOTINIC ACID.

nicotinamide adenine dinucleotide *See* NAD.

nicotinamide adenine dinucleotide phosphate (NADP) *See* NAD.

nicotine A colourless poisonous *alkaloid present in tobacco. It exerts its effect by binding to a class of acetylcholine receptors and stimulates the release of dopamine by neurons in the brain, thus stimulating the reward system and leading to addiction. It can be used as an insecticide. *See also* NEONICOTINOID.

nicotinic Describing one of the two main classes of *acetylcholine receptors, so called because the effect of acetylcholine on them can be mimicked by *nicotine. Nicotinic receptors are found at neuromuscular junctions in skeletal muscle and at postganglionic cells in ganglia of both divisions of the vertebrate autonomic nervous system. They are fast-acting receptors containing their own intrinsic *ion channel, which is directly activated by the receptor agonists. *Compare* MUSCARINIC.

nicotinic acid (niacin) A vitamin of the *vitamin B complex. It can be manufactured by plants and animals from the amino acid tryptophan. The amide derivative, **nicotinamide**, is a component of the coenzymes *NAD and NADP. These take part in many metabolic reactions as hydrogen acceptors. Deficiency of nicotinic acid causes the disease *pellagra in humans. Apart from tryptophan-rich protein, good sources are liver and groundnut and sunflower meals.

nictitating membrane A clear membrane forming a third eyelid in amphibians, reptiles, birds, and some mammals (but not humans). It can be drawn across the cornea independently of the other eyelids, thus clearing the eye surface and giving added protection without interrupting the continuity of vision.

nidation *See* IMPLANTATION.

nidicolous *See* ALTRICIAL SPECIES.

nidifugous *See* PRECOCIAL SPECIES.

ninhydrin A compound that reacts with amino acids to give a blue colour. Ninhydrin is commonly used in chromatography to analyse the amino-acid content of proteins.

nipple *See* MAMMARY GLANDS.

Nissl granules (Nissl bodies) Particles seen within the cell bodies of *neurons, consisting of rough endoplasmic reticulum and *polyribosomes. They are rich in RNA and stain strongly with basic dyes. They are named after F. Nissl (1860–1919), the German neurologist who discovered them.

nitrate A salt or ester of nitric acid. The salts contain the ion NO_3^-.

nitric oxide (nitrogen monoxide) A colourless gas, NO, that functions as a gaseous *second messenger in mammals and other vertebrates, especially in the cardiovascular and nervous systems, and as a neurotransmitter. It is produced in tissues from molecular oxygen and the amino acid arginine by the enzyme nitric oxide synthase and diffuses to neighbouring cells, where it stimulates formation of the intracellular messenger cyclic GMP. The activity of NO synthase, and hence production of NO, is regulated by a signal pathway involving binding of the neurotransmitter acetylcholine to its receptor on the cell surface, the production of the second messenger IP_3 (*see* INOSITOL), and release of calcium ions from the endoplasmic reticulum. The effects of nitric oxide include relaxation of smooth muscle and dilation of blood vessels. For example, NO induces the local

smooth muscle relaxation and vasodilation that enhance blood flow into the erectile tissues of the penis during sexual arousal. It also inhibits platelet aggregation and adhesion, acts as a neurotransmitter in the central nervous system, and may influence neuronal development. Certain cells of the immune system (macrophages and neutrophils) also produce nitric oxide, which is converted to the cytotoxic peroxynitrite anion ($^-$O-O-N=O). This is nonspecific activity against pathogens, including protist and metazoan parasites. (*see* INDUCIBLE NITRIC OXIDE SYNTHASE). Plants also produce NO as part of their immune response to attack by pathogens.

nitrification A chemical process in which nitrogen (mostly in the form of ammonia) in plant and animal wastes and dead remains is oxidized at first to nitrites and then to nitrates. These reactions are effected mainly by nitrifying archaea and bacteria, such as *Nitrosomonas* and *Nitrobacter* respectively. Unlike ammonia, nitrates are readily taken up by plant roots; nitrification is therefore a crucial part of the *nitrogen cycle. Nitrogen-containing compounds are often applied to soils deficient in this element, as fertilizer. *Compare* DENITRIFICATION.

nitrifying bacteria *See* NITRIFICATION.

nitrite A salt or ester of nitrous acid. The salts contain the ion NO_2^-.

nitrogen Symbol N. A colourless gaseous element that occurs in air (as dinitrogen, N_2, about 78% by volume) and is an essential constituent of proteins and nucleic acids in living organisms (*see* NITROGEN CYCLE).

nitrogenase An important enzyme complex that is present in those microorganisms that are capable of fixing atmospheric nitrogen (*see* NITROGEN FIXATION). Nitrogenase catalyses the conversion of atmospheric nitrogen into ammonia, which can then be used to synthesize nitrites, nitrates, or amino acids. The two main enzymes within the nitrogenase complex are **dinitrogenase reductase** and **dinitrogenase**.

nitrogen cycle One of the major cycles of chemical elements in the environment (*see* BIOGEOCHEMICAL CYCLE). Nitrates in the soil are taken up by plant roots and may then pass along *food chains into animals. Decomposing bacteria and archaea convert nitrogen-containing compounds (especially ammonia) in plant and animal wastes and dead remains back into nitrates, which are released into the soil and can again be taken up by plants (*see* NITRIFICATION). Though nitrogen is essential to all forms of life, the huge amount present in the atmosphere is not directly available to most organisms (*compare* CARBON CYCLE). It can, however, be assimilated by some specialized bacteria (*see* NITROGEN FIXATION) and is thus made available to other organisms indirectly. Lightning flashes also make some nitrogen available to plants by causing the combination of atmospheric nitrogen and oxygen to form oxides of nitrogen, which enter the soil and form nitrates. Some nitrogen is returned from the soil to the atmosphere by denitrifying bacteria (*see* DENITRIFICATION). See illustration.

nitrogen fixation A chemical process in which atmospheric nitrogen is assimilated into organic compounds in living organisms and hence into the *nitrogen cycle. The ability to fix nitrogen, by means of *nitrogenase enzymes, is limited to certain bacteria (e.g. *Azotobacter*, *Anabaena*), and, in some anaerobic environments, such as swamps, to methane-producing archaea (*see* METHANOGEN). Some bacteria (e.g. *Rhizobium*, *Bradyrhizobium*) are able to

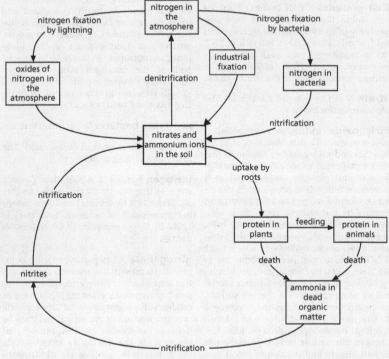

The nitrogen cycle

fix nitrogen in association with cells in the roots of leguminous plants, such as peas and beans, in which they form characteristic *root nodules (see BACTEROID); cultivation of legumes is therefore one way of increasing soil nitrogen. Certain nonleguminous plants are also hosts to nitrogen-fixing bacteria. For example, alder trees develop root nodules containing *Frankia*, a streptomycete-like organism. Various chemical processes are used to fix atmospheric nitrogen in the manufacture of *fertilizers. These include the Birkeland-Eyde process, the cyanamide process, and the Haber process.

nitrogenous base A basic compound containing nitrogen. The term is used especially of organic ring compounds, such as adenine, guanine, cytosine, and thymine, which are constituents of nucleic acids.

nitrogenous waste Any metabolic *waste product that contains nitrogen. *Urea and *uric acid are the most common nitrogenous waste products in terrestrial animals; freshwater fish excrete ammonia and marine fish excrete both urea and *trimethylamine oxide.

nitrogen oxides Oxides of nitrogen (NO_x), such as *nitric oxide (NO), dinitrogen oxide (N_2O), and nitrogen dioxide (NO_2), which are pollutants contributing to *acid rain. Nitrogen oxides are expelled in the emissions from car exhausts, aircraft, and factories. Nitric oxide is biologically important, with various roles in the body. High concentrations of nitrogen dioxide are irritating to the respiratory tract and can aggravate conditions such as asthma, while NO_x in general contribute to the formation of ground-level ozone. *See also* AIR POLLUTION.

nitrosamines A group of carcinogenic compounds with the general formula RR′NNO, where R and R′ are side groups with a variety of possible structures. Nitrosamines are a component of cigarette smoke and can be formed in the acidic environment of the stomach by the combination of nitrites (used as a food preservative) and amines; they also occur as contaminants of certain foods. Nitrosamines cause cancer in a number of organs, particularly in the liver, kidneys, and lungs. An example of a nitrosamine is N-nitrosodimethylamine, which has two methyl side groups (CH_3-).

NK cell *See* NATURAL KILLER CELL.

NMDA receptors *See* GLUTAMATE RECEPTOR.

NMR *See* NUCLEAR MAGNETIC RESONANCE.

nociception The detection by the body of injury or noxious stimuli, generally perceived as pain. The skin is richly supplied with nerve endings that respond to excessive heat or cold, tissue damage, or harmful chemicals and thus signal the body to take defensive action. Similar nerve endings serve muscles, joints, bones, and visceral organs. There are two principal types of nociceptive nerve fibres. **Aδ fibres** are medium-velocity myelinated fibres that are responsible for sensations of sharp stabbing pain caused by mechanical stimulation (e.g. pinpricks) and thermal stimuli (e.g. a hot stove). The cell bodies of such fibres lie in the dorsal ganglia, near the spinal cord, and the nerve impulses signalling pain are transmitted via relay neurons along the spinal cord to the brain. **C fibres** are similarly arranged but are slower unmyelinated fibres; they respond to chemical stimuli as well as thermal and mechanical stimuli, typically causing more prolonged aching or throbbing pain. Repeated stimulation of both Aδ and C fibres can lead to sensitization, in which the stimulus threshold for firing of the neurons is lowered, leading to increased pain. Also, tissue damage and inflammation cause the release of various substances, such as prostaglandins, serotonin, and bradykinin, that increase the sensitivity of nociceptive nerve fibres and thereby exacerbate the sensation of pain. Within the dorsal horn of the spinal cord the sensory 'pain' fibres relay their signals to other nerve fibres that ascend to the thalamus in the brain, from where the pain signals are transmitted widely to other brain areas, including the cortex, where a perception of localized pain is formed. The flow of nociceptive signals and hence consciousness of pain can be modified by signals from other sensory neurons. For example, stimulation of touch receptors in the skin by stroking can suppress the sensation of pain in nearby nociceptors, by activating inhibitory interneurons within the spinal cord. Mood or emotion can also affect the ability to feel pain. This is thought to involve descending neural pathways, originating in the midbrain, that end near the terminals of the nociceptive fibres in the dorsal horn of the spinal cord. Drugs such as aspirin and ibuprofen reduce prostaglandin synthesis and hence lower susceptibility to pain.

node 1. (in botany) The part of a plant stem from which one or more leaves arise. The nodes at the stem apex are very close together and remain so in species of monocotyledons that form bulbs. In older regions of the stem they are separated by areas of stem called **internodes. 2.** (in anatomy) A natural thickening, bulge, or distinct mass of tissue in an organ or part of the body. Examples are the **sinoatrial node** that controls the heartbeat (*see* PACEMAKER) and the *lymph nodes. **3.** (in phylogenetics) (**vertex**) A branch point in a *phylogram.

node of Ranvier *See* MYELIN SHEATH.

Nod factor *See* ROOT NODULE.

nodule 1. (in anatomy) A small node, such as a solid skin lesion up to 1 cm deep. **2.** (in botany) *See* ROOT NODULE.

nomad (in cytology) A cell that migrates or wanders from its site of formation. Certain types of *phagocytes are nomads.

Nomarski microscope (differential-interference contrast microscope) A type of light microscope that is useful for viewing live transparent unstained specimens, such as cells or microscopic organisms. The shadow-cast images give the illusion of depth to the outlines and surface features of organelles or other structures. An incident beam of plane-polarized light is split into parallel beams by a prism so that different parts of the beam pass through closely adjacent areas of the specimen. Slight differences in thickness and refractive index within the specimen cause interference between the beams as they exit the specimen and are recombined by a second prism: parts of the beam that are in phase will reinforce each other and produce a bright image, whereas parts that are out of phase will cancel each other out and produce a dark image. It is named after Polish-born physicist Georges Nomarski (1919-97). *Compare* PHASE-CONTRAST MICROSCOPE.

((())) SEE WEB LINKS
- Overview of Nomarski microscopy with interactive tutorials, from Olympus Microscopy Resource Center

noncoding DNA *See* JUNK DNA.

noncoding RNA (ncRNA) Any functional *RNA molecule that is not translated into protein. Noncoding RNAs include ribosomal RNA (rRNA); transfer RNA (tRNA); various forms of small ncRNA that are involved in *RNA interference, such as short interfering RNA (siRNA), micro RNA (miRNA), and *piwi-interacting RNA; *small nuclear RNA (snRNA) and *small nucleolar RNA (snoRNA); *long noncoding RNA; and *transfer-messenger RNA.

noncompetitive inhibition *See* INHIBITION.

noncyclic phosphorylation (noncyclic photophosphorylation) *See* PHOTOPHOSPHORYLATION.

nondisjunction The phenomenon that occurs when a pair of *homologous chromosomes do not separate in *meiosis but migrate to the same pole of the cell, resulting in an uneven number of chromosomes being present in the daughter cells. *Klinefelter's syndrome results from nondisjunction of the sex chromosomes.

nonreducing sugar A sugar that cannot donate electrons to other molecules and therefore cannot act as a reducing agent. Sucrose is the most common nonreducing sugar. The linkage between the glucose and fructose units in sucrose, which involves aldehyde and ketone groups, is responsible for the inability of sucrose to act as a *reducing sugar.

nonrenewable energy sources Sources of energy that use up the earth's finite mineral resources; these include *fossil fuels. Concern about the exhaustion of nonrenewable energy sources, together with the fact that burning fossil fuels contributes to *air pollution, carbon emissions, and the *greenhouse effect, has prompted increased use of **renewable energy resources**, which are not exhaustible. All depend directly or indirectly on the energy supplied by the sun, whether solar heating and photovoltaic cells, wind turbines, biofuels, or hydroelectric generators.

nonsense mutation A type of *point mutation that converts a codon normally specifying an amino acid to one of the *stop codons, thus signalling termination of translation and causing synthesis of the polypeptide chain to cease prematurely.

noradrenaline (norepinephrine) A hormone that is produced by the medulla of the *adrenal glands and also acts as a *neurotransmitter in the *sympathetic nervous system. Many of its general actions are similar to those of *adrenaline, but it is more concerned with maintaining normal body activity than with preparing the body for emergencies. *See* ADRENOCEPTOR.

norepinephrine *See* NORADRENALINE.

normalizing selection *See* STABILIZING SELECTION.

northern blotting *See* SOUTHERN BLOTTING.

nose The protuberance on the face of some vertebrates that contains the nostrils (*see* NARES) and part of the *nasal cavity. It therefore forms part of the olfactory system (*see* OLFACTION) and the external opening of the respiratory system.

nostrils *See* NARES.

notochord An elastic skeletal rod lying lengthwise beneath the nerve cord and above the alimentary canal in the embryos or adults of all chordate animals (*see* CHORDATA). Its function is to strengthen and support the body and act as a protagonist for the muscles. It is found in both adult and larval lancelets but in adult vertebrates it is largely replaced by the *vertebral column.

NPY *See* NEUROPEPTIDE Y.

nt *See* NUCLEOTIDE.

nucellus (*pl.* **nucelli**) The tissue that makes up the greater part of the ovule of seed plants. It contains the *embryo sac and nutritive tissue. It is enclosed by the integuments except for a small gap, the *micropyle. In certain flowering plants it may persist after fertilization and provide nutrients for the embryo.

nuclear-cytoplasmic ratio A measure of the size of a cell nucleus in relation to the cytoplasm. The nuclear-cytoplasmic ratio is often used as an index in the comparison of cells from normal and abnormal tissues. For example, cultured cancer cells show an increase in the nuclear-cytoplasmic ratio.

nuclear envelope The double membrane that separates the nucleoplasm (*see* NUCLEUS) of a cell from the cytoplasm. The membranes consist of *lipid bilayers that are separated by a **perinuclear space** (or **compartment**) some 20–40 nm wide. The outer membrane is continuous with the rough *endoplasmic reticulum and is structurally and functionally distinct from the inner membrane. The envelope is perforated at intervals by **nuclear pores**, typically about 100 nm in diameter, which provide a channel for the selective transfer of water-soluble molecules between the nucleus and the cytoplasm. Each nuclear pore is surrounded by a disc-shaped structure (**nuclear pore complex**) consisting of an octagonal arrangement of eight protein granules. Covering the internal surface of the envelope is the **nuclear lamina**, a mesh of intermediate filaments formed by proteins called *lamins. The lamina attaches to both the nuclear envelope and to the filamentous protein network that forms the **nuclear matrix** inside the nucleus; both the lamina and matrix maintain the shape of the nucleus.

nuclear factor κB (NFκB) A transcription factor involved in rapid cellular responses to stresses, such as injury or invading pathogens, and in many aspects of cell proliferation and survival, including programmed cell death (*apoptosis). It is found in the cells of organisms ranging from flies to mammals and has a key role in such disorders as systemic shock, inflammation, and cancer. The name derives from the finding that B cells require this factor in order to transcribe the κ light chain gene (hence *n*uclear *f*actor for κ light chain transcription in *B* cells). Many different extracellular signals can lead to activation of NFκB, notably signal pathways involving *Toll-like receptors, which mediate early immune responses to pathogens. NFκB is a heterodimer of two protein subunits, both of which have DNA-binding domains. In its inactive state in the cytoplasm it is bound by an inhibitor called IκB. In response to a signal the latter is phosphorylated, marking it for degradation and enabling the freed NFκB to migrate to the cell nucleus, where it 'switches on' the appropriate genes to produce the cellular response. Faulty control of NFκB signalling is a crucial factor in certain human tumours, because genes promoting cell proliferation remain switched on.

nucleariid Any of a group of unicellular protists that are the closest living relatives of fungi. Nucleariids are amoeboid with threadlike pseudopodia used for sensing and consuming prey, such as algae and cyanobacteria. Molecular evidence places them along with the fungi, animals, and certain other protists in the *opisthokonts.

nuclear magnetic resonance (NMR) The absorption of electromagnetic radiation (radio waves) by certain atomic nuclei placed in a strong and stable magnetic field. This results in a change of orientation of the nuclei, which respond to the magnetic field like miniature bar magnets. The main application of NMR is in a form of spectroscopy (**NMR spectroscopy**) used for chemical and biochemical analysis and structure determination. There are two methods of NMR spectroscopy. In continuous wave (CW) NMR, the sample is subjected to a strong magnetic field, which can be varied in a controlled way. As the field changes, absorption of radiation occurs at certain points; this produces oscillations in the field, which can be detected. Fourier transform (FT) NMR uses a fixed magnetic field and the sample is subjected to a high-intensity pulse of radiation covering a range of frequencies. The signal produced is analysed mathematically to give the NMR spectrum. The ^{1}H nucleus is the one commonly studied; other biochemically useful nuclei are ^{31}P, ^{13}C, ^{14}N, and ^{19}F, although these have lower natural abundance than hydrogen and produce weaker signals. The spectrum produced is characteristic of the molecule absorbing the radiation. NMR has been exploited as the basis of *magnetic resonance imaging.

nuclear pore *See* NUCLEAR ENVELOPE.

nuclear transfer (nuclear transplantation) A technique used in cloning animals in which a nucleus from a donor cell is inserted into an egg cell, which is then stimulated to develop as an embryo. The technique has been used successfully with various mammal species, most famously producing Dolly the sheep in 1997 (*see* CLONE). Previous to Dolly's birth, nuclear transfer had used cultured embryo cells in a relatively undifferentiated state. A single embryo cell is injected into an unfertilized egg cell, from which the chromosomes have been removed by micropipette. Fused recipient and donor cells are then stimulated by electrical pulses to begin dividing and form an embryo, like a normal fertilized egg cell. The embryo is then implanted into the uterus of a surrogate mother to continue its development. Dolly was the first mammal to be cloned from a fully differentiated adult body cell—**somatic cell nuclear transfer**. She demonstrated that it is possible to 'reprogram' such cells so that they can direct the development of a new individual. In Dolly's case, the donor cells were taken from a culture of sheep udder cells and starved into a state of quiescence in a low-nutrient medium. This was done to switch off all but essential genes and better mimic a natural fertilization.

There are several advantages in using adult body cells: cultures are easier to obtain and maintain; also, there is greater scope for genetically engineering such cells and screening them to select successfully modified cells. Nuclear transfer, using embryo cells or body cells, is now used increasingly to replicate elite animals in the livestock industry, to produce genetically engineered mammals for commercial use (e.g. goats that secrete human proteins in their milk), and to replicate endangered species. However, the failure rate is generally high, and even the few live clones produced often have congenital defects that shorten their lives. This shows that 'reprogramming' differentiated body cells poses formidable technical obstacles, particularly changing the pattern of epigenetic modifications of DNA so that the chromatin more closely resembles that of embryonic chromatin.

nuclease Any enzyme that breaks down nucleic acids to nucleotides. Nucleases are found in the small intestine. *See* DNASE; ENDONUCLEASE; EXONUCLEASE.

nucleic acid A complex organic compound in living cells that consists of a chain of *nucleotides. There are two types: *DNA (deoxyribonucleic acid) and *RNA (ribonucleic acid).

nucleic acid amplification test (NAAT) Any of various techniques that are used to detect specific target sequences of nucleic acid (RNA or DNA) in a sample. Such techniques are widely used in research, clinical medicine, and industry, e.g. to detect or assay the presence of infective organisms or viruses in samples of blood, tissue, food, or other material. Many such techniques utilize variants of the *polymerase chain reaction (PCR) to isolate and amplify the target sequence. Simpler and cheaper alternatives have

been devised without the heating and cooling cycles of PCR, to make nucleic acid testing in the field or clinic more feasible.

nucleic acid hybridization The association, *in vitro*, of two complementary nucleic-acid strands to form a hybrid double strand. *See* DNA HYBRIDIZATION.

nucleocytoplasmic large DNA virus (NCLDV; giant virus) Any of a diverse group of eukaryotic DNA viruses that have unusually large capsids, comparable in size with small bacteria, and contain large genomes, ranging in size from 100 kbp to 2.5 Mbp. Among the largest are *Mimivirus* and the related *Megavirus*, discovered in 2011, which measure 600 and 640 nm in diameter respectively and have a genome size of roughly 1.2 Mbp. However, these are outstripped by the *pandoraviruses, which are roughly 1000 nm long, and *Pithovirus*, which at 1500 nm long is the largest known virus. All four infect amoebas. Other NCLDVs include the phycodnaviruses (which infect algae), poxviruses, African swine fever virus, and iridoviruses, all of which infect metazoan animals. All are equipped with most of the genes needed for replication of their DNA and virion assembly: they either replicate in the cytoplasm of their host cells or pass to the cytoplasm after beginning their replication in the nucleus. Some regard their large genomes as evidence that NCLDVs have undergone reductive evolution from more complex cellular predecessors, and may represent ancestors of cellular life forms, whereas an alternative explanation is that they have acquired many of their genes by accretion from their hosts. In 2013 a new viral order, the *Megavirales*, was proposed to accommodate the NCLDVs on the basis of their shared genes, virion structure, and mode of replication. However, the relationship of NCLDVs to the other domains of life is contentious; for example, it has been proposed that the giant amoebal viruses are true microbes and represent a new domain of life.

nucleoid (nuclear region) The part of a cell of a bacterium (i.e. a prokaryotic *cell) that contains the genetic material *DNA and therefore controls the activity of the cell. It corresponds to the nucleus of the more advanced eukaryotic cells but is not bounded by a membrane.

nucleolar organizer A segment of a chromosome of a eukaryotic cell containing genes that encode ribosomal RNA. In the nondividing cell it is associated with the *nucleolus in the assembly of ribosomes. The nucleolar organizer consists of numerous tandem repeats of a single gene (*see* TANDEM ARRAY); each repeat is transcribed simultaneously to form multiple copies of a precursor RNA molecule containing all the larger ribosomal RNA subunits.

nucleolus (*pl.* **nucleoli**) A small dense round body within the nucleus of a nondividing eukaryotic cell that is the site of ribosome assembly. It forms around the *nucleolar organizer, which encodes most of the segments of ribosomal RNA. Ribosomal proteins migrate to the nucleolus from their assembly sites in the cytoplasm and are packaged into *ribonucleoproteins, which then return to the cytoplasm where they become mature ribosome particles. Two or more nucleoli may be present, depending on the species of organism and stage of the cell cycle.

nucleomorph A residual nucleus found in the plastids of certain single-celled algae. Nucleomorphs are the much-reduced nuclei of algal cells that were originally ingested by heterotrophic eukaryote cells as food; the algae retained their own plastid and the ability to photosynthesize inside the host cell. The ingested algae are thus secondary endosymbionts (*see* ENDOSYMBIONT THEORY). Nucleomorphs occur in chlorarachniophyte algae, whose plastids are derived from endosymbiotic green algae, and in cryptophytes, whose plastids are from red algae. The nucleomorph is located in the space between the second and third plastid membranes, corresponding to the cytosol of the engulfed endosymbiont. It typically contains some 300–500 genes, which are transcribed.

nucleoplasm (karyoplasm) The material contained within the *nucleus of a cell. The nucleoplasm is bound by the *nuclear envelope, which separates it from the cytoplasm.

nucleoprotein Any compound present in cells of living organisms that consists of a nucleic acid (DNA or RNA) combined with a protein. Chromosomes consist of nucleoprotein (DNA and proteins, mostly histones), as do ribosomes (RNA and protein). *See also* RIBONUCLEOPROTEIN.

nucleoside An organic compound consisting of a nitrogen-containing *purine or *pyrimidine base linked to a sugar (ribose or deoxyribose). An example is *adenosine. *Compare* NUCLEOTIDE.

nucleosome The fundamental unit of *chromatin, the material of which eukaryotic chromosomes are made. It consists of a core of *histone proteins around which are coiled about 160 base pairs of DNA. Consecutive nucleosomes, each 10 nm in diameter, are joined by uncoiled linker DNA, like beads on a string. This string is coiled into a 30 nm diameter solenoid, which undergoes further coiling in the fully condensed chromosome.

nucleotidase An enzyme that catalyses the breakdown of nucleotides. It is present in the epithelial cells of the small intestine and plays an important role in the digestion of proteins.

nucleotide Symbol nt. An organic compound consisting of a nitrogen-containing *purine or *pyrimidine base linked to a sugar (ribose or deoxyribose) and a phosphate group. *DNA and *RNA are made up of long chains of nucleotides (i.e. **polynucleotides**). *Compare* NUCLEOSIDE.

nucleus (*pl.* **nuclei**) 1. (in cytology) The large body embedded in the cytoplasm of all eukaryote *cells that contains the genetic material *DNA organized into *chromosomes. The nucleus functions as the control centre of the cell and is bounded by a double membrane (the *nuclear envelope). When the cell is not dividing, one or more *nucleoli are present in the nucleus and the chromosomal material (*chromatin) is dispersed in the nucleoplasm. In dividing cells the chromosomes become much shorter and thicker and the nucleolus disappears. The contents of the nucleus constitute the nucleoplasm. In certain protists there are two nuclei per cell, a **macronucleus** (or **meganucleus**) concerned with vegetative functions and a **micronucleus** involved in sexual reproduction. **2.** (in neuroanatomy) A cluster of nerve cell bodies in the central nervous system or in a nerve ganglion, usually having a shared function or anatomical features. **3.** (in physics) The core of an atom that contains most of its mass.

null hypothesis See SIGNIFICANCE. See also ERROR.

numerical taxonomy See TAXONOMY.

nut A dry single-seeded fruit that develops from more than one carpel and does not shed its seed when ripe. The fruit wall is woody or leathery. Many nuts are enclosed in a hard or membranous cup-shaped structure, the *cupule. The term nut is often loosely used of any hard fruit. For example, the walnut and coconut are in fact *drupes and the Brazil nut is a seed.

nutation The spiral movement of a plant organ during growth, also known as **circumnutation**. It is seen in climbing plants and helps the plant find a suitable support to twine around. Examples are the coiling movements of the shoot tips of runner beans and of the tendrils of sweet peas.

nutrient Any substance that is required for the nourishment of an organism, providing a source of energy or structural components. In animals nutrients form

part of the *diet and include the **major nutrients**, i.e. carbohydrates, proteins (see also ESSENTIAL AMINO ACID), and lipids (see also ESSENTIAL FATTY ACIDS), as well as vitamins and certain minerals (see ESSENTIAL ELEMENT). Plant nutrients, derived from carbon dioxide in the atmosphere and water (containing minerals) absorbed from the soil by the roots, are divided into *macronutrients and *micronutrients.

nutrition The process by which organisms obtain energy (in the form of food) for growth, maintenance, and repair. There are two main types of nutrition: *heterotrophic nutrition, employed by animals, fungi, and certain bacteria and archaea; and *autotrophic nutrition, found in most plants, bacteria, and archaea. Much less common are *mixotrophic organisms, which combine heterotrophic and autotrophic nutrition.

nyctinasty (sleep movements) *Nastic movements of plant organs in response to the changes in light and temperature that occur between day and night (and vice versa). Examples are the opening and closing of many flowers and the folding together of the leaflets of clover and other plants at night.

nymph The juvenile stage of *exopterygote insects, especially terrestrial species, such as grasshoppers, cockroaches, and earwigs; the juveniles of aquatic species (e.g. dragonflies, mayflies, stoneflies) are sometimes called **naiads** or **larvae**. The nymph resembles the adult except that the wings and reproductive organs are undeveloped. There is no pupal stage, the nymph developing directly into the adult. Compare LARVA.

objective The lens or system of lenses nearest to the object being examined through an optical instrument.

obligate anaerobes Organisms that are poisoned by the presence of free oxygen and obtain energy exclusively by either *fermentation or *anaerobic respiration.

occipital condyle A single or paired bony knob that protrudes from the occipital bone of the skull and articulates with the first cervical vertebra (the *atlas). In humans there is a pair of occipital condyles, one on each side of the *foramen magnum. Occipital condyles are absent in most fish, which cannot move their heads.

oceanic zone The region of the open sea beyond the edge of the continental shelf, where the depth is greater than 200 metres. *Compare* NERITIC ZONE.

ocellus (*pl.* **ocelli**) A simple eye occurring in insects and other invertebrates. It typically consists of light-sensitive cells and a single cuticular lens and provides information only about the intensity and direction of light.

ocular *See* EYEPIECE.

Odonata An order of *exopterygote insects containing the dragonflies and damselflies, most of which occur in tropical regions. Adult dragonflies have a pair of prominent *compound eyes, a compact thorax bearing two pairs of delicate membranous wings, and a long slender abdomen. They are strong fliers and prey on other insects, either in flight or at rest. The eggs are laid near or in water, and the newly hatched nymphs (naiads) are aquatic and resemble the adults, with rudimentary wings. They breathe through gills and feed on small aquatic animals. The nymph leaves the water for its final moult into the terrestrial adult.

odontoblast A cell that is responsible for producing the *dentine of vertebrate teeth. Odontoblasts are found around the lining of the *pulp cavity and have processes that extend into the dentine.

odorant A substance that is responsible for the perception of a smell and is thus detected by olfactory organs, such as the nose. *See* OLFACTION.

oedema The accumulation of *tissue fluid in the tissues of the body, causing swelling of the affected part. Localized oedema occurs during *inflammation. Generalized oedema can result from a variety of pathological conditions, including kwashiorkor (*see* MALNUTRITION) and heart or kidney failure; it can also occur as a side-effect of certain drugs and as a reaction to toxic chemicals.

oesophagus (gullet) (*pl.* **oesophagi** or **oesophaguses**) The section of the *alimentary canal that lies between the *pharynx and the stomach. It is a muscular tube whose function is to transfer food to the stomach by means of wavelike contractions (*peristalsis) along its length.

oestrogen One of a group of female steroid sex hormones, produced principally by the ovaries, that promote the onset of *secondary sexual characteristics (such as breast enlargement and development in women) and control the *oestrous cycle (*menstrual cycle in humans). **Oestradiol** is the most important; others include estrone and estriol, which are less active than oestradiol. Oestrogens are secreted at particularly high levels during ovulation, stimulating the uterus to prepare for pregnancy. They are used in *oral contraceptives (with *progestogens) and as treatment for various disorders of the female reproductive organs. Small amounts of oestrogens are produced by the adrenal glands and testes. Oestrogens are synthesized ultimately from cholesterol via intermediates that include testosterone and other male sex hormones (*see* ANDROGEN). These intermediates are converted to oestrogens by the enzyme aromatase, which is found in the ovaries and also in adipose tissue, an important source of oestrogens in postmenopausal women.

oestrous cycle (sexual cycle) The cycle of reproductive activity shown by most sexually mature nonpregnant female mammals except most primates (*compare* MENSTRUAL CYCLE). There are four phases:

(1) **pro-oestrus (follicular phase)**—*Graafian follicles develop in the ovary and secrete oestrogens;

(2) **oestrus (heat)**—ovulation normally occurs, the female Is ready to mate and becomes sexually attractive to the male;

(3) **metoestrus (luteal phase)**—*corpus luteum develops from ruptured follicle;

(4) **dioestrus**—*progesterone secreted by corpus luteum prepares uterus for implantation.

The length of the cycle depends on the species: larger mammals typically have a single annual cycle with a well-defined breeding season (they are described as **monoestrous**). The males have a similar cycle of sexual activity. Other species may have many cycles per year (i.e. they are **polyoestrous**) and the male may be sexually active all the time.

oestrus (heat) *See* OESTROUS CYCLE.

offset *See* RUNNER.

offspring (progeny) New individual organisms that result from the process of sexual or asexual reproduction. *See also* F_1; F_2.

oil Any of various viscous liquids that are generally immiscible with water. Natural plant and animal oils are either volatile mixtures of terpenes and simple esters (e.g. *essential oils) or are *glycerides of fatty acids.

oil-immersion lens *See* IMMERSION OBJECTIVE.

Okazaki fragment *See* DISCONTINUOUS REPLICATION.

oleaginous Producing or containing oil or lipids. Oleaginous microorganisms (e.g. yeasts), which normally contain 20–25% oil, are of interest in biotechnology as alternative sources of lipids or as possible sources for novel oils. The majority of the oils produced by oleaginous eukaryotic microorganisms are similar to plant oils.

olecranon process A bony process on the *ulna in the forelimb of vertebrates that projects beyond the joint between the humerus and the ulna.

oleic acid An unsaturated *fatty acid with one double bond, $CH_3(CH_2)_7CH{:}CH(CH_2)_7COOH$. Oleic acid is one of the most abundant constituent fatty acids of animal and plant fats, occurring in butterfat, lard, tallow, groundnut oil, soya-bean oil, etc. Its systematic chemical name is **cis-octadec-9-enoic acid**.

oleosome *See* SPHEROSOME.

olfaction The sense of smell or the process of detecting smells. This is achieved by receptors in **olfactory organs** (such as the *nose in vertebrates) that are sensitive to air- or water-borne chemicals. Stimulation of these receptors results in the transmission of information to the brain via the **olfactory nerve**. Fishes have olfactory receptors located in pits on the snout just in front of the eyes. Each pit is lined by a folded sensory epithelium, which detects chemicals in water flowing through the pit, via incurrent and excurrent openings (nares). Chemicals in the water are also detected by *taste buds located over the body surface. Insects smell airborne odorants using olfactory hairs on their antennae. *See* VOMERONASAL ORGAN.

• Tutorial on the sense of smell compiled by Tim Jacob of Cardiff University

olfactory lobe Either member of a pair of lobes in the forebrain, at the anterior end of the cerebrum. They contain the endings of the olfactory nerves (the first pair of cranial nerves) and are concerned with the sense of smell, being prominent in the dogfish and other animals that depend on this sense.

Oligocene The third geological epoch of the *Palaeogene period. It began about 34 million years ago, following the Eocene epoch, and extended for about 11 million years to the beginning of the Miocene epoch. The epoch was characterized by the continued rise of mammals; the first pigs, rhinoceroses, and tapirs made their appearance.

Oligochaeta In traditional classifications, a class of hermaphrodite annelid worms that bear only a few bristles (*chaetae). Oligochaetes are very abundant in freshwater and terrestrial habitats. The most familiar members of the class are the earthworm (*Lumbricus*) and the freshwater bloodworm (*Tubifex*). Molecular studies have revealed that the evolutionary relationships of annelids are more complex than traditional classifications would suggest, and many members of the Oligochaeta are now placed in the clade Sedentaria, along with some polychaetes and the leeches. *See* ANNELIDA.

oligodendrocyte *See* GLIA.

oligonucleotide A short polymer of *nucleotides.

oligosaccharide Any molecule that consists of several (2 to 20) *monosaccharides joined covalently by *glycosidic bonds. Oligosaccharides are components of *glycoproteins and occur, for example, on the surface of cells, where they play vital roles in cell recognition and adhesion.

oligotrophic Describing a body of water (e.g. a lake) with a poor supply of nutrients and a low rate of formation of organic matter by photosynthesis. *Compare* DYSTROPHIC; EUTROPHIC; MESOTROPHIC.

omasum (*pl.* **omasa**) The third of four chambers that form the stomach of ruminants. *See* RUMINANTIA.

OMIM (Online Mendelian Inheritance in Man) An online catalogue of human genetic disorders and their genes, originally based on the book *Mendelian Inheritance in Man* by Victor McKusick and colleagues at Johns Hopkins University. It contains descriptions of human inherited diseases, along with the chromosomal location(s) of causative genes and links to sequence data where known. The database is provided and updated by the US National Center for Biotechnology Information (NCBI). Each entry has a unique six-digit MIM number, with allelic variants denoted by a four-digit suffix following a decimal point.

ommatidium (*pl.* **ommatidia**) *See* COMPOUND EYE.

omnivore An animal that eats both animal and vegetable matter. Pigs, for example, are omnivorous. *Compare* CARNIVORE; HERBIVORE.

oncogene A dominant mutant allele of a cellular gene (a **proto-oncogene**) that disrupts cell growth and division and is capable of transforming a normal cell into a cancerous cell. Proto-oncogenes typically encode proteins involved in positive control of the cell division cycle, such as growth factor receptors, signal transduction proteins, and transcription factors. Mutations in these genes tend to relax control mechanisms and accelerate cell division, leading to the cell proliferation that is characteristic of cancer. Some oncogenic mutations cause inhibition of programmed cell death (*apoptosis), so that cancerous cells are less likely to be destroyed by the body's defences. Most oncogenic mutations arise during the lifetime of the individual and affect body cells (i.e. somatic cell mutations). However, in rare cases germ-line cells are affected, so that offspring might potentially inherit the oncogene. Certain oncogenes of vertebrates are derived from viruses (*see* ONCOGENIC). *Compare* TUMOUR-SUPPRESSOR GENE.

oncogenic Describing a chemical, organism, or environmental factor that causes the development of cancer. Some viruses are oncogenic to vertebrates, notably the *retroviruses (including the Rous sarcoma virus of chickens), avian sarcoma virus, mouse mammary tumour virus, simian virus 40, human papillomavirus, and human adenovirus. These viruses contain genes (known as *oncogenes) that become integrated into the host cell's DNA and are responsible for the transformation of a normal cell into a cancerous cell. *See also* GROWTH FACTOR.

oncotic pressure (colloid osmotic pressure) The part of the osmotic pressure produced by colloids (i.e. proteins and other large molecules) present within the blood vascular system. The vessel walls are relatively impermeable to these molecules. Oncotic pressure largely offsets the hydrostatic pressure that tends to drive water from blood vessels into the extracellular spaces and tissues. Hence, if oncotic pressure falls, there is increased risk of fluid accumulation and swelling in tissues (oedema).

one gene–one polypeptide hypothesis The theory that each *gene is responsible for the synthesis of a single *polypeptide. It was originally stated as the **one gene–one enzyme hypothesis** by the US geneticist George Beadle in 1945 but later modified when it was realized that genes also encoded nonenzyme proteins and individual polypeptide chains. It is now known that different polypeptides can be made from the same gene by *alternative splicing of the RNA transcript, and that some genes code for various types of RNA involved in control of gene expression and protein synthesis.

ontogeny The developmental course of an organism from the fertilized egg through to maturity. It has been suggested that 'ontogeny recapitulates *phylogeny', i.e. the stages of development, especially of the embryo, reflect the evolutionary history

of the organism. This idea is now discredited. *See* RECAPITULATION.

ontology A specification of the assumptions, terms, or concepts underlying a particular field of knowledge. For example, the Gene Ontology Consortium is an international collaboration between various databases in the field of genomics to standardize terminology used by researchers. It has developed three ontologies, each containing a controlled vocabulary for naming, respectively, cellular components, biological processes, and molecular functions. Standardization of terms in this way is vital for efficient searching of databases, particularly for devising and using automated search programs.

Onychophora A small phylum of caterpillar-like invertebrates, the velvet worms, that inhabit moist dark terrestrial habitats, such as forest litter and caves, in tropical and warm regions. The thin chitinous cuticle, which bears numerous papillae and sensory hairs giving it a velvety feel, is periodically moulted. The 110 or so known species, which include *Peripatus*, are generally small, with brownish bodies, although some are more brightly coloured. Sizes range from 14 mm to 200 mm in length, with females larger than males, and there may be from 14 to more than 40 pairs of unjointed hollow legs. Onychophorans capture prey, such as spiders and termites, by entangling them in a sticky secretion squirted from adhesive glands opening beside the mouth. The sexes are separate; some species lay eggs, whereas others are ovoviviparous or viviparous, the latter nourishing embryos internally via a placenta analogous to that of mammals. Onychophorans are closely related to arthropods and are thought to have descended from extinct marine forms that flourished in the Cambrian, such as *Aysheaia* and *Hallucigenia*, found in the *Burgess shale deposits. Such fossil forms are now sometimes placed with modern velvet worms in the phylum **Lobopodia**.

oocyte *See* OOGENESIS.

oogamy Sexual reproduction involving the formation and subsequent fusion of a large, usually stationary, female gamete and a small motile male gamete. The female gamete may contain nourishment for the development of the embryo, which is often retained and protected by the parent organism.

oogenesis The production and growth of the ova (egg cells) in the animal *ovary. Special cells (**oogonia**) within the ovary divide repeatedly by mitosis to produce large numbers of prospective egg cells (**primary oocytes**). When mature, these undergo meiosis, which halves the number of chromosomes. During the first meiotic division a **polar body** and a **secondary oocyte** are produced. At the second meiotic division the secondary oocyte produces a large haploid ootid, which matures into the ovum, and a second polar body. Oocytes may be present in the ovaries at birth and may represent the total number of eggs to be produced.

(((●))) SEE WEB LINKS

• Overview of oogenesis, including the development of follicles and the ovarian cycle, produced by Human Embryology

oogonium (*pl.* **oogonia**) **1.** The female sex organ (*gametangium) of algae and fungi. **2.** Any of the immature sex cells in the animal ovary that give rise to oocytes by mitotic divisions (*see* OOGENESIS).

Oomycota A phylum of fungus-like filamentous protists that includes the water moulds, downy mildews, and potato blight (*Phytophthora*), formerly classified as a class of fungi (Oomycetes). They are coenocytic and the cell wall is made of cellulose. Oomycotes are either saprotrophic or parasitic; they feed by extending hypha-like threads into the food source or host's body. Asexual reproduction is by means of flagellated *zoospores, which are released from a sporangium. Sexual reproduction involves the fusion of an antheridium and an oogonium and results in the production of a zygote, which can develop a wall of chitin and become a resistant oospore. They are classified as *stramenopiles.

oosphere (ovum; egg cell) The nonmotile female gamete in plants and some algae. In angiosperms (flowering plants) it is a cell in the *embryo sac of the ovule. In other plants it is situated in an *archegonium. In algae, such as *Fucus*, the oosphere is protected by an **oogonium** until it is shed into the water prior to fertilization. Many oospheres store food in the form of starch or oil droplets.

oospore A zygote that is produced as a result of *oogamy in certain algae and fungi. It contains food reserves, develops a protective outer covering, and enters a resting phase before germination. *Compare* ZYGOSPORE.

ootid *See* OOGENESIS.

open circulation *See* CIRCULATION.

open reading frame *See* READING FRAME.

operant conditioning *See* CONDITIONING.

operculum (*pl.* **opercula**) **1.** (in zoology) A lid or flap of skin covering an aperture, such as the gill slit cover of fish and larval amphibians and the horny calcareous operculum secreted by many gastropod molluscs, which closes the opening of the shell when the animal is inside. **2.** (in botany) The cone-shaped lid of the *capsule of mosses, which is forcibly detached to release the spores.

operon A functionally integrated genetic unit for the control of gene expression in bacteria, as proposed in the *Jacob-Monod hypothesis. Typically it comprises a closely linked group of **structural genes**, coding for protein, and adjacent loci controlling their expression—an **operator site** and a **promoter site**. The structural genes tend to encode enzymes concerned with a particular biochemical pathway. *Transcription of the structural genes is prevented by binding of a **repressor** molecule to the operator site. Another molecule, the *inducer, can bind to the repressor molecule, preventing it from binding to the operator and thus allowing the promoter site to bind the enzyme RNA polymerase, thereby initiating transcription. The repressor molecule is encoded by the **regulator gene**, which may be close to or distant from the operon. Some operons also have an **attenuator region** (*see* ATTENUATION) preceding the first structural gene, where transcription may either stall or proceed according to the amount of end-product in the cell. *See also* LAC OPERON.

opiate One of a group of drugs derived from **opium**, an extract of the poppy plant *Papaver somniferum* that depresses brain function (a **narcotic** action). Opiates include *morphine and its synthetic derivatives, such as *heroin and codeine. They are used in medicine chiefly to relieve pain, but the use of morphine and heroin is strictly controlled since they can cause drug dependence and tolerance. Opiates exert their effect by binding to cell receptors that bind *endorphins, the body's endogenous opioids, or natural painkillers.

opioid Any one of a group of substances that bind to *opioid receptors to produce pharmacological and physiological effects similar to those of morphine. Opioids are not necessarily structurally similar to morphine, although a subgroup of opioids, the *opiates, are morphine-derived compounds. The **endogenous opioids**, which occur naturally in the body, comprise the *endorphins, *enkephalins, and *dynorphins.

opioid receptor Any cell receptor molecule that binds an *opioid (i.e. as its ligand), such as morphine or an endogenous opioid (endorphin, enkephalin, or dynorphin). There are four types, labelled μ (mu), κ (kappa), δ (delta), and nociceptin (the latter has structural similarities to the other three 'classical' opioid receptors). All are *G-protein-coupled receptors, and their activation through binding of an appropriate ligand leads to analgesia (pain relief), hence their importance in medicine.

opisthokonts An assemblage of eukaryotic organisms that includes the kingdoms containing all animals and fungi, plus certain flagellated protists, such as the *DRIPs and choanoflagellates. Although the grouping is based chiefly on molecular systematics, members are characterized by having flagellated cells propelled by a single posterior flagellum, as in the sperm cells of mammals or the zoospores of chytrids (*see* CHYTRIDIOMYCOTA). This contrasts with the anterior flagella of flagellated cells in other eukaryotes.

opisthosoma The posterior section of the body (*see* TAGMA) of arachnids and other arthropods of the phylum *Chelicerata, which consists of those body segments that do not bear legs. *Compare* PROSOMA.

opportunistic Describing a species that can quickly exploit new resources as they arise, for example by rapidly colonizing a new environment. Such species characteristically exhibit *r* selection.

opsin The lipoprotein component of *rhodopsin, the light-sensitive pigment that occurs in the rod cells of the retina.

opsonin *See* OPSONIZATION.

opsonization The process in which certain *antibodies in the blood (known as **opsonins**) bind to the surface of an invading microorganism, which renders it more susceptible to phagocytosis by macrophages and neutrophils. *Acute-phase response proteins also act as opsonins, as do *complement proteins.

optical activity The ability of certain substances to rotate the plane of plane-polarized light as it passes through a crystal, liquid, or solution. It occurs when the molecules of the substance are asymmetric, so that they can exist in two different structural forms each being a mirror image of the other. The two forms are **optical isomers** or **enantiomers** (see illustration). The existence of such forms is also known as **enantiomorphism** (the mirror images being **enantiomorphs**). One form will rotate the light in one direction and the other will rotate it by an equal amount in the other. The two possible forms are described as *dextrorotatory or *laevorotatory according to the direction of rotation,

Optical activity: isomers of tartaric acid

D-form   L-form   *meso*-form

and the prefixes (+)- and (−)- are used, respectively, to designate the isomer, as in (+)-tartaric and (−)-tartaric acids. (The prefixes *d*- and *l*- are now obsolete.) An equimolar mixture of the two forms is not optically active. It is called a **racemic mixture** (or **racemate**) and designated by (+). In addition, certain molecules can have a **meso form** in which one part of the molecule is a mirror image of the other. Such molecules are not optically active.

Molecules that show optical activity have no plane of symmetry. The commonest case of this is in organic compounds in which a carbon atom is linked to four different groups. An atom of this type is said to be a **chiral centre**. Many naturally occurring compounds show optical isomerism and usually only one isomer occurs naturally. For instance, glucose is found in the dextrorotatory form. The other isomer, (−)-glucose, can be synthesized in the laboratory, but cannot be synthesized by living organisms.

optical fibre A glass fibre through which light can be transmitted with near total internal reflection and very little leakage through the sidewalls. In the **step-index fibre** a pure glass core, with a diameter between 6 and 250 μm, is surrounded by a glass or plastic cladding of lower refractive index. The cladding is usually between 10 and 150 μm thick. The interface between core and cladding acts as a cylindrical mirror at which total internal reflection of the transmitted light takes place. This structure enables a beam of light to travel through many kilometres of fibre. In the **graded-index fibre**, each layer of glass, from the fibre axis to its outer wall, has a slightly lower refractive index than the layer inside it. This arrangement also prevents light from escaping through the fibre walls by a combination of refraction and total internal reflection, and can be made to give the same transit time for rays at different angles.

Fibre-optic systems use optical fibres to transmit information, in the form of coded pulses or fragmented images (using bundles of fibres), from a source to a receiver. They are used, for example, in medical instruments (**endoscopes** or **fibrescopes**) to examine internal body cavities, such as the stomach and bladder.

optical isomers *See* OPTICAL ACTIVITY.

optical microscope *See* MICROSCOPE.

optic chiasm (optic chiasma) The point where some of the fibres from both *optic nerves cross over to the opposite side, forming an X-shaped structure. Sensory information from the right side of both the right and left eye is transmitted to the right primary visual cortex in the brain (via the **right lateral geniculate nucleus**), while information from the left side of each eye is carried via the left lateral geniculate nucleus to the left primary visual cortex. Additional neurons project from the primary visual cortex to connect to centres elsewhere in the cortex responsible for higher-order visual processing and integration. Thus each half of the brain receives sensory information from both eyes.

optic lobes *See* MIDBRAIN.

optic nerve The second *cranial nerve: a paired sensory nerve that runs from each eye to the brain. It is responsible for conveying visual stimuli received by the rods and cones in the retina to the brain for interpretation. *See also* OPTIC CHIASM.

optic vesicle The outgrowth from the forebrain of a vertebrate embryo that develops into the retina of the eye. When the optic vesicle comes into contact with the ectoderm covering the head of the embryo, these ectodermal cells become thicker: the ectoderm in this region invaginates and eventually becomes detached from the adjacent ectodermal cells to form the lens of the eye.

optimal foraging theory A theory, first formulated in 1966 by R. H. MacArthur and E. R. Pianka, stating that natural selection favours animals whose behavioural strategies maximize their net energy intake per unit time spent foraging. Such time includes both searching for prey and handling (i.e. killing and eating) it. The theory was originally devised in an attempt to explain why, out of the wide range of foods available, animals often restrict themselves to a few preferred types. The prediction is that an animal strikes a balance between two contrasting strategies: spending a long time (i.e. using more energy) searching for highly 'profitable' food items, or devoting minimal time (i.e. using less energy) to more common but less profitable food items. Various factors can cause animals to deviate from optimal foraging. For example, the risk of predation may force the animal to select less profitable food items in a relatively safe location, rather than opting for the energetically most efficient feeding strategy.

oral cavity *See* BUCCAL CAVITY.

oral contraceptive Any hormonal preparation taken in the form of a pill to prevent conception (*see* BIRTH CONTROL). The most common form is the combined pill, which contains an *oestrogen and a *progestogen. Both act to suppress ovulation, while the progestogen additionally causes changes in the viscosity of cervical mucus and alters the lining of the womb, both of which decrease the chances of fertilization should ovulation occur. The so-called 'minipill' contains only a progestogen and has fewer side effects than the combined pill. Emergency contraception (the so-called 'morning-after pill'), is designed to prevent pregnancy after unprotected sexual intercourse. There are several types. One contains the synthetic progestogen levonorgestrol, which delays ovulation and has to be taken within 72 hours of intercourse. Another type contains ulipristal acetate, which also delays ovulation, and is effective for up to five days following intercourse.

oral groove A ciliated channel found in certain protists and aquatic invertebrates down which food is directed into the mouth.

orbit (in anatomy) Either of the two sockets in the skull of vertebrates that house the eyeballs.

order (in taxonomy) A category used in the *classification of organisms that consists of one or several similar or closely related families. Similar orders form a class. Order names typically end in *-ales* in botany, e.g. Rosales (roses and orchard fruits), and in *-a* in zoology, e.g. Carnivora (flesh eaters).

Ordovician The second geological period of the Palaeozoic era, following the Cambrian and preceding the Silurian periods. It began about 485 million years ago and lasted for about 41 million years. The period was named by the British geologist Charles Lapworth (1842-1920) in 1879. *Graptolites, in deep-water deposits, are the dominant fossils. Other fossils include *trilobites, brachiopods, ectoprocts, gastropods, bivalves, echinoids, crinoids, nautiloid cephalopods, and the first corals.

(((∰))) SEE WEB LINKS

- A brief survey of the Ordovician period on the website of the University of California Museum of Paleontology

organ Any distinct part of an organism that is specialized to perform one or a number of functions. Examples are ears, eyes, lungs, and kidneys (in animals) and leaves, roots, and flowers (in plants). A given organ will contain many different *tissues.

organ culture The culture of complete living organs (**explants**) of animals and plants outside the body in a suitable culture medium. Animal organs must be small enough to allow the nutrients in the culture medium to penetrate all the cells. Whole plant roots and even root systems can be kept alive in such conditions for a considerable period of time. *See also* EXPLANTATION; SYNTHETIC BIOLOGY.

organelle A minute structure within a eukaryotic *cell that has a particular function. Examples of organelles are the nucleus, mitochondria, and lysosomes.

organic evolution The process by which changes in the genetic composition of populations of organisms occur in response to environmental changes. *See* ADAPTATION; EVOLUTION. *Compare* BIO-CHEMICAL EVOLUTION.

organism An individual living system, such as an animal, plant, or *microorganism, that is capable of reproduction, growth, and maintenance.

organizer An area of an animal embryo that causes adjacent areas of the embryo to develop in a certain way. The **primary organizer** (blastopore lip or archenteron roof) directs the overall development of the gastrula (*see* HENSEN'S NODE; SPEMANN'S ORGANIZER). A **secondary organizer** directs the development of a subsidiary part or tissue.

organ of Corti The sensory part of the *cochlea in the inner ear, which responds to sound. This organ, named after Italian anatomist A. G. G. Corti (1822–88), projects into the cochlear duct and consists of two membranes that run parallel to each other: sensory *hair cells, rooted in the **basilar membrane**, are in contact with the overlying **tectorial membrane** (see illustration). During the transmission of sound waves the basilar membrane vibrates, causing the sensory hairs to flex against the tectorial membrane; this results in the production of impulses, which are transmitted to the brain via the auditory nerve.

organogenesis The formation of organs during embryonic development. In animals this begins following the rearrangement of the cells at gastrulation, when the three germ layers are fully formed in their correct positions. Dividing cells of the *gastrula begin to differentiate and the rudimentary

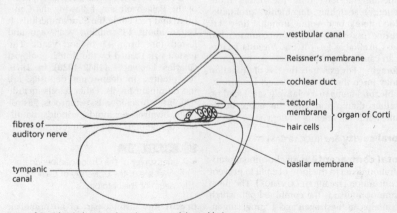

Organ of Corti (vertical section through one turn of the cochlea)

organs and organ systems begin to form. *See* DIFFERENTIATION; ECTODERM; ENDODERM; MESODERM.

orgasm The climax of sexual excitement in humans, which—in males—coincides with *ejaculation. A sense of physiological and emotional release is accompanied by a feeling of extreme pleasure.

Oriental region *See* FAUNAL REGION.

origin of life The process by which living organisms developed from inanimate matter, which is generally thought to have occurred on earth between 3500 and 4000 million years ago(mya). The oldest known fossils are *stromatolites, rocky masses formed by cyanobacteria as early as 3500 mya. Many hypotheses have been put forward, but it is supposed that the primordial atmosphere contained all the basic constituents of organic matter: ammonia, methane, hydrogen, and water vapour. These underwent a process of **chemical evolution** using a source of energy (e.g. from the sun and electric storms) to combine into ever more complex molecules, such as amino acids, proteins, and vitamins. For example, the 'primordial soup' theory, dating from the 1920s, proposes that in hot pools on oceanic shores, evaporation concentrated monomers such as amino acids to the extent that they underwent polymerization to form proteins. Eventually self-replicating nucleic acids, the basis of all life, could have developed. The very first organisms may have consisted of such molecules bounded by a simple membrane to form *protocells. One rival view, proposed in 2010, is that deep-sea hydrothermal vents may have provided the energy and catalytic conditions needed to sustain early life forms. Another alternative theory is that the chemical 'building blocks' of life, or even the earliest microorganisms, originated not on earth but were delivered from space by meteorites or other celestial bodies (*see* PANSPERMIA). *See also* PROTEINOID.

ornithine (Orn) An *amino acid, $H_2N(CH_2)_3CH(NH_2)COOH$, that is not a constituent of proteins but is important in living organisms as an intermediate in the reactions of the *urea cycle and in arginine synthesis.

ornithine cycle *See* UREA CYCLE.

orphan receptor A *receptor molecule in a cell whose ligand is unknown. When the ligand that binds to it is discovered, the receptor is termed 'adopted'.

orthogenesis An early theory of the nature of evolutionary change, which proposed that organisms evolve along particular paths predetermined by some factor in their genetic make-up. More recent understanding of selection pressure and other external forces that can be shown experimentally to affect the survival of organisms has proved the improbability of the theory.

orthologous Describing *homologous genes that are descendants of a single gene in a common ancestor. Thus, when a lineage splits to form two new species, any gene gives rise to two orthologues, which may subsequently diverge in their DNA sequence and function. *Compare* PARALOGOUS.

Orthonectida A phylum of microscopic wormlike animals that live as parasites inside the gonads of various marine invertebrates, including polychaete worms, bivalve molluscs, echinoderms, and flatworms. Some 45 species are known. The adult body typically measures less than 300 µm long and is covered in ciliated jacket cells; most internal organ systems are absent. Mating occurs between free-swimming males and females outside the host, and the fertilized eggs develop into ciliated larvae, which locate and enter the appropriate hosts via their genital ducts. Inside the gonad, each larval cell then develops and enlarges to contain multiple nuclei, which give rise to a new generation of tiny male and female adults inside a single cell. The presence of this amoeba-like plasmodium causes loss of fertility of the host gonad. Subsequently, the young orthonectids leave the host to become a new free-swimming sexual generation. Formerly regarded as closely related to *Dicyemida, the evolutionary affiliations of Orthonectida remain uncertain.

Orthoptera A large order of *exopterygote insects containing the grasshoppers, locusts, crickets, and—in some classification systems—the cockroaches (*see* DICTYOPTERA). They are characterized by enlarged hind legs modified for jumping and biting mouthparts and produce sounds by *stridulation. The crickets and long-horned grasshoppers (e.g. *Gryllus*, *Tettigonia*) have long threadlike antennae and stridulate by rubbing together modified veins on their forewings. The hearing organs are on the front legs. The short-horned grasshoppers and locusts (e.g. *Chorthippus*, *Locusta*) have short antennae and stridulate by rubbing pegs on the hind leg against a hardened vein on the forewing. The hearing organs are on the abdomen.

orthotropism The tendency for a *tropism (growth response of a plant) to be orientated directly in line with the stimulus concerned. An example is the vertical growth of main stems and roots in response to gravity (**orthogeotropism**). *Compare* PLAGIOTROPISM.

osculum (*pl.* **oscula**) **1.** The mouthlike aperture in the body wall of a sponge (*see* PORIFERA) through which water leaves the body cavity. **2.** Any of the suckers on the head (scolex) of a tapeworm, by which it attaches itself to the gut wall of its host. **3.** (in anatomy) A pore or tiny opening.

osmium tetroxide (osmium(IV) oxide) A yellow solid, OsO_4, made by heating osmium in air. It is used as a fixative in electron microscopy.

osmoconformer An animal whose body fluids are in osmotic balance with its environment. For many marine invertebrates the osmolarity and ionic concentrations of their body fluids are similar to those of the seawater in which they live. Such animals avoid the need to expend much energy on *osmoregulation. However, they generally maintain cell volumes by altering the concentration of intracellular *osmolytes. *Compare* OSMOREGULATOR.

osmole Symbol osmol. The *amount of substance that gives rise to or contains one mole of osmotically active particles when dissolved in water.

osmolyte Any compound that protects cells from desiccation by maintaining a high intracellular osmolarity (osmotic *concentration). Osmolytes are active in the process of *osmoregulation and are of particular use in the kidneys, where cells are exposed to highly concentrated fluids. Compounds that are known osmolytes include *polyols, amines (e.g. *trimethylamine), certain amino acids, and urea.

osmoreceptor Any of numerous neurons in the *hypothalamus of the vertebrate brain that respond to an increase in the osmolality of the blood. This results in the release of *antidiuretic hormone (ADH) and the subsequent conservation of water, thereby maintaining the *homeostasis of the body fluids.

osmoregulation The control of the water content and the concentration of salts in the body of an animal or protist (*see* OSMOREGULATOR). In freshwater species osmoregulation must counteract the tendency for water to pass into the animal by *osmosis. Various methods have been developed to eliminate the excess, such as *contractile vacuoles in protozoans, and *kidneys with well-developed glomeruli in freshwater fish. Marine vertebrates have the opposite problem: they prevent excessive water loss and enhance the excretion of salts by having kidneys with few glomeruli and short tubules (*see also* CHLORIDE SECRETORY CELL). Elasmobranch fishes also are equipped with salt-excreting *rectal glands. In terrestrial vertebrates the dangers of desiccation are reduced by the presence of long convoluted *renal tubules, which increase the reabsorption of water and salts. Marine birds and many reptiles remove excessive salt by means of *salt glands located in the head.

osmoregulator An animal that maintains a constant internal osmotic environment in spite of changes in its external environment. Vertebrates and some aquatic invertebrates, especially freshwater invertebrates, expend energy on *osmoregulation to maintain cell volumes and achieve optimum conditions for metabolism (i.e. homeostasis). *Compare* OSMOCONFORMER.

osmosis The net movement of water molecules from a dilute (*hypotonic) solution to a more concentrated (*hypertonic) solution through a differentially permeable membrane. The distribution of water in living organisms is dependent to a large extent on osmosis, water entering and leaving the cells through their partially permeable plasma membranes. The pressure required to stop the flow of pure water into a solution across a partially permeable membrane is a characteristic of the solution, and is called the **osmotic pressure**. In animal physiology osmosis is described in terms of osmolarity (*see* CONCENTRATION), with a net movement of water occurring from a region of low osmolarity to one of higher osmolarity. Plant physiologists use the concept of *water potential, so that water moves from an area of high (less negative) water potential to an area of low (more negative) water potential (*see also* PLASMOLYSIS; TURGOR). Animals have evolved various means to counteract the effects of osmosis (*see* OSMOREGULATION). *See also* ONCOTIC PRESSURE; TONICITY.

osmotic pressure *See* OSMOSIS.

osmotroph Any heterotrophic organism that obtains its nutrients by absorbing organic matter in solution from its surroundings. *Compare* PHAGOTROPH.

ossicle Any small bony or chitinous structure found in various skeletal parts of animals. The term ossicle usually refers to any of the chain of bones in the mammalian ear (*see* EAR OSSICLES).

ossification The process of *bone formation. It is brought about by the action of special cells called *osteoblasts, which deposit layers of bone in connective tissue.

In **intramembranous ossification** bones are formed directly in connective tissue (*see* MEMBRANE BONE); in **endochondral ossification** the bones are formed by the replacement of cartilage (*see* CARTILAGE BONE).

Osteichthyes In traditional classifications the class of vertebrates comprising the ray-finned fishes (subclass Actinopterygii), which comprise the bulk of living marine and freshwater fish species, and the lobe-finned fishes (subclass Sarcopterygii), which comprise the lungfishes (*see* DIPNOI) and *coelacanths. All have gills covered with a bony operculum, and a layer of thin overlapping bony *scales covers the entire body surface. Bony fish have a *swim bladder, which acts as a hydrostatic organ enabling the animal to remain suspended in the water at any depth. In the lungfishes this bladder acts as a lung. This traditional class is paraphyletic because it excludes amphibians, reptiles, and mammals, which all share a common ancestor with the bony fishes. Hence, Osteichthyes is now applied to the clade containing the bony fishes and their tetrapod relatives. *See also* DIPNOI; TELEOSTEI. *Compare* CHONDRICHTHYES.

osteoblast Any of the cells, found in *bone, that secrete collagen and other substances that form the matrix of bone (*see* OSTEOID). Osteoblasts are derived from **osteoprogenitor cells** in the bone marrow; they eventually become surrounded by bone within small cavities (lacunae) as *osteocytes. *See also* OSSIFICATION.

osteoclast Any of the cells, found on the surface of bone, that are involved in the breakdown of the bone matrix to enable the further development and remodelling of bone during growth and repair by *osteoblasts. *See also* PARATHYROID HORMONE; PERIOSTEUM.

osteocyte Any of the cells, found in bone, that perform the cellular activities, such as respiration and exchange of materials with the blood, that are required for maintenance of the bone tissue. They are derived from *osteoblasts.

osteoid A soft material, consisting mainly of *collagen, that is secreted by *osteoblasts and constitutes the uncalcified matrix of *bone. Osteoid is converted into hard bone matrix when it combines with calcium phosphate (hydroxyapatite) deposited from the blood (*see* OSTEONECTIN).

osteon *See* HAVERSIAN CANALS.

osteonectin A protein in bone that binds to collagen (*see* OSTEOID) and is involved in the formation of growth sites for calcium phosphate crystals, which are required for forming the hard bone matrix.

ostiole A pore in the fruiting body of certain fungi and algae, through which either spores or gametes are released. An ostiole occurs, for example, in the perithecium of ascomycete fungi (*see* ASCOCARP) and in the *conceptacle of brown algae.

ostium (*pl.* **ostia**) **1.** (in anatomy) A small opening, especially into a tube or hollow organ. **2.** Any one of the pores in the body wall of a sponge (*see* PORIFERA) through which water enters the body cavity. **3.** Any one of the apertures in the arthropod heart through which blood enters from the haemocoel.

Ostracoda A class of crustaceans comprising some 8000 known species. Ostracods are typically small (mostly <1 mm to 2 cm), with short oval bodies enclosed in a pair of hinged valves forming a carapace and resembling a tiny mussel, hence their common name of 'mussel shrimp'. The head bears five pairs of appendages, with another 1–3 pairs on the body. Most live in sediments on the floor of lakes or oceans.

otolith A gelatinous mass containing a high concentration of particles of calcium carbonate, which forms part of the *macula of the inner ear.

ouabain (G-strophanthin) A toxic glycoside derived from the seeds of certain African plants, such as *Strophanthus gratus*, that has been used historically as a drug to stimulate contractions of heart muscle in treating heart failure and other conditions. It specifically blocks the action of sodium/potassium ATPase (*see* SODIUM PUMP), a crucial membrane-bound transport protein that maintains differential ion concentrations between a cell's interior and the external environment. For this property, ouabain is used widely in biomedical research, and it is now known to occur naturally in the body at very low concentrations. The adrenal glands are thought to be the main source of endogenous ouabain in humans, which may play a role in sodium metabolism, e.g. modulating the effects of dietary salt intake on blood pressure.

outbreeding Mating between unrelated or distantly related individuals of a species. Outbreeding populations usually show more variation than *inbreeding ones and have a greater potential for adapting to environmental changes. Outbreeding increases the number of *heterozygous individuals, so disadvantageous recessive characteristics tend to be masked by dominant alleles.

outer ear (external ear) The part of the ear external to the *tympanum (eardrum). It is present in mammals, birds, and some reptiles and consists of a tube (the **external auditory meatus**) that directs sound waves onto the tympanum. In mammals it may include an external *pinna, which extends beyond the skull.

outgroup A species or higher taxon used in *systematics for comparison with a group of closely related species or taxa in order to assess whether particular characters shared by the group members are derived (*see* APOMORPHY) or ancestral (*see* PLESIOMORPHY). For the outgroup comparison to be valid, the outgroup must be outside the phylogenetic group under consideration but not too distantly related— ideally the *sister group—otherwise the comparison yields no useful information. For example, the reptiles, birds, and mammals are all amniotes known to comprise a *monophyletic group, the Amniota. Suppose that the evolutionary relations between several representatives of this group are being investigated, using reproductive physiology as one of a set of homologies (*see* HOMOLOGOUS). A dog and a kangaroo both bear live young (viviparity), whereas a sparrow, a turtle, and a crocodile all lay eggs (oviparity). Which is the ancestral trait and which is the derived trait? Outgroup comparison with frogs or fishes reveals that these anamniote fellow vertebrates exhibit oviparity: this is therefore inferred as the ancestral trait for the amniotes as a whole, and hence viviparity is a derived trait for mammals.

oval window (fenestra ovalis) A membrane-covered opening between the middle ear and the inner ear (*see* EAR), situated above the *round window. Vibrations of the tympanum are transferred across the middle ear by the *ear ossicles and transmitted to the inner ear by the oval window, which is connected to the third ear ossicle (stapes). *See also* VESTIBULAR CANAL.

ovarian cycle The cyclic series of events that lead to the release of mature egg cells (oocytes) from the ovary, i.e. ovulation, occurring in most female primates, including human females. It is controlled primarily by the level in the bloodstream of the gonadotrophin hormones follicle-stimulating hormone (FSH) and luteinizing hormone (LH) released by the anterior pituitary, which is stimulated by the hypothalamus. The ovarian cycle starts at the time of menstruation and is synchronized with the *menstrual cycle, which prepares the uterus to receive a fertilized egg. During the **follicular phase**, several primary oocytes begin to mature in the ovary, each surrounded by a cluster of cells forming a *Graafian or ovarian follicle. Initially the follicles release the sex hormone oestradiol at low levels; then the level rises to a peak, which triggers a surge in secretion of LH and FSH. This prompts the maturing follicle to move to the surface of the ovary and rupture, releasing the ovum into the *fallopian tube. In humans usually only a single follicle matures; the others disintegrate. The second half of the cycle is the **luteal phase**, during which the residual follicle forms the glandular *corpus luteum. This secretes progesterone and oestradiol, which promote thickening of the endometrium of the uterus, inhibit the release of gonadotrophin-releasing hormone (GnRH) by the hypothalamus, and hence reduce blood levels of FSH and LH (see illustration). If fertilization of the egg does not occur, continuing low levels of gonadotrophins cause the corpus luteum to degenerate and oestradiol levels to fall, thereby

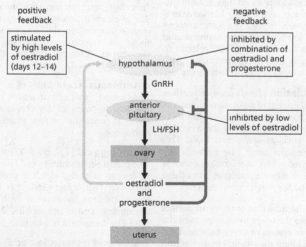

Ovarian cycle: positive and negative feedback in hormonal control

disinhibiting the hypothalamus and enabling another ovarian cycle to begin. In humans the ovarian cycle lasts about 28 days, like the menstrual cycle, with ovulation taking place around day 14.

ovarian follicle *See* GRAAFIAN FOLLICLE.

ovary 1. The reproductive organ in female animals in which eggs (ova) are produced. In most vertebrates there are two ovaries (in some fish the ovaries fuse together to form a single structure and in birds the left ovary only is functional). As well as eggs, they produce steroid hormones (*see* OESTROGEN; PROGESTERONE). In mammals each ovary is situated close to the opening of a *fallopian tube; it contains numerous follicles in which the eggs develop and from which they are released in a regular cycle. *See also* GRAAFIAN FOLLICLE; MENSTRUAL CYCLE; OOGENESIS; OVARIAN CYCLE; OVULATION; REPRODUCTIVE SYSTEM. **2.** The hollow base of the *carpel of a flower, containing one or more *ovules. After fertilization, the ovary wall develops into the fruit enclosing the seeds. In some species, the carpels are fused together to form a complex ovary.

overdominance The condition in which a heterozygous organism has greater fitness than the corresponding homozygotes. The hypothesis is that the heterozygous state leads to enhanced gene expression at a particular locus or set of loci relative to either homozygous alternative; this is proposed as an explanation of *hybrid vigour.

overpopulation The situation that arises when rapid growth of a population, usually a human population, results in numbers that cannot be supported by the available resources, such as space and food. This occurs when the birth rate exceeds the death rate, or when immigration exceeds emigration, or when a combination of these factors exists. *See* POPULATION GROWTH.

oviduct The tube that conveys an animal egg cell from the ovary to other parts of the reproductive system or to the outside. Eggs are passed along the oviduct by the action of muscles and cilia. *See* FALLOPIAN TUBE.

oviparity Reproduction in which fertilized eggs are laid or spawned by the mother and hatch outside her body. It occurs in most animals except marsupial and placental mammals. *Compare* OVOVIVIPARITY; VIVIPARITY.

ovipositor An organ at the hind end of the abdomen of female insects through which eggs are laid. It consists of a pair of modified appendages and is often long and piercing, so that eggs can be laid in otherwise inaccessible places. The sting of bees and wasps is a modified ovipositor.

ovoviviparity Reproduction in which fertilized eggs develop and hatch in the oviduct of the mother. It occurs in many invertebrates and in some fish and reptiles (e.g. the viper). *Compare* OVIPARITY; VIVIPARITY.

ovulation The release of an egg cell from the ovary, which in mammals is stimulated by *luteinizing hormone. The developing egg cell within its follicle migrates to the ovary surface; when mature, it is released from the follicle (which breaks open) into the body cavity, from where it passes into the oviduct. *See* OVARIAN CYCLE.

ovule The part of the female reproductive organs of seed plants that consists of the *nucellus, *embryo sac, and *integuments. The ovules of gymnosperms are situated on ovuliferous scales of the female cones while those of angiosperms are enclosed in the carpel. After fertilization, the ovule becomes the seed.

ovuliferous scale One of a group of large woody specialized leaves that form the female *cone of conifers and related trees. It bears the ovules, which develop into seeds.

ovum (egg cell) (*pl.* **ova**) **1.** (in zoology) The mature reproductive cell (*see* GAMETE) of female animals, which is produced by the ovary (*see* OOGENESIS). It is spherical, has a nucleus, is covered with an *egg membrane, and is not mobile. **2.** (in botany) The *oosphere of plants.

oxalic acid (ethanedioic acid) A crystalline solid, $(COOH)_2$, that is slightly soluble in water. Oxalic acid is strongly acidic and very poisonous. It occurs in certain plants, e.g. sorrel and the leaf blades of rhubarb.

oxaloacetic acid A compound, $HO_2CCH_2COCO_2H$, that plays an integral role in the *Krebs cycle. The anion, oxaloacetate, reacts with the acetyl group from acetyl coenzyme A to form citrate.

oxidase Any enzyme that catalyses *oxidation-reduction reactions that involve the transfer of electrons to molecular oxygen.

oxidation-reduction (redox) Originally, **oxidation** was simply regarded as a chemical reaction with oxygen. The reverse process—loss of oxygen—was called **reduction**. Reaction with hydrogen also came to be regarded as reduction. Later, a more general idea of oxidation and reduction was developed in which oxidation was loss of electrons and reduction was gain of electrons. This wider definition covered the original one and also applies to reactions that do not involve oxygen. However, it applies only to reactions in which electron transfer occurs—i.e. to reactions involving ions. It can be extended to reactions between covalent compounds by using the concept of **oxidation number** (or **state**). This is a measure of the electron control that an atom has in a compound compared to the atom in the pure element. An oxidation number consists of two parts:

(1) its sign, which indicates whether the control has increased (negative) or decreased (positive);

(2) its value, which gives the number of electrons over which control has changed.

The change of electron control may be complete (in ionic compounds) or partial (in covalent compounds). Oxidation is a reaction involving an increase in oxidation number and reduction involves a decrease. Thus in

$$2H_2 + O_2 \rightarrow 2H_2O$$

the hydrogen in water is $+1$ and the oxygen -2. The hydrogen is oxidized and the oxygen is reduced. Compounds that tend to undergo reduction readily are **oxidizing agents**; those that undergo oxidation are **reducing agents**. Redox reactions are fundamental to the metabolism of living organisms, particularly the oxidation of glucose via *glycolysis and the capture of energy via the Krebs cycle and the *electron transport chain. Coenzymes such as NAD^+ participate in these redox reactions as electron carriers.

oxidative burst (respiratory burst) The production of various highly reactive oxygen derivatives (*see* REACTIVE OXYGEN SPECIES) by certain cells, notably macrophages and neutrophils of the vertebrate immune system. These toxic compounds are used to kill bacteria engulfed by the macrophage, or secreted to attack larger parasites outside the macrophage. The enzyme NADPH oxidase catalyses the formation of superoxide anions (O_2^-), which are then converted by *superoxide dismutase to form hydrogen peroxide (H_2O_2), singlet oxygen, and hydroxyl radicals (OH·). Another phagocytic enzyme, myeloperoxidase, catalyses a reaction between chloride ions and hydrogen peroxide to yield hypochlorous acid (HOCl), which reacts with further hydrogen peroxide to yield more singlet oxygen. *Nitric oxide, and its peroxynitrite derivatives, are also produced. The net effect of this toxic cocktail is to disable the target cell by oxidizing crucial components of its cellular apparatus.

oxidative deamination A reaction involved in the catabolism of amino acids that assists their excretion from the body. An example of an oxidative deamination is the conversion of glutamate to α-ketoglutarate, a reaction catalysed by the enzyme glutamate dehydrogenase. *See* DEAMINATION.

oxidative decarboxylation The reaction in the *Krebs cycle in which oxygen, derived from two water molecules, is used to oxidize two carbon atoms to two molecules of carbon dioxide. The two carbon atoms result from the *decarboxylation reactions that occur during the Krebs cycle as the six-carbon compound citrate is converted to the four-carbon compound oxaloacetate.

oxidative phosphorylation A reaction occurring during the final stages of *aerobic respiration, in which ATP is formed from ADP and phosphate coupled to electron transport in the *electron transport chain. The reaction occurs in the mitochondria (*see* CHEMIOSMOTIC MECHANISM) and is the cell's principal method of storing the energy released by the oxidation of food. *See also* PHOSPHORYLATION; P/O RATIO.

oxidoreductase Any of a class of enzymes that catalyse *oxidation-reduction reactions, i.e. they are involved in the transfer of hydrogen or electrons between molecules. They include the *oxidases and *dehydrogenases.

oxygen Symbol O. A colourless odourless gaseous element. It is the most abundant element in the earth's crust (49.2% by weight) and is present in the atmosphere (28% by volume), mainly as dioxygen (O_2), with much smaller amounts of ozone (O_3). Atmospheric dioxygen is of vital importance for all organisms that carry out *aerobic respiration.

oxygen cycle The cycling of oxygen between the biotic and abiotic components of the environment (*see* BIOGEOCHEMICAL CYCLE). The oxygen cycle is closely linked to the *carbon cycle and the water cycle (*see* HYDROLOGICAL CYCLE). In the process of respiration oxygen is taken in by living organisms and released into the atmosphere, combined with carbon, in the form of carbon dioxide. Carbon dioxide enters the carbon cycle or is taken up by plants for *photosynthesis. During photosynthesis oxygen is evolved by the chemical splitting of water and returned to the atmosphere. In the upper atmosphere, ozone is formed from oxygen and dissociates to release oxygen (*see also* OZONE LAYER).

oxygen debt The physiological state that exists in a normally aerobic animal when insufficient oxygen is available for metabolic requirements (e.g. during a period of strenuous physical activity). To meet the body's increased demand for energy, pyruvate is converted anaerobically (i.e. by *fermentation, in the absence of oxygen) to lactic acid, which requires oxygen for its breakdown and accumulates in the tissues. When oxygen is available again lactic acid is oxidized in the liver, thus repaying the debt.

oxygen dissociation curve The S-shaped curve produced when the percentage saturation of haemoglobin with oxygen (i.e. the percentage of binding sites of haemoglobin that are occupied by oxygen molecules) is plotted against the partial pressure of oxygen (pO_2), which is a measure of the oxygen concentration in the surrounding medium. The steep rise of the curve indicates the high affinity of haemoglobin for oxygen: a small increase in pO_2 results in a relatively sharp increase in the percentage saturation of haemoglobin with oxygen. Therefore in the lungs, where the pO_2 is high, the blood is rapidly saturated with oxygen. Conversely, a small drop in pO_2 results in a large drop in percentage saturation of haemoglobin. Thus in tissues that utilize oxygen at a high rate, where the pO_2 is low, oxygen readily dissociates from haemoglobin and is released for use by the tissues. *See also* BOHR EFFECT.

Oxygen dissociation curve

oxygen-evolving complex (OEC) A small complex of proteins and manganese ions associated with the reaction centre of photosystem II (PSII) that splits (oxidizes) water molecules to yield oxygen during *photosynthesis. Essentially, this oxidation provides the low-energy electrons from water that are excited by PSII using the light energy trapped by chlorophyll pigments. The resultant high-energy electrons can then flow through the electron transport chain. PSII draws electrons singly from a cluster of four manganese ions held by each OEC. When this cluster has accumulated four positive charges, the OEC oxidizes two molecules of water, producing one molecule of oxygen and four protons. The OEC is located on the lumen side of the thylakoid membrane in the chloroplast and requires chloride ions to function. *See* PHOTOSYSTEMS I AND II.

oxygen sag curve The curve obtained when the concentration of dissolved oxygen in a river into which sewage or some other pollutant has been discharged is plotted against the distance downstream from the sewage outlet (see graph). Samples of water are taken at areas upstream and downstream from the sewage outlet. The presence of sewage reduces the oxygen content of the water and increases the *biochemical oxygen demand. This is due to the action of saprotrophic organisms that decompose the organic matter in the sewage and in the process use up the available oxygen.

oxyhaemoglobin *See* HAEMOGLOBIN.

oxyntic cell (parietal cell) Any of the cells in the gastric glands in the wall of the stomach that produce *hydrochloric acid, which forms part of the *gastric juice.

The oxyntic cells also produce intrinsic factor, which is involved in the absorption of vitamin B_{12} in the small intestine (*see* VITAMIN B COMPLEX).

oxytocin A hormone, produced by birds and mammals, that in mammals causes both contraction of smooth muscle in the uterus during birth and expulsion of milk from the mammary glands during suckling. Like *antidiuretic hormone, oxytocin is produced in the neurosecretory cells of the hypothalamus (*see* NEUROSECRETION) but is stored and secreted by the posterior pituitary gland. In some species the high circulating levels of oxytocin immediately after birth are associated with a critical period for the mother's bonding with her newborn young. *See also* NEUROPHYSIN.

ozonation The formation of ozone (O_3) in the earth's atmosphere. In the stratosphere, about 20–50 km above the surface of the earth, oxygen molecules (O_2) dissociate into their constituent atoms under the influence of *ultraviolet radiation of short wavelength (below about 240 nm). These atoms combine with oxygen molecules to form ozone (*see* OZONE LAYER). Ozone is also formed in the lower atmosphere (*troposphere) from nitrogen oxides and other pollutants by photochemical reactions (*see* PHOTOCHEMICAL SMOG). Through its oxidative damage to leaves and their photosynthetic apparatus, tropospheric ozone is increasingly a cause of impaired plant growth and reduced crop yields. It also poses a threat to human health by irritating the respiratory tract and triggering or exacerbating diseases such as asthma and lung diseases.

Oxygen sag curve

ozone layer (ozonosphere) A layer of the earth's atmosphere in which most of the atmosphere's ozone is concentrated. It occurs 15–50 km above the earth's surface and is virtually synonymous with the *stratosphere. In this layer most of the sun's ultraviolet radiation is absorbed by the ozone molecules, causing a rise in the temperature of the stratosphere and preventing vertical mixing so that the stratosphere forms a stable layer. By absorbing most of the solar ultraviolet radiation the ozone layer protects living organisms on earth. The fact that the ozone layer is thinnest at the equator is believed to account for the high equatorial incidence of skin cancer as a result of exposure to unabsorbed solar ultraviolet radiation. In the 1980s it was found that depletion of the ozone layer was occurring over both the poles, creating **ozone holes**. This is thought to have been caused by a series of complex photochemical reactions involving *nitrogen oxides produced from aircraft and, more seriously, *chlorofluorocarbons (CFCs) and halons. CFCs rise to the stratosphere, where they react with ultraviolet light to release chlorine atoms; these atoms, which are highly reactive, catalyse the destruction of ozone. Use of CFCs is now much reduced in an effort to reverse this human-induced damage to the ozone layer, and there are signs that the ozone holes are beginning to heal, although this process will likely take decades to complete. *See also* AIR POLLUTION.

(⊕) SEE WEB LINKS

• Description of the ozone layer, its depletion, and steps to protect it, produced by the US Environmental Protection Agency

P (parental generation) The individuals that are selected to begin a breeding experiment, crosses between which yield the *F_1 generation. In classical genetics studies, only pure-breeding (homozygous) individuals are selected for the P generation.

p21 A protein with a molecular mass of roughly 21 000 kDa that plays a crucial role in regulating the *cell cycle and enabling repair of damaged DNA before the cell proceeds to replication and division. Expression of the p21 gene is regulated by the product of the *tumour-suppressor gene p53, and occurs in response to DNA damage. The p21 protein binds to the cyclin-dependent kinase (CDK) present during the G_1 and S phases, thus inhibiting its activation by *cyclin. This pauses the cell cycle until the DNA is repaired, whereupon the p21 breaks down and the cell cycle resumes.

p53 *See* TUMOUR-SUPPRESSOR GENE.

pacemaker 1. (sinoatrial node) A small mass of specialized cardiac muscle cells in the mammalian heart, found in the wall of the right atrium near the opening for the vena cava. The cells initiate and maintain the heartbeat—their rhythmic electrical activity stimulates contractions of the atria, which in turn cause contractions of the ventricles (*see* ATRIOVENTRICULAR NODE). The electrical characteristics of the pacemaker cells result from the interplay of their various ion channels. These allow spontaneous and repeated generation of action potentials, which spread to neighbouring cardiac muscle cells via gap junctions between neighbouring cells. In particular, the change in polarity during each action potential depends on an influx of calcium ions (Ca^{2+}) through voltage-gated calcium channels, in contrast to the sodium ion

channels in other cardiac muscle cells, neurons, and skeletal muscle. The interval between the action potentials of the pacemaker cells governs the rate of heartbeat, and is regulated by the autonomic nervous system. The sympathetic neurotransmitter noradrenaline speeds up heartbeat, whereas acetylcholine, released by parasympathetic nerve endings, has the opposite effect. Similar pacemakers occur in the hearts of other vertebrates. **2.** An electronic or nuclear battery-charged device that can be implanted surgically into the chest to produce and maintain the heartbeat. These devices are used when the heart's own pacemaker is defective or diseased.

pachytene The period in the first prophase of *meiosis when paired *homologous chromosomes are fully contracted and twisted around each other.

Pacinian corpuscle (lamellated corpuscle) Any of certain sensory receptors that respond to pressure and vibration and are found in the subcutaneous layer of the skin, muscles, tendons, mesenteries, and joints of mammals. Each consists of a nerve ending surrounded by an oval capsule made of concentric layers of connective tissue. Pacinian corpuscles are fast adapting, and hence respond primarily to changes in pressure rather than sustained pressures. They are named after Italian anatomist Filipo Pacini (1812–83).

paedogenesis Reproduction by an animal that is still in the larval or pre-adult form. Paedogenesis is a form of *neoteny and is particularly marked in the axolotl, a larval form of the salamander, which retains its larval features owing to a lack of thyroid activity as an adaptation to life in an environment deficient in iodine, which is necessary for thyroid function. However, despite this it can breed, producing individuals like itself. If the thyroid hormone thyroxine is given, metamorphosis occurs.

paedomorphosis The evolutionary process in which larval or juvenile features of an ancestral organism are displaced to the adult forms of its descendants. It can arise by *neoteny or *progenesis. Paedomorphosis is thought to have occurred in the evolution of higher chordates from freeswimming larval tunicates, in which metamorphosis was eventually lost and sexual development accelerated until larval forms were capable of breeding.

PAGE (polyacrylamide gel electrophoresis) *See* GEL ELECTROPHORESIS.

PAI *See* PATHOGENICITY ISLAND.

pain Severe physical or mental discomfort or distress. Pain of a physical origin arises through detection of noxious stimuli or tissue injury by specific sensory neurons in the affected part of the body, which relay signals to the thalamus in the brain (*see* NOCICEPTION). However, perception of and susceptibility to pain are determined by higher centres in the cortex and can be influenced by such factors as emotional state.

pairing (synapsis) The close association between *homologous chromosomes that develops during the first prophase of *meiosis. The two chromosomes move together and a *synaptonemal complex of proteins forms between them, ensuring exact pairing of corresponding points along their lengths as they lie side by side. The resulting structure is called a **bivalent**.

pair-rule gene *See* SEGMENTATION GENES.

Palaearctic region *See* FAUNAL REGION.

palaeobotany The branch of *palaeontology concerned with the study of plants through geological time, as revealed by their *fossil remains (*see also* PALYNOLOGY). It overlaps with other aspects of plant study, including anatomy, ecology, evolution, and taxonomy.

Palaeocene The earliest geological epoch of the *Palaeogene period. It began about 66 million years ago, following the Cretaceous period, and extended for about 10 million years to the beginning of the *Eocene (the Palaeocene is sometimes included in the Eocene). It was named by the palaeobotanist W. P. Schimper in 1874. A major floral and faunal discontinuity occurred between the end of the Cretaceous and the beginning of the Palaeocene (*see* ALVAREZ EVENT): following the extinction of many reptiles the mammals became abundant on land. By the end of the epoch primates and rodents had evolved.

palaeoclimatology The study of climates of earlier geological periods. This is based largely on the study of sediments that were laid down during these periods and of fossils. The changes in the positions of the continents as a result of *continental drift and *plate tectonics complicate the study. Determination of historical perturbations in the climate provides vital information for the construction of accurate climate models that can predict the trajectory of future climate change.

palaeoecology The study of the relationships of *fossil organisms to each other and to their environments. It involves the study both of the fossils and of the surrounding rocks in which they are found. Trace fossils may provide information on the behaviour of the organism.

Palaeogene The older of the two geological periods of the *Cenozoic era, consisting of the *Palaeocene, *Eocene, and *Oligocene epochs. The Palaeogene began 66 million years ago and lasted until the start of the *Neogene period, 23 million years ago. It thus corresponds approximately to the first two-thirds of the *Tertiary period, a division that is no longer officially recognized.

Palaeolithic The Old Stone Age, lasting in Europe from about 2.5 million to 10 000 years ago, during which humans used primitive stone tools made by chipping stones and flints.

palaeomagnetic dating A technique for estimating the age of certain rocks based on determining the alignment of their magnetic field, which is fixed when the rocks are formed. The result can then be compared with the established timeline of the earth's geomagnetic reversals to give an estimate of the rock's age. It is useful when dating rock, particularly sedimentary material, that lacks the igneous intrusions required for radiometric *dating techniques.

palaeontology The study of extinct organisms, including their structure, environment, evolution, and distribution, as revealed by their *fossil remains. Palaeontological work also makes important contributions to geology in revealing stratigraphic relationships between rock strata and determining the physical appearance and climate of the earth during past geological ages. *See also* PALAEOBOTANY; PALAEOECOLOGY; PALAEOZOOLOGY.

Palaeozoic The first era of *Phanerozoic time. It follows the *Precambrian and is subdivided into the Early Palaeozoic, comprising the *Cambrian, *Ordovician, and *Silurian periods, and the Late Palaeozoic, comprising the *Devonian, *Carboniferous, and *Permian periods. It extended from about 541 million years ago to about 252 million years ago, when it was succeeded by the *Mesozoic era.

palaeozoology The branch of *palaeontology concerned with the study of animals throughout geological time, as revealed by their *fossil remains.

palate The roof of the mouth cavity of vertebrates, which separates the *buccal and nasal cavities. In mammals it is divided into two zones, the bony **hard palate** and the **soft palate**, and completely separates the buccal cavity from the air passage to enable simultaneous eating and breathing.

palea (*pl.* **paleae**) *See* LEMMA.

palindromic Describing a section of double-stranded DNA in which the sequence of bases on one strand is inverted and repeated on the other. Thus the following sequence is palindromic:

—ACTTGCAAGT—
—TGAACGTTCA—

Palindromic sequences are common in DNA and are the sites at which the DNA is cleaved by some *restriction enzymes. *See also* CRISPR.

palisade (**palisade mesophyll**) *See* MESOPHYLL.

pallium *See* CEREBRAL CORTEX.

palmitic acid (hexadecanoic acid) A saturated fatty acid, $CH_3(CH_2)_{14}COOH$. Glycerides of palmitic acid occur widely in plant and animal oils and fats.

palp An elongated sensory organ, usually near the mouth, in many invertebrates. Examples are the tactile head appendages of polychaete worms, the ciliated flap of tissue that produces feeding currents in bivalve molluscs, the distal part of the *mandibles of crustaceans, and the olfactory parts of the first and second *maxillae of some insects.

palynology (micropalaeontology) The study of fossil pollen and spores (**pollen analysis**) and various other *microfossils, such as coccoliths and dinoflagellates. Palynology is used in stratigraphy, palaeoclimatology, and archaeology. Pollen and spores are very resistant to decay and therefore their fossils are found in sedimentary rocks. They may be extracted by various methods, including boiling with potassium hydroxide solution, washing with strong oxidizing mixtures, and centrifuging repeatedly. Spores and pollen are classified according to shape, form of aperture, and both internal and external details of the exine (outer coat). They indicate the nature of the dominant flora, and therefore the climate and conditions of the period in which they lived.

PAMP (pathogen-associated molecular pattern) *See* IMMUNITY (sense 2).

pancreas (*pl.* **pancreases**) A gland in vertebrates lying between the duodenum and the spleen. Under the influence of the hormone *secretin it secretes **pancreatic juice** containing bicarbonate ions, which neutralize the acidity of chyme entering the duodenum from the stomach, and digestive enzymes or their precursors (*trypsin, *chymotrypsin, *amylase, *lipase, *nuclease, and *carboxypeptidase). This mixture is discharged into the duodenum via the pancreatic duct (*see* ACINUS). The pancreas also contains groups of cells—the *islets of Langerhans—that function as an *endocrine gland, producing the hormones *insulin and *glucagon, which regulate blood sugar levels.

pancreatic islets *See* ISLETS OF LANGERHANS.

pancreatin A mixture of digestive enzymes that have been extracted from the *pancreas.

pancreozymin *See* CHOLECYSTOKININ.

Pancrustacea A clade of *arthropods containing the crustaceans and insects. It has been established on the strength of molecular evidence showing the close evolutionary relationship between various lineages of the two groups.

pandemic An *epidemic that spreads worldwide. For example, a variant of the influenza A virus, H1N1, emerged in North America in 2009 and quickly spread to affect over 200 countries around the globe by August 2010. There was a low level of immunity among the human population, flu vaccines were ineffective, and up to 200 000 people are thought to have died.

Pandoravirus A genus of giant viruses first described in 2013. They infect amoebas and measure roughly 1×0.5 μm, making them larger than many bacteria, and they have unusually large genomes, of 1.9 and 2.5 million bases, with many unique genes of unknown function. The viral particles enter the host amoeba via phagocytic vacuoles and fuse with the vacuole membrane. Like other giant viruses, pandoraviruses are

regarded as *nucleocytoplasmic large DNA viruses, and may represent forms that originally evolved from cellular organisms.

Paneth cells *See* CRYPTS OF LIEBERKÜHN.

panicle A type of flowering shoot common in the grass family. The primary axis bears groups of *racemes and is itself racemose, as the youngest groups of flowers are at the top (e.g. oat). The term may be used loosely for any form of branched *racemose inflorescence; for example, the horse chestnut is a raceme of cymes. Both these arrangements are seen in the family Polygonaceae (docks and sorrels).

panmictic Describing a population in which mating is entirely random and any two (male and female) individuals are equally likely to mate. Random mating (or **panmixis**) is one of the assumptions of the *Hardy-Weinberg equilibrium but is probably uncommon in natural populations, in which spatial structuring and *assortative mating are usually evident.

panspermia The theory that the 'seeds of life' (in the form of microorganisms or their dormant spores) exist throughout the universe and are transported by meteorites or other small celestial bodies. According to this **strong panspermia theory**, when such a meteorite collides with a planet the microorganisms are released, potentially enabling them to colonize new sites where conditions are suitable for life. Proponents argue that this could be how life on Earth originated. A so-called **weak panspermia theory** posits that merely some of the essential building blocks of life, such as amino acids or nucleobases, were delivered in this way, and that these contributed to the emergence of the first life forms uniquely on earth. Our planet experienced intense meteor bombardment about 4 to 3.8 billion years ago and there is now strong evidence that meteorites can contain organic molecules; but whether and how these enabled life to begin nearly 3.5 billion years ago is contentious. *See also* ASTROBIOLOGY.

pantothenic acid A vitamin of the *vitamin B complex. It is a constituent of *coenzyme A, which performs a crucial role in the oxidation of fats, carbohydrates, and certain amino acids. Deficiency rarely occurs because the vitamin occurs in many foods, especially cereal grains, peas, egg yolk, liver, and yeast.

papain A protein-digesting enzyme (*see* PROTEASE) occurring in the fruit of the West Indian papaya tree (*Carica papaya*). It is used in biochemical research, as a digestant, in the manufacture of meat tenderizers, for preparing hides for tanning, in contact lens cleaning agents, and in various cosmetics and medicines.

paper chromatography A technique for analysing mixtures by *chromatography, in which the stationary phase is absorbent paper. A spot of the mixture to be investigated is placed near one edge of the paper and the sheet is suspended vertically in a solvent, which rises through the paper by capillary action carrying the components with it. The components move at different rates, partly because they absorb to different extents on the cellulose and partly because of partition between the solvent and the moisture in the paper. The paper is removed and dried, and the different components form a line of spots along the paper. Colourless substances are detected by using ultraviolet radiation or by spraying with a substance that reacts to give a coloured spot (e.g. ninhydrin gives a blue coloration with amino acids). The components can be identified by the distance they move in a given time. It is now primarily used as a teaching aid.

papilla (*pl.* **papillae**) Any cone-shaped protuberance projecting from the surface of an organ or organism. Papillae occur, for example, on the tongue, in the kidneys, and, in plants, on the surface of many petals.

papovavirus One of a group of *DNA-containing viruses, formerly members of the family *Papovaviridae*, that produce tumours in their hosts. They include the **papillomaviruses**, which produce nonmalignant tumours (such as warts) in all vertebrates and also certain types of cancer (e.g. cervical cancer) in humans. The **polyomaviruses** produce malignant tumours in certain classes of vertebrates (not including humans), and include the *SV40 virus.

pappus (*pl.* **pappi**) A group of modified *sepals, often in the form of a ring of silky hairs. For example, when the fruit of the dandelion matures a pappus of hairs persists at the top of a thin stalk forming a parachute-like structure, which serves to disperse the fruit.

parabiologist An individual who is trained in basic biological skills to enable them to support professional biologists, e.g. in a wildlife conservation programme. Training of indigenous people as parabiologists is increasingly seen as a key ingredient for the long-term success of conservation efforts. Skills might include wildlife monitoring, identification, sampling, and recording techniques, as well as knowledge of the principles of conservation.

parabiosis The surgical joining together of two organisms, or parts of organisms, so that they share a common circulation and can exchange hormones or other internally secreted substances. It is used, for example, as an experimental technique in insect endocrinology to show the effects on the adult of hormones produced by different larval stages.

paracellular pathway The route between cells. For example, substances can travel through epithelia by a paracellular pathway if the *tight junctions between constituent cells are not fully continuous (i.e. 'leaky'), as in the proximal tubule of a kidney nephron. Paracellular pathways lack any means of active transport, and substances can only move passively by simple diffusion. *Compare* TRANSCELLULAR PATHWAY.

paracrine Describing a type of cell signalling in which signal molecules affect only target cells located near the cell from which they are derived. Paracrine signals diffuse over short distances and typically act rapidly, e.g. as at a chemical synapse between two nerve cells. *Compare* AUTOCRINE. *See* SIGNAL TRANSDUCTION.

parallel evolution The development of related organisms along similar evolutionary paths due to strong selection pressures acting on all of them in the same way. It is debatable if the phenomenon really exists: many argue that all evolution is ultimately *convergent or divergent (*see* ADAPTIVE RADIATION).

parallelophyly The independent emergence of the same trait in two related lineages because of a genetic predisposition inherited from a common ancestor. Such a phenomenon could account for certain instances of apparent *homoplasy.

paralogous Describing *homologous genes that have arisen by duplication of an ancestral gene. The copies (**paralogues**) thus evolve side by side ('in parallel') in the genome of subsequent lineages, irrespective of speciation events, and hence display similarity of base sequence and, possibly, function. The class of homeotic genes known as *Hox* genes consists of 13 highly conserved paralogues, some of which have been identified in organisms as diverse as nematodes and mice. This is taken as evidence that they are the descendants of homeotic genes in an ancient common ancestor, possibly dating back to the Precambrian. *Compare* ORTHOLOGOUS.

paramorph A variant form within a species for which a more specific description does not exist because its taxonomic status cannot be determined.

paramutation *See* EPIGENETIC.

parapatric speciation A type of *speciation in which there is free exchange of genes between two populations of organisms living in directly adjacent but environmentally different habitats. Although individuals from the two populations can interbreed, this rarely occurs due to morphological, behavioural, or genetic differences, which are reinforced over time, leading eventually to reproductive isolation and formation of distinct species.

paraphyletic In systematics, describing a group of organisms that excludes one or more descendants of a particular single common ancestor. For example, the taxonomic group 'dinosaurs' is paraphyletic unless it includes the birds, which share the same common ancestor as extinct dinosaurs. In *cladistics such groups are regarded as invalid when constructing classification schemes, since cladists allow only *monophyletic groups, or clades, as a basis for taxonomic groupings. However, in evolutionary systematics paraphyletic groups, or **evolutionary grades**, are sometimes permitted in order to reflect biological similarities. *Compare* POLYPHYLETIC.

paraphysis (*pl.* **paraphyses**) Any of the sterile filaments of cells that are found around the sex organs of mosses (Bryophyta), brown algae (Phaeophyta), and certain fungi (the Ascomycota and Basidiomycota).

parapodium (*pl.* **parapodia**) *See* CHAETA; POLYCHAETA.

Paraquat Trade name for an organic herbicide, $1,1'$-dimethyl-$4,4'$-bipyridinium dichloride, used to control broadleaved weeds and grasses (*see* VIOLOGEN DYES). It is poisonous to humans, having toxic effects on the liver, lungs, and kidneys if ingested. Paraquat is not easily broken down and can persist in the environment adsorbed to soil particles. It is prohibited in the European Union. *See also* PESTICIDE.

parasexual cycle A type of reproduction in fungi that results in genetic recombination without the formation and fusion of gametes. It occurs in heterokaryotic mycelia, in which two haploid nuclei fuse and then divide by mitosis, during which a form of *crossing over occurs. During subsequent divisions chromosomes are progressively lost from the resultant diploid nucleus so that the haploid condition is restored. The asexual spores produced during the cycle will be genetically different from the original parent mycelium as a result of the crossing over that has occurred.

parasitism An association in which one organism (the **parasite**) lives on (**ectoparasitism**) or in (**endoparasitism**) the body of another (the *host), from which it obtains its nutrients. Some parasites inflict comparatively little damage on their host, but many cause characteristic diseases (these are, however, never immediately fatal, as killing the host would destroy the parasite's source of food; *compare* PARASITOID). Parasites are usually highly specialized for their way of life, which may involve one host or several (if the *life cycle requires it). They typically produce vast numbers of eggs, very few of which survive to find their way to another suitable host. **Obligate parasites** can only survive and reproduce as parasites; **facultative parasites** can also live as *saprotrophs. The parasites of humans include fleas and lice (which are ectoparasites), various bacteria, protists, and fungi (endoparasites causing characteristic diseases), and tapeworms (e.g. *Taenia solium*, which lives in the gut). *See also* HEMIPARASITE.

(⊕) SEE WEB LINKS

• Descriptions of human parasites feature in this website from the US Centers for Disease Control and Prevention

parasitoid An animal that lives in or on another animal (the **host**), which it consumes and eventually kills. Hence parasitoids may be regarded as intermediate between a true parasite (*see* PARASITISM) and a predator. Many wasps and other hymenopterans, and also some flies, are parasitoids for part of their life cycle, laying their eggs commonly in or on the larvae of other insects. The hatched parasitoids then use the tissues of their host for food during development into free-living adults. **Idiobiont parasitoids** paralyse or kill the host at the time of egg-laying. However, more common are **kainobiont parasitoids**, which allow the host to remain alive for extended periods. In some cases the parasitoid can prolong the development of the host larva and prevent it metamorphosing. *See also* HYPERPARASITE.

parasympathetic nervous system (PNS) Part of the *autonomic nervous system. Its nerve endings release acetylcholine as a *neurotransmitter and its actions tend to antagonize those of the *sympathetic nervous system. For example, the parasympathetic nervous system increases salivary gland secretion, decreases heart rate, and promotes digestion (by increasing *peristalsis)—'rest and digest'—while the sympathetic nervous system has opposite effects. However, in relation to reproductive function the PNS promotes erection of genitalia, thus complementing the action of the sympathetic system, which promotes ejaculation and vaginal contractions. Anatomically the preganglionic neurons of the PNS originate from the cranial nerves and from the lowest (sacral) region of the spinal cord, compared with the thoracic and lumbar regions of the spinal cord for the sympathetic division. Also, its ganglia lie close to the target organs, rather than alongside the spinal cord.

parathyroid glands Two pairs of *endocrine glands situated behind, or embedded within, the thyroid gland in higher vertebrates. They produce *parathyroid hormone, which controls the amount of calcium in the blood. *See also* C CELL.

parathyroid hormone (PTH; parathormone; parathyrin) A peptide hormone secreted by the *parathyroid gland in response to low levels of calcium in the blood. It acts to maintain normal blood levels of calcium by (1) increasing the number of *osteoclasts, which break down the bone matrix and release calcium into the blood; (2) increasing the reabsorption of calcium and magnesium ions in the kidney tubules, so that their concentration is maintained in the blood; (3) converting *vitamin D to its active form, which increases calcium absorption in the intestine. Parathyroid hormone acts in opposition to *calcitonin.

Parazoa *See* METAZOA.

parenchyma 1. A plant tissue consisting of roughly spherical relatively undifferentiated cells, frequently with air spaces between them. The cortex and pith are composed of parenchyma cells (*see* GROUND TISSUES). **2.** Loose *connective tissue formed of large cells. Its function is to pack the spaces between organs in some simple acoelomate animals, such as flatworms (Platyhelminthes).

parent 1. Either male or female partner that together produce offspring in the process of sexual reproduction. *See also* P. **2.** Denoting an organism or cell that gives rise to new organisms or cells, as by asexual reproduction or cell division.

parental care Any behaviour pattern in which a parent invests time or energy in feeding and protecting its offspring. Parental care is a form of *altruism since this type of behaviour involves increasing the *fitness of the offspring at the expense of the parents. The degree of parental care differs widely. For example, most species of fish show little or no parental care while humans and many other mammals care for their offspring until they reach adolescence.

parietal Of or relating to the wall of a body cavity or other structure.

parietal cell *See* OXYNTIC CELL.

parsimony *See* MAXIMUM PARSIMONY; PRINCIPLE OF PARSIMONY.

parthenocarpy The formation of fruit without prior fertilization of the flower by pollen. The resulting fruits are seedless and therefore do not contribute to the reproduction of the plant; examples are bananas and pineapples. Because this is regarded as a desirable trait, parthenocarpic cultivars of bananas, figs, breadfruit, and various other fruit crops are grown commercially. Plant hormones, such as gibberellins and auxins, can induce parthenocarpy, as can artificial fertilization using nonviable pollen.

parthenogenesis The development of an organism from an unfertilized egg. This occurs sporadically in many plants (e.g. dandelions and hawkweeds) and arthropods, and in some fishes, amphibians, and reptiles. In some species it is the main and sometimes only method of reproduction. For example, in some species of aphid, males are absent or very rare. The eggs formed by the females contain the full (diploid) number of chromosomes and are genetically identical. Variation is consequently very limited in species that reproduce parthenogenetically. *See* ARRHENOTOKY; THELYTOKY.

partially permeable membrane A membrane that is permeable to the small molecules of water and certain solutes but does not allow the passage of large solute molecules. This term is preferred to **semipermeable membrane** when describing membranes in living organisms. *See* OSMOSIS.

particulate inheritance The transmission from parent to offspring of separate units that determine characteristics. Gregor Mendel observed that *recessive characteristics, absent in the offspring of a cross in which only one parent possessed them, reappeared repeatedly in the progeny of subsequent crosses. This led him to formulate his theory of inherited 'factors' (now called *alleles) that retain their identity through succeeding generations (*see* MENDEL'S LAWS). *Compare* BLENDING INHERITANCE.

parturition The act of giving birth to young at the end of the *gestation period. In most mammals it is triggered by the secretion of ACTH by the fetal pituitary gland, which causes increased secretion of *corticosteroid hormones by the fetal adrenal glands. Cortisol produced by the fetus stimulates the activity of a placental enzyme that increases the level of oestradiol in the bloodstream, which in turn enhances the synthesis of prostaglandins by the uterus, leading to the muscular contractions of labour. Other key hormones include *oxytocin and *relaxin. In women this broad picture is made more complex by multiple paracrine, autocrine, and endocrine signals and various maternal and fetal control mechanisms.

pascal The *SI unit of pressure equal to one newton per square metre.

passive immunity *See* IMMUNITY.

passive transport *See* DIFFUSION.

Pasteur, Louis (1822–95) French chemist and microbiologist, who held appointments in Strasbourg (1849–54) and Lille (1854–57), before returning to Paris to the

Ecole Normale and the Sorbonne. From 1888 to his death he was director of the Pasteur Institute. In 1848 he discovered *optical activity, in 1860 relating it to molecular structure. In 1856 he began work on *fermentation, and by 1862 was able to disprove the existence of *spontaneous generation. He introduced *pasteurization (originally for wine) in 1863. He went on to study disease and developed vaccines against cholera (1880), anthrax (1882), and rabies (1885).

pasteurization The treatment of milk to destroy disease-causing bacteria, such as those of tuberculosis, typhoid, and brucellosis. Milk is heated to 65°C for 30 minutes or to 72°C for 15 minutes followed by rapid cooling to below 10°C. The method was devised by the French microbiologist Louis Pasteur (1822–95). *See also* FOOD PRESERVATION.

patch clamp technique A method for studying the activation of voltage-dependent transmembrane channels in neurons. A small piece of membrane containing only a few channels is isolated and microelectrodes are sealed to both sides of the membrane.

patella (kneecap) (*pl.* **patellae**) A small rounded movable bone that is situated in a tendon in front of the knee joint in most mammals (including humans). The function of the patella is to protect the knee.

pathogen Any disease-causing microorganism. Pathogens include viruses and many bacteria, fungi, and protists. *See* INFECTION.

pathogen-associated molecular pattern (PAMP) *See* PATTERN RECOGNITION RECEPTOR. *See also* IMMUNITY (2).

pathogenesis-related proteins (PR proteins) *See* SYSTEMIC ACQUIRED RESISTANCE.

pathogenicity island (PAI) A mobile genetic element in the chromosome of some bacteria that contains a cluster of genes responsible for the disease-causing properties of that particular bacterial strain or species. The element becomes integrated

into the host cell's chromosome following *lateral gene transfer from other bacteria by means of bacteriophage, transposons, or plasmids. The virulence genes of the PAI might, e.g. endow the bacterium with the ability to synthesize new enzymes or toxins, or make it resistant to antibiotics, and are significant in determining its potential to cause disease in higher organisms.

pathology The study of the changes in organs and tissues that are caused by or give rise to disease. This involves the examination of tissue samples, X-ray photographs, or other evidence taken from living patients or from cadavers. **Clinical pathology** applies these findings to clinical cases, particularly in the development of diagnostic tests and treatments. In **experimental pathology**, disease processes are studied using experimental animals, cell cultures, or other means.

(((●))) SEE WEB LINKS

• An Internet Pathology Laboratory providing an online resource for pathology students, from Florida State University College of Medicine

patristic Denoting similarity between organisms resulting from common ancestry. *Compare* HOMOPLASY.

pattern formation (patterning) The establishment of basic body symmetries and repetition of parts during development. In animal embryos, for example, pattern formation defines the body axes, the number and polarity of body segments, and other basic reference points to produce the framework in which tissues and organs can develop in the correct place and orientation. The process is controlled by the interaction of numerous *morphogens encoded by maternal and embryonic genes whose expression is intricately coordinated. *See* HOMEOTIC GENES; MATERNAL EFFECT GENES; SEGMENTATION GENES.

pattern recognition A branch of statistics used to classify data based upon several variables. This type of statistical analysis is particularly useful in interpreting biological data that consist of multiple observations on large numbers of individuals.

pattern recognition receptor (PRR) A protein that recognizes the pattern of certain molecules on the surface of invading pathogens, such as bacteria, fungi, or viruses, and binds to them, thereby eliciting an immune response by the host organism. Such receptors occur in both plants and animals, and may be membrane bound or present in plasma or other body fluids. The molecules recognized by them are called **pathogen-associated molecular patterns (PAMPs)**. An example is the spatial arrangement of mannose or fucose molecules on bacterial surfaces that matches the recognition sites on mannose-binding lectin, a PRR found in plasma. Coating by this *lectin makes the bacterium more susceptible to phagocytosis. *See also* IMMUNITY.

Pavlov, Ivan Petrovich (1849–1936) Russian physiologist, who became professor of physiology in St Petersburg in 1886. While working on the physiology of digestion he discovered that the mere sight of food stimulates the production of digestive juices. For this work he was awarded the 1904 Nobel Prize for physiology or medicine. Pavlov went on to demonstrate operant *conditioning in dogs and other animals. *See also* LEARNING (Feature).

PCR *See* POLYMERASE CHAIN REACTION.

PDGF *See* GROWTH FACTOR.

peat A mass of dark-brown or black fibrous plant debris produced by the partial decay of vegetation, such as mosses of the genus *Sphagnum*, in acidic wet places (*see* HYDROSERE). It may accumulate in depressions. When subjected to burial and hence pressure and heat, it may be converted to *coal. Peat is often extracted and used to improve soil and as a fuel. However, peat is a globally significant store of carbon, and peatlands are important for retaining water and as wildlife habitats. Hence, there is concern about the impact of extracting peat for horticulture and other uses.

peck order *See* DOMINANT.

pecten Any of various comblike structures in animals. The pecten in the eyes of birds is a heavily pigmented structure with a network of blood vessels attached to the optic nerve and projecting into the vitreous humour. It is thought to supply the retina with nutrients and oxygen, and maintain acid–base balance inside the eye. A simple form of this structure is found in the eyes of reptiles.

pectic substance *See* PECTIN.

pectin (pectic substance) A colloidal mixture of large polysaccharides consisting chiefly of galacturonic acid residues and typically of molecular mass 10 000 to 40 000 kDa. Pectin is an important constituent of the matrix of plant cell walls and of the *middle lamella between adjacent cell walls; it is also found in certain plant juices. Normally present in an insoluble form, in ripening fruits and in tissues affected by certain diseases it changes into a soluble form, which is evidenced by softening of the tissues. It is used in making jam as it forms a gel with sucrose.

pectoral fins *See* FINS.

pectoral girdle (shoulder girdle) The bony or cartilaginous structure in vertebrates to which the anterior limbs (pectoral fins, forelegs, or arms) are attached. In mammals it consists of two dorsal *scapulae (shoulder blades) attached to the backbone and two ventral *clavicles (collar bones) attached to the sternum (breastbone).

pedicel The stalk attaching a flower to the main floral axis (*see* PEDUNCLE). Some flowers, described as **sessile**, do not have a pedicel and arise directly from the peduncle.

pedigree A family tree depicting the number, gender, matings, and other traits of parents and their offspring in consecutive generations. In human clinical genetics a pedigree can provide valuable information about the inheritance of certain diseases, such as whether the abnormal *allele is dominant or recessive, and the probability that future offspring will also have the disease. Pedigrees are also important in breeding elite livestock, such as dogs or horses, particularly to identify potentially favourable matings.

pedipalp Either member of the second pair of appendages in arachnids, immediately posterior to the *chelicerae. In species with large chelicerae the pedipalps are typically adapted for walking or sensory functions; otherwise they are variously adapted for defence, killing prey, or digging.

pedology The science of the study of soils, including their origin and characteristics and their utilization.

peduncle 1. The main stalk of a plant that bears the flowers, which may be solitary or grouped in an *inflorescence. *Compare* PEDICEL. **2.** *See* BRACHIOPODA.

pelagic Describing organisms that swim or drift in a sea or a lake, as distinct from those that live on the bottom (*see* BENTHOS). Pelagic organisms are divided into *plankton and *nekton.

Pelecypoda *See* BIVALVIA.

pellagra A disease resulting from a deficiency of *nicotinic acid, which is characterized by dermatitis and mental disorder.

pellicle The thin, semirigid, transparent outer covering that protects and maintains the shape of certain protists, e.g. *Euglena*. In ciliated organisms, e.g. *Paramecium*, it consists of regular arrays of membranous sacs, called alveoli. The centre of each sac forms a depression, arising from which are one or two cilia. Beneath the pellicle are tracts of contractile filaments called kinetodesmata. These connect to the bases of the cilia, and to each other, to form a supporting cytoskeletal mesh that can alter the shape of the cell. Defensive organelles called trichocysts project between the alveoli. These can discharge a long thread with a harpoon-like tip. *See also* ALVEOLATES.

pelvic fins *See* FINS.

pelvic girdle (pelvis; hip girdle) The bony or cartilaginous structure in vertebrates to which the posterior limbs (pelvic fins or legs) are attached. The pelvic girdle articulates dorsally with the backbone; it is made up of two halves, each produced by the fusion of the *ilium, *ischium, and *pubis.

pelvis (*pl.* **pelvises** or **pelves**) **1.** *See* PELVIC GIRDLE. **2.** The lower part of the abdomen in the region of the pelvic girdle. **3.** A conical chamber in the *kidney into which urine drains from the kidney tubules before passing to the *ureter.

penicillin An *antibiotic derived from the mould *Penicillium notatum*; specifically it is known as **penicillin G** and belongs to a class of similar substances called penicillins. They are all active against a wide variety of bacteria, producing their effects by inhibiting cross-linking of peptidoglycans in the bacterial cell wall, especially Grampositive bacteria. Penicillins are used to treat a variety of infections caused by these bacteria, although many bacterial strains have developed resistance.

penis (*pl.* **penises** or **penes**) The male reproductive organ of mammals (and also of some birds and reptiles) used to introduce sperm into the female reproductive tract to ensure internal fertilization. It contains a duct (the *urethra) through which the sperms pass. The penis becomes erect during precopulatory activity, either by filling with blood or haemolymph or by the action of muscles, and can be inserted into the vagina (or cloaca). In mammals the urine also leaves the body through the penis. *See also* HEMIPENIS.

pentadactyl limb A limb with five digits, characteristic of tetrapod vertebrates (amphibians, reptiles, birds, and mammals). It evolved from the paired fins of primitive fish as an adaptation to locomotion on land and is not found in modern fish. The limb has three parts (see illustration): the upper arm or thigh containing one long bone, the forearm or shank containing two long bones, and the hand or foot, which contains a number of small bones. This basic design is modified in many species, according to the function of the limb, particularly by the loss or fusion of the terminal bones.

pentose A sugar that has five carbon atoms per molecule. *See* MONOSACCHARIDE.

pentose phosphate pathway (pentose shunt) A series of biochemical reactions that results in the conversion of glucose 6-phosphate to ribose 5-phosphate and generates NADPH, which provides reducing power for other metabolic reactions, such as synthesis of fatty acids. Ribose 5-phosphate and its derivatives are

upper arm
humerus
forearm
radius
ulna
wrist
carpals (actual number in man is reduced to 8)
palm
metacarpals
phalanges
fingers

A basic pentadactyl forelimb, as exemplified by the human arm

components of such molecules as ATP, coenzyme A, NAD, FAD, DNA, and RNA. In plants the pentose phosphate pathway also plays a role in the synthesis of sugars from carbon dioxide. In animals the pathway occurs at various sites, including the liver and adipose tissue.

PEP *See* PHOSPHOENOLPYRUVATE.

pepo (*pl.* **pepos**) *See* BERRY.

pepsin An enzyme that catalyses the breakdown of proteins to polypeptides in the vertebrate stomach. It is secreted as an inactive precursor, *pepsinogen.

pepsinogen The inactive precursor of the enzyme *pepsin. Pepsinogen is secreted into the stomach lumen by chief cells in gastric glands situated within the lining of the vertebrate stomach. In the lumen it is converted to pepsin by hydrochloric acid and also by the action of pepsin itself, in a self-perpetuating chain reaction.

peptidase *See* ENDOPEPTIDASE; EXOPEPTIDASE; PROTEASE.

peptide Any of a group of organic compounds comprising two or more amino acids linked by **peptide bonds**. These bonds are formed by the reaction between adjacent carboxyl (-COOH) and amino (-NH$_2$) groups with the elimination of water (see illustration). **Dipeptides** contain two amino acids, **tripeptides** three, and so on. *Polypeptides contain more than 20 and usually 100–300. Naturally occurring

humerus
ulna
radius

humerus
radius
ulna
carpals
metacarpals
phalanges
cloven hoof

humerus
carpals ulna
metacarpals
radius
phalanges

flipper of whale   foreleg of pig   wing of bat

The modified pentadactyl forelimb of various vertebrates

amino acid 1   amino acid 2   dipeptide   water

────── peptide bond

Formation of a peptide bond

oligopeptides (of less than 20 amino acids) include the tripeptide glutathione and the pituitary hormones antidiuretic hormone and oxytocin, which are octapeptides. Peptides also result from protein breakdown, e.g. during digestion.

peptide mapping (peptide fingerprinting) The technique of forming two-dimensional patterns of peptides (on paper or gel) by partial hydrolysis of a protein followed by electrophoresis and chromatography. The peptide pattern (or **fingerprint**) produced is characteristic for a particular protein and the technique can be used to separate a mixture of peptides.

peptide YY (pancreatic peptide YY; PYY) A hormone secreted by L-cells in the lining of the small intestine that acts as an appetite suppressant and counteracts the effects of the hormone *ghrelin. Active peptide YY consists of 36 amino acids and its release is stimulated by the presence of food in the digestive tract. It circulates in the bloodstream and binds to receptors in the brain to reduce appetite and promote satiety. It also acts on the stomach and intestine to slow the movement of digesta through the digestive tract. Levels in blood gradually fall following a meal and are at their lowest when fasting, e.g. overnight. It may have potential as an anti-obesity treatment.

peptidoglycan A macromolecule that is a component of the cell wall of bacteria; it is not found in eukaryotes. Consisting of chains of *amino sugars (N-acetylglucosamine and N-acetylmuramic acid) linked to a tripeptide (of alanine, glutamic acid, and lysine or diaminopimelic acid), it confers strength and shape to the cell wall. Archaea possess a similar polysaccharide,

pseudopeptidoglycan, with N-acetyltalosaminuronic acid instead of N-acetylmuramic acid.

peramorphosis Any outcome arising from evolutionary changes in developmental rates (*see* HETEROCHRONY) that involves the addition of new stages to the end of the ancestral development sequence. Peramorphic forms may arise either by *acceleration or by *hypermorphosis and conform to the theory of *recapitulation. In contrast, paedomorphic forms (*see* PAEDOMORPHOSIS), in which a new stage is added within the developmental sequence, do not exhibit recapitulation.

perception The interpretation of sensory information using both the raw data detected by the senses and previous experience. *Compare* SENSATION.

perennation The survival of biennial or perennial plants from one year to the next by vegetative means. In biennials and herbaceous perennials the aerial parts of the plant die down and the plants survive by means of underground storage roots (e.g. carrot), *rhizomes (e.g. couch grass, Solomon's seal), *tubers (e.g. dahlia), *bulbs (e.g. daffodil, snowdrop), or *corms (e.g. crocus, gladiolus). These **perennating organs** are also frequently responsible for *vegetative propagation. Woody perennials survive the winter by reducing their metabolic activity (e.g. by leaf loss in deciduous trees and shrubs).

perennial A plant that lives for a number of years. Woody perennials (trees and shrubs) have a permanent aerial form, which continues to grow year after year. Herbaceous (i.e. nonwoody) perennials have aerial shoots that die down each autumn and are replaced in spring by new shoots from an underground structure (*see* PERENNATION). Lupin and rhubarb are examples of herbaceous perennials. *Compare* ANNUAL; BIENNIAL; EPHEMERAL.

perfect (in botany) Describing a flower that has both male and female sex organs, i.e. both stamens and carpels. *Compare* IMPERFECT.

perforin A cytotoxic protein occurring in the lytic granules of *cytotoxic T cells. When released onto the surface of a target cell by an activated T cell, the perforin molecules polymerize to form a cylindrical pore through the target's plasma membrane, enabling entry of other cytotoxic molecules, such as **granzymes** and **granulysin**, which induce programmed cell death (*apoptosis).

perfusion techniques Methods of maintaining a live organ in isolation by circulating fluid containing essential nutrients and oxygen through and around the organ. Maintaining the circulation through an organ allows the delivery of nutrients to the tissues and the removal from them of toxins and waste products. Perfusion techniques are particularly useful in studying the metabolism of drugs in isolation in an intact organ, such as the liver or kidney.

perianth The part of a flower situated outside the stamens and carpels. In dicotyledons it consists of two distinct whorls, the outer of sepals (*see* CALYX) and the inner of petals (*see* COROLLA). In monocotyledons the two whorls are similar and often brightly coloured. In wind-pollinated flowers both whorls may be reduced or absent. In many horticultural varieties the number of perianth parts is multiplied, but the resulting 'double' flowers are often sterile.

pericardial cavity The cavity in vertebrates that contains the heart and is bounded by a membrane (the *pericardium). It is part of the *coelom.

pericardium (pericardial membrane) (*pl.* **pericardia**) The membrane that encloses the pericardial cavity, containing the vertebrate heart. The pericardium holds the heart in position while allowing it to relax and contract. It consists of two main parts: a tough outer fibrous layer (**fibrous pericardium**) and the more delicate **serous pericardium**, which consists of a double layer of *serous membrane, the inner layer being in close contact with the heart.

pericarp (fruit wall) The part of a fruit that develops from the ovary wall of a flower. The type of fruit that develops depends on whether the pericarp becomes dry and hard or soft and fleshy. The pericarp can be made up of three layers. The outer skin (**epicarp** or **exocarp**) may be tough and hard; the middle layer (**mesocarp**) may be succulent as in peach, hard as in almond, or fibrous as in coconut; and the inner layer (**endocarp**) may be hard and stony as in many *drupes, membranous as in citrus fruits, or indistinguishable from the mesocarp, as in many *berries.

perichondrium A dense layer of fibrous connective tissue that covers the surface of cartilage.

periclinal (in botany) Parallel to the surface of an organ or part. In periclinal cell division the plane of division is parallel to the surface of the plant body. *Compare* ANTICLINAL.

pericycle A plant tissue comprising the outermost layer of the root vascular tissue, lying immediately beneath the *endodermis. Lateral roots originate from the pericycle.

periderm *See* CORK CAMBIUM.

perigyny A floral arrangement in which the ovary is situated in a cup-shaped or flattened receptacle, from the margin of which the perianth and stamens arise (see illustration). The perianth and stamens are said to be **perigynous** with respect to the ovary, as seen in cherry and plum flowers. *Compare* EPIGYNY; HYPOGYNY.

perilymph The fluid of the *inner ear that fills the space between the bony labyrinth and the membranous labyrinth. *See also* COCHLEA. *Compare* ENDOLYMPH.

calyx

corolla

stamen

ovary

receptacle

Perigyny

perinuclear space (perinuclear compartment) *See* NUCLEAR ENVELOPE.

period *See* MENSTRUAL CYCLE.

periodontal membrane The membrane of connective tissue that surrounds the root of a *tooth and anchors it to its socket in the jawbone. Fibres of the periodontal membrane pass into the *cement covering the root, which provides a firm attachment.

periosteum (*pl.* **periostea**) The outer membrane that surrounds a bone. It contains connective tissue, capillaries, nerves, and a number of types of cell, including *osteoclasts. The periosteum plays an important role in bone repair and growth.

peripatric speciation A form of *allopatric speciation in which a founder population becomes established beyond the existing range of the main population and remains isolated because of difficult terrain or other factors. Such a founder population may involve just a few individuals or even a single fertilized female, and so has a much smaller gene pool compared with the main population. Also, it may be subject to different selection pressures. Together these factors cause potentially rapid evolutionary divergence, provided there is little or no interbreeding with the main population, so that in time it becomes a new species. *Compare* DICHOPATRIC SPECIATION.

Peripatus *See* ONYCHOPHORA.

peripheral nervous system (PNS) All parts of the nervous system excluding the *central nervous system. It consists of all the *cranial and *spinal nerves and their branches, which link the *receptors and *effectors with the central nervous system. *See also* AUTONOMIC NERVOUS SYSTEM.

periphyton The organisms, collectively, that live attached to the stems and leaves of freshwater plants.

periplasm (periplasmic space) The zone between the cytoplasmic membrane and the outer membrane in Gram-negative bacteria. It contains a thin layer of *peptidoglycan, has a gel like consistency, and contains various types of proteins concerned with the cell's metabolism. In Gram-positive bacteria the corresponding space is much narrower.

Perissodactyla An order of mammals having hoofed feet with an odd number of toes—the 'odd-toed ungulates'. They are all herbivores and include the tapirs, rhinoceros, and horses. The teeth are large and specialized for grinding. Cellulose digestion occurs in the caecum and large intestine. Fossils of the Eocene epoch, 60 million years ago, show that these animals were at that time already distinct from the cloven-hoofed *Artiodactyla (Cetartiodactyla).

peristalsis Repeated waves of smooth muscular contraction and relaxation that pass along the alimentary canal, forcing food contents along. It is brought about by contraction of the circular muscles of the gut wall in sequence. Similar muscular movements are used by burrowing worms for locomotion or to irrigate their burrows. *Compare* SEGMENTATION (sense 3).

peristome 1. A ring of toothlike structures around the opening of a moss *capsule. The teeth tend to bend and twist in dry weather, so opening the mouth of the capsule and allowing the spores to escape. In wet weather they close over the opening of the capsule. **2.** The area around the mouth in many invertebrates and some protists. It sometimes assists in food collecting. Examples are the spirally ciliated groove around the mouth of some ciliate protozoans and the first segment of the earthworm.

peritoneum (*pl.* **peritoneums** or **peritonea**) The thin layer of tissue (*see* SEROUS MEMBRANE) that lines the abdominal cavity of vertebrates and covers the abdominal organs. *See also* MESENTERY.

peritrichous 1. (of certain protists) Having cilia arranged around the mouth. **2.** (of bacteria) Having flagella arranged uniformly over the cell surface.

permafrost *See* TUNDRA.

permanent teeth The second and final set of teeth that mammals produce after shedding the *deciduous teeth. An adult human normally has 32 permanent teeth, consisting of incisors, canines, molars, and premolars (see illustration). These usually appear between the ages of approximately 6 and 21 years. *See also* DENTAL FORMULA; DIPHYODONT.

Permian The last geological period in the Palaeozoic era. It extended from the end of the Carboniferous period, about 299 million years ago, to the beginning of the Mesozoic era, about 252 million years ago. It was named by the British geologist Roderick Murchison (1792–1871) in 1841 after the Perm province in Russia. In some areas continental conditions prevailed, which continued into the following period, the Triassic. These conditions resulted in the deposition of the New Red Sandstone. During the period a number of animal groups became extinct, including the trilobites, tabulate and rugose corals, and blastoids (*see* MASS EXTINCTION). Amphibians and reptiles continued to be the dominant land animals and gymnosperms replaced ferns, clubmosses, and horsetails as the dominant plants.

(⊕) SEE WEB LINKS
- Overview of the Permian period from the University of California Museum of Paleontology

Upper jaw

Right side — Left side

| molars | pre-molars | canines | incisors | canines | pre-molars | molars |

Lower jaw

Permanent teeth of an adult human

pernicious anaemia *See* VITAMIN B COMPLEX.

peroxisome A small organelle (a type of *microbody) that is bounded by a single membrane and found in plant and animal cells. It contains enzymes that are involved in the transfer of hydrogen from various compounds to oxygen, thereby generating the highly toxic compound hydrogen peroxide (H_2O_2). Peroxisomal *catalase breaks down the H_2O_2 into water and oxygen, thus preventing it from damaging other cell components. Peroxisomes contain a vast array of enzymes and perform a range of functions, including oxidation of fatty acids and synthesis of phospholipids. They work in conjunction with other cell organelles, particularly mitochondria and, in plants, chloroplasts. They are responsible for neutralizing toxins absorbed from the blood, such as alcohol, especially in cells of the liver parenchyma and proximal convoluted tubule of the kidney. In plants, peroxisomes detoxify certain by-products of photosynthesis and oxidize glycolate (produced by *photorespiration) to glyoxylate, which can then be salvaged by a series of reactions involving mitochondria and chloroplasts as well as peroxisomes, ultimately to 3-phosphoglycerate, which can enter the Calvin-Bassham-Benson cycle.

persistent Describing a *pesticide or other pollutant that is not readily broken down and can persist for long periods, causing damage in the environment. For example, the herbicides *Paraquat and *DDT can persist in the soil for many years after their application.

pest Any of various organisms, such as fungi, insects, rodents, and plants, that harm crops or livestock or otherwise interfere with the wellbeing of human beings. **Weeds** are plant pests that grow where they are not wanted—often on cultivated land, where they compete with crop plants for space, light, nutrients, etc. Pests are controlled by the use of *pesticides and *biological control methods.

pesticide Any chemical compound used to kill pests that detract from agricultural

production or are in some way harmful to humans. Pesticides include **herbicides** (such as *glyphosate), which kill unwanted plants or weeds; **insecticides** (such as *pyrethrum), which kill insect pests; **fungicides**, which kill fungi; and **rodenticides** (such as *warfarin), which kill rodents. The main drawback with pesticides is that often they are nonspecific and may therefore be toxic to organisms that are not pests; they may also be nonbiodegradable, so that they persist in the environment and may accumulate in living organisms (*see* BIOACCUMULATION). Organophosphorus insecticides, such as malathion and parathion, are biodegradable but can also damage the respiratory and nervous systems in humans as well as killing useful insects, such as bees. They act as *anticholinesterases by inhibiting the action of the enzyme *cholinesterase. Organochlorine insecticides, such as dieldrin, aldrin, and *DDT, are very persistent and not easily biodegradable. *See also* NEONICOTINOIDS.

(⊕) SEE WEB LINKS
• Pesticides section of the UK's Health and Safety Executive website

PET *See* POSITRON EMISSION TOMOGRAPHY.

peta- Symbol P. A prefix used in the metric system to denote one thousand million million times. For example, 10^{15} metres = 1 petametre (Pm).

petal One of the parts of the flower that make up the *corolla. Petals of insect-pollinated plants are usually brightly coloured and often scented. Those of wind-pollinated plants are usually reduced or absent. Petals are considered to be modified leaves but their structure is simpler. Epidermal hairs may be present and the cuticle is often covered by lines or dots known as **nectar guides**, which direct insects to the *nectar.

petiole The stalk that attaches a *leaf blade to the stem. Leaves without petioles are described as **sessile**.

Petri dish A shallow circular flat-bottomed dish made of glass or plastic and having a fitting lid. It is used in laboratories chiefly for culturing bacteria and other microorganisms. It was invented by the German bacteriologist J. R. Petri (1852–1921).

petrification *See* FOSSIL.

Peyer's patches Numerous lymphoid organs found in the wall of the small intestine, which have a crucial role in protecting against infection. Each consists of a dome-shaped follicle containing chiefly B cells surrounded by smaller numbers of T cells. Within the surface of the follicle exposed to the gut lumen are specialized epithelial cells called **M cells** (multifenestrated cells). These have a microfolded luminal surface, and take up particles from the gut by endocytosis. The particles are then transferred through the M cells to the underlying lymphocytes for antigen processing. Within the follicle, *dendritic cells present antigen to naive lymphocytes, which become activated and leave Peyer's patches, returning to the blood via mesenteric lymph nodes and the thoracic duct. These then circulate in the blood and take up station as effector cells at various mucosal sites in the body, including the intestinal epithelium, primed in readiness for a subsequent encounter with the antigen. The organs were named after Swiss anatomist Johann K. Peyer (1653–1712).

pFc′ fragment *See* F(AB′)$_2$ FRAGMENT.

PFGE Pulsed-field gel electrophoresis. *See* GEL ELECTROPHORESIS.

PGD *See* PREIMPLANTATION GENETIC DIAGNOSIS.

pH *See* PH SCALE.

PHA *See* PHYTOHAEMAGGLUTININ.

Phaeophyta (brown algae) A phylum (or group, i.e. the phaeophytes) of protist *algae, now considered to be *stramenopiles, in which the green chlorophyll pigments are usually masked by the brown pigment fucoxanthin. Brown algae are usually marine (being abundant in cold water) and many species, such as the wracks (*Fucus*), inhabit intertidal zones. They vary in size from small branched filaments to ribbon-like bodies (known as kelps) many metres long.

phage *See* BACTERIOPHAGE.

phagemid *See* EXPRESSION VECTOR.

phagocyte A cell that is able to engulf and break down foreign particles, cell debris, and disease-producing microorganisms (*see* PHAGOCYTOSIS). Some protists and certain mammalian cells (e.g. *macrophages, *monocytes, and *neutrophils) are phagocytes. Phagocytes are important elements in the natural defence mechanism of most animals.

phagocytosis The process by which foreign particles invading the body or minute food particles are engulfed and broken down by certain animal cells (known as *phagocytes). It is a form of *endocytosis in which the plasma membrane of the phagocyte invaginates to capture the particle and then closes around it to form an acidified vesicle (**phagosome**). This then coalesces with a *lysosome, which contains enzymes that break down the particle, forming a **phagolysosome**. *Compare* PINOCYTOSIS.

phagotroph (macroconsumer) Any heterotrophic organism that feeds by ingesting organisms or organic particles, which are digested within its body. *Compare* OSMOTROPH.

phalanges (*sing.* **phalanx**) The bones that make up the *digits of the hand or foot in vertebrates. They articulate with the *metacarpals of the hand or with the metatarsals of the foot. In the basic *pentadactyl limb there are two phalanges for the first digit (the thumb or big toe in humans) and three for each of the others.

phanerophyte A plant life form in Raunkiaer's system of classification (*see* PHYSIOGNOMY). Phanerophytes are large shrubs and trees in which the overwintering (perennating) buds are located high above the ground. The buds are thus at risk of exposure to drought stress or frost, and such plants occur mainly in regions where frost and drought are uncommon, such as the tropics.

Phanerozoic The most recent eon of geological time, represented by rock strata containing clearly recognizable fossils. It comprises the *Palaeozoic, *Mesozoic, and *Cenozoic eras and has extended for about 541 million years from the beginning of the Cambrian period. *Compare* PROTEROZOIC.

pharate 'Cloaked': describing a larva or adult when inside the cuticle of the previous developmental stage. For example, a newly metamorphosed adult insect may remain 'cloaked' for some hours or even days within the pupal case until it receives the appropriate environmental cue that triggers emergence.

pharmacodynamics The study of the action of drugs on the body. *Compare* PHARMACOKINETICS.

pharmacogenomics (pharmacogenetics) The study of how genes affect the actions of drugs. The enormous growth in knowledge about human genetics arising from the *Human Genome Project, coupled with the rapid advance of computer systems to analyse the vast amounts of data, has revolutionized drug discovery and development. This approach, which combines *genomics and pharmacology, improves understanding of drug actions, suggests new potential drug molecules, and enables computer-based searches for likely drug targets. In addition, analysis of genetic data from individuals raises the prospect of drugs being tailormade to suit the genetic make-up of particular patients or groups of patients. This more precise targeting of drugs should make drugs more effective, with less risk of adverse side effects.

pharmacokinetics The movement of foreign substances, particularly drugs, throughout the body of an animal. Processes that influence the pharmacokinetics of a compound include uptake, distribution throughout the body tissues, the length of time the compound remains in the body, and its rate of clearance (e.g. by metabolism or excretion).

pharmacology The study of the properties of drugs and their effects on living organisms. Clinical pharmacology is concerned with the effects of drugs in treating disease.

pharynx (*pl.* **pharynges**) **1.** The cavity in vertebrates between the mouth and the *oesophagus and windpipe (*trachea), which serves for the passage of both food and respiratory gases. The presence of food in the pharynx stimulates swallowing (*see* DEGLUTITION). In fish and aquatic amphibians the pharynx is perforated by *gill slits. **2.** The corresponding region in invertebrates.

phase-contrast microscope A type of *microscope that is widely used for examining cells and tissues. It makes visible the changes in phase that occur when nonuniformly transparent specimens are illuminated. In passing through an object the light is slowed down and becomes out of phase with the original light. With transparent specimens having some structure diffraction occurs, causing a larger phase change in light outside the central maximum of the pattern. The phase-contrast microscope provides a means of combining this light with that of the central maximum by means of an annular diaphragm and a **phase-contrast plate**, which produces a matching phase change in the light of the central maximum only. This gives greater contrast to the final image, due to constructive interference between the two sets of light waves. This is **bright contrast**; in **dark contrast** a different phase-contrast plate is used to make the same structure appear dark, by destructive interference of the same waves.

phase I metabolism The first stage in the conversion of a foreign compound, such as a drug or toxin, into a form that can be eliminated by the body. Common reactions during this phase are oxidation, reduction, and hydrolysis; the resulting metabolites are chemically more reactive than the parent compound, enabling them to undergo the reactions of the second stage (*see* PHASE II METABOLISM).

phase II metabolism The second stage in adapting foreign compounds for elimination from the body (*compare* PHASE I METABOLISM). Phase II metabolism involves the addition of chemical groups (e.g. glycine or acetate), which usually makes the compound less toxic to body tissues and easier to excrete.

phase variation A form of *antigenic variation whereby certain bacteria can change their surface antigens to evade the adaptive immune response of their host. For example, successive generations of *Salmonella* bacteria can switch between two different proteins to construct their flagella. Immunoglobulins that are generated specific to one type of flagellar protein do not recognize the other type, and vice versa. Phase variation typically leads to successive waves of bacteria in the host and difficulty in eliminating the infection. The mechanism involves the inversion of segments of the bacterial genome that encode antigenic proteins, with different proteins expressed depending on the orientation of the segment relative to its promoter. Such inversions are heritable and reversible, depending on trigger stimuli encountered by successive generations.

phellem *See* CORK.

phelloderm *See* CORK CAMBIUM.

phellogen *See* CORK CAMBIUM.

phenetic Describing a system of *classification of organisms based on similarities and differences in as many observable characteristics as possible. A phenetic system does not aim to reflect evolutionary descent, although it may well do so. *Compare* PHYLOGENETIC.

phenocopy A *phenotype that is not genetically determined but mimics one that is. This occurs most commonly when environmental influences alone, such as diet, evoke a developmental trait that has a very similar genetic counterpart. For example, dietary deficiency of vitamin D causes the bone disease *rickets, and this form cannot be readily distinguished from genetically determined (nondietary) forms of the disease due to malabsorption or excessive loss of bone minerals.

phenology The study of the ways in which the timing and other aspects of periodic events, such as flowering in plants and breeding and migration in animals, are affected by climate and other environmental factors.

phenolphthalein A dye used as an acid-base indicator. It is colourless below pH 8 and red above pH 9.6. It is used in titrations involving weak acids and strong bases. It is also used as a laxative.

phenome A term used (by analogy with, e.g., genome or proteome) to mean essentially *phenotype in its broadest sense, i.e., all the measurable physical and biochemical attributes of an organism. Hence **phenomics** is the study of the interactions of genes, environment, and lifestyle to establish how these determine the phenotype and how these change throughout the life of the organism.

phenotype The observable characteristics of an organism. These are determined by its genes (*see* GENOTYPE), the dominance relationships between the *alleles, and by the interaction of the genes with the environment. *See also* NATURE AND NURTURE.

phenotypic plasticity The ability of an organism to adapt to changes in its environment by modifying its own development, form, behaviour, or other trait. Hence it depends on the capacity for a given genotype to respond to environmental factors and to adjust the phenotype accordingly. Plasticity is of great selective advantage in an unstable or unpredictable environment and can itself be inherited and evolve. Plants typically have marked phenotypic plasticity, showing considerable variation in, among other characters, mature height, seed number, and seed mass with different growing conditions, including factors such as light intensity, daylength, and temperature. Seasonal factors and the availability of food also affect the course of development in many animals. Social insects (such as bees) with the same genotype undergo different developmental pathways to become members of different castes, depending on diet and incubation temperature. Modification of an animal's behaviour through learning is another instance of the great importance of phenotypic plasticity.

phenylalanine *See* AMINO ACID.

phenylketonuria A genetic disorder in which there is disordered metabolism of the amino acid phenylalanine, leading to severe mental retardation of affected children. The disease is caused by the absence or deficiency of the enzyme phenylalanine hydroxylase, which results in the accumulation of phenylalanine in all body fluids. There are also high levels of the ketone phenylpyruvate in the urine, hence the name of the disease. The disease occurs in individuals who are homozygous for the defective recessive allele at the *PAH* gene on chromosome 12. Both parents of such individuals are thus heterozygous carriers of the allele. The advent of *gene probes has greatly aided accurate diagnosis, of both phenylketonurics and carriers.

pheophytin The first electron acceptor in the light-dependent reactions of *photosynthesis. It accepts an electron from the excited form of photosystem II (*see* PHOTOSYSTEMS I AND II), and passes it on via another acceptor, Q_A, to *plastoquinone. Pheophytin is a form of chlorophyll *a* in which the magnesium ion is replaced by two hydrogen ions. It participates in the crucial step of converting light energy to chemical energy.

pheromone (ectohormone) A chemical substance (*see* SEMIOCHEMICAL) emitted by an organism into the environment as a specific signal to another organism, usually of the same species. Pheromones play an important role in the social behaviour of certain animals, especially insects and mammals, and are also released by the hyphae of fungi to establish compatibility between mating types. They are used to attract mates, to mark trails, define territories, and to promote social cohesion and

coordination in colonies (*see* QUEEN SUBSTANCE). Pheromones are usually volatile organic acids or alcohols and can be effective at minute concentrations. The volatility depends on the nature and size of the molecule, and determines the role played by the pheromone. For example, a highly volatile pheromone will disperse rapidly and widely and serve effectively as a transient alarm signal, whereas a less volatile, more persistent pheromone is useful for marking territories or trails.

philopatry The tendency for an animal to stay in or return to its home area.

phloem (bast) A tissue that conducts food materials in vascular plants from regions where they are produced (called sources—notably the leaves) to regions, such as growing points and storage organs, where they are needed (called sinks). It consists of hollow tubes (sieve tubes) that run parallel to the long axis of the plant organ and are formed from elongated cells (*sieve elements) joined end to end and closely associated with *companion cells. The end walls of the sieve elements contain pores to allow passage of materials. In young plants and in newly formed tissues of mature plants **primary phloem** is formed by the activity of the *apical meristem (*see* PROTOPHLOEM; METAPHLOEM). In most plants **secondary phloem** is later differentiated by the vascular *cambium and this replaces the earlier formed phloem in older regions. *See also* MASS FLOW. *Compare* XYLEM.

phloem loading The processes by which the products of photosynthesis enter the phloem for transport to other tissues within the plant. The carbon fixed by photosynthesis is transported in phloem primarily as sucrose, formed by combination of glucose and fructose in the cytosol of photosynthetic cells (the source). From here the sucrose travels the short distance to the vicinity of the nearest phloem vessels, chiefly by diffusing from cell to cell via the *plasmodesmata. This **symplastic pathway** (*see* SYMPLAST) reaches the companion cells that lie alongside the phloem vessels (sieve elements) and can load sucrose into the phloem by simple diffusion through plasmodesmata. However, most sugar is loaded, via the **apoplastic pathway** (*see* APOPLAST), from the space between cells bordering the companion cells. The plasma membrane of the latter contains a sucrose/H^+ *cotransporter, which pumps sucrose molecules from the intercellular space into the companion cells, from which the sucrose diffuses into the sieve elements for transport to other parts of the plant where food is required (the sink). Energy is derived from the H^+ gradient across the companion cell membrane, which is maintained by hydrolysis of ATP to ADP. *Compare* PHLOEM UNLOADING.

phloem protein *See* P-PROTEIN.

phloem unloading The processes by which sugars and other substances leave the conducting vessels of the phloem and are delivered to their destination (the sink) in the plant. These processes mirror those responsible for *phloem loading in that both a symplastic pathway, involving intracellular diffusion via plasmodesmata, and an apoplastic pathway, via the intercellular space, are used variously, depending on the type of tissue and species of plant. Apoplastic unloading may have two possible routes. In one the sucrose diffuses from the phloem companion cells into the intercellular space, where it is split by an invertase enzyme into its component monosaccharides (glucose and fructose), which are actively taken up by the sink cells. In the other route the sucrose is actively pumped out of companion cells and subsequently diffuses into the sink cells. *See* SINK STRENGTH.

phloroglucinol A red dye (usually acidified with hydrochloric acid) that stains lignin in plant cells red.

phonotaxis (*pl.* **phonotaxes**) The movement of an organism in relation to a sound source. For example, females are often attracted by the courtship song of a potential mate (i.e. **positive phonotaxis**), or animals may flee from the sound of a predator (i.e. **negative phonotaxis**).

phoresy A method of dispersal in which an animal attaches itself to an animal of another species in order to be transported to a new site, where it releases itself, with little or no harm to the carrier. This method is adopted by various animals searching for new food sources, or by parasites when seeking a new host. For example, when their host dies, certain bird lice may attach themselves to blood-sucking flies in their effort to find a new avian host, and some insect parasitoids, which lay their eggs inside the eggs of other insects, attach themselves to adults of the host species to ensure immediate access to the eggs when the adult host lays them. Eggs may also be transferred in this manner. The Neotropical human bot fly, *Dermatobia hominis*, 'captures' a suitable carrier, such as a blood-sucking mosquito, to which it attaches 30 or so of its eggs. When the carrier insect reaches a human or bovine host, the bot fly eggs quickly hatch and the larvae release themselves from the carrier and invade the hair follicles of their host, where they develop.

Phoronida A phylum of marine invertebrates comprising a dozen or so species distributed widely. Phoronids, sometimes called 'horseshoe worms', are soft-bodied, sedentary wormlike creatures that live inside upright chitinous tubes in the soft sediment of shallow seas. Some live in colonies encrusting rocks or shells. They feed on suspended particles by means of a ciliated *lophophore of numerous tentacles, and have a U-shaped gut with the anus located alongside the mouth. Sexual reproduction results in free-swimming actinotroch larvae, which feed on plankton. Phoronids belong to the clade *Lophotrochozoa.

phosphagen A compound found in animal tissues that provides a reserve of chemical energy in the form of high-energy phosphate bonds. The most common phosphagens are *creatine phosphate, occurring in vertebrate muscle and nerves, and arginine phosphate, found in most invertebrates. During tissue activity (e.g. in muscle contraction) phosphagens give up their phosphate groups, thereby generating *ATP from

ADP. The phosphagens are then reformed when ATP is available.

phosphatase An enzyme that catalyses the removal of a phosphate group from an organic compound. *See* ALKALINE PHOSPHATASE.

phosphatide *See* PHOSPHOLIPID.

phosphatidylcholine *See* LECITHIN.

phosphodiester bond The covalent bond that links a phosphate group and a sugar group, by means of an oxygen bridge, in the sugar-phosphate backbone of a nucleic acid molecule. *See* DNA.

phosphoenolpyruvate (PEP) A three-carbon compound that is the substrate for carbon dioxide fixation during photosynthesis in C_4 plants (*see* C_4 PATHWAY). PEP is also an intermediate in *glycolysis (being the immediate precursor of pyruvate) and the *glyoxylate cycle.

phosphoglyceric acid (PGA; 3-phosphoglycerate) *See* GLYCERATE 3-PHOSPHATE.

phosphokinase *See* KINASE.

phospholipase Any of various enzymes that cleave particular bonds in the polar phosphate 'heads' of glycerophospholipids (*see* PHOSPHOLIPID). For example, phospholipase C (PLC) cleaves the phosphate-glycerol bond and is important in liberating

the second messengers inositol 1,4,5-trisphosphate (IP_3) and diacylglycerol (DAG) from phosphatidylinositol in cell plasma membranes (*see* INOSITOL). PLC plays a central role in *signal transduction; it is activated by protein tyrosine kinases, which themselves are activated by ligand binding to cell surface receptors. Phospholipase A_2 (PLA_2) is found in pancreatic juice and helps in the breakdown of ingested glycerophospholipids by cleaving the acyl residue (e.g. a fatty acid) from C2 of the glycerol.

phospholipid (phosphatide) One of a group of lipids having both a phosphate group and one or more fatty acids. **Glycerophospholipids** (or **phosphoglycerides**) are based on *glycerol; the three hydroxyl groups are esterified with two fatty acids and a phosphate group, which may itself be bound to one of a variety of simple organic groups (e.g. in *lecithin (phosphatidylcholine) it is choline; see formula). **Sphingolipids** are based on the alcohol sphingosine and contain only one fatty acid linked to an amino group. With their hydrophilic polar phosphate groups and long hydrophobic hydrocarbon 'tails', phospholipids readily form membrane-like structures in water. They are a major component of plasma membranes (*see* LIPID BILAYER).

phosphorus Symbol P. A nonmetallic element that is a major *essential element for living organisms. It is an important

Structure of the phospholipid lecithin (phosphatidylcholine), a phosphoglyceride constituent of tissues (especially bones and teeth) and of cells, being required for the formation of nucleic acids and energy-carrying molecules (e.g. ATP) and also involved in various metabolic reactions.

phosphorus cycle The cycling of phosphorus between the biotic and abiotic components of the environment (*see* BIOGEOCHEMICAL CYCLE). Inorganic phosphates (PO_4^{3-}, HPO_4^{2-}, or $H_2PO_4^-$) are absorbed by plants from the soil and bodies of water and eventually pass into animals through food chains. Within living organisms phosphates are built up into nucleic acids and other organic molecules. When plants and animals die, phosphates are released and returned to the abiotic environment through the action of bacteria. Small amounts move into the atmosphere as dust or sea spray. On a geological time scale, phosphates in aquatic environments eventually become incorporated into and form part of rocks; through a gradual process of erosion, these phosphates are returned to the soil, seas, rivers, and lakes.

The phosphorus cycle

Phosphorus-containing rocks are mined for the manufacture of fertilizers, which provide an additional supply of inorganic phosphate to the abiotic environment.

phosphorus:oxygen ratio *See* P/O RATIO.

phosphorylase *See* PHOSPHORYLATION.

phosphorylation The introduction of a phosphate group into a biomolecule in a reaction that is normally controlled by a **phosphorylase** enzyme. Phosphate is able to combine easily with inert organic compounds, making them chemically active. The first stage in many biochemical reactions is phosphorylation. The conversion of AMP and ADP to *ATP occurs by phosphorylation reactions in two main metabolic pathways, *oxidative phosphorylation, occurring in aerobic respiration, and *photophosphorylation, occurring in chloroplasts. The formation of other nucleotides also involves a phosphorylation reaction. **Substrate-level phosphorylation** occurs during *glycolysis; it generates two molecules of ATP per glucose molecule through reactions catalysed by water-soluble glycolytic enzymes located in the cytosol. An estimated 30% of cellular proteins are phosphorylated, representing the **phosphoproteome**. The phosphorylation of proteins plays a central role in most cellular activities and is essential for cell proliferation and development. Phosphorylation can alter a protein's function, activity, localization, and stability. The activity of many enzymes is controlled by phosphorylation: certain enzymes are activated when they are phosphorylated (*see* KINASE), while others are deactivated. Phosphorylation of these enzymes is under the control of hormones and other messengers.

photic zone *See* EUPHOTIC ZONE.

photoautotroph An autotrophic organism, such as a green plant or a photosynthetic bacterium, that synthesizes its organic materials from inorganic components using energy derived from the sun (solar energy) in the process of photosynthesis. *See* AUTOTROPHIC NUTRITION; PHOTOTROPH.

photoblastic Describing a seed whose germination is influenced by light. Seeds that are stimulated to germinate by light are described as **positively photoblastic**; seeds whose germination is inhibited by light are said to be **negatively photoblastic**. The response to light is mediated by *phytochrome-regulated production of the plant hormone *gibberellin.

photochemical smog A noxious smog produced by the reaction of *nitrogen oxides (e.g. from car exhausts, coal burning) with volatile hydrocarbons in the presence of ultraviolet light from the sun. The reaction is very complex and one of the products is ground-level *ozone. *See also* AIR POLLUTION.

photoheterotroph Any organism that uses energy derived from sunlight and relies on organic compounds in its environment as a source of carbon, instead of carbon dioxide, to manufacture its own organic compounds. For example, certain photosynthetic purple nonsulphur bacteria can use fatty acids, alcohols, carbohydrates, and other compounds as a carbon source in light conditions, while producing their ATP by *photophosphorylation. *See* PHOTOTROPH.

photoinhibition 1. The reduction in photosynthesis caused by exposure to abnormally high intensities of visible or ultraviolet light. Some degree of light-induced damage occurs normally in all photosynthesizing organisms, but repair mechanisms usually ensure that photosynthetic capability is not impaired. Very high light intensity magnifies the damage and can outstrip the repair mechanisms, leading to a decline in photosynthesis, which can be irreversible under extreme conditions. The damage is focused on photosystem II (PSII) in the reaction centres on the thylakoid membranes of the plastids (*see* PHOTOSYSTEMS I AND II). During photoinhibition, components of the reaction centre are degraded, causing disassembly of the PSII components. Damaged components are removed, and PSII is reassembled using newly synthesized components. 2. The (usually reversible) suppression of any biological process by exposure to light.

photolysis A chemical reaction produced by exposure to light or ultraviolet radiation. The photolysis of water, using energy from sunlight absorbed by chlorophyll, produces gaseous oxygen, electrons, and hydrogen ions and is a key reaction in *photosynthesis. *See* PHOTOPHOSPHORYLATION; PHOTOSYSTEMS I AND II.

photomicrography The use of photography to obtain a permanent record (a **photomicrograph**) of the image of an object as viewed through a microscope.

photomorphogenesis The development of plants under the influence of light. All the processes crucial to the growth and development of plants are triggered by light, including seed germination, stem elongation, chloroplast formation, and flowering. These light responses are mediated by various photoreceptor systems, including *phytochrome, which responds to red and far-red light, and *cryptochrome, which is sensitive to blue and green light. These photoreceptor molecules interact with cell signalling networks to regulate the expression of genes involved in development and also influence the production of plant hormones, notably gibberellins and brassinosteroids.

photonasty *See* NASTIC MOVEMENTS.

photoperiod *See* PHOTOPERIODISM.

photoperiodism The response of an organism to changes in day length (**photoperiod**). Such responses enable organisms to exploit favourable conditions associated with seasonal changes in climate and vegetation, for example to produce young when food is plentiful or to produce flowers when pollinating insects are abundant. Generally, a plant must experience a particular duration of unbroken darkness for the physiological switch for, say, nonflowering to flowering to take place. This is the **critical night length**, which varies between species and with latitude in the same species. Some actions are triggered when the night length falls below the critical threshold (long-day plants), whereas others are prompted when night length exceeds a critical threshold (short-day plants). Exposure to just a single requisite night length can trigger flowering in some species, whereas others require several successive exposures before flowering ensues. Moreover, some need an additional environmental stimulus, such as a period of cold, before the physiological switch is made. Plants have various photoreceptor molecules that are sensitive to light, including *phytochromes and *cryptochromes. It is known that in certain species these photoreceptor molecules interact with the daily fluctuations in levels of proteins responsible for the plant's innate circadian rhythm (*see* BIOLOGICAL CLOCK) and bring about photoperiodic changes in cell metabolism. In animals, photoperiodic control of breeding is controlled by *melatonin. *See* DAY-NEUTRAL PLANT; LONG-DAY PLANT; SHORT-DAY PLANT. *See also* DARK PERIOD.

photophore A gland or organ that is specialized for the production of light (*see* BIOLUMINESCENCE). Photophores are a common feature of invertebrates and fish living in the deep sea, often being arranged in lines or other patterns over the body surface to produce a characteristic display of light. They contain the light-producing chemicals or symbiotic bacteria responsible for the bioluminescence. The cells of many different tissues can be modified to form a photophore, including mucous glands in various fish and even the suckers of some deep-sea octopuses.

photophosphorylation The formation of ATP from ADP and inorganic phosphate using light energy in *photosynthesis (*compare* OXIDATIVE PHOSPHORYLATION). There are two pathways, noncyclic and cyclic photophosphorylation, which occur in the thylakoid membranes of the chloroplasts. **Noncyclic photophosphorylation** involves a linear flow of electrons derived from the *photolysis of water; the electrons are transferred through *photosystem II to photosystem I via an *electron transport chain (ETC) consisting of *plastoquinone, a *cytochrome complex, and *plastocyanin. The *cytochromes form the **cytochrome b_6–f complex** across the thylakoid membrane.

The ETC pumps hydrogen ions (H^+) into the thylakoid space, and these diffuse back to the stroma through the enzyme complex ATP synthase located in the thylakoid membrane, thus synthesizing ATP by the *chemiosmotic mechanism. Electrons continue to flow to photosystem I, which raises their energy levels sufficiently for the reduction of $NADP^+$ to NADPH by the enzyme $NADP^+$ reductase. This involves a second ETC but no production of ATP. Thus, ATP provides the energy, and NADPH the reducing power for the light-independent reactions of photosynthesis (*see* CALVIN–BASSHAM–BENSON CYCLE). In **cyclic photophosphorylation** the electrons from photosystem I that are raised to a higher energy level are recycled through the ETC back to photosystem I, and the resulting proton gradient across the thylakoid membrane again drives the phosphorylation of ADP to ATP by the enzyme ATP synthetase, but there is no production of NADPH. Some photosynthetic bacteria, such as the green sulphur bacteria and purple sulphur bacteria, have only a single photosystem, and cyclic electron flow is their only way of generating ATP. Cyclic electron flow also occurs in organisms with two photosystems, such as cyanobacteria and plants, and is an adaptation for protecting the photosynthetic machinery of plants against stress caused by high light intensity. In prokaryotes the components of photophosphorylation are arranged in folds of the plasma membrane.

photopic vision The type of vision that occurs when the cones in the eye are the principal receptors, i.e. when the level of illumination is high. Colours can be identified with photopic vision. *Compare* SCOTOPIC VISION.

photoprotection Protection of a plant's photosynthetic apparatus from the harmful effects of light. During periods of peak light intensity plants are able to utilize less than half the incoming energy. The surplus energy poses the risk of photooxidation, and the formation of highly reactive superoxide radicals that can destroy the cell's chlorophyll and many other cellular components. Much of the excess energy is trapped and dissipated as heat by carotenoids of the *xanthophyll cycle. Also, chloroplasts contain the enzyme *superoxide dismutase, which scavenges superoxide radicals.

photoreactivation *See* DNA PHOTOLYASE.

photoreceptor A sensory cell or group of cells that reacts to the presence of light. It usually contains a pigment that undergoes a chemical change when light is absorbed, thus stimulating a nerve. *See* EYE.

photorespiration A metabolic pathway that occurs in plants in the presence of light, in which ribulose bisphosphate carboxylase/oxygenase (*rubisco), the enzyme involved in carbon dioxide fixation with *ribulose bisphosphate, accepts oxygen, in place of carbon dioxide, resulting in the formation of a two-carbon compound, glycolate. Most of the fixed carbon represented by the glycolate can be salvaged by a series of reactions—the **glycolate pathway**—involving the peroxisomes and mitochondria, and returned to the chloroplasts. However, some of the carbon is lost as carbon dioxide. Unlike respiration, there is no production of ATP. In C_3 plants (*see* C_3 PATHWAY) photorespiration has the effect of reducing the rate of photosynthesis, as atmospheric oxygen can combine with rubisco. In C_4 plants (*see* C_4 PATHWAY) the effect of photorespiration is negligible as the affinity of phosphoenolpyruvate carboxylase for carbon dioxide is extremely high. As oxygen is a competitive inhibitor of rubisco, photorespiration will increase as oxygen concentration increases or as carbon dioxide concentration decreases.

photosynthate The products of *photosynthesis. The net output of the *Calvin-Bassham–Benson cycle is the three-carbon compound glyceraldehyde 3-phosphate, which is exported from the chloroplast to the cytosol of the cell. This undergoes condensation reactions to form the six carbon compounds fructose 1,6-bisphosphate and glucose 1-phosphate, which via a further condensation form the 12-carbon disaccharide sucrose. Some glyceraldehyde 3-phosphate can also be converted to amino acids and fats, but sucrose is the main form

in which carbohydrate is exported from photosynthesizing cells to the rest of the plant and hence the most abundant photosynthate.

photosynthesis The chemical process by which green plants and other *phototrophs synthesize organic compounds from carbon dioxide and water in the presence of sunlight. In plants and most algae it occurs in the *chloroplasts and there are two principal types of reactions. In the **light-dependent reactions**, which require the presence of light, energy from sunlight is absorbed by *photosynthetic pigments (chiefly the green pigment *chlorophyll) and used to bring about the *photolysis of water:

$$H_2O \rightarrow 2H^+ + 2e^- + \frac{1}{2}O_2$$

The electrons released by this reaction pass along a series of electron carrier molecules in an *electron transport chain; as they do so they lose their energy, which is used to convert ADP to ATP in the process of *photophosphorylation. The electrons and protons produced by the photolysis of water are used to reduce NADP:

$$2H^+ + 2e^- + NADP^+ \rightarrow NADPH + H^+$$

The ATP and NADPH produced during the light-dependent reactions provide energy and reducing power, respectively, for the ensuing **light-independent reactions** (formerly called the 'dark reaction'), which nevertheless cannot be sustained without the ATP generated by the light-dependent reactions. During these reactions carbon dioxide is reduced to carbohydrate in a metabolic pathway known as the *Calvin-Bassham–Benson cycle. Photosynthesis can be summarized by the equation:

$$CO_2 + 2H_2O \rightarrow [CH_2O] + H_2O + O_2$$

In photosynthetic bacteria the components of photosynthesis are arranged in folds of the plasma membrane and incorporate *bacteriochlorophyll as pigment. A form of photosynthesis also occurs in archaea of the genus *Halobacterium*, using a membrane-bound protein called *bacteriorhodopsin. Since virtually all other forms of life are directly or indirectly dependent on plants for food, photosynthesis is the basis for all life on earth. Furthermore virtually all the

Photosynthesis

atmospheric oxygen has originated from oxygen released during photosynthesis.

photosynthetic carbon reduction cycle See CALVIN–BASSHAM–BENSON CYCLE.

photosynthetic nitrogen use efficiency The ratio of a plant's rate of carbon assimilation to the concentration of organic nitrogen in its leaves: a measure of the efficiency with which the plant allocates nitrogen to its photosynthetic apparatus.

photosynthetic pigments The pigments responsible for the capture of light energy during the light-dependent reactions of *photosynthesis. In plants, algae, and cyanobacteria the green pigment *chlorophyll *a* is the **primary pigment**, absorbing blue and red light (*see* PHOTOSYSTEMS I AND II). The *carotenoids and various other pigments are *accessory pigments, absorbing light energy and

passing this on to the chlorophyll *a* molecules. Other photosynthetic bacteria use various types of *bacteriochlorophyll. *See also* BACTERIORHODOPSIN.

photosystems I and II The two systems of *photosynthetic pigments in the thylakoid membranes of chloroplasts that are involved in the light-dependent reactions of photosynthesis. Each photosystem comprises a complex of proteins spanning the thylakoid membrane. Attached to the proteins are multiple arrays of chlorophyll *a* and accessory pigments forming **light-harvesting complexes**, or **antenna pigments**, surrounding the photosystem's single **reaction centre**. Photons of light are trapped by the light-harvesting complexes and the excited pigment molecules rapidly transfer their energy via other molecules to a chlorophyll *a* molecule at the reaction centre, where the absorbed light energy is converted to chemical energy. The photons cause the release of energized electrons, or excitons, which behave in a quantum mechanical manner as both waves and particles. This behaviour contributes to the high efficiency with which light energy is transferred to the reaction centre (*see* QUANTUM BIOLOGY). In photosystem II a chlorophyll *a* molecule, known as P680, utilizes light of wavelength 680 nm; in photosystem I the chlorophyll *a* molecule, known as P700, absorbs light at a wavelength of 700 nm. Light energy is used in each reaction centre to raise electrons to higher energy levels to enable them to be taken up by electron acceptors. This causes P680 and P700 to become positively charged, or oxidized. The chlorophyll in photosystem II replaces its lost electrons with ones supplied by an associated complex of proteins called the *oxygen-evolving complex, which is responsible for the *photolysis of water:

$$2H_2O \rightarrow 4H^+ + 4e^- + O_2$$

The oxygen produced is given off as a gas, and the H^+ ions, together with the electrons from photosystem I, reduce $NADP^+$ (*see* PHOTOPHOSPHORYLATION).

phototaxis (*pl.* **phototaxes**) The movement of a cell (e.g. a gamete) or a unicellular organism in response to light. For example, certain algae (e.g. *Chlamydomonas*) can perceive light by means of a sensitive eyespot and move to regions of higher light concentration to enhance photosynthesis. *See* TAXIS.

phototroph Any organism that uses energy derived from the sun to manufacture organic compounds by photosynthesis. Most phototrophs are *photoautotrophs; a few bacteria are *photoheterotrophs.

phototropin Any of a class of photosensitive flavoproteins that are sensitive to blue light (wavelength 400–500 nm) and trigger reorientating growth of a plant towards a light source (i.e. *phototropism). The bound flavin mononucleotide acts as a chromophore, absorbing light and causing photoexcitation of a light-sensitive domain of the protein. This initiates autophosphorylation of the protein and triggers a signal cascade resulting in cell elongation. Phototropins also cause light-induced opening of stomata and movement of chloroplasts inside leaves. Light activates phototropins by inducing autophosphorylation.

phototropism (**heliotropism**) The growth of plant organs in response to light. Aerial shoots usually grow towards light, while some aerial roots grow away from light. The phototropic response is triggered by photosensitive molecules, such as *phototropins, and brought about by the differential distribution of the plant hormone *auxin in the illuminated part, which in turn causes differential growth of the shoot or root. *See* TROPISM.

phragmoplast A barrel-shaped array of short microtubules that organizes formation of the *cell plate in dividing plant cells. The phragmoplast is orientated parallel to the mitotic spindle and accumulates vesicles in the plane of the metaphase equator, midway between the daughter nuclei. The vesicles fuse, and their components form the new plasma membranes and cell walls of the daughter cells. *Compare* PHYCOPLAST.

phrenic nerve The nerve that forms a network in the diaphragm and is involved in the control of ventilation. *See* INSPIRATORY CENTRE.

pH scale A logarithmic scale for expressing the acidity or alkalinity of a solution. To a first approximation, the pH of a solution can be defined as $-\log_{10}c$, where c is the concentration of hydrogen ions in moles per cubic decimetre. A neutral solution at 25°C has a hydrogen-ion concentration of 10^{-7} mol dm^{-3}, so the pH is 7. A pH below 7 indicates an acid solution; one above 7 indicates an alkaline solution. pH stands for 'potential of hydrogen'. The scale was introduced by S. P. Sørensen (1868–1939) in 1909.

phycobiliprotein Any of a class of pigments, found chiefly in cyanobacteria and red algae, that act as *accessory pigments in photosynthesis, being attached to the *thylakoid membranes. Each comprises a protein bound to a **phycobilin**, a coloured tetrapyrrole prosthetic group (*see* PORPHYRIN). The main phycobiliproteins are *phycocyanin and *phycoerythrin. Within cells, phycobiliproteins form complex clusters called **phycobilisomes**.

phycobiont The algal or bacterial component of a *lichen.

phycocyanin An accessory photosynthetic pigment occurring mainly in cyanobacteria and red algae. It is a *phycobiliprotein, in which the pigmented prosthetic group is phycocyanobilin, which gives phycocyanin its blue colour.

phycoerythrin An accessory photosynthetic pigment occurring mainly in the red algae and cyanobacteria. It is a *phycobiliprotein, in which the pigmented prosthetic group is phycoerythrobilin, which gives phycoerythrin its red colour.

phycomycetes In older classification schemes, all primitive *fungi, many of which are found in water (e.g. the water moulds, which may be parasitic on fish) or in damp areas. Many are unicellular but those that form mycelia generally have hyphae lacking cross walls, which distinguishes them from the *Ascomycota and *Basidiomycota. They include the *Zygomycota.

phycoplast An array of microtubules that organizes cell division following mitosis in certain algae, e.g. *Chlamydomonas*. Unlike the *phragmoplast found in dividing cells of plants and many algae, the phycoplast consists of microtubules arranged perpendicular to the axis of the mitotic spindle and parallel to the plane of formation of the new cell wall. The new wall may form by infurrowing from the existing walls or by deposition of a *cell plate.

phyletic series A sequence of fossil forms that represent the succession of species in the evolution of a particular lineage. Often there are gaps in the fossil record, and phyletic series are incomplete. However, in some cases the fossil record is good, and a more complete phyletic series can be constructed. For example, most of the intermediate forms between the modern horse (*Equus*) and its dog-sized

ancestor, *Hyracotherium* (*Eohippus*) of the Eocene epoch, have been charted.

phyllode A flattened and widened petiole (leaf stalk) that looks like and functions as a leaf. In plants with phyllodes the true leaves are often tiny or do not develop. Australian and Pacific species of *Acacia* typically have phyllodes, whose vertical orientation is thought to protect the photosynthetic apparatus from excessive light. *Compare* CLADODE.

phyllotaxis (phyllotaxy) The arrangement of leaves on a plant stem. It is determined by the apical meristem of the growing shoot and is specific to each species. The leaves may be inserted up the stem in whorls (whorled phyllotaxy), in pairs (opposite phyllotaxy), or singly at each node. When arranged in pairs, the two leaves arise on opposite sides of the stem and are usually at right angles to the leaf pairs above and below them. Single leaves may be inserted alternately (alternate phyllotaxy) or in a spiral pattern (spiral phyllotaxy) up the stem. The most common arrangement in angiosperms is spiral phyllotaxy, which results in the minimum of shading of leaves by those above them. However, the shading afforded by opposite phyllotaxy may be beneficial where plants are subject to intense sunlight.

PhyloCode An initiative that started in 2000 to devise a set of rules for naming groups of organisms based explicitly on their evolutionary (i.e. phylogenetic) relationships, rather than on traditional rank-based codes of nomenclature. Its chief aim is to give each clade of organisms—i.e. an ancestor and all its descendants—a unique and unchanging name that will exist alongside rank-based classifications. Such a system will, it is claimed, avoid the ambiguities and confusion that arise due to continual taxonomic revisions and prevent different groups being given the same name. This will enhance the efficient storage and retrieval of the vast and ever-growing amounts of biological data drawn from across the entire tree of life.

phylogenetic Describing a system of *classification of organisms that aims to show their evolutionary history. *Compare* PHENETIC. *See* also Appendix 2 for a simplified Phylogenetic tree of the animal kingdom; and Appendix 3 for its counterpart for plants.

phylogenetic species concept (PSC; lineage species concept) The concept of a species as an irreducible group whose members are descended from a common ancestor and who all possess a combination of certain defining, or derived, traits (*see* APOMORPHY). Hence, this concept defines a species as a group having a shared and unique evolutionary history. It is less restrictive than the *biological species concept, in that breeding between members of different species does not pose a problem. Also, it permits successive species to be defined even if they have evolved in an unbroken line of descent, with continuity of sexual fertility. However, because slight differences can be found among virtually any group of organisms, the concept tends to encourage extreme division of species into ever-smaller groups.

phylogenetic tree 1. *See* EVOLUTIONARY TREE. **2.** *See* PHYLOGRAM.

phylogenomics The field of *bioinformatics that integrates knowledge about the evolutionary history of organisms (i.e. phylogeny) with structural and functional analysis of their genomes (i.e. genomics) and proteins (proteomics). A basic supposition is that well-characterized genes (or proteins) in one organism provide valuable insights into the likely sequence and function of homologous genes or proteins in related organisms. Conversely, comparisons of gene or protein sequences from different species can provide evidence of possible homology, and hence shared evolutionary origins. On a molecular level, the vast amounts of data from genomics projects are used to construct phylogenetic databases for genes and proteins; sequence data are correlated with experimental findings about protein structure and function and subject to statistical analysis. Sequence

data are used to construct phylogenetic trees (*phylograms) using principles of *maximum parsimony or *maximum likelihood. The validity and accuracy of such methods can be tested by experiments with living organisms—e.g. by allowing bacteriophage to evolve over multiple generations and lineages—or by computer simulations.

phylogeny The evolutionary history of an organism or group of related organisms. *Compare* ONTOGENY.

phylogeography A discipline first outlined in 1987 that studies the spatial distributions of genetic lineages among related species or within populations of a species. It seeks to describe not only when but also where species diverged, and how historical changes in topography, climate, continental drift, and other factors have determined the genetic makeup and distribution of their descendants. Changes in the composition of certain parts of the genome, for instance mitochondrial DNA and ribosome genes, provide the raw data for tracking evolutionary history (i.e. phylogeny) at different timescales, whether over millions of years for speciation events or thousands of years for changes in populations. The resulting insights enable modelling to predict effects of, for instance, climate change, habitat loss, and overexploitation by hunting, thus providing valuable information for species conservation.

phylogram (phylogenetic tree) A diagram representing the evolutionary relationships and distances between different groups or species of organisms using data obtained by phylogenetic methods, such as comparative morphology or sequence analysis of nucleic acids and proteins. The relationships are depicted by (usually) dichotomously branching lines to form a treelike pattern, with the lengths (or weights) of the branches proportional to the distance of the relationship in evolutionary terms. A **rooted tree** is one that specifies the oldest common ancestor and depicts the evolutionary splits in chronological sequence. An **unrooted tree** (sometimes called a **star diagram**) is

terminal node
(vertex)

polychotomous
internal node
internal node
(vertex)

root

branch
(edge)

Phylogram: rooted tree

Phylogram: unrooted tree

one that does not specify the oldest common ancestor, and thus the evolutionary origin: it indicates relationships but not evolutionary direction. As shown in the diagram, a tree consists of nodes (or vertices) connected by branches (or edges). In a rooted tree the earliest group to diverge is called the basal group, represented by D and its descendants E and F. If a node splits into three or more branches (as at C in the diagram) it is described as polychotomous, and the tree is said to be unresolved. A simple cladogram is similar to a rooted tree but shows only the relationships, not their evolutionary distance (*see* CLADISTICS).

phylum (*pl.* **phyla**) A category used in the *classification of organisms that consists of one or several similar or closely related classes. Examples of phyla are the Ascomycota, Bryophyta, and Chordata. In traditional classification schemes phyla are grouped into kingdoms, and for plants the

*division is sometimes used instead of the phylum.

physical map (in genetics) Any map that shows the arrangement of the material (i.e. the nucleoprotein) making up a chromosome or segment of a genome (*see* CHROMOSOME MAP) and the physical distance between various genetic sites (loci), ideally measured in nucleotides (nt) or base pairs (bp). There are several types of physical map, differing vastly in scale and detail. Moreover, a wide range of mapping techniques and strategies are available, the choice depending on the extent and detail required and existing knowledge of the organism's genome. The coarsest physical maps are ones depicting chromosome banding patterns, which are dark and light transverse bands obtained by staining entire chromosomes in mitosis. These **cytological maps** enable characterization of individual chromosomes and can reveal gross anomalies, such as missing or duplicated segments. The chromosomal location of a particular DNA base sequence of interest can be determined using a DNA probe and *fluorescence in situ hybridization (FISH). *Restriction mapping reveals the order and distance apart of sites cleaved by restriction enzymes; such sites are important genetic markers, used in pinpointing mutations. On a larger scale are **contig maps**; these show the order of cloned DNA segments taken from a *DNA library and fitted together to form a series of overlapping, or contiguous, segments, called a **contig**. Such segments are roughly on a gene-length scale. Once a contig has been correctly aligned, the base sequence of each component segment can be determined (*see* DNA SEQUENCING), and hence the overall sequence of the chromosomal DNA can be pieced together. Useful markers in establishing contigs are short unique sequences called *sequence-tagged sites. Genes that are expressed in a certain tissue or disease can be identified by searching databases for *expressed sequence tags, which are unique markers of complementary DNA. These can be used to construct a probe to identify a particular gene's location. Other useful markers in constructing physical maps include *variable number tandem repeats (minisatellite DNA) and *short tandem repeats (microsatellite DNA). The advent of fast, inexpensive DNA sequencing coupled with computerized analytical techniques and access to a fast-growing body of sequence information held on databases is rapidly changing approaches to genetic mapping. *Compare* LINKAGE MAP.

physiognomy (in ecology) The overall size and shape of an organism. Descriptions such as 'trees', 'shrubs', and 'herbs' are frequently used to characterize the general appearance of the vegetation of a region. Moreover, plant physiognomy can be broadly correlated with environmental conditions, so that regions of the world with similar climates tend to have a dominant vegetation of similar life forms. Various attempts have been made to refine this approach by defining physiognomic categories, or **life forms**, most notably by the Danish ecologist Christen Raunkiaer (1876–1960). His system of classifying life forms is based on the way in which plants survive harsh conditions, particularly the position of their perennating (or overwintering) buds in relation to the soil surface. He proposed five classes: *therophytes (annual plants); *geophytes (plants that produce underground perennating organs); *hemicryptophytes (herbaceous perennials); *chamaephytes (small shrubs); and *phanerophytes (large shrubs and trees).

physiological saline A liquid medium in which animal tissues may be kept alive for a few hours during experiments without pathological changes or distortion of the cells taking place. Such fluids are salt solutions that are isotonic with and have the same pH as the body fluids of the animal. A well-known example is **Ringer's solution**, formulated by the British physiologist S. Ringer (1835–1910), which is a mixture of sodium chloride, calcium chloride, sodium bicarbonate, and potassium chloride solutions formulated to match their concentrations in body fluids. In lactated Ringer's solution, sodium lactate is used instead of sodium bicarbonate.

physiological specialization The occurrence within a species of several forms that are identical in appearance but differ in physiology: these are termed **physiological races**. For example, many pathogenic fungi develop new physiological races in response to the strong selection pressure exerted when disease-resistant crop varieties are sown over large areas.

physiology The branch of biology concerned with the vital functions of plants and animals, such as nutrition, respiration, reproduction, and excretion.

physiome A complete description of the interplay of all processes that sustain the life of an organism or species and how these change in time and space—for instance on a daily basis, or with changing environmental conditions. Such a description integrates all aspects of an organism's biology, from its genes and proteins, through cellular metabolism and biochemistry, to anatomy. Attempts are underway to fulfil this goal for humans and other species by linking databases and computer models that focus on particular systems or networks, such as genomics, proteomics, and signal pathways, to form a single framework. Data from allied fields, such as pharmacology and pathology, are also incorporated. One ambition is to construct a virtual physiological human, as a means of facilitating understanding, predicting responses to drugs, diagnosing disease, etc.

physisorption *See* ADSORPTION.

phyto- Prefix denoting plants. For example, phytopathology is the study of plant diseases.

phytoaccumulation *See* PHYTOREMEDIATION.

phytoalexin An antimicrobial substance that is produced by a plant in response to infection by fungi or bacteria and helps to defend the plant by inhibiting the growth of invading microbes. Phytoalexins vary in their chemical nature; for example, in legumes they are mainly isoflavonoids, whereas in the potato family (Solanaceae) they are predominantly terpenes. Substances, such as small polysaccharides and proteins, that are produced by the invading pathogen act as *elicitors, triggering the manufacture of phytoalexins in the tissues surrounding the infection site. The production of phytoalexins is one component of a broader defensive reaction, known as *hypersensitivity (sense 2).

phytochelatin Any of a group of sulphur-rich peptides found in plants exposed to high concentrations of heavy metals (e.g. cadmium, zinc, lead, copper, mercury) and thought to play a role in detoxification. They are unusual peptides synthesized from *glutathione and with the general formula (γ-glutamic acid-cysteine)$_n$-glycine, where n is 2–11. They bind the metal ions in the cell cytoplasm and the resulting complex is transported to the cell vacuole, where they are sequestered by organic acids.

phytochrome Any of a class of plant pigments that can detect the presence/absence or quality of light, particularly of red and far-red wavelengths. Phytochromes are involved in regulating many processes that are linked to day length (photoperiod), such as seed germination and initiation of flowering, and growth responses to shading. Each consists of a light-detecting portion, called a *chromophore, linked to a small protein and exists in two interconvertible forms with different physical properties. In darkness phytochrome is virtually all in the form that absorbs red light (wavelength 660 nm), denoted P_r. When exposed to light, this P_r form is converted to the P_{fr} form, which absorbs light in the far-red region of the spectrum, at about 730 nm. Exposure to far-red light converts P_{fr} back to P_r. Because sunlight contains both red and far-red light, so long as the plant is exposed to light, the reversible interconversion between the two forms of phytochrome continues, reaching an equilibrium dependent on the relative amounts of red and far-red light. Conversion to the P_{fr} form exposes a nuclear localization signal sequence, so that it transfers from the cell cytosol to the nucleus. Here it interacts directly with some transcription factors and phosphorylates other proteins, triggering a signal cascade

that changes the activity of multiple genes. For example, P_{fr} stimulates the transcription of genes responsible for the production of plant hormones such as gibberellins that are involved in seed germination. In darkness, P_{fr} reverts over several hours to the inactive P_r ('red') form. See also PHOTOMORPHOGENESIS; PHOTOPERIODISM. Compare CRYPTOCHROME.

phytoecdysteroid See ECDYSONE.

phytogeography See PLANT GEOGRAPHY.

phytohaemagglutinin (PHA) Any of various plant-derived compounds that induce changes in lymphocytes normally associated with antigen challenge. These changes include cell enlargement, increased RNA and DNA synthesis, and, finally, cell division. Response to PHAs is used to test for competence of cell-mediated *immunity, for example in patients suffering chronic virus infections.

phytohormone See PLANT HORMONE.

phytomer A repeating structural unit, or 'building block', of higher plants. For example, in the shoot of a typical dicotyledonous plant a phytomer consists of a single node, the leaf or leaves arising from the node, the adjoining stem segment (internode), and the axillary bud(s) in the leaf axil(s).

phytoncide Any of various volatile antimicrobial organic compounds derived from trees, such as α-pinene and limonene, that are emitted as a means of defence against potential airborne pathogens. Human exposure to phytoncides, e.g. during 'forest bathing', is claimed to boost immune function, specifically the activity of *natural killer cells.

phytoplankton The photosynthesizing *plankton, consisting chiefly of microscopic green algae, diatoms, dinoflagellates, and cyanobacteria Near the surface of the sea there may be many millions of such organisms per cubic metre. Members of the phytoplankton are of great importance as they form the basis of food for all other forms of aquatic life, being the primary *producers. *Compare* ZOOPLANKTON.

phytoremediation The use of plants to decontaminate polluted land, water, or air. Different plant species that are able to grow on contaminated sites have evolved various ways of countering high concentrations of heavy metals, oil, solvents, or other toxic substances. In **phytotransformation** the plant takes up the toxin and degrades it by enzyme activity, thereby eliminating its toxic nature altogether. In **phytostabilization** the plant's root exudates or associated microorganisms precipitate the toxins (e.g. heavy metals) and make them harmless to other organisms. In **phytoaccumulation** (or **phytoextraction**) a plant takes up the contaminants and stores them in its tissues; the plant can then be harvested, thus removing the toxin with the plant. A good example is alpine pennycress, which is an effective accumulator of cadmium and zinc. Such plants are termed *hyperaccumulators.

phytostabilization *See* PHYTOR-EMEDIATION.

phytotelm A small body of water found on a plant. Examples include the contents of a pitcher plant or the small pool within a tree hole. Although often transitory, phytotelms may contain abundant life and represent well-demarcated animal communities that are very amenable to ecological study.

phytotransformation *See* PHYTOR-EMEDIATION.

pia mater The innermost of the three membranes (*meninges) that surround the brain and spinal cord of vertebrates. The pia mater lies immediately adjacent to the central nervous system, and the *choroid plexus, which secretes cerebrospinal fluid, is an extension of it.

pico- Symbol p. A prefix used in the metric system to denote 10^{-12}. For example, 10^{-12} farad $= 1$ picofarad (pF).

picornavirus One of a group of small RNA-containing viruses (*pico* = small; hence pico-RNA-virus) commonly present in the alimentary and respiratory tracts of vertebrates. They cause mild infections of these tracts but the group also includes the polioviruses, which attack the central nervous system causing poliomyelitis; and the aphthoviruses responsible for foot and mouth disease in cattle, sheep, and pigs.

pie chart A diagram in which percentages are shown as sectors of a circle. If x percent of the prey of a carnivore comprises species X, y percent species Y, and z percent species Z, a pie chart would show three sectors having central angles $3.6x°$, $3.6y°$, and $3.6z°$.

pigment A compound that gives colour to a tissue. Pigments perform a variety of functions. For example, *haemoglobin in vertebrate erythrocytes gives blood its characteristic red colour and enables the transport of oxygen throughout the body (*see* RESPIRATORY PIGMENT). Other biological pigments include *chlorophyll, a photosynthetic pigment in plants that is responsible for their green coloration; and *melanin, a brown pigment in animals that provides protection from ultraviolet light and can be used in camouflaging colorations. *See also* RHODOPSIN.

Pikaia *See* BURGESS SHALE.

pileus (*pl.* **pilei**) The umbrella-shaped cap of certain fungi, such as mushrooms. Spores are produced from *gills or pores on the lower surface.

piliferous layer The part of the root epidermis that bears *root hairs. It extends over a region about 4–10 mm behind the root tip. Beyond this the piliferous layer is sloughed off to reveal the hypodermis.

pilomotor Describing nerve fibres or the muscles innervated by them that can cause the position of the body hair to be altered. The *arrector pili is a pilomotor muscle.

Piltdown man Fossil remains, purported to have been found by Charles Dawson (1864–1916) at Piltdown, Sussex, in 1912, that were named *Eoanthropus dawsoni* and described as a representative of the true ancestors of modern humans. The skull resembled that of a human but the jaw was apelike. In 1953 dating techniques showed the specimen to be a fraud.

pilus (*pl.* **pili**) **1.** (in microbiology) A hairlike appendage projecting from the surface of a bacterium. Pili are found in both Gram-positive and Gram-negative bacteria and are typically 0.3–2.0 μm long; they fulfil a variety of functions, including adherence to surfaces, cell–cell interactions (e.g. between a pathogen and a host cell), *biofilm formation, and a form of motility called 'twitching', whereby the pili alternately extend and retract. Multiple subunits of the protein pilin make up each pilus, and variation in the nature of the subunits can alter the antigenic nature of the bacterium. A single cell may have numerous short pili, which are also called **fimbriae**. Bacteria containing a *sex factor can form one or several longer **sex pili**, which extend to contact another cell of the same species. The sex pilus then retracts to draw the two cells together so that DNA can be transferred in the process called *conjugation. **2.** (in anatomy) A hair.

pineal eye See MEDIAN EYE.

pineal gland An endocrine gland located between the two cerebral hemispheres and connected to the forebrain by a stalk. In humans and other mammals it produces the hormone *melatonin during the hours of darkness on a circadian (24-hour) cycle, according to timing signals received from two structures in the left and right hypothalamus, the suprachiasmatic nuclei (*see* BIOLOGICAL CLOCK). In many other vertebrates the equivalent structure is the pineal sac, which may be associated with a light-sensitive *median eye.

pinna (auricle) (*pl.* **pinnae**) The visible part of the *outer ear, present in some mammals. It is made of cartilage and its function is to channel sound waves into the external auditory meatus. In some species the pinna is movable and aids in detecting the direction from which a sound originates.

pinocytosis The process by which a living cell engulfs a minute droplet of liquid. It involves a mechanism similar to *phagocytosis. See ENDOCYTOSIS.

pipette A graduated tube used for transferring measured volumes of liquid. See also MICROPIPETTE.

piRNA See PIWI-INTERACTING RNA.

Pisces In some classifications, a superclass of the *Gnathostomata (jawed chordates) comprising the fishes (*compare* TETRAPODA). The whole body of a fish is covered with a tough, usually scaly, skin (*see* SCALES), which extends over the eye and contains pigments and sometimes slime glands. The circulatory system is a single circuit with blood passing through two sets of capillaries, one at the gills and the other in the body tissues. Three *semicircular canals are present in the inner ear. The sense of smell is particularly well developed and pressure waves are detected by the *lateral-line system. Fossils of fish date back to the Ordovician period, 505–438 million years ago. There are two classes of modern fish: *Chondrichthyes (cartilaginous fishes) and *Osteichthyes (ray-finned, or bony fishes).

(((⊕))) SEE WEB LINKS
• The Australian Museum Fish Site, an extensive site catering for fish enthusiasts at various levels

pistil The female part of a flower, consisting either of a single *carpel (**simple pistil**) or a group of carpels (**compound pistil**).

pit 1. A depression or cavity in the secondary wall of a plant cell that facilitates the movement of substances between adjacent cells. The equivalent structures in primary cell walls are called **primary pit fields**, areas where plasmodesmata are concentrated and pit development usually occurs. A pit comprises a **pit membrane**, consisting of the middle lamella plus the primary wall; and a **pit cavity**, the depression in the secondary wall. Pits usually occur in pairs (called **pit pairs**) on either side of the middle lamella between two adjacent cells. **2.** (coated pit) See ENDOSOME.

pitfall trap A simple trap for small invertebrate animals consisting of a tin that is placed in the ground with its rim at ground level. The trap, which contains some kind of bait, can be covered by a tile suspended above ground level by stones so that rain does not enter the tin.

pith 1. (medulla) The cylinder of *parenchyma tissue found in the centre of plant stems to the inside of the vascular tissue. It is light in weight and has been put to various commercial uses, notably the manufacture of pith helmets. **2.** (not in scientific usage) The white tissue below the rind of many citrus fruits. **3.** To destroy the central nervous system of an animal, especially a laboratory animal such as a frog, by severing its spinal cord.

Pithecanthropus See HOMO.

pituitary gland (pituitary body; hypophysis) A pea-sized endocrine gland attached by a thin stalk to the *hypothalamus at the base of the brain. It consists of two main lobes, the anterior and the posterior, separated in nonhumans by a much smaller intermediate lobe. The **anterior pituitary** (or **adenohypophysis**) secretes such hormones as *growth hormone, the *gonadotrophins, *prolactin, *thyroid-stimulating hormone, and *ACTH. Because these hormones regulate the growth and activity of several other endocrine glands, the anterior pituitary is often referred to as the **master endocrine gland**. Activity of the anterior pituitary itself is regulated by specific *releasing hormones produced by the hypothalamus (*see also* NEUROENDOCRINE SYSTEM). The **posterior pituitary** (or **neurohypophysis**) secretes the hormones *oxytocin and *antidiuretic hormone, and the intermediate lobe secretes *melanocyte-stimulating hormone. In humans the pituitary lacks a functionally discrete intermediate lobe.

piwi-interacting RNA (piRNA) Any of a group of small RNA molecules, typically 24–31 nucleotides long, that interact with piwi domain-containing proteins (named after the *Drosophila* mutant called P element-induced wimpy testis) or their homologues to form piRNA-induced silencing complexes. These silence transposons in the genome of germline and somatic cells, and are vital for normal formation of sperm and eggs in animals. They also take part in regulating gene expression in various tissues. At least 20 000 piRNAs are expressed in the human genome.

PKC *See* PROTEIN KINASE C.

placenta (*pl.* **placentae** or **placentas**) **1.** The organ in mammals and other viviparous animals by means of which the embryo is attached to the wall of the uterus. It is composed of embryonic and maternal tissues: extensions of the *chorion and *allantois grow into the uterine wall so that materials (e.g. oxygen, nutrients, and metabolic waste products) can pass between the blood of the embryo and its mother. There is, however, no direct connection between the maternal and embryonic blood: blood travels to the placenta from the embryo through the arteries of the umbilical cord and returns via the umbilical vein. The placenta is eventually expelled as part of the *afterbirth. **2.** A ridge of tissue on the ovary wall of flowering plants to which the ovules are attached. The arrangement of ovules on the placenta (**placentation**) is variable, depending on the number of carpels and whether they are free (*see* APOCARPY) or fused (*see* SYNCARPY).

Placentalia *See* EUTHERIA.

placode An area of thickening in the ectoderm of an early embryo that is the forerunner of a particular organ or structure. For example, the inner ear develops from the otic placodes, located on either side of the hindbrain. Likewise, hair follicles arise from individual epithelial placodes.

placoid scale (denticle) *See* SCALES.

Placozoa A phylum of simple aquatic animals containing just a single officially described species, *Trichoplax adhaerens*. This has a transparent round flattened body, between 0.2 and 3 mm in diameter, without head, tail, or appendages. It is covered in cilia, which it uses to crawl over surfaces, and it feeds by secreting enzymes from its ventral surface, part of which it may invaginate to form a temporary stomach. An adult comprises a few thousand cells of just four types, whose DNA content is the smallest of any animal. Placozoans can reproduce asexually by binary fission or budding, and sexual reproduction has been observed in laboratory cultures, although full details of the life cycle under natural conditions are unknown. The evolutionary relationships of placozoans with other animals, especially the sponges (*Porifera), remain speculative. For instance, some molecular studies of ribosomal RNA sequences have indicated that the Placozoa may be secondarily simplified descendants of more complex ancestors and are not closely related to the sponges. There is considerable genetic diversity in populations of placozoans from different sites around the world, indicating that the group may contain numerous species.

plagiotropism (diatropism) The tendency for a *tropism (growth response of a plant) to be orientated at an angle to the line of action of the stimulus concerned. For example, the growth of lateral branches and lateral roots is at an oblique angle to the stimulus of gravity (**plagiogeotropism**). *Compare* ORTHOTROPISM.

planarians *See* TURBELLARIA.

plankton The collection of *pelagic organisms that drift or float passively with the current in a sea or lake. Plankton includes many microscopic organisms, such as algae, protists, various animal larvae, and some worms, but also larger forms that are unable to swim effectively, such as jellyfish, *krill, and copepods. Plankton forms an important food source for many other members of the aquatic community and is divided into *zooplankton and *phytoplankton. *Compare* BENTHOS; NEKTON; NEUSTON.

plant 1. In traditional classifications, any living organism of the kingdom Plantae, which includes the various groups of algae and the land plants, or embryophytes. Plants are distinguished from other organisms by their life cycles, in which haploid male and female organisms develop from spores

produced by meiosis in the adult diploid organisms and produce gametes by mitosis; fertilization results in a diploid embryo that undergoes its early development in the haploid female. Most plants manufacture carbohydrates by *photosynthesis, in which simple inorganic substances are built up into organic compounds. The radiant energy needed for this process is absorbed by *chlorophyll, a complex pigment not found in animals. Plants also differ from animals in the possession of *cell walls (usually composed of *cellulose). Plants are immobile, as there is no necessity to search for food, and they respond slowly to external stimuli. For a classification of the plant kingdom, see Appendix 3. **2.** Any member of a supergroup of eukaryotic organisms—the Archaeplastida—recognized on the basis of recent molecular data. Besides land plants, this assemblage contains the single-celled and colonial green algae (*see* CHLOROPHYTA), the more distantly related red algae (*see* RHODOPHYTA), and some 13 species of single-celled algae—the *Glaucophyta—whose chloroplasts are similar to cyanobacteria.

(((●))) SEE WEB LINKS

● Open doors to the world of plants via the website of the Royal Botanic Gardens, Kew

plant geography (phytogeography) The study of the distribution of world vegetation, with particular emphasis on the influence of the environmental factors that determine this distribution.

plant hormone (plant growth regulator; phytohormone) Any of a number of organic chemicals that are synthesized by plants and regulate growth and development. They are usually made in a particular region, such as the shoot tip, and transported to other regions, where they take effect. *See* ABSCISIC ACID; AUXIN; BRASSINOSTEROID; CYTOKININ; ETHYLENE; GIBBERELLIN; JASMONATE; STRIGOLACTONE.

plantigrade Describing the gait of many mammals, including humans, in which the whole lower surface of the foot is on the ground. *Compare* DIGITIGRADE; UNGULIGRADE.

plant uncoupling protein *See* UNCOUPLING PROTEIN.

planula (*pl.* **planulae**) The ciliated free-swimming larva of many cnidarians, consisting of a solid mass of cells. It eventually settles on a suitable surface and develops into a *polyp.

plaque 1. A thin layer of organic material covering all or part of the exposed surface of a tooth. It contains dissolved food (mostly sugar) and bacteria. The bacteria in plaque metabolize the sugar and produce acid, which eats into the surface of the enamel of the tooth and eventually causes tooth decay (*dental caries). **2.** A clear area in a bacterial culture grown on an agar plate due to *lysis of the bacteria by a bacteriophage. **3.** A patch or differentiated zone on the surface of the skin or other organ. **4.** A fatty deposit that causes thickening of an arterial wall. *See* ATHEROSCLEROSIS.

plasma *See* BLOOD PLASMA.

plasma cell An effector cell of the immune system that secretes antibodies specific for a particular target antigen, such as one carried by an invading microorganism. Binding of the antibodies neutralizes the antigens (e.g. virus particles) or enhances the actions of other effector cells, such as phagocytosis by macrophages or neutrophils, in destroying the target. The plasma cells are produced by *B cells that become activated in response to encountering their specific antigen, in a process that requires participation by a helper *T cell. A single progenitor B cell can produce a clone consisting of thousands of plasma cells, each capable of secreting a trillion or so antibodies during a 4–5-day life.

plasmagel The specialized outer gel-like *cytoplasm of living cells (such as *Amoeba*) that move by extruding part of the cell (known as a *pseudopodium) in the direction of motion. A reversible conversion of plasmagel to the more fluid *plasmasol is involved in the continuous flow forward of

cytoplasm necessary for forming a pseudopodium (*see* AMOEBOID MOVEMENT). *See also* CYTOPLASMIC STREAMING.

plasma membrane (plasmalemma cell membrane) The partially permeable membrane forming the boundary of a cell. It consists mostly of protein and lipid (*see* LIPID BILAYER; FLUID MOSAIC MODEL) and plays various crucial roles in the cell's activities. A key task is to regulate the flow of materials into and out of the cell; this selectivity of traffic is accomplished, for example, by membrane proteins that act as *ion channels or *transport proteins. Other membrane proteins are receptors for signal molecules (e.g. hormones, growth factors, cytokines) arriving at the cell surface; they relay the signal to other components inside the cell. Supported by the cell's internal *cytoskeleton, the plasma membrane is the site of junctions with neighbouring cells (*see* CELL JUNCTION) and forms attachments to the *extracellular matrix, thus ensuring tissue integrity. In plants, fungi, bacteria, and many protists, it helps in assembling a cell wall or capsule on its outer surface. *Compare* NUCLEAR ENVELOPE.

plasma protein Any of the protein constituents of blood plasma. They comprise *albumins (about 60%), *globulins (e.g. antibodies and transport proteins—about 35%), and *fibrinogen (about 4%), with the remainder comprising chiefly various enzymes. They are significant contributors to maintaining the osmolarity of blood.

plasmasol The specialized inner sol-like *cytoplasm of living cells that move by producing *pseudopodia. *Compare* PLASMAGEL.

plasmid A structure in bacterial cells consisting of a looped DNA molecule that can exist and replicate independently of the chromosome. Plasmids provide genetic instructions for certain cell activities (e.g. resistance to antibiotic drugs). They can be transferred from cell to cell in a bacterial colony (*see* LATERAL GENE TRANSFER; SEX FACTOR). Plasmids are widely used as *vectors to produce recombinant DNA for *gene cloning.

plasmin (fibrinase; fibrinolysin) An enzyme, present in blood plasma, that breaks down a blood clot by destroying the fibrin threads of the clot and by inactivating factors involved in blood clotting, such as prothrombin and the *clotting factors. This occurs during *fibrinolysis. Plasmin is derived from an inactive precursor, *plasminogen.

plasminogen The inactive precursor of the enzyme *plasmin. Plasminogen is incorporated into blood clots and is converted to plasmin during *fibrinolysis in a reaction catalysed by the enzyme tissue plasminogen activator (TPA). TPA made with recombinant DNA technology is used to dissolve blood clots in patients suffering from cardiovascular disease or stroke.

plasmodesmata (*sing.* **plasmodesma**) Fine cytoplasmic strands that connect the *protoplasts of adjacent plant cells by passing through their cell walls. Plasmodesmata are cylindrical in shape (about 20–40 nm in diameter) and are lined by the plasma membrane of the two adjacent cells. The endoplasmic reticula of the two adjacent cells are connected by a narrower structure, the **desmotubule**, which runs through the centre of a plasmodesma. Plasmodesmata tend to occur in groups, forming distinct areas called **primary pit fields** (*see* PIT). They permit the passage between cells of substances including ions, sugars, amino acids, and macromolecules.

plasmodium (*pl.* **plasmodia**) A mass of cytoplasm containing many nuclei that comprises a stage in the life cycle of a plasmodial *slime mould. The plasmodium is the vegetative phase of the slime mould, often seen as a slimy layer on decaying wood; it is capable of slow amoeboid movement and feeds by engulfing particles of decaying organic matter.

plasmogamy Fusion of the cytoplasm of two (or more) cells. It precedes union of the nuclei in *fertilization and it occurs in *heterokaryosis.

plasmolysis The loss of water by *osmosis from a plant cell to the extent that the cytoplasm shrinks away from the cell wall.

This happens when the cell is placed in a solution that has a higher solute concentration than that of the cell sap, i.e. it has a lower (more negative) *water potential, since water always moves from an area of high water potential to an area of low water potential. *Compare* TURGOR.

plasticity *See* PHENOTYPIC PLASTICITY.

plastid An *organelle within a plant cell, often occurring in large numbers. Apart from the nucleus, plastids are the largest solid inclusions in a plant cell. For convenience they are classified into those containing pigments (*chromoplasts) and those that are colourless (*leucoplasts), although changes from one to the other frequently occur. Plastids develop from **proplastids**, colourless bodies found in meristematic and immature cells; they also arise by division of existing plastids. *See also* CHLOROPLAST; ENDOSYMBIONT THEORY.

plastocyanin A blue copper-containing protein that is found in chloroplasts and acts as an electron *carrier molecule in the light-dependent reactions of *photosynthesis (*see* PHOTOPHOSPHORYLATION). Plastocyanin consists of amino acid groups in association with a copper molecule which gives this compound a blue colour.

plastoglobulus (*pl.* **plastoglobuli**) A densely staining droplet found, often in large numbers, in plastids of plant cells. Plastoglobuli consist of lipid pigment and are especially prominent in coloured plastids (*chromoplasts), for example in ripening fruits. Plastoglobuli also occur in chloroplasts, but are masked by the green chlorophyll. When the chlorophyll breaks down as the leaves start to die in autumn, the pigmented plastoglobuli become apparent as the red or yellow 'fall' colours.

plastome The genome of a *plastid, such as that contained in the chloroplasts of a plant cell.

plastoquinone A quinone, found in chloroplasts, that functions as one of the *carrier molecules of the electron transport chain in the light-dependent reactions of

*photosynthesis. *See* PHOTOPHOSPHORY-
LATION.

plastron *See* CARAPACE.

plate count An estimate of the number of
viable cells in a culture. A plate count of a
bacterial culture is made by inoculating the
culture plate with a dilute solution of the
microorganisms and counting the number
of cells that appear in the resultant culture.

platelet (thrombocyte) A minute disc-
shaped cell fragment, 2–3 μm in diameter,
occurring in mammalian blood. Platelets
are formed as fragments of larger cells
(**megakaryocytes**) found in red bone marrow; they have no nucleus. They play an
important role in *blood clotting and
release *thromboxane A_2, serotonin and
other chemicals, which cause a chain of
events leading to the formation of a plug
at the site of the damage, thus preventing
further blood loss. There are typically
250 000 to 400 000 platelets per cubic millimetre in mammalian blood.

platelet-activating factor (PAF) A
phospholipid mediator of inflammation
produced by a variety of cell types. Originally named for its ability to activate and
aggregate platelets as part of the blood-
clotting mechanism, it is now known to
have a broad spectrum of biological effects.
PAF is one of the front-line agents produced by activated macrophages in the initial phase of an innate immune response to
infection. It contributes to the increased
permeability of local blood vessels, helping
to attract immune cells to the infection site,
and activates not only platelets but also
leucocytes, notably eosinophils, basophils,
and neutrophils, thereby amplifying the
inflammatory response. Its release from
mast cells during an allergic reaction contributes to constriction of the airways in
asthma and also to the reduction in blood
pressure during anaphylaxis.

platelet-derived growth factor *See*
GROWTH FACTOR.

plate tectonics The theory that the surface of the earth is made of lithospheric
plates, which have moved throughout

geological time resulting in the present-
day positions of the continents. The theory
explains the locations of mountain building
as well as earthquakes and volcanoes.
The rigid lithospheric plates consist of continental and oceanic crust together with the
upper mantle, which lie above the weaker
plastic asthenosphere. These plates move
relative to each other across the earth. Six
major plates (Eurasian, American, African,
Pacific, Indian, and Antarctic) are recognized, together with a number of smaller
ones. The plate margins coincide with
zones of seismic and volcanic activity.

A **constructive** (or **divergent**) plate margin occurs when two plates move away from
each other. It is marked by a mid-oceanic
ridge where basaltic material wells up from
the mantle to form new oceanic crust, in a
process known as **sea-floor spreading**.
The production of new crust at constructive
plate margins is compensated for by the
destruction of material along a **destructive**
(or **convergent**) plate margin. Along these
margins, which are also known as **subduc-
tion zones** and marked by an oceanic
trench, one plate (usually oceanic) is forced
to plunge down beneath the other (which
may be continental or oceanic). The crust
becomes partially melted and rises to form a
chain of volcanoes in the upper plate parallel to the trench. When two continental
plates collide the compression results in
the formation of mountain chains. A third
type of plate margin—the **transform plate
margin**—occurs where two plates are slip-
ping past each other.

Platyhelminthes In traditional classifications a phylum of acoelomate invertebrates
comprising the flatworms, characterized by
a flattened unsegmented body. The simple
nervous system shows some concentration
of cells at the head end. The mouth leads
to a simple branched gut without an
anus. Flatworms are hermaphrodite but
self-fertilization is unusual. Many species
are parasitic. The phylum contains the
classes *Turbellaria (planarians), *Trematoda (flukes), and *Cestoda (tapeworms).
Molecular evidence now suggests that the
turbellarians and trematodes are *paraphyletic groups, and that the majority of
flatworms are secondarily acoelomate,

belonging to a clade with the proposed name *Platyzoa. This distinguishes them from those platyhelminths traditionally placed in the order Acoelomorpha, which are primitively acoelomate and descended from an ancient animal lineage that is neither protostome nor deuterostome.

(⊕) SEE WEB LINKS

- Brief introduction to flatworms from the website of the University of California Museum of Paleontology

---

Platyzoa A proposed name for a clade of protostome animals based on molecular systematics and including the flatworms, rotifers, and certain other phyla.

pleiomorphism The existence of distinctly different forms during the life cycle of an individual, e.g. the caterpillar, pupa, and winged adult of a butterfly.

pleiotropic Describing an allele that has more than one effect in an organism. For example, the allele that causes the erythrocytes to have a distorted form in sickle-cell anaemia also causes these blood cells to rupture easily, thereby inducing anaemia.

Pleistocene The first epoch of the *Quaternary period. It extended from the end of the Pliocene, about 2.58 million years ago, to the beginning of the *Holocene (the present epoch), about 11 700 years ago. The Pleistocene is often known as the **Ice Age** as it was characterized by a series of glacials, in which ice margins advanced towards the equator, separated by interglacials when the ice retreated. *See also* ICE AGE.

plesiomorphy (ancestral trait) An evolutionary trait that is *homologous within a particular group of organisms but is not unique to members of that group—i.e. a derived trait (*see* APOMORPHY)—and therefore cannot be used as a diagnostic or defining character for the group. For example, vertebrae are found in zebras, cheetahs, and orang-utans, but the common ancestor in which this trait first evolved is so distant that the trait is shared by many other animals. Therefore, possession of vertebrae sheds no light on the phylogenetic

relations of these three species. *See also* SYMPLESIOMORPHY.

pleura (pleural membrane) (*pl.* **pleurae**) The double membrane that lines the thoracic cavity and covers the exterior surface of the lungs. It is a *serous membrane forming a closed sac, with a small space (the **pleural cavity**) between the two layers.

plexus (*pl.* **plexus** or **plexuses**) A compact branching network of nerves or blood vessels, such as the **brachial plexus**—a network of spinal nerves that supply branches to the forelimbs in vertebrates. *See also* CHOROID PLEXUS.

Pliocene The second epoch of the *Neogene period. Preceded by the Miocene and followed by the Pleistocene, it extended from about 5.3 million years ago to 2.58 million years ago. Mammals similar to modern forms existed during the epoch and the australopithecines (*see* AUSTRALOPITHECUS), hominins that coexisted with early humans, appeared.

plumule 1. (in zoology) A *down feather. 2. (in botany) The part of a plant embryo that develops into the shoot system. It consists of the stem apex and first leaves. In seedlings showing *epigeal germination the plumule grows above the soil surface together with the cotyledons; in seeds showing *hypogeal germination, the plumule alone emerges. *Compare* RADICLE.

pluripotent *See* STEM CELL.

pneumatophore 1. An aerial negatively geotropic root that acts as an organ of gaseous exchange. Pneumatophores are produced by mangroves and other plants that grow in waterlogged oxygen-deficient soils. *See* MANGROVE SWAMP. 2. The gas-filled float of certain colonial cnidarians of the class Hydrozoa, such as *Physalis* (Portuguese man-of-war).

pod *See* LEGUME.

podocyte A type of epithelial cell found in *Bowman's capsule. Podocytes possess major extensions resembling tentacles, each of which has numerous much finer fingerlike processes. These minor processes wrap around the blood capillaries of the glomerulus and are interwoven with those from other podocytes to form narrow slits. These slits (approximately 0.1 μm wide) form part of the filtration apparatus of the Bowman's capsule, permitting the passage of water and small molecules out of the blood plasma but acting as a barrier to blood cells and large molecules. Additional filtration is achieved by the walls of the glomerular capillaries, which have small openings (fenestrations), and the mesh of collagen fibres in the basement membrane between the capillaries and the podocytes.

pogonophoran Any marine tube-dwelling polychaete worm of the former phylum Pogonophora, now placed in the annelid family Siboglinidae. They are typically found in ocean sediment, some at great depths near hydrothermal vents or cold seeps. Many have a mass of feathery tentacles at the anterior end; hence their popular name of 'beard-worms'. Adults lack a digestive system and details of their nutrition are uncertain; they may rely on endosymbiotic bacteria utilizing inorganic or organic compounds as sources of energy.

poikilothermy The passive variation in the internal body temperature of an animal, which depends on the temperature of the environment. Most animals exhibit poikilothermy and are described as *ectotherms (exceptions are mammals, birds, some fishes and insects, and few reptiles). Although unable to maintain a constant body temperature, they can respond to compensate for very low or very high temperatures. For example, the tissue composition (especially cell osmotic pressure) can change to regulate the blood flow to peripheral tissues (and thus increase heat loss or heat absorption), and the animals can actively seek sun or shade. Seasonal changes in metabolism are usually under hormonal control. In particularly hot climates, ectotherms may undergo *aestivation to escape the heat. *Compare* HOMOIOTHERMY.

point mutation A change in the base sequence of DNA involving a single nucleotide, whether by *insertion, *substitution, *deletion, or *inversion. Point mutations

in coding regions of the genome will be transcribed into the corresponding messenger RNA. This can result in a misreading of the genetic code during the translation phase of protein synthesis and can change the order of amino acids making up a protein, which may or may not affect the function of that protein. *See also* INTERGENIC SUPPRESSOR; MISSENSE MUTATION. *Compare* CHROMOSOME MUTATION; SINGLE NUCLEOTIDE POLYMORPHISM.

polar body *See* OOGENESIS.

polarity 1. The property of a cell, tissue, or organism of being structurally and/or functionally different at opposite ends of its long axis. For example, plants consist of roots, which grow in the direction of the force of gravity, and stems, which grow away from the gravitational force (*see* GEOTROPISM). In developing insect embryos, the anterior-posterior polarity of larval body segments is determined by segment polarity genes. (*see* SEGMENTATION GENES). *See also* ANIMAL POLE. **2.** The property of molecules of having an uneven distribution of electrons, so that one part has a positive charge and the other a negative charge. Such **polar molecules** include *water.

polar molecule *See* POLARITY.

polar nuclei Two haploid nuclei in the centre of the *embryo sac of flowering plants. These nuclei fuse with a male gamete nucleus to form a triploid endosperm nucleus, which subsequently divides to form the endosperm. *See also* DOUBLE FERTILIZATION.

pollen The mass of grains containing the male gametes of seed plants, which are produced in large numbers in the *pollen sacs. The pollen grains of insect-pollinated plants may be spiny or pitted and are usually larger than those of wind-pollinated plants, which are usually smooth and light. The pollen grain contains the male *gametophyte generation; it contains two male nuclei: a *generative nucleus and a *tube nucleus. The wall of the mature pollen grain consists of the tough outer wall (**exine**) and the more delicate narrower

intine. The latter gives rise to the *pollen tube. *See also* POLLINATION.

pollen analysis *See* PALYNOLOGY.

pollen sac The structure in seed plants in which pollen is produced. In angiosperms there are usually four pollen sacs in each *anther; they contain the *microspore mother cells (microsporophytes). In gymnosperms variable numbers of pollen sacs are borne on the microsporophylls that make up the male *cone.

pollen tube An outgrowth of a pollen grain, which transports the male gametes to the ovule. It will only grow if the pollen grain is compatible with the female tissue. In angiosperms, the pollen grain is deposited on the stigma and the pollen tube grows down through the style and into the ovule. Germination of the pollen grain is initiated by imbibition of water from the stigma. Development of the pollen tube through the tissue of the style is guided by signal molecules secreted by the *synergids in the embryo sac. In some conifers, e.g. *Pinus* (pines), the pollen tube penetrates the *nucellus but does not develop further until the following year, when the female part of the plant is mature. *See also* EMBRYO SAC; FERTILIZATION; TUBE NUCLEUS.

pollex (*pl.* **pollices**) The innermost digit on the forelimb of a tetrapod vertebrate. It contains two phalanges (*see* PENTADACTYL LIMB) and in humans and higher primates it is the thumb, which is opposable (i.e. capable of facing and touching the other digits) and gives the hand greater manipulating ability. In some mammals a pollex is absent. *Compare* HALLUX.

pollination The transfer of pollen from an anther (the male reproductive organ) to a stigma (the receptive part of the female reproductive organ), either of the same flower (**self-pollination**) or of a different flower of the same species (**cross-pollination**). Cross-pollination involves the action of a pollinating agent to effect transfer of the pollen (*see* ANEMOPHILY; ENTOMOPHILY; HYDROPHILY). Transfer of pollen is followed by germination of the pollen grain and

growth of a *pollen tube through the style to deliver the male gametes to the ovule. *See also* FERTILIZATION; INCOMPATIBILITY.

pollutant Any substance, produced and released into the environment as a result of human activities, that has damaging effects on living organisms. Pollutants may be toxic substances (e.g. *pesticides) or natural constituents of the atmosphere (e.g. carbon dioxide) that are present in excessive amounts. *See* POLLUTION.

pollution An undesirable change in the physical, chemical, or biological characteristics of the natural environment, brought about by human activities. It may be harmful to human or nonhuman life. Pollution may affect the soil, rivers, seas, or the atmosphere (*see* AIR POLLUTION). Some *pollutants are **biodegradable** (e.g. sewage), which means that they can be rendered harmless by natural processes and need therefore cause no permanent harm if adequately dispersed or treated; others are **nonbiodegradable** (e.g. heavy metals in industrial effluents (*see* HEAVY-METAL POLLUTION) and DDT and other chlorinated hydrocarbons used as *pesticides), and these accumulate in the environment and may be concentrated in food chains. Other forms of pollution in the environment include noise (e.g. from jet aircraft, traffic, and industrial processes), thermal pollution (e.g. the release of excessive waste heat into lakes or rivers causing harm to wildlife), and light pollution (from street lights, buildings, etc., which can disorientate wildlife). Current pollution problems include high levels of carbon dioxide and other greenhouse gases in the atmosphere (*see* GREENHOUSE EFFECT); discarded plastic debris in watercourses and the oceans, harming aquatic and marine life; the disposal of radioactive waste; *acid rain; *photochemical smog; increasing levels of human waste; and pollution of inland and coastal waters by agricultural *fertilizers and sewage effluent, causing eutrophication (*see* EUTROPHIC). Attempts to contain or prevent pollution include strict regulations concerning factory emissions, the use of smokeless fuels, the banning of certain pesticides, greater use of renewable energy sources, the use of catalytic converters to cut pollutants in car exhausts, and the collection and recycling of plastic waste. *See also* CLIMATE CHANGE; SUSTAINABLE.

((∰)) SEE WEB LINKS
• The ten worst-polluted places in the world and the ten most-polluting industries; a report by Pure Earth (formerly the Blacksmith Institute) and Green Cross, Switzerland

polyacrylamide gel electrophoresis *See* PAGE.

polyandry A mating system in which a female mates with more than one male. In animals it is relatively rare compared to monogamous or polygynous systems, but occurs across a wide spectrum of animal groups, including certain human societies. In **simultaneous polyandry**, sperm from multiple males compete to fertilize eggs. For example, a honey bee queen will mate with several males during her courtship flight, and store their sperm throughout her life. In **sequential (serial) polyandry**, a female mates with a single male, leaving him for another male when her eggs have been fertilized and laid. The usual sexual *dimorphism is often reversed in polyandrous females, which tend to be larger than their male counterparts. In flowering plants, polyandry is the norm, with pollen from multiple donor plants being deposited on the same stigma, leading to fruits with different paternity. Exceptions are plants that perform self-fertilization or that receive pollen in discrete packages from a single donor. *Compare* POLYGYNY.

Polychaeta In traditional classifications a class of annelid worms in which each body segment has a pair of flattened fleshy lobes (**parapodia**) bearing numerous bristles (*chaetae). All polychaetes are aquatic and most of them are marine. They include the fanworms (*Sabella*), which construct tubes of sand, etc., in which they live; and the lugworms (*Arenicola*) and ragworms (*Nereis*), which burrow in sand or mud. Molecular studies have thrown the evolutionary

relationships of annelids into question and shown the polychaetes to be a paraphyletic group. *See* ANNELIDA.

polycistronic Describing a type of messenger *RNA that can encode more than one polypeptide separately within the same RNA molecule. Prokaryotic and chloroplast messenger RNAs are generally polycistronic. *Compare* MONOCISTRONIC.

polyembryony 1. The formation of more than one embryo in a plant seed. Often one embryo develops from the fertilized egg cell, while the others have formed asexually from other tissues in the ovule. **2.** The formation of more than one embryo from a single animal zygote. *Identical, or monozygotic, twins are produced in this way.

polygene Any of a group of genes influencing a quantitative trait, such as height in humans. *See* QUANTITATIVE INHERITANCE.

polygenic inheritance *See* QUANTITATIVE INHERITANCE.

polygyny A mating system in which a male mates with more than one female, but the female mates with only one male. There are numerous examples across the animal kingdom, including some human societies. Polygynous males tend to be larger than the females, with sometimes marked sexual *dimorphism, and compete with other males to secure access to females. Typically it is only the fittest, strongest males that mate and pass on their genes to the next generation, while weaker males are unsuccessful. The female's offspring will thus benefit by receiving paternal traits that will maximize their fitness. Care of the young is undertaken mainly or exclusively by the female, although she may profit from the availability of resources, such as optimal nest sites or foraging areas, that fall within the territory defended by the male. *Compare* POLYANDRY.

polymer A substance having large molecules consisting of repeated units (the monomers). There are a number of natural polymers, such as polysaccharides.

polymerase Any enzyme that catalyses the elongation of a polymeric molecule. **RNA polymerases** (types I to III) catalyse the synthesis of RNA using as a template either an existing DNA strand (**DNA-dependent RNA polymerase**) or an RNA strand. Type I is responsible for the synthesis of ribosomal RNA, type II for messenger RNA synthesis (*see* TRANSCRIPTION), and type III for transfer RNA synthesis. **DNA polymerases** catalyse the elongation of a new DNA strand during *DNA replication, using an existing DNA strand as template. **RNA-directed DNA polymerase** is more usually known as *reverse transcriptase.

polymerase chain reaction (PCR) A technique used to replicate a fragment of DNA so as to produce many copies of a particular DNA sequence. PCR can be employed as an alternative to *gene cloning as a means of amplifying genetic material for *DNA sequencing, and to provide the DNA fragments for cloning. The technique has also proved invaluable in forensic science, enabling amplification of minute traces of genetic material for *DNA profiling or for detecting *microsatellite DNA, and is used for a host of other applications, including *DNA fingerprinting and *DNA barcoding. The two strands of the DNA are separated by heating and short sequences of a single DNA strand (**primers**) are added, together with a supply of free nucleotides and a heat-stable DNA *polymerase (e.g. Taq polymerase obtained from *Thermus aquaticus*, a *hadobacterium that can withstand extreme heat, or Pfu polymerase from the archaean *Pyrococcus furiosus*). DNA polymerase uses single-stranded DNA as its template but requires a short section of double-stranded DNA to initiate the replication reaction. Hence, synthetic oligonucleotide primers are constructed so that their sequences are complementary to short regions flanking the 3′ region of DNA that is to be amplified; the oligonucleotides hybridize with the flanking sequences, forming the necessary double-stranded primers. In a series of heating and cooling cycles, the DNA sequence flanked by the primers doubles with each cycle and is thus rapidly amplified. **Reverse transcription PCR** (RT-PCR) is used for amplifying molecules of RNA, by initially converting the RNA into its complementary DNA molecule using the enzyme reverse transcriptase and then following the standard procedure for PCR. It can be used to analyse gene expression in samples taken from certain tissues or under particular conditions. **Real-time PCR** (RT-PCR) or **quantitative PCR** (qPCR) is a variant of the technique used for quantitatively estimating amounts of RNA or DNA in a sample. Essentially, it measures the rate of accumulation of DNA through successive cycles of PCR, on the basis that the more target molecules present in the starting sample, the faster is their subsequent amplification. All newly synthesized copies of the target are tagged with a fluorescent marker, and continuous monitoring of the amount of fluorescence indicates the rate of amplification and hence the amount of target molecule in the original sample. *See also* DIRECTED EVOLUTION; RANDOMLY AMPLIFIED POLYMORPHIC DNA.

((⊕)) SEE WEB LINKS
• Animated account of PCR from the DNA Learning Center

polymorphism The existence of two or more distinctly different forms (**morphs**) within a plant or animal species, particularly due to genetic differences. Two forms of genetic polymorphism exist: **transient polymorphism**, in which a particular form is in the process of spreading through a population, causing the relative proportion of phenotypes to alter; and **balanced polymorphism**, in which two or more forms coexist in a stable ratio within a population, each form possessing both advantageous and disadvantageous characteristics. An example of balanced polymorphism is the occurrence of **sickle-cell disease**, a genetic disease that principally affects Black populations of central Africa and is characterized by an abnormal form of the blood pigment haemoglobin (**haemoglobin S**) and sickle-shaped red blood cells. Three different types of individual occur in such populations: those who have two alleles (*AA*) for normal haemoglobin and therefore do not suffer from the disease; those with one normal and one abnormal allele (*AS*), who are described as having the **sickle-cell trait** and generally suffer no symptoms; and those with two abnormal alleles (*SS*), who suffer a chronic and eventually fatal form of anaemia. Normally such a harmful allele would have been eliminated from the population by the process of natural selection, but it is maintained in this case because people with the sickle-cell trait are resistant to a severe form of malaria endemic in central Africa. They exhibit heterozygote advantage (*see* BALANCING SELECTION). Environmental polymorphism is not due to genetic differences, an example being the *caste system of social insects, in which the different morphs—i.e. workers, drones, and queens—are determined by the larvae receiving different types of food. Many authorities prefer the term *polyphenism for this phenomenon. *Compare* MUTATION. *See also* RESTRICTION FRAGMENT LENGTH POLYMORPHISM; SINGLE NUCLEOTIDE POLYMORPHISM.

polynucleotide *See* NUCLEOTIDE.

polyol (polyhydric alcohol) Any compound that contains more than one alcohol group. The term is usually restricted to the sugar alcohols, which include such biologically important molecules as *myo*-*inositol *glycerol, **sorbitol** (common in the fruits of many plants), and **ribitol** (found in bacterial cell walls). Certain polyols act as *osmolytes in the kidney cells or function as *antifreeze molecules.

polyp The sedentary stage in the life cycle of the *Cnidaria, consisting of a cylindrical body fixed at one end to a firm base and having a mouth surrounded by a ring of tentacles at the other. Some polyps (e.g. *Hydra*) are single; others (e.g. the corals and *Obelia*) form colonies. Polyps typically reproduce asexually by budding to form either new polyps or *medusae. The latter reproduce sexually giving rise to new polyps. Sea anemones are solitary polyps that reproduce sexually to form new polyps.

polypeptide A *peptide comprising 20 or more amino acids. Polypeptides that constitute proteins may contain up to 1000 amino acids, typically 100 to 300. Shorter ones include certain antibiotics, e.g. gramicidin, and some hormones, e.g. *ACTH, which has 39 amino acids. The properties of a polypeptide are determined by the type, sequence, and three-dimensional arrangement of its constituent amino acids. One end, the C terminus, has a free carboxyl group, while the other end has a free amino group, the N terminus. *See* PROTEIN.

polypetalous Describing a flower in which the *corolla consists of separate petals. *Compare* GAMOPETALOUS.

polyphenism (environmental polymorphism) The occurrence within a species of distinct morphological types (**morphs**) in response to environmental cues. The morphs, which may overlap in space and/or time, are the result of particular signals in the animals' environment, such as nutrition, temperature, or day length, that interact with the animals' genes and 'switch' development of the young from one 'program' to another. The resultant morphs are better adapted to their environmental conditions or to fulfil a certain role in a social group. For example, the southern African butterfly *Bicyclus anynana* exhibits two distinct morphs: a wet season morph, with prominent eyespots on its hind wings; and a dry season morph that is brown with much reduced eyespots. Ambient temperature and levels of ecdysteroid hormones in the pupa determine which morph will emerge. Polyphenism occurring in successive generations exposed to seasonal changes in the environment is termed *cyclomorphosis. *Compare* POLYMORPHISM.

polyphyletic Describing a group of organisms that contains the descendants of two or more different ancestors, while excluding other descendants of a single ultimate common ancestor. Such a group may be constructed on the basis of certain shared traits, which may have evolved convergently (*see* CONVERGENT EVOLUTION), but it does not necessarily reflect any close evolutionary relationship, and is therefore rejected as a basis for phylogenetic classification. For example, mammals and birds share the trait of endothermy ('warm-bloodedness'), but the grouping 'endotherms' is polyphyletic because its members have quite different immediate ancestors and the defining trait evolved independently in each member. *Compare* MONOPHYLETIC; PARAPHYLETIC.

polyphyodont Describing a type of dentition in which the teeth are continuously shed and replaced during the lifetime of the animal. Sharks and frogs have a polyphyodont dentition. *Compare* DIPHYODONT; MONOPHYODONT.

polyploid Describing a nucleus that contains more than two sets of chromosomes (*see* DIPLOID) or a cell or organism containing such nuclei. For example, *triploid plants have three sets of chromosomes and **tetraploid** plants have four. Polyploidy is far more common in plants than in animals; many crops, in particular, are polyploid. For example, bread wheat is an allohexaploid ($6n$) containing two sets of chromosomes from each of three ancestral species, making six sets of chromosomes in total. Polyploidy can be induced chemically with *colchicine, e.g. in plant breeding to produce new hybrid species. *See also* ALLOPOLYPLOID; AUTOPOLYPLOID.

polyribosome (polysome) An aggregate of ribosomes in association with a single messenger RNA molecule during the *translation process of protein synthesis. Simultaneous translation by multiple ribosomes enables the cell rapidly to make numerous copies of a polypeptide. In eukaryotes, polyribosomes are attached to the surface of the rough endoplasmic reticulum and the outer membrane of the nucleus; in bacteria they are found free in the cytoplasm.

polysaccharide (glycan) Any of a group of carbohydrates comprising long chains of monosaccharide (simple sugar) molecules. **Homopolysaccharides** consist of only one type of monosaccharide; **heteropolysaccharides** contain two or more different types. Polysaccharides typically consist of

several hundred to several thousand monosaccharides joined by glycosidic linkages. They may have molecular weights of up to several million and are often highly branched. Major storage polysaccharides include *starch and *glycogen, whereas structural polysaccharides include *cellulose and *chitin.

polysepalous Describing a flower that possesses a calyx consisting of separate sepals. *Compare* GAMOSEPALOUS.

polysome *See* POLYRIBOSOME.

polyspermy The entry of several sperms into the egg during fertilization although only one sperm nucleus actually fuses with the egg nucleus. Physiological polyspermy occurs naturally in a wide range of animals, including arachnids, insects, cartilaginous fishes, urodele amphibians, reptiles, and birds—characteristically animals having yolky eggs. In other species polyspermy is pathological—it results in abnormal fertilization and death of the embryo. Hence the entry of a sperm into the egg triggers changes in the surface layer of the egg that prevent the entry of any additional sperms.

polysynaptic reflex A *reflex action that involves an electrical impulse being transferred from a sensory neuron to a motor neuron via at least one connecting neuron (**interneuron**) in the spinal cord. For example, stimulation of pain receptors in the skin initiates a **withdrawal reflex**, which involves several synapses with several motor neurons and results in the removal of the organism or part from the stimulus.

polyteny The condition of a chromosome, nucleus, or cell in which the DNA has repeatedly replicated, without subsequently separating (*see* ENDOMITOSIS). This results in a cablelike **giant chromosome**, consisting of up to 1000 parallel chromatids and often showing conspicuous transverse banding. Polyteny has been observed in the cells of certain insects of the order Diptera, notably in the salivary glands of *Drosophila*. Swellings (called **puffs**) have been observed in giant chromosomes at specific stages in the development of these insects. Puffs are associated with messenger RNA and are sites of gene transcription.

polythetic Describing a type of classification in which membership of a taxon is based on its constituent organisms sharing a large number of characteristics.

polytopic Describing a taxon that has arisen and exists in more than one geographical region.

polytypic Describing a species that exists in a variety of *forms or *subspecies that inhabit different geographical regions. *Compare* MONOTYPIC.

pome A type of fruit characteristic of apples and pears. The flesh of the fruit develops from the *receptacle of the flower, which completely encloses the fused carpels. After fertilization the carpels form the 'core' of the fruit, which contains the seeds. *See also* PSEUDOCARP.

pons (pons Varolii) (*pl.* **pontes**) A part of the hindbrain that links the medulla oblongata to the midbrain. As well as relaying impulses between different parts of the brain, it contains a group of neurons that help regulate breathing movements (*see* VENTILATION CENTRE). Also, certain pontine nuclei are active during REM *sleep. The pons is named after its discoverer, the Italian anatomist C. Varoli (1543–75).

population 1. A group of individuals of the same species that live within a defined area and interact with each other. The nature of a population is determined by such factors as density, *sex ratio, age structure, birth and death rates (*see* REPRODUCTION RATE), emigration, and immigration. The interaction of individuals is affected by how they are distributed within their habitat, i.e. their *dispersion. *Compare* COMMUNITY. **2.** The total number of individuals of a given species or other class of organisms in a defined area, e.g. the population of rodents in Britain. One

method of estimating the size of a population of mobile animals is the *mark-recapture method. *See also* METAPOPULATION. **3.** In statistics and experimental design, the complete set of entities from which a sample is drawn for study or experiment with the aim of making valid inferences about the population as a whole. *See* SAMPLING.

population bottleneck A situation in which a population suffers a steep decline in numbers owing to some adverse event, such as intensive hunting, habitat destruction, or an environmental disaster. This results in the survival of just a few individuals and consequent inbreeding, with the loss of many genetic variants, or alleles. This reduction in genetic diversity will persist in future generations even if the population recovers. Moreover, small populations are more likely to lose surviving rare alleles through *genetic drift. For example, the African cheetah (*Acinonys jubatus*) experienced such a bottleneck about 12 000 years ago, and the modern population retains a very low level of genetic diversity.

population dynamics The study of the fluctuations that occur in the numbers of individuals in animal and plant populations and the factors controlling these fluctuations. Changes in numbers depend on rates of births, deaths, emigration, and immigration (*see* POPULATION GROWTH). An important distinction is between controlling factors that are dependent on population density and have a stabilizing effect (e.g. food supply) and those that are independent of population density (e.g. catastrophes, such as flooding). Fluctuations of certain predators may follow those of their prey in fairly regular population cycles.

population genetics The study of the distribution of inherited *variation among a group of individuals of the same species, and of how their genetic makeup changes over time—the process of evolution. The potential for change depends on the sum total of alleles that are available to the organisms (the **gene pool**), which is determined by, e.g. the extent of genetic *polymorphism, mating patterns, migration between different populations, the *founder effect, and the rate of new mutations. The direction of genetic change depends primarily on the *selection pressures and the fitness of reproducing individuals as determined by their genotypes. In many cases alleles have little or no effect on fitness, and their frequencies may change in a random way (*see* GENETIC DRIFT). The idealized formulation of genotype frequency, given no selection, random mating, infinite population size, and no migration, is predicted by the *Hardy–Weinberg equilibrium.

population growth The increase in a population that occurs when the *birth rate is higher than the *death rate, or when immigration exceeds emigration, or when a combination of these factors is present. The **per capita growth rate** (r) is determined by the difference between the birth rate and the death rate. Under ideal conditions the population will achieve its maximum possible value of r, called the **intrinsic rate of increase** (r_{max}). A growth curve, obtained by plotting population size against time, is typically S-shaped (sigmoid) or J-shaped (see graphs). A sigmoid growth curve shows an initial phase of *exponential growth. The curve levels off when the environment has reached its **carrying capacity**, i.e. when food, space, and other conditions can support a given number of individuals without an increase in population numbers. A J-shaped growth curve shows an initial

phase of exponential growth that ceases abruptly, with a sudden decrease in population numbers. This decrease may be caused by a number of factors, such as the end of the life cycle of the prey or any other factor contributing to *environmental resistance that may suddenly take effect. *See also* BACTERIAL GROWTH CURVE; FECUNDITY; REPRODUCTION RATE; SURVIVORSHIP CURVE.

P/O ratio (phosphorus:oxygen ratio) The number of atoms of phosphorus (i.e. as phosphate) incorporated as ATP per molecule of oxygen (O_2) consumed during *oxidative phosphorylation in aerobically respiring cells. The yield of ATP from reduced coenzymes generated by the *Krebs cycle depends on how many hydrogen ions (H^+) are transported across the inner mitochondrial membrane for each molecule of NADH or $FADH_2$ entering the electron transport chain. It is now assumed that 10 H^+ ions are transported across for each molecule of NADH, and that four H^+ must re-enter the mitochondrial matrix via ATP synthase to produce one molecule of ATP. This equates to 2.5 ATP per NADH, and 1.5 ATP per molecule of $FADH_2$. These values give a net P/O yield of 31 ATP per glucose molecule. The picture is further complicated by which mechanism is used to transport cytosolic NADH (generated by glycolysis) into the mitochondrial matrix. The yield of 31 ATP applies only if the malate/aspartate shuttle is used, as in liver cells and heart cells; if an alternative

Sigmoid growth curve

J-shaped growth curve

Population growth curves

mechanism, known as the glycerol phosphate shuttle, is used, then the net yield is reduced to 29.5 ATP. The latter mechanism is found, for example, in muscle cells.

Porifera The phylum of marine and freshwater invertebrates that comprise the sponges, which live permanently attached to rocks or other surfaces. The body of a sponge is hollow and consists basically of an aggregation of cells between which there is little nervous coordination. The body is supported by an internal skeleton of spicules, whose composition defines the three groups of sponges. Flagellated cells (**choanocytes**) cause water to flow in through openings (**ostia**) in the body wall and out through openings (**oscula**) at the top; food particles are filtered from the water by the choanocytes. Glass sponges and demosponges have spicules made of silica, whereas calcareous sponges have spicules made of calcium carbonate. Demosponges are the most abundant group and have a skeleton consisting chiefly of a horny protein called spongin, as typified by the bath sponges. Various antibiotics and other potentially therapeutic compounds are produced by sponges, including possible anticancer drugs.

(((⊕))) SEE WEB LINKS
- Overview of the sponges, including their biology and classification, from the University of Michigan's Museum of Zoology Diversity Web

porin Any of a class of proteins that form water-filled channels across cell membranes. In the outer membrane of Gram-negative bacteria, most porins consist of three identical subunits that associate and extend fully through the membrane to form channels about 1 nm in diameter. These permit the entry and exit of hydrophilic low-molecular-weight substances; some contain a specific binding site and allow the passage only of certain substances. The pores can be opened and closed, which has significance for antibiotic resistance in pathogenic bacteria. Similar porin proteins occur in the mitochondrial outer membrane of eukaryotic cells, and in eukaryotic plasma membranes,

where they are responsible for controllable movements of water in or out of the cell associated with osmosis (*see* AQUAPORIN). **Nucleoporins** constitute the pores in the *nuclear envelope surrounding the nucleus.

porphyrin Any of a group of organic pigments characterized by the possession of a cyclic group of four linked nitrogen-containing rings (a **tetrapyrrole** nucleus), the nitrogen atoms of which are often coordinated to metal ions. Porphyrins differ in the nature of their side-chain groups. They include the *chlorophylls, which contain magnesium; and *haem, which contains iron and forms the *prosthetic group of haemoglobin, myoglobin, and the cytochromes.

Generalized structure of a porphyrin

portal vein (portal circulation; portal system) Any vein that collects blood from one network of capillaries and transports it directly to a second capillary network in another region of the body, without returning to the heart. *See* HEPATIC PORTAL SYSTEM.

positional cloning A technique for isolating and cloning a gene, for example a disease-associated gene, in the absence of any information about its product. It relies on the existence of suitable genetic markers (*see* MARKER GENE), whose inheritance can be traced along with the disease in question. With reference to an existing genetic map, the markers indicate the chromosomal region that contains the disease gene. DNA probes, based on DNA marker sequences in the map region, are used to select cloned segments from a *DNA library. A set of contiguous overlapping clones (contigs) is assembled by the process of *chromosome walking. The DNA sequence of the region can then be determined and analysed to identify stretches that might contain the sought-for disease gene. Having established the gene's base sequence, the structure and possible function of the corresponding protein can be deduced. Many important disease genes have been identified using this approach, including those for cystic fibrosis, Huntington's disease, and Friedreich's ataxia. *See also* PHYSICAL MAP.

positional information (in embryology) The suite of molecular cues that control *pattern formation in an early embryo.

positive feedback *See* FEEDBACK.

positive selection (directional selection) The process in which individuals at one extreme of a range of phenotypes are favoured, and the beneficial genetic variants (*alleles) carried by those individuals tend to become established in the population. Hence the overall make-up of the population shifts in a particular direction, often in response to changing environmental conditions. An example of positive selection is the increase in darker forms of the peppered moth (*Biston betularia*) that occurred in industrial areas, where the moths with darker wing coloration were better camouflaged than those with lighter wings against polluted tree trunks (*see* INDUSTRIAL MELANISM). The concept is central to the theory of *natural selection as first proposed by Charles Darwin and Alfred Russel Wallace in 1858. It is now thought that most mutations are selectively neutral and undergo genetic drift (*see* NEUTRAL THEORY OF MOLECULAR EVOLUTION), or are harmful and are eliminated by *purifying selection. *Compare* DISRUPTIVE SELECTION; STABILIZING SELECTION.

positive-sense (in genetics) Describing RNA or a single strand of DNA whose base sequence is the same as that required for translation into protein. By convention, the *coding strand of duplex DNA is the positive-sense (and nontranscribed) strand, whereas the complementary *anticoding strand is the *negative-sense (and transcribed) strand. Positive-sense single-stranded RNA viruses are the most abundant group of viruses, responsible for many plant diseases as well as the common cold. Their linear RNA genomes are translated directly into viral proteins by the host cell. *See* VIRUSES (feature).

positron emission tomography (PET) A noninvasive imaging technique that can produce three-dimensional pictures of certain biochemical changes within the body, such as areas of increased tissue metabolism. It is used for research in several areas, including neurology and pharmacology, and as a diagnostic technique in medicine, for example in detecting brain tumours. The technique is based on the detection of photons produced by the decay of short-lived radioactive isotopes, such as fluorine-18 in labelled fluorodeoxyglucose (*see* LABELLING), injected into the body. The isotopes emit positrons, which almost immediately collide with electrons to produce a pair of photons that travel in opposite directions. Simultaneously arriving photon pairs are detected by the scanner on either side of the body as flashes of light, and a computer calculates the point of their mutual origin within the body. By collating many thousands of such coincident flashes, the computer creates a map showing where the isotope is concentrated, which corresponds to locally increased blood flow or uptake as a consequence of elevated tissue activity. PET scans are often now integrated with images obtained by computerized *tomography or *magnetic resonance imaging so that the biochemical information can be correlated with anatomy of the site.

postcaval vein *See* VENA CAVA.

posterior 1. Designating the part of an animal that is to the rear, i.e. that follows when the animal is moving. In humans and bipedal animals (e.g. kangaroos) the posterior surface is equivalent to the *dorsal surface. **2.** Designating the side of a flower or axillary bud that faces towards the flower stalk or main stem, respectively. *Compare* ANTERIOR.

postreplicative repair A form of *DNA repair that fills large gaps in a new DNA strand resulting from failure of replication. It prevents gross chromosomal abnormalities and enables timely replication of DNA in the S phase of the *cell cycle. In **trans-lesion synthesis**, special polymerase enzymes synthesize a new DNA strand across the lesion that blocked the replicative polymerase. This is relatively prone to error but does restore genome stability and allow subsequent precision repair by, e.g., *mismatch repair. **Template switching** is an alternative mechanism that 'borrows' the information from a newly synthesized daughter strand as a template to synthesize a new strand that bypasses the lesion responsible for the gap. This is an error-free mechanism of postreplicative repair.

postsynaptic membrane The area of plasma membrane of the recipient neuron, muscle cell, or gland cell that bears receptors for the signals from a transmitting neuron at a *synapse.

post-tetanic potentiation *See* SYNAPTIC PLASTICITY.

post-transcriptional gene silencing *See* RNA INTERFERENCE.

post-transcriptional modification *See* RNA PROCESSING.

post-translational modification *See* TRANSLATION.

potassium Symbol K. A soft silvery metallic element that is an *essential element for living organisms. The potassium ion, K^+, is the most abundant cation in plant tissues, being absorbed through the roots and being used in such processes as protein synthesis. In animals the high intracellular concentration of K^+ relative to the extracellular fluid is a key component in establishing the *resting potential of neurons, and is maintained by the *sodium-potassium pump. Flows of potassium and sodium ions across the neuron plasma membrane

are responsible for the changes of electrical potential that accompany the transmission of nerve impulses. *See* POTASSIUM ION CHANNEL.

potassium–argon dating A *dating technique for certain rocks that depends on the decay of the radioisotope potassium-40 to argon-40, a process with a half-life of about $1.27 \times M\ 10^{10}$ years. It assumes that all the argon-40 formed in the potassium-bearing mineral accumulates within it and that all the argon present is formed by the decay of potassium-40. The mass of argon-40 and potassium-40 in the sample is estimated and the sample is then dated from the equation:

$$^{40}\mathrm{Ar} = 0.1102\ ^{40}\mathrm{K}(e^{\lambda t} - 1),$$

where λ is the decay constant and t is the time in years since the mineral cooled to about 300°C, when the $^{40}\mathrm{Ar}$ became trapped in the crystal lattice. The method is effective for micas, feldspar, and some other minerals.

potassium ion channel An *ion channel specifically for the passage of potassium ions (K^+) through cell membranes. There are several types. Ungated or 'leak' potassium channels are largely responsible for the *resting potential of cells. These are permanently open and allow K^+ ions to move in and out of the cell. Outward migration down the K^+ ion concentration gradient creates a net negative charge inside the cell. Voltage-gated K^+ channels play a key role in transmitting electrical signals in excitable cells, particularly neurons and muscle cells. During a nerve impulse, the sudden reversal of polarity (i.e. the *action potential) causes the voltage-gated K^+ channels to open, allowing excess positive charge, in the form of K^+ ions, to flood out of the cell thereby restoring the negative resting potential. Ligand-gated, or 'chemical' K^+ channels open in response to a chemical stimulus, and cause an excess negative charge inside the cell (i.e. hyperpolarization) by allowing more K^+ ions to leave than normal. They operate e.g. to reduce excitability of cells in parts of the brain during non-REM sleep.

p

See LIGAND-GATED ION CHANNEL; SODIUM ION CHANNEL; VOLTAGE-GATED ION CHANNEL. *See also* SODIUM-POTASSIUM PUMP.

potentiation The synergistic interaction of two substances or events to produce an effect that is greater than the effect of either substance or event in isolation. *See* SYNERGISM.

potometer An apparatus used to measure the rate of water loss from a shoot (*see* TRANSPIRATION) under natural or artificial conditions.

poxvirus One of a group of DNA-containing viruses, often enclosed in an outer membrane, that typically produce skin lesions in vertebrates. They include the viruses causing smallpox (variola) and cowpox (vaccinia) and also the myxoma virus responsible for myxomatosis. Some poxviruses produce tumours.

P-protein (phloem protein) A protein found in large amounts in the sap-conducting *sieve elements of phloem tissue in plants. It takes various forms in the mature sieve element, depending on plant species, ranging from a network of filaments to discrete crystalline bodies. One general property is its ability to form a gel, and it functions as a puncture repair substance, forming a plug at any site of damage in the sieve element, thus preventing loss of food materials being translocated by the phloem. In an intact functioning sieve element, the P-protein is mainly found lining the interior wall.

preadaptation An anatomical structure, physiological process, or behaviour pattern in an organism that is by chance highly suited to a new habitat to which the organism migrates or that improves the chance of the organism surviving a change in environmental conditions. For example, the lungs that developed in certain fish were probably originally a buoyancy aid before these fish began to adapt to a new environment on dry land. *See also* EXAPTATION.

Precambrian Describing the time from the formation of the earth, nearly 5 billion years ago, to the start of the Cambrian period, around 541 million years ago. The term 'Precambrian' is no longer used for a specific geological time interval, but remains as a general adjective. Precambrian time is now divided into three eons: *Hadean, *Archaean, and *Proterozoic, the latter extending to the start of the present (*Phanerozoic) eon. Fossils are rare, although *stromatolites indicate that there were flourishing populations of cyanobacteria and other bacteria. However, subsequent metamorphism of Precambrian rocks makes correlation of rocks and events extremely difficult. The largest areas of exposed Precambrian rocks are the shield areas, such as the Canadian (Laurentian) Shield and the Baltic Shield.

(((●))) SEE WEB LINKS

• Perspectives on Precambrian time provided by the website of the University of California Museum of Paleontology

precaval vein *See* VENA CAVA.

precipitin Any *antibody that combines with its specific soluble *antigen to form a precipitate. The term is sometimes applied to the precipitate itself. *See also* AGGLUTINATION.

precocial species Any species of bird in which hatching occurs at a relatively late stage of development. The young of such species are capable of locomotion and leave the nest soon after hatching; they are described as **nidifugous**. *Compare* ALTRICIAL SPECIES.

predation An interaction between two populations of animals in which one (the **predator**) hunts, captures, and kills the other (the **prey**) for food. Predator-prey relationships form important links in many food chains. They are also important in regulating population sizes of both predator and prey, especially when the predator relies on a single prey species. The term predation is also used, more loosely, for any feeding relationship in which an organism feeds on any other living organism (animal or plant).

predator An animal that obtains its food by *predation. All predators are *carnivores, although not all carnivores are predators.

pregnancy *See* GESTATION.

preimplantation genetic diagnosis (PGD) The screening of early embryos for disease-causing genes to enable the selection of 'healthy' embryos. The technique is used in conjunction with *in vitro* fertilization, which typically yields a number of embryos, and currently is approved in the UK for testing over 400 conditions. A single cell is removed from an eight-stage embryo and subjected to genetic testing; for example, it may be tested for a specific disease allele using either a gene probe and *fluorescence in situ hybridization or the *polymerase chain reaction. If the results are satisfactory, the embryo is implanted in the mother's uterus, and development proceeds. Removal of a single cell at this stage does not affect the embryo's subsequent development. PGD can help especially when couples who are being treated for fertility problems also have a history of genetic disease. However, use of PGD can be extended in nontherapeutic ways, such as choosing a baby's sex or selecting particular desired traits to produce so-called 'designer babies'. These highly controversial applications of PGD are prohibited in certain countries, including the UK. **Preimplantation tissue typing** is a related technique used to select embryos that are an exact tissue match for a living close relative, e.g. a sibling, who is suffering from a life-limiting blood disorder such as beta thalassaemia or Fanconi anaemia. The aim is for the healthy 'saviour sibling' to provide, in due course, stem cells that will help to treat the sick relative.

pre-messenger RNA *See* RNA PROCESSING.

premolar A broad ridged tooth in mammals that is situated behind the *canine teeth (when present) and in front of the *molars. Premolars are adapted for grinding and chewing food and are present in both the deciduous and *permanent teeth.

premotor cortex A region of cerebral cortex in the vertebrate brain situated immediately anterior and lateral to (in front of and to either side) the primary motor cortex. It integrates sensory information and controls the movements of muscles close to the body's vertical axis. The **supplementary motor area** lies between the premotor cortices. It is involved in planning complex movements, especially ones involving the hands.

premutation A gene variant (allele) that produces a normal individual but is predisposed to become a full mutation in subsequent generations. Genetic analysis of human families has revealed the involvement of premutations in several genetic diseases. For example, in the gene for Huntington's disease, normal individuals have a string of 6 to 39 CAG repeats near the start of the coding sequence. In individuals with the disease, this region extends to 36–180 CAG repeats. Individuals with CAG repeats in the low 30s have a premutation for Huntington's disease; this region of the gene is amplified during meiosis to become an abnormal allele, sufficient to produce the disease in that individual's offspring.

pre-replication complex *See* DNA REPLICATION.

presbyopia A loss of accommodation that normally develops in human eyes over the age of 45–50 years. Vision of distant objects remains unchanged but accommodation of the eye to near objects is reduced as a result of loss of elasticity in the lens of the eye. The defect is corrected by reading glasses using weak converging lenses.

pressure flow *See* MASS FLOW.

pressure potential Symbol Ψ_p. The component of *water potential due to the hydrostatic pressure that is exerted on water in a cell. In turgid plant cells it usually has a positive value as the entry of water causes the protoplast to push against the cell wall (*see* TURGOR). In xylem cells there is a negative pressure potential, or tension, as a result of transpiration. Water at atmospheric pressure has a pressure potential of zero.

presumptive Describing embryonic tissue that is not yet *determined but which will eventually develop into a certain kind of tissue by virtue of its position in the embryo.

presynaptic membrane The membrane of a neuron that releases neurotransmitter into the synaptic cleft between nerve cells (*see* SYNAPSE) or at a *neuromuscular junction.

prey An animal that is a source of food for a predator. *See* PREDATION.

Pribnow box (Pribnow–Schaller box; minus 10 sequence) A *consensus sequence of nucleotides—TATAAT—occurring in the *promoter region of prokaryote genes about 10 nucleotides before the start of transcription. The predominance of adenine and thymine bases means that hydrogen bonding between the two DNA strands in this region is relatively weak, enabling the strands to be separated more easily to permit transcription by RNA polymerase. *See also* TATA BOX.

prickle A hard sharp protective outgrowth, many of which may cover the surface of a plant. It contains cortical and vascular tissue and is not regarded as an epidermal outgrowth. *Compare* SPINE; THORN.

primary cell wall *See* CELL WALL.

primary consumer *See* CONSUMER.

primary growth The increase in size of shoots and roots of plants that results from the activity of the *apical (tip) meristems and subsequent expansion of the cells produced. The tissues thus produced are called **primary tissues** and the resultant plant parts constitute the **primary plant body**. *Compare* SECONDARY GROWTH.

primary immune response *See* IMMUNOLOGICAL MEMORY.

primary motor cortex *See* CEREBRUM.

primary producer *See* PRODUCER.

primary productivity The total amount of organic matter synthesized by the producers (e.g. green plants or phytoplankton) of an ecosystem. *See* PRODUCTIVITY.

primary somatosensory cortex *See* CEREBRUM.

primary structure *See* PROTEIN.

primase An enzyme that creates the short primer sequences needed for the *discontinuous replication of DNA. It is an RNA polymerase that synthesizes a short segment of RNA, consisting of 10 to 12 nucleotides, attached to the lagging strand of the unwound DNA molecule. Each primer sequence provides the free 3'-OH group required for the initiation of synthesis of an Okazaki fragment by DNA polymerase. These primers are subsequently removed before the Okazaki fragments are united into a continuous new DNA strand. *See also* PRIMOSOME.

Primates An order of mammals that includes the monkeys, apes, and humans. Primates evolved from arboreal insectivores, known as proto-primates, probably some 55–65 million years ago. Living primates are characterized by thumbs and big toes that are opposable (i.e. capable of facing and touching the other digits), which permits manual dexterity, and forward-facing eyes allowing *binocular vision. The brain, particularly the cerebrum, is relatively large and well-developed, accounting for the intelligence and quick reactions of these mammals. The young are usually produced singly and undergo a long period of growth and development to the adult form. *See also* HOMININ.

(⊕) SEE WEB LINKS
• Wide-ranging account of primate characteristics and taxonomy, compiled by Dennis O'Neil, Palomar College, California

primitive node (primitive knot) *See* HENSEN'S NODE.

primitive streak The longitudinal groove that develops in the *gastrula during the development of bird and mammal embryos. The cells in the primitive streak proliferate rapidly to form mesoderm cells, which migrate to the interior of the embryo.

primordial soup *See* ORIGIN OF LIFE.

primordium (*pl.* **primordia**) A group of cells that represents the initial stages in development of a plant organ. Root and shoot primordia are present in a young plant embryo while leaf primordia (or **leaf buttresses**) are seen as hornlike projections from the sides of the dome of the apical meristem, just below the shoot apex.

primosome A complex of enzymes and additional proteins that associates with the DNA molecule to initiate DNA replication in prokaryotes, such as *E. coli*. Its chief component is a *primase enzyme, responsible for synthesizing short single strands of RNA, which serve as primers for DNA polymerase (*see* POLYMERASE). The primosome also contains a **helicase** enzyme, which unwinds the two strands of the DNA in readiness for replication.

principle of parsimony The principle that the most acceptable explanation of an occurrence, phenomenon, or event is the simplest, involving the fewest entities, assumptions, or changes. In phylogenetics, for example, the preferred tree showing evolutionary relationships between species, molecules, or other entities is the one that requires the least amount of evolutionary change, that is, *maximum parsimony.

prion An abnormal form of a normal cell protein (PrP) found in the brain of mammals that is the agent responsible for the diseases scrapie in sheep, *bovine spongiform encephalopathy (BSE) in cattle, and *Creutzfeldt-Jakob disease in humans. Produced by mutation of the normal PrP gene, the abnormal prion protein induces the normal protein to fold incorrectly, causing it to form aggregates. These accumulate in the brain and progressively damage and destroy brain cells. Prion protein can be transmitted to other individuals of the same or closely related species, by injection or ingestion of infected tissue, and is transmissible between species that are not closely related, e.g. between cattle and humans.

Proboscidea The order of mammals that comprises the elephants. They are herbivorous, with a muscular trunk (*proboscis) used for drinking, bathing, and collecting food. The tusks are continuously growing upper incisors and the enormous ridged molar teeth are produced in sequence to replace worn teeth throughout life. The order, which evolved in the Eocene epoch, was formerly much larger and more widespread than it is today and included the extinct mammoths. Two species of modern elephants are generally recognized, the African elephant (*Loxodonta africana*) and Asian elephant (*Elephas maximus*), although there is evidence that there may be an additional African species, the forest elephant (*L. cyclotis*). The African elephant is now classed as 'vulnerable' and the Asian elephant as 'endangered', due to a combination of poaching and habitat loss.

proboscis (*pl.* **probosces** or **proboscides** or **proboscises**) **1.** A muscular and very flexible elongation of the nose, such as the trunk of an elephant, which has a fingerlike extremity and is capable of picking up and moving objects, taking in water, collecting food, etc. **2.** The elongated mouthparts of certain invertebrates, such as the two-winged flies (Diptera).

procambium (provascular tissue) (*pl.* **procambia** or **procambiums**) A plant tissue formed by the *apical meristems of shoots and roots. It consists of cells elongated parallel to the long axis of the plant. The procambium gives rise to the primary *vascular tissue.

procarcinogen The precursor of an active *carcinogen. The procarcinogen itself is not usually carcinogenic but is converted to the active carcinogen after it has been metabolized. For example, the drug diethylstilboestrol (a synthetic oestrogen no longer in clinical use) is metabolized to an epoxide intermediate, which can cause cervical cancer.

procaryote *See* PROKARYOTE.

prochlorophytes *See* CHLOROXYBACTERIA.

Proconsul A genus of extinct primates known from fossil remains, roughly 23 to 14 million years old, found in East Africa and assigned to four known species. They had a mix of apelike and monkey-like features, being quadrupeds but also tailless, and may be ancestral to the apes.

prodrug A drug that requires further metabolism in the liver before becoming biologically active. An example of a prodrug is the immunosuppressant azathioprine, which is metabolized in the body to the active compound 6-mercaptopurine.

producer An organism considered as a source of energy for those above it in a *food chain (i.e. at the next *trophic level). Photosynthetic *autotrophs, such as green plants and phytoplankton, which convert energy from sunlight into chemical energy, are notable examples of **primary producers**; herbivores are **secondary producers**, as they utilize energy from autotrophs and supply energy for carnivores. *Compare* CONSUMER.

productivity (production) (in ecology) The rate at which an organism, population, or community assimilates energy (**gross productivity**) or makes energy potentially available (as body tissue) to an animal that feeds on it (**net productivity**). The difference between these two rates is due to the rate at which energy is lost through excretion and respiration. Thus **gross primary productivity** is the rate at which plants (or other *autotrophs) assimilate energy, and **net primary productivity** is the rate at which energy is incorporated as cells or tissue. It is measured in kilojoules per square metre per year ($kJ\ m^{-2}\ y^{-1}$). In terrestrial plants, much of the net productivity is not actually available to *consumers, e.g. tree roots are not eaten by herbivores. *See also* ENERGY FLOW; SECONDARY PRODUCTIVITY.

profundal Occurring in or designating the deep-water zone of an inland lake. Light intensity, oxygen concentration, and (during summer and autumn) temperature are markedly lower than in the surface layer. *Compare* LITTORAL; SUBLITTORAL.

progenesis The maturation of gametes in an organism that is still otherwise at the juvenile stage of development (*see* PAEDOGENESIS). Progenesis can lead to *paedomorphosis.

progeny *See* OFFSPRING.

progesterone A hormone, produced primarily by the *corpus luteum of the ovary but also by the placenta, that prepares the inner lining of the uterus for implantation of a fertilized egg cell (*see* OVARIAN CYCLE). If implantation fails, the corpus luteum degenerates and progesterone production ceases accordingly. If implantation occurs, the corpus luteum continues to secrete progesterone, under the influence of *luteinizing hormone and *prolactin, for several months of pregnancy, by which time the placenta has taken over this function. During pregnancy, progesterone maintains the constitution of the uterus and prevents further release of eggs from the ovary. Small amounts of progesterone are produced by the testes. *See also* PROGESTOGEN.

progestogen One of a group of naturally occurring or synthetic hormones that maintain the normal course of pregnancy. The best known is *progesterone. In high doses progestogens inhibit secretion of *luteinizing hormone, thereby preventing ovulation, and alter the consistency of mucus in the vagina so that conception tends not to occur. They are therefore used as major constituents of *oral contraceptives. Synthetic progestogens are also known as **progestins**.

proglottid (proglottis) *See* CESTODA.

programmed cell death *See* APOPTOSIS.

progymnosperms An extinct group of plants that flourished in the mid- to late Devonian (360–350 million years ago) and contained the ancestors of modern gymnosperms (conifers and cycads). They were shrubs and trees, some up to 12 m tall, with frondlike leaves. Progymnosperms evolved a vascular cambium that was bifacial, capable of producing not only xylem on its inner face (as in the more primitive lycophytes and sphenophytes) but also phloem on its outer face. Moreover, as the cambium was pushed outwards during radial growth of the stem, the cambial cells were able to divide radially and so function indefinitely. These features permitted the growth of wider trunks, with more efficient vascular systems and stronger wood. Also, the first true simple leaves appear among certain members of the group, such as *Archaeopteris* (*see* TELOME THEORY). However, reproduction was by spores, not seeds. *See also* EUPHYLLOPHYTE.

prokaryote (procaryote) Any organism in which the genetic material is not enclosed in a cell nucleus. Prokaryotes consist exclusively of *archaea and *bacteria, which are now classified in separate *domains, *Archaea and *Bacteria. It is believed that eukaryotic cells (*see* EUKARYOTE) evolved as symbiotic associations of prokaryotes (*see* ENDOSYMBIONT THEORY).

prolactin (lactogenic hormone; luteotrophic hormone; luteotrophin) A hormone produced by the anterior pituitary gland in response to the release of prolactin-releasing hormone from the hypothalamus. In mammals it stimulates the mammary glands to secrete milk (*see* LACTATION) and the corpus luteum of the ovary to secrete the hormone *progesterone. Secretion of prolactin is increased by suckling. In birds prolactin stimulates secretion of crop milk by the crop glands.

proline *See* AMINO ACID.

prometaphase The period during the early part of *metaphase in which the chromosomes become attached to the spindle fibres and align themselves at right angles to the spindle poles in a plane through the centre of the cell.

promoter (in protein synthesis) The region of a DNA molecule that signals the start of *transcription. In eukaryote genes, the promoter typically has a core promoter very near to the transcription start site, often containing the base sequence TATA

(hence it is called the *TATA box). Other promoter elements are located 'upstream' of the core promoter and start site. Proteins called *transcription factors bind to the promoter and ensure that the enzyme responsible for transcription, RNA *polymerase, also binds correctly to the promoter with respect to the coding region of the gene. In prokaryotes and viruses, several genes can share the same promoter in units called *operons. Prokaryote promoters have their own characteristic base sequence, the *Pribnow box.

pronation Rotation of the lower forearm so that the hand faces backwards or downwards with the radius and ulna crossed. *Compare* SUPINATION.

proofreading (in genetics) A repair mechanism that helps to ensure faithful *DNA replication in living cells. It is a function of the enzyme DNA polymerase, which catalyses the replication process. This enzyme identifies and excises mismatched bases at the end of the growing strand, leaving the one free to accept the correct nucleotide instead, thereby restoring the correct complementary base sequence. *See also* DNA REPAIR.

propagation 1. (in botany) *See* VEGETATIVE PROPAGATION. 2. (in neurophysiology) The process whereby a nerve *impulse travels along the axon of a neuron or an action potential spreads throughout a muscle fibre (*see* NEUROMUSCULAR JUNCTION). *Compare* TRANSMISSION.

propagule Any cellular structure produced by an organism that is capable of dispersing and surviving in the environment before developing into a new individual. Examples are seeds, spores, and cysts. Plant cuttings are sometimes regarded as artificial propagules.

prophage The DNA of a temperate *bacteriophage following its incorporation into the host bacterium. The process of incorporation of the viral DNA is known as *lysogeny.

prophase The first stage of cell division, during which chromosomes contract and divide along their length (except for the centromeres) into chromatids. In *mitosis, the chromosomes remain separate from each other. In the first division of *meiosis, homologous chromosomes become paired (*see* PAIRING). By the end of first prophase the two chromosomes begin to move apart.

proplastid See PLASTID.

proprioceptor Any *receptor that is sensitive to movement, pressure, or stretching within the body. Proprioceptors occurring in muscles, tendons, and ligaments are important for the coordination of muscular activity and the maintenance of balance and posture. *See* MECHANORECEPTOR; STRETCH RECEPTOR.

prop root Any of the modified roots that arise from the stem of certain plants and provide extra support. Such stems are usually tall and slender and the prop roots develop at successively higher levels as the stem elongates, as in the maize plant. **Buttress roots**, which develop at the base of the trunks of many tropical trees, are similar but tend to have a more flattened appearance. **Stilt roots** are stouter than prop roots. Those formed at the base of the mangrove tree provide firm anchorage in the soft mud of the swamps.

prosencephalon See FOREBRAIN.

prosoma (*pl.* **prosomas** or **prosomata**) The anterior section of the body (*see* TAGMA) of arachnids and other arthropods of the phylum *Chelicerata, which consists of the fused head and thorax (cephalothorax) and bears the chelicerae and other appendages. *Compare* OPISTHOSOMA.

prostacyclin (prostaglandin I₂) An *eicosanoid normally produced from prostaglandin H_2 by endothelial cells lining the walls of blood vessels. It inhibits blood clotting by preventing the aggregation of platelets and causes widening of blood vessels (vasodilation). Hence, it is an antagonist of *thromboxane A_2. See PROSTAGLANDIN.

prostaglandin (PG) Any of a group of lipid-soluble organic compounds causing a range of physiological effects in animals. Prostaglandins have been detected in most body tissues, in some cases acting locally and in others having effects on more distant target tissues. They are synthesized within cell membranes from arachidonic acid, which is derived from membrane phospholipids by the action of phospholipase A. The arachidonic acid is converted by the enzyme *cyclo-oxygenase (prostaglandin endoperoxide synthase) to prostaglandin G_2, the precursor of all the various classes of prostaglandins, *prostacyclin, and *thromboxane A_2—collectively known as the **prostanoids**. PGs are short-lived signal molecules that bind to G-protein-coupled cell receptors. Among their array of effects, they cause the contraction of smooth muscle, e.g. in stimulating contractions of the uterus during parturition. Hence, natural and synthetic prostaglandins are used to induce abortion or labour in humans and domestic animals. Prostaglandins are also involved in inflammation, being released from macrophages and mast cells in response to injury or infection. Drugs such as *aspirin and ibuprofen exert their pain-relieving effects by blocking prostaglandin synthesis.

prostate gland A gland in male mammals that surrounds and opens into the urethra where it leaves the bladder. During ejaculation it secretes a slightly alkaline fluid into the *semen that activates the sperms and prevents them from sticking together. It also counteracts the acidity of the male and female genital tracts and contains prostaglandins, which stimulate contractions of the female reproductive tract.

prosthetic group A tightly bound nonpeptide inorganic or organic component of a protein. Prosthetic groups may be lipids, carbohydrates, metal ions, phosphate groups, etc. Some *coenzymes are more correctly regarded as prosthetic groups.

protamine Any of a group of proteins of relatively low molecular weight found in association with the chromosomal *DNA of vertebrate sperm cells. They contain a single polypeptide chain comprising about 67% arginine. Protamines serve in packaging the highly condensed DNA of the germ-cell chromosomes by replacing the histone proteins that normally associate with the DNA. Protamine sulphate is used therapeutically to treat heparin overdosage.

protandry 1. The condition in which the male reproductive organs (stamens) of a flower mature before the female ones (carpels), thereby ensuring that self-fertilization does not occur. Examples of protandrous flowers are ivy and rosebay willowherb. *Compare* PROTOGYNY; HOMOGAMY. *See also* DICHOGAMY. **2.** The condition in some hermaphrodite or colonial invertebrates in which the male gonads or individuals are sexually mature before the female ones. *Compare* PROTOGYNY.

protanopia *See* COLOUR BLINDNESS.

protease (peptidase; proteinase; proteolytic enzyme) Any enzyme that catalyses the hydrolysis of proteins into smaller *peptide fractions and amino acids, a process known as **proteolysis**. Examples are *pepsin and *trypsin. Several proteases, acting sequentially, are normally required for the complete digestion of a protein to its constituent amino acids. *See also* ACID PROTEASE.

proteasome A large complex of proteins found in the cytoplasm of cells that degrades cell proteins into small peptide units. Proteasomes participate in the disposal of damaged proteins and the removal of proteins that have reached the end of their functional life in the cell. They thus play a vital part both in the cell's response to stress and in the day-to-day running of the cell. A proteasome consists of a hollow cylinder of protease enzymes, each with its catalytic site facing the internal cylinder wall. Another protein complex guards the entrance to the cylinder. Proteins destined for degradation are usually tagged by a chain of small proteins, called *ubiquitins, which direct the target protein to unfold and enter the proteasome, where it is cleaved into peptides. Proteasomes are also involved in the production of peptides derived from viruses or other pathogens for presentation by cells in association with *MHC class I proteins.

protein Any of a large group of organic compounds found in all living organisms. Proteins comprise carbon, hydrogen, oxygen, and nitrogen and most also contain sulphur; molecular weights range from 6 to several thousand *kilodaltons (kDa). Protein molecules consist of one or several long chains (*polypeptides) of *amino acids linked via peptide bonds in a characteristic sequence. This sequence is called the **primary structure** of the protein. These polypeptides may undergo coiling (*see* ALPHA HELIX) or pleating (*see* BETA SHEET), the nature and extent of which is described as the **secondary structure**. The three-dimensional shape of the coiled or pleated polypeptides is called the **tertiary structure**, the functional unit of which is a *domain. Tertiary structure is stabilized by various interactions between the side chains of amino acids that are close together, such as *hydrogen bonds and van der Waals forces. Moreover, where two cysteine residues lie together, the sulfhydryl (–SH) groups on their side chains form covalent bonds, or *disulphide bridges, linking them together. **Quaternary structure** specifies the structural relationship of the component polypeptides in proteins that consist of two or more polypeptide chains, such as *haemoglobin and *collagen. *X-ray crystallography is a commonly employed technique to determine the three-dimensional structure of proteins.

Proteins may be globular or fibrous, with various intermediate forms. **Globular proteins** have compact rounded molecules and are usually water-soluble. Of prime importance are the *enzymes, proteins that catalyse biochemical reactions. Other globular proteins include the *antibodies, which combine with foreign substances in the body; the carrier proteins, such as haemoglobin; the storage proteins (e.g. *casein in milk and *albumin in egg white), and certain hormones (e.g. *insulin). Cell *receptors comprise another group of functionally vital proteins. **Fibrous proteins** are generally insoluble in water and consist of long coiled strands or flat sheets, which confer strength and elasticity. In this category are *keratin and collagen. Actin and myosin are the principal fibrous proteins of muscle, the interaction of which brings about muscle contraction. *Blood clotting involves the fibrous protein called fibrin.

When heated over 50°C or subjected to strong acids or alkalis, proteins lose their specific tertiary structure and may form insoluble coagulates (e.g. egg white). This usually inactivates their biological properties.

(⊕) SEE WEB LINKS

- Comprehensive survey of protein biochemistry, from the Virtual Library of Biochemistry, Molecular Biology and Cell Biology

protein blotting *See* WESTERN BLOTTING.

protein engineering The techniques used to alter the structure of proteins (especially enzymes) in order to improve their use to humans. This involves artificially modifying the DNA sequences that encode them so that, for example, new amino acids are inserted into existing proteins. Synthesized lengths of novel DNA can be used to produce new proteins by cells or other systems containing the necessary factors for *transcription and *translation. Alternatively, new proteins can be synthesized by **solid state synthesis**, in which polypeptide chains are assembled under the control of chemicals. One end of the chain is anchored to a solid support and the chemicals selectively determine which amino acids are added to the free end. The appropriate chemicals can be renewed during the process; when synthesized, the polypeptide is removed and purified. Protein engineering is used to synthesize enzymes (so-called 'designer enzymes') used in biotechnology. The three-dimensional tertiary structure of proteins is crucially important for their function, and this can be investigated using computer-aided modelling. *See also* DIRECTED EVOLUTION.

protein family A set of proteins that have evolved from a common ancestral protein and share an evolutionary history. Duplication of an ancestral gene gives rise to a duplicate protein that becomes available to evolution and can diverge in structure and function over evolutionary time. Other mechanisms of protein evolution, such as exon shuffling (*see* INTRON), can also be involved. The evolutionary histories linking members of a protein family can be deduced from the extent and molecular configurations of *homologous amino acid sequences and structures. Families are arranged in a hierarchy, with closely related proteins comprising subfamilies and related families grouped to form superfamilies.

protein kinase An enzyme that catalyses the transfer of a phosphate group from ATP to an intracellular protein, thereby affecting the biological activity of the protein (*see* KINASE). Protein kinases phosphorylate specific amino-acid residues of their target proteins, usually either serine, threonine, or tyrosine. They play an important role in increasing or decreasing enzyme activity and in transmitting signals from receptors on the cell surface (*see* SIGNAL TRANSDUCTION). The activity of the protein kinases is itself controlled by cyclic AMP, calcium ions, or other intracellular chemicals. It can be reversed by the action of phosphatase enzymes in the cell.

protein kinase C (PKC) Any of a family of proteins that act as serine/threonine kinases and are vital components of many intracellular signalling pathways. Inactive PKC is located in the cytoplasm, where it is activated by diacylglycerol (DAG) or other controllers (such as elevated levels of intracellular calcium ions), causing it to migrate to the plasma membrane. Here it phosphorylates serine and threonine residues on target proteins, which translocate to the nucleus and regulate gene transcription. PKC signalling pathways via *metabotropic receptors are crucial, for example, in the modulation of ion channel activity in nerve cells by such neurotransmitters as noradrenaline and GABA and also play a role in memory and learning.

proteinoid A protein-like substance formed by polymerization of amino acids under inorganic conditions, such as heating to over 140°C. In the 1970s it was discovered that proteinoids could also be formed by relatively mild heating (70°C) in the presence of certain inorganic catalysts (e.g. phosphoric acid). In water, proteinoids aggregate to form small round bodies called **proteinoid microspheres**, or 'protocells'. These have certain attributes of living cells, such as a differentially permeable filmlike outer layer, the ability to swell and shrink due to osmotic movements of water, and the capability for budding and binary fission. It has been proposed that such microspheres could have provided a suitable vehicle for the chemical components of life to evolve a primitive form of metabolism and pave the way for the emergence of the first living cells. *See* ORIGIN OF LIFE.

protein profiling The identification of individual proteins in a biological sample, often coupled with determination of their relative abundance. It is a crucial step in analysing the proteins synthesized by cells or tissues (*see* PROTEOMICS). Various techniques are applied, most of which employ gel or liquid electrophoresis in combination with mass spectrometry to separate the proteins and determine their relative amounts. Samples can be labelled differentially with isotopic or fluorescent tags to pinpoint differences in protein profiles when comparing samples—e.g. from normal tissue and from a tumour—to assess abnormalities.

protein sequencing The process of determining the amino-acid sequence of a protein or its component polypeptides. The technique traditionally used is Edman degradation (devised by Pehr Edman), in which the terminal amino-acid residues are removed sequentially and identified chromatographically. Each step is automated and the whole process can now be performed by a single machine—the sequenator. Large polypeptides must be cleaved into smaller peptides before sequencing. The identification of unknown protein samples, determination of protein structure, and detection of post-translational modifications, as well as amino acid sequencing now commonly

employ *mass spectrometry techniques, such as *matrix-assisted laser desorption/ionization-time of flight (MALDI-TOF). For large proteins, this can be complemented by peptide mapping, in which the polypeptide chain is cleaved at specific peptide links.

The results of chemical sequencing can often be compared with the amino-acid sequence deduced by *DNA sequencing. The gene coding for the protein under investigation may be found by screening a *DNA library, for example by *western blotting. However, the base sequence of the gene gives only the amino-acid sequence of the nascent protein, i.e. before post-translational modification. The sequence of the functional protein can only be found by chemical analysis.

protein synthesis The process by which living cells manufacture proteins from their constituent amino acids, in accordance with the genetic information carried in the DNA of the chromosomes. This information is encoded in messenger *RNA, which is transcribed from DNA in the nucleus of the cell (*see* GENETIC CODE; TRANSCRIPTION): the sequence of amino acids in a particular protein is determined by the sequence of nucleotides in messenger RNA. At the ribosomes the information carried by messenger RNA is translated into the sequence of amino acids of the protein in the process of *translation.

protein targeting The process whereby a newly synthesized protein is directed to its correct location within the cell. Protein targeting is determined by short sequences of amino acids in the protein (known as **signal sequences**), which direct it to the correct destination; for example to a mitochondrion or (in the case of a plant cell) a chloroplast. This movement can occur either during synthesis of the protein (co-translational) or after synthesis is complete (post-translational).

proteobacteria A large clade of Gram-negative bacteria that includes numerous free-living and pathogenic species. Proteobacteria share similarity in the sequence of their genes encoding ribosomal RNA, regarded as indicating evolutionary related-ness, and can be divided into five subgroups labelled alpha, beta, gamma, delta, and epsilon. Among the members of the clade are well-known human commensals and pathogens such as *Salmonella typhimurium* (responsible for gastrointestinal disease), *Helicobacter* sp. (causing gastric ulcers), *Vibrio cholera* (causing cholera), and *Escherichia coli*. The forms and lifestyles of proteobacteria are diverse, and include photoautotrophs, chemoautotrophs, and heterotrophs. *See also* SULPHUR BACTERIA.

proteoglycan One of a group of glyco-proteins found in the matrix of *connective tissue. Each consists of numerous carbohydrate chains covalently attached to a core protein; large complexes can form when hundreds of proteoglycans attach noncovalently to a single long polysaccharide, resulting in a fine web in which collagen fibres of the matrix are embedded. *See* GLYCOSAMINOGLYCAN.

proteolysis The enzymic splitting of proteins. *See* PROTEASE.

proteolytic enzyme *See* PROTEASE.

proteome The entire complement of proteins synthesized by a cell or organism at a given time. This can be determined by analysing protein constituents of cell contents using such techniques as gel electrophoresis, high-throughput liquid chromatography, microarrays, and mass spectrometry, coupled with automated database searching to identify proteins. A complementary approach is to infer the potential proteome from analysis of the genome (*see* GENOMICS). However, unlike the genome, the proteome is constantly changing due to the influence of intracellular and extracellular factors. *See* PROTEOMICS.

proteomics The study of the proteins synthesized by a particular cell or organism (*see* PROTEOME). This vast and rapidly expanding field is of fundamental significance to many areas of biology and medicine. It endeavours to determine what proteins a cell makes, how and when it makes them and in what quantities, how different proteins function, where

they function, and how they interact with other cell components, including other proteins. Moreover, proteomics seeks to discover the internal and external factors that influence the proteome, for example during development, disease, ageing, or evolution. *See* PROTEIN PROFILING. *See also* TRANSCRIPTOMICS.

Proterozoic The eon of geological time extending from the end of the *Archaean, about 2500 million years ago, to the start of the present eon (*see* PHANEROZOIC), about 541 million years ago. Life in the early Proterozoic was dominated by bacteria, which flourished in shallow seas and muds. They depended on a wide variety of metabolic strategies, including photosynthesis, which were crucial in determining the composition of the earth's atmosphere and oceans. The oldest eukaryotic fossils date from after the middle Proterozoic, about 1200 million years ago. These early protists are thought to have arisen through symbiotic associations of various prokaryotes (*see* ENDOSYMBIONT THEORY), probably on several independent occasions. Towards the end of this eon comes the first fossil evidence of multicellular animal life, the **Ediacaran fauna**, named after the Ediacara Hills in Australia but also found elsewhere. These rocks, dated to around 650 million years ago, reveal traces of soft-bodied fanlike or quiltlike creatures, perhaps unrelated to any modern forms, as well as animals resembling jellyfish and worms. The Proterozoic is divided into three eras: the Paleoproterozoic (2500–1600 mya), the Mesoproterozoic (1600–1000 mya), and the Neoproterozoic (1000–541 mya). Three periods of the Neoproterozoic are now recognized: the Tonian period (1000–720 mya), the *Cryogenian period (720–635 mya), and the Ediacaran period (635–541 mya), named after the characteristic faunal assemblage.

prothallus (*pl.* **prothalli**) A small flat-tened multicellular structure that represents the independent *gametophyte generation of clubmosses, horsetails, and ferns. In some of these plants a single prothallus bears both male and female sex organs. In others there are separate male and female prothalli.

prothoracic gland *See* ECDYSONE.

prothoracicotropic hormone (PTTH) An insect hormone, released by glands (corpora cardiaca) on either side of the brain, that stimulates release of the moulting hormone, *ecdysone, from the prothoracic glands. PTTH isolated from the moth *Manduca sexta* is a 30-kDa protein consisting of two identical peptide chains.

prothrombin (Factor II) One of the blood *clotting factors. It is the precursor of the enzyme thrombin, which catalyses the formation of the fibrin matrix of the blood clot. Prothrombin synthesis occurs in the liver and is dependent on adequate supplies of vitamin K. *See also* BLOOD CLOTTING.

protist Any eukaryotic organism that is essentially unicellular, colonial, or multicellular in form and lacks true cellular differentiation into tissues and organs. Protists include all eukaryotes except the plants, fungi, and animals. They display a diversity of forms, ranging from nonmotile or motile independent cells to leaflike or filamentous forms and sheets of cells. They can be autotrophic, saprotrophic, heterotrophic, or mixotrophic. Radical revisions to relationships between eukaryotic groups have been made in the light of molecular studies, and 'protist' is now purely a descriptive term. The kingdom Protista was originally proposed by Ernst Haeckel in 1866 to include the algae, bacteria, fungi, and protozoa; it was later restricted first to unicellular organisms, and then to protozoa, unicellular algae, and organisms then regarded as simple fungi.

protocell A spherical structure consisting of a double layer (bilayer) of fatty acid molecules surrounding a watery interior. Protocells can be made in the laboratory and demonstrate certain properties of living cells; primarily they are compartments in which biomolecules can become concentrated and interact to form more complex biomolecules. The bilayer mimics the plasma membrane in allowing small molecules such as sugars and nucleotides to enter, but preventing nucleic acids (DNA or RNA) from leaving. It is suggested that these properties illustrate how the earliest cells might have arisen from spontaneous assembly of fatty acids. *See also* ORIGIN OF LIFE; SYNTHETIC BIOLOGY.

Protoctista A former kingdom consisting of unicellular or simple multicellular organisms that possess nuclei and cannot be classified as animals, plants, or fungi. Molecular studies have prompted major revisions to the phylogeny of eukaryotic organisms, and the term 'protoctist' is no longer considered either descriptively useful or taxonomically valid.

protoderm A plant tissue formed by the *apical meristem of shoots and roots that subsequently gives rise to the epidermis.

protogyny 1. The condition in which the female reproductive organs (carpels) of a flower mature before the male ones (stamens), thereby ensuring that self-fertilization does not occur. Examples of protogynous flowers are plantain and figwort. *Compare* PROTANDRY; HOMOGAMY. *See also* DICHOGAMY. 2. The condition in hermaphrodite or colonial invertebrates in which the female gonads or individuals are sexually mature before the male ones. *Compare* PROTANDRY.

protonema (*pl.* **protonemata**) The first, usually filamentous, structure produced by a germinating moss or fern spore under dark conditions. The protonema of mosses bears buds that develop into the gametophyte plant; in ferns the protonema becomes the *prothallus.

protonephridium (*pl.* **protonephridia**) *See* NEPHRIDIUM.

proton pump A carrier protein or complex of proteins that transports protons (H^+) across biological membranes. Such pumps use energy, for example ATP, to establish a high concentration of protons on one side of the membrane compared to the other. This proton gradient is then exploited by the cell to drive various processes, including the transport of ions and small molecules across plasma membranes. Proton pumps are also involved in secretion of gastric acid by mucosal cells in the stomach lining, the uptake of mineral ions by plant root hairs, and the expansion of plant cell walls induced by auxins. The concept of a proton pump is central to the *chemiosmotic mechanism of ATP formation by an *electron transport chain in mitochondria and chloroplasts. Here, the pump is driven not by ATP hydrolysis but by electron transport, and the proton gradient that is set up represents a means of conserving the energy released by respiration, or captured by photosynthetic pigments. This energy is then stored as ATP by the enzyme *ATP synthetase. *See also* ACID GROWTH THEORY.

proto-oncogene *See* ONCOGENE.

protophloem The primary *phloem tissue that arises from the elongating regions of shoot or root apices. *Compare* METAPHLOEM.

protoplasm Formerly, the material comprising the living contents of a *cell, i.e. all the substances in a cell except large vacuoles and material recently ingested or to be excreted. The term is no longer used; the material of the cell is now referred to as the *cytoplasm, apart from the nucleoplasm inside the *nucleus.

protoplast (energid) The living unit of a cell, consisting of the nucleus and cytoplasm bounded by the plasma membrane. Protoplasts of bacterial and plant cells can be prepared by removing the cell wall; they are used to study the processes involved in cell metabolism and reproduction.

protostome An animal in which the mouth develops from the opening (blastopore) of the embryonic cavity (*see* ARCHENTERON). The name derives from Greek, meaning literally 'first mouth'. Other features typical of protostomes are spiral cleavage of the blastula, determinate development (i.e. the fate of cells is established at a very early embryonic stage), and schizocoelic formation of a coelom where one occurs (i.e. by formation of a cavity within a solid mass of mesoderm). Two main clades of protostomes are now recognized: *Ecdysozoa and *Lophotrochozoa (some authors propose a third, the *Platyzoa). *Compare* DEUTEROSTOME.

Prototheria A subclass of mammals—the monotremes—that lay large yolky eggs. It contains only the duckbilled platypus and the spiny anteater. After hatching, the young feed on milk from simple mammary glands inside a maternal abdominal pouch. In the anteater the eggs are also incubated in this pouch, while the platypus builds an underground nest. Adult monotremes have no true teeth. Their skeleton resembles that of a reptile, and although they are warm-blooded the body temperature is somewhat variable. They are believed to have diverged from other mammals about 180 million years ago.

protoxylem The primary *xylem tissue that is formed in the expanding region of roots and shoots before the shoot or root has completed the elongation process. In the secondary walls of these vessels the lignin is deposited in rings and spirals, which allows for further elongation. The protoxylem is succeeded by *metaxylem.

protozoa (*sing.* **protozoon** or **protozoan**) A name formerly given to a group of unicellular or acellular, usually microscopic *protists, now classified in various phyla (*see* APICOMPLEXA; CILIOPHORA; RHIZOPODA; ZOOMASTIGOTA). It is is no longer considered either descriptively useful or taxonomically valid.

Protura An order of tiny wingless insects that live in soil and leaf litter. There are about 800 known species. The slender elongate adults are usually 0.5 to 2.5 mm long and lack eyes and antennae; the front pair of legs are held forwards as sensors. The mouthparts, which are largely entognathous (concealed within folds), are adapted for feeding on decaying organic matter. The young resemble the adults, but three segments are added to the end of the abdomen during development. The relationship of proturans to other arthropod groups is controversial. Although some authorities place them in the subclass *Apterygota with the other wingless insects, other authors regard the Protura as a separate class within the superclass *Hexapoda, most closely affiliated to the *Collembola.

provascular tissue *See* PROCAMBIUM.

proventriculus (*pl.* **proventriculi**) **1.** The anterior part of the stomach of a bird, where digestive enzymes are secreted. Food passes from the proventriculus to the *gizzard. **2.** *See* GASTRIC MILL.

provirus The intermediate stage in the infection of a host cell by a virus, e.g. a *retrovirus, in which the viral genome is integrated into the DNA of the host, where it can undergo successive replications before being transcribed. In the case of retroviruses the single RNA strand of the virus is converted into double-stranded DNA by the enzyme *reverse transcriptase, then integrated into the host cell DNA and subsequently transcribed to form new RNA viruses. This integration also introduces retroviral oncogenes into the host, with the attendant risk of host-cell transformation to a cancer cell. A provirus, notably that of *HIV, can remain dormant for long periods before being transcribed. *See also* PROPHAGE.

proximal Denoting the part of an organ that is nearest to the organ's point of attachment. For example, the knuckles are at the proximal end of the fingers. *Compare* DISTAL.

proximal convoluted tubule (PCT; first convoluted tubule) The section of a *nephron situated between Bowman's capsule and the loop of Henle in the cortex of the vertebrate kidney. Reabsorption of most of the salt, water, and glucose from the *glomerular filtrate occurs in this tubule. Sodium ions (Na^+), glucose, and amino acids are all actively transported from the tubule by the cells lining the PCT and transferred to the extracellular fluid; water follows by osmosis, and bicarbonate ions and potassium ions also leave passively. At the same time hydrogen ions and ammonia are secreted into the lumen of the tubule, as are uric acid (in birds and reptiles) and drug metabolites, which are actively transferred from the blood

capillaries into the tubule cells. Movement of material both in and out of the tubule is facilitated by finger-like microvilli (*see* BRUSH BORDER) on the inner surface of the tubule, which increase its effective surface area.

PR protein (pathogenesis-related protein) *See* SYSTEMIC ACQUIRED RESISTANCE.

PRR *See* PATTERN RECOGNITION RECEPTOR.

pruning (in neurophysiology) *See* COMPETITION (sense 2).

PSC *See* PHYLOGENETIC SPECIES CONCEPT.

pseudoautosomal region *See* X INACTIVATION.

pseudocarp (false fruit) A fruit that incorporates, in addition to the ovary wall, other parts of the flower, such as the *receptacle. For example, the fleshy part of the strawberry is formed from the receptacle and the 'pips' on the surface are the true fruits. *See also* COMPOSITE FRUIT; POME; SOROSIS; SYCONUS.

Pseudociliata *See* DISCOMITOCHONDRIA.

pseudocoelomate Describing any invertebrate animal whose body cavity is a **pseudocoel**, a cavity between the gut and the outer body wall derived from a persistent blastocoel (*see* BLASTULA), rather than a true coelom. Pseudocoelomate animals include the Rotifera and Nematoda.

pseudogamy A form of *parthenogenesis in which the male gamete is required to activate development of the egg but fertilization does not occur. Generally, the sperm or pollen is from a member of the same species, although in some cases, for example in the fish *Poeciliopsis*, egg development is triggered by the sperm of a closely related species. Pseudogamous plants require pollination, and the endosperm undergoes fertilization even though the diploid embryo develops without it.

pseudogene A sequence of nucleotides in DNA that resembles a gene but has no apparent function. Although most are 'dead', many can be transcribed, and some transcripts have gained a role as non-coding RNAs that play a role in regulating functional genes. Pseudogenes are thought to arise by duplication of an existing gene through unequal crossing-over during meiosis, with, e.g., accompanying loss of the promoter or other flanking regions required for transcription. For example, the α- and β-globin gene clusters in humans contain several pseudogenes. **Processed pseudogenes** lack promoter and intron sequences and seem to have arisen by reverse transcription of a messenger RNA molecule and insertion of the DNA transcript into the chromosomal DNA at a new location, i.e. retrotransposition.

pseudoheart Any one of a series of contractile blood vessels in annelid worms that pump blood from the dorsal vessel to the ventral vessel.

pseudoparenchyma A tissue that superficially resembles plant parenchyma but is made up of an interwoven mass of hyphae (in fungi) or filaments (in algae). Examples of pseudoparenchymatous structures are the fruiting bodies (mushrooms, toadstools, etc.) of certain fungi and the thalli of certain red and brown algae.

pseudoplasmodium See SLIME MOULD.

pseudopodium (*pl.* **pseudopodia**) A temporary outgrowth of the cell of some protists (e.g. *Amoeba*), which serves as a feeding and locomotory organ. Pseudopodia may be blunt or threadlike, form a branching network, or be stiffened with an internal supporting rod. Phagocytic white blood cells also form pseudopodia to engulf invading bacteria. See AMOEBOID MOVEMENT.

pseudopregnancy A state resembling pregnancy that may occur in some mammals (e.g. rabbits and rodents) in which many of the phenomena of pregnancy are present but there is no fetus developing in the uterus. It is caused by an extended dioestrus (*see* OESTROUS CYCLE) in the absence of fertilization.

Psilotum See WHISK FERNS.

psychrophilic Describing an organism that lives and grows optimally at relatively low temperatures, usually below 15°C, and cannot grow above 20°C. Psychrophiles, which consist mainly of bacteria, algae, and fungi, are restricted to permanently cold climates. For example, the ice in polar regions often contains dense masses of algae. Such organisms are likely to be killed by even brief warming. Psychrophiles have various adaptations to enable survival at extremes of cold. These include a high content of unsaturated fatty acids in their plasma membranes, which maintain the membrane in a semifluid state. They also have enzymes that function optimally at relatively low temperatures. **Psychrotolerant** organisms live and grow best at higher temperatures (20–40°C), but are able to tolerate cold conditions. *Compare* MESOPHILIC; THERMOPHILIC.

Pteridophyta In traditional classification systems, a division of the plant kingdom that included ferns, horsetails, and clubmosses, i.e. the non-seed-bearing tracheophytes. In the light of recent molecular genetic evidence, this is no longer recognized (*see* FERNS).

Pteridospermales See CYCADOFILICALES.

pterobranch See HEMICHORDATA.

pterodactyls See PTEROSAURIA.

Pterosauria An extinct order of flying reptiles—the pterodactyls—that lived in the late Triassic, Jurassic, and Cretaceous periods (225–65 million years ago). Pterodactyls had beaked jaws and an elongated fourth finger that supported a membranous wing, with wingspans ranging from 50 cm to 13 m. They had long jointed tails, no feathers, and were capable of powered flapping flight.

Pterygota A subclass of insects that typically possess wings, although some are secondarily wingless. It comprises the *exopterygotes and the *endopterygotes. *Compare* APTERYGOTA.

PTH See PARATHYROID HORMONE.

PTTH *See* PROTHORACICOTROPIC HORMONE.

ptyalin An enzyme that digests carbohydrates (*see* AMYLASE). It is present in mammalian *saliva and is responsible for the initial stages of starch digestion.

puberty *See* ADOLESCENCE.

pubis (*pl.* **pubes**) One of the three bones that make up each half of the *pelvic girdle. It is the most anterior of the three pelvic bones. In mammals and many reptiles the pubes are united at a slightly movable joint, the **pubic symphysis**. *See also* ILIUM; ISCHIUM.

PubMed *See* MEDLINE.

puff (in genetics) *See* POLYTENY.

pufferfish Any fish belonging to the family Tetraodontidae, characterized by a rounded body that can be inflated as a defensive strategy when the animal is threatened by predators. Some species, including the Pacific fugu (*Takifugu rubripes*), contain lethal amounts of the neurotoxin **tetrodotoxin** in their internal organs. In spite of this, fugu is prepared by specially trained chefs and served as a delicacy in Japan and elsewhere. Pufferfish are remarkable for their highly compact genomes compared with other vertebrate species; for example, the fugu genome is about one-ninth the size of the human genome but is predicted to have more genes than humans. In 2002 this species became the second vertebrate species (after humans) to have a *draft sequence of its genome published; it is now used as a reference organism in comparative genomics.

pullulan A water-soluble polysaccharide composed of glucose units that are polymerized in such a way as to make it viscous and impermeable to oxygen. Pullulan is used in adhesives, food packaging, and moulded articles. It is derived from the fungus *Aureobasidium pullulans*.

pulmonary Of or relating to the lungs.

pulmonary artery The artery that conveys deoxygenated blood from the right ventricle of the heart to the lungs, where it receives oxygen.

pulmonary circulation The part of the circulatory system of birds and mammals that transports deoxygenated blood from the right side of the heart to the lungs and returns oxygenated blood to the left side of the heart. *Compare* SYSTEMIC CIRCULATION. *See* DOUBLE CIRCULATION.

pulmonary valve *See* SEMILUNAR VALVE.

pulmonary vein The vein that conveys oxygenated blood from the lungs to the left atrium of the heart.

pulp cavity The central region of a tooth, which is connected by a narrow channel at the tip of the root with the surrounding tissues. The pulp cavity contains the **pulp**—connective tissue in which blood vessels and nerve fibres are embedded, and it is lined with *odontoblasts, which produce the *dentine.

pulse (in physiology) A series of waves of dilation that pass along the arteries, caused by pressure of blood pumped from the heart through contractions of the left ventricle. In humans it can be felt easily where arteries pass close to the skin surface, e.g. at the wrist.

pulsed-field gel electrophoresis (PFGE) *See* GEL ELECTROPHORESIS.

pulvinus (*pl.* **pulvini**) A group of cells at the base of a leaf or leaflet in certain plants that, by rapidly losing water, brings about changes in the position of the leaves. In the sensitive plant (*Mimosa pudica*), the pulvinus is responsible for the folding of the leaves that occurs at nightfall or when the plant is touched or injured. *See* MOTOR CELL.

punctuated equilibrium A hypothesis, published in 1972 by N. Eldredge and Stephen J. Gould, proposing that in evolutionary history most change occurs very rapidly in short bursts lasting typically less than 100 000 years and is associated with speciation events. In between these speciation events are long periods (perhaps millions of years) of relative stasis, in which little evolutionary change occurs. This hypothesis, which contradicted the orthodox Darwinian view of evolution as a gradual and continuous process, prompted controversy and often heated debate. The authors based their hypothesis on studies of various fossil lineages (e.g. ammonite molluscs) in which forms intermediate between species are absent, citing this as evidence that speciation events are often so brief as not to be represented in the fossil record. Subsequent scrutiny of the evidence supports a pattern of punctuated equilibrium for some, but not all, lineages, so it cannot be regarded as universal. For example, the rodent lineage shows as much morphological change between speciation events as during speciation.

Punnett square A representation of all possible genotypes produced in the *F_2 generation of a Mendelian genetic cross; it was devised by R. C. Punnett. All possible male gametes are listed horizontally against a grid and all possible female gametes are listed vertically against the grid. The various combinations of these gametes are then written into the grid. See illustration at *dihybrid cross.

pupa (*pl.* **pupae**) The third stage of development in the life cycle of *endopterygote insects. During the pupal stage locomotion and feeding cease and *metamorphosis from the larva to the adult form takes place. There are three types of pupa. The commonest is the **exarate** or free pupa, in which the wings and other appendages are visible and movable. In the **obtect** type the wings are stuck to the body and immovable, as in the **chrysalis** of a butterfly or moth; and in the **coarctate** type an exarate pupa develops within a hard barrel-shaped **puparium**, as in the housefly and other Diptera.

pupil *See* IRIS.

pupillary reflex An alteration in the size of the pupil in response to a change in the light intensity. An increase in light intensity causes both pupils to contract (*see* IRIS) even if only one retina is directly stimulated. This is known as a *consensual reflex. Conversely, a decrease in light intensity causes dilation of the pupils.

pure line A population of plants or animals all having a particular feature that has been retained unchanged through many generations. The organisms are *homozygous and are said to 'breed true' for the feature concerned.

purifying selection (negative selection) The process whereby harmful genetic variants (*alleles) are eliminated from a population. For example, a new mutation that reduces the fitness of an individual is less likely to be transmitted to successive generations and the mutant trait will be quickly lost. Similarly, a change in environmental conditions may reduce the fitness of an existing trait and subject it to negative selection. *Compare* POSITIVE SELECTION. *See also* NATURAL SELECTION; NEUTRAL THEORY OF MOLECULAR EVOLUTION.

purine An organic nitrogenous base (see formula), sparingly soluble in water, that gives rise to a group of biologically important derivatives, notably *adenine and *guanine, which occur in *nucleotides and nucleic acids (DNA and RNA).

Purine

Purkyne cell (Purkinje cell) A type of nerve cell large numbers of which are found in the cortex of the *cerebellum. Purkyne cells are *multipolar neurons having large bodies from which arise several dendrons with highly branched *dendrites, which form a flat fan extending towards the surface of the cerebellum. They are named after the Czech physiologist J. E. Purkyne (1787–1869).

Purkyne fibres (Purkinje fibres) Modified fibres in the mammalian heart that originate in the *bundle of His and spread out in a network over the ventricles. Action potentials generated in the sinoatrial node (the *pacemaker of the heart) are conducted extremely rapidly through the ventricles, due to the extensive branching of the Purkyne fibres, causing both ventricles to contract almost simultaneously.

purple sulphur bacteria *See* SULPHUR BACTERIA.

putrefaction The microbial decomposition of organic matter, especially the anaerobic breakdown of proteinaceous material with the production of foul-smelling amines.

Pycnogonida *See* CHELICERATA.

pyknotic Describing a nucleus of a damaged cell that has decreased in volume and become darker due to some degree of condensation of the nuclear *chromatin. *See also* KARYORRHEXIS.

pyloric sphincter *See* SPHINCTER; STOMACH.

pyramidal cell A type of neuron found in the cerebral cortex of the brain. Pyramidal cells have a pyramid-shaped cell body and its dendrites extend from both the body and the axon.

pyramid of biomass A diagrammatic representation of the amount of organic material (*see* BIOMASS), measured in grams of dry mass per square metre (g m^{-2}), found in a particular habitat at ascending *trophic levels of a *food chain. Biomass decreases at each ascending level of the food chain. A pyramid of biomass is a more accurate representation of the flow of energy through a food chain than a *pyramid of numbers, but seasonal variations in the rate of turnover of the organisms at a particular level may result in higher or lower values for the amount of biomass sampled at a particular time than the average amount over the whole year. The best representation of energy flow in a food chain is a *pyramid of energy.

pyramid of energy A diagrammatic representation of the amount of energy, measured in kilojoules per square metre

Pyramid of energy

per year (kJ m^{-2} yr^{-1}), available at ascending *trophic levels of a *food chain in a particular habitat (see illustration). A pyramid of energy is the most accurate representation of the *energy flow through a food chain as it indicates how much energy is lost at each trophic level (through respiration, etc.). *See also* BOMB CALORIMETER. *Compare* PYRAMID OF BIOMASS; PYRAMID OF NUMBERS.

pyramid of numbers A diagramatic representation of the numbers of animals found in an area at ascending *trophic levels of a *food chain (see illustration). Because only a small proportion of the energy taken in by an organism is converted to tissue and is thus available to consumers at the next trophic level, the number of organisms that can be

Pyramid of numbers for a woodland food chain

supported at each level is generally much less than the number at the level that supplies its food (i.e. the level below). *See also* PYRAMID OF BIOMASS; PYRAMID OF ENERGY.

pyranose A *sugar having a six-membered ring containing five carbon atoms and one oxygen atom.

pyrenocarp *See* DRUPE.

pyrenoid A spherical protein body found in the *chloroplasts of many algae and the hornworts (*see* ANTHOCEROPHYTA). Pyrenoids are associated with the storage of starch: layers of starch are often found around them.

pyrethrum 1. Any of several plants of the genus Chrysanthemum, especially *C. cinerariifolium*, grown for its flowers and seed cases, which contain natural insecticidal compounds (**pyrethrins**). Various synthetic insecticides—the **pyrethroids**—are chemically similar to pyrethrins. Examples include permethrin and cypermethrin. The pyrethrins penetrate the insect's cuticle and are fast-acting, nontoxic to many animals (although fish are highly susceptible) and to plants, and readily biodegradable. **2.** An insecticidal preparation containing natural pyrethrins.

pyridoxine *See* VITAMIN B COMPLEX.

pyrimidine An organic nitrogenous base, sparingly soluble in water, that gives rise to a group of biologically important

Pyrimidine

derivatives, notably *uracil, *thymine, and *cytosine, which occur in *nucleotides and nucleic acids (DNA and RNA).

pyrrole An organic nitrogen-containing compound that forms part of the structure of *porphyrins.

Pyrrole

pyruvic acid (2-oxopropanoic acid) A colourless liquid organic acid, $CH_3COCOOH$. Pyruvate is an important intermediate compound in metabolism, being produced during *glycolysis and converted to acetyl coenzyme A, required for the *Krebs cycle. Under anaerobic conditions it is converted to lactate or ethanol *see* FERMENTATION.

PYY *See* PEPTIDE YY.

Q₁₀ *See* TEMPERATURE SENSITIVITY.

QTL *See* QUANTITATIVE TRAIT LOCUS.

quadrat An ecological sampling unit consisting of a small square area of ground within which all species of interest are noted or measurements taken. Quadrats may be spaced over a larger area to form an overall view when a total survey would be impracticable, or they may be used to sample along a *transect.

quadrate A paired bone in the upper jaw of bony fishes, amphibians, reptiles, and birds that articulates with the lower jawbone. It is absent in mammals, being reduced to a small bone (the incus) in the middle ear (*see* EAR OSSICLES).

quadriceps A group of *extensor muscles that cover the anterior surface and sides of the mammalian femur. These muscles join at the base and are connected to the tibia by a single tendon. *Muscle spindles in the quadriceps are responsible for the knee-jerk reflex (*see* STRETCH REFLEX).

qualitative variation *See* DISCONTINUOUS VARIATION.

quantitative inheritance (multifactorial inheritance; polygenic Inheritance) The determination of a particular characteristic, e.g. height or skin colour, by many genes (*quantitative trait loci), each having a small effect individually. Characteristics controlled in this way show *continuous variation.

quantitative structure-activity relationship (QSAR) A statistical algorithm that quantitatively defines the relationship between the chemical structure of a drug and its effect on an organism. QSAR studies are often used to predict the activity or toxicity of new drugs. Similar methods can be used to predict the metabolism of new drugs (**quantitative structure-metabolism relationships; QSMR**).

quantitative trait Any phenotypic trait that shows *continuous variation and can be measured quantitatively in terms of length, weight, concentration, test score, etc. Such traits, which include height, intelligence, and obesity, are determined by the cumulative effects of numerous genes at *quantitative trait loci.

quantitative trait locus (QTL) A region of a chromosome (genetic locus) containing one or more genes that contribute to determining a *quantitative trait. Typically, several or many such genes (polygenes), each having a small effect, act together to specify the trait.

quantitative variation *See* CONTINUOUS VARIATION.

quantum biology The study of biological phenomena in terms of quantum mechanics. Hypotheses about the role of quantum behaviour of subatomic particles in living processes date from the 1930s, but only in recent years has evidence emerged to substantiate them. For example, the dual wave-particle nature of excited electrons, or excitons, is now held to explain the high efficiency with which light energy is 'trapped' during photosynthesis in green plants, algae, and photosynthetic bacteria. The excitons are generated by photons of light being absorbed by chlorophyll molecules in the light-harvesting complex, or antenna pigments, of the photosynthetic apparatus of the cell (*see* PHOTOSYSTEMS I AND II). They rapidly transfer between multiple pigment molecules to reach the reaction centre, where the energy is absorbed to drive the chemical reactions of photosynthesis. The wavelike nature of an exciton enables it to simultaneously 'explore' various alternative routes to the reaction centre via the pigment molecules, and hence find the quickest pathway. Another phenomenon, called quantum tunnelling, accounts for the dramatic way in which enzymes can boost the rate of chemical reactions. Lightweight atoms, such as hydrogen, can transfer across seemingly insurmountable energy barriers by virtue of their wavelike nature, and effectively re-emerge instantaneously in another location. In this way the enzyme molecule is able to manipulate carbon–hydrogen bonds in ways that are energetically highly unfavourable. Quantum tunnelling is also proposed as a mechanism underlying the ability of olfactory receptors to detect very subtle differences in odorant molecules. Research suggests that quantum mechanical effects may occur in diverse other areas of biology, including bird navigation.

quantum hypothesis (in neurophysiology) The hypothesis, first formulated by Bernard Katz (1911–2003) and colleagues in the 1950s, that nerve impulses are transmitted across junctions (*synapses) in the nervous system by the release of discrete packets, or quanta, of neurotransmitter. It has since been verified that each quantum represents the contents of a single synaptic vesicle. The size of the vesicle, and hence the number of molecules of neurotransmitter it contains (i.e. the **quantum size**), is characteristic for any nerve cell terminal, although the number of vesicles discharged (i.e. the **quantum content**) can vary considerably. For example, at the *neuromuscular junction of a frog, each vesicle contains about 7000 molecules of the neurotransmitter acetylcholine, and discharge of about 200 such vesicles is needed to generate the end-plate potential that triggers contraction of the muscle.

quarantine A period of isolation imposed on an animal that moves from an area where particular diseases are prevalent to an area where those diseases are not prevalent. In the UK domestic animals or livestock entering the country are quarantined for the necessary incubation period (*see* INFECTION) in order to prevent the spread of a particular disease.

Quaternary The second and current period of the Cenozoic era, extending from about 2.58 million years ago, following the *Neogene period, to the present. The beginning of the Quaternary was based on the onset of a worldwide cooling. During the period four principal glacial phases occurred in Europe and North America, in which ice advanced towards the equator, separated by interglacials during which conditions became warmer and the ice sheets and glaciers retreated. The last glacial ended about 10 000 years ago.

quaternary structure *See* PROTEIN.

queen substance A *pheromone, *trans*-9-keto-2-decenoic acid, that is secreted by the mandibular salivary glands of a queen honeybee and inhibits the development of ovaries in the worker bees in the colony. The queen substance also affects the behaviour of the workers, preventing them from maintaining special brood cells for larvae that develop into new queens, and is a sexual attractant for drones during the queen's mating flights. As the queen ages, the secretion of queen substance diminishes: the workers then construct brood cells for future queens, which they feed exclusively with *royal jelly.

quinone Any of various compounds derived from benzene and containing C=O groups in an unsaturated ring. An example is *plastoquinone.

quorum sensing A density-dependent mechanism for communication and coordination among populations of bacteria in which the individual bacteria secrete and detect substances called **autoinducers**. These accumulate in the environment depending on the density of the population. Upon reaching a certain threshold concentration, the autoinducer triggers a specific response. Such mechanisms can regulate a wide range of bacterial activities, including bioluminescence, sporulation, virulence, and *biofilm formation. Some autoinducers are used for intraspecific communication, whereas others affect the behaviour of other species.

race 1. (in biology) A category used in the *classification of organisms that consists of a group of individuals within a species that are geographically, ecologically, physiologically, or chromosomally distinct from other members of the species. The term is frequently used in the same sense as *subspecies. **Physiological races**, for example, are identical in appearance but differ in function. They include strains of fungi adapted to infect different varieties of the same crop species. **2.** (in anthropology) A distinct human type possessing several characteristics that are genetically inherited. Traditionally, anthropologists used the concept of biological races such as Mongolian, Caucasian, and Ethiopian. However, this is now deprecated by many, who regard 'race' as reflecting primarily cultural and social characteristics rather than being rooted in genetic differences. Supporting this view is the finding that genetic variation within so-called 'racial' groups far outweighs the variation between groups.

raceme A type of *racemose inflorescence in which the main flower stalk is elongated and bears stalked flowers. An example is the lupin. *See also* PANICLE.

racemose inflorescence (indefinite inflorescence) A type of flowering shoot (*see* INFLORESCENCE) in which the growing region at the tip of the flower stalk continues to produce new flower buds during growth. As a result, the youngest flowers are at the top and the oldest flowers are at the base of the stalk. In a flattened inflorescence, the youngest flowers are in the centre and the oldest flowers are on the outside. Types of racemose inflorescence include the *capitulum, *catkin, *corymb, *raceme,

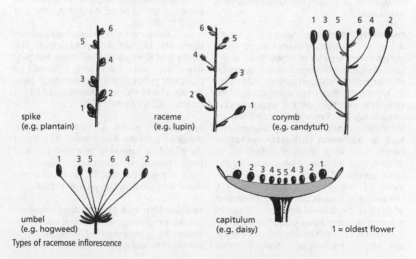

spike
(e.g. plantain)

raceme
(e.g. lupin)

corymb
(e.g. candytuft)

umbel
(e.g. hogweed)

capitulum
(e.g. daisy)

1 = oldest flower

Types of racemose inflorescence

*spadix, *spike, and *umbel (see illustration). *Compare* CYMOSE INFLORESCENCE.

rachis (rhachis) (*pl.* **rachides**) **1.** The main axis of a compound leaf or an inflorescence. **2.** The shaft of a *feather. **3.** The backbone.

rad *See* RADIATION UNITS.

radial symmetry The arrangement of parts in an organ or organism such that cutting through the centre of the structure in any direction produces two halves that are mirror images of each other. The stems and roots of plants usually show radial symmetry, while all animals belonging to the Cnidaria (e.g. jellyfish) and Echinodermata (e.g. starfish) are radially symmetrical—and typically sessile—in their adult form. The term **actinomorphy** is used to describe radial symmetry in flowers (e.g. a buttercup flower). *Compare* BILATERAL SYMMETRY.

radiation 1. (in physics) Energy travelling in the form of electromagnetic waves or photons. *See* ELECTROMAGNETIC SPECTRUM. **2.** (in physics) A stream of particles, especially alpha- or beta-particles from a radioactive source or neutrons from a nuclear reactor. *See also* BACKGROUND RADIATION. **3.** (in evolution) *See* ADAPTIVE RADIATION.

radiation damage Harmful changes that occur to inanimate materials and living organisms as a result of exposure to energetic electrons, nucleons, fission fragments, or high-energy electromagnetic radiation. In organisms this can cause changes to cells that alter their genetic structure (i.e. mutations), interfere with their division, or kill them. In humans, these changes can lead to **radiation sickness**, **radiation burns** (from large doses of radiation), or to long-term damage of several kinds, the most serious of which result in various forms of cancer (especially leukaemia). Ultraviolet radiation can cause distortion of DNA in the form of covalent linkage of adjacent thymine bases in one of the DNA strands. Unless corrected by the cell's *excision repair mechanism, these thymine

dimers interfere with DNA replication and may result in cancer.

radiation units Units of measurement used to express the activity of a radionuclide and the *dose of ionizing radiation.

The becquerel (Bq), the SI unit of activity, is the activity of a radionuclide decaying at a rate, on average, of one spontaneous nuclear transition per second. Thus 1 Bq = 1 s^{-1}. The former unit, the curie (Ci), is equal to $3.7 \times M\ 10^{10}$ Bq. The curie was originally chosen to approximate the activity of 1 gram of radium–226.

The gray (Gy), the SI unit of absorbed dose, is the absorbed dose when the energy per unit mass imparted to matter by ionizing radiation is 1 joule per kilogram. The former unit, the rad (rd), is equal to 10^{-2} Gy.

The sievert (Sv), the SI unit of dose equivalent, is the dose equivalent when the absorbed dose of ionizing radiation multiplied by the stipulated dimensionless factors is 1 J kg^{-1}. As different types of radiation cause different effects in biological tissue a weighted absorbed dose, called the **dose equivalent**, is used in which the absorbed dose is modified by multiplying it by dimensionless factors stipulated by the International Commission on Radiological Protection. The former unit of dose equivalent, the rem (originally an acronym for *r*oentgen *e*quivalent *m*an), is equal to 10^{-2} Sv.

In SI units, exposure to ionizing radiation is expressed in coulombs per kilogram, the quantity of X- or gamma-radiation that produces ion pairs carrying 1 coulomb of charge of either sign in 1 kilogram of pure dry air. The former unit, the roentgen (R), is equal to $2.58 \times M\ 10^{-4}$ C kg^{-1}.

radicle The part of a plant embryo that develops into the root system. The tip of the radicle is protected by a root cap and points towards the micropyle. On germination it breaks through the testa and grows down into the soil. *Compare* PLUMULE.

radioactive age The age of an archaeological or geological specimen as determined by a process that depends on a radioactive decay. *See* CARBON DATING;

FISSION-TRACK DATING; POTASSIUM-ARGON DATING; RUBIDIUM-STRONTIUM DATING; URANIUM-LEAD DATING.

radioactive tracing See LABELLING.

radioactive waste (nuclear waste) Any solid, liquid, or gaseous waste material that contains radionuclides (radioactive atomic nuclei). These wastes are produced in the mining and processing of radioactive ores, the normal running of nuclear power stations and other reactors, the manufacture of nuclear weapons, and in hospitals and research laboratories. Because high-level radioactive wastes can be extremely dangerous to all living matter and because they may contain radionuclides having half-lives of many thousands of years, their disposal has to be controlled with great stringency.

radioactivity The spontaneous disintegration of certain atomic nuclei accompanied by the emission of alpha-particles (helium nuclei), beta-particles (electrons or positrons), or gamma radiation (short-wavelength electromagnetic waves). **Natural radioactivity** is the result of the spontaneous disintegration of naturally occurring radioisotopes. The rate of disintegration is uninfluenced by chemical changes or any normal changes in their environment. However, radioactivity can be induced in many nuclides by bombarding them with neutrons or other particles. See also RADIATION UNITS.

radiobiology The branch of biology concerned with the effects of radioactive substances on living organisms and the use of radioactive tracers to study metabolic processes (see LABELLING).

((())) SEE WEB LINKS

• A tutorial providing an introduction to radiobiology from the University of Virginia Health Sciences Center

radiocarbon dating See CARBON DATING.

radiography The process or technique of producing images of an opaque object, traditionally on photographic film or on a fluorescent screen, by means of radiation (either particles or electromagnetic waves of short wavelength, such as X-rays and gamma-rays). The data are now generally converted into images (**radiographs**) by a computer, using electronic detectors such as flat panel screens to produce digital data for analysis, viewing, and storage. *Tomography employs numerous X-ray measurements processed by computer to build up a detailed picture of particular 'slices' of the body. Fluoroscopy is used to produce real-time moving images of internal organs, such as the heart; it is so named because traditionally it used a fluorescent screen to convert the X-rays into a visible image. See also AUTORADIOGRAPHY.

radioimmunoassay (RIA) A form of *immunoassay that uses a radioactively labelled antibody to bind to the target molecule.

radioisotope (radioactive isotope) An isotope of an element that is radioactive. See LABELLING.

radiolarian See RHIZARIA.

radiology The study and use of X-rays, radioactive materials, and other ionizing radiations for medical purposes, especially for diagnosis (**diagnostic radiology**) and the treatment of cancer and allied diseases (**therapeutic radiology** or **radiotherapy**).

radiometric dating (radioactive dating) See DATING TECHNIQUES; RADIOACTIVE AGE.

radiopaque (radio-opaque) Describing a medium that is opaque to X-rays and gamma rays. Examples are barium salts, used in diagnostic radiology of the digestive tract.

radiotherapy See RADIOLOGY.

radius (*pl.* **radiuses** or **radii**) (in anatomy) The smaller of the two bones in the lower section of the forelimb of a tetrapod vertebrate (*compare* ULNA). The radius articulates with some carpal bones and the ulna at the wrist and with the *humerus at the elbow. This sophisticated articulation of the radius enables humans (and some other animals) to twist the forearm (*see* PRONATION; SUPINATION).

radula (*pl.* **radulae**) A tonguelike organ of molluscs, consisting of a horny strip whose surface is studded with rows of horny teeth for rasping food. In some species it is modified for scraping or boring.

rainforest Any major terrestrial *biome in which trees form the dominant plants and annual rainfall is high (over 200 cm). Tropical rainforest is restricted to equatorial regions, such as the Amazon basin, central west Africa, and Southeast Asia. It is dominated by broadleaved evergreens and shows a very rich species diversity (*see* BIODIVERSITY). The leafy crowns of the trees typically form three layers of canopy, since the trees grow to different heights, which prevents much sunlight from reaching ground level. This limits the number of herbaceous plants and small shrubs that grow on the forest floor, but *epiphytes, vines, and creepers are abundant. The average temperature is about 27°C, which—together with the high humidity—encourages rapid decomposition of leaf litter, releasing minerals that replace those leached from the soil by the heavy rain. If the forest canopy is removed, the soil is destroyed rapidly due to leaching by rain. The soil of a rainforest, known as **latosol**, is acidic and typically red, due to the oxidation of iron oxide (Fe_2O_3) in the topsoil. Rainforests are thought to contain many undiscovered plant species that could be of benefit in the fields of medicine and biotechnology. The continued destruction of rainforest in many parts of the world, particularly in South America and Southeast Asia (*see* DEFORESTATION), will not only result in the loss of these and other species but also contributes to the *greenhouse effect and climate change.

r.a.m. *See* RELATIVE ATOMIC MASS.

Ramapithecus *See* SIVAPITHECUS.

ramus communicans (*pl.* **rami communicantes**) A structure containing nerve fibres that runs from a spinal nerve to a *ganglion of the sympathetic nervous system (*see* AUTONOMIC NERVOUS SYSTEM). The sympathetic ganglia form a series down each side of the vertebral column; each is connected to the ventral root of a spinal nerve by a pair of rami communicantes.

randomly amplified polymorphic DNA (RAPD) A DNA primer, made up of a random sequence of bases, used with the *polymerase chain reaction (PCR) to amplify segments of an organism's DNA. The primer effectively 'selects' all regions of DNA that happen to lie within inverted copies of the primer sequence. These regions, which will vary in length, are thus amplified by PCR and produce a series of bands when separated by electrophoresis. Because of variation in the sites of the primer sequences among individuals, the pattern of bands represents a genetic 'fingerprint' unique to each individual. This makes RAPD (pronounced 'rapid') PCR a useful technique for taxonomic studies, assessing kinship, or forensic investigation.

random sampling *See* SAMPLING.

rank (category) The position or status of a *taxon in a *classification hierarchy. Examples of ranks are the class, order, family, genus, and species.

RAPD *See* RANDOMLY AMPLIFIED POLYMORPHIC DNA.

raphe 1. A ridge on a seed marking the line of fusion between an anatropous (inverted) ovule and the *funicle. **2.** A groove on the valve of certain diatoms. **3.** Any other groove, ridge, or suture line marking the line of fusion between two originally separate structures.

raphide A needle-shaped crystal of insoluble calcium oxalate found in certain plants as a means of defence against browsing animals. The raphides occur in specialized cells (idioblasts) and in some cases may be forcefully expelled when the cell is disrupted, causing irritation to the mouth and digestive tract of the consumer.

rapid eye movement (REM) *See* SLEEP.

RAS protein A small *G protein that functions in cells as a molecular switch in signal transduction pathways regulating cell growth and other activities. In its active state it binds the nucleotide guanosine triphosphate (GTP), but because RAS has intrinsic GTPase activity, the GTP is rapidly converted to GDP, thereby returning RAS to its inactive state. Activation of RAS is mediated by a **guanine-nucleotide exchange factor** (GEF). When a surface receptor is activated, GEF transiently displaces GDP from RAS, enabling it to bind GTP. RAS exerts its effects by activating a cascade of sequentially activating protein kinases, including *MAP kinase, which ultimately regulates gene expression. RAS is encoded by the proto-oncogene *RAS* in humans, which is homologous to the rodent oncogene *ras* (named after *rat* sarcoma, and also found in a retrovirus). A gene mutation involving a single base change $(G \rightarrow T)$ is sufficient to transform the normal protein into an oncoprotein, which lacks the intrinsic GTPase activity and is permanently activated. It is associated with many human tumours.

rate-limiting step The slowest step in a metabolic pathway or series of chemical reactions, which determines the overall rate of the other reactions in the pathway. In an enzymatic reaction, the rate-limiting step is generally the stage that requires the greatest *activation energy or the transition state of highest *free energy.

Ratitae (Palaeognathae) A group comprising the flightless birds, including the ostrich, kiwi, cassowary, rhea, and emu, plus the extinct moa and elephant bird. They are typically large-bodied, have long legs, heavy bones and small wings, and lack a keel to the sternum. It is thought that these birds originated in the northern hemisphere during the Cretaceous period, between 115 and 105 million years ago, and their ancestors subsequently spread into the southern hemisphere. Following the mass extinction at the end of the Cretaceous, they diversified and independently lost the ability to fly; all except the kiwi then assumed the large body size. The partridge-like flighted tinamous of South America are also now regarded as ratites.

ray 1. (in optics) A narrow beam of radiation. **2.** (in botany) *See* MEDULLARY RAY. **3.** (in zoology) **a.** A bony or horny skeletal element in the fin of a bony fish (class *Osteichthyes). **b.** Any of an order (Rajiformes) of cartilaginous fishes with flattened bodies.

reabsorption (in excretion) *See* SELECTIVE REABSORPTION.

reactant *See* CHEMICAL REACTION.

reaction *See* CHEMICAL REACTION.

reaction centre *See* PHOTOSYSTEMS I AND II.

reaction norm The pattern of expression of a given phenotype across a range of values of some environmental variable. An example is the variation in yield of wheat with varying levels of nitrogen fertilizer applied to the soil. An accurate picture of a reaction norm is only attainable when using organisms of the same genotype, to exclude the influence of genetic variation. This can be accomplished in horticultural research, for example, by using clones of genetically identical plants. The reaction norms of different clones can then be compared. A typical pattern is a series of overlapping bell-shaped curves when plotting a graph of, say, grain yield versus nitrogen level. In natural populations, which are generally heterogeneous genetically, the interactions between genotype and environment are complex and make reaction norms much harder to discern. *Compare* PHENOTYPIC PLASTICITY.

reaction time (latent period) The period of time between the detection of a stimulus at a sensory receptor and the performance of the appropriate response by the effector organ. This delay is caused by the time taken for the impulse to travel across the synapses of adjacent neurons. The reaction time for a *reflex response, involving only a single linking synapse, is very short.

reactive oxygen species (ROS) Any of various chemical species that contain highly active oxygen, particularly the *free radical superoxide anion (O_2^-) and its derivatives, including hydrogen peroxide (H_2O_2), singlet oxygen (a metastable high-energy form of molecular oxygen), hydroxyl radical (OH·), and hypohalite ions (e.g. hypochlorite, OCl^-). Superoxide is produced normally as a by-product of aerobic respiration inside mitochondria; both it and other ROS are potentially harmful to living cells because of their highly reactive nature. They can damage DNA and other cell components and are implicated in various diseases (e.g. cancer, heart disease), hence cells have an array of mechanisms to remove them, including *superoxide dismutase and various *antioxidants. However, certain cells and tissues in both plants and animals use a process called the *oxidative burst to produce ROS in order to destroy invading pathogenic organisms, such as fungi and bacteria. *See also* PEROXISOME.

reading frame The 'reading' of bases in messenger RNA (or deduced from DNA) according to their correct triplet combinations for each coding unit (*codon) of the genetic code. Hence, the reading frame is established according to precisely where translation starts. For example, if translation starts one base either side of the correct base, an entirely different sequence of codons will be read, resulting in a faulty polypeptide or none at all. The hallmark of a functional gene is that it is transcribed to produce an **open reading frame** (ORF), consisting of a *start codon to pinpoint exactly where translation should start, a *stop codon to signal termination of translation, and typically a long sequence of codons that specify the constituent amino acids of the polypeptide (as well as *introns in most eukaryote genes).

readthrough The continuation of transcription of DNA beyond a normal stop signal, or terminator sequence, due to failure of RNA polymerase to recognize the signal. Readthrough can also occur in translation, when a mutation has converted a normal stop codon into one encoding an amino acid. This results in extension of the polypeptide chain until the next stop codon is reached, producing a so-called **readthrough protein**.

recapitulation The theory that some stages of evolution are repeated in the development of an individual organism, i.e. that *phylogeny is repeated in *ontogeny. It was proposed by the German biologist Ernst Haeckel (1834–1919). *See also* ACCELERATION.

Recent *See* HOLOCENE.

receptacle 1. (thalamus *or* **torus)** The tip of a flower stalk, which bears the petals, sepals, stamens, and carpels. The way the receptacle develops determines the position of the flower parts. It can be dilated and dome-shaped, saucer-shaped, or hollow and enclosing the gynoecium (*see* EPIGYNY; HYPOGYNY; PERIGYNY). In some plants it may become part of the fruit (*see* PSEUDOCARP). **2.** A swollen part of the thallus of some algae, e.g. *Fucus*, that bears the conceptacles in which the sex organs are situated.

receptive field A defined spatial distribution of stimuli that influences the activity of a particular neuron. For example, a sensory neuron relaying signals from a touch receptor will respond to stimuli affecting only the limited region of skin served by the receptor; similarly, in the eye, a given ganglion cell transmits signals to the brain exclusively from the photoreceptors to which it connects, within a defined area of the retina. Stimuli within the receptive field may either increase or decrease activity in the neuron, depending on the type of neuron and its location. Recording the activity of, say, a neuron in the striate cortex of the brain reveals how it responds selectively to spots of light or shapes sensed by the retina: stimuli in one part of the receptive field (the 'on' part) will increase the rate of firing of nerve impulses, whereas stimuli received in another part of the field (the 'off' part) will reduce the rate of firing. Auditory neurons in the brain also have their individual receptive fields, corresponding to sounds coming from a particular direction. Direction is determined by the differences (e.g. in timing or intensity) in the same sound arriving at each ear.

receptor 1. A cell or group of cells specialized to detect a particular stimulus and to initiate the transmission of impulses via the sensory nerves. The eyes, ears, nose, skin, and other sense organs all contain specific receptors responding to external stimuli (*see* EXTEROCEPTOR); other receptors are sensitive to changes within the body (*see* INTEROCEPTOR). *See also* BARORECEPTOR; CHEMORECEPTOR; ELECTRORECEPTOR; MAGNETORECEPTOR; MECHANORECEPTOR; OSMORECEPTOR; PROPRIOCEPTOR; THERMORECEPTOR. **2.** A protein that can bind with a specific ligand (i.e. a hormone, neurotransmitter, drug, or other chemical), thereby initiating a change within the cell. Some receptor proteins occur in the plasma membrane, typically spanning the membrane and having outward- and inward-facing sites. These bind large or polar ligands that cannot cross the plasma membrane. Other receptors are intracellular, located in the cytoplasm or inside the nucleus. They bind small or nonpolar ligands that are able to diffuse across the plasma membrane. Many intracellular receptors are transcription factors; when activated they move into the nucleus to regulate the expression of specific genes. Functionally, receptors fall into two main categories: *ionotropic receptors and *metabotropic receptors.

receptor-mediated endocytosis *See* ENDOCYTOSIS.

receptor potential *See* GRADED POTENTIAL.

recessive The *allele that is not expressed in the *phenotype when two different alleles are present in the cells of an organism. The aspect of a characteristic controlled by a recessive allele only appears when two such alleles are present, i.e. in the **double recessive** condition. *Compare* DOMINANT.

recipient An individual who receives tissues or organs of the body from another (the *donor).

reciprocal cross A *cross reversing the roles of males and females to confirm the results obtained from an earlier cross. For example, if the pollen (male) from tall plants is transferred to the stigmas (female) of dwarf plants in one cross, the reciprocal cross would use the pollen of dwarf plants to pollinate the stigmas of tall plants.

recognition sequence *See* RESTRICTION ENZYME.

recombinant DNA DNA that contains genes from different sources, especially when combined by the techniques of *genetic engineering rather than by breeding experiments. Genetic engineering is therefore also known as **recombinant DNA technology**. Recombinant DNA is formed during *gene cloning or in the creation of *genetically modified organisms.

recombination The rearrangement of segments of genes or other DNA sequences. It occurs when reproductive cells (gametes) are formed, consequent to the *independent assortment of parental sets of chromosomes and exchange of chromosomal material (*see* CROSSING OVER) that occur during *meiosis. This results in offspring that have a combination of characteristics different from that of their parents. Recombination is also crucial to generating the huge diversity of receptors found on cells of the immune system and their corresponding antibodies. Hence during B-cell development a recombinase enzyme splices together random segments from a large gene cluster to create functional *immunoglobulin genes for the light and heavy chains of the B-cell receptors. Because of the vast number of possible combinations, each progenitor cell and its clone of daughter cells is likely to have a unique antigen specificity. A similar mechanism accounts for T-cell diversity. Recombination can also be induced artificially by *genetic engineering techniques.

recombinational repair A form of *DNA repair that fills large gaps in a new DNA strand resulting from damage due to, e.g., UV irradiation. In *E. coli*, in which this repair process is best understood, thymine dimers in a damaged template strand cause the DNA polymerase to skip several hundred bases, leaving a large gap in the newly synthesized daughter strand and the template strand unpaired in this region. The repair is performed primarily by the action of an enzyme, the RecA protein (encoded by the *recA* locus), which fills in the gap by appropriating a complementary segment of DNA from the other, undamaged, newly synthesized duplex (double-stranded) DNA. The enzyme coats the unpaired template strand region and insinuates it into the other DNA duplex. Here it pairs with the complementary region of its original sister strand (the other template strand), displacing the newly formed strand. The two ends of this paired region are nicked by an endonuclease enzyme, hence freeing the original damaged region with its complementary 'filler' strand. The strand from which the 'filler' has been taken is then itself filled in by DNA polymerase I. Although the thymine dimer remains in the original damaged strand, it can be excised at a later stage. Most importantly, the integrity of the DNA is maintained. Many of the enzymes and processes of postreplicative repair are also involved in *recombination. See also* POSTREPLICATIVE REPAIR.

recon The smallest unit of DNA in which crossing over and recombination can occur. A recon consists of two pairs of nucleotides.

reconciliation ecology The branch of ecology concerned with designing new wildlife habitats and enhancing existing ones in areas that are already exploited by humans for e.g. agriculture, housing, industry, or mining. The aim is to promote biodiversity alongside human land use—in other words, to reconcile the two. Such a strategy can create greater connectivity between remaining areas of pristine habitat, e.g. by establishing wildlife corridors, besides increasing the value of local *ecosystem services. Reconciliation ecology is multifaceted and tailored to the requirements of each particular area.

(⊕) SEE WEB LINKS

• Living landscapes projects of UK wildlife trusts

recruitment 1. (in neurophysiology) The activation of extra motor neurons in order to bring about an increased response to a stimulus that is present at an even intensity. **2.** (in ecology) The attracting of beneficial organisms, e.g. predatory animals, by plants as a means of defending the plant against attack from pests. Such an attack may, for example, trigger the release of volatile chemicals that attract the attention of beneficial parasitic wasps; these may then mount a counterattack against caterpillars eating the leaves of the plant.

rectal gland A gland found near the rectum of elasmobranch fishes that removes excess sodium chloride (NaCl) from the blood. It consists of numerous blind-ending tubules surrounded by blood capillaries; the tubules drain into a duct that opens into the intestine. *Sodium-potassium pumps in the cells lining the tubules create a concentration gradient of Na^+ ions across the serosal (blood) surface of the tubules, which in turn drives a cotransport system that moves chloride ions (Cl^-) from the blood into the tubule cells. The resulting raised intracellular concentration of Cl^- ions causes them to diffuse from the tubule cells into the duct lumen, and the consequent electrochemical gradient causes Na^+ ions to follow between the cells. Water also diffuses passively into the lumen, creating a solution that has the same osmolarity as the blood but containing a much higher concentration of NaCl. *See also* SALT GLAND.

rectum (*pl.* **rectums** or **recta**) The muscular portion of the *alimentary canal between the *colon and the *anus. Its main function is the storage of *faeces and their elimination via the anus.

recycling 1. The recovery and processing of materials after they have been used, which enables them to be reused. For example, used paper, cans, and glass can be broken down into their constituents, which form the raw materials for the manufacture of new products. **2.** The continual movement of essential elements between the biotic (living) and abiotic (nonliving) components of the environment. *See* CARBON CYCLE; NITROGEN CYCLE; OXYGEN CYCLE; PHOSPHORUS CYCLE; SULPHUR CYCLE.

red algae *See* RHODOPHYTA.

red blood cell *See* ERYTHROCYTE.

red fluorescent protein (RFP) Any of several proteins that emit red fluorescent light and are used as markers in cell biology and cell imaging microscopy (*see* FLUORESCENT PROTEIN). The first to be widely used, termed DsRed, was derived from the coral animal *Discosoma striata*. Following excitation it takes several hours to emit red light, after going through a green phase, which can cause confusion if *green fluorescent protein is being used at the same time. Subsequent mutated forms of DsRed, such as DsRed-Express, have improved brightness and given more rapid maturation of the red fluorescence. Another commercially available RFP is HcRed, derived from the coral *Heteractis crispa*.

redox *See* OXIDATION-REDUCTION.

Red Queen hypothesis A hypothesis, proposed by L. M. Van Valen in the early 1970s, that describes how the *coevolution of competing species creates a dynamic equilibrium, in which the probability of extinction remains fairly constant over time. Hence, evolution is seen neither as 'progressive'—with a species' chances of survival improving over time—nor as 'escalatory'—with increasing vulnerability to extinction over time. Instead, as one species evolves improvements that make it more competitive, its competitors experience selection pressures that force them to evolve in order to keep pace with it. Ones that lag too far behind will become extinct. The hypothesis is named after the remark made by the Red Queen in Lewis Carroll's *Through the Looking Glass*: "Here, you see, it takes all the running you can do, to keep in the same place."

red tide A sudden often toxic proliferation of marine phytoplankton, notably dinomastigotes, that colours the sea red, brown, or yellowish due to the high concentration of the organisms' photosynthetic accessory pigments. Some dinomastigotes, such as *Gonyaulax*, produce potent toxins, which may kill fish and invertebrates outright or accumulate in the food chain, posing a hazard to humans eating shellfish and other seafood. These phytoplankton blooms may be related to nutrient-rich inputs from the land, or upwelling oceanic waters, and are initiated by the activation of cystlike forms lying on the seabed. *See also* ALGAL BLOOM.

reducing sugar A monosaccharide or disaccharide sugar that can donate electrons to other molecules and can therefore act as a reducing agent. The possession of a free ketone (–CO–) or aldehyde (–CHO) group enables most monosaccharides and disaccharides to act as reducing sugars. Reducing sugars can be detected by *Benedict's test. *Compare* NONREDUCING SUGAR.

reduction *See* OXIDATION-REDUCTION.

reduction division *See* MEIOSIS.

reflex An automatic and innate response to a particular stimulus. A reflex response is extremely rapid. This is because it is mediated by a simple nervous circuit called a **reflex arc**, which at its simplest involves only a receptor linked to a sensory neuron, which synapses with a motor neuron (supplying the effector) in the spinal cord, without the sensory and motor signals having to travel to and from the brain. Such reflexes are known as **monosynaptic spinal reflexes**; an example is the *stretch reflex. Other spinal reflexes involve more than one synapse (*see* POLYSYNAPTIC REFLEX); an example is the **withdrawal reflex** of the hand from a painful stimulus (such as fire). **Cranial reflexes** are mediated by pathways in the cranial nerves and brain; examples are the blinking and swallowing reflexes. *See also* CONDITIONING.

reflex action An automatic movement produced in response to a stimulus (*see* REFLEX).

reflex arc *See* REFLEX.

reforestation The replanting of trees on areas of land where forests have been cleared by felling or burning (*see* DEFORESTATION) or by natural means. Reforestation is particularly important in countries, such as Brazil, where large areas of forest have been destroyed by deforestation, although planted forest has much less species diversity (*see* BIODIVERSITY) than the original forest. It also helps to counteract global emissions of carbon dioxide, by fixing the gas as plant material, and to stabilize the exchange of water vapour with the atmosphere and prevent shifts in rainfall patterns. Hence reforestation can play a part in slowing global warming and *climate change due to the *greenhouse effect.

refractory period The period after the transmission of an impulse in a nerve or muscle in which the membrane of the axon or muscle fibre regains its ability to transmit impulses (*see* ACTION POTENTIAL). The **absolute refractory period** is the interval immediately following an action potential during which the sodium ion channels are inactivated and no stimulus, however large, will trigger a further impulse. This period can last from about 0.4 to 2 ms, depending on the type of neuron. During the ensuing **relative refractory period**, the ion channels continue to reactivate, but an action potential can be triggered only if the stimulus is abnormally large.

regeneration The growth of new tissues or organs to replace those lost or damaged by injury. Many plants can regenerate a complete plant from a shoot segment or a single leaf, this being the basis of many horticultural propagation methods (*see* CUTTING). The capacity for regeneration in animals is less marked, although some cnidarians and annelids reproduce asexually by *fragmentation followed by regeneration. Some planarians and sponges can regenerate whole organisms from small pieces, and crustaceans (e.g. crabs), echinoderms (e.g. brittlestars), and some reptiles and amphibians can grow new limbs or tails (*see* AUTOTOMY), but in mammals regeneration is largely restricted to wound

healing and, in certain cases, regrowth of damaged nerve fibres.

regma (*pl.* **regmata**) A dry fruit that is characteristic of the geranium family. It is similar to the *carcerulus but breaks up into one-seeded parts, each of which splits open to release a seed.

regulation (in embryology) The processes involved in animal embryonic development that counteract any abnormalities that may arise in the different developmental stages. Regulative (or regulation) embryos or eggs can compensate for the removal of sections of the embryo or egg at an early stage of development so that subsequent development is not affected; in such embryos the direction of development of the cells is not *determined until cleavage is well advanced. Regulation also includes **twinning**: the formation of two embryos from the cleavage of a single embryo (*see* TWINS).

regulator Any organism that can maintain a constant *internal environment largely independently of the external environment. This is generally achieved by homeostatic mechanisms (*see* HOMEOSTASIS). Regulators tend to occupy habitats in which environmental conditions are variable.

regulator gene *See* OPERON.

regulatory enzyme Any enzyme that is involved in controlling the different metabolic pathways in the cell by switching them on or off. Regulatory enzymes exist in active and inactive forms; they include *allosteric enzymes and those enzymes whose activity is controlled by *kinases.

regulatory genes Genes that control development by regulating the expression of structural genes responsible for the formation of body components. Some encode *transcription factors, which are proteins that interact with regulatory sites of other genes causing activation or repression of developmental pathways. Many other regulatory genes produce *noncoding RNAs, which control gene expression in various ways; some degrade target messenger RNA molecules or block their translation; others block transcription of the structural genes by altering chromatin structure (*see* CHROMATIN REMODELLING). Much of development in quite different organisms, such as mammals and insects, is controlled by genes that are structurally very similar, thought to have descended from genes in ancient common ancestors. Prime examples are the *homeotic genes, such as the *Hox* genes of mammals. Genomes also contain numerous **regulatory sequences**, such as *enhancers, *promoters, and terminators, which are nontranscribed regions that play crucial roles in gene expression.

reinforcement (in animal behaviour) Increasing (or decreasing) the frequency of a particular behaviour through *conditioning, by arranging for some biologically important event (the **reinforcer**) always to follow another event. In instrumental conditioning an **appetitive reinforcer**, or **reward** (e.g. food), given after a response made by the animal, increases that response; an **aversive reinforcer**, or **punishment** (e.g. an electric shock) decreases the response.

Reissner's membrane The membrane in the *cochlea of the inner ear that separates the *vestibular canal from the cochlear duct. It is named after the German anatomist Ernst Reissner (1824–78).

relative abundance *See* ABUNDANCE.

relative atomic mass (atomic weight; r.a.m.) Symbol A_r. The ratio of the average mass per atom of the naturally occurring form of an element to 1/12 of the mass of a carbon-12 atom.

relative density (r.d.) The ratio of the density of a substance to the density of some reference substance. For liquids or solids it is the ratio of the density (usually at 20°C) to the density of water (at its maximum density). This quantity was formerly called **specific gravity**.

relative growth rate A measurement of the *productivity of a plant, defined as the

increase in dry mass per unit of plant mass over a specified period of time.

relative molecular mass (molecular weight) Symbol M_r. The ratio of the average mass per molecule of the naturally occurring form of an element or compound to 1/12 of the mass of a carbon-12 atom. It is equal to the sum of the relative atomic masses of all the atoms that comprise a molecule.

relative refractory period *See* REFRACTORY PERIOD.

relaxin A mammalian protein hormone produced by the corpus luteum in both nonpregnant and pregnant females and by the placenta during the terminal stages of pregnancy. In males it is synthesized by the prostate and released in seminal fluid. Among its many biological effects, it relaxes the pubic symphysis and dilates the cervix of the uterus, thereby aiding parturition.

release factor (RF) A protein that terminates *translation of messenger RNA (mRNA) during protein synthesis and releases the completed polypeptide chain from the ribosome. It recognizes a *stop codon in the mRNA and attaches to the A site of the ribosome (the binding site for an aminoacyl tRNA molecule), blocking further elongation of the polypeptide chain. Then the mRNA is released and the ribosome complex dissociates, releasing the polypeptide. Prokaryotes have two principal release factors, RF-1 and RF-2, whereas eukaryotes have just one, eRF.

release-inhibiting hormone (RIH) A hormone that inhibits the secretion of another hormone. The hypothalamus produces several hormones that inhibit the release of hormones by the anterior lobe of the pituitary (adenohypophysis). They include **MSH-inhibiting hormone** (which inhibits melanocyte-stimulating hormone), **prolactin-inhibiting hormone**, and *somatostatin. These RIHs are released by neurosecretory cells into the hypothalamo-hypophyseal portal blood vessels, which convey the hormones through the pituitary stalk into the anterior pituitary.

releaser *See* SIGN STIMULUS.

releasing hormone (releasing factor)
A hormone that is produced by the hypothalamus and stimulates the release of a hormone from the anterior *pituitary gland into the bloodstream. Each hormone has a specific releasing hormone; for example, thyrotrophin-releasing hormone stimulates the release of *thyroid-stimulating hormone.

relict A group of organisms that survives as a remnant of a formerly much larger group, in terms of either taxonomic diversity (**evolutionary relict**) or geographical distribution (**geographical relict**). The term can be applied to species, genera, other taxa, or to populations or even to entire communities.

relictual (in systematics) Describing features that have been inherited relatively unchanged from ancient ancestors. For example, the chloroplasts of modern land plants are very similar to those of green algae, even though the two groups diverged from a common ancestor some 470 million years ago. *See* PLESIOMORPHY.

rem *See* RADIATION UNITS.

REM sleep *See* SLEEP.

renal Of or relating to the *kidney. For example, the **renal artery** and **renal vein** convey blood towards and away from the kidney, respectively.

renal capsule *See* BOWMAN'S CAPSULE.

renal tubule Any of the sections of a *nephron of the vertebrate kidney that are concerned with reabsorption of water and solutes from the *glomerular filtrate. *See also* PROXIMAL CONVOLUTED TUBULE; LOOP OF HENLE; DISTAL CONVOLUTED TUBULE.

renaturation The reconstruction of a protein or nucleic acid that has been *denatured such that the molecule resumes its original function. Some proteins can be renatured by reversing the conditions (of temperature, pH, etc.) that brought about denaturation.

renin A proteolytic enzyme (*see* PROTEASE) that is involved in the formation of the hormone *angiotensin, which raises blood pressure, in part by stimulating the release of the hormone *aldosterone from the adrenal cortex (the **renin-angiotensin-aldosterone system; RAAS**). Renin is secreted into the blood by juxtaglomerular cells (*see* JUXTAGLOMERULAR APPARATUS) of the kidney under the control of the sympathetic nervous system; its release also occurs in response to a fall in blood-sodium levels and to falling blood pressure. Renin catalyses cleavage of the circulating precursor angiotensinogen to produce angiotensin I, precursor of the active hormone. The net effect is to raise blood pressure, blood volume, and the *glomerular filtration rate.

rennin (chymosin) An enzyme secreted by cells lining the stomach in mammals that is responsible for clotting milk. It acts on a soluble milk protein (**caseinogen**), which it converts to the insoluble form *casein. This ensures that milk remains in the stomach long enough to be acted on by protein-digesting enzymes.

Renshaw cell A type of inhibitory interneuron found in feedback circuits in the central nervous system that control the level of excitation of motor neurons. The Renshaw cells receive input from side branches (collaterals) of motor axons and in turn form inhibitory synapses with the motor-neuron cell bodies. Strychnine blocks transmission at the glycinergic inhibitory synapses of Renshaw cells, causing convulsions and death from paralysis of respiratory muscles. These cells are named after the US neurophysiologist B. Renshaw (1911–48).

repetitive DNA DNA whose base sequence is repeated many times throughout the genome of an organism. It is common in eukaryotes, accounting for about half of the total DNA in mammals, for example, and can be divided into various types. Some is involved in maintaining the structure of the chromosomes, but a significant proportion remains of uncertain function, and is probably *junk DNA. **Highly**

repetitive DNA typically consists of short (usually <100 bp, but sometimes up to 500 bp) sequences that are repeated hundreds or thousands of times in tandem at any one site—**simple sequence DNA**. It includes the *short tandem repeats (STRs), which form the heterochromatin flanking the centromere and are used as genetic markers in *DNA profiling. Tandemly repeated short sequences also occur at each chromosome tip (telomeric DNA). Both types are important for maintaining chromosome structure. **Moderately repetitive DNA** comprises longer sequences repeated typically 10 to 1000 times and more evenly dispersed throughout the genome. Among these are multiple copies of particular genes or gene sequences; these may be members of *gene families or duplicates of genes encoding histones, transfer RNAs, and ribosomal RNAs, which often form tandem arrays. But by far the biggest proportion of repetitive DNA in many eukaryote organisms consists of copies of mobile genetic elements called *transposons. In humans, for example, the various types of transposon, notably *SINEs and *LINEs, account for around 45% of the entire genome. Other distinct types of repetitive DNA lie both in noncoding introns within genes and between genes, where they may act as 'spacer' DNA. Among these are *variable number tandem repeats (VNTRs), sequences of 5–100 nucleotides repeated hundreds or thousands of times at numerous sites within the genome. Centrifugation of the total DNA separates out repetitive DNA in distinct bands according to its base composition, and hence buoyant density. These bands are thus seen as 'satellites' distinct from the band containing the bulk of genomic DNA. There are several categories, including *satellite DNA, minisatellite DNA (*see* VARIABLE NUMBER TANDEM REPEAT), and *microsatellite DNA.

replacing bone *See* CARTILAGE BONE.

replication *See* DNA REPLICATION.

replicon A DNA sequence that is replicated as a unit from a single initiation site (origin of replication). The genome of a bacterium or a virus comprises a single replicon; eukaryotes contain a number of replicons on each chromosome.

repolarization The restoration of the *resting potential in neurons or muscle fibres following the passage of a nerve impulse. Repolarization is brought about by diffusion of potassium ions out of the neuron and by active elimination of sodium ions (*see* SODIUM PUMP).

reporter gene A gene that is used to 'tag' another gene or DNA sequence of interest, such as a *promoter. Expression of the reporter is easily monitored and permits the function or whereabouts of the 'target' sequence to be tracked. It helps pinpoint which cells contain the tagged gene, for instance among a population of genetically engineered bacteria, or shows the varying degrees of expression of the tagged gene within different tissues of the body. For example, the β-galactosidase gene (*lacZ*) is a common reporter gene whose activity can be detected on indicator plates by causing a colour change in a dye. One use of reporter genes is to investigate the function of 'foreign' promoters in *transgenic organisms. The promoter is inserted in a *vector 'upstream' of the reporter gene, and the vector is allowed to transfect the organism. How well and in what tissues the promoter functions can then be assessed by expression of the reporter. Another striking example, obtained from the jellyfish *Aequorea victoria*, is the reporter gene that encodes *green fluorescent protein (GFP). Tissues in which the gene is expressed emit a green fluorescent light, making them easily identifiable.

repressor A protein that can prevent the expression of a gene. *See* OPERON.

reproduction The production of new individuals more or less similar in form to the parent organisms. This may be achieved by a number of means (*see* SEXUAL REPRODUCTION; ASEXUAL REPRODUCTION) and serves to perpetuate or increase a species.

reproduction rate A measure of the *fecundity of a cohort of females in a population. It can be expressed as the average number of daughters born, assuming age-specific birth rates, throughout the reproductive life of the females. The **gross reproduction rate** assumes that none of the cohort will die during their reproductive life, whereas the **net reproduction rate** accounts for the expected proportion of deaths during that time (i.e. mortality). Hence, a net reproduction rate greater than 1 means that the population will maintain or increase its size—for example, a rate of 2 would indicate a doubling of the population size within a generation. Conversely, a rate of less than 1 means the population size will decline.

reproductive system The organs that are involved in the process of *sexual reproduction in an organism. The reproductive system of a flowering plant is found in the *flower and consists of the stamens (male

Human male reproductive system

Human female reproductive system

organs) and carpels (female organs). In mammals the reproductive system consists of the testes, epididymis, sperm duct, and penis in the male and the ovaries, fallopian tubes, and uterus in the female (see illustration).

Reptilia The class that contains the first entirely terrestrial vertebrates, which can live in dry terrestrial habitats as their skin is covered by a layer of horny scales, preventing water loss. They breathe atmospheric oxygen by means of lungs assisted by respiratory movements principally involving the ribs (there is no diaphragm). Reptiles are cold-blooded (*see* POIKILOTHERMY) but behavioural patterns make it possible for them to maintain a fairly even body temperature throughout the day. Fertilization is internal and the majority of reptiles lay eggs on land. These eggs have a porous shell to provide protection from desiccation and allow gas exchange. In some reptiles the eggs are retained within the body of the mother until the young are ready to hatch, thereby greatly reducing juvenile mortality (*see* OVOVIVIPARITY).

The class includes the modern crocodiles, lizards and snakes (*see* SQUAMATA), and tortoises and turtles, as well as many extinct forms, such as the *dinosaurs and *Pterosauria. This grouping is paraphyletic because it excludes the birds, descendants of *theropod dinosaurs. Hence, contemporary classification schemes generally include birds as a constituent group of the reptiles.

(⊕) SEE WEB LINKS
• This Amniota page from the Tree of Life project shows the phylogenetic relationships of reptile groups, birds, and mammals

rescue effect In ecology, the saving from extinction of a small subpopulation of organisms by the immigration of individuals from another subpopulation. *See* METAPOPULATION.

residual volume The amount of air remaining in the lungs after maximum expiration, which cannot be expelled from the lungs voluntarily. An average human has a residual volume of about 1–1.2 litres. *See also* VITAL CAPACITY.

resin A naturally occurring mixture of hydrocarbons secreted by many trees (especially conifers) into ducts or canals. Resins comprise chiefly terpenes, and are found either as brittle glassy substances or dissolved in essential oils. Their functions are similar to those of gums and mucilages, i.e. protecting the plant from predators such as wood-boring insects. *Compare* PHYTONCIDE.

resistance 1. (in microbiology) The degree to which pathogenic microorganisms remain unaffected by antibiotics and other drugs. Resistance to antibiotics is a growing and serious problem in human and veterinary medicine, to the extent that some strains of pathogen are now susceptible to very few drugs. Genes for antibiotic resistance are often carried on *plasmids or *transposons, which can spread across species barriers. **2.** (in ecology) **a.** The degree to which a *pest can withstand the effects of a pesticide. It depends on the selection and spread within a pest population of genes that confer the ability to destroy, or minimize the effects of, a pesticide. **b.** *See* ENVIRONMENTAL RESISTANCE. **3.** (in immunology) The degree of *immunity to infection that an animal possesses.

resistance protein *See* R PROTEIN.

resistance response A long-term response to any stimulus that potentially threatens the wellbeing of an organism. An example of a resistance response is the release of *ACTH (adrenocorticotrophic hormone) in response to stress. This hormone stimulates the production of mineralocorticoid hormones (e.g. *aldosterone), which promote the excretion of hydrogen ions produced in excess as a result of the increased metabolic activity resulting from the *alarm response. Resistance responses diminish as normal conditions resume.

resolving power A measure of the ability of an optical instrument to form separable images of close objects or to separate

close wavelengths of radiation. In optics the resolving power determines the minimum distance between two points that can be separated; the greater the resolving power, the better the resolution.

respiration 1. (cellular or **tissue respiration)** The metabolic process in which substances, usually organic compounds, are broken down to simpler products with the release of energy, which is incorporated into special energy-carrying molecules (*see* ATP) and subsequently used for other metabolic processes. In plants, animals, fungi, and many protists, respiration requires oxygen, and carbon dioxide is an end product. It involves a series of enzyme-catalysed reactions that release the energy of the chemical bonds in a food source and store it as ATP, the cell's usable source of fuel. It can be divided into four stages. In the first, *glycolysis, glucose is partially oxidized to pyruvate. This stage uses the coenzyme NAD^+ as the electron acceptor and produces some ATP by substrate-level *phosphorylation; it does not require oxygen. In the second stage, the pyruvate produced by glycolysis is oxidized to form two molecules of acetyl coenzyme A. In the third stage, the acetyl groups enter the *Krebs cycle in which they are completely oxidized via a sequence of reactions that yields carbon dioxide, a high-energy compound called GTP, and the reduced coenzymes NADH and FADH2. In the final stage, these reduced coenzymes donate electrons to a series of protein complexes and associated molecules called an *electron transport chain (ETC or respiratory chain). This further releases energy in small increments and generates ATP by the *chemiosmotic mechanism. In aerobic organisms, oxygen is the final electron acceptor, and the end products are water and carbon dioxide. In anaerobic bacteria and archaea, a range of other molecules or ions can be utilized as the terminal electron acceptor in the ETC, and various end products are formed. In cells of eukaryotic organisms, all stages except glycolysis take place in the mitochondria. Biochemists often reserve the term 'cellular respiration' for stages two to four as outlined above.

r

Compare FERMENTATION. **2. (external respiration)** The exchange of oxygen and carbon dioxide between the body tissues and the environment (*see* VENTILATION). In many animals, the exchange of gases takes place at *respiratory organs (e.g. *lungs in air-breathing vertebrates) and is assisted by *respiratory movements (e.g. breathing). In plants, oxygen enters through pores on the plant surface and diffuses through the tissues via intercellular spaces or dissolved in tissue fluids.

( ) SEE WEB LINKS

• Overview of cellular respiration, with illustrations, from About.com.

respiratory burst *See* OXIDATIVE BURST.

respiratory chain *See* ELECTRON TRANSPORT CHAIN.

respiratory movement The muscular movement that enables the passage of air to and from the lungs or other *respiratory organs of an animal. The mechanism of the movement varies with the species. In insects abdominal muscles relax and contract rhythmically to encourage the flow of air through the *tracheae. In amphibians air is drawn into the lungs by a pumping action of the muscles in the floor of the mouth. **Breathing** in mammals involves the muscle of the *diaphragm and the *intercostal muscles between the ribs. Contraction of these muscles lowers the diaphragm and raises the ribs, so that the lungs expand and air is drawn in (*see* INSPIRATION). Relaxation has the opposite effect and forces air out during *expiration. *See also* VENTILATION CENTRE.

respiratory organ Any animal organ across which exchange of carbon dioxide and oxygen takes place. The surface membranes of such organs are always moist, thin, and well supplied with blood. Examples are the *lungs of air-breathing vertebrates, the *gills of fish, the *tracheae of insects, and the *lung books of arachnids.

respiratory pigment 1. A coloured compound that is capable of reversibly binding with oxygen at high oxygen concentrations and releasing it at low oxygen concentrations. Most such pigments are present in the blood (**blood pigments**), transporting oxygen within the circulatory system from the *respiratory organs to the tissues of the body; an exception is *myoglobin, which occurs in the muscles. In vertebrates and many invertebrates the respiratory pigment is *haemoglobin, contained in the erythrocytes (red blood cells). Besides haemoglobins, invertebrate respiratory pigments include *haemocyanin, haemerythrin, and chlorocruorin. **2.** Any of the proteins involved in cellular respiration as components of an *electron transport chain. They include the *cytochromes.

respiratory quotient (RQ) The ratio of the volume of carbon dioxide produced by an organism during respiration to the volume of oxygen consumed. It varies according to the nature of the food being metabolized. For aerobic respiration under 'normal' metabolic conditions, the RQ is usually about 1.0 for carbohydrates, about 0.7 for fats and other lipids, and roughly 0.8 for proteins.

respirometer Any device that measures an organism's oxygen uptake. Simple respirometers consist of a chamber (in which the organism is placed) connected to a *manometer. Carbon dioxide is chemically removed from the chamber so that only oxygen uptake is measured. Human oxygen consumption is generally measured by a device known as a **spirometer**, which can also be used to measure depth and frequency of breathing.

response The physiological, muscular, or behavioural activity that can be elicited by a *stimulus.

resting potential The difference in electrical potential that exists across the plasma membrane of an excitable cell, such as a neuron, when it is in a stable, nonexcited state, i.e. its resting *membrane potential. It is the result of differences in concentrations of negatively and positively charged ions and molecules between the cytosol and the extracellular fluid, and the differential permeability of the resting plasma membrane to certain ions. A principal mechanism for establishing the resting potential is the *sodium-potassium pump. This actively exports sodium (Na$^+$) ions and imports potassium ions (K$^+$). Other cations, notably protons (H$^+$) and calcium ions (Ca^{2+}), may also be actively pumped out of the cell. These movements of cations create a net negative charge inside the cell, i.e. an electrical gradient is established across the plasma membrane. However, the K$^+$ ions concentrated inside the cell also experience a concentration, or chemical, gradient, with a tendency to diffuse out of the cell through K$^+$ leak channels in the plasma membrane. But the positive K$^+$ ions are simultaneously attracted in the opposite direction by the negatively charged cell interior (due chiefly to negatively charged proteins and other large molecules). Hence, an equilibrium is established by the action of the two gradients combined, i.e. the **electrochemical gradient**; the electrical charge difference between interior and exterior at this equilibrium point is the cell's resting potential. Its value is measured by inserting a *microelectrode into the cell, and measuring the potential difference across the plasma membrane. The resting potential of neurons is generally –60 to –70 mV. *Compare* ACTION POTENTIAL; GRADED POTENTIAL.

restriction enzyme (restriction endonuclease) A type of enzyme that can cleave molecules of foreign DNA at a particular sequence of (usually four to six) bases called a **restriction site**, or **recognition sequence**. Restriction enzymes are produced by many bacteria and protect the cell by cleaving (and therefore destroying) the DNA of invading viruses. The bacterial cell is protected from attack by its own restriction enzymes by modifying the bases of its DNA during replication. Restriction enzymes are widely used in the techniques of genetic engineering (*see* DNA LIBRARY; DNA PROFILING; DNA SEQUENCING; GENE CLONING; RESTRICTION MAPPING).

restriction fragment length polymorphism (RFLP) The occurrence of different cleavage sites for *restriction enzymes in the DNA of different individuals of the same species. Cleavage of DNA from different individuals with restriction enzymes thus produces differing sets of restriction fragments. The deletion of existing restriction sites or the creation of new ones is the result of random base changes in the noncoding stretches of DNA (*introns) between genes. RFLPs have provided geneticists with a powerful set of genetic markers for mapping the genome (*see* RESTRICTION MAPPING), for identifying particular genes and mutations (*see* GENE TRACKING), and for characterizing organisms (*see* DNA FINGERPRINTING; RIBOTYPING).

restriction mapping A technique for determining the sites at which a length of DNA (e.g. from a chromosome) is cleaved by *restriction enzymes. By cleaving the DNA with various such enzymes, both individually and in combination, and analysing the resultant number and size of fragments by electrophoresis, a **restriction map**, indicating the order of restriction sites in the original DNA, can be deduced. This can then be integrated with a classical *linkage map. Gene deletions or rearrangements that alter the restriction sites can be detected as changes in the pattern of fragments obtained. This may be used, for instance, to diagnose certain genetic abnormalities in the fetus. The fragments are separated by *gel electrophoresis and identified using specific *gene probes, as in the *Southern blotting technique. The absence of a certain fragment in a fetal DNA digest can be diagnostic of a pathological change in the fetal gene containing the corresponding restriction site. *See also* PHYSICAL MAP.

restriction point *See* CELL CYCLE.

resurrection biology *See* DE-EXTINCTION.

reticular activating system *See* BRAINSTEM.

reticular formation *See* BRAINSTEM.

reticulate 1. Marked with or having the form of a network; e.g. a reticulate leaf is one with a network of veins. **2.** A form of evolution in which DNA from unrelated organisms is combined to form a new and distinct line of organisms. There are two principal mechanisms.

In hybridization—which occurs naturally in many plants as well as some fish, amphibians, and invertebrates—usually individuals of two different species mate to form a genetically new species, or to re-form an ancestral lineage. The other is *lateral gene transfer, which is common in bacteria and introduces new traits to existing strains or species. Reticulate evolution is typically represented in phylogenetic trees by a union, instead of a separation, of two branches—a depiction called a **reticulation**.

reticuloendothelial system *See* MONONUCLEAR PHAGOCYTE SYSTEM.

reticulum (*pl.* **reticula**) The first of four chambers that form the stomach of ruminants. Like its companion chamber, the *rumen, it acts essentially as a fermentation vat, housing a dense population of bacteria and protists that digest the plant material ingested by the animal. *See* RUMINANTIA.

retina (*pl.* **retinas** or **retinae**) The light-sensitive membrane that lines the interior of the eye. The retina consists of two layers. The outer layer (**pigment epithelium**) is pigmented, which prevents the back reflection of light and consequent decrease in visual acuity. The inner layer contains

pigment epithelium

rod

cone

horizontal cell

bipolar cell

amacrine cell

ganglion cell

to optic nerve

Structure of the retina

nerve cells, blood vessels, and two types of light-sensitive cells (*rods and *cones). Light passing through the lens stimulates individual rods and cones, which generates nerve impulses that are transmitted through **bipolar** and **ganglion cells** to the optic nerve, and hence to the brain, where the visual image is formed. Information can also be transferred horizontally within the retina via a network of **horizontal** and **amacrine cells** (see illustration). The retinal neurons are supported by glia called *Müller cells.

SEE WEB LINKS

• Anatomy of the retina, prepared by members of the John Moran Eye Center, University of Utah

retinal (retinene) *See* RHODOPSIN; VITAMIN A.

retinol *See* VITAMIN A.

retrotransposon A type of *transposon found in the DNA of various organisms, including yeast, *Drosophila*, and mammals, that multiplies by forming copies of itself using a mechanism similar to that of retroviruses. It undergoes transcription to RNA, then creates a DNA copy of the transcript with the aid of the enzyme *reverse transcriptase. This DNA copy can then reintegrate into the cell's genome, leaving the original version *in situ*. Retrotransposons, like retroviruses, can cause cancer in vertebrates by inserting in the genome near growth-promoting genes, causing the latter to be abnormally expressed. Retrotransposons account for a large proportion of the moderately *repetitive DNA found in eukaryote genomes, the most abundant being the *LINEs and *SINEs.

retrovirus An RNA-containing virus that converts its RNA into DNA by means of the enzyme *reverse transcriptase; this enables it to become integrated into its host's DNA as a *provirus. Some retroviruses can cause cancer in animals: they contain *oncogenes (cancer-causing genes), which are activated when the virus enters its host cell and starts to replicate. Also, their insertion into the host genome may activate

growth-promoting host genes abnormally. The special properties of retroviruses make them useful as *vectors for inserting genetic material into eukaryotic cells. The best-known retrovirus is *HIV, responsible for AIDS in humans.

SEE WEB LINKS

• Series of tutorials about retroviruses and their applications, compiled by the Nolan Lab of Stanford University

reverse genetics Any approach to genetic investigation that aims to find the function for some known protein or gene. It contrasts with the more traditional *forward genetics approach, in which an unknown gene is sought for a known function (identified by the effect of a mutation). For example, analysis of gene sequences reveals open reading frames, which are the hallmarks of functional genes (*see* READING FRAME). Reverse genetics methods can be used to discover the function of such genes. For example, the gene can be cloned, subjected to mutation, and then reinserted into the organism (e.g. a bacterium or yeast cell) to see what effect the mutation has on function. A similar approach can be taken starting with a protein of unknown function. The amino-acid sequence can be back-translated into genetic code, a DNA probe constructed for part of the DNA sequence, and the relevant gene selected from a *DNA library of the organism.

reverse transcriptase An enzyme, occurring in *retroviruses, that catalyses the formation of double-stranded DNA using the single RNA strand of the viral genome as template. This enables the viral genome to be inserted into the host's DNA and replicated by the host. Reverse transcriptase is thus an RNA-directed DNA *polymerase. The enzyme is used in genetic engineering for producing *complementary DNA from messenger RNA, e.g. in determining levels of gene expression in the technique called RT-PCR (*see* POLYMERASE CHAIN REACTION). *See also* RNA SEQUENCING.

reversion (in genetics) The restoration of a mutation to the wild-type genotype or phenotype by a second mutation. The gene or organism affected in this way is called a **revertant**.

rewilding The large-scale management of an ecosystem to reinstate natural processes and reduce human intervention so that ultimately 'nature can take care of itself'. It typically involves reintroduction of indigenous species that previously filled key ecological roles in maintaining the historical 'wild' landscape or seascape. Coupled with this is a return to more natural patterns of drainage and less intensive farming. For example, the Rewilding Europe programme focuses initially on rewilding nine areas of Europe, with reintroduction of species such as the European bison, wild horse, wolf, beaver, and tauros (a modern version of the extinct aurochs; *see* DE-EXTINCTION).

RF *See* RELEASE FACTOR.

RFLP *See* RESTRICTION FRAGMENT LENGTH POLYMORPHISM.

RFP *See* RED FLUORESCENT PROTEIN.

R_F value (in chromatography) Retardation factor: the distance travelled by a given component divided by the distance travelled by the solvent front. For a given system at a known temperature, it is a characteristic of the component and can be used to identify components. For example, the photosynthetic pigments of an organism and the metabolites of a drug excreted in the urine can be identified by their R_F values in thin-layer or paper chromatography.

Rh *See* RHESUS FACTOR.

rhabdom *See* COMPOUND EYE.

rhachis *See* RACHIS.

rhesus factor (Rh factor) An *antigen whose presence or absence on the surface of red blood cells forms the basis of the rhesus *blood group system. (The factor was first recognized in rhesus monkeys.) Most people possess the Rh factor, i.e. they are rhesus positive (Rh+). People who lack the factor are Rh−. If Rh+ blood is given to an Rh− patient, the latter develops anti-Rh

antibodies. Subsequent transfusion of Rh+ blood results in *agglutination, with serious consequences. Similarly, an Rh− pregnant woman carrying an Rh+ fetus may develop anti-Rh antibodies in her blood; these will react with the blood of a subsequent Rh+ fetus, causing anaemia in the newborn baby.

rhizarians A clade of eukaryotic protists comprising the testate (shelled) amoebas. Although constructed chiefly on the basis of molecular systematics, the group features threadlike (filose) pseudopodia. Members include radiolarians with their intricately sculpted siliceous skeletons; the foraminiferans, characterized by pore-covered shells made of sand or organic materials; and the cercozoans, such as the euglyphid amoebas. Some authorities now regard the rhizarians as members of the SAR supergroup, along with the *alveolates and *stramenopiles.

rhizoid 1. One of a group of delicate and often colourless hairlike outgrowths found in certain algae and the gametophyte generation of bryophytes and ferns. They anchor the plant to the substrate and absorb water and mineral salts. **2.** A modified hypha found in certain fungi that serves to anchor the fungus to the substrate, and in some cases to secrete digestive enzymes and absorb the solubilized nutrients.

rhizome A horizontal underground stem. It enables the plant to survive from one growing season to the next and in some species it also serves to propagate the plant vegetatively. It may be thin and wiry, as in couch grass, or fleshy and swollen, as in *Iris*. Compact upright underground stems, as in rhubarb, strawberry, and primrose, are often called **rootstocks**.

Rhizopoda In older classifications, a phylum of protists that contains the amoebas and cellular *slime moulds. *See also* AMOEBA; PROTOZOA.

rhizosphere The zone immediately surrounding the actively growing region of a plant root. Typically 1–2 mm thick, it consists of a *biofilm of water and soluble substances derived from the plant, soil constituents, and a community of fungi, bacteria, and other microorganisms that interact with each other and with the plant. The rhizosphere has a profound influence on the growth and survival of the plant, notably in helping it to absorb nutrients from the soil and inhibiting root pathogens, and on soil composition and structure. The root releases or actively secretes a wide range of soluble substances, including amino acids, proteins, sugars, polysaccharides, and vitamins, as well as the layer of slime (*mucigel) that helps to protect and lubricate the growing root tip. These substances attract, and are exploited as food by, microflora (fungi, bacteria, etc.) and microfauna (e.g. nematodes, mites, and snails), which together form a complex community in the rhizosphere. Some microorganisms are beneficial to the plant, for example by mineralizing soil nutrients to make them available to the plant or by producing vitamins, antibiotics, or signal molecules that encourage plant growth. Also, conditions in the rhizosphere encourage the germination of spores of fungi that form mutualistic *mycorrhizas with plant roots; in some cases the rhizospere can inhibit the growth of competing plant species. The community of organisms found in any particular rhizosphere can be characterized by direct analysis of their genetic material derived from a suitable sample, using the techniques of *metagenomics.

Rhodophyta (red algae) A phylum (or group, i.e. the rhodophytes) of *algae that are often pink or red in colour due to the presence of the pigments *phycocyanin and *phycoerythrin. Members of the Rhodophyta may be unicellular or multicellular; the latter form branched flattened thalli or filaments. They are commonly found along the coasts of tropical areas. Sexual reproduction is by means of carpospores (*see* CARPOGONIUM). Red algae are now regarded as members of the eukaryote assemblage Archaeplastida: *see* PLANT (sense 2).

rhodopsin (visual purple) The light-sensitive pigment found in the *rods of the vertebrate retina. It consists of a protein component, **opsin**, linked to a nonprotein *chromophore, **retinal** (or **retinene**), a derivative of *vitamin A. Light falling on the rod is absorbed by the retinal, which changes its form and separates from the opsin component; this initiates the transmission of a nerve impulse to the brain via an intracellular signalling cascade involving the G protein *transducin. The great sensitivity of rhodopsin allows vision in dim light (night vision). *See also* DARK ADAPTATION.

rhombencephalon *See* HINDBRAIN.

Rhombozoa *See* DICYEMIDA.

rhynchocoel *See* NEMERTEA.

rhyniophytes A group of extinct vascular land plants that flourished in the early Devonian (416–380 million years ago). They had short upright aerial axes ('stems'), several centimetres tall, that arose from rhizomes or corms and branched dichotomously to form two equal growing points. The 'naked' branches lacked leaves and ended in multicellular spore-forming organs (sporangia). Rhyniophytes, such as *Cooksonia* and *Rhynia* (named after fossil-rich chert deposits at Rhynie in Scotland), were the earliest true vascular plants, having a solid mass of xylem vessels in the centre of the stem. In the 1980s, evidence emerged of the coexistence of similar gametophyte forms alongside the sporophyte rhyniophytes. This prompted speculation that these might have been alternative forms of the same species and that subsequent reduction of the gametophytes could have led to evolution of the predominant sporophyte forms of modern vascular plants (*see* TRANSFORMATION HYPOTHESIS). *Compare* TRIMEROPHYTES; ZOSTEROPHYLLOPHYTES.

rhytidome *See* BARK.

rib One of a series of slender curved bones that form a cage to enclose, support, and protect the heart and lungs (*see* THORAX). Ribs occur in pairs, articulating with the *thoracic vertebrae of the spinal column at the back and (in reptiles, birds, and mammals) with the *sternum (breastbone) in front. Movements of the rib cage, controlled by *intercostal muscles between the ribs, are important in breathing (*see* RESPIRATORY MOVEMENT).

ribbon worm *See* NEMERTEA.

riboflavin *See* VITAMIN B COMPLEX.

ribonuclease *See* RNASE.

ribonucleic acid *See* RNA.

ribonucleoprotein (RNP) Any complex, or conjugate, of protein and RNA. Ribonucleoproteins take part in many key cellular activities, notably transcription of DNA into RNA, processing of the RNA transcript, and translation of the mature RNA during protein synthesis. *Ribosomes, responsible for transcription, are prime examples of RNPs. Certain RNPs are restricted to the nucleus whereas others are found in both the nucleus and the cytoplasm. The most common RNP occurring in the nucleus is **heterogeneous nuclear RNP (hnRNP)**, which consists of protein bound to the primary transcript of DNA (*see* TRANSCRIPTION). It may be associated with **small nuclear RNP (snRNP)**, which is involved in the removal of intron sequences from the primary transcript to form messenger RNA, which eventually leaves the nucleus (*see* RNA PROCESSING). A third type, **small cytoplasmic RNP (scRNP)**, is found in the cytoplasm; examples include signal recognition particles—complexes of a single RNA and several proteins that help newly synthesized polypeptides to enter the endoplasmic reticulum (*see* SIGNAL HYPOTHESIS). *See also* HETEROGENEOUS NUCLEAR RNA.

ribose A *monosaccharide, $C_5H_{10}O_5$, rarely occurring free in nature but important as a component of *RNA (ribonucleic acid). Its derivative **deoxyribose**, $C_5H_{10}O_4$, is equally important as a constituent of *DNA (deoxyribonucleic acid), which carries the genetic code in chromosomes.

ribosomal RNA *See* RIBOSOME; RNA.

ribosome A small spherical body within a living cell that is the site of *protein synthesis. Ribosomes consist of two subunits, one large and one small, each of which comprises a type of RNA (called **ribosomal RNA, or rRNA**) and numerous proteins. In eukaryotes, large and small subunits are assembled at the *nucleolus and then exit the nucleus, via the nuclear pores, to become fully functional ribosomes in the cytoplasm. During *translation, the ribosome subunits assemble around the messenger RNA; the ribosome also has three binding sites for transfer RNA molecules. Ribosomes are described in terms of their sedimentation coefficients (i.e. their rates of sedimentation in an ultracentrifuge), which are measured in Svedberg units (symbol S). The prokaryote (70S) ribosome comprises a 50S (large) subunit and a 30S (small) subunit; the eukaryote (80S) ribosome has large 60S and small 40S subunits. The eukaryote large ribosomal subunit comprises three different rRNA molecules plus 49 different proteins; the small subunit contains a single rRNA and 33 proteins. Usually there are many ribosomes in a cell, either attached to the *endoplasmic reticulum or nuclear envelope or free in the cytoplasm. During protein synthesis they are associated with messenger RNA as *polyribosomes in the process of *translation.

riboswitch A region of an RNA molecule, usually a messenger RNA (mRNA), that can bind a specific metabolite and 'switch off' further transcription or translation of the mRNA. It acts as a mechanism in bacterial cells for regulating the concentration of a metabolite by controlling the synthesis of, for example, an enzyme or transport protein involved in synthesis or uptake of the metabolite. When levels of the metabolite are high, it binds to the riboswitch and alters the folding pattern (secondary structure) of the RNA helix such that further transcription of the mRNA is halted; when levels fall, and no metabolite is bound, the riboswitch adopts an alternative structure and transcription is able to proceed.

ribotyping The characterization of isolates of microbial organisms such as yeast or bacteria based on variations in the base sequence of their ribosomal RNA genes. It is used e.g. to identify organisms causing disease, for phylogenetic or geographical population studies, or to track the source of microbial contaminants in food. The

DNA is extracted from a sample of a pure culture of the organism and digested by an appropriate *restriction enzyme; the resulting restriction fragments are separated by gel electrophoresis and blotted onto a nylon membrane. A radiolabelled DNA probe is then added to bind to the ribosomal DNA fragments, thus revealing the pattern of ribosomal restriction fragments that is characteristic for the isolate. This 'fingerprint' can then be recorded, digitized, and stored in a database for analysis and comparison with other ribotypes.

ribozyme (catalytic RNA) Any RNA molecule that can catalyse changes to its own molecular structure. Self-splicing *introns, found in some ribosomal RNA and protein-coding genes in organelle genomes, are examples of ribozymes (*see* RNA PROCESSING). RNA in the large subunit of ribosomes also has ribozyme activity, in catalysing the formation of a peptide bond between the incoming amino acid and the end of the growing polypeptide chain—the peptidyltransferase reaction (*see* TRANSLATION). The spliceosome, responsible for removing introns and splicing exons during RNA processing, contains small nuclear RNA molecules that catalyse the splicing reactions. Ribozymes have properties very similar to *viroids and *virusoids. They can also copy themselves to some extent, and it is proposed that the earliest life forms relied on RNA both to store genetic information and to catalyse chemical reactions—the so-called 'RNA world' hypothesis.

ribulose A ketopentose sugar (*see* MONOSACCHARIDE), $C_5H_{11}O_5$, that is involved in carbon dioxide fixation in photosynthesis as a component of *ribulose bisphosphate.

ribulose bisphosphate (RuBP) A five-carbon sugar that is combined with carbon dioxide to form two three-carbon intermediates in the first stage of the light-independent reactions of *photosynthesis (*see* CALVIN CYCLE). The enzyme that mediates the carboxylation of ribulose bisphosphate, ribulose bisphosphate carboxylase/oxygenase (*see* RUBISCO), is also involved in *photorespiration.

rickets A childhood condition caused by decalcification of bone, resulting in deformed bones. Rickets is associated with chronic deficiency of *vitamin D or calcium and with disorders that cause poor phosphate reabsorption from the kidney *nephrons.

rickettsia A very small coccoid or rod-shaped Gram-negative bacterium belonging to an order (Rickettsiales) of the phylum Proteobacteria. With one exception, rickettsias are obligate parasites, being unable to reproduce outside the cells of their hosts. Rickettsias can infect such arthropods as ticks, fleas, lice, and mites, through which they can be transmitted to vertebrates, including humans. The group includes the causal agents of Q fever, Rocky Mountain spotted fever, and forms of typhus. The only genus that can be grown in culture outside host cells is *Bartonella* (*Rochalimaea*), which includes *B. quintana*, the causal agent of trench fever.

rigor mortis The stiffening of the body of an animal after death, due to a temporary rigidity of the muscles. This condition arises because ATP, which is no longer synthesized after death, is required to break down the bridges that form between actin and myosin filaments in muscle tissue during contraction.

RIH *See* RELEASE-INHIBITING HORMONE.

Ringer's solution *See* PHYSIOLOGICAL SALINE.

ring species Two species with a looped or ringlike distribution pattern, for example circumpolar, which comprises a series of interbreeding forms that are intermediate between the two species. The latter occur where the two ends of the 'ring' meet. Ring species thus demonstrate how the evolution of differences among the different populations or subspecies leads eventually to the formation of new species.

RISC (RNA-induced silencing complex) *See* MICRORNA; RNA INTERFERENCE.

ritualization An evolutionary process in which the form or context of an action is altered because it comes to play a role in social communication. For example, many *courtship and greeting ceremonies in animals include ritual food presentation (though the quantities of food may be negligible), derived from the action of feeding the young.

river continuum concept The concept of a river as an ecosystem whose character changes continually between its source and its mouth according to the nature of its energy inputs. The headwaters are likely to be narrow, fast flowing, and shaded by trees and other vegetation, so that virtually all the energy enters in the form of leaves, twigs, and other debris from the surroundings. The fauna is dominated by detritivores and filter feeders. Further downstream, as the river broadens and becomes less shaded, it is colonized by algae and plants, which contribute significant energy to the community and are exploited by grazers. Towards the mouth, increased sediment loading reduces light penetration and *in situ* photosynthesis may decline. Energy, in the form of biomass and detritus, is constantly flowing downstream, hence the energetics of any particular section of the river are influenced by events upstream. The result is a longitudinal continuum of ecosystem structure, with certain predictable properties.

RMR *See* ROOT MASS RATIO.

RNA (ribonucleic acid) A complex organic compound (a nucleic acid) in living cells that is concerned with *protein synthesis. In some viruses, RNA is also the hereditary material. Most RNA is synthesized in the nucleus and then distributed to various parts of the cytoplasm. An RNA molecule consists of a long chain of *nucleotides in which the sugar is *ribose and the bases are adenine, cytosine, guanine, and uracil (see illustration; *compare* DNA). **Messenger RNA (mRNA)** is responsible for carrying the *genetic code transcribed from DNA to specialized sites within the cell (known as *ribosomes), where the information is translated into protein composition (*see* RNA PROCESSING; TRANSCRIPTION; TRANSLATION). **Ribosomal RNA (rRNA)** is present in *ribosomes; it is single-stranded but helical regions are formed by *base pairing within the strand. **Transfer RNA (tRNA, soluble**

RNA, sRNA) is involved in the assembly of amino acids in a polypeptide chain being synthesized at a ribosome. Each tRNA is specific for an amino acid and bears a triplet of bases complementary with a triplet on mRNA (*see* ANTICODON; ELONGATION). RNA can associate with proteins to form complexes called *ribonucleoproteins. *See also* ANTISENSE RNA; NONCODING RNA.

SEE WEB LINKS
- Animated account of the roles of RNA in protein synthesis: part of the website DNA from the Beginning

RNAase *See* RNASE.

RNA interference (RNAi; post-transcriptional gene silencing) The ability of double-stranded RNA to interfere with, or suppress, the expression of a gene with a corresponding base sequence. The phenomenon occurs in many types of organisms, including plants, fungi, and animals. The RNA transcript of a noncoding RNA gene folds back on itself to form double-stranded RNA, which is normally a rarity in cells; this is cut into fragments by a ribonuclease enzyme called *Dicer. There are essentially two pathways by which gene silencing is accomplished: one involves short, single-stranded *microRNAs (miRNAs), encoded by the genome, which function in controlling gene expression by binding to complementary base sequences on messenger RNAs (mRNAs); the other uses *short interfering RNAs (siRNAs), which target potentially 'rogue' mRNAs (e.g. derived from infecting viruses, or transposons) for destruction. In either case, processing by Dicer results in a short (typically 22-nucleotide) single strand of RNA being incorporated into an assemblage of proteins called the **RNA-induced silencing complex (RISC)**. The RISC binds to the target RNA and either blocks translation (in the case of a RISC-miRNA combination) or triggers cleavage and degradation of the mRNA (with a RISC-siRNA combination). RNA interference is now used as a powerful and versatile experimental tool to suppress particular genes, as a form of *knockout. *See also* PIWI-INTERACTING RNA.

Detail of molecular structure of the sugar–phosphate backbone. The carbon atoms of ribose are numbered 1' (1 'prime') to 5'. Each ribose unit is attached via its 5' carbon atom to a phosphate group, and via its 1' carbon atom to a base.

Single-stranded structure of RNA

adenine (A)

guanine (G)

cytosine (C)

uracil (U)

The four bases of RNA

Molecular structure of RNA

RNA polymerase *See* POLYMERASE.

RNA processing (post-transcriptional modification) The modification of newly formed RNA transcripts to produce functional messenger RNA (mRNA) molecules. It occurs in the nucleus of eukaryotic cells and involves removal of the noncoding stretches (*introns) from the primary transcript and splicing together of the discontinuous coding sequences (*exons) (i.e. **gene splicing**). The primary transcript, or **pre-messenger RNA** (pre-mRNA),

associates with other small RNA molecules and proteins (*see* HETEROGENEOUS NUCLEAR RNA), and the leading (5') end is capped with a special nucleotide (7-methylguanosine), the **5' cap**. When transcription is terminated, a portion of the tail (3') end of the transcript is removed and replaced by a stretch of 100–250 adenine-containing nucleotides, forming the **poly(A) tail**. Splicing of pre-mRNA is performed by a complex particle called a **spliceosome**, which is about the size of a ribosome and comprises small RNA molecules and proteins. It binds

RNA INTERFERENCE

The 1990s were remarkable for a series of advances in genetics and molecular biology, but perhaps the most amazing was the discovery of an unexpected genetic control mechanism, the phenomenon of RNA interference (RNAi).

The work of Fire and Mello

By the mid-1990s it was known that RNA is capable of inhibiting the activity of genes, particularly when the RNA base sequence matches that of the gene. But the work of US molecular geneticists Andrew Fire (1959–) and Craig Mello (1960–) proved to be crucial in unravelling the precise mechanism. The focus of their experiments was the nematode *Caenorhabditis elegans*, specifically a gene that affected muscle action. They injected worms with RNA whose base sequence matched that of the target gene. When 'silencing' of the target gene occurred, it caused a muscle defect in the worm and consequent twitching movements.

Key findings

- Effective silencing required double-stranded RNA molecules (dsRNAs).
- One strand of the dsRNA (the antisense strand) had to possess a base sequence complementary to that of the target gene's mRNA and be capable of binding to it.
- Silencing was specific, affecting only the target gene mRNA.
- Silencing involved 'mature' mRNAs, and hence probably took place in the cytoplasm.
- The dsRNA seemed to act as some form of catalyst.
- The target mRNA disappeared from the cytoplasm, so was probably degraded.

For their discoveries concerning RNAi, Fire and Mello were jointly awarded the 2006 Nobel Prize for physiology or medicine.

Dicer and RISC

It is now established that dsRNA precursors formed in the nucleus are cut into shorter fragments (typically 21 or 22 nucleotides) in the cytoplasm by a protein called Dicer.

One strand associates with an assembly of proteins called the RNA-induced silencing complex (RISC). This binds to a complementary base sequence on its target mRNA and causes silencing.

MicroRNA (miRNA) and short interfering RNA (siRNA)

There are two basic pathways leading to RNA silencing: the miRNA pathway and the siRNA pathway, as shown in the diagram.

miRNA pathway

- Uses small RNA molecules encoded by the cell's DNA as means of regulating gene expression.
- A double-stranded hairpin precursor (pre-miRNA) is trimmed by Dicer to form miRNA.
- The miRNA is incorporated into a RISC.
- miRNA binds imperfectly to target mRNA causing suppression of translation into protein, but not degradation of the mRNA.

siRNA pathway

- dsRNA precursors originate from various sources, including virus infection, introduced transgenes, and transposons.
- Dicer trims precursors to form siRNAs, which are incorporated in RISCs.
- siRNA–RISC binds to its target mRNA sequence completely, triggering cleavage and degradation of the mRNA.

Functions of RNAi

- Regulatory mechanism for an estimated 50% of all protein-coding genes (in mammals).
- Helps protect cells against certain viruses by targeting viral RNA for destruction.
- Helps to silence potentially disruptive transposons in the genome by destroying RNA copies arising from transposon replication.

Applications

RNAi is a precise and efficient tool for *knockout of specific genes when studying gene function in experimental organisms (although, unlike *genome editing, its effects are temporary). It also has potential for new forms of targeted but reversible gene therapy.

sequentially to sites adjoining each intron and bends the intron into a loop structure (**lariat**). The lariat is cut out, and the cut ends of the pre-mRNA molecule are joined together. The mechanism known as *alternative splicing produces variant proteins from the same gene by excluding certain exons from the mature mRNA, depending on the type of tissue or the developmental stage of the cell. In some organisms, the

RNA molecules themselves catalyse splicing of the introns (*see* RIBOZYME). The fully processed mRNA molecule remains associated with RNAs and proteins while it is exported from the nucleus to the ribosomes for *translation in the cytoplasm.

(((( ))) SEE WEB LINKS

- Animated account of RNA processing: part of the website DNA from the Beginning

RNase (ribonuclease; **RNAase**) Any enzyme that catalyses the cleavage of nucleotides in RNA. Each RNase has a specificity for a different cleavage site. For example, RNase A is a digestive enzyme secreted by the pancreas that hydrolyses phosphodiester bonds in the nucleotide chain. Other RNases are active at the cellular level, for instance in modifying ribosomal RNA after transcription. RNase P cuts the 5′ end of precursors of transfer RNA (tRNA) molecules.

RNA sequencing (RNA-seq) A technique used to study *gene expression by determining the nature and quantity of RNAs in a sample. It exploits the increasing availability of inexpensive next-generation *DNA sequencing to produce an accurate and quantitative assay of the RNA transcripts in a given tissue, and is replacing *DNA microarray analysis for many applications. There are several variants of the technique, focusing on different aspects of the RNA transcriptome, including messenger RNA (mRNA), ribosomal RNA, and noncoding RNA, but all employ the same essential steps. Firstly the RNAs are isolated from the tissue sample and cut into small, similar-sized fragments. These are then converted to double-stranded complementary DNA (cDNA) molecules using reverse transcriptase and DNA polymerase. The cDNA fragments are sequenced and analysed by computer. This can then be used to match the transcripts with the genome sequence when known or to assemble the short sequences into longer contiguous sequences from scratch (*de novo* assembly). Compared to microarrays, RNA sequencing provides a more comprehensive and sensitive assay of gene expression and can identify, e.g., alternatively spliced transcripts from the same gene. Moreover, unlike microarrays it does not require a probe based on pre-existing sequence information.

RNA splicing *See* RIBOZYME; RNA PROCESSING.

RNP *See* RIBONUCLEOPROTEIN.

rod A type of light-sensitive receptor cell present in the *retinas of vertebrates. Rods contain the pigment *rhodopsin and are essential for vision in dim light. Each consists of a narrow cylindrical outer segment, which detects light; an inner segment containing mitochondria; a cell body containing the nucleus; and a terminal synaptic body, which transmits signals to an adjacent neuron. The rhodopsin molecules are organized within a stack of membranous discs in the outer segment. Rods are not evenly distributed on the retina, being absent in the *fovea and occupying all of the retinal margin. *See also* DARK ADAPTATION. *Compare* CONE.

Rodentia An order of mammals characterized by a single pair of long curved incisors in each jaw. These teeth are specialized for gnawing: they continue growing throughout life and have enamel only on the front so that they wear to a chisel-shaped cutting edge. Rodents often breed throughout the year and produce large numbers of quickly maturing young. The order includes the squirrels, beavers, rats, mice, and porcupines.

roentgen The former unit of dose equivalent (*see* RADIATION UNITS). It is named after the discoverer of X-rays, W. K. Roentgen (1845–1923).

root 1. (in botany) The part of a vascular plant that grows beneath the soil surface in response to gravity and water. It anchors the plant in the soil and absorbs water and mineral salts. Unlike the stem, it never produces leaves, buds, or flowers and usually does not contain chlorophyll. The *radicle (embryonic root) may give rise either to a

Section through the tip of a plant root

tap root system with a single main **tap root** from which **lateral roots** develop, or a **fibrous root system**, with many roots of equal size. Adventitious roots arise from the stem—whether above ground or underground—and these are common in monocots such as grasses and cereals, replacing the primary root as the plant grows. The *apical meristem at the root tip gives rise to a protective sheath, the *root cap, and to the primary tissues of the root. The vascular tissues usually form a central core (see illustration). This distinguishes roots from stems, in which the vascular tissue often forms a ring. A short distance behind the root tip **root hairs** develop from the epidermis and greatly increase the surface area for absorption of water and minerals (*see* RHIZOSPHERE). Beyond this, lateral roots develop.

Roots may be modified in various ways. Some are swollen with food to survive the winter, as in the carrot. Certain plants, such as orchids, have absorptive aerial roots; others, such as ivy, have short clasping roots for climbing. The roots of leguminous plants, such as beans and peas, contain *root nodules, which have an important role in nitrogen fixation. Other modifications include *prop roots, stilt roots, and buttress roots, which support the plant; and *pneumatophores. The roots of most plants form associations with fungi called *mycorrhizae, which greatly enhance the plant's ability to absorb nutrients from the soil and to communicate with neighbouring plants. **2.** (in dentistry) The portion of a *tooth that is not covered with enamel and is embedded in a socket in the jawbone. Incisors, canines, and premolars have single roots; molars normally have several roots. **3.** (in anatomy) The point of origin of a nerve in the central nervous system. There are two roots for every *spinal nerve (*see* DORSAL ROOT; VENTRAL ROOT).

root cap (calyptra) A cone-shaped structure that covers the root tip and develops as a result of cell division by a meristem at the root apex (*see* CALYPTROGEN). It protects the root tip as it grows between the soil particles. The cells are constantly worn away by friction and are replaced by the meristem. *See also* MUCIGEL.

root hair *See* ROOT.

root mass ratio (RMR) The dry mass of the roots of a plant divided by the total dry mass of the entire plant, typically measured in grams per gram or kilograms per kilogram. This reflects the proportion of resources that a plant allocates to its root system, which varies with such factors as species, growing conditions, and season.

root nodule A swelling on the roots of certain plants, especially those of the family Fabaceae (Leguminosae), that contains bacteria (notably *Rhizobium*) capable of fixing atmospheric nitrogen into ammonia, which is subsequently converted to nitrates and amino acids (*see* BACTEROID; NITROGEN FIXATION). Plants that possess root nodules increase soil fertility by increasing the nitrate content of the soil. The practice of *crop rotation will normally include the cultivation of a leguminous species. Legumes attract rhizobia by releasing flavonoids from their roots into the soil. These also trigger synthesis by the bacteria of Nod factors, which cause certain cells in the root to divide and form a primary nodule meristem, which gives rise to the root nodule. Certain nonleguminous plants, such as alder (*Alnus*) and bayberry (*Myrica*), also form root nodules, although the bacteria are filamentous actinomycetes (*Frankia*), not rhizobia.

root pressure The pressure that forces water, absorbed from the soil, to move through the roots and up the stem of a plant. This pressure can be demonstrated by cutting a stem, from which water will exude. A *manometer can be attached to a plant stem to measure the root pressure. Root pressure is believed to be due to both the osmosis of water from the soil into the root cells, and the active pumping of salts into the *xylem tissue, which maintains a concentration gradient along which the water will move. *See also* TRANSPIRATION.

rootstock *See* RHIZOME.

ROS *See* REACTIVE OXYGEN SPECIES.

Rotifera A phylum of microscopic (0.04–2.00 mm) *pseudocoelomate aquatic animals belonging to the *Lophotrochozoa and characterized by a crown of cilia at the head end. These are used in locomotion and in some species for feeding: the crown resembles a rotating wheel when the cilia are beating. Rotifers possess jaws and are covered with a layer of chitin (the **lorica**). There is no circulatory system and gas exchange occurs across the body surface. Some rotifers reproduce by *parthenogenesis, existing only as females for many generations. Indeed, molecular analysis of asexual rotifers of the clade Bdelloidea indicates that they have maintained exclusively female lines for over 50 million years.

(((●))) SEE WEB LINKS

• Basic descriptions and striking images of rotifers by Wim van Egmond

roughage *See* (DIETARY) FIBRE.

rough endoplasmic reticulum (rough ER) *See* ENDOPLASMIC RETICULUM.

Roundup *See* GLYPHOSATE.

round window (fenestra rotunda) A membrane-covered opening between the middle ear and the inner ear (*see* EAR), situated below the *oval window. Pressure waves transmitted through the perilymph in the *cochlea are released into the middle ear through the round window, which bulges out into the middle-ear cavity.

roundworms *See* NEMATODA.

royal jelly A protein-rich food substance that is secreted from the hypopharyngeal and mandibular glands of nurse honeybees and fed to the newly hatched larvae. For the first three days after hatching, larvae that will develop into worker bees are fed on royal jelly, after which their diet changes to pollen and nectar. However, larvae that will develop into queen bees continue to be fed exclusively on royal jelly throughout the larval stage. This special diet contains an ingredient that silences a particular gene,

Dnmt3, that encodes an enzyme involved in epigenetic silencing of hundreds of genes. When these genes are inactive, the bee larva develops into a worker; but when Dnmt3 is itself silenced, these genes are switched on, leading to the development of a queen with all the necessary attributes, such as functional ovaries, a larger body, and sexual behaviours.

R protein (resistance protein) Any intracellular plant protein that acts as a receptor for a specific effector molecule released into the cell by invading bacteria, fungi, or other pathogens. Binding of an effector to its R protein receptor triggers a signal transduction pathway that activates effector-triggered immunity (*see* IMMUNITY sense 2) to combat the infection. R proteins are encoded by resistance (*R*) genes, while their corresponding effectors (or virulence factors) are encoded by Avirulence (*Avr*) genes in the respective pathogen. For example, a rust fungus that carries a particular *Avr* gene will induce an effective immune response in a variety of wheat having the corresponding *R* gene. Hence the desire of plant breeders to create crop varieties carrying *R* genes.

RQ *See* RESPIRATORY QUOTIENT.

rRNA *See* RNA.

r selection A type of selection that favours organisms with a high *biotic potential (*r* value). Organisms that are *r* selected (*r* strategists) are able to colonize a habitat rapidly, utilizing the food and other resources before other organisms are established and begin to compete. The *r* strategists tend to be relatively small organisms with short life spans (e.g. bacteria) and often live in temporary or unstable environments; characteristically their survival depends on their ability to produce large numbers of offspring rather than on their ability to compete. *Compare* K SELECTION. *See also* POPULATION DYNAMICS; SURVIVAL CURVE.

RT-PCR *See* POLYMERASE CHAIN REACTION.

rubidium-strontium dating A method of dating geological specimens based on the decay of the radioisotope rubidium-87 into the stable isotope strontium-87. Natural rubidium contains 27.85% of rubidium-87, which has a half-life of $4.7 \times M$ 10^{11} years. The ratio $^{87}Rb/^{87}Sr$ in a specimen gives an estimate of its age (up to several thousand million years).

rubisco (ribulose bisphosphate carboxylase/oxygenase) The enzyme that catalyses the crucial step in the *Calvin-Bassham-Benson cycle of photosynthesis, namely the incorporation of a molecule of carbon dioxide with a molecule of ribulose bisphosphate, forming two molecules of phosphoglycerate. This reaction fixes gaseous carbon dioxide in the form of organic carbon, providing a source of energy and other materials for plants and all their dependent organisms. However, rubisco has a dual activity, being able to catalyse the reaction between oxygen and ribulose bisphosphate, splitting the latter into one molecule of phosphoglycerate and one molecule of phosphoglycolate. This is the wasteful process of *photorespiration, which competes with photosynthesis. Reflecting its key role in life, rubisco is the most abundant of all natural proteins.

RuBP *See* RIBULOSE BISPHOSPHATE.

ruderal Describing a plant species that is characteristic of land with a high level of disturbance (e.g. resulting from environmental disaster or human activity) but rich in resources (water, nutrients, etc.). Such species tend to quickly colonize bare patches and reproduce rapidly, but are poor competitors against more robust species. *See also* CSR STRATEGIES. *Compare* FUGITIVE SPECIES.

Ruffini's capsule (Ruffini's corpuscle) A specialized nerve ending, found in skin and in joint capsules and ligaments, that responds to pressure and stretching, providing information about certain types of touch stimuli and movements of joints and limbs. Named after Italian histologist Angelo Ruffino (1864–1929), it consists of a highly branched unmyelinated nerve ending within a spindle-shaped capsule, which is filled with a mesh of collagen fibres. The

nerve endings weave around the collagen fibres and are activated by movement of surrounding tissues.

rumen (*pl.* **rumens** or **rumina**) The second of four chambers that form the stomach of ruminants. It is the main fermentation chamber, housing a dense and complex population of anaerobic bacteria and protists that digest the plant material ingested by the animal. *See* RUMINANTIA.

Ruminantia A suborder of hooved mammals (*see* ARTIODACTYLA) comprising the sheep, cattle, goats, deer, and antelopes. They are characterized by a four-chambered stomach (see illustration). The oesophagus empties into the first and most forward chamber, the **reticulum**. This communicates freely with the second and largest chamber, the **rumen**. Here food is fermented by the large population of anaerobic bacteria and protists; cellulose and other normally indigestible plant materials are broken down by the microbial enzymes, such as *cellulase. Periodically the animal regurgitates material from the rumen and chews it before swallowing it once again; this process, known as 'chewing the cud', helps to break down fibrous food. The digesta enters the third chamber, the **omasum**, from the reticulum. Here water and some nutrients are absorbed, before the contents pass to the fourth and final chamber, the **abomasum**, which functions rather like a normal stomach,

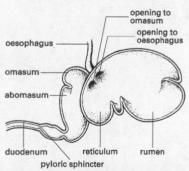

Ruminantia: section of the stomach of a ruminant

secreting acidic abomasal juice and digestive enzymes from its walls.

runner A stem that grows horizontally along the soil surface and gives rise to new plants either from axillary or terminal buds. Runners are seen in the creeping buttercup and the strawberry. **Offsets**, e.g. those of the houseleek, are short runners.

rusts A group of parasitic fungi of the phylum *Basidiomycota. Many of these species attack the leaves and stems of cereal crops: characteristic rust-coloured streaks of spores appear on infected plants. The life cycles of some rusts may be complex; many form a number of different types of spore and some require two different host plants. *Compare* SMUTS.

saccharide *See* SUGAR.

Saccharomyces An industrially important genus of yeasts. *S. cerevisiae*, of which there are at least 1000 strains, is used in baking (*see* BAKER'S YEAST), brewing, and wine making; it is also used in the production of various proteins and other compounds in biotechnology, including industrial alcohol, and as a model organism for experimental studies in cell biology and genetics. Other yeasts used in the production of beer include *S. uvarum* (or *carlsbergensis*); it is distinguished from *S. cerevisiae* by its ability to ferment the disaccharide melibose using α-galactosidase, an enzyme not produced by *S. cerevisiae*. *S. boulardii* is used as a probiotic dietary supplement for the prevention and treatment of diarrhoeal diseases.

saccharose *See* SUCROSE.

sacculus (saccule) (*pl.* **sacculi**) A fluid-filled chamber of the *inner ear from which the *cochlea arises in reptiles, birds, and mammals. It bears patches of sensory epithelium concerned with balance (*see* MACULA).

sacral vertebrae The vertebrae that lie between the lumbar and the caudal vertebrae in the *vertebral column. The function of the sacral vertebrae is to articulate securely with the *pelvic girdle, and they are usually fused to form a single bone (the **sacrum**) to provide a firm support. The number of sacral vertebrae varies from animal to animal. Amphibians have a single sacral vertebra, reptiles have two, and mammals have three or more.

safranin A stain used in optical microscopy that colours lignified tissues, cutinized tissues, and nuclei red and chloroplasts pink. It is used mainly for plant tissues, in conjunction with a green or blue counterstain.

sagittal Describing a section through an organism that is bisected in the longitudinal plane.

salicylic acid (1-hydroxybenzoic acid) A naturally occurring carboxylic acid, HOC_6H_4COOH, that acts as a plant hormone. Its production is induced by pathogen infection, and its methylated derivative, methyl salicylate, moves through phloem to other parts of the plant. There it is reconverted to salicylic acid, which initiates the generalized state of heightened immunity known as *systemic acquired resistance. Salicylic acid also plays a part in senescence and seed germination. When extracted from plants it is used in making aspirin and in the foodstuffs and dyestuffs industries.

saline Describing a chemical compound that is a salt, or a solution containing a salt. *See also* PHYSIOLOGICAL SALINE.

salinization An increase in the salt content of the soil (or of a body of fresh water), which can lead to the retardation of plant growth and eventually renders the soil infertile. This problem is particularly acute in hot areas, where water readily evaporates from the soil, and it may also occur as a consequence of *irrigation. Salts dissolved in irrigation water build up in the soil as the water evaporates, making the water potential of the soil water more negative. This reduces the gradient of water potential between the soil water and the roots, reducing the rate of water uptake by the roots.

saliva A watery fluid secreted by the *salivary glands in the mouth. Production of saliva is stimulated by the presence of food in the mouth and also by the smell or thought of food. Saliva contains mucin, which lubricates food and eases its passage into the oesophagus, and in some animals salivary *amylase (or ptyalin), which begins the digestion of starch. *Lysozyme, an antimicrobial enzyme, also occurs in saliva, helping to protect the teeth and oral cavity against the bacteria in food. The saliva of insects is rich in digestive enzymes, and that of bloodsucking animals contains an anticoagulant.

salivary glands Glands in many terrestrial animals that secrete *saliva into the mouth. In humans there are three pairs: the sublingual, submandibular, and the submaxillary glands. The salivary gland cells of some insect larvae produce giant chromosomes (*see* POLYTENY), which are widely used in the study of genetics and protein synthesis.

Salmonella A genus of rod-shaped Gram-negative bacteria that inhabit the intestine and cause disease (**salmonellosis**) in humans and animals. They are aerobic or facultatively anaerobic, and most are motile. Salmonellae can exist for long periods outside their host, and may be found, for example, in sewage and surface water. Humans may become infected by consuming contaminated water or food, especially animal products, such as eggs, meat, and milk, or vegetables that have been fertilized with contaminated manure. The bacteria can also be transmitted from human or animal carriers by unhygienic food preparation. The genus is divided into two species, *S. bongori* and *S. enterica*; the latter can be divided into six subspecies, each of which has various serotypes—there are over 2500 *Salmonella* serotypes. For example, the serotypes Enteritidis and Typhimurium of *S. enterica* cause gastroenteritis, whereas the serotypes Typhi and Paratyphi cause typhoid fever and paratyphoid fever, respectively.

salt A compound formed by reaction of an acid with a base, in which the hydrogen of the acid has been replaced by metal or other positive ions. Typically, salts are crystalline ionic compounds, such as Na^+Cl^- (*sodium chloride).

salt gland 1. Either of a pair of glands found in the nasal region of most marine birds and many reptiles that remove excess salt (sodium chloride) from the blood and excrete it via the nostrils. Typically it consists of numerous lobes surrounding a central canal; each lobe contains many secretory tubules arranged radially around a duct, which drains into the canal. Blood capillaries run alongside the tubules in a countercurrent arrangement, and salt is actively pumped across the tubule wall from the blood into the lumen of the salt gland, using a mechanism similar to that of the *rectal gland of elasmobranch fishes. However, the resulting solution has an osmolarity greater than that of the blood. **2.** A gland found in certain plants, especially 'salt-loving' plants (halophytes) or desert plants, that excretes accumulated sodium chloride onto the leaves.

samara A dry single-seeded indehiscent fruit in which the fruit wall hardens and extends to form a long membranous wing-like structure that aids dispersal. Examples are ash and elm fruits. The sycamore fruit is a double samara and technically a *schizocarp. *See also* ACHENE.

sampling In statistics and experimental design, the selection of a small group of entities to represent a larger *population of entities. In **random sampling**, a chance process is used so that each individual of a population has an equal chance of being selected as part of the sample. In **stratified sampling**, the population is divided into subgroups or strata (e.g. male and female), and the number of individuals chosen from each corresponds to the relative size of that group within the population. In **systematic sampling**, individuals are chosen at fixed intervals; for example, every tenth animal in a population. In **sampling with replacement**, each individual chosen is returned to the general pool before the next selection is made (with the effect that some items may be selected more than once).

Sanger, Frederick (1918–2013) British biochemist who worked at Cambridge

University and the Medical Research Council. He was awarded two Nobel Prizes for chemistry. The first was for his discovery of the amino-acid sequence in bovine insulin, which enabled insulin to be synthesized. His second Nobel Prize was for discovering the sequence of 5400 nucleotides in a strand of viral DNA. His technique for sequencing nucleotides has been widely applied (*see* DNA SEQUENCING). The Sanger Institute, Cambridge, was founded in 1992 and named in his honour. It is one of the world's leading genomics research establishments.

sap 1. The sugary fluid that is found in the phloem tissue of plants. Sap is the medium in which carbohydrates, produced in photosynthesis, and other organic molecules are transported and stored in plants. It typically comprises up to 90% sucrose along with variable proportions of amino acids, mineral ions, hormones, and other small molecules. **2. (cell sap)** The fluid that is contained in the *vacuoles of plant cells. It is a solution of organic and inorganic compounds, including sugars, amino acids, salts, pigments, and waste products.

saponin Any of a class of *glycosides, found widely in plants, that have detergent properties and form a lather when shaken with water. They are especially concentrated in the soapwort (Saponaria officinalis), whose foliage was formerly boiled and used as a soap substitute. Chemically saponins consist of a sugar group (e.g. glucose) linked to a steroid or triterpene group (a **sapogenin**). Their presence in plants is thought to act as a deterrent to herbivores—they are bitter-tasting and cause gastric irritation if ingested. They are also highly toxic to fish. If injected into the bloodstream they disrupt red cells, through their effects on plasma membranes. Some saponins are extracted commercially from plants such as yuccas and quillaja and may have beneficial effects on health.

saprotroph (saprobe; saprobiont) Any organism that feeds by absorbing nutrients from dead organic matter. Most saprotrophs are bacteria and fungi. Saprotrophs

are important in *food chains as they bring about decay and release nutrients for plant growth. *Compare* PARASITISM.

sapwood (alburnum) The outer wood of a tree trunk or branch. It consists of living *xylem cells, which both conduct water and provide structural support. *Compare* HEARTWOOD.

SAR A *supergroup of eukaryote protists named after the initial letters of its constituent clades: *stramenopiles, *alveolates, and *rhizarians. It is tentatively proposed on the basis of possible homologies revealed by whole-genome DNA sequencing of representatives of the three clades.

sarcolemma The contractile membrane that surrounds a muscle fibre. It represents the plasma membrane of fused embryonic muscle cells that develop to form a single muscle fibre. *See also* TRANSVERSE TUBULES.

sarcoma (*pl.* **sarcomas** or **sarcomata**) *See* CANCER.

sarcomere Any of the functional units that make up the myofibrils of a *skeletal muscle fibre. Each sarcomere is bounded by two membranes (*Z lines), which provide the points of attachment of *actin filaments; another membrane (the **M band** or **line**) is the point of attachment of the

Structure of a sarcomere

*myosin filaments. The sarcomere is divided into various bands reflecting the arrangement of the filaments (see illustration). During muscle contraction the actin and myosin filaments slide over each other (*see* SLIDING FILAMENT THEORY) and the length of the sarcomere shortens: the Z lines are drawn closer together and the I and H bands become narrower.

sarcoplasm The cytoplasm of a muscle fibre. Sarcoplasm contains chemicals that are required for muscle contraction, including glycogen, ATP, and phosphocreatine. In addition the sarcoplasm of active muscles tends to be rich in mitochondria.

sarcoplasmic reticulum The specialized endoplasmic reticulum found in the fibres of skeletal and cardiac muscle, which forms a network of membrane-lined cavities surrounding the contractile myofibrils that run through the fibres. The release of calcium ions from the sarcoplasmic reticulum into the cytosol following stimulation from an action potential (*see* TRANSVERSE TUBULES) causes subsequent contraction of the myofibrils (*see* SARCOMERE). The calcium ions are immediately returned to the sarcoplasmic reticulum by *calcium ion pumps in the membrane.

satellite DNA The proportion of the DNA of a eukaryotic cell that consists of copies of nucleotide sequences, often in large numbers. For example, in humans these sequences can vary in length from 5 to over 300 bp, which are repeated in tandem 10^5 to 10^6 times. It occurs mainly around the centromeres and telomeres of the chromosomes. The highly repetitive nature of this DNA fraction gives it a distinctive base composition, and consequently when samples of DNA are centrifuged it forms so-called 'satellite bands' quite separate from the band representing the bulk of the cell's DNA. *See* REPETITIVE DNA. *Compare* MICROSATELLITE DNA; VARIABLE NUMBER TANDEM REPEATS.

saturated 1. Denoting a compound consisting of molecules that have only single bonds (i.e. no double or triple bonds). Saturated compounds can undergo substitution reactions but not addition reactions. *See* FATTY ACID. *Compare* UNSATURATED. **2.** Describing a cellular transport system in which all the carrier molecules for a particular substance are fully loaded, so that the rate of transport across the plasma membrane is maximal. **3.** Describing an enzyme molecule or system in which the active sites are fully occupied and the rate of reaction is determined primarily by how quickly the catalytic reaction takes place. It occurs when the substrate is present in excess, and as soon as each active site releases the product(s), another substrate molecule occupies it. **4.** Describing a soil in which all the pores are filled with water.

savanna *See* GRASSLAND.

scala (*pl.* **scalae**) Any one of three fluid-filled canals of the *cochlea in the inner ear: the **scala media** (cochlear duct), **scala tympani** (tympanic canal), and **scala vestibuli** (vestibular canal).

scales The small bony or horny plates forming the body covering of fish and reptiles. The wings of some insects, notably the Lepidoptera (butterflies and moths), are covered with tiny scales that are modified cuticular hairs.

In fish there are three types of scales. **Placoid scales (denticles)**, characteristic of cartilaginous fish, are small and tooth-like, with a projecting spine and a flattened base embedded in the skin. They are made of *dentine, have a pulp cavity, and the spine is covered with a layer of enamel. Teeth are probably modified placoid scales. **Cosmoid scales**, characteristic of lungfish and coelacanths, have an outer layer of hard **cosmin** (similar to dentine) covered by modified enamel (**ganoine**) and inner layers of bone. The scale grows by adding to the inner layer only. In modern lungfish the scales are reduced to large bony plates. **Ganoid scales** are characteristic of primitive ray-finned fishes, such as sturgeons. They are similar to cosmoid scales but have a much thicker layer of ganoine and grow by the addition of material all round. The scales of modern teleost fish are reduced to thin bony plates.

In reptiles there are two types of scales: horny epidermal **corneoscutes** sometimes fused with underlying bony dermal **osteoscutes**. The feathers of birds are now thought to have evolved from the scales of their reptile ancestors.

scanning electron microscope *See* ELECTRON MICROSCOPE.

scanning probe microscopy Any of several microscopic techniques that are based on measuring the interaction between a very sharp-tipped probe and the surface of the sample. The probe repeatedly scans the sample surface in a systematic way (or the sample is moved beneath a static probe) and records the surface topography; deflections of the tip can be caused by various types of forces, including mechanical contact, electrostatic forces (as in *atomic force microscopy), van der Waals forces, capillary forces, and magnetic forces (in magnetic force microscopy). The resulting data are processed by a computer to produce images, with resolutions of the order of a fraction of a nanometre. Widely used in chemistry, such techniques are now commonly used in biology to study biomolecules and cell surfaces at the nanometre scale. The most common variations are atomic force microscopy and *scanning tunnelling microscopy.

(⊕)) SEE WEB LINKS

• Elegant overview compiled by the Chemistry Department of Queen Mary University of London

scanning tunnelling microscopy (STM) A variation of *scanning probe microscopy that relies on the quantum mechanical phenomenon of electrons 'tunnelling' through barriers by virtue of their wavelike properties. It detects the very small current created by tunnelling electrons when a fine-tipped metal conductor is scanned over the surface of a sample. The amount of current flowing depends on the distance between the metal probe and the surface. If the apparatus is operated in constant-current mode, feedback from a current sensor is used to alter the position of the probe relative to the surface to maintain a constant current. The deflections of the probe (or the sample) then correspond to the surface topography. The data are processed by a computer and a grey-scale map of the surface is produced. Although most suitable for imaging substances that are electrical conductors, STM is now being used for studying organic systems and biological materials at the nanometre scale, by first applying the nonconductive material as a thin film to a conductive substrate, such as graphite.

scapula (shoulder blade) (*pl.* **scapulae** or **scapulas**) The largest of the bones that make up each half of the *pectoral (shoulder) girdle. It is a flat triangular bone, providing anchorage for the muscles of the forelimb and an articulation for the *humerus at the *glenoid cavity. It is joined to the *clavicle (collar bone) in front.

scavenger An animal that feeds on dead organic matter. Scavengers (such as hyenas) may feed on animals killed by predators or they may be *detritivores.

Schiff's reagent A reagent used for testing for aldehydes and ketones; it consists of a solution of fuchsin dye that has been decolorized by sulphur dioxide. Aliphatic aldehydes restore the pink immediately, whereas aromatic ketones have no effect on the reagent. Aromatic aldehydes and aliphatic ketones restore the colour slowly. It is named after the German chemist Hugo Schiff (1834–1915).

schizocarp A dry indehiscent fruit formed from carpels that develop into separate one-seeded fragments called **mericarps**, which may be dehiscent, as in the *regma, or indehiscent, as in the *cremocarp and *carcerulus.

schizogeny The localized separation of plant cells to form a cavity (surrounded by the intact cells) in which secretions accumulate. Examples are the resin canals of some conifers and the oil ducts of caraway and aniseed fruits. *Compare* LYSIGENY.

Schwann cell (neurilemma cell) A cell that forms the *myelin sheath of nerve fibres (axons) of the peripheral nervous system. Each cell is responsible for a given stretch (called an **internode**) along any particular axon and typically serves several neighbouring axons simultaneously; adjacent internodes are separated by small gaps (**nodes of Ranvier**) where the axon is bare. During its development the cell wraps itself around the axon, so the sheath consists of concentric layers of Schwann cell membrane. These cells are named after the German physiologist Theodor Schwann (1810–82).

scintillation counter A type of particle or radiation counter that makes use of the flash of light (scintillation) emitted by an excited atom falling back to its ground state after having been excited by a passing photon or particle. The scintillating medium is usually either solid or liquid and is used in connection with a photomultiplier, which produces a pulse of current for each scintillation. The pulses are counted to enable the radioactivity of the source to be calculated. The distribution of a radiolabelled compound, such as a drug, in an organism can be determined in this way by testing tissue samples from different organs after drug administration.

scion *See* GRAFT.

sclera (*pl.* **scleras** or **sclerae**) *See* SCLEROTIC.

sclereid A type of *sclerenchyma cell that is shorter than a *fibre; its lignified walls typically contain branched *pits. Sclereids provide support and the hard texture typical of seed coats and nut shells. Clumps of them also occur in pear fruits, giving them their characteristic gritty texture.

sclerenchyma A plant tissue whose cell walls have become impregnated with lignin. Due to the added strength that this confers, sclerenchyma plays an important role in support; it is found in the stems and also in the midribs of leaves. The cell walls contain *pits, enabling the exchange of substances between adjacent cells. Mature sclerenchyma cells are dead, since the lignin makes the cell wall impermeable to water and gases, and so they occur in parts of the plant that are no longer growing lengthways. Sclerenchyma cells take the form of *fibres or *sclereids. *Compare* COLLENCHYMA; PARENCHYMA.

sclerophyllous Describing scrub or woodland in which the plants have small leathery evergreen leaves. Such leaves are an adaptation to conserve water, and sclerophyllous vegetation is characteristic of Mediterranean-type climates, which have summer drought and occur in California, South Australia, South Africa, and Chile, as well as the Mediterranean region itself.

scleroprotein Any protein that forms long filaments, such as the *collagens and *keratins. They are typically insoluble in aqueous solutions, tend to clump together in aggregates, and perform various structural roles.

sclerotic (sclera) The tough external layer of the vertebrate eye. At the front of the eye, the sclera is modified to form the *cornea.

sclerotium (*pl.* **sclerotia**) **1.** In certain fungi, such as *Sclerotium* and *Claviceps*, a compact mass of fused hyphae produced as a resistant resting structure when conditions are unfavourable. **2.** In plasmodial slime moulds, an irregular compact mass of cellular components formed to survive unfavourable conditions.

sclerotization A process in which proteins undergo natural hardening, or 'tanning', as in conversion of the relatively soft elastic larval protein into the rigid cuticle of the adult insect. A similar process occurs following moulting (*see* ECDYSIS). Enzymatically generated orthoquinones react with the free amino groups of the cuticular proteins to form cross-linkages, thereby producing a hard rigid covering consisting of structural proteins called sclerotins.

scolex (*pl.* **scolices**) *See* CESTODA.

scorpions *See* ARACHNIDA.

scotopic vision The type of vision that occurs when the rods in the eye are the principal receptors, i.e. when the level of illumination is low. With scotopic vision colours cannot be identified. *Compare* PHOTOPIC VISION.

SCP *See* SINGLE-CELL PROTEIN.

scRNP *See* RIBONUCLEOPROTEIN.

scrotum (*pl.* **scrota** or **scrotums**) The sac of skin and tissue that contains and supports the *testes in most mammals. It is situated outside the body cavity and allows sperm to develop at the optimum temperature, which is slightly lower than body temperature.

scurvy A disease caused by deficiency of *vitamin C, which results in poor collagen formation. Symptoms include anaemia, skin discoloration, and tooth loss. Scurvy was a common disease among sailors in the 16th–18th centuries, when no fresh food was available on long sea voyages.

scutellum (*pl.* **scutella**) The tissue in a grass seed that lies between the embryo and the endosperm. It is the modified cotyledon of grasses, being specialized for the digestion and absorption of the endosperm.

Scyphozoa *See* CNIDARIA.

SDS-PAGE (sodium dodecyl sulphate-polyacrylamide gel electrophoresis) *See* GEL ELECTROPHORESIS.

seasonal isolation *See* ISOLATING MECHANISM.

seaweeds Large multicellular *algae living in the sea or in the intertidal zone. They are commonly species of the *Chlorophyta, *Phaeophyta, and *Rhodophyta.

sebaceous gland A small gland occurring in mammalian *skin. Its duct opens into a hair follicle, through which it discharges *sebum onto the skin surface.

sebum The substance secreted by *sebaceous glands onto the surface of the *skin. It is a complex mixture of lipids, including fatty acids and cholesterol, that protects, lubricates, and waterproofs the skin and hair and helps prevent desiccation. It is colonized by certain bacteria, notably *Propionibacterium acnes*, which may serve an immune function.

second Symbol s. The SI unit of time equal to the duration of 9 192 631 770 periods of the radiation corresponding to the transition between two hyperfine levels of the ground state of the caesium-133 atom.

secondary cell wall *See* CELL WALL.

secondary consumer *See* CONSUMER.

secondary growth (secondary thickening) The increase in thickness of plant shoots and roots through the activities of the vascular *cambium and *cork cambium. It is seen in most dicotyledons and gymnosperms but not in monocotyledons. The tissues produced by secondary growth are called **secondary tissues** and the resultant plant or plant part is the **secondary plant body**. *Compare* PRIMARY GROWTH.

secondary immunological response *See* IMMUNOLOGICAL MEMORY.

secondary metabolite A substance produced by living organisms that is not essential to vital processes such as respiration, nutrition, and growth, but is produced for a particular purpose, often in response to environmental stresses. Examples include the antibiotics synthesized by certain bacteria, and the numerous defensive chemicals produced by plants to resist herbivores or insect pests (*see* ALKALOID; COUMARIN; GLYCOSIDE; TANNIN).

secondary productivity The rate of biomass formation or energy fixation by heterotrophic organisms, such as grazers and decomposers (*see* HETEROTROPHIC NUTRITION). These derive all their energy from photosynthetic plants and other autotrophs, either directly or indirectly, and their productivity determines the number of *trophic levels and the lengths of the *food chains within an ecosystem—both of which are likely to be increased by greater secondary productivity. *See* PRODUCTIVITY.

secondary sexual characteristics External features of a sexually mature animal that, although not directly involved in copulation, are significant in reproductive behaviour. The development of such features is controlled by sex hormones (androgens or oestrogens); they may be seasonal (e.g. the antlers of male deer or the body colour of male sticklebacks) or permanent (e.g. breasts in women or facial hair in men). In humans they develop during *adolescence.

secondary structure *See* PROTEIN.

secondary thickening *See* SECONDARY GROWTH.

second convoluted tubule *See* DISTAL CONVOLUTED TUBULE.

second messenger A chemical within a cell that is responsible for initiating the response to a signal from a chemical messenger (such as a hormone, neurotransmitter, or growth factor) that cannot enter the target cell itself, for example because it is not lipid-soluble and is therefore unable to cross the plasma membrane. A common second messenger is *cyclic AMP; the signal for its formation within the cell by the enzyme *adenylate cyclase is transmitted from hormone receptors on the cell surface by a *G protein. Inositol 1,4,5-trisphosphate (*see* INOSITOL), diacylglycerol, calcium ions, and nitric oxide are other examples of second messengers.

secretin A hormone produced by S cells in the lining of the anterior *duodenum in response to the presence of hydrochloric acid and partially digested food (chyme) entering the small intestine from the stomach. Secretin circulates in the bloodstream and causes the pancreas to secrete alkaline pancreatic juice and stimulates bile production in the liver. Secretin, whose function was first demonstrated in 1902, was the first substance to be described as a hormone.

secretion 1. The manufacture and discharge of specific substances into the external medium by cells in living organisms. (The substance secreted is also called the secretion.) Secretory cells are often specialized and organized in groups to form *glands. The substances produced may be released

directly into the blood (**endocrine secretion;** *see* ENDOCRINE GLAND) or through a duct (**exocrine secretion;** *see* EXOCRINE GLAND). Secretions can be classified according to the manner of their discharge. **Merocrine (eccrine) secretion** occurs without the secretory cells sustaining any permanent change; in **apocrine secretion** the cells release a secretory vesicle incorporating part of the secretory cell membrane; and **holocrine secretion** involves the disruption of the entire cell to release its accumulated secretory vesicles. Substances destined for secretion are prepared and packaged into membranous vesicles by the *Golgi apparatus inside the cell. **2.** The process by which a substance is pumped out of a cell against a concentration gradient. Secretion has an important role in adjusting the composition of urine as it passes through the *nephrons of the kidney.

secretion vector *See* EXPRESSION VECTOR.

Sedentaria A clade of annelids proposed on the basis of recent molecular evidence. The sedentarians tend to be burrowing or tube-dwelling forms, in contrast to the more active *Errantia, and include the leeches and earthworms. *See* ANNELIDA.

seed The structure in angiosperms and gymnosperms that develops from the ovule after fertilization. Occasionally seeds may develop without fertilization taking place (*see* APOMIXIS). The seed contains the *embryo and nutritive tissue, either as *endosperm or food stored in the *cotyledons. Angiosperm seeds are contained within a *fruit that develops from the ovary wall. Gymnosperm seeds lack an enclosing fruit and are thus termed **naked**. The seed is covered by a protective layer, the *testa. During development of the testa the seed dries out and enters a resting phase (dormancy) until conditions are suitable for germination.

Annual plants survive the winter or dry season as seeds. The evolution of the seed habit enabled plants to colonize the land, since seed plants do not depend on water for fertilization (unlike the lower plants).

seed coat *See* TESTA.

seed ferns *See* CYCADOFILICALES.

seed leaf *See* COTYLEDON.

seed plant Any plant that produces seeds. Most seed plants belong to the phyla *Anthophyta (flowering plants) or *Coniferophyta (conifers); others include the *Cycadophyta, the Ginkgophyta (gingkos), and the Gnetophyta (*see* GYMNOSPERM).

segmentation 1. *See* METAMERIC SEGMENTATION. **2.** *See* CLEAVAGE. **3.** Describing a type of movement occurring in the stomach and small intestine in which regions of the gut wall undergo alternating ringlike contraction and relaxation. This produces a back-and-forth movement of the gut contents, breaking them up and mixing them with digestive juices to promote digestion and absorption of nutrients. *Compare* PERISTALSIS.

segmentation genes Any of three classes of genes that determine the pattern of development of the early insect embryo. They have been extensively studied in the fruit fly *Drosophila* and are key to determining the number and polarity of the body segments in the adult fly. They are expressed in a sequential cascade normally commencing about 3 hours after fertilization. First to be 'switched on' are the **gap genes**, so named because mutations of these genes result in missing portions, or gaps, in the body plan of the larva. They are differentially expressed along the anteroposterior axis of the embryo according to the gradient of mRNAs derived from *maternal effect genes in the mother's ovary and define broad longitudinal zones in the embryo. The gap genes in turn regulate expression of **pair-rule genes**, which refine these zones into units corresponding to two body segments each. Mutation in these results in the loss of alternate segments. Finally, the **segment polarity genes** define the boundaries and polarity of each body segment of the embryo; mutations in these cause the loss of part of the affected segment(s) and its replacement with its mirror image—e.g. two anterior half segments orientated back to back. Following completion of the segmentation

sequence, subsequent cell fate and organ development within each segment is regulated by the expression of *Hox* genes (*see* HOMEOTIC GENES).

segregation The separation of pairs of *alleles during the formation of reproductive cells so that they contain one allele only of each pair. Segregation is the result of the separation of *homologous chromosomes during *meiosis. *See also* MENDEL'S LAWS.

Selachii In some classifications, a subclass or superorder of the Chondrichthyes (cartilaginous fishes), containing the sharks. Their sharp teeth develop from the toothlike placoid *scales (denticles) and are rapidly replaced as they wear out.

selectin *See* CELL ADHESION MOLECULE; LECTIN.

selection The process by which one or more factors acting on a population produce differential mortality and favour the transmission of specific characteristics to subsequent generations. *See* ARTIFICIAL SELECTION; BALANCING SELECTION; DISRUPTIVE SELECTION; NATURAL SELECTION; POSITVE SELECTION; NEGATIVE SELECTION; SEXUAL SELECTION; STABILIZING SELECTION.

selection coefficient Symbol s or t. A measure of the sum of the forces acting to oppose the reproductive success of a particular organism, phenotype, or genotype. It varies between 0 and 1 and is related to *fitness (W) by the equation: $W = 1 - s$. Hence, as the selection coefficient increases, fitness decreases.

selection pressure The extent to which organisms possessing a particular characteristic are either eliminated or favoured by environmental demands. Hence, selection pressure can be negative or positive. It indicates the degree of intensity of *natural selection. *See also* SELECTION.

selective breeding *See* BREEDING.

selective reabsorption The absorption of some of the components of the *glomerular filtrate back into the blood as the filtrate flows through the *nephrons of the kidney. Glucose, amino acids, and salts can be reabsorbed against a concentration gradient and their transport across the nephron into the capillaries requires energy (*see* ACTIVE TRANSPORT). Other components, such as ammonia and urea, are secreted rather than absorbed (*see* SECRETION), while certain ions, including potassium and bicarbonate, can either leave passively (*see* PROXIMAL CONVOLUTED TUBULE) or be actively taken up (*see* DISTAL CONVOLUTED TUBULE) by the tubules according to the omolarity of the interstitial fluid surrounding the tubules, which reflects the overall ionic balance throughout the body.

self-fertilization *See* FERTILIZATION.

selfish DNA Regions of DNA that apparently have no function and are replicated alongside the organism's genes to be passed from parent to offspring. *Transposons are good examples; certain other types of *repetitive DNA also have 'selfish' characteristics, as do nonfunctional relict genes and *pseudogenes. Selfish DNA is so called as it seemingly exists only to pass copies of itself from one generation to another; it does so by acting like a 'molecular parasite', using the organism in which it is contained as a survival machine. This is known as the **selfish DNA theory**. The greatest amounts of apparently selfish DNA are found in vertebrates and higher plants. Selfish DNA may indeed exist because the cell has no way of halting its increase in the genome as long as it poses no selective disadvantage. Alternatively, its function may simply be unrecognized. Transposons make up some 45% of the human genome, accounting for a large proportion of *junk DNA. The term was coined by British evolutionary biologist Richard Dawkins (1941–) in his book *The Selfish Gene* (1976).

self-pollination *See* POLLINATION.

self-splicing *See* RIBOZYME.

self-sterility The condition found in many hermaphrodite organisms in which male and female reproductive cells produced by the same individual will not fuse

to form a zygote, or if they do, the zygote is unable to develop into an embryo. In plants this is usually termed **self-incompatibility** (*see* INCOMPATIBILITY).

Seliwanoff's test A biochemical test to identify the presence of ketonic sugars, such as fructose, in solution. It was devised by the Russian chemist F. F. Seliwanoff. A few drops of the reagent, consisting of resorcinol crystals dissolved in equal amounts of water and hydrochloric acid, are heated with the test solution and the formation of a red precipitate indicates a positive result.

semaphorin One of a class of proteins that act as guidance molecules during the development of nerve cells, immune cells, blood vessels, bone, and other tissues. There are several classes, some occurring in invertebrates and others in vertebrates; some semaphorins are secreted by cells, whereas others remain bound to the plasma membrane. However, all semaphorin molecules have a characteristic region called a sema domain. Class 3 semaphorins act as short-range cues to guide the *growth cone at the tip of elongating nerve fibres along appropriate pathways through tissue. They may either attract the growth cone or repel it, depending on the particular semaphorin and the type of nerve cell. The action of semaphorins is effected by binding to any of various cell surface receptors.

semelparity The strategy of reproducing only once during a lifetime, after which death is inevitable, even if the reproductive event is unfruitful. Semelparous organisms include most annual and biennial plants and representatives from most invertebrate animal taxa; vertebrate examples include the Pacific salmon and Atlantic eel. *Compare* ITEROPARITY.

semen A slightly alkaline fluid (pH 7.2–7.6) containing sperm and various secretions that is produced by a male mammal during copulation and is introduced into the body of the female by *ejaculation. Spermatozoa are produced by the *testes and the secretions by the *prostate gland, *seminal vesicles, and *Cowper's glands. Semen also contains enzymes that activate the sperm after ejaculation.

semicircular canals The sense organ in vertebrates that is concerned with the maintenance of physical equilibrium (sense of balance). It occurs in the *inner ear and consists of three looped canals set at right angles to each other and attached to the *utriculus. The canals contain a fluid (**endolymph**) that flows in response to movements of the head and body. A swelling (*ampulla) at one attachment point of each canal contains sensory cells that respond to movement of the endolymph in any of the three planes. These sensory cells initiate nervous impulses to the brain.

semiconservative replication The generally accepted method of *DNA replication, in which the two strands of the DNA helix separate and free nucleotides pair with the exposed bases on the single chains to form two new DNA molecules, each containing one original and one newly synthesized strand of DNA. *Compare* CONSERVATIVE REPLICATION; DISPERSIVE REPLICATION.

semilunar valve Either of two *valves in the heart, found in the pulmonary artery (**pulmonary valve**) and in the aorta (**aortic valve**), that prevent the backflow of blood into the right and left ventricles from the pulmonary artery and the aorta, respectively, thus maintaining blood flow in a single direction. Each semilunar valve consists of three pockets, which close by filling with blood when the force of contraction that expels the blood from the heart diminishes.

seminal receptacle *See* SPERMATHECA.

seminal vesicle 1. A pouch or sac in many male invertebrates and lower vertebrates that is used for storing sperm. **2.** One of a pair of glands in male mammals that secrete a liquid component (**seminal fluid**) of *semen into the vas deferens. This secretion is alkaline, which neutralizes the acidic conditions in the female genital tract, and contains fructose, used by the sperm as a

source of energy. In humans, seminal fluid is thickened by the presence of mucus and the protein *fibrinogen; it accounts for 60% of the volume of semen.

seminiferous tubules *See* TESTIS.

semiochemical A chemical that affects the behaviour of an organism. Such chemicals include *pheromones, which are used for communication between members of the same species, and *allelochemicals, which act as chemical signals between members of different species. Terrestrial organisms may respond to chemical signals both in the air and on the ground, using their senses of smell or taste; whereas aquatic organisms are influenced very largely by chemicals dissolved or dispersed in their watery medium. Synthetic semiochemicals, especially artificial sex pheromones, are now widely used to control populations of agricultural and horticultural pests, notably insect pests. For example, they can be used to lure male insects to field traps, or to raise the general level of sex attractants in the area and so confuse males and thereby disrupt mating.

semipermeable membrane *See* PARTIALLY PERMEABLE MEMBRANE.

senescence The changes that occur in an organism (or a part of an organism) between maturity and death, i.e. ageing. Characteristically there is a deterioration in functioning as the cells become less efficient in maintaining and replacing vital cell components. In animals this results in a decline in physical ability and, in humans, there is also often a reduction in mental ability. Not all the parts of the body necessarily become senescent at the same time or age at the same rate. For example, in deciduous trees the shedding of senescent leaves in the autumn is a normal physiological process. Various theories have been advanced to explain why ageing occurs. The **telomere theory of ageing** stems from the finding that the tips (*telomeres) of chromosomes in body cells shorten with each round of chromosomal replication and cell division. It is suggested that eventually essential genetic information at the chromosomal extremities is lost, thus disrupting vital cellular functions and causing cell death. An alternative account is the **mitochondrial theory of ageing**, which proposes that over an individual's lifetime, mutations accumulate in *mitochondrial DNA to the extent that there is an age-related decline in the formation of ATP by oxidative phosphorylation. More recently it has been proposed that changes in the epigenetic modifications of DNA, and of the histone proteins that package DNA in chromosomes, could be linked to age-related functional deficits (*see* CHROMATIN REMODELLING). For example, there is a gradual loss of DNA methylation with increasing age, which might disrupt patterns of gene expression. Similar changes are associated with certain cancers. Other proposed mechanisms of ageing in humans include decline in immunological function, cross-linking of proteins or attack by free radicals, both of which lead to tissue damage, or simply the wear and tear of components (rather like an old car).

((⊕)) SEE WEB LINKS

- Examination of the causes and consequences of ageing, compiled by João Pedro de Magalhães of Harvard University

sensation The raw data detected by the *senses. For example, red is a colour sensation. *Compare* PERCEPTION.

sense organ A part of the body of an animal that contains or consists of a concentration of *receptors that are sensitive to specific stimuli (e.g. sound, light, pressure, heat). Stimulation of these receptors initiates the transmission of nervous impulses to the brain, where sensory information is analysed and interpreted. Examples of sense organs are the *eye, *ear, *nose, *taste buds, and *mechanoreceptors.

senses The faculties that enable animals to perceive information about their external environment or about the state of their bodies in relation to this environment (*see* SENSE ORGAN; VISION; HEARING; BALANCE; OLFACTION; TASTE; TOUCH). Specific

*receptors are sensitive to pain, temperature, chemicals, etc.

sensillum (*pl.* **sensilla**) Any of various hairlike or peglike sense organs found in insects and other arthropods, comprising a cluster of basal receptor cells whose dendrites extend inside a sheath and hairshaft. They respond to a wide range of chemical and mechanical stimuli, acting as chemoreceptors of both 'smell' and 'taste', touch receptors, mechanoreceptors to detect movement at joints in the skeleton, and auditory organs. They may protrude from the surface, for example as *setae, or they may be embedded in the cuticle. All the constituent cells of any one sensillum are derived from a single mother cell; they may include receptor cells responsive to several different classes of stimuli. The antennae, in particular, bear large numbers of olfactory sensilla. Chemicals enter the hairshaft via one or numerous pores and elicit nerve impulses in the dendrites. For example, males of the silkworm moth (Bombyx mori) have in each antenna about 17 000 sensilla that respond to the female sex pheromone; each sensillum has some 3000 pores, giving sensitivity to just a few molecules of this vital odour.

sensitive period *See* IMPRINTING (1).

sensitivity (irritability) One of the fundamental properties of all organisms: the capacity to detect, interpret, and respond to changes in the environment (e.g. the stimuli of light, touch, chemicals, etc.). Multicellular animals have specialized *sense organs and *effector organs for this purpose; in unicellular organisms, which lack a nervous system, the reception of and response to a stimulus occur in the same cell.

sensitization 1. (of a cell) The alteration of the integrity of a plasma membrane resulting from the reaction of specific *antibodies with *antigens on the surface of the cell. In the presence of *complement, the cell ruptures. **2.** (of an individual) Initial exposure to a specific antigen such that re-exposure to the same antigen causes a severe immune response, as in *allergies or *anaphylaxis. **3.** (in animal behaviour) The process whereby an animal gives a stronger than normal response to a stimulus after previous application of a different stimulus. For example, if the tail of a sea slug (*Aplysia*) is given an electric shock, a subsequent tap on the animal's siphon will cause a marked withdrawal of the siphon that is greater than that without a preceding tail shock. *Compare* DISHABITUATION.

sensory cell *See* RECEPTOR.

sensory neuron A nerve cell (*see* NEURON) that transmits information about changes in the internal or external environment to the central nervous system. Sensory neurons are of two types. **Somatic sensory neurons** occur in peripheral nerves in the skin, skeletal muscle, joints, and bones. **Visceral sensory neurons** are located in sympathetic and parasympathetic nerves in the heart, lungs, and other organs.

sepal One of the parts of a flower making up the *calyx. Sepals are considered to be modified leaves with a simpler structure. They are usually green and often hairy but in some plants, e.g. monk's hood, they may be brightly coloured.

sepsis Infection of the soft tissues or blood by pathogenic microorganisms, arising usually after these have entered the body through a skin wound. It results in destruction of the tissues by the pathogens or their *toxins.

septum (*pl.* **septa**) Any dividing wall in a plant or animal. Examples are the septa that separate the chambers of the heart.

sequence analysis The process of characterizing sequences of biomolecules, particularly the nucleotides of nucleic acids (DNA or RNA) or the amino acids of proteins. Once the order of nucleotides of, say, a genome fragment has been established by *DNA sequencing, the sequence data can be analysed using computer software. This will automatically identify such features as open *reading frames, *promoters, *enhancers, and *repetitive DNA

and translate any putative coding sequences into corresponding amino acid sequences. The unknown sequence is compared with existing sequence data held on any of numerous databases. Likely homology is revealed by its degree of *alignment with other DNA sequences, which will provide clues about its evolutionary relationships with other biomolecules (*see* PHYLOGENOMICS) and possible membership of a *protein family. Moreover, coding sequences that correlate with particular functional *domains in the corresponding protein can be identified. *See also* BIOINFORMATICS.

sequence database A database containing the sequences of biomolecules, which may be the sequences of nucleotides for nucleic acids (DNA or RNA) or of amino acids for proteins. The data also include relevant *annotations. *See* GENBANK; INTERNATIONAL NUCLEOTIDE SEQUENCE DATABASE COLLABORATION; UNIPROT.

sequence-tagged site (STS) A unique short sequence of DNA, typically 200 to 500 nucleotides long, that can be used as a tag, or marker, in physical mapping of cloned DNA segments. An STS is derived by sequencing a short length of a particular clone, then using the sequence data to design primers for the *polymerase chain reaction (PCR). Using the primers, PCR will amplify the sequence in every clone containing it. Clones that have a unique sequence in common must overlap. By repeating the process for different STSs, the clones can be aligned into a series of overlapping segments (contigs). *See* PHYSICAL MAP.

sere A complete *succession of plant communities, which results in the climax community. A sere is composed of a series of different plant communities that change with time. These communities are known as **seral stages** or **seral communities**.

serine *See* AMINO ACID.

seroconversion The stage in an immune response when antibodies to the infecting agent are first detected in the bloodstream. For example, people infected with HIV typically seroconvert about 4–6 weeks following the initial infection, when antibodies against viral proteins are first produced.

serology The laboratory study of blood serum and its constituents, particularly *antibodies and *complement, which play a part in the immune response.

serotonin (5-hydroxytryptamine; 5-HT) A lipid-soluble amine (see formula) that functions as both a neurotransmitter in the central nervous system and a signal molecule in various roles. It is synthesized from the amino acid tryptophan and occurs chiefly in enterochromaffin cells in the gut lining, in blood platelets, and in the raphe nuclei of the brainstem. In the gut, serotonin triggers responses to various mechanical or chemical stimuli, by stimulating the release of acetylcholine from neurons in the gut wall, which in turn promotes peristaltic movements of the gut musculature and secretion from glands. Serotonin also has more general effects on the activity of smooth muscles, particularly on the contraction or relaxation of blood vessel walls. Serotonergic (serotonin-releasing) neurons project from the raphe nuclei to virtually all parts of the brain and also modulate transmission in the spinal cord. Serotonin plays a key role in arousal, mood, aggression, and the sleep-wake cycle; reduced serotonin levels are associated with various disorders, including depression, migraine, and obsessive-compulsive disorder. Several types of antidepressive drugs act by enhancing serotonin levels in the brain, including fluoxetine (Prozac); this is a selective serotonin reuptake inhibitor (SSRI) that slows the reuptake of serotonin by neurons and prolongs its residence time in the synaptic cleft. Platelets take up serotonin from the bloodstream and release it at sites of blood clots, where it causes blood vessels to contract. Serotonin binds to a large family of serotonin (5-HT) receptors, which are coupled to diverse intracellular signal pathways. Most are *metabotropic receptors, but some are coupled directly to ion channels (ionotropic). Different tissues have different types of 5-HT receptors, whereby serotonin elicits different effects. For example, serotonin binding to 5-HT$_2$ receptors in large blood vessels causes vasoconstriction and an increase in blood pressure, whereas by binding to 5-HT$_1$ receptors in the endings of sympathetic nerves it reduces the release of noradrenaline, which tends to reduce blood pressure. Serotonin is also found in mushrooms and in certain nuts, fruits, and vegetables. *See also* LYSERGIC ACID DIETHYLAMIDE.

serous membrane (serosa) A tissue consisting of a layer of *mesothelium attached to a surface by a thin layer of connective tissue. Serous membrane lines body cavities that do not open to the exterior; the *peritoneum, *pleura, and serous *pericardium are examples.

serpin (serine protease inhibitor; serine proteinase inhibitor) Any of a large family of proteins, found in all groups of organisms, that are characterized by irreversibly inhibiting serine proteases (i.e. proteases with a serine residue at the active site). Serpins, which can occur both inside and outside cells, are involved in a wide range of metabolic processes, including regulation of complement activation, blood coagulation, tumour suppression, inflammation, and blood vessel formation (angiogenesis). The most abundant serpin in human plasma in α_1-antitrypsin, which inhibits trypsin and, most significantly, elastase released by leucocytes. Some serpins lack inhibitory activity, including angiotensinogen, the precursor of *angiotensin.

Sertoli cells (sustentacular cells) Cells that line the seminiferous tubules in the *testis. Named after Italian histologist Enrico Sertoli (1842–1910), Sertoli cells protect the *spermatids and convey nutrients to both the developing and mature spermatozoa. They also produce a hormone, **inhibin**, which

Serotonin

can inhibit *follicle-stimulating hormone and thereby regulate production of spermatozoa.

serum (*pl.* **sera** or **serums**) *See* BLOOD SERUM.

sessile 1. Describing animals that live permanently attached to a surface, i.e. sedentary animals. Many marine animals, e.g. sea anemones and limpets, are sessile. **2.** Describing any organ that does not possess a stalk where one might be expected. For example, the leaves of the oak (Quercus robur) are attached directly to the twigs.

seta (*pl.* **setae**) **1.** A bristle or hair in many invertebrates. Setae are produced by the epidermis and consist either of a hollow projection of cuticle containing all or part of an epidermal cell (as in insects) or are composed of chitin (as in the *chaetae of annelid worms). **2.** *See* SPOROGONIUM.

set point A particular value for a physiological variable, such as blood pressure or heart rate, that the body maintains to ensure *homeostasis of its internal environment. Any fluctuation above or below the set point is detected by sensors, which provide feedback to control centres; these in turn initiate hormonal or nervous signals to make physical or biochemical changes in effectors (cells or organs) that will restore the variable to its set point. For example, human body temperature has an average set point of 37°C, which is monitored by sensors in the hypothalamus in the brain. A rise in temperature elicits cooling responses, such as sweating and dilation of surface blood vessels, whereas a fall below the set point causes shivering and constriction of surface vessels. Set points can themselves change, depending on time of day, season, or other factors, such as hibernation.

sewage Waste matter from industrial and domestic sources that is dissolved or suspended in water. Raw (untreated) sewage is a pollutant. It has a high content of organic matter (notably faeces and nitrogenous waste) and therefore provides a rich source of food for many decomposers (bacteria, fungi) and *detritivores, some of which are pathogenic to humans. The release of raw

(untreated) sewage into a river causes eutrophication (*see* EUTROPHIC); there is a sudden increase in the *biochemical oxygen demand (BOD), as the organisms that feed on sewage proliferate and use up the available dissolved oxygen in the river. Oxygen-sensitive organisms, such as fish, will die. Certain organisms can proliferate in particular concentrations of sewage, depending on their tolerance, and can be used as markers of the extent to which a river is polluted by sewage. For example, *Tubifex* worms are able to tolerate high concentrations of sewage.

Sewage can be treated before release. This involves a number of stages, including filtration, sedimentation, and microbial degradation (notably by *methanogens). When most of the solid waste has been removed, the remaining liquid (**effluent**) is discharged into rivers, etc. During sedimentation, particulate organic matter accumulates at the bottom of large tanks. This material, known as **sludge**, is periodically removed, further decomposed by microorganisms (*see* ACTIVATED SLUDGE PROCESS), and then sold as fertilizer or dumped.

Sewall Wright effect *See* GENETIC DRIFT. It is named after US statistician Sewall Wright (1889–1988).

sex chromosome A chromosome that operates in the sex-determining mechanism of a species. Many animals have two different types of sex chromosome. For example, in mammals there is a large X chromosome and a much smaller Y chromosome; the female has two matching X chromosomes, making it the *homogametic sex, whereas the male has one X chromosome and one Y chromosome, making it the *heterogametic sex. In other animal groups, such as butterflies and moths, birds, and reptiles, the situation is reversed: males have two matching W chromosomes, while females have one W chromosome and one Z chromosome. Sex chromosomes carry genes governing the development of sex organs and secondary sexual characteristics. They also carry other genes unrelated to sex (*see* SEX LINKAGE). A few plants have distinct sex chromosomes, one example being the maidenhair tree (*Gingko biloba*), in which male and female trees are distinguished by their X and Y chromosomes. *See* SEX DETERMINATION; TESTIS-DETERMINING FACTOR; X INACTIVATION.

sex determination The process by which the distinction between males and females is established in a species. It is usually under genetic control. Equal numbers of males and females are produced when sex is determined by *sex chromosomes or by a contrasting pair of alleles. In mammals, development of male or female sex organs depends on hormonal signals from the gonads of the early embryo. Embryos having a Y chromosome produce a *testis-determining factor, which switches embryonic development to the male pathway. Otherwise, the embryo develops as a female. In some species (e.g. bees) females develop from fertilized eggs and males from unfertilized eggs. This does not produce equal numbers. Environmental factors can also be crucial in governing the sex of developing individuals. For example, temperature can affect the sex ratios of broods of certain turtles. High incubation temperatures (>28°C) produce a preponderance of males, while lower temperatures (<26°C) give rise to more females.

sex factor A *plasmid found in the cytoplasm of certain bacteria that is responsible for initiating *conjugation and gene transfer. The sex factor in *Escherichia coli* is known as the **F factor**; it can be transferred from a donor cell (F⁺) to a recipient (F⁻) by conjugation, enabled via formation of a sex *pilus. Occasionally the F factor may become integrated with the chromosome of an F⁺ bacterium to form a **high-frequency recombinant** (Hfr) cell. This enables the part of the host chromosome nearest to the F factor to be transferred to the recipient by conjugation.

sex hormones Steroid hormones that control sexual development and regulate the functioning of the reproductive system. The most important are the *androgens and *oestrogens.

sex linkage The tendency for certain inherited characteristics to occur far more frequently in one sex than the other. For example, red-green colour blindness and *haemophilia affect men more often than women. This is because the genes governing normal colour vision and blood clotting occur on the X *sex chromosome. Women have two X chromosomes. If one carries an abnormal allele it is likely that its effects will be masked by a normal allele on the other X chromosome. However, men only have one X chromosome and any abnormal alleles therefore will not be masked. *See also* CARRIER.

sex ratio The ratio of the number of females to the number of males in a *population. Because the mortality rates in the two sexes may be different, the sex ratios in different age classes may differ.

sexual cycle *See* MENSTRUAL CYCLE; OESTROUS CYCLE.

sexual dimorphism *See* DIMORPHISM.

sexual intercourse (coitus; copulation; mating) The process by which spermatozoa from a male are deposited in the body of a female during *sexual reproduction. In mammals the penis of the male becomes erect and stiff as its tissues become filled with blood, enabling it to be inserted into the vagina of the female. Thrusting movements of the penis result in *ejaculation, in which *semen, containing spermatozoa, is deposited in the vagina. The males of many invertebrates deposit their sperm in packets called **spermatophores**, which the female must then locate to transfer the sperm into her reproductive tract. In some cases the male assists in this process.

sexually transmitted disease (STD; sexually transmitted infection, STI) Any disease that is passed from one individual to another during sexual intercourse or other types of sexual activity. They include gonorrhoea, caused by the bacterium *Neisseria gonorrhoeae*; syphilis, due to infection by the bacterium *Treponema pallidum*; genital herpes, which is caused by a herpesvirus; chlamydia, caused by the bacterium *Chlamydia trachomatis*; trichomoniasis, caused by the parasitic protist *Trichomonas vaginalis*; and *AIDS, resulting from infection with *HIV, a retrovirus. Pubic lice and scabies mites are also transmitted by close sexual contact. The transmission of sexually transmitted diseases can be reduced by limiting the number of sexual partners and by the use of condoms (*see* BIRTH CONTROL), which reduces the risk of contact with body fluids that harbour the microorganisms that cause these diseases.

SEE WEB LINKS

• Extensive account of STDs from NHS Choices

sexual reproduction A form of reproduction that involves the fusion of two reproductive cells (*gametes) in the process of *fertilization. Normally, especially in animals, it requires two parents, one male and the other female. However, most plants bear both male and female reproductive organs and self-fertilization may occur, as it does in hermaphrodite animals. Gametes are formed by *meiosis, a special kind of cell division in the parent reproductive organs that both reassorts the genetic material and halves the chromosome number. Meiosis thus ensures genetic variability in the gametes and therefore in the offspring resulting from their subsequent fusion. Sexual reproduction, unlike *asexual reproduction, therefore generates variability within a species. However, it depends on there being reliable means of bringing together male and female gametes, and many elaborate mechanisms have evolved to ensure this.

sexual selection The means by which individuals with certain characteristics have a greater likelihood of obtaining mates compared with others of their kind. In **intersexual selection**, or **mate choice**, members of one sex select members of the opposite sex with which to mate. For example, females often choose to mate with the male that gives the best courtship display and therefore has the brightest coloration, loudest mating call, etc. In **intrasexual selection** members of one sex compete among themselves for access to mates. This might entail aggressive displays

or encounters between competing males to establish dominance and defence of territory or a group of females. Intersexual selection is assumed to have given rise to certain *secondary sexual characteristics, particularly of male animals, resulting in sexual *dimorphism. A theory of sexual selection was first advanced by Charles *Darwin to explain the evolution of apparently nonadaptive features, exemplified by the extravagant tail of the peacock. These features would be inherited by its male offspring and would thus tend to become exaggerated down the generations. It is speculated that secondary sexual features, such as long tail feathers, indicate to the female that the male is vigorous and healthy and thus has good genes to transmit to her offspring.

SH2 domain (Src homology 2 domain) A functional unit of a protein, typically comprising 60–100 amino acids, that forms a binding site for another protein. It is named after the prototype of such domains in members of the *Src tyrosine kinase family and is found in many intracellular signalling proteins.

SH3 domain *See* SRC TYROSINE KINASE.

shared derived trait *See* SYNAPOMORPHY.

shikimic acid pathway The key metabolic pathway in the biosynthesis of the aromatic amino acids—tyrosine, phenylalanine, and tryptophan—which occurs in plants, bacteria, and fungi, but not in animals. This series of reactions makes these amino acids available to animals, for which they are *essential amino acids, required in the diet for proteins and as precursors for a wide range of aromatic molecules. The initial step of the pathway is the condensation of erythrose-4-phosphate (from the *pentose phosphate pathway) and phosphoenolpyruvate (from *glycolysis). The product of this is then cyclized and reduced to form the intermediate compound shikimate, with the phenolic ring structure characteristic of the aromatic amino acids. Combination with a further molecule of phosphoenolpyruvate produces chorismate, from which alternative pathways lead to either phenylalanine/tyrosine or to trypto-phan. An enzyme responsible for a step in the conversion of shikimate to its immediate derivative, 3-enolpyruvyl-shikimate-5-phos-phate, is inhibited by the herbicide *glypho-sate, thus blocking the biosynthesis of aromatic amino acids by the plant.

Shine-Dalgarno sequence A sequence of five to nine (typically seven) nucleotides preceding the *start codon in prokaryotic messenger RNA (mRNA) that is recognized by the ribosome as the correct site for bind-ing the mRNA molecule prior to the start of translation. The sequence (AGGAGGU) binds a complementary sequence on the 16S ribosomal subunit, helping to form a stable complex between the ribosome and mRNA. The role of this sequence was first proposed by John Shine (1946–) and Lynn Dalgarno (1935–).

shivering See THERMOGENESIS.

shoot The aerial part of a vascular plant. It develops from the *plumule and consists of a stem supporting leaves, buds, and flowers.

short-day plant A plant in which flower-ing can be induced or enhanced by long nights, usually of more than 12 hours dur-ation. Examples are strawberry and chrysan-themum. See PHOTOPERIODISM. Compare DAY-NEUTRAL PLANT; LONG-DAY PLANT.

short interfering RNA (siRNA) Any of several types of small RNA molecules that are produced in cells from double-stranded (ds) precursors by the action of the protein *Dicer. This cuts the precursor to form a single-stranded RNA, typically 21 or 22 nucleotides long, which is incorporated into an RNA-induced silencing complex (RISC) that binds to sites with the comple-mentary sequence on messenger RNA (mRNA) molecules, thereby triggering deg-radation of the mRNA. This mechanism of *RNA interference probably evolved as a means of combating infection by RNA vir-uses, which replicate via double-stranded RNA intermediates. dsRNA can also be introduced by transfection of cells; this is performed experimentally for short-term

silencing of particular genes using siRNAs designed to bind to specific mRNAs. The base sequence of siRNAs matches perfectly that of their binding site and triggers cleav-age of the target RNA, whereas *micro-RNAs, which are normally encoded by the cell to control gene expression, often bind their targets imperfectly and merely sup-press translation of the mRNA.

short interspersed element See SINE.

short sequence repeat (SSR) See MICROSATELLITE DNA.

short-sightedness See MYOPIA.

short tandem repeats (STRs) Short sequences of nucleotides in DNA, 2–6 bp long, that are repeated in tandem at par-ticular loci throughout the genome and constitute *microsatellite DNA. The num-ber of repeats at any one site varies typically between five and thirty. STRs are widely used as the basis for *DNA profiling because they are highly variable and their small size makes them readily amenable to detection and amplification by the *polymerase chain reac-tion. See also REPETITIVE DNA.

shotgun cloning A method for cloning the entire genome of an organism in which the DNA is broken into small fragments at random using restriction enzymes. Each fragment is then incorporated with a clon-ing vector and introduced into a host organism. The fragments are maintained as a *DNA library. See GENE CLONING.

shotgun sequencing A strategy for sequencing large DNA fragments or entire genomes based on sequencing and order-ing small random fragments, typically 20–300 kbp long, obtained by *shotgun cloning. It relies primarily on computa-tional power to assemble the numerous and overlapping sequence reads of cloned fragments, obtained by repeated rounds of fragmentation and sequencing, into con-tiguous sequences, or contigs. Coalescence of the contigs yields a fully assembled whole-genome sequence. By sidestepping the laborious phase of constructing phys-ical maps of clones, shotgun sequencing is potentially faster than other approaches, especially for prokaryote genomes. However, genomes containing large amounts of repetitive DNA and the lack of a reference genome can prove problematical. It was employed by US geneticist Craig Venter's Celera Genomics company to produce a draft sequence of the human genome, from some 27 million different reads, in just three years and is now widely used to characterized microbial communities using the techniques of *metagenomics.

shoulder girdle See PECTORAL GIRDLE.

shunt vessel A blood vessel that links an artery directly to a vein, allowing the blood to bypass the capillaries in certain areas. Shunt vessels can control blood flow by constriction and dilation. In *endotherms the shunt vessels dilate in response to cold, thereby cutting off the blood flow to the extremities and preventing heat loss.

siblings Individuals that have both parents in common.

sibling species (cryptic species) Two or more groups of organisms that cannot be readily distinguished by their appearance or other traditional taxonomic criteria but whose members cannot interbreed successfully. For example, many North American species of cricket, occupying the same geographical range, can be distinguished only by their song.

sickle-cell disease See POLYMORPHISM.

sieve element A type of plant cell occurring within the *phloem. Sieve elements combine to form a series of tubes (**sieve tubes**) connecting the leaves, shoots, and roots in a fine network. Food materials are transported from one element to another via perforations termed **sieve areas** or **sieve plates** (*see* MASS FLOW). Sieve elements contain little cytoplasm and no nucleus, but are supported in their metabolic activities by adjacent *companion cells in angiosperms and by albuminous cells in gymnosperms. *See also* P-PROTEIN.

sievert The SI unit of dose equivalent (*see* RADIATION UNITS). It is named after the Swedish physicist Rolf Sievert (1896–1966).

sieve tube A tube within the *phloem tissue of a plant, composed of joined *sieve elements.

sigma factor A protein in *prokaryote cells that binds to RNA polymerase and also to the promoters of a specific set of genes, thereby directing the polymerase to transcribe those genes. Different sigma factors are expressed depending on which sets of genes are to be transcribed. *See* TRANSCRIPTION.

sigmoid growth curve See POPULATION GROWTH.

signal hypothesis A hypothesis to explain how ribosomes become attached to membranes within cells in order to deliver the appropriate proteins to cell organelles, such as mitochondria and chloroplasts, or transport proteins outside the cell membrane. It proposes that the leading end of the nascent polypeptide chain consists of a **signal peptide.** This sticks out from the ribosome and is recognized by a ribonucleoprotein particle called a **signal recognition particle (SRP)**. When the complex of ribosome and SRP encounters a membrane, the SRP binds to a **docking protein (signal recognition particle receptor)** on the membrane surface. Synthesis of the polypeptide, which has hitherto been stalled, now resumes, and the polypeptide (or fully formed protein) passes into the membrane, where the signal peptide is removed by a signal peptidase enzyme. Once translation is completed, the ribosome dissociates and is freed from the membrane. It is thought that the signal sequence tags the protein for insertion at particular sites, by interacting with membrane-bound glycoproteins (**signal sequence receptors**). If the signal sequence is not the correct one, the ribosome is released before delivering its protein. The hypothesis, which was formulated in the early 1970s by workers including Günter Blobel (1936–2018) and César Milstein (1927–2002), is now widely accepted.

signal transducer and activator of transcription (STAT) See JANUS KINASE.

signal transduction Any mechanism by which binding of an extracellular signal molecule to a cell-surface receptor triggers a response inside the cell. The mechanism depends on the type of signal molecule (e.g. hormone, paracrine, or autocrine signals), but it often involves changes in concentration of a *second messenger (e.g. cyclic AMP, inositol trisphosphate, calcium ions) within the cell, which in turn can amplify the initiating signal and create a cascade of interactions that affect numerous cell activities. Many receptors are associated with *G proteins, which act to turn signal transduction pathways on and off. Other important components of signal transduction include *protein kinases, which activate enzymes by transferring a phosphate group from ATP. See also IONOTROPIC RECEPTOR; METABOTROPIC RECEPTOR.

(((•))) **SEE WEB LINKS**

- Free downloads of posters for cell signal pathways from Cambridge-based bioreagents company Abcam

significance In research and statistical analysis, the degree to which an observed difference between two test groups or conditions is unlikely to have occurred by chance alone. In testing for significance, an initial assumption is made that there is no meaningful difference between the groups or conditions under investigation (the **null hypothesis**). Various statistical procedures are then applied, and a result indicating that there is a probability of less than 5% (P value <0.05) that the difference occurred by chance is usually sufficient to overturn the null hypothesis: the observed difference is **statistically significant**. Some tests are **parametric tests**, based on the assumption that observations will be distributed in a normal or Gaussian *distribution, where 95% of observations lie within two *standard deviations of the mean (**Student's t test** to compare means). Nonparametric tests (**Mann-Whitney U tests**) make no assumptions about distribution patterns. See also STANDARD ERROR OF THE MEAN.

sign stimulus (releaser) The essential feature of a stimulus, which is necessary to elicit a response. For example, a red belly (characteristic of courting male sticklebacks) is the sign stimulus necessary to provoke an attack from a rival male; even a very crude model fish is attacked if it has a red undersurface. Similarly, territorial fighting in male robins is triggered by the sign stimulus of a red breast; a bunch of red feathers is enough to induce territorial behaviour. See also STIMULUS FILTERING.

silent mutation An alteration in the genetic code that has no apparent effect on the phenotype of an organism.

silicula (silicle or silicule) (pl. **siliculae**) A type of *capsule formed from a bicarpellary ovary. It is longitudinally flattened and divided lengthwise into two cavities (**loculi**). It is broader than a *siliqua. Examples include the fruits of Alyssum and candytuft.

siliqua (silique) (pl. **siliquae** or **siliquas**) A type of *capsule formed from a bicarpellary ovary. It resembles a *silicula but is longer than it is broad; an example is the fruit of the wallflower. See also LOMENTUM.

silk A material produced by the silk glands of spiders, some insects, and certain other invertebrates (e.g. centipedes). It is exuded as a liquid by a protruding appendage, the *spinneret, but quickly hardens after leaving the gland. Silk is composed of α-keratin crystals embedded in a rubbery matrix of amino-acid chains, giving the material its flexibility and strength. A spider typically has several silk glands each producing a different type of silk, with properties determined by the nature of the amino-acid matrix. The spider switches from one gland to another to produce the silk appropriate for the task. For example, the silk used for wrapping prey is softer and distinct from structural silk used for the main fibres of a capture web. The silk may also be coated with a lipid waterproofing layer, as well as fungicides and bactericides to prevent attack from microorganisms. Spider silk has outstanding toughness and elasticity and on an equal-weight basis is stronger than steel or Kevlar, but it is completely biodegradable. Hence it continues to inspire attempts to produce synthetic equivalents through *bioengineering, possibly using some form of recombinant DNA technology.

Silurian A geological period of the Palaeozoic era following the Ordovician period and extending until the beginning of the Devonian period. It began about 444 million years ago and lasted for about 25 million years. The Silurian was named by Roderick Murchison (1792–1871) after an ancient British tribe that inhabited South Wales, where he observed rocks of this period. The majority of Silurian life was marine but during the later part of the period primitive plants began to make their appearance on land. Trilobites and graptolites became less common, brachiopods were numerous and varied, crinoids became common for the first time, and corals also increased. The only known vertebrates during the Silurian were primitive fish; the first jawed fish appeared later in the period. The Caledonian orogeny (mountain-building period) reached its peak towards the end of the Silurian.

(⊕) SEE WEB LINKS

- Brief survey of the Silurian period on the website of the University of California Museum of Paleontology

SINE (short interspersed element) Any of a class of dispersed moderately *repetitive DNA found in eukaryotes, consisting of numerous copies ($>10^5$) of relatively short (<500 bp) sequences scattered throughout the genome. SINEs are not translated into proteins, occur mostly in introns, and are nonautonomous *retrotransposons, which require a reverse transcriptase from another source (probably a *LINE) in order to replicate via an RNA intermediate. The most notable example in humans and other primates is the *Alu family.

single-cell protein (SCP) Protein produced by microorganisms, such as algae, bacteria, filamentous fungi, and yeasts, that is extracted for use as a component of human and animal foods. Growing demand for protein is focusing attention both on expanding

the range of substrates used to grow the microbial cultures that yield SCP, including food waste and crop residues, and on streamlining production systems for photosynthetic algae that require carbon dioxide and light as raw materials to make SCP.

single circulation The type of circulatory system that occurs in fishes, in which the blood passes only once through the heart in each complete circuit of the body *Compare* DOUBLE CIRCULATION.

single-locus sequence typing A variant of *multilocus sequence typing that involves sequencing a specific locus of one gene to characterize a sample of DNA.

single nucleotide polymorphism (SNP) A variation in the base sequence occurring at any given single position in the genome (for example A instead of C) that is found in more than 1% of the population. It thus differs from a *point mutation only in its greater frequency. SNPs can be found in all parts of the genome, including structural genes, regulatory regions, and noncoding DNA. In the human genome overall, about 10 million SNPs are thought to occur, making them exceptionally useful as *molecular markers. Some are known to be linked to disease-causing alleles. They can be detected by various techniques, including RNA sequencing, polymerase chain reaction, or microarrays. *See* INTERNATIONAL HAPMAP PROJECT.

sink strength The relative ability of a nonphotosynthesizing organ or tissue—so-called sink sites (*see* MASS FLOW)—to acquire and use the products of photosynthesis (i.e. photoassimilates) produced elsewhere in the plant. Sink strength is one factor determining the rate of flow of photoassimilate (chiefly sucrose) from leaves via *phloem to sink organs such as grains, fruits, or tubers. The strength of these sinks is determined by their size and physiological ability to maintain an effective concentration gradient of sucrose between phloem and tissue cells. The latter depends on the efficiency of the sucrose transporter, a cellular protein that actively unloads sucrose from phloem, and of enzymes like invertase, which converts sucrose into glucose and fructose.

sinoatrial node *See* PACEMAKER.

sinus A saclike cavity or organ in an animal, such as the *sinus venosus.

sinusoid A tiny blood vessel or blood-filled space in an organ. Sinusoids replace capillaries in certain organs, notably the liver; they allow more direct contact between the blood and the tissue it is supplying.

sinus venosus A thin-walled chamber of the heart of vertebrate embryos that conveys deoxygenated blood to the single atrium. It is retained in adult fishes but in higher vertebrates it becomes incorporated into the left atrium.

siphonaceous (siphonous) Describing algae and other protists in which the thallus is coenocytic (*see* COENOCYTE) and often takes the form of a tubular structure.

Siphonaptera An order of secondarily wingless insects comprising the fleas. The body of a flea is laterally compressed and bears numerous backward-directed spines. Fleas live as blood-sucking ectoparasites of mammals and birds, having mouthparts adapted to piercing their host, injecting saliva to prevent clotting, and sucking up the blood. The long bristly legs can transmit energy stored in the elastic body wall to leap relatively long distances (over 300 mm horizontally). Apart from causing irritation, fleas can transmit disease organisms, most notably bubonic plague bacteria, which can be carried from rats to humans by the rat flea (*Xenopsylla cheopis*). The whitish wormlike legless larvae feed on organic matter. After two moults the larva spins a cocoon and undergoes metamorphosis into the adult.

Siphunculata (Anoplura) An order of secondarily wingless insects comprising the sucking lice: blood-sucking ectoparasites of mammals, with piercing and sucking mouthparts forming a snoutlike proboscis. They constitute an irritating pest to humans and domestic animals and can transmit diseases, including typhoid. The human louse (*Pediculus humanus*) exists

in two forms, or ecotypes: the head louse (*P. humanus capitis*) and the body louse (*P. humanus corporis*).

siRNA *See* SHORT INTERFERING RNA.

sirtuin Any of a family of proteins that characteristically bind *NAD and enzymatically remove acetyl groups from a range of targets, notably histones, transcription factors, and other enzymes. This deacetylase activity is thought to play a significant role in the post-transcriptional modification of proteins. Evidence suggests that sirtuins play numerous roles in cells, including stress responses, DNA repair, energy metabolism, ageing, and tumour formation. Controversy surrounds the proposition that sirtuins mediate the health benefits of calorie-restricted diets, such as increased longevity, and that sirtuins are activated by the compound resveratrol, found in red wine.

sister species (sister group) Either of the two descendant species or groups formed when one lineage splits during evolution. Hence, the sister species (or sister group) is the one most closely related to any given species (or group), since both share an ancestral species (or group) not shared by any other species (or group). In classification systems based on the principles of *cladistics, sister species are always grouped together. Moreover the sister group is the prime choice for use in *outgroup comparison.

site-directed mutagenesis A technique used in genetic engineering for introducing particular changes to the base sequence of a gene at a specific site. It allows precise and specific mutations to be made and is done, for example, to modify the amino-acid sequence of the protein expressed by the gene to investigate how such a change affects the protein's structure and function, or in *protein engineering. First, the gene of interest is cloned and made available as single-stranded DNA—a widely used vector for this purpose is the bacteriophage M13. Then an artificial oligonucleotide, containing perhaps 20–30 nucleotides, is constructed containing the desired change in base sequence. This is allowed to hybridize with the complementary (apart from the mutation site) single-stranded DNA and is then extended at either end by the enzyme DNA polymerase using the single-stranded DNA as template. The two strands, each with their vector, are then separated and cloned, and clones containing the mutant gene are selected. *Compare* GENOME EDITING.

Site of Special Scientific Interest *See* SSSI.

SI units Système International d'Unités:
the international system of units now recommended for all scientific purposes. A coherent and rationalized system of units derived from the **m.k.s. units** (a metric system based on the metre, kilogram, and second), SI units have superseded *c.g.s. units and *Imperial units. The system has seven **base units** and two **dimensionless units** (formerly called **supplementary units**), all other units being derived from these nine units. There are 22 derived units with special names. Each unit has an agreed symbol (a capital letter or an initial capital letter if it is named after a scientist, otherwise the symbol consists of one or two lower-case letters). Decimal multiples of the units are indicated by a set of prefixes; when possible a prefix representing 10 raised to a power that is a multiple of three should be used. *See* Appendix 1.

Sivapithecus (*Ramapithecus*) A genus of
extinct primates that lived about 12–8 million years ago. Fossil remains of sivapithecines have been found in India and Pakistan, the Near East, and East Africa. Early discoveries of jaw fragments suggested that they chewed from side to side and had fairly short muzzles, both of which are humanoid features. However, subsequent finds, including a complete jaw, were not hominoid, and sivapithecines are now regarded as ancestral to the Asian great apes (e.g. orang-utans), not the hominins. *See also* DRYOPITHECUS; AUSTRALOPITHECUS.

skeletal muscle (striated muscle; voluntary muscle) Muscle that is generally

Structure and action of a skeletal muscle

Figure labels (left diagram): tendon; epimysium; bundle of muscle fibres; single muscle fibre; nucleus; myofibril; light band; dark band; sarcomere unit containing contractile proteins

Figure labels (right diagrams): origin; biceps (flexor) contracted; triceps (extensor) relaxed; insertion; tendon; arm flexed; biceps (flexor) relaxed; triceps (extensor) contracted; arm extended attached to the skeleton and under the control of the will, although it is also involved in involuntary movements such as breathing and maintaining posture. An individual muscle consists of bundles of long **muscle fibres**, each bounded by a *sarcolemma and containing *sarcoplasm, *sarcoplasmic reticulum, and many nuclei. The whole muscle is covered with a strong connective tissue sheath (**epimysium**) and attached at each end to a bone by inextensible *tendons. Running through each fibre are smaller fibres (**myofibrils**) having alternate light and dark bands, which contain protein filaments responsible for the muscle's contractile ability and give the muscle its typical striped appearance under the microscope. The functional unit of a myofibril is the *sarcomere. The end of the muscle that is attached to a nonmoving bone is called the **origin** of the muscle; the end attached to a moving bone is the **insertion**. As a muscle contracts, it becomes shorter and fatter, moving one bone closer to the other. Since a muscle cannot expand, another muscle (the **extensor**) is required to move the bone in the opposite direction and stretch the first muscle (known as the **flexor**). The flexor and extensor are described as **antagonistic muscles**.

The speed, strength, and endurance of contractions of a particular muscle are determined by the number of muscle fibres that are contracting at any one time, and the proportions in the muscle of *slow-twitch fibres and *fast-twitch fibres. See illustration.

skeleton The structure in an animal that provides mechanical support for the body, protection for internal organs, and a framework for anchoring the muscles. The skeleton may be external (*see* EXOSKELETON) or internal (*see* ENDOSKELETON). Both types require *joints to allow *locomotion. The skeleton of higher vertebrates consists of a system of *bones (*see* APPENDICULAR SKELETON; AXIAL SKELETON). Soft-bodied animals have a *hydrostatic skeleton.

⊕ SEE WEB LINKS
- The e-Skeletons Project site explores the comparative skeletal anatomy of humans and other primates

skin The outer layer of the body of a vertebrate. It is composed of two layers, the *epidermis and *dermis, with a complex nervous and blood supply. The skin may bear a variety of specialized structures, including *hair, *scales, and *feathers. This skin has an important role in protecting the body from mechanical injury, water loss, and the entry of harmful agents (e.g. disease-causing bacteria; *see* IMMUNITY). It is also a sense organ, containing receptors sensitive to pain, temperature, and pressure (*see* MEISSNER'S CORPUSCLES; MERKEL'S DISC; PACINIAN CORPUSCLE). In warm-blooded animals it helps regulate body temperature by means of hair, fur, or feathers and *sweat glands.

skull The skeleton of the head. In mammals it consists of a *cranium enclosing the brain and the bones of the face and jaw. All the joints between the individual bones of the skull are immovable (*see* SUTURE) except for the joint between the mandible (lower jaw) and the rest of the skull. There is a large opening (**foramen magnum**) at the base of the skull through which the spinal cord passes from the brain.

SLA *See* SPECIFIC LEAF AREA.

sleep A readily reversible state of reduced awareness and metabolic activity that occurs periodically in many animals. Usually accompanied by physical relaxation, the onset of sleep in humans and other mammals is marked by a change in the electrical activity of the brain, which is recorded by an *electroencephalogram as waves of low frequency and high amplitude (**slow-wave sleep**). This is interspersed by short bouts of high-frequency low-amplitude waves (similar to wave patterns produced when awake) associated with restlessness, dreaming, and rapid eye movement (REM); this is called **REM** (or **paradoxical**) **sleep** and is often accompanied by an increased pulse rate and dilation of the pupils. Humans typically experience four or five cycles of non-REM and REM sleep per night, with the latter accounting for around 20% of the total. Four stages can be distinguished in each phase of non-REM sleep, with stages 3

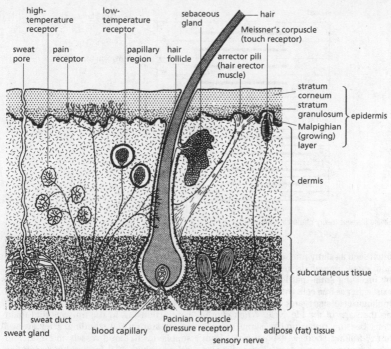

Structure of mammalian skin

and 4 considered to be the deepest, slow-wave sleep. Several regions of the brain are involved in sleep and arousal, especially the reticular formation of the *brainstem and certain pontine nuclei (*see* PONS). Sleep is essential for life, but understanding of its function remains elusive. It may be necessary for repair and maintenance of neural connections in the brain or for consolidating memories and learning. The latter is supported by the finding that individuals have impaired memory of tasks learned immediately before a period of sleep deprivation.

sleep movements *See* NYCTINASTY.

sliding filament theory A widely accepted model for the mechanism of muscle contraction in which the *actin and *myosin filaments of striated muscle slide over each other to shorten the length of the muscle fibres (*see* SARCOMERE).

Myosin-binding sites on the actin filaments are exposed when calcium ions bind to *troponin molecules in these filaments. This allows bridges to form between actin and myosin, which requires ATP as an energy source. Hydrolysis of ATP in the heads of the myosin molecules causes the heads to change shape and bind to the actin filaments. The release of ADP from the myosin heads causes a further change in shape and generates mechanical energy that causes the actin and myosin filaments to slide over one another (see illustration).

slime 1. (in plant physiology) *See* MUCIGEL. **2.** (in microbiology) *See* GLYCOCALYX (sense 2).

slime mould Any of various small simple eukaryotic organisms that live in damp terrestrial habitats and superficially resemble fungi, to which they are unrelated. They are

Sliding filament theory of muscle contraction

often seen as slimy masses on rotting wood and show *amoeboid movement, feeding by ingesting small particles of food. They exist either as free cells (**myxamoebas**) or as multinucleate aggregates of cells depending on the stage of the life cycle. When conditions become unfavourable, slime moulds form fruiting bodies (sporangia), from which spores are released. These disperse and subsequently germinate into small amoebas, thereby completing the life cycle. The classification of slime moulds has long been contentious. Modern molecular systematics identifies several distinct groups. **Plasmodial slime moulds** live as 'giant cells' (*see* PLASMODIUM), formed by the fusion of individual flagellated cells and containing many nuclei. They fall into two groups: Myxogastria and Protostelia. **Cellular slime moulds** live mainly as separate amoeboid cells but aggregate to form a cellular swarm called a **pseudoplasmodium**, in which the plasma membranes of individual cells are retained. There are two groups: the Dictyosteliida, which, like the plasmodial slime moulds, are included in the *amoebozoans; and the acrasid slime moulds (Acrasidae), which belong to the *excavates. Another unrelated group, the Labyrinthulomycota, consists of the **slime nets**, protists that secrete filaments along which the cells glide.

slow-twitch fibre (type I fibre) A type of skeletal muscle fibre that produces relatively slow, low-tension contractions and is highly resistant to fatigue. Slow-twitch fibres can produce lengthy and sustained contractions, as in the muscles that maintain posture when standing. They are well supplied with blood vessels and mitochondria, produce ATP by aerobic respiration, carry substantial reserves of fuel in the form of glycogen and fat, and contain large amounts of the oxygen-storing protein *myoglobin; hence they are the chief component of 'red' muscle. Athletes whose muscles contain a high proportion of slow-twitch fibres are generally suited to endurance events, such as long-distance running or cycling. *Compare* FAST-TWITCH FIBRE.

slow virus Formerly, a quasi-viral or subviral agent thought to be responsible for the diseases scrapie in sheep and bovine spongiform encephalopathy (BSE) in cattle. These diseases are now generally regarded as being caused by abnormal *prion proteins.

slow-wave sleep *See* SLEEP.

sludge *See* SEWAGE.

small cytoplasmic ribonucleoprotein (scRNP) *See* RIBONUCLEOPROTEIN.

small intestine The portion of the *alimentary canal between the stomach and the large intestine. It is subdivided into the *duodenum, *jejunum, and *ileum. It plays an essential role in the final digestion and absorption of food.

small nuclear ribonucleoprotein (snRNP) *See* RIBONUCLEOPROTEIN.

small nuclear RNA (snRNA) Any of a class of small RNA molecules, typically less than 200 nt in length, that function with other RNAs and proteins in splicing introns from pre-messenger RNA following transcription of DNA in cells. *See* RNA PROCESSING.

small nucleolar RNA (snoRNA) Any of a class of small RNA molecules, generally 60–300 nt in length, that guide enzymes to modify the chemical structure of ribosomal RNAs, either by adding a methyl group or by changing the nucleoside uridine to pseudouridine. They are associated chiefly with the *nucleolus.

small ubiquitin-related modifier *See* SUMO.

smell *See* OLFACTION.

smooth endoplasmic reticulum (smooth ER) *See* ENDOPLASMIC RETICULUM.

smooth muscle (involuntary muscle) Muscle whose activity is under the control of the *autonomic (involuntary) nervous system. Smooth muscle comprises long spindle-shaped cells without striations and thus has a 'smooth' rather than a striated appearance. These cells occur singly, in groups, or as sheets in the skin, around hair follicles, and in the digestive tract, the respiratory tract, the urinogenital tract, and the circulatory system. The cells contract slowly in spontaneous rhythms or when stretched; they may show sustained contraction (tonus) for long periods without fatigue. The thin (actin) and thick (myosin) filaments of smooth muscle interdigitate but are not organized into regular bands, as in the *sarcomeres of skeletal muscle. Instead, the thin filaments are anchored to *dense bodies within the cell or to attachment plaques on the inside surface of the plasma membrane. **Single-unit smooth muscles** are typically small elongated cells that taper at each end. They are connected by *gap junctions, which transmit spontaneous (myogenic) electrical depolarizations rapidly from cell to cell, thereby stimulating a wave of contraction, as in the peristaltic contractions of the intestinal tract. By contrast, the cells of **multi-unit smooth muscles** act independently and contract only when stimulated by autonomic nerves, as in the walls of blood vessels. Neurotransmitter is released from swellings (called varicosities) in the autonomic axons, which each activate numerous muscle cells. *Compare* CARDIAC MUSCLE; SKELETAL MUSCLE.

smuts A group of parasitic fungi of the phylum *Basidiomycota. Many of these species attack the ears of cereal crops, replacing the grain by a mass of dark spores. *Compare* RUSTS.

snakes *See* SQUAMATA.

SNAP *See* SNARE.

SNARE (soluble NSF attachment receptor) Any of a family of proteins that facilitate fusion between intracellular transport vesicles and the plasma membrane (or other target membrane) and are essential for secretion of cellular materials by *exocytosis. Each has a characteristic SNARE motif of 60–70 amino acids, which can associate with similar motifs in other SNAREs to form a SNARE complex. SNARE proteins are vital for normal release of neurotransmitter from synaptic vesicles at the *active zones of synapses. Essentially, SNARE proteins in the vesicle membrane (v-SNARES) form a complex with other SNARE proteins in the target membrane (t-SNARES), which leads to docking of the vesicle, fusion of the membranes, and formation of a fusion pore, allowing the contents of the vesicle to escape. The principal v-SNARE is **synaptobrevin** (belonging to a family called vesicle-associated membrane proteins, or VAMPs). The first step involves activation of the SNARE proteins by *N*-ethylmaleimide-sensitive factor (NSF), soluble NSF attachment protein

S

(**SNAP**), and ATP. Then, following tethering of the vesicle to its target membrane by a protein called RAB1, synaptobrevin associates with two t-SNARES, called **syntaxin** and SNAP-25 (synaptosome-associated protein of 25 kDa), to form a SNARE complex; this causes the vesicle to become irreversibly docked with the target membrane. A further protein, **synaptotagmin**, is recruited; this is sensitive to the presence of calcium ions (Ca^{2+}) and is thought to act as a switch mechanism, regulating the completion of membrane fusion in response to the influx of Ca^{2+} ions through calcium channels, and subsequent dissociation of the SNARE complex. Tetanus and botulinum toxins are proteases that degrade components of the SNARE complex, hence their potent effects on nerve function.

snoRNA See SMALL NUCLEOLAR RNA.

SNP See SINGLE NUCLEOTIDE POLYMORPHISM.

snRNA See SMALL NUCLEAR RNA.

snRNP See RIBONUCLEOPROTEIN.

social behaviour Any behaviour exhibited by a group of animals that interact with each other. Social behaviour ranges from moving as a herd in order to minimize the effects of predators to performing designated roles in highly organized societies. For example, within a colony of bees specific tasks, including tending the larvae, foraging for food, and controlling the temperature within the colony by wing fanning, are performed by different individuals (see CASTE; SUPERORGANISM). The application of evolutionary theory to social behaviour is termed **sociobiology**. This field is concerned primarily with genetically determined aspects of behaviour and their adaptive significance in terms of natural selection and reproductive success. See also EUSOCIAL.

sodium Symbol Na. A soft silvery element that is a major *essential element required by animals. It is important in maintaining the *acid-base balance and in controlling the volume of extracellular fluid and functions in the transmission of nerve impulses (see SODIUM PUMP; SODIUM ION CHANNEL).

sodium chloride (common salt) A colourless crystalline solid, NaCl, that is soluble in water. Sodium chloride has a key role in biological systems in maintaining electrolyte balances. It is used as a food preservative (see FOOD PRESERVATION). See also OSMOREGULATION.

sodium fluoride A crystalline compound, NaF. It is highly toxic but in very dilute solution (less than 1 part per million) it is used in the *fluoridation of water for the prevention of tooth decay on account of its ability to replace OH groups with F groups in the material of dental enamel.

sodium ion channel An *ion channel through a cell membrane specifically to allow the passage of sodium ions (Na^+). Voltage-gated sodium channels are crucial in producing the rapid reversal in electrical polarity (i.e. the *action potential) of excitable cells that occurs during passage of a nerve impulse. As an impulse arrives at any point, the sodium channels open in response to slight depolarization of the plasma membrane, allowing Na^+ to flood into the cell and causing the action potential, which travels further along the neuron to transmit the nerve impulse. Within a millisecond, the sodium channels close, and *potassium ion channels open to allow potassium ions (K^+) to enter the cell, thereby restoring the negative polarity of the resting potential at that point in the cell. Ligand-gated sodium channels also occur. For example, in the postsynaptic membrane of an excitatory synapse or neuromuscular junction, such channels open in response to a neurotransmitter (e.g. acetylcholine in the peripheral nervous system) to cause depolarization. In the cone cells of the retina, ligand-gated sodium channels close in response to a chemical signal generated by light; the consequent hyperpolarization leads to release of neurotransmitter by the cone cell and generation of a nerve impulse in connecting neurons. See LIGAND-GATED ION CHANNEL;

VOLTAGE-GATED ION CHANNEL. *See also* SODIUM-POTASSIUM PUMP.

sodium-potassium pump (Na⁺/K⁺ ATPase) A membrane *transport protein that exchanges sodium ions (Na⁺) for potassium ions (K⁺) across the plasma membrane of a eukaryotic cell. It exports three Na⁺ ions for every two K⁺ ions imported. This maintains the differential concentrations of each ion across the plasma membrane so that the cell normally has a low Na⁺ and a high K⁺ concentration relative to the extracellular fluid. This differential is vital to cellular function, e.g. in establishing the *resting potential of a neuron. The process requires energy in the form of ATP, being a form of *active transport.

softwood *See* WOOD.

soil The layer of unconsolidated particles derived from weathered rock, organic material (*humus), water, and air that forms the upper surface over much of the earth and supports plant growth. The formation of soil depends on the parent material (i.e. the original material from which the soil is derived), the climate and topography of the area, the organisms present in the soil, and the time over which the soil has been developing. Soils are often classified in terms of their structure and texture. The structure of a soil is the way in which the individual soil particles are bound together to form aggregates or peds. The structure types include platy, blocky, granular, and crumbs. The texture of a soil denotes the proportion of the various particle sizes that it contains. The four main texture classes are sand, silt, *clay, and *loam, of which loams are generally the best agricultural soils as they contain a mixture of all particle sizes. A number of distinct horizontal layers can often be distinguished in a vertical section (profile) of soil—these are known as **soil horizons**. Four basic horizons are common to most soils: an uppermost A horizon (or **topsoil**) containing the organic matter and living organisms forming a complex ecosystem; an underlying B horizon (or **subsoil**), which contains little organic material and

is strongly leached; a C horizon consisting of weathered rock; and a D horizon comprising the bedrock. Topsoil contains a vast assemblage of organisms, including bacteria, protists, fungi, algae, earthworms, nematodes, and insects. These decompose and recycle dead organic matter, forming the humus and ultimately making nutrients available in the form of inorganic minerals for plant roots to absorb. They also contribute to the physical and chemical nature of the soil; for example, earthworms clump particles together, thereby improving aeration.

soil erosion The removal and thinning of the soil layer due to climatic and physical processes, such as high rainfall, which is greatly accelerated by certain human activities, such as *deforestation. Soil erosion can lead to a loss of agricultural land and if unchecked, eventually results in *desertification. Careful soil husbandry and appropriate cropping practices are essential to prevent soil erosion. Measures include the planting of shelter belts to reduce wind erosion of soil; terracing and contour ploughing to retain soil on hill slopes; the use of cover crops where otherwise the soil surface would be bare, e.g. between crops; and minimal cultivation to help maintain the network of plant roots that is vital for stabilizing the soil.

sol A *colloid in which small solid particles are dispersed in a liquid continuous phase.

solenocytes *See* FLAME CELLS.

solute The substance dissolved in a solvent in forming a *solution.

solute potential Symbol Ψ_s. The component of *water potential that is due to the presence of solute molecules. It always has a negative value as solutes lower the water potential of the system.

solution A homogeneous mixture of a liquid (the *solvent) with a gas or solid (the **solute**). In a solution, the molecules of the solute are discrete and mixed with the molecules of solvent. There is usually some interaction between the solvent and solute molecules. For example, when sodium chloride dissolves in water the sodium ions attract polar water molecules, with the negative oxygen ion pointing towards the positive Na^+ ion. This is called **hydration**.

solvent A liquid that dissolves another substance or substances to form a *solution. **Polar solvents** are compounds, such as water, in which there is some separation of charge in the chemical bonds. These solvents are capable of dissolving ionic compounds or covalent compounds that ionize. **Nonpolar solvents** are compounds, such as benzene, that do not dissolve ionic compounds but will dissolve nonpolar covalent compounds.

soma (*pl.* **somata** or **somas**) *See* CELL BODY.

somatic 1. Relating to all the cells of an animal or plant other than the reproductive cells. Thus a somatic *mutation is one that is not heritable. **2.** Relating to organs and tissues of the body other than the gut and its associated structures. The term is applied especially to skeletal muscles, the sense organs, and the nervous system. *Compare* VISCERAL.

somatic cell hybridization *See* CELL FUSION.

somatic cell nuclear transfer (SCNT) *See* NUCLEAR TRANSFER.

somatic hypermutation A mechanism by which activated B cells can generate antibodies of increased specificity and with improved antigen binding over the course of an adaptive immune response. Special adaptations of these cells give rise to a relatively high rate of enzyme-induced base changes to the DNA of genes encoding the variable regions of both heavy and light chains of the antibody molecules. Such mutations are confined to clones of B cells generated by a T-cell-mediated response to the presence of specific antigen (*see* HELPER T CELL) and do not affect germ-line cells, so cannot be inherited. Over the course of several weeks, B cells that accumulate beneficial mutations producing better

antigen binding receive signals that promote their proliferation, whereas their counterparts with deleterious mutations can no longer bind antigen and die. Consequently, the efficiency of antigen binding of the B-cell population, and hence the overall effectiveness of the antibody response, increase over time. *Compare* SOMATIC RECOMBINATION.

somatic recombination A mechanism for generating diversity of antibodies in B cells and of receptors in T cells (*see* T-CELL RECEPTOR). The variable regions of an antibody's light (V_L) and heavy (V_H) chains, which determine the molecule's binding properties for antigen, are encoded by genes that are assembled from either two (in the case of V_L) or three (for V_H) gene segments. Both V_L and V_H chains are joined to an additional constant region (C_L and C_H, respectively), encoded by a further set of genes. The mature V_L gene is made up of a **V gene segment** and a **J** (or **joining**) **gene segment**. For example, the human λ light chain genetic locus spans 1050 kbp on chromosome 22 and comprises several clusters of gene segments encoding constant, variable, and joining regions of the light chain. There are thought to be about 29–33 functional V gene segments, 4–5 functional J gene segments, and 4–5 constant gene segments. So there are potentially of the order $30 \times 5 = 150$ different combinations of V and J gene segments, which means that 150 different mature VL genes can be made. Adding to this the 230 possible combinations for the second type of light chain (κ light chains, encoded by a gene cluster on chromosome 2) makes 380 different light chains. Heavy chain genes contain a third gene segment, called a **diversity** (or **D_H**) **gene segment**, and possible combinations of the V_H, D_H, and J_H segments found at the heavy chain locus on chromosome 14 total roughly 6000. Therefore, altogether there are theoretically 1.9×10^6 (i.e. 320×6000) different combinations of the several different antibody regions due to somatic recombination. This is one way in which B cells generate their vast repertoire of antibodies. The recombination mechanism, called

V(D)J recombination, is a multistep enzymatic process requiring the products of two genes that are expressed only in developing lymphocytes, namely the **recombination-activating genes** *RAG1* and *RAG2*. Diversity of T-cell receptors is generated by a similar mechanism involving the rearrangement of gene segments encoding variable regions of the α and β chains. Further diversity is generated by additional mechanisms, namely D-J recombination, which causes variation in amino acid sequence at the D-J junction of the heavy chain, and *somatic hypermutation. All these mechanisms together account for potentially something of the order of 10^{13} different antibodies—sufficient to match the great diversity of potential antigens.

somatic sensory neuron *See* SENSORY NEURON.

somatomedin *See* INSULIN-LIKE GROWTH FACTOR.

somatosensory cortex *See* CEREBRUM.

somatostatin (growth hormone inhibiting hormone; GHIH) A hormone, secreted by the hypothalamus, that inhibits the release of *growth hormone from the anterior pituitary gland (*see* RELEASE-INHIBITING HORMONE). The secretion of somatostatin is stimulated by various factors, including very high blood glucose levels, which result from the effect that growth hormone has on glucose metabolism. It is also produced by the D cells of the *islets of Langerhans in the pancreas, where it can inhibit the release of glucagon and insulin.

somatotrophin *See* GROWTH HORMONE.

somite Any of a series of segmented blocks of tissue in animal embryos that develop from the mesoderm. In vertebrate embryos they lie on the dorsal side of the notochord and give rise to the vertebral column, ribs, dermis, and striated muscle. In invertebrates showing *metameric segmentation the somites give rise to the body segments.

sonicator A device used to break down cells using ultrasonic waves. Ultrasonic waves are transmitted through a metal rod, which is placed in a suspension of the cells. The vibrations cause the plasma membranes to rupture and the cellular contents are released into the surrounding medium.

Sonic hedgehog (SHH) A protein that has a crucial role in patterning and development of tissues in vertebrates, particularly of the nervous and skeletal systems. It is homologous to **Hedgehog protein**, which performs a similar function in the fruit fly *Drosophila*—Hedgehog was so named because mutation of its gene in *Drosophila* produced flies bearing spiky denticles resembling the spines of a hedgehog. Sonic hedgehog, named after a video game character, is one of three proteins belonging to the Hedgehog family in vertebrates—the others being Desert hedgehog (DHH) and Indian hedgehog (IHH). It acts as a *morphogen, influencing the fate of embryonic cells in the vicinity according to its source and concentration. For example, it is released by the notochord and induces the formation of floor plate cells and motor neurons along the ventral midline of the spinal cord. Patterning of developing limbs is also regulated by SHH released from an organizer group of cells (**the zone of polarizing activity**, or **ZPA**) at the base of each limb bud. The concentration of SHH conveys positional information about posterior-anterior polarity of the limb for developing limb structures.

soredium (*pl.* **soredia**) A small body, consisting of a core of algal cells surrounded by fungal hyphae, that functions as a structure of vegetative propagation in *lichens. Soredia, which range from 25 to 100 μm in diameter, are released like spores and are dispersed in air currents. *Compare* ISIDIUM.

sorosis (*pl.* **soroses**) A type of *composite fruit formed from an entire inflorescence spike. Mulberry and pineapple fruits are examples.

sorus (*pl.* **sori**) **1.** Any of the spore-producing structures on the undersurface of a fern frond, visible as rows of small brown dots. **2.** A reproductive area on the thallus of some algae, e.g. *Laminaria.* **3.** Any of various spore-producing structures in certain fungi.

SOS response A suite of metabolic mechanisms that are invoked in a bacterial cell in response to damage by ultraviolet light, certain chemicals, or other mutagens. Such agents interfere with DNA replication, causing large gaps in newly synthesized DNA strands. Hence the main functions of the SOS response are to fill these gaps and maintain the integrity of the cell's DNA and to suspend cell division—and hence further rounds of DNA replication—until repairs have been made. In *E. coli* the key to this response is the RecA protein. This is involved in *postreplicative repair, which is one of the SOS repair mechanisms. Moreover, activation of RecA by DNA damage causes it to cleave another protein, LexA, which is a repressor for several DNA repair genes. Inactivation of LexA causes these genes to become activated, and their products are instrumental in filling gaps in damaged DNA. However, some of the SOS repair mechanisms are error-prone, introducing mutations in the nucleotide sequences, and it is these repair mechanisms that are the direct cause of mutations in such circumstances. In cells infected with lambda prophage, activation of RecA protein causes derepression of the prophage genes, leading to vegetative growth of the phage and ultimately host cell lysis. Hence the phage has evolved a means of detecting the SOS response, and is able to 'escape' from a damaged host.

Southern blotting A chromatographic technique for isolating and identifying specific fragments of DNA, such as the fragments formed as a result of DNA cleavage by *restriction enzymes. The mixture of fragments is subjected to electrophoresis through an agarose gel, followed by denaturation to form single-stranded fragments. These are transferred, or 'blotted', onto a nitrocellulose filter where they are immobilized in their relative positions. Specific *gene probes labelled with a radioisotope or fluorescent marker are then added. These hybridize with any complementary fragments on the filter, which are subsequently revealed by autoradiography or a fluorescence detector. The technique was devised by US biologist E. M. Southern (1938–). A similar technique for detecting RNA fragments is called **northern blotting**, by analogy. *See also* WESTERN BLOTTING.

spacer DNA Nonrepetitive DNA that lies between transcribed sequences, particularly tandemly repeated RNA genes or regulatory sequences. It is also a feature of many recombinant DNA constructs produced artificially for use in, e.g., genetic engineering and genome editing.

spadix (*pl.* **spadices**) A flowering shoot (a type of *spike) with a large fleshy floral axis bearing small, usually unisexual, flowers. It is protected by a large petal-like bract, the **spathe**, and is characteristic of plants of the family Araceae (e.g. calla lily).

spathe *See* SPADIX.

spatial summation *See* SUMMATION.

special creation The belief, founded in the scriptures of many religions, that every species was individually created by God in the form in which it exists today. In Christianity and Judaism this belief accords with the Book of Genesis. Such a view was the generally accepted explanation of the origin of life until the advent of *Darwinism. The idea still holds sway in certain religious communities, e.g. among fundamentalist Christians in the USA. However, special creation is contradicted by fossil evidence and genetic studies, and the pseudoscientific arguments of **creation science** cannot stand up to logical examination.

specialization 1. Increasing *adaptation of an organism to a particular environment. **2.** *See* PHYSIOLOGICAL SPECIALIZATION.

speciation The development of one or more species from an existing species. It occurs when *sympatric or *allopatric populations diverge so much from the parent population that interbreeding can no longer occur between them. *See* DOBZHANSKY-MULLER MODEL. *See also* DICHOPATRIC SPECIATION; PARAPATRIC SPECIATION.

species 1. A group of organisms that resemble each other more than they resemble members of other groups and cannot be subdivided into two or more species. The precise definition of what constitutes a species differs depending on which species concept is applied. According to the *biological species concept, a species comprises a group of individuals that can usually breed among themselves and produce fertile offspring. However, many other species concepts have been proposed, including the *phylogenetic species concept and various *typological species concepts. Typically, a species consists of numerous local populations distributed over a geographical range. Within a species, groups of individuals become reproductively isolated because of geographical or behavioural factors (*see* ISOLATING MECHANISM), and over time may evolve different characteristics and form a new and distinct species. **2.** A rank, or category, used in the *classification of organisms. Similar species are grouped into a genus, and a single species may be subdivided into *subspecies or *races. *See also* BINOMIAL NOMENCLATURE.

species diversity *See* BIODIVERSITY.

species evenness The relative *abundance of different species in a particular sample, community, or area of habitat. A community with greater evenness equates to less variation in abundances of the species present, and to greater species diversity, and hence *biodiversity. *Compare* SPECIES RICHNESS.

species richness The number of species of organisms present in a particular sample, community, or area of habitat. It is one measure of species diversity, and hence of *biodiversity. *Compare* SPECIES EVENNESS.

specific leaf area (SLA) The total area of the leaves of a plant divided by the dry mass of the leaves, typically measured in square centimetres per gram or square metres per kilogram. A high value for SLA is typical of fast-growing plants with high photosynthetic capability, whereas a slow-growing or drought-resistant plant would typically have a low SLA. *See also* LEAF AREA RATIO.

spectral karyotyping A form of *chromosome painting used in clinical medicine particularly for rapidly identifying the origin of *accessory chromosomes or exchanges/insertions between chromosomes within the full metaphase set of chromosomes (i.e. the karyotype) of an individual. Following the principles of *fluorescence in situ hybridization (FISH), the technique uses specific probes to label each pair of homologous chromosomes with a distinct combination of fluorescent markers. Spectral analysis of the resulting multicolour image reveals the abnormalities, which can often be readily visualized.

spectrin A protein that forms part of the fibrous network (*cytoskeleton) underlying the plasma membrane of erythrocytes and other cell types. Spectrins essentially consist of two helical polypeptide subunits that associate to form long flexible molecules. These are arranged in a polygonal mesh 'glued' together by short actin filaments and attached to the plasma membrane by ankyrin proteins. This mesh is involved in maintaining cell shape, particularly the biconcave shape of erythrocytes, and also serves as a platform for membrane channels, receptors, and transporters. Spectrins are also found in heart muscle cells closely associated with the contractile fibres.

spectroscopy The study of methods of producing and analysing *spectra using spectroscopes and other instruments. Spectroscopy is employed in a wide range of biological research areas, such as biochemistry and toxicology, for the identification of metabolites and other compounds of biological significance. *See* MASS SPECTROMETRY.

spectrum (*pl.* **spectra**) A range of electromagnetic energies arrayed in order of increasing or decreasing wavelength or frequency. The **emission spectrum** of a body or substance is the characteristic range of radiations it emits when it is heated, bombarded by electrons or ions, or absorbs photons. The **absorption spectrum** of a substance is produced by examining,

S

through the substance and through a spectroscope, a continuous spectrum of radiation. The energies removed from the continuous spectrum by the absorbing medium show up as black lines or bands; with a substance capable of emitting a spectrum these are in exactly the same positions in the spectrum as the emission lines and bands would occur in the emission spectrum.

Emission and absorption spectra may show a **continuous spectrum**, a **line spectrum**, or a **band spectrum**. A continuous spectrum contains an unbroken sequence of frequencies over a relatively wide range; it is produced by incandescent solids, liquids, and compressed gases. Line spectra are discontinuous lines produced by excited atoms and ions as they fall back to a lower energy level. Band spectra (closely grouped bands of lines) are characteristic of molecular gases or chemical compounds. Absorption spectra of chlorophylls and other photosynthetic pigments are important in the study of photosynthesis. *See* ACTION SPECTRUM.

Spemann's organizer (Spemann–Mangold organizer; dorsal lip) A cluster of cells in the dorsal mesoderm that determines the development of the spinal cord and brain in the early amphibian embryo. Discovered by the German embryologist Hans Spemann (1869–1941) and his graduate co-worker Hilde Mangold (1898–1924), it forms near the lip of the blastopore following gastrulation and releases signal proteins (including follistatin, noggin, and chordin) that diffuse to overlying ectoderm. These signals block the action of the growth factor bone morphogenetic protein 4 (BMP4) and induce cells to differentiate into nervous tissue instead of epidermis, forming the neural plate and subsequently the neural cord. *Compare* HENSEN'S NODE.

sperm 1. The male gamete of an animal or plant. *See* ANTHEROZOID. **2.** A single *spermatozoon. **3.** Spermatozoa, collectively.

spermatheca (seminal receptacle) (*pl.* **spermathecae**) A sac or receptacle in

some female animals (e.g. many insects) or hermaphrodite animals (e.g. earthworms) in which sperm from the mate is stored until the eggs are ready to be fertilized, e.g. when environmental conditions are appropriate. Sperm are released from the spermatheca to fertilize eggs as they travel through the reproductive tract before being passed out of the body.

spermatid A nonmotile cell, produced during *spermatogenesis, that subsequently differentiates into a mature spermatozoon. Four spermatids are formed after two meiotic divisions of a primary *spermatocyte and therefore contain the haploid number of chromosomes.

spermatium (*pl.* **spermatia**) A type of nonmotile male sex cell that is produced by the red algae (Rhodophyta) and by certain fungi (e.g. the *rusts).

spermatocyte A diploid cell in the testis that divides by meiosis to give rise to four *spermatids (*see* SPERMATOGENESIS). A **primary spermatocyte** develops from a *spermatogonium. Two **secondary spermatocytes** result from the first meiotic division of a primary spermatocyte; each of these produces two spermatids after the second meiotic division.

spermatogenesis The series of cell divisions in the testis that results in the production of spermatozoa. Within the seminiferous tubules of the testis germ cells grow and divide by mitosis to produce *spermatogonia. These divide by mitosis to produce *spermatocytes, which divide by meiosis to produce *spermatids. The spermatids, which thus have half the number of chromosomes of the original germ cells, then develop into spermatozoa.

spermatogonium (*pl.* **spermatogonia**) Any of the diploid cells in the walls of the seminiferous tubules of the testis that give rise to the primary *spermatocytes. *See also* SPERMATOGENESIS.

Spermatophyta In traditional classifications, a division of the plant kingdom containing plants that reproduce by means of

*seeds. In modern systems seed plants are grouped into separate phyla, the most important of which are the *Anthophyta and *Coniferophyta.

spermatozoid *See* ANTHEROZOID.

spermatozoon (sperm) (*pl.* **spermatozoa**) The mature mobile reproductive cell (*see* GAMETE) of male animals, which is produced by the testis (*see* SPERMATOGENESIS). It consists of a head section containing a *haploid nucleus and an *acrosome, which allows the sperm to penetrate the egg at fertilization; a middle section containing *mitochondria to provide the energy for movement; and a tail section, which contains a single *flagellum (sense 2) that lashes to drive the sperm forward.

sperm competition Competition between sperm from different males to reach and fertilize the egg cell of a single female. Sperm competition can occur among rodents in which a male mates a number of times with the same female, with a rest period between successive matings during which the sperm journeys towards the egg. If a second male mates with the female during a rest period its own sperm may disrupt the movement of sperm from the first male and succeed in fertilizing the egg cell. Certain animals in which sperm competition is possible have evolved features to minimize this interference. For example, in moths and butterflies the male cements the opening of the female genitalia after mating, thereby preventing further matings with other males. An ingenious

mechanism operates in the fly *Johannseniella nitida*, in which the female eats a copulating male except for his genitalia, which remain in the body of the female and prevent further mating.

Sphenophyta (Arthrophyta) *See* HORSETAILS.

spherosome (oleosome) A small spherical organelle found in the cytoplasm of plant cells. Up to 1 mm in diameter and bounded by a single membrane, it synthesizes and stores lipids.

sphincter A specialized muscle encircling an opening or orifice. Contraction of the sphincter tends to close the orifice. Examples are the **anal sphincter** (round the opening of the anus) and the **pyloric sphincter** (at the lower opening of the *stomach). A precapillary sphincter is a ring of smooth muscle that surrounds the entrance to a capillary bed and by contracting restricts the entry of blood.

sphingolipid *See* PHOSPHOLIPID.

spiders *See* ARACHNIDA.

spike A type of *racemose inflorescence in which stalkless flowers arise from an undivided floral axis, as in plantain and *Orchis*. In the family Gramineae (Poaceae; sedges and grasses) the flowers are grouped in clusters called **spikelets** (see illustration), which may be arranged to form a compound spike (as in wheat). *See* GLUME; LEMMA.

awn

lemma
palea } enclosing floret

glume

Spike: structure of a spikelet

spinal column *See* VERTEBRAL COLUMN.

spinal cord The part of the vertebrate central nervous system that is posterior to the brain and enclosed within the *vertebral column. It consists of a hollow core of *grey matter (H-shaped in cross section) surrounded by an outer layer of *white matter; the central cavity contains *cerebrospinal fluid. The white matter contains numerous longitudinal nerve fibres organized into distinct tracts: **ascending tracts** consist of sensory neurons, conducting impulses towards the brain; **descending tracts** consist of motor neurons, transmitting impulses from the brain. Paired *spinal nerves arise from the spinal cord.

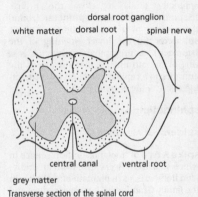

Transverse section of the spinal cord

spinal nerves Pairs of nerves that arise from the *spinal cord (*compare* CRANIAL NERVES). In humans there are 31 pairs (one from each of the vertebrae). Each nerve arises from a *dorsal root and a *ventral root and contains both motor and sensory fibres (i.e. they are mixed nerves). The spinal nerves form an important part of the *peripheral nervous system.

spinal reflex *See* REFLEX.

spindle A structure formed from *microtubules in the cytoplasm during cell division that moves chromatids (*see* MITOSIS) or chromosomes (*see* MEIOSIS) diametrically apart and gathers them in two clusters at opposite ends (poles) of the cell. Broad

in the middle and narrowing to a point at either pole, its construction is directed by a microtubule-organizing centre, the *centrosome. In the preliminary stages of cell division the centrosome divides, and the two daughter centrosomes move to opposite poles. Each organizes three sets of microtubules (or 'spindle fibres'): the first set consists of a tuft of fibres, called the *aster, radiating towards the cell periphery; the second set extends towards the centre of the cell to attach to chromatids or chromosomes; the third set also extends through the centre to overlap with its counterpart from the opposite pole at the **spindle equator**, the region halfway between the poles. The spindle becomes fully formed by metaphase, when the chromatids are attached to spindle fibres via their centromeres and lie at the spindle equator. During anaphase this set of fibres shortens and hauls the attached chromatids towards the corresponding pole of the cell. Also, the overlapping fibres at the equator actively engage and slide past each other to elongate the entire spindle. *See also* MUSCLE SPINDLE.

spindle attachment *See* CENTROMERE.

spine 1. *See* VERTEBRAL COLUMN. **2.** A hard pointed protective structure on a plant that is formed through modification of a leaf, part of a leaf, or a stipule. The edge of the holly leaf is drawn out into spines, but in cacti the whole leaf is modified as a spine. *Compare* PRICKLE; THORN.

spinneret A small tubular appendage from which *silk is produced in spiders and some insects. Spiders have four to six spinnerets on the hind part of the abdomen, into which numerous silk glands open. The silk is secreted as a fluid and hardens on contact with the air. Various types of silk are produced depending on its use (e.g. for webs, egg cocoons, etc.). The spinnerets that produce the cocoons of insects are not homologous with those of spiders. For example, the spinneret of the silkworm is in the pharynx and the silk is produced by modified salivary glands.

spinocerebellar tracts Pathways of neurons that transfer sensory information from the spinal cord to the cerebellum of the brain.

spiracle 1. A small paired opening that occurs on each side of the head in cartilaginous fish. It is the reduced first *gill slit, its small size resulting from adaptations of the skeleton for the firm attachment of the jaws. In modern teleosts (bony fish) the spiracle is closed up. In tetrapods the first gill slit develops into the middle ear cavity. **2.** Any of the external openings of the *tracheae along the side of the body of an insect.

spirillum (*pl.* **spirilla**) Any rigid spiral-shaped bacterium. Generally, spirilla are Gram-negative (*see* GRAM'S STAIN), aerobic, and highly motile, bearing flagella either in tufts or singly. They occur in soil and water, feeding on organic matter.

spirochaete Any nonrigid corkscrew-shaped bacterium that moves by means of flexions of the cell, produced by a series of rotatory axial fibrils underlying the cell's outer sheath. Most spirochaetes are Gram-negative (*see* GRAM'S STAIN), anaerobic, and feed on dead organic matter. They are particularly common in sewage-polluted waters. Some, however, cause disease in humans and other animals; *Treponema pallidum*, the agent of syphilis, is an example.

spirometer *See* RESPIROMETER.

spleen A vertebrate lymphoid organ, lying behind the stomach, in which worn-out red blood cells (erythrocytes) are collected and disposed of. It forms part of the peripheral lymphoid system and is a site where antigen-presenting cells stimulate lymphocytes to mount adaptive immune responses against invading pathogens. The spleen consists chiefly of red pulp comprising numerous sinuses where erythrocytes are stored and old ones destroyed. Their remnants are ingested and degraded by macrophages. Interspersed among the red pulp are areas of white pulp. These contain distinct zones of lymphocytes—B cells and T cells—arranged around central arterioles.

spliceosome *See* INTRON; RNA PROCESSING.

splicing *See* INTRON; RNA PROCESSING.

sponges *See* PORIFERA.

spongy bone *See* BONE.

spongy mesophyll *See* MESOPHYLL.

spontaneous generation The discredited belief that living organisms can somehow be produced by nonliving matter. For example, it was once thought that microorganisms arose by the process of decay and even that vermin spontaneously developed from household rubbish. Controlled experiments using sterilized media by Pasteur and others finally disproved these notions. *Compare* BIOGENESIS. *See also* BIOPOIESIS.

sporangiophore The simple or branched stalk that bears one or more sporangia.

sporangium (*pl.* **sporangia**) A reproductive structure in plants that produces asexual spores. A **megasporangium** produces megaspores, which give rise to the female gametophyte; in seed plants it is represented by the *ovule. A **microsporangium** produces microspores, which give rise to the male gametophyte; it is represented in seed plants by the *pollen sac. *See also* SPOROPHYLL.

spore A reproductive cell that can develop into an individual without first fusing with another reproductive cell (*compare* GAMETE). Spores are produced by plants, fungi, bacteria, and some protists. A spore may develop into an organism resembling the parent or into another stage in the life cycle, either immediately or after a period of dormancy. In plants showing *alternation of generations, spores are formed by the *sporophyte generation and give rise to the *gametophyte generation. In ferns, the rows of brown reproductive structures on the undersurface of the fronds are spore-producing bodies.

spore mother cell (sporocyte) A diploid cell that gives rise to four haploid spores by meiosis. *See also* MEGASPORE MOTHER CELL; MICROSPORE MOTHER CELL.

sporocyte *See* SPORE MOTHER CELL.

sporogonium (*pl.* **sporogonia**) The *sporophyte generation in mosses and liverworts. It is made up of an absorptive **foot**, a stalk (**seta**), and a spore-producing **capsule**. It may be completely or partially dependent on the *gametophyte.

sporophore (fructification) The aerial spore-producing part of certain fungi; for example, the stalk and cap of a mushroom.

sporophyll A leaf that bears *sporangia (spore-producing structures). In ferns the sporophylls are the normal foliage leaves, but in other plants the sporophylls are modified and arise in specialized structures such as the strobili (cones) of clubmosses, horsetails, and gymnosperms and the flower of angiosperms. Most plants produce spores of two different sizes (small **microspores** and large **megaspores**). The sporophylls bearing these are called **microsporophylls** and **megasporophylls** respectively.

sporophyte The generation in the life cycle of a plant that produces spores. The sporophyte is *diploid but its spores are *haploid. It is either completely or partially dependent on the *gametophyte generation in mosses and liverworts, but is the dominant plant in the life cycle of clubmosses, horsetails, ferns, and seed plants. *See also* ALTERNATION OF GENERATIONS.

Sporozoa *See* APICOMPLEXA.

sport An individual organism that shows the effects of a mutation. The term is usually applied to atypical forms of plants in horticulture.

squalene An intermediate compound formed in the synthesis of cholesterol; it is a hydrocarbon containing 30 carbon atoms. The immediate oxidation of squalene to squalene 2,3-epoxide is the last common step in the synthesis of *sterols in animals, plants, and fungi.

Squamata An order of reptiles comprising the lizards and snakes. They are thought to have appeared at the end of the Triassic period, about 200 million years ago, and have invaded a wide variety of habitats. Most lizards have four legs and a long tail, eardrums, and movable eyelids. Snakes are limbless reptiles that lack eardrums; the eyes are covered by transparent immovable eyelids and the articulation of the jaws is very loose, enabling a wide gape to facilitate swallowing prey whole. In both snakes and lizards, males possess paired *hemipenes for copulation.

squamous epithelium A type of *epithelium that is made up of flattened cells and therefore presents a small distance over which substances have to pass. This type of epithelium is found lining the alveoli in the lungs and Bowman's capsule in the kidney.

Src tyrosine kinase Any of a family of proteins that can add a phosphate group to specific tyrosine residues of other proteins; they are homologous to the Src protein (pronounced 'Sark') encoded by the viral oncogene v-*src*, which can produce tumours in chickens. Commonly associated with cell receptors, these kinases are involved in signal transduction pathways controlling cell division and proliferation and antigen recognition by lymphocytes. They contain the prototypical *SH2 domain (Src homology 2 domain) and the **SH3 domain**, which binds to proline-rich regions in other proteins.

SRY protein *See* TESTIS-DETERMINING FACTOR.

SSSI (Site of Special Scientific Interest) The legal designation for an area of land in England, Scotland, or Wales that has been identified by Natural England, Scottish National Heritage, or Natural Resources Wales as being of special interest because of its flora, fauna, or geological or physiographical features. Such sites are protected from development activities and funds are available for their conservation and management. There are over 9000 SSSIs in Britain; similar sites in Northern Ireland are designated Areas of Special Scientific

Interest (ASSIs) by the Northern Ireland Environment Agency.

stabilizing selection (normalizing selection) *Natural selection that acts to maintain the constancy of a species over successive generations. It involves selection against the extremes of the range of phenotypes for a particular characteristic. For example, babies whose birth weight is substantially below or above the average of 3.6 kg historically have a greater mortality than babies of average birth weight (although medical advances have now greatly reduced this pattern of selection in humans). *Compare* DIRECTIONAL SELECTION; DISRUPTIVE SELECTION. *See also* BALANCING SELECTION.

staining A technique in which cells or thin sections of biological tissue that are normally transparent are immersed in one or more coloured dyes (**stains**) to make them more clearly visible through a microscope. Staining heightens the contrast between the various cell or tissue components. Stains are usually organic salts with a positive and negative ion. If the colour comes from the negative ion (organic anion), the stain is described as **acidic**, e.g. *eosin. If the colour comes from the positive ion (organic cation), the stain is described as **basic**, e.g. *haematoxylin. **Neutral stains** have a coloured cation and a coloured anion; an example is *Leishman's stain. Cell constituents are described as being **acidophilic** if they are stained with acidic dyes, **basophilic** if receptive to basic dyes, and **neutrophilic** if receptive to neutral dyes. **Vital stains** are used to colour the constituents of living cells without harming them (*see* VITAL STAINING); **nonvital stains** are used for dead tissue.

Counterstaining involves the use of two or more stains in succession, each of which colours different cell or tissue constituents. **Temporary staining** is used for immediate microscopical observation of material, but the colour soon fades and the tissue is subsequently damaged. **Permanent staining** does not distort the cells and is used for tissue that is to be preserved for a considerable period of time.

Electron stains, used in the preparation of material for electron microscopy, are described as **electron-dense** as they interfere with the transmission of electrons. Examples are salts of heavy metals, such as lead hydroxide, phosphotungstic acid (PTA), uranyl acetate (UA), and osmic acid.

stamen One of the male reproductive parts of a flower. It consists of an upper fertile part (the *anther) on a thin sterile stalk (the **filament**). Stamens represent the plant's microsporophylls, producing microspores that develop into pollen grains. *See* SPOROPHYLL.

staminode A sterile stamen.

standard deviation (SD) In statistics, a measure of the dispersion of data about the mean. For a set of values $a_1, a_2, a_3, \ldots a_n$, the mean m is given by $(a_1 + a_2 + \ldots + a_n)/n$. The **deviation** of each value is the absolute value of the difference from the mean $\#|m - a_1\#|$, etc. The standard deviation is the square root of the mean of the squares of these values, i.e. $\sqrt{[(\#|m-a_1\#|^2 + \ldots \#|m-a_n\#|^2)/n]}$. A large standard deviation indicates that data points vary across a wide range of values, whereas a small SD indicates the opposite. *See also* DISTRIBUTION; SIGNIFICANCE.

standard error of the mean (SEM) The *standard deviation within a sample of observations (e.g. of plant sizes) divided by the square root of the number of observations comprising that sample. Whereas the standard deviation of a sample provides a measure of the scatter of the data, the SEM measures the extent to which the *means of several different samples taken from the same *population would vary. It is therefore an indication of the accuracy of the sample mean as an estimate of the mean of the population as a whole.

standard metabolic rate (SMR) The rate of energy metabolism required to maintain a non-growing, non-active, ectothermic animal. It is measured at a specified temperature, because the SMR of ectotherms varies according to the environmental temperature. The SMR of aquatic animals such as fishes and shrimps can be calculated from their oxygen consumption, e.g. by monitoring changes in the oxygen concentration of the water in a specially designed aquarium. *Compare* BASAL METABOLIC RATE.

standing biomass *See* STANDING CROP.

standing crop The total amount of living material in a specified population at a particular time, expressed as *biomass (**standing biomass**) or its equivalent in terms of energy. The standing crop may vary at different times of the year; for example, in a population of deciduous trees between summer and winter.

stapes (stirrup) (*pl.* **stapes** or **stapedes**) The third of the three *ear ossicles of the mammalian middle ear.

Staphylococcus A genus of spherical nonmotile Gram-positive bacteria that occur widely as saprotrophs or parasites. The cells occur in grapelike clusters. Many species inhabit the skin and mucous membranes, and some cause disease in humans and animals. *S. aureus* infection can lead to boils and abscesses in humans; this species also produces *toxins that irritate the gastrointestinal tract and result in staphylococcal food poisoning. Certain strains are resistant to antibiotics, and infection with these is very difficult to treat. For example, some strains of methicillin-resistant *S. aureus* (**MRSA**) are now resistant to nearly all antibiotics and pose a grave threat to patients, causing lethal infections of the skin, lungs, and blood.

starch A *polysaccharide consisting of various proportions of two glucose polymers, *amylose and *amylopectin. It occurs widely in plants as a carbohydrate energy store, occurring as granules in chloroplasts and other plastids, especially in the storage cells of roots, tubers, seeds, and fruits. Starch is therefore a major energy source for animals. When digested it ultimately yields glucose. Starch granules are insoluble in cold water but disrupt if heated to form a gelatinous solution.

star diagram *See* PHYLOGRAM.

Starling's law Either of two physiological principles first described by the British physiologist Ernest H. Starling (1866–1927). **Starling's law of the heart** states that the more the muscle fibres of the heart are stretched at the end of diastole, the stronger their contraction during the following systole part of the *cardiac cycle. This principle generally operates under normal conditions and ensures that the amount of blood pumped out of the heart (cardiac output) automatically adjusts to the amount of venous blood returning to the heart. This relationship produces an S-shaped curve (**Starling's curve**) when stroke volume is plotted against ventricular pressure at the end of diastole. **Starling's law of the capillaries** describes the movement of water and small solutes between the capillaries and the surrounding intercellular space, as well as how blood volume is maintained in capillaries. At the arterial end of the capillaries, blood pressure is relatively high, and fluid is forced out through the capillary walls by hydrostatic pressure. Towards the venous end, blood pressure falls, and much of the fluid returns to the capillaries due to the inwardly directed osmotic pressure created by the large proteins retained in the capillaries. There is roughly a net 10% loss of fluid from the capillary bed, but this eventually returns to the venous blood via the lymphatic vessels.

start codon (initiation codon) The triplet of nucleotides on a messenger *RNA molecule (*see* CODON) at which the process of *translation is initiated. In eukaryotes the start codon is AUG (*see* GENETIC CODE), which codes for the amino acid methionine; in bacteria the start codon can be either AUG, coding for *N*-formyl methionine, or GUG, coding for valine. *Compare* STOP CODON.

startle display A response by an animal to discovery by a predator in which the potential victim exposes some previously hidden markings, for example an eyespot, in order to surprise its attacker. This may enable the victim to escape or divert the attentions of the predator elsewhere. Some camouflaged insects use this as a second line of defence, flicking their forewings forwards when disturbed to reveal a pair of vividly coloured eyespots on the hindwings.

STAT (signal transducer and activator of transcription) *See* JANUS KINASE.

stationary phase 1. *See* BACTERIAL GROWTH CURVE. **2.** *See* CHROMATOGRAPHY.

statoblast A chitinous bud that is produced asexually by moss animals and other members of the *Bryozoa. It stores food and is able to remain dormant for long periods, withstanding drought and extreme temperatures. Statoblasts may be liberated and dispersed after the death of the parent zooid, and subsequently 'germinate' to produce a zooid capable of founding a new colony.

statocyst (otocyst) A balancing organ found in many invertebrates. It consists of a fluid-filled sac lined with sensory hairs and contains granules of calcium carbonate, sand, etc. (**statoliths**). As the animal moves the statoliths stimulate different hairs, giving a sense of the position of the body or part of it. The *semicircular canals in the ears of vertebrates act on the same principle and have a similar function.

statocyte Any plant cell that contains *statoliths.

statolith 1. *See* STATOCYST. **2.** A membrane-bound group of starch grains (*see* AMYLOPLAST) that according to the starch statolith hypothesis acts as a sensor to gravity. Starch statoliths are found in specialized cells (statocytes) in the root cap at the growing tip, and in the tissues close to the vascular bundles in shoots. Under the influence of gravity the statoliths sink to the bottom of the cell and influence the release of the plant hormone *auxin (*see* GEOTROPISM). Their mechanism of action in triggering the transport of auxin across the plasma membrane is not understood. One theory is that the statoliths exert pressure on membrane systems inside the cell (e.g. the endoplasmic reticulum), which in turn potentially affects the opening or closing of ions channels and ultimately leads to differential growth. Alternatively, the weight of the statoliths may mechanically 'stretch' the upper side and compress the lower side of the root cells, causing the root tip to grow downwards.

stearic acid (octadecanoic acid) A solid saturated *fatty acid, $CH_3(CH_2)_{16}COOH$, that occurs widely (as *glycerides) in animal and vegetable fats.

stele The vascular tissue (i.e. *xylem and *phloem) of *tracheophyte plants, together with the endodermis and pericycle (when present). The arrangement of stelar tissues is very variable. In roots the stele often forms a solid core, which better enables the root to withstand tension and compression. In stems it is often a hollow cylinder separating the cortex and pith. This arrangement makes the stem more resistant to bending stresses. Monocotyledons and dicotyledons can usually be distinguished by the pattern of their stelar tissue. In monocotyledons the vascular bundles are scattered throughout the stem whereas in dicotyledons (and gymnosperms) they are arranged in a circle around the pith.

stem The part of a plant that usually grows vertically upwards towards the light and supports the leaves, buds, and reproductive structures (see illustration). The leaves

Transverse section through a herbaceous stem

develop at the *nodes and side or branch stems develop from buds at the nodes. The stems of certain species are modified as bulbs, corms, rhizomes, and tubers. Some species have twining stems; others have horizontal stems, such as *runners. Another modification is the *cladode. Erect stems may be cylindrical or angular; they may be covered with hairs, prickles, or spines and many exhibit secondary growth and become woody (*see* GROWTH RING). In addition to its supportive function, the stem contains *vascular tissue that conducts food, water, and mineral salts between the roots and leaves. It may also contain chloroplasts and carry out photosynthesis.

stem cell A cell that is not differentiated itself but can undergo unlimited division to form other cells, which either remain as stem cells or differentiate to form specialized cells. For example, stem cells in the bone marrow divide to produce daughter cells that differentiate into various types of immune cell (e.g. monocytes, lymphocytes, mast cells). Also, stem cells in the intestine continually divide to replace cells sloughed off the gut lining. **Embryonic stem cells**, such as those taken from an early human embryo, are capable of differentiating into all of the various tissue cells found in a fully developed individual—they are described as **totipotent**. Stem cells that can differentiate into a more limited range of tissues are described as **pluripotent**. Cultures of such cells have the potential to provide replacement tissues and organs for medical use, including transplantation and the treatment of various diseases, such as type 1 diabetes, Parkinson's disease, and Huntington's disease. However, ethical concerns have led to tight controls on research using human embryonic stem cells in many countries, including the USA and UK. Such objections can be overcome by techniques that turn tissues cells into stem cells. For example, adult skin cells can be harvested from a patient, cultured, and genetically manipulated so that they are effectively dedifferentiated. One technique uses a retroviral cloning vector to introduce copies of crucial stem cell regulatory genes into cultured skin fibroblasts. The fibroblasts are deprogrammed by these 'master genes' to form **induced pluripotent stem cells (iPS cells)**. These can then be induced to differentiate into various tissues, with the great advantage that they are accepted as 'self' tissue by the patient's immune system. Meristem tissue from plants contains stem cells from which new plantlets can be grown using the technique of *micropropagation. Commercial cloning of plants also makes use of the totipotent *callus tissue that forms over wounds. *See also* HAEMOPOIETIC TISSUE; MERISTEM.

(((●))) SEE WEB LINKS
- Website of the International Stem Cell Forum, which promotes good practice and progress in stem cell research

steno- A prefix denoting narrowness. For example, **stenohaline** aquatic organisms can tolerate only small variations in the salinity of the water. *Compare* EURY-.

stereocilium (*pl.* **stereocilia**) A relatively short, nonmotile cilium, some 20 to 300 of which form an array on the apical surface of a sensory *hair cell. The stereocilia are usually associated with a single, much longer *kinocilium but lack the 9 + 2 arrangement of internal microtubules characteristic of flagella. Instead, their shaft contains numerous fine longitudinal actin filaments. Bending of the kinocilium causes the array of stereocilia to bend en masse, thereby changing the pattern of impulses in the adjacent sensory neuron.

stereoisomerism The existence of chemical compounds (**stereoisomers**) that have the same molecular formulae and functional groups, but differ in the arrangement of groups in space. Optical isomerism is one form of this (*see* OPTICAL ACTIVITY).

sterigma (*pl.* **sterigmata**) A small spore-bearing stalk found in certain fungi. In the Basidiomycota, sterigmata develop as outgrowths of a basidium, each bearing a basidiospore at its tip.

sterile 1. (of organisms) Unable to produce offspring. *See also* HYBRID; INCOMPATIBILITY; SELF-STERILITY; STERILIZATION. **2.** (of objects, food, etc.) Free from contaminating microorganisms. *See* STERILIZATION.

sterilization 1. The process of destroying microorganisms that contaminate food, wounds, surgical instruments, etc. Common methods of sterilization include heat treatment (*see* AUTOCLAVE; PASTEURIZATION) and the use of *disinfectants and *antiseptics. **2.** The operation of making an animal or human incapable of producing offspring. Men are usually sterilized by tying and then cutting the *vas deferens (**vasectomy**); in women the operation often involves permanently blocking the fallopian tubes by means of clips (**tubal occlusion**). *See also* BIRTH CONTROL.

sternum (breastbone) (*pl.* **sternums** or **sterna**) **1.** A shield-shaped or rod-shaped bone in terrestrial vertebrates, on the ventral side of the thorax, that articulates with the *clavicle (collar bone) of the pectoral girdle and with most of the ribs. It is absent in fish, and in birds it bears a *keel. **2.** The ventral portion of each segment of the exoskeleton of arthropods.

steroid Any of a group of lipids derived from a saturated compound called cyclopentanoperhydrophenanthrene, which has a nucleus of four rings (see formulae). Some of the most important steroid derivatives are the steroid alcohols, or *sterols, such as cholesterol. Other steroids include the *bile acids, which aid digestion of fats in the intestine; the sex hormones (*androgens and *oestrogens); and the *corticosteroid

hormones, produced by the adrenal cortex. *Vitamin D is also based on the steroid structure.

sterol Any of a group of *steroid-based alcohols having a hydrocarbon side-chain of 8–10 carbon atoms. Sterols exist either as free sterols or as esters of fatty acids. Animal sterols (**zoosterols**) include *cholesterol and lanosterol. The major plant sterol (**phytosterol**) is beta-sitosterol, while fungal sterols (**mycosterols**) include *ergosterol.

sticky end (cohesive end) A single unpaired strand of nucleotides protruding from the end of a double-stranded DNA molecule. It is able to join with a complementary single strand, e.g. the sticky end of another DNA molecule, thus forming a single large double-stranded molecule. Sticky ends provide a means of annealing segments of DNA in many applications, e.g. in packaging of *vectors in genetic engineering, and in hybridizing primers with target DNA in the *polymerase chain reaction.

stigma 1. The glandular sticky surface at the tip of a *carpel of a flower, which receives the pollen (*see* POLLINATION). In insect-pollinated plants the stigmas are held within the flower, whereas in wind-pollinated species they hang outside it. **2.** *See* EYESPOT.

stilt root *See* PROP ROOT.

stimulus (*pl.* **stimuli**) Any change in the external or internal environment of an organism that provokes a physiological or behavioural response in the organism. Specific *receptors are sensitive to particular

steroid nucleus   cholesterol (a sterol)   testosterone (an androgen)
Steroid structure

types of stimuli, such as sounds, smells, pain, injury, visual stimuli, or temperature.

stimulus filtering The process of separating useful sensory information from the many thousands of stimuli present in the external environment, so that only potentially useful information is sent to the brain. Many sensory organs are adapted so that they receive stimuli only within a certain range. For example, the human eye only detects colours in the visible region of the spectrum and not in the ultraviolet or infrared regions. Another method of stimulus filtering involves *sign stimuli, which are more important than other stimuli in eliciting a response and, once detected, may lead to other stimuli being ignored.

stipe 1. The stalk that forms the lower portion of the fruiting body of certain fungi, such as mushrooms, and supports the umbrella-shaped cap. **2.** The stalk between the holdfast and blade (**lamina**) of certain brown algae, notably kelps.

stipule An outgrowth from the petiole or leaf base of certain plants. Those of the garden pea are leaflike photosynthetic organs. The stipules of the lime tree are scalelike and protect the winter buds, whereas those of the false acacia (*Robinia*) are modified as spines.

STM *See* SCANNING TUNNELLING MICROSCOPY.

stock *See* GRAFT.

stolon A long aerial side stem that gives rise to a new daughter plant when the bud at its apex touches the soil. Plants that multiply in this way include blackberry and currant bushes. Gardeners often pin down stolons to the soil to aid the propagation of such plants. This process is termed **layering**.

stoma (*pl.* **stomata**) **1.** (in botany) A pore, large numbers of which are present in the epidermis of leaves (especially on the undersurface) and young shoots. Stomata function in the exchange of carbon dioxide and oxygen between the plant and the atmosphere and are the principal route of evaporative water loss. Each stoma is bordered by two semicircular *guard cells, whose movements (due to changes in water content) control the size of the aperture and hence the movement of gases and water vapour. The density of stomata is determined both by genetics and by the environmental conditions to which the plant is exposed during its development; some plants may have a stomatal density of up to 20 000 per square centimetre. A decline in stomatal density in some species over the past 100 years is thought to be associated with the marked increase in atmospheric levels of carbon dioxide during that time. The term stoma is also used to mean both the pore and its associated guard cells. *See* STOMATAL CONDUCTANCE. **2.** (in anatomy) Any small opening or pore, especially an artificial opening connecting an internal cavity or organ with the body surface.

stomach The portion of the vertebrate *alimentary canal between the oesophagus and the small intestine. It is a muscular organ, capable of dramatic changes in size and shape, in which ingested food is stored and undergoes preliminary digestion. Cells lining the stomach produce *gastric juice, which is thoroughly mixed with the food by muscular contractions of the stomach. The resultant acidic partly digested food mass (*chyme) is discharged into the *duodenum through the pyloric *sphincter for final digestion and absorption. Some herbivorous animals (*see* RUMINANTIA) have multichambered stomachs from which food is regurgitated, rechewed, and swallowed again.

stomatal conductance The rate at which water vapour passes through the stomata of a plant per unit leaf area, typically measured in millimoles per square metre per second. It varies between plants, depending on the distribution density, size, and pore thickness of the stomata, and in the same plant over time according to the difference in vapour pressure between the inside of the plant and the external environment and the degree of opening of the stomatal pores.

stomium (*pl.* **stomia**) A region of thin-walled cells in certain spore-producing structures that ruptures to release the spores. For example, in the sporangium of the fern *Dryopteris* the stomium ruptures when the annulus dries out.

stop codon (termination codon) The triplet of nucleotides on a messenger *RNA molecule (*see* CODON) at which the process of *translation ends. It is recognized by proteins called *release factors, which bind to the A site of the ribosome. This effectively stops the formation of a polypeptide chain at that point. The three stop codons are UGA, UAA, and UAG (*see* GENETIC CODE). *Compare* START CODON.

storage compound *See* FOOD RESERVES.

stramenopiles An assemblage of eukaryotic organisms, sometimes regarded as constituting the phylum (division) Heterokontophyta, proposed on the basis of recent evidence from molecular systematics. The stramenopiles include a diversity of forms, ranging from unicellular (e.g. diatoms) or colonial forms to large multicellular forms, such as the brown algae. Many have hitherto been classified in separate phyla, including the diatoms (*see* BACILLARIOPHYTA), brown algae (*see* PHAEOPHYTA), oomycetes (*see* OOMYCOTA), golden-brown algae (*see* CHRYSOPHYTA), and yellow-green algae (*see* XANTHOPHYTA). Stramenopiles are characterized by having, at some stage in their life cycles, cells with flagella of two distinct types: one smooth flagellum and one *tinsel flagellum with hairlike projections. Other features are the presence in certain groups of chlorophylls *a* and *c* but the absence of chlorophyll *b*, use of the polysaccharide chrysolaminarin as a food reserve, and a double layer of endoplasmic reticulum enveloping the chloroplast membranes. The stramenopiles, along with the *alveolates and *rhizarians, are now widely regarded as constituting the *SAR supergroup.

stratification 1. The arrangement of the components of an entity in layers (**strata**). Stratification is a feature of sedimentary rocks and *soils. It is also seen in *stratified epithelium, and thermal stratification can occur in some lakes (*see* THERMOCLINE). **2.** The practice of placing certain seeds between layers of peat or sand and then exposing them to low temperatures for a period, which is required before they will germinate. *See* VERNALIZATION.

stratified epithelium *Epithelium that is made up of a number of layers of platelike (squamous) cells and is present in areas of the body that are subject to wear and tear. Stratified squamous epithelium containing *keratin forms the outer layer of the skin; stratified epithelial cells that do not contain keratin are found in moist areas, such as the mouth and vagina. A **pseudostratified epithelium**, e.g. the ciliated cells of the mucous membrane lining the respiratory tract, consists of a single layer of cells whose nuclei are arranged at varying heights within the layer, suggestive of a truly stratified epithelium.

stratified sampling *See* SAMPLING.

stratosphere The layer of the earth's atmosphere that lies above the troposphere and extends to about 50 km above the earth's surface. The temperature within the stratosphere remains fairly constant but can rise in the upper regions of this layer due to absorption of ultraviolet radiation by ozone. *See* OZONE LAYER.

stratum corneum The layers of dead keratinized cells that form the outermost layers of mammalian *epidermis. The stratum corneum provides a water-resistant barrier between the external environment and the living cells of the *skin.

Streptococcus A genus of spherical Gram-positive bacteria occurring widely in nature, typically as chains or pairs of cells. Many are saprotrophic and exist as usually harmless commensals inhabiting the skin, mucous membranes, and intestine of humans and animals. Others are parasites, some of which cause diseases. There are several systems of classifying streptococcal species. The Lancefield system, devised by US microbiologist Rebecca Lancefield (1895–1981) in the 1920s, identifies groups serologically according to unique carbohydrate antigens in their cell wall. There are 20 Lancefield groups, designated A to V; the common pathogen *S. pyogenes*, responsible for 'strep throat' and scarlet fever, is a group A streptococcus, while *S. agalactiae* is group B, and so on. However, some species lack a Lancefield antigenic marker, notably *S. pneumoniae*, which is responsible for pneumococcal infections ranging from bronchitis to septicaemia and meningitis. Another classification system depends on the pattern of breakdown (haemolysis) of blood cells when the bacterium is incubated on blood agar. Alpha-haemolytic bacteria, e.g. *S. pneumoniae*, cause only a partial haemolysis, whereas beta-haemolytic bacteria, e.g. *S. pyogenes*, lyse the blood cells completely, forming a clear zone around the bacterial colonies.

streptomycin *See* ACTINOBACTERIA; ANTIBIOTICS.

streptophyte Any of a clade of green plants including the land plants (embryophytes) and the closely related multicellular freshwater algae, i.e. the stoneworts and coleochaetophytes. All characteristically retain the egg inside the parent, possess a *phragmoplast to organize cell plate formation during cell division, and have cytoplasmic connections (*plasmodesmata) between adjacent cells. In some classifications the streptophytes are synonymous with the charophytes.

stress protein *See* HEAT-SHOCK PROTEIN.

stretch receptor A specialized cell or group of cells found in muscles (*see* MUSCLE SPINDLE) or tendons that is sensitive to mechanical stress. When muscles or tendons are stretched, the strain is registered by special sensory nerve cells and converted (transduced) into nerve impulses, which are transmitted to the central nervous system. *See* GOLGI TENDON ORGAN; MECHANORECEPTOR.

stretch reflex (myotatic reflex) The *reflex initiated when a muscle is stretched; an example is the **knee-jerk reflex**. Stretching of a muscle causes impulses to be generated in the *muscle spindles. These impulses are transmitted by sensory neurons to the spinal cord, where the sensory neurons synapse with motor neurons; these initiate contraction of the same muscle so that it returns to its original length. Since the reflex action involves the transmission of impulses across only one set of synapses, the response is rapid and described as **monosynaptic**. *See* GOLGI TENDON ORGAN.

striated muscle *See* SKELETAL MUSCLE.

stridulation The production of sounds by insects rubbing one part of the body against another. The parts of the body involved vary from species to species. Stridulation is typical of the Orthoptera (grasshoppers, crickets, cicadas), in which the purpose of the sounds is usually to bring the sexes together, although they are also used in territorial behaviour, warning, etc.

strigolactone Any of a class of *plant hormones derived from carotenoids that function chiefly in controlling apical dominance, attracting mycorrhizal fungi to roots, and promoting seed germination. Strigolactones modulate the flow of auxins from a plant's growing shoot tips down the stem, thereby suppressing the growth of axillary buds further down the stem. This effect is antagonized by *cytokinins moving up the stem from the root, permitting the growth of axillary buds nearest the base of the stem. Strigolactones released by the roots into the soil stimulate the growth of hyphae of mycorrhizal fungi (*see* MYCORRHIZA) towards the roots—the first stage in establishing a symbiotic association with the root system. In an example of opportunistic evolution, the germination of seeds of the parasitic witchweed plant (*Striga* spp.), a serious pest of cereal crops in Africa, is also triggered by strigolactones.

strobilus (*pl.* **strobili**) **1.** A type of *composite fruit that is formed from a complete inflorescence. It produces *achenes enclosed in bracts and when mature becomes cone-shaped. The hop fruit is an example. **2.** *See* CONE.

stroke volume The volume of blood ejected at each stroke of the heart during *systole. Human stroke volumes generally lie in the range 50–100 mL, with a typical value for a normal person at rest being 70 mL; this increases to a maximum of 120 mL during strenuous exercise. *See* CARDIAC OUTPUT.

stroma (*pl.* **stromata**) Tissue that forms the framework of an organ; for example, the tissue of the ovary that surrounds the reproductive cells, or the gel-like matrix that surrounds the grana of *chloroplasts.

stromatolite A rocky column or mound consisting of layers of trapped sediment and precipitated limestone formed by the unchecked growth of millions of *cyanobacteria. Stromatolites are found only in areas where other organisms that would normally keep down the bacterial numbers cannot survive, such as extremely salty bays. Such photosynthetic bacteria were abundant during the Proterozoic and Archaean eons, from as early as 3500 million years ago. They began to generate oxygen, which ultimately led to the dominance of life forms dependent on aerobic respiration as a means of obtaining energy. The white rings of fossilized microorganisms found in rocks of this age are the remains of stromatolites.

strontium Symbol Sr. A soft yellowish metallic element. The isotope strontium-90 is present in radioactive fallout (half-life 28 years), and can be metabolized with calcium so that it collects in bone.

structural gene *See* OPERON.

strychnine A colourless poisonous crystalline alkaloid found in certain plants.

STS *See* SEQUENCE-TAGGED SITE.

style The stalk of a carpel, between the stigma and the ovary. In many plants it is elongated to aid *pollination.

subarachnoid space The space between the *arachnoid membrane and the *pia mater, two of the membranes (*meninges) that surround the brain and spinal cord. It is filled with *cerebrospinal fluid.

subclavian artery A paired artery that passes beneath the collar bone (clavicle) and branches to supply blood to the arm. The left subclavian artery arises from the aorta; the right from the innominate artery.

subcutaneous tissue The tissue that underlies the *dermis (*see* SKIN). It is made up of loose fibrous *connective tissue, muscle, and fat (*see* ADIPOSE TISSUE), which in some animals (e.g. whales and hibernating mammals) forms an insulating layer or an important food store.

suberin A mixture of waxy substances, similar to *cutin, present in the thickened cell walls of many trees and shrubs, particularly in mature cork cells. The deposition of suberin (**suberization**) provides a protective water-impermeable layer. *See also* CASPARIAN STRIP.

sublittoral 1. Designating or occurring in the shallow-water zone of a sea, over the continental shelf and below the low tide mark. **2.** Designating or occurring in the zone of a lake below the littoral zone, to a depth of 6–10 metres.

submucosa A layer of *areolar connective tissue that lies beneath the *mucosa, for example in the stomach and intestine.

subsoil *See* SOIL.

subspecies A group of individuals within a *species that breed more freely among themselves than with other members of the species and resemble each other in more characteristics. Reproductive isolation of a subspecies may become so extreme that a new species is formed (*see* SPECIATION). Subspecies are sometimes given a third Latin name, e.g. the mountain gorilla, *Gorilla gorilla beringei* (*see also* BINOMIAL NOMENCLATURE).

substance P A *neuropeptide comprising 11 amino-acid residues that is found widely in tissues, especially in the nervous system and gut. It mediates the effects of the parasympathetic nervous system by stimulating contractions of the gut, causing dilation of blood vessels (vasodilation), and increasing the flow of saliva from salivary glands. Substance P is also released as a neurotransmitter within the spinal cord by sensory neurons responding to pain and other noxious stimuli (*see* NOCICEPTION).

substitution (in genetics) A *point mutation in which one base pair in the DNA sequence is replaced by another. In **transition** mutations a pyrimidine base (i.e. thymine or cytosine) is replaced by another pyrimidine base or a purine base (adenine or guanine) is replaced by another purine base; whereas in **transversion** mutations a pyrimidine base is replaced by a purine base, or vice versa. A substitution thus alters the base sequence of the genetic code; however, the effect on the gene product depends on the nature of the change. A **synonymous substitution** changes the codon to another that codes for the same amino acid, whereas a **nonsynonymous substitution** creates a codon for a different amino acid, which may or may not affect the functioning of the protein. Sickle-cell anaemia is an example of a substitution mutation in which thymine is replaced by adenine in the triplet coding for the sixth amino acid in the β-chain of haemoglobin.

substrate 1. (in biochemistry) The substance upon which an *enzyme acts in biochemical reactions. **2.** (in biology) The material on which a sedentary organism (such as a barnacle or a plant) lives or grows. The substrate may provide nutrients for the organism or it may simply act as a support.

substrate-level phosphorylation *See* PHOSPHORYLATION.

subtilisin Any of a group of protein-digesting enzymes (*see* PROTEASE) produced by *Bacillus subtilis* and related species. Subtilisins have broad specificities and work over a wide range of pH values. **Subtilisin Carlsberg** is an alkaline protease, derived from *Bacillus licheniformis*, that is used widely in detergents. Roughly 30–40% of the peptide bonds in casein, a milk protein, are hydrolysed by subtilisin Carlsberg. Homologous proteins are also found in plants, such as melon and tomato.

succession (in ecology) The sequence of communities that develops in an area from the initial stages of colonization until a stable mature **climax community** is achieved. Many factors, including climate and changes brought about by the colonizing organisms, influence the nature of a succession; for example, after many years shrubs produce soil deep enough to support trees, which then shade out the shrubs. Change in communities caused by abiotic factors, such as the gradual silting up of a pond, is termed **allogenic succession**; change brought about primarily through species' interactions (i.e. biotic factors), such as competition, is called **autogenic succession**. Autogenic succession can be either primary or secondary. **Primary succession** occurs on newly exposed substrate that has not previously supported a living community, such as a volcanic lava flow; **secondary succession** occurs where the vegetation of an area has been partially or completely removed, for example by fire or cultivation, but soil and seeds remain. In **cyclical succession** the climax community depends on a pattern of recurring disturbances that maintain conditions for its survival. Fire-adapted forests are prime examples, where the mature trees have adapted to withstand periodic low-intensity fires. The fires clear understorey scrub and supply nutrients to the soil (in the form of ash) for germinating tree seeds. *See also* SERE.

succinate A salt of succinic (butanedioic) acid, $HOOC(CH_2)_2COOH$, a four-carbon fatty acid. Succinate occurs in living organisms as an intermediate in metabolism, especially in the *Krebs cycle.

succulent A plant that conserves water by storing it in fleshy leaves or stems. Succulents are found either in dry regions or in areas where there is sufficient water but it is not easily obtained, as in salt marshes. Such plants are often modified to reduce water loss by transpiration. For example, the leaves of cacti are reduced to spines.

See also CRASSULACEAN ACID METABOLISM; XEROPHYTE.

succus entericus (intestinal juice) The alkaline secretion produced by glands in the wall of the intestine. It consists of the viscous, alkaline, mucoid secretion of *Brunner's glands in the first part of the duodenum, combined with water electrolytes secreted by cells of the *crypts of Lieberkühn and additional mucus secreted by goblet cells in both small and large intestine. It helps to counteract the highly acidic and proteolytic chyme entering the small intestine from the stomach, thus protecting the duodenum from damage, and serves to lubricate passage of the digesta through the intestine. It was formerly believed to contain digestive enzymes, but duodenal enzymes are now known to be confined to cells of the brush border.

sucker (turion) A shoot that arises from an underground root or stem and grows at the expense of the parent plant. Suckers can be dug up with a portion of root attached and used to propagate a plant. If, however, a plant is grafted onto a different rootstock, as many roses are, any suckers will be of the wild rootstock, rather than the ornamental scion, and must be removed.

sucrase A carbohydrate-digesting enzyme, produced in the brush border of the small intestine, that breaks down the disaccharide sucrose into the monosaccharides glucose and fructose.

sucrose (cane sugar; beet sugar; saccharose) A sugar comprising one molecule of glucose linked to a fructose molecule. It occurs widely in plants and is the principal form in which carbohydrate is exported from photosynthetic cells in leaves to other parts of the plant. It is particularly abundant in sugar cane and sugar beet (15–20%), from which it is extracted and refined for table sugar. If heated to 200°C, sucrose becomes caramel.

sugar (saccharide) Any of a group of water-soluble *carbohydrates of relatively low molecular weight and typically having a sweet taste. The simple sugars are called *monosaccharides. More complex sugars (oligosaccharides) comprise between two and 20 monosaccharides linked together: *disaccharides contain two, trisaccharides three, and so on. The name is often used to refer specifically to *sucrose (cane or beet sugar).

sulpha (sulfa) drugs *See* SULPHONAMIDES.

sulphonamides Organic compounds containing the group $-SO_2.NH_2$. The sulphonamides are amides of sulphonic acids. Many have antibacterial action and are also known as **sulpha drugs**, including sulfadiazine, $NH_2C_6H_4SO_2NHC_4H_3N_2$, sulfathiazole, $NH_2C_6H_4SO_2NHC_5H_2NS$, and several others. They act by preventing bacteria from reproducing and are used to treat a variety of bacterial infections, especially of the gut and urinary system.

sulphur Symbol S. A yellow nonmetallic element that is an *essential element in living organisms, occurring in the amino acids cysteine and methionine and therefore in many proteins. It is also a constituent of various cell metabolites, e.g. *coenzyme A. Sulphur is absorbed by plants from the soil as the sulphate ion (SO_4^{2-}). *See* SULPHUR CYCLE.

sulphur bacteria Any of various groups of unrelated bacteria that utilize sulphur, sulphide, or sulphate in their metabolism. The anaerobic photoautotrophic **green sulphur bacteria** have specialized cellular organelles called chlorosomes, which are essentially sacs containing bacteriochlorophylls and accessory pigments, making them highly efficient at photosynthesis in low light conditions. They oxidize sulphides as electron donors and produce elemental sulphur as a by-product. The photosynthetic pigments in **purple sulphur bacteria** (gamma proteobacteria) give them a range of colours from purple to orange; they also can oxidize hydrogen sulphide and yield sulphur as a by-product:

$$2H_2S + CO_2 \rightarrow H_2O + CH_2O + 2S$$

The chemoautotrophic **sulphur-oxidizing bacteria**, such as *Thiobacillus*, derive energy from the oxidation of sulphur or its compounds (e.g. sulphide), producing sulphate (SO_4^{2-}). The anaerobic heterotrophic **sulphate-reducing bacteria**, such as *Desulfovibrio*, require sulphate for respiration, deriving energy from its reduction to sulphur or hydrogen sulphide. The aerobic **filamentous sulphur bacteria** (e.g. *Beggiatoa*), which belong to the phylum Myxobacteria (gliding bacteria), can grow by oxidizing sulphides to sulphates.

sulphur bridge *See* DISULPHIDE BRIDGE.

sulphur cycle The cycling of sulphur between the biotic (living) and abiotic (nonliving) components of the environment (*see* BIOGEOCHEMICAL CYCLE). Most of the sulphur in the abiotic environment is found in rocks, although a small amount is present in the atmosphere as sulphur dioxide (SO_2), produced by combustion of fossil fuels, and volatile organic sulphur compounds such as dimethyl sulphide, produced by phytoplankton in the oceans. Sulphate (SO_4^{2-}), derived from the weathering and oxidation of rocks and also found in oceanic sediment, is taken up by plants and phytoplankton and incorporated into sulphur-containing proteins. In this form sulphur is passed along food chains to animals. Decomposition of dead organic matter and faeces by anaerobic sulphate-reducing bacteria returns sulphur to the abiotic environment in the form of hydrogen sulphide (H_2S). Hydrogen sulphide can be converted back to sulphate or to elemental sulphur by the action of different groups of photosynthetic and sulphide-oxidizing bacteria (*see* SULPHUR BACTERIA). Elemental sulphur becomes incorporated into rocks.

sulphur dioxide (sulphur(IV) oxide) A colourless liquid or pungent gas, SO_2, formed by sulphur burning in air. Sulphur dioxide is produced by burning fossil fuels in power stations; it dissolves in water to give a mixture of sulphuric and sulphurous acids. *See also* ACID RAIN; AIR POLLUTION.

summation 1. (in neurophysiology) The combined effect of the changes in electric potential elicited in one or more postsynaptic membranes by the transmission of impulses at *synapses or *neuromuscular junctions. If the *excitatory postsynaptic potential is sufficiently large, it will trigger an action potential in the postsynaptic neuron or muscle fibre. Summation occurs when one or a few postsynaptic potentials

The sulphur cycle

alone are insufficient to elicit a response in the postsynaptic cell; it may consist of the effect of two or more potentials evoked simultaneously at different synapses on the same postsynaptic cell (**spatial summation**) or in rapid succession at the same synapse (**temporal summation**). *Inhibitory postsynaptic potentials also exhibit summation, causing hyperpolarization of the recipient cell and making it less likely to produce an action potential in response to excitatory input. **2.** *See* SYNERGISM.

SUMO (small ubiquitin-related modifier) Any of a family of proteins with structural similarity to *ubiquitin that are attached to other proteins within cells and modify their function or change their cellular location. Attachment of SUMO—**SUMOylation**—occurs chiefly in the nucleus as a post-translational modification and generally prolongs a protein's life. It can, for example, determine how a protein controls transcription or whether it moves from the nucleus to the cytosol.

supercoiling A form of DNA in which the double helix is further twisted about itself, forming a tightly coiled structure. This is the form generally adopted by DNA in nature, and enables it to condense sufficiently to be packaged into living cells (*see* CHROMATIN). In **negative supercoiling** the DNA is twisted about an axis in a direction opposite to that of the clockwise turns of the (right-handed) helix; this decreases the number of turns of one helix around the other. In **positive supercoiling** the twist of the supercoils is in the same direction as that of the double helix, which increases the number of turns of one helix around the other. Supercoiling must be temporarily removed when DNA replication takes place, and the degree of supercoiling can affect gene transcription. Changes in supercoiling are performed by *topoisomerase enzymes.

supercooling The cooling of a liquid below its normal freezing point without changing from a liquid to a solid. Water is relatively amenable to supercooling, and many organisms living in cold environments exploit this property to avoid freezing of their body water and consequent tissue damage. Supercooling of extracellular fluids requires first the exclusion of potential *ice-nucleating agents, such as food particles, that could promote the formation of ice crystals. Hence, animals preparing for winter frequently void their gut contents. Second, the animal manufactures *antifreeze molecules, such as glycerol, which depress the freezing point. These measures can be extremely effective in achieving supercooling.

supergene A cluster of tightly linked genes that affect the same trait and are inherited apparently as a single unit. For example, supergenes are known to determine shell colour and patterning in terrestrial snails, and wing shape and colouring essential to mimicry in certain butterflies. This tight linkage is advantageous since it ensures that the appropriate combination of alleles for the character concerned is virtually always transmitted intact to the animal's offspring. Any rare recombinants will be selected against. A supergene can arise through *inversion of part of a chromosome; crossing over within the inversion produces unbalanced recombinant chromatids, and hence zygotes containing these are nonviable. Thus, with respect to the inversion, only nonrecombinant zygotes are generally produced, giving a very low apparent rate of recombination between the loci in this region.

supergroup An assemblage of organisms constituting a primary clade of a *domain. It is thus effectively the highest-ranking taxon after domain and has superseded *kingdom in many classifications. The eukaryotes are currently placed in four hypothetical supergroups, chiefly on the basis of molecular studies of their genomes. These are the *excavates, *SAR, Archaeplastida (*see* PLANT sense 2), and *unikonts.

superior Describing a structure that is positioned above or higher than another structure in the body. For example, in flowering plants the ovary is described as superior when located above the other organs of the flower (*see* HYPOGYNY). *Compare* INFERIOR.

supernormal stimulus 1. (in animal behaviour) A stimulus that produces a more vigorous response than the normal stimulus eliciting that particular response. For example, a female herring gull will brood a giant egg in preference to its own eggs, which are smaller. A supernormal stimulus is an exaggerated *sign stimulus. **2.** (in neurophysiology) A stimulus that is more intense than a normal stimulus and is capable of inducing a response in a nerve fibre during the relative *refractory period.

supernumerary chromosome *See* ACCESSORY CHROMOSOME.

superorganism An association or colony of individuals that collectively functions like a single organism. The concept is best illustrated by the colonies of *eusocial insects, such as ants, bees, wasps, and termites, which are characterized by high levels of self-organization, cooperation, division of labour, and communication. *Castes of individuals are evident, whose members are specialized to perform certain tasks, and some members (e.g. worker castes) forgo reproduction to tend to the needs of reproductive members. More loosely, the term 'superorganism' is used in many contexts, e.g. to describe an individual human and their associated microbial community (*microbiome) or human society in general. Even planet Earth can be viewed as a superorganism according to the *Gaia hypothesis.

superoxide dismutase (SOD) A widely distributed enzyme that removes the superoxide radical ($O_2\cdot$), with the formation of molecular oxygen and hydrogen peroxide, in the following reaction:

$$O_2\cdot + O_2\cdot + 2H^+ \rightarrow O_2 + H_2O_2$$

The hydrogen peroxide formed is removed by the action of *catalase. The superoxide anion, which damages tissues, is a *free radical formed by the partial reduction of molecular oxygen and a normal by-product of aerobic respiration in mitochondria. It is

also produced during the metabolic breakdown of various toxins (including drugs and chemical poisons) and as part of the immune response to the presence of bacteria and virus-infected cells. *See* OXIDATIVE BURST.

super-resolution microscopy *See* CONFOCAL FLUORESCENCE MICROSCOPY.

supination Rotation of the lower forearm so that the hand faces forwards or upwards with the radius and ulna parallel. *Compare* PRONATION.

supplementary units *See* SI UNITS.

suprachiasmatic nucleus *See* BIOLOGICAL CLOCK; HYPOTHALAMUS.

supramolecular adhesion complex (SMAC) *See* IMMUNOLOGICAL SYNAPSE.

suprarenal glands *See* ADRENAL GLANDS.

surface tension Symbol γ. The property of a liquid that makes it behave as if its surface is enclosed in an elastic skin. The property results from intermolecular forces: a molecule in the interior of a liquid experiences a force of attraction from other molecules equally from all sides, whereas a molecule at the surface is only attracted by molecules below it in the liquid. The surface tension is defined as the force acting over the surface per unit length of surface perpendicular to the force. It is measured in newtons per metre. It can equally be defined as the energy required to increase the surface area by one square metre, i.e. it can be measured in joules per metre squared (which is equivalent to $N\,m^{-1}$).

The surface tension of water is very strong, due to the intermolecular hydrogen bonding, and is responsible for the formation of drops, bubbles, and **meniscuses** (the curved surfaces of columns of liquid), as well as the rise of water in a capillary tube (**capillarity**), the absorption of liquids by porous substances, and the ability of liquids to wet a surface. Capillarity is very important in plants as it contributes to the transport of water, against gravity, within the plant.

surfactant (surface active agent) A substance, such as a detergent, added to a liquid to increase its spreading or wetting properties by reducing its *surface tension.

survivorship curve (survival curve) A graph that shows the relationship between the survival and age of a population of a particular species. Survival can be represented as the percentage of individuals of the original population that have survived after a specified time. Some types of organism, notably those that produce many offspring (e.g. bacteria), have a low survival rate in the early stages of development (line A on graph; *see* R SELECTION). Other organisms, which have a low rate of reproduction, tend to be long-lived and their population numbers are maintained over a long period (line B on graph; *see* K SELECTION). This type of survivorship curve is typical of a human population in a developed country.

suspension culture A type of culture in which intact cells are maintained in suspension in the culture medium so that they are distributed evenly within it. *Compare* MONOLAYER CULTURE.

suspensor The chain of cells that anchors a plant embryo in the surrounding gametophyte tissue. In flowering plants the suspensor attaches the embryo to the embryo

sac and extends to push the embryo into the endosperm.

suspensory ligaments *See* LENS.

sustainable Describing a system whose current state or conditions do not jeopardize its ability to function to its full potential over the long term. In terms of economic and social development, one of the earliest and most enduring definitions comes from the Brundtland Report of 1987, which describes 'sustainable development' as 'development which meets the needs of the present, without compromising the ability of future generations to meet their own needs'. In ecology, the concept is often used to assess the impact of human activities on ecosystems at various scales, from local to global. Criteria of ecological sustainability can thus vary, but include measures such as projected effects on biodiversity, habitats, resources (e.g. water, soil, air, minerals), and the climate. It is widely accepted that on current trends many aspects of human behaviour and lifestyles are unsustainable, and are damaging crucial components of the global ecosystem and its ability to provide *ecosystem services. Chief areas of concern are human population growth, resource depletion, *pollution, misuse of the land and seas, and *climate change. Moreover, these underlying impacts

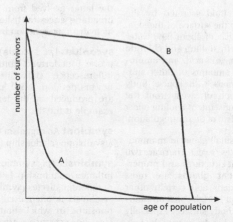

Survival curves

frequently have adverse social and political consequences, such as social instability, poverty, war, famine, and mass migration, which compound the problems. One of the United Nations' Millennium Development Goals is to 'ensure environmental sustainability' by integrating the principles of sustainable development into the policies of member states. This remains a distant prospect, although such principles are steadily gaining wider appreciation and acceptance. *See also* ECOLOGICAL FOOTPRINT.

(⊕) SEE WEB LINKS
- UN Sustainable Development Knowledge Platform

suture The line marking the junction of two body structures. Examples are the immovable joints between the bones of the skull and, in plants, the seam along the edge of a pea or bean pod.

SV40 A DNA-containing virus belonging to the *papovavirus group that is much used in cancer research for its ability to induce *transformations in cells. It was originally isolated from monkeys—hence its name, which is an abbreviation of **Simian Virus 40**.

swallowing *See* DEGLUTITION.

swamp *See* HYDROSERE.

sweat The salty fluid secreted by the *sweat glands onto the surface of the skin. In humans it consists of about 99% water plus minerals (chiefly sodium and chloride ions), lactate, urea, uric acid, and ammonia, among small amounts of other substances. In mammals with sparse body hair the evaporation of sweat from the skin surface and consequent cooling effect is a major mechanism of thermoregulation.

sweat gland A small gland in mammalian skin that secretes *sweat. There are two types with different structures and properties. **Eccrine sweat glands** are most numerous in humans and certain other primates; they occur in high concentrations over much of the body surface, especially on the palms of the hands, soles of the feet,

upper torso, and forehead. The gland lies in the subcutaneous layer of the skin and discharges sweat directly to the skin surface through a narrow coiled duct. **Apocrine sweat glands** discharge their secretion into hair follicles. In horses and marsupials they produce a watery secretion that functions in cooling the body surface, whereas in other mammals, particularly ones with a body covered by dense hair or fur, they produce an oily secretion that mixes with sebum as it passes alongside the hair shaft to reach the skin surface. Here the odourless secretion is subject to microbial action that can produce smelly substances, some of which may act as pheromones. In humans, apocrine sweat glands are confined to the armpits, groin, and around the nipples.

swim bladder (air bladder; gas bladder) An air-filled sac lying above the alimentary canal in bony fish that regulates the buoyancy of the animal. Air enters or leaves the bladder either via a pneumatic duct opening into the oesophagus or stomach or via capillary blood vessels, so that the specific gravity of the fish always matches the depth at which it is swimming. This makes the fish weightless, so less energy is required for locomotion. In lungfish it also has a respiratory function. The lungs of tetrapods are homologous with the swim bladder, and it is now thought that the latter evolved from the lungs of air-breathing ancestral fishes and developed its hydrostatic function by specialization.

syconium (*pl.* **syconia**) A type of *composite fruit formed from a hollow fleshy inflorescence stalk inside which tiny flowers develop. Small *drupes, the 'pips', are produced by the female flowers. An example is the fig.

symbiont An organism that is a partner in a symbiotic relationship (*see* SYMBIOSIS).

symbiosis (*pl.* **symbioses**) A long-term intimate relationship between individuals of different species (**symbionts**). The term symbiosis is often used to describe relationships in which both species benefit (*see* COOPERATION; MUTUALISM), but it may be used for other close associations, such as *commensalism, *inquilinism, and *parasitism. Many symbioses are obligatory (i.e. the participants cannot survive without the interaction); for example, a lichen is an obligatory symbiotic relationship between an alga or a cyanobacterium and a fungus. *See also* ENDOSYMBIONT THEORY.

symmetry Regularity in the arrangement of the parts of an organism. *See* BILATERAL SYMMETRY; RADIAL SYMMETRY.

symmorphosis A hypothesis, proposed by Ewald Weibel and Charles Richard Taylor in 1981, postulating that biological systems adhere to an 'economy of design' giving a close match between their various structural and functional parameters. Hence, no single parameter in the system has unnecessary excess capacity, beyond the requirements of the system. The hypothesis was tested initially by analysing the mammalian respiratory system. Here it was found that, except for the lungs, the structures of the oxygen-transfer system, including blood, heart, muscle capillaries, and mitochondria, are well matched to the functional capacity of the system. However, components often serve in more than one physiological system; for example, the blood and blood vessels are also parts of the excretory system. Therefore, apparent spare capacity in one system might be needed for another system.

sympathetic nervous system The division of the *autonomic nervous system whose activation broadly causes arousal and prepares the body for 'fight or flight'. It thus antagonizes the effects of the *parasympathetic nervous system. Preganglionic sympathetic neurons arise on either side of the body from the thoracic and lumbar regions of the spinal cord, connect with ganglia arranged in chains alongside the spinal cord, and release acetylcholine. Some preganglionic neurons connect with visceral ganglia, such as the coeliac ganglion or inferior mesenteric ganglion. The postganglionic sympathetic neurons release mainly noradrenaline as a neurotransmitter (smaller amounts of adrenaline are also released).

The actions of the sympathetic nervous system tend to antagonize those of the *parasympathetic nervous system, thus achieving a balance in the organs they serve. For example, the sympathetic nervous system decreases salivary gland secretion, increases heart rate, and constricts blood vessels, while the parasympathetic nervous system has opposite effects.

sympathetic tone The condition of a muscle when the *tone is maintained predominantly by impulses from the *sympathetic nervous system.

sympatric Describing groups of similar organisms that, although in close proximity and theoretically capable of interbreeding, do not interbreed because of differences in behaviour, flowering time, etc. *See* ISOLATING MECHANISM. *Compare* ALLOPATRIC.

symphysis (*pl.* **symphyses**) A *joint that is only slightly movable; examples are the joints between the vertebrae of the vertebral column and that between the two pubic bones in the pelvic girdle. The bones at a symphysis articulate by means of smooth layers of cartilage and strong fibres.

symplast All the *protoplasts in plants and their interconnecting cytoplasmic channels—the *plasmodesmata. This effectively forms a continuous system of cytoplasm bounded by the plasma membranes of the cells. The movement of water through the symplast is known as the **symplastic pathway**. It is one of the routes by which water travels across the root cortex from the root hairs and is the only means by which water crosses the *endodermis. *Compare* APOPLAST.

symplesiomorphy An ancestral trait that is shared by two or more modern groups. Because of their ancient origin, symplesiomorphies are not usually helpful in assessing more recent evolutionary relationships within a larger group. For example, the simple leaves of some modern flowering plants are probably inherited from simple-leaved ancestors. But this does not mean that all simple-leaved modern plants are closely related. *See* PLESIOMORPHY.

sympodium (*pl.* **sympodia**) The composite primary axis of growth in such plants as lime and horse chestnut. After each season's growth the shoot tip of the main stem stops growing (sometimes terminating in a flower spike); growth is continued by the tip of one or more of the lateral buds. *Compare* MONOPODIUM.

symporter A *transport protein that couples the movement of a substance across a cell membrane with the simultaneous movement of ions in the same direction. The substance is 'pulled' uphill against its concentration gradient by the flow of the ions down their pre-existing concentration gradient. Hence, it is a form of active transport, powered by the energy stored in the ionic concentration gradient, which is maintained by cellular mechanisms. For example, kidney tubule cells contain a sodium/glucose symporter that exploits the gradient of sodium ions (Na^+) across the plasma membrane; for every two Na^+ that flow inwards it imports one glucose molecule. *Compare* ANTIPORTER; UNIPORTER.

synapomorphy (shared derived trait) A trait that is present in the ancestor of a group and in all members of the group. It is thus evidence of the common ancestry of the group. For example, mammary glands are a synapomorphy of mammals. Such traits define the strictly *monophyletic groups, or clades, which are the basis of cladistic classification systems (*see* CLADISTICS). *Compare* APOMORPHY; PLESIOMORPHY.

synapse The junction between two adjacent neurons (nerve cells), i.e. between the axon ending of one (the **presynaptic neuron**) and the dendrites of the next (the **postsynaptic neuron**). In **chemical synapses** the swollen tip of the axon of the presynaptic neuron, called the **synaptic knob** (or **bouton**), contains vesicles of *neurotransmitter substance. At a synapse, the membranes of the two cells (the **pre-** and **postsynaptic membranes**) are in close contact, with only a minute gap (the **synaptic cleft**) between them. A nerve *impulse is transmitted across the synapse by the release from the presynaptic membrane of neurotransmitter (*see* SNARE), which diffuses across the synaptic cleft to the postsynaptic membrane. This may trigger the propagation of the impulse from the dendrite along the length of the postsynaptic neuron. Depending on their type, most neurons give rise to multiple synapses with other neurons or effector organs

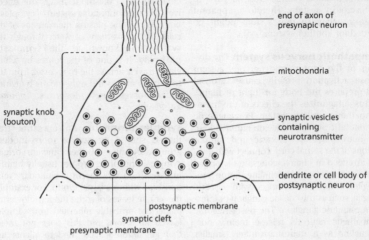

end of axon of presynapic neuron

mitochondria

synaptic knob (bouton)

synaptic vesicles containing neurotransmitter

dendrite or cell body of postsynaptic neuron

postsynaptic membrane

synaptic cleft

presynaptic membrane

Structure of a synapse

(e.g. muscles) and themselves receive inputs from many hundreds or thousands of synapses. The signals received or sent can be excitatory or inhibitory (*see* SUMMATION). Much less common are **electrical synapses**, in which ions flow directly from one neuron to another via *gap junctions. These are found, for example, in the heart and in certain parts of the vertebrate central nervous system. Such an arrangement ensures virtually instantaneous transmission of impulses. During development many superfluous synapses formed in the early embryo are eliminated by a process called **synaptic pruning**. This ensures that only functionally relevant and active synapses persist into adulthood. *See also* ACTIVE ZONE; EXCITATORY POSTSYNAPTIC POTENTIAL. *Compare* NEUROMUSCULAR JUNCTION.

SEE WEB LINKS

• Multilevel summary of synaptic function, sponsored by the Canadian Institutes of Health Research

synapsid (stem mammal) Any of a clade of tetrapod animals that contains the modern mammals, their ancestors, and extinct relatives. Synapsids are characterized by an opening, the temporal fenestra, on each side of the skull behind the eye socket, thought originally to provide a passage for cheek muscles to operate closure of the jaw. Synapsids first appear in the fossil record during the late Carboniferous period, over 300 mya, and had hairless bodies with a reptilian appearance, such as *Dimetrodon*, which had a prominent spine sail along its back. The Permian saw the emergence of larger herbivorous and carnivorous forms (*see* THERAPSID), but diversity was much reduced in the mass extinction of the late Permian. During the Triassic, synapsids became more mammal-like, with hairy bodies and a more upright gait, though they still laid eggs. These gave rise to the first true mammals during the Jurassic period.

synapsis *See* PAIRING.

synaptic cleft *See* SYNAPSE.

synaptic knob *See* SYNAPSE.

synaptic plasticity (neuronal plasticity) Change in the efficacy or connections of the junctions (synapses) between neurons in the nervous system. It is a crucial process that underlies modification of an animal's behaviour during development and in response to previous activity or experience, including learning and memory. Various mechanisms produce changes in the efficacy of synapses, which can range in duration from fractions of a second to days or weeks. For example, the arrival of impulses in rapid succession at a motor end plate causes an augmented postsynaptic potential lasting perhaps 100–200 ms (*see* FACILITATION). If a motor neuron is stimulated at high frequency for a long period (i.e. tetanic stimulation), it is followed by a period of reduced synaptic efficacy (**synaptic depression**), due to depletion of synaptic vesicles in the presynaptic nerve endings. Following recovery from synaptic depression, a synapse often displays **post-tetanic potentiation**, in which the amplitude of the response is increased relative to that obtained by a comparable stimulus before tetanic stimulation. The activity of synapses is also modulated by various agents, including neurotransmitters, released into the vicinity by other nerve endings or present in the bloodstream. For example in the sluglike gastropod mollusc *Aplysia*, enhanced sensitivity to a stimulus (behavioural sensitization) is triggered by the release of serotonin. This facilitates the release of neurotransmitter at certain synapses, thereby increasing the amplitude and duration of the postsynaptic potentials for several minutes. In such processes as learning and memory it is thought that **long-term potentiation** (LTP) and **long-term depression** (LTD) play important roles in strengthening or inhibiting existing neural circuits of the brain, with effects lasting for hours, days, or longer. Studies on mammalian brains have revealed that two types of *glutamate receptors, the AMPA and NMDA receptors, play crucial roles in both LTP and LTD. For example, the long-lasting increase in the response of a postsynaptic cell that typifies LTP involves a series of action potentials arriving simultaneously at both the presynaptic cell and at other synapses nearby. This activates NMDA receptors in the postsynaptic membrane and triggers the recruitment of AMPA receptors, thereby stably increasing the size of the *excitatory postsynaptic potential.

synaptobrevin See SNARE.

synaptonemal complex A ribbon-like complex of proteins that binds together the members of each homologous pair of chromosomes during *pairing (synapsis) in prophase of the first division of *meiosis. Formation of the complex begins at the telomeres, which at this stage are attached to the nuclear envelope, and proceeds along the length of the chromosomes, 'zipping' them together so that matching regions are aligned.

synaptotagmin See SNARE.

syncarpy The condition in which the female reproductive organs (*carpels) of a flower are joined to each other. It occurs, for example, in the primrose. Compare APOCARPY.

syncytium (pl. **syncytia**) A group of animal cells in which cytoplasmic continuity is maintained. For example, the cells of striated muscle form a syncytium. In some syncytia the cells remain discrete but are joined together by cytoplasmic bridges. For example, developing male germ cells in the testes of mammals are linked throughout the process of differentiation until an individual sperm is released into the lumen of the seminiferous tubule (see SPERMATOGENESIS). Compare PLASMODIUM.

synecology The study of ecology at the level of the *community. A synecological study aims to investigate the relationships between different species that form a community and their interactions with the surrounding environment. Synecology involves both *biotic and *abiotic factors. Compare AUTECOLOGY.

synergids The two nuclei in the *embryo sac of flowering plants that are closely associated with the oosphere, or egg cell, to form the **egg apparatus**. The synergids

secrete signal molecules that diffuse through the style tissue and guide the developing pollen tube towards the embryo sac; they may also cause the tip of the pollen tube to burst, releasing the male gametes into the embryo sac.

synergism (summation) 1. The phenomenon in which the combined action of two substances (e.g. drugs or hormones) produces a greater effect than would be expected from adding the individual effects of each substance. See also POTENTIATION. **2.** The combined action of one muscle (the **synergist**) with another (the **agonist**) in producing movement. Compare ANTAGONISM.

syngamy See FERTILIZATION.

synomone See ALLELOCHEMICAL.

synonymous substitution See SUBSTITUTION.

synovial membrane The membrane that lines the ligament surrounding a freely movable joint (such as that at the hip or elbow). It secretes a fluid (**synovial fluid**) that lubricates the layers of cartilage forming the articulating surfaces of the joint.

syntaxin See SNARE.

synteny The conservation of chromosomal segments containing groups of homologous genes in the same order and orientation among different species. Chromosomal regions containing such genes are termed **syntenic blocks**, and identifying them in related species is helpful in determining whether similar genes are truly *orthologous or not. Syntenic blocks can contain from a few up to several thousand genes, and syntenic maps have been compiled to show how these blocks are distributed between the chromosomes of different species. For example, the long arm of mouse chromosome 1 contains a large region that it shares with human chromosome 2. Hence, similar genes localized to these regions are likely to share a common ancestor.

synthesis The formation of chemical compounds from more simple compounds. See BIOSYNTHESIS.

synthetic biology A field of biology that draws on principles of engineering to create new biological parts, devices, or systems, and to redesign or improve natural systems. Underlying synthetic biology is the capability to make DNA polymers with specific base sequences that behave as genes. These synthetic polymers consist of various small DNA sequences that are 'stitched' together by homologous recombination to yield functional genes with the required properties, e.g. multigene transcription units with promoters, ribosome-binding sites, and terminators. The corresponding proteins with desired properties can then be assembled when the engineered gene is matched with the necessary molecular apparatus. A common strategy is to introduce the engineered genes into a modified bacterium, such as E. coli. Entirely synthetic cells, e.g. based on *protocells, are another option. Synthetic biology is becoming a vast new biotechnology sector, with a host of applications using cellular 'factories' designed to produce new materials, drugs, and vaccines, harness light energy via photosynthesis, degrade toxic chemicals, act as biosensors, or even function as biological computers, among others. Much interest centres on the potential of DNA to permanently store digital data. Files in digital format, including images, text, and music, comprise bits of information in the form of 1s and 0s. Pairs of this binary code can be assigned to one of the four bases that constitute the nucleotides of the DNA strands—A, C, G, and T. A DNA molecule is then assembled whose bases correspond to the sequence of digits in the data file. Subsequently the information can be retrieved by 'reading' the coding strand of the relevant DNA molecule by *DNA sequencing. DNA is a high-density storage medium, is very durable if stored correctly, and can easily be replicated using existing technology such as PCR. At present the main obstacle is the slow speed of synthesizing and reading the DNA compared with, say, a hard drive.

synthetic theory See NEO-DARWINISM.

syphilis See SEXUALLY TRANSMITTED DISEASE.

syrinx The sound-producing organ of a bird, situated at the lower end of the trachea where it splits into the bronchi. It has a complex structure with a number of vibrating membranes.

systematics The study of the diversity of organisms and their natural relationships. It is sometimes used as a synonym for *taxonomy. The term **biosystematics** (or **experimental taxonomy**) describes the experimental study of diversity, especially at the species level. Biosystematic methods include breeding experiments, field work, biochemical work (known as **chemosystematics**), and cytotaxonomy. See also MOLECULAR SYSTEMATICS.

systematic sampling See SAMPLING.

Système International d'Unités See SI UNITS.

systemic acquired resistance (SAR) A generalized state of enhanced immunity to infection demonstrated by plants following an initial localized injury. The hypersensitive response of the plant to local infection (see HYPERSENSITIVITY) produces certain substances, such as methylsalicylic acid, that over a period of hours to days evoke resistance throughout the plant, even in parts quite distant from the original injury site (see SYSTEMIC SIGNALLING). Signal molecules, such as systemins and *jasmonates, are disseminated throughout the plant and cause the increased expression of genes encoding **pathogenesis-related (PR) proteins**. These include enzymes capable of degrading microbial cell walls (e.g. *chitinases and glucanases), *defensins, and other proteins that block microbial enzymes (e.g. protease inhibitors). *Salicylic acid (the precursor of the analgesic drug aspirin) is also a well-known signal molecule in plant defences. See also IMMUNITY (sense 2).

systemic arch A paired blood vessel in the embryos of tetrapods that carries blood from the aorta to the trunk and hind limbs. It is derived from the fourth *aortic arch. Adult amphibians and reptiles retain both arches, while birds and mammals only have one.

systemic circulation The part of the circulatory system of birds and mammals that transports oxygenated blood from the left ventricle of the heart to the tissues in the body and returns deoxygenated blood from the tissues to the right atrium of the heart. *Compare* PULMONARY CIRCULATION. *See* DOUBLE CIRCULATION.

systemic signalling The dissemination of chemical signals throughout a plant in response to injury or infection of part of the plant. The signals trigger a systemic defence response that helps the plant to combat spread of the infection or restrict further injury by herbivorous animals. For example, insects chewing on a leaf of a tomato plant elicit the release of a systemic signal that is transported throughout the plant and causes the plant to express genes encoding proteinase inhibitors that interfere with the insect's digestion. Foremost among these systemic signals are the *jasmonates and a group of peptides called **systemins**. *See also* SYSTEMIC ACQUIRED RESISTANCE.

systems biology An approach that seeks to study organisms as complete systems—networks of interacting genes, biomolecules, and biochemical reactions, i.e. the *interactome. It thus attempts to integrate all relevant structural and functional information, rather than focusing on, say, just one particular gene or protein at a time. This involves amassing and organizing data obtained from genomics, proteomics, and other areas of bioinformatics and managing and analysing the data to identify patterns, formulate hypotheses, and ultimately create computer models that will enable accurate predictions of cellular and organismal responses. Such models could have a radical impact on medicine and biology in the future. For example, aberrations in a particular cancer patient could enable treatment tailored to that individual's particular form of cancer.

systems ecology The use of the techniques of systems analysis, such as mathematical modelling, to study the complex interactions and relationships of ecological systems.

systole The phase of the heartbeat during which the ventricles of the heart contract to force blood into the arteries. *Compare* DIASTOLE. *See* BLOOD PRESSURE.

tachycardia An increase in heart rate. The heart rate is controlled by the opposing actions of the sympathetic nervous system, which accelerates the heartbeat; and the parasympathetic nervous system, which slows it down. The heart rate is normally in the range 60–100 beats per minute (bpm). During exercise there is an increase in sympathetic activity and the heart rate increases to over 100 bpm. Following exercise, at rest, the heart rate returns to normal. However, sustained tachycardia at rest may indicate a form of abnormal heart rhythm (arrhythmia). A slow heart rhythm of less than 60 bpm at rest (i.e. **bradycardia**) is common in athletes but may also indicate arrhythmia.

tachymetabolism Metabolism that is sustained at a relatively high rate. Characteristic of 'warm-blooded' animals (*see* ENDOTHERM), it generates the heat required to maintain the body temperature of such animals at a constant high level. A tachymetabolic animal typically has a metabolic rate at least five times that of an ectothermic ('cold-blooded') animal of equivalent size. *Compare* BRADYMETABOLISM.

TACK A supergroup of *archaea named after the initials of four of its member phyla: the *Th*aumarchaeota, *A*igarchaeota, *Cr*enarchaeota, and *K*orarchaeota. The phylogenetic identity and relationships of these and other prokaryote lineages within the TACK supergroup have been established largely on the basis of *metagenomic analysis of environmental samples of DNA taken from sediments, seawater, and other sources. TACK members are regarded as being more closely related to each other than to the other major groups of archaea, the Euryarchaetoa and the DPANN supergroup.

tactic movement *See* TAXIS.

tagma (*pl.* **tagmata**) A section of the body of an arthropod that is formed by the fusion of mesodermal *somites and has a distinct function and structure. The basic tagmata are the head, thorax, and abdomen, but the form of the tagmata (known as **tagmosis**) varies between arthropod groups, each group having its characteristic tagmosis. For example, many crustaceans have a *cephalothorax and abdomen, while arachnids have a *prosoma and *opisthosoma.

tagmosis *See* TAGMA.

taiga (boreal forest) The world's largest terrestrial *biome, consisting mainly of evergreen coniferous forests (chiefly pine, fir, and spruce), which occurs across subarctic North America and Eurasia and at high altitudes in temperate regions. In certain parts, such as northeastern Siberia, deciduous conifers and broadleaved trees, such as larch and birch, are dominant. Over the northermost taiga the ground is permanently frozen within about one metre of the surface, which prevents water from filtering down to deeper levels in the soil. This means that bogs may form in depressions. For at least six months of the year temperatures are below freezing but there is a short growing season lasting 3–5 months. Where average annual temperatures are higher, sunlight may melt the frozen soil during the summer. The soil in taiga areas is acidic and infertile. *Compare* TUNDRA.

TALEN Transcription activator-like effector nuclease. An engineered protein consisting of a nuclease enzyme and a DNA-binding module based on a family of proteins (transcription activator-like effectors) occurring naturally in bacteria of the genus *Xanthomonas*. They are used in *genome editing to introduce changes to the DNA of cells at specific sites.

tandem array (in genetics) A series of copies of a gene arranged in tandem along a chromosome. *Nucleolar organizers, for example, can contain up to 250 copies of a single ribosomal RNA (rRNA) gene in tandem. Genes for histone proteins also occur in tandem arrays. Such arrays ensure that large amounts of the gene product are synthesized by the cell.

tannin One of a diverse group of complex phenolic compounds commonly found in leaves, unripe fruits, and the bark of trees. Tea, chocolate, and wine are among the common tannin-rich foods. Tannins help defend the plant against pathogens and herbivorous animals by virtue of their astringent taste. They also reduce digestibility of the plant material by forming complexes with proteins, carbohydrates, and other nutrients, which causes satiety in the grazing animal. Some tannins have commercial uses, notably in the production of leather and ink.

tapetum (*pl.* **tapeta**) **1.** A membranous layer, such as the thin sheet of nerve fibres in the wall of the lateral ventricles in the brain. **2.** A reflecting layer, containing crystals of guanine, in the *choroid of the eye of many nocturnal vertebrates. It reflects light back onto the retina, thus improving vision and causing the eyes to shine in the dark. **3.** A layer of nutritive or nurse cells in the anther of a flowering plant, or in the sporangium of a fern, that contributes to the development and maturation of the pollen grains or spores.

tapeworms *See* CESTODA.

taphonomy The study of the biological, chemical, and physical processes that change organisms after death, leading ultimately to their preservation as fossils in rock. The initial phase in this process is removal or decay of the organism's soft parts by scavengers or microbes. Thereafter the remaining hard parts, such as bones

and shells, may undergo disarticulation and fragmentation, and are often moved from the site of death, for example by water currents or the wind. Exposure to such movements causes abrasion against other solid particles, and consequent erosion of sharp edges. Burial underneath sediment is common, resulting in flattening or collapse of shell or skeletal cavities, depending on their mechanical strength. The chemical components of the remains may also change; for example, the calcium carbonate of a shell may be converted from aragonite to calcite. Also, concretions of carbonates often form with shell cavities, sometimes protecting them from collapse. The physical and chemical processes occurring after burial of an organism's remains are termed **diagenesis**. Knowledge of all these processes, which may take thousands or millions of years, enables a fuller and more accurate interpretation of *fossil remains.

tap root *See* ROOT.

tarsal (tarsal bone) One of the bones that form the ankle (*see* TARSUS) in terrestrial vertebrates.

tarsus (*pl.* **tarsi**) The ankle (or corresponding part of the hindlimb) in terrestrial vertebrates, consisting of a number of small bones (**tarsals**). The number of tarsal bones varies with the species: humans, for example, have seven.

taste 1. (gustation) The sense that enables the different substances in solution (**tastants**) to be detected and distinguished (*see* TASTE BUD). **2.** The flavour of a substance.

(⊕) SEE WEB LINKS

• Tutorial on the sense of taste, compiled by Tim Jacob of Cardiff University

taste bud A small sense organ in most vertebrates, specialized for the detection of taste. In terrestrial animals taste buds are concentrated on the upper surface of the *tongue—the human tongue has 5000 to 10 000, found mostly on the sides of the papillae. Each taste bud contains a cluster of receptor cells (**gustatory receptors**) whose ends converge on a tiny opening, the taste pore. Microvilli project from the apical surface of the receptor cells into the taste pore to receive the substances (tastants) that elicit the taste stimulus. The receptor cells are surrounded by supporting cells and basal cells, which generate new receptor cells. Each receptor cell is specific for one of five classes of tastant: sweet, salt, bitter, sour, and umami (the latter is from the Japanese for 'delicious' and is a savoury or meaty taste evoked by certain amino acids and the flavour enhancer monosodium glutamate). Binding of a tastant with a receptor protein in a receptor cell causes the release of neurotransmitter onto associated sensory neurons, which transmit nerve impulses to the brain. The mechanism of signal transduction varies according to the type of tastant. For example, salty and some sour tastants bind to *ionotropic receptors and cause the opening of ion channels that allow the influx of sodium ions, which leads to depolarization of the receptor cell, whereas some bitter tastants bind to G protein-coupled *metabotropic receptors that stimulate intracellular signalling pathways, e.g. involving the release of calcium ions from stores.

In fishes the distinction between taste and smell is less clear; taste buds can be found in lips, gill rakers, the oral cavity, pharynx, oesophagus, and distributed over the surface of the body and on the fins. All provide information about the surrounding water or substrate. They also have receptors in olfactory pits on the head. Insects have taste receptors on their mouthparts and feet. These take the form of hollow hairlike projections of the cuticle (sensillae) inside which are the dendrites of several sensory cells.

TATA box (Hogness box) A sequence of nucleotides that serves as the main recognition site for the attachment of RNA polymerase in the *promoter region of eukaryotic genes. Located at around 25 nucleotides before the start of transcription, it consists of the seven-base *consensus sequence TATAAAA, and is analogous to the *Pribnow box in prokaryotic promoters.

TATA-box-binding protein (TBP) A protein that binds to the *TATA box in the promoter of eukaryotic genes and is necessary for initiation of transcription. It bends the DNA molecule and associates with many other transcription factors to assemble a pre-initiation complex, thereby enabling binding of RNA polymerase.

taxis (taxic response; tactic movement) The movement of a cell (e.g. a gamete) or a microorganism in response to an external stimulus. Certain microorganisms have a light-sensitive region that enables them to move towards or away from high light intensities (positive and negative *phototaxis respectively). Many bacteria move in response to chemical stimuli (**chemotaxis**); a specific example is **aerotaxis**, in which atmospheric oxygen is the stimulus. Taxic responses are restricted to cells that possess cilia, flagella, or some other means of locomotion. The term is usually not applied to the movements of higher animals. *See also* GEOTAXIS. *Compare* KINESIS; TROPISM.

taxon (*pl.* **taxa**) Any named taxonomic group of any *rank in the hierarchical *classification of organisms. Thus the taxa Papilionidae, Lepidoptera, Hexapoda, and Arthropoda are named examples of a family, order, class, and phylum, respectively.

taxonomy The study of the theory, practice, and rules of *classification of living and extinct organisms. The naming, description, and classification of a given organism draws on evidence from a number of fields. **Classical taxonomy** is based on morphology and anatomy. **Cytotaxonomy** compares the size, shape, and number of chromosomes of different organisms. **Numerical taxonomy** uses mathematical procedures to assess similarities and differences and establish taxonomic groups. Modern taxonomy generally classifies organisms according to their evolutionary history (i.e. phylogeny), and valid taxa should reflect this by being *monophyletic, i.e. containing only an ancestral form and all its descendants. *See also* SYSTEMATICS.

TCA cycle *See* KREBS CYCLE.

T cell (T lymphocyte) Any of a population of *lymphocytes that are the principal agents of cell-mediated *immunity. T cells are derived from the bone marrow but migrate to the thymus to mature (hence *T cell*). After leaving the thymus the mature T cells continually circulate between lymph nodes and the bloodstream as naive T cells. Eventually a naive T cell is presented with antigen by an *antigen-presenting cell (e.g. dendritic cell or macrophage) within lymphoid tissue and becomes 'armed' to act rapidly as an **effector T cell** when it subsequently encounters its specific antigen. Subpopulations of T cells play different roles in the immune response and can be characterized by their surface antigens (*see* CD). *Helper T cells carry the *CD4 antigen on their surface, and recognize foreign antigens provided these are presented by cells (such as macrophages and B cells) bearing *MHC class II proteins. The helper T cell binds to its target cell by means of *T-cell receptors. Helper T cells are essential in the majority of infections for stimulating B cells to proliferate and differentiate into clones of antibody-producing plasma cells. They also ensure that bacteria ingested by macrophages do not survive and grow inside the host cells.

*Cytotoxic T cells, which carry the *CD8 antigen, recognize foreign antigen bound with *MHC class I protein on the surface of virus-infected cells and destroy the cell by releasing cytolytic proteins. **Regulatory T cells** (T_{reg} or **suppressor T cells**) are important in regulating the activity of other lymphocytes and are crucial in maintaining tolerance to self tissues. *See also* IMMUNOLOGICAL MEMORY.

T-cell receptor A protein molecule found attached to the plasma membrane of a T cell that recognizes foreign antigen displayed on the surface of host cells and activates an immune response by the T cell. A single T cell typically has about 30 000 such receptors. Each is similar, but not identical, to one arm of a Y-shaped antibody molecule (i.e. a *Fab fragment). It consists of an α chain and a β chain (corresponding to the light and heavy chains of a Fab fragment), each of which has an outer variable region (designated V_α or V_β) and an inner constant region (C_α or C_β). The V_α and V_β regions together form the single antigen-binding site. Each chain has a short hinge region close to the membrane, where the two chains are linked by a disulphide bond, and a transmembrane region terminating in a short cytoplasmic tail inside the cell. The two major classes of T cells are characterized by the nature of the *coreceptors that operate in conjunction with the T-cell receptor. The subset carrying CD4 coreceptors recognize MHC class II proteins, whereas cells carrying CD8 coreceptors recognize MHC class I. In either case, both the coreceptor and T-cell receptor bind to the complex of MHC molecule and processed antigen fragment on the surface of an *antigen-presenting cell or infected cell, thereby triggering the intracellular signal pathway that activates the cell. The great diversity of potential antigens is matched by the formidable array of T-cell receptors, each with a slightly different chemical structure and hence different binding properties. This diversity is generated by *somatic recombination in T-cell progenitors (*see* THYMOCYTE), in a manner similar to that responsible for producing the vast number of different antibodies.

TDF *See* TESTIS-DETERMINING FACTOR.

tectorial membrane *See* ORGAN OF CORTI.

tectum (*pl.* **tecta**) *See* MIDBRAIN.

teeth *See* DECIDUOUS TEETH; PERMANENT TEETH; DENTITION; TOOTH.

tegmentum (*pl.* **tegmenta**) *See* MIDBRAIN.

Teleostei The largest group (superorder or infraclass) of the *Osteichthyes (bony fish), containing about 20 000 species. Teleosts have colonized an extensive variety of habitats and show great diversity of form. The group includes the eel, seahorse, plaice, and salmon. They have been the dominant fish since the Cretaceous period (about 70 million years ago).

telocentric *See* CENTROMERE.

telomere The end of a chromosome, which consists of tandemly repeated short sequences of DNA that perform the function of ensuring that each cycle of *DNA replication has been completed. Each time a cell divides some sequences of the telomere are lost; eventually (after 60–100 divisions in an average cell) the cell dies (the **telomere theory of ageing** (*see* SENESCENCE) is based on this phenomenon). Thus for a limited period telomeres provide protection for the genes located at the ends of the chromosome. Also, proteins associated with telomeric DNA prevent the unequal ends of the replicated DNA strands from activating repair pathways that might cause the cell cycle to stall or even lead to death of the cell. Human telomeres consist of the nucleotide sequence TTAGGG repeated hundreds to thousands of times, and some 50–200 nucleotides are lost during each replication cycle. In yeasts, protists, and the stem cells and germ cells of higher organisms, the chromosomes are kept at their appropriate lengths by the action of **telomerase**, which catalyses the addition of lost telomeric sequences. Telomerase is an enzyme consisting of RNA and protein that is inactive in normal cells of higher organisms. Its presence in tumours is linked to the uncontrolled multiplication of cancer cells.

telome theory The theory that the leaves (megaphylls) of ferns and seed plants evolved by the modification of terminal branches (**telomes**) of stems. It envisages that firstly, instead of the primitive equal (dichotomous) branching of the stem, there developed a main axis with lateral side branches. Next, each lateral branch system was restricted to one plane, instead of forming a three-dimensional pattern. Lastly, the spaces between the telomes in each lateral branch system were gradually filled in by webbing consisting of thin sheets of photosynthetic parenchymatous tissue. There is fossil evidence for this sequence of events occurring in certain *trimerophytes and *progymnosperms, notably in trees of the genus *Archaeopteris*, of the late Devonian period. Megaphylls are unique to the *euphyllophyte group of vascular land plants.

telophase A stage of cell division. In *mitosis the chromatids that separated from each other at *anaphase collect at the poles of the spindle. A nuclear membrane forms around each group, producing two daughter nuclei with the same number and kind of chromosomes as the original cell nucleus. In the first telophase of *meiosis, complete chromosomes from the pairs that separated at first anaphase form the daughter nuclei. The number of chromosomes in these nuclei is therefore half the number in the original one. In the second telophase, daughter nuclei are formed from chromatids (as in mitosis). In some organisms, the first telophase of meiosis is abbreviated or omitted (*see* INTERKINESIS).

temperature inversion An abnormal increase in air temperature that occurs in the troposphere, the lowest level of the earth's atmosphere. This can lead to pollutants becoming trapped in the troposphere (*see* AIR POLLUTION).

temperature sensitivity (Q^{10}) A measure of how the rate of a reaction or metabolic process varies with temperature. It is given by the formula:

$$Q_{10} = \frac{R_T}{R_{T-10}}$$

where R_T is the reaction rate at temperature T, and R_{T-10} is the reaction rate at a temperature 10°C lower. Values of Q_{10} are typically between 2 and 3 for biological processes (a Q_{10} of 2 means that the rate doubles for a temperature rise of 10°C).

template Any molecule that acts as a pattern for the synthesis of a new molecule. For example, the two nucleotide chains of a DNA molecule can separate and each acts as a template for the synthesis of the complementary chain (*see* DNA REPLICATION).

temporal summation *See* SUMMATION.

tendon A thick strand or sheet of fibrous *connective tissue that attaches a muscle to a bone. Tendons consist of *collagen fibres and are therefore inelastic: they ensure that the force exerted by muscular contraction is transmitted to the relevant part of the body to be moved.

tendril A slender branched or unbranched structure found in many climbing plants. It may be a modified stem, leaf, leaflet, or petiole. Tendrils respond to contact with solid objects by twining around them (*see* THIGMOTROPISM). The cells that touch the object lose water and decrease in volume in comparison to the outer cells, thus causing the tendril to curve.

tentacle Any of the soft flexible appendages in aquatic invertebrate animals that are used principally for feeding. Water flows over the tentacles, which are able to capture food and direct it to the oral aperture. Tentacles are possessed by many cnidarians, some echinoderms (including sea cucumbers), and by cephalopod molluscs, in which the tentacles bear rows of suckers.

tera- Symbol T. A prefix used in the metric system to denote one million million times. For example, 10^{12} volts = 1 teravolt (TV).

teratogen Any environmental factor that acts on a fetus to cause congenital abnormality. Examples include ionizing radiation (e.g. X-rays), nutritional deficiencies, drugs (e.g. thalidomide), toxic chemicals, and virus infections (e.g. rubella).

tergum (*pl.* **terga**) The protective cuticle found on the dorsal surface of body segments in arthropods.

termination codon *See* STOP CODON.

terminator (termination sequence) A sequence of bases in DNA that signals the termination of *transcription of a gene or operon and release of the RNA transcript from the RNA polymerase. Prokaryotes have two different mechanisms of termination: **rho-dependent termination** involves binding of a protein transcription factor called rho to a specific site on the newly formed transcript. Meanwhile, the RNA polymerase stalls at a site downstream on the DNA molecule called a **transcription stop point**. Rho then moves along the transcript to the RNA polymerase and causes the transcript to

dissociate from the template strand of the DNA. In **rho-independent termination**, the end of a gene is marked by a particular sequence called an inverted repeat, followed by a string of adenine (A) bases. This causes the RNA transcript to form a hairpin loop structure followed by a string of uracil (U) bases. The hairpin loop causes the RNA polymerase to stall, and the weak binding between the Us on the RNA and the As on the DNA template strand enables release of the RNA transcript. In eukaryotes, the mechanism of termination varies according to the type of RNA polymerase. Specific terminator sequences have been identified for types I and III RNA polymerases, whereas termination for RNA polymerase type II, responsible for transcribing protein-coding genes, is more complex, with cleavage of the pre-messenger RNA occurring before the polymerase has stopped transcribing.

terpenes A group of unsaturated hydrocarbons present in plants (*see* ESSENTIAL OIL). Terpenes consist of isoprene units, $CH_2:C(CH_3)CH:CH_2$. Monoterpenes have two units, $C_{10}H_{16}$, sesquiterpenes three units, $C_{15}H_{24}$, diterpenes four units, $C_{20}H_{32}$, etc. Derivatives of terpenes, known as **terpenoids**, include the *plant hormones abscisic acid and gibberellin, and the carotenoid and chlorophyll pigments used in photosynthesis.

territory A fixed area that an animal or group of animals defends against intrusion from others of its species by various types of **territorial behaviour**. Outside the territory (which may contain food sources, hiding places, and nesting sites) others are not threatened (*compare* HOME RANGE). Many mammals indicate their territory boundaries with scent markings, while birds sing territorial songs that repel would-be intruders. Animals in neighbouring territories normally respect each other's boundaries, which reduces overt *aggression. Some animals are territorial only at certain times of the year, usually the breeding season (*see* COURTSHIP; LEK).

Tertiary Formerly, the older geological period of the Cenozoic era (*compare* QUATERNARY). It began about 65 million years ago, following the Cretaceous period, and extended to the beginning of the Quaternary, 2.58 million years ago. The Tertiary period was characterized by the rise of the modern mammals and the development of shrubs, grasses, and other flowering plants. It is no longer officially recognized as a division of geological time, having been replaced by the *Palaeogene and *Neogene periods.

tertiary consumer See CONSUMER.

tertiary structure See PROTEIN.

testa (seed coat) (*pl.* **testae**) The lignified or fibrous protective covering of a seed that develops from the integuments of the ovule after fertilization. *See also* HILUM; MICROPYLE.

test cross A mating (cross) made to identify hidden *recessive alleles in an individual of unknown genotype. This individual is crossed with one that is *homozygous for the allele being investigated (i.e. a homozygous recessive). The homozygous recessive individual may be the parent of the individual being investigated (*see* BACK CROSS).

testis (testicle) (*pl.* **testes**) The reproductive organ in male animals in which spermatozoa are produced. In vertebrates there are two testes; as well as sperm, they produce steroid hormones (*see* ANDROGEN). In most animals the testes are within the body cavity but in mammals, although they develop within the body near the kidneys, they come to hang outside the body cavity in a *scrotum. Most of the vertebrate testis is made up of a mass of **seminiferous tubules**, lined with *Sertoli cells, in which the sperms develop (*see* SPERMATOGENESIS). It is connected to the outside by means of the *vas deferens. *See* REPRODUCTIVE SYSTEM.

testis-determining factor (TDF; SRY protein) A protein that plays a crucial role in sex determination in mammals. It is encoded by the *SRY* (sex reversal on Y) gene on the Y chromosome, and switches embryonic development from the default female pathway to the male pathway, by driving testis formation. The male developmental cascade is then consolidated by secretion of the male sex hormone testosterone by the testes.

testosterone The principal male sex hormone. *See* ANDROGEN.

tetanus 1. (*pl.* **tetani**) A powerful sustained contraction of a skeletal muscle resulting from the summation of a series of rapid muscular contractions (**twitches**) that are induced by repeated stimulation of the muscle. **2.** A disease caused by the bacterium *Clostridium tetani*, which generally enters the body via an open wound. A toxin produced by the bacterium irritates the nerves, which induce muscle spasms that begin in the jaw muscles (giving the disease the informal name of **lockjaw**).

tetracosactide *See* ACTH.

tetrad A group of four *haploid cells formed at the end of the second division of *meiosis.

tetraploid Describing a nucleus, cell, or organism that has four times ($4n$) the haploid number (n) of chromosomes. *See also* POLYPLOID.

Tetrapoda In some classifications, a superclass of jawed chordates (*Gnathostomata) comprising all vertebrate animals with four limbs, i.e. the amphibians, reptiles, birds, and mammals. The skeleton of the limbs of all tetrapods is based on the same five-digit pattern (*see* PENTADACTYL LIMB).

tetraspore Any one of four spores produced by meiosis in an independent diploid sporophyte (**tetrasporophyte**) in the life cycle of certain red algae. The tetraspores are produced within a **tetrasporangium** and germinate to produce the sexual haploid generation. The tetrasporophyte develops from a *carpospore.

tetrodotoxin *See* PUFFERFISH.

thalamus (*pl.* **thalami**) **1.** (in anatomy) Either of a pair of structures in the vertebrate *forebrain that lie above the hypothalamus and form part of the *limbic system. Each relays sensory information to the cerebral cortex on the same side of the brain, such as inputs from the retina of the opposite eye via the optic nerves. The thalamus acts as a 'gatekeeper' for the neuromodulatory inputs of neurons in the *brainstem, and thus regulates the wakefulness of neurons in the cortex. Thalamocortical neurons project from the thalamus to all parts of the cortex and maintain wakefulness by releasing excitatory neurotransmitters, such as glutamate and aspartate, and maintaining a characteristic spontaneous rhythmic electrical activity. At the onset of sleep, inputs from the brainstem hyperpolarize the thalamocortical neurons, which switch to an oscillating pattern of electrical activity that synchronizes cortical neurons to the same pattern—one characteristic of sleep when monitored by an electroencephalogram. This pattern essentially disconnects the cortex from the outside environment. Parts of the thalamus, along with other structures of the limbic system, are involved in the generation and expression of emotions. **2.** (in botany) *See* RECEPTACLE.

thalassaemia A blood disease caused by a *substitution mutation in the DNA sequence that controls the synthesis of the α- or β-chains of *haemoglobin. Thalassaemia results in anaemia and growth retardation.

thallus (*pl.* **thalli**) A relatively undifferentiated vegetative body with no true roots, stems, leaves, or vascular system. It is found in the algae, fungi, mosses, liverworts, and lichens, and in the gametophyte generation of clubmosses, horsetails, and ferns.

T$_H$ cell *See* HELPER T CELL.

theca (*pl.* **thecae**) *See* CAPSULE.

thelytoky The phenomenon occurring in the reproduction of certain animals in which fertilized eggs give rise to males and unfertilized eggs to females. It is found among aphids and certain other insects, and in some mites. The males are diploid, whereas females are haploid and transmit only the maternal genome—their production represents thelytokous *parthenogenesis. *Compare* ARRHENOTOKY.

therapeutic half-life (in pharmacology) The time taken for half the dose of a drug to be excreted: used to calculate the most effective and nontoxic dosing intervals. It can be determined by administering a therapeutic dose of the drug labelled with a radioisotope (*see* LABELLING) and measuring the time for half of it to be excreted in the urine.

therapeutic index (therapeutic window) The therapeutic usefulness of a drug, expressed as a ratio of the maximum nontoxic dose and the minimum effective dose of the drug. The therapeutic index is commonly used as a measure of the range of the dose considered to be safe. However, due to the variation in individual response to a given drug, the therapeutic index cannot be calculated accurately.

therapsid Any of a group of *synapsid animals that includes the immediate ancestors of modern mammals. The therapsids flourished during the Permian period, first appearing in the fossil record around 280 million years ago (mya). Although they still resembled reptiles in some respects, mammal-like features were increasingly apparent, such as teeth specialized for different functions. They diversified in form, with both carnivores and herbivores occupying a wide range of habitats. For example, the plant-eating dicynodonts had horny beaklike jaws with teeth much reduced except for a pair of prominent tusklike upper canines. However, many therapsids disappeared during the end-Permian mass extinction, and it was their successors, the cynodonts, that dominated the Triassic and from which the earliest true mammals evolved about 200 mya.

thermal denaturation *See* DENATURE.

thermal hysteresis protein *See* ANTIFREEZE MOLECULE.

thermocline A steep temperature gradient that exists between the relatively warm surface layer and the much colder deep layer of a body of water such as a lake or ocean. In a lake the thermocline occurs in the middle zone (the **metalimnion**) and gives rise to thermally induced vertical stratification of the water. The metalimnion lies between the relatively warm **epilimnion** above and the cold **hypolimnion** below. The thermocline may be short-lived, especially in shallow lakes where wind action can mix the water from different levels. However, it can exist for most of the summer period in temperate lakes and sometimes nearly all year in tropical lakes. A thermocline can speed up the process of eutrophication by preventing the diffusion of oxygen from the epilimnion to the hypolimnion (*see* EUTROPHIC). The depth and temperature range of oceanic thermoclines vary with season and latitude, typically being deepest and strongest during the summer months and in tropical latitudes.

thermogenesis The production of heat within tissues to raise body temperature or as an adaptive response (**adaptive thermogenesis**) to 'burn off' excess food energy intake. Heat production occurs especially in birds and mammals, animals that maintain their temperature within a narrow range (i.e. *endotherms), but is also found in some 'cold-blooded' vertebrates and invertebrates. There are two types of thermogenesis, of which the more familiar is **shivering**. This involves repeated rapid contractions of antagonistic sets of skeletal muscles, which produce little net movement so that most of the chemical energy (in the form of ATP) is converted to heat rather than mechanical work. Many endothermic insects, such as bees and moths, perform synchronous contraction of their flight muscles to generate heat before taking off in cold conditions. **Nonshivering thermogenesis** takes place in *fat cells (adipose tissue) and involves the breakdown of stored fat to generate heat in situ instead of its being transported to the liver for conversion to ATP. This process is activated by the sympathetic nervous system and is accomplished in two ways: nonproductive cyclical active transport of ions across the fat-cell plasma membrane and uncoupling of electron transport from ATP synthesis within the fat-cell mitochondria (*see* UNCOUPLING PROTEIN). The net result is release of energy from existing ATP and oxidation of fat to produce heat instead of

ATP. Certain mammals have deposits of a special adipose tissue called *brown fat that is adapted to provide the body with bursts of intense heat. Stimulation of brown fat oxidation enables rapid warming during the arousal of hibernating animals, for example. Brown fat deposits are also present in human babies and other neonate mammals, to help protect them against hypothermia, and in adult humans, where they provide a means for cold-stimulated thermogenesis. The latter ability is associated with maintaining a lean body mass. Another form of adaptive thermogenesis is the ability to dissipate excess food energy as heat, instead of converting it to fat. In humans, such an ability can enable an individual to maintain food intake without becoming overweight, and seems to have a strong degree of genetic control. It may have evolved as an adaptation when humans consumed large quantities of energy-rich but nutrient-poor foods. The mechanisms involved are significant for our understanding of the causes and treatment of human obesity.

thermogenin *See* UNCOUPLING PROTEIN.

thermography A medical technique that makes use of the infrared radiation from the human skin to detect an area of elevated skin temperature that could be associated with an underlying cancer. The heat radiated from the body varies according to the local blood flow, thus an area of poor circulation produces less radiation. A tumour, on the other hand, has an abnormally increased blood supply and is revealed on the **thermogram** (or **thermograph**) as a 'hot spot'.

thermonasty *See* NASTIC MOVEMENTS.

thermoneutral zone The range of environmental temperatures over which the heat produced by a 'warm-blooded' animal (*see* ENDOTHERM) remains fairly constant. Hence, it is the range in which the animal is 'comfortable', having neither to generate extra heat to keep warm nor expend metabolic energy on cooling mechanisms, such as panting. Animals adapted to cold environments tend to have broader thermoneutral zones than ones living in hot environments.

thermophilic Describing an organism that lives and grows optimally at extremely high temperatures, typically over 40°C. The majority are prokaryotes, such as the archaea found in hot springs and in undersea hydrothermal vents (**thermoacidophils**); some of these (called **hyperthermophiles**) thrive at temperatures above 80°C and can even survive at temperatures above the boiling point of water. Some eukaryotes, especially certain protists and fungi, are capable of surviving temperatures up to about 60°C. Thermophiles have various adaptations in order to thrive at such high temperatures. For example, their proteins and nucleic acids have structural modifications that give them much greater heat stability, so that the cell machinery is able to function. Also, the chemical makeup of their cell membranes is adapted, for example by inclusion of lipids rich in saturated fatty acids. *Compare* MESOPHILIC; PSYCHROPHILIC.

thermoreceptor Any specialized free nerve ending (*see* RECEPTOR) that is sensitive to temperature. Mammals have several types of thermoreceptors in the skin and tongue, each specific for detecting changes within a certain temperature range. Some respond when the skin is warmed (i.e. above body temperature) and others when the skin is cooled (i.e. below body temperature). Curiously, the same receptors also respond to chemicals that are perceived as 'hot' or 'cool' when tasted, such as cayenne pepper or menthol. In many cases the response of the thermoreceptor involves the opening of *TRP protein ion channels in the cell's plasma membrane, which triggers the influx of calcium ions and the initiation of a signal to the brain via sensory neurons. Thermoreceptors are also located at various sites inside the body, including the large blood vessels, viscera, abdominal wall, spinal cord, and brain. Information from both external and internal thermoreceptors is conveyed to the body's 'thermostat' in the *hypothalamus, where it is integrated to enable the appropriate responses.

thermoregulation Regulation of body temperature by any means, whether physiological or behavioural. Some animals, particularly mammals and birds, can maintain a fairly constant internal body temperature (*see* HOMOIOTHERMY), whereas in others the body temperature varies with the temperature of the environment (*see* POIKILOTHERMY). In mammals body temperature is regulated by the *hypothalamus in the brain, in conjunction with the autonomic nervous system. The hypothalamus contains temperature sensors and also receives input from *thermoreceptors in the skin and viscera, besides other information about the body (e.g. its state of wakefulness and activity). Temperature is controlled by homeostatic mechanisms that maintain it at a **set point**. When cooling is detected, the hypothalamus triggers responses such as constriction of blood vessels in the skin to conserve body heat and shivering to generate additional heat. When the body is too warm, the skin blood vessels are dilated and the ventilation rate may increase (i.e. panting). Seasonal responses, such as changes in thickness of body hair between summer and winter, also contribute to thermoregulation (*see* ACCLIMATION). *See also* HETEROTHERM; THERMOGENESIS.

therophyte A plant life form in Raunkiaer's system of classification (*see* PHYSIOGNOMY). Therophytes are annual plants that complete their life cycle in a short period when conditions are favourable and survive harsh conditions as seeds. They are typically found in deserts and other arid regions.

Theropoda A clade of saurischian ('lizard-hipped') dinosaurs whose descendants include the modern birds. It comprised bipedal carnivores, including *Tyrannosaurus rex*, which was over 12 m long and 4.5 m tall. Shared traits included hollow, thin-walled bones, three extended clawed digits on both the hands and feet, and a furcula ('wishbone'). Some fossil theropods had feathers very similar to those of modern birds, including *Archaeopteryx*, which lived about 150 million years ago.

thiamin(e) *See* VITAMIN B COMPLEX.

thigmotropism (haptotropism) The growth of an aerial plant organ in response to localized physical contact. For example, when a tendril of sweet pea touches a supporting structure, it curves in the direction of the support and coils around it. The coiling is the result of differential growth of cells on opposite sides of the tendril. *See* TROPISM.

thin-layer chromatography A technique for the analysis of liquid mixtures using *chromatography. The stationary phase is a thin layer of an absorbing solid (e.g. alumina) prepared by spreading a slurry of the solid on a plate (usually glass) and drying it in an oven. A spot of the mixture to be analysed is placed near one edge and the plate is stood upright in a solvent. The solvent rises through the layer by capillary action carrying the components up the plate at different rates (depending on the extent to which they are absorbed by the solid). After a given time, the plate is dried and the location of spots noted. It is possible to identify constituents of the mixture by the distance moved in a given time. The technique needs careful control of the thickness of the layer and of the temperature.

thoracic cavity The space within the *thorax, which in vertebrates contains the heart, lungs, and rib cage.

thoracic duct The main collecting vessel of the *lymphatic system, running longitudinally in front of the backbone. The thoracic duct drains its lymph into the superior vena cava.

thoracic vertebrae The *vertebrae of the upper back, which articulate with the *ribs. They lie between the *cervical vertebrae and the *lumbar vertebrae and are distinguished by a number of articulating facets for attachment of the ribs. In humans there are 12 thoracic vertebrae.

thorax (*pl.* **thoraces** or **thoraxes**) The anterior region of the body trunk of animals. In vertebrates it contains the heart and lungs within the rib cage. It is particularly well-defined in mammals, being separated from the *abdomen by the *diaphragm. In insects the thorax is divided into an anterior **prothorax**, a middle **mesothorax**, and a posterior **metathorax**, each of which bears a pair of legs; the hindmost two segments also both carry a pair of wings. In other arthropods, especially crustaceans and arachnids, the thorax is fused with the head to form a **cephalothorax**.

thorn A hard side stem with a sharp point at the tip, replacing the growing point. In some plants the development of thorns and subsequent suppression of the growing points may be a response to dry conditions. Examples are the thorns of gorse and hawthorn. *Compare* PRICKLE; SPINE.

thread cell (nematoblast; cnidoblast) A specialized cell found only in the ectoderm of the *Cnidaria. It contains a **nematocyst**, a fluid-filled sac within which lies a long hollow coiled thread. When a small sensory projection (**cnidocil**) on the surface of the thread cell is touched, e.g. by prey, the thread is shot out and adheres to the prey, coils round it, or injects poison into it. Numerous thread cells on the tentacles of jellyfish produce their sting.

threat display *See* AGONISTIC BEHAVIOUR.

threatened species *See* ENDANGERED SPECIES.

threonine *See* AMINO ACID.

threshold (in physiology) The minimum intensity of a stimulus that is necessary to initiate a response.

thrombin An enzyme that catalyses the conversion of fibrinogen to fibrin. *See* BLOOD CLOTTING; PROTHROMBIN.

thrombocyte *See* PLATELET.

thromboplastin (tissue factor; Factor III) A membrane glycoprotein expressed on the surface of damaged tissue cells that initiates the cascade of reactions leading to formation of a blood clot. It forms a complex with Factor VIIa in the presence of phospholipid and calcium ions; this complex converts Factor X to Xa, which in turn converts *prothrombin to *thrombin. *See* BLOOD CLOTTING.

thrombosis (*pl.* **thromboses**) The obstruction of a blood vessel by a mass of blood cells and fibrin (**thrombus**), which can result from excessive *blood clotting, fragmentation of an atherosclerotic plaque (*see* ATHEROSCLEROSIS), or inflammation of the blood vessel.

thromboxane A$_2$ (TXA$_2$) An eicosanoid, related to the *prostaglandins, that promotes blood clotting and causes constriction of blood vessels (vasoconstriction). It is released by platelets when these are activated by local tissue damage and attracts other platelets to the injury site, forming a platelet plug. Its action thus antagonizes that of *prostacyclin. TXA$_2$ is spontaneously converted to the inactive TXB$_2$.

thylakoid Any of the flattened saclike membranous structures that are stacked on top of one another to form the grana (*see* GRANUM) within the stroma of plant *chloroplasts. Chlorophyll and other photosynthetic pigments are situated in the thylakoid membranes, which are the site for the light-dependent reactions of *photosynthesis (*see also* PHOTOPHOSPHORYLATION). The photolysis of water (*see* PHOTOSYSTEMS I AND II) occurs in the **thylakoid space** between the thylakoid membranes. Thylakoid membranes also occur in photosynthetic prokaryotes such as cyanobacteria.

thymidine A nucleoside consisting of one thymine molecule linked to a D-deoxyribose sugar molecule.

thymine A *pyrimidine derivative and one of the major component bases of *nucleotides and the nucleic acid *DNA.

thymocyte An immature or developing lymphocyte in the thymus that is the forerunner of a T cell. Progenitor cells enter the thymus from bone marrow to become thymocytes, which proliferate and undergo distinct developmental stages marked by changes in their CD surface proteins

(*see* CD). During maturation they express on their surface a precursor T-cell receptor (pre-T-cell receptor) and both CD8 and CD4 proteins (**double-positive cells**). At the same time they slowly migrate from the outer cortex of the thymus to the inner medullary region. Some 98% of thymocytes express surface receptors that fail to recognize self antigen, and these cells undergo apoptosis and die; only the 2% of cells capable of distinguishing self antigen survive to maturity in the medulla. Here they become **single-positive cells**, expressing exclusively CD4 or CD8 surface proteins, and are exported from the thymus in the bloodstream to become helper T cells and cytotoxic T cells, respectively. *See also* SOMATIC RECOMBINATION.

thymus (*pl.* **thymi**) An organ, present only in vertebrates, that is concerned with development of *lymphoid tissue, particularly the white *blood cells involved in cell-mediated *immunity (*see* T CELL; THYMOCYTE). In mammals it is a bilobed organ in the region of the lower neck, above and in front of the heart. The thymus undergoes progressive shrinkage (involution) throughout life, starting after the first 12 months. Haemopoietic stem cells from the bone marrow migrate to the thymus, attracted by chemotactic factors, and begin to divide and differentiate to form the many subpopulations of T cells. As their progeny cells migrate through the thymus from its cortex to medulla, they interact with thymic 'nurse cells' and with each other and are influenced by various extracellular proteins and thymic peptide hormones (e.g. thymosin and thymopoietin). All these factors help to promote the differential expression of surface antigens and development of distinctive immunological competences.

thyrocalcitonin *See* CALCITONIN.

thyroglobulin (TGB) A glycoprotein, made in the thyroid gland, that consists of about 5000 amino acids, some of which are tyrosine residues. TGB is the precursor of the thyroid hormones, thyroxine and triiodothyronine. Iodine binds to the tyrosine residues in thyroglobulin, which is then hydrolysed into **iodotyrosines** that combine to form triiodothyronine (T_3) or thyroxine (tetraiodothyronine or T_4).

thyroid gland A bilobed endocrine gland in vertebrates, situated in the base of the neck. It secretes two iodine-containing **thyroid hormones, thyroxine (T_4**; see formula) and **triiodothyronine (T_3)**, which are formed in the gland from *thyroglobulin. They control the rate of many metabolic processes in the body, including energy metabolism, blood pressure, heart rate, muscle tone, and digestive function; they also influence physical development and activity of the nervous system. Growth and activity of the thyroid are controlled by *thyroid-stimulating hormone, secreted by the anterior *pituitary gland. Adequate amounts of dietary iodine are required for the body to synthesize sufficient T_3 and T_4, and iodine deficiency causes enlargement of the thyroid (i.e. 'goitre'). The thyroid gland also contains **C cells**, which secrete *calcitonin.

(⊕) SEE WEB LINKS

• Overview of thyroid disease and function from MedicineNet

thyroid-stimulating hormone (TSH; thyrotrophin; thyrotropin) A hormone, secreted by the anterior pituitary gland, that controls the synthesis and secretion of the two thyroid hormones, thyroxine and triiodothyronine, in the *thyroid gland. The secretion of thyroid-stimulating hormone is controlled by thyrotrophin-releasing hormone (TRH) from the hypothalamus. The release of TRH depends on many factors, including the levels of TSH, glucose, and thyroxine in the blood and the rate of metabolism in the body.

thyrotrophin-releasing hormone (TRH) *See* THYROID-STIMULATING HORMONE.

thyroxine (T_4) The principal hormone of the *thyroid gland. *See also* THYROGLOBULIN.

Thysanura The former name for an order of insects containing the bristletails, silverfish, and firebrats. Its use is now deprecated, although it is still commonly found in popular texts. The silverfish and firebrats are now placed in the order Zygentoma, in the subclass *Apterygota. These wingless insects typically have slender bodies with long threadlike antenna and a segmented tail-like extension to the abdomen bordered on either side by the elongated paired cerci. They are mostly detritivores living in litter or under bark, although some species are found in deserts and others are familiar inhabitants of human dwellings. These domestic species feed on food debris, paper, cotton, and similar materials. Bristletails are now placed in the order Archaeognatha. They are mostly nocturnal, feeding on algae, lichens, and other plant material and sheltering in litter or beneath bark during the day. They can spring some distance by arching the thorax and flexing the abdomen.

tibia (*pl.* **tibiae**) **1.** The larger of the two bones of the lower hindlimb of terrestrial vertebrates (*compare* FIBULA). It articulates with the *femur at the knee and with the *tarsus at the ankle. The tibia is the major load bearing bone of the lower leg. **2.** The fourth segment of an insect's leg, which is attached to the femur.

ticks *See* ACARINA.

tidal volume The volume of air taken in or expelled by an animal breathing normally at rest during each cycle of *ventilation.

$$HO-\bigcirc-O-\bigcirc-CH_2-\overset{\overset{\displaystyle NH_3^+}{|}}{\underset{\underset{\displaystyle H}{|}}{C}}-COO^-$$

Thyroxine

The average human has a tidal volume of approximately 500 cm^3.

tight junction (zonula occludens) The region between the plasma membranes of two adjacent cells that are so closely positioned that there is no intercellular space between them. This type of junction fuses cells together and provides a selective barrier to the diffusion of substances between cells. Tight junctions are common between cells of the intestinal epithelium, where they act to seal off the intestinal lumen from the intercellular fluid. *See also* CELL JUNCTION.

Tiktaalik An extinct genus of lobe-finned fish whose fossils display features of both fish and four-legged (tetrapod) animals; hence its nickname 'fishapod'. It was first discovered at a site in the Canadian Arctic and dated to 375 million years ago, some 10 million years earlier than the first known tetrapods. Like its fish relatives, *Tiktaalik* had prominent pectoral and pelvic fins, gills, and lungs, and was covered in scales. However, it also possessed adaptations suggestive of an amphibious lifestyle, such as a flattened head with eyes on top and a functional neck. A full set of ribs and robust limbs enabled it to breathe air and prop itself up in shallow waters. Also, the bones of the front fins have the same pattern, including a wrist, found in all limbed animals.

tiller A shoot that develops at the base of a plant stem from an axillary bud. Tillers are often produced in response to injury of the main stem, as occurs when a tree is lopped.

time-lapse photography A form of photography used to record a slow process, such as plant growth. A series of single exposures of the object is made on film at predetermined regular intervals. The film produced can then be projected at normal cine speeds and the process appears to be taking place at an extremely high rate.

tinsel flagellum A type of eukaryotic *flagellum (sense 2) with numerous hair-like projections (**mastigonemes**) along the shaft. They occur in certain protists, particularly the fungus-like oomycotes and hyphochytrids. The mastigonemes vary in nature: some are solid protein rods; others are tubular. They increase the power generated by the flagellum.

Ti plasmid *See* AGROBACTERIUM TUMEFACIENS.

tissue A collection of cells of similar structure organized to carry out one or more particular functions. For example, in animals nervous tissue is specialized to perceive and transmit stimuli. An organ, such as a lung or kidney, contains many different types of tissues.

tissue culture The growth of the tissues of living organisms outside the body in a suitable culture medium. Culture (or nutrient) media contain a mixture of nutrients either in solid form (e.g. in *agar) or in liquid form (e.g. in *physiological saline). Tissue culture is invaluable for gaining information about factors that control the growth and differentiation of cells, e.g. in testing drugs, cloning, and the production of *stem cells. Culture of plant tissues such as *callus has resulted in the regeneration of complete plants, enabling commercial propagation (e.g. of orchids) and—through culture of meristem tissues—the production of virus-free crop plants. *See also* EXPLANTATION; MICROPROPAGATION; TISSUE ENGINEERING.

tissue engineering The creation of synthetic or semisynthetic tissue that can be used instead of human tissue in surgery or for research. Different kinds of tissue have been developed or are currently being researched, including skin, bone, cartilage, cornea, and spinal tissue, using various combinations of biopolymers, cultured cells, and growth factors. For example, the first such product to gain approval for clinical use was a form of artificial skin consisting of a thin sheet of collagen gel infiltrated with two layers of cultured human cells—keratinocytes on the outer surface to form the 'epidermis', and fibroblasts on the inner surface to form the 'dermis'. More rigid tissues, such as synthetic bone and cartilage, are typically based on a biopolymer scaffold, which is seeded with cultured bone cells or cartilage cells to secrete the natural tissue material. The scaffold can be inserted in situ, for example at a fracture site, or used to construct a new body part entirely *in vitro*. For successful colonization and growth by cultured cells or the body's own cells, the scaffold is treated with appropriate growth factors; for example, bone cells require a substance called bone morphogenetic protein. Another strategy is the manipulation of stem cells in three-dimensional cultures by sequentially exposing them to a suite of signalling molecules (*morphogens) that mimic the factors determining their development into a particular tissue or organ in the body. Tissue engineering requires detailed knowledge of cell culture techniques, biomolecules, and the biomechanical properties of natural tissues.

(()) SEE WEB LINKS

• Introduction to tissue engineering from the US National Institutes of Biomedical Imaging and Bioengineering

tissue fluid (interstitial fluid) The fluid that bathes the body cells of animals, consisting of water, ions, and dissolved gases and food substances. It is formed when blood is ultrafiltered (*see* ULTRAFILTRATION) from the capillaries into the intercellular spaces. The pressure in the arterial capillaries causes most components of the blood to pass across the capillary walls; blood cells and most of the plasma proteins are retained in the capillaries. The tissue fluid surrounds the body cells, facilitating the exchange of nutrients and waste materials. At the venous end of the capillaries, the tissue fluid is drawn into the capillaries by *osmosis. In animals with an open *circulation, such as insects and other arthropods, the haemolymph serves also as the tissue fluid.

tissue plasminogen activator (TPA)
See PLASMINOGEN.

tissue typing The process of identifying the major histocompatibility (MHC) antigens in humans encoded by the *HLA system in both the recipient and the potential donor before organ transplantation. If the donor and recipient tissue types do not match exactly, the transplanted organ will be rejected. HLA typing is imprecise, and individuals who are shown on testing to be HLA identical very rarely have identical MHC genotypes. This follows from the complexity and polymorphism of the alleles determining *histocompatibility. Even a graft from an HLA-identical sibling will provoke rejection (albeit more slowly), because of differences between non-MHC proteins that act as minor histocompatibility antigens. Only grafts between identical twins are likely to be free of rejection problems.

titre 1. The number of infectious virus particles present in a suspension. **2.** A measure of the amount of *antibody present in a sample of serum, given by the highest dilution of the sample that results in the formation of visible clumps with the appropriate antigen (*see* AGGLUTINATION).

TLR *See* TOLL-LIKE RECEPTOR.

T lymphocyte *See* T CELL.

tmRNA *See* TRANSFER-MESSENGER RNA.

TNF *See* TUMOUR NECROSIS FACTOR.

toads *See* AMPHIBIA.

tobacco mosaic virus (TMV) A rigid rod-shaped RNA-containing virus that causes distortion and blistering of leaves in a wide range of plants, especially the tobacco plant. It is transmitted by insects when they feed on plant tissue. TMV was the first virus to be discovered. *See* VIRUS (Feature).

tocopherol *See* VITAMIN E.

tolerance 1. The ability of an organism to withstand extreme variations in environmental conditions, such as drought. **2.** The build-up of *resistance to drugs or other chemicals (such as pesticides), which occurs after prolonged use or application. Increasingly large doses of the chemical are required to produce the desired effect in the organism. **3. (immunological tolerance)** The phenomenon by which the cells of the immune system are constrained from mounting an immune response against 'self' tissues. During their development and maturation, lymphocyte precursors (i.e. precursors of both *B cells and *T cells) undergo a series of selection processes to ensure that they are capable of recognizing the body's own tissue markers, particularly the *histocompatibility proteins, and that they do not respond to the wide range of other 'self' antigens when the latter are combined with these marker proteins. Any precursor lymphocytes that fail these selection procedures are eliminated, ensuring that only tolerant clones are produced (*see* THYMOCYTE). The term 'tolerance' also embraces failure of the immune response in an animal exposed to a foreign antigen to which an immune response would normally be mounted. This commonly follows exposure to the antigen during fetal life, presumably when the immune system is developing self-tolerance.

Tollens reagent A reagent used in testing for aldehydes, named after German chemist B. C. G. Tollens (1841–1918). It is made by adding sodium hydroxide to silver nitrate to give silver(I) oxide, which is dissolved in aqueous ammonia (giving the complex ion $[Ag(NH_3)_2]^+$). The sample is warmed with the reagent in a test tube. Aldehydes reduce the complex Ag^+ ion to metallic silver, forming a bright silver mirror on the inside of the tube (hence the name **silver-mirror test**). Ketones give a negative result.

Toll-like receptor (TLR) Any of a family of receptor proteins associated with innate immune responses by lymphocytes, macrophages, and other cells (*see* IMMUNITY). They bind certain constituents of invading pathogens, such as bacterial lipopolysaccharide or peptidoglycan, and activate a signal pathway inside the cell that triggers the release of cytokines and chemokines, such as tumour necrosis factor α. TLR activation also leads to the surface expression by *dendritic cells and macrophages of co-stimulatory molecules that are essential for activating naive T cells, which can then initiate adaptive immune responses. Toll signalling involves the activation of a transcription factor, *nuclear factor κB, which consequently migrates from the cytoplasm to the nucleus, where it activates gene transcription. TLR proteins in humans and other vertebrates are homologous with the Toll protein originally identified because of its role in dorsoventral patterning in embryological development of the fruit fly *Drosophila*. Similar proteins participate in plant immunity against viruses, which suggests an ancient origin for TLR-based defence mechanisms.

tomography The use of X-rays to photograph a selected plane of a human body with other planes eliminated. The CT (computerized tomography) **scanner** is a ring-shaped X-ray machine that rotates through 180° around the horizontal patient, making numerous X-ray measurements every few degrees. The vast amount of information acquired is built into a three-dimensional image of the tissues under examination by the scanner's own computer. The patient is exposed to a dose of X-rays only a fraction of that used in a normal diagnostic X-ray. *See also* POSITRON EMISSION TOMOGRAPHY.

tone (tonus) The state of sustained tension in muscles that is necessary for the maintenance of posture. In a tonic muscle contraction, only a certain proportion of the muscle fibres are contracting at any given time; the rest are relaxed and recovering for subsequent contractions. The fibres involved in tone contract more slowly than the fast fibres used for rapid responses by the same muscle. The proportions of slow and fast fibres depend on the function of the muscle.

tongue A muscular organ of vertebrates that in most species is attached to the floor of the mouth. It plays an important role in manipulating food during chewing and swallowing and in terrestrial species it bears numerous *taste buds on its upper surface. In some advanced vertebrates the tongue is used in the articulation of sounds, particularly in human speech.

tonicity 1. The effective osmotic pressure of a solution relative to another solution from which it is separated by a membrane permeable only to water molecules. It is determined by the concentration of non-penetrating solutes in the solution relative to that in the other solution. Hence, the tonicity of a solution surrounding a cell determines whether the solution will cause water to flow in or out of the cell across the plasma membrane. A **hypertonic solution** will cause water to flow out of the cell, whereas a **hypotonic solution** will cause water to flow into the cell. If the solution is **isotonic**, there will be no net flow of water across the cell membrane. *See* OSMOSIS. **2.** The degree of muscle *tone.

tonoplast (vacuole membrane) The single membrane that bounds the *vacuole of plant cells. It contains proteins that transport solutes, including sugars and amino acids, from the cytoplasm into the vacuole.

tonsil A mass of *lymphoid tissue, several of which are situated at the back of the mouth and throat in higher vertebrates. In humans there are the **palatine tonsils** at the back of the mouth, **lingual tonsils** below the tongue, and **pharyngeal tonsils** (or **adenoids**) in the pharynx. They are concerned with the production of *lymphocytes and form part of the *gut-associated lymphoid tissue protecting the gut against infection.

tonus *See* TONE.

tooth Any of the hard structures in vertebrates that are used principally for biting

Section through an incisor tooth

and chewing food but also for attack, grooming, and other functions. In fish and amphibians the teeth occur all over the palate, but in higher vertebrates they are concentrated on the jaws. They evolved in cartilaginous fish as modified placoid *scales, and this is reflected in their structure: a body of bony *dentine with a central *pulp cavity and an outer covering of *enamel on the exposed surface (**crown**). The portion of the tooth embedded in the jawbone is the **root** (see illustration).

In mammals there are four different types of teeth, specialized for different functions (*see* CANINE TOOTH; INCISOR; MOLAR; PREMOLAR). Their number varies with the species (*see* DENTAL FORMULA). *See also* DECIDUOUS TEETH; PERMANENT TEETH.

top carnivore *See* CONSUMER.

topoisomer An isomer of, typically, a large complex molecule that is distinct from other similar isomers by virtue of its topology, such as degree of twisting or interlocking of ring structures. For example, topoisomers of DNA may be distinguished according to the degree of *supercoiling of the double helix.

topoisomerase An enzyme that creates or changes the state of a *topoisomer, particularly a topoisomer of DNA. Such enzymes change the way in which DNA is packaged in living cells by altering the degree of *supercoiling of the DNA molecules. For example, DNA gyrase, a topoisomerase found in *E. coli*, introduces negative supercoiling by breaking both strands of the DNA double helix and passing another stretch of the double helix through the gap, which is then sealed. This creates a twist in the circular DNA of the bacterial chromosome. Other topoisomerases can remove supercoiling by introducing a temporary break in one DNA strand and passing the other strand through it.

topsoil *See* SOIL.

tornaria (*pl.* **tornariae** or **tornarias**) A ciliated larva produced by some acorn worms (*see* HEMICHORDATA).

torpor *See* HETEROTHERM.

torus (*pl.* **tori** or **toruses**) *See*
RECEPTACLE.

total peripheral resistance *See* BLOOD
PRESSURE.

totipotent 1. Describing differentiated
plant cells that, when isolated, have the
ability to develop into an entire new plant
if provided with the suitable growing
medium. **2.** Describing embryonic cells or
*stem cells at a stage before their fate is
irreversibly determined, when they have
the ability to develop into any differentiated
cell given the appropriate stimulation.

touch The sense that enables the pressure
exerted by objects and substances, and
hence their surface texture, to be perceived.
Touch receptors occur in the *skin, being
concentrated in sensitive areas such as the
tips of the finger in humans. Generally, they
consist of nerve endings (dendrites) encapsulated in connective tissue and located at
different depths in the skin according to the
type of stimulus they detect. *Meissner's
corpuscles respond to changes in pressure,
whereas *Merkel's discs provide information about sustained pressure. *Ruffini's
capsules and *Pacinian corpuscles respond
to vibrating stimuli of varying frequencies,
while naked dendrites wind around the
base of hairs to sense hair movements.

toxicogenomics The study of the toxic
effects of drugs or other substances on patterns of gene expression in particular cells
or tissues. It is thus a synthesis of toxicology
and genomics, particularly *transcriptomics,
and is aimed at identifying potential new
drugs more efficiently and revealing the genetic basis of toxic reactions to compounds.

toxicology The science of the study of
poisons. Originally developed by Paracelsus
(1493–1541), toxicology is concerned with
the investigation of the deleterious effects of
all foreign substances (*xenobiotics) on living
organisms.

toxin A poison produced by a living organism, especially a bacterium. **Endotoxins** are
lipopolysaccharides derived from the outer
membrane of Gram-negative bacteria, such

as strains of Escherichia and Salmonella.
They are released only when the bacterial
cell dies or disintegrates and are seldom
fatal. **Exotoxins** are secreted by a bacterial
cell into the surrounding medium and are
generally highly toxic to the host. For
example, *botulinum toxin produced by
Clostridium botulinum is one of the most
poisonous substances known. Eukaryotes
can also produce toxins. Toxic 'algal blooms'
of dinoflagellates release various neurotoxins
into seawater, poisoning fish and shellfish
and also humans who consume affected seafood (*see* DINOMASTIGOTA). *Ergot poisoning
is the result of eating grain contaminated
with toxins produced by the fungus *Claviceps
purpurea*. In the body a toxin acts as an
*antigen, producing an immune response.
High-affinity IgG and IgA antibodies bind to
the receptor-binding sites of secreted bacterial exotoxins, thereby neutralizing them by
preventing them from binding to receptors
on host cells. Such antibodies are called
neutralizing antibodies.

trace element *See* ESSENTIAL ELEMENT.

trace fossil *See* FOSSIL.

trachea (*pl.* **tracheae** or **tracheas**)
1. (windpipe) The tube in air-breathing
vertebrates that conducts air from the
throat to the *bronchi. It is strengthened
with incomplete rings of cartilage. **2.** An
air channel in insects and most other terrestrial arthropods. Tracheae occur as
ingrowths of the body wall. They open to
the exterior by **spiracles** and branch into
finer channels (**tracheoles**) that terminate
in the tissues (*see also* AIR SAC). In some
larger and more active insects, pumping
movements of the abdominal muscles
cause air to be drawn into and out of the
tracheae. Also, by opening and closing different spiracles during certain phases of
ventilation, the insect causes the directional flow of air through the tracheae, as
in the locust.

tracheid A type of cell occurring within
the *xylem of conifers, ferns, and related
plants. Tracheids are elongated and their
walls are usually extensively thickened by
deposits of lignin. Water flows from one

tracheid to another through unthickened regions (pits) in the cell walls. *Compare* VESSEL ELEMENT.

tracheole *See* TRACHEA.

tracheophyte Any plant that has elaborate tissues, including *vascular tissue; a conspicuous *sporophyte generation; and complex leaves with waterproof cuticles. Tracheophytes include plants of the phyla Psilophyta, *Lycophyta, *Sphenophyta, Filicinophyta, *Coniferophyta, and *Anthophyta. In traditional classification systems these were regarded as classes of the division Tracheophyta.

tracing (radioactive tracing) *See* LABELLING.

transaminase An enzyme that catalyses the transfer of an amino group from one molecule to another. Transaminases play an important role in the synthesis of amino acids (*see* TRANSAMINATION).

transamination A biochemical reaction in amino acid metabolism in which an amine group is transferred from an amino acid to a keto acid to form a new amino acid and keto acid. The coenzyme required for this reaction is pyridoxal phosphate.

transcellular pathway The route through cells. Substances moving across epithelia are mostly forced to pass through the epithelial cells, rather than passing between the cells (the *paracellular pathway). This means that both the mucosal surface (facing the exterior) and the serosal surface (facing the interior) of the cells can act as 'gatekeepers', keeping out certain substances, while transporting others from one side to another, often against a concentration gradient.

transcriptase *See* REVERSE TRANSCRIPTASE.

transcription The process in living cells in which the genetic information of *DNA is transferred to a molecule of messenger *RNA (mRNA) as the first step in *protein synthesis (*see also* GENETIC CODE). Transcription takes place in the cell nucleus or (in prokaryotes) in the nuclear region. It involves the action of RNA *polymerase enzyme in assembling the nucleotides necessary to form a complementary strand of mRNA from the DNA template (*see also* PROMOTER), and (in eukaryote cells) the subsequent removal of the noncoding sequences from this primary transcript (pre-mRNA; *see* RNA PROCESSING) to form a functional mRNA molecule. The process begins with **initiation**, the binding of RNA polymerase to a sequence of bases 'upstream' of the coding sequence called the *promoter, which orients the polymerase to the correct DNA strand—the template strand—and in the correct 'downstream', or 5' to 3', direction. Binding of the enzyme is assisted by *transcription factors and additionally, in prokaryotes, by *sigma factors. The two DNA strands unwind in the vicinity of the enzyme, and transcription begins at an initiation site in the promoter region. In the **elongation** phase, the chain of RNA nucleotides complementary to the base sequence of the template DNA is assembled by the polymerase from a pool of free ribonucleoside triphosphates (NTPs). The polymerase moves along the DNA template, adding nucleotides to the 3' end of the growing RNA chain. Formation of the covalent bonds linking the nucleotides is driven by hydrolysis of two of the three phosphate groups attached to each incoming NTP. As elongation proceeds, the DNA double helix rewinds behind and unwinds in front of the polymerase, so that only about 10 base pairs are unwound at any time. Elongation continues until RNA polymerase reaches a **termination** sequence, where transcription stops and the pre-mRNA transcript is released (*see* TERMINATOR). In eukaryotes, the extent of DNA packaging in *chromatin must be modified before a region of DNA can be transcribed. At sites of active transcription the normally supercoiled chromatin structure is decondensed and loops out, consisting of naked DNA or individual linked nucleosomes. The term 'transcription' is also applied to the assembly of single-stranded DNA from an RNA template by the enzyme *reverse transcriptase. *Compare* TRANSLATION.

transcription factor Any of a group of proteins that work synergistically to regulate gene activity by increasing or decreasing the binding of RNA polymerases to the DNA molecule during the process of *transcription. This is achieved by the ability of the transcription factors to bind to the DNA molecule (*see* DNA-BINDING PROTEINS). Transcription factors contain *finger domains, which are often in repeated sequences called **multifinger loops**. In eukaryotes **general transcription factors** are active in the transcription of all genes. First, several bind to the promoter site near to the transcription start site and ensure the correct positioning of RNA polymerase II with respect to the coding region of the gene. Binding of additional transcription factors and other proteins forms the **initiation complex**, and enables RNA polymerase II to unwind the DNA double helix and begin RNA synthesis. However, efficient transcription of many genes requires **regulatory transcription factors**, which determine whether the gene is switched 'on' or 'off' by binding to regulatory sites called control elements. **Proximal control elements** are located near to the promoter, whereas **distal control elements**, grouped together to form *enhancers, may be located many thousands of base pairs from the coding region. The binding of activators or repressors to these control elements is crucial to regulating transcription. Another mechanism involves the recruitment of chromatin-modifying proteins by activators or repressors. These add or remove acetyl groups on the histone proteins of chromatin near the promoter, thus promoting or silencing transcription. These mechanisms in eukaryotes are crucial in ensuring that genes are expressed in a tissue-dependent manner and, during development, that genes are expressed at the appropriate time and place within the embryo. Hundreds of transcription factors may be involved in controlling just one gene. Many cellular signal transduction pathways, triggered by the binding of a substance to its receptor on the cell surface, ultimately lead to the phosphorylation and activation of transcription factors in the nucleus, and hence alterations in gene expression. *See also* OPERON; SIGMA FACTOR.

transcriptome The full complement of RNA transcripts of the genes of a cell or organism. The types and relative abundance of different transcripts, i.e. the messenger RNAs (mRNAs), can be obtained by analysing cell contents using oligonucleotide *microarrays. Alternatively, the RNA is isolated from cells, fragmented, reverse transcribed into complementary DNA, and subject to *shotgun sequencing (*see* RNA SEQUENCING). Either method potentially can provide a 'snapshot' of the expression pattern of the cell's genes. *See* TRANSCRIPTOMICS.

transcriptomics The study of the RNA transcripts of a cell, tissue, or organism (i.e. the *transcriptome). Transcriptomics is concerned with determining how the transcriptome, and hence pattern of gene expression, changes with respect to various factors, such as type of tissue, stage of development, hormones, drugs, or disease. It complements and overlaps with *proteomics.

transducin A *G protein that acts as a second messenger in the response to light by vertebrate photoreceptors. When a photon of light strikes a molecule of rhodopsin (the visual pigment in the rod cells of the retina of the eye), it activates transducin. This sets in train a cascade of events within the rod cell that amplifies the signal and evokes a response. The transducin activates numerous molecules of phosphodiesterase, which hydrolyses the signalling molecule *cyclic GMP (cGMP). The resultant lowered intracellular concentration of cGMP causes closure of sodium channels in the rod-cell membrane that carry the 'dark current'—inwardly flowing sodium ions; however, outflow of potassium ions persists, the rod cell hyperpolarizes (its internal electric potential becomes more negative), and its release of neurotransmitter to the relay neuron is reduced. When the light stimulus ceases, cGMP is regenerated by guanylate cyclase, the sodium channels reopen, transducin is deactivated, and neurotransmitter release returns to resting levels. Hence, the transducin pathway can produce a response to the smallest possible light stimulus—a single photon—a measure of its exquisite sensitivity.

transduction 1. The transfer of genetic material from one bacterial cell to another by means of a *bacteriophage. **2.** The conversion of stimuli detected in the *receptor cells into electric impulses, which are transported by the nervous system. *See also* SIGNAL TRANSDUCTION.

transect A straight line across an expanse of ground along which ecological measurements are taken, continuously or at regular intervals. Thus an ecologist wishing to study the numbers and types of organisms at different distances above the low-tide line might sample at five-metre intervals along a number of transects perpendicular to the shore.

***trans* fatty acid** An unsaturated fatty acid in which one or more of the double bonds are in the *trans* configuration. In most naturally occurring unsaturated fatty acids the double bonds are in the *cis* configuration, in which the hydrocarbon chain is effectively 'kinked' at each double bond. Hydrogenation of natural oils to 'harden' them in food manufacturing to extend the shelf life of products such as cakes, biscuits, and pastries. A consequence of this is the conversion of *cis* double bonds to the *trans* configuration, in which the hydrocarbon chain is not kinked. Studies have shown that *trans* fatty acids are harmful to health, by lowering levels of 'good' cholesterol (in high-density lipoproteins, or HDLs), and health authorities recommend minimizing their intake in the diet.

transfection 1. (in genetic engineering) Any process by which eukaryotic cells take up foreign DNA in a form that is replicated and passed to daughter cells, thereby changing the genetic constitution of the cell line or organism concerned. It is synonymous with *transformation in bacteria. Various techniques have been devised, depending on the nature of the recipient cell. The most popular are the use of *liposomes, electroporation, microinjection, and *biolistics. *See* GENETICALLY MODIFIED ORGANISMS (Feature). **2.** (in microbiology) The uptake by a bacterium of DNA from a bacterial virus (bacteriophage). *Compare* TRANSFORMATION.

transferase Any of a class of enzymes that catalyse the transfer of a group of atoms from one molecule to another.

transfer cell A type of plant cell specialized for the short-distance high-volume transport of materials. Transfer cells have numerous knobs and ridges on the inner surface of the primary cell wall. These greatly increase the surface area of the plasma membrane, which follows the contours of the protuberances. The expanded plasma membrane can accommodate a large number of *transport proteins, which are responsible for transporting materials in or out of the cell. Transfer cells are found mainly in salt-secreting glands and in regions where photosynthate (sugar) is loaded or unloaded into sieve elements of phloem.

transfer-messenger RNA (tmRNA) A type of *RNA known only from bacteria and some bacteriophages that combines features of both transfer RNA (tRNA) and messenger RNA (mRNA). It contains a tRNA domain and a short coding sequence, and serves to rescue ribosomes that have stalled during protein synthesis due to a faulty mRNA molecule. The tmRNA dissociates the unfinished polypeptide from the ribosome, enabling the latter to be recycled. It also tags both the polypeptide and the defective mRNA for degradation.

transferrin A plasma protein (a β-*globulin) that can bind to iron and transport it to the liver for storage (*see* FERRITIN) or to the cells of bone marrow, where it is used for the formation of haemoglobin. One transferrin molecule can bind to two iron ions.

transfer RNA *See* RNA.

transformation 1. A permanent heritable change in a cell, particularly a bacterial cell, that occurs as a result of its acquiring foreign DNA. For example, nonvirulent bacterial cells can be transformed into virulent forms if cultured in a medium containing killed virulent bacteria. In cloning, the host cells are transformed by the insertion of recombinant DNA. *See* GENE CLONING. **2.** The conversion of a normal cell into a malignant cell (*see* CANCER), which is brought about by mutation of genes and consequent cellular changes, sometimes through the action of *carcinogens or *oncogenic viruses.

transformation hypothesis A hypothetical explanation of how the sporophyte came to be the more prominent form in the life cycles of most vascular plants. It postulates that early vascular plants had gametophyte and sporophyte generations that were very similar in appearance, with prominent upright branching forms. Over time, the gametophytes became smaller and simpler, whereas the sporophyte got progressively more elaborate in form. Eventually, the gametophytes became retained within the sporophyte—the situation in modern seed plants. Hence, the modern sporophyte evolved through the transformation of an existing form. *Compare* INTERPOLATION HYPOTHESIS.

transforming growth factor beta (TGF-β) Any of several related proteins, including the multifunctional transforming growth factor beta-1 (TGF-β1). The latter controls many cell functions, including proliferation, differentiation, bone remodelling, collagen formation, and immune cell development. TGF-β1 is secreted by diverse cells, and numerous cells have specific receptors for it. It positively and negatively regulates many other growth factors. Somatic mutations in the TGF-β1 gene are associated with certain cancers. TGFs give their name to a large superfamily of proteins, including *bone morphogenetic proteins, that act as *cytokines for a wide range of cell types in animals, both during development and in adults. They regulate cell growth, division, adhesion, and differentiation, cell migration, and programmed cell death (apoptosis), and stimulate or repress other growth factors. *Inhibin is a TGF family member that regulates sperm formation and maturation of ovarian follicles.

Transforming growth factor alpha (TGF-α) is a member of the *epidermal growth factor family of cytokines.

transgene A gene that is taken from one organism and inserted into the germ line of another organism, either of the same or a different species, so that it is replicated as part of the genome and present in all the recipient's cells. The resulting organism is described as *transgenic. *See* GENETICALLY MODIFIED ORGANISMS.

transgenic Describing an organism whose genome incorporates and expresses genes from an unrelated organism, particularly one of another species. Transgenic individuals are created by genetic engineering, using suitable *vectors to insert the desired foreign gene into the fertilized egg or early embryo of the host. For example, the gene for rat growth hormone can be inserted into fertilized mouse eggs to produce mice with cells that produce rat growth hormone. Transgenic organisms offer considerable commercial potential; *see* GENETICALLY MODIFIED ORGANISMS (Feature). *Compare* GENOME EDITING.

transient polymorphism *See* POLYMORPHISM.

transition (in genetics) *See* SUBSTITUTION.

transition zone *See* HYPOCOTYL.

translation The process in living cells in which the genetic information encoded in messenger *RNA (mRNA) in the form of a sequence of nucleotide triplets (*codons) is translated into a sequence of amino acids in a polypeptide chain during *protein synthesis (see illustration). Translation takes place on *ribosomes in the cell cytoplasm. Initiation of the process involves the formation of an initiation complex and binding of the small ribosomal subunit to the mRNA (*see* INITIATION FACTOR). In prokaryotes binding takes place at the *Shine–Dalgarno sequence, whereas in eukaryotes the small subunit binds to the 5′ cap at the end of the mRNA molecule then moves to the start codon. The ribosome moves along the

(a)

(b)

(c)

The stages of translation in protein synthesis

mRNA 'reading' each codon in turn. Molecules of transfer RNA (tRNA), each bearing a particular amino acid, are brought to their correct positions along the mRNA molecule: base pairing occurs between the bases of the codons and the complementary base triplets of tRNA (*see* ANTICODON; WOBBLE). In this way amino acids are assembled in the correct sequence to form the polypeptide chain (*see* ELONGATION). Translation is terminated at a stop codon, which binds a *release factor. Several ribosomes can perform translation simultaneously on the same mRNA molecule (*see* POLYRIBOSOME), thus increasing the rate of protein synthesis. Following their release from the ribosome, most assembled polypeptides undergo some form of chemical modification to produce a fully functional protein. These **post-translational modifications** can take place in the *endoplasmic

reticulum and might include cleavage of certain amino acids; cutting or joining of polypeptides; addition of sugars to form glycoproteins; or addition of phosphate groups.

translocation 1. (in botany) The movement of minerals and chemical compounds within a plant. There are two main processes. The first is the uptake of soluble minerals from the soil and their passage upwards from the roots to various organs by means of the water-conducting vessels (*xylem). The second is the transfer of organic compounds, synthesized by the leaves, both upwards and downwards to various organs, particularly the growing points. This movement occurs within the *phloem tubes. *See also* MASS FLOW. **2.** (in genetics) A type of *chromosome mutation in which a section of one chromosome is broken off and becomes attached to another chromosome, resulting in a loss of genetic information from the first chromosome.

transmembrane domain A functional region of a protein that spans the phospholipid bilayer of a biological membrane, such as the plasma membrane of a cell. Integral membrane proteins typically comprise two or more such domains, alternating with intracellular and extracellular domains arranged on either side of the membrane. Amino acids of the transmembrane domains interact with the fatty acyl groups of the membrane phospholipids, thereby anchoring the protein in the membrane.

transmission 1. (in neurophysiology) The one-way transfer of a nerve *impulse from one neuron to another across a *synapse. *See also* NEUROTRANSMITTER. *Compare* PROPAGATION. **2.** (in medicine) The spread of an *infection from person to person. This can occur in various ways, such as close contact with an infected person, including sexual contact (*see* SEXUALLY TRANSMITTED DISEASE); contact with a *vector or a *carrier of the disease; consuming food or drink contaminated with the infecting microorganism; and breathing in contaminated droplets of moisture, produced by coughing and sneezing.

transmission electron microscope *See* ELECTRON MICROSCOPE.

transmitter *See* NEUROTRANSMITTER.

transpiration The loss of water vapour by plants to the atmosphere. It occurs mainly from the leaves through pores (stomata) whose primary function is gas exchange. The water is replaced by a continuous column of water (and dissolved nutrients) moving upwards from the roots within the *xylem vessels. The flow of this column of water is known as the **transpiration stream**, which is maintained by *root pressure and a combination of cohesive and adhesive forces in the xylem vessels according to the **cohesion-tension theory** (*see* COHESION). *See also* POTOMETER.

transplantation *See* GRAFT.

transport protein A protein that penetrates or spans a cell membrane to permit the passage of a substance through the membrane. Some transport proteins form pores, or *channels, through which particular ions or molecules can pass. These channel proteins are often gated, enabling them to open and close in response to signals; categories include *ligand-gated ion channels and *voltage-gated ion channels. Other types of transport protein bind the substance on one face of the membrane, then change shape so that the substance is carried by the protein through the membrane to be released at the other face. These include the *uniporters, which transport just one substance, and the *cotransporters, which transport two or more different substances. Transport proteins often require energy to drive the transport process; this is provided by hydrolysis of ATP or by an existing concentration gradient. *See* ACTIVE TRANSPORT.

transposon (transposable genetic element) A mobile genetic element, known informally as a 'jumping gene', that can become integrated at many different sites in the genome, either by moving from place to place or by producing copies of itself that insert elsewhere in the genome. The simplest types are known as **insertion sequences**, typically found in bacteria and consisting of some 700–2500 base pairs and with numerous short repeated nucleotide sequences at either end. Larger and more complex are the **composite transposons**, which consist of a central portion, possibly containing functional genes, flanked by insertion sequences at either end. Transposons were first discovered by the US geneticist Barbara McClintock (1902–92) in maize in the 1940s and have since been found in other eukaryotes and in prokaryotes. They can disrupt gene expression or cause deletions and inversions, and hence affect both the genotype and phenotype of the organisms concerned. However, most eukaryotic transposons are *retrotransposons, which form RNA copies of themselves in a manner similar to the replication of retroviruses. Transposons account for a sizable proportion of the *repetitive DNA in eukaryotes—making up around 45% of the human genome and 85% of the maize genome. *See also* ALU FAMILY.

transverse tubules (T tubules) A series of infoldings of the plasma membrane (sarcolemma) of skeletal muscle fibres that envelop each myofibril at the Z line (*see* SARCOMERE). The T tubules transfer action potentials from the sarcolemma to the *sarcoplasmic reticulum, which releases calcium ions (Ca^{2+}) into the cytosol causing contraction of the muscle fibres. This release is accomplished by the action of two types of receptors: dihydropyridine (DHP) receptors in the wall of the T tubule; and ryanodine receptors in the sarcoplasmic reticulum membrane, which act as Ca^{2+} channels. Each DHP receptor is physically linked to a ryanodine receptor. An action potential arriving in the T tubule triggers a change in shape of the DHP receptors. This causes the associated ryanodine receptor channels to open, allowing Ca^{2+} to flow out of the sarcoplasmic reticulum into the sarcoplasm. This affects the binding properties of the proteins *troponin and tropomyosin, leading to the interaction of actin and myosin filaments that produces muscle contraction. In the absence of further action potentials, calcium pumps in the sarcoplasmic reticulum membrane remove the Ca^{2+} from the sarcoplasm and restore the resting state of the muscle fibre.

transversion (in genetics) *See* SUBSTITUTION.

T$_{reg}$ Abbreviation for regulatory T cell. *See* T CELL.

Trematoda In traditional classifications, a class of parasitic flatworms (*see* PLATYHELMINTHES) comprising the flukes, such as *Fasciola* (liver fluke) and *Schistosoma mansoni* (blood fluke). Flukes have suckers and hooks to anchor themselves to the host and their body surface is covered by a protective cuticle. The whole life cycle may either occur within one host or require one or more intermediate hosts to transmit the infective eggs or larvae (*see* CERCARIA; MIRACIDIUM). *Fasciola hepatica*, for example, undergoes larval development in a land snail (the intermediate host) and infects sheep (the primary host) when contaminated grass containing the larvae is swallowed.

trial-and-error learning *See* LEARNING.

Triassic The earliest period of the Mesozoic era. It began about 252 million years ago, following the Permian, the last period of the Palaeozoic era, and extended until about 201 million years ago when it was succeeded by the Jurassic. It was named, by F. von Alberti in 1834, after the sequence of three divisions of strata that he studied in central Germany—Bunter, Muschelkalk, and Keuper. The Triassic rocks are frequently difficult to distinguish from the underlying Permian strata and the term **New Red Sandstone** is often applied to rocks of the Permo-Triassic. During the period marine animals diversified: molluscs were the dominant invertebrates—ammonites were abundant and bivalves replaced the declining brachiopods. Reptiles were the dominant vertebrates and included turtles, phytosaurs, dinosaurs, and the marine ichthyosaurs.

triazines A group of organic compounds based on a heterocyclic ring containing three carbon atoms and three nitrogen atoms and including several commonly used agricultural herbicides, such as simazine and atrazine. The triazine herbicides work by inhibiting electron transport of photosynthesis. They are absorbed primarily through the plant root and bind to a particular site in photosystem II, thereby preventing binding of *plastoquinone, a key electron acceptor in the electron transport chain. Certain crop plants are resistant to triazines—for example, maize roots contain an enzyme that renders them inactive—hence their usefulness as selective weedkillers. Attempts have been made to introduce the resistance gene into other crop species, to broaden the applicability of triazine herbicides. However, these efforts have been hampered because the target gene is part of the chloroplast genome, not the nuclear genome. Also, many weeds have now evolved triazine resistance.

tribe A category used in the *classification of plants and animals that consists of several similar or closely related genera within a family. For example the Bambuseae, Oryzeae, Paniceae, and Aveneae are tribes of grasses.

tricarboxylic acid cycle *See* KREBS CYCLE.

triceps (*pl.* **triceps**) A muscle that runs parallel to the humerus and works antagonistically with the *biceps, causing the arm to extend. *See* ANTAGONISM; SKELETAL MUSCLE.

trichogyne *See* CARPOGONIUM.

trichome A hairlike projection from a plant epidermal cell. Examples include root hairs and the stinging hairs of nettle leaves.

Trichoplax *See* PLACOZOA.

trichromatic theory *See* COLOUR VISION.

tricuspid valve (right atrioventricular valve) A valve, consisting of three flaps, situated between the right atrium and the right ventricle of the mammalian heart. When the right ventricle contracts, forcing blood into the pulmonary artery, the tricuspid valve closes the aperture to the atrium, thereby preventing any backflow of blood. The valve reopens to allow blood to flow from the atrium into the ventricle. *Compare* BICUSPID VALVE.

triglyceride (triacylglycerol) An ester of glycerol (propane-1,2,3-triol) in which all three hydroxyl groups are esterified with a fatty acid. Triglycerides are the major constituent of fats and oils and provide a concentrated food energy store in living organisms as well as cooking fats and oils, margarines, etc. Their physical and chemical properties depend on the nature of their constituent fatty acids. In **simple triglycerides** all three fatty acids are identical; in **mixed triglycerides** two or three

different fatty acids are present. During digestion, triglycerides are hydrolysed by lipase enzymes to a mixture of fatty acids and mono- and diglycerides. These lipid-soluble products pass through the epithelial cell membranes in the small intestine, where the triglycerides are resynthesized and packaged into *chylomicrons for transfer into the lacteals of the intestinal villi.

triiodothyronine (T$_3$) A hormone secreted by the *thyroid gland. *See also* THYROGLOBULIN.

trilobite An extinct marine arthropod belonging to the class Trilobita (some 4000 species), fossils of which are found in deposits dating from the Precambrian to the Permian period (590–250 million years ago). They disappeared during the later Permian mass extinction. Trilobites were typically small (1–7 cm long); the oval flattened body comprised a head (covered by a semicircular dorsal shield) and a thorax and abdomen, which were protected by overlapping dorsal plates with a raised central part and flattened lateral portions, presenting a three-lobed appearance. The head bore a pair of antenna-like appendages and a pair of compound eyes; nearly all body segments bore a pair of Y-shaped (biramous) appendages—one branch for locomotion and the other fringed for respiratory exchange. Trilobites were bottom-dwelling scavengers.

trimerophytes A group of extinct land plants that flourished during the latter part of the Devonian period (380–360 million years ago), members of which evolved into the ancestors of seed plants and ferns. Among the earliest *euphyllophytes, they were robust herbs and small shrubs, some up to 3 m tall, and showed significant changes in branching pattern compared with their predecessors, the *rhyniophytes. Instead of equal (*dichotomous) branching, one stem tended to be dominant and give rise to smaller lateral branches, some bearing spore-forming organs and others acting as leaves. Moreover, the branching pattern became more regular. This trend culminated in such forms as *Pertica* *quadrifolia*, a plant with an obvious main trunk and numerous small lateral branches.

trimethylamine A nitrogenous compound, $(CH_3)_3N$, that acts as an *osmolyte in the cells of the kidney. This compound can also be converted to **trimethylamine oxide**, which is excreted by many marine fish and gives them their characteristic odour.

triose A sugar molecule that contains three carbon atoms. *See* MONOSACCHARIDE.

triplet code *See* CODON; GENETIC CODE.

triploblastic Describing an animal having a body composed of three embryonic cell layers: the *ectoderm, *mesoderm, and *endoderm. Most multicellular animals are triploblastic; the ctenophores, cnidarians, and placozoans, which are *diploblastic, are exceptions.

triploid Describing a nucleus, cell, or organism that has three times ($3n$) the haploid number (n) of chromosomes (*see also* POLYPLOID). Triploid organisms are normally sterile because one-third of their chromosomes will lack partners, and so are unable to pair during *meiosis. This can be useful to plant breeders, for example in banana cultivation: sterile triploid bananas can be propagated asexually and will not contain any seeds.

trisomy The condition of a nucleus, cell, or organism in which one of the pairs of homologous chromosomes has gained an additional chromosome, resulting in a chromosome number of $2n + 1$ (*see* ANEUPLOID). Trisomy is the cause of a number of human genetic abnormalities, including *Down's syndrome; Patau's syndrome, in which there is an extra chromosome 13 (**trisomy 13**); and Edwards' syndrome, in which there is an extra chromosome 18 (**trisomy 18**).

tritanopia *See* COLOUR BLINDNESS.

tritiated compound *See* LABELLING.

TRK (Trk) *See* NEUROTROPHIN RECEPTOR.

tRNA *See* RNA.

trochanter 1. Any of several bony knobs on the femur of vertebrates to which muscles are attached. **2.** The second segment of an insect's leg, between the *coxa and the *femur.

trochophore The pelagic planktonic larva of polychaete worms, some molluscs, and certain other invertebrates. It is top-shaped and usually has two bands of cilia encircling the body.

trophic cascade The pattern of effects of a single consumer species on other trophic levels in a food chain. Many different patterns can be described, involving predators, herbivores, and primary producers such as plants and algae. However, a common pattern stems from the effects of a predator controlling the abundance of its herbivore prey, which in turn impacts lower down the food chain. For example, off the Aleutian Islands and southeastern Alaska, the sea otter (*Enhydra lutris*) feeds on sea urchins, which in turn graze the holdfasts of large algae (kelp). At sites where otters have historically been absent due to hunting by humans, the overabundance of sea urchins leads to sparse kelp beds, and relative paucity of other organisms that live in these beds. When otters recolonize such sites, urchin populations are controlled and the kelp beds recover, benefiting the entire ecosystem.

trophic efficiency A measure of the efficiency of energy flow between trophic levels in a *food chain. It is the percentage of energy from a trophic level that is used by the organisms of the next trophic level for growth and reproduction. Hence, it does not include energy lost via respiration and faeces, nor material from the previous trophic level that is not consumed. Trophic efficiencies can range from 5% to 20%, depending on the type of ecosystem, with an average of about 10%. This means that considering just three trophic levels, only 1% of net primary production is available to a secondary consumer.

trophic level The position that an organism occupies in a *food chain. For example, green plants (which obtain their energy directly from sunlight) are the primary *producers; herbivores are primary *consumers (and secondary producers). A carnivore that eats only herbivores is a secondary consumer and a tertiary producer. Many animals feed at several different trophic levels.

trophoblast The layer of epithelial cells that forms the outer wall of a blastocyst (*see* BLASTULA). It embeds the blastocyst into the wall of the uterus (*see* IMPLANTATION) and contributes to the *chorion.

tropic 1. Describing a hormone that controls the activity of another endocrine gland. **2.** Of or relating to a *tropism.

tropism The directional growth of a plant organ in response to an external stimulus, such as light, touch, or gravity. Growth towards the stimulus is a **positive tropism**; growth away from the stimulus is a **negative tropism**. *See also* GEOTROPISM; HYDROTROPISM; ORTHOTROPISM; PHOTOTROPISM; PLAGIOTROPISM; THIGMOTRO-PISM. *Compare* NASTIC MOVEMENTS; TAXIS.

tropomyosin A protein found in the *actin filaments in muscles. The molecule consists of two elongated strands that run along the length of the filament. When the muscle is at rest, the tropomyosin molecule covers the binding site of the actin molecule, where interaction with myosin occurs. On contraction of the muscle, the tropomyosin is displaced by another protein, *troponin, allowing the interaction of actin with myosin. *See* SLIDING FILAMENT THEORY.

troponin A complex of three polypeptide chains that are found at regular intervals along the length of an *actin filament. During muscle contraction, troponin binds to calcium ions, displacing *tropomyosin and exposing the binding site on the actin filament. This allows the interaction of actin and myosin to occur. *See* SLIDING FILAMENT THEORY.

troposphere The lowest level of the earth's atmosphere, extending from the earth's surface to a height of about 10 km (its thickness varies from 7 km at the poles to 28 km at the equator). Within the troposphere temperature falls with increasing height, although *temperature inversions can occur.

TRP protein (transient receptor potential protein) Any of a large family of proteins that form various ion channels in the outer membrane of cells. They occur in both vertebrate and invertebrates and are involved in sensory perception for a wide range of stimuli, including temperature, pain, taste, and mechanical force. Binding of, say, a sour substance to TRP channels in a *taste bud receptor cell will trigger transient opening of the channel, allowing the entry of calcium ions; this prompts release of neurotransmitter, which causes depolarization of adjacent sensory neurons, thereby sending a signal to taste centres in the brain.

trumpet cell An elongated cell, large numbers of which are joined end to end to transport photosynthate (sugar) through the bodies of brown algae. The end walls are perforated with holes forming a sieve, analogous to the arrangement of sieve elements in *phloem. However, trumpet cells and sieve elements are not homologous structures, but an example of convergent evolution.

trypsin An enzyme that digests proteins (*see* ENDOPEPTIDASE; PROTEASE). It is secreted in an inactive form (**trypsinogen**) by the pancreas into the duodenum. There, trypsinogen is acted on by an enzyme (**enterokinase**) produced in the brush border of the duodenum to yield trypsin. The active enzyme plays an important role in the digestion of proteins in the anterior portion of the small intestine. It also activates other proteases in the pancreatic juice (*see* CARBOXYPEPTIDASE; CHYMOTRYPSIN).

trypsinogen *See* TRYPSIN.

tryptophan *See* AMINO ACID.

TSH *See* THYROID-STIMULATING HORMONE.

T tubules *See* TRANSVERSE TUBULES.

tube feet *See* ECHINODERMATA.

tube nucleus (vegetative nucleus) One of the two nuclei produced when the haploid nucleus of a *pollen grain divides by mitosis (*compare* GENERATIVE NUCLEUS). The tube nucleus, which lies inside the tube cell, controls growth of the *pollen tube.

tuber A swollen underground stem or root in certain plants. It enables the plant to survive the winter or dry season and is also a means of propagation. A **stem tuber**, such as the potato, forms at the end of an underground stem. Each tuber represents several nodes and internodes. The following season several new plants develop from the terminal and axillary buds (eyes). **Root tubers**, such as those of the dahlia, are modified food-storing adventitious roots and may also give rise to new plants.

tubicolous Describing invertebrate animals that live in tubes that they have themselves constructed. For example, the polychaete fanworms (*Sabella*) build tubes of sand particles.

tubulin A protein of which the *microtubules of cells are formed.

Tullgren funnel A device used to remove and collect small animals, such as insects, from a sample of soil or leaf litter. The sample is placed on a coarse sieve fixed across the wide end of a funnel and a 100-watt light bulb, in a metal reflector, is placed about 25 cm above the funnel. The heat from the bulb dries and warms the sample, causing the animals to move downwards and fall through the sieve into the funnel, which directs them into a collecting dish or tube below. The dish can contain water or alcohol to prevent the animals from escaping.

tumour *See* NEOPLASM.

tumour necrosis factor (TNF) Either of two proteins that act as *cytokines, eliciting numerous responses in virtually all types of cells by activating cellular signalling pathways and promoting the expression of genes. They are involved in many aspects of the body's immune defences, notably in their ability to destroy tumour cells; they also assist in the killing of virally infected cells by natural killer cells. TNF-α is a glycoprotein produced chiefly by monocytes and macrophages, whereas TNF-β (or lymphotoxin) is produced by T cells. Both bind to either of the cell-surface receptors—tumour necrosis factor receptors I and II (TNFR-I and TNFR-II). These proteins give their name to a family of related proteins that typically regulate aspects of immune responses and cell death.

tumour-suppressor gene (anti-oncogene) Any of a class of genes that suppress the development of cancers in living cells. The products of tumour-suppressor genes are typically involved in monitoring replication of DNA and progress of the cell cycle and in promoting repair of damaged DNA. Hence, mutation and consequent loss of activity of such genes leads to the formation of tumours. For example, development of the childhood cancer of retinal cells, known as retinoblastoma, is caused by loss of the chromosome segment containing the tumour-suppressor gene *RB*. This codes for a protein, retinoblastoma protein, that restrains cell proliferation. The RB tumour-suppressor protein can also be inactivated by an oncoprotein encoded by an adenovirus; this is the mechanism by which such DNA tumour viruses cause cancer. Another common tumour-suppressor gene is the p53 gene, mutations of which are linked to about half of all human cancers. Its product, a 53-kDa protein, acts as a key transcription factor for genes that encode cell cycle-inhibiting proteins. If genetic damage has occurred, the p53 protein halts the cell cycle to allow repair of the faulty DNA. If the damage is severe, it can trigger cell death.

tundra A terrestrial *biome characterized by a lack of trees and a permanently frozen subsoil. Tundra lies to the north of the *taiga in North America and Eurasia; the vegetation is dominated by grasses, sedges, lichens, mosses, heathers, and low shrubs. The growing season, which occurs during the warmest part of the year when the average daily mean temperature is about 10°C, lasts only 2–4 months, during which the topsoil thaws to a depth of 30 cm, allowing roots to penetrate it. However, below this level the soil is permanently frozen (**permafrost**); water cannot filter through the soil and may lie in surface depressions during the growing season. Global warming is now affecting the ecology and economy of tundra regions dramatically. By the mid-21st century, the area of permafrost in the northern hemisphere is predicted to decline by around 20–35%, and by 2080 the depth of thawing is expected to increase by 30–50%. This will release large amounts of carbon that hitherto have been 'locked' in the frozen soil, thereby exacerbating the warming process. Already, buildings and other structures erected on permafrozen ground are subsiding and craters are forming, causing disruption of infrastructure and dislocation of communities. Many Arctic lakes are draining away and disappearing, with profound impacts on river flows, aquatic biodiversity, and fisheries. *Compare* TAIGA.

tunicates *See* UROCHORDATA.

Turbellaria In traditional classifications, a class of free-living flatworms (*see* PLATYHELMINTHES) comprising the planarians, which occur in wet soils, fresh water, and marine environments. Their undersurface is covered with cilia, used for gliding over stones and weeds. Planarians can also swim by means of undulations of the body.

turgor The condition in a plant cell or other cell with a semirigid cell wall (i.e. cells of archaea, bacteria, fungi, and some protists) when the force causing water to enter the cell by *osmosis is balanced by the pressure of the fluid cell constituents (i.e. the protoplast) pushing against the cell wall—i.e. the **turgor pressure**. A state of turgor thus prevents further water from entering the cell. The turgor pressure is thus comparable in magnitude to the *pressure potential, which is the pressure exerted by the cell on the watery cell contents; the pressure potential is significantly positive when the cell is turgid. Turgidity assists in maintaining the rigidity of plants; a decrease in turgidity leads to *wilting. *See also* WATER POTENTIAL. *Compare* PLASMOLYSIS.

turion 1. A winter bud, covered with scale leaves and mucilage, that is produced by certain aquatic plants, such as frogbit. Turions become detached and remain dormant on the pond or lake bottom during the winter before developing into new plants the following season. **2.** *See* SUCKER.

Turner's syndrome A genetic disorder of women caused by the absence of the second *sex chromosome (such women are XO, rather than the normal XX). It is characterized by a lack of ovaries and menstrual cycle. Affected women are sterile and lack secondary sexual characteristics, although the external genitalia are present. The syndrome is named after the US endocrinologist H. H. Turner (1892–1970), who first described it.

twins Two individuals born to the same mother at the same time. Twins can develop from the same egg (*see* IDENTICAL TWINS) or from two separately fertilized eggs (*see* FRATERNAL TWINS).

two-hybrid screening *See* YEAST TWO-HYBRID SCREEN.

tylose A balloon-like extension of a parenchyma cell that protrudes into the lumen of a neighbouring xylem vessel or tracheid through a *pit in the cell wall. Tyloses form most commonly in older woody tissue, possibly in response to injury; they may eventually block the vessels and thus help prevent the spread of fungi and other pathogens within the plant. Tyloses may become filled with tannins, gums, pigments, etc., giving heartwood its dark colour, and their walls can remain thin or become lignified.

tympanic cavity *See* MIDDLE EAR.

tympanum (tympanic membrane; eardrum) (*pl.* **tympanums** or **tympana**) **1.** In mammals, the membrane that separates the *outer ear from the *middle ear. It vibrates in response to sound waves and transmits these vibrations via the *ear ossicles of the middle ear to the site of hearing (the *cochlea of the *inner ear). In amphibians and some reptiles there is no external ear and the tympanum is exposed at the skin surface. **2.** In insects, an analogous structure that forms part of the **tympanal organ**, used for detecting sound. Paired tympanal organs can occur at one of various locations on the insect body, including the thorax, abdomen, or legs, depending on species. The tympanum lies on the body surface stretched over an enlarged air channel (*trachea) and is associated with sensory *hair cells, which detect vibrations of the tympanum and send nerve impulses to the central nervous system.

type I fibre *See* SLOW-TWITCH FIBRE.

type II fibre *See* FAST-TWITCH FIBRE.

type specimen The specimen used for naming and describing a *species or subspecies. If this is the original specimen collected by the author who named the species it is termed a **holotype**. The type specimen is not necessarily the most characteristic representative of the species. The term **type** is also applied to any taxon selected as being representative of the rank to which it belongs. For example, the genus *Solanum* (potato) is said to be the type genus of the family Solanaceae.

typological species concept The concept of a species as a group whose members share certain characteristics that distinguish them from other species. This Aristotelian concept was applied to the natural world by the early taxonomists, but by the late 19th century was being supplanted by other concepts, notably the *biological species concept. These could better account for the many cases in which species appear to be virtually indistinguishable (*see* SIBLING SPECIES) or where intermediate phenotypes

occur due to hybridization. However, taxonomists must use a typological approach when attempting to classify exclusively asexual organisms (*see* AGAMOSPECIES). *See also* PHYLOGENETIC SPECIES CONCEPT.

tyramine A biologically active amine, derived from tyrosine, that mimics the effects of adrenaline, causing increased heart activity and raising blood pressure. It occurs naturally in body tissues and can also be found in ergot, mistletoe, putrefied animal tissue, and cheese. Certain antidepressant drugs, called monoamine oxidase inhibitors (MAO inhibitors), prevent the normal metabolism of tyramine, leading to dangerously high blood pressure. Hence people prescribed such drugs are advised not to eat cheese.

tyrosine *See* AMINO ACID.

tyrosine kinase Any of a large family of proteins that catalyse the phosphorylation of a tyrosine residue of a protein by ATP. They are components of numerous signalling pathways inside cells, notably ones regulating cell growth and differentiation. There are two main types. **Receptor tyrosine kinases** are cell surface receptors having intrinsic tyrosine kinase activity. There are around 20 classes in humans; examples include the insulin receptor, *fibroblast growth factor receptors, *neurotrophin receptors, and *epidermal growth factor receptors. **Nonreceptor tyrosine kinases**, such as *Src tyrosine kinases, occur wholly in the cytosol but can associate with surface receptors to activate intracellular signal pathways.

ubiquinone (coenzyme Q) Any of a group of related quinone-derived compounds that serve as electron carriers in the *electron transport chain reactions of cellular respiration. Ubiquinone molecules have side chains of different lengths in different types of organisms but function in similar ways. They are small hydrophobic molecules and the only components of the electron transport chain that are not proteins. They move freely within the inner mitochondrial membrane.

ubiquitin A small protein (consisting of 76 amino acid residues), found universally in eukaryotes, that tags proteins destined for degradation by *proteasomes or alters the protein's function, location, or trafficking. It forms a covalent bond with lysine residues in an ATP-dependent reaction termed **ubiquitination** or **ubiquitinylation**. Any of the seven lysine residues of a ubiquitin molecule can also form bonds with other ubiquitin molecules, giving rise to chains. The particular lysine residues involved, and the polyubiquitin chain length, are crucial in determining whether ubiquitination tags proteins for delivery to proteasomes or signals other modifications to the protein that affect its properties or fate in the cell. Ubiquitin is important in both the normal life of the cell and in a cell's response to stress; it is considered to be a *heat-shock protein. *See also* SUMO.

(⊕) SEE WEB LINKS
- Explore the structure and function of ubiquitin on the Protein Data Bank website

UCP *See* UNCOUPLING PROTEIN.

ulna (*pl.* **ulnae** or **ulnas**) The larger of the two bones in the forearm of vertebrates (*compare* RADIUS). It articulates with the outer carpals at the wrist and with the humerus at the elbow.

ultimobranchial bodies *See* C CELL.

ultracentrifuge A high-speed centrifuge used to measure the rate of sedimentation of colloidal particles or to separate macromolecules, such as proteins or nucleic acids, from solutions. Ultracentrifuges are electrically driven and are capable of rotation speeds up to 150 000 rpm.

ultradian rhythm A biological rhythm whose period is less than a day. Numerous physiological processes undergo rhythmical changes of less than 24 hours' duration, such as appetite, heart rate, blood pressure, and growth hormone secretion. *Compare* CIRCADIAN RHYTHM; INFRADIAN RHYTHM. *See* BIORHYTHM.

ultrafiltration The process in which hydrostatic pressure causes water and small dissolved molecules and ions to move across a membrane against a *concentration gradient. Ultrafiltration is responsible for the formation of *tissue fluid and *glomerular filtrate from blood. In both these processes the ultrafiltered fluid has the same composition as the plasma except that it does not contain blood cells or large protein molecules.

ultramicroscope A form of microscope that uses a sheet of laser light projected from either side to illuminate a thin horizontal plane through a specimen. This means there is no interference from out-of-focus light above or below the illuminated plane. Light from the specimen is focused by an objective lens positioned above and passes through a filter and a tube lens to a camera, which records the image. The specimen is moved stepwise vertically through the light sheet, and the resultant stack of images is used to construct a three-dimensional picture of the object, with micrometre resolution. Specimens are prepared by various fluorescent staining techniques, and fluorescence is thus confined to the illuminated plane. Solid specimens are first subject to a clearing process to make them transparent.

ultramicrotome *See* MICROTOME.

ultrasonics The study and use of pressure waves that have a frequency in excess of 20 000 Hz and are therefore inaudible to the human ear. Ultrasound is used in medical diagnosis, particularly in conditions such as pregnancy, in which X-rays could have a harmful effect.

ultrastructure The submicroscopic, almost molecular, structure of living cells, which is revealed by the use of an electron microscope.

ultraviolet microscope A *microscope that has quartz lenses and slides and uses *ultraviolet (UV) radiation as the illumination. The use of shorter wavelengths than the visible range enables the instrument to resolve smaller objects and to provide greater magnification than the normal optical microscope. Also, greater contrast can be achieved in some cases where there is differential absorption of UV light by different parts of the specimen. Digital imaging means that the final image can be viewed directly through the eyepieces or recorded electronically.

ultraviolet radiation (UV) Electromagnetic radiation having wavelengths between that of violet light and long X-rays, i.e. between 400 nanometres and 4 nm. In the range 400–300 nm the radiation is known as the **near** ultraviolet. In the range 300–200 nm it is known as the **far ultraviolet**. Below 200 nm it is known as the **extreme ultraviolet** or the **vacuum ultraviolet**, as absorption by the oxygen in the air makes the use of evacuated apparatus essential. The sun is a strong emitter of UV radiation

but only some reaches the surface of the earth. Ultraviolet radiation is classified in three ranges: UV-A (320–400 nm), UV-B (290–320 nm), and UV-C (230–290 nm). All UV-C and some UV-B radiation is absorbed by the *ozone layer in the atmosphere, so that 95% of UV reaching the earth's surface is UV-A. The shorter the wavelength, the more harmful but less penetrating to skin; however, all UV radiation is considered potentially harmful, causing burning, pigmentation (tanning), premature ageing of skin, and skin cancers. The risk of skin cancers has been increased by the depletion of the ozone layer. UV radiation is used clinically in the treatment of certain skin complaints, such as psoriasis. It is also used to induce *vitamin D formation in patients who are allergic to vitamin D preparations.

Most UV radiation for practical use is produced by various types of mercury-vapour lamps. Ordinary glass absorbs UV radiation and therefore lenses and prisms for use in the UV are made from quartz.

umbel A type of *racemose inflorescence in which stalked flowers arise from the same point on the flower axis, resembling the spokes of an umbrella. An involucre (cluster) of bracts may occur at the point where the stalks emerge. This arrangement is characteristic of the family Umbelliferae (Apiaceae; e.g. carrot, hogweed, parsley, parsnip), in which the inflorescence is usually a compound umbel.

umbilical cord The cord that connects the embryo to the *placenta in mammals. It contains a vein and two arteries that carry blood between the embryo and placenta. It is severed after birth to free the newly born animal from the placenta, and shrivels to leave a scar, the navel, on the animal.

uncoupling protein (UCP) Any of a family of proteins that uncouple electron transport from *oxidative phosphorylation in mitochondria, so that the energy derived from aerobic tissue respiration is released chiefly as heat instead of being used to make ATP. The archetype is the UCP found in *brown fat, a type of adipose tissue associated with newborn or hibernating mammals. Called UCP1, or thermogenin, it enables oxidation of fat reserves within the tissues to generate heat (*see* THERMOGENESIS), which is vital to combat cold stress in small mammals or when awakening from hibernation. It works by promoting the leakage of protons back into the mitochondrial matrix across the inner mitochondrial membrane, thus abolishing the electrochemical gradient necessary for ATP production according to the *chemiosmotic theory. Similar UCPs occur in a wide range of eukaryotic organisms, including plants, fungi, and protists. **Plant uncoupling proteins** are thought mainly to control the production of *reactive oxygen species in mitochondria and to regulate levels of respiration in response to stress. *See also* ALTERNATIVE RESPIRATORY PATHWAY.

undernourishment *See* MALNUTRITION.

undulipodium (*pl.* **undulipodia**) A term sometimes used to designate a eukaryotic *flagellum (sense 2) or a *cilium (which have the same structure), to emphasize the distinction between these structures and the *flagellum (sense 1) of a bacterium.

ungulate A herbivorous mammal with hoofed feet (*see* UNGULIGRADE). Ungulates are grouped into two orders: *Artiodactyla and *Perissodactyla.

unguligrade Describing the gait of ungulates (e.g. horses and cows), in which only the tips of the digits (i.e. the hooves) are on the ground and the rest of the foot is off the ground. *Compare* DIGITIGRADE; PLANTIGRADE.

unicellular Describing tissues, organs, or organisms consisting of a single cell. For example, the reproductive organs of some algae and fungi are unicellular. Unicellular organisms include archaea, bacteria, protists, and certain fungi and algae. *Compare* ACELLULAR; MULTICELLULAR.

unikonts A *supergroup of eukaryote organisms that contains animals, fungi, amoebas, and some non-amoeboid protists.

The relationships of these diverse groups have been proposed largely on the basis of molecular studies of their genomes. They comprise two large groups: the *amoebozoans and the *opisthokonts.

unipolar neuron A *neuron that has one main process, the axon, extending from its cell body. Unipolar neurons include many sensory neurons and many vertebrate motor neurons and interneurons. *Compare* BIPOLAR NEURON; MULTIPOLAR NEURON.

uniporter A *transport protein that moves molecules across a membrane in one direction only down their concentration gradient. *Compare* ANTIPORTER; SYMPORTER.

UniProt (Universal Protein Resource) An international consortium that maintains several databases serving as a central repository for information about protein sequences and functions. Funded chiefly by the US National Institutes of Health, it is a crucial resource for molecular biologists worldwide. UniProt was established in 2002 to pool the database resources of three organizations: the European Bioinformatics Institute (EBI), the Swiss Institute of Bioinformatics (SIB), and the Protein Information Resource (PIR) of the US National Biomedical Research Foundation. Hitherto these organizations had separately hosted the databases TrEMBL (translated EMBL), Swiss-Prot, and Protein Sequence Database, respectively. The formation of UniProt enabled these databases to be amalgamated and streamlined to form the basis for the UniProt Knowledgebase and associated databases.

Uniramia (Mandibulata) In some classifications, a phylum or subphylum of *arthropods that contains the classes *Chilopoda (centipedes), *Diplopoda (millipedes), Pauropoda and Symphyla (centipede-like animals), and *Hexapoda (insects). Members are characterized by *uniramous appendages and a single pair of antennae.

uniramous appendage A type of appendage that is characteristic of insects and other members of the *Uniramia. It consists of an unbranched series of segments (*see* COXA; TROCHANTER; FEMUR). *Compare* BIRAMOUS APPENDAGE.

unisexual Describing animals or plants with either male or female reproductive organs but not both. Most of the more advanced animals are unisexual but plants are often *hermaphrodite. Flowers that contain either stamens or carpels but not both are also described as unisexual. *See also* MONOECIOUS; DIOECIOUS.

unit A specified measure of a physical quantity, such as length, mass, time, etc., specified multiples of which are used to express magnitudes of that physical quantity. For scientific purposes previous systems of units have now been replaced by *SI units.

Universal Protein Resource *See* UNIPROT.

unsaturated Denoting a compound having double or triple bonds in its molecules. Unsaturated compounds can undergo addition reactions as well as substitution. *See* FATTY ACID. *Compare* SATURATED.

upper critical temperature (critical thermal maximum) The maximum body temperature that can be tolerated by an organism. When body temperature exceeds the upper critical point, cellular components and process become disrupted, and the organism eventually dies. For most animals the upper critical temperature lies in the range 30–45°C, while few plants can survive leaf temperatures in excess of 50°C, exceptions being desert plants such as agaves and cacti, which can tolerate temperatures of 60°C or more. Some specialized prokaryotes, notably certain archaea, live near deep-sea vents at temperatures exceeding 100°C (*see* THERMOPHILIC). *Compare* LOWER CRITICAL TEMPERATURE.

upregulation An increase in the sensitivity of a cell to a chemical substance, such as a hormone, signal molecule, or drug, due to an increase in the density of cell-surface receptors for that molecule. Upregulation of a gene entails an increased rate of transcription and enhanced expression of its

effect on the cell or tissue, e.g. through an increased concentration of RNA or protein. The converse, **downregulation**, reduces the cell's sensitivity or gene expression.

uracil A *pyrimidine derivative and one of the major component bases of *nucleotides and the nucleic acid *RNA.

uranium–lead dating A group of *dating techniques for certain rocks that depends on the decay of the radioisotopes uranium–238 to lead–206 (half-life 4.5×10^9 years) or the decay of uranium–235 to lead–207 (half-life 7.1×10^8 years). One form of uranium–lead dating depends on measuring the ratio of the amount of helium trapped in the rock to the amount of uranium present (since the decay $^{238}U \rightarrow {}^{206}Pb$ releases eight alpha-particles). Another method of calculating the age of the rocks is to measure the ratio of radiogenic lead (^{206}Pb, ^{207}Pb, and ^{208}Pb) present to nonradiogenic lead (^{204}Pb). These methods give reliable results for ages of the order 10^7–10^9 years.

urea (carbamide) A white crystalline water-soluble solid, $CO(NH_2)_2$. Urea is the major end product of nitrogen excretion in mammals and other *ureotelic animals, being synthesized by the *urea cycle. Urea is synthesized industrially from ammonia and carbon dioxide for use in urea-formaldehyde resins and pharmaceuticals, as a source of nonprotein nitrogen for ruminant livestock, and as a nitrogen fertilizer.

urea cycle (ornithine-urea cycle) The series of biochemical reactions that converts ammonia, which is highly toxic, and carbon dioxide to the much less toxic *urea during the excretion of metabolic nitrogen derived from the deamination of excess amino acids. In *ureotelic vertebrates, these reactions take place in the liver. The cycle requires energy in the form of ATP, and involves the addition of two –NH2 groups and carbon dioxide to ornithine to form arginine, via the intermediates citrulline and arginosuccinate. The enzyme arginase then catalyses the removal of urea from arginine, and ornithine is re-formed. In mammals the urea is ultimately excreted in solution in *urine. For elasmobranch fishes urea is an important component of body fluids, helping to maintain their osmolarity slightly above that of seawater. It is also their main nitrogenous waste product, being excreted across the gills.

ureotelic Describing animals that excrete nitrogenous waste in the form of *urea. Most terrestrial vertebrates are ureotelic, converting ammonium ions formed during the breakdown of amino acids into urea by the *urea cycle in the liver. In teleost fishes and many invertebrates, urea is formed from uric acid via the uricolytic pathway. *Compare* AMMONOTELIC; URICOTELIC.

ureter The duct in vertebrates that conveys urine from the *kidney to the *bladder.

urethra The duct in mammals that conveys urine from the *bladder to be discharged to the outside of the body. In males the urethra passes through the penis and is joined by the *vas deferens; it therefore also serves as a channel for sperm.

uric acid The end product of purine breakdown in most mammals, birds, terrestrial reptiles, and insects and also (except in mammals and elasmobranch fishes; *see* UREA) the major form in which metabolic nitrogen is excreted. Being fairly insoluble, uric acid can be expelled in solid form, which conserves valuable water in arid environments. Bird droppings are a mixture of faeces and white uric acid. In humans small quantities of uric acid are formed as the end product of nucleic acid metabolism. The accumulation of uric acid in the synovial fluid of joints causes gout.

uricotelic Describing animals that excrete nitrogenous waste in the form of *uric acid or the closely related guanine. Uricotelic animals include birds, reptiles, and most terrestrial arthropods. *Compare* AMMONOTELIC; UREOTELIC.

uridine A nucleoside consisting of one uracil molecule linked to a D-ribose sugar molecule. The derived nucleotide uridine diphosphate (UDP) is important in carbohydrate metabolism.

urinary system The collection of organs and tissues that perform *osmoregulation and *excretion. The mammalian urinary system consists of two *kidneys each linked to the bladder by a ureter.

urine The aqueous fluid formed by the excretory organs of animals for the removal of metabolic waste products. In higher animals, urine is produced by the *kidneys, stored in the *bladder, and excreted through the *urethra or *cloaca. Apart from water, the major constituents of urine are one or more of the end products of nitrogen metabolism—ammonia, urea, uric acid, and creatinine. It may also contain various inorganic ions, the pigments urochrome and urobilin, amino acids, and purines. Precise composition depends on many factors, especially the habitat of a particular species: aquatic animals produce copious volumes; terrestrial animals need to conserve water and produce much less (about 1.0–1.5 litres per day in humans).

uriniferous tubule *See* NEPHRON.

Urochordata (Tunicata) A clade of marine nonvertebrate chordates in which the body is typically enclosed in a protective **tunic** of a cellulose-like material and the notochord and dorsal hollow nerve cord are evident only in the free-swimming tadpole-like larval stage. The pharynx is specialized for filter feeding (*see* ENDOSTYLE). There are three classes: the sessile Ascidiacea (sea squirts), the pelagic Thaliacea (salps), and the pelagic Larvacea (in which the larval form is retained into adulthood).

uterine cycle *See* MENSTRUAL CYCLE.

uterus (womb) (*pl.* **uteri**) The organ of female mammals in which the embryo develops. Paired in most mammals but single in humans, it is situated between the bladder and rectum and is connected to the *fallopian tubes and to the *vagina. The lining (*see* ENDOMETRIUM) shows cyclical changes (*see* MENSTRUAL CYCLE; OESTROUS CYCLE) associated with egg production (*see* OVARIAN CYCLE) and provides a thick spongy layer in which the fertilized egg becomes embedded. The outer wall of the uterus (**myometrium**) is thick and muscular; by contracting, it forces the fully grown fetus through the vagina to the outside.

utriculus (utricle) (*pl.* **utriculi**) A chamber of the *inner ear from which the *semicircular canals arise. It bears patches of sensory epithelium concerned with detecting changes in the direction and speed of movement (*see* MACULA).

UV *See* ULTRAVIOLET RADIATION.

vaccination *See* IMMUNIZATION.

vaccine A liquid preparation of treated disease-producing microorganisms or their products used to stimulate an immune response in the body and so confer resistance to the disease (*see* IMMUNIZATION). Vaccines are administered orally or by injection (**inoculation**). They take the form of dead viruses or bacteria that can still act as antigens, live but weakened microorganisms (*see* ATTENUATION), specially treated *toxins, or antigenic extracts of the pathogenic microorganism. For **recombinant DNA vaccines**, such as the hepatitis B vaccine, the gene encoding a suitable antigen, such as a protein subunit, is isolated from the pathogen and inserted into the genome of a host cell, such as the bacterium *E. coli* or yeast (*see* GENE CLONING). These recombinant cells are cultured so that they produce multiple copies of the subunit antigen, which is extracted and purified as a vaccine. Another strategy under trial employs **DNA vaccines**, which essentially introduce the genes for specific pathogenic antigens into body cells, so that the individual's own cells synthesize the antigens and display them on the cell surfaces. The pathogen's genes are inserted into a plasmid for delivery into the body by various methods. In principle, these vaccines should evoke an immune response with no risk of disease, as well as being relatively easy to make. Animal tests have shown their effectiveness in stimulating immunity against a range of diseases, but their clinical effectiveness remains to be established.

vacuole A membrane-bound vesicle within the cytoplasm of a living *cell that is filled with air, water or other liquid, *cell sap, or food particles. Vacuoles are derived from the endoplasmic reticulum and Golgi apparatus, and form part of the *endomembrane system of the cell. Plant vacuoles perform various functions, including storage of food reserves or waste products; in some cases these are toxic or unpalatable substances that deter browsing animals. Vacuoles in cells of petals or other flower parts may contain the colourful anthocyanin pigments that attract pollinators, whereas in seeds vacuoles store proteins for use by the germinating embryo. As plant cells mature smaller vacuoles coalesce to form usually one large central vacuole bounded by a single-layered membrane (**tonoplast** or **vacuole membrane**). This central vacuole occupies most of the cell's interior. It is filled with cell sap, including most of the cell's inorganic ions, and it swells by absorbing water from the cytoplasm. The pressure exerted by the vacuole is transmitted through the cytoplasm to the cell wall; in a growing cell this *turgor pressure helps the cell wall to expand, whereas in a mature cell it stiffens the wall, thus lending support to the plant. Animal cells usually have several small vacuoles. *See also* CONTRACTILE VACUOLE.

vacuole membrane (tonoplast) *See* VACUOLE.

vagal tone The effect produced on the heart by the parasympathetic nerve fibres (which are carried in the *vagus nerve) controlling the heart rate. The parasympathetic nerve fibres slow the heart rate, e.g. from approximately 70 beats per minute to 60 beats per minute. Hence they counteract the effects of the *sympathetic nerve fibres. *See* PARASYMPATHETIC NERVOUS SYSTEM.

vagina (*pl.* **vaginas** or **vaginae**) The tube leading from the uterus to the outside. Sperm are deposited in the vagina during copulation and the fully developed fetus is born through it. In a number of mammals the vagina may be sealed when the animal is not sexually receptive and only open during oestrus. Its lining produces mucus, which prevents friction and the entry of infective organisms. In humans it is surrounded externally by the *vulva.

vagus nerve The tenth *cranial nerve: a paired nerve that supplies branches to many major internal organs. It carries motor nerve fibres to the heart, lungs, and viscera and sensory fibres from the viscera.

valine *See* AMINO ACID.

valve 1. Any of various structures for restricting the flow of a fluid through an aperture or along a tube to one direction. Valves in the heart (*see* BICUSPID VALVE; SEMILUNAR VALVE; TRICUSPID VALVE), veins, and lymphatic vessels consist of two or three flaps of tissue (**cusps**) fastened to the walls. The cusps are flattened to the walls to allow the normal passage of blood or lymph, but a reverse flow causes them to block the vessel or aperture, so preventing further backflow. **2.** Any of the parts that make up a capsule or other dry fruit that sheds its seeds. **3.** One of the two halves of the cell wall of a diatom. **4.** Either of the two hinged portions of the shell of a bivalve mollusc.

van der Waals interactions Weak interactions between nonpolar molecules or regions of molecules produced by transient changes in positive and negative charges arising from random variations in electron distribution. They influence how certain biologically important molecules behave when in close proximity (e.g. an enzyme and substrate) and help determine the three-dimensional structure of large biomolecules such as proteins and nucleic acids. They are named after the Dutch physicist Johannes van der Waals (1837–1923).

variable In statistics and research, any characteristic relating to an individual or group that can take a number of different values. **Categorical** or **qualitative variables** are descriptive characteristics, such

as sex or species; **quantitative variables** relate to a numerical scale and are subdivided into **discrete variables**, found only at fixed points (e.g. number of offspring), and **continuous variables**, found at any point on a scale (e.g. weight). *See also* CONFOUNDING; CORRELATION.

variable number tandem repeats (VNTR; minisatellite DNA) A family of genetic loci found in eukaryotes consisting of short (15–100 bp) sequences of DNA repeated in tandem arrays; in humans these arrays are typically 1–5 kb long. The alleles for any particular locus all have the same sequence but differ as to how many times the sequence is repeated. VNTR loci contribute to *repetitive DNA and have proved valuable in *DNA profiling. The VNTR sequences can be released intact from a DNA sample using restriction endonuclease enzymes and identified using *gene probes with *Southern blotting. Alternatively, the VNTR alleles can be amplified by the *polymerase chain reaction, separated by gel electrophoresis, and the resultant patterns compared without the need for special gene probes. Since each VNTR locus typically has many different alleles, the likelihood of two individuals having identical sets of alleles for even a few such loci is very remote. For forensic purposes, VNTRs have now been supplanted by *short tandem repeats.

variation The differences between individuals of a plant or animal species. Variation may be the result of environmental conditions; for example, water supply and light intensity affect the height and leaf size of a plant. Differences of this kind, acquired during the lifetime of an individual, are generally not transmitted to succeeding generations since the genes are not affected (*see* PHENOTYPIC PLASTICITY). However, environmental factors may lead to *epigenetic changes arising during the life of an organism, and these can sometimes be transmitted to offspring. Although these alter the phenotype, they do not affect the genotype and may be reversed subsequently. **Genetic variation**, due to differences in genetic constitution, is inherited (*see* CONTINUOUS VARIATION;

DISCONTINUOUS VARIATION). The most important sources of genetic variation are *mutation and *recombination (*see also* CROSSING OVER). It is also increased by *outbreeding. Wide genetic variation improves the ability of a species to survive in a changing environment, since the chances that some individuals will tolerate a particular change are increased. Such individuals will survive and transmit the advantageous genes to their offspring.

variegation The occurrence of differently coloured patches, spots, or streaks in plant leaves, petals, or other parts, due to absence of pigment or different combinations of pigment in the affected area of the part. Variegation may be brought about by infection, for example *tobacco mosaic virus infection, or by differences in pigmentation genes contained in the plastids (e.g. chloroplasts) of cells of the variegated part (*see* CHIMAERA). Hence such traits show *cytoplasmic inheritance. Lack of the photosynthetic pigment chlorophyll in the affected patches will reduce the plant's growth potential.

variety A category used in the *classification of plants and animals below the *species level. A variety consists of a group of individuals that differ distinctly from but can interbreed with other varieties of the same species. The characteristics of a variety are genetically inherited. Examples of varieties include breeds of domestic animals. *See also* CULTIVAR. *Compare* SUBSPECIES.

varve dating (geochronology) An absolute *dating technique using thin sedimentary layers of clays called **varves**. The varves, which are particularly common in Scandinavia, have alternate light and dark bands corresponding to winter and summer deposition. Most of them are found in the Pleistocene series, where the edges of varve deposits can be correlated with the annual retreat of the ice sheet, although some varve formation is taking place in the present day. By counting varves it is possible to establish an absolute time scale for fossils up to about 20 000 years ago.

vasa recta A network of thin-walled blood capillaries that branch from the efferent arterioles leaving the *glomerulus in juxtamedullary *nephrons of the kidney in mammals and birds. The vasa recta forms a series of interconnected U-shaped loops around the *loop of Henle and its vessels eventually coalesce into venules that ultimately drain into the renal vein. Water from the kidney medulla is reabsorbed by the vasa recta during the formation of urine.

vascular bundle (fascicle) A long continuous strand of conducting (vascular) tissue in tracheophyte plants that extends from the roots through the stem and into the leaves. It consists of *xylem and *phloem, which are separated by a *cambium in plants that undergo secondary thickening. *See* VASCULAR TISSUE.

vascular cambium *See* CAMBIUM.

vascular plants All plants possessing organized *vascular tissue. *See* TRACHEOPHYTE.

vascular system 1. A specialized network of vessels for the circulation of fluids throughout the body tissues of an animal. All animals, apart from simple invertebrate groups, possess a vascular system to enable the passage of a fluid (i.e. blood or haemolymph) around the body to transport respiratory gases, nutrients, excretory products, and other metabolites into and out of the cells. In vertebrates it consists of a muscular *heart, which pumps blood through major blood vessels (*arteries) into increasingly finer branches until in the *capillaries it is in intimate contact with tissues (*see* MICROCIRCULATION). It then returns to the heart via another network of vessels (the *veins). This closed system of *circulation also enables a stable *internal environment for tissue function (*see* HOMEOSTASIS), the transmission of chemical messengers (*hormones) around the body, and a means of defending the body against pathogens and damage via the immune system. Small invertebrates, such as arthropods and many molluscs, have an open circulation system filled with haemolymph, which fulfils many of the same functions as blood (*see* HAEMOCOEL).

A **water vascular system** is characteristic of the *Echinodermata. **2.** The system of *vascular tissue in plants.

- Interactive round-up of the human cardiovascular system, presented by the Texas Heart Institute

vascular tissue (**vascular system**) The tissue that conducts water and nutrients through the plant body in higher plants (*tracheophytes). It consists of *xylem and *phloem. Since the xylem and phloem tissues are always in close proximity to each other, distinct regions of vascular tissue can be identified (*see* VASCULAR BUNDLE). The possession of vascular tissue has enabled the higher plants to attain a considerable size and dominate most terrestrial habitats.

- Animations of how the conducting tissues of plants develop (Michigan State University)

vas deferens (*pl.* **vasa deferentia**) One of a pair of ducts carrying sperm from the testis (or *epididymis) to the outside, in mammals through the *urethra.

vas efferens (*pl.* **vasa efferentia**) Any of various small ducts carrying sperm. In reptiles, birds, and mammals they convey sperm from the seminiferous tubules of the testis to the *epididymis; in invertebrates they carry sperm from the testis to the vas deferens.

vasoactive intestinal peptide *See* VIP.

vasoconstriction The reduction in the internal diameter of blood vessels, especially arterioles or capillaries. The constriction of arterioles is accomplished by smooth muscle fibres of the arteriole walls controlled by the action of nerves of the autonomic nervous system, and by certain hormones; it results in an increase in *blood pressure.

vasodilation (**vasodilatation**) The increase in the internal diameter of blood vessels, especially arterioles or capillaries. The vasodilation of arterioles is mediated by the relaxation of smooth muscle fibres of the arteriole walls due primarily to autoregulation following local changes in the tissue environment, and indirectly also to the influence of the *autonomic nervous system. Low oxygen concentrations and high carbon dioxide concentrations in a tissue cause local relaxation of systemic capillary smooth muscle, thereby widening local capillaries and increasing local blood supply (the opposite applies in the lungs, where low O_2 causes vasoconstriction). Inhibition of sympathetic nervous stimulation by parasympathetic fibres also causes vasodilation, as do the parasympathetic neurotransmitters acetylcholine, substance P, and vasoactive intestinal polypeptide. *See* BLOOD PRESSURE.

vasomotor centre The region of the *medulla oblongata of the brain that mediates the diameter of the blood vessels (via the action of the *vasomotor nerves) and hence controls blood pressure (*see* VASOCONSTRICTION; VASODILATION).

vasomotor nerves The nerves of the *autonomic nervous system that control the diameter of blood vessels. **Vasoconstrictor nerves** of the sympathetic division decrease the diameter and cause *vasoconstriction. Inhibition of these by nerves of the parasympathetic division leads to relaxation of smooth muscle in blood vessel walls, as does direct stimulation by vasodilator nerves (*see* VASODILATION).

vasopressin *See* ANTIDIURETIC HORMONE.

V(D)J recombination *See* SOMATIC RECOMBINATION.

vector 1. An animal, usually an insect, that passively transmits disease-causing microorganisms from one animal or plant to another or from an animal to a human. *Compare* CARRIER. **2.** (**cloning vector**) A vehicle used in *gene cloning to insert a foreign DNA fragment into the genome of a host cell. For bacterial hosts several different types of vector are used: *bacteriophages, *artificial chromosomes, *plasmids, and their hybrid derivatives, *cosmids. The foreign DNA is spliced into the vector using specific *restriction enzymes and *DNA ligases to cleave the vector DNA and join the foreign DNA to the two ends created (**insertional vectors**). In some phage vectors, part of the viral genome is enzymically removed and replaced with the foreign DNA (**replacement vectors**). *Retroviruses can be effective vectors for introducing recombinant DNA into mammalian cells. In plants, derivatives of the tumour-inducing (Ti) plasmid of the crown gall bacterium, *Agrobacterium tumefaciens*, are used as vectors. *See also* EXPRESSION VECTOR.

vegetal pole *See* ANIMAL POLE.

vegetative propagation (vegetative reproduction) 1. A form of *asexual reproduction in plants whereby new individuals develop from specialized multicellular structures (e.g. *tubers, *bulbs) that become detached from the parent plant. Examples are the production of strawberry plants from *runners and of gladioli from daughter *corms. Artificial methods of vegetative propagation include grafting (*see* GRAFT), *budding, and making *cuttings. **2.** Asexual reproduction in animals, e.g. budding in *Hydra*.

vein 1. A blood vessel that carries blood towards the heart. Most veins carry deoxygenated blood (the *pulmonary vein is an exception). The largest veins are fed by smaller ones, which are formed by the merger of *venules. Veins have thin walls and a relatively large internal diameter. *Valves within the veins ensure that the flow of blood is always towards the heart. *Compare* ARTERY. **2.** A vascular bundle in a leaf (*see* VENATION). **3.** Any of the tubes of chitin that strengthen an insect's wing.

velamen (*pl.* **velamina**) A whitish spongy sheath of dead empty cells that surrounds the aerial roots of epiphytic plants, such as certain orchids. It absorbs any surface water on the roots.

velum (*pl.* **vela**) **1.** A membranous or veil-like structure, such as certain sheets of white matter in the brain. **2.** *See* ANNULUS; VELAMEN.

velvet worms *See* ONYCHOPHORA.

vena cava (*pl.* **venae cavae**) Either of the two large veins that carry deoxygenated blood into the right atrium of the heart. The **precaval vein** (**anterior** or **superior vena cava**) receives blood from the head and forelimbs; the **postcaval vein** (**posterior** or **inferior vena cava**) drains blood from the trunk and hindlimbs.

venation 1. The arrangement of veins (vascular bundles) in a leaf. The leaves of dicotyledons have a central main vein (midrib) with side branches that themselves further subdivide to form a network (**net** or **reticulate venation**). The leaves of monocotyledons have parallel veins (**parallel venation**). **2.** The arrangement of the veins in an insect's wing, which is often important in classification.

venter The swollen base of an *archegonium, in which the egg cell (oosphere) develops.

ventilation The process by which a continuous exchange of gases is maintained across respiratory surfaces. Often called external *respiration, this is achieved by *respiratory movements; in air-breathing vertebrates it is movement of air into and out of the lungs (*see also* AIR SAC; EXPIRATION; INSPIRATION; TRACHEA; VENTILATION CENTRE). The **minute ventilation rate** (or **minute respiration rate**) of an animal is the volume of air breathed per minute, i.e. *tidal volume × number of breaths per minute. It can be measured with the aid of a *respirometer.

ventilation centre Two paired groups of neurons (i.e. *nuclei, sense 2) in the *medulla oblongata of the brain that control the process of breathing (i.e. *ventilation). The partial pressure of carbon dioxide in the blood and the pH of the blood are monitored by chemoreceptors in the arteries and on the surface of the medulla. These include the *carotid bodies in the carotid arteries and the **aortic bodies** in the wall of the aorta close to the heart. The ventilation centre responds to an increase in the amount of carbon dioxide in the blood by increasing the rate of breathing. The nuclei comprising the ventral respiratory group have both inspiratory and expiratory neurons and set the basic respiratory rhythm, while the dorsal respiratory group (*see* INSPIRATORY CENTRE) alters the basic rhythm in response to the body's oxygen requirements. The pontine respiratory centre, located in the pons, fine-tunes breathing during activities such as speaking, exercise, and sleeping. Furthermore, conscious centres within the cerebral cortex can override the involuntary control mechanisms, e.g. during singing or holding of breath. Changes in the rate and depth of breathing are achieved by nerve impulses sent from the ventilation centre to the muscles of the ribs and diaphragm. A negative feedback circuit prevents overexpansion of the lungs.

ventral Describing the surface of a plant or animal that is nearest or next to the ground or other support, i.e. the lower surface. In bipedal animals, such as humans, it is the forward-directed (*anterior) surface. *Compare* DORSAL.

ventral aorta The artery in vertebrate embryos that carries blood from the ventricle of the heart to the *aortic arches. In adult fish it branches into afferent branchial arteries supplying the gills; in adult tetrapods it is represented by the ascending part of the *aorta. *Compare* DORSAL AORTA.

ventral root The part of a *spinal nerve that leaves the spinal cord on the ventral side and contains motor fibres. *Compare* DORSAL ROOT. *See* SPINAL CORD.

ventricle 1. A chamber of the *heart that receives blood from an *atrium and pumps it into the arterial system. Amphibians and fish have a single ventricle, but mammals, birds, and reptiles have two, pumping deoxygenated blood to the lungs and oxygenated blood to the rest of the body, respectively. **2.** Any of the four linked fluid-filled cavities in the brain of vertebrates. One of these cavities is in the *medulla oblongata, two are in the cerebral hemispheres (*see* CEREBRUM), and the fourth is in the posterior part of the *forebrain. The ventricles contain cerebrospinal fluid filtered from the blood by the *choroid plexus.

venule A small blood vessel that receives blood from the capillaries and transports it to a vein.

vermiform appendix *See* APPENDIX.

vernalization The promotion of flowering by exposure of a plant to low temperatures. For example, winter cereals will not flower unless subjected to a period of chilling early in their development. Winter cereals are therefore sown in the autumn for flowering the following year. However, if germinating seeds are artificially vernalized they can be sown in the spring for flowering the same year. Biennial plants, such as carrot (*Daucus carota*), will remain in their nonflowering rosette form until subjected to cold treatment. For vernalization to be effective, the plant tissue must be actively metabolizing and supplied with carbohydrate (i.e. energy) and oxygen. In biennials, perception of the cold stimulus is confined to the shoot apex, and cold treatment of other parts of the plant is ineffective. Studies in thale cress (*Arabidopsis thaliana*) have shown that flowering is suppressed by high concentrations of a protein called FLC (Flowering locus C). Levels of FLC fall when the plant is exposed to prolonged cold, mediated by other proteins that cause *chromatin remodelling of the FLC gene. This cessation of FLC production enables flowering genes to be switched on and flowers to be produced as the vernalized plant develops during the ensuing spring. However, this mechanism does not seem to be conserved in other species, such as wheat.

vertebra (*pl.* **vertebrae**) Any of the bones that make up the *vertebral column. In mammals each vertebra typically consists of a main body, or **centrum**, from which arises a **neural arch** through which the spinal cord passes, and **transverse processes** projecting from the side. There are five groups of vertebrae, specialized for various functions and varying in number with the species. In humans, for example, there are 7 *cervical vertebrae, 12 *thoracic

V

vertebrae, 5 *lumbar vertebrae, 5 fused *sacral vertebrae, and 5 fused *caudal vertebrae (forming the *coccyx).

vertebral column (backbone; spinal column; spine) A flexible bony column in vertebrates that extends down the long axis of the body and provides the main skeletal support. It also encloses and protects the *spinal cord and provides attachment for the muscles of the back. The vertebral column consists of a series of bones (*see* VERTEBRA) separated by discs of cartilage (*intervertebral discs). It articulates with the skull by means of the *atlas vertebra, with the ribs at the *thoracic vertebrae, and with the pelvic girdle at the sacrum (*see* SACRAL VERTEBRAE).

vertebrate Any one of a large group of animals comprising all chordates (*see* CHORDATA) that possess a backbone made of vertebrae (*see* VERTEBRAL COLUMN). Vertebrates include the fishes, amphibians, reptiles (including birds), and mammals. Characteristically in vertebrates, the vertebral column replaces the *notochord during development as the principal supporting structure.

vesicle A small, usually fluid-filled, membrane-bound sac within the cytoplasm of a living cell. Vesicles occur, for example, as part of the *Golgi apparatus, as *lysosomes, or as *microbodies. *Compare* VACUOLE.

vessel 1. (in botany) A tube within the *xylem composed of joined *vessel elements. Vessels facilitate the efficient movement of water from the roots to the shoots and leaves of a plant. **2.** (in zoology) Any of various tubular structures through which substances are transported, especially a blood vessel or a lymphatic vessel.

vessel element A type of cell occurring within the *xylem of flowering plants, many of which, end to end, form water-conducting **vessels**. Vessel elements are frequently very broad and have side walls thickened by deposits of lignin over most of the surface area. However, the end walls are broken down to provide connections with the cells both above and below them,

forming a continuous hollow 'pipe'. *Compare* TRACHEID.

vestibular apparatus The part of the inner ear that is responsible for balance. The vestibular apparatus is continuous with the cochlea. It consists of the three *semicircular canals, which detect movements of the head (*see* AMPULLA), and the *utriculus and *sacculus, which detect the position of the head (*see* MACULA). *See* EAR.

vestibular canal A canal in the *cochlea of the inner ear that connects with the *oval window. It contains *perilymph, through which pressure waves are transmitted from the oval window via *Reissner's membrane to the endolymph within the cochlear duct. *See also* ORGAN OF CORTI.

vestibule A chamber that leads to a body cavity or that links one cavity to another. For example, a vestibule leads from the vulva into the *vagina in the mammalian reproductive system.

vestigial organ Any part of an organism that has diminished in size during its evolution because the function it served decreased in importance or became totally unnecessary. Examples are the human appendix and the wings of the ostrich.

viable count A measure of the number of living cells within a culture.

vibrio Any comma-shaped bacterium. Generally, vibrios are Gram-negative (*see* GRAM'S STAIN), motile, and aerobic. They are widely distributed in soil and water and while most feed on dead organic matter some are parasitic, e.g. *Vibrio cholerae*, the causal agent of cholera.

vicariant event The formation of a physical barrier that splits an existing population into two or more subpopulations and prevents or impedes breeding and hence free interchange of genetic material. Over time, such events lead to the splitting of the original species into two distinct species (*see* ALLOPATRIC). They include the uplifting of mountain ranges, shifts in the course of rivers, or the separation of continental land masses via movement of tectonic plates.

villus (*pl.* **villi**) A microscopic outgrowth from the surface of some tissues and organs, which serves to increase the surface area of the organ. Numerous villi line the interior of the small intestine. Their shape may vary from finger-like (in the *duodenum) to spadelike (in the *ileum). Intestinal villi are specialized for the absorption of soluble food material: each contains blood vessels and a lymph vessel (*see* LACTEAL).

Chorionic villi occur on the chorion of the mammalian placenta, where they increase the surface area for the exchange of materials between the fetal and maternal blood.

viologen dyes (bipyridylium dyes) A group of organic compounds based on two linked pyridine rings and including certain agricultural and horticultural herbicides, notably *Paraquat and **Diquat**. These generally kill all nonwoody plants and are used as nonselective weedkillers; they are active ingredients in various contact and systemic herbicides. They act by interfering with electron transport in photosynthesis and by generating superoxide anions, which damage chloroplasts and other cell components. These herbicides are also highly toxic to animals, and their use is strictly controlled.

VIP (vasoactive intestinal peptide) A widely distributed 28-amino-acid *neuropeptide that acts as a neurotransmitter or regulator of cell function in many organs and tissues, including the central and peripheral nervous systems, heart, lungs, kidneys, and immune system. It also acts as a peptide hormone, being secreted by endocrine cells of the upper part of the small intestine in response to the entry of partially digested food from the stomach. VIP, along with *secretin, stimulates the pancreas to produce a thin watery secretion containing bicarbonate. This raises the pH in the intestine in preparation for secretion of pancreatic enzymes. VIP also inhibits gastric secretion. Other physiological effects

of VIP include widening of blood vessels (vasodilation), increased cardiac output, widening of the airways (bronchodilation), and relaxation of smooth muscle. In nervous tissue it can function as a *cotransmitter. For example, parasympathetic neurons supplying the salivary gland secrete both acetylcholine and (at high-frequency stimulation) VIP, causing vasodilation and secretion of saliva. VIP also modulates the function of immune cells, such as macrophages and helper T cells, and has an anti-inflammatory role.

virion *See* VIRUS.

viroid Any of various small naked single-stranded RNA molecules that infect plant cells and cause disease. Smaller than viruses, the 30 or so known viroids are not enclosed in a protein coat of any kind: they generally consist of fewer than 400 nucleotides, do not contain any genes, but do have *ribozyme activity. The circular RNA strand undergoes extensive base pairing within itself, forming a double-stranded structure that mimics DNA and is replicated by the host cell's enzymes. The progeny viroids spread to other cells, establishing a wider infection; in some cases they infect the seeds and hence the new plants following germination. Pathogenicity is thought to be due to RNA interference of the host's genes induced by viroid-specific small interfering RNAs. Viroids include many commercially important disease agents, such as coconut cadang-cadang, citrus exocortis, and potato spindle tuber viroid. The agents are also spread by contaminated tools, insects, or rubbing by large animals. *Compare* VIRUSOID.

virology The scientific study of *viruses. *See* MICROBIOLOGY.

(((●))) SEE WEB LINKS
• Website for All the Virology on the WWW, a major portal for virology resources via the Internet

virulence The disease-producing ability of a microorganism. *See also* PATHOGEN.

virulence factor Any of various molecules that facilitate invasion of host cells by bacteria, viruses, fungi, or other pathogens or help them to evade host defences. Typically proteins located on the cell surface of the pathogen or secreted by the pathogen allow adherence to or penetration of host cells. Moreover, the polysaccharide capsules of bacteria such as pneumococci prevent phagocytosis by host immune cells. Various bacterial *toxins are also virulence factors, as are the pili (see PILUS sense 1) via which cell attachment can be accomplished.

virus A particle that is too small to be seen with a light microscope or to be trapped by filters but is capable of independent metabolism and reproduction within a living cell. Outside its host cell a virus is completely inert. A mature virus (a **virion**) typically ranges in size from 20 to 600 nm in diameter. It consists of a core of nucleic acid (DNA or RNA) surrounded by a protein coat (see CAPSID). Some (the **enveloped viruses**) bear an outer envelope consisting of proteins and lipids. Inside its host cell the virus initiates the synthesis of viral proteins and undergoes replication. The new virions are released when the host cell disintegrates. Viruses are parasites of animals, plants, and some bacteria (see BACTERIOPHAGE). Viral diseases of animals include the common cold, influenza, AIDS, herpes, hepatitis, polio, and rabies (see ADENOVIRUS; ARBOVIRUS; HERPESVIRUS; HIV; MYXOVIRUS; PAPOVAVIRUS; PICORNAVIRUS; POXVIRUS); some viruses are also implicated in the development of cancer (see RETROVIRUS). Plant viral diseases include various forms of yellowing and blistering of leaves and stems (see TOBACCO MOSAIC VIRUS). *Antiviral drugs are effective against certain viral diseases and *vaccines (if available) provide protection against others. See also INTERFERON; NUCLEOCYTOPLASMIC LARGE DNA VIRUS; PANDORAVIRUS. See Feature.

virusoid A subviral particle that infects plants and consists of a small naked single-stranded circular RNA molecule. The virusoid is introduced to its host cell by an infecting plant virus, in which it is carried as a passenger. The virusoid nucleic acid is replicated by the cell's RNA polymerase, in the form of a precursor comprising tandem repeats of the virusoid structure. These repeats are cut out and the ends joined together in a self-splicing mechanism, in which the RNA acts as a *ribozyme. Compare VIROID.

visceral Relating to the internal organs (the **viscera**) that lie in the coelomic cavities of animals, i.e. in the thoracic and abdominal cavities of mammals. Compare SOMATIC.

visceral sensory neuron See SENSORY NEURON.

vision The sense that enables perception of objects in the environment by means of the *eyes.

visual acuity Sharpness of vision: the ability of the eye to distinguish between objects that lie close together. This hinges on the ability of the eye to focus incoming light to form a sharp image on the retina. Visual acuity depends on the *cone cells, which are most densely packed in the *fovea, close to the centre of the retina, and are therefore in the optimum position to receive focused light. In addition, each cone cell synapses with a single bipolar cell in the *retina and is thus able to send a separate signal, via the optic nerve fibres, to the brain.

visual cortex (striate cortex) The region of the *cerebral cortex of the brain where sensory information from the eyes is interpreted. See also RECEPTIVE FIELD.

visual purple See RHODOPSIN.

vital capacity The total amount of air that can be exhaled after maximum inspiration. The vital capacity of an average human is typically 4.5–5 litres; in trained male athletes it can be 6 litres or more. However, some air always remains in the lungs (see RESIDUAL VOLUME). Hence, the total lung capacity is the sum of vital capacity and residual volume.

VIRUSES

The first evidence that certain diseases could be caused by agents smaller than bacteria was published in 1892 by Russian microbiologist Dmitri Ivanowski, who showed that an agent capable of passing through a filter for bacteria could cause tobacco mosaic disease in plants. A few years later, Dutch microbiologist Martinus Beijerinck (1851–1931) confirmed that this agent had certain properties of living things and called it a 'filterable virus'.

Subsequently many other diseases of plants and animals were shown to be caused by these agents, which came to be called simply viruses. The term bacteriophage (often shortened to phage) for bacterial viruses was coined in 1917 by French scientist F. d'Herelle. Today thousands of different viruses have been discovered, and a wealth of knowledge exists about their structure and mode of life. Virus-infected cells have proved extremely useful experimental systems for studying aspects of cell metabolism, including DNA replication and protein synthesis, and viruses are commonly used as vectors for cloning DNA.

Structure

Virus particles (virions) come in various shapes and sizes, ranging from about 20 nm (e.g. parvoviruses) to 1500 nm for *Pithovirus*, a *nucleocytoplasmic large DNA virus. Essentially they fall into two structural classes—helical and icosahedral—depending on how the viral nucleic acid and protein subunits of the capsid are arranged.

viral nucleic acid (RNA)

protein subunit

In rod-shaped viruses, such as the tobacco mosaic virus, the protein subunits form a helical array and the nucleic acid (RNA) runs along an internal groove.

protein subunits of capsid

nucleic acid core

protein fibre

The capsid of an icosahedral virus, such as human adenovirus, takes the form of a polyhedron with 20 triangular faces (an icosahedron). The projecting protein fibres attach to surface receptors of their host cell in the initial phase of infection.

icosahedral head

helically arranged protein subunits of tail

tail fibres

tail

base plate

Some bacteriophages, such as the T4 phage, have a more complex structure. The capsid is equipped with a hollow helical tail and tail fibres. After the tail fibres attach to the surface of a host cell, the tail contracts and injects the viral DNA into the host bacterium.

V

Baltimore classification

A common method of classifying viruses is based on the nature of their genetic material (DNA or RNA) and how they convert that genetic information into viral messenger RNA (mRNA). It was devised by US biologist David Baltimore (1938–). The viral mRNA strand that is translated into viral proteins by the host cell is always denoted the positive (+) sense strand, and the template strand from which it is copied is denoted as the negative (−) sense strand.

Class	Genetic material	Genome	Replication method	Examples
I	DNA	one double-stranded DNA (dsDNA) molecule	host cell enzymes transcribe viral DNA into viral mRNA	adenoviruses, herpesviruses, poxviruses
II	DNA	one positive- or negative-sense single-stranded DNA (ssDNA) molecule	viral ssDNA is copied in host cell into dsDNA, which is copied into viral mRNA	parvoviruses
III	RNA	10–12 dsRNA molecules	negative strands of genomic RNAs copied into positive mRNAs	reoviruses (e.g. rotavirus)
IV	RNA	single positive-sense RNA strand	genomic positive strand can act as mRNA; further copies are made via synthesis of a negative RNA strand	polioviruses, togaviruses, astroviruses, caliciviruses
V	RNA	one or several positive-sense RNA strands	genomic RNA is template for viral mRNAs	orthomyxoviruses (e.g. influenza virus), paramyxoviruses, rhabdoviruses (e.g. rabies virus)
VI	RNA	two identical positive-sense RNA strands	the enzyme reverse transcriptase copies a genomic RNA strand into a single negative DNA strand, and then into dsDNA, which becomes integrated into the host genome. The host transcribes this into positive viral RNA	retroviruses (e.g. HIV)
VII	DNA	circular, part dsDNA, part	host transcribes viral DNA into mRNA, some of which ssDNA encodes viral proteins; some becomes 'pregenomic RNA', which is copied by reverse transcriptase inside the virion into viral DNA	hepadnaviruses (e.g. hepatitis B virus)

vital staining A technique in which a harmless dye is used to stain living tissue for microscopical observation. The stain may be injected into a living animal and the stained tissue removed and examined (**intravital staining**) or the living tissue may be removed directly and subsequently stained (**supravital staining**). Microscopic organisms, such as protists, may be completely immersed in the dye solution. Vital stains include trypan blue, vital red, and Janus green, the latter being especially suitable for observing mitochondria.

vitamin One of a number of organic compounds required by living organisms in relatively small amounts to maintain normal health. There are some 14 generally recognized major vitamins: the water-soluble *vitamin B complex (containing 9) and *vitamin C and the fat-soluble *vitamin A, *vitamin D, *vitamin E, and *vitamin K. Most B vitamins and vitamin C occur in plants, animals, and microorganisms; they function typically as *coenzymes. Vitamins A, D, E, and K occur only in animals, especially vertebrates, and perform a variety of metabolic roles. Animals are unable to manufacture many vitamins themselves and must have adequate amounts in the diet. Foods may contain vitamin precursors (called **provitamins**) that are chemically changed to the actual vitamin on entering the body. Many vitamins are destroyed by light and heat, e.g. during cooking.

vitamin A (retinol) A fat-soluble vitamin that cannot be synthesized by mammals and other vertebrates and must be provided in the diet. Green plants contain precursors of the vitamin, notably carotenes, that are converted to vitamin A in the intestinal wall and liver. It is also present in dairy products. The aldehyde derivative of vitamin A, **retinal**, is a constituent of the visual pigment *rhodopsin. Deficiency affects the eyes, causing night blindness, **xerophthalmia** (dryness and thickening of the cornea), and

eventually total blindness. Vitamin A is also important for cell differentiation and growth, especially in maintaining the integrity of epidermal and mucosal surfaces, which act as a barrier against infection. The competence of macrophages, which are crucial to innate immunity, requires vitamin A, as does the proper utilization of stored iron in the manufacture of red blood cells (haemopoiesis); vitamin A deficiency can result in iron deficiency anaemia and defective immunity.

vitamin B complex A group of water-soluble vitamins that characteristically serve as components of *coenzymes. Plants and many microorganisms can manufacture B vitamins but dietary sources are essential for most animals. Heat and light tend to destroy B vitamins.

    Vitamin B$_1$ (thiamin(e)) is a precursor of the coenzyme thiamine pyrophosphate, which functions in the Krebs cycle and carbohydrate metabolism. Deficiency leads to *beriberi in humans and to polyneuritis in birds. Good sources include whole-grain or fortified cereals, beans, peas, and nuts.

    Vitamin B$_2$ (riboflavin) occurs in green vegetables, yeast, liver, and milk. It is a constituent of the coenzymes *FAD and FMN, which have an important role in the metabolism of all major nutrients as well as in the oxidative phosphorylation reactions of the *electron transport chain. Deficiency of B$_2$ causes inflammation of the tongue and lips, mouth sores, and conjunctivitis.

    Vitamin B$_6$ (pyridoxine) is widely distributed in cereal grains, yeast, liver, milk, etc. It is a constituent of a coenzyme (pyridoxal phosphate) involved in amino acid metabolism. Deficiency can cause anaemia, dermatitis, fatigue, and other signs.

    Vitamin B$_{12}$ (cyanocobalamin or cobalamin) is manufactured only by microorganisms and natural sources are entirely of animal origin. Liver is especially rich in it. One form of B$_{12}$ functions as a coenzyme in a number of reactions, including the oxidation of fatty acids and the synthesis of DNA. It also works in conjunction with *folic acid (another B vitamin) in the synthesis of the amino acid methionine and it is required for normal production of red blood cells.

VITAMINS

1897	Dutch physician Christiaan Eijkman (1858-1930) cures beriberi in chickens with diet of whole rice.
1906-07	British biochemist Sir Frederick Hopkins demonstrates existence of accessory dietary elements essential for growth.
1912	Polish-born US biochemist Casimir Funk (1884-1967) extracts antiberiberi factor (an amine) from rice husks and coins the term 'vitamine' (vital amine; later changed to 'vitamin').
1913	US biochemist Elmer McCollum (1879-1967) discovers and names vitamin A (retinol) and names antiberiberi factor vitamin B.
1920	McCollum names antirachitic factor vitamin D.
1922	US embryologist Herbert Evans (1882-1971) discovers vitamin E (tocopherol).
1926	German chemist Adolf Windaus (1876-1959) discovers that ergosterol is converted to vitamin D in the presence of sunlight.
1931	German chemist Paul Karrer (1889-1971) determines the structure of (and synthesizes) vitamin A.
1932	Hungarian-born US biochemist Albert Szent-Györgyi (1893-1986) and US biochemist Charles King (1896-1986) independently isolate vitamin C (ascorbic acid).
1933	Polish-born Swiss chemist Tadeus Reichstein (1897-1996) and British chemist Walter Haworth (1883-1950) independently synthesize vitamin C. US chemist Roger Williams (1893-1988) discovers the B vitamin pantothenic acid.
1934	Danish biochemist Carl Dam (1895-1976) discovers vitamin K.
1935	Karrer and Austrian-born German chemist Richard Kuhn (1900-67) independently synthesize vitamin B_2 (riboflavin).
1937	US chemist Robert Williams (1886-1965) synthesizes vitamin B_1 (thamine).
1938	Karrer synthesizes vitamin E.
	Kuhn isolates and synthesizes vitamin B_6 (pyridoxine).
1939	Dam and Karrer isolate vitamin K.
1940	Szent-Györgyi and US biochemist Vincent Du Vigneaud (1901-78) discover 'vitamin H' (the B vitamin biotin).
	Roger Williams determines the structure of pantothenic acid.
	US biochemist Edward Doisey synthesizes vitamin K.
1948	US biochemist Karl Folkers (1906-97) isolates vitamin B_{12} (cyanocobalamin).
1956	British chemist Dorothy Hodgkin (1910-94) determines the structure of vitamin B_{12}.
1971	US chemist Robert Woodward (1917-79) and Swiss chemist Albert Eschenmoser (1925-) synthesize vitamin B_{12}.

Vitamin B$_{12}$ can only be absorbed from the gut in the presence of a glycoprotein called **intrinsic factor**; lack of this factor or deficiency of B$_{12}$ results in pernicious anaemia.

Other vitamins in the B complex include *nicotinic acid, *pantothenic acid, *biotin, and *lipoic acid. *See also* CHOLINE; INOSITOL.

vitamin C (ascorbic acid) A colourless crystalline water-soluble vitamin found especially in citrus fruits and green vegetables. Most organisms synthesize it from glucose, but humans and other primates and various other species must obtain it from their diet. It functions as a scavenger of free radicals within cells and in extracellular fluid and blood plasma, and is needed for the synthesis of *collagen, a protein found in tendons, skin, and other connective tissues. Deficiency leads to *scurvy and delayed wound healing. Vitamin C is readily destroyed by heat and light.

vitamin D A fat-soluble vitamin occurring in the form of two steroid derivatives: **vitamin D$_2$ (ergocalciferol**, or **calciferol**), found in yeast; and **vitamin D$_3$ (cholecalciferol**), which occurs in animals. Vitamin D$_2$ is formed from a steroid by the action of ultraviolet light and D$_3$ is produced by the action of sunlight on a cholesterol derivative in the skin. Fish-liver oils are the major dietary source. The active form of vitamin D (**calcitriol**) is manufactured in the kidneys in response to the secretion of *parathyroid hormone, which occurs when blood calcium levels are low. It causes increased uptake of calcium from the gut and promotes calcium reabsorption by the kidneys, thereby increasing the supply of calcium for bone synthesis. Vitamin D deficiency causes *rickets in growing animals and softening of bones (osteomalacia) in mature animals. Both conditions are characterized by weak deformed bones.

vitamin E (tocopherol) A fat-soluble vitamin, consisting of several closely related compounds, that is the main antioxidant of cell membranes and other lipid-rich tissue components. Deficiency leads to a range of disorders in different species, including muscular dystrophy, liver damage, and infertility.

Good sources are cereal grains and green vegetables. Vitamin E prevents the oxidation of unsaturated fatty acids in cell membranes by removing lipid peroxyl radicals, so maintaining their structure (*see* ANTIOXIDANTS).

vitamin K Any of several related fat-soluble compounds, including **vitamins K$_1$ (phylloquinone)** and **K$_2$ (menaquinone)**, that act as coenzymes in the gamma-carboxylation of glutamate amino acid residues in certain proteins. Such a reaction is essential for the function of various proteins involved in blood coagulation (e.g. prothrombin and certain other coagulation factors) and bone metabolism. Deficiency of vitamin K, which leads to extensive bleeding, is rare because a form of the vitamin is manufactured by intestinal bacteria. Green vegetables and egg yolk are good sources.

vitelline membrane *See* EGG MEMBRANE.

vitreous humour The colourless jelly that fills the space between the lens and the retina of the vertebrate eye.

viviparity 1. (in zoology) A form of reproduction in animals in which the developing embryo obtains its nourishment directly from the mother via a *placenta or by other means. Viviparity occurs in some insects and other arthropods, in certain fishes, amphibians, and reptiles, and in the majority of mammals. *Compare* OVIPARITY; OVOVIVIPARITY. **2.** (in botany) **a.** A form of *asexual reproduction in certain plants, such as the onion, in which the flower develops into a budlike structure that forms a new plant when detached from the parent. **b.** The development of young plants on the inflorescence of the parent plant, as seen in certain grasses and the spider plant.

VNTR *See* VARIABLE NUMBER TANDEM REPEATS.

vocal cords A pair of elastic membranes that project into the *larynx in air-breathing vertebrates. Vocal sounds are produced when expelled air passing through the larynx vibrates the cords. The pitch of the sound produced depends on the tension of the cords, which is controlled by muscles and cartilages in the larynx.

voltage clamp An experimental apparatus for measuring the flow of ions through membrane *channels in cells. As the flow of ions amounts to a flow of current this can easily be measured. Typically a cell is placed in solution with three electrodes, two intracellular and one extracellular. One of the intracellular electrodes is used to measure voltage across the membrane relative to the extracellular electrode; the second intracellular electrode passes current into the cell. When the membrane becomes either less negative (depolarization) or more negative (hyperpolarization) these effects can be detected by the electrodes.

voltage-gated ion channel Any *ion channel that opens and closes in response to changes in electrical potential across the cell membrane in which the channel is situated. There are several types of voltage-gated channel, each allowing the selective passage of a particular ion. Two types are especially important in transmitting *action potentials along axons: voltage-gated sodium channels and voltage-gated potassium channels. The sodium channels open rapidly in response to initial depolarization of the axon plasma membrane, allowing sodium ions (Na^+) to flood in. Depolarization also triggers less rapid opening of the potassium channels, which permits outflow of potassium ions (K^+), thus acting to restore the membrane potential to its resting state. Voltage-dependent calcium channels also carry some of the depolarizing current in some cells. The sodium channel protein has positively charged voltage-sensing regions, which move towards negative charges on the outer surface of the membrane when the latter becomes depolarized. This opens the channel, allowing passage of sodium ions. Within a millisecond of channel opening, the voltage-sensing region returns to its original location, and a channel-inactivating segment moves to block the channel and allow the channel protein to revert to its resting state. *See also* CALCIUM ION CHANNEL; POTASSIUM ION CHANNEL; SODIUM ION CHANNEL.

voluntary Controlled by conscious thought. *See* SKELETAL MUSCLE. *Compare* INVOLUNTARY.

voluntary muscle *See* SKELETAL MUSCLE.

volutin Polyphosphate granules that are used as food reserves by certain bacteria, including the cyanobacteria.

vomeronasal organ An organ of smell found in many terrestrial vertebrates, but not apparent in humans. Vomeronasal organs are typically paired blind-ending sacs opening into the oral or nasal cavities. They are lined with olfactory epithelium, whose cells express receptor proteins for airborne chemicals. The vomeronasal organs tend to respond especially to *pheromones used in chemical signalling between members of the same species, and their receptor proteins belong to different protein families than those expressed by the nasal epithelium. Also, the axons of vomeronasal receptors project to a different site in the brain from the nasal olfactory axons.

vulva The female external genitalia, comprising in women two sets of skin folds surrounding the opening to both the vagina and (just in front) the urethra. The thicker, outer folds are the labia majora, and the thinner, inner folds are the labia minora; the latter meet in front to partially enclose the erectile *clitoris. *See* LABIUM.

waggle dance *See* DANCE OF THE BEES.

Wallace, Alfred Russel (1823–1913) British naturalist, who in 1848 went on an expedition to the Amazon, and in 1854 travelled to the Malay Archipelago. There he noticed the differences between the animals of Asia and Australasia and devised *Wallace's line, which separates them. This led him to develop a theory of *evolution through *natural selection, which coincided with the views of Charles *Darwin; their theories were presented jointly to the Linnaean Society in 1858.

Wallace's line An imaginary line that runs between the Indonesian islands of Bali and Lombok and represents the separation of the Australian and Oriental faunas. It was proposed by A. R. *Wallace, who had noted that the mammals in Southeast Asia are different from and more advanced than their Australian counterparts. He suggested this was because the Australian continent had split away from Asia before the better adapted placental mammals evolved in Asia. Hence the isolated Australian marsupials and monotremes were able to thrive while those in Asia were driven to extinction by competition from placental mammals. *See also* ZOOGEOGRAPHY.

Wallerian degeneration The characteristic sequence of changes that occurs in a vertebrate sensory or motor peripheral neuron after its axon has been severed. The cut length of axon now separated from the cell body degenerates, and associated Schwann cells dedifferentiate and proliferate and—with invading macrophages—phagocytose the remnant axon and myelin sheath. The cell body swells, the nucleus moves from a central to a more peripheral location, and the prominently staining endoplasmic reticulum (*see* NISSL GRANULES) becomes dispersed. Within hours, new axonal sprouts emerge from the axon stump and start growing towards their target within the basal lamina that ensheathed the original axon. The process is named after British physiologist Augustus Waller (1816–70).

warfarin 3-(alpha-acetonylbenzyl)-4-hydroxycoumarin: a *coumarin derivative used as a synthetic *anticoagulant both therapeutically in clinical medicine and, in lethal doses, as a rodenticide (*see* PESTICIDE).

warm-blooded animal *See* ENDOTHERM.

warning coloration (aposematic coloration) The conspicuous markings of an animal that make it easily recognizable and warn would-be predators that it is a poisonous, foul-tasting, or dangerous species. For example, the yellow-and-black striped abdomen of the wasp warns of its sting. *See also* MIMICRY.

waste product Any product of metabolism that is not required for further metabolic processes and is therefore excreted from the body. Common products include *nitrogenous wastes (such as *urea, uric acid, and ammonia), carbon dioxide, and *bile.

water A colourless liquid, H_2O. In the gas phase water consists of single H_2O molecules in which the H-O-H angle is 105°. The structure of liquid water is still controversial; hydrogen bonding of the type $H_2O\ldots$H-O-H imposes a high degree of structure (*see* HYDROGEN BOND) and current models supported by X-ray scattering studies have short-range ordered regions, which are constantly disintegrating and re-forming. This ordering of the liquid state is sufficient to make the density of water at about 0°C higher than that of the relatively open-structured ice; the maximum density occurs at 3.98°C. This accounts for the well-known phenomenon of ice floating on water and the contraction of water below ice, a fact of enormous biological significance for all aquatic organisms. The collective hydrogen bonding also explains the tendency of water molecules to stay together—the phenomenon of **cohesion**, which is vital for the uptake of water and dissolved nutrients by plants—and also the **adhesion** of water molecules to vessel walls. Another consequence of the hydrogen bonding is the high specific heat of water and its large latent heat of evaporation. This means it acts as a very effective 'heat bank', so that oceans and other large water bodies have a moderating influence on climate by slowly absorbing or emitting heat. Similarly, water is a very effective means of evaporative cooling of surfaces, whether of a forest or an athlete. Water is a powerful *solvent for both polar and ionic compounds; molecules or ions in solution are frequently strongly hydrated. Pure liquid water is very weakly dissociated into H_3O^+ (hydroxonium) and OH^- (hydroxyl) ions by self-ionization and consequently any compound that increases the concentration of the positive ion, H_3O^+, is acidic and compounds increasing the concentration of the negative ion, OH^-, are basic (*see* ACID). The phenomena of ion transport in water and the division of materials into **hydrophilic** (water loving) and **hydrophobic** (water hating) substances are central features of almost all biological chemistry.

water cycle *See* HYDROLOGICAL CYCLE.

water potential Symbol Ψ. The difference between the chemical potential of the water in a biological system and the chemical potential of pure water at the same temperature and pressure. It is manifested as a force acting on water molecules in a solution separated from pure water by a membrane that is permeable to water molecules only and can be expressed as the sum of the *solute potential and the *pressure potential:

$$\psi = \psi_s + \psi_p$$

Water potential is measured in kilopascals (kPa). The water potential of pure water is zero; aqueous solutions of increasing concentration have increasingly negative values. Water tends to move from areas of high (less negative) water potential to areas of low (more negative) water potential. *Osmosis in plants is described in terms of water potential. In soils and other extracellular systems, another factor, called *matric potential, can contribute significantly to water potential. *See also* TURGOR.

water vascular system *See* ECHINO-DERMATA.

Watson, James Dewey (1928-) US biochemist, who moved to the Cavendish Laboratory, Cambridge, in 1951 to study the structure of *DNA. In 1953 he and Francis *Crick announced the now accepted two-stranded helical structure for the DNA molecule. In 1962 they shared the Nobel Prize for physiology or medicine with Maurice Wilkins (1916–2004), who with Rosalind Franklin (1920–58) had made X-ray diffraction studies of DNA.

Watson-Crick model The double-stranded twisted ladder-like molecular structure of *DNA as determined by James *Watson and Francis Crick at Cambridge, England, in 1953. It is commonly known as the **double helix**.

wax (in biology) Esters of fatty acids, usually having a protective function. Examples are the beeswax forming part of a honeycomb and the wax coating on some leaves, fruits, and seed coats, which acts as a protective water-impermeable layer supplementing the functions of the cuticle. The seeds of a few plants contain wax as a food reserve.

weed *See* PEST.

Weismannism The theory of the **continuity of the germ plasm** published by August Weismann (1834–1914) in 1886.

It proposes that the contents of the reproductive cells (sperms and ova) are passed on unchanged from one generation to the next, unaffected by any changes undergone by the rest of the body. It thus rules out any possibility of the inheritance of acquired characteristics, and has become integral to neo-Darwinian theory. However, Weismann's dogma is challenged by the finding that *epigenetic modifications of *chromatin acquired during an individual's lifetime can be transmitted to their offspring.

Wernicke's area An area of the brain located in the posterior temporal lobe of the cerebral hemisphere and concerned with understanding language, whether as speech or written words. It is named after the German neurologist Carl Wernicke (1848–1905), who identified its crucial role in comprehending meaning; patients with lesions in this area cannot understand simple instructions, and produce confused or incoherent speech. Wernicke's area receives nervous inputs from the auditory and visual cortices and is thought to associate these inputs with speech sounds, which are then articulated by *Broca's area. Wernicke's and Broca's areas form part of a complex language-processing network that occurs most commonly in the left cerebral hemisphere (in 90% of right-handed and 70% of left-handed persons).

western blotting (protein blotting) An *immunoassay for determining very small amounts of a particular protein in tissue samples or cells. The sample is subjected to electrophoresis on SDS-polyacrylamide gel to separate constituent proteins. The resultant protein bands are then 'blotted' onto a polymer sheet. A radiolabelled or fluorescently labelled antibody specific for the target protein is added; this binds to the protein, which can then be detected by autoradiography or a fluorescence detector. A variation of this technique is used to screen bacterial colonies containing cDNA clones in order to isolate those colonies expressing a particular protein. The name is derived by analogy to that of *Southern blotting.

whalebone (baleen) Transverse horny plates hanging down from the upper jaw on each side of the mouth of the toothless whales (*see* CETACEA), forming a sieve. Water, containing plankton on which the whale feeds, enters the open mouth and is then expelled with the mouth slightly closed, so that food is retained on the baleen plates.

whales *See* CETACEA.

whey *See* CURD.

whisk ferns Primitive *tracheophyte plants represented by the extant genus *Psilotum*. Along with the sole species of the related genus *Tmesipteris* and numerous extinct forms that flourished in the Devonian period, they make up the family Psilotaceae. Whisk ferns have rhizoids rather than roots; the stems, which show *dichotomous branching, may be naked or bear scalelike leaves. Recent molecular genetic evidence shows that they belong to the clade of *monilophytes, which also contains the *horsetails and true *ferns.

white blood cell *See* LEUCOCYTE.

white matter Part of the tissue that makes up the central nervous system of vertebrates. It consists chiefly of bundles of nerve axons enclosed in their whitish *myelin sheaths. *Compare* GREY MATTER.

white muscle A type of muscle tissue consisting of glycolytic *fast-twitch fibres specialized for rapid contraction. In fish white muscles are used for fast swimming movements and for escape reactions. They lie deeper in the body than the red muscles used for slow swimming and are arranged in a helical pattern rather than parallel with the body axis. This arrangement produces marked body curvature when they contract.

wild type Describing the form of an *allele possessed by most members of a population in their natural environment. Wild-type alleles are usually *dominant.

wilting The condition that arises in plants when more water is lost by evaporation than is absorbed from the soil. This causes the cells to lose their *turgor and the plant structure droops. The plant hormone abscisic acid (ABA) accumulates in leaves and signals the *guard cells to close leaf pores (stomata) and reduce the loss of water vapour. Plants can normally recover from wilting if water is added to the soil, but permanent wilting and possible death can result if the plant does not have access to water for a long period of time. In certain plants wilting is important as a mechanism to avoid overheating: when the leaves droop they are taken out of direct contact with the sun's rays. When the sun sets the plant can begin to transpire at the normal rate and the cells of the leaves regain their turgor.

windpipe *See* TRACHEA.

wind pollination *See* ANEMOPHILY.

wing *See* FLIGHT.

withdrawal reflex *See* POLYSYNAPTIC REFLEX; REFLEX.

WNT protein Any of a family of proteins that are involved in signal transduction pathways crucial in regulating cell proliferation, differentiation, and survival, both during embryological development and in later life. The components of the WNT signal pathway show marked similarity in a wide range of organisms, both invertebrates and vertebrates, signifying the pathway's fundamental role and ancient origin. The prototypical family member is encoded by the *wingless* gene (named after the effects of its mutation) in the fruit fly *Drosophila*. The Wingless protein (WG) plays a vital part in establishing the segment polarity of the embryonic fly (*see* SEGMENTATION GENES). It binds to the receptor protein **Frizzled**, which through the mediation of other pathway components, including the protein Dishevelled, causes the accumulation of β-catenin (*see* CATENIN) in the cytosol. This migrates to the nucleus, where it interacts with the transcription factor TCF (T-cell factor) to activate transcription of target genes. The homologous gene in mammals, *Wnt1*, acts through a similar pathway. WNT1 protein binds to receptors of the Frizzled family and is involved in development of the central nervous system. Other members of the WNT family have roles in the patterning of various tissues.

wobble The occurrence of nonstandard base pairing between the anticodon of certain *transfer RNA (tRNA) molecules and codons of messenger RNA (mRNA) during *translation of the genetic message that directs protein synthesis in living cells. The bases in positions 3 and 2 of the anticodon (corresponding to positions 1 and 2 of the codon) observe strictly standard pairing, whereas the base in position 1 of the anticodon can 'wobble' somewhat in its pairing. Hence, the cell does not need a unique tRNA for each of the 61 codons that encode amino acids, but requires only about 40 or so tRNAs, to ensure that each codon binds to a tRNA carrying the correct amino acid. *See* GENETIC CODE.

Wolff's law The principle that *bone undergoes changes in density and shape in response to the mechanical stresses imposed on it. Physical activity is essential to maintain healthy bones, whereas inactivity results in loss of mineral substance and weakening of bones, as evidenced by astronauts exposed to zero gravity for long periods. Mechanical loading creates hydrostatic pressure gradients within fluid-filled spaces in the bone tissue. These internal fluid pressures stimulate osteocytes and osteoblasts to promote bone metabolism and matrix production, thereby remodelling the bone. The law is named after German anatomist Julius Wolff (1836–1902).

womb *See* UTERUS.

wood The hard structural and water-conducting tissue that is found in many perennial plants and forms the bulk of trees and shrubs. It is composed of tough-walled secondary *xylem and associated cells, such as fibres. The wood of angiosperms is termed **hardwood**, e.g. oak and mahogany, and that of gymnosperms **softwood**, e.g. pine and fir. New wood is added to the outside of the old wood each growing season by divisions of the vascular cambium (*see* GROWTH RING). Only the outermost new wood (*sapwood) functions in water conduction; the inner wood (*heartwood) provides only structural support.

Woronin body A rounded granular body, bounded by a double membrane, found in the hyphae of filamentous ascomycote fungi. One or more Woronin bodies are closely associated with the pore in each cross-wall (septum) of the hyphae. If the hypha is injured, the Woronin body is swept by the flow of cytoplasm to plug the pore and minimize further damage. It is named after the Russian mycologist Mikhail Woronin (1838–1903).

wort *See* BREWING.

xanthophyll A member of a class of oxygen-containing *carotenoid pigments, which provide the characteristic yellow and brown colours of autumn leaves.

xanthophyll cycle A cyclic series of interconversions involving certain xanthophylls, which plays a vital role in the *photoprotection of plant chloroplasts by quenching reactive oxygen species (ROS) such as hydrogen peroxide and singlet oxygen. Under intense illumination, violaxanthin is converted, via an intermediate, to zeaxanthin. The latter acts as a quenching agent by dissipating excess energy from chlorophyll as heat, thereby avoiding possible damage to the photosynthetic apparatus. In dim light, the zeaxanthin is reconverted to violaxanthin, so that all incident light energy is used for photosynthesis. Xanthophylls also 'mop up' ROS formed by other plant stressors such as drought, heat, and salinity.

Xanthophyta A phylum of mostly freshwater eukaryotic organisms, traditionally known as yellow-green algae, that possess carotenoid pigments (including xanthins), which are responsible for their colour, in addition to chlorophylls. Xanthophytes occur in a variety of forms—unicellular, colonial, filamentous, and siphonaceous; motile cells have two unequal-sized flagella. Storage products are oil and the polysaccharide **chrysolaminarin**. Xanthophytes are included in a eukaryotic assemblage known as the *stramenopiles.

X chromosome *See* SEX CHROMOSOME.

xenobiotic Any substance foreign to living systems. Xenobiotics include drugs, pesticides, and carcinogens. *Detoxification of such substances occurs mainly in the liver.

xeric Denoting conditions characterized by an inadequate supply of water. Xeric conditions exist in arid habitats, extremely cold habitats, and in salt marshes. Certain plants are adapted to live in such conditions. *See* HALOPHYTE; XEROPHYTE.

xeromorphic Describing the structural modifications of certain plants (*xerophytes) that enable them to reduce water loss, particularly from their leaves and stems.

xerophthalmia *See* VITAMIN A.

xerophyte A plant that is adapted to live in conditions in which there is either a scarcity of water in the soil, or the atmosphere is dry enough to provoke excessive transpiration, or both. Xerophytes have special structural (**xeromorphic**) and functional modifications, including swollen water-storing stems or leaves (*see* SUCCULENT) and specialized leaves that may be hairy, rolled, or reduced to spines or have a thick cuticle to lower the rate of transpiration. Another common adaptation of xerophytes is *crassulacean acid metabolism, which enables the uptake of carbon dioxide at night, when temperatures are lowest, and closure of leaf pores (stomata) during the daytime to conserve water Examples of xerophytes are desert cacti and many species growing on sand dunes and exposed moorlands. Some *halophytes have xeromorphic features. *Compare* MESOPHYTE; HYDROPHYTE.

X inactivation The inactivation of all but one of the X chromosomes in cells of female mammals to prevent an abnormal dose of X-linked genes compared with cells of male individuals, which carry a single X chromosome. Hence, in humans, one of the two X chromosomes is normally selected for inactivation and assumes a highly condensed state as a *Barr body. X inactivation is controlled by a genetic locus called the **X inactivation centre**. This includes the *XIST* gene, which is transcribed as long noncoding RNAs (lncRNAs). These coat the inactivated chromosome, causing it to become condensed as heterochromatin, as in a Barr body, and hence largely inactivated. Some genes in the 'silenced' chromosome remain active, particularly in the **pseudoautosomal region**, a region near one of the telomeres. *See also* SEX CHROMOSOME.

X-ray crystallography The use of X-ray diffraction to determine the structure of crystals or molecules, such as nucleic acids. The technique involves directing a beam of X-rays at a crystalline sample and recording the diffracted X-rays on a photographic plate or other detector, such as a transducer that can create a digital image. The diffraction pattern consists of a pattern of spots on the image, and the crystal structure can be worked out from the positions and intensities of the diffraction spots. X-rays are diffracted by the electrons in the molecules and if molecular crystals of a compound are used, the electron density distribution in the molecule can be determined.

X-rays Electromagnetic radiation of shorter wavelength than ultraviolet radiation and longer wavelength than gamma radiation. The range of wavelengths is 10^{-11} m to 10^{-9} m. X-rays can pass through many forms of matter and they are therefore used medically and industrially to examine internal structures. X-rays are produced for these purposes by an X-ray tube.

xylem A tissue that transports water and dissolved mineral nutrients in vascular plants. In flowering plants it consists of hollow **vessels** that are formed from cells (*vessel elements) joined end to end. The end walls of the vessel elements are perforated to allow the passage of water. In nonflowering vascular plants, such as conifers and ferns, the constituent cells of the xylem are called *tracheids. In young plants and at the shoot and root tips of older plants **primary xylem** is formed by the apical meristems (*see* PROTOXYLEM; METAXYLEM). In plants showing secondary growth this xylem is replaced in most of the plant by **secondary xylem**, formed by the vascular *cambium. The walls of the xylem cells are thickened with lignin, the extent of this thickening being greatest in secondary xylem. Xylem contributes greatly to the mechanical strength of the plant: *wood is mostly made up of secondary xylem. *See also* FIBRE. *Compare* PHLOEM.

xylenes *See* DIMETHYLBENZENES.

X

YAC *See* (YEAST) ARTIFICIAL CHROMOSOME.

Y chromosome *See* SEX CHROMOSOME.

yeast artificial chromosome (YAC)
See ARTIFICIAL CHROMOSOME.

yeasts Any single-celled fungus. The 'true'
or budding yeasts constitute the subphylum Saccharomycotina of the phylum
*Ascomycota. They occur as single cells or
as groups or chains of cells; true yeasts
reproduce asexually by *budding and sexually by producing ascospores. Yeasts of the
genus *Saccharomyces* ferment sugars; they
are used in the baking and brewing industries (*see* BAKER'S YEAST), and in genetics as
experimental model organisms.

yeast two-hybrid screen A technique
for detecting an interaction between two
proteins by means of expression of a
*reporter gene in yeast cells. It can be
used in high-throughput screening of
libraries of genes or potential genes (open
*reading frames) from any species (e.g.
mouse or human) to identify pairs of
genes that encode interacting proteins.
Essentially, a plasmid carrying a gene for
the query protein, termed the 'bait', is
introduced to numerous yeast cells, each
of which contains a plasmid carrying a particular gene from the gene library, termed
the 'fish'. The plasmids are constructed

such that both the bait and the fish proteins
are expressed in the cell as hybrid proteins,
in which the bait or fish protein is fused
to a second protein involved in activating
expression of a reporter gene. Binding of
the bait and fish proteins inside the yeast
nucleus allows activation of the reporter
gene, and the yeast colony containing
those particular genes can be pinpointed
and analysed further.

yellow body *See* CORPUS LUTEUM.

yolk The food stored in an egg for the use
of the embryo. It can consist mainly of
protein (**protein yolk**) or of phospholipids
and fats (**fatty yolk**). The eggs of oviparous
animals (e.g. birds) contain a relatively
large yolk.

yolk sac One of the protective membranes
surrounding the embryos of birds, reptiles,
and mammals (*see* EXTRAEMBRYONIC
MEMBRANES). The embryo derives nourishment from the yolk sac via a system of
blood vessels. In birds and reptiles the
yolk sac encloses the yolk; in most mammals a fluid replaces the yolk.

Y organ Either of a pair of glands in the
head of certain crustaceans that secrete
the hormone *ecdysone and play a role in
the moulting process.

zeatin A naturally occurring *cytokinin that was first identified in 1963 in the grains of the maize plant (*Zea mais*).

Zika virus A mosquito-borne flavivirus that usually causes mild symptoms in humans, including fever, skin rashes, conjunctivitis, muscle and joint pain, malaise, and headache. However, infection during pregnancy can result in brain abnormalities of the fetus, notably microcephaly (an abnormally small skull). Zika virus can also trigger Guillain-Barré syndrome—a more severe disease of the peripheral nerves, possibly causing breathing difficulties. Infection is usually from a mosquito bite, although the virus may also be transmitted by sexual intercourse. The virus was first isolated in a rhesus monkey in the Zika forest of Uganda in 1947.

zinc Symbol Zn. A blue-white metallic element that is a trace element (*see* ESSENTIAL ELEMENT) required by living organisms. It functions as the prosthetic group of a number of enzymes and in plants is required for auxin production. Certain plant species can accumulate very high concentrations, and flourish on land heavily contaminated with zinc (*see* HYPERACCUMULATOR).

zinc finger A structural motif characteristic of certain proteins that bind to DNA, notably *transcription factors. It consists of a finger-like fold of amino acids at the base of which lie two cysteine and two histidine residues. These four residues bind a single zinc ion in a tetrahedral array. Each finger typically interacts with three base pairs in the nucleic acid molecule. Unlike other DNA-binding motifs, such as the *helix-turn-helix and the *leucine zipper, the zinc finger is also found in proteins that bind to RNA, such as RNA-directed RNA polymerase. Artificially constructed proteins containing arrays of zinc fingers are used as *genome editing tools. The amino acid composition of the zinc finger DNA-binding sites is engineered to bind to a target site within the genome of a cell. A nuclease enzyme, introduced with the zinc finger protein as part of an editing module, then cuts both DNA strands at the target site. This can result in gene deletion or integration of newly introduced DNA at the target site, following DNA repair by the cell.

zinc finger nuclease An engineered endonuclease enzyme that is used in *genome editing to make targeted cuts in both strands of a DNA molecule. It consists of a short chain of zinc finger modules, which recognize and bind to specific base sequences in the target DNA, coupled to a DNA-cleaving domain. The double-strand breaks stimulate the cell's own DNA repair mechanisms so that strands are rejoined following, e.g. insertion of a new stretch of DNA or deletion of an existing segment.

Z line (Z band) The thin membrane that separates adjacent *sarcomeres in the fibres of skeletal (striated) muscle. It is visible under an electron microscope as a thin dark line in the middle of a lighter band. The centre point of each thin (actin) filament is anchored to the Z line and exerts a pull on the membrane when the fibre contracts.

zona pellucida A layer of *glycoprotein that surrounds the plasma membrane of a mammalian egg cell. It develops as a jelly coat around the primary oocyte and is surrounded by the *granulosa cells.

zonation The distribution of the different species of a community into separate zones, which are created by variations in the environment. A clear example of zonation occurs on a rocky shore, where different species of seaweed (*Fucus*) occupy different zones, according to their ability to withstand desiccation. For example, the species found in the splash zone, which is never completely submerged in water, is better adapted to exposure than those found in zones lower down the shore, where they are submerged for longer periods. Animals, particularly stationary species, such as barnacles, also exhibit zonation on a rocky shore; as with the seaweeds, this may depend on the ability of different species to withstand desiccation. Competition between species may also contribute to zonation.

zone fossil *See* INDEX FOSSIL.

zone of polarizing activity (ZPA) *See* SONIC HEDGEHOG.

zonula adherens *See* ADHERENS JUNCTION.

zonula occludens *See* TIGHT JUNCTION.

zoogeography The study of the geographical distributions of animals. The earth can be divided into several *faunal regions separated by natural barriers, such as oceans, deserts, and mountain ranges. The characteristics of the fauna of each region are believed to depend particularly (but not wholly) on the process of *continental drift and the stage of evolution reached when the various land masses became isolated. For example Australia, which has been isolated since Cretaceous times, has the most primitive native mammalian fauna, consisting solely of marsupials and monotremes. *See also* WALLACE'S LINE.

zooid An individual member of a colony of invertebrate animals, especially an individual of the phylum *Bryozoa.

zoology The scientific study of animals, including their anatomy, physiology,

biochemistry, genetics, ecology, evolution, and behaviour.

Zoomastigota (Zoomastigina) In older classifications, a phylum of parasitic or free-living heterotrophic *protists that possess one or more *flagella (sense 2) for locomotion.

zoonosis (*pl.* **zoonoses**) An infectious disease of nonhuman vertebrates that can be transmitted to humans. Rabies and anthrax are well-known examples, and certain midges and the tsetse fly act as carriers for a variety of nematode-worm zoonoses.

zooplankton The animal component of *plankton. All major animal phyla are represented in zooplankton, as adults, larvae, or eggs; some are just visible to the naked eye but most cannot be seen without magnification. Examples include copepods, shrimps, *krill, other small crustaceans, jellyfish, and tunicates. Near the surface of the sea there may be many thousands of such animals per cubic metre.

zoosporangium (*pl.* **zoosporangia**) *See* ZOOSPORE.

zoospore A spore that possesses one or more flagella and is therefore motile. Released from a sporangium (called a **zoosporangium**), zoospores are produced by many algae and certain protists, such as the potato blight (*Phytophthora infestans*) and other members of the *Oomycota.

zosterophyllophytes (zosterophylls) A group of extinct vascular plants that lived in the Devonian period. They were small herbs, similar in many ways to the *rhyniophytes; for example, members of the principal genus, *Zosterophyllum*, grew about 15 cm tall in swampy areas and had naked stems that branched dichotomously (equally). However, other members evolved lateral branches and scalelike outgrowths. Moreover, the arrangement of their spore-forming organs and xylem indicates that they represent a line of evolution, separate from the rhyniophytes, that led to the lycophytes (e.g. clubmosses) but not to the seed plants.

zwitterion (ampholyte ion) An ion that has a positive and negative charge on the same group of atoms. Zwitterions can be formed from compounds that contain both acid groups and basic groups in their molecules. For example, the amino acid glycine has the formula $H_2N.CH_2.COOH$. However, under neutral conditions, it exists in the different form of the zwitterion $^+H_3N.CH_2.COO^-$, which can be regarded as having been produced by an internal neutralization reaction (transfer of a proton from the carboxyl group to the amino group). Glycine therefore has some properties characteristic of ionic compounds, e.g. a high melting point and solubility in water. In acid solutions, the positive ion $^+H_3NCH_2$ COOH is formed. In basic solutions, the negative ion $H_2NCH_2COO^-$ predominates. The name comes from the German *zwei*, two.

zygomorphy See BILATERAL SYMMETRY.

Zygomycota A phylum of saprotrophic or parasitic fungi that includes the bread mould (*Mucor*). Their hyphae lack cross walls and they can reproduce asexually by sporangiospores formed within a *sporangium or sexually by means of *zygospores.

zygospore A zygote with a thick resistant wall, formed by some algae and fungi

(*see* ZYGOMYCOTA). It results from the fusion of two gametes, neither of which is retained by the parent in any specialized sex organ (such as an oogonium). It enters a resting phase before germination. *Compare* OOSPORE.

zygote A fertilized female *gamete: the product of the fusion of the nucleus of the ovum or ovule with the nucleus of the sperm or pollen grain. *See* FERTILIZATION.

zygotene The second phase of the first *prophase of meiosis, in which *pairing (synapsis) of homologous chromosomes takes place. Intimate contact is made between identical regions of homologues, in a process involving proteins and DNA organized to form a **synaptonemal complex**.

zymase An extract of brewer's yeast containing a mixture of glycolytic enzymes (*see* GLYCOLYSIS) that can perform fermentation.

zymogen Any inactive enzyme precursor that, following secretion, is chemically altered to the active form of the enzyme. For example, the protein-digesting enzyme *trypsin is secreted by the pancreas as the zymogen trypsinogen. This is changed in the small intestine by the action of another enzyme, enterokinase, to the active form.

Appendix 1. SI units

Table 1.1. Base and dimensionless SI units

Physical quantity	Name	Symbol
length	metre	m
mass	kilogram	kg
time	second	s
electric current	ampere	A
thermodynamic temperature	kelvin	K
luminous intensity	candela	cd
amount of substance	mole	mol
*plane angle	radian	rad
*solid angle	steradian	sr

*dimensionless units

Table 1.2. Derived SI units with special names

Physical quantity	Name of SI unit	Symbol of SI unit
frequency	hertz	Hz
energy	joule	J
force	newton	N
power	watt	W
pressure	pascal	Pa
electric charge	coulomb	C
electric potential difference	volt	V
electric resistance	ohm	Ω
electric conductance	siemens	S
electric capacitance	farad	F
magnetic flux	weber	Wb
inductance	henry	H
magnetic flux density (magnetic induction)	tesla	T
luminous flux	lumen	lm
illuminance	lux	lx
absorbed dose	gray	Gy
activity	becquerel	Bq
dose equivalent	sievert	Sv

Table 1.3. Decimal multiples and submultiples to be used with SI units

Submultiple	Prefix	Symbol	Multiple	Prefix	Symbol
10^{-1}	deci	d	10	deca	da
10^{-2}	centi	c	10^2	hecto	h
10^{-3}	milli	m	10^3	kilo	k
10^{-6}	micro	μ	10^6	mega	M
10^{-9}	nano	n	10^9	giga	G
10^{-12}	pico	p	10^{12}	tera	T
10^{-15}	femto	f	10^{15}	peta	P
10^{-18}	atto	a	10^{18}	exa	E
10^{-21}	zepto	z	10^{21}	zetta	Z
10^{-24}	yocto	y	10^{24}	yotta	Y

Table 1.4. Conversion of units of SI units

From	To	Multiply by
in	m	2.54×10^{-2}
ft	m	0.3048
sq. in	m^2	6.4516×10^{-4}
sq. ft	m^2	9.2903×10^{-2}
cu. In	m^3	1.63871×10^{-5}
cu. ft	m^3	2.83168×10^{-2}
l(itre)	m^3	10^{-3}
gal(lon)	l(itre)	4.546 09
miles/hr	$m\ s^{-1}$	0.477 04
km/hr	$m\ s^{-1}$	0.277 78
lb	kg	0.453 592
$g\ cm^{-3}$	$kg\ m^{-3}$	10^3
lb/in^3	$kg\ m^{-3}$	$2.767\ 99 \times 10^4$
dyne	N	10^{-5}
poundal	N	0.138 255
lbf	N	4.448 22
mmHg	Pa	133.322
atmosphere	Pa	$1.013\ 25 \times 10^5$
hp	W	745.7
erg	J	10^{-7}
eV	J	$1.602\ 10 \times 10^{-19}$
kW h	J	3.6×10^6
cal	J	4.1868

Appendix 2. Simplified phylogenetic tree of the animal kingdom

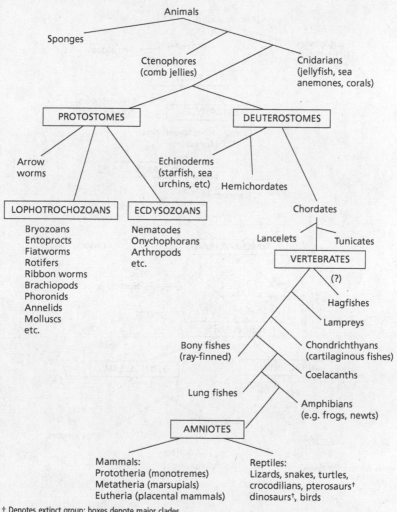

† Denotes extinct group; boxes denote major clades.

Appendix 3. Simplified phylogenetic tree for plants

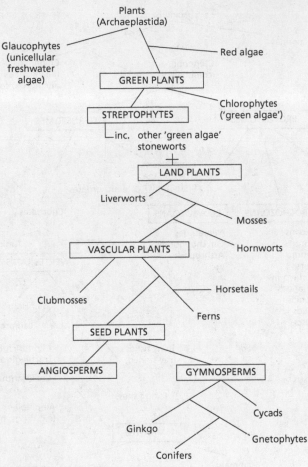

Boxes denote major clades.

Appendix 4. Geological time scale

millions of years ago	Eon	Era	Period	Epoch	millions of years ago
	Phanerozoic	Cenozoic	Quaternary	Holocene	
				Pleistocene	2.58
			Neogene	Pliocene	
				Miocene	23
			Palaeogene	Oligocene	
				Eocene	
				Palaeocene	66
		Mesozoic	Cretaceous		145
			Jurassic		201
			Triassic		252
		Palaeozoic	Permian		299
			Carboniferous		359
			Devonian		419
			Silurian		444
			Ordovician		485
			Cambrian		541
541	Proterozoic	Neoproterozoic	Ediacaran		541
			Cryogenian		635
			Tonian		720
		Mesoproterozoic	Stenian		1000
			Ectasian		
			Calymmian		1600
		Palaeoproterozoic	Statherian		
			Orosirian		
			Rhyacian		
			Siderian		2500
2500	Archaean				2500
4000					4000
4500			Hadean		4500

Appendix 5. Navigating the body

The ability to describe accurately the location of parts of the body in relation to the whole is extremely useful for the biologist and essential in medicine. This entails knowledge of various directional terms and imaginary planes based on particular body axes or lines of orientation. These are applied somewhat differently depending on whether the animal normally stands on two legs (e.g. a human) or four legs (e.g. a dog). Such a stance is the **anatomical position**—a reference position that is always used in anatomical descriptions.

Anatomical term	Relates to . . .	
	in bipedal animals (e.g. human)	in quadrupeds (e.g. dogs)
abdominal	anterior trunk below diaphragm	ventral trunk behind diaphragm
anterior	front, or in front of	= ventral
caudal	—	tail or tail end
cephalic	head	head
cervical	neck	neck
coronal	1. crown of head	crown of head
	2. coronal (frontal) plane	
cranial	cranium of skull	head or front end of body
crural	leg	leg
distal	farthest, or away from the trunk or point of attachment	
dorsal	back or top	back
dorsal plane	= frontal plane	= frontal plane
facial	face	face
femoral	thigh	thigh or femur
frontal	1. forehead	1. forehead
	2. frontal (coronal) plane	2. frontal (dorsal) plane
inferior	lower, or towards the feet	—
lateral	side, or farthest from median plane	
lumbar	lower back	loin
medial	centre, or nearest to the median plane	
palmar	palm of hand	back of front foot
plantar	sole of foot	back of hind foot
posterior	back, or to the rear of	= dorsal
proximal	nearest, or towards the trunk or point of attachment	
rostral	—	nose or front of head
sagittal	a front-to-back direction	a top-to-bottom direction
superior	upper, or towards the head	—
transverse	at right angles to the long axis of the body	
ventral	front of the body	undersurface of the body

Appendix 6. Model organisms and their genomes

Scientific name	Description/ common name	Type of organism	Genome size (Mb)	Haploid no. of chromosomes	No. of genes	Website
Escherichia coli K-12	nonpathogenic reference strain of E. coli	eubacterium	4.6	1	4576	https://ecocyc.org/
Dictyostelium discoideum		amoebozoan cellular slime mould	34	6	13 243	http://dictybase.org/
Saccharomyces cerevisiae	budding yeast used in baking and brewing	fungus	12.1	16	6600*	https://www.yeastgenome.org/
Neurospora crassa	red bread mould	fungus	40*	7	9820*	https://www.broadinstitute.org/ fungal-genome-initiative/ neurospora-crassa-genome-project
Zea mays	maize (corn)	monocotyledonous flowering plant of the grass family	2000–3000*	10 (including supernumerary (B) chromosomes)	~40 000	https://www.maizegdb.org/
Arabidopsis thaliana	thale cress	dicotyledonous flowering plant of the mustard family	125	5	27 873	https://www.arabidopsis.org/
Caenorhabditis elegans		nematode	97	6 autosomes+X sex chromosome	17 800*	https://www.ncbi.nlm.nih.gov/ IEB/Research/Acembly/index. html?worm

Drosophila melanogaster	fruit fly	dipteran insect	16.5	3 autosomes+X and Y sex chromosomes	15 185	http://flybase.org/
Danio rerio	zebrafish	teleost fish	1762	25	25 591	https://www.sanger.ac.uk/science/data/zebrafish-genome-project
Gallus domesticus	chicken	bird	1050	38 autosomes+Z and W sex chromosomes	18 346	https://www.ncbi.nlm.nih.gov/genome/?term=gallus%20gallus
Mus musculus	laboratory mouse	mammal	2632	19 autosomes+X and Y sex chromosomes	22 630	http://www.informatics.jax.org/
Homo sapiens	human	mammal	3000	22 autosomes+X and Y sex chromosomes	20 376	https://www.ncbi.nlm.nih.gov/genome/guide/human/
Trypanosoma brucei	protozoan	parasite	35	11 (+ many small ones)	8747	https://www.sanger.ac.uk/resources/downloads/protozoa/trypanosoma-brucei.html

* Approximate or estimated number of open reading frames (ORFs) or protein-coding genes

Appendix 7. Major mass extinction of species

Extinction event(s)	Date (millions of year ago)	Organisms most affected	Estimate of percentage of species made extinct	Cause(s)
Late Cambrian (series)	c. 488	trilobites, brachiopods, conodonts (primitive toothed vertebrates), ?soft-bodied arthropods	?	?change in sea level
Late Ordovician	c. 444	echinoderms, brachiopods, trilobites, ostracods, nautiloids	70–85	glaciation and fall in sea levels
Late Devonian (series)	c. 360	cephalopods, corals, brachiopods, bryozoans, echinoderms, trilobites, ammonites, agnathans, armoured fishes	70–83	?global cooling and reduced oxygen levels in deeper waters
Late Permian (Permo-Triassic; PTr)	251	corals, crinoids, ammonites, brachiopods, bryozoans, trilobites, land plants, insects, terrestrial vertebrates	<95	?volcanic activity with consequent global warming and changes in marine environment
Late Triassic	200	brachiopods, ammonites, bivalve and cephalopod molluscs, marine reptiles, conodonts, labyrinthodonts (primitive amphibians), insects	80	?climatic changes due to continental drift
Cretaceous-Tertiary (K–T)	65	dinosaurs, flying reptiles, ammonites, fish, brachiopods, planktonic organisms, plants	75–85	?meteorite collision (Alvarez event)

Appendix 8. Nobel prizewinning contributions to biology

Year	Nobel prize	Name of prizewinner(s)	Nationality	Nature of work or discovery
1901	Physiology or Medicine	Emil Adolf von Behring (1854–1917)	German	Developed a diphtheria antitoxin based on serum derived from immune individuals
1902	Physiology or Medicine	Ronald Ross (1857–1932)	British	Established that malaria parasites are transmitted by mosquitoes
1904	Physiology or Medicine	Ivan Petrovich Pavlov (1849–1936)	Russian	Investigated importance of sight and smell of food in stimulating the digestive system
1905	Physiology or Medicine	Robert Koch (1843–1910)	German	Discovered the bacterium responsible for tuberculosis
1906	Physiology or Medicine	Camillo Golgi (1843–1926)	Italian	Identified fundamental aspects of nervous system organization
		Santiago Ramón y Cajal (1852–1934)	Spanish	
1907	Physiology or Medicine	Charles Louis Alphonse Laveran (1845–1922)	French	Identified the protozoan responsible for malaria
1908	Physiology or Medicine	Ilya Ilyich Mechnikov (1845–1916)	Russian	Discovered phagocytosis
		Paul Ehrlich (1854–1915)	German	Studied antisera and immunity
1909	Physiology or Medicine	Emil Theodor Kocher (1841–1917)	Swiss	Studied the physiology, pathology, and surgery of the thyroid gland
1910	Physiology or Medicine	Albrecht Kossel (1853–1927)	German	Identified the chemical nature of cell components, particularly proteins and nucleic acids
1913	Physiology or Medicine	Charles Robert Richet (1850–1935)	French	Discovered anaphylaxis
1914	Physiology or Medicine	Robert Bárány (1876–1936)	Austro-Hungarian	Investigated the physiology of the inner ear and devised the Bárány test for diagnosing disease of the vestibular apparatus

Appendix 8 (cont.)

Year	Nobel prize	Name of prizewinner(s)	Nationality	Nature of work or discovery
1915	Chemistry	Richard Martin Willstätter (1872–1942)	German	Determined key aspects of chemical nature of chlorophyll and other plant pigments
1919	Physiology or Medicine	Jules Bordet (1870–1961)	Belgian	Discovered the immune component alexin, later called complement
1920	Physiology or Medicine	Schack August Steenberg Krogh (1874–1949)	Danish	Discovered physiological mechanism controlling capillary diameter in the blood vascular system
1922	Physiology or Medicine	Archibald Vivian Hill (1886–1977)	British	Discovered that heat produced following muscle contraction indicated oxygen consumption
		Otto Fritz Meyerhof (1884–1951)	German	Showed that lactic acid produced by contracting muscles was subsequently converted to glycogen by aerobic reactions
1923	Physiology or Medicine	Frederick Grant Banting (1891–1941)	Canadian	Discovered insulin
		John James Richard Macleod (1876–1935)	Canadian	
1927	Chemistry	Heinrich Otto Wieland (1877–1957)	German	Characterized the nature of bile acids
1928	Chemistry	Adolf Otto Reinhold Windhaus (1876–1959)	German	Identified key aspects of sterol chemistry and link with vitamin D
1929	Physiology or Medicine	Christiaan Eijkman (1858–1930)	Dutch	Identified cure for beriberi
		Frederick Gowland Hopkins (1861–1947)	British	Discovered vitamins
1930	Physiology or Medicine	Karl Landsteiner (1868–1943)	Austrian	Discovered the ABO system of human blood groups
1930	Chemistry	Hans Fischer (1881–1945)	German	Determined the chemical nature of haem
1931	Physiology or Medicine	Otto Heinrich Warburg (1883–1970)	German	Identified the enzymes involved in cell respiration

Appendix 8 (cont.)

Year	Nobel prize	Name of prizewinner(s)	Nationality	Nature of work or discovery
1932	Physiology or Medicine	Charles Scott Sherrington (1857–1952)	British	Studied nervous control and integration of muscle reflexes
		Edgar Douglas Adrian (1889–1977)	British	Investigated principles of nervous signalling based on impulse frequency
1933	Physiology or Medicine	Thomas Hunt Morgan (1866–1945)	US	Established chromosomes as the physical basis of genetic linkage
1935	Physiology or Medicine	Hans Spemann (1869–1941)	German	Discovered an embryonic organizer
1936	Physiology or Medicine	Henry Hallett Dale (1875–1968)	British	Discovered that acetylcholine is a chemical transmitter of nerve signals
		Otto Loewi (1873–1961)	Austrian	
1937	Physiology or Medicine	Albert von Szent-Györgyi (1893–1936)	Hungarian	Discovered fundamental components of cellular respiration
		Walter Norman Haworth (1883–1950)	UK	Discovered ring structures of sugars and synthesized vitamin C (ascorbic acid)
		Paul Karrer (1889–1971)	Swiss	Determined the structure of carotene and synthesized vitamins A and B2 (riboflavin)
1938	Physiology or Medicine	Corneille Jean François Heymans (1892–1968)	Belgian	Determined the role of the carotid sinus in regulating heart rate and blood pressure
1938	Chemistry	Richard Kuhn (1900–67)	German	Determined the structures of vitamins A and B2 and synthesized vitamin B6 (pyridoxine)
1943	Physiology or Medicine	Henrik Carl Peter Dam (1895–1976)	Danish	Discovered and characterized vitamin K
		Edward Adelbert Doisy (1893–1986)	US	
1944	Physiology or Medicine	Joseph Erlanger (1874–1965)	US	Identified different classes of nerve fibres according to their conducting velocity
		Herbert Spencer Gasser (1888–1963)	US	

Appendix 8 (cont.)

Year	Nobel prize	Name of prizewinner(s)	Nationality	Nature of work or discovery
1945	Physiology or Medicine	Alexander Fleming British (1881–1955)	British	Discovered, isolated, and purified penicillin
		Ernst Boris Chain (1906–79)	British	
		Howard Walter Florey (1898–1968)	Australian	
1946	Physiology or Medicine	Hermann Joseph Müller (1890–1967)	US	Discovered that X-rays cause a high rate of mutations
1947	Physiology or Medicine	Carl Ferdinand Cori (1896–1984)	US	Discovered how glycogen is broken down and resynthesized
		Gerty Theresa Cori (1896–1957)	US	
		Bernardo Alberto Houssay (1887–1971)	Argentina	Studied the effects of pituitary hormones on blood glucose
1948	Chemistry	Arne Wilhelm Kaurin Tiselius (1902–71)	Swedish	Developed electrophoresis as a technique for separating proteins and confirmed the existence of different classes of serum proteins
1950	Physiology or Medicine	Edward Calvin Kendall (1886–1972)	US	Identified the structure and biological effects of the adrenocortical hormones
		Tadeus Reichstein (1897–1996)	Swiss	
		Philip Showalter Hench (1896–1965)	US	
1953	Physiology or Medicine	Hans Adolf Krebs (1900–81)	British	Discovered the citric acid cycle (Krebs cycle)
		Fritz Albert Lipmann (1899–1986)	US	Discovered coenzyme A and established its importance in intermediary metabolism
1955	Physiology or Medicine	Axel Hugo Theodor Theorell (1903–82)	Swedish	Established mechanism of action of oxidative enzymes
1955	Chemistry	Vincent du Vigneaud (1901–78)	US	Synthesized the hormone oxytocin

Appendix 8 (cont.)

Year	Nobel prize	Name of prizewinner(s)	Nationality	Nature of work or discovery
1957	Chemistry	Alexander R. Todd (1907–97)	British	Synthesized the purine and pyrimidine bases of nucleic acids and also various coenzymes, including FAD, ADP, and ATP
1958	Physiology or Medicine	George Wells Beadle (1903–89)	US	Formulated the one gene-one enzyme hypothesis (now known as the one gene-one polypeptide hypothesis)
		Edward Lawrie Tatum (1909–75)	US	
		Joshua Lederberg (1925–2008)	US	Discovered genetic recombination and conjugation in bacteria
1958	Chemistry	Frederick Sanger (1918–2013)	British	Determined the amino acid sequence of bovine insulin
1959	Physiology or Medicine	Severo Ochoa (1905–93)	US	Discovered enzymes that catalyse the formation of RNA and DNA from their respective nucleotides
		Arthur Kornberg (1918–2007)	US	
1960	Physiology or Medicine	Medicine Frank Macfarlane Burnet (1899–1985)	Australian	Discovered acquired immunological tolerance
		Peter Brian Medawar (1915–87)	British	
1961	Physiology or Medicine	Georg von Békésy (1899–1972)	US	Established the physical mechanism of hearing within the cochlea of the inner ear
1961	Chemistry	Melvin Calvin (1911–97)	US	Determined the reactions of carbon assimilation during photosynthesis (the Calvin cycle)
1962	Physiology or Medicine	Francis Harry Compton Crick* (1916–2004)	British	Discovered the chemical structure of DNA and its significance for the transfer of genetic information*
		James Dewey Watson (1928–)	US	
		Maurice Hugh Frederick Wilkins (1916–2004)	New Zealand-British	

Appendix 8 (cont.)

Year	Nobel prize	Name of prizewinner(s)	Nationality	Nature of work or discovery
1962	Chemistry	Max Ferdinand Perutz (1914–2002)	British	Determined the structure of the protein myoglobin using X-ray crystallography
		John Cowdery Kendrew (1917–97)	British	
1963	Physiology or Medicine	John Carew Eccles (1903–97)	Australian	Discovered how ionic movements are intrinsic to nerve cell excitability
		Alan Lloyd Hodgkin (1914–98)	British	
		Andrew Fielding Huxley (1917–2012)	British	
1964	Physiology or Medicine	Konrad Bloch (1912–2000)	US	Discovered crucial steps in cholesterol synthesis
		Feodor Lynen (1911–79)	German	Identified key role of coenzyme A in fatty acid metabolism
1965	Physiology or Medicine	François Jacob (1920–2013)	French	Formulated the operon model of gene regulation
		Jacques Monod (1910–76)	French	
		André Lwoff (1902–94)	French	Determined the mechanism by which bacterial cells infected with bacteriophages undergo lysogeny
1966	Physiology or Medicine	Peyton Rous (1879–1970)	US	Discovered that certain viruses can cause cancer in animals
1967	Physiology or Medicine	Ragnar Granit (1900–91)	Swedish	Discovered fundamental aspects of the neurophysiology of vision
		Haldan Keffer Hartline (1903–83)	US	
		George Wald (1906–97)	US	
1968	Physiology or Medicine	Robert W. Holley (1922–93)	US	Elucidated the genetic code and its role in protein synthesis
		Har Gobind Khorana (1922–2011)	US	
		Marshall W. Nirenberg (1927–2010)	US	

Appendix 8 (cont.)

Year	Nobel prize	Name of prizewinner(s)	Nationality	Nature of work or discovery
1969	Physiology or Medicine	Alfred D. Hershey (1908–97)	US	Established that DNA is the genetic material of bacteriophages
		Max Delbrück (1906–81)	US	Demonstrated genetic recombination between viruses
		Salvador E. Luria (1912–91)	US	
1970	Physiology or Medicine	Bernard Katz (1911–2003)	British	Discovered the nature of certain neurotransmitters and the mechanism of their storage, release, and inactivation at synapses
		Ulf von Euler (1905–83)	Swedish	
		Julius Axelrod (1912–2004)	US	
1970	Chemistry	Luis F. Leloir (1906–87)	Argentinian	Discovered the role of sugar nucleotides in glycogen synthesis
1971	Physiology or Medicine	Earl W. Sutherland (1915–74)	US	Discovered cyclic adenosine monophosphate and demonstrated its importance as a second messenger in cell signaling
1972	Physiology or Medicine	Gerald M. Edelman (1929–2014)	US	Determined the chemical structure of antibodies
		Rodney R. Porter (1917–85)	British	
1972	Chemistry	Christian B. Anfinsen (1916–95)	US	Established that amino acid sequence alone determines the biological activity of enzymes
		Stanford Moore (1913–82)	US	Identified the chemical groups contributing to the active site of the enzyme ribonuclease
		William H. Stein (1911–80)	US	
1973	Physiology or Medicine	Karl von Frisch (1886–1982)	German	Demonstrated certain basic aspects of individual and social behaviour in animals under natural conditions
		Konrad Lorenz (1903–89)	Austrian	
		Nikolaas Tinbergen (1907–88)	British	

Appendix 8 (cont.)

Year	Nobel prize	Name of prizewinner(s)	Nationality	Nature of work or discovery
1974	Physiology or Medicine	Albert Claude (1899–1983)	Belgian	Discovered certain cell components, including lysosomes and ribosomes
		Christian de Duve (1917–2013)	Belgian	
		George E. Palade (1912–2008)	US	
1975	Physiology or Medicine	David Baltimore (1938–)	US	Discovered the role of the enzyme reverse transcriptase during infection by RNA viruses
		Howard Martin Temin (1934–94)	US	
		Renato Dulbecco (1914–2012)	US	Established the concept of virus-induced transformation of normal cells into cancer cells
1977	Physiology or Medicine	Roger Guillemin (1924–)	US	Discovered peptide hormones in the brain
		Andrew V. Schally (1926–)	US	
		Rosalyn Yalow (1921–2011)	US	Developed radioimmunoassay for peptide hormones
1978	Physiology or Medicine	Werner Arber (1929–)	Swiss	Discovered restriction enzymes and their applications in molecular genetics
		Daniel Nathans (1928–99)	US	
		Hamilton O. Smith (1931–)	US	
1978	Chemistry	Peter D. Mitchell (1920–92)	British	Formulated the chemiosmotic theory of biochemical energy transfer
1979	Physiology or Medicine	Alan M. Cormack (1924–98)	US	Developed computerized tomography
		Godfrey N. Hounsfield (1919–2004)	British	
1980	Chemistry	Paul Berg (1926–)	US	Pioneered recombinant DNA techniques
		Walter Gilbert (1932–)	US	Developed techniques for sequencing nucleic acids
		Frederick Sanger (1918–2013)	British	

Appendix 8 (cont.)

Year	Nobel prize	Name of prizewinner(s)	Nationality	Nature of work or discovery
1980	Physiology or Medicine	Baruj Benacerraf (1920–2011)	US	Determined the genetic basis of histocompatibility antigens on body cells and their significance in immune mechanisms and tissue transplantation
		Jean Dausset (1916–2009)	French	
		George D. Snell (1903–96)	US	
1981	Physiology or Medicine	Roger W. Sperry (1913–94)	US	Identified the main functional specializations of right and left cerebral hemispheres
		David H. Hubel (1926–2013)	US	Made key insights into the structural and functional organization of the visual cortex
		Torsten N. Wiesel (1924–)	Swedish	
1982	Physiology or Medicine	Sune K. Bergström (1916–2004)	Swedish	Discovered key aspects of the nature, metabolism, and biological actions of prostaglandins and related substances
		Bengt I. Samuelsson (1934–)	Swedish	
		John R. Vane (1927–2004)	British	
1982	Chemistry	Aaron Klug (1926–2018)	British	Developed crystallographic electron microscopy to determine the structure of protein-nucleic acid complexes
1983	Physiology or Medicine	Barbara McClintock (1902–92)	US	Discovered mobile genetic elements (transposons)
1984	Physiology or Medicine	Niels K. Jerne (1911–94)	Danish	Developed theories explaining antibody specificity and diversity
		Georges J. F. Köhler (1946–95)	German	Described the hybridoma technique for producing monoclonal antibodies
		César Milstein (1927–2002)	Argentine-British	

Appendix 8 (cont.)

Year	Nobel prize	Name of prizewinner(s)	Nationality	Nature of work or discovery
1985	Physiology or Medicine	Michael S. Brown (1941–)	US	Described the importance in cholesterol metabolism of cell surface receptors for low-density lipoproteins
		Joseph L. Goldstein (1940–)	US	
1986	Physiology or Medicine	Stanley Cohen (1922–2020)	US	Discovered, respectively, epidermal growth factor and nerve growth factor
		Rita Levi-Montalcini (1909–2012)	Italian-US	
1987	Physiology or Medicine	Susumu Tonegawa (1939–)	Japanese	Demonstrated how genetic recombination in immune cells produces the diversity of antigen receptors and antibodies
1988	Chemistry	Johann Deisenhofer (1943–)	German	Determined the three-dimensional structure of a photosynthetic reaction centre
		Robert Huber (1937–)	German	
		Hartmut Michel (1948–)	German	
1989	Physiology or Medicine	J. Michael Bishop (1936–)	US	Discovered that viral oncogenes are derived from normal cellular genes
		Harold E. Varmus (1939–)	US	
1989	Chemistry	Sidney Altman (1939–)	Canadian-US	Discovered the catalytic properties of RNA
		Thomas R. Cech (1947–)	US	
1991	Physiology or Medicine	Erwin Neher (1944–)	German	Studied the operation of singleion channels in cell membranes
		Bert Sakmann (1942–)	German	
1992	Physiology or Medicine	Edmond H. Fischer (1920–)	Swiss-US	Discovered the importance of protein phosphorylation as a regulatory mechanism for cellular functions
		Edwin G. Krebs (1918–2009)	US	

Appendix 8 (cont.)

Year	Nobel prize	Name of prizewinner(s)	Nationality	Nature of work or discovery
1993	Physiology or Medicine	Richard J. Roberts (1943–)	British	Discovered 'split genes', consisting of exons and intervening noncoding introns
		Phillip A. Sharp (1944–)	US	
1993	Chemistry	Kary B. Mullis (1944–)	US	Invented the polymerase chain reaction technique
		Michael Smith 1932–2000	Canadian	Developed the technique of site-directed mutagenesis
1994	Physiology or Medicine	Alfred G. Gilman (1941–2015)	US	Discovered G proteins and their role in cell signaling
		Martin Rodbell (1925–98)	US	
1995	Physiology or Medicine	Edward B. Lewis (1918–2004)	US	Identified the genes that control the development of Drosophila embryos
		Christiane Nüsslein Volhard (1942–)	German	
		Eric F. Wieschaus (1947–)	US	
1995	Chemistry	Paul J. Crutzen (1933–)	Dutch	Made fundamental contributions to understanding of ozone chemistry and the threat posed to the ozone layer by human-derived chemicals
		Mario J. Molina (1943–)	US	
		F. Sherwood Rowland (1927–2012)	US	
1996	Physiology or Medicine	Peter C. Doherty (1940–)	Australian	Discovered the mechanism by which cytotoxic T cells recognize virus-infected cells
		Rolf M. Zinkernagel (1944–)	Swiss	
1997	Physiology or Medicine	Stanley B. Pruisner (1942–)	US	Discovered prions
	Chemistry	Paul D. Boyer (1918–2018)	US	Determined the structure and mechanism of ATP synthetase
		John E. Walker (1941–)	British	
		Jens C. Skou (1918–)	Danish	Discovered sodium/potassium ATPase (the sodium pump) in cell membranes

Appendix 8 (cont.)

Year	Nobel prize	Name of prizewinner(s)	Nationality	Nature of work or discovery
1998	Physiology or Medicine	Robert F. Furchgott (1916–2009)	US	Discovered that nitric oxide is a key signalling molecule in the cardiovascular system
		Louis J. Ignarro (1942–)	US	
		Ferid Murad (1936–)	US	
1999	Physiology or Medicine	Günter Blobel (1936–2018)	US	Formulated the signal hypothesis whereby proteins are tagged with peptide 'address labels' to specify their destination within the cell
2000	Physiology or Medicine	Arvid Carlsson (1923–2018)	Swedish	Demonstrated the mechanism of action of dopamine as a neurotransmitter in the brain
		Paul Greengard (1925–2019)	US	
		Eric R. Kandel (1929–)	US	Discovered how changes in nerve synapses are the basis of learning and memory
2001	Physiology or Medicine	Leland H. Hartwell (1939–)	US	Identified key genes and proteins involved in regulating the cell cycle
		R. Timothy Hunt (1943–)	British	
		Paul M. Nurse (1949–)	British	
2002	Physiology or Medicine	Sydney Brenner (1927–2019)	British	Identified crucial genes that regulate organ development and programmed cell death (apoptosis)
		H. Robert Horvitz (1947–)	US	
		John E. Sulston (1942–)	British	
2003	Chemistry	Peter Agre (1949–)	US	Discovered water channels in cell membranes
		Roderick MacKinnon (1956–)	US	Determined the spatial conformation of the potassium ion channel

Appendix 8 (cont.)

Year	Nobel prize	Name of prizewinner(s)	Nationality	Nature of work or discovery
2004	Physiology or Medicine	Richard Axel (1946–)	US	Discovered the family of genes encoding olfactory receptors and how olfactory signals are received by the brain
		Linda B. Buck (1947–)	US	
2004	Chemistry	Aaron Ciechanover (1947–)	Israeli	Discovered the process of ubiquitin-mediated protein degradation in living cells
		Avram Hershko (1937–)	Israeli	
		Irwin Rose (1926–2015)	US	
2006	Physiology or Medicine	Andrew Z. Fire (1959–)	US	Discovered RNA interference
		Craig C. Mello (1960–)	US	
2006	Chemistry	Roger D. Kornberg (1947–)	US	Determined the molecular basis of transcription in eukaryotic cells
2007	Physiology or Medicine	Mario R. Capecchi (1937–)	US	Developed the technique of using embryonic stem cells to create mouse strains carrying targeted gene modifications
		Martin J. Evans (1941–)	British	
		Oliver Smithies (1925–2017)	US	
2008	Chemistry	Osamu Shimomura (1928–2018)	Japanese	Discovery and development of the green fluorescent protein, GFP
		Martin Chalfie (1947–)	US	
		Roger Y. Tsien (1952–)	US	
2008	Physiology or Medicine	Harald zur Hausen (1936–)	German	Discovery of human papilloma viruses causing cervical cancer
		Françoise Barré-Sinoussi (1947–)	French	Discovery of human immunodeficiency virus
		Luc Montagnier (1932–)	French	

Appendix 8 (cont.)

Year	Nobel prize	Name of prizewinner(s)	Nationality	Nature of work or discovery
2009	Chemistry	Venkatraman Ramakrishnan (1952–)	Indian	Studies of the structure and function of ribosomes
		Thomas A. Steitz (1940–)	US	
		Ada E. Yonath (1939–)	Israeli	
2009	Physiology or Medicine	Elizabeth H. Blackburn (1948–)	Australian	Discovery of how chromosomes are protected by telomeres and the enzyme telomerase
		Carol W. Greider (1961–)	US	
		Jack W. Szostak (1952–)	British	
2010	Physiology or Medicine	Robert G. Edwards (1925–2013)	British	Development of *in vitro* fertilization
2011	Physiology or Medicine	Bruce A. Beutler (1957–)	US	Discoveries concerning the activation of innate immunity
		Jules A. Hoffmann (1941–)	Luxembourg	Discovery of the dendritic cell and its role in adaptive immunity
		Ralph M. Steinman (1943–2011)	Canadian	
2012	Physiology or Medicine	John B. Gurdon (1933–)	British	Discovered how to genetically reprogram mature cells into pluripotent stem cells
		Shinya Yamanaka (1962–)	Japanese	
2012	Chemistry	Robert J. Lefkowitz (1943–)	US	Discovered a family of cell receptors called G-protein-coupled receptors
		Brian K. Kobilka (1955–)	US	
2013	Physiology or Medicine	James E. Rothman (1950–)	US	Elucidated mechanisms of intracellular trafficking of transport vesicles
		Randy W. Schekman (1948–)	US	
		Thomas C. Südhof (1955–)	German	
2014	Physiology or Medicine	John O'Keefe (1939–)	US-British	Discovered the cells that constitute a positioning system in the brain
		May-Britt Moser (1963–)	Norwegian	
		Edvard I. Moser (1962–)	Norwegian	

Appendix 8 (cont.)

Year	Nobel prize	Name of prizewinner(s)	Nationality	Nature of work or discovery
2016	Physiology or Medicine	Yoshinori Ohsumi (1971–)	Japanese	Discovered mechanisms for autophagy
2017	Physiology or Medicine	Jeffrey Hall (1945–)	US	Discoveries of molecular mechanisms controlling the circadian rhythm
		Michael Rosbash (1944–)	US	
		Michael Young (1949–)	US	
2018		James P. Allison (1948–)	US	Discovery of cancer therapy by inhibition of negative immune regulation
		Tasuku Honjo (1942–)	Japanese	
2019	Physiology or Medicine	Tasuku Honjo (1942–)	Japan	Discoveries of how cells sense and adapt to oxygen availability
		William G. Kaelin Jr. (1957–)	US	
		Sir Peter J. Ratcliffe (1954–)	UK	
		Gregg L. Semenza (1956–)	US	
2020	Physiology or Medicine	Harvey J. Alter (1935–)	US	Discovery of Hepatitis C virus
		Michael Houghton	UK	
		Charles M. Rice (1952–)	US	
2020	Chemistry	Emmanuelle Charpentier (1968–)	French	Development of a method for genome editing
2021	Medicine	Jennifer Doudna (1964–)	US	Discoveries of receptors for temperature and touch
		David Julius (1955–)	US	
		Ardem Patapoutian (1967–)	Lebanon	

Note: The Physiology or Medicine prizewinners listed above have been selected for their contributions to biology; they do not represent an exhaustive list of prizewinners

* Rosalind Franklin (1920–58), British X-ray crystallographer, was also a key contributor to discovering the structure of DNA, but Nobel Prizes are not awarded posthumously

Appendix 9. Evolution

Sources of information about this fundamental theory in modern biology

Subject	Source
Charles Darwin: co-author (with A.R. Wallace) of the theory of evolution of species by natural selection that bears his name—Darwinism	**Darwin online**: a compendium of Darwin's works, including books, articles, and other publications, plus a wealth of information about his life and legacy. **Charles Darwin and Evolution: 1809–2009**: website created by Christ's College, Cambridge, to celebrate the bicentenary of Darwin's birth.
Alfred Russel Wallace: sometimes overlooked but a key figure in formulating the theory of natural selection, and a contributor to many other areas of science	**The Alfred Russel Wallace Website**: lots of information about Wallace's life and works, plus news of events and other activities of the Wallace Fund.
Evolutionary theory and evidence, including natural selection and the modern synthesis	**Human Evolution**: lively introduction to human evolution from the Natural History Museum, London. **History of Evolutionary Thought**: extensive survey of various strands of scientific thought since the 1700s that have led to our current understanding of evolution. Site created by the University of California Museum of Paleontology. **PBS Evolution**: comprehensive learning resource for students, teachers, and the general public, including a searchable multimedia library, plus web activities. Devised by WGBH Educational Foundation and Clear Blue Skies Production, Inc. **Tree of Life**: contemporary interactive illustration of Darwin's concept of the 'Tree of Life' hosted by the Wellcome Trust. **The Simpsons**: watch Homer evolve from a single cell to a couch potato in 1.29 (on YouTube).

Appendix 10. Useful websites

The following is a selection of authoritative, quality-controlled websites that provide free information on essential biological topics. To access any of these sites, go to the dictionary's web page at www.oxfordreference.com/page/biol.

Access Excellence—The National Health Museum
A portal providing links to numerous learning and teaching resources, ranging from specialist journals to lively graphics and videos.

American Society for Microbiology
Podcasts and videos provide a wide-ranging introduction to current topics in microbiology.

BioMed Central
Publisher of some 300 open-access peer-reviewed journals in biology, medicine, and health.

Botanical Society of America
Provides links to a wide range of resources, including talks, videos, and blogs.

Cells Alive
An educational site that presents pictures of all different types of cells and some fun videos.

DOE Genomes Program
The US Department of Energy hosts this site describing the history and achievements of the Human Genome Project and other projects sponsored by the DOE, including the Genomic Science programme.

EMBO Journal
Access to tables of contents and abstracts from the well-known academic journal.

Fishbase
An international consortium runs this large online database. The records (currently over 32 000 species) include taxonomic information, distribution, recent species status on the IUCN RedList of Threatened Species, size and growth parameters, diet composition, trophic levels, and other biological features of marine and freshwater fishes of the world.

GRID-Arendal
Provides a wealth of maps, statistics, and graphics on environmental topics.

Human Proteome Organization
A starting point for insights into the pioneering and rapidly developing world of proteomics. HUPO is the international body responsible for consolidating and coordinating the work of national and regional proteomics groups.

Met Office
Climate science section of the UK Met Office website: provides in-depth information on climate change issues, including monthly bulletins and the latest observations.

Nature Magazine Online
The online weekly journal that offers cutting-edge news articles and features, lively debate columns, and information on the latest science research, as well as links to numerous publications and databases.

Public Library of Science
A nonprofit organization publishing scientific and medical literature as a freely available public resource, including *PLoS Biology*, *PLoS Genetics*, and *PLoS Medicine*.

PubMed Central
The US National Institutes of Health (NIH) free digital archive of biomedical and life sciences journal literature.

Royal Botanic Gardens, Kew
Besides an introduction and special features about the UK's premier plant collection, there is a wealth of information about horticulture, plant conservation, and other plant sciences.

Society for Conservation Biology
An information site from the international professional organization dedicated to promoting the scientific study of the phenomena that affect the maintenance, loss, and restoration of biological diversity.

Wildlands Network
A site produced by the leading North American conservation movement, highlighting aspects of its work and providing links to other related resources.

ZSL (Zoological Society of London)
This site contains links to London Zoo, Whipsnade Wild Animal Park, and the ZSL's research division, the Institute of Zoology.